Solid-State Chemistry of Inorganic Materials

MATERIALS RESEARCH SOCIETY
SYMPOSIUM PROCEEDINGS VOLUME 453

Solid-State Chemistry of Inorganic Materials

Symposium held December 2–5, 1996, Boston, Massachusetts, U.S.A.

EDITORS:

Peter K. Davies
University of Pennsylvania
Philadelphia, Pennsylvania, U.S.A.

Allan J. Jacobson
University of Houston
Houston, Texas, U.S.A.

Charles C. Torardi
DuPont Company
Central R&D, Experimental Station
Wilmington, Delaware, U.S.A.

Terrell A. Vanderah
NIST, Ceramics Division
Gaithersburg, Maryland, U.S.A.

PITTSBURGH, PENNSYLVANIA

This material is based upon work supported by the National Science Foundation under Grant No. DMR-9612039. Any opinions, findings, and conclusions or recommendations expressed in this materials are those of the author (s) and do not necessarily reflect the views of the National Science Foundation.

Single article reprints from this publication are available through
University Microfilms Inc., 300 North Zeeb Road, Ann Arbor, Michigan 48106

CODEN: MRSPDH

Published by:

Materials Research Society
9800 McKnight Road
Pittsburgh, Pennsylvania   15237
Telephone (412) 367-3003
Fax (412) 367-4373
Website: http://www.mrs.org/

Library of Congress Cataloging in Publication Data

Solid–state chemistry of inorganic materials : symposium held December
    2–5, 1996, Boston, Massachusetts, U.S.A. / editors, Peter K. Davies,
    Allan J. Jacobson, Charles C. Torardi, Terrell A. Vanderah
        p. cm—(Materials Research Society symposium proceedings ; v. 453)
        Includes bibliographical references and index.
        ISBN 1-55899-357-6
        1. Solid state chemistry—Congresses. 2. Inorganic compounds—
    Congresses. 3. Semiconductors—Congresses. I. Davies, Peter K.
    II. Jacobson, Allan J. III. Torardi, Charles C. IV. Vanderah, Terrell A.
    V. Series: Materials Research Society symposium proceedings ; v. 453.
QD478.S6337 1997   97-8050
541′.0421—dc21   CIP

Manufactured in the United States of America

CONTENTS

*Invited Paper

PART III: <u>MATERIALS SYNTHESIS</u>

*Invited Paper

PART IV: <u>THEORY</u>

*Invited Paper

PART V: <u>OPTICAL PROPERTIES</u>

PART VI: <u>ELECTRONIC AND MAGNETIC PROPERTIES</u>

*Invited Paper

*Invited Paper

PART VII: <u>DIELECTRICS AND FERROELECTRICS</u>

*Invited Paper

PART VIII: <u>SOLID-STATE IONICS</u>

PART IX: <u>SURFACES AND INTERFACES</u>

*Invited Paper

PREFACE

During the last 50 years the importance and scope of solid state chemistry has grown in response to the continuing challenge to understand, control, and predict the structures and properties of solids at the atomic level and to synthesize new compounds with enhanced physical response. The many successes in the preparation of materials with unique electronic, optical, magnetic and catalytic properties are a clear testament to the vitality and importance of solid state chemistry to materials research. Today, researchers in this field can be found in numerous relevant areas including materials science and engineering, ceramics, chemistry, chemical engineering, mineralogy/geology and condensed matter physics. The inherently multidisciplinary nature of solid state chemistry provides the intellectual basis for its continuing vitality; however, a disadvantage results in that this nontraditional discipline has not been consistently and/or substantially represented in national meetings. Solid state chemistry is now recognized as a core activity in materials research; therefore, the Materials Research Society meetings appear to provide the most appropriate multidisciplinary forum to present and discuss important new advances. This symposium was deliberately organized with two purposes: to initiate a series of regularly occurring symposia in this field, hosted by the MRS; and to attract researchers from all relevant fields to highlight recent advances in solid state chemistry and to discuss their impact on the development and application of inorganic materials.

Symposium R, "Solid State Chemistry of Inorganic Materials," was one of the largest and most comprehensive meetings on this topic held under the auspices of a major United States professional society. The symposium, which received over 250 submissions, consisted of three and one-half days of oral sessions and required three evenings for presentation of the 124 poster papers. The gathering was enhanced by a strong international component, with many of the 19 invited speakers traveling from abroad. The technical presentations, although highly interdisciplinary and directed towards many different classes of solids, reflected the shared challenge to understand, control, and predict the structures and properties of inorganic solids. The papers included in this proceedings volume encompass the subjects of the technical sessions which include chalcogenides, synthetic methods and reactivity, theory and modeling, electronic structure, optical properties, electronic and magnetic properties, dielectrics and ferroelectrics, solid-state ionics and surfaces and interfaces. The organizers would like to thank all of the participants whose collective attendance and effort resulted in a lively, highly successful, and very well attended symposium. Many participants contributed their time onsite to expeditiously review the manuscripts for this volume. We especially thank our international attendees who contributed large blocks of time from their busy schedules to travel and participate in this meeting. Last, but certainly not least, the organizers recognize and appreciate that the success of this meeting was critically enhanced by the financial support received from the National Science Foundation, the Texas Center for Superconductivity at the University of Houston, the DuPont Company, and Exxon Research and Engineering Company.

Peter K. Davies
Allan J. Jacobson
Charles C. Torardi
Terrell A. Vanderah

February 1997

Materials Research Society Symposium Proceedings

ABOUT CHEMICAL BONDS IN MISFIT LAYER CHALCOGENIDES

J. ROUXEL and A. MEERSCHAUT
Institut des Matériaux de Nantes, UMR CNRS N° 110-Université de Nantes,
BP 32229, 2 rue de la Houssinière, 44322 NANTES Cedex 3 - France.

ABSTRACT

On the basis of crystal structures and electronic properties, a critical discussion of the nature of chemical bonds in misfit layered chalcogenides is given. It concerns also the various mechanisms of non-stoichiometry that are present in these incommensurate structures. Misfit chalcogenides with a LnY slab opposed to TY_2 ones (T = Ti, V, Cr, Nb, Ta) appear as infinite two-dimensional intercalation compounds with an electron donation from LnY to TY_2. In case of PbY or SnY slabs a more complex mechanism involving coupled substitutions in both slabs seems to prevail.

INTRODUCTION

All solids of low-dimensional character can be "incommensurate" by a misfit effect if they contain two or more interpenetrating sublattices. It concerns not only one- and two- dimensional solids themselves for which the stacking of different slabs or fibers provides the sublattices, but also 3-D structures showing tunnels. The latter can be regarded as the negative of one-dimensional structures. Filling the tunnels by various chemical species generates the second sublattice.

Indeed, the two or more sublattices may show periods that coincide in some direction(s) but are incommensurate in, at least, one direction, i.e. their ratio is given by an irrational number in that case [1]. For example a one dimensional misfit structure is usually composed of separate columns of different kinds or of similar columns enclosed in tunnels provided by a given framework. Incommensurability comes from the ratio of periodicities along the direction of columns. The $Ba_{1+x}Fe_2S_4$ ($0.062 \leq x \leq 0.143$) phases [2] are composed of chains of edge-sharing $[FeS_4]$ tetrahedra separated by Ba^{2+} columns. Depending on x the Ba-Ba distances vary and govern the periodicity along the column, thus justifying the name of infinitely adaptive series [3]. The chimney-ladder phases TX_{2-x} [4] present chimneys in a tetragonal framework provided by a transition metal T, in which X (Si, Ge, Ga) is arranged on helices (the ladders).

Layered misfit structures show an alternate stacking of two different slabs. The misfit character arises from the mismatch between the intralayer periodicities of the two layers nets. A very good illustration is provided by the $(MY)_{1+x}TY_2$ phases (M = Ln, Sn, Pb, Bi, Sb ; Y = S, Se ; T = Ti, V, Cr, Nb, Ta ; $0.08 < x < 0.28$). The great number of phases that have been prepared and characterized so far, the variety of structural arrangements and properties, allow to introduce a discussion about the origin of their stability. Why do these phases exist preferentially to the two binaries MY and TY_2 ? Along with a critical discussion of structural types, electrical properties and non-stoichiometry, this will be the subject of the present paper.

I - Preparation and Structural features

The misfit chalcogenides $(MY)_{1+x}(TY_2)$ are obtained from the elements or the binary (MY/M_2Y_3 and TY_2) chalcogenides when available, by heating in evacuated quartz ampoules at temperatures ranging from 900°C (Pb, Sn derivatives) to 1100°C (lanthanides derivatives). A slight excess of sulfur or selenium is used. In case of sulfides another way which is particularly convenient for the rare earth derivatives consists in treating the corresponding ternary oxydes in

Mat. Res. Soc. Symp. Proc. Vol. 453 © 1997 Materials Research Society

flowing argon/H$_2$S gas at 1300-1500°C. The only way to get (LnS)$_{1+x}$CrS$_2$ phases is to start from LnCrO$_3$ + Ln$_2$O$_3$ in appropriate ratio. Iodine is a good transport agent to facilitate crystal growth (<5mg/cm^3). Chlorine has also been used. It is formed in-situ by thermal decomposition of (NH$_4$)$_2$PbCl$_6$. Nice layered black crystals are obtained.

In the structure two types of slabs alternate along a common axis which gives the $\bar{c}$ direction. MY slabs, also named "Q" slabs, are a distorted (100) double slice of the NaCl structure type. TY$_2$ slabs, also named "H" slabs, consists of edges sharing octahedra building TiS$_2$, VS$_2$, CrS$_2$ layers, or edges sharing trigonal prisms building NbS$_2$ or TaS$_2$ layers. Among the **a** and **b** in-plane parameters of each sublattice, the **b** parameters have always been found identical so far, but the ratio of the **a** parameters is an irrational number (figure 1). Consequently the full diffraction pattern of both subsystems is given by a scattering vector $\vec{S}$ determined by four reciprocal lattice vectors :

$$\vec{S} = H\vec{a}_1^* + K\vec{a}_2^* + L\vec{a}_3^* + M\vec{a}_4^*$$

in which

$$\vec{a}_1^* = \vec{a}_{TY_2}^*, \quad \vec{a}_2^* = \vec{b}_{TY_2}^* = \vec{b}_{MY}^*, \quad \vec{a}_3^* = \vec{c}_{TY_2}^* = \vec{c}_{MY}^*, \quad \vec{a}_4^* = \vec{a}_{MY}^*$$

HKL0 reflections with H ≠ 0 and 0KLM reflections with M ≠ 0 correspond to TY$_2$ and MY subsystems, respectively. The 0KL0 reflections are common to both subsystems. HKLM reflections of the whole crystal appear as satellites due to the mutual modulation of the two subsystems, the wave vector of the modulation acting on the TY$_2$ subsystem being $\vec{a}_4$, and that of the MY subsystem being $\vec{a}_1$. Structure determinations have to be made in (3+1) dimensional space groups [5]. However one can also use a composite approach based on the structure determination of each sublattice in a first step, then using 0KL0 reflections to fix their relative position [6]. Satellites have been rarely measured and introduced in the refinement processes. However, as the effect of the modulation also influences the intensity of the main reflections, the modulation is consequently accessible without using satellites, although with larger standard deviations in its parameters.

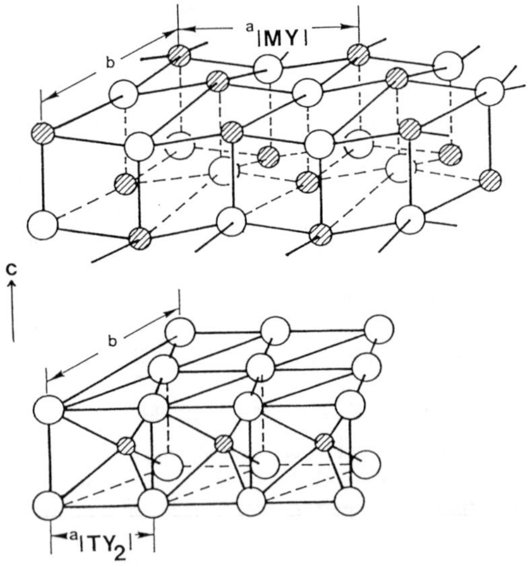

Figure 1 - Schematic representation of a misfit layer structure type with MY and TY$_2$ slabs stacked along the $\bar{c}$ direction ; incommensurability is observed along $\bar{a}$.

Phases with two or three adjacent TY_2 slabs have been made [7-13]. They show a van der Waals gap between successive TY_2 slabs. This allows to practice intercalation chemistry [14] and also to exfoliate the structure [15]. Other classical features of parent binary dichalcogenides such as the presence of extra T atoms in van der Waals gap have also been observed [12].

With two successive $(Nb-Ta)Y_2$ slabs a 2H stacking mode (periodicity of two slabs along $\vec{c}$) was expected in the misfit compounds like in pristine dichalcogenides. This is what was indeed observed in all the bilayer selenides derivatives as represented in case of $(PbSe)_{1.12}(NbSe_2)_2$ in figure 2a. Surprisingly the sulfides show 2/3 of a 3R arrangement for the two $(Nb-Ta)S_2$ slabs (in the pure dichalcogenides the 3R arrangement implies 3 slabs). In addition two polymorphs of $(PbS)_{1.14}(NbS_2)_2$ were found (fig. 2b,c). Both show a partial population of the van der Waals gap by extra niobium atoms leading to the chemical formulation $(PbS)_{1.14}[(NbS_2)Nb_{0.098}(NbS_2)]$. One form is orthorhombic and shows a reverse orientation of successive 3R-type $(NbS_2)_2$ blocks. In the other one, with monoclinic symmetry, the orientation is identical from one block to the other.

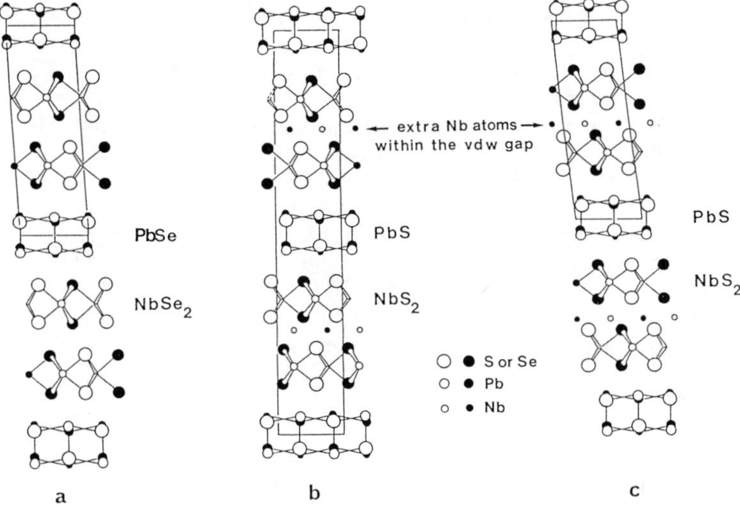

Figure 2 - Projection along the misfit direction of the structures of misfit bilayer compounds :
a) (PbSe)$_{1.12}$(NbSe$_2$)$_2$ with a 2H-stacking mode
b) - c) orthorhombic and monoclinic forms for (PbS)$_{1.14}$(NbS$_2$)$_2$; both have a 2/3 part of a 3R arrangement of NbS$_2$ slabs. Extra niobium atoms are present between adjacent NbS$_2$ slabs (within the van der Waals gap).

In a trilayer misfit compound like $(SnS)_{1.17}(NbS_2)_3$ [16] the NbS_2 part shows a 3R stacking mode.

Other important feature concerns the possible preparation of commensurate phases. This is achieved through appropriate substitutions in the Q slabs or through combined substitutions and rotations of Q with respect to H. Substituting Bi by Ca in the Q slabs of $[(Bi/Ca)S]_{1.20}TiS_2$ [17] allows to modify the ratio of **a** parameters and to reach commensurability. $[(Nb_{1-y}La_y)S]NbS_2$ phases [18] also are commensurate. They show a rotation around the **c** axis by 45° of the $(Nb_{1-y}La_y)S$ slab with respect to the NbS_2 one (the **a** parameter has been changed by a factor of

$\sqrt{2}$). The smaller radius of Nb^{3+} compared to La^{3+} (0.721 and 1.032 Å respectively) allows to play with cell dimensions.

Finally a misfit compound with two Q slabs for one H slab (Franckeite type) has recently been prepared [19]. It has the composition $[(Pb,Sb)S]_{2.13}NbS_2$. The presence of antimony at the interface between adjacent Q slabs has been necessary to stabilize the structure [20].

II - Electrical Properties

Misfit layered chalcogenides show interesting magnetic properties which associate low-dimensionality and incommensurability [21-24]. They will not be discussed in this contribution. We will focus on electrical properties because they will feed the discussion on chemical bonds and stability of misfit compounds.

The particular geometrical and chemical features of layered misfit chalcogenides suggest high anisotropy of the conductivity, localization effects due to the modulation of the structure, superconductivity and eventually charge density waves instability in relation with $(Nb/Ta)Y_2$ slabs. Some of these features are present. The largest interest is presently devoted to the superconducting properties with two very important points which are anisotropy and progressive stacking of superconducting layers. For example $(PbS)_{1.14}NbS_2$ is superconducting [25] below $T_c = 2.475$ K. It is a type II superconductor. The temperature dependence of H_{c_2} is linear as expected from the Ginzburg-Landau theory with $dH_{c_2\perp}/dT = 1.6 kG/K$ and $dH_{c_2//}/dT = 17.2$ kG/K (// and $\perp$ mean respectively parallel and perpendicular to the layers). The anisotropy of H_{c_2} near T_c (figure 3) is 8.75 to be compared with 2.9-3 for pure NbS_2. Contrary to the monolayer phase, $(PbS)_{1.14}(NbS_2)_2$ does not show superconductivity down to 1.2 K [26]. This is to be related to the 3R stacking mode of the NbS_2 slabs. It is known that $3R-NbS_2$ is not superconducting either. With a 2H stacking mode $(PbSe)_{1.14}(NbSe_2)_2$ shows a superconducting transition below 3.64 K [8,26]. In $2H-NbSe_2$ itself the transition is at 7.2 K. In the mono layered phase $(PbSe)_{1.14}NbSe_2$ Tc is observed at 2.05 K. In the trilayered phase $(PbSe)_{1.14}(NbSe_2)_3$ it reaches the value of 4.8 K [10]. The structure of bulk $NbSe_2$ is progressively rebuilt between PbSe slabs. One can measure the contribution of one slab and the influence of their stacking mode.

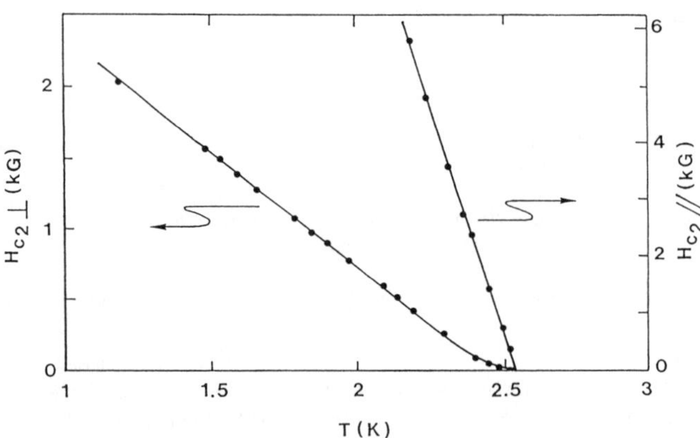

Figure 3 - Anisotropy of H_{c_2} in $(PbS)_{1.14}NbS_2$ as a function of temperature.

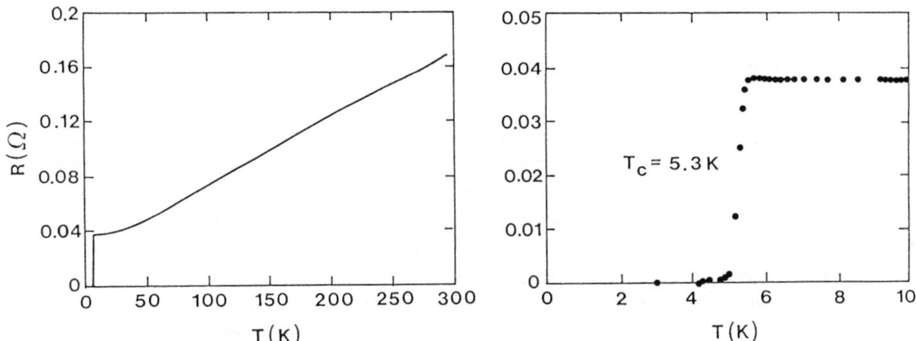

Figure 4 - Variation of resistance of (LaSe)₁.₁₄(NbSe₂)₂ as a function of temperature.

Superconducting transitions can be sharp (figure 4 shows the transition at 5.3 K observed in the case of $(LaSe)_{1.14}(NbSe_2)_2$ [11]. Sometimes they are rather broad. In that case we could observe with the help of HREM the simultaneous presence of monolayered, bilayered and trilayered domains in the same crystal.

Misfit layer compounds with $(Nb/Ta)Y_2$ slabs are all metallic. The in-plane resistivity ρ_{ab} is about $10^{-6}\Omega m$ (same for compounds with TiS_2 sandwiches) [1d]. The anisotropy $\rho_{//c}/\rho_{ab}$ is closed to 50 with rare earth derivatives [27,28], but much bigger for Pb or Sn derivatives where it can reach for example 250 at 300 K and 800 at 4 K for $(SnS)_{1.17}NbS_2$ [29]. Measurements of Hall (R_H) and Seebeck (α) coefficients (both being positive for misfit compounds when M = Ln) have been made as well as photoelectron spectroscopy studies [30-37]. The rare earth phases show about 0.1 hole/T in the d_{z^2} conduction band of the TY_2 slab which means that 0.9 electrons have been transferred from the LnY to the TY_2 slab (the d_{z^2} band can accommodate $2e^-$ and is already half filled in pristine Nb/TaY_2 slabs). In the case of SnS or PbS Q-slabs an increased population of the d_{z^2} band is also observed. A band structure calculation on $(SnS)_{1.17}NbS_2$ was made in a commensurate approach by taking a value of 5/3 for the ratio of **a** parameters (which corresponds to $(SnS)_{1.20}NbS_2$ composition). It shows [38] an additional electronic population of 0.36 e^-/Nb. Measurements also agree with a d_{z^2} band more than half filled. $(MS)_{1+x}TiS_2$ phases (M = Sn, Pb) are also metallic [39,40]. A population of the mostly $3d$-t_{2g} band of TiS_2 slabs by 0.4-0.7 electrons is deduced. R_H and α are negative.

A similar electron transfer to the VS_2 slab is observed in case of $(PbS)_{1.12}VS_2$. With averaged V-V distances of 3.256 (2×) and 3.294 (4×) [41] a metallic behaviour is expected in relation with a partial population of the t_{2g} band of VS_2 slabs and the correlative existence of holes in the valence band of PbS. Indeed $Na_{0.3}VS_2$ which shows similar VS_2 slabs [42,43] with V-V distances of 3.288 Å and $Na_{0.95}VS_2$ with V-V distances of 3.346 Å are metallic. With longer distances (3.566 Å) $NaVS_2$ is a semi-conductor. Clearly localization occurs for shorter metal-metal distances in the misfit compounds. This is to be related to the distribution of potential experienced by the electrons due to the relative modulation of atomic positions in one slab by the other. In addition the holes in the valence band of PbS should not contribute or be absent due to defects (see & IV). $(LnS)_{1+x}VS_2$ phases also show semi-conducting behaviour [44-47]. A semi-metallic phase has been observed in addition to the semi-conducting one in case of samarium [46].

The rare earth misfit compounds $(LnS)_{1+x}CrS_2$ are semi-conductors. An optical gap of 1.2-1.3 eV was reported for Y, Ge, Dy, Ho, Er derivatives [48]. This will be discussed now in & III.

III - Electronic properties and non stoichiometry

We have already found three types of non-stoichiometry which are expressed through (i) the 1+x value that comes from the incommensurability of the structure, (ii) the presence of extra T atoms between successive TY_2 slabs in multilayer phases, (iii) the possible substitution of M in the Q slabs leading eventually to commensurate situations. There is, in addition, another form of non-stoichiometry which was found as a consequence of the semi-conducting behaviour of $(LnS)_{1+x}CrS_2$ phases.

Magnetic measurements in $(GdS)_{1.27}CrS_2$ [49] as well as chemical shifts of the K absorption edge and XPS binding energies of core levels in the La, Nd, Gd phases [50] show that chromium is at the +3 oxidation state. It means that one electron has been transferred per chromium atom from LnS slab to CrS_2 one. One can assume that, like in VS_2 layer, Cr-Cr distances are too long to allow an electronic delocalization in that slab. However assuming the LnS layer to be $Ln^{3+}S^{2-}e^-$ with one electron delocalized in the 5d band like in rare earth monochalcogenides, x electrons would remain in that band (1+x are initially present) and lead to metallic behaviour. At that point a very precise chemical analysis was made on a single crystal of the Gd-Cr phase with the help of a CAMECA SX50 electron microprobe. It appears that there are vacancies in the GdS layer (they were already suspected on the basis of chemical refinements). The exact chemical formulation is $(Gd_{0.91}\square_{0.09}S)_{1.27}CrS_2$. The number of vacancies is such that there is an exact charge balance between Gd^{3+}, Cr^{3+} and S^{2-} [51]. A similar result was recently found for $(La_{0.95}\square_{0.05}Se)_{1.21}VSe_2$ [52]. The question is open whether it is a general situation or not in case Q slabs populated with M = Ln and T = Nb, Ta.

The presence of vacancies can be checked with the help of appropriate chemical substitutions. Let us consider for example $(La_{0.94}\square_{0.06}S)_{1.2}CrS_2$; it is possible to substitute Sr^{2+} for La^{3+} in the Q slab [53,54]. However the process stops where one lanthanum is left per chromium atom. At that moment all the vacancies are filled up with Sr^{2+} cations. The formulation can be written $(LaS)(SrS)_{0.20}CrS_2$. The process is driven by the necessity to get a transfer of one electron per chromium atom. Sr^{2+} cations act as inert species, just filling the vacancies [55].

IV - Electronic properties and chemical bonds in the misfit compounds

We have just shown above that there is an electronic transfer of one electron from LaS to CrS_2 layers in the corresponding misfit compound. $(LnS)_{1+x}CrS_2$ phases appear as an infinite two dimensional intercalation compounds. They can be compared to $NaCrS_2$ which shows the same structure as $NaTiS_2$ the intercalation compound of sodium between TiS_2 slab. There is no layered CrS_2 with CdI_2 structure. The reason is the instability of Cr^{4+} in presence of sulfur. The corresponding d levels of chromium are too deeply engaged in the sp band of sulfur. They would be filled up at the expense of this band at the top of which holes would appear. These holes could associate and give anionic pairs leading to a CrS_2 with a pyrite or a marcasite structure, like for MnS_2 or FeS_2 just next to the right. This is not observed maybe due to the instability (distortions) that would be attached to the presence of a Cr^{2+} (d^4) cation. Cr^{3+}, on contrary, is highly stable in octahedral coordination (t_{2g}^3). When giving one electron to the CrS_2 slab we are indeed refilling the top of the sp band, neutralizing holes and preventing the formation of anionic associations [56]. This is the reason of the stabilization of CrS_2 slabs in $NaCrS_2$ as well as in the misfit compounds.

It is of interest to see how band structure calculations confirm all these results. Such calculation cannot be made in 3+1 dimensions. There are consequently two approaches (i) a composite approach in which we make calculations for each sublattice and adjust the two Fermi levels. This is an approach that was largely used in previous discussions just above where mention was made for example of a d_{z^2} conduction band of NbS_2 slabs or t_{2g} levels of TiS_2 ones. When doing that we just express the equality of chemical potentials between A and B in chemical compound, A being the "Q" slab and B the "H" one. Another approach is the commensurate approach. Calculations are made in a bigger unit cell close to commensurability. For $(LaS)_{1+x}CrS_2$ the x value is nearly 0.2. It is possible to form a superstructure by using 3 LaS units and 5 CrS_2 units along the incommensurate $\bar{a}$ direction. The calculations were made [57,55] using the T.B-LMTO-ASA method [58-61]. Figure 5 shows the total spin projected density of states of $(LaS)(SrS)_{0.20}CrS_2$.

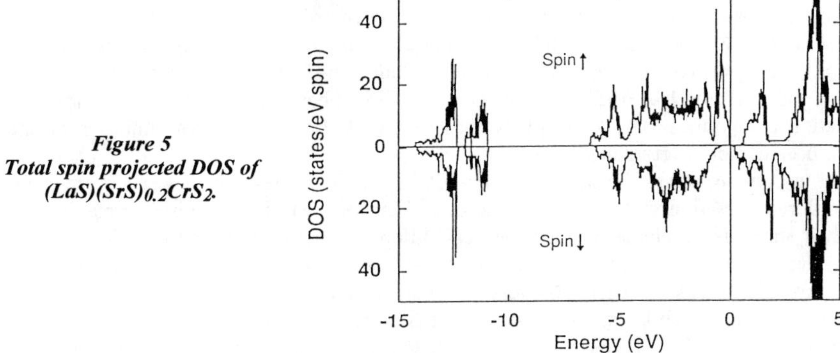

Figure 5
Total spin projected DOS of
$(LaS)(SrS)_{0.2}CrS_2$.

With increasing energy we find successively the 3s sulfur states around -13 eV, then the 3p S states hybridized with Cr d states from -6 to -1 eV. La and Sr are fully ionized to the +3 and +2 states (no contribution below E_F, La 4f bands being 4 eV above E_F for example). Figure 6 shows the spin projected D.O.S. of the Cr 3d orbitals. A magnetic moment of 3 μ_B is calculated from the difference of the spin densities (↑) and (↓) in agreement with the magnetic measurements. The three Cr 3d electrons are fully polarized on t_{2g} (↑). One electron has clearly been transferred to chromium.

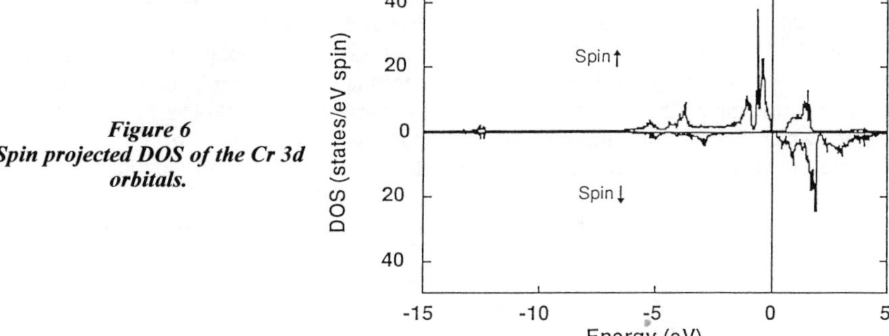

Figure 6
Spin projected DOS of the Cr 3d
orbitals.

E_F falls in an energy gap in agreement with the semi-conducting properties. This gap is open between t_{2g} ($\uparrow$) and t_{2g} ($\downarrow$) orbital set of chromium. Any additional electron would occupy t_{2g} ($\downarrow$) and destabilize the structure. Thus $(LaS)_{1.20}CrS_2$ without vacancies and with an increased electronic transfer does not exist.

Calculations made on $(LaSr_{0.2})S_{1.20}$ and CrS_2 separated slabs show the same features except that LaS components are pushed up towards higher energies in the misfit compound. The charge transfer which occurs when the misfit compound forms depopulates La states and fills Cr t_{2g} states. It results in a stabilization of the whole structure. The question of the presence of some covalent bonding between La and the S atoms of the CrS_2 slabs can be raised up with respect to what has been found for $(SnS)_{1.17}NbS_2$ [62] and $(SnS)_{1.2}TiS_2$ [63]. Charge density and Electron Localization Function (ELF) calculations were made on $(LaS)(SrS)_{0.2}CrS_2$. The charge around the S atoms of LaS slab are found spherical which agrees with an ionic behaviour. No significant charge density is found between the S atoms of the CrS_2 layer and La in LaS slabs. The lone pairs of the S atoms in the CrS_2 layer even point towards the free space but not to the La atoms. However interaction between these pz orbitals and Ln empty states which have some antibonding character is not to be totally excluded [64]. It has been discussed in layered dichalcogenides themselves [65]. Clearly $(LnS)_{1+x}CrS_2$ phases are infinite two dimensional intercalation compounds. This concept extends to all the LnS derivatives although vacancies have not been depicted yet in all systems.

Let us come finally to the case of the $[(Sn/Pb)S]_{1+x}TS_2$ phases. They clearly correspond to a very different situation with respect to the one discussed just above. This was manifested all along their study, for example through the very different "charge transfer to the TY_2 slabs" or the much higher anisotropy ratios of conductivity. But the most important feature is the fact that they do not exist in case of CrS_2 H-layers. It means that the electronic transfer, if any, is not sufficient to stabilize such layers. We think that it is rather difficult to propose a charge transfer with PbS or SnS Q-slabs. At least Pb has never been found at a higher oxidation state than 2+ in the presence of sulfur and could not account for the increased population of 0.30 to 0.60 e⁻ found in the conduction bands of H-slabs. Again very precise analysis of single crystals with the help of the XS50 Castaing microprobe shows that some T element is present in Q-slabs at the 3+ oxidation state, the charge equilibrium being realized through an equivalent reduction of T in the H-slab [66] :

$$M^{2+} \text{ (Q slab)} + T^{4+} \text{ (H) replaced by } T^{3+} \text{ (Q)} + T^{3+} \text{ (H)}$$

Each Nb^{3+} for example in an NbS_2 H-slab brings one additional electron and explains the increased population of the corresponding conduction band. There is no need to consider a 4+ oxidation state of Sn or Pb. Stability of the whole structure comes from the fact that we have Q-slabs and H-slabs that are positively and negatively charged, respectively.

The large difference in size of M^{2+} and T^{3+} cation in the Q-slab should result in a segregation of rows parallel to the $\bar{b}$ direction, according to a minimization of interfacial energies. As in some minerals where this is known (cylindrite for example [1a]) the proportion according to which these rows follow each other along $\bar{a}$ may have to do with semi-commensurate or incommensurate states of the structure. It may add another cause of incommensurability to purely geometric factors associated to the relative size of the Q and H slabs as shown in LnS derivatives. But the undulations may also fix the position of one slab with respect to the other.

REFERENCES

1. For detailed description see (a) E. Mackovicky and B.G. Hyde, Struct. Bonding **46**, p.101 (1981), (b) <u>Incommensurate sandwiched layered compounds</u>, A. Meerschaut Ed. (Mat. Sci. Forum **100**, Trans Tech. Pub. 1992), (c) J. Rouxel, A. Meerschaut, G.A. Wiegers, J. Alloys Comp. **229**, p.144-157 (1995), (d) G.A. Wiegers, Prog. Solid State Chem. Elsevier Science Ltd **24**, p. 1 (1996).

2. I.E. Grey, J. Solid State Chem. **11**, p. 128 (1973).

3. J.S. Anderson, J. Chem. Soc., Dalton, p. 1107 (1973).

4. H. Novotny, W. Jeitschko and E. Parthe, Acta Cryst. **22**, p. 417 (1967).

5. S. van Smaalen, <u>Incommensurate sandwiched layered compounds</u>, A. Meerschaut Ed. (Mat. Sci. Forum **100**, Trans Tech. Pub. 1992)

6. G.A. Wiegers, A. Meetsma, R.J. Haange, S. van Smaalen, J.L. de Boer, A. Meerschaut, P. Rabu and J. Rouxel, Acta Cryst. **B46**, p. 324-332 (1990).

7. A. Meerschaut, L. Guemas, C. Auriel and J. Rouxel, Eur. J. Solid State Inorg. Chem. **27**, p. 557 (1990).

8. C. Auriel, A. Meerschaut, R. Roesky and J. Rouxel, Eur. J. Solid State Inorg. Chem. **29**, p. 1079 (1992).

9. A. Meerschaut, C. Auriel and J. Rouxel, J. Alloys Comp. **183**, p. 129 (1992).

10. Y. Oosawa, Y. Gotoh, J. Akimoto, T. Tsunoda, M. Sohma, Jap. J. Appl. Phys. **3IL**, p. 1096 (1992).

11. R. Roesky, A. Meerschaut, J. Rouxel and J. Chen, Z. Anorg. Allg. Chem. **619**, p. 117 (1993).

12. C. Auriel, A. Meerschaut and J. Rouxel, Mat. Res. Bull. **28**, p. 675 (1993).

13. Y. ren, A. Meetsma, G.A. Wiegers and S. van Smaales, Acta Cryst. **B52**, p. 389 (1996).

14. - L. Herman, J. Morales, J. Pattanayak and J.L. Tirado, J. Solid State Chem. **100**, p. 262 (1992).
    - C. Barriga, P. Lavela, J. Morales, J. Pattanayak and J.L. Tirado, Chem. Mat. **4**, p. 1021 (1992).

15. P. Bonneau, J.L. Mansot and J. Rouxel, Mat. Res. Bull. **28**, p. 757 (1993).

16. L.M. Hoistad, A. Meerschaut, P. Bonneau and J. Rouxel, J. Solid State Chem. **114**, p. 435 (1995).

17. Hung Yi-Chung and H. Shiou-Jyh, Inorg. Chem. **32**, p. 5427 (1993).

18. R. Roesky, A. Meerschaut, A. van der Lee and J. Rouxel, Mat. Res. Bull. **29(11)**, p. 149 (1994).

19. A. Lafond, A. Meerschaut, Y. Moëlo and J. Rouxel, C.R. Acad. Sc. Paris, serie IIB, **322**, p. 165 (1996).

20. A. Lafond, A. Nader, S. Perrin, Y. Moëlo, A. Briggs, A. Meerschaut, P. Monceau and J. Rouxel, J. Alloys Comp., to be submitted.

21. K. Suzuki, N. Kojima, T. Ban and I. Tsujikawa, J. Phys. Soc. Japan **59(1)**, p. 266 (1990)

22. O. Pena, P. Rabu and A. Meerschaut, J. Phys. Cond. Matter. **3**, p. 9929 (1991).

23. T. Terashima, and N. Kojima, J. Phys. Soc. Japan **61(9)**, p. 3303 (1992).

24. K. Suzuki, T. Kondo, M. Iwasaki and T. Enoki, Jpn J. Appl. Phys. **32(32-3)**, p. 341 (1993).

25. A. Smontara, P. Monceau, L. Guemas, A. Meerschaut, P. Rabu and J. Rouxel, J. Fizika **21(3)**, p. 201 (1989).

26. P. Monceau, J. Chen, O. Laborde, A. Briggs, C. Auriel, R. Roesky, A. Meerschaut and J. Rouxel, Physica B **194-196**, p. 2361 (1994).

27. T. Terashima, Thesis, University of Kyoto, Japan (1993).

28. T. Terashima and N. Kojima, J. Phys. Soc. Japan **63(2)**, p. 658 (1994).

29. C.M. Fang, J. Baas and G.A. Wiegers, (to be published)

30. G.A. Wiegers, A. Meetsma, R.J. Haange and J.L. de Boer, J. Less Commun Metals **68**, p. 347 (1991).

31. A.R.H.F. Ettema, G.A. Wiegers, C. Haas and T.S. Turner, Physica Scripta **T41**, p. 265 (1992).

32. A.R.H.F. Ettema, C. Haas and T.S. Turner, Phys. Rev. **B47**, p. 12794 (1993).

33. A.R.H.F. Ettema and C. Haas, J. Phys. : Condens. Matter. **5** p. 3817 (1993).

34. A.R.H.F. Ettema, S. van Smaalen, C. Haas and T.S. Turner, Phys. Rev. **B49**, p. 10585 (1994).

35. Y. Ohno, Phys. Rev. **B48(8)**, p. 5515 (1993).

36. Y. Ohno, J; Phys. : Condens. Matter. **4**, p. 7815 (1992).

37. Y. Ohno, Phys. Rev. **B44(3)**, p. 1281 (1991).

38. C.M. Fang, A.R.H.F. Ettema, C. Haas, G.A. Wiegers, H. van Leuken and R.A. de Groot, Phys. Rev. **B52(4)**, p. 2336 (1995).

39. G.A. Wiegers and R.J. Haange, Eur. J. Solid State Inorg. Chem. **28**, p. 1071 (1991).

40. C. Auriel, A. Meetsma, C. Deudon, J. Baas, G.A. Wiegers, P. Monceau and C. Chen, Eur. J. Solid State Inorg. Chem. **32**, p. 947 (1995).

41. M. Onoda, K. Kato, Y. Gotoh, Y. Oosawa, Acta Cryst. **B46**, p. 487 (1990).

42. G.A. Wiegers, R. van der Meer, H. van Heiningen, H.J. Kloosterboer and A.J.A. Alberink, Mat. Res. Bull. **9**, p. 1266 (1993).

43. C.F. van Bruggen and G.A. Wiegers, J. Solid State Chem. **27**, p. 9 (1979).

44. Y. Gotoh, M. Goto, K. Kawaguchi and Y. Oosawa, Mat. Res. Bull. **25**, p. 307 (1990).

45. N. Cho, S. Kikkawa, F. Kanamaru, Y. Takeda, O. Yamamoto, H. Kido and T. Hoshikawa, Solid State Ionics **63-65**, p. 696 (1993).

46. T. Kondo, K. Suzuki and T. Enoki, Solid State Commun. **84**, p. 999 (1992).

47. K. Suzuki, T. Kondo, T. Enoki and S. Bandow, Synthetic Metals **55-57**, p. 1741 (1993).

48. T. Takahashi, S. Osaka and O. Yamada, J. Phys. Chem. Solids **43**, p. 1131 (1973).

49. A. Lafond and A. Meerschaut, Mat. Res. Bull. **28**, 979 (1993).

50. T. Murugesan, S. Ramesh, J. Gopalakrishnan and C.N.R. Rao, J. Solid State Chem. **38**, p. 165 (1981).

51. J. Rouxel, Y. Moëlo, A. Lafond, F.J. DiSalvo, A. Meerschaut and R. Roesky, Inorg. Chem. **33**, p. 3358 (1994).

52. Y. Ren, J. Baas, A. Meetsma, J.L. de Boer and G.A. Wiegers, Acta Cryst. **B52**, p. 398 (1996).

53. T. Nishikawa, Y. Yasni and M. Sato, J. Phys. Soc. Japan **63(9)**, p. 3218 (1994).

54. Y. Yasui, T. Nishikawa, Y. Kobayashi, M. Sato, T. Nishioka and M. Kontani, J. Phys. Soc. Japan **64(10)**, p. 3890 (1995).

55. L. Cario, D. Johrendt, A. Lafond, C. Felser, A. Meerschaut and J. Rouxel, Phys. Rev. B submitted (1996).

56. J. Rouxel, Chem. Eur. J. **2(9)**, p. 1053 (1996).

57. C. Fang, Thesis University of Groningen (The Netherlands) (1996).

58. O.K. Anderson, Phys. Rev. **B12**, p. 3060 (1975).

59. O.K. Anderson and O. Jepsen, Phys. Rev. Letters **53**, p. 2571 (1984).

60. TB LMTO - ASA 46 Program : a program for ab initio band calculations. G. Krier, O. Jepsen, A. Burkhardt and O.K. Anderson, Max-Planck-Institut, Germany (1994).

61. O. Jepsen and O.K. Anderson, Z. Phys. **B97**, p. 35 (1995).

62. M.C. Fang, A.R.H. Ettema, C. Haas and G.A. Wiegers, Phys. Rev. **B52**, p. 2336 (1995).

63. M.C. Fang, R.A. de Groot, G.A. Wiegers and C. Haas, J. Phys. Cond. Matter. **8**, p. 1663 (1996).

64. Preliminary calculations (F. Boucher and J. Rouxel, to be published) are in favor of such interaction. When writing this paper the authors have just been aware of similar proposition : S.P. Abramov, to be published in J. Alloys Comp. (1997).

65. E. Sandré, R. Brec and J. Rouxel,, J. Phys. Chem. Solids **50(8)**, p. 801 (1989).

66. Y. Moëlo, A. Meerschaut, J. Rouxel and C. Auriel, Chem. Mater. 7, p. 1771 (1995).

Thermoelectric Properties and Electronic Structure of BaBiTe$_3$

Duck-Young Chung[a], Stéphane Jobic[b], Tim Hogan[c], Carl R. Kannewurf[c], Raymond Brec[b], Jean Rouxel[b] and Mercouri G. Kanatzidis*[a], [b]

[a]Department of Chemistry and Center for Fundamental Materials Research, Michigan State University, East Lansing, MI 48824. (b) Institut des Matériaux de Nantes, 2 Rue de la Houssinière, 44072 Nantes Cedex 03, France (c) Department of Electrical Engineering and Computer Science, Northwestern University, Evanston, IL 60208

ABSTRACT

The compound BaBiTe$_3$ was prepared by the reaction of Ba/Bi/Te at over 700 °C either in K$_2$Te$_4$ or BaTe$_3$ flux and was recrystallized in a Ba/BaTe$_3$ flux. The black rod-shaped polycrystalline material crystallizes in the orthorhombic space group P2$_1$2$_1$2$_1$ with a=4.6077(2) Å, b=17.0437(8) Å, c=18.2997(8) Å. Its structure is made of interdigitating columnar anionic [Bi$_4$Te$_{10}$(Te$_2$)]$_{\infty}$ "herring-bone" shaped segments which arrange into layers with Ba^{2+} ions between them. The electrical conductivity, thermopower, thermal lattice conductivity, infrared absorption properties of this material suggest it is a narrow gap semiconductor.

INTRODUCTION

Interest in Group 15 chalcogenides is increasing because of their promising thermoelectric properties.[1,2] Bi$_2$Te$_3$ and its solid solutions are currently the leading thermoelectric materials for room temperature cooling applications. This reflects the fact that this material exhibits the unusual combination of simultaneously high electrical conductivity, high thermoelectric power and low thermal conductivity. The origin of this phenomenon is not entirely understood. In most materials high electrical conductivity is characterized by low thermoelectric power and vice versa. Therefore, it is difficult to construct a robust theoretical framework to help search for new thermoelectric materials. Nevertheless, new materials incorporating bismuth have not been extensively explored during the past three decades, and thus we have embarked in a chemically based approach to prepare and study the properties of structural and chemical derivatives of Bi$_2$Te$_3$. The variation in properties of such materials could reveal interesting trends from which we could benefit in constructing a better theoretical framework for understanding and designing efficient thermoelectric materials. We have examined the Ba/Bi/Te system in detail and report on the synthesis, structure, electronic structure, spectroscopic properties, thermal conductivity, and electrical conductivity and

Mat. Res. Soc. Symp. Proc. Vol. 453 © 1997 Materials Research Society

thermoelectric power of BaBiTe$_3$. This material has one of the lowest κ_L (lattice thermal conductivity) values known for chalcogenides and for thermoelectric materials.

DISCUSSION

BaBiTe$_3$ was prepared with a flux technique using K$_2$Te$_4$ and mixed BaTe$_x$/Cs$_2$Te$_x$ fluxes. The full details of the synthesis are reported elsewhere.[4]

The structure of BaBiTe$_3$ is layered (Figure 1) with [BiTe$_3$] slabs sandwiching Ba cations in pseudo-trigonal prismatic sites. The Ba^{2+} cations are surrounded by nine Te atoms which form a distorted tri-capped trigonal prism. The Coulombic interactions between Ba^{2+} and the nine Te atoms in its coordination polyhedron seem to be important in stabilizing this structure type. We point out that BaBiTe$_3$, BaBiSe$_3$ and BaSbTe$_3$ are all isostructural,[5,6] while the strontium compounds[6,7] are not, probably because a smaller cation such as Sr^{2+} cannot accommodate nine Te atoms in its immediate coordination environment. The size of alkaline earth metal appears to play an important role in determining the structure type of ternary Group 15 chalcogenide compounds.

The coordination sphere of Bi is best described as distorted octahedral. Each slab contains parallel, rod-shaped columnar [Bi$_4$Te$_{10}$]$_\infty$ segments with a NaCl sub-structure which are made of edge-sharing BiTe$_6$ octahedra, see Figure 2.

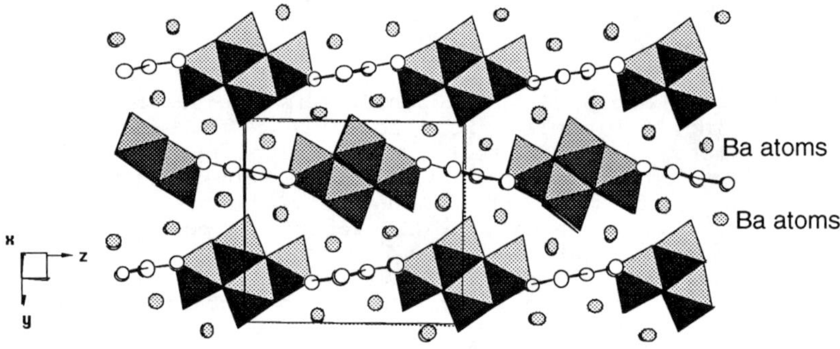

Figure 1. The layered BaBiTe$_3$ structure viewed down the a-axis. The non-bonded atoms are Ba.

This structure type exhibits short Te-Te contacts between [Bi$_4$Te$_{10}$]$_\infty$ segments in the same layer. The BaBiTe$_3$ contains polytelluride fragments, as can be deduced by assigning a 2+ and a 3+ formal oxidation state to Ba and to Bi respectively. The average formal oxidation state of Te is -1.667 suggesting the presence of Te-Te bonding. This leads to two possible descriptions of the compound. In the first, the Bi/Te slabs can be thought of as composed of columnar [Bi$_4$Te$_{10}$]$_\infty$ segments alternating with parallel "zig-zag" Te$_n{}^{n-}$ "chains" with Te---Te contacts see Figure 3. In the second, the slabs can be thought of as composed of interdigitating columnar $[Bi_4Te_{10}(Te_2)]_\infty$ "herring-bone" shaped segments. Within the would-be zig-zag Te_n^{n-} "chains" the Te-Te distance is 3.170(2) Å. The Te-Te distance between Te atoms in the would-be "zig-zag" "chain" and Te atoms bound to an adjacent $[Bi_4Te_{10}]$ segment at 3.098(2) Å. Because this distance is 0.072 Å shorter than the distance inside the would-be chain one is inclined to prefer the interdigitated "herring-bone" description of the structure; this is supported by theoretical calculations.[4] Such short contacts were found to be bonding in polymeric $TiTe_2$ and $IrTe_2$ with a charge transfer from the Te^{2-} sp band to cation d-levels.[8]

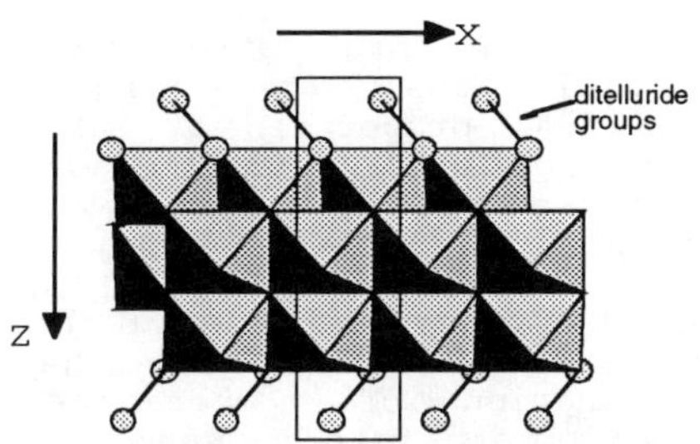

Figure 2. The infinite columnar $[Bi_4Te_{10}]_\infty$ rod which makes up the layers in $BaBiTe_3$.

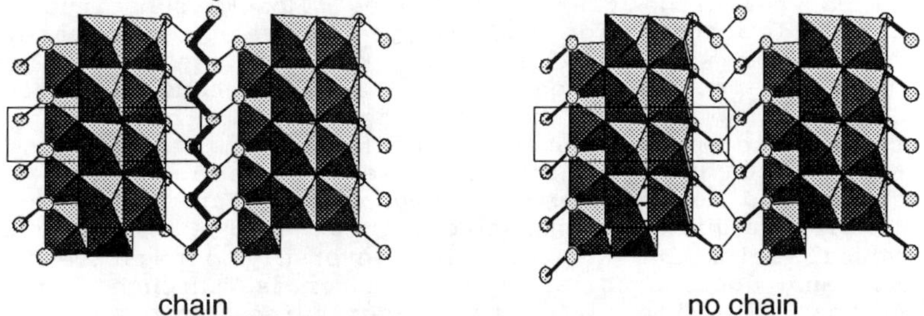

Figure 3. The extended framework of a layer viewed down the b-axis showing the two alternative extremes in bonding description regarding the polytelluride part of the structure. (left) $[Bi_4Te_{10}]_\infty$ segments alternating with parallel "zig-zag" Te-chains (right) interdigitating columnar $[Bi_4Te_{10}(Te_2)]_\infty$ "herring-bone" shaped segments.

In order to evaluate the electronic structure of the material we performed extended Hückel band structure calculations. A +3 oxidation state was assigned to bismuth. Hence its electronic configuration should be $6s^2 6p^0$ with the two remaining electrons acting as a lone pair (which of course becomes somewhat delocalized through interactions with the nearest neighbors). The band dispersion along the high symmetry axis calculated with the Extended Hückel method is shown in

Figure 4. In the [-10. eV, -6. eV] range, six bands are observed, the four lowest based on the so-called Bi lone pairs hybridized with tellurium orbitals, the two highest based on Te orbitals associated with the interface of the $[Bi_4Te_{10}(Te_2)]_\infty$ "herring-bone" shaped segments, with a slight contribution of the Bi-s orbitals. Although the Fermi level crosses slightly two partially filled bands along the Γ-X axis suggesting a metallic behavior, we have to keep in mind that such calculations do not yield accurate band positions and in borderline cases, like the one dealt with here, it is difficult to ascertain the electrical properties of the phase e.g. metallic or semiconductor.

Charge Transport Properties.

The nature of the electronic band structure suggests strongly that the material will have interesting charge transport properties. The electrical properties of $BaBiTe_3$ were measured from single crystal samples. The conductivity at room temperature was found between 40 and 50 S/cm with a weak negative temperature coefficient consistent with a metal-like or semimetallic material, see Figure 5. Qualitatively, the temperature dependence is similar to that of Bi_2Te_3.[1c,9] The nature of the carriers was probed with thermoelectric power measurements as a function of temperature. We find it interesting that we can obtain both n-type and p-type samples of this material based on preparation method. N-type material forms from Te-rich cesium polytelluride flux, and p-type forms from potassium polytelluride flux. In Cs_2Te_x flux Te substitution of Bi atoms could account for the n-type nature of the material. Given the similar size of K^+ and Ba^{2+}, one may speculate that in K_2Te_x flux some substitution of Ba^{2+} by K^+ atoms renders the compound p-type by creating holes in the valence band. In both cases the room temperature Seebeck coefficient is quite large, ~-200 μV/K and ~+200 μV/K respectively, see Figure 5. These values decrease and tend towards zero at lower temperatures. The p-type samples exhibit a kink of unknown origin at ~215 K and an unusual sign reversal below 185 K. The sign reversal to negative values is smooth and continuous as the data cross through zero and suggests that electrons at low temperatures are either more numerous or more mobile than holes. This behavior is not observed in the n-type samples where the Seebeck coefficient decreases but remains negative.

Undoubtedly, the thermopower data reflect the substantial complexity of the electronic band structure at the Fermi level of this material, as it is suggested by the band structure calculations. Certainly, the band structure in Figure 4 is consistent with the simultaneous occurrence of holes and electrons as carriers. Despite the metal-like temperature dependence of the electrical conductivity and thermopower one cannot unequivocally conclude that $BaBiTe_3$ is a metal since the magnitude of the Seebeck coefficients is too large for a typical metal. The overall behavior is similar to the parent binary phase Bi_2Te_3, albeit the ternary derivative has a lower electrical conductivity. Therefore, at this stage neither the extended Hückel electronic structure calculations nor the charge transport measurements are sufficient to distinguish between a truly metallic behavior from a semimetallic or narrow gap semiconductor behavior. Optical absorption measurements however, are of considerable help in this regard. Diffuse reflectance spectroscopy on BaBiTe$_3$ at room temperature revealed the presence of absorptions at 0.28 eV and 0.42 eV which we assign to an energy band-gap, see Figure 6. The transition at 0.42 eV is assigned to excitations between other more widely spaced bands.

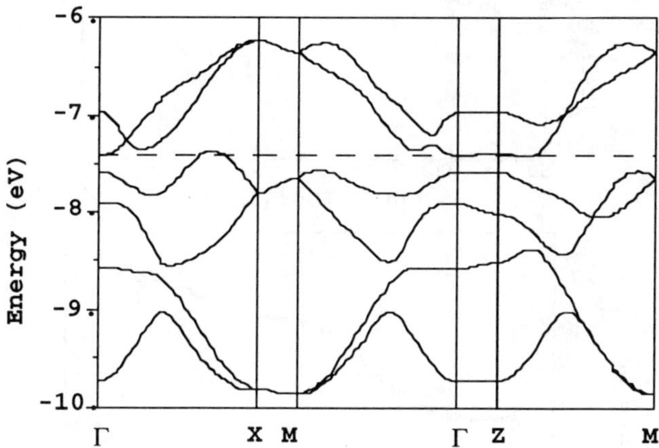

Figure 4. Calculated electronic band dispersion of a $[Bi_4Te_{12}]_n^{8n-}$ layer. The Fermi level is represented by the dashed line. $\Gamma=(0,0,0)$, $X=(1/2,0,0)$, $Z=(0,0,1/2)$, $M=(1/2,0,1/2)$.

We also performed self-consistent *ab-initio* band structure calculations within the TB-LMTO-ASA method.[10] All the atoms were taken into account. In contrast to the extended Hückel calculations, which are not suitable for the estimation of band gaps, the more rigorous TB-LMTO-ASA calculations suggest an opening of a small indirect gap (0.3 eV), although the magnitude of this gap is most likely incorrect. This model is more consistent with the measured transport properties. In the 70 - 300 K temperature range, the "exhaustion regime" might be reached (in a narrow gap semiconductor) explaining the metal-like properties with a low conductivity. According to the temperature, the different nature of the charge carriers could originate from the differential depopulation of the valence and conduction bands and from the greatly differing mobilities of the holes and electrons. At low temperatures hole localization may leave electrons as the only carriers and to n-type conductivity. That the materials can be doped both n- and p-type implies the occurrence of donor and acceptor levels in the gap. The acceptor levels may be based on a slight excess of Bi situated on Te sites, while the donor levels would suggest the occurrence of a slight excess of Te on Bi sites. Such a non-stoichiometry is common among heavy chalcogenides and should be probed further because it could permit the tuning of the properties of BaBiTe3 and the achievement of good thermoelectric performance.

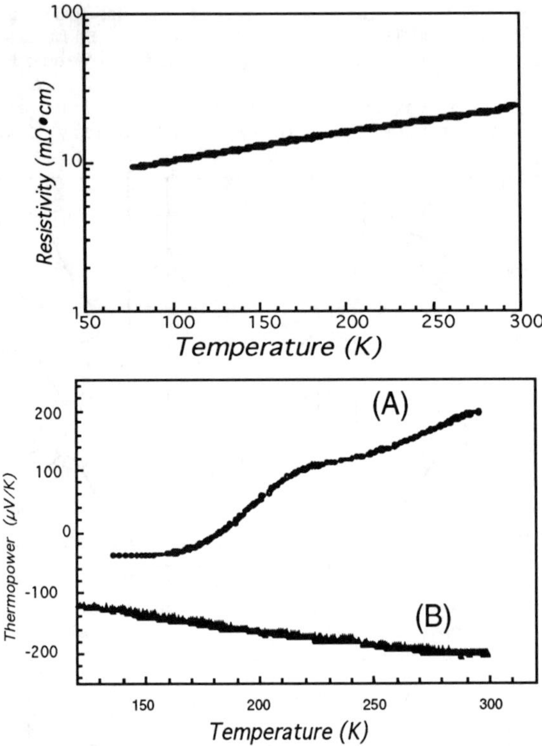

Figure 5. (Top) Variable temperature electrical resistivity data for single crystals of p-type BaBiTe3. (bottom) Variable temperature thermoelectric power data for (A) p-type and (B) n-type BaBiTe3.

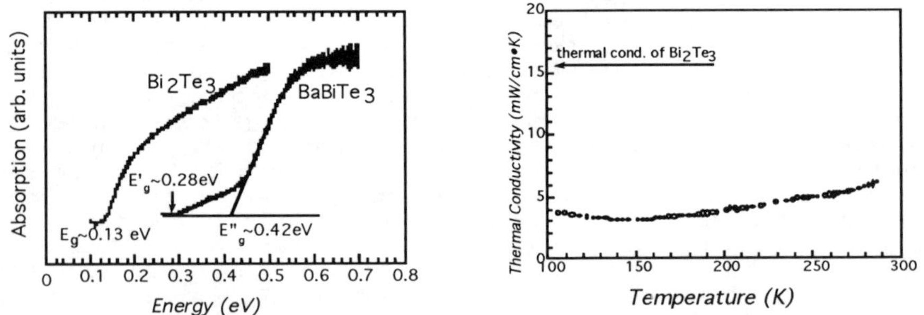

Figure 6. (Left) Infrared absorption spectra of BaBiTe3 and Bi2Te3. The semiconductor energy gaps are indicated in the spectra. (Right) Thermal conductivity data for p-type BaBiTe3.

Thermal conductivity.

Thermoelectric applications demand a very low thermal conductivity in the temperature range of interest. Therefore, the thermal conductivity of $BaBiTe_3$ was measured as a function of temperature, Figure 6. The data show a remarkably low thermal conductivity with respect to that of Bi_2Te_3 (κ_L ~1.6 W/m-K)[11]. Using the measured values of the electrical resistivity in conjunction with the Wiedemann-Franz law[12], we estimate the maximum possible value of the electronic thermal conductivity contribution to be below 1% of the total thermal conductivity of $BaBiTe_3$. Thus, essentially all heat in the compound is carried by lattice phonons. Hence, at least from the perspective of thermal transport, the $BaBiTe_3$ satisfies one of the key requirements for a useful thermoelectric material, and in fact, it possesses one of the lowest thermal conductivity values ever reported for a potential thermoelectric material. The very low thermal conductivity of $BaBiTe_3$ implies that the modification of the structure and composition of Bi_2Te_3 by the incorporation of $BaTe_x$ or A_2Te_x (A=alkali metal) into it is a correct approach in lowering its lattice thermal conductivity. If ways could be found to enhance the electrical conductivity and, at the same time, preserve or even increase the already large thermopower values a promising thermoelectric material might be obtained. To achieve this we need additional information regarding the transport properties including carrier concentrations and mobilities. Then optimization of these properties could be accomplished by controlling accurately the stoichiometry of these materials. This work is currently underway.

In summary, $BaBiTe_3$ is a narrow gap semiconductor with an indirect gap and a very low lattice thermal conductivity which makes it attractive for further studies as a thermoelectric material. This work suggests that this is possible by K doping in Ba sites (p-type) and Te doping of Bi sites (n-type). This could greatly increase the potential thermoelectric performance of this material. The complex electronic structure of $BaBiTe_3$ at the Fermi level is responsible for the large values of thermoelectric power, and suggests that large changes could be observed in the thermoelectric properties of this material with the application of a relatively small pressure.

Acknowledgments. Financial support from the Office of Naval Research (contract # N00014-94-1-0935, MGK and CRK) is gratefully acknowledged. At NU this work made use of Central Facilities supported by NSF through the Materials Research Center (DMR-91-20521). MGK thanks the "Pays de la Loire" region for support during a sabbatical stay in Nantes, 1996.

References

1. (a) *CRC Handbook of Thermoelectrics*, Rowe, D. M., Eds.; CRC press, Inc., 1995 (b) Kaibe, H.; Tanaka, Y.; Sakata, M.; Nishida, I. *J. Phys. Chem. Solids*, **1989**, *50*, 945-950 (c) Jeon, H. -H.; Ha, H. -P.; Hyun, D. -B.; Shim, J. -D. *J. Phys. Chem. Solids*, **1991**, *4*, 579-585
2. (a) Kanatzidis, M. G.; McCarthy, T. J.; Tanzer, T. A.; Chen, L. -H.; Iordanidis, L.; Hogan, T.; Kannewurf, C. R.; Uher, C.; Chen, B. *Chem. Mater.*, **1996**, *8*, 1465-1474. (b) McCarthy, T. J.; Ngeyi, S. -P; Liao, J. -H.; DeGroot, D. C.; Hogan, T.; Kannewurf, C. R.; Kanatzidis, M. G. *Chem. Mater.*, **1993**, *5*, 331-340. (c) McCarthy, T. J.; Tanzer, T. A.; Kanatzidis, M. G. *J. Amer. Chem. Soc.*, **1995**, *117*, 1294-1301.
3. (a) Testardi, L. R.; Bierly, J. N. Jr.; Donahoe, F. J. *J. Phys. Chem. Solids*, **1962**, *23*, 1209 (b) Champness, C. H.; Chiang, P. T.; Parekh, P. *Can. J. Phys.*, **1965**, *43*, 653-659 (c) Yim, W. M.; Fitzke, E. V. *J. Electrochem. Soc.*, **1968**, *115*, 556-560
4. Chung, D.-Y.; Jobic, S.; Hogan, T.; Kannewurf C. R.; Brec, R.; Rouxel, J.; Kanatzidis, M. G. submitted.
5. Volk, K. ; Cordier, G. ; Cook, R. ; Schäfer, H. *Z. Naturforsch.*, **1980**, *35b*, 136-140
6. (a) Cook, R.; Schäfer, H. *Studies in Inorganic Chemistry*, **1983**, *3*, 757-760. (b) Cook, R.; Schäfer, H. *Rev. Chim. Miner.*, **1982**, *19*, 19-27
7. Chung, D.-Y.; Kanatzidis, M. G. to be submitted.
8. (a) Canadell, E.; Jobic, S.; Brec, R.; Rouxel, J.; Whangbo, M. -H. *J. Solid State Chem.*, **1992**, *89*, 189-199 (b) Jobic, S.; Deniard, P.; Rouxel, J.; Jouanneaux, A.; Fitch, A. N. *Z. anorg. allg. Chem.*, **1991**, *598/599*, 199-215
9. Buist, R. J. *CRC Handbook of Thermoelectrics*, Rowe, D. M., Eds.; CRC press, Inc.: 1995, p143.
10. (a) Andersen, O. K.; Jepsen, O. *Phys. Rev. Lett.*, **1984**, *53*, 2571-2574 and references therein. (b) Krier, G.; Jepsen, O.; Burkhardt, A.; Andersen, O. K. *The TB-LMTO-ASA program*, version 45; Max-Planck-Institut Für Festkörperforschung: Heisenbergstr. 1, D-70569 Stuttgart, Germany
11. Encyclopedia of Materials Science and Engineering, *"Thermoelectric Semiconductors"*; Cambridge, Mass., MIT Press: Oxford, Pergamon Press: 1986, p4968
12. Kittel, C. *Introduction to Solid State Physics*, 6th Eds.; John Wiley & Sons, Inc.: 1986, p150

SYNTHESIS AND STRUCTURE OF NOVEL TERNARY MANGANESE TELLURIDES: $MMnTe_2$ (M = Li, Na), $Na_3Mn_4Te_6$, AND $Na_3Mn_{4.7}Te_6$

Chwanchin Wang*, Joonyeong Kim** and Timothy Hughbanks**
*Present address: NEC Research Institute, 4 Independence Way, Princeton NJ 08540
**Department of Chemistry, Texas A&M University, College Station, Texas 77843-3255

ABSTRACT

The synthesis and structure of four ternary manganese tellurides are reported. All the title compounds form layered structures wherein $MnTe_4$ tetrahedra are the fundamental building blocks. In two compounds $MMnTe_2$ (M = Li, Na), $MnTe_4$ tetrahedra share three corners in the formation of $[MnTe_{3/3}Te]^{1-}$ layers, the fourth Te is bound to only one Mn and formally bears a negative charge. $MnTe_4$ tetrahedra share both corners and edges in the formation of $[Mn_4Te_6]^{3-}$ and $[Mn_{4.7}Te_6]^{3-}$ layers in $Na_3Mn_4Te_6$ and $Na_3Mn_{4.7}Te_6$, respectively. Structural relationships between these ternary tellurides and oxides will be discussed.

INTRODUCTION

Low dimensional transition metal chalcogenides have been extensively studied because of their unusual physical properties, interesting structural chemistry, and many applications.[1-5] Until recently, investigation of the first-row ternary transition metal chalcogenides had mostly concentrated on the exploration of sulfides and selenides. These materials often show interesting magnetic and semiconducting properties and display rich structural chemistry.[6-8] Recently, the investigation of ternary M-Mn-Te (M = alkali metals) systems have produced two structure types: one (Na_6MnTe_4) with isolated $[MnTe_4]$ tetrahedra[9] and the other (A_2MnTe_2; A = K, Rb, Cs) with one-dimensional $^1_\infty[MnTe_{4/2}]$ chains.[10] During our investigation of quaternary M-Mn-Zr-Te systems (M = alkali metals), we were able to isolate four new layered compounds, $MMnTe_2$ (M = Li, Na), $Na_3Mn_4Te_6$ and $Na_3Mn_{4.7}Te_6$. Herein, we report the synthesis and structure of these new manganese tellurides.

EXPERIMENT

All operations were carried out under an inert-gas atmosphere. Elemental starting materials were Zr (99.2%, including 4.5 wt. % Hf, Johnson Matthey), Na (99.99%, Johnson Matthey), Mn (99.99%, Johnson Matthey), and Te (99.99%, Johnson Matthey). NaCl (> 99.0%, Aldrich) was sublimed at least twice before use. All the title compounds were synthesized by the use of Nb tubes that were in turn sealed in an evacuated (~ 10^{-4} Torr) silica tubes. $NaMnTe_2$ was originally prepared in a reaction designed to yield a quaternary Na-Mn-Zr-Te compound; a 0.3 gram mixture of Na, Mn, Zr, and Te at a ratio of 1:2:2:3 combined with NaCl (NaCl:Zr = 1:1) was finely ground in placed in a reaction vessel as described above. The temperature of the reaction vessel was raised to 550 °C over 12 hrs and held at 550 °C for 2 days. It was then ramped uniformly to 1000 °C over 2 days and maintained at 1000 °C for 21 days. The reaction vessel was then cooled at a rate of 5 °C/h to 300 °C and quenched to room temperature. Orange-red, plate-like crystals were observed as a very minor product. $LiMnTe_2$ was prepared by mixing elemental Li, Mn, and Te in stoichiometric proportions. The temperature of the reaction vessel was uniformly raised to 450 °C over 10 hrs, maintained at 450 °C for one day, uniformly increased to 750 °C over 10 hrs, held at 750 °C for 100 hrs, and finally cooled to room temperature at a rate of 2 °C/h. This yielded a mixture of $LiMnTe_2$ and Li_2Te as indicated by the powder diffraction pattern. The composition of this 2-phase mixture was established for all elements using Atomic Absorption spectroscopy. Microprobe analysis was used to independently determine Mn and Te composition. Various locations within samples subject to microprobe analysis showed either no detectable manganese (Li_2Te) or Mn:Te = 1:2 ($LiMnTe_2$). The lithium content (x) in Li_xMnTe_2 was determined to be 0.97 after the Mn:Te ratio was used to determine the phase ratio (Li_2Te:Li_xMnTe_2 = 40:60).

Mat. Res. Soc. Symp. Proc. Vol. 453 © 1997 Materials Research Society

Quantitative synthesis of $NaMnTe_2$ was attempted by loading a 1:1:2 ratio of Na:Mn:Te. The temperature of the reaction vessel was uniformly raised to 600 °C over 2 days, maintained at 600 °C for 3 days, then cooled to room temperature at a rate of 3 °C/h. Instead of $NaMnTe_2$, more than 90% reaction product was determined to be $Na_3Mn_4Te_6$. When the same reaction was attempted at 700 °C, under otherwise identical reaction conditions as described for the synthesis of $Na_3Mn_4Te_6$, $Na_3Mn_{4.7}Te_6$ was the major product. An independent synthesis of $Na_3Mn_{4.7}Te_6$ by mixing elemental Na, Mn, and Te in the proportion 3:4.7:6 resulted in the formation of $Na_3Mn_{4.7}Te_6$ in greater than 90% yield.

RESULTS

After the discovery of $NaMnTe_2$, attempts to synthesize it in silica tubes over a temperature range of 400 to 700°C resulted solely in the formation of Na_6MnTe_4. We therefore carried all subsequent reactions in Nb tubes. Attempts to find a rational synthesis of $NaMnTe_2$ by mixing elements in stoichiometric proportions led to the discovery of $Na_3Mn_4Te_6$ and $Na_3Mn_{4.7}Te_6$. When rational synthesis of $NaMnTe_2$ proved difficult, we found that $LiMnTe_2$ could be directly prepared by mixing elements in stoichiometric proportions. A mixture of Li_2Te and $LiMnTe_2$ were observed in the product. The composition of this mixture was determined by dual use of Atomic absorption spectroscopy and microprobe analysis and showed 40:60 molar mixture of Li_2Te and $LiMnTe_2$. Discrepancies between "loaded" compositions and those observed in products may reflect some involvement of Nb tubes in the reactions — perhaps by formation of binaries and or ternaries of Nb, Mn, and/or Te. More careful attention to possible reactions occuring in the tube walls is needed.

The structure of $NaMnTe_2$ is shown in Figure 1.[11] Fairly regular [MnTe4] tetrahedra are the building blocks of the structure. Each MnTe4 tetrahedron shares three of its Te atoms in one ab plane to form two dimensional $[MnTe_{3/3}Te]^{1-}$ layers. The most unusual feature of the structure is asymmetry of Te binding to manganese; half are 3-coordinate and half are terminal. This produces an unprecedented, polar layered structure. The Mn-Te$_{terminal}$ bond distance (2.768(5) Å) is found to be slightly shorter than other three Mn-Te bond distances (2.784(2) Å). These layers are then held together by Na atoms to complete the three dimensional structure. The Na atom is in a slightly irregular octahedral environment having Na-Te bond distances, $3 \times$ 3.33(1) Å and 3×3.21 (1) Å. The shortest Te-Te contact is 4.519 Å. It is notable that $NaMnTe_2$ is the only compound that contains exclusively Mn(III) among known manganese tellurides.

The structure of $NaMnTe_2$ can be viewed as an inverse-$ScCuS_2$-type structure.[6] In the former, Mn(III) occupies the tetrahedral site and Na(I) sits on the octahedral site, while Sc(III) occupies the octahedral site and Cu(I) sits on the tetrahedral site in the latter. A high pressure form of $AgAlS_2$ was the only example reported to be isostructural with $ScCuS_2$.[12] Another structural similarity can also be drawn between wurtzite and $NaMnTe_2$. If the sodium ions in $NaMnTe_2$ are moved from octahedral to the appropriate tetrahedral sites between the $[MnTe_{3/3}Te]^{1-}$ layers, a wurtzite derivative structure results. A preliminary structure determination of $LiMnTe_2$ shows that it contains same $[MnTe_2]^{1-}$ layered structure as the sodium compound. However, owing to the low scattering factor of Li atoms, it has uncertain whether $LiMnTe_2$ is isostructural with $NaMnTe_2$ or, perhaps, has a wurtzite-like structure.

Attempts to rationally synthesize $NaMnTe_2$ at 600 °C led to the discovery of a novel compound, $Na_3Mn_4Te_6$. The projection of $Na_3Mn_4Te_6$ structure approximately down the a axis is shown in Figure 2.[13] Again, fairly regular [MnTe4] tetrahedra are the building blocks of the structure. Mn-Te bond distances within the tetrahedron range between 2.793 and 2.806 Å – comparable to those of Na_6MnTe_4[9] and K_2MnTe_2.[10] The [MnTe4] tetrahedron shares three corners with three other tetrahedra in the formation of two dimensional network (Figure 3) which is reminiscent of the 3636 Kagomé net. The two-dimensional $[Mn_4Te_6]^{3-}$ layers propagate parallel to the ab plane and are composed of two such Kagomé networks by flipping one net by 180° and fusing two nets together by sharing edges of MnTe4 tetrahedra as shown in Figure 3. These layers are then held together by Na ions to complete the structure. There are two crystallographically unique Na ions, each of which is coordinated to six Te centers (Na-Te

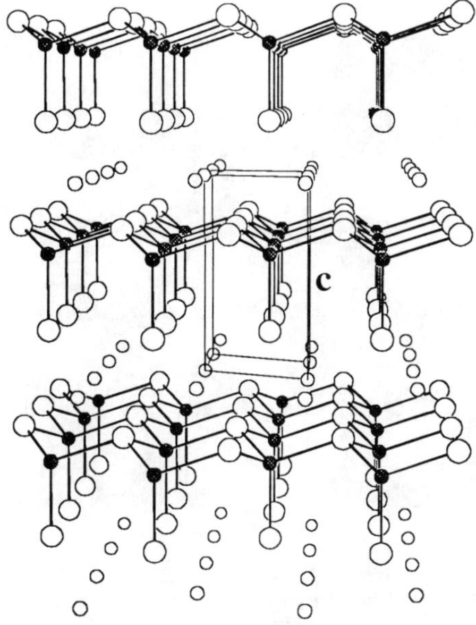

Figure 1. The structure of NaMnTe$_2$ viewed with the c-axis vertical. Mn atoms are shown as black circles, Te atoms as large open circles, and Na as small open circles.

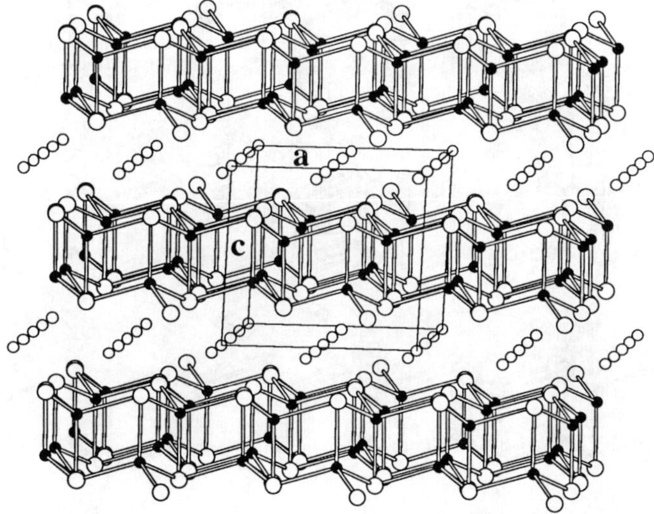

Figure 2. The structure of Na$_3$Mn$_4$Te$_6$ viewed with the c-axis vertical. Mn atoms are shown as black circles, Te atoms as large open circles, and Na as small open circles.

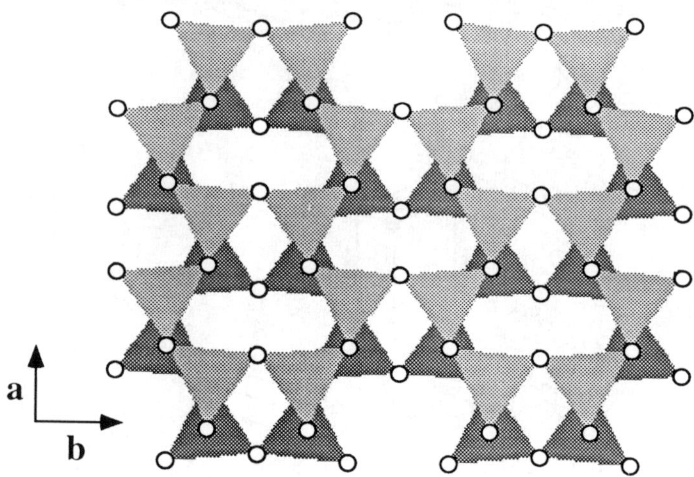

Figure 3. Polyhedron presentation of two-dimensional $[Mn_4Te_6]^{3-}$ layers viewed on the ab plane. $MnTe_4$ tetrahedra are shown and Te atoms are shown as open circles.

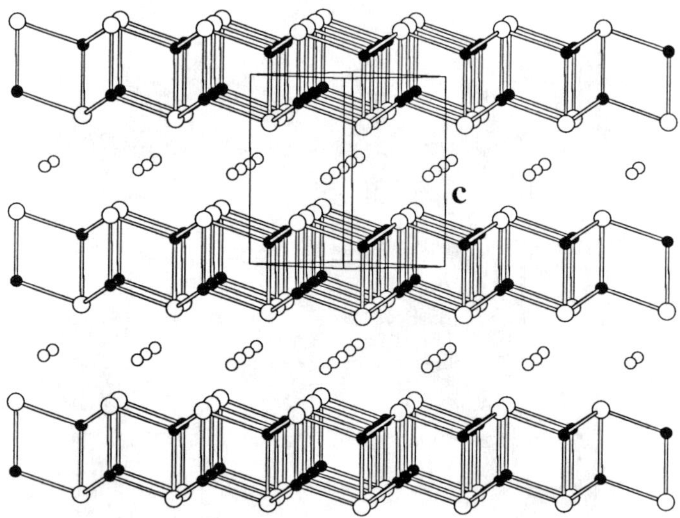

Figure 4. The structure of $Na_3Mn_{4.7}Te_6$ projected approximately along the c axis. Mn atoms are shown as black circles, Te as large open circles, and Na as small open circles.

distances range from 3.113 to 3.397 Å). $Na_3Mn_4Te_6$ is presumably a Mn(III)/Mn(II) mixed valence compound.

It is worthwhile noting that a similar Kagomé net is also observed in B_2O_3.[14] Unlike fusion of two nets found in $Na_3Mn_4Te_6$, the nets in B_2O_3 stack such that vertices of the BO_4 tetrahedra are shared and a three-dimensional network results. Nevertheless, the structures are quite similar, *when viewed in projection*. In Figure 3, it is apparent that there are four empty tetrahedral sites within each Mn_8Te_{12} repeat unit of the $[Mn_4Te_6]^{3-}$ layers. If these empty tetrahedral sites are filled up with Mn atoms, one obtains the hypothetical compound, $NaMn_2Te_2$ (in which all Mn would have a formal +1.5 oxidation state).

In an attempt to rationally synthesize $NaMnTe_2$ at 700 °C, we discovered another new compound, $Na_3Mn_{4.7}Te_6$. $Na_3Mn_{4.7}Te_6$ is found to adopt a defective La_2O_3 ($\equiv O_3La_6O_6$) structure type;[15] a projection of the structure is shown in Figure 4.[16] $[MnTe_4]$ tetrahedra are once again the building blocks of the structure (Mn-Te bond distances range between 2.749 and 2.808 Å). These tetrahedra share three vertices with tetrahedra in the same half of the layer (as in $NaMnTe_2$) and three edges with tetrahedra in the other half of the layer to form two dimensional $[Mn_{4.7}Te_6]^{1-}$ layers that are once again interspersed with Na ions to complete the three dimensional structure. There is one crystallographically unique Na atom in the structure which is exclusively coordinated to six Te atoms (Na-Te bond distances: $6 \times 3.255(2)$ Å). Mn atoms partially and randomly occupy 78% of the tetrahedral sites and have an average formal charge close to +2.

These new compounds, $MMnTe_2$ (M = Li, Na), $Na_3Mn_4Te_6$, and $Na_3Mn_{4.7}Te_6$, are all structurally related. As was also mentioned above, a complete filling of the tetrahedral sites with Mn ions within the double-layers of $Na_3Mn_4Te_6$ results in the composition "$NaMn_2Te_2$". This hypothetical compound can also be viewed as parent compound of $MMnTe_2$ (M = Li, Na) and $Na_3Mn_{4.7}Te_6$. Each layer in $NaMn_2Te_2$ has a [Te-Mn-Mn-Te] stacking sequence. If approximately one quarter of the Mn atoms are randomly removed from the $[Mn_2Te_2]$ layer, layered $Na_3Mn_{4.7}Te_6$ ($\equiv NaMn_{1.56}Te_2$) results. On the other hand, if 1/3 of the Mn atoms are removed in a specific ordered fashion to yield $[Te-Mn_{2/3}-Mn_{2/3}-Te]$, we obtain $Na_3Mn_4Te_6$ ($\equiv NaMn_{1.33}Te_2$). This is reminiscent of $Na_2Mn_2S_3$ in which 1/3 of the Mn atoms are removed with a different ordering (zigzag).[17] Finally, if all atoms from one Mn layer removed, the unusual polar $[MnTe_2]^{1-}$ layer is formed.

CONCLUSION

The layered compounds $MMnTe_2$ (M = Li, Na), $Na_3Mn_4Te_6$ and $Na_3Mn_{4.7}Te_6$ have been synthesized and characterized by the use of single-crystal and powder x-ray diffraction, microprobe analysis, and Atomic absorption spectroscopy. Layered $MMnTe_2$ (M = Li, Na) are of particular interest since they represent an unprecedented, polar layered structure. This may provide a unique opportunity of these compounds can be ion-exchanged and the intrinsically polar environment can be exploited. Other than their interesting structures, these compounds may display interesting magnetic properties since all three compounds exhibit different oxidation states, Mn(III) in $MMnTe_2$ (M = Li, Na), mixed valence Mn(III)/Mn(II) in $Na_3Mn_4Te_6$, and Mn(II) in $Na_3Mn_{4.7}Te_6$. Magnetic studies are currently underway.

ACKNOWLEDGMENTS

This research was generously supported by the National Science Foundation through grant DMR-9215890 and Texas Advanced Research program grant 010366-097 and was carried out at Texas A&M. The R3m/v single-crystal X-ray diffractometer and crystallographic computing system were purchased from funds provided by the National Science Foundation (CHE-8513273).

REFERENCES

1. F. Hulliger, in <u>Structural Chemistry of Layer-type Phases</u>, edited by F. Levy, <u>Physics and Chemistry of Materials with Layered Structures</u> (Reidel, Dordrecht, Holland, 1976), vol. 5.
2. J. Rouxel, R. Brec, Ann. Rev. Mater. Sci. **16**, 137 (1986).
3. J. Rouxel, <u>Crystal Chemistry and Properties of Materials with Quasi-one-dimensional Structures</u> (Reidel, Dordrecht, Holland, 1986).
4. R. Schöllhorn, in <u>Progress in Intercalation Research,</u> edited by W. Müller-Warmuth and R. Schöllhorn (Kluwer, Netherlands, 1994).
5. M. S. Whittingham and L. B. Ebert, <u>Applications of Intercalation Compounds</u> (Reidel, Dordrecht, Holland, 1979).
6. J. P. Dismukes, R. T. Smith, J. Phys. Chem. Solids **32**, 913 (1971).
7. W. Bronger, C. Herudek, J. Huster, D. Schmitz, Z Anorg. Allg. Chem **619**, 243 (1993).
8. W. Bronger, U. Ruschewitz, P. Müller, J. Alloys and Comp. **218**, 22 (1995).
9. W. Bronger, H. Balk-Hardtdegen, Z. Anorg. Allg. Chem. **574**, 89 (1989).
10. W. Bronger, H. Balk-Hardtdegen, D. Schmitz, Z. Anorg. Allg. Chem. **574**, 99 (1989).
11. Crystal data for $NaMnTe_2$: trigonal, space group P3m1 (No. 156), $a = 4.5630(6)$ Å, $c = 7.542(2)$ Å, $V = 135.99(6)$ Å^3, $Z = 1$, $D_{calc} = 4.068$ g cm^{-3}, $R = 0.034$, $R_w = 0.080$.
12. K. J. Range, G. Engert, A. Weiss, Z. Naturforsch. Teil B: Anorg. Chem., Org. Chem. **29B**, 186 (1974).
13. Crystal data for $Na_3Mn_4Te_6$: monoclinic, space group C2/m (No. 12), $a = 8.274(4)$ Å, $b = 14.083(6)$ Å, $c = 7.608(6)$ Å, $\beta = 91.97(4)°$, $V = 886.0(9)$ Å^3, $Z = 2$, $D_{calc} = 3.952$ g cm^{-3}, $R = 0.077$, $R_w = 0.1296$.
14. C. T. Prewitt, R. D. Shannon, Acta Crystallogr. Sect. B. **24B**, 869 (1968).
15. P. Villars and L. D. Calvert, <u>Pearson's Handbook of Crystallographic Data for Intermetallic Phases</u> (American Society for Metals, Ohio, ed. 2nd., 1991), vol. 1-4.
16. Crystal data for $NaMnTe_2$: trigonal, space group P$\bar{3}$m1 (No. 164), $a = 4.4973(8)$ Å, $c = 7.638(2)$ Å, $V = 133.77(5)$ Å^3, $Z = 1$, $D_{calc} = 4.517$ g cm^{-3}, $R = 0.044$, $R_w = 0.060$.
17. K. Klepp, P. Böttcher, W. Bronger, J. Solid State Chem., **47**, 301 (1983).

CRYSTAL GROWTH AT LOW TEMPERATURES - SOLVOTHERMAL SYNTHESIS AND CRYSTAL STRUCTURES OF TWO MERCURY TELLURIDES

JING LI *, ZHEN CHEN *, JAMAL L. KELLEY *, DAVIDE M. PROSERPIO **
*Department of Chemistry, Rutgers University, Camden, NJ 08102, jingli@crab.rutgers.edu
**Dipartimento di Chimica Strutturale e Stereochimica Inorganica, Universita' di Milano, 20133 Milano, Italy, davide@csmtbo.mi.cnr.it

ABSTRACT

Two mercury tellurium compounds $[Mn(en)_3]_2Hg_2Te_9$ and $\{[Mn(en)_3]_2Cl_2\}(Hg_2Te_4)$ have been synthesized at 180°C by solvothermal reactions using ethylenediamine as solvent. Single crystal X-ray diffraction analyses show that both compounds crystallize in monoclinic system, space group $P2_1/c$ (no. 14) with four formula units in the unit cell. The cell dimensions are: a = 12.019(1) Å, b = 28.731(4) Å, c = 18.267(4) Å, β = 98.04(1)°, V = 6246(2) Å^3 for $[Mn(en)_3]_2Hg_2Te_9$ and a = 15.720(3) Å, b = 16.812(3) Å, c = 15.224(3) Å, β = 114.31(1)°, V = 3668(1) Å^3 for $\{[Mn(en)_3]_2Cl_2\}(Hg_2Te_4)$. $[Mn(en)_3]_2Hg_2Te_9$ contains pseudo one-dimensional chains of weakly bound $(Hg_2Te_9)^{4-}$ Zintl anions. The structure of $\{[Mn(en)_3]_2Cl_2\}(Hg_2Te_4)$ contains one-dimensional chains of Zintl anion $^1_\infty[(Hg_2Te_4)^{2-}]$ formed by linking five-membered rings Hg_2Te_3 through bridging Te atoms. In both structures ethylenediamine incorporates into the manganese complex cation as bidentate ligand.

INTRODUCTION

During our recent efforts in synthesizing new solid-state chalcogenide materials, we have developed a solvothermal route in which single crystals are grown at temperatures below 200°C [1-5]. By utilizing an organic or inorganic solvent under pressure and at temperatures below its critical temperature, such a route enables the reaction to take place at a much lower temperature than what is typically used in a solid-state synthesis. This allows for the preparation of materials that may not be accessible by other higher temperature techniques. Several new tellurides containing chalcophilic metals (e.g. Hg[1,2], In[3], Sb[4], and Sn[5]) have been synthesized using ethylenediamine as solvent. The structures of these compounds vary from molecular tellurometalates [1, 5] to extended one- or two-dimensional framework [2-4]. Here we present the synthesis and structure characterization of two new mercury tellurides, $[Mn(en)_3]_2Hg_2Te_9$ and $\{[Mn(en)_3]_2Cl_2\}(Hg_2Te_4)$.

EXPERIMENT

Chemicals

Mercury (I) chloride (99.5%, Fisher Scientific), manganese (II) chloride (97%, Alfa AESAR), Te (99.8%, Aldrich Chemical Company) and binary alkali-metal precursors (A_2Te, A = Li, Rb) were used as starting materials. A_2Te were prepared by direct combination of stoichiometric amount of alkali-metal and Te in liquid ammonia. Ethylenediamine (99%, anhydrous, Fisher Scientific) was the solvent used in all reactions.

Mat. Res. Soc. Symp. Proc. Vol. 453 © 1997 Materials Research Society

Synthesis of [Mn(en)$_3$]$_2$Hg$_2$Te$_9$ (**I**)

Single crystals of **I** were grown from a solvothermal reaction containing the following reactants: 0.075 g (0.25 mmol) of Rb$_2$Te, 0.059 g (0.125 mmol) of Hg$_2$Cl$_2$, 0.032 g (0.25 mmol) of MnCl$_2$ and 0.096 g (0.75 mmol) of Te. The starting materials were weighed and mixed in a glove box under inert argon atmosphere. A thick-wall Pyrex tube (9 mm OD, ca. 5-6 inches long) was used as the reaction container. Approximately 5.5 mmol of ethylenediamine (en) was added to the mixture. The tube was then sealed under vacuum ($\sim 10^{-3}$ torr) after the liquid was condensed by liquid nitrogen. The sample was heated in an oven at 180°C for 7 days. After cooled to room temperature, the reaction product was washed with 35% and 95% ethanol respectively and dried with diethyl ether. Dark, columnar crystals of **I** were isolated. Microprobe analysis was performed on the selected crystals using Kevex EDAX (energy dispersive analysis by x-ray) on a Hitachi S-2400 scanning electron microscope and gave a Mg/Hg/Te ratio in agreement with the molecular formula. Powder x-ray diffraction analysis of the sample also indicated the existence of HgTe and Te.

Synthesis of {[Mn(en)$_3$]$_2$Cl$_2$}(Hg$_2$Te$_4$) (**II**)

Column-like crystals of **II** were isolated from a similar reaction using the same solvent. The starting materials included 0.052 g (0.25 mmol) of Li$_2$Te, 0.059 g (0.125 mmol) of Hg$_2$Cl$_2$, 0.032 g (0.25 mmol) of MnCl$_2$ and 0.096 g (0.75 mmol) of Te. The sample was prepared in the same way as described above for **I**. The same heating and isolation scheme was used. EDX analysis gave the approximate composition in agreement with the molecular formula. Powder x-ray diffraction analysis on the final product indicated a ca. 40% yield. Both HgTe and Te were also present in the mixture.

Crystal Structure Determination

A black crystal (0.11 x 0.10 x 0.06 mm) of **I** and a red crystal (0.19 x 0.06 x 0.06 mm) of **II** were mounted on glass fiber in air on an Enraf-Nonius CAD4 automated diffractometer, and 25 intense reflections [17°<2θ<23°] were centered using graphite-monochromated Mo-Kα radiation (0.71073Å). Least-squares refinement of their setting angles resulted in the unit-cell parameters reported in Table I, together with other details associated with data collection and refinement. Data were collected using the ω-scan method with a scan interval of 1.1°, within the limits 6°<2θ<40° (-11 ≤ h ≤ 11, 0 ≤ k ≤ 27, 0≤ l ≤ 17) for **I** and 6°<2θ<42° (0 ≤ h ≤ 15, -16 ≤ k ≤ 0, -15 ≤ l ≤ 13) for **II**. The diffracted intensities were corrected for Lorentz and polarization effects and decay (**I** 66% and **II** 26%). An empirical absorption correction based on ψ-scans was applied to all data (transmission range **I** 1.0-0.285 and **II** 1.0-0.526, respectively). The structure was solved by direct methods with SIR92 [6] and refined by full-matrix least-squares on F_o^2 using 5772 (**I**) and 3890 (**II**) significant [$I > 1\sigma(I)$] independent reflections. Anisotropic thermal displacements were assigned to all the heavy atoms (Hg, Mn, Te, Cl). The hydrogen atoms were located in ideal positions. The final difference electron density map showed no features with a height greater than 1.5 electron. For **I**, one of the two independent (Hg$_2$Te$_9$)$^{4-}$ anions was disordered over a center of symmetry, with each atom having an occupancy of 0.5 except for Te(15). The assignment of atom positions to the disordered anion or its inverted image was made with reference to the geometry of the ordered one. All calculations were performed using SHELX-93 [7]. Crystal structure drawings were produced with SCHAKAL [8]. The final atomic coordinates and isotropic equivalent displacement parameters are listed in Table II for the anions in both structures. Selected bond distances are reported in Table III.

Table I. Crystallographic Data for **I** and **II**.

$[Mn(en)_3]_2Hg_2Te_9$ (**I**)	fw 2028.08	monoclinic	$P2_1/c$ (No.14)
a = 12.019(1) Å	b = 28.731(2) Å	c = 18.267(2) Å	β = 98.04(1)°
V = 6246(2) Å^3	Z = 6	D_c = 3.222 g cm^{-3}	μ = 14.148 mm^{-1}
R^a [$I > 1\sigma(I)$]	R1 0.041	wR2^b 0.095	
$\{[Mn(en)_3]_2Cl_2\}\{(Hg_2Te_4)\}$ (**II**)	fw 1240.92	monoclinic	$P2_1/c$ (No.14)
a = 15.720(3) Å	b = 16.812(3) Å	c = 15.224(3) Å	β = 114.31(1)°
V = 3668(1) Å^3	Z = 4	D_c = 2.632 g cm^{-3}	μ = 12.310 mm^{-1}
R^a [$I > 1\sigma(I)$]	R1 0.041	wR2^c 0.087	

^{a} R1 = $\Sigma\,||F_o| - |F_c||\,/\,\Sigma\,|F_o|$, ^{c} wR2 = $[\Sigma\,w\,(F_o^2 - F_c^2)^2\,/\,\Sigma\,wF_o^4\,]^{1/2}$
^{b} Weighting: $w = 1/[\,\sigma^2(F_o^2) + (0.0592P)^2 + 26.5036P]$, where $P = (F_o^2 + 2F_c^2)/3$
^{c} $w = 1/[\,\sigma^2(F_o^2) + (0.0438P)^2 + 38.7536P]$

RESULTS

The structure of **I** contains polymeric chains of weakly bound $(Hg_2Te_9)^{4-}$ anions separated by the $[Mn(en)_3]^{2+}$ cations. Two perpendicular views of the chain along the b axis are shown in Figure 1 (top and middle). The anionic structure of **II** is similar to that of $(Hg_2Te_9)^{4-}$ found in $[Fe(en)_3]_2Hg_2Te_9$ [1] (Figure 1, bottom). Both, formally written as $[(Hg^{2+})_2(Te^{2-})(Te_2^{2-})(Te_3^{2-})_2]$, consists of a five-membered ring Hg_2Te_3 and two bent (Te_3^{2-}) that bond to the two Hg atoms in the 1 and 3 positions of the ring (see Figure 1). The coordination of the Hg atoms are trigonal planar in both structures. The Te···Te interatomic distances between the two neighboring $(Hg_2Te_9)^{4-}$ ions are also similar, 3.384(14)Å, 3.523(3)Å in **I** and 3.488(2)Å in $[Fe(en)_3]_2Hg_2Te_9$. Both are considerably shorter than the sum of the van der Waals radii of 4.12Å [9], indicating a weak secondary interaction [10]. The main difference between **I** and $[Fe(en)_3]_2Hg_2Te_9$ is that one third of the $(Hg_2Te_9)^{4-}$ unit in **I** have the two $(Te_3)^{2-}$ arms in the "trans" position, whereas all $(Te_3)^{2-}$ pairs are in "cis" position in the $[Fe(en)_3]_2Hg_2Te_9$ structure. The Hg-Te and Te-Te distances in **I** are comparable with those found in $[Fe(en)_3]_2Hg_2Te_9$, as well as in other mercury telluride compounds $(Hg_4Te_{12})^{4-}$ [11-12], $(Hg_2Te_5)^{2-}$ [11], $(Hg_3Te_7)^{4-}$ [13], and $(Hg_2Te_4)^{2-}$ [13]. All contain a common structure motif of five-membered Hg_2Te_3 ring. The two independent cations $[Mn(en)_3]^{2+}$ exist as one conformer *lel$_3$* (or $\Delta\lambda\lambda\lambda$ and its enantiomer) [14], and the crystal structure of manganese tris-ethylenediamine has never been reported previously.

Compound **II** contains one-dimensional polymeric chains of anionic $_\infty^1[(Hg_2Te_4)^{2-}]$ (Figure 2) running along c and cationic coordination complex $\{[Mn(en)_3]_2Cl_2\}^{2+}$. The chain is built up by connecting Hg_2Te_3 five-membered rings through bridging Te(4) and is isostructural to $(Et_4N)_2Hg_2Te_4$ [13]. The latter was isolated in moderate yield via the extraction of the intermetallic material with nominal composition $K_2Hg_2Te_3$ prepared from the melt. The Hg coordination in this compound is again trigonal planar. The Hg-Te and Te-Te bond distances are

Table II. Atomic coordinates and s.o.f. for the anions **I** (top) and **II** (bottom). U(eq) is defined as one third of the trace of the orthogonalized U_{ij} tensor.

Atom	x	y	z	U(eq)	s.o.f.
Hg(1)	0.42702(8)	0.23251(4)	0.04958(5)	0.0420(3)	1
Hg(2)	0.42757(8)	0.09795(4)	0.04072(5)	0.0415(3)	1
Te(1)	0.27527(13)	0.16576(7)	0.00415(9)	0.0457(5)	1
Te(2)	0.65286(14)	0.21361(6)	0.06939(11)	0.0537(5)	1
Te(3)	0.65391(13)	0.11850(6)	0.06730(10)	0.0519(5)	1
Te(4)	0.38554(13)	0.32165(6)	0.09641(9)	0.0401(5)	1
Te(5)	0.16839(13)	0.31904(6)	0.13122(9)	0.0426(5)	1
Te(6)	0.18433(13)	0.32504(6)	0.28202(9)	0.0379(5)	1
Te(7)	0.39748(13)	0.00416(6)	0.05982(9)	0.0379(5)	1
Te(8)	0.18065(13)	-0.00238(6)	0.09390(9)	0.0423(5)	1
Te(9)	0.19744(13)	-0.00300(6)	0.24499(9)	0.0389(5)	1
Hg(3A)	0.5685(3)	0.0609(2)	0.5048(3)	0.0446(12)	0.50
Hg(3B)	0.5206(3)	0.0705(2)	0.4913(3)	0.0433(11)	0.50
Te(10)	0.3263(3)	0.0207(2)	0.4673(2)	0.0458(10)	0.50
Te(11)	0.3340(3)	0.06317(14)	0.4918(2)	0.0440(10)	0.50
Te(12)	0.7196(3)	0.02674(14)	0.5439(2)	0.0530(10)	0.50
Te(1B)	0.5641(7)	0.1630(3)	0.4707(9)	0.048(2)	0.50
Te(1A)	0.6039(6)	0.1536(3)	0.4788(8)	0.042(2)	0.50
Te(2A)	0.8198(9)	0.1565(10)	0.4424(6)	0.047(3)	0.50
Te(2B)	0.7827(9)	0.1576(10)	0.4405(6)	0.048(3)	0.50
Te(15)	0.77920(14)	0.15762(7)	0.29064(9)	0.0489(5)	1
Hg(1)	0.75168(6)	0.17132(6)	0.64550(6)	0.0489(3)	1
Hg(2)	0.76436(6)	0.29480(5)	0.42797(6)	0.0513(3)	1
Te(1)	0.74142(12)	0.32227(10)	0.59091(11)	0.0686(5)	1
Te(2)	0.76682(10)	0.05581(9)	0.51961(10)	0.0511(4)	1
Te(3)	0.78742(10)	0.14186(9)	0.37591(10)	0.0504(4)	1
Te(4)	0.76167(11)	0.40928(8)	0.30214(10)	0.0510(4)	1

comparable with those reported for $(Et_4N)_2Hg_2Te_4$. The two independent cations $[Mn(en)_3]^{2+}$ exist as two different conformers *lel₃* and *lel₂ob* (or $\Delta\lambda\lambda\lambda$ and $\Delta\lambda\lambda\delta$ and theirs enantiomers) [14]. The two cations form a quasi one-dimensional chain along **c** by hydrogen bonds between chlorine and some N-H (Cl···N 3.34-3.54 Å). This is the first example of a crystal structure of $[M(en)_3]^{n+}$ complexes containing two different conformers.

Hg(1)-Te(1)	2.696(2)	Hg(1)-Te(2)	2.742(2)	Hg(1)-Te(4)	2.767(2)
Hg(2)-Te(1)	2.693(2)	Hg(2)-Te(7)	2.748(2)	Hg(2)-Te(3)	2.760(2)
Hg(3A)-Te(10)	2.680(6)	Hg(3A)-Te(1A)	2.749(9)	Hg(3A)-Te(11)	2.795(5)
Hg(3B)-Te(10)	2.722(6)	Hg(3B)-Te(1B)	2.746(10)	Hg(3B)-Te(12)	2.753(5)
Te(2)-Te(3)	2.733(3)	Te(11)-Te(12)	2.720(6)	Te(4)-Te(5)	2.772(2)
Te(7)-Te(8)	2.769(2)	Te(1B)-Te(2B)	2.764(10)	Te(1A)-Te(2A)	2.770(10)
Te(5)-Te(6)	2.740(2)	Te(8)-Te(9)	2.739(2)	Te(2A)-Te(15)	2.746(11)
Te(2B)-Te(15)	2.731(10)	Te(4)-Te(1B)	3.384(14)	Te(4)-Te(1A)	3.685(12)
Te(7)-Te(7)	3.523(3)	Mn(1)-N (ave.)	2.26(2)	Mn(2)-N (ave.)	2.28(2)
Hg(1)-Te(1)	2.655(2)	Hg(1)-Te(4)	2.691(2)	Hg(1)-Te(2)	2.806(2)
Hg(2)-Te(1)	2.689(2)	Hg(2)-Te(4)	2.704(2)	Hg(2)-Te(3)	2.758(2)
Te(2)-Te(3)	2.751(2)	Mn(1)-N (ave.)	2.27(2)	Mn(2)-N (ave.)	2.28(3)

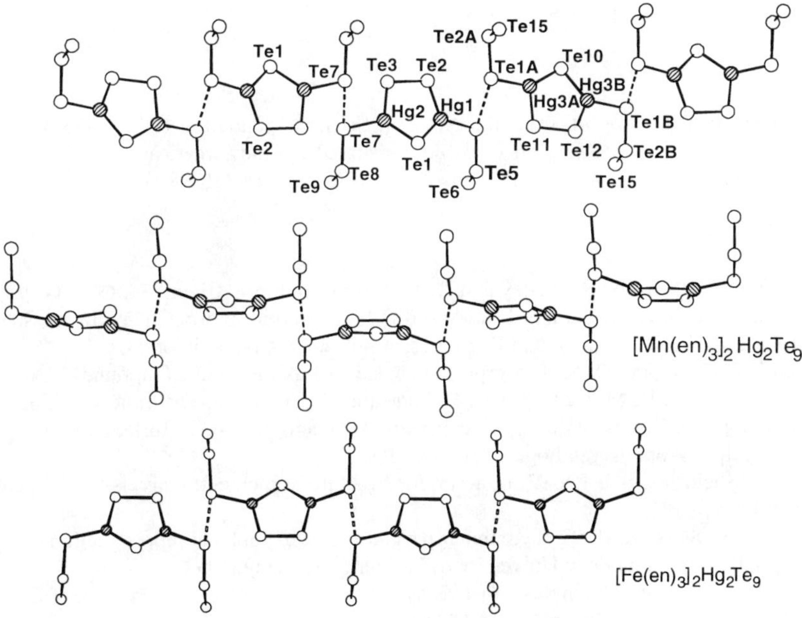

Figure 1. Two perpendicular views of the polymeric chain $(Hg_2Te_9)^{4-}$ in $[Mn(en)_3]_2Hg_2Te_9$ (**I**) (top and middle, only one ordered model is shown), and in $[Fe(en)_3]_2Hg_2Te_9$ (bottom). The cross-hatched circles are Hg atoms and the open circles Te.

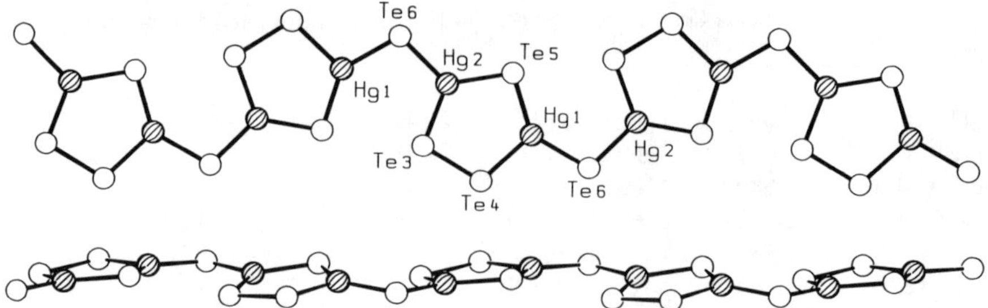

Figure 2. Two perpendicular views of the polymeric chain $^1_\infty[(Hg_2Te_4)^{2-}]$ in $\{[Mn(en)_3]_2Cl_2\}(Hg_2Te_4)$ (**II**). The cross-hatched circles are Hg atoms and the open circles Te.

CONCLUSIONS

Two new mercury telluride compounds have been synthesized via unconventional low temperature method. Single crystal X-ray diffraction analysis shows both structures contain Hg_2Te_3 five-membered rings that form 1-D chains through weak Te$\cdots$Te interactions (as in **I**) or bridging Te (as in **II**). The cation $[Mn(en)_3]^{2+}$ has been structurally characterized for the first time, in two different conformers.

ACKNOWLEDGMENTS

The financial support from the National Science Foundation (DMR-9553066) and the Donors of the Petroleum Research Fund administrated by the American Chemical Society are greatly appreciated. JL is a Henry Dreyfus Teacher-Scholar 1994-1998.

REFERENCES

1.   J. Li, B. G. Rafferty, S. Mulley and D. M. Proserpio, Inorg. Chem. **34**, 6417 (1995).
2.   J. Li, Z. Chen, K.-C. Lam, S. Mulley and D.M. Proserpio, Inorg. Chem., in press.
3.   J. Li, Z. Chen, T. J. Emge and D. M. Proserpio, Inorg. Chem., in press.
4.   J. Li, Z. Chen, and D. M. Proserpio, to appear in J. Alloys and Compounds.
5.   J. Li, Z. Chen, D. M. Proserpio, T. J. Emge and Y. Tan, to appear in Inorg. Chim. Acta.
6.   A. Altomare, G. Cascarano, C. Giacovazzo, A. Guagliardi, M.C. Burla, G. Polidori, and M. Camalli, J. Appl. Crystallogr. **27**, 435 (1994).
7.   G. M. Sheldrick, SHELX-93: program for structure refinement: University of Goettingen: Germany, 1994.
8.   E. Keller, SCHAKAL 92: a computer program for the graphical representation of crystallographic models: University of Freiburg: Germany, 1992.
9.   A. Bondi, J. Phys. Chem. **68**, 441 (1964).
10. N. W. Alcock, Adv. Inorg. Chem. **15**, 1 (1972).
11. R. C. Haushalter, Angew. Chem., Int. Ed. Engl. **24**, 433 (1985).
12. K. -W. Kim and M. G. Kanatzidis, Inorg. Chim. Acta **224**, 163 (1994).
13. S. S. Dhingra, C. J. Warren, R. C. Haushalter and A. B. Bocarsly, Chem. Mater. **6**, 2382 (1994).
14. Y. Saito, <u>Inorganic Molecular Dissimmetry</u>, Springer-Verlag, Berlin, 1992, pp. 56-59.

THE SYNTHESIS OF TRANSITION METAL SULFIDES AND SULFO-SALT CRYSTALS IN HYDROTHERMAL BRINES

J. W. KOLIS, M. B. KORZENSKI
Department of Chemistry, Clemson University, Clemson SC 29634
KJOSEPH@Clemson.edu

ABSTRACT

A new procedure for the synthesis of transition metal sulfosalts in hydrothermal brines is presented. Large, well formed crystals of metal antimony sulfides can be prepared under appropriate conditions in supercritical brines. A new compound prepared by this method, $MnSb_2S_4$, is structurally characterized.

INTRODUCTION

In the last several years we have been exploring the use of supercritical fluids as a route to solid state compounds [1,2]. We are particularly interested in the synthesis of metal chalcogenides and their derivatives. We found that supercritical amines such as ammonia or ethylenediamine are excellent media for the preparation of alkali metal chalcogenide complexes of coinage metals like $Cs_2Ag_7S_4$ [3] and $K_2Ag_{12}Se_7$ [4]. This reaction type was readily extended to the more complex quaternary species containing group 15 elements as well, leading to a great many complexes with the general formula $A_wM_xQ_yE_z$ where A is an alkali metal, M is a coinage metal usually Ag or Cu, Q is P, As or Sb, and E is S, Se or Te. In virtually all cases, the coinage metal is monovalent and the main group framework is valence precise. Thus the compounds are invariably closed shell semiconductors with bandgaps between 1.8 and 2.8 eV. In an attempt to form more electronically and magnetically interesting metal chalcogenide solids in supercritical amines, open shelled higher valent transition metals such as Fe(II), Yb(III) and Mn(II) were added in place of the monovalent alkali and coinage metals. However, a rather severe and obvious limitation crops up immediately in that the amine solvent coordinates strongly to the di- or trivalent metal center. This effectively sequesters the metal center from the main group framework, leading to complexes with formulas like $[Mn(NH_3)_6][Sb_6S_{10}]$ and $[Yb(NH_3)_8][Ag(S_4)_2]$. The metal complex acts as a templating counterion around which the main group framework forms, leading to a classical Werner complex encapsulated within a Zintl-type structure [5]. The overall structure is often quite complex and the framework is usually quite open to accommodate the large cationic coordination compound.

It is clear that the ideal supercritical solvent will be sufficiently polar to solubilize the anions, but a somewhat less basic ligand than ammonia, so as not to sequester the metal cations. An obvious solvent which meets these criteria is water. Initially it may appear to have a potential drawback in that it may hydrolyze the main group chalcogenide anions, leading to formation of metal oxides. However, this is not generally the case, and none of the main group sulfides or selenides that we investigated underwent appreciable hydrolysis in supercritical water. However, we did find that the metals do not react with main group solids in pure supercritical water. Thus we turned our attention to geothermal brines.

Geochemists have long known that a wide variety of metal sulfides and oxides can be transported throughout the Earth's crust by hydrothermal fluids [6]. Detailed investigation by a number of workers has revealed that these fluids are superheated brines with high concentrations of alkali and alkaline earth salts of halides, carbonates, hydroxide and sulfides [7]. These brines serve as mineralizers to solubilize a wide variety of metal sulfides to a surprising extent. For example galena, PbS, can be dissolved to nearly 1000 ppm in concentrated chloride brines at 350°C via the following equilibria [8].

$$PbS + Cl^- \text{-------> } PbCl^+ + S^{2-}$$

Mat. Res. Soc. Symp. Proc. Vol. 453 © 1997 Materials Research Society

It is postulated that concentration in excess of 100 ppm of metal cations are sufficient to mobilize metals ions sufficiently to achieve transport and crystal growth [9]. With recent technological advances, it is now possible to simulate the conditions of natural geothermal brines in the laboratory [10]. Thus we took advantage of the insights obtained by geochemists to mobilize normally intractable solids, and use these brines as solvents to prepare large single crystals of both known and new metal sulfosalts. As a test case, we have selected the reactions of stibnite Sb_2S_3 with various transition metals. In this paper we present several convenient preparations of several new and known metal antimony sulfides in hydrothermal brines, and relate these methods to the naturally occurring processes in geothermal brines. In many cases, these preparations are substantial improvements over any existing synthetic methods.

EXPERIMENTAL

Reagents. Chemicals in this work were used as purchased from traditional vendors and were of the highest available commercial grade. Water was distilled from potassium permanganate and potassium hydroxide. All solids were manipulated under an argon atmosphere in a Vacuum Atmosphere Dri-Lab glovebox. A detailed synthesis is given below. All other syntheses were performed under nearly identical conditions using stoichiometries given in Table 1.

Berthierite, $FeSb_2S_4$ (1). An amount of 0.0160 g (0.286 mmol) of Fe, 0.0970 g (0.286 mmol) Sb_2S_3, and 0.25 g of NaCl were added to a fused quartz ampoule, and 0.6 mL of distilled water was then added. The ampoule was subsequently flame-sealed in vacuo ($\approx 10^{-3}$ Torr). The ampoules were then placed in a high pressure autoclave and counter-pressured with 3000 psi of argon using an Air Driven Gas Compressor by Autoclave Engineers. The autoclave was then heated in a tube furnace for 3 days at 375°C. The products were filtered and washed repeatedly with distilled water and acetone to give approximately 90% yield of the black metallic-looking needles.

X-RAY Crystallography A suitable single crystal was sealed in epoxy and mounted on a Rigaku AFC7R diffractometer. Data was collected at room temperature and the structure was solved in routine fashion using the teXsan suite of programs [11]. Data was corrected for absorption using psi scans and all atoms were refined anisotropically. Crystallographic Data for $MnSb_2S_4$: formula weight, 426.68; crystal system, orthorhombic; space group, Pnma; a = 11.456, b = 3.819, c = 14.355, V = 628.1Å^3; Z=4; ρ= 4.512 g/cm^3; μ = 0.608; λ = 0.71073; 2θ range = 3.5-55°; T (K) = 295; R= 2.91%, R_w= 2.93%; total reflns = 1227; obsd. reflns. = 668; parameters = 43.

RESULTS AND DISCUSSION

Stibnite reacts well with a number of transition metals in hydrothermal brines to form well formed single crystal of metal antimony sulfides. The type and concentration of mineralizer definitely plays an important role in the reactivity of the solids. In most cases the reactants are unchanged in pure water heated to hydrothermal temperatures. However upon addition of sodium chloride or sodium carbonate, smooth reaction occurred in many cases quantitatively, with formation of very high quality crystals in sizes often exceeding several millimeters. We envision the role of the small anion is to attack the stibnite, displacing sulfide, which then attacks more stibnite, creating a polynuclear main group anion [12]. The resultant anion can attack a metal cation.

$$Sb_2S_3 + xs\ Cl^- \longrightarrow SbCl_4^- + S^{2-}$$

$$Sb_2S_3 + S^{2-} \longrightarrow Sb_2S_4^{2-}$$

$$Sb_2S_4^{2-} + M^{2+} \longrightarrow MSb_2S_4 \quad (M = Fe,\ Mn)$$

Clearly the reactions are more complex than these above, but they serve to illustrate the role of the halide in generating and transporting the main group anions. These reactions all have precedence in the geochemical literature, and are frequently invoked to account for the transportation and deposition of metal sulfide minerals. The metal cations are either produced by oxidation of the metal by superheated water, or by attack of mineralizer on any metal sulfide produced in solution. We have determined that superheated water will oxidize active metals such as iron to metal oxides or hydroxides.

Table I. Reactants, products and conditions for synthesis of typical metal sulfosalts in hydrothermal brines.

Charge	Mineralizer	Rxn Conditions	Products	% Yield
Fe, Sb_2S_3	NaCl or Na_2CO_3	375°C, 3300 psi	$FeSb_2S_4$	>90%
FeS, Sb	" "	" "	" "	>90%
Fe, Sb, S	" "	" "	$FeSb_2S_4$, FeS	30%, 50%
Mn, Sb_2S_3	" "	" "	$MnSb_2S_4$	>90%
MnS, Sb	" "	" "	" "	>90%
Mn, Sb, S	" "	" "	$MnSb_2S_4$, MnS	30%, 50%
12 Cu, 2 Sb_2S_3, 7 S	NaCl	375°C, 3300 psi	$Cu_{12}Sb_4S_{13}$	100%
3 Cu, Sb, 4 S	"	" "	Cu_3SbS_4	100%
3 Cu, Sb, 3 S	"	" "	Cu_3SbS_3	>90%
Cu, Sb, 2 S	"	" "	$CuSbS_2$	>90%
5 Ag, Sb, 4 S	NaCl	375°C, 3300 psi	Ag_5SbS_4	>90%
3 Ag, Sb, 3 S	"	" "	Ag_3SbS_3	100%
Ag, Sb, 2 S	"	" "	$AgSbS_2$	>90%

In the case of the silver and copper antimony sulfides, all of the species described are known to occur naturally as minerals. Note that their identity can be controlled to a very high degree by the original stoichiometry of the starting materials. It is our experience that Ag and Cu are generally more reactive toward the various sulfides than the earlier transition elements. In the case of Mn and Fe, they form only one metal antimony sulfide, but it can be formed from a variety of routes. Stibnite reacts with the metal powder to form the sulfosalt in quantitative yield, while antimony metal reacts with the metal sulfide to form the same product. The reaction of the pure elements is somewhat less efficient in that the product is formed in low yield, and is apparently in competition with the formation of the metal sulfide as well.

In virtually all cases the quality of the product was high, with very well formed crystals being produced in excellent yield (Table I). It is well known that supercritical fluids have some unique properties particularly low viscosity, which generally enhance crystal growth. These properties lead to vastly improved routes to these metal sulfosalts. For example the best procedure reported in the literature for tetrahedrite $Cu_{12}Sb_4S_{13}$ calls for heating the elements at 400°C for 180 days [13]. Preparations of the other compounds listed here are invariably impure when made by traditional melt techniques.

In several cases, full crystal structures were performed on single crystals where only poor structures were reported in the literature previously. In all cases, the composition of the bulk of the sample was confirmed by powder diffraction. The iron compound $FeSb_2S_4$ has been observed as a naturally occurring mineral called Berthierite, and its structure [14] and magnetic properties [15] have been reported previously. However we found that manganese reacts similarly to iron and forms a new compound $MnSb_2S_4$, which was characterized by single crystal x-ray diffraction.

$MnSb_2S_4$ crystallizes in the orthorhombic space group Pnma and is isostructural with the known mineral Berthierite (Figure 1).

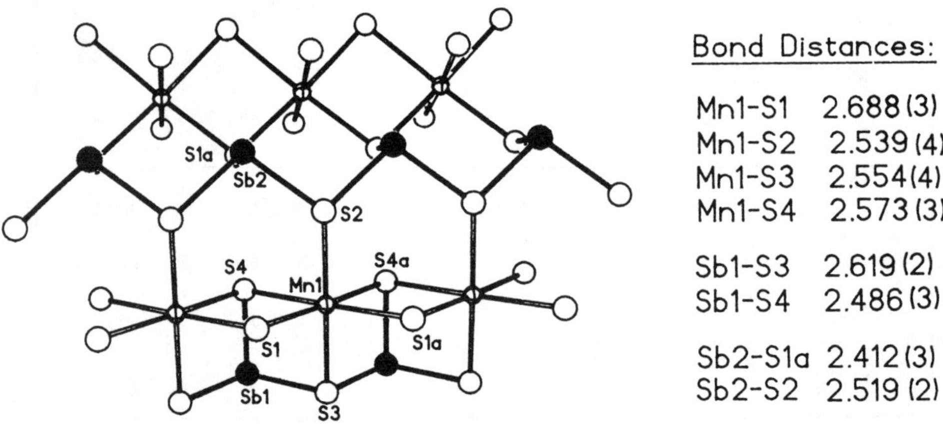

Figure 1. View of a fragment of MnSb2S4, showing the bridging (SbS$_2$)$_\infty^-$ chains and face sharing metal sulfide octahedra. The open spheres are S atoms, crossed spheres are Mn atoms, and filled spheres are Sb atoms.

The two dimensional layered structure is made up of seven unique atoms, one Mn, two Sb and four S atoms, all of which lie in mirror planes. The manganese atoms coordinated by six sulfur atoms in a slightly distorted octahedral coordination environment with Mn-S distances ranging from 2.539(4) to 2.688(3), and angles no smaller than 86.8(1)°. The Mn-S distances are substantially longer than Mn-S distances in coordination compounds, which are typically less than 2.43 Å [16]. The distances are comparable to the distances in MnS_2, which is abnormally long relative to other first row pyrites [17]. These long M-S distances have also been observed in the Berthierite analog. The octahedra are trans edge-shared by the terminal sulfides (S1 and S4) of $(SbS_2)_\infty^-$ chains. The bridging sulfides (S3) of the $(SbS_2)_\infty^-$ chains are also bound to Mn atoms of adjoining chains to create a complex two dimensional corrugated layered structure (Figure 2). All Sb-S distances within the individual layers range from 2.486(3) to 2.619(2)Å, which are comparable to those found in sulfosalt systems. Due to the stereochemically active lone pair on each pyramidal Sb(III), long Sb···S interactions of 2.889(4) Å are observed between each antimony and sulfur atoms in neighboring layers. This gives rise to a set of channels down the b axis of 28 membered rings with alternating Mn-S-Sb-S-Mn sequences.

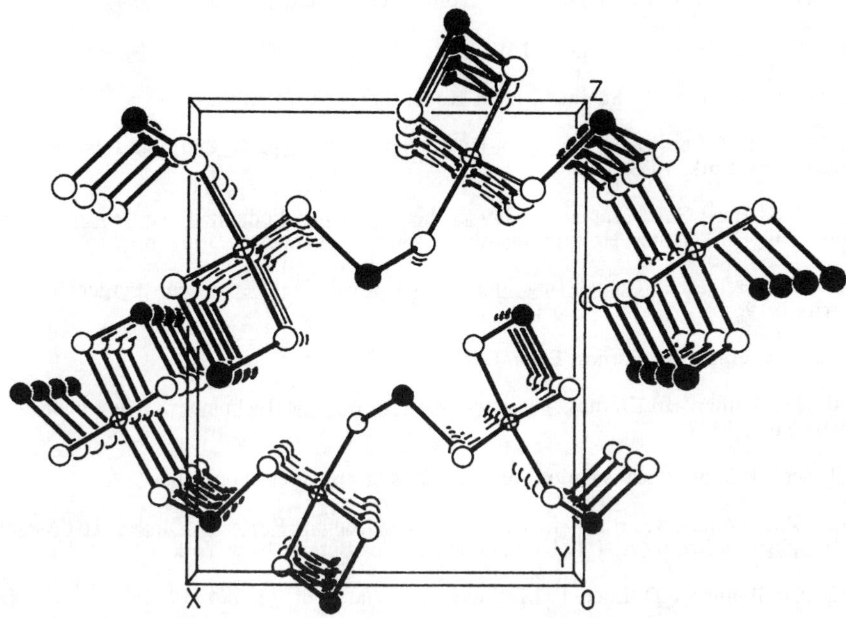

Figure 2. View of the unit cell of $MnSb_2S_4$ down the y axis showing the corrugated layers which interact with each other via long Sb···S interactions.

CONCLUSIONS

A number of metal sulfosalts can be prepared easily in hydrothermal brines. The presence of mineralizer is crucial in these preparations. Preliminary work on other systems shows that the identity of the mineralizer has a profound effect on the nature and composition of the products. Although many of these compounds have been reported, they are known only as minerals or their preparations are inconvenient. The compounds generally form as large well formed single crystals, which is significant in that several of these compounds (e.g. pyrargerite, Ag_3SbS_3) have properties of interest to materials scientists. To demonstrate the utility of the method a new compound $MnSb_2S_4$ has been prepared and structurally characterized.

Acknowledgments: We are indebted to the National Science Foundation (CHE-9102548) and to the Petroleum Research Fund, administered by the ACS, for support of this work, and to George Schimek for help with the crystallography.

REFERENCES

1. G.L. Schimek, W.T. Pennington, P. T. Wood, J.W. Kolis, J Solid State Chemistry 123, 277 (1996).

2. P.T.Wood, G.L. Schimek, J.W. Kolis, Chem .Mater. 8, 721, (1996).

3. P.T.Wood, W.T. Pennington, J.W. Kolis, Inorg. Chem. 33, 1556 (1994).

4. P.T.Wood, W.T. Pennington, J.W. Kolis, J.W. J. Am. Chem. Soc. 114, 9233 (1992).

5. D. M. Young, G. L. Schimek, J. W. Kolis, Inorg. Chem. in press.

6. Hubert L. Barnes Geochemistry of Hydrothermal Ore Deposits 2nd Ed., John Wiley and Sons, New York, 1979.

7. H.L. Barnes, G.K. Czamanske, in Geochemistry of Hydrothermal Ore Deposits 1st Ed., edited by H.L. Barnes, Holt Reinhardt and Winston, New York, 1967, p. 334.

8. H.C. Helgesen, Complexing and Hydrothermal Ore Deposition, Pergamon Press, Oxford, 1964.

9. D.A. Crerar, H.L. Barnes, Econ. Geol. 71, 772 (1976).

10. G.C. Ulmer, H.L.Barnes, Hydrothermal Experimental Techniques, Wiley-Interscience, New York, 1987.

11. See ref. 2 for typical solutions of single crystal structures in our labs.

12. V.I. Popolitov, B.N. Litvin, in Crystallization Processes under Hydrothermal Conditions, edited by A.N. Lobachev, Consultants Bureau, New York, 1973, p 73.

13. B.J. Skinner, F.D. Luce, E. Mackovicky, Econ. Geol. 72, 924 (1972).

14. M.J. Beurger, T. Hahn Am. Mineral. 40, 226 (1955).

15. M. Wintenberger, G. Andre, Physica B 162, 5 (1990)

16. D. Coucouvanis, P.R. Patil, M.G., Kanatzidis, B. Detering, N.C. Baenziger Inorg. Chem. 24, 24, (1985).

17. A.F. Wells, Structural Inorganic Chemistry 5th Ed. Clarendon Press, Oxford, 1984, p. 759.

Part II
Synthesis and Reactivity

CRYSTALLIZATION OF PEROVSKITES FROM SOLUTIONS

I. MAURIN, P. BARBOUX and J. P. BOILOT
Laboratoire de Physique de la Matière Condensée, CNRS URA 1254D
Ecole Polytechnique 91128 Palaiseau Cedex France

ABSTRACT

$LaMnO_3$ and $PbZrO_3$ perovskites have been synthesized through coprecipitation and sol-gel techniques. After the drying and calcination steps, amorphous mixtures of oxides have been obtained. They are unstable towards phase separation. Therefore, their homogeneity and reactivity strongly depend on the parameters of the calcination, such as the temperature and heating rate. In all cases, the perovskite structure crystallizes around 600°C through nucleation- or diffusion-limited kinetics. The reaction rate depends on the history of the material. A mechanistic interpretation of the reaction, based on the similarity between the structures of PbO, ZrO_2, Mn_2O_3 and La_2O_3 is proposed.

INTRODUCTION

Solution techniques have been largely investigated in the last ten years for the low temperature synthesis of multielement oxides. They offer a mean to control the homogeneity of the elements through the reaction of molecular precursors such as alkoxides in the sol-gel process or metallo-organics in the Metallo-Organic Deposition process (MOD)[1]. These concepts of "better homogeneity through chemistry" may be generalized to other techniques such as the polymeric route, where homogeneity is maintained through a viscous chelating polymer, or more simply to the coprecipitation techniques in which a polychelating reagent traps the different metal cations together through a rapid and random precipitation process (citrates, oxalates ...) [2, 3].

However, even if obtained after the condensation or precipitation steps, it is not clear that the homogeneity is maintained along the following reaction steps of drying and calcination. During these reactions, the residual organics or bridging hydroxyl groups are removed and the reaction progressively yields a mixture of pure oxides. This mixture may be metastable or unstable as compared to phase separation [4]. Therefore, there will be a competition between the crystallization of the multielement phase and phase partitioning.

It is clear that the kinetics of cationic diffusion plays an important role in the phase separation. In systems of cations having low charge and high mobility, phase separation may always be expected. This is the case for the $YBa_2Cu_3O_7$ system as an example. Unless ultra-rapid heating rate is used, phase separation occurs before the temperature onset of crystallization is reached [5]. On the contrary, diffusion limited crystallization of metastable phases will be observed in systems with high valent cations such as ZrO_2-Al_2O_3 or TiO_2-Al_2O_3 [6,7]. In these systems, short range ordering and the crystallization of a metastable phase will be preferred to long-range phase partitioning.

As a result, solution techniques may lead to a large decrease of the reaction temperature. Whereas usual solid state reactions go to completion within a few hours at 900°C for $PbTiO_3$ and $BaTiO_3$, they can be obtained at temperatures as low as 300°C for $PbTiO_3$ or even from the room temperature solution for $BaTiO_3$ [8, 9]. But, these appear to be the most favorable cases. For other perovskite systems such as $PbZrO_3$, the solution methods hardly decrease the reaction temperature down to 600°C. In this last case, X-ray absorption experiments indicate that the coordination of Zr

Mat. Res. Soc. Symp. Proc. Vol. 453 © 1997 Materials Research Society

in the amorphous material obtained from the solution is eight-fold [4]. Transformation to a six fold coordination may be the local barrier to the crystallization of the perovskite structure. Or is it due to a phase separation of the PbO-ZrO_2 mixture during the calcination or at higher temperatures?

In order to answer to this question and to gain any information on the mechanism of the rearrangements leading to the crystalline structure from the amorphous material, we have compared the synthesis of two perovskites $LaMnO_3$ and $PbZrO_3$. The comparison appears interesting because all the crystalline structures of the single oxides involved in these systems relate to the fluorite system (like ZrO_2) by removing one quarter (for La_2O_3 and Mn_2O_3) or one half of the anions (for PbO) (Fig. 1). Metastable mixed phases could be easily obtained or, in the opposite case, the kinetics of phase separation could be easily studied through the cationic diffusion across a single anionic lattice.

Recently, the $LaMnO_3$ system (partly substituted in the La site by divalent cations) has been largely investigated for its large magnetoresistance effect [10] as well as for application in fuel cells based on its high oxygen mobility [11]. The $PbZrO_3$ perovskite is antiferroelectric and has been used in the $PbZr_xTi_{1-x}O_3$ solid solution (0<x<1) for ferroelectric thin films and ceramics [8].

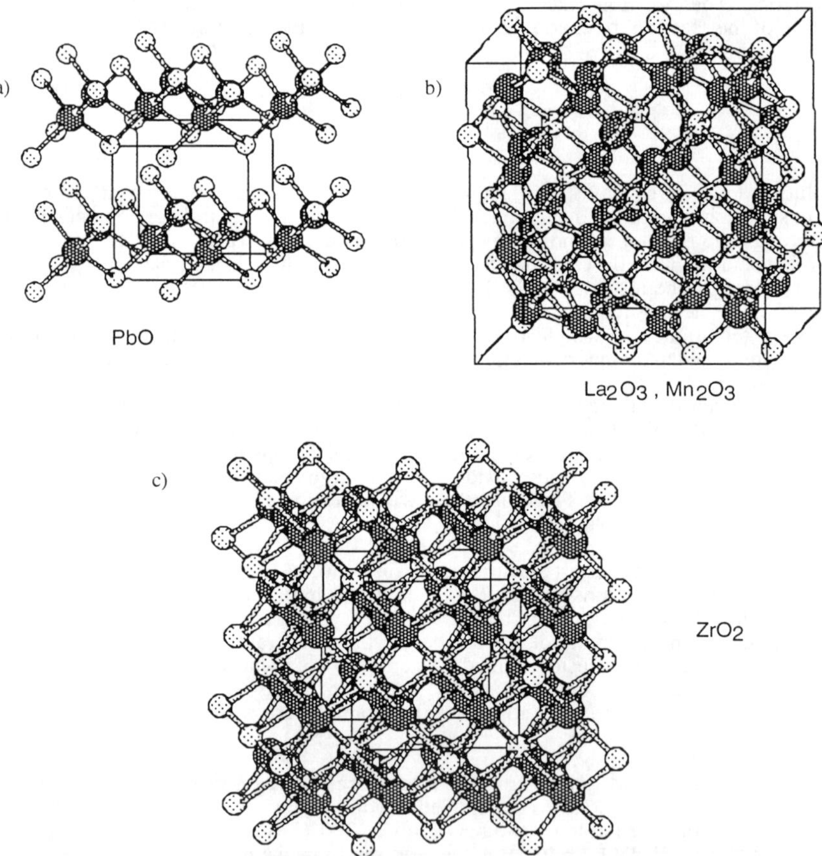

Figure 1: Comparison of the structures of a) PbO, b) La_2O_3 or Mn_2O_3 and c) ZrO_2.

EXPERIMENTAL

Synthesis of LaMnO$_3$

Pechini method: LaMnO$_3$ has been synthesized with a method derived from the citrate-ethylene-glycol process similar to previous works [3, 12, 13]. Typically, La(NO$_3$)$_3$.6H$_2$O and Mn(NO$_3$)$_2$.6H$_2$O were dissolved in the appropriate ratio (0.5 M). Ethylene glycol and citric acid were added to the concentration 0.85 M and 1M respectively. After 4 hours at boiling point of water, a viscous polymer is obtained through esterification. Further drying at 120°C yields a dark brown resin. Its calcination under oxygen is rough as shown by the TGA-DTA experiment (Fig. 2a).

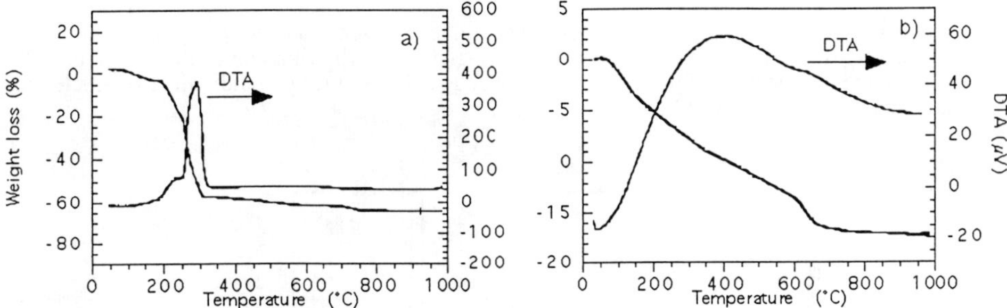

Figure 2: Thermal analysis of LaMnO$_3$ precursor powders obtained by a) the citrate method; b) the hydroxide coprecipitation. (5°C/min. in oxygen)

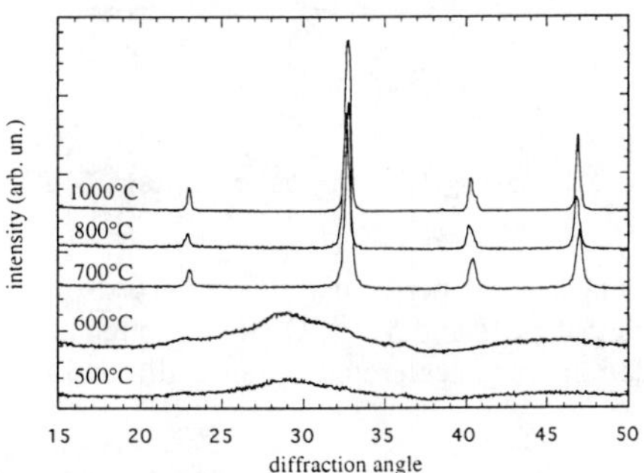

Figure 3: Synthesis of LaMnO$_3$ by the citrate method after calcination for 2 hours at different temperatures.

After a very rough calcination which yields a 58% weight loss at 350°C, a residual weight loss occurs in two steps between 400°C and 800°C. It corresponds to the decomposition of one CO$_3$ group per LaMnO$_3$ formula unit. Therefore, the amorphous material obtained at 400°C corresponds to the chemical formula LaMnO$_2$CO$_3$. Indeed, upon a too rough calcination at 500°C (large amount calcined in oxygen leads to explosive reaction), crystalline La$_2$O$_2$CO$_3$ is clearly observed in the X-ray diffraction pattern. In similar experiments in which smaller amounts of powders have been calcined and probably better thermalized by the crucible during the exothermic decomposition, an amorphous material is observed instead (Fig. 3). In any case, after calcination at 700°C for 2 hours the perovskite structure is obtained.

In complete agreement with the work of M. Verelst et al. [13], we observe, after crystallization at 700°C, a large non stoichiometry since the iodometric titration of the oxidation state of Mn indicates an oxygen composition of $LaMnO_{3.2}$. This non-stoichiometry has been analyzed by Roosmalen et al. as caused by cationic La and Mn vacancies better than by an oxygen excess [12].

The same method has been used for thin film deposition by the spin-coating method. In order to adapt the rheology to spin-coating, the solution was mixed with a viscous organic polymer obtained by the reaction of hexamethylenetetramine and acetylacetone in acetic acid (1/1/10)[14]. Smooth films have been deposited onto a (100) MgO substrate. After crystallization at 700°C for 15 hours, they exhibit a strong preferential orientation with the (001) direction of the pseudo-cubic $LaMnO_3$ structure perpendicular to the substrate [Fig. 4]. This indicates a strong alignment of the LaO planes of the perovskite structure with the MgO rocksalt structure.

The alignment of a film has two possible origins. It can be due to a thermal stress induced upon cooling by the different thermal expansion coefficients of the film and of the substrate or to a substrate-oriented nucleation of the film. In this last case, the texturing would indicate a nucleation-controlled crystallization of $LaMnO_3$.

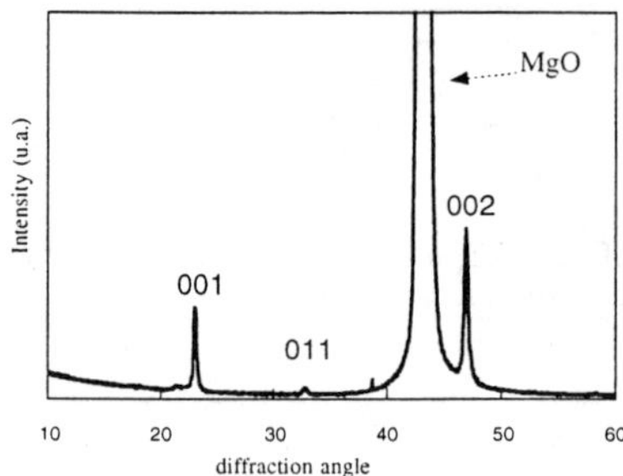

Figure 4: Diffraction pattern of a $LaMnO_3$ film deposited onto a MgO substrate (pseudo-cubic indexation).

Hydroxide route: Precipitation was performed under very strong basic conditions in order to overcome the different acidities of La^{3+} and Mn^{2+} [15]. Although at such a pH, Mn^{2+} is readily oxidized by air to Mn^{3+}, oxydation was accelerated by an additional oxidizer. The process is as follows:

One ml of H_2O_2 (30%) is added to 25 ml of a 50 weight % aqueous solution of NaOH. A solution of $La(NO_3)_3. 6H_2O$ (4.33 g =0.01 mole) and $Mn(NO_3)_2. 6H_2O$ (2.87 g=0.01 mole) in 20 ml water is slowly added to this solution. A brown precipitate is immediatly obtained. Filtration and thorough washing are performed on a Buchner funnel. The powder is dried at 120°C overnight.

By X-ray diffraction, the resulting powder exhibits a poorly crystalline pattern which may corresponds to a Mn-La mixture isostructural with the pure $La(OH)_3$ structure. The weight loss observed in the TGA (Fig. 2b) at 650°C is 18%, in good agreement with that expected for a mixture of $La(OH)_3$-$Mn(OH)_3$ (18.6%).

This precipitate is metastable towards dehydration since heating the precipitate for one day in its mother liquor at 150°C in a sealed tube yields a separation and a mixture of well crystallized $La(OH)_3$ and Mn_2O_3, as observed by X-ray diffraction. This is also in good agreement with the weight loss at 600°C (11.1%)

The crystalline precipitate decomposes upon heating through progressive dehydrations (Fig. 2b) and becomes amorphous after heating at 400°C or 500°C for one hour (Fig. 5). The broad diffraction pattern obtained in the first moments of the calcination at 400°C or 500°C may be an average structure of the La_2O_3 and Mn_2O_3 structure containing the cations with some degree of homogeneity. However, we have used X-ray absorption experiments at the La L_{III}-edge and Mn-K edge at the LURE synchrotron in Orsay (France). We failed to observe any Mn in the vicinity of La and reciprocally. However, second neighbors may not be observed in disordered materials and these results do not forbid that Mn and La can be homogeneously mixed.

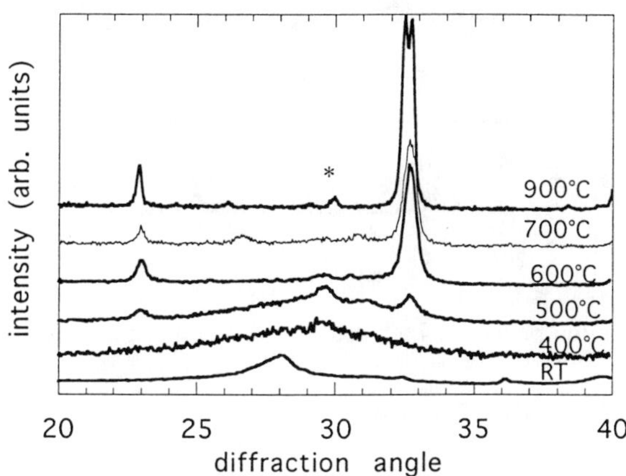

Figure 5: X-ray diffraction pattern of $LaMnO_3$ by the hydroxide route heat-treated at different temperatures for 12 hours in air. Peak marked with * corresponds to traces of $La_2O_2CO_3$.

In the temperature range between 200°C and 500°C broad exothermic rearrangements are observed in the differential thermal analysis (Fig. 2b). These exothermic reactions may be associated to the progressive phase separation and cationic rearrangements. The onset of formation of the $LaMnO_3$ perovskite structure appears to be around 500°C as shown by X-ray diffraction.

<u>Synthesis of PbZrO3</u>

The synthesis has been adapted from a previously described method [4, 16]. Anhydrous lead acetate (32.5 g=0.1 mole) obtained by drying $Pb(CH_3COO)_2$. $3H_2O$ at 150°C for 4 hours is dissolved upon heating in 100 ml of a 1M solution of $Zr(OPr)_4$ in propanol. The reaction is performed in a flow of dry nitrogen. The mixture is refluxed for 2 hours after dissolution. Then, the solvent and by-products are removed upon vacuum distillation at 80°C. The resulting polymeric material is dissolved in 150 ml of propanol (concentration of Zr: 0.66M).Then, 50 ml of the alkoxide solution is hydrolyzed upon a very slow addition of 2.4 ml of HNO_3 1M in water dispersed in 10 ml propanol ($H_2O/Zr=4$). The resulting fine precipitate is filtered and dried at 120°C in air.

After an endothermic weight loss (7-10%) corresponding to residual water and OH groups an amorphous material is obtained at 400°C. At 500°C, it rapidly crystallizes to a poorly crystalline material reminiscent of a fluorite structure (Fig. 6c), intermediate between the crystalline patterns of PbO (orthorhombic, Fig. 6a) and ZrO_2 (tetragonal, Fig. 6b) obtained separately by the precipitation of pure alkoxides in a similar process [4]. This pattern must be related to an intergrowth of the PbO and ZrO_2 structures. An homogeneous statistical mixture would yield a sharper peak at a position intermediate between PbO and ZrO_2, whereas a macroscopic mixture would yield a superposition of pattern 6a and 6b. From the peak width, the intergrowth domain length may be of 5 nm. Upon longer heating, the perovskite forms at 500°C within days.

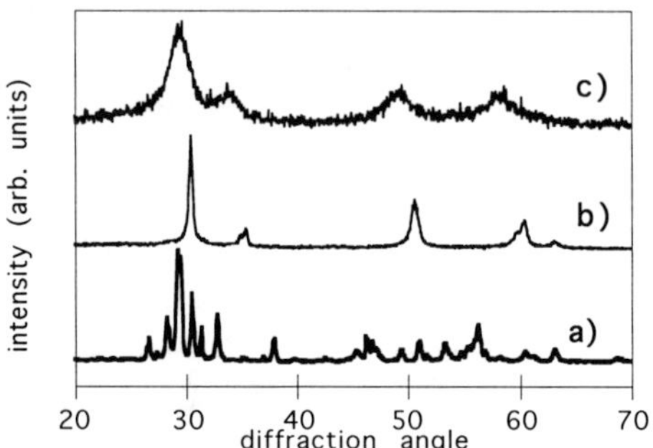

Figure 6: X-ray diffraction pattern of PbO and ZrO_2 crystallized at 400°C and c) Pb-Zr mixture annealed at 500°C for 1 hour.

KINETICS OF CRYSTALLIZATION

For the kinetics studies, the samples, previously calcined at 400°C for 12 hours have been directly placed in the hot furnace at the appropriate temperature and quenched at various times. The final weight loss is observed within the first few minutes of the reation. This indicates that the elimination of the last organics, CO_2 and hydroxyls may not be the limiting step in the process either. A mixture of oxides is readily obtained. Note that the weight losses observed in the thermal analysis of Fig. 2 are not indicative of the decomposition temperatures in static conditions.

The transformation reaction can be analyzed through the Avrami equation of kinetics:

$$\frac{dy}{dt} = K(1-y)t^{n-1}$$

Here y is the fraction of the crystallized phase determined from the integrated intensity of the x-ray diffraction for the (011) peaks at $2\theta = 32°$) . When plotting

$$Ln(-Ln(1-y))=nLn(t)+ Ln\ K/n$$

a straight line should be obtained in which the time exponent n yields information on the nucleation and growth rate [16,17].

For the $LaMnO_3$ system, the crystallization kinetics observed at 600°C in the citrate and the hydroxide method are shown in Figure 7. The slope of the straight lines observed in the first times of the reaction indicates a time exponent n=0.9. This may correspond to a nucleation-controlled reaction (without growth) in which a value of n=1 is expected [17, 18].

Nucleation-controlled reactions are often observed in divided powders since crystallization occurs as soon as the nuclei attain the size of the grain. No growth is necessary. Such nucleation-controlled crystallization has been observed for PZT powders obtained by the spraying of nitrates. The small size of the grains (16nm) can explain the control by a nucleation process [19].

In our experiments performed below 650°C, the size of the diffraction domain is constant throughout the crystallization (20nm) and independent of the temperature (Pechini or hydroxide). However, for these two systems we still have two different crystallization rates. This cannot be understood if only a nucleation mechanism is observed. The difference may come from the extent of the phase separation before crystallization occurs. In the Pechini method, the exothermic calcination favored a larger phase separation and a lower nucleation rate.

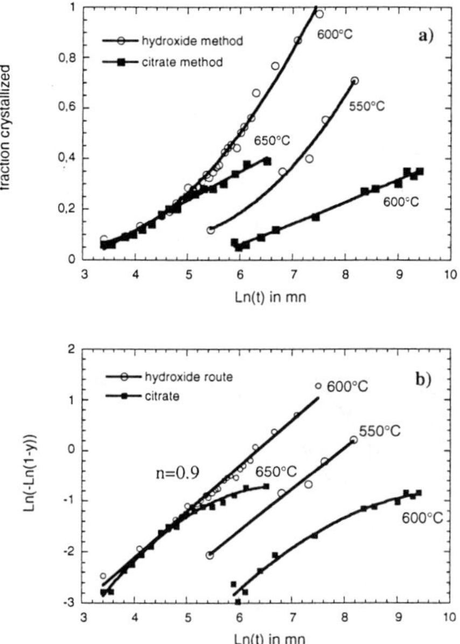

Figure 7: Kinetics of crystallization of LaMnO$_3$ at 600°C and 550°C for the hydroxide method and at 600°C and 650°C for the citrate method.

The crystallization of the LaMnO$_3$ phase from the citrate method is much slower than for the coprecipitated hydroxides as indicated by Fig. 7a. Moreover, the Avrami analysis (Fig. 7b) does not apply correctly if we use only one mechanism throughout the whole crystallization. At short times, the reaction follows a nucleation-limited with a time-proportional crystallization. With longer times, the rate decreases indicating that the reaction is controlled by another mechanism, with a time exponent lower than 1. This may be related to a diffusion at the interface between the La- and Mn-rich regions.

The analysis of the Pb-Zr system through the Avrami method is also not straightforward. The linear fit is fair and yields a time exponent between 1.2 and 1.8 depending on the temperature (Fig. 8b). At 500°C , the value of 1.15 may indicate a nucleation-limited reaction whereas at 600°C, the time exponent is n= 1.54 which approaches n= 3/2 expected for a 3D-diffusion controlled crystallization.

The perosvkite crystallizes with a constant domain size of 100 nm. This is much larger than the elementary particles (5 nm) of the precipitate analyzed by transmission electron microscopy [4].

Therefore, we believe that a diffusion-controlled step is involved in the crystallization of this phase and this implies some microscopic heterogeneities.

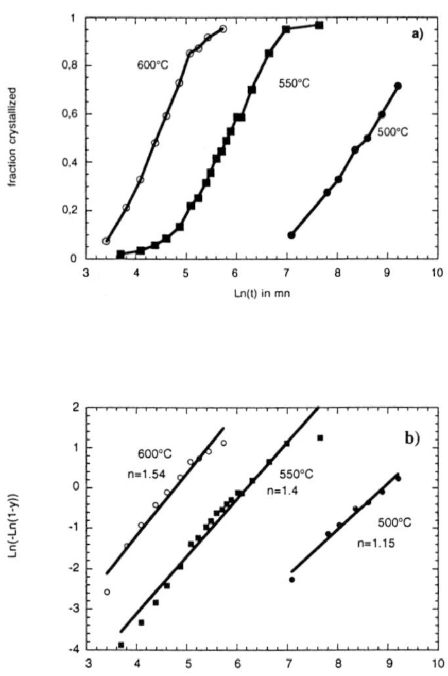

Figure 8: Kinetics of crystallization of PbZrO3 from sol-gel method.

DISCUSSION

We propose a mechanism for the crystallization of these systems. The main step of the crystallization must be the transformation of the oxygen sublattice. We recall that the starting disordered fluorite structure can be considered as a structure with oxygen placed on the cubic lattice of a cation-deficient CsCl structure. The perovskite will form by a cooperative shift to a cubic face centered packing of oxygens (also including the A cation) which is found in the perovskite. This reaction may be of the displacive type similar to the transformation from the CsCl-structure to a NaCl-structure [20]. This change of anionic lattice could be the limiting step of the nucleation.

A limited cationic separation can explain the different reaction rates of nucleation observed in the different systems. It will affect the value of K since the oxygen sublattice will only switch to the perovskite sublattice when the two A and B-type cations are locally present in the same ratio to stabilize the structure. Therefore, the problem can be considered as the competition between two mechanisms: 1) a phase separation that occurs between the cations in the fluorite lattice; 2) a local transformation of the anionic structure from a structure in which the interaction between the cations is repulsive to an anionic structure in which the interaction may be considered as attractive because it only takes place in the regions where the cationic ratio A/B is near the appropriate stoichiometry. This is summarized in the Figure 9. In this Figure, the energy of the amorphous system has been computed using enthalpy and entropy of the simplest statistical model [20] with:

$$F = E - TS$$

$$S = -Nk\left[c\ln(c) + (1-c)Ln(1-c)\right]$$

$$E = \frac{1}{2}NZ\left[Uc(1-c) + cU_{BB} + (1-c)U_{AA}\right]$$

where F, S and E are the free energy, the configuration entropy and the enthalpy of the system. c is the local concentration of the component B. U_{BB}, U_{AA} and U_{AB} are the interaction energies for the couples of particles AA, BB and AB respectively. $U = 2U_{AB} - U_{AA} - U_{BB}$. Z is the coordination between cations.

Figure 9 is drawn with a value of U positive (repulsive) in the fluorite lattice and strongly negative in the perovskite one. The U value was chosen so that the perovskite can only form in the cationic ratio A/B between 0.9 and 1.1.

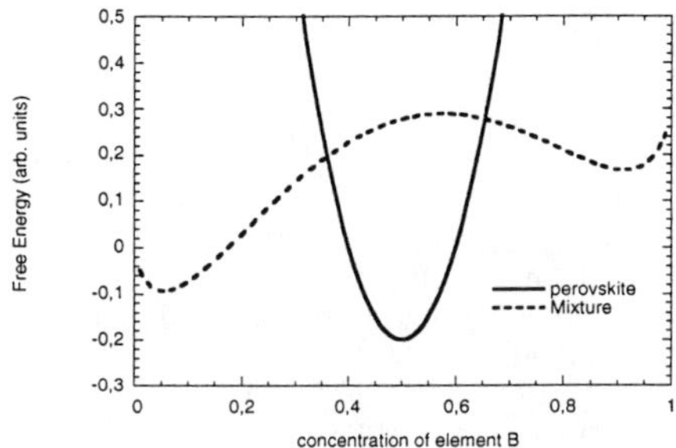

Figure 9: Energy of the system LaMnO$_3$ as a function of the concentration of B cations depending on the structure type (perovskite and fluorite).

The energy of the starting mixture is at its maximum when the cations are homogeneously mixed, in the domain of the perovskite phase. However, due to the similarity between the crystalline structure of the starting oxides, the repulsive interaction may be very small. At the temperatures of the synthesis the system may be found near the miscibility domain. It has been shown that after the rapid relaxation due to the spinodal decomposition, a slow growth of the heterogeneous domains occurs (α t$^{-1/3}$ [22]). Therefore, the phase separation may be limited to domains having a few repeat units and the crystallization is mostly limited by nucleation. The importance of the dispersion states of the cations has a very important effect on the activation energy for the formation of PbZrO$_3$ system too [23]. This probably indicates the change of mechanism from diffusion to nucleation -limited reactions.

The difference of reaction mechanisms between the two systems (La,Mn) and (Pb,Zr) may have two origins. First, a larger phase separation is probably observed in the Pb-Zr system whose oxides present much different structures than in the La-Mn case. Besides, Roosmalen et al. have shown a broad solubility range for the perovskite that may form with the La/Mn ratio in the range 0.8 to 1.2 [23, 24]. Therefore, the perovskite phase may nucleate even in regions where the cationic stoichiometry is not fully achieved.

The difference between the citrate and the hydroxide precipitation can be related to the highly exothermic calcination of the organics and nitrates. This decomposition may cause a phase separation as observed in the Pb-Zr systems containing a large number of residual organics.

CONCLUSION

Phase separation occurs in some multielement systems obtained by calcination of solution precursors. Although formidable, this phenomenon may be limited. It is slow at low temperatures because of the low diffusion kinetics. At higher temperatures, near the miscibility domain, the activity of the system may be also low enough to restrict the separation. Therefore, in systems such as those we have presented here, phase separation does not play an important role in the development of the multielement phase. The difficult cooperative nucleation of the perovskite anionic sublattice from a fluorite-related sublattice is probably the limiting factor.

The confinement of the mixture into small scale systems may also limit the phase separation. In order to eliminate the interfacial energy in small confined systems rapid interdiffusion has been observed to yield an amorphous homogeneous mixture. Recent examples have been shown in methods such as the physical evaporation of superlattices [26] or solution deposition of alternative layers of TiO_2-ZrO_2 systems [27], or mecanosynthesis [28]. In all these systems, the crystallization has been shown to be controlled by nucleation.

We believe that such nucleation controlled reactions are important from the material point of view. In the case of thin films deposited onto crystalline substrates, they will allow a better texturing from a nucleation induced at the contact with the subtrate. This has been the case, here, for the $LaMnO_3$ films deposited onto a MgO substrate.

REFERENCES

[1] R. W. Vest, Ferroelectrics 102 (1990) 53
[2] P. A. Lessing, Ceramic Bulletin 68 (1989) 1002.
[3] Pechini, US Patent N° 3 330 697 Jul 11, 1967.
[4] P. Barboux, F. Ribot, P. Griesmar and L. Mazerolles J. Solid State Chem. 117 (1995) 343.
[5] J.W. Cahn, Trans. Met. Soc. of AIME, 242 (1968) 166.
[6] F.F Lange, M.L. Balmer and C.G. Levi, J. of Sol-Gel Science and Technology 2 (1994) 317.
[7] L. Coury-Bonhomme, N. Lequeux, C. Mussotte and P. Boch, J. Sol Gel Science and Technology, 2 (1994) 371.
[8] C.D. Chandler, C. Roger and M. Hampden-Smith, Chem. Rev. 1993 93 1205.
[9] F. Chaput and J.P. Boilot, J. Mater. Sci. Lett. 6 (1987) 1110.
[10] R. Von Helmolt, J. Wecker, B. Holzapfel, L. Schultz, and K. Samwer, Phys. Rev. Lett. 71 (1993) 2331.
[11] R. Koc and H.U. Anderson, J. Materials Science 27 (1992) 5837.
[12] J.A.M. Van Roosmalen, E.H. P. Cordfunke and R.B. Helmholdt, J. of Solid State Chemistry 110 (1994) 100.
[13] M. Verelst, N. Rangavittal, N.R. Rao and A. Rousset, J. of Solid State Chem. 104 (1993) 74.
[14] I. Valente, Thesis of the University Pierre et Marie Curie, Paris (1989).
[15] C.F. Baes and E. Maessmer, "The hydrolysis of cations", Wiley, New York (1976).
[16] S. P. Faure, P. Barboux, P. Gaucher and J. Livage, J. Mat. Chem. 2 (1992) 713.
[17] H. Tanaka, N. Koga, A.K. Galwey, J. Chem. Educ. 72 (1995) 251.
[18] M. Avrami, J. Chem. Physics, 9 (1941) 177.
[19] Y. Yoshikawa and K. Tsuzuki, J. Am. Ceram. Soc. 73 (1990) 31.
[20] M.J. Buerger, Phase Transformations in Solids,, J. Wiley and Sons, New York (1951) 183-211
[21] R. Becker, Ann. Phys. 32 (1938) 128.
[22] J.D. Gunton, M. San Miguel and P.S. Sahni, Phase transitions , Vol. 8, Academic Press, London, (1983) pp. 267-466.
[23] J. Takahashi, N. Kakuta and A. Ueno, Mat. Res. Bull. 26 (1991) 243.
[24] J.A.M. Roosmalen,, E.H.P. Cordfunke and J.P.P. Huijsmans, Solid State Ionics 66 (1993) 285.
[25] J.A.M. Roosmalen, P. Van Vlandeeren, E.H.P. Cordfunke, W.L. Ijdo and D.J.W. Ijdo, J. of Solid State Chemistry 114 (1995) 516.
[26] L. Fister and D.C. Johnson, J. Am. Chem. Soc. 114 (1992) 4639.
[27] H. Shin, M. Argawai, M. DeGuire and A. Heuer, J. Am. Ceram. Soc. 79 (1996) 1975.
[28] D. Michel, F. Faudot, E. Gaffet, L. Mazerolles, J. Amer. Ceram. Soc, 76 (1993) 2884.

ON THE STRUCTURES OF $Mo_2O_5(OCH_3)_2$ AND $Mo_2O_5(OCH_3)_2 \cdot 2CH_3OH$

E. M. M^{C}CARRON III[*], R. L. HARLOW I, Z. G. LI, C. SUTO[†], Y. YUEN[†]
Central Research and Development, The DuPont Company,
Experimental Station, 356/341, Wilmington, DE 19880-0356, USA

ABSTRACT

The reaction of molybdenum trioxide dihydrate, $MoO_3 \cdot 2H_2O$, with methanol produces the title compounds. That these molybdenum oxy-methoxides decompose with liberation of CH_2O suggests that they represent exquisite models for selective oxidation of methanol to formaldehyde over molybdate catalysts. While a number of physical techniques have been employed to elucidate certain of their structural aspects, unfortunately, the detailed structures of these materials remain unknown. Problems, both physical and crystallographic, encountered in attempts to determine these structures are discussed. Also discussed is the recent progress made by employing the complimentary nature of electron and powder diffraction techniques.

INTRODUCTION

Interest in understanding the mechanism of the selective oxidation of methanol to formaldehyde over molybdate catalysts [1] naturally led to attempts to synthesize model compounds, i.e., compounds which would hopefully mimic certain aspects of the heterogeneous catalytic process [2]. $Mo_2O_5(OCH_3)_2$ and the closely related $Mo_2O_5(OCH_3)_2 \cdot 2CH_3OH$ [3] are such compounds. Analogous to the catalytic process, these oxy-methoxides were observed to decompose with liberation of CH_2O. While the results of numerous physical studies allowed us to propose a reasonable structure model for $Mo_2O_5(OCH_3)_2$ [2], the inability to either grow single crystals or to index the x-ray powder pattern prohibited structural confirmation. In this paper we make use of the complimentary nature of electron and powder diffraction techniques to determine the unit cell parameters of the two molybdenum oxy-methoxides. Moreover, this new structural information, coupled with insights gleaned from detailed studies of reaction chemistry, has allowed us to propose a new structure for these materials which is in agreement with all the physical data collected to date – with the peculiar exception of the x-ray structure factors themselves.

EXPERIMENTAL

Reaction chemistry was carried out as previously described [3]. A JEM-200EX (at 200 kV accelerated voltage) microscope, equipped with Gatan 1024x1024 CCD camera, was used to obtained electron diffraction patterns on these samples which were directly deposited on conventional TEM grids. High-resolution, x-ray powder diffraction patterns were obtained at the National Synchrotron Light Source which is a DOE User Facility located at Brookhaven National Laboratory, Upton, NY. The samples were mounted in capillaries and the patterns collected at beamlines X3b1 and X7a. In either case, an Si(111) monochromator, a Ge(200) analyzer, and slits on the order of 1x8 mm were used in conjunction with a scintillation counter to achieve the highest possible resolution and signal to noise ratio.

---

[*] mccarron@esvax.dnet.dupont.com
[†] NSF-sponsored Summer Research Program in Solid State Chemistry (CS: '95 and YY: '96)

Mat. Res. Soc. Symp. Proc. Vol. 453 © 1997 Materials Research Society

RESULTS

Reaction Chemistry

In earlier papers [2,3], it was reported that the reaction of yellow crystalline $MoO_3 \cdot 2H_2O$ with neat CH_3OH produced two white crystalline compounds, $Mo_2O_5(OCH_3)_2$ and $Mo_2O_5(OCH_3)_2 \cdot 2CH_3OH$. In the present study, not only these two compounds but also a third compound, white amorphous $MoO_3 \cdot CH_3OH$, was found. Compound formation is controlled by the reaction temperature:

$$MoO_3 \cdot 2H_2O + CH_3OH \xrightarrow{0°C} MoO_3 \cdot CH_3OH + 2\ H_2O \qquad (1)$$

$$2\ MoO_3 \cdot 2H_2O + 4\ CH_3OH \xrightarrow{30°C} Mo_2O_5(OCH_3)_2 \cdot 2CH_3OH + 5\ H_2O \qquad (2)$$

$$2\ MoO_3 \cdot 2H_2O + 2\ CH_3OH \xrightarrow{60°C} Mo_2O_5(OCH_3)_2 + 5\ H_2O \qquad (3)$$

Furthermore, the reaction of the hydrate with methanol to yield the methoxides was found not to be limited to just the yellow mono- and di-hydrates, but to occur with both the white monohydrate $MoO_3 \cdot H_2O$ and white hemihydrate $MoO_3 \cdot 1/2H_2O$ as well. Finally, conversion of the intermediate $Mo_2O_5(OCH_3)_2 \cdot 2CH_3OH$ to the methoxide can be accomplished simply by gentle heating [3]:

$$Mo_2O_5(OCH_3)_2 \cdot 2CH_3OH \xrightarrow{150°C} Mo_2O_5(OCH_3)_2 + 2\ CH_3OH \qquad (6)$$

Diffraction Data - Unit Cells

Table I: Unit cell data.

	$Mo_2O_5(OCH_3)_2 \cdot 2CH_3OH$	$Mo_2O_5(OCH_3)_2$
System:	monoclinic	monoclinic
Space Group:	$P2_1/c$	$P2_1/m$ or $P2_1$
a:	8.0535(6)	7.600(2) Å
b:	11.6648(11)	6.208(2) Å
c:	6.2910(4)	3.953(1) Å
β:	98.918(6)	102.55(1)°
V:	583.85	182.0 Å³
Interlayer Spacing:	11.6648(11)	7.418(2) Å

Partial cell information, obtained from selected area electron diffraction (SAED) patterns, was used to guide the indexing and to confirm the results. Lattice parameters for crystalline $Mo_2O_5(OCH_3)_2 \cdot 2CH_3OH$ and $Mo_2O_5(OCH_3)_2$ were determined from peak positions in their synchrotron diffraction patterns using the indexing program TREOR. They are listed in Table I.

<u>Diffraction Data - Structure Refinement</u>

The synchrotron diffraction patterns of $Mo_2O_5(OCH_3)_2 \cdot 2CH_3OH$ and $Mo_2O_5(OCH_3)_2$ are generally characterized by anisotropically broadened peaks which not only hindered the unit cell determinations but also suggested that structure determinations would not be possible. This has proven true for $Mo_2O_5(OCH_3)_2 \cdot 2CH_3OH$. In the case of $Mo_2O_5(OCH_3)_2$, however, a Patterson map based on intensities obtained from a LeBail extraction, contained peaks which could easily be interpreted as one Mo atom situated on a mirror plane in space group $P2_1/m$. Subsequent Fourier maps suggested that the Mo was five coordinate with 3 oxygen atoms in the mirror plane and two, one each above and below the plane, in the direction of the c-axis. The packing which results from this model shows that the structure consists of sheets of corner-shared Mo octahedra which stack along the a-axis. Attempts to locate the methyl group have not been successful although there is sufficient space between the sheets to accommodate this moiety. The fit between the observed and calculated intensities based on this model is quite good: the present Bragg R value is approximately 0.08.

DISCUSSION

Table II lists physical observations made in previous studies on the methoxides along with structural inferences taken from them. Based primarily upon these inferences and in the absence of any unit cell data, a structure for $Mo_2O_5(OCH_3)_2$ was proposed [2] which is almost certainly wrong. The structure consisted of stacked double layers of $MoO_{5/2}(OCH_3)_{1/1}$ corner sharing octahedra (ReO_3-type). While this structure fits the requirements of low dimensionality and a single carbon environment, it does not pass a simple color test. Mo(VI) oxide and oxides containing covalently bound adducts, which are based on an ReO_3-type corner sharing through oxygen, appear intensely yellow colored (e.g., β-MoO_3 [4], $MoO_3 \cdot 2H_2O$ [5], and $MoO_3 \cdot C_5H_5N$ [6]). A white coloration, on the other hand, is indicative of the edge-sharing double chain motif (as shown in Figure 2a) found in α-MoO_3 itself [7], $MoO_3 \cdot H_2O$ [8,9], and $MoO_3 \cdot 1/2H_2O$ [10]. Note particularly, that in the case of the hydrolysis of white $Mo_2O_5(OCH_3)_2$ the product is the white hemihydrate [3] – further supporting this conjecture. Thus, any new structure proposed for $Mo_2O_5(OCH_3)_2$ ought to include this edge-sharing double chain structural element.

Table II. Physical characteristics of $Mo_2O_5(OCH_3)_2$ and structural inferences drawn.

<u>Observation</u>	<u>Structural Information</u>	<u>Reference</u>
stoichiometry, $MoO_{5/2}X$	*<3D*	[3]
plate-like morphology	*2D*	[4]
white color	*$MoO_{3/3}O_2$-type chains*	[3]
single line ^{13}C-NMR spectrum	*1 type of C atom*	[4]

$Mo_2O_5(OCH_3)_2$ has a plate-like morphology. Electron diffraction patterns taken normal to the plate dimension confirm that the plane of the plate contains the b- and c-axes. Based upon the moderate heating conditions necessary for its conversion to methoxide (Equation 4), it appears reasonable to conclude that the intermediate $Mo_2O_5(OCH_3)_2 \cdot 2CH_3OH$ ought to be made up of the same basic structural units as the end-product itself. Thus, one might expect to find some correlation between the lattice parameters of these two materials. The unit cell data is listed in Table I. Indeed, one finds that:

$$b_{methoxide} \approx c_{intermediate}$$
$$c_{methoxide} \approx 2\, a_{intermediate}$$
leaving $\quad a_{methoxide} \quad$ and $\quad b_{intermediate} \quad$ as the unique axes.

Furthermore, as a check of the validity of the unit cells and also the underlying assumption that the structural units are in fact the same, one might expect the following relationship to hold true:

$$2^* \,(V_{methoxide} + 2\,V_{methanol}^{\dagger}) \approx V_{intermediate}$$
$$2\,(182.0 + 2(52.4))\ \text{\AA}^3 \approx V_{intermediate}$$
$$573.6\ \text{\AA}^3 \approx 584.9\ \text{\AA}^3.$$

The agreement to within 2% is indeed quite remarkable and confirms the soundness of the above arguments. Finally, if it is assumed that the $Mo_2O_5(OCH_3)_2$ structure is actually layered, in accord with both the platy-habit of the crystallites and the molybdenum atoms positions taken from the structural refinement, then the a-axis of the methoxide and the b-axis of the intermediate become the axes associated with the layer stacking direction and are thus a measure of the interlayer spacing (IS). If one now takes the volume of a methanol molecule$\dagger$ and simply calculates a spherical diameter (sd), one has, as yet another check of this developing structure model, the following relationship:

$$IS_{methoxide} + sd_{methanol} \approx IS_{intermediate}$$
$$7.42 + 4.64\ \text{\AA} \approx IS_{intermediate}$$
$$12.06\ \text{\AA} \approx 11.66\ \text{\AA}$$

which is quite reasonable considering that the calculation is most likely an overestimation, since it does not allow for any preferred orientation (of a non-spherical methanol molecule) nor for any nestling of the layers.

It is now possible to combine this crude model of the methoxide structures (i.e., stacked layers of composition $Mo_2O_5(OCH_3)_2$ and separated by single layers of methanol molecules in the case of $Mo_2O_5(OCH_3)_2 \cdot 2CH_3OH$) with the reaction chemistry outlined above to propose a structure for the layers themselves.

The fact that a yellow (containing corner-shared octahedral layers) hydrate reacts to form an amorphous white (presumably containing edge-shared octahedral chains) methanolate $MoO_3 \cdot CH_3OH$ (Equation 1) implies that a rather substantial structural rearrangement must necessarily take place. Moreover, that it takes place at relatively low temperature agrees with the observed lack of crystallinity. One speculates in the absence of diffraction data that white $MoO_3 \cdot CH_3OH$ might be a simple structural analogue of white $MoO_3 \cdot H_2O$, the structure of which is shown in Figure 1a.

---

* The factor 2 is associated with the doubling of a-axis of the intermediate relative to the c-axis of the methoxide.
$\dagger$ The volume of a methanol molecule (52.4 $\text{\AA}^3$) was calculated from its crystal structure [12].

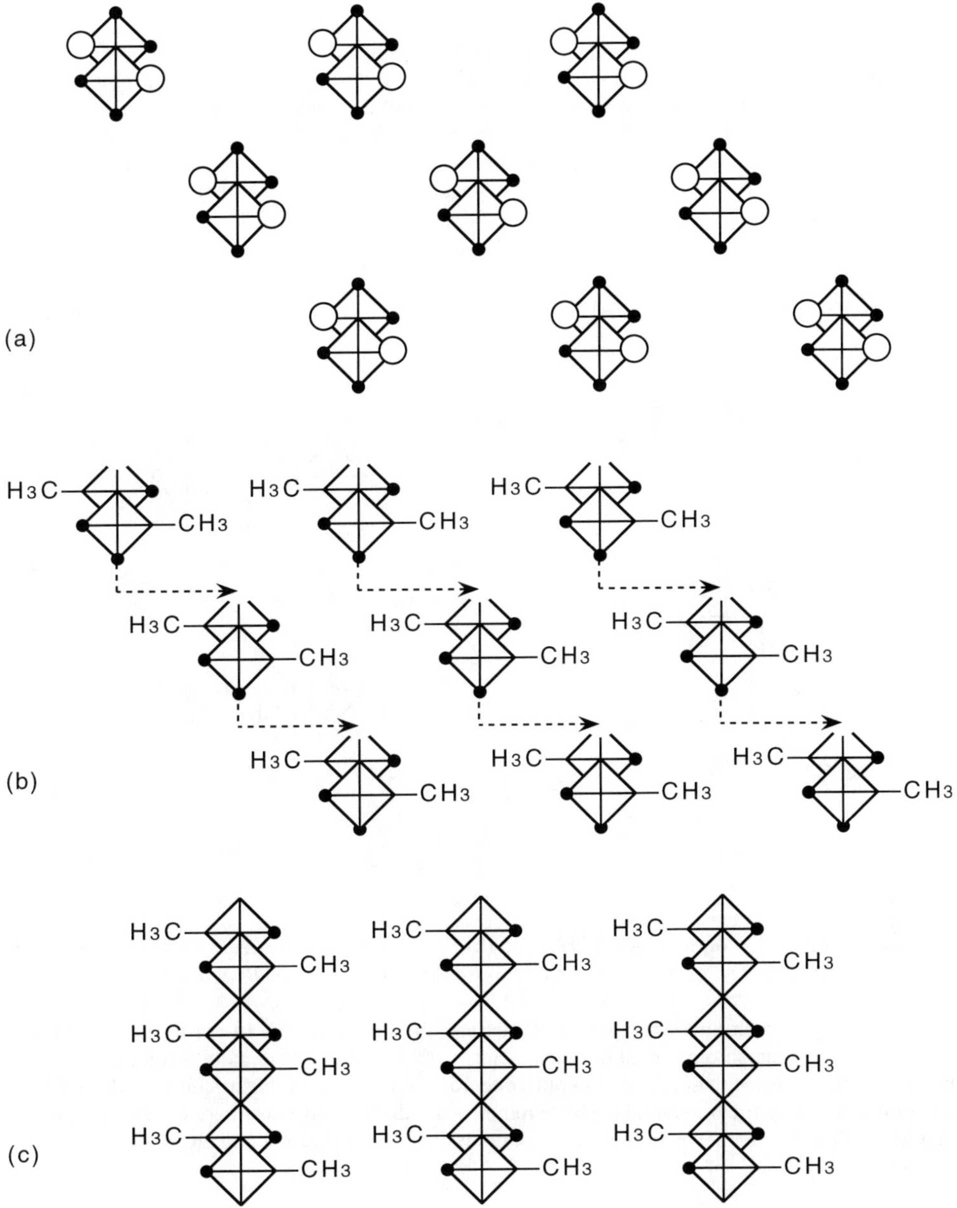

Figure 1. Conversion of (a) MoO$_3$·CH$_3$OH via (b) H$_2$O elimination to give (c) the proposed structure of Mo$_2$O$_5$(OCH$_3$)$_2$ (*open and filled circles represent methanol and singularly coordinated molybdenyl oxygen, respectively*).

The loss of water from amorphous $MoO_3 \cdot CH_3OH$ necessary to give the crystalline dimethoxides appears to be a thermally, though relatively mildly so, activated process (Equations 2 and 3). Figures 1 and 2 show how simple water loss involving two alcoholic protons and a molybdenyl oxygen (filled circle directed toward the top of the page in Figure 1)[1] can generate a layered structure for $Mo_2O_5(OCH_3)_2$ in fair agreement with the observed unit cell parameters. Taking an average *trans*-MoO_6 octahedral distance (*trans*-d_{oct}) to be ~3.9Å[2], one can generate the following lattice parameters normal to the layer stacking direction for comparison to those found in $Mo_2O_5(OCH_3)_2$ (see Figure 2b):

$$3\ trans\text{-}d_{oct} = 11.7\ \text{Å} \qquad \approx \qquad 2\ b_{\ methoxide} = 12.42\ \text{Å}$$
$$trans\text{-}d_{oct} = 3.9\ \text{Å} \qquad \approx \qquad c_{\ methoxide} = 3.95\ \text{Å}$$

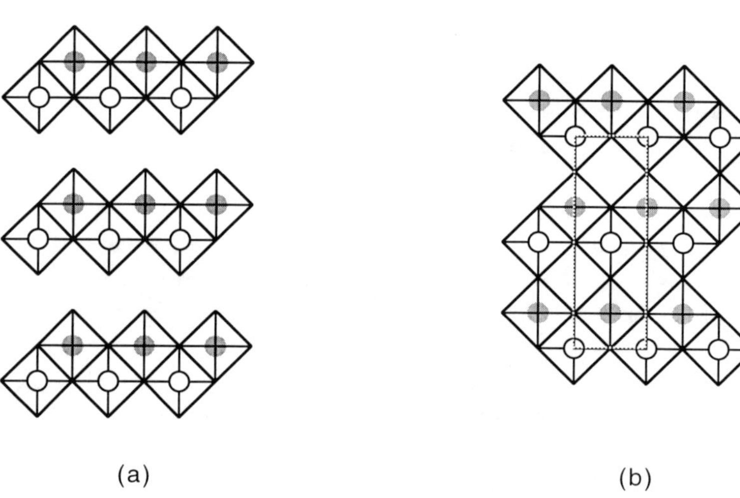

(a)   (b)

Figure 2. Conversion of (a) $MoO_3 \cdot CH_3OH$ to (b) $Mo_2O_5(OCH_3)_2$ viewed normal to the chain axis (*methyl groups represented by circles; proposed unit cell outlined*).

That this seemingly reasonable structural model does not fit with the structural refinement of the synchrotron data is an anathema. The five-coordinate molybdenum suggested by the structural refinement is found in only one other solid state oxide structure K_4MoO_5 [11] and in this structure the bipyramidal MoO_5 moieties are discrete. It must therefore be concluded that the truth lies — somewhere in all this tangle of data. Fortunately, there are still avenues yet to explore.

---

[1] Had the molybdenyl oxygen at right angles been chosen instead, the formation of the resultant one dimensional structure would necessitate massive structural rearrangement – at odds with both the platy crystal habit and the relatively low temperature of the transformation (Equation 3).
[2] The value of 3.9Å for *trans*-d_{oct} was derived by averaging the intralayer lattice parameters for α-MoO_3 [7].

56

CONCLUSIONS

Based upon numerous physical measurements, reaction chemistry, and a determination of lattice parameters, structures have been proposed for the compounds $Mo_2O_5(OCH_3)_2$ and $Mo_2O_5(OCH_3)_2 \cdot 2CH_3OH$. Both structures consist of layers of composition $Mo_2O_5(OCH_3)_2$ and, in the case of the methanolate, separated by a single layer of methanol molecules. The $Mo_2O_5(OCH_3)_2$ layers are postulated to contain the edge-shared octahedral chains as found in α-MoO_3 itself. However, the conundrum that this reasonable model does not mesh with the model determined using synchrotron diffraction data continues to perplex and confound. We invite any attempt at resolution.

ACKNOWLEDGMENTS

As 1996 represents the tenth anniversary of this puzzle, we wish to acknowledge all that have long suffered it: B. Baker, J. C. Calabrese, W.-H. Cheng, U. Chowdhry, M. K. Crawford, D. B. Chase, W. E. Farneth, T. P. Feist, A. Ferretti, L. E. Firment, S. C. Freilich, R. P. Groff, F. Hong, K. J. Karel, J. R. Long, C. J. Machiels (*in memoriam*), F. Ohuchi, J. B. Parise, A. W. Sleight, R. H. Staley, C. C. Torardi, and A. J. Vega; and the technical assistance of T. Q. Hawk, K. E. Lehman, and M. W. Sweeten.

REFERENCES

1. W.-H. Cheng, U. Chowdhry, A. Ferretti, L. E. Firment, R. P. Groff, C. J. Machiels, E. M. McCarron III, F. Ohuchi, R. H. Staley, and A. W. Sleight, *Proceedings of the 2nd Symposium of IUCCP of Dept. Chemistry, Texas A&M*, 165 (1984); "Heterogeneous Catalysis", B. L. Sharpiro, Ed., Texas A&M University Press, College Station, Texas.

2. E. M. McCarron III and A. W. Sleight, *Polyhedron* **5**, 129 (1986).

3. E. M. McCarron III, R. H. Staley and A. W. Sleight, *Inorganic Chemistry* **23**, 1043 (1984).

4. E. M. McCarron III, *J. Chem. Soc., Chem. Commun.*, 336 (1986).

5. B. Krebs, Acta Cryst. B **28**, 2222 (1972); J. R. Günter, *J. Solid State Chemistry* **5**, 354 (1972); and references therein.

6. J. W. Johnson, A. J. Jacobson, S. M. Rich, and J. F. Brody, *J. Amer. Chem. Soc.* **103**, 5246 (1981).

7. L. Kihlborg, *Arkiv Kemi* **21**, 357 (1963); and references therein.

8. I. Böschen and B. Krebs, Acta Cryst. B **30**, 1795 (1975).

9. H. R. Oswald, J. R. Günter, and E. Dubler, *J. Solid State Chemistry* **13**, 330 (1975).

10. P. Bénard, L. Seguin, D. Louër, & M. Figlarz, *J. Solid State Chemistry* **108**, 170 (1994).

11. T. Betz and R. Hoppe, *J. Less-Common Metals* **105**, 87 (1985).

12. JCPDS-ICDD, card # 12-874; G. R. S. Krishna Murti, *Indian J. Physics* **33**, 458 (1959).

SOL-GEL SYNTHESIS OF THE STRONTIUM-COPPER OXYCARBONATE SUPERCONDUCTOR $Sr_2CuO_2(CO_3)_{1-x}(BO_3)_x$

AUDREY J. BABCOCK, ALEXANDER R. PICO AND CATHERINE J. PAGE*
Department of Chemistry, University of Oregon, Eugene, Oregon 97403

ABSTRACT

We have developed an ambient-pressure sol-gel synthetic route to superconducting borate-doped $Sr_2CuO_2(CO_3)$ using polyether alkoxide precursors. In our sol-gel preparation, the starting solutions contain strontium and copper alkoxide complexes in 2-(2-methoxy-ethoxy)ethanol. Boron is incorporated into the solution by using an aqueous boric acid solution for hydrolysis. Dried gels were examined by x-ray diffraction and thermal gravimetric analysis. By experimenting with various firing sequences and atmospheres we have established a successful route for reproducibly preparing relatively pure $Sr_2CuO_2CO_3$ and boron-doped phases. Final products were characterized by x-ray diffraction, elemental analysis and magnetic susceptibility. Samples of nominal composition $Sr_2CuO_2(CO_3)_{0.85}(BO_3)_{0.15}$ are superconducting with a T_c(onset) of ~25K. Subsequent treatment at 1100°C and 3 GPa for one hour increased T_c(onset) to ~35K.

INTRODUCTION

A series of alkaline earth-copper oxycarbonates and related intergrowth structures containing infinite CuO_2 layers have recently been discovered, some of which exhibit high-temperature superconductivity [1-10]. The 'parent' compound of these materials is tetragonal $Sr_2CuO_2(CO_3)$, in which alternating carbonate and $(CuO_2)_\infty$ layers are separated by layers of strontium cations [1-2].

$Sr_2CuO_2(CO_3)$ can be prepared by conventional solid state synthesis (~1000°C) under ambient pressure (flowing mixtures of O_2 and CO_2, or partially enclosed conditions which inhibit escape of CO_2) [1-4]. These compounds do not exhibit superconductivity. To drive the system superconducting, holes must be introduced into the $(CuO_2)_\infty$ layers by adjusting the charge on the Sr-(CO_3)-Sr layers. To date, this has been accomplished in one of two ways. The first involves substitution of excess Cu for ~ 10% of the C in the $(Sr,Ba)_2CO_3$ layer [2,3,5] (e.g. $(Sr_{0.55}Ba_{0.45})_2Cu_{1+y}O_{2+2y+\delta}(CO_3)_{1-y}$ prepared using hot isostatic pressing has a T_c onset of ~40K). The second approach to introducing holes is to substitute BO_3^{3-} groups for some of the carbonate ions in the Sr_2CO_3 layers [6,7] (e.g., when prepared by conventional synthesis under ambient pressure $Sr_2CuO_2(CO_3)_{0.85}(BO_3)_{0.15}$ is reported to have a T_c of ~35K [6]). Borate-doped $Sr_2CuO_2(CO_3)$ is the parent compound of a series of oxycarbonate superconductors with the formula $Sr_2(Ca,Sr)_{n-1}Cu_n(CO_3)_{1-x}(BO_3)_xO_y$ (n = 1, 2, 3) which can be prepared under pressure [7]. In this case the superconducting critical temperatures have been reported to be 50K, 105K and 115K for n = 1, 2, and 3, respectively.

Previously, we have demonstrated that high-quality $YBa_2Cu_3O_{7-\delta}$ can be prepared via sol-gel synthesis at relatively low temperatures [11-14] (~700°C *vs* 900-950°C needed for

Mat. Res. Soc. Symp. Proc. Vol. 453 © 1997 Materials Research Society

conventional solid state synthesis). Alkoxide sol-gel synthesis involves the hydrolytic condensation of homogeneous solutions of metal alkoxide complexes which have the appropriate metal stoichiometry of the desired solid state product [15-18]. The condensed 'gel' can generally be fired at moderate temperatures to form oxide glasses or crystalline phases. Because diffusion barriers in the sol-gel process are significantly reduced compared to those of conventional solid state synthesis, high temperatures may not be needed to produce a phase of interest. Indeed, it is sometimes possible to access metastable phases that cannot be prepared at high temperature using this synthetic route [19]. Another advantage of this synthetic method is the ability to shape precursor gels into useful forms before processing to obtain the brittle ceramic materials.

Our sol-gel synthetic route employed for the synthesis of $YBa_2Cu_3O_{7-\delta}$ makes use of soluble bimetallic barium-copper 2-(2-methoxy)ethoxyethoxide (MEE) complexes that we discovered previously [11-14]. Use of this type of soluble bimetallic complex is critical since most copper alkoxides are insoluble and therefore cannot be used in alkoxide sol-gel synthesis. After addition of the yttrium component, transmission electron microscopy (TEM) showed the particles in solution are smaller than $\sim16\text{Å}$, indicating that true molecular-scale mixing is achieved in the precursor solutions.

We have also found that MEE complexes of other highly electropositive elements including strontium solubilize copper alkoxides, presumably by forming soluble bimetallic complexes. Solutions containing strontium-copper MEE complexes with Sr:Cu ratios of 2:1 were employed in the sol-gel synthesis of the title compound.

We report here the alkoxide sol-gel synthesis of the parent compound $Sr_2CuO_2(CO_3)$ and its superconducting boron-doped isostructural analog $Sr_2CuO_2(CO_3)_{0.85}(BO_3)_{0.15}$ We embarked on this project to see whether this synthetic route could produce these superconducting materials at relatively low temperatures without high pressure. We also have examined whether high pressure treatment improves the superconducting properties of material prepared at ambient pressure.

EXPERIMENTAL

Preparation of our polyether alkoxide solutions has been modified slightly from our previously reported method [11-14], although inert-atmosphere handling and preparation of precursor materials are the same. Strontium metal was dissolved in dry 2-(2-methoxy)ethoxy-ethanol, and a stoichiometric amount of copper methoxide was added, resulting in a clear royal blue solution after several hours of stirring. Resulting solutions were on the order of 0.04 M in metals. Hydrolysis was induced by the addition of either pure water or an aqueous boric acid solution in a 100:1 molar ratio of H_2O to metals (50:1 also worked well). In the case of the boric acid solution addition, a quantitative amount of boric acid was dissolved in the total amount of water before addition so that the amount of boron in the hydrolyzed solution was 0.15 times the number of moles of copper. Hydrolyzed solutions were poured into Petri dishes, loosely covered and dried over low heat ($\sim50°C$) for $\sim$ 2 days.

Dried gels were fired in alumina boats under a flowing atmosphere of either a 90% O_2/10% CO_2 mixture or using a premixed 95% oxygen/5% carbon dioxide directly from a cylinder (this latter atmosphere source gave more reproducible results). The firing program used involved ramping at 5°/min. from room temperature to the pyrolyis temperature (generally 925°C), a soak at this temperature, and slow cooling by turning off the furnace and leaving the sample in place to cool to room temperature. High-pressure, high-temperature firings were done using a piston-cylinder apparatus for which the sample was sealed in a gold capsule placed in a calcium fluoride pressure cell with graphite heater tubes.

Dried gels and fired products were characterized by powder x-ray diffraction (XRD) using a Scintag XDS-2000 θ–θ powder diffractometer. Thermal gravimetric measurements were made using a DuPont Instruments 951 Thermogravimetric Analyzer (TGA) at a ramp rate of 2°C/minute under flowing 95% O_2/5% CO_2. Magnetic susceptibility measurements were made with a SQUID magnetometer at fields of 3 and 5 G.

RESULTS AND DISCUSSION

Thermogravimetric analysis (TGA) of the dried gels showed a large mass loss from ~25-200°C, followed by another small sharp mass loss at ~840°C, as shown in a representative TGA (Fig. 1). The first mass loss is due to loss of residual water and organic content. The smaller mass loss at ~840°C may be due to some CO_2 loss associated with decomposition of $SrCO_3$.

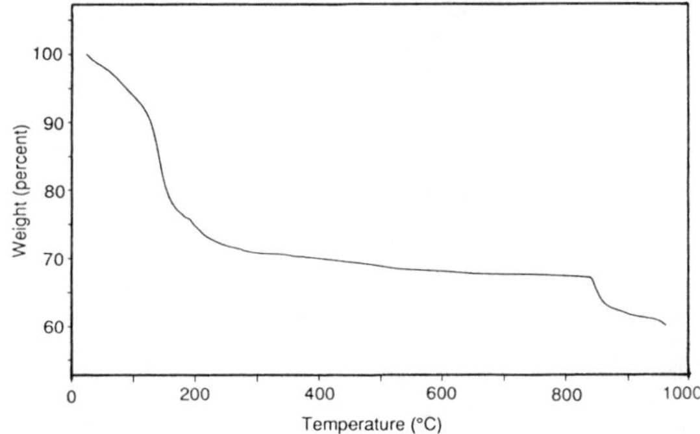

Figure 1. TGA of a dried gel (containing boron) heated in flowing 95% O_2/5% CO_2.

Dried gels fired to 900-925°C under flowing 90%/10% or 95%/5% O_2/CO_2 atmosphere produced materials that appeared by X-ray diffraction to be phase-pure $Sr_2CuO_2(CO_3)$ (boron-doping does not alter the structure since borate anions are very nearly the same size as carbonate anions). The firing atmosphere was found to be critical for obtaining the desired phase. When flowing pure O_2 was used as the firing atmosphere (at 925°C), the major phase produced was orthorhombic Sr_2CuO_3. When a flowing pure CO_2 firing atmosphere is used, copper metal and $SrCO_3$ are formed. Even a 50%/50% mixture of O_2/CO_2 yields predominantly $SrCO_3$. We found it is necessary to use an O_2/CO_2 mixture with the CO_2 percentage between ~2.5% and ~10% to produce the desired oxycarbonate phase. Firing temperatures are also important. At temperatures >950°C the orthorhombic Sr_2CuO_3 apparently becomes more stable and mixtures of this phase and the oxycarbonate are formed. At temperatures less than 850°C, $SrCO_3$ is the only crystalline product. When dried gel is fired at 875°C $SrCO_3$ is initially formed (e.g. after 10 hours), but with extended firing (~30 hr) at 875°C this is converted entirely to the desired $Sr_2CuO_2CO_3$ phase.

A diffraction pattern for a sample of nominal composition $Sr_2CuO_2(CO_3)_{0.85}(BO_3)_{0.15}$ prepared by firing at 925°C for 10 hours under 90%O_2/10%CO_2 is shown in Figure 2(a). Elemental analysis of this sample gave a composition close to the nominal composition, with a slightly higher boron concentration (corresponding to ~18-20% boron doping). [Analysis: Calculated for $Sr_2CuO_2(CO_3)_{0.85}(BO_3)_{0.15}$: Sr, 53.00% ; Cu, 19.22%; C, 3.09%; B, 0.49%. Found: Sr, 52.81%; Cu, 19.48%; C, 3.32%; B, 0.61%.] This sample was found to be superconducting, with a T_c onset of ~25K, as shown by the magnetic susceptibility in Fig. 2(b). The Meissner effect for this sample was very small (<2%), probably due to a small volume of superconducting material in the sample and perhaps also to the small particle size. The broad transition also suggests inhomogeneity in terms of boron doping and/or oxygen content.

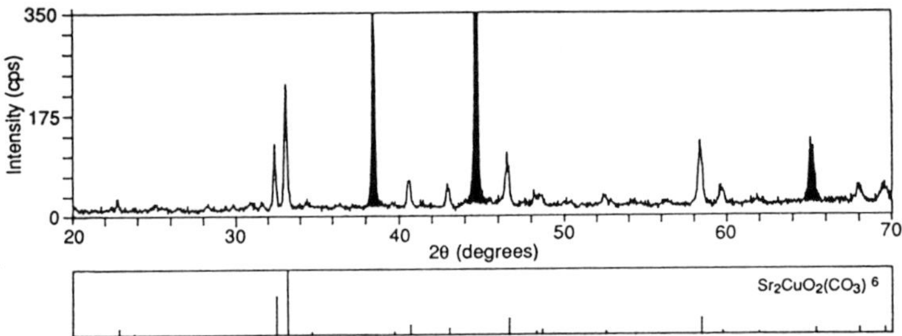

Figure 2. (a) XRD of $Sr_2CuO_2(CO_3)_{0.85}(BO_3)_{0.15}$ fired at 925°C under 90% O_2/10% CO_2. The filled black lines are from the aluminum XRD substrate.

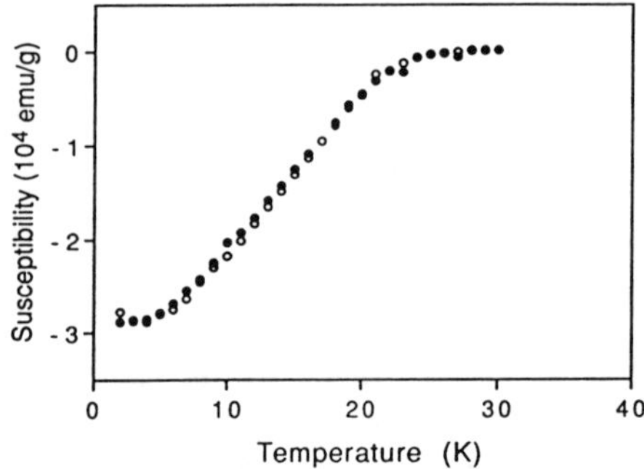

Figure 2. (b) Magnetic susceptibility of $Sr_2CuO_2(CO_3)_{0.85}(BO_3)_{0.15}$ measured at 3 G. Filled circles represent zero field cooled data; open circles are field cooled data.

To investigate whether pressure treatment would improve the superconducting properties, we subjected the same sample in a sealed gold capsule to a high-temperature, high-pressure treatment at 1100°C and 3GPa for one hour. (Note, however, that we did not include $KClO_4$ in the capsule to provide an oxidizing agent, as is often done in high-pressure treatments of copper oxide superconductors.) The x-ray diffraction pattern after the treatment is shown in Figure

3(a). Note that the major difference in the XRD of this sample with respect to the XRD of the sample before high pressure treatment (Fig. 2(a)) is that the diffraction peak intensities have changed. The (004) and (008) lines are now much more prominent, suggesting preferred orientation and perhaps a higher aspect ratio of the crystallites. Magnetic susceptibility measurements of this sample show a higher T_c (~35K) and a larger Meissner effect (54%) for this sample after pressure treatment, as shown in Figure 3(b).

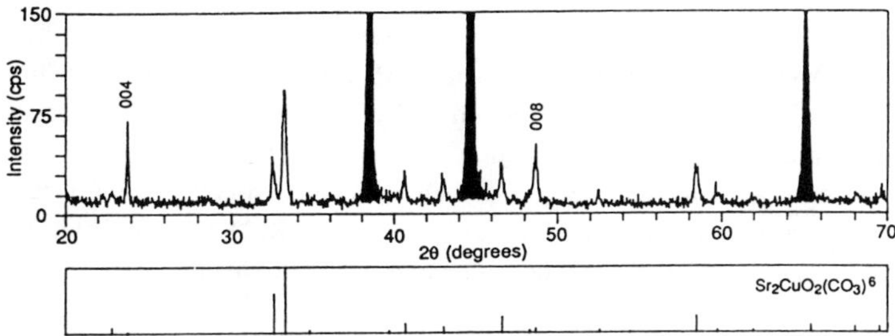

Figure 3. (a) XRD of $Sr_2CuO_2(CO_3)_{0.85}(BO_3)_{0.15}$ after treatment at 1100°C and 3GPa. The filled black lines are from the aluminum XRD substrate.

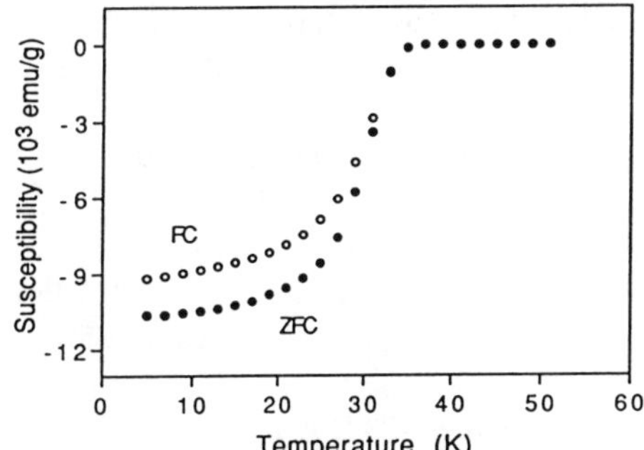

Figure 3. (b) Magnetic susceptibility of $Sr_2CuO_2(CO_3)_{0.85}(BO_3)_{0.15}$ after treatment at 1100°C and 3GPa (field=5G). Note the difference in temperature and susceptibility scales relative to Fig. 2(b).

CONCLUSIONS

In summary, we have developed an all-alkoxide sol-gel synthesis at ambient pressure to $Sr_2CuO_2(CO_3)_{0.85}(BO_3)_{0.15}$ which appears to be relatively pure by XRD. Magnetic susceptibility measurements indicate that while some of the sample superconducts at ~25K, the material is not a bulk superconductor, and the breadth of the observed superconducting transition suggests the sample is inhomogeneous with respect to boron and/or oxygen content. High temperature, high pressure treatment of the material prepared at ambient pressure produces higher quality, more crystalline material that superconducts at ~35K with a sharper transition and a much larger Meissner fraction characteristic of bulk superconductivity. We are currently optimizing the firing parameters for the ambient-pressure sol-gel synthesis to see if better material can be prepared without high pressure treatment, and are simultaneously exploring different high-pressure treatments to optimize superconducting properties of these materials. In addition, we are now using this synthetic route to prepare other (higher-Tc) analogs in the $Sr_2(Ca,Sr)_{n-1}Cu_n(CO_3)_{1-x}(BO_3)_xO_y$ family [7].

ACKNOWLEDGMENTS

We gratefully acknowledge support for AJB and ARP through the University of Oregon Chemical Physics Institiute NSF-REU Program (CHE 9531402). We thank Professor Dana Johnston, Ms. Jennifer Pickering, Mr. Brandon Schwab and Mr. Mike Dewey of the University of Oregon Geology Department for use of and help with the high-pressure equipment. Dr. Peter Klavins and Professor Robert Shelton (Physics Department, UC Davis) are gratefully acknowledged for providing the magnetic susceptibility measurements.

REFERENCES

1. Y. Miyazaki, et al., Physica C **191**, p. 434 (1992).
2. K. Kinoshita, T. Yamada, Nature **357**, p. 313 (1992).
3. K. Kinoshita, T. Yamada, Jpn. J. Appl. Phys. **31**, p. L832 (1992).
4. A. R. Armstrong, P. P. Edwards, J. Solid State Chem. **98**, p. 432 (1992).
5. F. Izumi, et al., Physica C **196**, p. 227 (1992).
6. M. Uehara, H. Nakata, J. Akimitsu, Physica C **216**, p. 453 (1993).
7. M. Uehara, et al., Physica C **229**, p. 310 (1994).
8. J. Akimitsu, H. Nakata, M. Uehara, J. Supercond. **7**, p. 19 (1994).
9. C. Greaves, M. Al-Mamouri, P. Slater, P. P. Edwards, Physica C **235-240**, p. 158 (1994).
10. B. Raveau, C. Michel, B. Mercey, J. F. Hamet, M. Hervieu, J. Alloys Compd. **229**, p. 134 (1995).
11. C. S. Houk, C. J. Page, Adv. Mater. **8**, p. 173 (1996).
12. C. S. Houk, G. A. Burgoine, C. J. Page, Chem. Mater. **7**, p. 173 (1995).
13. C. S. Houk, G. A. Burgoine, C. J. Page, Mater. Res. Soc. Symp. Proc. **346**, p. 29 (1994).
14. C. J. Page, C. S. Houk, G. A. Burgoine, Mater. Res. Soc. Symp. Proc. **271**, p 155 (1992).
15. H. Schmidt, J. Non-Cryst. Solids **100**, p. 51 (1988).
16. R. C. Mehrotra, J. Non-Cryst. Solids **100**, p. 1 (1988).
17. H. Reuter, Advanced Materials **3**, p. 258 (1991).
18. G. R. Lee, J. A. Crayston, Advanced Materials **5**, p. 434 (1993).
19. L. Bonhomme-Coury, N. Lequeux, S. Mussotte, P. Boch, Sol-Gel Sci. Tech. **2**, p. 371 (1994).

ORDERING IN LITHIUM AND SODIUM INSERTED $W_{18}O_{49}$

A.MARTINEZ DE LA CRUZ[1], LETICIA M. TORRES-MARTINEZ[1], F. GARCIA-ALVARADO[2], E. MORAN[3] AND M.A. ALARIO-FRANCO[3].

[1] Facultad de Ciencias Químicas, Universidad Autónoma de Nuevo León, Apartado Postal 1864, Monterrey,N.L., MEXICO.

[2] Facultad de Ciencias Experimentales y Técnicas, Universidad San Pablo-CEU, Urb. Montepríncipe, Apd. Correos 67, Boadilla del Monte, Madrid, SPAIN.

[3] Departamento de Química Inorgánica, Facultad de Ciencias Químicas, Universidad Complutense, Madrid 28040, SPAIN.

ABSTRACT

The insertion of alkali metals in $W_{18}O_{49}$ is governed by size of the cations. In this way, lithium insertion ($r^{VI} = 0.76$ Å) seems to be optimal and a maximum of 23.5 lithium per formula can be reached before the irreversible reduction of $W_{18}O_{49}$. On the other hand, when sodium is inserted ($r^{VI} = 1.02$ Å) an order of magnitude decrease in the amount inserted is observed and a maximum of only 1.8 sodium per formula can be intercalated. On the basis of the different phases that we have previously detected by electrochemical methods in the $Li / W_{18}O_{49}$ and $Na / W_{18}O_{49}$ systems, we have started a structural characterization by electron diffraction techniques. These studies have revealed some lithium and sodium ordering states for certain compositions; this can be explained taking into account the different types of tunnels present in $W_{18}O_{49}$.

INTRODUCTION

When WO_3 is reduced, different structural families are formed with a complicated structural arrangement and stoichiometry WO_{3-x}. In the region of the phase diagram poor in oxygen, $W_{18}O_{49}$ (this is WO_{3-x}, $x = 0.28$) exists as a stoichiometric and very stable phase. The reduction of WO_3 to $WO_{2.72}$ produces a framework of polyhedra WO_6 and WO_7 which form three type of tunnels as shown in Figure 1.

Due to its quite open structure, different authors have used $W_{18}O_{49}$ as a host in lithium insertion reactions [1,2]. Nevertheless the different phases formed, $Li_xW_{18}O_{49}$, have only been characterized by X-Ray diffraction. Recently, we have started an electrochemical study of lithium insertion in this material coupled with the microstructural characterization by electron microscopy and diffraction [3,4]. In this way, for a sample with composition $Li_{17}W_{18}O_{49}$ lithium ordering was observed. The location of the inserted lithium in this phase is not easy to predict, due to the large number of available sites, 14 per unit cell: 1 hexagonal, 6 triangular and 7 quadrangular. In any case, for $x_{max} = 23.5$, lithium multioccupancy in at least two tunnels type (quadrangular and hexagonal) seems necessary.

Based in previous work where it has been shown that *hexagonal tungsten bronze* type-phases intercalate monovalent cations larger than lithium (Na^+ or K^+) but only in the hexagonal tunnels [5] we have also performed electrochemical sodium insertion in $W_{18}O_{49}$ [6]. Taking into account that the sodium ion is almost 40 % larger than a lithium ion, it was expected that sodium insertion would be more *selective* than that of lithium. In fact, the maximum amount of sodium inserted, $y_{max} = 1.8$, suggests that sodium is incorporated in $W_{18}O_{49}$ by occupying the unique hexagonal tunnel present in the oxide framework. The different compositions, $Na_yW_{18}O_{49}$, were characterized by electron diffraction and a simple model of occupancy can be proposed correlating the observed composition with a posible structure.

Mat. Res. Soc. Symp. Proc. Vol. 453 © 1997 Materials Research Society

EXPERIMENTAL

The synthesis of $W_{18}O_{49}$ was carried out following the procedure previously described [7]. The synthesis of the lithiated phases, $Li_xW_{18}O_{49}$, were carried out by chemically indirect reactions with n-butyllithium in n-hexane. In a typical experiment a few grams of $W_{18}O_{49}$ were mixed with the organometallic reagent in differents molar ratios. After stirring for a week, the solvent was extracted and the powder washed several times with dry hexane. Lithium contents were determined by Inductively Coupled Plasma (ICP) as previously described [3].

Samples of composition $Na_yW_{18}O_{49}$ were obtained by the discharge of Swagelok electrochemical cells with configuration Na / $NaClO_4$ 1 mol dm^{-3} in PC / $W_{18}O_{49}$, where PC = Propylene Carbonate [6]. Different sodium cells were discharged in potentiostatic conditions (scan rate of 10 mV/h) and the experiments were interrupted for predetermined values of y in $Na_yW_{18}O_{49}$. The sodium content was determined by the evaluation of the total charge transfered in each step potential supposing that side reactions do not occur. The microstructural characterization of the inserted phases was done by electron diffraction using an electron microscope JEOL 2000FX.

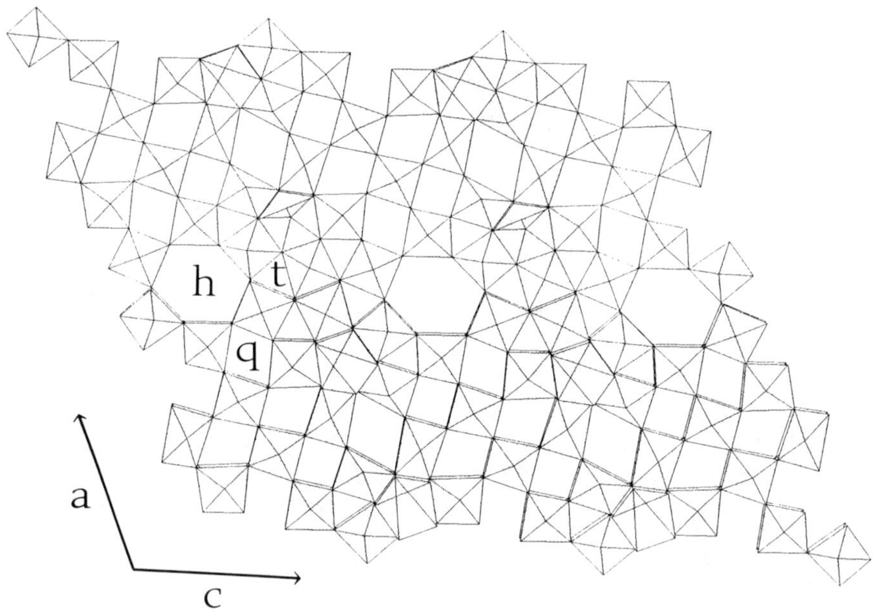

Figure 1. Schematic representation of the $W_{18}O_{49}$ structure projected along the b direction. Tunnels running along that direction are labelled as h (hexagonal), q (quadrangular) and t (triangular).

RESULTS AND DISCUSSIONS

Chemical analysis of the lithiated samples indicates the formation of $Li_xW_{18}O_{49}$, with x = 0.4, 2, 5, 9, 10 and 17. In all cases $Li_xW_{18}O_{49}$ can easily be related to the $W_{18}O_{49}$ structure. It can be deduced that the $W_{18}O_{49}$ framework is maintained since, in spite of a certain loss of crystallinity, the six patterns can be indexed using the $W_{18}O_{49}$ cell. A shift of the peaks to lower angles indicates that the insertion reaction ocurrs expanding the cell volume [3].

Figure 2 shows a typical electron diffraction pattern of a sample with a relatively small amount of lithium (x ~ 5). This type of diagram was frequently observed for different crystals with compositions between $0 < x \leq 5$. In agreement with the X-Ray diffraction results, all electron diffraction patterns showed that the $W_{18}O_{49}$ cell is maintained. Some streaking, also seen in the diagram, indicates the presence of some disorder in the crystal.

However, the situation for the samples with x = 9, 10 and 17 is very different. Figure 3 show some electron diffraction patterns of $Li_{10}W_{18}O_{49}$ along [1 -1 1] and [1 -3 1] zone axes. In this case, the observed extra spots cannot be indexed on the basis of the $W_{18}O_{49}$ cell. A new cell with parameters 2a x 2b x 2c can be proposed that can account for the presence of the extra spots. It is most likely that the new cell is related with a lithium ordering within the framework oxide. Similar results were observed for the composition x = 9.

We have already reported a similar situation for x = 17, where also a double cell allowed us to explain the extra diffraction effects [3,4]. On the basis of these results, it would appear that lithium *ordering* in $W_{18}O_{49}$ seems to ocurr for samples with high lithium content ($x \geq 9$). Taking into account that $W_{18}O_{49}$ has 14 tunnels per formula and that, experimentally, one can insert up to approximately 23.5 lithium per formula, a multioccupancy of some of the tunnels is necesary to justifity such an inserted amount.

Figure 2. Electron diffraction pattern of $Li_5W_{18}O_{49}$ showing disorder along the [1 0 -1] zone axis.

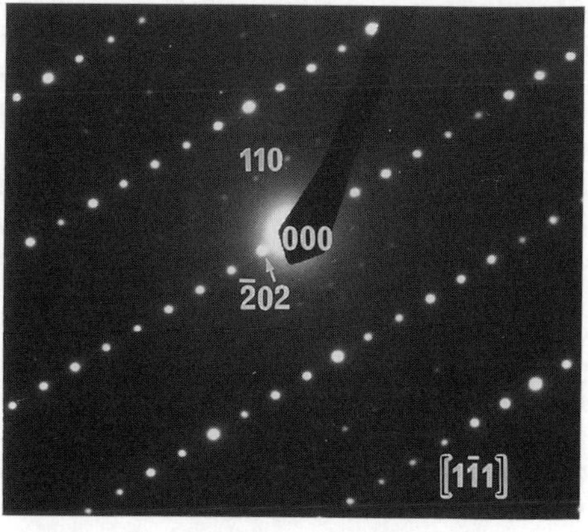

Figure 3a. Electron diffraction pattern of $Li_{10}W_{18}O_{49}$ along [1 -1 1] zone axis. This indexing has been based on the cell 2a x 2b x 2c.

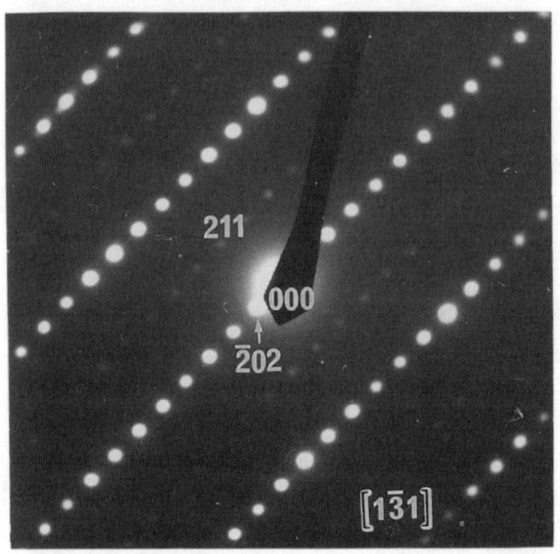

Figure 3b. Electron diffraction pattern of $Li_{10}W_{18}O_{49}$ along the [1 -3 1] zone axis. This indexing has been made based on the cell 2a x 2b x 2c.

Although it is very difficult to predict the maximum occupancy of lithium in each tunnel, some considerations can be made. In early works, for related structures, there are evidences that lithium atoms preferentially occupy the triangular positions [8-10]. If this is also the case in here, when lithium insertion begings in $W_{18}O_{49}$ six different holes can accomodate the first lithium ions and an ordering of this cation has not been detected.

For high lithium content, quadrangular and hexagonal tunnels accomodate one or more lithium. In this situation lithium ordering is more likely and many models of ocupation can be proposed for the compositions between 9 and 17 lithium per formula. Nevertheless for the maximum lithium content, 23.5 lithium per formula, a model of occupation can be suggested :

tunnel type	tunnel / $W_{18}O_{49}$ cell	occupancy / tunnel	Total
triangular	6	1	6
quadrangular	7	2	14
hexagonal	1	3	3

The insertion of sodium, a larger cation than lithium, leads to a decrease an order of magnitude lower than the amount of cation inserted in $W_{18}O_{49}$. So that when sodium is inserted electrochemically in $W_{18}O_{49}$ a maximum of 1.8 sodium can be reached [6]. Having in mind the ionic radius of Na^+ (r^{VI} = 1.02 Å) and the maximum amount intercalated, we propose that sodium insertion proceeds by the occupancy of the unique hexagonal tunnel present in $W_{18}O_{49}$ cell. On the other hand, since the triangular tunnels are very small to accomodate sodium, the incorporation of sodium could happen in the quadrangular tunnels but it would lead at least to a minimum composition of $Na_7W_{18}O_{49}$. In agreement with this considerations, Takusagawa et al. [9] showed that sodium atoms occupy preferably tunnels larger than the quadrangular ones.

When sodium is inserted in $W_{18}O_{49}$ the framework is maintained, as in the lithium case, with only a certain loss of crystallinity [6]. In this case however, the shift of the peaks is almost negligible. This is not surprising if we recall the small amount of sodium that can be inserted. On the other hand, the insertion of sodium in the larger tunnel should produce a minimum structural distortion. A microstructural characterization was carried out for two samples with composition $Na_yW_{18}O_{49}$, where y = 0.45 and 0.9. Figure 4 shows some electron diffraction patterns for $Na_{0.45}W_{18}O_{49}$ (along the [1 -1 0] and [1 -3 0] zone axes). Again the extra spots observed can only be indexed on the basis of a new supercell. Here, it corresponds to the dimensions a x b x 2c with respect to the $W_{18}O_{49}$ cell. This means that sodium is ordered only along the c-axis. Different crystals were observed and the a and b-axis were never observed to be doubled.

These results are in good agreement with the model suggested, where if sodium atoms occupy the hexagonal tunnels, sodium ordering may exist for y = 0.5 in at least one of the three directions. On the other hand, insertion of 0.5 sodium atoms in quadrangular tunnels leading to a situation of ordering highly unlikely since the number of quadrangular tunnel per unit cell is quite large.

The incorporation of 1 sodium per formula must proceed by the total occupancy of the unique hexagonal tunnel. In this case, there is sodium ordering but as every hexagonal tunnel is occupied there is no superstructure and the unit cell of $W_{18}O_{49}$ is maintained. The electron diffraction patterns for this sample is in agreement with this reasoning. In this way, all patterns can be easily indexed on the basis of the parent oxide cell. In fact evidence of superstructure were never found in the many crystals investigated having this composition.

More data are obviously necessary to confirm the two models of occupation of lithium and sodium in the three type of tunnels of $W_{18}O_{49}$. In this way different composition are now being investigated to fully scan the systems $Li_xW_{18}O_{49}$ and $Na_yW_{18}O_{49}$.

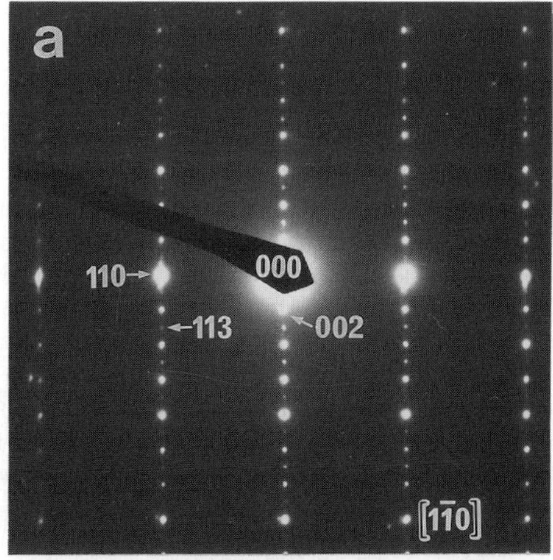

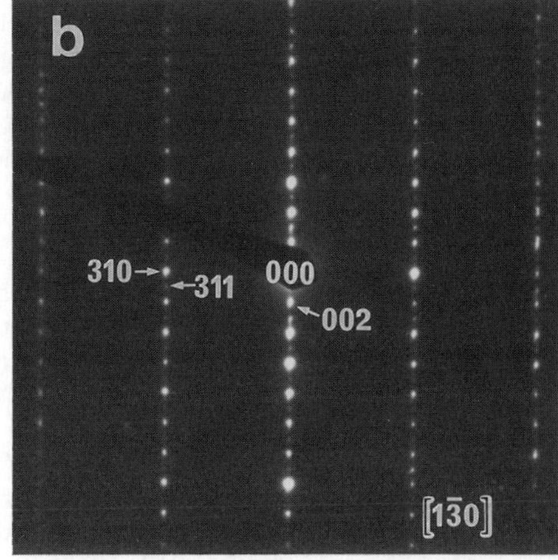

Figure 4. Electron diffraction patterns of $Na_{0.45}W_{18}O_{49}$ along the a) [1 -1 0] and b) [1 -3 0] zone axes. The indexing has been made based on the cell a x b x 2c.

CONCLUSIONS

The insertion of alkali metals in $W_{18}O_{49}$ is clearly governed by the size of the alkali cations. While lithium insertion ($r^{VI} = 0.76$ Å) leads to a maximum composition of $Li_{23.5}W_{18}O_{49}$, when sodium is inserted, only $Na_{1.8}W_{18}O_{49}$ can be reached ($r^{VI} = 1.02$ Å). Nevertheless, X-Ray diffraction studies have shown that the $W_{18}O_{49}$ cell is maintained in both cases with only a certain loss of crystallinity. On the other hand, an electron microscopy study showed new structural features. Six lithiated samples with compositions related to 0.4, 2, 5, 9, 10 and 17 were synthetized and characterized by electron diffraction techniques.

For samples with x = 0.4, 2, 5 the electron diffraction patterns are in agreement with the X-Ray diffraction and only some streaking indicates a certain disorder in the crystal. On the contrary, samples with lithium content x = 9, 10 and 17 showed extra spots in their electron diffraction patterns that can only be indexed on the basis of a new cell 2a x 2b x 2c.

Taking into account the ionic radius of sodium and the maximum insertion in $W_{18}O_{49}$ we propose that sodium insertion proceeds by the occupancy of the unique hexagonal tunnel present in the unit cell. The microstructural characterization of two samples with composition $Na_{0.45}W_{18}O_{49}$ where sodium is ordered along c-axis and $Na_{0.9}W_{18}O_{49}$ where ordering is not observed seems to confirm this occupancy model.

On the basis of these results we believe that both lithium and sodium ordering can be related with the filling of the largest tunnels. Lithium ordering for high cation content suggest that this order can be associated with the incorporation of one or more lithium in quadrangular and hexagonal tunnels. On the other hand, sodium insertion seems to be more *selective* and a simple model where only the hexagonal tunnels is occupied can explain the results obtained by the electron diffraction study.

ACKNOWLEDGMENTS

We thank Dr. A. Kuhn for his kind help in the electron microscopy study. AMC wants to thank UANL (MEXICO) for making possible the stay at Universidad Complutense. The authors would like to thank CICYT (SPAIN) grant MAT95-0809 for financial support and the Centro de Microscopía Electrónica (UCM) for technical assistance. FGA also thanks Universidad San Pablo-CEU for supporting the project 5/96.

REFERENCES

1.- Kent H. Cheng and M. Stanley Whittingham, Solid State Ionics **1**, 151 (1980).
2.- S.A. Kay, D. Phil. Thesis, University of Oxford (1986).
3.- A. Martínez de la Cruz, F. García-Alvarado, E. Morán, M.A. Alario-Franco and Leticia M. Torres-Martínez, J. Mater. Chem. **5**, 513 (1995).
4.- A. Martínez de la Cruz, Leticia M. Torres-Martínez, F. García-Alvarado, E. Morán and M.A. Alario-Franco, Mat. Res. Soc. Proc. Vol. **369**, 131 (1995).
5.- E. Banks and A. Goldstein, Inorg. Chem. **7**, 967 (1968).
6.- A. Martínez de la Cruz et al., to be published.
7.- A. Magnéli, Ark. Kemi **1**, 213 (1949).
8.- B.O. Marinder, T. Hörlin and A. Magnéli, Mat. Res. Bull. **14**, 387 (1979).
9.- F. Takusagawa and R.A. Jacobson, J. Solid St. Chem. **18**, 163 (1976).
10.- S.C. Abrahams, P.B. Jamieson and J.L. Bernstein, J. Chem. Phys. **54(6)**, 2355 (1971).

REACTIVITY OF CHRYSOTILE ASBESTOS IN ACIDS: MECHANISM OF TRANSFORMATION TO SILICON DIOXIDE HEMIHYDRATE UPON LEACHING OF MAGNESIUM

Anne VAILLANCOURT* **, Georges DENES* and Raymond LE VAN MAO**
*Laboratory of Solid State Chemistry and Mössbauer spectroscopy, and **Catalysis Laboratory, Laboratories for Inorganic Materials, Department of Chemistry and Biochemistry, Concordia University, Montreal, Quebec, H3G 1M8, Canada, gdenes@vax2.concordia.ca

ABSTRACT

Acid leaching chrysotile asbestos results in a stepwise loss of magnesium which reduces the strain locally. This results in a partial unwiding of the structural sheets which makes possible further leaching. Strong acids are used for a coarse brutal attack of the chrysotile structure, whereas weak acids allow a fine control of the magnesium leaching degree required to have the adequate properties of the partially leached asbestos which is used for the preparation of zeolites with enhanced catalytic properties. The final material is not carcinogenic. The leached materials were characterized by atomic absorption spectrometry, bulk density and BET specific surface area measurements, and X-ray diffraction.

INTRODUCTION

Large deposits of chrysotile asbestos are found in many places around the world. Its high thermal and chemical stability had resulted in many practical applications, particularly for fire insulation. Unfortunately, its fibrous texture makes it highly carcinogenic, and this has resulted in a ban of its use in many countries. We recently developped a treatment to make it harmless and we designed a safe practical use of the treated asbestos. The treatment consists in acid leaching, which removes the magnesium, destroys the fibrous structure and gives non-carcinogenic silicon dioxide hemihydrate $SiO_2.0.5H_2O$, which we use as a source of silica for the synthesis of zeolites [1]. By controlling carefully the final stages of magnesium leaching, the catalytic properties of the zeolite obtained from the partially leached chrysotile are enhanced compared to those of zeolites prepared from classical sources of silica [2].

In this work, we have investigated the detailed mechanism of the acid leaching process by use of atomic absorption spectrometry, X-ray diffraction, bulk density and BET specific surface area measurements. Chrysotile asbestos $Mg_3Si_2O_5(OH)_4$ is made of an infinite silica sheet covered with a layer of magnesium which is octahedrally coordinated to four hydroxyl groups and two oxygen atoms bridging to silicon. The large size of the magnesium containing layers as compared to the size of the silicon containing layers makes highly strained sheets, which bend and wind around themselves to give hollow fibers which are carcinogenic [3]. The behavior upon leaching can be explained based on these particular structure and texture.

Mat. Res. Soc. Symp. Proc. Vol. 453 © 1997 Materials Research Society

EXPERIMENTAL PROCEDURES

The chrysotile asbestos used in this work was short fiber commercial grade originating from the batch 7TF12 which was mined in Asbestos, Quebec. We showed recently that this chrysotile is orthochrysotile-2Ocl (JCPDS No. 22-1162), with an average fiber diameter of 156Å, and that it is very strongly stressed in the b direction of the unit-cell, due to sheet bending, which is caused by the large size of the octahedrally coordinated magnesium, compared to the smaller size of the tetrahedrally coordinated silicon [4].

Elemental analysis was carried out by atomic absorption spectrometry (AAS) on a Perkin-Elmer 503 absorption spectrometer. The experimental chemical composition (in wt% dried oxides) of chrysotile asbestos is: 41.6% SiO_2, 50.2% MgO, 8.0% Fe_2O_3, 0.1% Al_2O_3 and 0.1% Na_2O. From the analytical results, the degree of advancement of leaching, i.e. the "Magnesium Leaching Degree" (MLD) was calculated as follows:

$$MLD(\%) = 100[(MgO)_i - (MgO)_f]/(MgO)_i \qquad (1)$$

where $(MgO)_i$ and $(MgO)_f$ are the magnesium oxide content of the samples before and after leaching, respectively.

Leaching was performed according to a two step process designed in our laboratories [1, 2]. The fibers were first subjected to a strong leaching by a mineral acid (HCl or H_2SO_4) at various acid concentrations and various temperatures from $20^\circ C$ to $80^\circ C$, for various periods of time. Then, the acid was diluted with cold water, the solid was filtered, washed with water and dried at $120^\circ C$. In a second step, mild leaching was performed by using a weak acid, such as oxalic acid (OXA) and acetic acid (ACA).

X-ray powder diffraction was performed on a Philips X-ray powder diffractometer that has been automated by Sietronics. The Ni-filtered K_α peak of copper was used [$\lambda(K_{\alpha 1}=$ 1.54051Å). Diffraction pattern were run at the angular velocity of $1^\circ(2\theta)/min$ with a step size of $0.02^\circ(2\theta)$. NaCl was used as an internal standard for line position, line intensity and linewidth. The chrysotile degree of crystallinity (DC) was calculated as follows:

$$DC(\%) = 100 I_f/I_i \qquad (2)$$

where I_i and I_f are the normalized intensities of a well defined Bragg peak (or group of peaks) of chrysotile before and after leaching, respectively. Data accumulation and processing were carried out by using the SIE112 software of Sietronics. The JCPDS powder database and the Micro Powder Diffraction Search-Match (μPDSM) of Fein Marquat were used for phase identification.

Bulk density measurements were carried out by displacement in CCl_4 (Archimedean method) by using a special home made apparatus which allows wetting the degassed sample with CCl_4 while still under vacuum (error < 2%).

Specific surface area measurements were performed on a classical multipoint BET volumetric apparatus at liquid nitrogen temperature, after outgassing the samples at $350^\circ C$ for 3 hours under a vacuum of 10^{-4} torr.

RESULTS AND DISCUSSION

Figure 1 gives the chrysotile degree of crystallinity DC versus the normality of the strong acid. Up to normality 1N, DC increases to 115%. This may look strange. It does not mean that the crystallinity of chrysotile increases. It is due to the presence of two impurities, brucite $Mg(OH)_2$ (JCPDS No. 7-0239) and pyroaurite $Mg_6Fe_2CO_3(OH)_{16}.4H_2O$ (JCPDS No. 25-0521), which are present in substantial amounts in the the chrysotile used. These impurities are basic and therefore dissolve at acid normalities that are too low for leaching chrysotile. It results a lower sample mass for the same amount of crystalline chrysotile, thus a higher apparent DC. All the results presented below have been corrected for this, by recalculating DC, based on DC = 100% at 1N normality. Above 1N, leaching of chrysotile becomes significant, and it results in a decrease of DC, which reaches zero at ca. 7N. The final material, at DC = 0, contains no magnesium and is amorphous silicon dioxide hemihydrate $SiO_2.0.5H_2O$, with sometimes a minor peak for α-quartz. It should be noticed that the nature of the acid, HCl or H_2SO_4 does not affect the results.

Figure 2 shows the relationship between DC and the magnesium leaching degree MLD. The loss of crystallinity with increasing MLD is expected. In a first step, DC decreases rapidly to about 40% at MLD = 50 %. This probably corresponds to the dissolution of "brucite-type Mg", not to be confused with the free brucite $Mg(OH)_2$. "Brucite-type Mg" makes up 2/3 of the total Mg content of chrysotile and has the same coordination as Mg in free brucite. The fact it is coordinated by OH groups only (no Mg-O-Si bridges) and OH groups are at the surface of the wound sheet, make it more prone to acidic attack. It is not understood at this point why DC decreases faster in HCl. From MLD = 50% to 80%, DC changes little. This probably corresponds to the removal of some of the "skeletal Mg", which bridges to Si. This requires unwinding the sheets, which are still held together by Si-O bonds, and thus minimized the loss of crystallinity. When enough skeletal Mg has been removed (MLD > 80%), structural collapse takes place and DC falls rapidly to zero.

The bulk density versus MLD is shown on figures 3 and 4. Four regions, A to D, can be identified, in terms of how fast the density changes with increasing MLD. In regions A (MLD $\cong$ 0 - 45%) and C (MLD $\cong$ 65 - 90%), the density is nearly constant. On the other hand, in regions B (MLD $\cong$ 45 - 65%) and D (MLD $\cong$ 90 - 100%), it decreases rapidly. Region A (< 1N) corresponds to the dissolution of the basic impurities and minor attack on the outer surface of the fibres. Since these impurities have a density similar to that of chrysotile, their disappearance from the sample has little influence on the density of the samples. Region B (1 - 2.5N) corresponds to the leaching of a large part of the "brucite-type magnesium" creating many Mg vacancies. Since the volume changes little whereas the mass decreases, the density decreases rapidly. In region C (2.5 - 4.5N), there is significant removal of "skeletal Mg" with partial structural collapse, which eliminates the Mg vacancies created, thereby decreasing the volume, which compensates for the loss of mass, resulting in a nearly constant density. In region D (> 4.5N), the most internal "skeletal Mg" is being lost with a complete structural collapse. The density decreases rapidly to 1.9 g/cm^3, which we assign to amorphous silicon hemihydrate, $SiO_2.0.5H_2O$. This value is lower than those of all

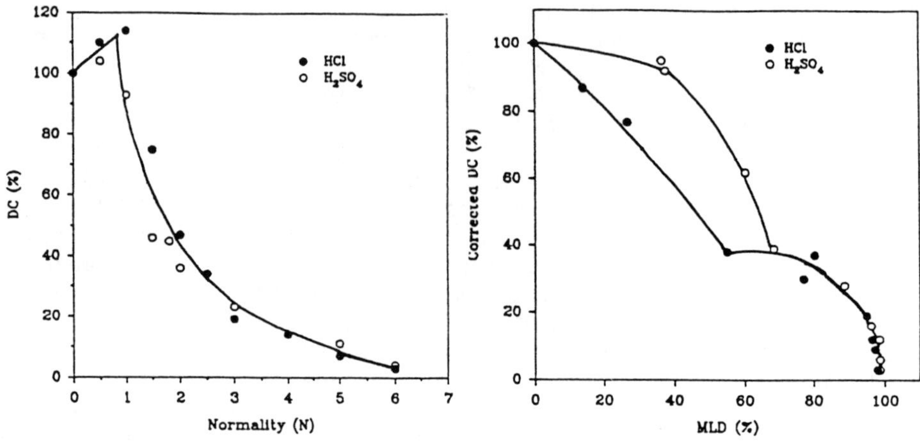

Figure 1: Chrysotile degree of crystallinity DC versus the normalty of the acid used for leaching at 80°C for 24 hours.

Figure 2: Degree of crystallinity DC versus MLD (4 hours at 80°C)

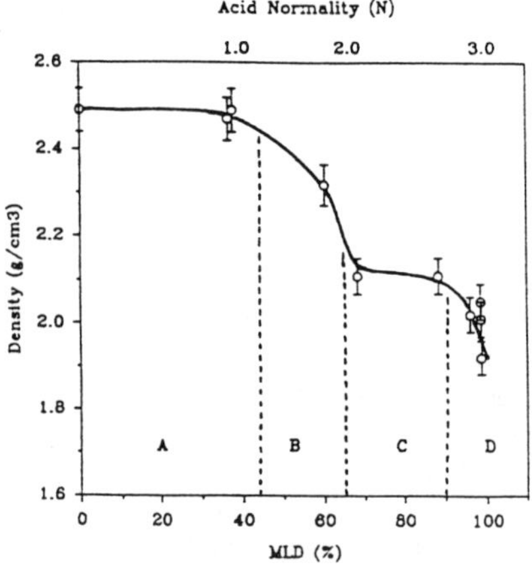

Figure 3: Bulk density versus MLD (H_2SO_4, 4 hours at 80°C).

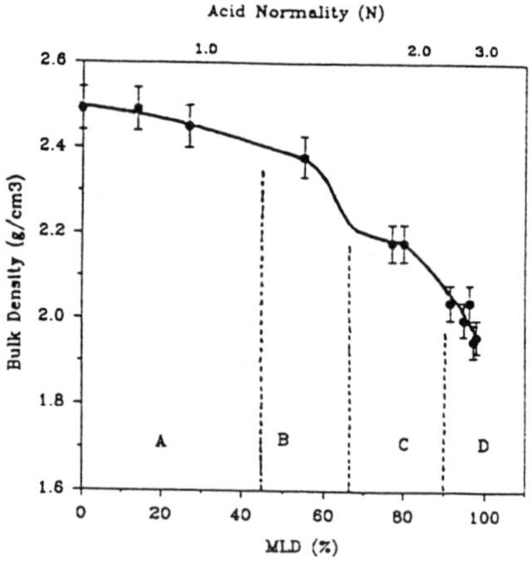

Figure 4: Bulk density versus MLD (HCl, 4 hours at 80°C).

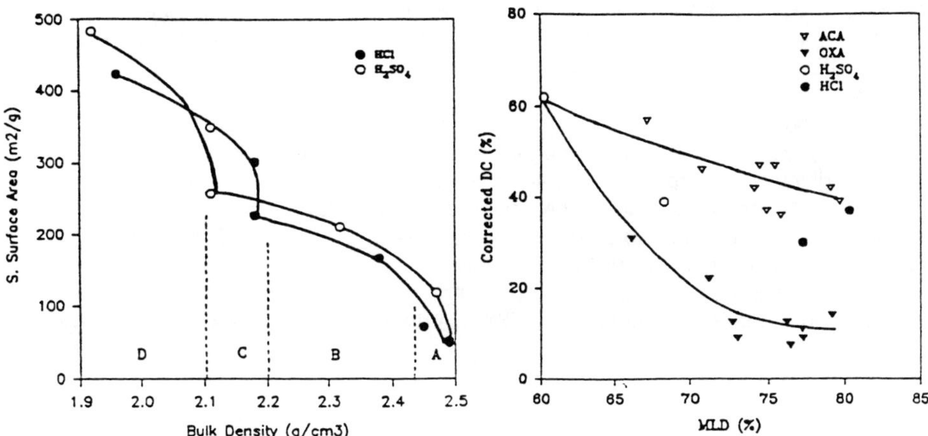

Figure 5: BET specific surface area versus bulk density.

Figure 6: Degree of crystallinity versus MLD for secondary leaching with weak acids (4 hours at 80°C).

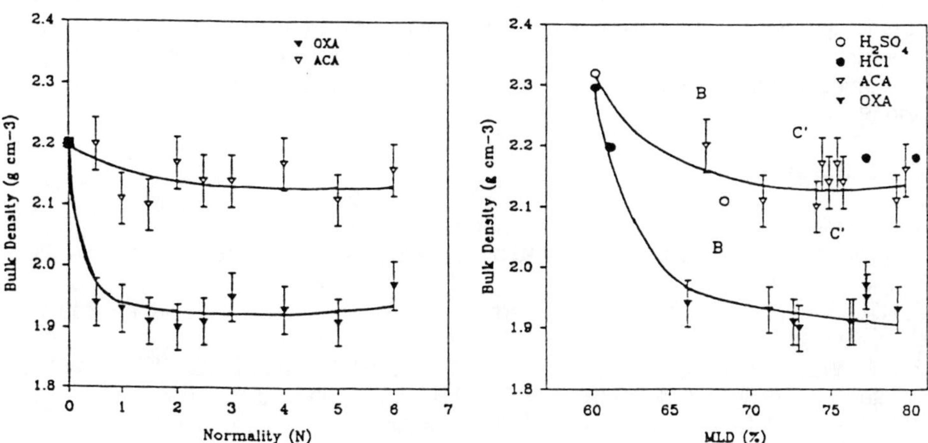

Figure 7: Bulk density versus acid normality in secondary leaching.

Figure 8: Bulk density versus MLD in secondary leaching.

polymorphs of crystalline SiO_2. This is undoubtely due to its amorphous state and the presence of water.

The BET specific surface area varies with the bulk density in a similar fashion, i.e. the same four regions can be identified for both acids (fig. 5). The surface area increases when magnesium is leached out, creating vacancies, and when the structure collapses to give $SiO_2.0.5H_2O$, which is made of very small particles and therefore has a high surface area.

Figures 6 to 8 show the variation of DC and the bulk density with secondary leaching by weak acids. Starting values after partial primary leaching with strong acids are also given for comparison. Only minor smooth changes are observed for acetic acid (ACA), when oxalic acic (OXA) gives an initial sudden change followed by little further change. This probably reflects the relative acid strength of the two weak acids. Although the two weak acids are probably unable to remove efficiently the skeletal magnesium, their weakness has the advantage of providing a controllable fine tuning of the leaching process, i.e. some sort of "chimie douce", in contrast with the brutal strong acid attack.

CONCLUSION

The above observations show that acid leaching chrysotile asbestos dissolves very rapidly the basic impurities, brucite and pyroaurite. Then, acid attack on the outer layers of the fibers occur on the hydroxyl side, which is the convex side, and is therefore exposed to the acidic solution. "Brucite-type magnesium" is removed preferentially, with no major structural reorganization. The final step involves the removal of "skeletal magnesium", which results in a structural collapse with formation of amorphous silicon dioxide hemihydrate. The bulk density and the BET surface area vary in agreement with these steps. Soft leaching with weak acids makes possible the fine tuning of DC, MLD and the surface area, which is very useful for optimizing the properties of the zeolites (A, X and ZSM-5) prepared by using the leached chrysotile 5 and 6].

REFERENCES

1. G. Denes, P. Kipkemboi, R. Le Van Mao and A. Vaillancourt, in <u>Production and Processing of Fine Particles</u>, Edited by A.J. Plumpton (Pergamon Press, New York, 1988), p. 599-607.
2. R. Le Van Mao, P. Kipkemboi, P. Levesque, A. Vaillancourt and G. Denes, Zeolites **9**, 405 (1989).
3. A.P. Middleton, in <u>Asbestos</u>, Vol. 1, edited by L. Michaels and S.S. Chissicks (Wiley-Interscience, New York, 1979), p.266; A.A. Hodgson, ibid., p. 75-79.
4. G. Denes, R. Le Van Mao and A. Vaillancourt, in <u>Thin Films: Stresses and Mechanical Properties V</u>, edited by S.P. Baker, C.A. Ross, P.H. Townsend, C.A. Volkert and P.Børgesen (Mat. Res. Soc. Symp. Proc. **356**, Pittsburg, PA, 1995), p. 105-110.
5. R. Le Van Mao, B. Sjiariel and J. Dunnigan, Zeolites **11**, 804 (1991).
6. R. Le Van Mao, P. Lévesque, B. Sjiariel and P.H. Bird, Can. J. Chem. **63**, 3464 (1985).

V_2O_5 AND β-Cu$_x$V$_2$O$_5$ THIN FILMS GROWN BY MOCVD

A. GLEIZES, J.-M. BUFFORN, C. SALZMAN, E. VIRETTE, F. SENOCQ
E.N.S. Chimie de Toulouse (I.N.P.T.), Laboratoire "Cristallochimie, Réactivité et Protection des Matériaux", 118 route de Narbonne, F-31077 Toulouse cedex 4, France, gleizes@cict.fr

ABSTRACT

Thin films of V_2O_5 and β-Cu$_x$V$_2$O$_5$ have been grown by MOCVD using VO(O^{i}Pr)$_3$ and Cu(tmhd)$_2$ (tmhd = tetramethylheptanedionato) as precursor molecules. Films were grown on chips of Si$_3$N$_4$ coated silicon wafers in a cold wall reactor using a H.F. heater. A mixture of helium and oxygen was used as a reactive carrier gas, and depositions were performed at low pressure. The films were examined by SEM, characterized by XRD, and analyzed by EDS and EMPA techniques. All the films proved to be carbon free.

For V_2O_5, the substrate temperature was varied from 450 to 630°C. The films are highly c-axis oriented for both ends of the temperature range, and less oriented for intermediate temperatures. For 630°C, the XRD pattern consists almost entirely of reflections (001) and (002).

For β-Cu$_x$V$_2$O$_5$, the substrate temperature was varied from 450°C to 650°C. Pure β-phase films with x varying from 0.25 to 0.55 have been obtained above 500°C, by using well chosen gas phase composition. Morphology and texture depend dramatically on the temperature. The most oriented films exhibit a strong anisotropy of electrical surface conductivity.

INTRODUCTION

The layered structure of V_2O_5 has long been used as an intercalation host for small cations, giving rise to vanadium oxides bronzes (VOB) with a wide range of applications including for instance catalysis, rechargeable batteries, and optical devices. The host structure is made of sheets generated from edge and corner sharing VO$_4$ square pyramids, stacked along the [001] direction of an orthorhombic lattice (scheme 1). Since Wadsley's pioneering work in 1955 [1], VOB have been extensively investigated [2 and references therein]. The copper family [3] comprises three phases. α-Cu$_x$V$_2$O$_5$ (x < 0.02) results from copper atoms intercalated between the layers of the V_2O_5 structure. ε-Cu$_x$V$_2$O$_5$ (x > 0.85) has also a layered structure though different from V_2O_5. β-Cu$_x$V$_2$O$_5$ (0.26 < x < 0.64) is characterized by tunnels parallel to the [010] direction of a monoclinic lattice (scheme 2).

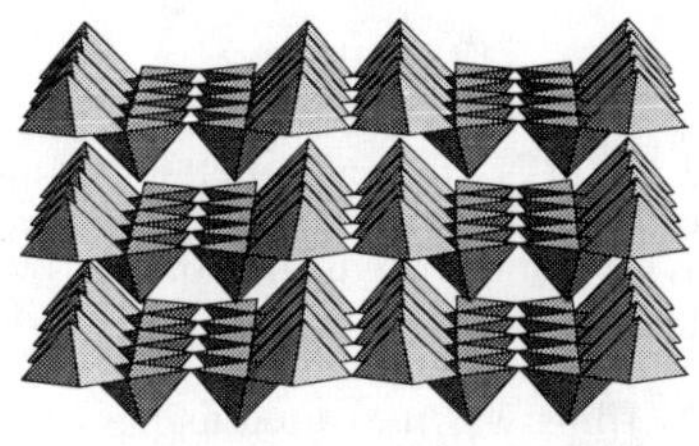
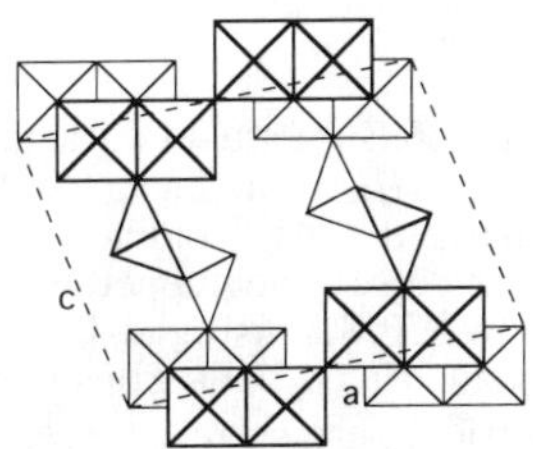

Scheme 1   Scheme 2

For their applications, V_2O_5 and VOB are often shaped into films. Such films have been made by evaporation, sputtering, electro-deposition, sol-gel technology and spin coating followed by chemical or electrochemical introduction of small cations. So far no CVD processing has been reported for these materials. Formation of V_2O_5 was observed by Takahashi, Kanamori, Hashimoto, Moritani and Masuda while using MOCVD to prepare VO$_2$ films [4]. We undertook to use the MOCVD technique to grow thin films of V_2O_5 and β-Cu$_x$V$_2$O$_5$. This VOB was retained because of our experience in the MOCVD of copper containing compounds.

Mat. Res. Soc. Symp. Proc. Vol. 453 © 1997 Materials Research Society

EXPERIMENT

<u>Preparation of precursors</u>: Vanadium(V) tris(isopropoxy)oxide, $VO(O^iPr)_3$ was prepared from V_2O_5 (Aldrich, purity 98%) and ^iPrOH (Prolabo, purity 99%, distillation on $SnCl_2$ prior to use) according to the method described in [5]. It was purified by distillation (65°C, 2 torr). Copper bis(2,2,6,6-tetramethyl-3,5-heptanedionate, $Cu(tmhd)2$, was prepared from copper acetate (Aldrich, purity 97%) and tetramethylheptanedione (Aldrich, 95%)) according to a method adapted from [6]. The crude product was recrystallized from methanol.

<u>Chemical Vapor Deposition (CVD) Apparatus</u>: CVD experiments were carried out in a horizontal cold-wall reactor consisting of a quartz tube (inside diameter = 46 mm), and a steel susceptor heated by a high-frequency generator. The susceptor was placed on a quartz support inclined at about 7° to the direction of the gas flow. A thermocouple plugged into the susceptor allowed regulation of the temperature. The substrates were chips of Si_3N_4-coated silicon wafer laying on the susceptor. Before use they were degreased in boiling acetone and air dried. Helium was used as a carrier gas. Oxygen was added to the precursor/helium mixture at the entrance of the furnace. Gas flows were controlled through mass-flowmeters. Total pressure was measured independently of the gas composition and automatically monitored using a capacitance manometer and a throttle valve control sytem. Experimental parameters were fixed or varied as reported in Table I:

TABLE I. Experimental conditions for the MOCVD of V_2O_5 and of $Cu_xV_2O_5$.

Parameters	V_2O_5	$Cu_xV_2O_5$
Total pressure (torr)	25	5 - 50
Substrate temperature (°C)	450 - 630	450 - 650
$VO(O^iPr)_3$ temperature (°C)	25	35
$Cu(tmhd)_2$ temperature (°C)	-	115 - 130
He flow on $VO(O^iPr)_3$ (sccm)	12	20
He flow on $Cu(tmhd)_2$ (sccm)	-	30 - 50
O_2 flow (sccm)	50	0 - 130
Duration (hr)	2 - 6	4

<u>Characterization</u>

The films were submitted to grazing incidence ($\Omega = 2°$) and/or θ-θ (2θ scanning range of 5-80°) X-ray diffraction on a SEIFERT XRD 3000TT diffractometer, using back-monochromatized, CuK_α radiation. SEM observations and analyses were performed with a JEOL 6400 microscope equipped with an EDS analyser (Link Analytical). Analyses were also performed by EPMA using a CAMECA microprobe (accelerating voltage of 10kV)and $Cu_{0.40}V_2O_5$ and $Cu_{0.60}V_2O_5$ powder samples as standards.

The electrical surface resistance R_s of $Cu_xV_2O_5$ films was measured using a four-probe technique. R_s was deduced from the applied tension U (V) and the measured intensity I (A) according to the formula: $R_s = (\pi/\ln2)(U/I)$.

RESULTS

<u>CVD of V_2O_5</u>

Chemical vapor depositions of V_2O_5 were carried out at 450, 500, 550, 600 and 630°C, using the conditions reported in Table I. A large excess of oxygen was used in order to ensure complete oxidation of vanadium.

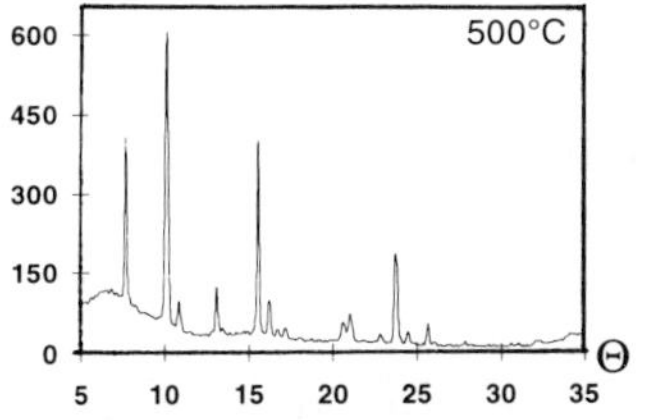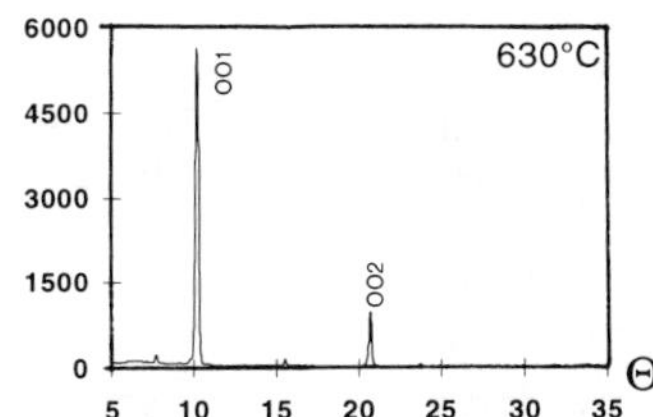

Fig. 1 - XRD patterns of V_2O_5 films grown at 500°C and 630°C. Indexation from [7].

The films proved to be carbon free on EDS analysis. The SEM showed the deposits obtained at T > 550°C to be made of a thin layer of small, platelike crystals (about a few µm in size) covered with a non-adhesive, hairy layer of whiskers up to 0.4 mm in length. For T ≤ 550°C, the surface of the films consisted only of small crystals whose average size decreased from a few µm (550°C) to 0.5 µm (450°C).

XRD patterns showed lines relevant to V_2O_5 only, with some marked changes in relative intensities from one diagram to the other one. The patterns have five lines in common, namely (001), (101), (110), (310)-(400), (002). For 630°C (Figure 1), only lines (001) and (002) showed prominently, thus indicating a highly textured film with the V_2O_5 layers parallel to the surface of the substrate. The films grown at 600, 550 and 500°C were less oriented (Figure 1), but the one grown at 450°C was more oriented than them. The degree of orientation is most likely to be also related to the duration of deposition which was 4 hours at 450°C, only 2 to 3 hours for 500, 550 and 600°C, and 6 hours for 630°C.

CVD of $Cu_xV_2O_5$

A series of experiments were conducted with the substrate temperature fixed at 550°C and by varying the pressure and the ratio of molar fractions $\chi(O_2)/\chi(V)$ (V stands for $VO(O^iPr)_3$). When oxygen was not used, the deposits comprised metal copper or copper oxides and vanadium oxides different from V_2O_5. Their colors varied from lustreless black to shining copper-red according to the metal copper content. When oxygen was added to the vapor phase, the color of the deposit varied from silver grey to shining bluish dark.

The results are reported on the diagram of Figure 2. Deposits grown at 5 torr were made of copper oxides in a matrix of V_2O_4. Some of them contained $Cu_{11}O_2(VO_4)_6$. β-$Cu_xV_2O_5$ was obtained at higher pressures (25, 30, 50 torr), either pure (marked o in Figure 2) or mixed with V_6O_{13} (o∗) or with $Cu_2V_2O_7$ (o+). For P = 25 torr, increasing $\chi(O_2)/\chi(V)$ favored the formation of $Cu_2V_2O_7$, and for a molar ratio of 5000 the deposit contained only this phase. Pure β-$Cu_xV_2O_5$ formed for the lowest $\chi(O_2)/\chi(V)$ ratios.

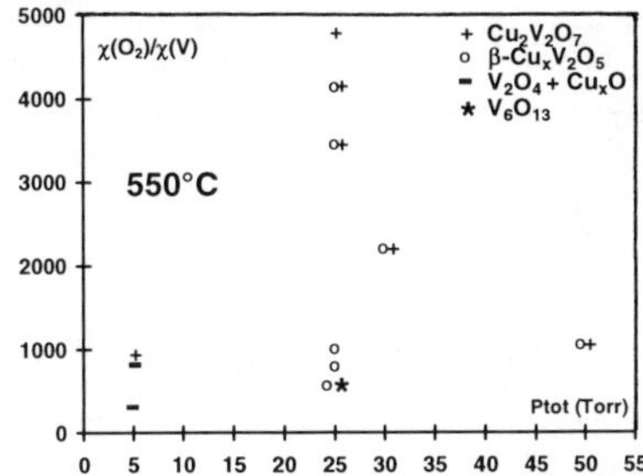

Fig. 2 - Phases obtained as a function of total pressure and molar ratio $\chi(O_2)/\chi(V)$.

In a second series of experiments, the substrate temperature was varied from 450 to 650°C while keeping the molar fractions nearly constant. Their values were choosen so as to maximize chances to deposit pure $Cu_xV_2O_5$ from the former study at 550°C. The pressure was maintained at 25 torr. These experiments are presented in Table II.

TABLE II. Substrate temperatures, precursor and O_2 molar fractions, and measured x values for the films of $Cu_xV_2O_5$.

Samples	T (°C)	$10^4\,\chi(Cu)$	$10^4\,\chi(V)$	$\chi(O_2)$	$\chi(Cu)/\chi(V)$	$\chi(O_2)/\chi(V)$	x
A	450	1.3	6.3	0.5	0.21	794	>0.2[a]
B	500	1.4	6.3	0.5	0.22	794	>0.2[b]
C1	550	1.6	8.4	0.5	0.19	595	0.40
C2	550	1.4	5.3	0.5	0.26	943	0.54
D	600	1.6	6.3	0.5	0.25	794	>0.2[c]
E1	650 (2hr)	2.2	6.3	0.5	0.35	794	0.28
E2	650 (4hr)	2.2	6.3	0.5	0.35	794	0.25

[a]: EPMA not reliable due to film morphology

[b]: mixture of α and β phases

[c]: one extra unidentified line on XRD pattern

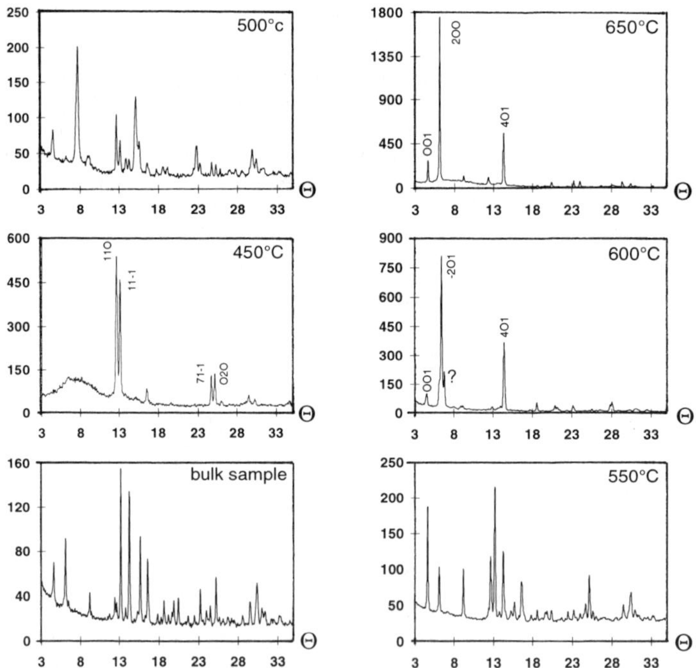

Fig. 3 - XRD patterns of films of β-$Cu_xV_2O_5$. Indexation from [8].

From XRD (Figure 3), β-$Cu_xV_2O_5$ proved to be the only phase deposited at 450, 550 and 650°C. The film grown at 500°C proved to be a mixture of β-$Cu_xV_2O_5$ and α-$Cu_xV_2O_5$ (or V_2O_5). The film grown at 600°C showed one extra diffraction line which could not be attributed definitely.

Copper contents x were determined by EPMA. Neither EPMA or EDS showed significant presence of carbon.

The substrate temperature proved to influence both the morphology and the texture of the films dramatically. In the SEM, the deposits appeared to be made of small crystals whose shapes and sizes varied with the temperature. For 450°C, they were made of a few μm large, acicular crystals forming radiating groups, hence showing a wide range of orientations about a normal to the surface of the substrate. This morphology precluded reliable electron microprobe analysis. For 650°C, crystals were several tens of μm large, elongated plates, nearly parallel to the surface of the substrate. For intermediate temperatures of 500, 550 and 600°C, intermediate shapes, sizes and textures were observed. Deposits grown at 650°C also comprised roughly spherical nodules with diameters up to 15 μm, having about the same composition as the crystals.

XRD patterns (Figure 3) confirm the tendency to texturation on increasing temperature. They have five lines in common, namely (001), (200), (002), (401) and (40-5) with different relative intensities from one pattern to the other one. It should be noticed that the strongly oriented films obtained at 600 and 630°C do not show the same prominent lines. In both cases the tunnels of the β-phase structure are parallel to the surface of the substrate. The pattern of the film made of acicular crystals forming radiating groups (450°C) shows four prominent lines different from those at 600 and 630°C.

The surface electrical resistance R_S was measured in three directions: i) parallel to the gas flow direction in the reactor (o in Figure 4), ii) perpendicular to this direction (∗), and iii) halfway between (+). No measurement was made on the film grown at 450°C because of its morpholy (*vide supra*). R_S is isotropic for the films grown at 500°C and 550°C, and markedly anisotropic at 600 and 650°C, in accordance with the orientation effect observed by XRD. It is difficult to correlate x and R_S since x is known for only two of the samples submitted to electrical measurements. Consistently enough, the highest resistance (*ca.* 2000 Ω) corresponds to the lowest copper content, $x = 0.25$ (650°C), and the lowest resistance (*ca.* 200 Ω) to the highest copper content, $x = 0.45$ (550°C).

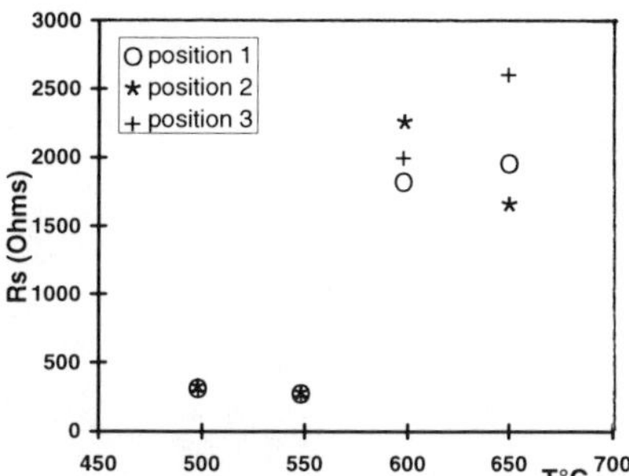

Fig. - 4. Surface electrical resistance measured in three directions on β-$Cu_xV_2O_5$ films.

CONCLUSIONS

This preliminary work has demonstrated the feasability of growing films of V_2O_5 and β-$Cu_xV_2O_5$ by using the MOCVD technique at low pressure. A large excess of oxygen in the vapor phase is needed. At temperature $\geq$ 600°C the films are highly oriented. The layers of V_2O_5 and the tunnels of β-$Cu_xV_2O_5$ tend to be parallel to the surface of the substrate. More investigations are needed to control the copper content.

AKNOWLEDGEMENT

We thank the CNRS for supporting this work. We are grateful to Mrs. M. Reversat and Dr. D. Oqab for performing SEM, and to Mr. de Parceval for performing EMPA at Institut National des Sciences Appliquées (INSA) de Toulouse.

REFERENCES

1. A. D. Wadsley, Acta Crystallogr. **8**, 695 (1955).

2. J.-M. Savariault, E. Deramond and J. Galy, Z. Krist. **209**, 405 (1994).

3. A. Casalot, A. Deschanvres, P. Hagenmuller and B. Raveau, Bull. Soc. Chim. Fr., 1730 (1965). J. Galy, A. Casalot, M. Pouchard and P. Hagenmuller, Bull. Soc. Fr. Minéral. Cristallogr. **XC**, 544 (1967). A. Casalot, Thèse d'Etat, Bordeaux (1968). J. Galy, J. Darriet, A. Casalot, J. B. Goodenough, J. Solid State Chem. **1**, 339 (1970). J. Galy, D. Lavaud, A. Casalot, P. Hagenmuller, J. Solid State Chem. **2**, 531 (1970).

4. Y. Takahashi, M. Kanamori, H. Hashimoto, Y. Moritani and Y. Masuda, J. Mater. Sci. **24**, 192 (1989).

5. A. Lachowitz, W. Höbold and K. H. Thiele, Z. anorg. allg. Chem. **418**, 65 (1975).

6. G. S. Hammond, D. C. Nonhebel, C. S. Wu, Inorg. Chem. **2**, 73 (1963).

7. V_2O_5, orthorhombic, Pmmn, a = 11.516(4), b = 3.5656(4), c = 4.3727(4) Å, after R. Enjalbert and J. Galy, Acta Crystallogr. **C42**, 1467 (1986).

8. $\beta\text{-}Cu_{0.4}V_2O_5$, monoclinic, C2/m, a = 15.17, b = 3.62, c = 10.13 Å, β = 108.0° after A. Casalot, A. Deschanvres, P. Hagenmuller and B. Raveau, Bull. Soc. Chim. Fr., 1730 (1965).

CVD-GROWTH OF THIN-FILM LAYERED Se-CARBON COMPOUNDS

L. GRIGORIAN, S. FANG, G. SUMANASEKERA, A.M. RAO, L. SCHRADER[1], and P.C. EKLUND
Department of Physics and Astronomy and Center for Applied Energy Research, University of Kentucky, Lexington, KY 40506, grigorian@alpha.caer.uky.edu
[1] University of the South, Sewanee, TN 37383

ABSTRACT

Oriented films of layered Se-carbon compounds are grown on Ni substrates in evacuated sealed quartz tubes. Results of X-ray diffraction, Raman scattering and the c-axis electrical transport studies are reported. A discussion of possible models for the carbon -Se interaction is presented.

INTRODUCTION

Traditionally, graphite intercalation compounds (GIC) are produced by exposing a pre-existing graphitic host, such as highly oriented pyrolytic graphite (HOPG), to the vapor or solution of the intercalant which then must diffuse into the host graphite [1, 2]. This kinetic limitation makes it impractical to apply traditional intercalation techniques to insert a number of potentially interesting intercalants, e.g., Se, Te, Ge, Si, etc.

A chemical vapor deposition (CVD) approach to produce GICs has been recently reported [3, 4]. They used a hydrocarbon (e.g., benzene) vapor to supply carbon atoms for growth of the graphitic host, mixed with a vapor containing the guest species. An inert carrier gas (argon) flow transported the vapors of reagents to the hot zone of the furnace where the hydrocarbon molecules decomposed due to high temperature and catalytic action of a transition metal (e.g., Ni) substrate. The guest species were captured between growing graphitic carbon layers forming a series of new GICs [3, 4]. Formation of the graphitic host and the intercalation occured at the same time, so that the diffusion kinetics was no longer a limiting factor. The CVD-grown graphite and GIC films were reported to be oriented with the c-axis of the microcrystals aligned normal to the substrate. Some of these materials exhibited very high c-axis thermoelectric power, $S=150$-900 $\mu V/K$ and thermoelectric figure of merit, $Z=S^2 \sigma / \kappa$, where σ and κ are respectively the c-axis electrical and thermal conductivities. For Se- and Te-GICs, ZT >30 [3], that is more than an order of magnitude larger than that of any previously known thermoelectric material. If this result is verified, these materials would represent a major breakthrough in the area of thermoelectrics.

However, very little structural and spectroscopic data were reported for the Sharp Corp. samples, and the physical origin of the anomalous thermoelectric properties was not investigated. In this work, we present the results on synthesis and characterization of oriented films of Se-carbon layered compounds grown by a *closed-tube* CVD method. Results of X-ray diffraction (XRD), Raman scattering, c-axis electrical resistivity and thermoelectric power measurements on these films are reported and discussed. In our closed-tube system, the reactor volume is being filled with H_2 gas released with the decomposition of the hydrocarbon. This may affect the reaction in several possible ways, e.g., through reaction with Se vapor to form H_2Se, H-termination of dangling carbon bonds in the growing graphite film, etc. This aspect makes our approach different from the open-tube CVD process used by the Sharp Corp. group where all the gaseous by-products and unreacted precursors are removed from the film growth zone by carrier gas stream.

SYNTHESIS AND EXPERIMENTAL DETAILS

The anthracene powder and Se shots were loaded in one end of a quartz tube, the Ni substrate in the other end, and the tube was sealed under a vacuum of 10^{-3} Torr. The Ni-containing end was introduced into a pre-heated furnace to allow the Ni substrate to warm-up first, while the reagents were kept at $20°C$. After the Ni substrate's temperature reached $900°C$, the tube was pushed further into the furnace to a pre-determined position so that anthracene and Se were heated to 350

Mat. Res. Soc. Symp. Proc. Vol. 453 © 1997 Materials Research Society

and 500°C, respectively. The Se-carbon film starts growing on the Ni substrate as anthracene decomposes. To terminate the growth process and anneal the film, the reagent-containing end of the tube was pulled out of the furnace, while the sample-containing end was kept at 900°C. The annealing usually took ~20 hours. After annealing, the tube was allowed to cool to the room temperature over 2-3 hours. The chemical composition and quality of the films could be controlled through the temperatures of the reagents and that of the Ni substrate, as well as the molar ratio of Se:C in the precursors. It is important to synchronize the initiation of the sublimation processes to ensure that both reagents reach the Ni substrate simultaneously. If either vapor reached the substrate first, the Se-carbon films were found to contain nickel selenide or pyrolytic graphite impurities. By manipulating the temperature at the Se and hydrocarbon source, predominantly single-phase Se-carbon films could be grown. Adhesion of the CVD-grown films to the Ni substrates was observed to be better if the film deposition rate was lower than 0.1 micron per hour.

X-ray diffraction patterns of the CVD-grown films were taken using a Rigaku diffractometer (Cu Kα). Raman scattering spectra were measured with an Ar laser (514.5 nm). DC electrical resistivity was measured in the direction perpendicular to the surface (i.e., parallel to the c-axis) using a 4 silver paint probe geometry. Thickness of the films was determined by observing the film cross-section with a scanning electron microscope (Hitachi). Thermoelectric power measurements were carried out with an experimental set-up described elsewhere [6].

RESULTS

<u>X-ray Diffraction</u>

Without Se present in the reaction tube, high quality graphitic films were grown at T_{Ni} =900°C, where T_{Ni} is temperature of the Ni substrate. Comparison with a commercial HOPG (Union Carbide) sample shows that our film exhibits sharper XRD reflections (a typical full width at half maximum (FWHM) in our films is 0.19° vs 0.28° in HOPG). Strongly enhanced (00L) reflections indicate that the film is preferentially oriented with its c-axis perpendicular to the substrate. The XRD data also show that the film has an A-B-A-B-type graphitic crystal structure with the interplanar separation of 3.365 Å, which is only slightly higher than the separation of 3.354 Å observed in the HOPG, and significantly less than 3.40 Å observed for turbostratic graphite [1].

When Se vapor was also present in the reaction tube, the XRD patterns changed completely due to the incorporation of Se layers into the crystal structure of the graphitic host (Fig.1). Usually, the Se-containing samples exhibited XRD patterns which could be interpreted as arising from a mixture of various stage Se-carbon layered compounds with $n = 2 - 5$, as inferred from the values of the c-axis repeated distance (I_c) calculated from the experimentally observed (00L) reflections. In addition to the relatively sharp reflections (FWHM -0.5°) identified with various stage Se-carbon phases, many of the samples exhibited broad (FWHM ~10-15°) background coming from poorly crystallized graphitic carbon. We were able to grow films of a Se-carbon layered compound which exhibited only the reflections expected for a pure stage-3 compound (Fig.1). The (00L) XRD intensities for the stage-3 Se-carbon films were analyzed within the framework of the model developed in ref.[7]:

$$I_{00L} \text{ (calc)} = K\,L\,P\,A\, |\, F_{00L}\,|^2 \qquad (1),$$

where I_{00L} (calc) are the calculated intensities of the (00L) reflections, F_{00L} is the structure factor, K is a scale factor, and L, P and A are the angle-dependent Lorentz, polarization and absorption factors, respectively. The structure factor is calculated according to:

$$F_{00L} = \Sigma\, N_j f_j\, \exp(2\pi i\, z_j\, L)\, \exp(-B_j\, \lambda^{-2}\, \sin^2\theta_L) \qquad (2),$$

where N_j is atomic density in the j-th atomic layer, f_j is the atomic scattering factor, z_j is the layer coordinate and B_j is the Debye-Waller factor. The XRD intensities calculated using the eq.(1) and (2) were compared to the experimental data using a least-square fit procedure until an optimal set of

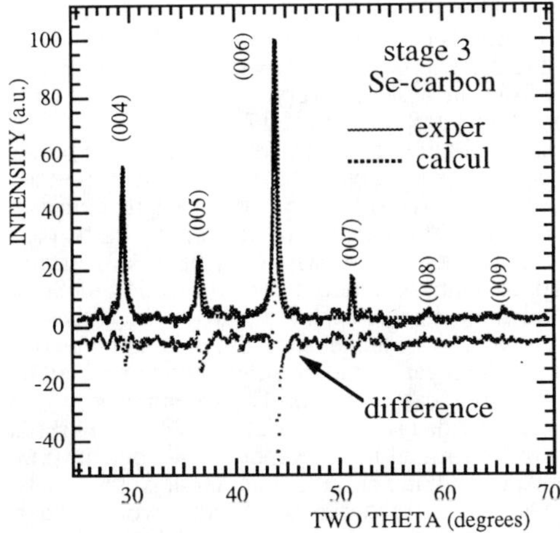

Fig.1. X-ray diffraction pattern of the CVD-grown stage-3 Se-carbon film oriented with it's **c**-axis perpendicular to the surface. The difference curve was downshifted for clarity.

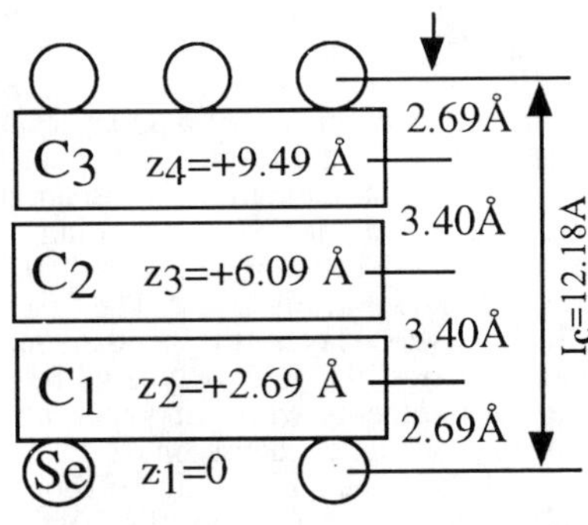

Fig.2a. The ionic structural model for the stage-3 Se-carbon layered materials, showing stacking of carbon (C) and Se layers along the effective **c**-axis. This model was used for the calculation of the X-ray diffraction pattern shown in the Fig.1.

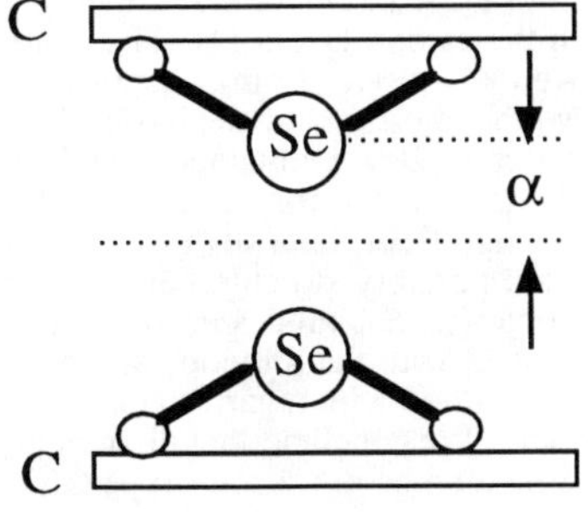

(b). The pendant model

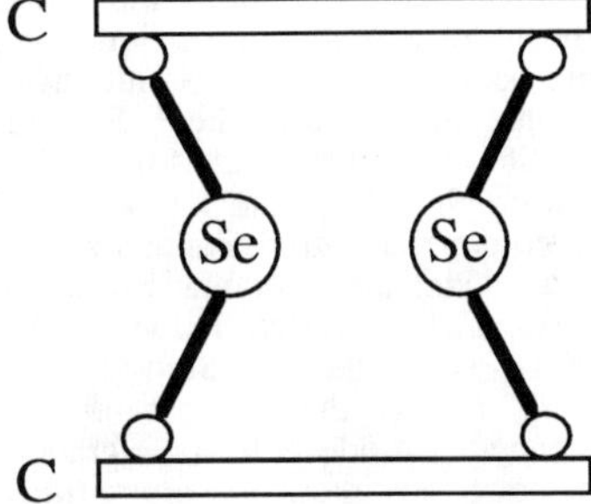

(c). The bridging model

Fig.2b and c. The alternative structural models for covalent Se-carbon bonding. The carbon atoms (small circles) are depicted as protruding out of the graphitic carbon layers (rectangles) to show local distortions.

parameters was found, yielding the following parameters: I_c=12.18 Å (stage 3), corresponding to a C-C interlayer spacing of 3.40 Å, a Se layer thickness of 1.98 Å and ratio of Se:C=1:24. The resulting layer coordinates are summarized in Fig.2a. The C and Se layers were presumed to be planar, and the thickness of a Se layer obtained from the best fit is very close to the size of a Se^{2-} ion (1.98 Å) and significantly larger that the size of a neutral Se atom (1.17 Å).

Alternative structural models assuming covalent bonding between Se and carbon can also be discussed. Covalent C-Se-C bonding might be anticipated given the very small difference in electronegativity between Se and carbon. Two possible ways to bond Se atoms to graphitic carbon layers are shown in Fig.2 b, c. The first way is to have Se covalently bonded to two carbon atoms within the *same* layer (the pendant model, Fig.2b), similar to what has been suggested for "graphite oxide" [11]. This type of Se-C bonding could be realized due to reaction of Se with aromatic fragments while in vapor phase, and then the benzene rings with the attached Se "pendants" would condense onto Ni substrate into a pendant-model film. Since Se is bonded only to one of the adjacent bounding carbon layers, the Se layer would presumably be expected to be displaced from the central position between the bounding carbon layers. The second way is to have bridging linkages with carbon atoms in adjacent layers (the bridging model, Fig.2c). The bridging, as well as the ionic, geometries are not likely to be formed in the vapor phase, but have to be formed during the film growth process. In both the bridging and the ionic models, the Se atoms (ions) are forced to be in the center between the bounding carbon layers. In other words, the layer coordinate of Se is $z_1=\pm\alpha$ for the pendant model, and $z_1 = 0$ for the bridging and the ionic models. The value of α depends on the choice of the specific carbon atoms to which the Se bonds and the induced puckering in the carbon layer. Having both + and - signs reflects the fact that statistically both "down" and "up"-bonded Se is equally possible. Structural analysis of the (00L) XRD data shows that at α >0.15 Å, the misfit between theory and experiment diverges sharply. This result suggests that a central or near-central position is most likely for the Se, suggesting the pendant model is not favored. However, it is not entirely clear if a 0.15 Å displacement is large enough to support the pendant model. This would depend on the extent of attendant puckering in C layers.

<u>Raman Scattering Spectra</u>

The pure CVD graphite film exhibited Raman spectrum characterized by bands at 1353 and 1582 cm⁻¹. The 1353 cm⁻¹ band is due to disorder-induced first order scattering and it's intensity has been correlated with the finite basal plane crystallite size in the carbon layers [1]. The sharp strong peak at 1582 cm⁻¹ is identified with the Raman-allowed E_{2g} mode. Using the ratio of integrated intensities of the 1353 and the 1582 cm⁻¹ bands we have estimated the basal-plane crystallite size to be about 180 Å which is comparable to the value of 200 Å obtained for CVD-grown graphite [5]. The Se-carbon films exhibit an intense band at 1358 cm⁻¹ and a poorly-resolved doublet with peaks at 1590 and 1617 cm⁻¹ (Fig.3). The significant intensity of the 1358 cm⁻¹ band indicates a dramatic decrease in basal crystallite size and possibly structural disorder in the hexagonal sp^2 carbon network. The doublet peaks are identified with E_{2g} modes in the interior (1590 cm⁻¹) and the bounding (1617 cm⁻¹) carbon layers. Consistent with the previous acceptor-type GIC results, the Raman bands in the Se-carbon film were up-shifted with regard to the high-frequency E_{2g} mode of the pure graphite film observed at 1582 cm⁻¹. We conclude that these high-frequency bands in our Se-carbon films are consistent with the formation of an acceptor-type GIC.

<u>Transport Measurements</u>

The resistivity (R) of the Se-carbon films was found to be very weakly T-dependent and ~100 times higher as compared to the non-intercalated reference samples (CVD-grown graphitic film and HOPG) from 77 to 300 K (Fig.4). The dR/dT <0 for the stage-3 film and >0 for the mixed-stage sample. Typically, the in-plane R decreases significantly upon intercalation with both donors and acceptors [9, 10]. However, the change in the c-axis R has been shown to be sensitive to the type of an intercalant: it *decreases* dramatically for donor-type GICs, but *increases* for acceptor-type GICs, and the increase was found be as high as several orders of magnitude, depending on the particular intercalant and the stage index [9, 10]. Thus, the preliminary electrical resistivity data for our Se-carbon films are also in a good agreement with traditional acceptor-type GIC behavior.

An alternative explanation for the increased R is puckering of carbon layers induced by C-Se-C covalent bond, leading to local distortions of the in-plane conjugation. A similar situation has been found in the case of graphite oxide, where O atoms are covalently bound to carbon atoms in a fashion similar to our pendant model (Fig.2b), resulting in puckered graphite layers and insulating behavior [11]. In our case, the damage to the the c-axis conduction should be less severe, as the concentration of Se in our samples is quite low: the carbon-to-intercalant ratio is 24:1 for Se vs 1:2 for O in the graphite oxide.

We have carried out preliminary measurements of c-axis thermopower (S) in a few Se-carbon films at 290 K in order to determine the sign of charge carriers and test the potential of these samples as thermoelectric materials. All the measured films exhibited S >0 indicating a p-type conductivity consistent with the assumption of an acceptor-type GIC. The largest c-axis S value observed in our Se-carbon films so far was 45 mV/K, higher than any previous in-plane or c-axis value in any GICs or other carbon-based material [1, 2, 12], but still much lower than that claimed by the Sharp Corp. (900 μV/K [3]). The other key factor in the thermoelectric figure of merit Z is the ratio of electrical to thermal conductivities, σ/κ, and this ratio may be improved significantly with Se intercalation in spite of the observed decrease in conductivity. Indeed, the total thermal conductivity, κ_{total} is contributed mainly by an electronic (κ_e) and a lattice (κ_l) terms: $\kappa_{total} = \kappa_e + \kappa_l$ [12]. According to the Wiedemann-Franz law, $\kappa_e \propto \sigma$ [12], and the

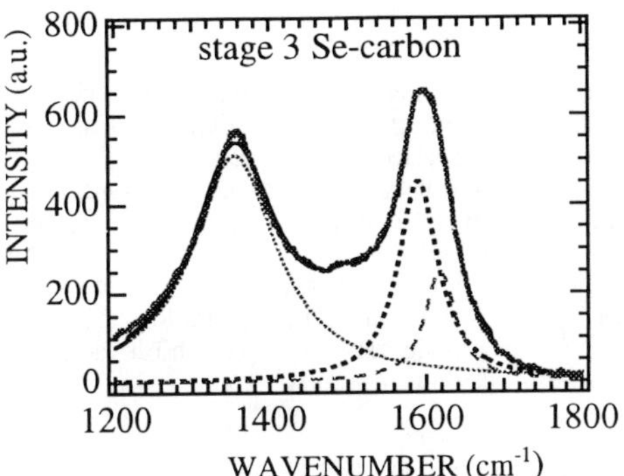

Fig.3. Raman scattering spectra of the CVD-grown stage-3 Se-carbon film (the fitting frequencies are indicated in the text).

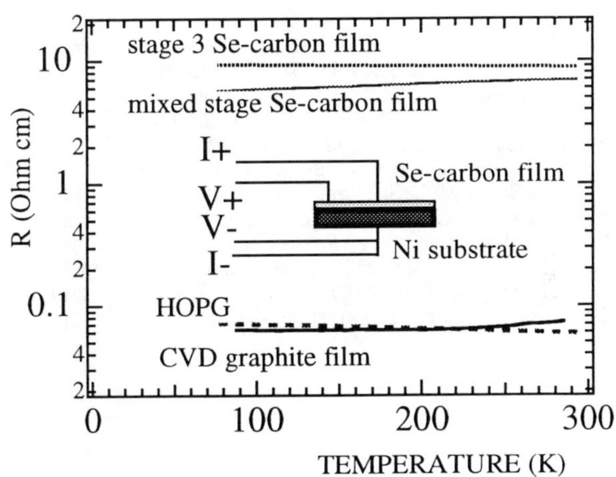

Fig.4. Resistivity (c-axis) vs temperature for the HOPG slab, the CVD graphitic film, stage-3 and mixed-stage (2 + 3) Se-carbon films.

decrease in conductivity may be compensated with a corresponding decrease in κ_e. On the other hand, the introduction of high atomic mass Se layers into the graphitic crystal structure is expected to suppress κ_l resulting in an overall improvement of the σ/κ ratio and therefore in Z.

SUMMARY

In summary, we have shown that homogeneous oriented submicron films of Se-carbon layered compounds can be grown by CVD method in sealed quartz tubes. Our experimental data on X-ray diffraction and photoelectron spectroscopy, as well as Raman scattering and the c-axis electrical transport measurements can be reasonably explained with either the covalent Se-carbon bonding models, or the ionic model assuming the formation of an acceptor-type Se-GIC with electron transfer from carbon to Se. The weight of the experimental evidence indicate that a GIC classification for these Se-carbon layer compounds is probably correct, although further work on better quality films will be necessary to clarify the ionic vs. covalent nature of the Se-C bonding. The CVD-grown Se-carbon films exhibit the largest thermoelectric power observed so far among the carbon-based compounds.

ACKNOWLEDGEMENTS

We would like to thank G. Dresselhaus, M.S. Dresselhaus and G. Mahan for useful comments. This work was funded in part by Grants from ARPA # MDA 972-95-1-0021 (G.S., S.F., P.C.E. and A.M. Rao) and NSF EPSCOR # OSR 9452895 (L.G. and L.S.)

REFERENCES

1. M.S. Dresselhaus and G. Dresselhaus, *Advances in Physics* **30** (1981) 139.
2. S. Solin and H. Zabel, Eds. *Graphite Intercalation Compounds*, Springer-Verlag, Berlin, 1992.
3. H. Nakaya, et al., European Patent Application #92301388.2 (1992).
4. Y. Yoshimoto, et al., US Patent #5,273,778 (1993).
5. M. Yudasaka, et al., *Appl. Phys. Lett.* **64** (1994) 842.
6. P.C. Eklund and A.K. Mabatah, *Rev. Sci. Instr.* **48** (1977) 775.
7. M.H. Boca, M.L. Saylors, D.S. Smith and P.C. Eklund, *Synth. Metals* **6** (1983) 39.
8. G.Malmsen, I.Thoren, S.Hogberg, J.E. Bergmark and S.E. Karlsson, *Phys. Scr.* **3** (1971) 96.
9. A.R. Ubbelohde, *Proc. Royal Soc. A* **321** (1972) 445.
10. G.Foley, C. Zeller, E.R. Falardean and F.L. Vogel, *Solid State Commun.* **24** (1977) 371.
11. G.Hennig, in *Progress in Inorganic Chemistry*, ed.F.Cotton, v.1, Intersci. Publ., NY, 1959.
12. C. Kittel, "Introduction to Solid State Physics", John Wiley and Sons, Inc., New York, 1986.

INFLUENCE OF HYDROGEN INCORPORATION AND OXYGEN REMOVAL ON SUPERCONDUCTING $YBa_2Cu_3O_{7-\delta}$

T. HIRATA
National Research Institute for Metals, 1-2-1, Sengen, Tsukuba, Ibaraki 305, hirata@monokusa.nrim.go.jp

ABSTRACT

The influence of hydrogen incorporation and oxygen removal on superconducting $YBa_2Cu_3O_{7-\delta}$ is discussed in comparison with their respective effects. Both hydrogen incorporation and oxygen removal induce the orthorhombic (O)-to-tetragonal (T) transition and antiferromagnetic ordering as well. However, there exist no consistent changes in the c-axis for the O-to-T transition and the superconducting transition temperature T_c upon hydrogen incorporation in $YBa_2Cu_3O_{7-\delta}$; on the other hand, the c-axis certainly increases in the corresponding transition driven by oxygen nonstoichiometry and T_c decreases with increasing δ in $YBa_2Cu_3O_{7-\delta}$. Plots of T_c vs. the lattice parameter c reveal no significant suppression in T_c upon hydrogenation than oxygen removal. It is concluded that the Cu-O bond distances or the Cu-Cu plane spacings along the c-axis change somewhat differently in two cases; it is believed that the charge-carrier concentration is reduced less significantly upon hydrogenation than oxygen removal in $YBa_2Cu_3O_{7-\delta}$, giving rise to the aforementioned finding.

INTRODUCTION

Recently, a great deal of attention has been paid to hydrogen in high T_c-superconductors (HTSC)[1], because the structure as well as physical properties of HTSC could be influenced by hydrogenation; hydrogen occupies some interstitial sites in HTSC, forming either dilute solid solution or hydrogen concentrated phase(s). A striking structural change is that an orthorhombic (O)-to-tetragonal (T) transition takes place at $x\approx0.8$ in $H_xYBa_2Cu_3O_{7-\delta}$ [2]. The lattice parameter c decreases both in the orthorhombic and tetragonal phases such that a=b=c/3 is attained at x=1 in $H_xYBa_2Cu_3O_{7-\delta}$ [3]. However, it should be noticed that the c-axis does not always decrease upon hydrogenation for $YBa_2Cu_3O_{7-\delta}$. For example, all the lattice parameters of $H_xEuBa_2Cu_3O_{7-\delta}$ increase up to x=0.5, and the O-to-T transition is induced at x=0.8 and the orthorhombic phase vanishes for x>1 [4]. No consistent changes of the c-axis raise a question whether hydrogen induced O-to-T transition can be identified with the corresponding transition driven by oxygen removal in $YBa_2Cu_3O_{7-\delta}$, where the c-axis certainly expands with increasing δ [5,6].

Turning attention to the superconducting transition temperature T_c, there exists no conformity in the changes of T_c upon hydrogen incorporation for $YBa_2Cu_3O_{7-\delta}$ as we will see later, while T_c obviously decreases with increasing δ in $YBa_2Cu_3O_{7-\delta}$ [7-9]. Hydrogen incorporation causes a noticeable increase of the normal state resistivity for $YBa_2Cu_3O_{7-\delta}$ [10], implying that the material becomes less conducting due to the reduction in the charge-carrier concentration; this is corroborated by a decrease in the inverse Hall coefficient R_H^{-1} with increasing x for $H_xYBa_2Cu_3O_{7-\delta}$ thin films [11]. The charge-carrier concentration is also controlled in terms of oxygen nonstoichiometry for $YBa_2Cu_3O_{7-\delta}$, which alters the nominal copper valence and hence determines T_c's. Interestingly enough, both hydrogen incorporation and oxygen removal induce antiferromagnetic ordering in $YBa_2Cu_3O_{7-\delta}$; adding one hydrogen has the same effect as removing one oxygen for the superconductor to antiferromagnet transition. In the context of similar/dissimilar influence of hydrogenation and oxygen removal on superconducting $YBa_2Cu_3O_{7-\delta}$, no consistent changes in the c-axis and T_c upon hydrogen incorporation urge us on the Cu-O bond distances or the Cu-Cu plane spacings along the c-axis; note that these bond distances/plane spacings are responsible for the charge-carrier concentration in the Cu(2)-O plane, which plays an important role in the size of T_c.

RESULTS

Orthorhombic-to-Tetragonal Transition

For $H_xYBa_2Cu_3O_{7-\delta}$ (δ=0.09), the lattice parameters b and c decrease as a function of x whereas the a-axis increases with x up to x=0.8, where the O-to-T transition is induced [2]; the c-axis remains unchanged while the a-axis increases with further increasing x. We note a slight increase in the a and c

Mat. Res. Soc. Symp. Proc. Vol. 453 © 1997 Materials Research Society

axes with x up to x=1.0 for tetragonal $YBa_2Cu_3O_6$ [12]. As the O-to-T transition proceeds, the orthorhombicity (b-a)/(b+a) decreases [13,14]. Dauo et al. have found that all the lattice parameters of $H_xEuBa_2Cu_3O_{7-\delta}$ increase up to x=0.5 until the O-to-T transition is induced at x=0.8 [4]. On the other hand, Harrington et al. [3] emphasize that the c-axis generally decreases upon hydrogen incorporation for $YBa_2Cu_3O_{7-\delta}$ (δ=0.07). Obviously, the c-axis holds the key when comparing with the O-to-T transition driven by oxygen nonstoichiometry for $YBa_2Cu_3O_{7-\delta}$.

Superconducting Transition Temperature T_c

Table I compiles the values of T_c before and/or after hydrogen incorporation for $YBa_2Cu_3O_{7-\delta}$ and related compounds. For our later argumentation, the corresponding lattice parameters c are also listed along with the orthorhombic distortion p=(b-a); it should be recalled that p is correlated with T_c [29,30].

Antiferromagnetic Ordering

It is of interest that antiferromagnetic (AFM) ordering is induced upon hydrogen incorporation as well as oxygen removal for $YBa_2Cu_3O_{7-\delta}$. The Mössbauer spectra of $H_xYBa_2Cu_3O_{7-\delta}$ (x=0.1 to 1.3) doped with 2 at% [57]Fe indicate the induction of antiferromagnetism at the Cu(2) sites; this is substantiated by a well defined magnetic sextet in the Mössbauer spectrum of $H_{1.3}YBa_2Cu_3O_{7-\delta}$, reflecting that iron substituted at the Cu(2) sites orders antiferromagnetically [21]. In addition, the muon-spin-rotation (μSR) technique has indicated that AFM ordering occurs at x=0.5 and saturates at x=1.2 for $H_xYBa_2Cu_3O_{7-\delta}$ [31,32]. Here, it should be noticed that AFM ordering is also induced with increasing δ, i.e. oxygen removal in $YBa_2Cu_3O_{7-\delta}$ [33,34]. The x dependence of the Knight shift and linewidth for $H_xYBa_2Cu_3O_7$ or $H_xYBa_2Cu_3O_{6.7}$ is compared with their behavior as a function of δ for $YBa_2Cu_3O_{7-\delta}$ in the [89]Y nuclear magnetic resonance line [35,36]. It turns out that hydrogen incorporation and oxygen removal exert similar influence on these parameters; the superconductor to antiferromagnet transition occurs at H/O ratio$\approx$1. It is reported that the long-range AFM ordering tends to increase with x in $H_xYBa_2Cu_3O_7$ with another evidence-the [63]Cu31.5MHz line in the Cu NQR spectra broadens with x and disappears at x=1 for $H_xYBa_2Cu_3O_7$ [37,38], and the ^{1}H line in ^{1}H-NMR spectra of $H_{0.11}YBa_2Cu_3O_7$ splits as well [39].

Relationship between T_c and Cu-O Bond Length or Cu-Cu Plane Spacing; Relevance to the Charge-Carrier Concentration

It is established that T_c decreases with increasing δ in $YBa_2Cu_3O_{7-\delta}$ [8,9]. With increasing δ in $YBa_2Cu_3O_{7-\delta}$ [9,40], the Cu(2)-O(1) bond length increases whereas the Cu(1)-O(1) bond length shortens to eventually decrease the Cu(2)-Cu(2) plane spacing, where the Cu(2)-O(1) bond length corresponds to the distance from the Cu(2)-O plane to the apical oxygen O(1), which bridges the conducting Cu(2)-O plane to the charge reservoir Cu(1)-O chain. Miceli et al. [41] have found that T_c linearly decreases with the Cu(1)-O(1) bond length's reduction for $YBa_2Cu_{3-x}Co_xO_{7-\delta}$. The changes in the Cu-O bond length are associated with the charge transfer among the ions, which shift the nominal Cu valence, i.e. the charge-carrier concentration. There exists good correlation between T_c and the charge-carrier concentration-hole content in the p-type high-T_c superconductors [42-44]. Consequently, it is of significance to know how hydrogen incorporation affects the Cu valence in the Cu(2)-O plane relative to oxygen removal in $YBa_2Cu_3O_{7-\delta}$. Within the relationship between T_c and the Cu-O bond length or the Cu-Cu plane spacing which is relevant to the c-axis, however, it is required to inspect the change in the c-axis upon hydrogen incorporation relative to oxygen removal; in fact, T_c changes linearly with d-spacing between Cu-O sheets along the c-axis in high T_c cuprates with the perovskite structure within a range of d=3-4 Å [45].

DISCUSSION

For our subsequent discussion, it is a premise that no oxygen loss takes place upon hydrogenation, and that hydrogen is distributed homogeneously in $YBa_2Cu_3O_{7-\delta}$. No oxygen loss occurs under mild hydriding conditions (T=573 K, P_{H2}<760 torr) for $YBa_2Cu_3O_{7-\delta}$ [46,47]; on the other hand, the hydrogen-depth profile of $H_xYBa_2Cu_3O_{7-\delta}$ thin films [11,20] assures a uniform concentration of hydrogen at ~100 nm, expect for its enhancement near the surface.

Under these presupposition, let us discuss the changes in the Cu-O bond length or the Cu-Cu plane spacing along the c-axis upon hydrogenation and oxygen removal for YBa$_2$Cu$_3$O$_{7-\delta}$. Harrington et al. [3] have proposed that the Cu-Cu plane spacings change somewhat differently upon hydrogen incorporation and oxygen removal, namely the Cu(2)-Cu(2) plane spacing increases whereas the Cu(2)-Cu(1) plane spacing decreases based on the fact that the c-axis decreases upon hydrogenation for YBa$_2$Cu$_3$O$_{7-\delta}$ as shown in Fig.1. On the other hand, the opposite applies to oxygen removal, in accordance with the trend in the experimentally determined Cu-O bond distances and the expanded c-axis as well [9,40].

However, no compilation of data in Table I allows a definite conclusion that the c-axis generally decreases upon hydrogenation for YBa$_2$Cu$_3$O$_{7-\delta}$. It depends on the occupation site(s) for hydrogen as well as hydrogen concentration whether the c-axis increases or decreases upon hydrogenation for YBa$_2$Cu$_3$O$_{7-\delta}$. Two interstitial sites are candidated for hydrogen accommodation in orthorhombic YBa$_2$Cu$_3$O$_{7-\delta}$. The octahedral interstitial site (H$_I$) is surrounded by four Ba and two Cu(1) atoms in the Cu(1)-O plane, and another site H$_{II}$ is surrounded by four Y and two Cu(2) atoms (labelled H in Fig.1). Based on the lattice parameter changes, Fujii et al. [2] consider that hydrogen preferentially occupies the site H$_I$ up to x=0.8, increasing the a-axis while the b-and c-axes are decreased for H$_x$YBa$_2$Cu$_3$O$_{7-\delta}$; beyond x=1.0, hydrogen occupies the site H$_{II}$ to further increase the a-axis whereas the c-axis remains unchanged. However, it is difficult to account for why the c-axis does not increase if hydrogen occupies the site H$_{II}$. When the site H$_{II}$ is occupied with hydrogen, we conjecture that the Cu(2)-Cu(2) plane spacing increases and both the [Cu(1)-O(1)] and [Cu(2)-O(1)] bond lengths shorten so as to cause a decrease in the Cu(2)-Cu(1) plane spacing, in accordance with Ref.[3]; it is most probable that some competitive change between these bond lengths/plane spacings determines whether the c-axis decreases or increases upon hydrogenation.

It is of particular importance that the Cu-O bond length or the Cu-Cu plane spacing changes somewhat differently upon hydrogenation and oxygen removal for YBa$_2$Cu$_3$O$_{7-\delta}$, which causes no significant suppression in T$_c$ upon hydrogen incorporation than oxygen removal. In Fig.2, we present the plots of T$_c$ vs. the lattice parameter c upon hydrogen incorporation and oxygen removal for YBa$_2$Cu$_3$O$_{7-\delta}$. No significant depression in T$_c$ is noticeable with increasing the c-axis upon hydrogen incorporation than oxygen removal prior to the O-to-T transition. This fact is also supported by the orthorhombic distortion p=(b-a)≈0.06 Å (see Table I), which is within the regime of relatively high T$_c$≈80 K in the p dependence of T$_c$ for YBa$_2$Cu$_3$O$_{7-\delta}$ [29,30].

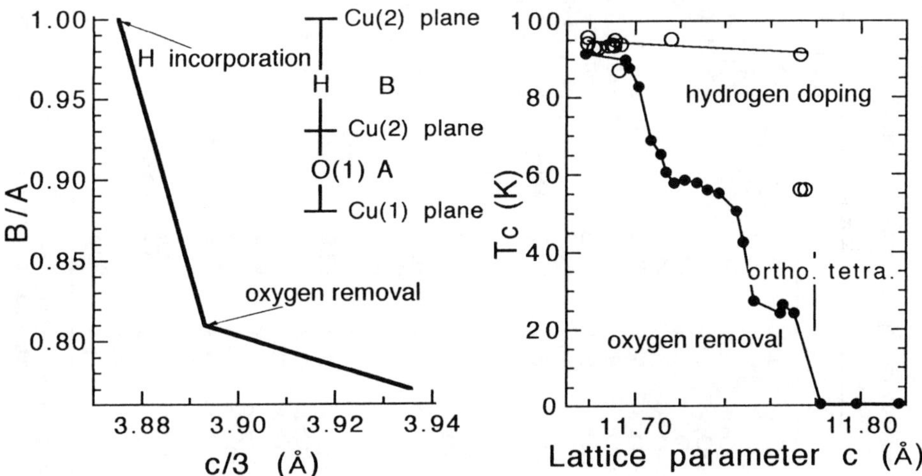

Fig.1 The relative Cu plane spacing as a function of the lattice parameter c/3 upon oxygen removal and hydrogen incorporation for YBa$_2$Cu$_3$O$_{7-\delta}$ after Ref.[3]; the apical oxygen O(1) and a possible candidate site (labelled H) for hydrogen occupation are shown as well.

Fig.2 Plots of T$_c$ vs. the lattice parameter c upon hydrogen incorporation and oxygen removal for YBa$_2$Cu$_3$O$_{7-\delta}$. o; hydrogen incorporation. •; oxygen removal after Ref. [9].

Table I Superconducting transition temperatures T_c's before and after hydrogen incorporation for $YBa_2Cu_3O_{7-\delta}$ and related compounds; the corresponding lattice parameter c are also compiled along with the orthorhombic distortion b-a.

Materials		T_c (K)	ΔT_c (K)	H concentration	c (Å)	Δc (Å)	b-a (Å)	Ref.
$H_xYBa_2Cu_3O_{6.91}$	bulk	95	±	x<2.0	11.640	±		15
$H_xRBa_2Cu_3O_7{}^a$	bulk	92	91(-)	x≤0.38	?	±		10
$H_xYBa_2Cu_3O_7$	bulk	91	92.5(+)	0.14<x<5.0	?			16
$H_xYBa_2Cu_3O_7$	bulk	92	±	x=0.05, 0.20 0.60	?	±		17
$H_xYBa_2Cu_3O_7$	bulk	93	±	?	11.6995	11.6819(-)	0.066	18
$H_xYBa_2Cu_3O_7$	bulk	93	63(-)	?	11.6995	11.6919(-)	0.054	19
$H_xYBa_2Cu_3O_{6.91}$	bulk	92	±	x<2.0	11.70	-		2
$H_xYBa_2Cu_3O_7$	film	?	-	x=0.08-0.41	?			20
$H_xYBa_2Cu_3O_7$	film	88~89	38(-)	x≈0.6	11.75	13.65(+)		11
$H_xYBa_2Cu_3O_7{}^b$	film	73	-	0<x<1.3	11.66	11.72(+)		21
$H_xYBa_2Cu_3O_{7-\delta}$	film	90.5	86(-)	x<0.3				22
$H_xEuBa_2Cu_3O_{7-\delta}$	bulk	96	-	x=0.23,0.54,0.79,1.05	11.715	+		4
$H_xYBa_2Cu_3O_{6.9}$	bulk	93	-	0<x<0.8	?			23
$H_xYBa_2Cu_3O_{6.7}$	bulk	70		x=1.0	11.705	11.713(+)	0.041	24
$H_xYBa_2Cu_3O_7$	bulk	91.2	+		11.6848	11.673(-)	0.058	14
$H_xYBa_2Cu_3O_{7-\delta}{}^c$	bulk	90	87(-)	x=0.32	11.7085	11.6929(-)		25
$D_xYBa_2Cu_3O_{7-\delta}$	bulk	90	-	x=0.22	11.739	+		26
$H_xYBa_2Cu_3O_{7-\delta}$	bulk	92	±	0≤x≤1.52; δ≈0.07	11.690	-		3
$H_xYBa_2Cu_3O_{7-\delta}$	bulk	92.0	93.5(+)	x=0.13	11.6828	11.6882(+)	0.064	27
$H_xYBa_2Cu_3O_{7-\delta}$	bulk	92.0	93.5(+)	x=0.19	11.6828	11.6907(+)	0.061	27
$H_xYBa_2Cu_3O_{7-\delta}$	bulk	92.0	93.8(+)	x=0.38	11.6828	11.6900(+)	0.062	27
$H_xYBa_2Cu_3O_{7-\delta}$	bulk	92.0	93.8(+)	x=0.49	11.6828	11.6937(+)	0.060	27
$H_xErBa_2Cu_3O_{7-\delta}$	bulk	94.8	94.2(-)	x=0.30	11.6732	11.679(+)	0.060	28
$H_xErBa_2Cu_3O_{7-\delta}$	bulk	94.8	95.8(+)	x=0.48	11.6732	11.679(+)	0.063	28
$H_xDyBa_2Cu_3O_{7-\delta}$	bulk	93.9	95.0(+)	x=0.35	11.689	11.691(+)	0.065	28
$H_xHoBa_2Cu_3O_{7-\delta}$	bulk	89.5	93.0(+)	x=0.38	11.676	11.684(+)	0.061	28
$H_xGdBa_2Cu_3O_{7-\delta}$	bulk	93.6	95.2(+)	x=0.30	11.708	11.716(+)	0.058	28
$H_xNdBa_2Cu_3O_{7-\delta}$	bulk	90.3	91.0(+)	x=0.25	11.7572	11.773(+)	0.055	28
$H_xNdBa_2Cu_3O_{7-\delta}$	bulk	90.3	56(-)	x=0.43	11.7572	11.773(+)	0.053	28
$H_xNdBa_2Cu_3O_{7-\delta}$	bulk	90.3	56.0(-)	x=0.56	11.7572	11.775(+)	0.050	28

a; R=Y, Er, Eu. b; doped with 2 at.% ^{57}Fe, T_c decreases at ~40 K/mol atom-hydrogen up to x≈1.0. c; single crystal. + indicates an increase in T_c whereas - does a decrease in T_c and ? represents no specification.

Within the relationship between T_c and the hole concentration in the p-type cuprate superconductors [42-44], the aforementioned fact indicates no significant reduction of the hole concentration upon hydrogenation than oxygen removal for $YBa_2Cu_3O_{7-\delta}$. The effective Cu-valence $(2+n_h)$ or the hole concentration n_h was determined to be $(2+n_h)=2.25-0.35x$ and $(2+n_h)=2.02-0.35x$ for $YBa_2Cu_3O_{6.91}$ and $YBa_2Cu_3O_{6.54}$, respectively by an iodometric technique [15]. It is believed that no hole concentration decreases considerably upon hydrogenation than oxygen removal for $YBa_2Cu_3O_{7-\delta}$ [40]; in fact, there is evidence that T_c of $H_xYBa_2Cu_3O_{6.9}$ only slightly decreases with x [23].

CONCLUSIONS

The influence of hydrogen incorporation and oxygen removal on superconducting $YBa_2Cu_3O_{7-\delta}$ is discussed in comparison with their respective effects. Both hydrogen incorporation and oxygen removal induce the orthorhombic-to-tetragonal transition and antiferromagnetic ordering as well. The present work highlights that the Cu-O bond length or the Cu-Cu plane spacing along the c-axis changes somewhat differently upon hydrogenation and oxygen removal for $YBa_2Cu_3O_{7-\delta}$. This causes no significant suppression in T_c upon hydrogen incorporation, along with another evidence that the orthorhombic distortion $p=(b-a) \approx 0.06$ Å is within the regime of relatively high $T_c \approx 80$ K in the p dependence of T_c for $YBa_2Cu_3O_{7-\delta}$. Within the relationship between T_c and the hole concentration of the p-type cuprate superconductors, it is believed that the effective Cu valence or the hole concentration is not greatly reduced upon hydrogenation than oxygen removal for $YBa_2Cu_3O_{7-\delta}$.

ACKNOWLEDGMENTS

The author greatly thanks Dr. Y. Asada for his critical comments by reading the manuscript.

REFERENCES

1.   T. Hirata, physica status solidi (a), **156**, 227(1996).
2.   H. Fujii, H. Kwanaka, W. Ye, S. Orimo and H. Fukuba, Jpn. J. Appl. Phys. **27**, L525(1988).
3.   I. Harrington, C. Korn, S.D. Goren, H. Shaked and G. Kimmel, Physica C **226**, 255(1994).
4.   J.N. Daou, J.P. Burger and P. Vajda, J. Less-Common Met. **172-174**, 425(1991).
5.   J.M. Tarascon, P. Barboux, B. G. Bagley, L.H. Greene, W.R. Mckinnon and G.W. Hull, in Chemistry of High-temperature Superconductors, edited by. D.L. Nelson, M.S. Whittingham and T. F. George (American Chem. Soc., Washington DC, 1987), p.198.
6.   J.M. Tarascon, L.H. Greene, B. Bagley, W.R. Mckinnon, P. Barboux and G.W. Hull, in Novel Superconductivity, edited by S.A. Wolf and V.Z. Kresin (Plenum Press 1987), p.705.
7.   L.H. Greene and B.G. Bagley, in Solid State Physics, Advances in Research and Applications, Vol.42, edited by H. Ehrenreich and D. Turnbull (Academic Press Inc. New York, 1989), p.509.
8.   R.J. Cava, B. Batlogg, C.H. Chen, E.A. Rietman, S.M. Zahurak and D. Werder, Phys. Rev. B **36**, 5719(1987).
9.   J.D. Jorgensen, B.W. Veal, A.P. Paulikas, L.J. Nowicki, G.W. Crabtree, H. Claus and W.K. Kwok, Phys. Rev. B **41**, 1863(1990).
10.   M. Nicolas, J.N. Daou, I. Vedel, P. Vajda, J.P. Burger, J. Lesueur and L. Dumoulin, Solid State Commun. **66**, 1157(1988).
11.   G. Dortmann, J. Erxmeyer, S. Blässer, J. Steiger, J. Paatsch, A. Weidinger, H. Karl and B. Stritzker, Phys. Rev. B **49**, 600(1994).
12.   S.D. Goren, C. Korn, H. Shakel, E. Rössler, H.M. Vieth, K. Lüders, Physica C **140-144**, 223(1994).
13.   A.M. Balagurov, G.M. Mironova, L.A. Rudnickij and V.Ju. Galkin, Physica C **172**, 331(1990).
14.   D. Fruchart, J. L. Soubeyroux, D. Tran Qui, C. Pique, C. Rillo, F. Lera, V. Orera, J. Flokstra and D.H.A. Blank, J. Less-Common Met. **157**, 233(1990).
15.   V.V. Sinitsyn, I.O. Bashkin, E.G. Ponyatovskii, V.M. Prokopenko, R.A. Dilanyan, V.Sh. Shekhtman, M.A. Nevedomskaya, I.N. Kremenskaya, N.S. Sidorov, R.K. Nikolaev and Zh.D. Sokolovskaya, Sov. Phys. Solids State. **31**, 2056(1989).
16.   C.Y. Yang, X.-Q. Yang, S.M. Heald, J.J. Reilly, T. Skotheim, A.R. Moodenbaugh and M. Suenaga, Phys. Rev. B **36**, 8798(1987).

17. I. Natkaniec, A.V. Belushkin, J. Brankowski, E.A. Goremychkin, J. Mayer, I.L. Sashin, V. K. Fedotov, A.I. Kolesnikov, I.O. Bashkin, V.V. Sinicyn, E.G. Ponyatovskii, Phyica C **162-164**, 1369(1989).

18. I. Garrote, E. Morán, M.A. Alario-Franco, J.M. Rojo and J. Sanz, J. Mater. Chem. **5**. 1171(1995).

19. T. Kato, K. Usami, J. Kuniya and S. Mastuda, Jpn. J. Appl. Phys. **27**, L1104(1988).

20. J. Erxmeyer, G. Dortmann, J. Steiger, O. Boebel and A. Weidigner, J. Less-Common Met. **172-174**, 419(1991).

21. I. Felner, B. Brosh, S.D. Goren, C. Korn and V. Volterra, Phys. Rev. B **43**, 10368(1991).

22. A. Richter, G. Irmer, K. Kessler, M. Panzner and K. Herzog, J. Alloys and Compounds. **187**, 59(1992).

23. T. Takabatake, W. Ye, S. Orimo, T. Tamegai and H. Fujii, Physica C **162-164**, 65(1989).

24. V. G. Hadjiev, M.V. Abrashev, M.N. Iliev and L.N. Bozukov, Physica C **171**, 257(1990).

25. D. Porath, A. Grayevsky, N. Kaplan, D. Shaltiel, U. Yaron and E. Walker, J. Alloys and Compounds. **204**, 79(1994).

26. T. Kamiyama, S. Tomiyoshi, H. Omori, H. Yamuchi, T. Kajitani, T. Matsunaga and H. Yamamoto, Physica B **148**, 491(1987).

27. J.J. Reilly, M. Suenaga, J.R. Johnson, P. Thomson and A.R. Moodenbaugh, Phys. Rev. B **36**, 5694(1987).

28. J.R. Johnson, M. Suenaga, P. Thomson and J.J. Reilly, Zeitschrift für Physikalische Chemie (N. F). **163**, 721(1989).

29. A. Kulpa, A.C.D. Chaklader, N.R. Osborne, G. Roemer and B. Sullivan and D.Ll. Williams, Solid State Commun. **71**, 265(1989).

30. Y. Asada, E. Nakada, A. Miyazaki and T. Yokokawa, J. Jpn. Inst. Met. **55**, 105(1991).

31. C. Nidermayer, H. Glückler, R. Simon, A. Golnik, M. Rauer, E. Recknagel, A. Wedinger, J. I. Budnick, W. Paulus and Schöllhorn, Phys. Rev. B **40**, 11386(1989).

32. H. Glückler, A. Weidinger, A. Golnik, C. Nidermayer, M. Rauer, R. Simon, E. Recknagel, J. I. Budniek, W. Paulus and R. Schöllhorn, Physica C **162-164**, 149(1989).

33. H. Alloul, M. Mendels, G. Collin and P. Monod, Phys. Rev. Lett. **61**, 746(1988).

34. G. Balakrishnan, R. Dupree, I. Farnan, D. Mckpaul and M.E. Smith, J. Phys. C: **21**, L847(1988).

35. S.D. Goren, C. Korn, V. Volterra, H. Riesemeier, E. Rössler, M. Schaefer, H.M. Vieth and K. Lüders, Phys. Rev. B **42**, 7949(1990).

36. S.D. Goren and C. Korn, C. Perrin, W. Hoffmann, H.M. Vieth and K. Lüders, Phys. Rev. B **52**, 3091(1995).

37. S.D. Goren, C. Korn, H. Riesemeier and K. Lüders, Phys. Rev. B **47**, 2821(1993).

38. H. Lütgemeier, S. Schmenn, H. Schone, Yu. Baikov, I. Felner, S. Goren and C. Korn, J. Alloys and Compounds. **219**, 29(1995).

39. S.D. Goren, C. Korn, V. Volterra, M. Schaefer, H. Riesemeier, E. Rössler, H. Stenschke, H. M. Vieth and K. Lüders, Solid State Commun. **70**, 279(1989).

40. A.W. Hewat, Physica B **180 & 181**, 369(1992).

41. P.F. Miceli, J.M. Tarascon, L.H. Greene, P. Barboux, F.J. Rotella and J.D. Jorgensen, Phys. Rev. B **37**, 5932(1988).

42. H. Zhang and H. Sato, Phys. Rev. Lett. **70**, 1697(1993).

43. J.L. Tallon, C. Bernhard, H. Shakel, R.L. Hitterman and J.D. Jorgensen, Phys. Rev. B **51**, 12911(1995).

44. S. Tanaka and Y. Motoi, Phys. Rev. B **52**, 85(1995).

45. C. Kawabata and T. Nakanishi, J. Phys. Soc. Jpn. **59**, 3835(1990).

46. I. Garrote, E. Morán, M.A. Alario-Franco, J.M. Rojo and J. Sanz, Physica C **235-240**, 385(1994).

47. T. Kubokawa, Y. Kumaki, K. Fujii, Y. Tatsumi, H. Taimatsu and H. Kaneco, Jpn. J. Appl. Phys. **33**, L507(1994).

SYNTHESIS OF NEW ALKALI-METAL-INTERCALATED LAYERED-SILICATE COMPOUNDS AND THEIR MAGNETIC PROPERTIES

N. WADA*, HIDEAKI OKUI*, YASUYUKI OMURA*, AKIHIKO FUJIWARA**,
HIROYOSHI SUEMATSU**, YOUICHI MURAKAMI***
*Faculty of Engineering, Toyo University, Kawagoe City, Japan 350, nwada@krc.eng.toyo.ac.jp
**Department of Physics, University of Tokyo, Tokyo, Japan 113
***National Laboratory for High Energy Physics, Tsukuba, Japan 305

ABSTRACT

Single crystals and powders of vermiculite are dehydrated and then intercalated with alkali metals (K and Rb) using a two-temperature-zone furnace. The samples, originally transparent, exhibit metallic silver color upon intercalation. Our x-ray diffraction experiments indicate that the c-axis lattice constant becomes smaller after the intercalation process. A SQUID magnetometer is used to measure the magnetization of the samples as functions of temperature and magnetic field. Evidence for Pauli paramagnetism and enhancement of ferromagnetism are found in the Rb- and K-intercalated compounds, in addition to Curie paramagnetism due to iron impurities in the original vermiculite.

INTRODUCTION

Layered silicates have been extensively studied for the last several decades, because of their interesting physical, chemical properties and industrial applications. It is noted that expandable layered silicates are particularly intriguing, since they can be easily cation-exchanged, hydrated and intercalated with even large neutral or ionic molecules. The application of pillared layered silicates to catalysis is also noteworthy [1].

Zeolites, that are quite similar to layered silicates in the chemical composition, have been also well studied and the subject of alkali-metal intercalation [2-4]. The alkali-metal-intercalated zeolite compounds exhibit unusual properties, possibly due to their small-metallic-particle-size or charge-transfer effects. Appearance of a ferromagnetic phase at low temperature is surprising [4]. Since, as far as we know, no one has ever tried to intercalate layered silicates with alkali metals, we decided to synthesize the intercalation compounds. Below, we discuss the synthesis of the layered-silicate intercalation compounds and the structural and magnetic properties studied by synchrotron x-ray diffraction and SQUID.

EXPERIMENTAL

Vermiculite, that is available in a large crystalline form, is used as a starting material for the intercalation process. The samples are Llano vermiculite, one of the Clay Minerals Society's source clays, and have been well characterized by many researchers. The Chemical composition for Llano vermiculite [5] has been obtained as $(Si_{5.72}Al_{2.28})(Mg_{5.88}Al_{0.10}Fe_{0.03}Ti_{0.02})O_{20}(OH)_4 Mg_{0.93} \cdot m(H_2O)$, where the number of water molecules m can vary depending on its hydration state.

The original crystals contain hydrated Mg cations as interlamellar ions and can be easily cation-exchanged by immersing in a salt solution. We used RbCl and KCl water solutions for ion exchange. After being completely ion-exchanged, a single-crystal sample (several mm in size) or a powder was placed in a pyrex or quartz tube and evacuated by a vacuum pump. Interlamellar water

95

Mat. Res. Soc. Symp. Proc. Vol. 453 © 1997 Materials Research Society

molecules were taken out by heating the sample to ~200 °C while evacuating. Then, a chunk of alkali metal was inserted into the glass tube in a glove bag filled with argon gas. The glass tube was evacuated and sealed. The oven had two temperature zones; the sample side was typically heated to 300 °C while the alkali metal side was 200 °C. Overnight, one may find that a single crystal looks metallic silver or a powder looks dark gray.

For synchrotron-x-ray-diffraction experiments, we transferred the powder sample to a thin quartz tube (0.3 mm in diameter and 0.01 mm in thickness) in an Ar-gas-filled glove bag and then sealed the tube off. The x-ray experiments were done at the Photon Factory of the National Laboratory for High Energy Physics in Japan. The measurements were conducted at room temperature using an imaging plate. It was found from the diffraction pictures that the powder samples were not well powdered; the diffraction rings were not too uniform. It was also found that small crystals tend to align themselves because vermiculite is a layered material. We will come back to discuss about these facts later.

Magnetization measurements were conducted using a SQUID magnetometer (MPMS, Quantum Design). We measured both temperature dependence and magnetic-field dependence of the samples.

RESULTS AND DISCUSSIONS

X-ray-diffraction patterns obtained from dehydrated Rb cation-exchanged vermiculite and Rb doped vermiculite are shown in Figure 1. Both samples were powders. The original diffraction patterns contained broad features originated from the quartz tube. The intensities from the quartz tube were simply subtracted from the original diffraction patterns.

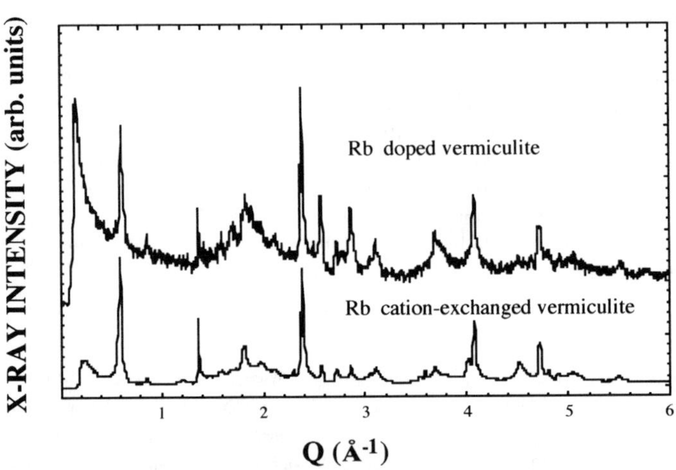

Fig. 1. Room-temperature synchrotron-radiation-x-ray-diffraction patterns obtained from Rb cation-exchanged vermiculite and Rb doped vermiculite. Diffraction intensities from the quartz tubes are subtracted from the original diffraction patterns.

The relative Bragg peak intensities appear to differ significantly. Since the observed relative intensities should also depend on the crystalline size and preferred crystal orientation of the powder samples as mentioned above, we may not quickly draw a conclusion that Rb atoms were really introduced to the interlayer spacing from this fact only [6]. Here, we emphasize the fact that the Bragg peak positions shifted slightly but definitely. The c-axis lattice constant of Rb cation-exchanged vermiculite can be estimated to be 10.51 Å from the out-of-plane (001) peak at $Q \approx 0.6$ Å^{-1}, whereas that of Rb doped vermiculite is estimated to be 10.25 Å. The peaks at $Q \approx 1.36$ Å^{-1} can be assigned to be in-plane (200). The a-axis constants for Rb cation-exchanged and Rb doped vermiculite are similar and are estimated to be 9.24 Å. The "shrinkage" of the lattice upon Rb and K intercalation was also confirmed by neutron-diffraction experiments [7].

We speculate that charge transfer from the intercalated alkali-metal atoms to the host silicate layer may be responsible for this shrinkage of the lattice. Another possibility is that a small amount of water molecules attached to the interlamellar ions may allow the lattice to be somewhat larger than the completely dehydrated state that may be realized during intercalation. We think that this possibility is rather unlikely because the Bragg peak widths are almost the same before and after the intercalation; if a small amount of water molecules were present, the c-axis stacking would become more random.

Also, we note that the broad feature at $Q \approx 1.9$ Å^{-1} appears to be more pronounced in the pattern of doped vermiculite and might suggest that the amount of in-plane Rb atoms increased after intercalation. The width of the feature suggests that the intercalant layers were considerably disordered.

Figure 2 shows temperature dependence of the magnetization at $B = 30,000$ gauss for Rb cation-exchanged and Rb doped vermiculite. Since Rb cation-exchanged vermiculite could be slightly hydrated if any, and since it was not possible to estimate how much extra Rb atoms are intercalated to the host material, the values for magnetization might be off by some amount. But the following arguments should not be affected too much by this uncertainty.

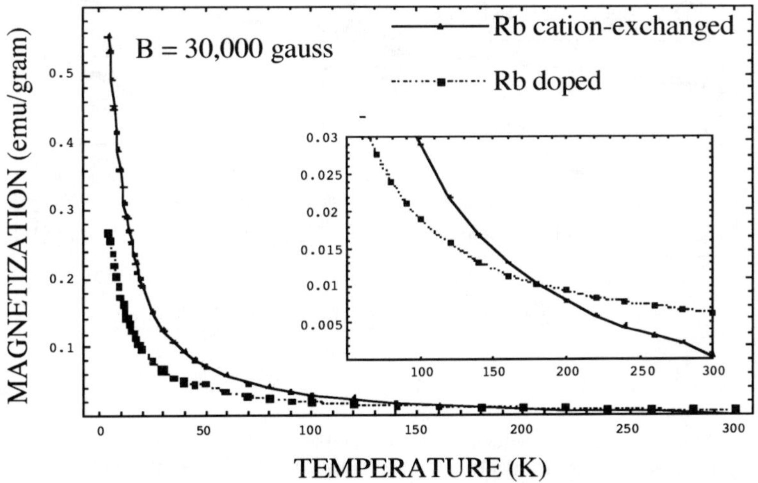

Fig. 2. Magnetization of Rb cation-exchanged and doped vermiculite as a function of temperature. The applied field was 30,000 gauss. The insert displays an expansion of the higher-temperature data.

It is noticed from the figure that for a given temperature, the measured magnetization value of Rb ion-exchanged vermiculite is about twice larger than that of Rb doped vermiculite at lower temperatures. Because of the C/T dependence observed, we may say that the number of free spins responsible for Curie paramagnetism is far less in the doped vermiculite than in the cation-exchanged vermiculite. The free Curie spins originate from iron impurities in the sample. To avoid any confusions caused by the naturally-occurring samples, we used exactly the same powder to measure cation-exchanged and doped data; we first measured the dehydrated, undoped sample, then intercalated it with Rb atoms, and again measured the doped sample. Thus, we can confidently state that this large change in Curie paramagnetism is caused by intercalation. We speculate that charge transfer might be responsible for this reduction in the free spin density, e.g., screening due to metallic layers.

As can be seen in the insert in Fig. 2, the magnetization curves cross at $T \approx 180$ K. The magnetization of Rb doped vermiculite is larger than that of Rb cation-exchanged vermiculite above this temperature. The question is what causes this behavior. Simple Curie picture cannot be used to explain this finding.

To discuss further about this net positive magnetization of the doped sample observed at higher temperatures, we depict magnetic-field dependence of both Rb doped and cation-exchanged vermiculite samples in Fig. 3.

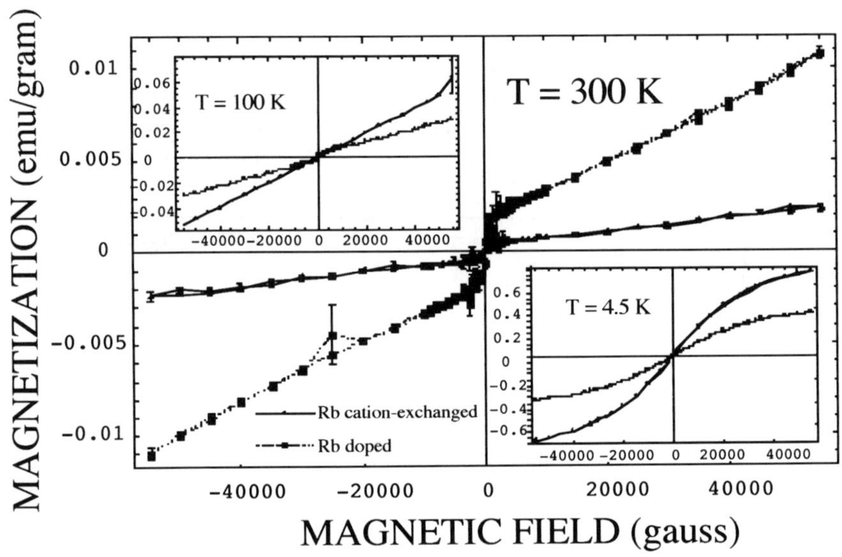

Fig. 3.   Magnetic-field dependence of magnetization taken from Rb cation-exchanged and Rb doped vermiculite powder samples. The data were taken at $T = 300$ K, while the insets were taken at $T = 100$ K and 4.5 K as indicated. The error bars, which are usually much smaller that the symbols, are also shown here for all the data points.

The magnetization data taken at 300 K shows a weak ferromagnetic behavior at low magnetic fields for both doped and cation-exchanged samples.The magnitudes of the ferromagnetic component, however, are not the same; the doped vermiculite sample exhibited a more pronounced ferromagnetic behavior as seen from the figure. This observation has been confirmed by conducting the same experiments on several K- and Rb-intercalated samples. It is not clear at this moment what causes the increase of the ferromagnetic component upon intercalation.

It is also seen from Fig. 3 that the magnetization slope of Rb doped vermiculite is much steeper at higher magnetic fields than that of the cation-exchanged sample. This fact cannot be explained by Curie paramagnetism, since the amount of free spins in Rb doped vermiculite is much smaller than that of Rb cation-exchanged vermiculite as discussed above. We speculate that this difference in the slopes might be due to Pauli paramagnetism.

The inserts in Fig. 3 are also shown to demonstrate that the magnitude of magnetization of Rb doped vermiculite is much smaller than that of Rb cation-exchanged vermiculite at lower temperatures as expected from the discussions above (see Fig. 2).

CONCLUSIONS

For the first time, intercalation of alkali-metals to layered silicates has been succeeded by vapor transport. We have examined their structures by synchrotron-x-ray diffraction and found that the new material indeed has a different structure from the original. The change is not drastic but definite. The c-axis lattice constant actually becomes smaller upon intercalation. The relative Bragg intensities also change upon intercalation, but because of the sample's nature, we could not draw a decisive conclusion on the intercalation from this point of view.

The magnetization measurements conducted on doped and undoped samples indicate that intercalation caused a drastic reduction in the Curie paramagnetization, and an increase in the ferromagnetization. In addition, there seems to be a Pauli paramagnetism component produced by doping. It would be quite interesting if one could really make the layered silicates metallic by intercalation. Further confirmation is needed.

Though characterization of layered silicates is relatively difficult, we believe that this intercalation compound system may become a quite exciting field of research in the near future.

ACKNOWLEDGEMENT

We thank Prof. Shoji Yamanaka of Hirosima University for preparing vermiculite powders and valuable discussions.

REFERENCES

1. T.J. Pinnavaia, Science **220,** p. 365 (1983).

2. P.P Edwards, L.J. Woodal, P.A. Anderson, A.R. Armstrong, and M. Slaski, Chemical Society Reviews, 1993, p.305.

3. B.Xu and L. Kevan, J. Phys. Chem.**9 6**, p.2642 (1992)

4. T. Goto, Y. Nozue, and T. Kodaira, Materials Science and Engineering **B19,** p.48 (1993).

5. P.G.Slade, C. Dean, P.K. Schultz, and P.G. Self, Clays Clay Miner. **3 5**, 177 (1987).

6. Preliminary x-ray diffraction experiments on alkali-metal-intercalated artificial fluor-tetrasilicic mica (this mica has little impurities and also expandable, and is available as a completely randomly oriented powder) indicated that both x-ray positions and relative intensities change upon intercalation; H. Okui, Y. Omura, N.Wada, and Y. Murakami, to be published.

7. W.A. Kamitakahara, N.Wada, H. Okui, and Y. Omura, to be published.

Part III

Materials Synthesis

EXAMPLES OF HYDROTHERMAL TITRATION AND REAL TIME X-RAY DIFFRACTION IN THE SYNTHESIS OF OPEN FRAMEWORKS

J.B. PARISE[*,**], K. TAN[*], P. NORBY[§,**], Y. KO[*] AND C. CAHILL[**]
* Earth and Space Sciences, State University of New York, StonyBrook NY 11794-2100
** Department of Chemistry, State University of New York, StonyBrook NY 11794-2100
§ Department of Chemistry, Brookhaven National Laboratory, Upton NY 11973

ABSTRACT

Hydrothermal titration (HTT) techniques represent a new synthetic strategy for the production of open frameworks. As an example, initial digestion of amorphous GeS_2 in the presence of tetraethyl ammonium hydroxide (TEAOH) at 100°C is followed by injection of Ag^+ into this mixture to condense the $[Ge_4S_{10}]^{4-}$ clusters and form a framework consisting of 8-ring channels bounded by alternating GeS-clusters and Ag^+. The rational use of this apparatus depends upon determining the timing of injection. Insight into the changes occurring during reactant digestion is provided by real time synchrotron X-ray powder diffraction. Changes in the the system dimethylamine-Sb-S have been followed in real time at different temperatures. The transformation from mixtures of DMA, Sb and S to the known open phase DMA-SbS-SB8 was followed by observing monochromatic X-ray scattering detected on an imaging plates over a 8 hr period with one continuous exposure.

INTRODUCTION

Open frameworks consisting of corner linked polyhedral elements are capable of occluding organic molecules and therefore are of interest for their potential as selective agents for molecular sieving, catalysis, ion exchange, ion conduction and host-guest chemistry [1-3]. The most common of these frameworks are the aluminosilicates [1, 4-6]. However, the range of materials capable of forming open frameworks has increased greatly over the past decade. Exploratory synthesis aimed at expanding the chemistries capable of forming open frameworks and the types of frameworks which form from known chemistries is worthwhile. For example, following their description by Krebs [7], synthetic routes to knit isolated tetrahedral Sn- and Ge-sulfide clusters into frameworks have been found. Berdard et al. [8-10] suggested main group metal sulfides might form a novel family of open framework materials and this was accomplished using two procedures: self-condensation of the clusters in the presence of organic amines and condensation from solution using transition metal ions along with the organic amines [8-28]. There is evidence that both the organic amines and the transition metal act as structure directing agents. Another desirable structural feature of the sulfides is their tendancy to cluster, promoting the formation of large channels (Figure 1) in corner-connected networks of these clusters.

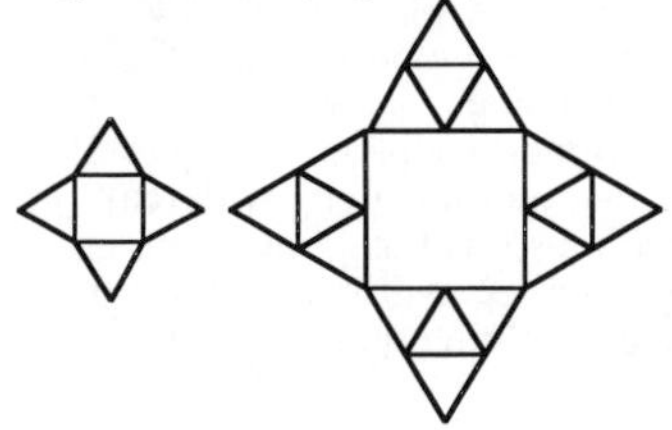

Figure 1. Schematic illustration of the increase in pore size when similarly sized structural elements cluster (right) and corner link.

The techniques used in the synthesis, modification and exploration for new framework materials is a variation of low temperature hydrothermal synthesis [4, 29-32]. Typically, reactive starting materials, such as amorphous oxides and metal sulfide, are mixed to form gels or slurries; these mixtures also contain alkali metals, alkaline earth metals and/or tertiary amines, believed to behave as templates, or reaction modifiers, and serving to direct the crystallization towards a specific framework type [30-32]. The mixture is then treated hydrothermally, typically between room temperature and 200°C. This technique has been used as the default procedure for the synthesis of these materials. In the case of the metal sulfides mineralizers, such as S^{2-}, HS^-, HCO_3^- and other alkaline solutions [33] dissolve GeS_2, SnS_2 and Sb_2S_3 above 100°C [34] and the formation of aqueous metal complexes with sulfide and bisulfide ligands has been proposed to account for observed metal solubility [35]. Proper control of pH and the concentration of the individual components leads to the isolation of pure phases of these clusters; further condensation to knit them together to form frameworks was also predicted [7, 36]. The condensation of pre-formed clusters from solution using transition metals and transformations from one framework to another with aging time have recently been observed [16, 22, 28]. This raises the possibility that redirection of the reaction at high temperature might be a fruitful strategy for the production of new materials. Such a hydrothermal titration [37] allows introduction of structure directing agents to reactants aged at autogenous pressures. Knowledge of the transformation history of the reactants provides appropriate cues for the redirection of the synthesis.

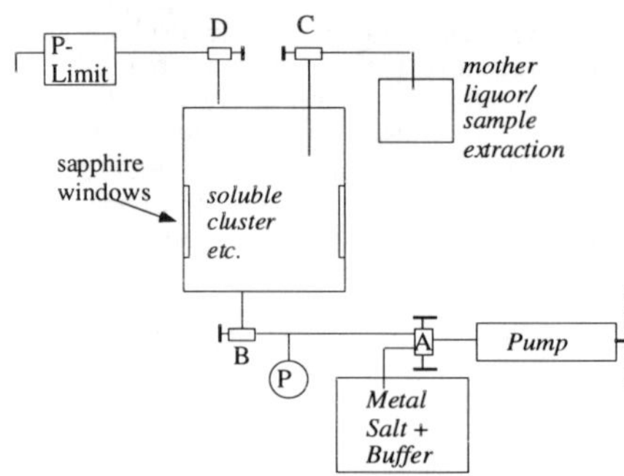

Figure 2 Schematic representation of a high-temperature hydrothermal titrator (HHT). The designators A, B, C, D are high pressure valves and P is a pressure gauge. In a typical experimental run, precursor material, such as are required to produce soluble clusters for example, are introduced into the reactor. This reactor can be equipped with sapphire windows to allow radiation access. The reactor is secured and heated externally . The contents can be sampled through valve "C". Further reactants can be loaded into the pump through valve "A", pressure generated above the autogenous pressure in the reactor, and then these can be loaded into the reactor through valve "B".

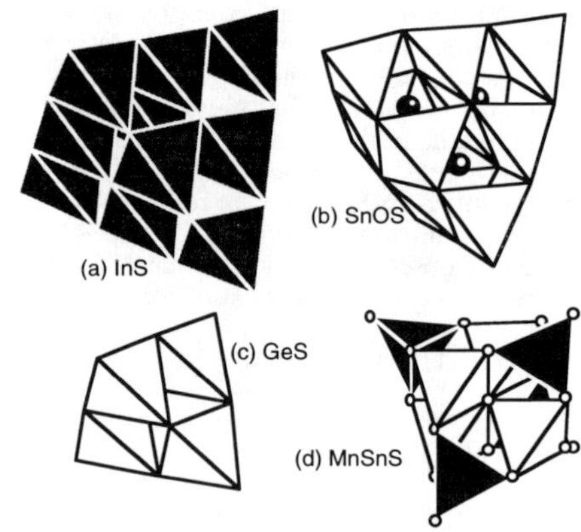

Figure 3. Four examples of super-tetrahedral clusters which can be used to construct open frameworks either by self condensation or through the use of transition metal to "knit" these cluster together (see Figure 4 for example); (a) $[In_{10}S_{20}]^{10-}$ [38], (b) $[Sn_{10}S_{20}O_4]^{8-}$ [14, 22], (c) $[Ge_4S_{10}]^{4-}$ and (d) $[Mn_4Sn_4S_{17}]^{10-}$ [28]

As empirical knowledge of the chemistry involved in molecular sieve synthesis improves [30-32] the possibilty of redirecting reactions toward desired products becomes viable. This intervention can be carried out quite simply using a hydrothermal titration apparatus (Figure 2). In this publication we describe initial experiments with hydrothermal titration along with real time diffraction experiments designed to guide the use of the apparatus. Although the systems described here are open framework sulfides, the technique is applicable to hydrothermal reactions occurring in the 0 - 0.2 GPa and 100 - 600°C ranges for the synthesis of microporous oxides.

EXPERIMENTS AND RESULTS

Several metal sulfide clusters are available from the literature to attempt the construction of open frameworks (Figure 1) and we have also synthesized several over the past few years (Figure 3). We have applied the hydrothermal titration (HTT) technique based upon the principles described above in the condenstation of GeS-clusters by Ag^+ ions added after the solution containing the cluster was at pressure and temperature (Figure 2 [37]).

Synthesis and structural characterization of a novel open framework [(C$_8$H$_{20}$N][AgGe$_2$S$_5$] (TEA-AgGS-SB1)

The reactions involving amorphous germanium sulfide as a starting material are good candidates for hydrothermal titration; depending upon conditions, germanium sulfide clusters can condense about Cu or Mn to form at least three structure types [9, 24, 25, 39-41]. One of these frameworks contains close to linear S-Cu$^+$-S linkages between [Ge$_4$S$_{10}$]$^{4-}$ clusters and its structure suggests similar chemistry might be possible with Ag$^+$. Formation of a new framework might then be induced by a hydrothermal titration reaction; injecting a solution containing Ag$^+$ after the isolated [Ge$_4$S$_{10}$]$^{4-}$ clusters formed at temperature and pressure [36]. We report here the synthesis using hydrothermal titration of a novel germanium sulfide condensed about silver, [C$_8$H$_{20}$N][AgGe$_2$S$_5$], (1) and its characterization using synchrotron X-ray powder diffractometry.

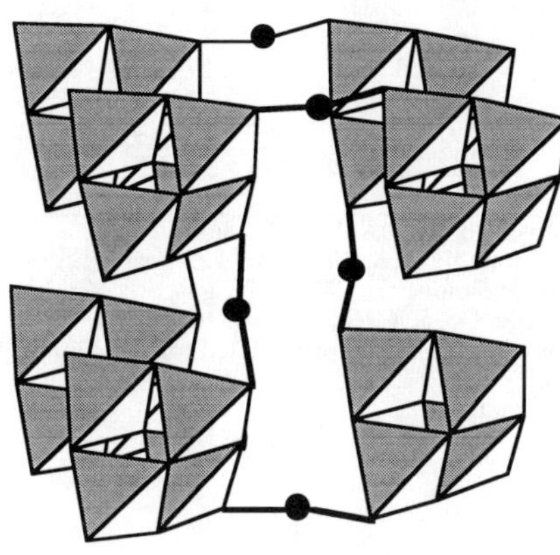

Figure 4. Tetrahedral clusters of [Ge$_4$S$_{10}$]$^{4-}$ anions joined by two-coordinated Ag$^+$ ions (solid circles) in the structure of 1. For the sake of clarity the [TEA]$^+$ ions occupying the cavities within the structure are omitted.

A homogeneous powder of 1 was produced by at first heating a slurry of 0.2 g of amorphous GeS$_2$, 0.1 g of 1M tetraethylammonium (TEA) chloride and 2.5 ml of 40% TEA-carbonate solution to 100°C in a Pyrex$^®$-lined stainless steel reactor (Figure 2). The temperature of this solution was stabilized and held at 100°C for 10 minutes. Then 8 ml of 1% silver acetate solution was injected using a high pressure pump (Figure 2) and the mixture heated at 130°C overnight. The product was a micro-crystalline powder and included flecks of an unidentified black material, amorphous to X-rays which was assumed to be Ag$_2$S. After hand-separation of the black material, The product was washed with water and ethanol solution in an ultrasonic cleaner. The yield was 90% based on GeS$_2$. Although the intensity distribution in the powder X-ray diffraction pattern was unique, closer inspection suggested a structural relationship with the previously described TEA-CuGS-SB1 [40] and the new material was designated TEA-AgGS-SB1 in accordance with established nomenclature [9]. All peaks in the pattern were indexed on the basis of the tetragonal unit cell ascribed to TEA-CuGS-SB1 [40]. Electron-probe microanalysis, carried out using a Cameca Camebax instrument, indicated the presence of Ag, Ge and S in a 1:2:5 molar ratio. The composition is consistent with the refined crystal structure, which was determined from data collected using a position sensitive detector [42] at beamline X7A of the National Synchrotron Light Source (NSLS). A pattern suitable for Rietveld refinement was obtained by combining in 0.01° steps data from the central 2° of the detector, which was stepped at 0.25° intervals from 5 to 35° in 2θ. The starting model for Rietveld [43] analysis, including the position of the [TEA]$^+$ ion, was taken from the single crystal structure analysis of TEA-CuGS-SB1 [40]. The Ag compound is isostructural with this material with [Ge$_4$S$_{10}$]$^{4-}$ clusters linked to form a three dimensional framework (Figure 4). The position of the organic template, which was refined using soft constraints, is similar in the structures of both Cu and AgGS-SB1. Details of the refined model are presented in Table 1.

The major difference between the Cu- and Ag-forms of the framework (Figure 4) concern the linkage provided by the transition metal (M) between [Ge$_4$S$_{10}$]$^{4-}$ tetrahedral clusters. Although the S-M-S angles are comparable (172 and 169° for the Cu- and Ag-materials, respectively) the M-S distance is longer for the Ag- (2.440(8) Å), than for the Cu-framework

(2.134(2) Å), consistent with the larger radius for Ag^+. The geometry of the clusters is similar to those found in comparable frameworks. For **1** the average Ge-S bond length is 2.235Å while it is 2.213 and 2.220Å in the isostructural framework TEA-CuGS-SB1 and in another Ag^+-containing material Dab-AgGS-SB2; the later compound is obtained in a static hydrothermal experiment [44]. The distribution of Ge-S distances in the M^+GS-frameworks studied is bimodal, with those involved in bonding between the clusters and silver (Figure 1) shortened with respect to those bonded within the cluster. The average Ge-S distances are 2.15 and 2.26Å, respectively.

Table I. Fractional atom coordinates and isotropic displacement parameters for TEA-AgGS-SB1*

atom	x	y	z	U_{iso} x 10^2
Ge(1)	0.42969(32)	0.07038(28)	0.1342(5)	4.4(2)
Ag(1)	0.37005(28)	0.25	0.125	7.8(1)
S(1)	0.5	0.0	0.0131(15)	3.7(4)
S(2)	0.3566(5)	0.1321(7)	0.0185(8)	4.1(4)
S(3)	0.5010(6)	0.1476(7)	0.2433(12)	3.1(5)
N(1)[¶]	0.25	-0.1061(16)	-0.125	12.3(9)[§]
C(1)[¶]	0.3166(4)	-0.1537(16)	-0.1105(16)	12.3(9)[§]
C(2)[¶]	0.3288(14)	-0.1780(18)	0.0109(20)	12.3(9)[§]
C(3)[¶]	0.2402(11)	-0.0580(16)	-0.0173(6)	12.3(9)[§]
C(4)[¶]	0.1682(15)	-0.0182(15)	-0.0230(33)	12.3(9)[§]

*$[C_8H_{20}N][AgGe_2S_5]$, tetragonal *I* $\bar{4}2d$, a = 18.0050(4)Å, c = 11.0765(4)Å, V=3590.8(3)Å^3, Z=8, synchrotron radiation, λ=0.64830(3) Å, powder sample in 0.2 mm glass capillary, incident beam slits set to 10 x 0.5 mm, position sensitive detector operating in escape mode with 42 psi Kr gas, at beamline X7A, NSLS; 497 reflections in pattern, Ag anisotropic, Ge, S, C, N, isotropic (C, N constrained equal); no absorption correction performed; anomalous dispersion calculated for Ag, Ge, S; Rietveld refienemnt R_{wp} = 3.63, R_p = 1.93, $R_{Bragg}(F^2)$ = 10.05%, χ^2 = 2.36.
∞ equivalent isotropic parameter as 1/3 trace of diagonalized matrix
¶ soft constraints used for TEA (overall weight 0.1): for C-C 1.54(1); for N-C 1.48(1)Å
§ thermal parameters for TEA constrained to be equal

<u>*In-situ* Time Resolved Powder Diffraction</u>

Knowledge of the structural changes occuring during synthesis will be useful in directing hydrothermal titration. As an example of this approach we describe preliminary *in situ* synchrotron X-ray powder diffraction experiments on a model antimony sulfide system [27]. Although no injection was carried out during these studies, structural changes occuring in the reaction, as a function of time and temperature, were monitored. This provides a basis for deciding when and at what temperature to carry out HHT injection experiments in the laboratory.

Crystallization of antimony sulfide in the presence of organic templates has produced many open frameworks [11, 12, 26, 27, 45-50] (Figure 5). However, the products obtained are often poorly crystallized and a mixture of phases in one batch is very common. In some cases, structures are found to depend on the temperature and heating history. For example, using dimethylamine (DMA) to direct the crystallization produces at least three distinct structure types as determined by single crystal and powder X-ray diffractometry [27]. When the temperature is maintained at 100 °C [27] only needle shaped crystals are obtained. After three days at 200 °C, only plate shaped crystals of a material consisting of connected SbS-chains forms (Figure 5). At conditions itermediate between these, either mixtures of these two materials or a third phase are produced.

The mechanism of formation and transformation involved in the DMA-SbS system are not yet clear. The slurries of reactants used for the synthesis of antimony sulfides are heterogeneous. They can be a mixture of crystalline Sb_2S_3 and an amine solution saturated by gaseous H_2S or elemental Sb and S plus template solution. The digestion of crystalline starting materials is expected to be critical to the nucleation of open structures. Knowledge of this procedure may provide insights into the chemistry before and during the crystallization of open framework structures. Analysis of the kinetics of phase formation or phase transformation requires real time study since quench experiments may not necessarily provide an accurate picture of what is happening during these procedures.

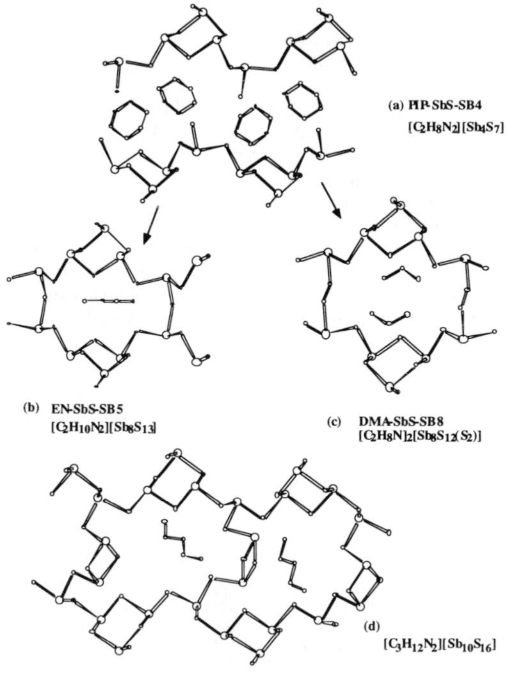

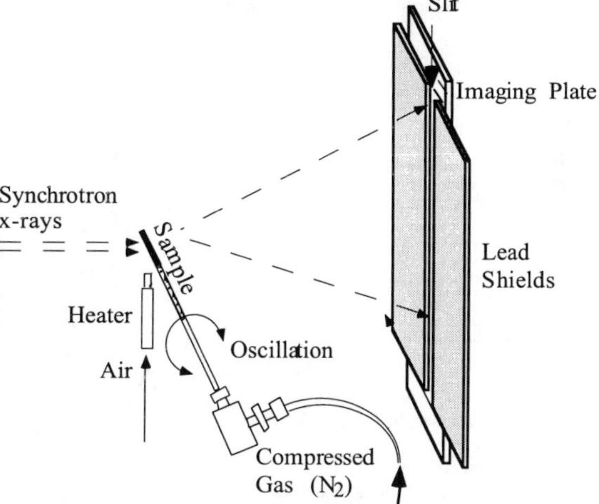

Figure 5. The structural relationship among antimony sulfides templated by (a) piperazinium (SBSpip or PIP-SbS-SB4) [11], (b) enthylenediamomnium (EN-SbS-SB5) [12], (c) DMA-SbS-SB8 [27]and (d) *n*-proplydiammonium [50]. The common structural element in these materials is the SbS_3-

In-situ time resolved synchrotron powder diffraction has been developed to study a variety of chemical kinetic processes when crystalline materials are involved [51-56]. Powder diffraction serves as a powerful tool for the identification of crystalline phases. It also has advantages for determining crystallinity and for crystal structure refinement. With an area detector, such as an imaging plate, a powder diffraction pattern can be obtained in minutes when a synchrotron X-ray radiation source is used. This makes possible the study of time-dependent chemical reactions and the instrumental setup at X7B beamline of the National Synchrotron Light Source (NSLS) at the Brookhaven National Laboratory (BNL) has made possible studies such as the kinetics of crystallization, transformation and degradation along with determination of crystalline or amorphous intermediate phases [53, 54, 56-58].

Instrumental setup

For time resolved powder diffraction (Figure 6), a 3 mm wide slit, alligned perpendicular to the sample contained in a thin walled quartz capillary (o.d. = 0.7 mm), is placed in front of a 20×40 cm Fuji imaging plate (IP). The plate is translated continually behind the slit exposing a strip of the full powder diffraction pattern from the sample. An air heater is placed below the capillary and used to control the temperature. Typical beam conditions include a wavelength of about 1Å (calibrated with LaB_6 powder before each run)

Figure 6 Schematic representation of the experimental set-up for time-resolved experiments at the X7B beamline at the NSLS.

and a beamsize is 2×1 mm. The sample consists of a homogeneous mixture of fine powders of Sb and S, which were packed into the capillary. Dimethylamine (DMA) solution was then injected and the open end of the capillary was then fitted into the head of a specially designed goniometer head [57], which was mounted on a Huber diffractometer and connected to a compressed gas tank to apply a pressure of 300 psi, to compensate for the calculated vapor pressure of water at 200 °C [57] (Figure 6).

Typically (Figures 7) the temperature was raised to 100°C over 30 min, held there for 1 hr, ramped to 200°C in 30 min and finally held at 200°C for 3 hr. The scan speed was adjusted to perform the experiment in one exposure of the imaging plate (IP), which was scanned using a BAS2000 IP reader. An IP is read at 100 µm pixel resolution to produce an image 2048×4096 pixels represented by a Cartesian vector, $P(x_i, y_j)$. Assuming the IP scans along the y-axis in the experiments (Figure 6), the intensity distribution in pixels of $P(x_i, y_j)$ ($i = 1$ to 4096) represents a powder diffraction pattern at time $t_j = (y_j-y_0)/V$ (y_0 is the pixel position at the beginning of the scan and V is the scan speed in pixels/sec.). From a scanner, the diffraction intensity that each pixel recorded is scaled to a digitized relative intensity $U(x_i, y_i)$. Suppose there is a diffraction peak located at $P(m,n)$ with FWHM of $2l$, the intensity of the peak can be integrated as:

$$H_p^*(m,n) = \sum_{i=m-l}^{i=m+l} U(i,n) - 1/2[1/5 \sum_{i=m-l-5}^{i=m-l} U(i,n) + 1/5 \sum_{i=m+5}^{i=m+l+5} U(i,n)] \tag{1}$$

The terms in parenthesis in equation (1) represent an averaged background intensity.

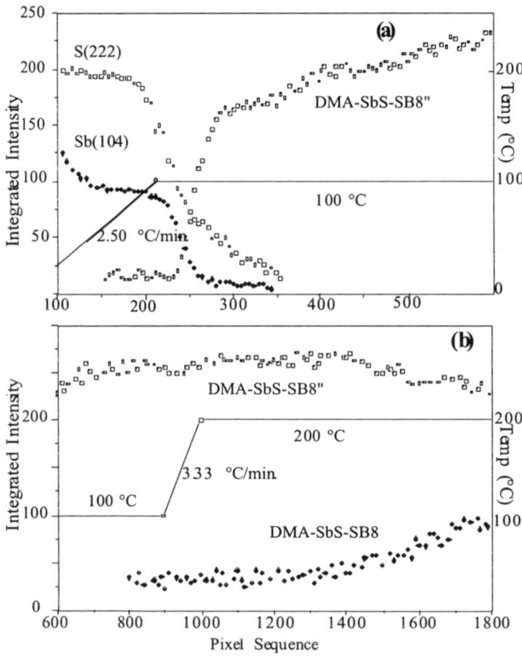

Figure 7 Summary of the phases reacting, forming and transforming during the hydrothermal synthesis of DMA-SbS. The bottom scale represents pixel sequence accross the imaging plate (IP) which is translated behind a slit (Figure 6) at a rate of 3.75 pixel/min. The resolution of the IP reader is 100 µm/pixel. Superimposed is the temperature sequence used during the experiment and the integrated intensities of selected reflections for the phases indicated. The internsities were integrated according to the methodology given in equations (1) and (2). The starting materials for the reaction were elemental antimony and sulfur placed in a capillary (Figure 6) with dimethylammonium solution.

In practice, $H_p(m,n)$ is smoothed to improve statistics:

$$H_p(m,n) = \sum_{j=n-2}^{j=n+2} H_p^*(m,j) \qquad (2)$$

Using this methodology, integrated intensities of selected reflections representative of a particular phase present during digestion are plotted in Figure 7. From these changes in relative intensity the hydrothermal synthesis of the DMA-SbS is divided into 3 steps: digestion of crystalline reactants, crystallization of an as yet uncharacterized open structure and transformation to a second structure, DMA-SbS-SB8 described recently by Tan *et al.* [27].

Digestion of crystalline reactants The room temperature powder diffraction data (Figure 7) shows the characteristic peaks for a mixture of elemental antimony and sulfur (orthorhombic α-S_8). With an increase in temperature, the intensities of these peaks gradually decreased and finally disappeared when the temperature was held at 100 °C (Figure 6). Using equations 1 and 2, the integrated intensities of reflections of Sb(104) and S(222) were extracted as a function of time to estimate the mass variation of these two elements during this heating period (Figure 7). It is obvious that the dissolution of Sb and S was not proportional, suggesting that they did not react directly with one another. From Figure 7, it is clear that antimony reacted first. It is known that Sb can react with water upon heating to form antimony trioxide [59]

$$2Sb + 3H_2O \text{ ---> } Sb_2O_3 + 3H_2 \qquad (3)$$

This reaction has also been observed in a hydrothermal experiment in which a mixture of antimony and water was sealed into a pyrex® tube and heated in a hydrothermal bomb at 100 °C. The reaction produced a small yield of fine needle-shaped crystals of Sb_2O_3 mixed with a predominance of unreacted Sb powder. Antimony oxide was not observed in the final products from a hydrothermal synthesis of antimony sulfide since the compound can dissolve in bases to give solutions of antimonates [59]

$$Sb_2O_3 + 6OH^- \text{ ---> } 2SbO_3^{3-} + 3H_2O \qquad (4)$$

The dimethylamine solution used in the real time experiments was basic and reactions (3) and (4) could account for the decline of the reflection intensity (about 20%) from crystalline antimony during the heating while the intensity of the diffraction pattern of sulfur remained unchanged. The reaction involving Sb seemed to reach equilibrium quickly within the sealed tube as evidenced by the flattening of the curve plotting the intensity of the Sb(104) reflection (Figure 7).

At about 80 °C, sulfur began to react (Figure 7) but did not react with antimony directly. Although the reaction of sulfur in an amine solution is complicated, especially during heating, it is known that sulfur reacts with organic amines to form N,N'-polythiobisamines [60]

$$2RR'NH + S_n \text{ ---> } (RR'N)_2S_{n-1} + H_2S \qquad (5)$$

For the DMA template R and R' are methyl groups and n is 8. Since the dimethylamine solution is basic, the S_8 ring can also be activated by nucleophilic attack. In both reactions, the ring is opened and sulfur chains or chain compounds are formed; free radicals may also be involved. All reactions of S_8 require the initial attack of the cyclic species to open the ring. Other reactions involved in the system may include base-induced redox disproportion of the S_8 to sulfide and thiosulfate/sulfate [16].

$$S_8 + 18H_2O \longrightarrow 4S^{2-} + 2S_2O_3^{2-} + 12H_3O^{+1} \qquad\qquad (6)$$

$$S_8 + 24H_2O \longrightarrow 6S^{2-} + 2SO_4^{2-} + 16H_3O^{+1} \qquad\qquad (7)$$

In the previous experiments, a predominance of $S_2O_3^{2-}$ and small amounts of SO_4^{2-} and SO_3^{2-} species have been detected in amine/sulfur solution [26] or in the mother liquor [16]. These species are not expected to contribute to the nucleation of open structures.

When the temperature was raised to about 100 °C, antimony underwent further reaction. This likely involved the polysulfide species, with cleaving of S-S bonds in the sulfur spesies to form various antimony polysulfides and sulfide complexes which are believed to be the precursor species of open framework structures. The reaction also accelerated the dissolution of S and eventually Sb as well (Figure 7).

Crystallization of DMA-SbS-SB". After about 50% of the sulfur and 30% of the antimony were dissolved (Figure 7), a new powder diffraction pattern appeared, indicating the crystallization of a new phase at 100 °C. The pattern for this phase was similar to that of the layered structure DMA-SbS-SB8', a unique phase found in the DMA-SBS system and recently described by Tan *et al.* [27]. The first peak of the new phase was located at $2\theta = 3.75$ ° with a *d*-spacing of 15.24 Å, different to that measured for the low angle peak of DMA-SbS-SB8' (17.193 Å). The nucleation of -SB8" was probably kinetically controlled and a model for the precipitated particles was used to fit experimental data and thereby extract kinetic parameters. Assuming that particle growth was limited by diffusion, that the nucleation was random and that the crystal growth occurred at a constant rate, the percentage of the crystallization could be expressed as [29, 61]

$$P_t\big/P_f = 1 - \exp(-k(t - t_0)^n) \qquad\qquad (8)$$

where P_t and P_f are the percentages of crystallization at time t and at the final time, respectively. Suppose that the diffraction intensity from the DMA-SbS-SB8" is proportional to the mass percentage of this material which crystallized, then:

$$I = C(1 - \exp(-k(t - t_0)^n)) \qquad\qquad (9)$$

where C is a proportionality constant. From an imaging plate the recorded intensity is

$$U(x_i, y_j) = B(x_i, y_j) + C^*I(x_i, y_j) \qquad (y_j = y_0 + Vt_j) \qquad\qquad (10)$$

$$\text{or } U(x_i, y_j) = B(x_i, y_j) + C^*C(x_i)(1 - \exp(-kt_j^n)) \qquad\qquad (11)$$

where B is the background intensity and C^* is a constant.

Since the first peak, located at $2\theta = 3.75$ ° ($\lambda = 0.99726\text{Å}$) was characteristic of DMA-SbS-SB8", its variation in intensity was calculated using Formula 1 (Figure 8). Applying Formula 11 to fit this curve through a least squares procedure, the kinetic parameters n and k were obtained; $n = 1.896$ and $k = 0.016$ min.$^{-n}$. The index n was related to the shape of particles [61]. In this case, it was very close to 2, indicating that the particles were rod-shaped. This morphology was consistent with that of the needle crystals of DMA-SBS-SB8', suggesting a similarity between DMA-SbS-SB8' and DMA-SbS-SB8". After its crystallization at 100 °C, DMA-SbS-SB8" was stable without a change in its powder diffraction pattern.

Transformation of DMA-SbS-SB8". About 100 minutes after the temperature was increased to 200 °C, a set of new reflections gradually appeared corresponding to the known diffraction pattern for DMA-SbS-SB8 [27]. Because of the time limitation of the *in-situ* experiment, the completion of the transformation was not observed. However, prolonged heating, of up to 3 weeks, in static bomb experiments [27] produce only the DMA-SbS-SB8 phase indicating this is

the stable phase under hydrothermal conditions at 200°C. Another possible reaction involving
DMA-SbS-SB8" was the formation of Sb_2S_3 upon loss of liquid. This was observed in
experiment in which liquid was lost from a burst capillary.

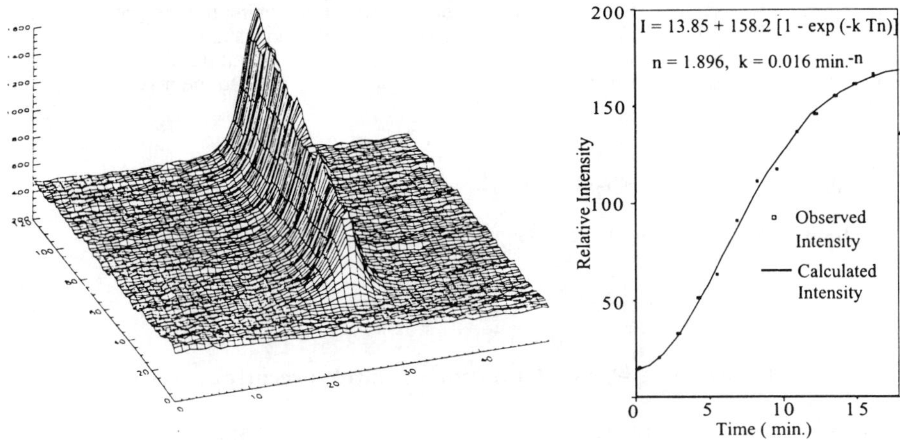

Figure 8 Variation in the first peak in the synchrotron X-ray powder diffraction pattern for
DMA-SbS-SB8". The peak is shown to the left as a function of time (min.). The integrated
intensity of this peak (dots) during crystal growth (Figure 7) and the theoretical fitting to the
integrated intensities is shown to the right.

The possible reaction schemes involving the DMA-SbS system in the hydrothermal synthesis
are summarized in Figure 9. The first step involves digestion or mineralization of crystalline Sb
and S and consequent reaction of polysulfide species with Sb to form antimony sulfide
precursors. Condensation to form the first open framework compound (SB8") then occurs. This
condensation (Figure 7) was rapid. As discussed in a previous publication [27], the structural
backbone of the first open framework might consist of SbS-double chains, formed from $Sb_4S_7^{2-}$
single chains connected through multiple (S_n) bridges (Figure 5). The number of sulfur atoms in
the bridge could change depending on experimental conditions and could be controlled by the
degree of S-excess of precursor species. Therefore, a variety of open structures, such as DMA-
SbS-SB8' found in the laboratory experiment or DMA-SbS-SB8" detected in the *in-situ*
experiments could be produced. This group of compounds have similar needle-morphologies and
this can be observed in the static experiments or predicted from the kinetic parameters for DMA-
SbS-SB8" derived from the variation in diffraction peak width in the *in-situ* experiments (Figure
8).
Both of the precursor open structures (-Sb8' and -SB8") are unstable at 200°C (Figure 7) and
above this temperature gradually transformed to DMA-SbS-SB8. The results of earlier
autogenous hydrothermal experiments [27] indicate the structural transformation can take up to 3
days. Since this procedure was accompanied by the elimination of excessive S atoms in the
precursor open structures, the transformation might also occur in several steps. Therefore, the
powder diffraction patterns recorded during the transformation may show only some of the
characteristic peaks of DMA-SbS-SB8 [27]. Intensities of some other reflections may be weaker
than that calculated according to the structural model of DMA-SbS-SB8. For example, the (002)
reflection, which represents the repeat of the double chains in the structure of DMA-SbS-SB8
[27], was very weak. While preferred orientation effects might also account for this observation,
variation in the connection between Sb_4S_7-single chains in the precursor open structures may lead

to disorder along [001]; such disorder might arise from a variation in the (S_n) bridges or the packing of the double chains in the early stage of the transformation (Figure 5).

In recent reports of real-time *in-situ* studies of hydrothermal synthesis of microporous tin sulfides, a three-stage mechanistic pathway was also proposed for the formation of these open structures [62]. Some results from these experiments are similar to those observed in this study on antimony sulfide system. However, monochromatic diffraction using imaging plates at the X7B beamline allows the accurate and reliable determination of changes in integrated peak intensity. In this way the complexity of the reaction pathway is revealed during the hydrothermal synthesis; for example the prereaction of antimony prior to the involvement of the sulfur (Figure 7).

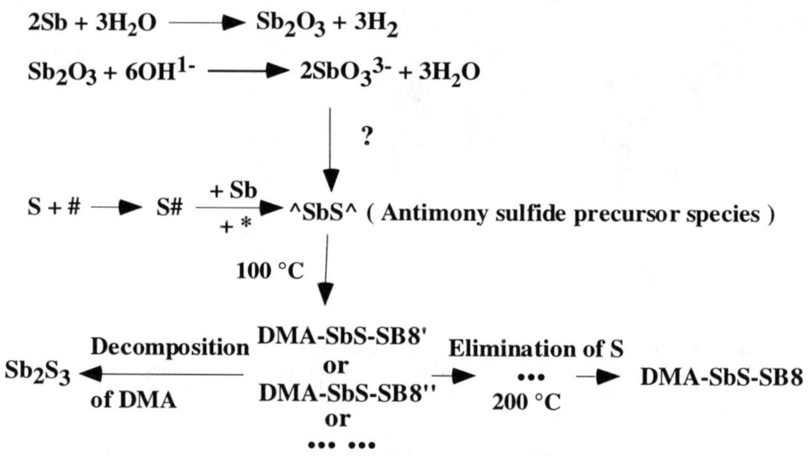

$$2Sb + 3H_2O \longrightarrow Sb_2O_3 + 3H_2$$

$$Sb_2O_3 + 6OH^{1-} \longrightarrow 2SbO_3^{3-} + 3H_2O$$

$$\downarrow \ ?$$

$$S + \# \longrightarrow S\# \xrightarrow[+\ *]{+\ Sb} \text{^SbS^} \ (\text{ Antimony sulfide precursor species })$$

$$100\ ^\circ C \downarrow$$

$$Sb_2S_3 \xleftarrow[\text{of DMA}]{\text{Decomposition}} \begin{array}{c}\text{DMA-SbS-SB8'}\\ \text{or}\\ \text{DMA-SbS-SB8''}\\ \text{or}\\ \bullet\bullet\bullet\ \bullet\bullet\bullet\end{array} \xrightarrow[200\ ^\circ C]{\text{Elimination of S}} \bullet\bullet\bullet \longrightarrow \text{DMA-SbS-SB8}$$

Figure 9. Schematic of the possible reaction pathways for the system DMA-Sb-S. The symbols #, * and ^ represent species in solution. For example, S# implies a possible polysulfide species.

CONCLUSIONS

Hydrothermal titration is a powerful new methodology for exploratory synthesis, especially when the production of such kinetically stablized phases as open frameworks are desired. The technique differs from static experiments, in which all reactants are sealed into a bomb and the reaction allowed to proceed unperturbed, by activating reactants and producing precursor materials prior to precipitation

Preliminary *in-situ* experiments on the DMA-SbS system provide insights into the chemistry of digestion of crystalline starting materials and structural formation and transformation. Such data are required for the rational design of synthetic strategies for the hydrothermal titration apparatus. Further investigation of this system requires combination with other techniques such as EXAFS or spectroscopy to focus on microscopic analysis to complement the X-ray diffraction studies reported here. This will be especially important in the initial digestion stage of the reactions where non-crystalline materials are produced.

The combination of these techniques at synchrotron light sources is now possible with the present generation of hydrothermal titration apparatus. The use of a glass capillary was necessitated by the low energies achievable at bending magnet sources. The apparatus used in the laboratory (Figure 2) will be usable directly at higher energy beamlines such as X17 at the NSLS and at the Advanced Photon Source. This will allow observation of changes as materials are injected and extracted from the reaction.

ACKNOWLEDGMENTS

This work was supported by NSF grant DMR-9413003 to JBP. Work at the National Synchrotron Light Source (NSLS) was supported under contract DE-AC02-76CH00016 at beamlines X7A and X7B. The NSLS is supported by the US Department of Energy, Divisions of Material Sciences and Chemical Sciences. We are grateful for the efforts of Drs. Jonathan C. Hanson and David Cox for their help and encouragement on beamlines X7B and X7A respectively.

REFERENCES

1. R. M. Barrer, Zeolites and Clay Minerals as Sorbents and Molecular Sieves, Academic Press, London, 1978, pp. 12
2. N. Herron, Y. Wang, M. M. Eddy, G. D. Stucky, D. E. Cox, K. Moller, and T. Bein, J. Am. Chem. Soc. **11**, 530 (1989).
3. G. A. Ozin, A.Kuperman, and A. Stein, Angew. Chem Int. Ed. Engl. **28**, 359 (1989).
4. R. M. Barrer, Zeolites **1**, p. 130 (1981).
5. D. W. Breck, Zeolite Molecular Sieves, Krieger, Malabar,FL, 1984, pp. 115
6. W. M. Meier, and D. H. Olson, Atlas of Zeolite Structure Types, Butterworths, London, 1987, pp. 100
7. B. Krebs, D. Voelker, and K. Stiller, Inorg. Chem. Acta L101 (1982).
8. R. L. Bedard, L. D. Vail, S. T. Wilson, and E. M. Flanigen, **U. S. Patent 4,880,761**, (1989).
9. R. L. Bedard, S. T. Wilson, L. D. Vail, J. M. Bennett, and E. M. Flanigen in Zeolites: Facts, Figures, Future. Proceedings of the 8th International Zeolite Conference, edited by P. A. J. a. R. A. v. Santen (Elsevier 49, city, 1989), p. 375.
10. R. L. Bedard, L. D. Vail, S. T. Wilson, and E. M. Flanigen, **U. S. Patent 4,933,068**, (1990).
11. J. B. Parise, and Y. Ko, Chem. Mat. **4**, 1446 (1992).
12. K. Tan, Y. Ko, and J. B. Parise, Acta Cryst. **C50**, 1439 (1994).
13. J. B. Parise, Ko, Y., Rijssenbeek, J., Nellis, D. M., Tan, K., Koch, S., J. Chem Soc., Chem. Commun. 527 (1994).
14. J. B. Parise, and Y. Ko., Chem. Materials **6**, 718 (1994).
15. Y. Ko, C. Cahill, and J. B. Parise, J. Chem. Soc., Chem. Commun. 69 (1994).
16. T. Jiang, G. A. Ozin, and R. L. Bedard, Adv. Mater. **6**, 860 (1994).
17. P. T. Wood, W. T. Pennington, and J. W. Kolis, Inorg. Chem. **33**, 1556 (1994).
18. H. Ahari, G. A. Ozin, R. L. Bedard, S. Petrov, and D. Young, Adv. Mater. **6**, 370 (1995).
19. Y. Ko, K. Tan, D. M. Nellis, S. Koch, and J. B. Parise, J. Solid State Chem. **114**, 506 (1995).
20. D. M. Nellis, Y. Ko, K. Tan, S. Koch, and J. B. Parise, J. Chem Soc., Chem. Commun. 541 (1995).
21. P. Enzel, G. S. Hederson, G. A. Ozin, and R. L. Bedard, Adv. Mater. **7**, 64 (1995).
22. J. B. Parise, Y. Ko, K. Tan, D. M. Nellis, and S. Koch, J.Solid State Chem. **117**, 219 (1995).
23. K. Tan, Y. Ko, and J. B. Parise, Acta Cryst. **C51**, 398 (1995).
24. C. L. Bowes, W. U. Huynh, S. J. Kirkby, A. Malek, G. O. Ozin, S. Petrov, M. Twardowski, D. Young, R. L. Bedard, and R. Broach, Chem. Mater. **8**, 2147 (1996).
25. K. Tan, Y. Ko, J. B. Parise, and A. Darovsky, Chem. Mater. **8**, 448 (1996).
26. Y. Ko, K. Tan, J. B. Parise, and A. Darovsky, Chem. Mater. **8**, 493 (1996).
27. K. Tan, Y. Ko, J. B. Parise, J. H. Park, and A. Darovsky, Chem. Mater. **8**, 2510 (1996).
28. J. B. Parise, D. M. Nellis, and Y. Ko, in preparation (1996)
29. R. M. Barrer, Hydrothermal Chemistry of the Zeolites, Academic Press., London, 1982, p. 43
30. M. E. Davis, and P. F. Lobo, Chem. Mater. **4**, 756 (1992).
31. M. E. Davis, Chemtech. 22 (1994).

32. S. I. Zones, M. M. Olmstead, and D. S. Santilli, J. Am. Chem. Soc. **114**, 4195 (1992).
33. D. J. Vaughan, and J. R. Craig, <u>Mineral Chemistry of Metal Sulfides</u>, Cambridge University Press, Cambridge, 1978, p. 57
34. R. C. Weast, Eds., Handbook of Chemistry and Physics, CRC press, Cleveland, 1974.
35. L. R. Gardner, Geochim. Cosmochim. Acta **38**, 1297 (1974).
36. B. Krebs, Angew. Chem. Int. Ed. Engl. **22**, 113 (1983).
37. M. A. A. Schoonen, and H. L. Barnes, Geochem. Cosmochem. Acta **55**, 3491 (1991).
38. C. L. Cahill, and J. B. Parise, in preparation (1996)
39. O. M. Yaghi, Z. Sun, D. A. Richardson, and T. L. Groy, J. Am. Chem. Soc. **116**, 807 (1994).
40. K. Tan, A. Darovsky, and J. B. Parise, J. Am. Chem. Soc. **117**, 7039 (1995).
41. C. L. Cahill, and J. B. Parise, submitted (1996)
42. G. C. Smith, Synch. Rad. News **4**, 24 (1991).
43. H. M. Rietveld, J. Appl. Crystallog. **2**, 65 (1969).
44. J. B. Parise, and K. Tan, J. Chem. Soc., Chem. Comm. 1687 (1996).
45. W. S. Sheldrick, and H. J. Haüsler, Z.anorg.allg.Chem **561**, 149 (1988).
46. J. B. Parise, J. Chem. Soc., Chem. Commun. 1553 (1990).
47. J. B. Parise, Science **251**, 293 (1991).
48. T. J. MaCarthy, and M. G. Kanatzidis, Inorg. Chem. **33**, 1205 (1994).
49. X. Wang, and F. Liebau, J. Solid State Chemistry 385 (1994).
50. X. Wang, Eur. J. Solid State Inorg. Chem. **32**, 303 (1995).
51. J. Anwar, and P. Barnes, Phase Transitions **39**, 3 (1992).
52. P. Barnes, S. M. Clark, D. Hausermann, E. Henderson, and C. H. Fentiman, Phase Trans. **39**, 117 (1992).
53. K. Ståhl, and J. Hanson, J. Applied Cryst. **27**, 543 (1994).
54. P. Norby, A. N. Christensen, and J. C. Hanson in <u>Studies in Surface Science and Catalysis</u>, edited by J. Weitkamp, H. G. Karge, H. Pfeifer and W. Hölderich (Elsevier 84, city, 1994), p. 179
55. A. N. Christensen, and M. S. Lehmann, J. Solid State Chem. **51**, 196 (1984).
56. A. N. Christensen, P. Norby, and J. C. Hanson, J. Solid State Chem. **114**, 556 (1995).
57. P. Norby, Mat. Sci. Forum **228-231**, 147 (1996).
58. P. Norby, A. Darovsky, J. C. Hanson, and I. Meshhovsky, in preparation
59. Gmelin-Institute, Eds., Gmelis Handbuch der Anorganischen Chemie, Antimon, Teil A,B, 18, Verlag Chemie, GMBH., Weinheim, Bergstrasse, 1949.
60. F. A. Cotton, and G. Wilkinson, <u>Advanced Inorganic Chemistry</u>, John Wiley & Sons, New York, 1988, p 189
61. D. Turnbull in <u>Phase Changes</u>, Edited by F. Seitz and D. Turnbull, 3, Academic Press Inc., New York, 1956, p. 225.
62. R. J. Francis, S. J. Price, J. S. O. Evans, S. O'Brien, D. O'Hare, and S. M. Clark, Chem. Mater. **8**, 2102 (1996).

HYDROTHERMAL SYNTHESIS OF NOVEL VANADIUM OXIDES

M. STANLEY WHITTINGHAM[*], ELIZABETH BOYLAN, RONGJI CHEN, THOMAS CHIRAYIL, FAN ZHANG and PETER Y. ZAVALIJ
Chemistry Department and Materials Research Center, State University of New York at Binghamton, Binghamton, NY 13902-6000, USA

ABSTRACT

Extending our prior work on tungsten and molybdenum oxides, we have found that a wide variety of vanadium oxides can be prepared using hydrothermal methods. These include a number of layer compounds as well as cluster complexes. The starting reaction medium usually contained vanadium pentoxide, an alkali containing compound such as LiOH, an organic template such as tetramethylammonium, and the pH of the whole was controlled by the addition of acid. Reaction temperature was 150°C to 200°C, and time was up to 3 days. A new lithium vanadium oxide, which has the simplest structure of any layered vanadium oxide, was formed. The lithium could be readily removed leading to a new form of vanadium dioxide. This vanadium oxide was also capable of intercalating a variety of other ionic and molecular species. Several other new vanadium oxides containing the TMA cation were also formed; one of these $TMAV_3O_7$ readily absorbed oxygen to form $TMAV_3O_8$. Addition of zinc or iron to the reaction medium caused the formation of layer structures containing double V_2O_5 layers; for iron the TMA was retained in the structure whereas for zinc the TMA was excluded. Changing the organic entity resulted in other new structures, for example methylamine and dimethylamine gave tetragonal structures.

INTRODUCTION

There has been much interest in the synthesis and characterization of new materials by soft chemistry approaches over the last decade [1]. Such materials might be expected to exhibit new structures and different properties than known materials as they are formed under conditions of kinetic rather than thermodynamic control. One particularly successful synthesis approach [2] is hydrothermal under mild conditions, that is no higher than 200°C. Though extensively utilized for many years for the synthesis of zeolites and phosphates, only in the last six years has serious focus been placed on transition metal oxides. Tungsten oxides form a range of tunnel structures [3-8], as well as Keggin clusters .[9]. The well-known hexagonal molybdenum oxides were also formed hydrothermally [10, 11], in addition to a new layered molybdenum oxide containing the tetramethylammonium ion, TMA [12]. The TMA cation was found to be particularly adept in forming new structures [13, 14], and it is not necessarily retained in the structure formed[15]. A wide range of new vanadium oxide structures have subsequently been formed using this cation [16-21]. In the last year, a number of new vanadium oxide structures containing other organic cations have been reported [22-28]. Layered manganese oxides have also been formed in our laboratory by the hydrothermal decomposition of aqueous permanganate solutions. In the case of the alkali metal permanganates birnessite type phases were formed [29, 30]. However, when TMA permanganate is decomposed in the presence of nickel ions the compound $NiMnO_3$ is formed; it is not the known ilmenite form [31] but a new structure with very different magnetic properties [32].

This paper is concerned with the formation of vanadium oxides under mild hydrothermal conditions, and will discuss in particular the role of pH and organic

Mat. Res. Soc. Symp. Proc. Vol. 453 © 1997 Materials Research Society

cation in determining the structure formed, and the reactivity of the formed vanadium oxides for both redox and intercalation reactions. Primary consideration will be given to the TMA cation. In addition, the impact of adding a second transition metal such as iron, manganese, or zinc will be described.

EXPERIMENTAL

In a typical synthesis the reactants V_2O_5, $N(CH_3)_4OH$, and $LiOH$ in a molar ratio of 1:2:1 in water were acidified with acetic acid and heated at 185°C in a Teflon lined Parr reactor for 3 days; typically 20 gm of reactants were dissolved in 70 ml of water. The iron vanadium oxide was formed by heating $FeCl_3$, V_2O_5 and TMAOH in aqueous solution at pH of 3-6 in a 1:1:2.8 molar ratio, for 2.5 days at 165°C. In the case of the zinc compound a 1:1:1 aqueous solution of $ZnCl_2$, V_2O_5 and TMAOH was heated at 165°C for 2.5 days; the initial pH was 4.7. The methylamine complex was formed by heating V_2O_5 and methylamine in water at a pH of 3.5 for 4 days at 200°C.

X-ray powder diffraction was performed using $CuK\alpha$ radiation on a Scintag $\theta-\theta$ diffractometer. The TGA data was obtained on a Perkin Elmer model TGA 7, the FTIR on a Perkin Elmer 1600 series, the Electron Microprobe on a JEOL8900, and the chemical analysis was obtained on an ARL Spectrospan-7 DCP Atomic Emission Spectrometer. The degree of reduction of the vanadium oxide by lithium was determined by reaction with n-butyl lithium.

RESULTS AND DISCUSSION

<u>Tetramethylammonium Ion</u>

When the reaction medium contained the TMA cation a variety of materials were precipitated from solution. X-ray diffraction of the products formed showed that at least six different phases were formed depending on the initial pH. These are shown in Fig. 1. At the highest pH, 10, the known Li_3VO_4 is formed [33]; in this structure all the metal atoms are in tetrahedral sites. As the pH is lowered the new phase $N(CH_3)_4V_3O_7$ is formed [19]; this phase is also formed in the absence of lithium ions. At a pH of about 5 a new layered vanadium dioxide compound $Li_xV_{2-\delta}O_{4-\delta} \bullet H_2O$ is found, as reported earlier [15, 17, 18]. Lowering the pH further leads to the formation of another tetramethylammonium vanadate, $N(CH_3)_4V_4O_{10}$ whose structure we reported recently.[15, 16] At the highest acidities two new compounds are formed, both of which contain tetramethylammonium ions and have string like morphologies unlike the platelets of the other TMA compounds; the strands shown in the figure are around a μm in cross-section and up to several hundred μm in length. They appear to have layer structures with repeat distances of 11.5 Å and 19.1Å at pHs of 3.3 and 2.6 respectively. We are presently determining their structures from powder data.

The trivanadate, $N(CH_3)_4V_3O_7$, has a lath-like appearance and it's structure is related to that recently reported for $(DABCO)V_6O_{14}$, [25, 27] but has some significant differences. The structure contains zig-zag strings of edge sharing VO_5 square pyramids; these strings are joined together by VO_4 tetrahedra. The structure has lattice parameters of a = 18.478 Å, b = 6.552 Å, c = 8.427 Å and β = 91.1°, with space group $P2_1/n$. The tetramethylammonium ions reside between the vanadium oxide layers, as indicated in Fig. 1. The sheets are almost planar in direct contrast to those of the DABCO hexavanadate $C_6H_{14}N_2V_6O_{14} \bullet H_2O$ which are highly puckered, leading to a longer a and a shorter c axis, a = 19.20 Å, c = 7.547 Å and β = 111.3°, whereas the dimension along the chain remains almost unchanged, b = 6.625 Å [25] . On heating to around 200°C and prior to the loss of the organic cations at 350°C the structure takes up

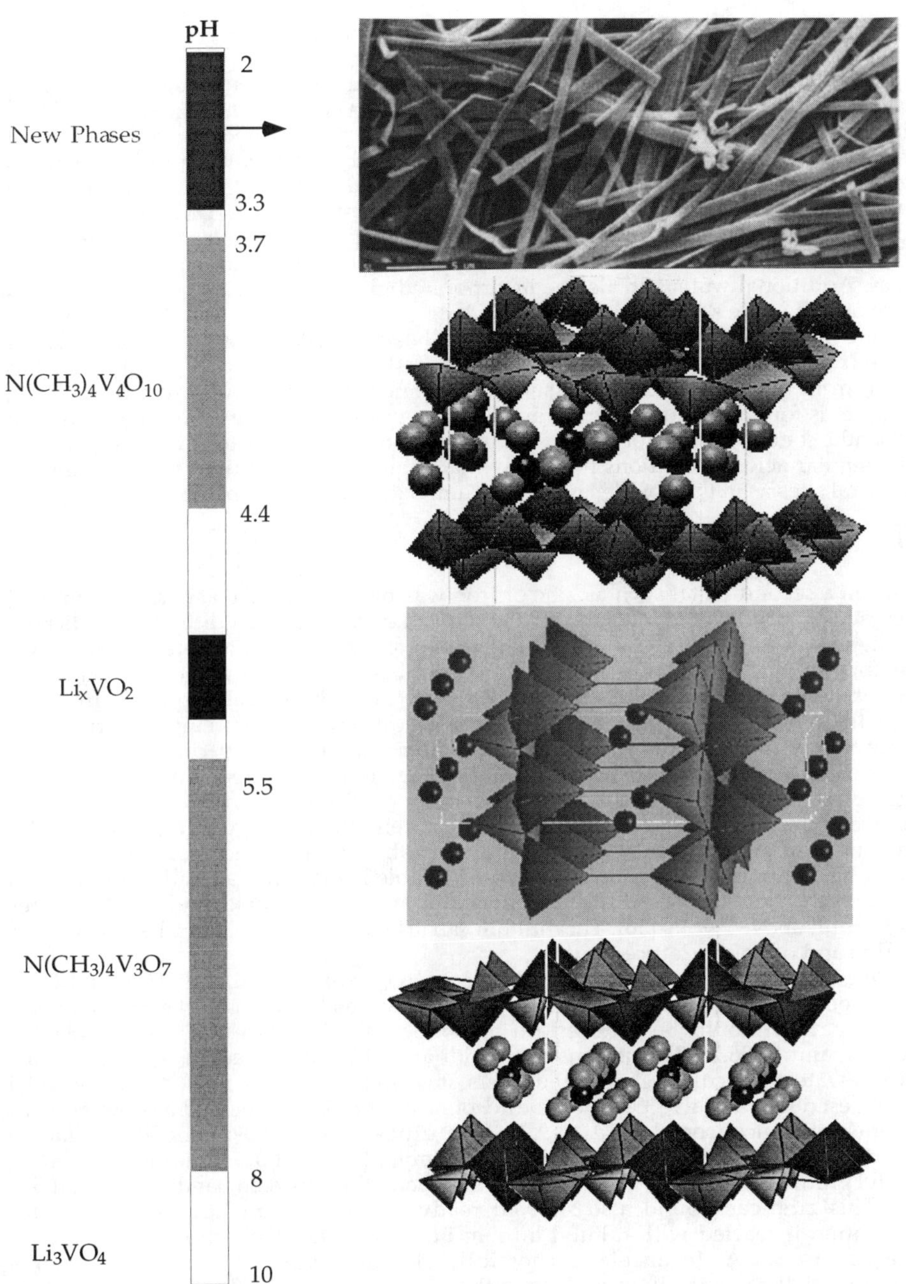

Fig. 1 Some compounds formed on reaction of V$_2$O$_5$, N(CH$_3$)$_4$OH, and LiOH as a function of pH. Top is electron micrograph of phase formed at pH=3.3 with strings about 1 micron across.

an additional oxygen atom to give the fully oxidized phase $N(CH_3)_4V_3O_8$. This appears to have a triclinic structure.

In the tetravanadate, $N(CH_3)_4V_4O_{10}$, all the vanadium is present in square pyramids just as in the layered vanadium dioxide, $Li_xV_{2-\delta}O_{4-\delta}\bullet H_2O$. In this dioxide the TMA cations are not retained in the structure but it is not formed in the absence of tetramethylammonium ions. This dioxide readily and reversibly undergoes dehydration and redox reactions with lithium [18], and all the lithium can be removed in an electrochemical cell. In addition, other solvents such as DMSO are readily incorporated into the structure as are alkylamines expanding the lattice just as in the layered disulfides. The alkylamines can form both interleaved monolayer and bilayer phases. Additional water can also be incorporated giving two layers of water between the vanadium oxide sheets.

Some trends in the structures of figure 1 can be seen. Li_3VO_4 has all the vanadium in tetrahedral sites. As the pH is lowered the first phase formed $N(CH_3)_4V_3O_7$ has vanadium in both tetrahedral and square pyramidal sites. At still lower pHs all the vanadium is found in square pyramids. The structures of the phases formed under the most acidic conditions are not known but by comparison with cluster complexes formed under similar acidic conditions it is likely that some of the vanadium will be found in octahedral sites as in the double layer vanadium oxides discussed below.

<u>Double V_2O_5 Structures with Zinc and Iron</u>

When a second metal such as zinc or iron was present in the reaction medium [34] the crystals formed were large, up to 500 μm on an edge, and x-ray diffraction indicated that the repeat unit was significantly larger than for the compounds described above. Thus for iron and zinc the lattice repeat was 13.1 Å and 10.4 Å respectively and the former only dropped to 12.9 Å on heating to 150°C to drive out all the water from the lattice. In both these compounds the structure is best described as being constructed of double layers of V_2O_5 following Galy [35]. The iron compound contains ≈10% by weight of TMA which is lost by 430°C on heating, and the composition is tentatively described as $[N(CH_3)_4]_zFe_yV_2O_{5+\delta}\bullet nH_2O$ where $y \approx 0.1$ and z and n are around 1/6. As noted above the lattice contracts very little on dehydration, suggesting that the TMA ions act as pillars and the water fills the available vacant space. Unlike the TMA containing compounds described above, this double layer material readily reacts with lithium both chemically and electrochemically and takes up close to 1 lithium per formula unit above 2 volts [36]. This lithium can also be removed electrochemically, and cycled in and out of the lattice.

The zinc compound had the formula, $Zn_{0.4}V_2O_{5.4}\bullet nH_2O$ and both TGA and FTIR showed no evidence for the presence of organic cations between the layers. It's x-ray powder pattern could be indexed on a monoclinic cell with a=11.74Å, b=3.62Å, c=10.50Å, and β=96.7°. This is consistent with a double layer sheet of V_2O_5 with zinc ions and water molecules between the layers. It is almost certainly related to the ν and δ phases described by Galy [35]. The product of heating to 150°C showed a contraction in the repeat distance from 10.43Å to 9.23Å; on exposure to atmospheric air the lattice expanded again. The FTIR spectrum was very similar to that of the iron compound with vanadyl group at 1008 cm^{-1}, and the other main vanadium-oxygen bands at 777 and 530 cm^{-1}. This zinc compound also showed ready reactivity with lithium. Thus, after dehydration it reacted with n-butyl lithium in hexane and it's lattice repeat distance decreased to 8.62Å. In an electrochemical cell initial voltage was over 3.5 volts suggesting that the vanadium is close to the +5 oxidation state. The lithium insertion then proceeded in a number of stages, reminiscent of that of V_2O_5 itself. These steps were maintained during the subsequent charge and discharge cycles, showing that the structure of this material did not change during the redox reactions.

<u>Methylamine and Dimethylamine</u>

When the hydrothermal reaction was carried out in the presence of methylamine or dimethylamine shiny black square platelike crystals, as shown in figure 2 were formed. These have the formula $(amine)_{0.55}V_2O_{5-\delta}$, where the vanadium oxidation state is 4.35 and appear to have a tetragonal unit cell with a=6.19 Å and c=8.98 Å with space group P 4/mmm. The details of the structure are presenly being elucidated. The structure is readily reduced by the intercalation of lithium ions. On heating in oxygen or nitrogen, V_2O_5 and V_6O_{13} are the principal reaction products respectively.

Fig. 2. Electron micrograph showing vanadium oxide square platelets. (Bar is 5μm long)

CONCLUSIONS

A wide range of new vanadium oxides have been synthesized hydrothermally using the tetramethylammonium ion as a structure directing molecule. The structure formed depends critically on the pH of the reaction medium. Several of these vanadium oxides readily undergo redox reactions, and the dioxide is a ready host for intercalation reactions. Addition of a second transition or post-transition metal, such as zinc or iron leads to the formation of double sheet V_2O_5 structures. These double sheet structures are good hosts for lithium ions even when the tetramethylammonium ions are present, unlike the single sheet $TMAV_3O_7$ and $TMAV_4O_{10}$. Other organic species, such as methylamine, lead to different structures.

ACKNOWLEDGEMENTS

We thank the Department of Energy, through Lawrence Berkeley Laboratory, and the National Science Foundation through grant DMR-9422667 for partial support of this work.

REFERENCES

1.   J. Rouxel, M. Tournoux, and R. Brec, ed. *Soft Chemistry Routes to New Materials - Chimie Douce-*. Materials Science Forum, Vol. 152-153. 1994, Trans Tech Publications: Switzerland.

2. M. S. Whittingham, Current Opinion in Solid State and Materials Science, **1** (1996) 227.
3. J. R. Günter, M. Amberg, and H. Schmalle, Mat. Res. Bull., **24** (1989) 289.
4. K. P. Reis, A. Ramanan, and M. S. Whittingham, Chem. Mater., **2** (1990) 219.
5. K. P. Reis and M. S. Whittingham, J. Solid State Chem., **91** (1991) 394.
6. K. P. Reis, A. Ramanan, W. Gloffke, and M. S. Whittingham. *Synthesis, Diffusion and Ion-Exchange in Open Structure Sodium Tungstates and YBaCuTungstates.* in *Solid State Ionics II.* **210** (1991) 473. Boston, MA: Materials Research Society.
7. K. P. Reis, A. Ramanan, and M. S. Whittingham, J. Solid State Chem., **96** (1992) 31.
8. K. P. Reis, E. Prince, and M. S. Whittingham, Chem. Mater., **4** (1992) 307.
9. P. Zavalij, J. Guo, M. S. Whittingham, R. A. Jacobson, V. Pecharsky, C. Bucher, and S.-J. Hwu, J. Solid State Chem., **123** (1996) 83.
10. J. Guo, P. Zavalij, and M. S. Whittingham, Eur. J. Solid State Chem, **31** (1994) 833.
11. J.-D. Guo, P. Zavalij, and M. S. Whittingham, J. Solid State Chem, **117** (1995) 323.
12. J. Guo, P. Zavalij, and M. S. Whittingham, Chem. Mater., **6** (1994) 357.
13. M. S. Whittingham, J. Li, J. Guo, and P. Zavalij. *Hydrothermal Synthesis of New Oxide Materials using the Tetramethyl Ammonium Ion.* in *Soft Chemistry Routes to New Materials.* **152-153** (1993) 99. Nantes, France: Trans Tech Publications Ltd.
14. Y. J. Li and M. S. Whittingham, Solid State Ionics, **63** (1993) 391.
15. M. S. Whittingham, J. Guo, R. Chen, T. Chirayil, G. Janauer, and P. Zavalij, Solid State Ionics, **75** (1995) 257.
16. P. Zavalij, M. S. Whittingham, E. A. Boylan, V. K. Pecharsky, and R. A. Jacobson, Z. Kryst., **211** (1996) 464.
17. T. Chirayil, P. Zavalij, and M. S. Whittingham, Solid State Ionics, **84** (1996) 163.
18. T. Chirayil, P. Zavalij, and M. S. Whittingham, J. Electrochem. Soc., **143** (1996) L193.
19. T. A. Chirayil, P. Y. Zavalij, and M. S. Whittingham, Chem. Commun., (1996) in press.
20. L. F. Nazar, Personal Communication, (1996)
21. E. A. Boylan, T. Chirayil, J. Hinz, P. Zavalij, and M. S. Whittingham, Solid State Ionics, **90** (1996) 1.
22. D. Riou and G. Férey, J. Solid State Chem., **120** (1995) 137.
23. D. Riou and G. Férey, Inorg. Chem., **34** (1995) 6250.
24. D. Riou and G. Férey, J. Solid State Chem., **124** (1996) 151.
25. L. F. Nazar, B. E. Koene, and J. F. Britten, Chem. Mater., **8** (1996) 327.
26. Y. Zhang, C. J. O'Connor, A. Clearfield, and R. C. Haushalter, Chem. Mater., **8** (1996) 595.
27. Y. Zhang, R. C. Haushalter, and A. Clearfield, J. Chem. Soc., Chem. Commun., (1996) 1055.
28. Y. Zhang, J. R. D. DeBord, C. J. O'Connor, R. C. Haushalter, A. Clearfield, and J. Zubieta, Angew. Chem. Int. Ed. Engl., **35** (1996) 989.
29. R. Chen, P. Zavalij, and M. S. Whittingham, Chem Mater, **8** (1996) 1275.
30. R. Chen, T. Chirayil, and M. S. Whittingham, Proceedings of the 10th International Symposium on Solid State Ionics, Singapore, December 1995. Solid State Ionics, **86-88** (1996) 1.
31. M. Pernet, J. C. Joubert, and B. Ferrand, Solid State Communications, **16** (1975) 503.
32. R. Chen, P. Y. Zavalij, and M. S. Whittingham, Chem. Mater., (1997) in press.
33. R. D. Shannon and C. Calvo, J. Solid State Chem., **6** (1973) 538.
34. F. Zhang, P. Y. Zavalij, and M. S. Whittingham, Mater. Res. Bull., (1997) in press.
35. J. Galy, J. Solid State Chem., **100** (1992) 229.
36. F. Zhang, P. Y. Zavalij, and M. S. Whittingham, This volume, (1997)

NEW LOW TEMPERATURE PHASES OF TUNGSTEN OXYNITRIDE STABILIZED BY DOPING WITH Fe, Co, AND Ni

F. CHEN, J. AITKEN, S. PARASHER
Rider University, Department of Chemistry,
2083 Lawrenceville Rd, Lawrenceville, NJ 08648

ABSTRACT

Several low temperature tungsten-containing oxynitride phases were isolated and stabilized by doping with Fe, Co, and Ni. These Phases have the rocksalt structure and form around 420 °C by reacting MWO_4 (M=Fe, Co, Ni) with dry ammonia. They served as intermediates in the formation of the hexagonal ternary nitrides (MWN_2). X-ray powder diffraction analyses of these phases indicate that $W_{0.62}(NO)$ could be the parent structure type in each case. The incorporation of group VIIIB elements in the $W_{0.62}(NO)$ structure may play an important role in the formation and stabilization of these low temperature phases.

INTRODUCTION

Although nitrides exhibit a variety of properties which have wide applications in technology [1], they have not received as much attention as many other solid-state materials because they are more challenging and costly to synthesize. In addition, many nitrides have unfavorable free energies of formation [2] and the inductive effect predicts that it will be difficult to form nitrides which do not contain highly electropositive elements [3].

There are few routes to synthesize ternary nitrides. We have chosen to study the less explored method of ammonolysis of ternary oxides. We feel that this route would be the most economical and practical because of the availability of oxide precursors and the use of low temperatures and ambient pressures. Moreover, the oxide precursor method is advantageous because the elements are already mixed at atomic level making diffusion lengths shorter and reaction time faster [2]. Furthermore, this low temperature synthesis can lead to "metastable phases" not obtainable at high temperatures.

We believe that ammonolysis of ternary oxides will yield many new nitrides in the future. However, at this point, only a few nitrides have been made via ammonolysis of their oxide precursors [3]. Our approach to finding more suitable precursors has been to study the phase formation process using known compounds and to use that knowledge to search for new nitrides. One of the group of known compounds we have chosen for study are three new ternary transition metal nitrides, MWN_2 (M=Fe, Co, Ni). They can be synthesized by reaction of the corresponding tungstate precursors with flowing NH_3 at 600-800 °C and all crystallized in a hexagonal structure with a slight difference in the anion stacking sequence or cation coordination [4,5]. We performed a systematic study under various reaction temperatures and times to monitor the phase formation progress of these nitrides.

Here we report three low temperature phases of tungsten oxynitrides stabilized by doping with Fe, Co, and Ni which were discovered as the intermediate phases during our study on the phase formation of this group of nitrides.

Mat. Res. Soc. Symp. Proc. Vol. 453 © 1997 Materials Research Society

EXPERIMENT

Synthesis:

Precursors. Oxide precursors were prepared using a simple wet chemistry method [4]. For example, to prepare a precursor with a Co:W ratio of 1.5:1.0, the following method was used. 400 mL of a 0.17M solution of $CoCl_2$ was added drop-wise to a 150 mL solution of 0.3 M $Na_2WO_4 \cdot 2H_2O$. This was performed slowly with constant stirring. The precipitate was filtered by vacuum and washed with two 30 mL portions of H_2O and one 30 mL portion of ether. The product was allowed to dry overnight in a desiccator. The compound was further dried at 150 °C for 24 hours. At this point, the amorphous $CoWO_4$ was obtained . To confirm the product, a small portion was placed in an alumina boat and put into a hinged furnace with a quartz tube flow through reactor in an atmosphere of nitrogen at 700 °C for 12 hours. Well crystallized $CoWO_4$ (JCPD # 15-867) was confirmed using powder X-ray diffraction analysis.

Ammonolysis. Ammonolysis of the amorphous transition metal (Fe, Co, Ni) tungsten oxide were conducted. (We use the amorphous oxide because we feel that it will be more susceptible to ammonia treatment, however, experiments using both crystalline and amorphous precursors have shown no significant difference so far.) The sample was placed in an alumina boat and put into a hinged furnace with a quartz tube flow through reactor in an atmosphere of dried ammonia. The temperature was increased from room temperature to the desired temperature at 3 °C/min and held there for 12-48 hours at which point the sample was quenched to room temperature by turning the furnace off and immediately opening it. The ammonia continued to pass over the sample until it has reached room temperature.

X-ray Diffraction

X-ray diffraction (XRD) patterns of the powdered samples, both oxide precursors and the products of ammonolysis, were obtained in the 5-80° 2θ range with CuKα radiation using a Rigaku DMAX/B X-ray diffractometer operating at 45 KV and 40 mA.

Chemical Analysis

CAMECA SX50 EPMA (Electron probe microanalysis) was used for the preliminary elemental analyses of these phases. In order to detect the nitrogen content, a LPC1 multilayer synthetic crystal (W-Si) with 2d spacing 60 Å was used for this particular purpose. A modified Kjeldahl method was also used for the nitrogen analysis to confirm the EPMA results.

RESULTS

A cubic phase was first observed from a X-ray diffraction pattern taken of a sample after ammonolysis of $CoWO_4$ around 420-430 °C. Subsequently, we also found similar phases in samples obtained using $FeWO_4$ or $NiWO_4$ as the precursors. However, the Co and Ni phases always contain observable amounts of Co or Ni metal. In order to isolate the desired phase in pure form, a 3M HCl solution proved to be effective in dissolving the metal impurities. X-ray diffraction patterns of all three purified phases are shown on Figure 1.

Although these phases resemble $W_{0.62}(NO)$, WN, and β-W_2N, the 2θ value of their corresponding diffraction peaks are noticeably shifted from those of the reference (Fig 1).

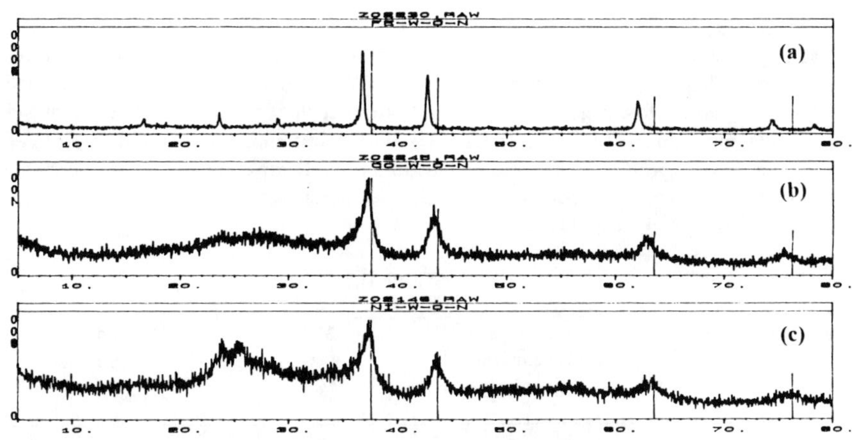

Fig.1 XRD patterns of low temperature M-W-O-N phases (a) M=Fe (b) M=Co (c) M=Ni. The vertical lines represent peaks given in JCPD file (#25 -1254) for $W_{0.62}(NO)$.

In order to quantify these observations, NBS silicon was used as a 2θ standard for accurate peak positions. The NRCVax Powder-Pattern Lattice Parameter Refinement Routine in the computer program Dispow was used to index these peaks. The structure was confirmed to be cubic with a larger lattice parameter than either $W_{0.62}(NO)$, WN, and β-W_2N (except for the Ni-W-N-O phase which has almost the same lattice parameter as $W_{0.62}(NO)$). These phases have consistent relative intensities in their XRD patterns which are in close agreement with those of $W_{0.62}(NO)$ (Table I). Therefore, we suggest that $W_{0.62}(NO)$ may be their parent phase.

Table I   Comparison of $W_{0.62}(NO)$, WN, and β-W_2N XRD Intensities with the New Low Temperature Cubic Phase

h k l	$W_{.62}(NO)^a$ (a = 4.138 Å) I/I_o	WN^b (a = 4.1196 Å) I/I_o	β-W_2N^c (a = 4.126 Å) I/I_o	Low temperature phase* (a - see Table II) I/I_o
1 1 1	100	75	100	100
2 0 0	68	100	47	73
2 2 0	45	75	33	40
3 1 1	46	75	44	37

JCPD # a. 25-1254; b. 3-1031; c. 25-1257.   * Co-W-O-N was used as an example

Table II gives the results of chemical analysis, the calculated chemical formulae and the observed lattice constants for these phases. Analysis for metal is considered accurate to ±3 atomic% while values for O and N may be uncertain to 5-8 atomic percent. The results showed that within experimental error the N:O ratio was essentially 1 and this conclusion has been used in calculating the chemical formulae given in Table II. Since only a trace of nickel (< 1 wt%) was found in the Ni-W-N-O phase from the analysis, and the lattice parameter of the phase is almost the same as that of $W_{0.62}(NO)$, it is uncertain if the nickel is incorporated in the phase or just simply present as a separate low level metal impurity which is below the XRD detection limits. Therefore, the chemical formula of Ni-W-N-O has not been determined and is not listed in the table. However, experiments indicate that the presence of the trace nickel is critical for the formation of the cubic phase at such a low temperature (<430 °C) [7]. The literature data of $W_{0.62}(NO)$ [6] are also included in Table II for comparison. Figure 2 shows the relationship between lattice parameter change and the degree of incorporation of Fe, Co, and Ni in the $W_{0.62}(NO)$ structure.

Table II

Lattice parameter (Å)	Formula (a) (according to the analyses)	M weights %	Method of preparation
4.233 ± 0.005	$Fe_{20}W_{22}(ON)_{58}$	16.6	$FeWO_4$ 430 °C 48 h NH_3
4.176 ± 0.005	$Co_4W_{27}(ON)_{70}$	3.1	$CoWO_4$ 425 °C 24 h NH_3
4.138 ± 0.0009	Ni-W-O-N (b)	< 1	$NiWO_4$ 412 °C 24 h NH_3
4.138	$W_{62}(ON)_{100}$	0	paratungstate 700 °C 48 h NH_3

(a) The formula has been written based on atomic percentage - the preliminary results from electron probe microanalysis.

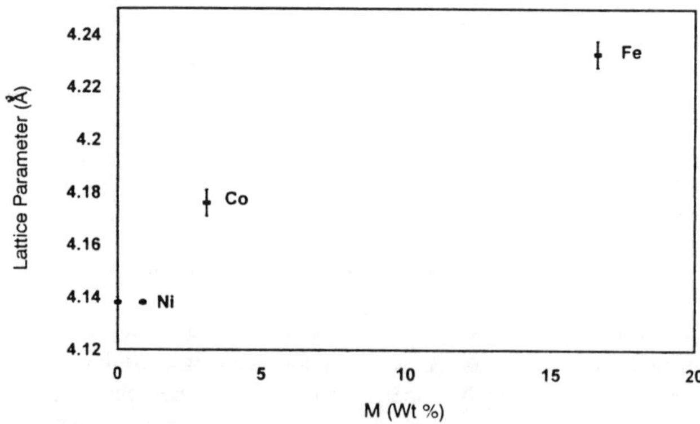

Fig.2   Lattice parameter changes with the degree of incorporation of Fe, Co, Ni in the $W_{0.62}(NO)$ structure

DISCUSSION

Three low temperature phases of transition metal (Fe, Co, Ni) containing tungsten oxynitrides were isolated during our study. The salient points of this study which yield information allowing us to further understand synthetic nitride chemistry are summarized below.

(1) To our knowledge, these phases may be the only tungsten containing nitrides/oxynitrides synthesized and stabilized at temperatures below 430 °C; most of known tungsten containing binary or ternary nitrides/oxynitrides are formed and stabilized at temperatures ranging 600-900 °C.

(2) The X-ray powder diffraction patterns of these phases were indexed with a face-centered cubic unit cell, and indicated that $W_{0.62}(NO)$ could be the parent structure in each case. The incorporation of group VIIIB elements in the $W_{0.62}(NO)$ structure may play an important role in the formation and stabilization of these low temperature phases. If this is true, further study may help us to develop more controllable low temperature synthetic routes using the strategy of low level doping with suitable transition metal element(s) .

(3) The variation of the lattice parameters and chemical analysis of the compounds confirmed that the incorporation of group VIIIB elements into these phases decreases in the order of Fe > Co > Ni. This is in agreement with the trend of the free energies of formation of the binary nitride of these elements. This observation may help us to select alternate appropriate chemical systems in searching for new ternary nitrides.

oxy = oxynitride (+ Co, or Ni)
mix = oxynitride + nitride (+ Co, or Ni)
nitride = nitride (+ Co, or Ni)

Fig. 3   The progression of the phase formation of ternary nitride M-W-N

$FeWN_2$

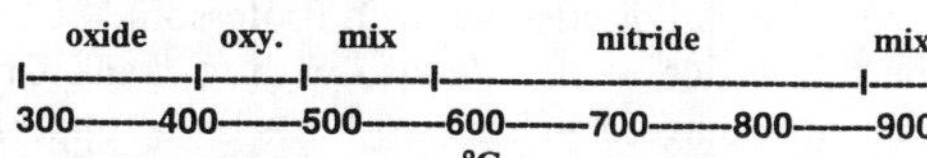

$Co_xW_yN_z$

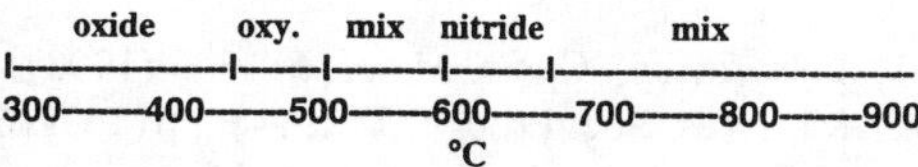

$Ni_xW_yN_z$

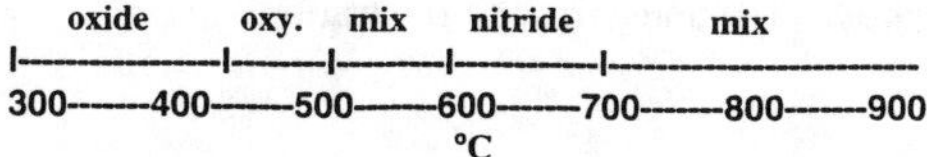

(4) Figure 3 shows the the detailed phase progression for a group of hexagonal ternary nitrides M-W-N (M=Fe, Co, Ni). The three reported low temperature cubic phases were found and isolated as the intermediate phases during this study. This is the first evidence of the <u>two step reaction sequence</u> in the formation of these nitrides synthesized by an ammonolysis process. Previously it has been proposed that the ternary oxide is directly converted to the ternary nitride via ammonolysis.   The ternary oxide was thought to serve as a template, ultimately determining the structure of the nitride.

Oxide --------- > Nitride

However, our experiments show that the oxide is first converted to an intermediate or intermediates. The intermediate(s), cubic oxynitride in this case, subsequently reacts with NH_3 to form final hexagonal nitride.   We believe, however, that the structure of both oxide and the intermediate(s) play a role in determining what the structure of nitride will be [7] or determining whether or not they can be converted to the pure nitride.

Oxide --------- > Intermediate(s) --------- > Nitride

Therefore, these low temperature oxynitride phases also provide key information to help us learn more about reaction progress during the ammonolysis, and to aid in finding additional oxide precursor systems.

CONCLUSION

Three low temperature transition metal tungsten oxynitrides with a rocksalt structure were found as the intermediates in the synthesis of hexagonal M-W-N (M=Fe, Co, Ni). These are the only known tungsten containing nitrides/oxynitrides that can be synthesized and stabilized at temperatures below 430 °C. In addition to the fact that these phases themselves may have interesting properties, this finding also is the first evidence to support the concept of a two step reaction sequence for the formation of hexagonal M-W-N (M=Fe, Co, Ni) via ammonolysis. A further detailed study on these phases is in progress.

ACKNOWLEDGMENTS

The authors express our appreciation for very helpful discussions with Professors H.C. zur Loye, F.J. DiSalvo and W.H. McCarroll on this nitride work. We also wish to thank Dr.E. Vicenzi for performing EPMA analysis at Princeton University. This work has been supported by a grant from the Research Corporation.

REFERENCES

1.   F.J. DiSalvo, Science **247**, 649 (1990).
2.   S.H. Elder, F.J. DiSalvo, L. Topor, and A. Navrotsky, Chem. Mater. **5**, 1545 (1993).
3.   J.D. Houmes, D.S. Bem, H.C. zur Loye, Mat. Res. Soc. Proc. **327** (1994) pp. 153-164.
4.   D.S. Bem and H.C. zur Loye, J. Solid State Chem. **104**, 467 (1993).
5.   P.S. Herle, N.Y. Vasanthacharya, M.S. Hegde, and J. Gopalakrishnan, J. Alloys and Compounds **217**, 22 (1995).
6.   R. Kiessling and L. Peterson, Acta Metallurgica **2**, 675 (1954).
7.   F.Chen, J. Aitken, S. Parasher, J. Truszkowski, manuscript in preparation.

SYNTHESIS AND STRUCTURE OF A METAL-ORGANIC SOLID HAVING THE CADMIUM(II) SULFATE NET

O. M. YAGHI, HAILIAN LI, M. O'KEEFFE
Department of Chemistry and Biochemistry, Goldwater Center for Science and
Engineering, Arizona State University, Tempe, AZ 85287-1604, oyaghi@asu.edu

ABSTRACT

Reaction of silver(I) nitrate and hexamethylenetetramine (HMTA) gives crystals of
$Ag_2(HMTA)(NO_3)_2$, which was formulated by elemental microanalysis and a single crystal
x-ray study. Its $Ag_2(HMTA)$ open-framework is the first example of a decorated $CdSO_4$
net. The relative orientation of the tetrahedral HMTA building blocks in this structure point
to numerous opportunities toward constructing novel chiral and polar porous frameworks.

INTRODUCTION

Soluble organic,[1] metal-organic,[2] organometallic,[3] and inorganic[4] molecular
building blocks have been linked using addition copolymerization and condensation
polymerization reactions to yield open-framework solids. Generally, these structures are
held together by attractive and directional forces such as hydrogen bonding and metal-
ligand coordination. The demonstrated ability to manipulate and exploit these forces
toward the design of porous solids, coupled to the vast number of molecular building units
that are available as starting materials have yielded framework structures of diverse
topology and architecture. For example, structures that adopt topologies related to the
archetypes diamond, [4,5] rutile,[6] α-$ThSi_2$,[2d,e] PtS,[7] and quartz[8] have been reported. The
structures of these solids are conceptually related to the assembled analogues by simply
replacing their constituent atoms and/or bonds with metal ions and large molecular entities
such as clusters, metal complexes or organic ligands to yield less dense frameworks. This
report shows how silver(I) ions combined with the *tetra*-monodentate ligand,
hexamethylenetetramine (HMTA), produce the first example of a three-dimensional, four-
connected (3-D, 4-C) open-framework solid that is topologically related to $CdSO_4$.[9]

EXPERIMENTAL

A 70 mL aqueous solution of silver nitrate (0.280 g, 1.65 mmol) was combined at
room-temperature with another 10 mL aqueous solution of hexamethylenetetramine (0.100
g, 0.71 mmol). The resulting clear colorless mixture was stirred for 1 min. and allowed to
stand in an uncovered container for 2 days. The large colorless crystals that appeared at the
bottom of the container were isolated from the small amount of needle-like crystals at the
surface of the solution and washed with 1:1 solution mixture of water and ethanol to give
0.061 g (18 % yield) of $Ag_2(HMTA)(NO_3)_2$. This compound is stable in air, and insoluble
in ethanol, chloroform and tetrahydrofuran, however, it slowly dissolves in water,
acetonitrile and *N, N*-dimethylformamide. Anal. Calcd for $C_6H_{12}N_6O_6Ag_2$: C, 15.01; H,
2.52; N, 17.51; Ag, 44.95. Found: C, 14.96; H, 2.51; N, 17.26; Ag, 44.97. The

Mat. Res. Soc. Symp. Proc. Vol. 453 ©1997 Materials Research Society

homogeniety of the bulk product was confirmed by comparison of the observed and calculated x-ray powder diffraction patterns. The calculated pattern was produced using (SHELXTL-XPOW) program with the single crystal data. Single crystals of $Ag_2(HMTA)(NO_3)_2$ were analyzed at $20\pm1\ ^\circ C$, tetragonal, space group $P4_122$ (no. 142) with $a = 6.751$ (1) Å, $c = 26.240$ (5) Å, $V = 1195.9$ (4) Å^3, $Z = 4$, $\rho_{calcd} = 2.666$ g. cm^{-3}, $\mu_a(MoK\alpha) = 3.31$ mm^{-1} and R_1 (unweighted, based on F) = 0.027. Atomic coordinates for nonhydrogen atoms in crystalline $Ag_2(HMTA)(NO_3)_2$ are listed below.

Atomic Coordinates for Nonhydrogen Atoms in Crystalline $Ag_2(N_4C_6H_{12})(NO_3)_2$

Atom Type	Fractional Coordinates			Equivalent Isotropic Thermal Parameter,
	10^4x	10^4y	10^4z	B, Å^2 x 10
Ag_1	5000	5263(1)	0	21(1)
Ag_2	567(1)	9433(1)	1250	26(1)
N_1	1830(6)	3668(5)	-38(1)	14(1)
N_2	142(6)	1120(5)	464(1)	14(1)
N_3	4858(7)	8269(7)	855(1)	22(1)
O_1	3449(9)	7472(8)	631(2)	43(1)
O_2	4493(7)	9677(7)	1151(1)	31(1)
O_3	6576(8)	7710(9)	786(2)	43(1)
C_1	1923(7)	2389(7)	418(1)	15(1)
C_2	1634(7)	2395(7)	-496(1)	15(1)
C_3	0	4899(9)	0	14(1)
C_4	0	-111(8)	0	15(1)

RESULTS AND DISCUSSION

Reacting aqueous solutions of silver(I) nitrate and HMTA in approximately 2 : 1 mole ratio yields large grey rectangular crystals of $Ag_2(HMTA)(NO_3)_2$, which was formulated by elemental microanalysis and structurally characterized by an x-ray single crystal study. This compound was found to have an extended network composed of the building units shown in Figure 1.

The asymmetric unit consists of two crystallographically distinct silver atoms (Ag1 and Ag2) with each lying on a 2-fold-axis, one adamantane-like HMTA unit (N1 and N2), and one nitrate anion (N3). Each silver (I) is bound to two symmetry equivalent HMTA units (Ag1-N1, N1a = 2.462 (5) Å and Ag2-N2b, N2d = 2.373 (3) Å) and two nitrate anions with each being shared by two symmetry inequivalent silver atoms. Here, Ag1

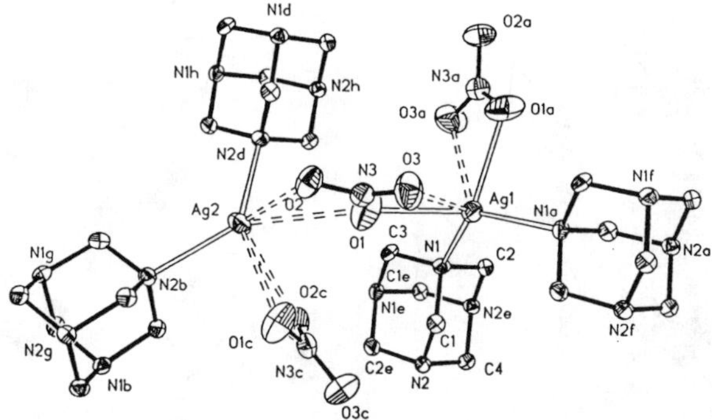

Figure 1. Perspective drawing of the coordination environment for the silver atoms in the solid-state structure of $Ag_2(HMTA)(NO_3)_2$ with nonhydrogen atoms represented by thermal vibration ellipsoids drawn to encompass 50 % of their electron density. Hydrogen atoms have been omitted for purpose of clarity. Atoms labelled with an additional letters *a-h* are related to atoms without such labels by the following symmetry operations: *a*, 1-x, y, -z; *b*, -y, 1-x, _-z; *c*, 1-y, 1-x, _-z; *d*, x, 1+y,z; *e*, -x, y, -z; *f*, 1+x, y, z; *g*, -y,1+x, _-z; and *h*, -x, 1+y, -z.

forms two strong (Ag1-O1, O1a = 2.462 (5) Å) and two weak (Ag1-O3, O3a = 2.849 (5) Å) bonding interactions to the nitrate anions N3 and N3a.[10] The anion N3 is shared with Ag2, through intermediate (Ag2-O2 = 2.668 (5) Å) and weak (Ag2-O1 = 2.859 (5) Å) bonds, which also form between Ag2 and a second anion (N3c) that is shared with a different Ag1. Excluding consideration of weak Ag-O interactions, each silver atom has a distorted tetrahedral coordination (N1-Ag1-O1 = 85.5 (2)°, N1-Ag1-O1a = 105.4 (2)°, N1-Ag1-N1a = 126.6 (2)°, and N2d-Ag2-O2 = 90.3 (1)°, O2-Ag2-N2b = 123.6 (2)°, N2b-Ag2-N2d = 129.7 (2)°) due to the steric demands of the HMTA units. The bond length and angle values of HMTA in this structure remain unchanged upon its coordination to the metal center as they are coincident with those observed for pure HMTA.[11] In the present structure, each HMTA is bridged by a silver(I) center to form a 3-D, 4-C network shown in Figure 2a.

It is worth noting that the nitrate anions point to the center of a 2-D channel system having rectangular pores with aperture dimension of nearly 9 × 18 Å, that is constricted at the center to 4 × 18 Å as shown in Figure 2b.

Attempts to exchange the nitrates by immersing solid samples of $Ag_2(HMTA)(NO_3)_2$ in alcoholic solutions containing excess potential guest anions such as BF_4^-, PF_6^-, and $CF_3SO_3^-$ at room-temperature were unsuccessful. Simple anions such as Cl⁻ and Br⁻ gave new crystalline phases as observed by x-ray powder diffraction data obtained on the products. It appears that the integrity of the framework is compromised upon mobilizing the nitrates due to their close and multiple bonding interactions to the silver ions. This is in contrast to the anion exchange behavior of the Ag(4,4'-bpy)˜NO₃ solid,[2a] where the process occurs reversibly and without destruction of its framework. Here, each nitrate ion bonds weakly to silver (Ag-O = 2.78 (1) and 2.83 (2) Å) using one oxygen to give loosely bound anions that can be mobilized without affecting the structural organization of the host framework.

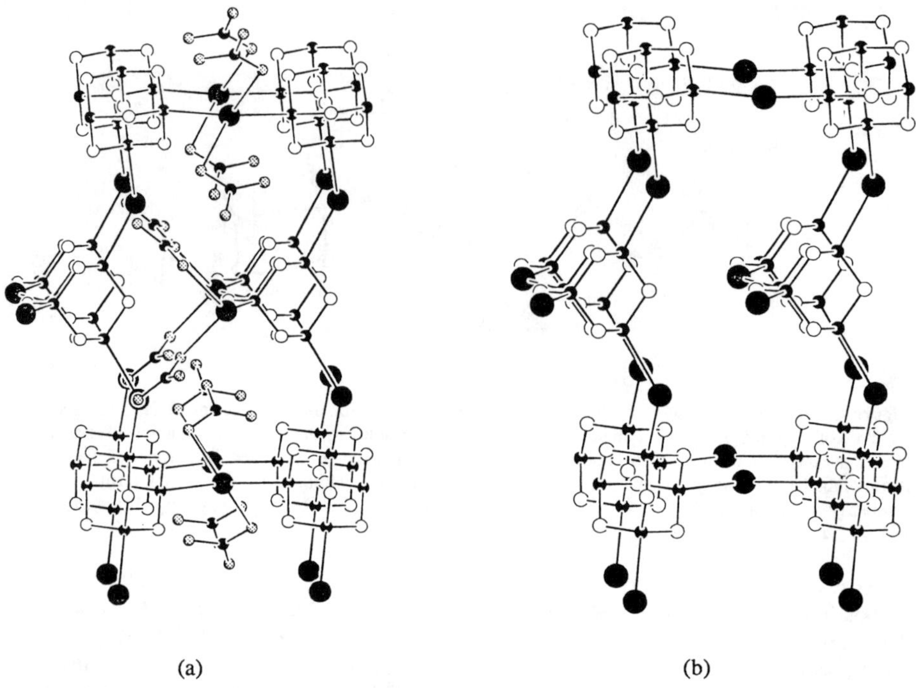

(a)   (b)

Figure 2. Perspective view for the solid-state structure of $Ag_2(HMTA)(NO_3)_2$ (a) The nitrogen atoms (small dark spheres) of HMTA are bound to silver atoms (large dark spheres) which are coordinated to two nitrate ligands (small dark spheres for nitrogen and shaded spheres for oxygen). (b) The same view is shown without the nitrate ions.

The novelty of the $Ag_2(HMTA)$ framework structure lies in its relationship to the 4-C net of $CdSO_4$, which is revealed upon comparison of Figure 2b and 3. The $CdSO_4$ structure and the derived net are shown in Figure 3a and b, respectively. Replacing every vertex in the net of $CdSO_4$ (Fig. 3b) by a tetrahedron of vertices (4 N of HMTA) gives the $Ag_2(HMTA)$ framework, in which the tetrahedra are linked by silver atoms—shown as lines in Figure 3c. This process has been referred to as *decorating*. [9a,12] The two simplest 4-C nets (only two vertices in the topological repeat unit) are those of C in diamond, and Cd and S in $CdSO_4$. Numerous[5] decorated diamond networks exist such as the porous Mn-Ge-S framework of the $MnGe_4S_{10} \cdot 2TMA$,[4] where Ge_4S_6 and MnS_4 tetrahedra alternately replace the carbon vertices in diamond. On the other hand, the $Ag_2(HMTA)$ network is the first example of a decorated $CdSO_4$ net which is interesting because, unlike other decorated nets reported in the literature where the derived nets are unique, for this structure an infinite series of nets related as polytypes can be derived. The simplest of such possibilities have chiral $P4_122$ (Figure 3c) (or the enantiomorphic form) and polar $Ama2$ (Figure 3d) symmetries. The $Ag_2(HMTA)$ structure has the first symmetry and the HMTA groups are arranged in helices around the 4_1 axes.

130

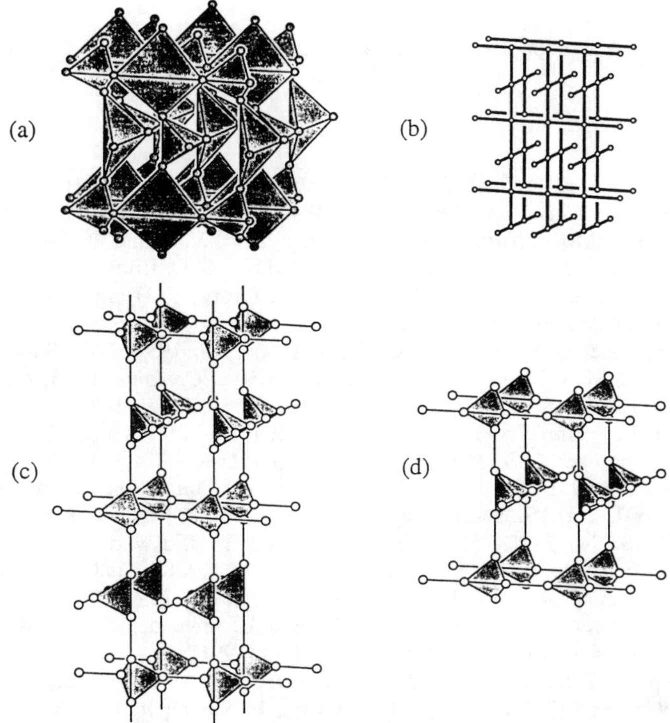

Figure 3. A schematic illustration of the structure of $CdSO_4$ represented as CdO_4 and SO_4 tetrahedra (a) and its net represented having vertices (open spheres) corresponding to Cd and S atoms and edges (solid lines) corresponding to —O— links (b), including its two simplest decorated nets having chiral (c) and polar (d) symmetries with the tetrahedra of vertices and the lines connecting them representing the HMTA building blocks and Ag(I) ions, respectively.

The importance of the anion in the formations of metal-organic networks is shown by the fact that in the presence of PF_6^-, the silver ions and HMTA building units assemble to give crystalline $Ag(HMTA)^- PF_6(H_2O)$. This framework is based on the $SrSi_2$ net, in which the vertices are alternately occupied by Ag and HMTA.[13] A 3-D, 3-C network in crystalline $Ag(HMTA)NO_3$[14] is produced by the combination of Ag(I) and HMTA (unreported quantities) in the presence of NO_3^- at room-temperature yields. To our knowledge the extended 3-D, 4-C network present in $Ag_2(HMTA)(NO_3)_2$ is the first of its kind in the metal coordination chemistry of HMTA. Given the simplicity of the building units used in the assembly of this solid, it is our current objective to examine in detail and elaborate on the anion-framework interactions and their connection to the resulting structure.

ACKNOWLEDGEMENT

This work was supported by the National Science Foundation (Grant CHE-9522303).

REFERENCES

(1)   For example: (a) K. Endo; T. Sawaki; M. Koyanagi; K. Kobayashi; H. Masuda; Y. Aoyama *J. Am. Chem. Soc.* **1995**, *117*, 8341-8352. (b) D. Venkataraman; S. Lee; J. Zhang; J. S. Moore *Nature* **1994**, *371*, 591-593. (c) X. Wang; M. Simard; J. D. Wuest *J. Am. Chem. Soc.* **1994**, *116*, 12119-12120. (d) O. Ermer; L. Lindenberg *Helv. Chim. Acta* **1991**, *74*, 825-877. (e) R. M. Barrer; V. H. Shanson *J. Chem. Soc. Chem. Commun.* **1976**, 333-334.

(2)   For example: (a) O. M. Yaghi; H. Li *J. Am. Chem. Soc.* **1996**, *118*, 295-296. (b) F. Robinson; M. J. Zaworotko *J. Chem. Soc., Chem. Commun.* **1995**, *23*, 2413-2414. (c) O. M. Yaghi; G. Li; H. Li *Nature* **1995**, *378*, 703-706. (d) O. M. Yaghi; H. Li *J. Am. Chem. Soc.* **1995**, *117*, 10401-10402. (e) G. B. Gardner; D. Venkataraman; J. S. Moore; S. Lee *Nature* **1995**, *374*, 792-795. (f) D. Venkataraman; G. B. Gardner; S. Lee; J. S. Moore *J. Am. Chem. Soc.* **1995**, *117*, 11600-11601. (g) S. Subramanian; M. J. Zaworotko *Angew. Chem. Int. Ed. Engl.* **1995**, *34*, 2127-2129. (h) M. Fujita; Y. J. Kwon; O. Sasaki; K. Yamaguchi; K. Ogura *J. Am. Chem. Soc.* **1995**, *117*, 7287-7288. (i) Batten, S. R.; Hoskins, B. F.; Robson, R. *J. Am. Chem. Soc.* **1995**, *117*, 5385-5386. (j) T. Iwamoto *in Inclusion Compounds*, Vol. 5, eds. J. L. Atwood, J. E. D. Davies, and D. D. MacNicol, Oxford University Press, Oxford (1991).

(3)   For example: (a) J. Lu; W. T. A. Harrison; A. J. Jacobson *Angew. Chem. Int. Ed. Engl.* **1995**, *34*, 2557-2559. (b) P. Schwarz; E. Siebel; R. D. Fischer; D. C. Apperley; N. A. Davies; R. K. Harris *Angew. Chem. Int. Ed. Engl.* **1995**, *34*, 1197-1199. (c) S. B. Copp; S. Subramanian; M. J. Zaworotko *Angew. Chem. Int. Ed. Engl.* **1993**, *32*, 706-709.

(4)   O. M. Yaghi; Z. Sun; D. A. Richardson; T. L. Groy *J. Am. Chem. Soc.* **1994**, *116*, 807-808.

(5)   M. J. Zaworotko *Chem. Soc. Rev.* **1994**, 283-288.

(6)   S. R. Batten; B. F. Hoskins; R. Robson *J. Chem Soc. Chem. Commun.* **1991**, 445.

(7)   B. F. Abrahams; B. F. Hoskins; D. M. Michail; R. Robson *Nature* **1994**, 369, 727-729.

(8)   (a) S. C. Abrahams; J. L. Bernstien; R. Liminga *J. Chem. Phys.* **1980**, *73*, 4585-4590. (b) B. F. Hoskins; R. Robson; N. V. Y. Scarlett *Angew. Chem. Int. Ed. Engl.* **1995**, *34*, 1203-1204.

(9)   (a) M. O'Keeffe *Z. Krist.* **1991**, *196*, 21-37. (b) K. Aurivillius; C. Stalhandske *Z. Krist.* **1980**, *153*, 121-129.

(10)   R. J. Lancashire In *Comprehensive Coordination Chemistry*; G. Wilkinson, Ed; Pergamon: Oxford, U.K., 1987; Vol. 5 pp 777-851.

(11)   L. N. Becka; D. W. J. Cruickshank *Proc. Roy. Soc. A*, **1963**, *273*, 435-454.

(12)   (a) M. O'Keeffe; S. T. Hyde *Z. Krist.* **1996**, *211*, 73-78. (b) M. O'Keeffe; B. G. Hyde *Crystal Structures I. Patterns and Symmetry*, Mineral Soc. Am.: Washington DC, 1996.

(13)   L. Carlucci; G. Ciani; D. M. Proserpio; A. Sironi *J. Am. Chem. Soc.* **1995**, *117*, 12861-12862.

(14)   A. Michelet; B. Viossat; P. Khodadad; N. Rodier *Acta Cryst.* **1981**, *B37*, 2171-2175.

HYDROTHERMAL SYNTHESIS OF VANADIUM OXIDES

THOMAS CHIRAYIL, PETER ZAVALIJ AND M. STANLEY WHITTINGHAM
Chemistry Department and Materials Research Center, State University of New
York at Binghamton, Binghamton, NY 13902-6000, USA

ABSTRACT

The pH of the reaction media is a critical factor in the formation of new
vanadium oxides with the tetramethylammonium (TMA) ion via the conventional
hydrothermal method. New vanadium oxides; $TMAV_3O_7$, $Li_{0.6}V_{2-\delta}O_{4-\delta}.H_2O$ and
$Li_{0.6}V_{2-\delta}O_{4-\delta}$, $TMAV_4O_{10}$, and two more new vanadium phases with d spacings of
11.5Å and 19.1Å are formed as the pH is varied with acetic acid. The synthesis and
characterization of $TMAV_3O_7$, a new vanadium oxide with tetramethylammonium
ions residing between its layers is discussed. The intercalation of alkylamines,
DMSO, and water into $Li_{0.6}V_{2-\delta}O_{4-\delta}.H_2O$ is described. Microwave hydrothermal
synthesis is a new and alternate soft chemistry technique used to accelerate the
synthesis of $TMAV_4O_{10}$. This method also aided in the formation of a new cluster
compound, $[Li(H_2O)_4]_2TMA(V_{10}O_{28}).4H_2O$.

INTRODUCTION

Extensive research is currently underway to find promising candidates for
cathode materials in advanced lithium batteries [1]. Vanadium oxides are prime
candidates because of their high free energy of reaction with lithium and their
relatively low cost. Conventional hydrothermal synthesis is a soft chemistry
technique that has resulted in the synthesis of many new, metastable, open or
layered vanadium oxides with the tetramethylammonium ion [2, 3].

By changing the pH of the reaction media with acetic acid, different vanadium
oxide compounds with new structures are observed. At pH 10, a known compound
Li_3VO_4 is formed [4]. Lowering the pH into the range 9-5.5 results in a new
vanadium oxide, $TMAV_3O_7$ [5]. In the weakly acidic regime of pH 5.5-4.4, the new
lithium vanadium dioxides, $Li_{0.6}V_{2-\delta}O_{4-\delta}.H_2O$ [6] and $Li_{0.6}V_{2-d}O_{4-d}$ [7], are formed.
Intercalation reactions with $Li_{0.6}V_{2-\delta}O_{4-\delta}.H_2O$ showed that an additional water,
DMSO and alkylamines can be used to expand its layers. At a still lower pH range,
4.4-3.7, the first vanadium oxide with an organic cation in the layers, $TMAV_4O_{10}$,
was observed [8, 9]. At high acidities, pH range 3.3-2.0, two more new vanadium
phases with layered structures are formed. The vanadium oxides at pH 3.3 and 2.6
has d spacings of 11.5Å and 19.1Å respectively. Both compounds contain TMA ions.

Hydrothermal reactions can be enhanced in a number of ways, including by
providing heating through microwaves. Komarneni et al. [10] used microwave
heating of the hydrothermal reactor in catalyzing the synthesis of crystalline unary
oxides such as TiO_2, ZrO_2, and Fe_2O_3 and binary oxides such as $KNbO_3$ and $BaTiO_3$ as
well as Ti and Zr phosphates. We use the microwave hydrothermal synthesis to
accelerate the formation of the layered vanadium oxides as well as to synthesize
new ones. The time period required for synthesis, via the conventional
hydrothermal is of the order of several days, the use of microwave radiation reduces
this to the order of a few hours.

Mat. Res. Soc. Symp. Proc. Vol. 453 © 1997 Materials Research Society

EXPERIMENTAL

The new vanadium oxide, $TMAV_3O_7$, was prepared hydrothermally by mixing V_2O_5 powder with 25 weight % of tetramethylammonium hydroxide in aqueous solution from Alfa Chemicals in a 1:2 molar ratio. The pH of the reaction mixture was lowered into the range 9-6.2 with dilute acetic acid. The reaction mixture was transferred to a 125-ml Parr bomb, sealed, and reacted hydrothermally in the oven for 3 days at 185°C. The lithiated version of this new vanadium oxide can also be made by adding LiOH to the reaction mixture and heating under the same conditions. The resulting crystals were filtered, washed and dried in air. The crystals were characterized by powder x-ray diffraction, TGA, FTIR, DCP, electron microprobe and also by its charge and discharge cycle as a cathode in a lithium secondary battery.

Intercalation reactions with DMSO and alkylamines such as butylamine, hexylamine, octylamine, and decylamine were conducted with $Li_{0.6}V_{2-\delta}O_{4-\delta}\cdot H_2O$ at room temperature. Intercalating species and the vanadium compound were placed in a vial undisturbed for 6-7 days before being characterized by powder x-ray diffraction and TGA.

Microwave hydrothermal synthesis was performed in a MDS-2000 microwave oven by CEM Corporation. V_2O_5 + TMA + LiOH were mixed in a 1:2:1 molar ratio, acidified with nitric acid to pH 3.7, before heating. The reaction time varied from 30-60 minutes.

RESULTS AND DISCUSSION

TMAV$_3$O$_7$

The powder x-ray patterns of $TMAV_3O_7$ and $Li_xTMAV_3O_7$ are similar. The lithium in the layers did not affect the layer spacing of 9.2Å. The x-ray pattern of this new vanadium oxide is shown in figure 1. The DCP analysis on the lithiated compound indicated 0.1 lithium per complex thus giving the general formula $Li_{0.1}TMAV_3O_7$.

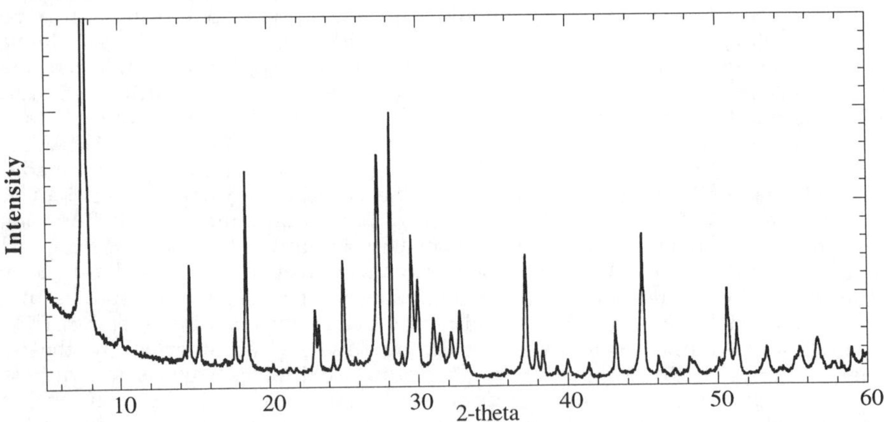

Figure 1. Powder x-ray pattern of $TMAV_3O_7$

The layered structure of this new vanadium compound is confirmed from the electron microprobe image shown in figure 2 below.

Figure 2. Electron microprobe images of $TMAV_3O_7$ crystallites

Indexing of the powder pattern using the crystallographic programs package CSD showed that $TMAV_3O_7$ has a monoclinic symmetry with space group P21/n. The lattice parameters are a=18.481Å, b=6.552Å c=8.427Å and β is 91.132°. The structure of this new vanadium oxide is shown in figure 3. $TMAV_3O_7$ structure contains zig-zag strings of edge sharing VO_5 square pyramids and the strings are joined together by VO_4 tetrahedra. The tetramethylammonium ions reside between the layers of the vanadium oxide [5].

Figure 3. Structure of $TMAV_3O_7$, from [5].

Thermogravimetric analysis, TGA, of $TMAV_3O_7$ shows an increase in weight around 200°C. The weight increased about 4% on heating to 260°C. The weight increase can be attributed to the gain of one oxygen to give a fully oxidized phase "$TMAV_3O_8$". The TGA pattern of $TMAV_3O_7$ is shown in figure 4. Powder x-ray diffraction showed the decrease in d-spacing from 9.2Å to 8.6Å upon heating to 260°C. The oxidation state of vanadium in both TMAV$_3$O$_7$ and "TMAV$_3$O$_8$" was determined by chemical analysis and was found to be 4.38 and 5.0 respectively. This confirmed that TMAV$_3$O$_7$ was fully oxidized when heated to 260°C under oxygen.

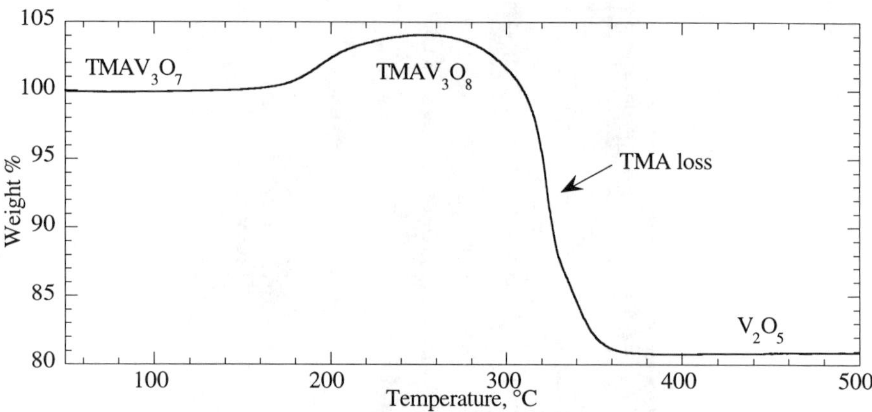

Figure 4. TGA graph of TMAV$_3$O$_7$. At 260°C, the compound is fully oxidized to "TMAV$_3$O$_8$". Further heating shows the loss of TMA and the formation of V$_2$O$_5$

<u>Intercalation of alkylamines, DMSO and additional H$_2$O into Li$_{0.6}$V$_{2-\delta}$O$_{4-\delta}$.H$_2$O</u>

After its first report, Li$_{0.6}$V$_{2-\delta}$O$_{4-\delta}$.H$_2$O, several experiments were done on this compound that showed some interesting results. Li$_{0.6}$V$_{2-\delta}$O$_{4-\delta}$.H$_2$O can be heated to

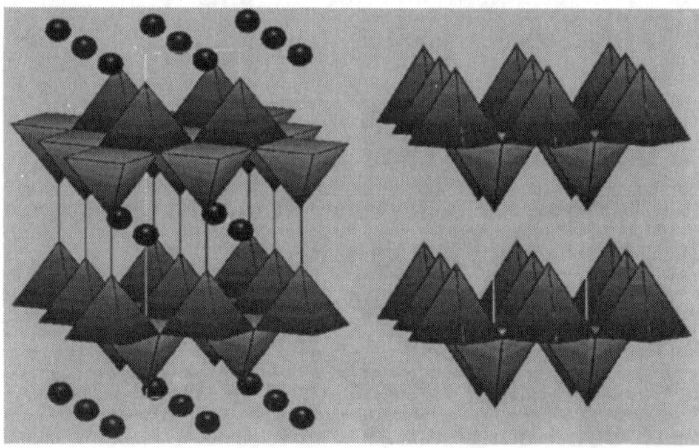

Figure 5. Structures of Li$_{0.6}$V$_{2-\delta}$O$_{4-\delta}$.H$_2$O(left) and Li$_{0.6}$V$_{2-\delta}$O$_{4-\delta}$ (right), from [7].

remove the water giving Li$_{0.6}$V$_{2-\delta}$O$_{4-\delta}$. The structures of the hydrated and dehydrated phases are shown in figure 5. By electrochemically removing the lithium by charging the Li$_{0.6}$V$_{2-\delta}$O$_{4-\delta}$ cathode in a lithium secondary battery, a new "VO$_2$" can be formed [7]. Intercalation of many solvents into Li$_{0.6}$V$_{2-\delta}$O$_{4-\delta}$.H$_2$O layers was found to be possible. The intercalation of an additional water increased the d-spacing from

7.89Å to 10.88Å. The intercalation of DMSO also increases the layer spacing to 12.99Å. Several alkylamines can also be intercalated into the vanadium oxide layers. The change in the d-spacing of $Li_{0.6}V_{2-\delta}O_{4-\delta}.H_2O$ with different alkylamines is shown in table 1. The intercalation of alkylamines often resulted in the formation of two phases due to the different stacking arrangements of alkyl chains.

Table 1. Lattice Spacings for Alkylamine Intercalation Compounds of $Li_{0.6}V_{2-\delta}O_{4-\delta}$

Compound	d-spacing (Å)	d-spacing (Å)
$Li_{0.6}V_{2-\delta}O_{4-\delta}.H_2O$	7.89	-
$Li_{0.6}V_{2-\delta}O_{4-\delta}.H_2O$ + Butylamine	16.20	13.84
$Li_{0.6}V_{2-\delta}O_{4-\delta}.H_2O$ + Hexylamine	20.92	16.29
$Li_{0.6}V_{2-\delta}O_{4-\delta}.H_2O$ + Octylamine	23.80	21.22
$Li_{0.6}V_{2-\delta}O_{4-\delta}.H_2O$ + Decylamine	29.22	23.60

Microwave Hydrothermal Synthesis

The use of microwaves is advantageous because it can increase the kinetics by one or two orders of magnitude over the conventional method. $TMAV_4O_{10}$ is a layered vanadium oxide that can be made in 30 minutes by this new method, whereas 2-3 days are required for its formation via the conventional hydrothermal process. A new cluster compound, $[Li(H_2O)_4]_2TMA_4(V_{10}O_{28}).4H_2O$, is precipitated out of the solution following several days at ambient temperature after the microwave hydrothermal synthesis for one hour. The structure of this new compound is shown in figure 6, and contains layers comprising decavanadate clusters joined together by hydrogen bonds from the $Li(H_2O)_4$ tetrahedra. The tetramethyl-ammonium ions and water reside between these sheets [11].

Fig. 6. Structure of a new vanadium cluster, $[Li(H_2O)_4]_2TMA_4(V_{10}O_{28}).4H_2O$, from [12]

CONCLUSIONS

The conventional hydrothermal method is a rich technique that has yielded many new vanadium oxides with the tetramethylammonium ion. Within the last few years, after the report of our first new vanadium oxide, $TMAV_4O_{10}$, several other organic molecule/vanadium oxide complexes have been reported [13], but so far only the tetramethylammonium ion has been found to form more than one structure type. The formation of $TMAV_3O_7$ depends critically on the pH of the reaction medium. $TMAV_3O_7$ can be fully oxidized to "$TMAV_3O_8$" by heating it under oxygen to 260°C. The structures of "$TMAV_3O_8$" and the two new vanadium phases formed at high acidities are currently being determined. We were able to intercalate alkylamines, DMSO and an additional H_2O into the layers of $Li_{0.6}V_{2-\delta}O_{4-\delta}.H_2O$. A new "$VO_2$" is formed electrochemically by driving the lithium out. Microwave heating of the hydrothermal reactor was used to decrease the time period in the synthesis of $TMAV_4O_{10}$, and was instrumental in the formation of a new cluster compound. Further studies of this method on metal oxides as well as titanium phosphates are currently being pursued.

ACKNOWLEDGMENTS

We thank the Department of Energy through Lawrence Berkeley Laboratory and the National Science Foundation DMR-9422667 for their support of this project.

REFERENCES

1. J-M. Tarascon and editor, Solid State Ionics, **69** (1994)
2. M. S. Whittingham, J. Guo, R. Chen, T. Chirayil, G. Janauer, and P. Zavalij, Solid State Ionics, **75** (1995) 257.
3. L. F. Nazar, Personal Communication, (1996)
4. R. D. Shannon and C. Calvo, J. Solid State Chem., **6** (1973) 538.
5. T. A. Chirayil, P. Y. Zavalij, and M. S. Whittingham, Chem. Commun., (1996) in press.
6. T. Chirayil, P. Zavalij, and M. S. Whittingham, Solid State Ionics, **84** (1996) 163.
7. T. Chirayil, P. Zavalij, and M. S. Whittingham, J. Electrochem. Soc., **143** (1996) L193.
8. P. Zavalij, M. S. Whittingham, E. A. Boylan, V. K. Pecharsky, and R. A. Jacobson, Z. Kryst., **211** (1996) 464.
9. E. A. Boylan, T. Chirayil, J. Hinz, P. Zavalij, and M. S. Whittingham, Solid State Ionics, **90** (1996) 1.
10. S. Komarneni, Q. H. Li, and R. Roy, J. Mater. Chem., **4** (1994) 1903.
11. P. Zavalij, T. Chirayil, M. S. Whittingham, V. K. Pecharsky, and R. A. Jacobson, Acta Cryst, **C52** (1996)
12. M. S. Whittingham, R. Chen, T. Chirayil, and P. Zavalij, Solid State Ionics, **93** (1997) in press.
13. D. Riou and G. Ferey, J. Solid State Chem., **120** (1995) 137; Inorganic Chem., **34** (1995) 6250.; J. Solid State Chem., 1996, **124** (1996) 151.; Y. Zhang, C. J. O'Connor, A. Clearfield, and R. C. Haushalter, Chem Mater., **8** (1996) 595

HIGH PRESSURE SYNTHESIS OF SOLID STATE MATERIALS

PAUL F. McMILLAN*, **
* Materials Research Science and Engineering Center, Dept. of Chemistry and Biochemistry,
Arizona State University, Tempe, AZ 85287-1604 pmcmillan@asu.edu
** Additional collaborators are listed in the text

ABSTRACT

Within the ASU MRSEC, we are using a combination of high pressure synthesis techniques (multi-anvil, piston cylinder), combined with *in situ* studies in diamond anvil cells, and *ab initio* calculations, to prepare a range of new materials at high pressure, and to study their physical properties under extreme conditions of pressure and temperature. Work in progress includes experiments on nitrides, particularly glasses and crystalline materials based in the P_3N_5-HPN_2-PON system, silicate and germanate perovskites and related vanadates and niobates, and high hardness materials in the B-C-N-O system. Some of our recent results are presented.

INTRODUCTION

During the past four years, we have established a Materials Research Group (MRG) in the synthesis and study of new solid state materials under high pressure conditions. The work involved the research groups of co-investigators C.A. Angell, J.R. Holloway, M. O'Keeffe, W.T. Petuskey, O.F. Sankey and G.H. Wolf, combining backgrounds in solid state chemistry and physics, geochemistry, and materials engineering. This research is being continued as a thrust area in our newly formed Materials Research Science and Engineering Center (MRSEC), with the addition of J. Kouvetakis, O. Yaghi (materials chemistry), R. Hemley, H.-K. Mao (high pressure mineralogy and physics), K. Leinenweber (solid state chemistry), and J. Tyburczy (geophysics), to the research team. Our principal goals are to synthesize new solids under high pressure conditions, to investigate the effect of high pressure on materials properties, and generally to explore the use of the pressure variable in designing materials synthesis routes. The work integrates theory and experiment, and diamond cell and multi-anvil techniques, combining approaches from solid state chemistry and physics, materials science, and high pressure geochemistry and geophysics. On-going work in three areas of interest is summarized below.

NITRIDE GLASSES AND P-N FRAMEWORKS

One of the initial goals within the MRG was to use high pressure to stabilize nitride materials in a molten state (due to the increased a_{N_2} in the melt), and to quench glasses containing N^{3-} as the only network-forming anion. This was carried out by T. Grande, and a range of pure nitride glasses were prepared in the system Li_3N-Ca_3N_2-P_3N_5, at pressures of 5-15 kbar and ~1000°C, using double capsule techniques (Ta or W inside Pt) [1-4]. The glasses were optically transparent (usually red in colour, but occasionally pale green-yellow), and they combined high hardness with high refractive index ($n \approx 2$), and high softening temperature (> 750°C) (Fig. 1).

Synthesis of the P_3N_5 precursor was accomplished via ammonolysis of $PNCl_2$ to give first low density, X-ray amorphous P-N-H, which was successively pyrolyzed to first PN(NH) (phospham), then P_3N_5 (P. Coffman) [5,6]. The amorphous P-N-H could be vitrified to give a bulk glassy material by high pressure treatment at 500-750°C (L. Hilt, R. Pacalo) [7]. We also investigated the high pressure behaviour of the crystalline PN(NH), which has a cristobalite-like structure at ambient conditions, as well as the related material PON [8,9].

High pressure treatment of PON at high temperature (500°C at 5 GPa) yields an ordered quartz phase, which can be quenched to room pressure (R. Pacalo) [10]. Heating at higher pressure (10 GPa) appears to give a further structure related to quartz, which is still not well characterized. Unlike PON, heating PN(NH) at 500°C and 5 GPa allows a distorted

Mat. Res. Soc. Symp. Proc. Vol. 453 © 1997 Materials Research Society

(monoclinic) cristobalite-like phase to be recovered [11]. The structures and properties of these materials are under investigation [10,11].

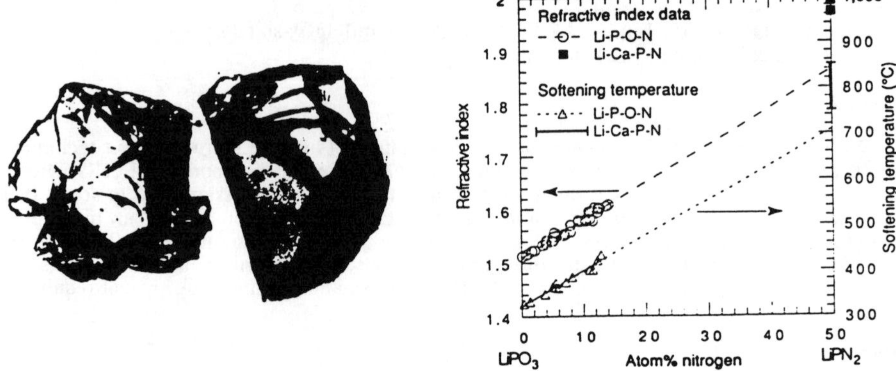

Figure 1. Pure nitride glasses in the system Li_3N-Ca_3N_2-P_3N_5 prepared by T. Grande by quenching the melts at high pressure [1-4], viewed in transmitted light in an optical microscope. The glass samples at left are ~1 mm in dimension. The glasses are clear and reddish in colour. The dark edges and cracks are due to the high refractive index of these glasses (n≈2). At right, the refractive index and softening temperature for glasses in the Li(Ca)PO3-Li(Ca)PN2 system are plotted against nitrogen content [ref. 1].

Compression of PON (and PN(NH)) at ambient temperature results in displacive transformations within the cristobalite type structure, but the open tetrahedral framework is retained to the highest pressures (70 GPa) (K. Kingma) [8,9]. This is quite unlike the isostructural SiO_2 polymorphs, which undergo transitions to phases with Si in high coordination, or pressure-induced amorphization [12-14]. It is quite likely that the high pressure stability of the PON tetrahedral framework reflects the reluctance of P to take VI-fold coordination by O or N, presumably associated with the difficulty of N or O to be 3-coordinated by P atoms, due to unsatisfied bond valence sums and P...P non-bonded repulsion (Fig. 2). The result is that PON and PN(NH) provide useful materials for exploring the limits of compressional behaviour in tetrahedral framework structures, to very high density [8,9].

Figure 2. A bond valence argument for the reluctance of phosphorus to take high coordination in oxides and nitrides. In low density SiO_2 polymorphs (top), the Si-O bond valence is 4/4 = 2/2 = 1, for these structures with tetrahedral Si and 2-coordinate oxygen. The resulting valence sums at Si and O are exactly satisfied (4 and 2). In stishovite, Si is octahedrally coordinated and O is 3-coordinate, so the bond valence is 1/3, and the valence sums are again exactly satisfied. In the case of PON (P^{5+}, O^{2-}, N^{3-}), a compound with octahedral P would have bond valencies 5/6 = 0.833. Summing these around a 3-coordinate N atom gives a valence sum of 2.5, siginificantly lower than 3 required by nitrogen. The corresponding sum around oxygen is likewise significantly greater than 2. If such an octahedrally coordinated $P(O_3N_3)$ could be made, it should have an intrinsic tendency to ferroic behaviour, from the inherent site asymmetry and unsatisfied bond valencies, which would result in structural distortion and/or electronic polarization.

DECOMPRESSED PEROVSKITES AND RELATED PHASES

Our work on oxide perovskites prepared at high pressure had its roots in geophysics studies. At high pressure, above 23 GPa, $MgSiO_3$ transforms to the perovskite structure, and it has been proposed that this phase constitutes the bulk of the Earth's lower mantle [15]. This material can be recovered to room pressure, where it exists in a non-polar orthorhombic structure. The analogous $CaSiO_3$ crystallizes in the cubic perovskite structure above 10 GPa, but amorphizes upon decompression [16,17]. Theoretical studies within the MRG, combining first principles calculations and molecular dynamics simulations (A. Chizmeshya, M. Hemmati, P. Poole, J. Shao), demonstrated that the amorphization was triggered by softening of a zone center optic mode associated with the ferroelectric distortion of the SiO_6 octahedra, followed by collapse of the resulting tetragonally distorted structure [18] (Fig. 3). This work has recently been complemented by accurate LAPW calculations [19]. The metastable $MgSiO_3$ perovskite transforms to the amorphous phase with silicon in tetrahedral coordination upon heating to 423 K at ambient pressure [20]. Analogous behaviour is observed for $CaGeO_3$ perovskite, synthesized above 6.5 GPa, when it is heated above 823 K at room pressure [21]. From the combination of these observations, it seemed that high pressure silicate and germanate perovskites, synthesized at high pressure and decompressed metastably to room pressure, might provide a range of materials with interesting ferroic behaviour, due to the inherent instability associated with the low-lying ferroelectric optic modes [18]. This possibility had already been suggested by G. Samara [22,23].

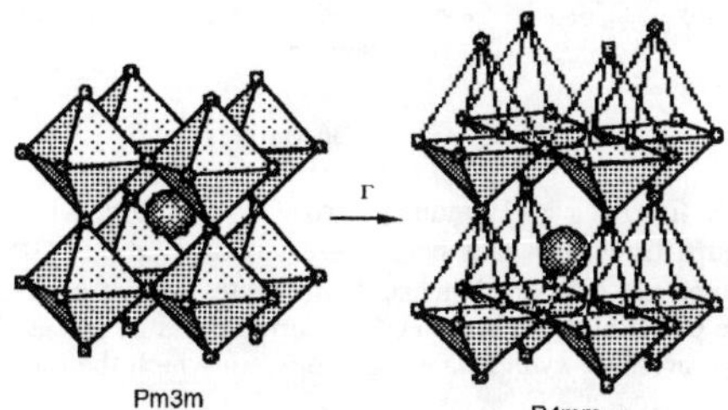

Fig. 3. A zone centre ferroic instability encountered for $CaSiO_3$ perovskite on decompression (ref. 18)

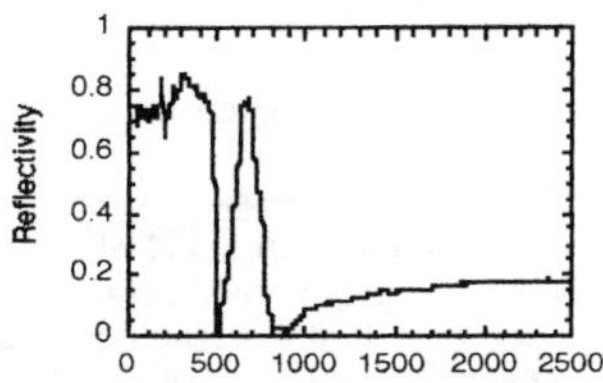

Fig. 4. IR reflectivity of $SrGeO_3$ perovskite at ambient pressure. Note the high reflectivity as $\omega \to 0$, associated with a large value of ε_0 (refs. 26,28).

The dielectric constant of decompressed $CaGeO_3$ perovskite inferred from infrared reflectivity studies is quite normal ($\varepsilon_0 \sim 25$) [24]. $SrGeO_3$ transforms to a garnet phase above 1 GPa, and to perovskite at 5 GPa [25]. We have synthesized $SrGeO_3$ perovskite and estimated its dielectric constant from far-IR reflectivity measurements and capacitance (100 kHz) measurements on powdered compacts (A. Grzechnik, H. Hubert) (Fig. 4). The dielectric constant obtained by both techniques is $\varepsilon_0 \sim 100$ [26]. We have further investigated the low- and high-temperature behaviour of this material, including its back-transformation first to garnet and amorphous germanate then to low pressure tetrahedral phases, and have investigated the instability by first principles calculations [27]. We have also investigated the relationships between structure, properties and metastability in perovskite solid solutions prepared along the $SrTiO_3$-$SrGeO_3$ [28] and $CaSiO_3$-$CaTiO_3$ joins [29]. A new ordered perovskite compound (Ca_2SiTiO_6) was obtained in this study.

In a related series of studies, we have investigated the high pressure synthesis ad behaviour of alkali metavanadates (A. Grzechnik), which might give rise to useful non-linear optic materials. Ambient pressure $LiVO_3$ has a tetrahedral chain structure. On pressurization, it undergoes a sequence of displacive phase transitions, and eventually adopts the $LiNbO_3$ structure above 7 GPa [30]. This phase is not recoverable, as it reverts to low pressure tetrahedral structures on decompression below ~4.5 GPa (Fig. 5). However, $LiNbO_3$-$LiVO_3$ solid solutions containing up to 20% $LiVO_3$ component can be recovered from syntheses at 8 GPa [31,32]. The dielectric constant decreases slightly with vanadium substitution. High pressure treatment of α-$NaVO_3$, followed by *in situ* FTIR spectroscopy, reveals no pressure-induced amorphization, as had been previously reported. Instead, a transition to an octahedrally coordinated phase occurs at ~7 GPa, probably with the $LiNbO_3$ structure [33]. This material reverts to the low pressure structure on decompression.

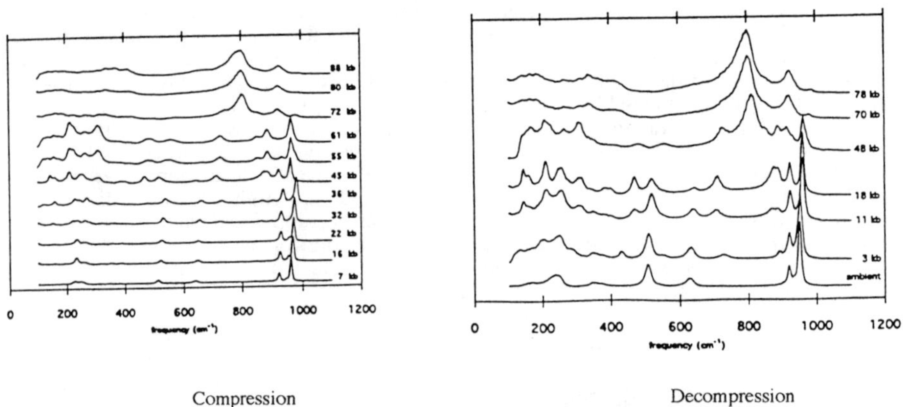

Compression   Decompression

Figure 5. Results of compression experiments on $LiVO_3$ followed by Raman spectroscopy [ref. 30]. At left, the low pressure structures with tetrahedral vanadate groups transform to the $LiNbO_3$ structure at 7.2 GPa. On decompression (at right), the high pressure phase is retained to ~4.5 GPa. This study illustrates the use of diamond cell experiments to follow materials *in situ*, and evaluate phase changes which occur during pressure release. This is particularly important in evaluating the results of high pressure *synthesis* experiments, in which the recovered materials may not represent the phases present at high pressure.

COVALENT MATERIALS

In this area of research, we have focused our attention on materials with high hardness in the B-N-C-O system (H. Hubert, L. Garvie). Here, the characterization of these light element containing materials is greatly facilitated by interaction with the electron microscopy group of P. Buseck at ASU, particularly in electron energy loss spectroscopy (L. Garvie), as well as TEM imaging and diffraction (B. Devouard).

In the B-O system, there had been previous reports of phases with B_2O stoichiometry prepared at high pressure with both the graphite (layered) and ordered diamond structures. We carried out a series of *ab initio* calculations to investigate the relative energetics of these structures [34]. It was found that the B/O ordering pattern on the diamond lattice previously

144

suggested was unstable to exfoliation of the lattice (into BO and B layers). We followed up this theoretical work by an experimental study in attempts to reproduce the prior syntheses. These were unsuccessful: instead, boron suboxide (nominally B_6O) was obtained in runs beginning with mixtures of B and B_2O_3 [35] (Fig. 6). Since this material itself is known to have extremely high hardness (competitive with cubic BN), the focus of the study switched to this phase.

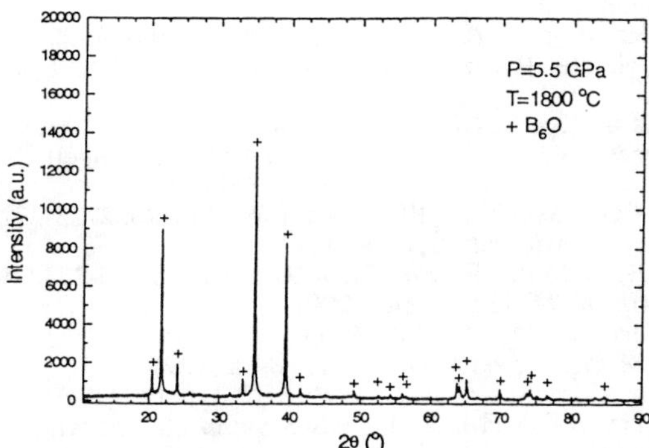

Figure 6. Powder X-ray diffraction of B_6O prepared from a mixture of $B+B_2O_3$ at ~5.5 GPa and 1800°C (refs. 36,37).

In these studies, the molten $B-B_2O_3$ (with some $B(OH)_3$ component) acts as a flux from which large grains (several tens of µm) of B_6O could be obtained, which is useful for better characterization of this important material. In addition, high pressure increases the activity of oxygen in the charge, so that the recovered material is much closer to stoichiometric B_6O in composition [36-37]. In some runs, nano-rods of B_6O_{1-x} are observed which have grown via liquid-solid-liquid (LSL) processing in the $B-B_2O_3$ flux. Even more interesting, at one particular pressure, near-perfect icosahedra of B_6O up to several tens of µm in dimension grow in the charge, via Mackay packing rather than conventional crystal growth [37]. These have potential applications in high-wear environments. On-going work focuses on related materials in the B-C-N-O system, grown at high pressure [38].

ACKNOWLEDGEMENTS

The research carried out within the Materials Research Group was supported by NSF grants DMR-9121570 and -9542864, and by Arizona State University. The MRSEC is supported by a cooperative agreement (DMR-9632635) with the National Science Foundation

REFERENCES

[1] Grande T, Holloway JR, McMillan PF, Angell CA. *Nature* **369**, 43-45 (1994).
[2] Grande T, Jacob S, Holloway JR, McMillan PF, Angell CA. *J. Non-Cryst. Solids* **184**, 151-154 (1995).
[3] McMillan PF, Angell CA, Grande T, Holloway JR. U.S. Patent No. 5,455,211 (1995).
[4] McMillan PF, Angell CA, Grande T, Holloway JR. U.S. Patent App. Ser. No. 08/447,534.
[5] Coffman P, Ph.D. dissertation, Chemistry and Biochemistry, ASU (1996).
[6] Coffman P, Pacalo RE, McMillan PF, Petuskey WT, in prep.
[7] Pacalo RE, McMillan PF, Petuskey WT, Coffman P, Hilt L, in prep.

[8] Kingma KJ, Pacalo REG, McMillan PF (1996) In "US-Japan Conference on High Pressure-Temperature Research", Syono Y, Manghnani MH, eds, in press.

[9] Kingma KJ, Pacalo REG, McMillan PF (1996) J. Solid State Chem., submitted.

[10] Pacalo RE, McMillan PF, Petuskey WT, Coffman P, in prep.

[11] Pacalo RE, McMillan PF, Petuskey WT, Parise J, Coffman P, in prep.

[12] Kingma KJ, Hemley RJ, Mao H-K, Veblen DR. *Phys Rev Lett* **70**, 3927-3930 (1993).

[13] Kingma KJ, Meade C, Hemley RJ, Mao H-K, Veblen DR. *Science* **259**, 666-669 (1993).

[14] VerHelst-Voorhees M, Yarger JL, Diefenbacher J, Poe BT, McMillan PF, Smith DJ, Wolf GH. *Phys Rev Lett*, submitted

[15] Hemley RJ, Cohen RE. *Ann Rev Earth Planet Sci* **20**, 553-600 (1992).

[16] Mao HK, Chen LC, Hemley RJ, Jephcoat AP, Wu Y, Bassett WA, *J Geophys.Res* **94**, 17889-17894 (1989).

[17] Kanzaki M, Stebbins JF, Xue X, *Geophys Res Lett* **18**, 463-466 (1991).

[18] Hemmati M, Chizmeshya A, Wolf GH, Poole PH, Shao J, Angell CA. *Phys Rev B* **51**, 14841-14848 (1995).

[19] Chizmeshya A, Wolf GH, McMillan, PF *Geophys Res Lett* **23**, 2725-2728 (1996).

[20] Durben DJ, Wolf GH *Am Mineral* **77**, 890-893 (1992).

[21] Durben DJ, Wolf GH, McMillan PF *PhysChem Minerals* **18** 215-223 (1991).

[22] Samara GA *J Phys Chem.* **94**, 1127-1134 (1990)

[23] Samara GA *Ferroelectrics* **117**, 347-372 (1991)

[24] Lu R, Hofmeister AM *Phys Chem Minerals* **21**, 78-84 (1994)

[25] Shimizu Y, Syono Y, Akimoto S *High Temp Press* **2**, 113-120 (1970)

[26] Grzechnik A, Petuskey WT, McMillan PF. MRS Symp Thermodynamics and Kinetics of Phase Tansformations (1996).

[27] Grzechnik A, Chizmeshya A, Wolf GH, McMillan, PF, in prep.

[28] Grzechnik A, Hubert H, McMillan P, Petuskey W Proc. Symp. "Integrated Ferroelectrics", in press (1996).

[29] Leinenweber K, Grzechnik A, Voorhees M, Navrotsky A, Yao N, McMillan P *Phys Chem Minerals*, in press (1996).

[30] Grzechnik A, McMillan PF. *J. Phys. Chem. Solids* **56**, 159-164 (1995).

[31] Grzechnik A, Ph.D. dissertation, Chemistry and Biochemistry, ASU (1996)..

[32] Grzechnik A, McMillan PF. *J. Solid State Chem*, submitted (1996).

[33] Grzechnik A, McMillan PF. *Solid State Commun* **99**, 869-871 (1996).

[34] Grumbach MP, Sankey OF, McMillan PF. *Phys. Rev. B* **52**, 15807-15811 (1996).

[35] Hubert H, Garvie LAJ, Leinenweber K, Buseck P, Petuskey WT, McMillan PF *MRS Symp Covalent Ceramics III* (1996).

[36] Hubert H, Garvie LAJ, Buseck P, Petuskey WT, McMillan PF *J. Solid State Chem*, submitted (1996).

[37] Hubert H, Garvie LAJ, Devouard B, Buseck P, Petuskey WT, McMillan PF *J. Solid State Chem*, submitted (1996).

[38] Hubert H, Garvie LAJ, Devouard B, Buseck P, Petuskey WT, McMillan PF *Science*, submitted (1996).

[38] Hubert H, Garvie LAJ, Devouard B, Buseck P, Petuskey WT, McMillan PF *MRS Symp* (poster at this meeting) (1996).

EARLY TRANSITION METAL OXIDES AND A PINCH OF SALT: NEW SYNTHETIC ROUTES TO TITANATES AND NIOBATES

M. J. GESELBRACHT, R. J. SCAROLA, D. INGRAM, C. GREEN, AND J. H. CALDWELL
Department of Chemistry, Reed College, Portland, OR 97202

ABSTRACT

Molten salts have traditionally been used as recrystallizing solvents for the crystal growth of mixed metal oxides. In recent years, many examples of the direct preparation of mixed metal oxides from molten salt solutions have been reported. We have been exploring the use of chloride melts as a reaction media to prepare complex early transition metal oxides. Specific examples of new synthetic routes to interesting titanates and niobates will be presented. In one case, we have prepared a series of layered niobate perovskite solid acids from molten salts at temperatures well below those traditionally used in solid state syntheses. In the second case, we have discovered a new synthetic route to a poorly characterized reduced calcium titanate that is otherwise very difficult to make. The synthesis and characterization of these two classes of compounds will be described.

INTRODUCTION

Many interesting oxide materials have been prepared by traditional solid state synthesis techniques. However, very high temperatures are often required to overcome the slow diffusion rates of solid state reactants. We have been exploring the chemistry of early transition metal oxides in molten chloride fluxes to find better routes to materials with interesting properties. In this paper, we will present new synthetic routes involving molten salt fluxes to two different types of oxides, namely the layered perovskite $KCa_2Nb_3O_{10}$ and the reduced titanate $CaTi_2O_4$.

Layered perovskites such as $KCa_2Nb_3O_{10}$ have captured much attention recently due to their potential use as solid acid catalysts [1]. The proposed structure of $KCa_2Nb_3O_{10}$ [2] is shown in Figure 1. The structure consists of perovskite-type slabs, separated along the c axis by a layer of potassium ions. Adjacent perovskite slabs are not stacked directly on top of each other but are displaced by half a unit cell along the a direction to produce a trigonal prismatic coordination of the potassium ions. The interlayer potassium ions exhibit ion-exchange properties and can be exchanged for protons to produce $HCa_2Nb_3O_{10}$, which acts as a solid Brønsted acid [3].

The layered perovskites are typically prepared by solid state reaction of the constituent carbonates and oxides at temperatures ranging from 1100–1400°C (*e.g.* eq. 1). While

$$K_2CO_3 + 4\,CaCO_3 + 3\,Nb_2O_5 \longrightarrow 2\,KCa_2Nb_3O_{10} + 5\,CO_2 \qquad (1)$$

this method is successful, it also presents several disadvantages: First, the best samples require several stages of repeated grinding and firing. Second, the alkali metal carbonates are fairly volatile at these high temperatures; to compensate, a small excess is typically added after the initial firing step. Third, the solid state synthesis generally does not lead to crystal formation. While a few layered perovskites have been crystallized by slow cooling in alkali metal sulfate fluxes from 1300°C [2,4], detailed crystallographic information is lacking for most of the phases due to the difficulty of growing high quality single crystals. To overcome some of these disadvantages, we have investigated synthetic routes in molten chloride fluxes that lead to $KCa_2Nb_3O_{10}$.

We are also interested in utilizing molten chloride fluxes in the synthesis of reduced titanates. In particular, we have sought a new synthetic route to $CaTi_2O_4$, which was first prepared and structurally characterized in 1956 by Bertaut and Blum [5]. They grew crystals of $CaTi_2O_4$ by electrolysis of TiO_2 at Pt electrodes in molten $CaCl_2$. A similar method was used by Bright et al. to grow crystals of $CaTi_2O_4$ electrochemically starting from $CaTiO_3$ in molten $CaCl_2$ [6]. However neither of these previous reports provide detailed experimental information and attempts to repeat this work in our lab have been unsuccessful [7]. We report here a much simpler route to $CaTi_2O_4$ that can be used to prepare large samples for further study.

Mat. Res. Soc. Symp. Proc. Vol. 453 © 1997 Materials Research Society

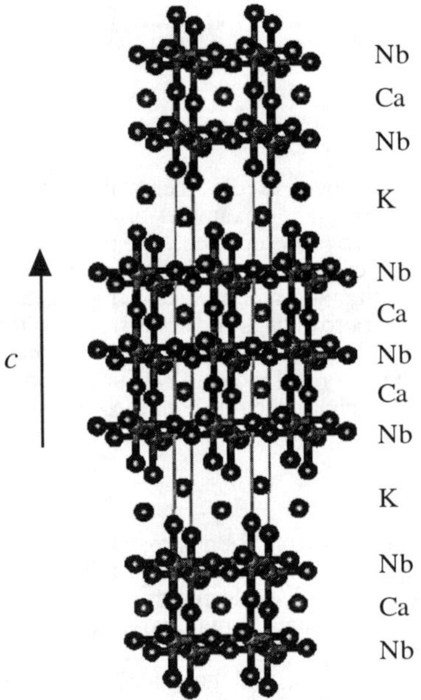

Figure 1. Idealized view of the proposed structure (ref. 2) of $KCa_2Nb_3O_{10}$. Perovskite-type slabs, composed of three corner-shared NbO_6 octahedra, are shifted relative to one another along the c direction. This displacement results in a trigonal prismatic coordination of the potassium ions located between the perovskite layers.

EXPERIMENT

All materials were characterized by X-ray powder diffraction on a Scintag XDS 2000 diffractometer using copper K_α radiation. For refinement of lattice parameters, silicon powder was added to diffraction samples as an internal standard.

Molten Salt Synthesis of $KCa_2Nb_3O_{10}$

Stoichiometric amounts of the reactants, typically 0.0927 g (0.671 mmol) K_2CO_3, 0.269 g (2.68 mmol) $CaCO_3$, and 0.535 g (2.01 mmol) Nb_2O_5, were ground with 5.00 g (67.1 mmol) KCl. The resultant mixture was fired in an alumina crucible at 900°C for 24 hours. After cooling to room temperature, the layered perovskite product was isolated by dissolving the KCl in deionized water and filtering the mixture to yield a white polycrystalline solid.

The temperature dependence of the molten salt synthesis was investigated by firing the reaction mixture for 24 hours at temperatures ranging from 800-1200°C. The products at 900°C as a function of time were monitored by loading several identical crucibles in the furnace at one time, raising the temperature of the furnace to 900°C (requiring approximately 20 minutes), and then subsequently removing the crucibles from the hot furnace after soak times ranging from 30 minutes to 72 hours. The solid state products were then isolated by dissolving the KCl in water and characterized by X-ray powder diffraction.

Synthesis of CaTi$_2$O$_4$

Anhydrous CaCl$_2$ was subjected to further drying by heating at 150°C under vacuum for 6-12 hours. Equimolar quantities of Ti metal powder and CaTiO$_3$ (typically 0.154 g Ti and 0.438 g CaTiO$_3$) were ground together with 5.00 g dried CaCl$_2$ in a nitrogen-filled glove bag, and the mixture was poured into an alumina combustion boat. The reaction boat was covered with a larger alumina boat inverted over the top and quickly inserted into a tube furnace at 120°C through which nitrogen or argon gas was flowing. After purging the tube furnace with inert gas for 30 minutes, the temperature was raised from 120°C to 1000°C at 30°/min, held at 1000°C for 6 hours, and then cooled to 700°C at 2°/min. The furnace was then turned off and allowed to cool to room temperature. The reaction boat was removed, and the CaCl$_2$ flux was dissolved in deionized water by sonicating the boat in an ultrasonic cleaner. Unreacted CaTiO$_3$ could usually be separated from the other insoluble products by carefully decanting off the supernatant, which contained the fine CaTiO$_3$ powder in suspension. Adding more deionized water and repeating the sonication and decanting steps several times until the supernatant was clear, dramatically improved the separation. Examining the remaining insoluble products under a low magnification microscope (1–7x magnification) revealed the predominant bluish-black crystalline needles of CaTi$_2$O$_4$ as well as polycrystalline clumps of TiO. The TiO impurities were separated from the CaTi$_2$O$_4$ crystals by sieving the products through a series of steel sieves of varying mesh size. The composition and purity of each fraction was monitored by examination under the microscope and X-ray powder diffraction.

RESULTS

Molten Salt Synthesis of KCa$_2$Nb$_3$O$_{10}$

The reaction of stoichiometric amounts of K$_2$CO$_3$, CaCO$_3$, and Nb$_2$O$_5$ in molten KCl readily formed the layered perovskite, KCa$_2$Nb$_3$O$_{10}$, at temperatures above 850°C. A typical X-ray powder diffraction pattern obtained from a sample fired in KCl at 950°C for 48 hours is shown in Figure 2. The reflections were indexed according to an orthorhombic unit cell, and the refined lattice parameters (a = 3.855(1) Å, b = 3.8740(7) Å, c = 29.444(5) Å) agreed with those reported in the literature for samples prepared by the traditional solid state synthesis route [3].

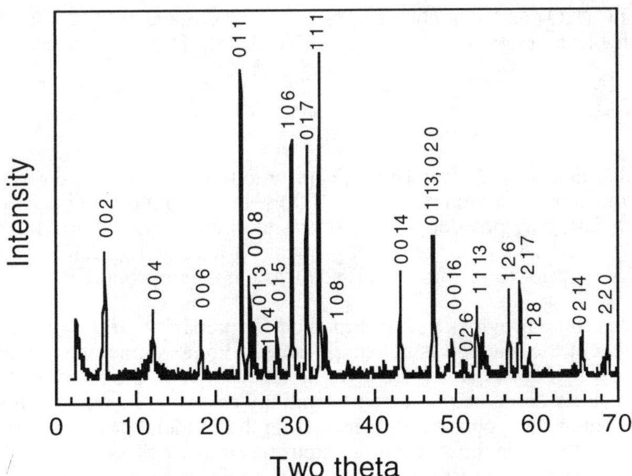

Figure 2. X-ray diffraction pattern of KCa$_2$Nb$_3$O$_{10}$ prepared by reaction in molten KCl at 950°C for 48 hours. The reflections are indexed according to an orthorhombic unit cell (a = 3.855(1) Å, b = 3.8740(7) Å, c = 29.444(5) Å).

In order to find the best synthesis conditions, a number of parameters of the molten salt synthesis were varied. First of all, mixtures of KCl and $CaCl_2$ were considered for the molten salt flux, as temperatures below the melting points of either pure KCl or pure $CaCl_2$ could be accessed. However, the layered perovskite phase was only stable in pure KCl, readily decomposing to $Ca_2Nb_2O_7$ in melts containing $CaCl_2$. Thus, the successful molten salt synthesis of $KCa_2Nb_3O_{10}$ requires KCl.

The reaction temperature was also varied. After 24 hours, the layered perovskite was the dominant product at all temperatures above 900°C. At 850°C, $KCa_2Nb_3O_{10}$ formed, but products also contained a considerable amount of $Ca_2Nb_2O_7$. When the reaction temperature was lowered to 800°C, the products were not very crystalline, and the reflections for $KCa_2Nb_3O_{10}$ were very weak. The conclusions of this study indicate that the optimal reaction temperature for the molten salt synthesis is 900°C or above.

In order to learn more about the chemistry of the molten salt synthesis, reactions were quenched from 900°C after various time intervals, and the products were characterized by X-ray diffraction. Reflections for the layered perovskite phase were clearly evident after as little as 30 minutes, although $Ca_2Nb_2O_7$ and $KNbO_3$ were also present. By 90 minutes, the reflections for $KNbO_3$ had disappeared and $KCa_2Nb_3O_{10}$ and $Ca_2Nb_2O_7$ remained. After 14 hours, new impurities appeared to grow in; the most intense reflections for $KCa_2Nb_3O_{10}$, $Ca_2Nb_2O_7$, and $CaNb_2O_6$ were all of equal intensity in the diffraction patterns. These impurities persisted in the products until reaction times of 24 hours, when the layered perovskite became the dominant phase. Little change occurred between 24 and 72 hours of reaction at 900°C. Therefore, the optimal conditions for preparing large samples of $KCa_2Nb_3O_{10}$ by the molten salt synthesis route involve firing the reactants for 24 hours at 900°C in molten KCl.

To underscore the role of the molten chloride flux in the synthesis of $KCa_2Nb_3O_{10}$, we investigated the traditional solid state reaction at 900°C. X-ray diffraction of the products revealed only weak reflections for the layered perovskite at this temperature; $Ca_2Nb_2O_7$ and $K_{5.75}Nb_{10.85}O_{30}$ were the major products. In our opinion, the molten salt synthesis is a better route to $KCa_2Nb_3O_{10}$. Although this method does not produce single crystals, it does lower the reaction temperature by 300°C and eliminates the need for intermediate grinding steps and for the addition of excess K_2CO_3.

Synthesis of CaTi$_2$O$_4$

Combining $CaTiO_3$ and Ti metal powder in a $CaCl_2$ flux at 1000°C for 6 hours led to the formation of bluish-black crystalline needles of $CaTi_2O_4$ (eq. 2). The successful isolation of

$$CaTiO_3 \ + \ Ti \ \xrightarrow[\text{flowing } N_2 \text{ or Ar}]{CaCl_2 \ / \ 1000°C} \ CaTi_2O_4 \qquad (2)$$

$CaTi_2O_4$ depended critically on drying the $CaCl_2$ prior to the reaction. When equimolar quantities of $CaTiO_3$ and Ti metal powder were reacted at 1000°C in $CaCl_2$ that had been dried overnight in an oven at 130°C, a dark gray powder was obtained. Despite the dark color, the diffraction pattern of this powder identified $CaTiO_3$ as the major product. If instead, the $CaCl_2$ was dried overnight under vacuum at 150°C prior to reaction, $CaTi_2O_4$ was the major product identifiable in the X-ray diffraction pattern.

Although this route provides a simple method for preparing large samples of crystalline $CaTi_2O_4$, the product of the reaction is not single-phase. For example, when reaction (2) was conducted under flowing N_2 gas, a surface layer of TiN formed in the reaction boat. The formation of this impurity could be prevented by initially covering the reaction mixture in the boat with a thin layer of pure $CaCl_2$ powder, thus insulating the Ti metal powder from direct contact with the N_2 gas. The other two most common impurities in the $CaTi_2O_4$ product were unreacted $CaTiO_3$ and TiO. Attempts to adjust the stoichiometry of the reactants to obtain a single phase product have so far been unsuccessful as small amounts of adventitious oxygen in the system lead to ready oxidation of the Ti metal powder. However, because the $CaTi_2O_4$ typically formed as crystalline needles, whereas the $CaTiO_3$ and TiO were polycrystalline powders, pure samples of $CaTi_2O_4$ could be prepared by careful physical separation techniques. The X-ray diffraction pattern of a typical sample of purified $CaTi_2O_4$ is shown in Figure 3 along with the calculated diffraction pattern based on the structural model from single crystal diffraction data. The purity of this sample can be measured by the fact that all of the reflections in the diffraction pattern can be assigned to $CaTi_2O_4$. The refined lattice parameters ($a = 9.744(3)$ Å, $b = 9.989(3)$ Å, $c = 3.149(1)$ Å) agree well with those previously reported in the literature [5,6].

Due to the similarity in composition of $CaTi_2O_4$ to the superconducting lithium titanate, $LiTi_2O_4$ ($T_c = 13.7$ K) [8], we are very interested in characterizing the electronic and magnetic properties of $CaTi_2O_4$. The lithium titanate differs from $CaTi_2O_4$ in that $LiTi_2O_4$ crystallizes in the spinel structure rather than a tunnel structure, and the average titanium oxidation state is +3.5 in $LiTi_2O_4$ as compared to +3 in $CaTi_2O_4$. The analogous magnesium compound, $MgTi_2O_4$,

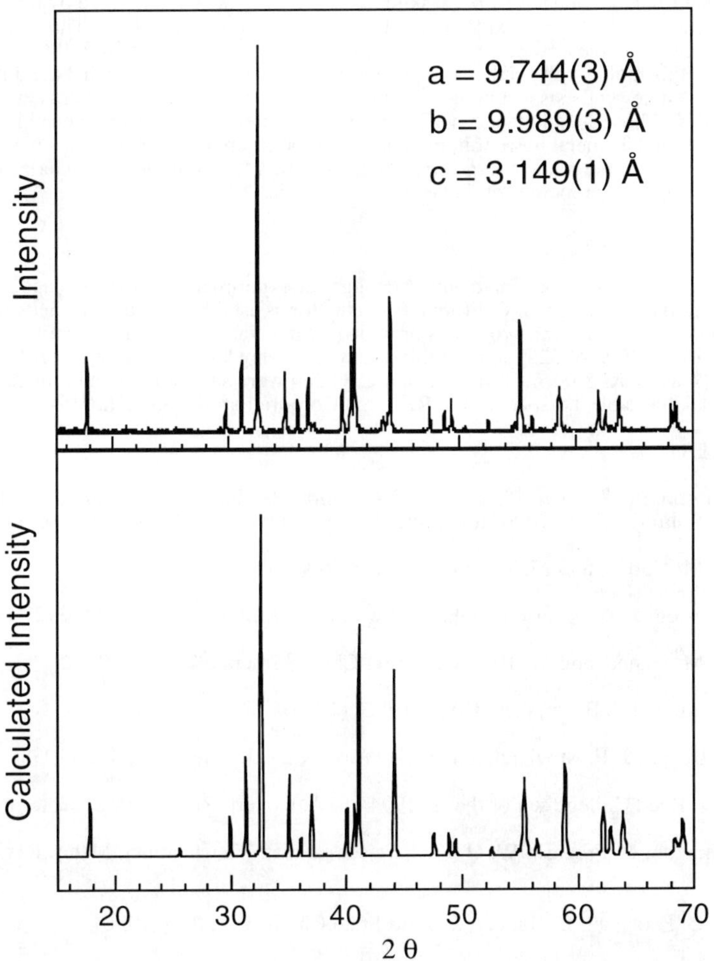

Figure 3. X-ray powder diffraction pattern of a sample of $CaTi_2O_4$ purified by physical separation techniques (top) and the calculated X-ray diffraction pattern (bottom) based on the structural model refined from single crystal diffraction data (ref. 5).

crystallizes in the spinel structure with an average titanium oxidation state of +3 and was recently reported to be a low resistive semiconductor [9]. We are interested in understanding the relationships between structure, composition, and properties in these related materials. The molten salt synthesis route to $CaTi_2O_4$ now provides a good route to large samples of crystalline material for further study. Preliminary magnetism studies indicate that $CaTi_2O_4$ is paramagnetic from room temperature down to 1.8 K. Thus, $CaTi_2O_4$ does not appear to superconduct at these temperatures.

CONCLUSION

We have discovered two new synthesis routes to interesting titanates and niobates using molten chloride fluxes. The layered perovskite, $KCa_2Nb_3O_{10}$, was synthesized by reacting the constituent metal carbonates and oxides in molten KCl for 24 hours at 900°C. This new preparative route offers the advantage of lowering the reaction temperature by 300°C. We are currently using this molten salt synthesis to prepare other layered perovskites that are difficult to prepare by solid state synthesis techniques and to explore the preparation of new materials in this family of oxides. The reduced titanate, $CaTi_2O_4$, was also prepared from a molten chloride flux by reacting $CaTiO_3$ and Ti metal powder in molten $CaCl_2$ for 6 hours at 1000°C. This reaction provides a simple, reliable route to large crystals that can be used to further characterize the electronic and magnetic properties of $CaTi_2O_4$.

ACKNOWLEDGMENTS

We would like to thank L. Macdonald and M. Rogge for experimental assistance, and A. M. Stacy and B. Reiser at the University of California Berkeley for assistance with the magnetism measurements. Funding for this work was provided by the National Science Foundation ILI program (Grant #DUE-9352625) and a William and Flora Hewlett Foundation Award of Research Corporation (Award #C-3488). D. Ingram and C. Green were supported by a Partners in Science, M.J. Murdock Charitable Trust Award of Research Corporation (Award #HS0354).

REFERENCES

1. A. J. Jacobson, in *Chemical Physics of Intercalation II*, edited by P. Bernier, J. E. Fischer, S. Roth, and S. Solin (NATO ASI Series B **305**, Plenum Press, New York, 1993) pp.117-139.

2. M. Dion, M. Ganne, and M. Tournoux, Mater. Res. Bull. **16**, 1429 (1981).

3. A. J. Jacobson, J. W. Johnson, and J. T. Lewandowski, Inorg. Chem. **24**, 3727 (1985).

4. M. Dion, M. Ganne, and M. Tournoux, Rev. Chim. Minerale **21**, 92 (1984).

5. E. F. Bertaut and P. Blum, Acta Cryst. **9**, 121 (1956).

6. N. F. H. Bright, J. F. Rowland, and J. G. Wurm, Can. J. Chem. **36**, 492 (1958).

7. W. Breyer, Reed College senior thesis, 1994. P. A. Covert, Reed College senior thesis, 1995.

8. D. C. Johnston, H. Prakash, W. H. Zachariasen, and R. Viswanathan, Mater. Res. Bull. **8**, 777 (1973).

9. H. Hohl, C. Kloc, and E. Bucher, J. Solid State Chem. **125**, 216 (1996).

NEW ONE-DIMENSIONAL COMMENSURATE AND INCOMMENSURATE STRUCTURAL FORMS OF SR-CO OXIDE

R. CHRISTOFFERSEN, A.J. JACOBSON, S.L. HEGWOOD, L. LIU
Department of Chemistry, University of Houston, Houston, TX, 77204-5641

ABSTRACT

New forms of Sr-Co oxide with one-dimensional structures related to mixed-layer hexagonal perovskites have been synthesized and characterized by TEM. Crystals of $Sr_5Co_4O_{12}$ grown from molten KOH flux and oxidized under slow cooling have structures based on a 3/2 ratio of mixed $[Sr_3CoO_6]$ and $[Sr_3O_9]$ layers. This is the first known structure with layers of this type stacked in a non-integer ratio, yielding chains of face-sharing octahedral and trigonal prismatic Co-sites with a $3 + 1$ sequence along the c-axis. In oxygen-deficient $Sr_5Co_4O_{12-x}$, the $\{110\}$ structural modulation of the stoichiometric 5:4 phase becomes rotated to an irrational orientation, possibly in association with vacancy ordering, forming an incommensurate superstructure. For the compound $Sr_6Co_5O_{15}$, previously known to have a 1/1 ratio of $[Sr_3CoO_6]$ and $[Sr_3O_9]$ layers with a $4 + 1$ octahedral/trigonal prismatic site sequence, introduction of excess oxygen atoms leads to formation of a commensurate rhombohedral superlattice in which c is doubled relative to the stoichiometric 6:5 phase.

INTRODUCTION

The occurrence of unique physical, electronic and magnetic properties in inorganic structures with one-dimensional atomic arrangements is well-known, with charge-density wave phenomena in metals and metallic conductors being perhaps the most widely studied example [1]. In oxides, one-dimensional structures are not particularly common, but an increasing number of linear structures are now being found within a sub-family of the derivative phases related to the hexagonal perovskites [2]. In these structures a proportion of the $[AX_3]$ cation-oxygen layers of normal perovskite are replaced by $[A_3A'X_6]$ layers in which a single A' cation substitutes for oxygen triplet-groups within the $[AX_3]$ layer. Stacking these layers in an hexagonal, rather than cubic, sequence produces chains of face-sharing octahedral and trigonal prismatic sites along the c-axis whose relative proportions depend on the ratio of the two layer types. The resulting structural series can be described by the general formula $A_{3n+3}A'_nB_{3+n}O_{9+6n}$ where n is the ratio of $[A_3A'O_6]$ to $[A_3O_9]$ ($=[AX_3]$) layers [2].

Recently, Harrison et al. [3] showed that the so-called low-temperature hexagonal phase in the Sr-Co oxide system is $Sr_6Co_5O_{15}$, an n=1 member of the above series with $[Sr_3O_9]$ and $[Sr_3CoO_6]$ layers stacked 1:1 in alternate fashion. The chains of face-sharing coordination polyhedra in this structure have a sequence of 4 octahedral and 1 trigonal prismatic site, the octahedra most likely containing Co^{4+} and the trigonal prisms Co^{2+}, although other charge-ordering schemes are possible. The structure's c-axis repeat contains 3 $[Sr_3CoO_6]$ layers and 3 $[Sr_3O_9]$ layers for a total of 6 cation-oxygen layers. This layer sequence will be referred to here as 6H' due to its similarity to the 6H perovskite polytype [4].

To date $Sr_6Co_5O_{15}$ has remained the only known form of Sr-Co oxide with an hexagonal, rather than cubic, perovskite-related structure. However, in the course of synthesis experiments designed to refine our understanding of $Sr_6Co_5O_{15}$, we have uncovered three additional hexagonal perovskite-related forms of Sr-Co oxide and a TEM study of these is reported here. Two of the forms have structures and metal-oxygen stoichiometries that have not been seen previously within the hexagonal perovskite-related family of compounds.

EXPERIMENTAL PROCEDURES

Samples were synthesized in powder form from mixed-oxide and citrate-gel precursors, and as mm-size single crystals grown from molten KOH. The mixed-oxide synthesis route followed the procedures outlined in [3], with the exception that a starting composition of exactly $Sr_6Co_5O_{15}$ was used and the final powder was re-ground and re-heated to 875°C multiple times to ensure ho-

153

mogeneity. As a means to make homogeneous samples of $Sr_6Co_5O_{15}$ more rapidly, a citrate-gel Pechini-type route [5] was adapted from procedures reported by [6,7]. For this method, Sr and Co were delivered into a pre-formed citrate gel using a nitric acid solution prepared from Co metal and $SrCO_3$. The mixture was then dried and calcined in the same manner as [6,7], initially heated at 1060°C for 10 hours, and then annealed at 875°C for 8 hours to produce the final Sr-Co oxide powder. The powder itself was shown by SEM examination to consist of loosely aggregated "microcrystals", 50-100 μm long, with well-developed prismatic forms.

The molten KOH flux technique was adapted from procedures used to grow Ba-Ni oxide single crystals by [8]. Starting chemicals of reagent grade $SrCO_3$ (Alfa 99.99%), Co_3O_4 (Aldrich 99.995%), and KOH (EM) were ground in a Sr:Co:K molar ratio of 1:0.83:8. The mixture was heated for 10 minutes at 750°C in an alumina crucible to calcine the carbonate and then cooled to room temperature. It was then re-heated in a box furnace for 10 min at 1100°C to melt the mixture, and then slowly cooled by shutting off the furnace power. After washing with water the sample yielded shiny, black prismatic crystals 1 to 3 mm long.

The crystals within the Pechini-route powder and the KOH-flux sample were large enough to analyze individually by electron probe microanalysis (EPMA). EPMA was carried out on a JEOL SXA-8600 electron probe micro-analyzer using Co metal (Co), $SrTiO_3$ (Sr) and CuO_2 (O) for analytical standards. Raw analyses were corrected using $\phi(\rho z)$ procedures [9]. It should be noted that, due to oxygen's inherently lower X-ray yield, microprobe analyses for O have larger error limits (as much as 5-10% relative) compared to those for heavier elements.

Transmission electron microscope (TEM) observations of the samples, primarily for the purpose of studying their electron single-crystal diffraction patterns, were performed utilizing a JEOL 2000FX analytical TEM operating at 200 keV. The mixed-oxide and Pechini-route samples were prepared for TEM study by suspending finely ground powders in high-purity acetone and transferring drops of the suspension to a holey-carbon support film on a 200 mesh Cu grid. For the KOH-flux material it was possible to make TEM samples from individual single crystals by crushing them between glass slides, and transferring the fragments to a holey-carbon grid.

High-resolution TEM (HRTEM) observations of the KOH-flux crystals utilized a JEOL 4000EX high-resolution TEM operating at 400 keV. HRTEM image simulations were performed using the Cerius [10] multislice program on a SiliconGraphics Indy workstation. Routine X-ray powder diffraction characterization of the samples utilized a Scintag Model XDS 2000 diffractometer. Thermogravimetric analysis (TGA) was carried out using a TA Instruments 2950 TGA system.

RESULTS

<u>As-Synthesized Samples</u>

Selected-area [100] (hexagonal cell) electron diffraction patterns for the as-synthesized powders and flux-grown single crystals are shown in Figs. 1 and 2. Although electron diffraction patterns from several zone axis orientations were studied, the [100] orientation reveals most of the key geometrical differences between the sample's reciprocal lattices.

For the mixed-oxide sample (Fig. 1a), the reciprocal cell dimensions and pattern of allowed reflections in [100] and other diffraction patterns agree with the rhombohedral 6H' structure previously reported for $Sr_6Co_5O_{15}$ by [3]. The stoichiometry of the sample studied by [3] was well constrained by the neutron refinement and the X-ray powder diffraction data for the present sample are consistent with this stoichiometry as well.

The Pechini-route microcrystals were shown by microprobe analyses to have uniform Sr:Co proportions tightly grouped around 6:5, with O contents on a formula basis ranging from $Sr_6Co_5O_{15}$ to $Sr_6Co_5O_{15.5}$, with an average of $Sr_6Co_5O_{15.25}$. The electron diffraction results (Fig. 1b) show the crystals to have superlattice reflections along $c*$ at positions relative to the 6H' diffraction patterns of $00l$, $l=1/2$ n, with overall pattern indexing being consistent with a rhombohedral, 2 x $c_{6H'}$, or 12H', superlattice.

EPMA of the KOH-flux crystals showed them to be Co-depleted relative to the mixed-oxide and Pechini samples, with a consistent Co/Sr ratio of 0.80±0.01. Together with the O analyses a stoichiometry of $Sr_5Co_4O_{12-x}$ is indicated, with x=0.2-0.7. Although for reasons already

Figure 1. Selected-area [100] electron diffration patterns for (a) $Sr_6Co_5O_{15}$ synthesized by mixed-oxides, (b) $Sr_6Co_5O_{15+x}$ Pechini-route "microcrystals", (c) $Sr_5Co_4O_{12-x}$ single crystals slow-cooled in pure O_2. Arrows show S vector.

discussed, O contents measured by microprobe have large associated errors, replicate analyses of the same grains using different O standards were carried out and confirm that the KOH-flux crystals have less than 12 oxygens in the above formula unit.

Electron diffraction patterns of the KOH-flux crystals have both incommensurate satellite reflections and commensurate main spots, the latter corresponding to reflections expected from a basic two-layer, or 2H', substructure (Fig. 2a). The vector, S, describing the satellite positions is normal to [100] but has an irrational orientation in the b*-c* plane, making an angle α of approximately 41.2° with b* (Fig. 2b). It should be noted that for values of α such that S lines up with rows of 2H' main reflections, a commensurate superlattice of the 2H' sub-lattice results, as in the case of the 6H' and 12H' structures.

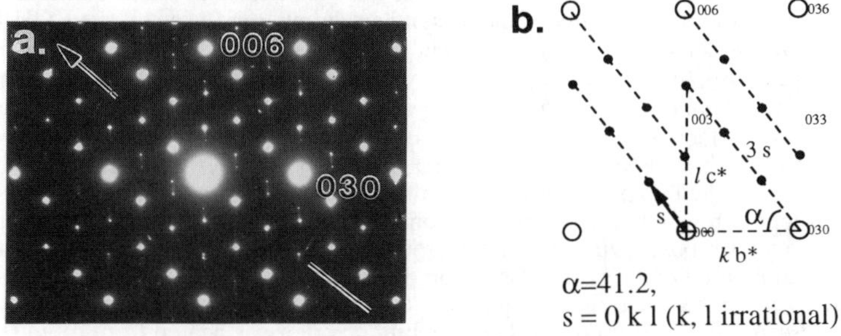

$\alpha = 41.2,$
$s = 0\ k\ l\ (k, l\ \text{irrational})$

Figure 2. (a) Selected-area [100] electron diffraction pattern for $Sr_5Co_4O_{12-x}$ single crystals grown from molten KOH flux. Line shows orientation of satellite reflections. (b) Diagram showing orientation and angle of satellite vector S.

The nature of the incommensurate modulation in the KOH-flux crystals was studied by high-resolution TEM imaging along [100], the zone axis that is perpendicular to S and hence parallel to the plane of the modulation. For the 6H' structure a [100] projection results in superposition of the octahedral/trigonal prismatic columns of Co sites over their adjacent columns of Sr atoms, in projection forming dense (010) layers of metal cations separated by less dense O layers. As shown in Fig. 3a the metal cation and O layers appear in [100] high-resolution images as prominent vertical (010) fringes, with the S-modulation appearing as a wave-like undulation in the contrast of these fringes oriented approximately 49° from (010).

Because the projected positions of the Co and Sr atoms within the (010) layers have separations below the microscope resolution, the exact nature of the modulation in terms of atomic displacements or site occupancies within the Co or Sr columns is not easily determined. In certain images, we have noted *en echelon* arrangments of short, diagonal, rows of bright dots forming strips oriented perpendicular to S (Fig. 3b). The dots run between 2 to 3 adjacent (010) fringes, and may diagonally correlate cation sites of similar type in adjacent columns of Sr or Co. A small relative rotation of such diagonal site correlations from their normal orientation in the commensurate parent structure may be the geometrical explanation for the incommensurate modulation and we are presently investigating this possibility with detailed image simulations.

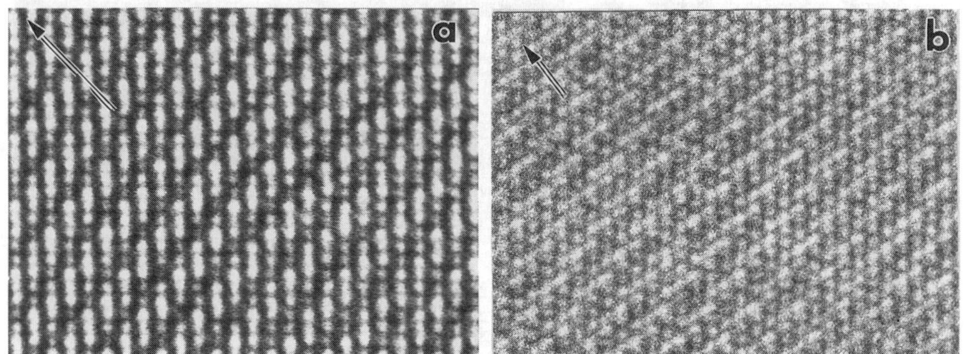

Figure 3. High-resolution TEM images for [100] orientation in incommensurate $Sr_5Co_4O_{12-x}$. Incommensurate modulation vector shown by arrow. (a) Defocus = -104 nm, (b) Defocus = -40 nm.

<u>Post-Synthesis Annealing and TGA Experiments</u>

One annealing experiment and one TGA experiment were conducted to make an initial assessment of the temperature and pO_2 stability of the as-synthesized phases. To see if the KOH-flux crystals were incommensurate due to oxygen deficiency relative to a possible stoichiometric composition such as $Sr_5Co_4O_{12}$, the crystals were initially heated unground at 875°C in pure O_2, and then step-cooled to room temperature in pure O_2 over a 2 month period. After this experiment the crystals had a dull gray surface patina but their interiors remained black and shiny with no optically visible decomposition. Within TEM samples prepared from the same crystal, crystal fragments were found to have either the incommensurate or commensurate 6H' structures, or a third commensurate superstructure in which, relative to its incommensurate position, the satellite vector is rotated 2.7° closer to the b* axis, giving superlattice reflections with exact 1/5 spacings relative to the first main reflection along $c*$ (Fig. 1c). Based on the first of these reflections being first-order the new superlattice would have a c-repeat equal to 5/6 that of the 6H' structure, making it 5H'.

In the TGA experiment, a portion of the 12H' Pechini powder was heated from room temperature to 875°C under artificial air (79%N_2/21%O_2) at 2°C/minute, and then cooled to room temperature at this same rate. A net 0.3% weight decrease due to O loss was measured. Scanning electron microscope observations showed the microcrystals in the powder to be exactly the same size and shape before and after the TGA run, suggesting that O loss occurred with minimal recrystallization of the sample. Assuming the crystals had initial compositions that were slightly O-rich as indicated by microprobe, the oxygen loss measured by TGA would correspond to changing the sample's average stoichiometry from $Sr_6Co_5O_{15.25}$ to $Sr_6Co_5O_{15}$. Consistent with this change, electron diffraction showed the majority of material in the TGA residue to have the 6H' structure, together with a lesser amount of 12H' material and some 5H' material.

DISCUSSION

<u>5H' and Incommensurate Structures</u>

The proportion of distinct A and B cation sites and oxygen in $Sr_5Co_4O_{12}$ would make it an n=1.5 member of the hexagonal perovskite-type structural series discussed above, with a 2:3 proportion of [A_3O_9] (=[Sr_3O_9]) layers to [$A_3A'O_6$] (=[Sr_3CoO_6]) layers. To our knowledge such a non-integer member of the mixed-layer hexagonal perovskite derivative structures has not previously been synthesized. A simple layer sequence for $Sr_5Co_4O_{12}$ based on a 2:3 mixed-layer proportion is shown in Fig. 4, together with the sequence for the $Sr_6Co_5O_{15}$ 6H' phase.

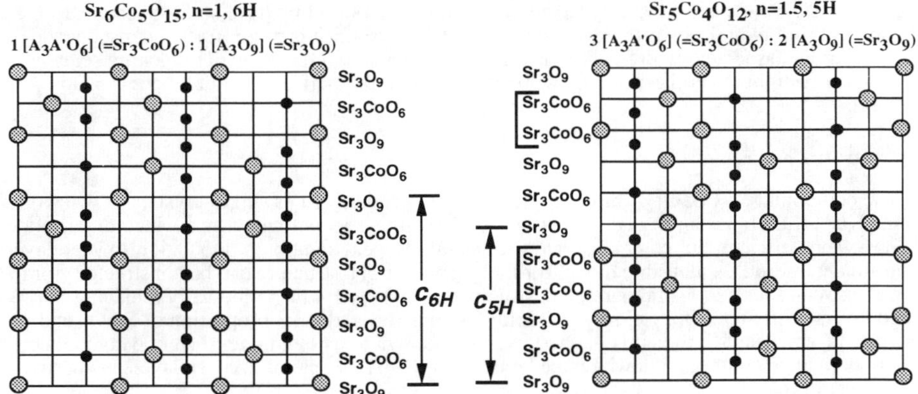

Figure 4. Schematic [1-10] structural projections showing stacking sequence of layers containing Sr (large, gray) and Co (small, black) in $Sr_6Co_5O_{15}$ (left) and $Sr_5Co_4O_{12}$ (right).

The sequence for $Sr_5Co_4O_{12}$ is produced if the $Sr_6Co_5O_{15}$ sequence is interrupted every 5 layers by replacing a $[Sr_3O_9]$ layer with a $[Sr_3CoO_6]$ layer and then continuing the sequence (Fig. 4). The resulting double $[Sr_3CoO_6]$ layers reduce the Co/Sr ratio from 5/6 to 4/5 because there are two octahedral B-sites between an $[Sr_3O_9]$ and $[Sr_3CoO_6]$ layer, but only one between two successive $[Sr_3CoO_6]$ layers. The sequence of sites along the polyhedral chains is changed from 4 octahedra + 1 trigonal prism in the 6:5 structure to 3 + 1 in the 5:4 structure.

The 5:4 model structure in Fig. 4 results in a c-axis repeat that is 5/6 of that for the 6H' structure, hence matching the c-axis periodicity we observe for our 5H' phase by electron diffraction. We do not know by direct microanalysis whether our crystal fragments with the 5H' structure are $Sr_5Co_4O_{12}$, but the fact that they form as an oxidation product of the oxygen-deficient incommensurate phase, $Sr_5Co_4O_{12-x}$, whose Sr/Co ratio is known, is strong circumstantial evidence that they have this composition. (The reason why oxidized crystals of the incommensurate phase contain regions of 6H', 6:5 structure along with 5H' is not fully understood, but it may be due to localized, partial decomposition of the crystals at lower temperature.) Moreover, the diffraction patterns of the 5H' and incommensurate phases are nearly the same, differing only by a small (2.7°) relative rotation of **S**. This suggests that the 5H' structure is the basic commensurate atomic arrangement from which the incommensurate structure is derived, and since the latter is cation-deficient, a role for vacancy or charge ordering in producing the incommensurate modulation would seem to be a strong possibility. We are presently using the 5H' structure as a basis for HRTEM image simulations to see if underlying features in our experimental images of the incommensurate structure reflect the 5H' stacking model.

6H' and 12H' Structures

Like the incommensurate and 5H' structures, the 6H' and 12H' structures have the same Sr/Co ratio but different oxygen contents and different structures. Our TGA results show that, similar to the incommensurate and 5H' phases, the 12H' and 6H' structures can be experimentally converted one into the other, only by reduction rather than oxidation. Unlike the former structures, however, the "oxygen-deficient" phase in this case (6H') is the one with the stoichiometric composition, while the non-stoichiometric 12H' phase has an oxygen-excess.

The electron diffraction patterns of the 6H' and 12H' structures, and hence their respective superstructures, show a greater degree of relative difference, i.e. a fully doubled c-axis, than what is seen between the incommensurate and 5H' structures. Because of the need to maintain a Sr:Co stoichiometry of 6:5, our attempts to explain the 12H' structure by altering the sequence, but not the ratio, of the Sr_3O_9 and Sr_3CoO_6 mixed layers in the 6H' structure have so far been unsuccessful. Our working hypothesis is therefore that the 12H' superlattice is more likely a consequence of defect ordering associated with the presence of excess oxygen in the basic 6H' stacking sequence. Work to confirm this hypothesis, along with evaluation of alternative structural models, is in progress.

CONCLUSIONS

The compound $Sr_5Co_4O_{12-x}$ has been successfully synthesized as single-crystals grown from molten KOH flux. It has an hexagonal perovksite derivative superstructure that is incommensurate when oxygen deficient, but commensurate with a c-axis repeat composed of 5 cation-oxygen layers when stoichiometric. A plausible model for this 5-layer or 5H' structure can be constructed from the n=1 mixed-layer perovskite-derivative structure of $Sr_6Co_5O_{15}$ by periodically replacing Sr_3O_9 mixed layers with Sr_3CoO_6, yielding a structure with n=1.5 and a 2:3 proportion of Sr_3O_9 and Sr_3CoO_6 layers. The 5H' model is the first example known to us of a mixed-layer hexagonal perovskite-derivative compound in which the $[A_3O_9]$ and $[A_2A'O_6]$ cation-oxygen layers occur in a non-integer ratio.

A new 12H' superstucture based on a doubled c-axis was found for the 6H', n=1, rhombohedral structure of $Sr_6Co_5O_{15}$. Electron microprobe analyses and TGA results indicate that the 12H' phase contains excess oxygen relative to stoichiometric $Sr_6Co_5O_{15}$, and ordering of defects associated with this oxygen is a possible explanation for the cell doubling.

ACKNOWLEDGMENTS

A portion of the TEM work for this study was conducted at the Center for High Resolution Electron Microscopy at Arizona State University and access to this facility is gratefully acknowledged. We thank the Texas Center for Superconductivity and the Robert A.Welch Foundation for financial support of this work. This research was also supported in part by the MRSEC program of the National Science Foundation under Award Number DMR-9632667.

REFERENCES

1. R.E. Thorne, Phys. Today, **49**, 42-47, (1996).
2. J. Darriet and M.A. Subramanian, J. Mater. Chem., **5**, 543-552 (1995).
3. W.T.A. Harrison, S.L. Hegwood, and A.J. Jacobson, J. Chem. Soc., Chem. Commun., 1953-1954, (1995).
4. L. Katz and R. Ward, Inorg. Chem., **3**, 205-207 (1964).
5. M.P. Pechini, U.S. Pat. No. 3 330697 (11 July 11 1967).
6. L.-W. Tai and P.A. Lessing, J. Mater. Res., **7**, 502-510 (1992).
7. L.-W. Tai and P.A. Lessing, J. Mater. Res., **7**, 511-521 (1992).
8. J. A. Campa, E. Gutierrez-Puebla, M.A. Monge, I. Rasines and C. Ruiz-Valero, J. Solid State Chem., **108**, 230-235 (1994).

9. G.F. Bastin, F.J.J. van Loo and H.J.M. Heijliger, *Evaluation of the Use of Gaussian ϕ (ρz) Curves in Quantitative Electron Probe Microanalysis: A New Optimization*, (University of Technology, Eindhoven, The Netherlands).
10. M.R. Stapleton, Molecular Simulations LTD, Cambridge, UK, 1992.

THE SYSTEM Li$_3$SbO$_4$ - CuO

M. A. CASTELLANOS R.*, S. TRUJILLO T.*, J. M. S. SKAKLE **, A. R. WEST **

* Universidad Nacional Autónoma de México, Facultad de Química, México, DF, 04510, México
** University of Aberdeen, Department of Chemistry, Meston Walk, Aberdeen, UK.

ABSTRACT

Phase formation in the system Li$_3$SbO$_4$-CuO has been studied by solid state chemistry reaction, at temperatures between 650° and 1150°C. Two new complex oxides were found and characterized by X-ray powder diffraction (XRD), Li$_3$CuSbO$_5$ (O) and Li$_3$Cu$_2$SbO$_6$ (S). The latter exhibits a new partially ordered rock salt structure and shows a low level of electronic conductivity [1]. The crystal structure of Li$_3$SbO$_4$ was determined [2].

INTRODUCTION

Previously, we have studied the formation of new Cu-complex oxides, Li$_2$CuZrO$_4$, Li$_2$CuHfO$_4$ [3],   Li$_{2/3(1-X)}$Ti$_{1/3(1-X)}$ Cu$_X$O; $0.125 \leq x \leq 0.175$ [4], Li$_{(3-3X)}$Ta$_{(1-X)}$Cu$_{4X}$O; $0 \leq x \leq 1$ [5], and Li$_2$CuSnO$_4$ [6]. These were synthesized as part of a program to obtain new Cu-containing rock salt structures.

This report gives some preliminary results, at temperatures below 1150°C, on the phase diagram of the binary system Li$_3$SbO$_4$-CuO. Two new phases O and S were found. Both compositions on this join have the required 1:1 stoichiometry for the rock salt structure. The crystal structure of Li$_3$Cu$_2$SbO$_6$, with a spacegroup C2/c, is a partially ordered rock salt structure related to the ordered structure of Li$_2$TiO$_3$ by additional cation order: Ti sites are occupied by Cu and Sb in ordered fashion; Li sites contain Li/Cu in two sets, but with non-statistical occupancy, and Li. In Li$_2$TiO$_3$, the cation distribution alternates between layers of Li and layers of Li and Ti , whereas in S all the layers contain a mixture of cations. Bond lengths show a Jahn-Teller elongation of Cu-O octahedra consistent with magnetic susceptibility data that indicate paramagnetic Cu^{2+} ions. S is a very modest electronic semiconductor [1]. The powder pattern of Li$_3$SbO$_4$ was examined because, although the XRD data were collected on a high resolution instrument, the pattern could not be adequately indexed on the previously described tetragonal unit cell. The crystal structure of Li$_3$SbO$_4$ was reinvestigated, it presents a space group P2/c, with an ordered rock salt structure type, and distorted Sb-O octahedra, which form zig-zag chains running parallel to c in a monoclinic supercell. Attempts to transform the monoclinic Li$_3$SbO$_4$ structure to a cubic rock salt disordered form were useless [2].

Mat. Res. Soc. Symp. Proc. Vol. 453 © 1997 Materials Research Society

EXPERIMENTAL

Synthesis

was used direct from the bottle as drying was found unnecessary. The CuO was dried at 700°C prior to use. Several compositions were prepared basically in two steps. First, stock quantities of Li_3SbO_4 were prepared by reacting stoichiometric mixtures of Li_2CO_3 and Sb_2O_5, in Pt crucibles in a muffle furnace.

Initial temperatures were 650°C for 4h to expel CO_2, followed by 1000°C for 24h. In the second step, the CuO oxide was mixed in the appropriate stoichiometry with Li_3SbO_4; several compositions were weighed out in 5 - 7g amounts, mixed and ground with an agate mortar and pestle, using acetone as vehicle, dried and reacted in Pt crucibles. For each phase optimum conditions of temperature, and time, were found by trial and error. All of them were air -cooled to room temperature.

Care was required to ensure that lithia loss by volatilization was not a serious problem; this was confirmed by separate weight loss checks. The synthesis conditions including only the final firing and product colors can be described as follows.

$$Sb_2O_5 + 3Li_2CO_3 \xrightarrow[24\,h]{1000°C} 2Li_3SbO_4 + 3CO_2\uparrow \qquad \text{White}$$

$$Li_3SbO_4 + CuO \xrightarrow[12\,h]{1000°C} Li_3CuSbO_5 \qquad \text{Green fluorescent}$$

(or from Li_2CO_3 , Sb_2O_5, CuO at 1050°C/17h)

$$Li_3SbO_4 + 2CuO \xrightarrow[85\,h]{1000°C} Li_3Cu_2SbO_6 \qquad \text{Green olive}$$

(or from Li_2CO_3 , Sb_2O_5, CuO at 1025°C/12h)

Reaction products were analyzed by X-ray powder diffraction using a SIEMENS D5000 diffractometer with CuK_a radiation. Si was added as an internal standard for accurate d-spacing measurements, in the O phase. Intensities were taken as peak heights on the diffractometer traces.

RESULTS

The results of heating experiments on a number of different starting compositions, on the join Li_3SbO_4 - CuO, given in Fig. 1, were studied experimentally. For the synthesis of pure samples of the new compounds, it was essential to avoid loss of lithia by volatilization. It was found early on that the two steps method described above, usually gave a more pure product than did single stage reaction of the three components (Li_2CO_3, CuO, Sb_2O_5).

S and O required temperatures 1000 °C for their synthesis. Powder X - ray diffraction showed that the two phases were highly crystalline single phases. The X - ray diffraction pattern of O is displayed in Fig. 2 and its diffraction data in Table I.

There was no evidence for any solid solution formation with either O or S. They appear to be lineal phases. Li_3SbO_4 and CuO do not seem to form solid solution at least at 5 and 95 mol % of CuO. Instead, the phase diagram, Fig. 1, shows areas of compositions that yielded mixtures of O and Li_3SbO_4, O plus S, and S and CuO. The area of each of these regions are marked on the phase diagram. In the region rich in CuO, at temperatures above 1000°C, the reversible formation of cuprite, Cu_2O, was observed.

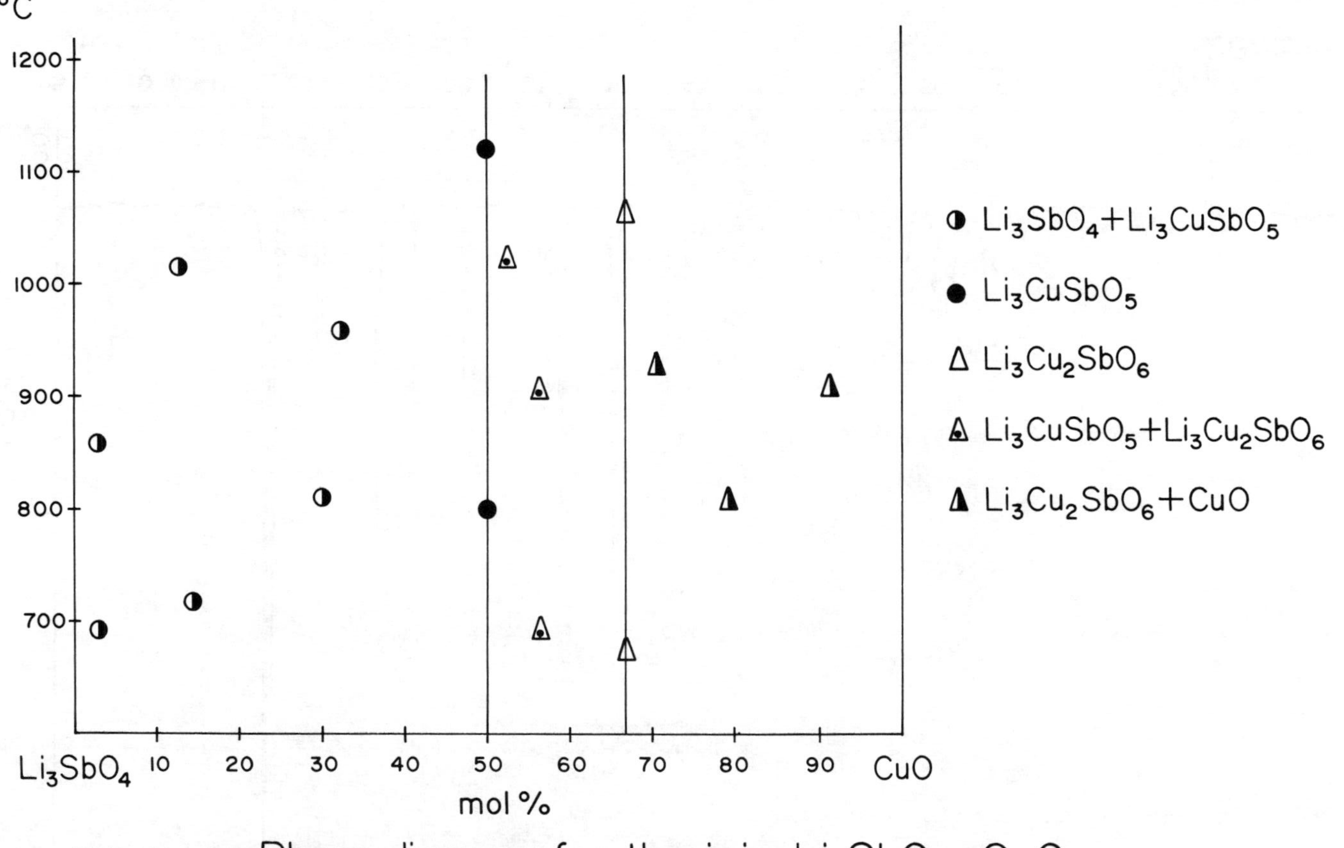

FIG. No. 1 Phase diagram for the join Li_3SbO_4-CuO

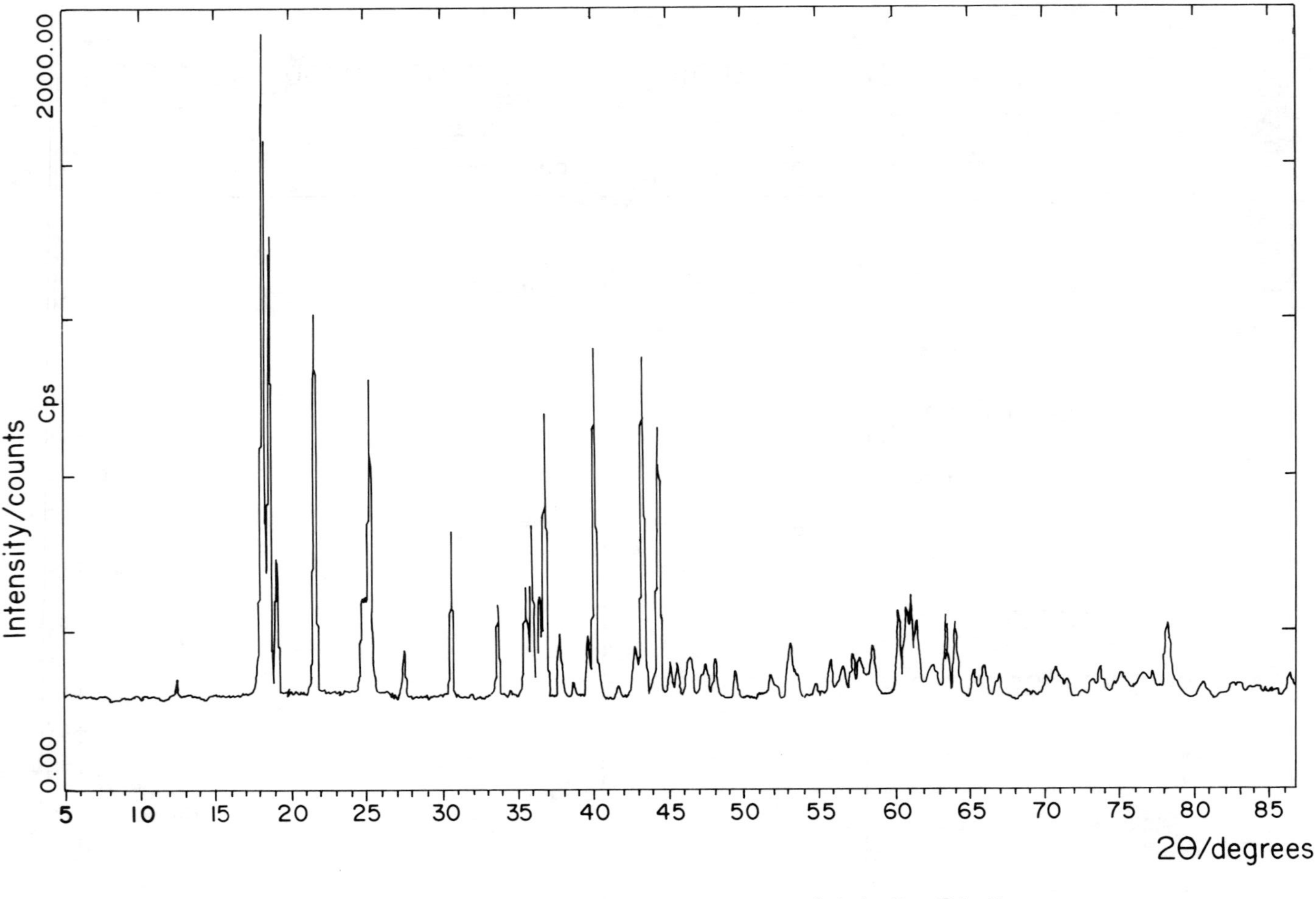

FIG. No. 2   X-ray diffaction pattern of Li_3CuSbO_5

Table I. Observed °2θ, d-spacings (Å) and intensities for the X-ray diffraction pattern of Li_3CuSbO_5

N	°2θ	d	I_{rel}
1	12.363	7.1534	5.69
2	18.114	4.8931	100.00
3	18.514	4.7885	49.20
4	19.014	4.6636	18.03
5	21.549	4.1204	46.03
6	22.771	3.9019	7.65
7	24.599	3.6160	13.11
8	24.843	3.5810	16.55
9	25.151	3.5378	33.68
10	25.383	3.5060	11.31
11	25.546	3.4841	6.66
12	27.367	3.2562	10.10
13	28.433	3.1365	45.39
14	28.689	3.1091	6.81
15	30.485	2.9299	28.28
16	30.696	2.9102	7.00
17	31.448	2.8423	5.88
18	33.569	2.6674	16.21
19	35.412	2.5327	17.99
20	35.624	2.5181	8.81

5 and 95 mol % of CuO. Instead, the phase diagram, Fig. 1, shows areas of compositions that yielded mixtures of O and Li_3SbO_4, O plus S, and S and CuO. The area of each of these regions are marked on the phase diagram. In the region rich in CuO, at temperatures above 1000°C, the reversible formation of cuprite, Cu_2O, was observed.

CONCLUSIONS

The formation of two new phases have been found in this system, Li_3CuSbO_5 (O) and $Li_3Cu_2SbO_6$ (S). The determination of the crystal structure of O is in progress whereas that of S has been fully determined by Rietveld refinement In both phases were used different starting materials and the products of reaction were the same. As a consequence of this study the crystal structure of Li_3SbO_4 was also determined. The phase relations among all of these compounds can be observed from the phase diagram .

ACKNOWLEDGMENTS

We thank L. Baños for obtaining the X-ray diffraction data and M. Guadalupe Calderón C. for correcting the manuscript.

REFERENCES

1. J. M. S. Skakle, M. A. Castellanos R., S. Trujillo Tovar, and A. R. West; J. Solid State Chemistry, submitted Aug 26, 1996.

2. J. M. S. Skakle, M. A. Castellanos R., S. Trujillo Tovar, S. M. Fray and A. R. West; J. Mater. Chem., in press.

3. M. Castellanos, M. Chavez Martinez, A. R. West, Z. Kristallogr. 190,161-169 (1990).

4. J. V. Padilla Saldivar, B. Sc. thesis, Universidad Nacional Autónoma de México, 1994.

5. M. C. Salcedo L., B. Sc. thesis, Benemérita Universidad Autónoma de Puebla, México,1995.

6. De León G. A. M., M. Sc. thesis, Universidad Nacional Autónoma de México, 1993.

SYNTHESIS OF $GaSr_2Ca_{n-1}Cu_nO_{2n+3}$ $(n = 2 - 5)$ PREPARED BY SOLID-STATE REACTION METHOD

E. CHAVIRA, E. AGUILA, L. BAÑOS, and O. NAVARRO.
Instituto de Investigaciones en Materiales, U.N.A.M., Apartado Postal 70-360, 04510, México D.F., MEXICO, chavira@servidor.unam.mx

ABSTRACT

We present some experimental results of the family of compounds with a general formula $GaSr_2Ca_{n-1}Cu_nO_{2n+3}$ $(n = 2, 3, 4, 5)$. This system has been synthesized using the solid-state reaction method at ambient pressure. We have found that the reaction temperature for these compounds is $930 \pm 3^{\circ}C$; the samples melt just above this temperature. The powder diffraction data were analyzed assuming a single-phase and the obtained structure is different to the one prepared at high pressure [1,2]. However, comparing the powder patterns we found that the compounds are isostructural to the $Al - Sr - Ca - Cu - O$ system prepared at high pressure [3]. For $n = 2, 3, 4$ and 5 a shift of the (105) peak has been observed, which is due to the increases in the number of Ca and $Cu - O_2$ planes in the crystalline structure. Notice that to obtain a single-phase for the above compounds, high pressure is not necessary.

INTRODUCTION

The copper oxide compounds with superconducting properties [4-7] have a structure characterized by the presence of $Cu - O_2$ layers. These layers are bonded in the third dimension by metal-oxygen layers (MO) containing large and strongly ionic metal ions, which act as charge reservoirs in the structural arrangement of $MO - CuO_2 - MO$ blocks. Some of these oxide compounds have the highest critical temperature observed up today.

A family of compounds $GaSr_2Ca_{n-1}Cu_nO_{2n+3}$ $(n = 3, 4)$ was studied by Takayama-Muromachi et al. [1] at high-pressure annealing and were assumed to have orthorhombic lattices. A series of related compounds $AlSr_2Ca_{n-1}Cu_nO_{2n+3}$ $(n = 4, 5)$ was prepared also at high-pressure by Isobe et al. [3], obtaining a tetragonal cell. An interesting feature found in the Al-based compounds is their relatively high-T_c; the highest record for the $AlSr_2Ca_3Cu_4O_{11}$ composition is $T_c = 110K$ [3]. These facts motivate us to explore the structure of similar compounds such as the Ga-based system. In the present study, we successfully prepared the compounds with a general formula $GaSr_2Ca_{n-1}Cu_nO_{2n+3}$ $(n = 2, 3, 4, 5)$ at ambient pressure. From X-ray data for the stable phase of this system, tetragonal cell is observed which is isostructural to the Al-based compounds.

EXPERIMENT

Samples were synthesized using the method of solid state reaction. Prior to weighing, the starting materials $(Ga_2O_3, SrCO_3, CaCO_3$ and $CuO)$ were dehydrated, and the purity was confirmed by X-ray powder diffraction (XRD). The reagents were mixed in an agate mortar and fired in high-density alumina crucibles to obtain mixtures having the following nominal compositions: $GaSr_2CaCu_2O_7$ (1:2:1:2), $GaSr_2Ca_2Cu_3O_9$ (1:2:2:3),

Mat. Res. Soc. Symp. Proc. Vol. 453 © 1997 Materials Research Society

$GaSr_2Ca_3Cu_4O_{11}$ (1:2:3:4) and $GaSr_2Ca_4Cu_5O_{13}$ (1:2:4:5). The decomposition of the oxysalts were carried out in situ at $860°C$ for 12h. In order to determine if the carbonates were present or if intermediates phases were formed we tested the samples by weighing and using XRD. After the heat treatment at $860°C$, the temperature of the mixtures were increasing gradually by steps of $10°C$ to allow solid state reaction to continue. Slow cooling till room temperature with intermediate grinding was done for each step. In preliminary synthesis experiments we found that the lower reaction temperature is $900°C$, but the best results are obtained around $900°C - 930°C$ during 48h. Increasing the temperature $20°C$ or $30°C$ above $930°C$, depending on composition, causes the samples to melt.

Phase identification of the resulting polycrystalline samples was performed by means of X-ray powder diffraction with $Cu\ K_{\alpha 1}$ radiation, using a SIEMENS D5000 diffractometer equipped with a graphite monochromator on the counter side and an auto-divergence-slit system. This equipment is able to detect up to a minimum of 1% of impurities. The d-spacing and the lattice constants were determined by a least-squares method, for accurate measurements KCl ($a = 6.2931Å$) was added as an internal standard.

RESULTS AND DISCUSSION

Single-phase of homogeneous samples were obtained for the $GaSr_2Ca_{n-1}Cu_nO_{2n+3}$ ($n = 2, 3, 4, 5$) system at ambient pressure. The structure of the stable phase in this system is tetragonal, i.e., it is different from the orthorhombic structure reported by Takayama-Muromachi et al. [1] and Khasanova et al. [2] for the same system (samples with $n = 3$, 4) prepared at high pressure. However, comparing the powder pattern of the $GaSr_2Ca_2Cu_3O_9$ compound (see Figure 1), it is observed that the diffractogram is isostructural to the one for $AlSr_2Ca_2Cu_3O_9$ composition, which has a tetragonal cell and was obtained at high pressure [3].

Comparing the spectrum given in Figure 1 for the $GaSr_2Ca_2Cu_3O_9$ compound with the one reported by Isobe et al. [3] for $AlSr_2Ca_2Cu_3O_9$ composition, we note that some of the lines in the Ga-based diffractogram were shifted to higher angles (2Θ), due to the substitution of Al-cation by Ga-cation. In Table 1, we show the principal lines of the powder X-ray pattern for $GaSr_2Ca_2Cu_3O_9$ compound. In this Table we have indexed only the d-values of the Al-based system obtained by Takayama-Muromachi et al. [1].

$d_{obs.}(Å)$	$d_{calc.}(Å)$	hkl	I	$d_{obs.}(Å)$	$d_{calc.}(Å)$	hkl	I
4.9183	4.9223	0 0 1	21	2.0314	2.0354	0 0 9	33
4.8860	4.8900	0 0 3	1	1.9434	1.9474	2 0 0	57
4.0701	4.0741	0 0 4	27	1.7970	1.8010	0 0 10	37
3.4433	3.4473	0 0 5	26	1.7827	1.7867	0 0 11	54
3.1204	3.1244	1 0 3	23	1.6282	1.6322	2 1 5	38
2.9619	2.9659	0 0 6	62	1.4997	1.5037	2 1 6	28
2.8111	2.8151	1 0 4	75	1.4038	1.4078	1 1 11	25
2.5870	2.5910	1 1 0	100	1.3755	1.3795	2 2 0	31
2.3938	2.3978	1 0 5	35				

Table 1. X-ray powder diffraction data of the $GaSr_2Ca_2Cu_3O_9$ composition.

A set of representative X-ray diffraction patterns are shown in figure 2 for $n = 4$ and

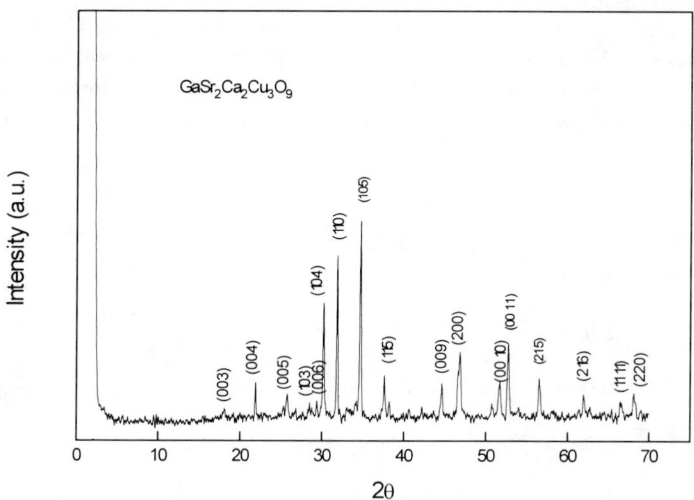

Figure 1: X-ray powder patterns for Ga-1:2:2:3. Main peaks are assigned to a tetragonal cell with a=3.8948A and c=19.6537A.

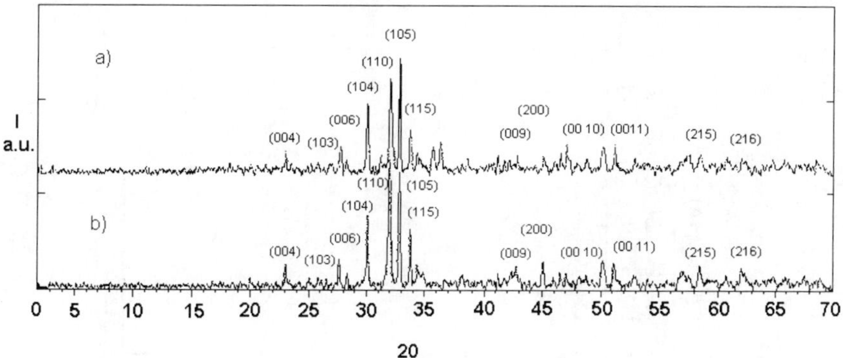

Figure 2: X-ray diffraction patterns for a) Ga-1:2:3:4, and b) Ga-1:2:4:5.

5 in the $GaSr_2Ca_{n-1}Cu_nO_{2n+3}$ system. The intensity for $n = 4$ is normalized to the most intense (105) peak, the same has been done for $n = 5$ but here the most intense peak is the (110) peak. Increasing the n value shifts each peak slightly implying that the tetragonal symmetry is preserved. In general, for $n = 2, 3, 4$ and 5, there is a shift of the peaks due to the increases in the number of Ca and $Cu - O_2$ planes in the crystalline structure.

The lattice parameters for $GaSr_2Ca_{n-1}Cu_nO_{2n+3}$ system are plotted as a function of n in figure 3 and figure 4. Comparing these lattice parameters with those of the

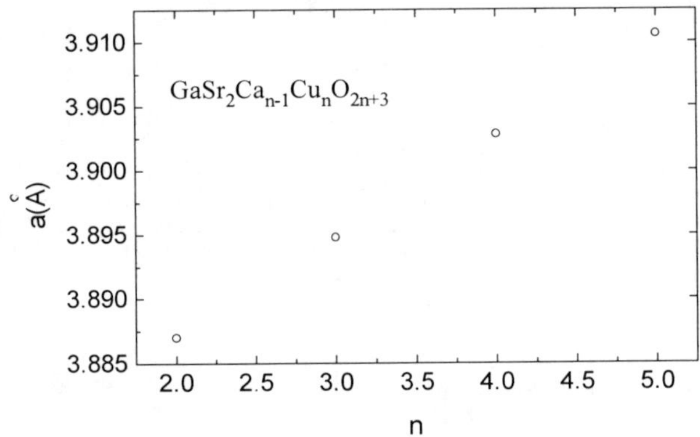

Figure 3: Variation of a-lattice parameter with n in Ga-based system.

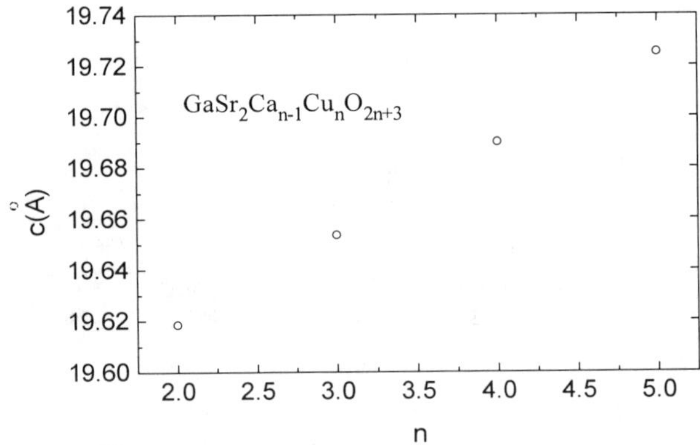

Figure 4: Variation of the c-lattice parameter with n in Ga-based system

$AlSr_2Ca_{n-1}Cu_nO_{2n+3}$ compound for $n = 4$ ($a = 3.839\AA$ and $c = 17.72\AA$) and for $n = 5$ ($a = 3.845\AA$ and $c = 20.87\AA$) [3], we observe that $a = 3.903\AA$ ($n = 4$) is longer whereas $c = 19.690\AA$ is much longer, but in $n = 5$, $a = 3.911\AA$ is slightly longer and $c = 19.725\AA$ is shorter. The latter behavior could be attributable to a replacement of the Ga ion in the $Cu - O_2$ planes. However, to make definite determination more experimental work is necessary.

The change in the cell volume is represented in figure 5. As Ca and Cu content increases, the volume increases monotonically.

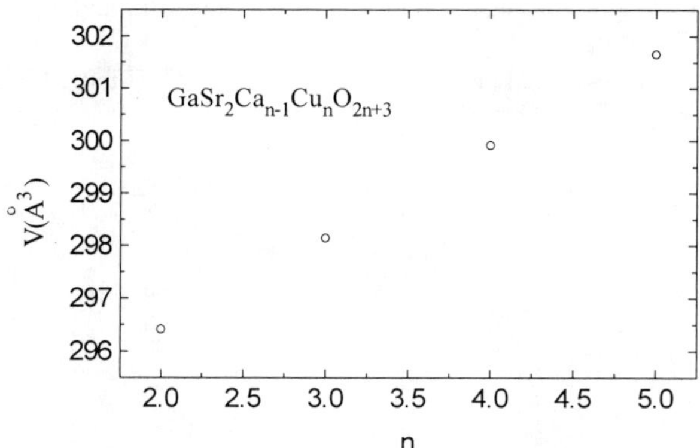

Figure 5: Change of the cell volume vs. n content.

CONCLUSIONS

We established the thermodynamic stability of the $GaSr_2Ca_{n-1}Cu_nO_{2n+3}$ system ($n = 2, 3, 4$, and 5), synthesized at ambient pressure with a reaction temperature of $930 \pm 3°C$. By X-ray powder diffraction for a single phase of this Ga-system, it is determined that this system is isostructural to the one for $AlSr_2Ca_2Cu_3O_9$ composition, which has a tetragonal cell and was obtained at high pressure. We can say that to obtain a single-phase for the above Ga compounds, high pressure is not necessary.

ACKNOWLEDGMENTS

We want to thank R. Escamilla and L. F. Pérez for technical support. This work was partially supported by CONACyT 2661P-A9507, and by the DGAPA-UNAM Grants IN102196 and IN-104595.

REFERENCES

[1] E. Takayama-Muromachi and M. Isobe, Jpn. J. Appl. Phys. **33** (1994) L1399.

[2] N.R. Khasanova, F. Izumi, E. Takayama-Muromachi, and A. W. Hewat, Physica C **258** (1996) 227.

[3] M. Isobe, T. Kawashima, K. Kosuda, Y. Matsui, and E. Takayama-Muromachi Physica C **234** (1994) 120.

[4] J. G. Bednorz and K. A. Muller, Z, Phys. B **64** (1986) 189.

[5] M. H. Wu, J. R. Aschburn, C. J. Tory, P. H. Hor, R. C. Meng, L. Gao, Z. H. Huang, Y. Q. Wang ans C. W. Chu, Phys. Rev. Lett. **58** (1987) 908.

[6] C. Michel, J. Provost, F. Deslandes, B. Raveau, J. Beille, R. Cabanel, P. Lejay, A. Sulpice, J. L. Tholerance, R. Tournier, B. Chevallier, G. Demazeau, and J. Etourneau, Z. Phys. B **68** (1987) 417.

[7] S. S. Parkin, V. Y. Lee, E. M. Engler, A. I. Nazzal, T. C. Huang, G. Gorman, R. Savoy, and R. Beyer, Phys. Rev. Lett. **60** (1988) 2539.

CONTROLLING DEFECTS IN DOUBLE-LAYER CUPRATES BY CHEMICAL MODIFICATIONS

P. A. SALVADOR *, K. B. GREENWOOD §, K. OTZSCHI §, J. W. KOENITZER §, B. M. DABROWSKI ‡, K. R. POEPPELMEIER §, T. O. MASON *†
* Materials Science and Engineering Department, Northwestern University, 2225 N. Campus Dr., Evanston, IL 60208-3108
§ Department of Chemistry, Northwestern University, Evanston IL 60208
‡ Department of Physics, Northern Illinois University, DeKalb, IL 60115
† Author to whom correspondence should be addressed.

ABSTRACT

In-situ high temperature electrical conductivity and thermopower have been measured simultaneously on a number of ordered perovskite-like oxides containing double $CuO_{4/2}$ sheets. Equilibrium measurements have been conducted as a function of oxygen partial pressure, temperature and chemical substitution in order to understand the relationships between the chemical architecture and the transport and defect properties. Data for $LaBa_2Cu_2NbO_8$ and $LaCa_2Cu_2GaO_7$ are presented and compared with those of known triple perovskite superconductors, $Y_{1-x}Ca_xSr_2Cu_2GaO_7$ and $YBa_2Cu_3O_{7-\delta}$, and several quadruple perovskites, $Ln'Ln''Ba_2Cu_2M_2O_{11}$ (Ln = Lanthanide, Y; M = Sn, Ti). These materials belong to a general family of superconductors which are constructed from similar 'active' layers (double perovskite blocks of square-pyramidal copper-oxygen sheets), and interleaved with fixed valence cations in perovskite-like 'conditioning' layers. Similarities in the transport properties of the non-superconducting and superconducting materials at elevated temperatures are illustrated, and the amount and types of defects, including carrier concentrations, are correlated with the internal chemistry and inner architecture of each material.

INTRODUCTION

We have been investigating the relationships between the inner architecture, internal chemistry, and physical properties of layered copper oxides, and how these relationships can be used to rationally design materials with specific properties [1-4]. This communication reports on these relationships in superconductors and potential superconductors with triple and quadruple perovskite structures (see Figure 1). Emphasis is on the influence that the chemistry of the constituent cations have on the physical properties of these materials.

The structures of high-temperature superconducting cuprates can be constructed by stacking distinct structural elements along a single axis, leading to highly two-dimensional structural features and properties. The structural element which is the 'active' layer for superconductivity is a perovskite block containing copper as the sole B-cation, of stoichiometry $(ACuO_{3-x})_m$ (where m is an integer denoting the number of perovskite blocks in this layer). This 'active' layer (AL) is interleaved by a variety of other structural elements (including perovskite, fluorite, and rock-salt slabs) which we denote as the 'conditioning' layer (CL), owing to its influence over the superconducting properties via charge transfer mechanisms, internal stresses, and defect incorporation. One of the earliest and best known superconductors is $YBa_2Cu_3O_{7-x}$ ($T_c \approx 92$ K [5, 6]), which has a triple oxygen-deficient perovskite structure, as illustrated in Figure 1(a). The AL of this material is comprised of the now familiar double-perovskite block with closely-spaced, facing sheets of copper square pyramids. Interleaved between this AL block is a single oxygen-deficient perovskite CL, also containing copper as the B-cation, in a coordination varying from linear (x = 1) to square planar(x = 0) and a valence state nominally ranging from +1 to +3.

In order to investigate the underlying properties of this double perovskite AL motif, without complications from the CL (such as nonstoichiometry or valence changes) we have investigated the crystal and defect chemistry of structurally related materials having perovskite-like CLs which contain fixed valence cationic constituents. The structures of several of these materials are illustrated in Figure 1(b-d), where it is immediately apparent that the AL copper coordination is quite similar in each compound. Figure 1(b) depicts the structure of $Y_{1-x}Ca_xSr_2Cu_2GaO_7$ ($T_c \approx 40K$) [7] and $LaCa_2Cu_2GaO_7$ [8], which have a single oxygen-deficient perovskite-like CL, with gallium in tetrahedral coordination. Figure 1(c) depicts the structure of $LaBa_2Cu_2NbO_8$, which has a single perovskite-like CL having niobium in octahedral coordination [9, 10]. Figure 1(d) illustrates the

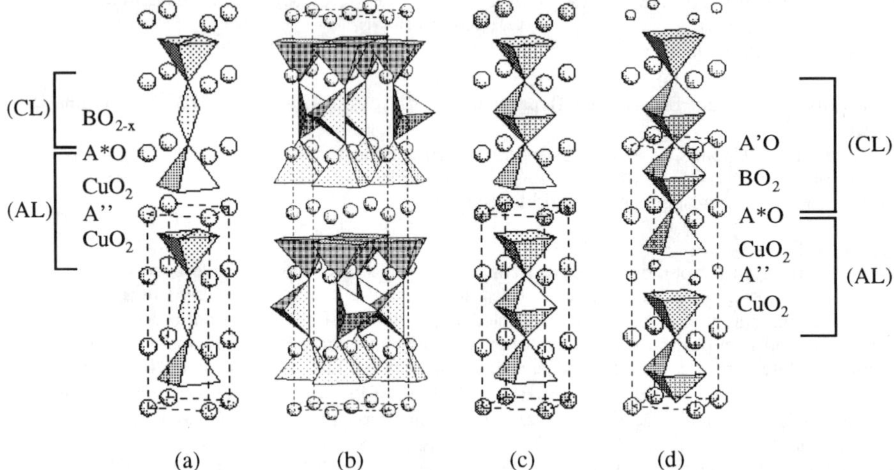

Figure 1. Polyhedral representations of several 'pure'-perovskite layered cuprates: (a) YBa$_2$Cu$_3$O$_{7-x}$ [6], (b) $LnAE_2$Cu$_2$GaO$_7$ [7], (c) $LnAE_2$Cu$_2$NbO$_8$ [10], and (d) $Ln'Ln''$Ba$_2$Cu$_2$Ti$_2$O$_{11}$ [2]. In all cases, the spheres denote the A-site perovskite cations (lanthanides and Y {Ln}; alkaline earths {AE}), the B-cations are centered in the polyhedra, and oxygen are at the apexes of the polyhedra. (AL) denotes the 'active' layer for superconductivity, and in these materials is comprised of a double oxygen deficient block of copper-based perovskite units. (CL) denotes the 'conditioning' layer which is comprised of one or two perovskite type blocks with various B-cations and coordination polyhedra.

structure of $Ln'Ln''$Ba$_2$Cu$_2$Ti$_2$O$_{11}$ (Ln = lanthanide, Y) [2, 11, 12], which has a CL that consists of a double perovskite block having titanium in octahedral coordination. Materials which adopt the latter two structures have not exhibited superconductivity, despite their similarities to known superconductors.

We have previously reported on the structural features and physical properties of several of these materials, where it is apparent that both the AL and CL play an important role in determining the ultimate physical properties [1, 2, 4, 7, 11, 13-15]. In this communication we report high temperature electrical characterization of LaCa$_2$Cu$_2$GaO$_7$ and LaBa$_2$Cu$_2$NbO$_8$, which are compared with prior results for YBa$_2$Cu$_3$O$_{7-x}$, Y$_{1-x}$Ca$_x$Sr$_2$Cu$_3$$_2GaO_{7-x}$, and La$_2Ba_2Cu_2Ti_2O_{11}$. Chemical control over the stable defect species is possible with site-selective cationic substitutions in the AL and CL blocks, and have important consequences for the superconducting properties.

EXPERIMENTAL SECTION

All samples were prepared by solid state reaction of appropriate stoichiometric mixtures of LnO$_{1.5}$ (Ln = lanthanide, Y), AECO$_3$ (AE = Ca, Sr, Ba), CuO, TiO$_2$, Ga$_2$O$_3$, and Nb$_2$O$_5$, all of which are of purity above 99.99%. Synthesis conditions are similar to those described elsewhere [2, 3, 8, 10]. Powder X-Ray diffraction patterns were collected with a Rigaku diffractometer using CuKα radiation to confirm phase-purity [8, 10]. Simultaneous 4-point DC conductivity and thermopower measurements were performed on select samples using methods described elsewhere [1, 4, 16].

RESULTS

High temperature electrical conductivity and thermopower measurements of LaCa$_2$Cu$_2$GaO$_7$ and LaBa$_2$Cu$_2$NbO$_8$ were measured over a range of temperatures and oxygen pressures (see Figure 2 for exact conditions). The thermopower is plotted as a function of the natural logarithm of the conductivity ('Jonker' plots [17])in Figure 2 for these two materials. Also included in Figure 2 are data of the triple perovskites YBa$_2$Cu$_3$O$_{7-x}$ and YSr$_2$Cu$_2$GaO$_7$, and the quadruple perovskite La$_2$Ba$_2$Cu$_2$Ti$_2$O$_{11}$. The details of this analysis technique and its applicability to the parent insulating compounds of high-T_c cuprates are given elsewhere [1, 4, 13, 15, 17]. The position of the data on the 'Jonker' curve, the portion of the curve traced out as a function of chemical substitutions or oxygen partial pressure, and the size of the curve are the preeminent features which reveal the nature of a material's internal chemistry [4]. Moreover, the transport parameters can be determined from data spanning only a small region of the Jonker curve. The characteristic shape of this curve is demonstrated by the dashed lines through the data for $La_2Ba_2Cu_2Ti_2O_{11}$ [1] and $LaBa_2Cu_2NbO_8$ in Figure 2. Only the positive thermopower axis is included in these plots, owing to the p-type behavior of these layered cuprates. Superconducting compositions have values near the high conductivity intercept, as illustrated by $YBa_2Cu_3O_{7-x}$ in oxygen at 650 °C.

Both $LaCa_2Cu_2GaO_7$ and $LaBa_2Cu_2NbO_8$ exhibit p-type semiconductivity (positive thermopowers) at elevated temperatures, similar to the other materials displayed in Figure 2. The data for $LaCa_2Cu_2GaO_7$ fall on the extrinsic 'leg' of the Jonker curve, having the characteristic linear dependence of the thermopower on the natural logarithm of the conductivity (at any temperature). This implies that holes are the dominant carriers over the entire experimental range. In contrast, the data for $LaBa_2Cu_2NbO_8$ falls on the intrinsic side (dual carrier) of the pear, indicating that electrons are also contributing to the electrical properties of this material, albeit as minority carriers. The thermal generation of carriers in the niobate is evidenced in the temperature-dependent lower conductivity intercept, also observed for $La_2Ba_2Cu_2Ti_2O_{11}$. The temperature-independent high conductivity intercepts, for both $LaCa_2Cu_2GaO_7$ and $LaBa_2Cu_2NbO_8$, are consistent with band conduction, or small polaron conduction with activation energies $\approx kT$.

Appreciable portions of the Jonker curve are traced out for both $LaCa_2Cu_2GaO_7$ and $LaBa_2Cu_2NbO_8$ as a function of oxygen pressure. The data for $LaCa_2Cu_2GaO_7$ traces out a large portion of the extrinsic p-type leg, in stark contrast to that for the isostructural material, $YSr_2Cu_2GaO_7$. Furthermore, $LaCa_2Cu_2GaO_7$ has a much lower thermopower and higher conductivity, indicating that the carrier concentration is much greater than that in $YSr_2Cu_2GaO_7$. The absolute magnitude of the thermopower in $LaCa_2Cu_2GaO_7$, which is nominally undoped, is more similar to the 10% calcium doped compound, than to the undoped compound in the $Y_{1-x}Ca_xSr_2Cu_2GaO_7$ system. While the extrinsic carrier concentration in $LaCa_2Cu_2GaO_7$ (as

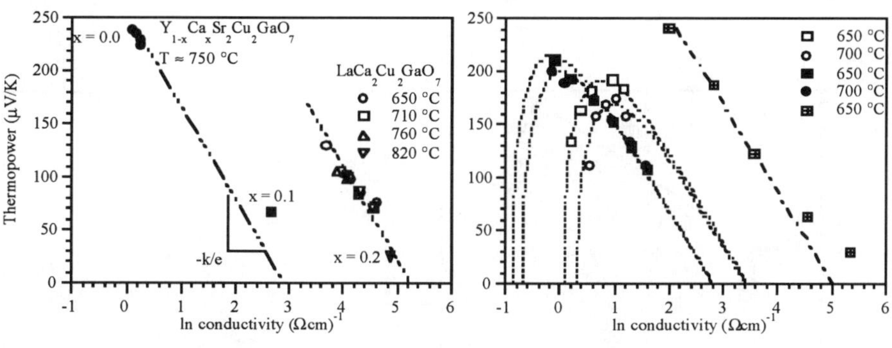

(a)   (b)

Figure 2. Plots of the thermopower versus the natural logarithm of the conductivity for several layered cuprates. Compositions are denoted in (a), while in (b), the unfilled symbols are for $LaBa_2Cu_2NbO_8$, filled symbols for $La_2Ba_2Cu_2Ti_2O_{11}$ (from Ref. [1]), and cross-square symbols for $YBa_2Cu_3O_{7-x}$ (from Ref. [13]). The highest conductivity data point, for any composition and temperature, corresponds to an oxygen pressure of 1 atm. Subsequent points correspond to decreases, by an order of magnitude, of oxygen partial pressure moving counterclockwise around the curve. The -k/e (k is Boltzman's constant and e the unit of electronic charge) slope seen in all these curves that expected for an extrinsic p-type semiconductor.

reflected in the extent of the curve traversed) is strongly dependent on the oxygen pressure, it is weakly dependent on the temperature, as was also observed in the extrinsic data of $La_2Ba_2Cu_2Ti_2O_{11}$. These observations are consistent with intercalation of oxygen between the square pyramidal sheets in the AL layer, resulting in oxidation of the CuO_2^{2-} sheets [1]. In contrast to this AL intercalation, the large dependence of the electrical properties on the oxygen partial pressure for $YBa_2Cu_3O_{7-x}$, and on temperature (not shown) [13, 18], is associated with intercalation of oxygen in the CL, illustrating that the chemical differences of the B-site CL cation can be used to control the defect population of this layer.

We can further demonstrate the difference between the isostructural systems, $LaCa_2Cu_2GaO_7$ and $YSr_2Cu_2GaO_7$, by plotting a modified ('reduced') thermopower value as a function of the 'normalized' doping level (described below). A simple mathematical factoring of the thermopower equation, for an extrinsic p-type semiconductor, results in an expression directly proportional to the logarithm of the hole concentration. This expression, termed the reduced thermopower, is given by:

$$Q_R = -(\frac{e}{2.303k})Q = \log p - \log Ne^A \tag{1}$$

where Q is the thermopower, p is the hole concentration, N is the density of states, and A is the transport constant. The 'normalized' doping level is the carrier concentration, or aliovalent doping concentration, per copper plane per formula unit. Thus, in the $Y_{1-x}Ca_xSr_2Cu_2GaO_7$ series, the 'normalized' doping level is x/2. Superconducting layered cuprates exhibit very similar 'normalized' transport properties. Because the reduced thermopower values are quite similar as a function of normalized dopant concentrations, we can infer that the normalized values of the density of states and transport constant (the term on the far right of Eq. 1) are also similar in the parent semiconducting cuprates[15, 19, 20]. Furthermore, we can plot an 'expected' value line for the reduced thermopower versus dopant concentration (based on analysis of data for $Y_{1-x}Ca_xSr_2Cu_2GaO_7$ [15] and $La_{2-x}AE_xCuO_{4\pm\delta}$ [19, 20]), and thereby estimate the dopant levels in layered cuprates when unknown. Examples are given for several materials in Figure 3.

The calcium doped $Y_{1-x}Ca_xSr_2Cu_2GaO_7$ series (Figure 3a) lies on the 'expected' value line of extrinsic layered cuprates [15]. The carrier concentrations in the nominally undoped parent compounds, which exhibit extrinsic semiconducting behavior, can be estimated. $LaCa_2Cu_2GaO_7$ has

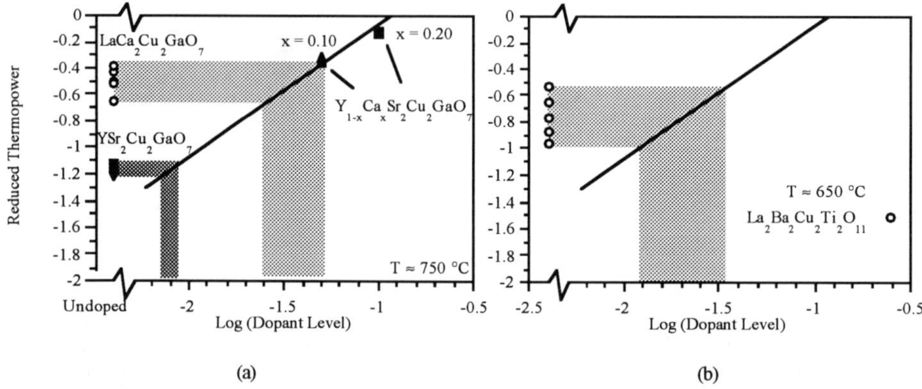

Figure 3. Plots of the reduced thermopower versus the logarithm of the normalized dopant concentration for several perovskite layered cuprates (see text for references). The dopant level is normalized to the number of copper layers per formula unit. The line of slope 1 corresponds to the 'expected' value for layered cuprates (see text) and is identical in each plot.

an oxygen pressure-dependent carrier concentration which varies from 0.025-0.05 as the oxygen pressure is varied from 10^{-4} to 1 atm. This is consistent with 0.025-0.050 excess oxygen incorporated in the AL, as has been observed in the related material (identical AL), $La_2CaCu_2O_{6\pm\delta}$, when partial lanthanum occupancy of the A" site occurs [21]. In contrast, the normalized doping level in undoped $YSr_2Cu_2GaO_7$ system is only 0.007-0.008, much smaller in both absolute magnitude and in the variation as a function of oxygen partial pressure. This is consistent with the small yttrium cation occupying the A" site in the AL, excluding interstitial oxygen in this region.

Chemical control over the defect species in the active layer can also be demonstrated in the quadruple perovskites by site-selective substitutions on the A" site. When the large lanthanides reside on the A" site, as in $Ln_2Ba_2Cu_2Ti_2O_{11}$ (Ln = La, Eu), oxygen intercalation has been observed [1], as is illustrated in Figure 3b. The approximate carrier concentrations of $La_2Ba_2Cu_2Ti_2O_{11}$ vary from 0.01-0.03 holes per copper plane, corresponding to 0.01-0.03 extra oxygen in the active layer. Recently, appreciable suppression of this intercalation mechanism has been demonstrated in $LaYBa_2Cu_2Ti_2O_{11}$ [4], in which the smaller yttrium cation preferentially occupies the A" site. Chemical control over ionic defects in the active layer is possible through appropriate chemical design.

Returning to the Jonker behavior of $LaBa_2Cu_2NbO_8$, the beginning of a transition from intrinsic to extrinsic behavior is observed as the oxygen partial pressure is increased. This behavior is similar to that of the quadruple perovskites, $La_2Ba_2Cu_2Ti_2O_{11}$ and $Eu_2Ba_2Cu_2Ti_2O_{11}$ [1]. However, the quadruple perovskites exhibit a greater degree of extrinsic behavior. Owing to this transitional behavior, a complete Jonker 'curve' can be fit to these data and apparent band gaps determined for these two temperatures. The values for the gap energies (assuming phonon scattering is dominant [1, 13]) are 0.33 eV (650 °C) and 0.30 eV (700 °C). These gap energies are consistent the values observed for the quadruple perovskites, and are intermediate between $La_2Ba_2Cu_2Ti_2O_{11}$ and $La_2Ba_2Cu_2Sn_2O_{11}$ [1]. The in-plane Cu-O bond lengths are approximately 1.97 Å and 2.01 Å, respectively for the latter two materials, while they are $\approx$ 1.99 Å in $LaBa_2Cu_2NbO_8$. This suggests that the band gaps are inversely scaling with the in-plane bond length for this AL structural motif. The triple perovskites which have a gallium based CL are also in agreement with this trend, although the exact gap energies cannot be determined since we lack data in the intrinsic region. However, the large extrinsic thermopower value observed for $YSr_2Cu_2GaO_7$ requires that the band gap be larger than any of the materials which exhibit intrinsic behavior at the same temperature, consistent with its in-plane bond length of $\approx$ 1.94 Å [7]. The CL exerts a strong influence over the in-plane bond length, through the size-mismatch of the different structural blocks, and therefore the gap energies can be controlled by choosing the CL block to correspond to a specific bond length. CLs which lead to smaller bond lengths result in larger gaps, indicating a greater splitting of the Hubbard bands in the parent insulating materials as the copper-oxygen overlap becomes greater. Furthermore, the smaller bond length materials stabilize higher carrier concentrations, as demonstrated by the high carrier concentrations in the $LaCa_2Cu_2GaO_7$, which has an average in-plane bond length of $\approx$ 1.92 Å [8].

CONCLUSIONS

The high-temperature electrical property data for $LaCa_2Cu_2GaO_7$ and $LaBa_2Cu_2NbO_8$ are extremely similar to those of other layered cuprates [1, 4, 13-15, 18]. Both materials exhibit electrical properties which are significantly affected by variations in the oxygen partial pressure. Owing to the preference of the CL cationic species to have fixed valence states and coordination environments, the CL in both of these materials can be considered oxygen-stoichiometric, and therefore cannot account for this oxygen partial pressure sensitivity. The dependence of the electrical properties on the oxygen pressure and temperature is consistent with intercalation of oxygen between the double copper oxygen sheets of this AL motif. Moreover, the band gaps of these semiconducting materials scale inversely as the in-plane copper-oxygen bond length.

The preeminent concern of this investigation is understanding the factors which influence oxidation of the CuO_2^{2-} sheets in layered cuprates. Several aspects of the internal chemistry, such as the intercalation behavior, oxidation behavior, and defect chemistry, are closely linked to features of the inner architecture, such as constituent cations/anions, coordination environments, internal stresses, and bond lengths. The ultimate superconducting properties are to a large extent controlled by the ability of a structure to maintain a single coordination environment around the 'active' layer copper-oxygen sheets (see Figure 1) and to support an appropriate amount of charge carriers in these sheets. Intercalation of oxygen in the AL is deleterious to superconductivity as it perturbs the 2D copper-oxygen sheets. This can be chemically controlled by the preferential ordering of small cations on the A-site between these sheets. The CL-AL mismatch directly affects the band gap, or Hubbard splitting, of the copper-oxygen sheets and therefore has a direct influence on the ability of these sheets to be doped. Smaller bond-lengths are more amenable to oxidation and thereby more amenable to superconductivity. Thus, the realization of superconductivity in non-superconducting layered cuprates, such as the series $Ln'Ln''AE_mCu_2Ti_mO_{5+3m}$ [3] , requires attention to the inner architecture of both the AL and CL.

ACKNOWLEDGMENTS

This work was supported by the National Science Foundation (Award No. DMR-91-20000) through the Science and Technology Center for Superconductivity and made use of MRL Central Facilities supported by the National Science Foundation, at the Materials Research Center of Northwestern University (Award No. DMR-9120521). One of the authors (KO) is partly supported by a Research Fellowship of the Japan Society for the Promotion of the Science for Young Scientists.

REFERENCES

1. P. A. Salvador, L. Shen, T. O. Mason, K. B. Greenwood and K. R. Poeppelmeier, J. Solid State Chem. **119**, p. 80 (1995).
2. K. B. Greenwood, G. M. Sarjeant, K. R. Poeppelmeier, P. A. Salvador, T. O. Mason, B. Dabrowski, K. Rogacki and Z. Chen, Chem. Mater. **7**, p. 1355 (1995).
3. K. D. Otzschi, K. R. Poeppelmeier, P. A. Salvador, T. O. Mason, H. Zhang and L. D. Marks, J. Am. Chem. Soc. **118**, p. 8951 (1996).
4. P. A. Salvador, T. O. Mason, K. Otzschi, K. B. Greenwood, K. R. Poeppelmeier and B. Dabrowski, J. Am. Chem. Soc. (Submitted-1996).
5. M. K. Wu, J. R. Ashburn, C. J. Torng, P. H. Hor, R. L. Meng, L. Gao, Z. J. Huang, Y. Q. Wang and C. W. Chu, Phys. Rev. Lett. **58**, p. 908 (1987).
6. M. A. Beno, L. Soderholm, I. D. W. Capone, D. G. Hinks, J. D. Jorgensen, J. D. Grace, I. K. Schuller, C. U. Segre and K. Zhang, Appl. Phys. Lett. **51**, p. 57 (1987).
7. J. T. Vaughey, J. P. Thiel, E. F. Hasty, D. A. Groenke, C. L. Stern, K. R. Poeppelmeier, B. Dabrowski, D. G. Hinks and A. W. Mitchell, Chem. Mater. **3**, p. 935 (1991).
8. J. W. Koenitzer, K. B. Greenwood, K. R. Poeppelmeier, P. A. Salvador, T. O. Mason and B. Dabrowski, (Manuscript In Preparation).
9. N. Murayama, E. Sudo, K. Kani, A. Tsuzuki, S. Kawakami, M. Awano and Y. Torii, Jpn. J. Appl. Phys. **27**, p. L1623 (1988).
10. C. Greaves and P. R. Slater, Physica C **161**, p. 245 (1989).
11. M. T. Anderson, K. R. Poeppelmeier, J. P. Zhang, H.-J. Fan and L. D. Marks, Chem. Mater. **4**, p. 1305 (1992).
12. A. Gormezano and M. T. Weller, J. Mater. Chem. **3**, p. 771 (1993).
13. M.-Y. Su, C. E. Elsbernd and T. O. Mason, J. Am. Ceram. Soc. **73**, p. 415 (1990).
14. M.-Y. Su, C. E. Elsbernd and T. O. Mason, Physica C **160**, p. 114 (1989).
15. G. W. Tomlins, N.-L. Jeon, T. O. Mason, D. A. Groenke, J. T. Vaughey and K. R. Poeppelmeier, J. Solid State Chem. **109**, p. 338 (1994).
16. A. Trestman-Matts, S. E. Dorris and T. O. Mason, J. Am. Ceram. Soc. **66**, p. 589 (1983).
17. G. H. Jonker, Philips Res. Repts **23**, p. 131 (1968).
18. M.-Y. Su, S. E. Dorris and T. O. Mason, J. Solid State Chem. **75**, p. 381 (1988).
19. L. Shen, P. A. Salvador and T. O. Mason, J. Am. Ceram. Soc. **77**, p. 81 (1994).
20. L. Shen, P. A. Salvador and T. O. Mason, J. Phys. Chem. Solids **57**, p. 1311 (1996).
21. A. Fuertes, X. Obradors, J. M. Navarro, P. Gómez-Romero, N. Casañ-Pastor, F. Pérez, J. Fontcuberta, C. Miravitlles, J. Rodríguez-Carvajal and B. Martínez, Physica C **170**, p. 153 (1991).

STABILITY OF IONIC TIN(II) IN A CHLORIDE FLUORIDE MATRIX: THE UNEXPECTED $Ba_{1-x}Sn_xCl_{1+y}F_{1-y}$ SOLID SOLUTION CRYSTALLIZING IN THE BaClF STRUCTURE

Georges DENES, Julie BRATIGNY, David LE ROUX and Abdualhafeed MUNTASAR
Laboratory of Solid State Chemistry and Mössbauer spectroscopy, Laboratories for Inorganic Materials, Department of Chemistry and Biochemistry, Concordia University, Montreal, Quebec, H3G 1M8, Canada, gdenes@vax2.concordia.ca

ABSTRACT

We have prepared a wide $Ba_{1-x}Sn_xCl_{1+y}F_{1-y}$ solid solution ($0<x<0.25$, $-0.20<y<0.20$) that has the BaClF structure without any sign of lattice distortion or ordering of the cations or of the anions other than that already present in BaClF. Furthermore, ^{119}Sn Mossbauer spectroscopy shows that tin(II) is present in this material in the form of Sn^{2+} ions instead of being covalently bonded Sn(II) with a stereoactive lone pair of electrons, which is nearly always the case when the anions are small and strongly electronegative, like for F and Cl. The tin situation is similar to that observed in $SnCl_2$.

INTRODUCTION

In a fluoride or a chloride matrix, divalent tin nearly all the time forms covalent bonds to the surrounding halogens. The tin(II) valence orbitals are hybridized, and the tin non-bonding electron pair (lone pair) is located on one of the hybrid orbitals. Such a lone pair is called *stereoactive* since it occupies one of the apeces of the polyhedron of coordination of tin, taking approximately the same room as a fluoride or an oxide ion in the solid structure [1]. The lone pair distorts the polyhedron of coordination in the first coordination sphere, similarly to the case of individual molecules, in agreement with the VSEPR rules [2]. However, in a solid, one must take into account the atoms beyond the first sphere of coordination. The tin(II) coordination would be a simple highly symmetric polyhedron (e.g. an octahedron, with F or O) if the lone pair did not distort it. In practice, the lone pair keeps some number of potential neighbors away, outside bonding range, and the actual coordination of tin(II) is lowered [1, 3]. For large and more mildly electronegative anions, such as Br, I, Se and Te, tin(II) is present in the form of the Sn^{2+} stannous ion and it forms ionic bonding to its neighbors. In such a case, the tin valence orbitals are not hybridized, the lone pair is present on the 5s unhybridized orbital and it is said to be *non-stereoactive* since it has a spherical shape, unless it is distorted by the non-uniform electric field gradient from the neighboring ions. Since this distortion is usually very small, tin has a fairly regular coordination, e.g. octahedral in $CsSnBr_3$, which is a cubic perovskite [4]. One of the very rare cases of a non-stereoactive lone pair for a chloride is $SnCl_2$ [5].

BaClF has the PbClF structure, which is derived from the fluorite type BaF_2 by ordering the two types of anions along the *c* axis of the unit-cell. In BaF_2, the unit-cell is cubic, the space group Fm3m, Ba is in cubic coordination whereas F is tetrahedrally coordinated (fig. 1a). In BaClF, Ba and F occupy the same relative positions as in BaF2. To the contrary, the large size of Cl^- anions forces them to move by a vector (1/4 1/4 0) relative to the fluoride positions in BaF_2 in such a way as to form an axial Ba-Cl

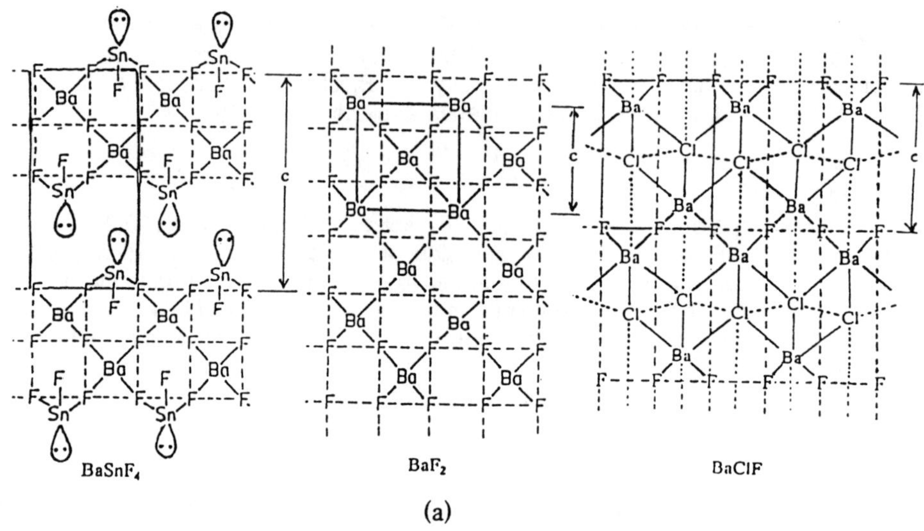

(a)

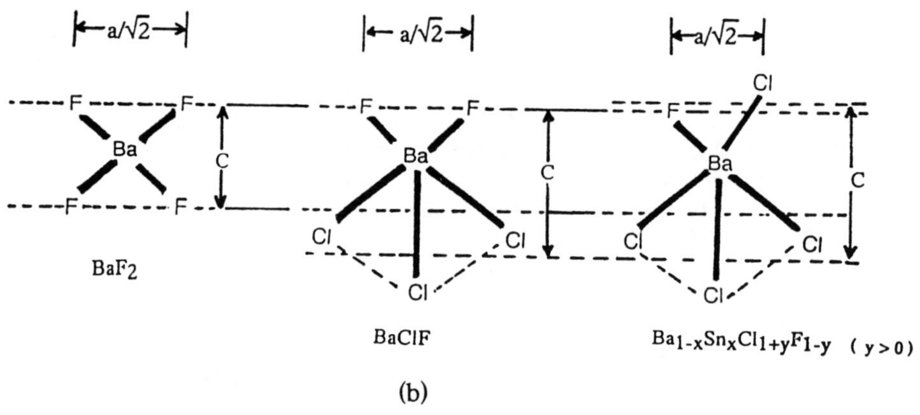

(b)

Figure 1: (a) Structures of $BaSnF_4$, BaF_2 and $BaClF$, (b) increase of c in $Ba_{1-x}Sn_xCl_{1+y}F_{1-y}$ with $y > 0$

bond parallel to the **c** axis. In addition, since the chloride ions are too bulky to fit in the same plane, they shift along the **c** axis, to form a corrugated sheet, in contrast with the flat fluoride ion planes (fig. 1a). The coordination is $BaF_4ClCl'_4Cl''$, where Cl forms a strong axial bond; there are four equatorial Ba-Cl' bonds a little longer, and Cl" forms a very weak axial interaction with Ba through the midlle of the fluoride square. All the chloride ions are crystallographically identical. The structure of $BaSnF_4$ is also derived from BaF_2, also by ordering, however here is ordering between the two kinds of metals (fig. 1a). The structure of BaF_2 is isotropic, contrary to $BaSnF_4$ which is bidimensional and very strongly anisotropic, due to the stereoactivity of the tin lone pairs, which form cleavage planes. On the other hand, BaClF has a thre-dimensional structure and no cleavage planes.

EXPERIMENTAL PROCEDURES

The following reactants were used in the synthesis: SnF_2 (99%) from Ozark Mahoning, $BaCl_2.2H_2O$ (analytical grade) from American Chemicals, BaF_2 (99%) from Allied chemicals and Dye Corporations, HF in 40% aqueous solution from Mallinckrodt, and doubly distilled water. Elemental analysis for Ba and Sn was carried out by atomic absorption spectrometry (AAS) on a Perkin-Elmer 503 absorption spectrometer. Fluorine analysis was carried out by use of a specific fluoride ion electrode from Orion and chloride was titrated by argentometry. X-ray powder diffraction was performed on a Philips X-ray diffractometer that has been automated by Sietronics. The Ni-filtered K_α peak of copper was used [$\lambda(K_{\alpha 1}$= 1.54051Å)]. Diffraction patterns were run at the angular velocity of $1°(2\theta)$/min with a step size of $0.02°(2\theta)$. LiF was used as an internal standard for line position, line intensity and linewidth. Mössbauer spectroscopy was carried out using an Elscint diving system, a scintillation detector equipped with a (Tl)NaI crystal, a Pd foil to absorbs the source X-ray lines, and a Tracor-Northern 7200 multichannel analyzer equipped withy a built-in single channel analyzer and an amplifier. The nominal 10mCi $Ca^{119m}SnO_3$ source was supplied by Amersham. Commercial α-SnF_2 and $CaSnO_3$ were used for calibration. Isomer shifts are referenced to $CaSnO_3$ at ambient temperature. All measurments were carried out in the constant acceleration mode, at ambient temperature, in the velocity range ±8mm/s.

RESULTS AND DISCUSSION

Mixing aqueous solutions of SnF_2 and $BaCl_2$ results in the precipitation of a white solid. The materials obtained are a function of the Ba/Sn ratio and how the two solutions are mixed. When the barium chloride solution is added to the SnF_2 solution, $BaSn_2F_6$ and other unidentified phases are obtained for (Ba/Sn)<0.33. For 0.33<(Ba/Sn)<0.83, several new phases are obtained, two which have been analyzed, $BaSn_2Cl_2F_4$ and $Ba_3Sn_3Cl_4F_8.2.6H_2O$. For (Ba/Sn)>0.83, a phase which has the diffraction pattern of BaClF is obtained. The same BaClF type diffraction pattern was also obtained from solid state reactions between appropriate amounts of BaF_2, $BaCl_2$ and SnF_2, heated at 350°C, 500°C or 600°C in copper tubes under argon. This gave a $Ba_{1-x}Sn_xCl_{1+y}F_{1-y}$ solid solution in the range 0<x<0.25, -0.20<y<0.20. These phases

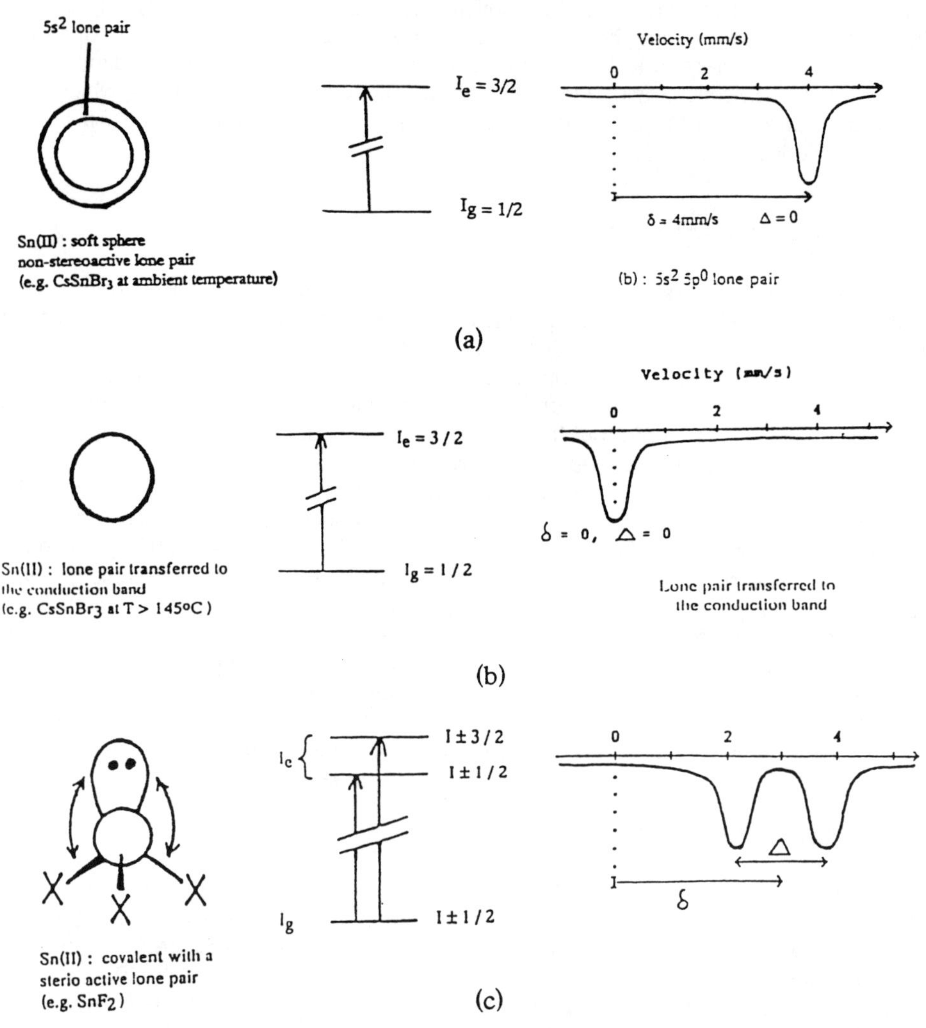

Figure 2: Tin-119 Mössbauer spectroscopy versus the electronic state of tin(II).

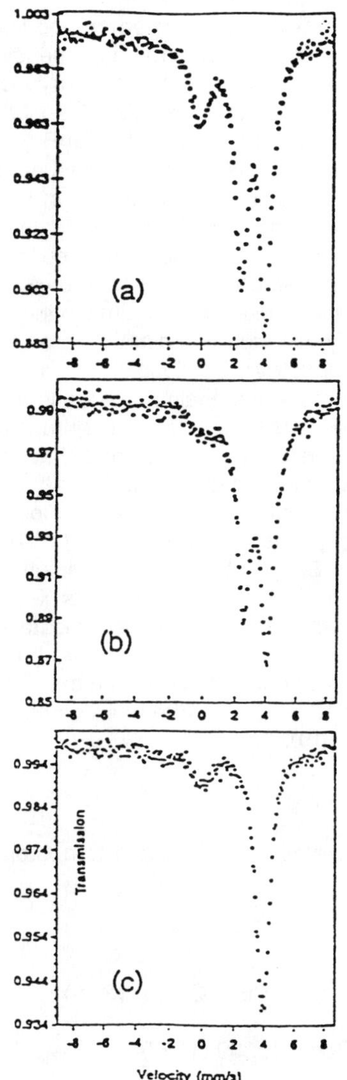

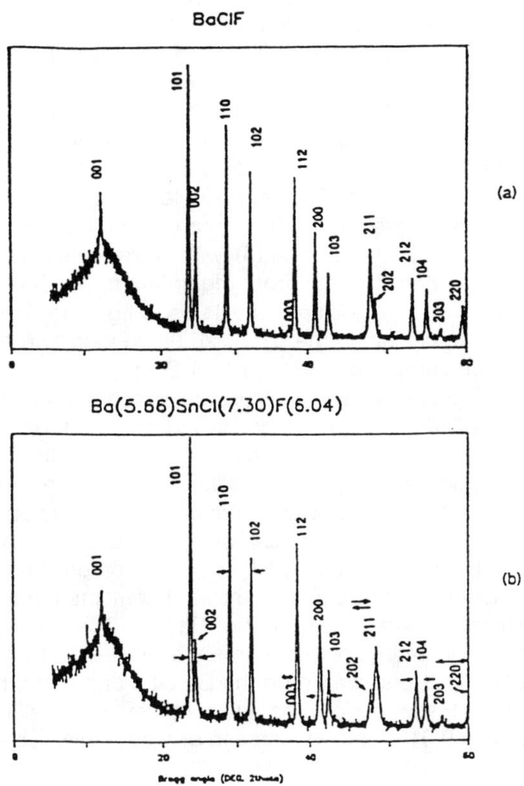

Figure 4: X-ray diffraction patterns of BaClF and $Ba_{0.85}Sn_{0.15}Cl_{1.10}F_{0.90}$

Figure 3: Mössbauer spectrum of :
(a) $BaSn_2Cl_2F_4$,
(b) $Ba_3Sn_3Cl_4F_8 \cdot 2.6H_2O$,
(c) $Ba_{0.85}Sn_{0.15}Cl_{1.10}F_{0.90}$

were studied by ^{119}Sn Mössbauer spectroscopy, in order to check the tin situation in each phase.

The ^{119}Sn Mössbauer spectrum is strongly dependent on the tin electronic state (fig. 2). For tin(II), the stannous ion Sn^{2+} gives a single line at ca. 4 mm/s (fig. 2a). The high isomer shift is due the high s electron density in the 5s orbital, located around the tin ion. The absence of electric field gradient (e.f.g.) around tin arises from the sperical shape of the 5s orbital and the regular coordination that results from the sphericity of the tin ion. This gives no splitting of the tin-119 nuclear levels, and a single 1/2 --> 3/2 transition upon absorption of a γ-photon gives a single line. In the case of $CsSnBr_3$, a transition from semiconducteur to metal occurs at 145°C, and this gives the Mössbauer spectrum of figure 2b, i.e. a single line at $\delta \approx$ 0mm/s. This is due to the transfer of the lone pair to the conduction band of the solid, i.e. away from the tin nucleus. In the case of covalently bonded tin(II) with a stereoactive lone pair, the axial lone pair pulls s electron density away from the tin atom, which results in a smaller isomer shift (δ= 2.5-3.5mm/s). In addition, the large e.f.g. due to both the presence of p and often d electron density in the hybridized lone pair, and a highly distorted coordination, creates a large quadrupole splitting Δ= 1-2mm/s. The resulting spectrum is a doublet (fig. 2c).

Figure 3 shows the Mossbauer spectrum for some of the materials prepared in the $BaCl_2/SnF_2$ system. For $BaSn_2Cl_2F_4$ and $Ba_3Sn_3Cl_4F_8.2.6H_2O$, a large quadrupole doublet is observed, which shows that tin(II) is covalently bonded and has a stereoactive lone pair. In addition, the weak peak at ca. 0mm/s is due to tin(IV). This is SnO_2 which forms on the surface of the particles and is present in much lower amount that the intensity of its line suggests, due to its very high recoil-free fraction [6]. The spectrum of $Ba_{0.85}Sn_{0.15}Cl_{1.10}F_{0.90}$, a composition of the $Ba_{1-x}Sn_xCl_{1+y}F_{1-y}$ solid solution obtained by precipitation, shows it contains the Sn^{2+} ion. Figure 4 compares its X-ray diffraction pattern with that of BaClF. There is no line splitting which would indicate a lower symmetry, or superstructure reflections that would show ordering. Therefore both metals are disordered on the Ba site, and the extra y Cl are disordered with the (1-y)F on the F site. Arrows on figure 4b show a small line shift, which are due to a decrease of *a* by 0.91% and an increase of *c* by 0.97% (see figure 1b).

ACKNOWLEDGMENTS

The Natural Science and Engineering Research Council of Canada and Concordia University are acknowledged for financial support.

REFERENCES

1. J. Galy, G. Meunier, S. Andersson and S. Aström, J. Solid State Chem. **13**, 142 (1975).
2. R.J. Gillespie and R. Nyholm, Quater. Rev. Chem. Soc. **11**, 339 (1957).
3. I.D. Brown, J. Solid State Chem. **11**, 214 (1974).
4. J. Barrett, S.R.A. Bird, J.D. Donaldson and J. Silver, J. Chem. Soc. (A) 3105 (1971).
5. N. Greenwood and T.C. Gibb, <u>Mössbauer Spectroscopy</u>, Chapman and Hall, London (1971).
6. G. Denes, J. Solid State Chem. **77**, 54 (1988).

Part IV
Theory

A PRIORI PREDICTION OF CRYSTAL STRUCTURES

I.D. BROWN
Brockhouse Institute for Materials Science, McMaster University, Hamilton, Ontario, Canada
L8S 4M1, idbrown@mcmaster.ca

ABSTRACT

The arrangement of atoms in a crystal is determined by two factors, the bonding preferences of individual atoms (giving rise to short range order), and the translational symmetry operations of the space group (giving rise to long range order). Chemical rules can be used to determine which atoms are bonded, and hence the maximum possible symmetry of the formula unit. Space group theory is then used to find the space groups that are compatible with this symmetry. In favourable cases, using the principle of maximum symmetry, the structure can be completely determined, but in all cases the analysis yields insights into the restrictions that determine what crystal structures might be possible.

INTRODUCTION

It should be possible, in principle, to predict the crystal structure of an inorganic solid from a knowledge of only its chemical formula. Despite nearly a century of attempts this can still be only done in a few cases, and then only in an *ad hoc* manner. There is no systematic way of developing the crystal structure from a knowledge of only the chemical formula.

To understand the problem of modelling inorganic structures, it is useful to break the process into three steps. The first is to find the topology of the structure, that is, the arrangement of the atoms relative to each other. The second is to determine the geometry, that is, to find the exact positional coordinates of the atoms, and the third is to select which of several possible structures is the most stable under any given set of conditions.

Of these three steps, only the second, the refinement of the geometry, can be performed with any reliability. There are several methods for determining accurate coordinates for the atoms once their general location is known, the most popular being the use of two-body potentials which are able to position atoms with an accuracy of 1 pm [1].

The third step requires a knowledge of the free energy of the crystal to an accuracy which is currently at or beyond the limit of our calculations. While this is an interesting part of the problem of structure prediction, it will not be discussed further here.

The first step, the problem of predicting the topology of the structure, is the subject of this paper. The paper explores how the topology is constrained by the laws of chemistry and crystallography. For compounds such as $SrTiO_3$, these constraints lead directly to a single structure, but even in the large number of cases where the structure is not uniquely determined, such as $MgSiO_3$ discussed below, analysis of the constraints provides insight into the factors that limit the possible topologies.

DESCRIPTION OF THE METHOD

The topology of a crystal structure can be represented by a network consisting of an infinite array of atoms linked together by chemical bonds. Since the network must conform to the symmetry of one of the 230 possible 3-dimensional space groups, it can be factored into two parts, the symmetry operators of the space group and the topology of the asymmetric unit. If

Mat. Res. Soc. Symp. Proc. Vol. 453 © 1997 Materials Research Society

these can both be separately found, the infinite bond network can be generated.   Thus the problem of predicting the topology is reduced to two steps, the determination of the topology of the asymmetric unit (related to the short range order), and the determination of the space group (related to the long range order). The short range order is determined by the laws of chemistry, the long range order by the laws of space group theory.

In practice it is better to start, not with the asymmetric unit, but with the formula unit, since the shape and contents of the asymmetric unit depend on the space group which is initially unknown.   In general, the formula unit comprises several asymmetric units related by non-translational symmetry elements. The space group must then be one that contains the symmetry elements of the formula unit.

Where more than one matching space group exists the choice between them is made using the principle of maximum symmetry which states that:

In modelling any structure, symmetry is broken only when required by chemical or crystallographic constraints.

In other words, always choose the highest possible symmetry.

The short range order is determined by the bonds that are formed between neighbouring atoms, that is, by the coordination number and bonding preference of each atom. Average coordination numbers have been determined for most cations [2], and the choice of bonded neighbours is determined by the valence matching principle which states that cations and anions will tend to bond to each other when they have the same bonding strength, where 'bonding strength' is defined as the atomic valence (or oxidation state) divided by the average coordination number [3]. These constraints, combined with the principle of maximum symmetry, allow one to generate possible graphs for the bond network of one formula unit [3].

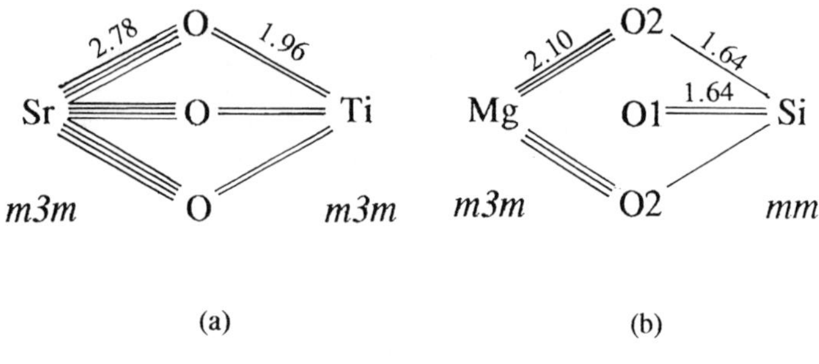

Figure 1

Bond graphs for (a) SrTiO$_3$ and (b) MgSiO$_3$. Also shown are the bond lengths predicted using the network equations described in [3] and the maximum possible site symmetries of the cations

Two such graphs, one for SrTiO$_3$ and one for MgSiO$_3$, are illustrated in Figure 1. They represent a version of the bond network which has been collapsed to one formula unit by extracting the formula unit from the infinite network and reconnecting the symmetry equivalent bonds that are broken during this process. Note that multiple lines in the bond graph do not indicate double or triple bonds as they would in the bond graph of an organic molecule. Rather each line of a multiplet represents a separate bond between different, but chemically equivalent, atoms in the infinite bond network. When properly drawn, the bond graph gives the correct local environment around each atom, including the coordination number and the chemical character of the bonded neighbours, but it does not, of course, contain any information about the long range order.

Using the valence matching principle and the principle of maximum symmetry, a unique graph can be drawn which, providing the graph is not too complex, usually corresponds to the observed structure. The most likely space group is then the highest symmetry space group that has Wyckoff positions that match in both symmetry and multiplicity the atoms of the bond graph. The structure is completely determined in those cases where the bond graph can be correctly drawn and can be mapped into a space group of high symmetry.

The first step in finding a space group is to determine the highest possible site symmetry of each atom in the bond graph. Since these site symmetries will ultimately correspond to the site symmetries of a Wyckoff position of a space group [4], only the 32 crystallographic point groups need be considered. For example, the Ti atom in Fig. 1a and the Mg atom in Fig. 1b both form six bonds to chemically equivalent O atoms. Both could, therefore, occupy sites of octahedral symmetry belonging to point group m3m. The Sr atom (Fig. 1a) forms 12 bonds to chemically equivalent O atoms and so could form a cuboctahedron with a site symmetry of m3m. However the highest site symmetry that the Si atom of Fig. 1b can have is mm. Even if the bonds are arranged tetrahedrally, the four O atoms to which Si is bonded belong to two chemically distinct species, so the $\bar{4}$-, 3- and 2-fold axes of the tetrahedral symmetry are lost.

Even though one can assign possible site symmetries to each of the atoms, they may not be able to occupy sites with these symmetries in the crystal. A powerful constraint on the crystallographic symmetry is supplied by the relationship between the order, O_w, and the multiplicity, m_w, of a Wyckoff position, namely:

$$m_w O_w = O_s \tag{1}$$

where O_s is the order of the space group. The order of a Wyckoff position is the number of times the symmetry operations of the space group transform the site into itself. Those symmetry elements that do not transform the site into itself generate symmetry related sites equal in number to the multiplicity of the Wyckoff position. The product of the order and the multiplicity is, therefore, just the total order, O_s, of the space group, i.e., the number of images generated by all the symmetry elements of the space group which is the same as the multiplicity of the general position.

An important corollary to Eqn. 1 is that atoms having the same multiplicity must have crystallographic site symmetries of the same order. This condition is satisfied, for example, by the Sr and Ti atoms in Fig. 1a, both of which appear in the formula unit with a multiplicity of 1 and which can, therefore, both have the same site symmetry m3m, but it is not satisfied for the two cations in Fig. 1b. Again, both Mg and Si have multiplicities of 1 but the order of m3m is 48 while that of mm is only 4. The crystallographic symmetry of Mg must be a subgroup of m3m that retains the octahedral geometry but which has an order of only 4. The candidate site symmetries are $\bar{4}$, 4, 222, 2/m and mm, but the first three of these can be eliminated using the following argument. With a site symmetry of mm, Si lies on the intersection of two mutually perpendicular mirror planes. O1 must lie on one of these mirror planes and O2 on the other. Since the multiplicity of O2 is twice that of Si, the order of its site symmetry is 2 and its point group can therefore only be m. The site that Mg occupies must, therefore, be one in which each of its six O2 ligands has site symmetry m. The only candidates from the above list are 2/m and mm. Similarly, O1 will have a site symmetry of order 4, and the only candidates that contain at least one mirror plane are also 2/m and mm. An analysis of the bond graph, combined with the principle of maximum symmetry, shows that one should look for a space group that has one site of multiplicity 1 and symmetry mm (for Si), two sites of multiplicity 1 and symmetry 2/m or mm (for Mg and O1) and one site of multiplicity 2 and symmetry m for O2 [*]. If a satisfactory space group cannot be found, it is necessary either to lower the symmetry or to adopt a different bond graph.

There are only 15 space groups that match the above requirements, thus considerably reducing the task of looking for a structure, but the search is not trivial. The low symmetry means that the structure will have between 5 and 8 free crystallographic parameters that must be chosen in such a way as to produce a 3-dimensionally connected network in which the bonds have the predicted lengths without bringing non-bonded atoms into too close contact or producing structures that are too open. It is easy to show why some of the space groups are unlikely to produce stable structures, but an exhaustive study cannot be done systematically. Interestingly, $MgSiO_3$ does not crystallise in any of these 15 space groups, but adopts the pyroxene structure with a bond graph that includes two formula units. It would be difficult to predict this *a priori*, but the analysis shows why $MgSiO_3$ does not crystallise wiht a high symmetry structure.

By contrast, determining the structure of $SrTiO_3$ is simple. A space group is sought with two Wyckoff positions having site symmetry m3m and one Wyckoff position of triple the multiplicity. Only 3 space groups have sites with m3m symmetry: Im3m, Fm3m and Pm3m and only one of these, Pm3m, has sites with the right multiplicity. Two structures are possible (with Sr and Ti interchanged), but only one of these, the perovskite structure, matches the bond graph. All atom coordinates are fixed by symmetry and only one parameter, the length of the cubic cell needs to be determined. However, this single crystallographic parameter must match two chemically determined parameters, the Sr-O and the Ti-O bond lengths, and such a choice is not, in general, possible. By chance, for $SrTiO_3$ the fit is exact, and the predicted bond lengths of 2.78 and 1.96 Å respectively both lead to a predicted cell dimension of 3.92 Å, close to the observed value of 3.90 Å. In cases where the fit is not exact, the bonds must be strained with the bonds to one cation being stretched and the bonds to the other compressed. These large strains lead to the instabilities that are a characteristic feature of perovskite structures [6]. They result in displacive phase transitions to structures with reduced symmetry and a range of fascinating physical properties such as ferroelectricity and superconductivity.

CONCLUSIONS

Combining chemical and crystallographic constraints leads to a better understanding of the crystal structures that can be formed by a given compound. In some cases the crystal structure can be directly derived, but in all cases, this approach gives insights into the factors that limit the possible structures. A more complete account of this approach will be forthcoming [7].

ACKNOWLEDGEMENTS

I would like to thank Profs H.Burzlaff and R.Galiulin for stimulating discussions and the Natural Science and Engineering Research Council of Canada for a research grant.

REFERENCES

1. C.W.Burnham, Am. Mineral. **75**, p.443 (1990).
2. I.D.Brown, Acta Cryst. **B44**, p.545 (1988).
3. I.D.Brown, Acta Cryst. **B48**, p.553 (1992).
4. International Tables for Crystallography, Vol A. D Reidel Publishing Co. Dordrecht, 1983.
5. The multiplicities given in International Tables [4] need to be reduced by eliminating the contributions of the translational symmetry elements since these cannot generate any Wyckoff positions. Normally this means dividing the multiplicity given in International Tables by the multiplicity of the Wyckoff (a) position.
6. A.M.Glazer, Acta Cryst. **B28**, p.3384 (1972).
7. I.D.Brown, (1997) Submitted to Acta Cryst.

LONG-RANGE ORDER EFFECTS IN FERROELECTRIC Pb(Zr$_{1/2}$Ti$_{1/2}$)O$_3$

G. SÁGHI-SZABÓ, Ronald E. COHEN
Carnegie Institution of Washington, 5251 Broad Branch Rd., N.W., Washington, DC 20015

ABSTRACT

The local orbital extension of the Linearized Augmented Planewave (LAPW+LO) method within the local density and general gradient approximations was used to optimize internal coordinates and calculate total energies of Pb(Zr$_{1/2}$Ti$_{1/2}$)O$_3$ (PZT) superlattices with B site cations ordered along the [001] and [111] directions. Ferroelectric structures and bond-length distributions similar to those obtained from experimental data were found for all three investigated chemically ordered phases. The Ti atom sits in an off-center position of a slightly distorted TiO$_6$ octahedron, the structure of which is mostly independent of the chemical ordering pattern. Coupling between the Ti and Zr containing octahedra results in highly distorted ZrO$_6$ units. This relatively high energy part of the structure could decrease the stability of these perovskite compounds against zone-boundary rotations of the BO$_6$ octahedra. Polar, zone-center only distortions result in lower energy [111] ordered superstructures when compared to the [001] ordered structure. The lowest energy chemically ordered PZT is the one with B-site cations ordered along the [111] direction and has *I4mm* symmetry. Our total energy results and a simple statistical model predict a wide miscibility gap PZT at the morphotropic composition is therefore likely a multiphase or metastable material which may be responsible for its sensitivity to synthesis and annealing conditions.

INTRODUCTION

Many of the technologically important ferroelectrics are oxides with a perovskite or perovskite related structure. Lead Zirconate Titanate (PZT) is a well known member of this class of materials, mainly due to its excellent ferroelectric, dielectric and piezoelectric properties. Chemical ordering has not been considered as a possibly important feature of PZT. Here we show, however, that the energetics of ordering, ferroelectricity, other structural relaxations, and exsolution are all on the same low energy scale, comparable with thermal energies at or below synthesis temperatures.

B-site chemical order has a subtle effect on many properties of A(B'B")O$_3$ ferroelectric perovskites.[1] Relaxor behavior can result from long-range order with a short coherence length, i.e. a diffuse phase transition, whereas both local atomic disorder and long coherence lengths result in normal ferroelectric (FE) or antiferroelectric (AFE) behavior.[2] Ignoring covalency, the degree of order depends on the differences in size and charge between the two B-site ions; generally speaking, the larger the difference the more favored the ordered state is. Normal ferroelectrics are best characterized by a sharp first or second order change in permittivity about their Curie temperature, weak frequency dependence of permittivity and strong anisotropy of light scattering, while relaxor ferroelectrics have very broad permittivity maxima with strong frequency dependence and weak remnant polarization. Relaxors also show very weak anisotropy to light scattering and no x-ray line splitting due to their pseudocubic structure. 1:1 ordering along [111] is very common in Pb containing perovskites with the formula of Pb(B'B")O$_3$, nevertheless, long-range chemical ordering has not been observed experimentally

Mat. Res. Soc. Symp. Proc. Vol. 453 © 1997 Materials Research Society

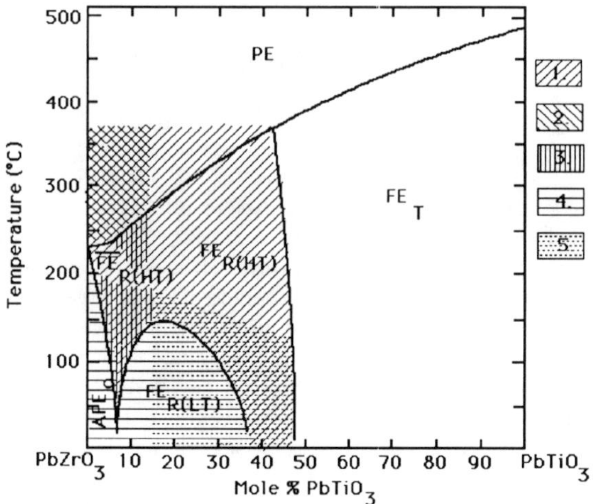

Figure 1. Phase diagram of the PZT solid solution (after Jaffe *et al.*[4] and Viehland *et al.*[9]) Regions: (1) disordered R-type oxygen tilt; (2) disordered M-type oxygen tilt; (3) locally ordered M-type tilt; (4) ordered R-type tilt; (5) constricted polarization state. The upper bounds of the shaded regions are the highest temperatures investigated. The boundaries between disordered fields may not be sharp.

in pure PZT.[3] This result can be attributed to the small size difference between Zr and Ti and to their identical valence electronic structure.

The phase diagram of the $PbZrO_3$ (PZ) - $PbTiO_3$ (PT) solid solution system (Fig. 1) was determined first by Jaffe, Cook and Jaffe[4] from x-ray diffraction measurements. There is only one high temperature cubic paraelectric phase in the experimental phase diagram of PZT. The low temperature PT-PZ diagram contains regions corresponding to several phase transitions. Compositions for $x>0.52$ in $PbZr_{1-x}Ti_xO_3$ show only one low temperature ferroelectric phase with space group *P4mm*. The phase diagram near pure $PbZrO_3$ is considerably more complicated. Pure $PbZrO_3$ (PZ) at low temperatures occurs in a complex antiferroelectric orthorhombic structure. Pulsed neutron scattering experiments show, that atoms in PZ are significantly displaced locally from the average sites determined by crystallographic analysis.[5] Near the ferroelectric-paraelectric transition temperature of the high Zr-content PZT, the structure of the material is less well established. Transmission electron spectroscopy studies[6] revealed an intermediate ferroelectric phase with rhombohedral symmetry. Other investigations showed an antiferroelectric (AFE) tetragonal phase[7] of $PbZr_{1-x}Ti_xO_3$ for $x<0.03$. The antiferroelectric phase disappears when $x>0.07$. Compositions of $PbZr_{1-x}Ti_xO_3$ for $0.07<x<0.35$ show a phase transition below the Curie temperature between ferroelectric rhombohedral phases with space groups of *R3m* and *R3c*. This latter transition involves changes in the tilt of the oxygen octahedra around the [111] directions.[8] Dielectric spectroscopic studies of coupling between the various tilted states and the polarization revealed the presence of dielectric dispersion at the *R3m* - *R3c* boundary.[9]

The phase diagram of the solid solution shows a nearly temperature independent morphotropic phase transition for compositions near $PbZr_{0.55}Ti_{0.45}O_3$. The most widely used piezoelectric ceramic is PZT with compositions near this morphotropic boundary. At that concentration the piezoelectric moduli, the remnant polarization of polycrystalline ceramics and the material's dielectric susceptibility have their maxima. When PZT is prepared by solid-solid reaction of constituent oxides (PbO, ZrO_2 and TiO_2), there is always a coexistence region of tetragonal and rhombohedral phases, with the tetragonal phase being on the Ti-rich side of the phase boundary.[10] In this region of unknown width, the lattice parameters of each phase do not change with varying composition, but only the relative amounts of the two phases change according to the lever rule.[11] Single phase samples at the morphotropic phase boundary without compositional fluctuations at the B-site were successfully created using a solution technique.[12] Even this single phase material transforms to the two phase (*R3m* and *P4mm*) ceramics with the lapse of time after poling. The computationally simplest of systems which has a composition close to that of the morphotropic boundary is $Pb(Zr_{1/2}Ti_{1/2})O_3$, the stoichiometry which was investigated in this study.

First-principle density functional theory (DFT) methods have been used effectively to contribute to the microscopic understanding of the basic principles underlying ferroelectricity. Ferroelectricity in $BaTiO_3$ and $PbTiO_3$ systems was studied[13,14] in great detail by self-consistent total energy calculations using the full-potential linearized planewave (LAPW) method. The electronic structure of $PbTiO_3$ has been extensively investigated in order to determine the underlying potential surface of this ferroelectric material.[15] It has been shown that the hybridization between the O *2p* and Ti *3d* states is the interaction responsible for the ferroelectric instability. Large, tetragonal strain was found to stabilize the tetragonal ferroelectric distortion in $PbTiO_3$.[13,14] The ultra-soft-pseudopotential method[16] was applied to the analysis and comparison of the properties of a series of perovskite compounds. Perovskite based materials have also been thoroughly investigated within the local density approximation (LDA) using a local orbital extension of the general potential LAPW method (LAPW+LO).[17-20] The linear-response approach to density-functional calculations within the local density approximation was applied to the calculation of phonon spectra, Born effective charges and dielectric constants for various semiconductors.[21,22]

Ferroelectric instabilities in perovskites are very sensitive to volume. Since LDA usually gives volumes that are too small by several percent, the ferroelectric instability is suppressed at LDA optimized volumes in some materials. Results of the calculations based on the LDA are in best agreement with experimental data when obtained at experimental lattice parameters. Calculations within the generalized gradient approximation (GGA)[23] have yielded ferroelectric parameters very similar to those of the LDA calculations, but the GGA lattice parameters are closer to the experimental results, and thus the well-depths are more accurate at optimized volumes. All of these works have demonstrated that sufficiently high quality DFT results can account for the overall nature of the potential energy surface in perovskite type ferroelectric materials, including depth and shape of soft mode displacement wells together with hybridization effects.

METHOD

Total energy calculations presented here were performed within the LDA and GGA[24] using the full-potential LAPW+LO method.[25] The calculations are fully self-consistent without the use of shape approximations for the charge density or potential. The Hedin and Lundqvist[26] exchange parametrization was used in the calculations. The PW91[23] density functional was applied to obtain the GGA results. LAPW sphere radii of 2.3, 1.7, 1.7 and 1.6 a.u. were used for

Pb, Zr, Ti and O, respectively. Pb *5d, 6s, 6p*, Zr *4s, 4p, 4d, 5s*, Ti *3s, 3p, 3d, 4s* and O *2s* and *2p* orbitals were treated as valence orbitals. The LAPW+LO method includes local orbitals (LO) in addition to the normal LAPW basis to allow treatment of all the valence bands in a single energy window and allow greater variational freedom. Local orbitals were used to include the semi core Zr *4s, 4p* and Ti *3s, 3p* states with the valence bands as well as to help the relaxation of the linearization of the Pb *5d*, Zr *4d*, Ti *3d* and O *2s, 2p* states. There is no pseudopotential approximation, core states of the atoms were calculated self-consistently in the crystal potential and were treated fully relativistically, while valence states were treated semirelativistically. The special points method[27-30] was applied for Brillouin-zone samplings. The results of the LDA calculations were checked for convergence with respect to the number of k-points and the plane wave cutoff energy. A 6x6x6 mesh and $Rk_{max}=8$ was used to calculate the LDA results, however, calculations with a smaller 4x4x4 mesh and $Rk_{max}=7$ result in the same relative and relaxation energies to within several tenths of a mRy. No separate convergence tests were performed within the GGA.

In this study, the internal coordinates of $PbTiO_3$ were optimized at the experimentally determined lattice parameters using analytical LAPW-LDA and LAPW-GGA Hellman-Feynmann forces[31] with Pulay's incomplete-basis-set correction.[32] Total energy calculations were performed on the [111] and [001] ordered phases of PZT as well as on the two end members of the PZT solid solution, PZ and PT. B site cation ordering in PZT along the [001] direction results in a tetragonal structure with *P4/mmm* symmetry. It is important to realize that the tetragonal structure is a simple consequence of the chemical ordering pattern and is not the result of either ferroelectric instability or long-range ordering of local atomic displacements, whereas the experimental tetragonal and rhombohedral structures result from unstable zone-center polar phonon modes. Introducing ferroelectric distortion along the [001] direction further reduces the symmetry of the system to *P4mm*. Ordering of B-site cations along the [111] direction generates a phase with *Fm3m* space group. In its cubic, undistorted form, this latter ordering corresponds to the rocksalt type arrangement of the B site cations. Ferroelectric atomic relaxation leads to either a structure with *R3m* or *I4mm* space groups, depending on whether the atoms move along [111] or [001] axis, respectively. The smallest ordered $Pb(Zr_{1/2}Ti_{1/2})O_3$ unit cells contain 10 atoms, corresponding to two formula units.

RESULTS

PbTiO₃

Our optimum lattice constant for the high symmetry cubic phase - 3.885 Å- is in good agreement with previous calculations. The displacement of the Ti atom relative to the Pb atom was found to be 0.186 Å in the relaxed structure with experimental lattice parameters. This result is very close to the relative Ti position corresponding to the minimum of the Ti-Pb displacement well found in an earlier study.[13] Table I contains the unrelaxed and refined coordinates in terms of the experimental tetragonal lattice parameters as well as the corresponding total energies. The calculated energy within the LDA decreases upon ferroelectric relaxation by 19.7 mRy per five-atom formula unit.

In addition, total energies within the GGA were determined for the unrelaxed and relaxed tetragonal $PbTiO_3$ using experimental lattice parameters for both structures as well as the internal coordinates from the LDA calculations for the latter one. Forces calculated within the

GGA using the LDA optimized structure were small, and show that internal coordinates are very close to their GGA optimum values; no significant energy decrease is expected from further internal relaxation within the GGA. However, the GGA relaxation energy, 13.8 mRy per five-atom unit cell, differs significantly from the value obtain within the LDA.

PbZrO$_3$

The electronic structure of PbZrO$_3$ has been recently determined and analyzed using the LAPW-LO method within the LDA.[19] It was found that the lowest energy (not fully relaxed) orthorhombic antiferroelectric (AFE) PZ structure lies 0.274 eV (20.1 mRy/ five-atom formula unit) below the cubic perovskite structure. This relaxation energy, which includes both internal and external structural relaxation energies, was subtracted from the calculated total energy of the cubic perovskite structure in order to determine the total energy of the pure PbZrO$_3$ end point of the PZT phase diagram. The lattice constant of the cubic perovskite PZ structure was optimized using Rk$_{max}$=8.0 and a 6x6x6 k-point mesh in the LDA calculations, and Rk$_{max}$=7.0 with a smaller, 4x4x4, k-point mesh in the GGA calculations. The calculated lattice parameters of the cubic perovskite structure are 4.11 Å and 4.14 Å within the LDA and the GGA, respectively. We obtained the final values of -49473.7198 Ry within the LDA and -49513.4999 Ry within the GGA for the total energy of the orthorhombic AFE phase using the earlier mentioned procedure.

TABLE I. Internal atomic coordinates, LDA and GGA total energies (Ry units) of the *unrelaxed* and relaxed species.

	Pb(Zr$_{1/2}$Ti$_{1/2}$)O$_3$ *P4mm*[a]	Pb(Zr$_{1/2}$Ti$_{1/2}$)O$_3$ *I4mm*[b]	Pb(Zr$_{1/2}$Ti$_{1/2}$)O$_3$ *R3m*[c]	PbTiO$_3$ *P4mm*[d]
Pb	*(0 0 0)* (0 0 0)	*(¾ ¼ ½)* (0.783 0.283 ½)	*(¼ ¼ ¼)* (0.271 0.271 0.271)	*(0 0 0)* (0 0 0)
Pb			*(¾ ¾ ¾)* (0.768 0.768 0.768)	
Zr	*(0 0 ½)* (0 0 0.533)	*(0 0 0)* (0 0 0)	*(0 0 0)* (0 0 0)	
Ti	*(½ ½ ¼)* (½ ½ 0.238)	*(½ ½ 0)* (0.497 0.497 0)	*(½ ½ 0)* (½ ½ 0)	*(½ ½ ½)* (½ ½ 0.455)
O$_1$	*(½ ½ ¾)* (½ ½ 0.735)	*(¼ ¼ 0)* (0.239 0.239 0)	*(¼ ¼ ¾)*(0.245 0.245 0.714)	*(½ 0 ½)* (½ 0 0.374)
O$_2$	*(½ ½ 0)* (½ ½ 0.948)	*(¾ ¾ 0)* (0.721 0.721 0)	*(¼ ¾ ¾)*(0.230 0.730 0.730)	*(½ ½ 0)* (½ ½ 0.879)
O$_3$	*(½ ½ ½)* (½ ½ 0.475)	*(¼ ¼ ½)*(0.225 0.225 0.514)		
O$_4$	*(½ 0 ¼)* (½ 0 0.185)			
O$_5$	*(½ 0 ¾)* (½ 0 0.708)			
E(LDA)$_{unrel}$	-46730.4436	-46730.4460	-46730.4478	-43987.2205
E(LDA)$_{rel}$	-46730.4727	-46730.4749	-46730.4748	-43987.2403
E(GGA)$_{unrel}$	-46767.5715[e]	-46767.5707[e]	-46767.5724[e]	-44021.7380[e]
E(GGA)$_{rel}$	-46767.5930[e]	-46767.5935[e]	-46767.5936[e]	-44021.7518[e]

Total energies shown are per five-atom formula unit for all phases, including PZT unit cells containing ten atoms. Unrelaxed and optimized coordinates above are given in terms of the following lattice vectors (Bohr units, a$_i$=(x$_i$ y$_i$ z$_i$)):

[a] a$_1$=(7.616 0 0), a$_2$=(0 7.616 0), a$_3$=(0 0 15.722), Rk$_{max}$=8.0, 6x6x4 k-mesh

[b] a$_1$=(0 7.615 7.861), a$_2$=(0 -7.615 7.861), a$_3$=(-7.615 0 7.861), Rk$_{max}$=8.0, 6x6x6 k-mesh

[c] a$_1$=(7.696 7.696 0), a$_2$=(0 7.696 7.696), a$_3$=(7.696 0 7.696), Rk$_{max}$=8.0, 6x6x6 k-mesh

[d] a$_1$=(7.377 0 0), a$_2$=(0 7.377 0), a$_3$=(0 0 7.846), Rk$_{max}$=8.0, 6x6x6 k-mesh

[e] GGA results were obtained using Rk$_{max}$=7.0 and a less dense k-point mesh, 4x4x4 for Pb(Zr$_{1/2}$Ti$_{1/2}$)O$_3$(*R3m*), Pb(Zr$_{1/2}$Ti$_{1/2}$)O$_3$(*I4mm*) and PbTiO$_3$, 4x4x2 for the Pb(Zr$_{1/2}$Ti$_{1/2}$)O$_3$(*P4mm*) structures

<u>Ordered PZT phases</u>

Lattice parameter optimizations performed on cubic perovskite $Pb(Zr_{1/2}Ti_{1/2})O_3$ phases with ordered B-site cations in the unrelaxed, high symmetry positions, gave identical results, 4.01 Å for both [111] and [001] ordered structures, showing no direct volume effect accompanying the two different ordering schemes at this level of approximation.

Total energy calculations for the [001] ordered *P4mm* and [111] ordered *I4mm* structures were performed at the experimentally determined lattice parameters of the microcrystalline $Pb(Zr_{1/2}Ti_{1/2})O_3$ phase with *P4mm* symmetry as well as at a volume obtained from the cubic perovskite optimization, with the c/a ratio set to 2.0 in the latter case. The volume of the [111] ordered *R3m* phase was kept equal to that of the experimental tetragonal phase in one set of the calculations, while in the other set, the volume of the 10 atom unit cell was set equal to the volume of the optimized high symmetry perovskite structure. For both *R3m* sets, the small rhombohedral distortion was neglected.

All of the symmetry allowed internal atomic relaxations were determined using the method described under the section about $PbTiO_3$. The optimizations were performed with the parameter Rk_{max} set to 7.0 and with a smaller k-point mesh, 4x4x4 for the *R3m, I4mm* and 4x4x2 for the *P4mm* structures. The optimization procedure was halted when the absolute values of gradients reached a value smaller than $1x10^{-4}$ Ry/Bohr. The internal coordinates determined this way were input into a new set of calculations, in which the total energies corresponding to these optimized structures were recalculated with a larger k-point mesh and Rk_{max}. These latter values were determined during convergence tests. The same internal parameters, sets optimized within the LDA, were used to determine total energies within the GGA. The smaller set of k-point mesh and Rk_{max} (4x4x4 or 4x4x2 and 7.0) was applied to calculate GGA forces and total energies. Force calculations performed within the LDA with the denser mesh and more basis functions, as well as GGA forces show that atomic positions obtained using the smaller set within the LDA calculations were already sufficiently converged. Further relaxation of internal coordinates in the former two cases would not have lead to any significant energy decrease. Table I contains the LDA and GGA total energies corresponding to the unrelaxed and relaxed structures as well as the calculated final internal coordinates. The relaxation energies, the energy difference between the unrelaxed and computationally optimized structures, are summarized in Table II.

DISCUSSION

Ferroelectric instability results from a delicate balance between short-range repulsions which would favor the occupancy of high symmetry sites and long-range dipolar forces favoring ferroelectric distortion. Covalent bonding between oxygens and B-site cations are crucial to this distortion since they decrease the short-range repulsive forces, allowing the displacement of minima on the potential surface from the high symmetry, often cubic, sites.

Spontaneous strain and volume effects also play an important role in ferroelectric phase transitions. It is known, that ferroelectric instability could be suppressed at optimized volumes obtained by LDA calculations. $PbZr_{1/2}Ti_{1/2}O_3$ still shows significant ferroelectric distortion at optimized volumes which are altogether nearly 5% smaller than the experimentally determined values. Changing the c/a ratio from 2.03 to 2.00 at constant, 911.8 $Bohr^3$ volume in the [111] ordered *I4mm* PZT phase decreases the total relaxation energy by 13% from 28.8 mRy to 25.0 mRy per five-atom formula unit (based on calculations with Rk_{max}=7.0 and 4x4x4 k-point mesh).

If the c/a ratio change is accompanied with volume compression from 911.8 Bohr^3 to 871.0 Bohr^3, the relaxation energy drops to 23.1 mRy per five-atoms, an additional 7% decrease. We obtain a similar, 9% decrease in the relaxation energy if we uniformly compress by the same amount the volume of the [111] ordered PZT phase with R3m symmetry. In the case of the [001] ordered PZT with *P4mm* symmetry, the former decrease in volume and c/a ratio reduces the relaxation energy by nearly 24%.

TABLE II. Relaxation energies per 5-atom unit cell in units of mRy (PZT=$PbZr_{1/2}Ti_{1/2}O_3$)

	$PbTiO_3$ *P4mm*	$PbZrO_3$ *Pbam*	$PbZr_{1/2}Ti_{1/2}O_3$ *P4mm*	$PbZr_{1/2}Ti_{1/2}O_3$ *I4mm*	$PbZr_{1/2}Ti_{1/2}O_3$ *R3m*
GGA[b]	13.8	-	21.5	23.3	21.2
LDA[c]	19.7	20.1[a]	29.1	28.9	27.0
LDA[d]	-	-	21.1	23.1	24.3

[a] LDA result includes external (strain) and internal AFE relaxation energy from Ref. 19

[b] Rk_{max}=7.0, 4x4x4 k-point mesh for PT(P4mm), PZT(I4mm), PZT(R3m) and 4x4x2 mesh for PZT(P4mm), experimental lattice parameters

[c] Rk_{max}=8.0, 6x6x6 k-point mesh for PT(P4mm), PZT(I4mm), PZT(R3m) and 6x6x4 mesh for PZT(P4mm), experimental lattice parameters

[d] Rk_{max}=7.0, 4x4x4 k-point mesh for PT(P4mm), PZT(I4mm), PZT(R3m) and 4x4x2 mesh for PZT(P4mm), lattice parameters were optimized for the cubic perovskite structure

The relaxation energy, or the energy difference between the ferroelectrically relaxed phases and phases with atoms at the high symmetry points, can be decomposed into two parts. First is the structural relaxation due to the size difference between the Zr and Ti atoms. The remaining part is due to the energy change accompanying polar zone-center instability. The latter was a significant element in the total relaxation energy even at compressed volumes. It accounts for almost 50% of the relaxation energy in the compressed (871.0 Bohr^3) [001] ordered *P4mm* PZT and for about 28% in the [111] ordered PZT phase with *R3m* symmetry having same volume. We found that the ferroelectric part of the relaxation energy is strongly volume dependent. In the *R3m* and *I4mm* symmetry phases, the 911.8 Bohr^3 to 871.0 Bohr^3 uniform compression of the unit cells (at c/a=2.0 ratio) decreases the depth well of the ferroelectric instability by more than 40%.

The BO_6 octahedron can be regarded as the building block of the perovskite structure. We can compare the theoretical interatomic distances, or bond lengths, obtained from our structural optimizations to those calculated using effective ionic radii.[33] For TiO_6 and ZrO_6, the sum of Zr-O ionic radii is 2.12 Å, whereas the sum of Ti-O radii equals to 2.00 Å. The ideal O-O distances therefore in the octahedra would be 3.00 Å in ZrO_6 and 2.83 Å in TiO_6. In the optimized *P4mm* structure (Fig. 2) the Ti atom is displaced along the [001] axis by 0.2 Å from the center of a slightly distorted octahedra. The Ti-O distances are split into 1.77, 2.17 and 2.02 Å. The O-O distance is very close to the sum of ionic radii, varying from 2.80 Å to 2.85 Å. These distances and distance distributions within the TiO_6 octahedron are very similar to those obtained from theoretical and experimental $PbTiO_3$ structures.

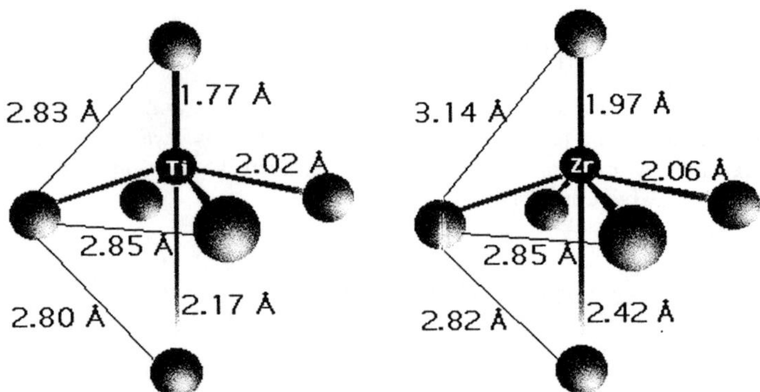

Figure 2. Relaxed structures of the TiO_6 and ZrO_6 octahedra in the [001] ordered PZT.

The ZrO_6 octahedron in the *P4mm* structure is more distorted than the TiO_6 octahedron. The Zr atom is shifted in the same direction as the Ti atom by more than 0.2 Å. The distribution of Zr-O distances is large and are split into 1.97, 2.06 and 2.42 Å. The distribution of the O-O distances, 2.82, 2.85 and 3.14 Å, is similar to that obtained from a structure of $PbZrO_3$ proposed by Fujishita and Hoshino.[34] Pulsed neutron scattering studies of the local atomic structure of PZT[35] also favor this distribution as opposed to the very widely distributed atomic distances calculated from the structure proposed by Jona *et al.*[36] In general, the distortion of the BO_6 octahedra and the bond-length distributions are similar in the other phases. However, the high distortion of the ZrO_6 octahedron, also found in the other two ordered structures, might be overemphasized by the small cell and assume symmetry for our computations, in which only zone-center distortions were allowed. In the real crystal, these high energy units can further relax by local or long-range zone-boundary tilts or rotations of the oxygen octahedra. The presence of R-type and M-type oxygen tilt structures were revealed by experimental techniques in the phase diagram of PZT with Zr/Ti ratios between 95/5 and 55/45.

Calculations within the GGA decrease every relaxation energy by approximately 5 mRy per five-atom formula unit when compared to those obtained from LDA calculations. We have adjusted the LDA PZT relaxation energy for comparisons with the GGA results by similarly decreasing it from 20 to 15 mRy per 5-atom formula unit, since we have no GGA results for the ground state PZ structure. We find that at zero K, the mechanical mixture of fully relaxed $PbTiO_3$ and $PbZrO_3$ phases has the lowest energy when compared to the energies of each ordered phase. The lowest energy relaxed, ordered PZT is the tetragonal body centered phase with B cations ordered along [111]. This phase with randomly ordered B-sites and a ferroelectric distortion along [001] gives a material with overall *P4mm* symmetry, which is in good agreement with the symmetry of the lowest energy low temperature PZT phase corresponding to the $Pb(Zr_{1/2}Ti_{1/2})O_3$ stoichiometry obtained by crystallographic methods. Calculations within both the LDA and GGA give almost identical total energy for the [111] ordered PZT phase with *R3m* symmetry. The [001] ordered, relaxed PZT phase with *P4mm* symmetry is the highest in energy based on the LDA, lying approximately 2 mRy above the other two phases, while the result of the calculation within the GGA is less than 1 mRy apart from the total energies of the *R3m* and *I4mm* phases.

Next we consider the thermodynamics of ordering and exsolution at finite temperature. Long-range order of B-site cations has not been observed experimentally in undoped PZT.

Theoretically, short range order could present in the PZT solid solution, since atoms of different sizes and slightly different charges substitute for each other. The simplest statistical approximation completely neglects the possible short-range ordering in the material. Since the calculated final total energies of all three ferroelectric low temperature phases as well as the energies of the two high temperature cubic ordered phases are very similar, the assumption of randomly distributed B-site cations seems to be valid both below and above the ferroelectric-paraelectric transition temperature.

As a first approximation, we used the simplest type of non-ideal model (symmetrical regular solution) to obtain the molar free energies of mixing of our binary regular solid solution. There are two positions, Zr and Ti, in the $Pb_2(ZrTi)O_6$ system at which mixing can occur. The non-ideal contribution to the free energy was treated as a symmetrical function of the composition, and was assumed to be pressure and temperature independent. This approximation completely neglects certain terms in the free energy, e.g. by assuming that the entropy of mixing is not altered by non-ideal interactions and that no long-range ordering or clustering of atoms occur.

As a preliminary result, we found a wide miscibility gap in the PZT phase diagram using the energy data of the low temperature ferroelectric PZT. PT and PZ phases. The gap closes at around 750 °C and 1300 °C based on the GGA and LDA calculations, respectively. The best estimate, 750 °C, is below, but in the range of sintering temperatures used in PZT synthesis,[37] and details of the sintering procedure can have significant effects on the structural and ferroelectric properties of PZT ceramics. Immiscibility and incipient ordering, whether equilibrium or due to disequilibrium and sample history may be important parameters in PZT ceramics, as well as grain size and texture.

Highly oriented PZT films have been prepared using layer-by-layer deposition techniques.[38-40] These low temperature methods can create films oriented along various axes depending on the substrate used in the deposition process. So far, however, there has been no evidence found either for long range ordering or for superlattices in these layers. Our results are consistent with this observation, namely that epitaxially grown films do not seem to maintain axial B-site cation ordering even when grown as such.[41]

CONCLUSIONS

Ab initio electronic structure studies of ferroelectrics allow us to investigate the underlying principles of ferroelectricity. Our results help us to understand the energetics and forces that drive the ferroelectric transitions, therefore helping us interpret the sensitivity of the perovskite ferroelectrics to preparation conditions, crystal defects and chemical fluctuations.

The overall differences between the various chemically ordered PZT phases is small, leading to similar total energies and comparable distribution of interatomic distances. Our calculations confirm the importance of volume effects and spontaneous strain. We find that tetragonal strain and larger volumes stabilize ferroelectric long-range ordered structures. Increased pressure leads to smaller relaxation energies between the high symmetry cubic and relaxed phases as well as to a significant decrease of the well depth attributed to polar, ferroelectric instabilities. Using our computed total energies and a simple model, we find a miscibility gap. The small energy difference between the ferroelectric ordered structures indicates that PZT is most likely disordered. Near these temperatures, also because of the small differences between the various ordered and disordered phases, there could be an abundance of available microstructures, which in turn could explain some of the observed sensitivity of ferroelectric PZT to fluctuations in synthesis conditions (composition, temperature and processing time) near the morphotropic phase boundary.

ACKNOWLEDGMENTS

This work is supported by the Office of Naval Research. The authors would like to thank L.W. Finger, H. Krakauer, I.I. Mazin and D.J. Singh for helpful discussions.

REFERENCES

[1] N. Setter and L. E. Cross, J. Mater. Sci. **15**, 2478 (1980).

[2] I-W. Chen, P. Li and Y. Wang, J. Phys. Chem. Solids, **57**, 1525 (1996).

[3] C. A. Randall, A. S. Bhalla, T. R. Shrout and L. E. Cross, Ferroelec. Lett., **11**, 103 (1990).

[4] B. Jaffe, W. R. Cook, and H. Jaffe, *Piezoelectric Ceramics,* (Academic Press, London, 1971); M. E. Lines and A. M. Glass, *Principles and Applications of Ferroelectrics and Related Materials,* (Clarendon Press, Oxford, 1977).

[5] S. Teslic, T. Egami and D. Viehland, J. Phys. Chem. Solids, **57**, 1537 (1996).

[6] A. M. Glazer, K. Roleder and J. Dec, Acta Cryst. B**49**, 846 (1993).

[7] R. W. Whatmore and A. M. Glazer, J. Phys. C **12**, 1505 (1979).

[8] C. A. Randall, M. G. Matsko, W. Cao and A. S. Bhalla, Solid State Commun. **24**, 769 (1977).

[9] D. Viehland, J.-F. Li, X. Dai and Z. Xu, J. Phys. Chem. Solids **57**, 1545 (1996).

[10] W. Cao and L. E. Cross, Phys. Rev. B **47**, 4825 (1993).

[11] P. Ari-Gur and L. Benguigui, Solid State Commun. **15**, 1077 (1974).

[12] K. Kakegawa, J. Mohri, S. Shirasaki and K. Takahashi, J. Am. Ceram. Soc. **65**, 515 (1982).

[13] R. E. Cohen and H. Krakauer, Phys. Rev. B **42**, 6416 (1990).

[14] R. E. Cohen and H. Krakauer, Ferroelec. **136**, 65 (1992).

[15] K. M. Rabe and U. V. Waghmare, Ferroelec. **164**, 15 (1995).

[16] R. D. King-Smith and D. Vanderbilt, Phys. Rev. B **49**, 5828 (1994).

[17] D. J. Singh and L. L. Boyer, Ferroelectrics **136**, 95 (1992).

[18]D. J. Singh, Ferroelectrics **164**, 143 (1995).

[19]D. J. Singh, Phys. Rev. B **52**, 12559 (1995).

[20]D. J. Singh, Phys. Rev. B **53**, 176 (1996).

[21]C.-Z. Wang, R. Yu and H. Krakauer, Phys. Rev. Letters **72**, 368 (1994).

[22]R. Yu and H. Krakauer, Phys. Rev. B **49**, 4467 (1994).
[23] J. P. Perdew and Y. Wang, Phys. Rev. B **45**, 13244 (1992).

[24] D. J. Singh, Ferroelectrics **164**, 143 (1995).

[25]D. J. Singh, Planewaves, *Pseudopotentials and the LAPW Method*, (Kluwer Academic Publishers, Boston, 1994), p. 56.

[26]L. Hedin and B. I. Lundquist, J. Phys. C **4**, 2064 (1971).

[27]A. Baldereschi, Phys. Rev. B **7**, 5212 (1973).

[28]D. J. Chadi and M. L. Cohen, Phys. Rev. B **8**, 5747, (1973).

[29]H. J. Monkhorst and J. D. Pack, Phys. Rev. B **13**, 5188 (1976).

[30]H. J. Monkhorst and J. D. Pack, Phys. Rev. B **16**, 1748 (1977).

[31] R. Yu, D. Singh and H. Krakauer, Phys. Rev B **43**, 6411 (1991).

[32] P. Pulay, Mol. Phys. **17**, 197 (1969).

[33]R. D. Shannon and C. T. Prewitt, Acta Cryst., 925 (1969); R. D. Shannon, Acta Cryst. A**32**, 751 (1976).

[34]H. Fujishita and S. Hoshino, J. Phys. Soc. Jpn. **53**, 226 (1984).

[35]S. Teslic, T. Egami and D. Viehland, J. Phys. Chem. Solids, **57**, 1537 (1996).

[36]F. Jona, G. Shirane, F. Mazzi and R. Pepinski, Phys. Rev. **105**, 849 (1957).

[37] P. G. Lucuta, F. Constantinescu and D. Barb, J. Am. Ceram. Soc. **68**, 533 (1985).

[38]A. Masuda, Y. Yamanaka, M. Tazoe, Y. Yonezawa, A. Morimoto and T. Shimizu, Jpn. J. Appl. Phys. **34**, 5154 (1995).

[39] K. Akoi, Y. Fukuda, K. Numata and A. Nishimura, J. Appl. Phys. **34**, 746 (1995)

[40] Y. Sotome, J. Senzaki, S. Morita, S. Tanimoto, T. Hirai, T. Ueno, K. Kuroiwa and S. Tanimoto, Jpn. J. Appl. Phys. **33**, 4066 (1994).

[41] T. Hirai, K. Teramoto, T. Goto and Y. Tarui, Jpn. J. Appl. Phys. **34**, 539 (1995).

FIRST PRINCIPLES COMPUTER SIMULATION OF
THE DEFECT CHEMISTRY OF RUTILE TiO$_2$

I.DAWSON and P.D. BRISTOWE
Department of Materials Science and Metallurgy, University of Cambridge, Pembroke Street,
Cambridge, CB2 3QZ, UK

J.A. WHITE and M.C. PAYNE
Cavendish Laboratory(TCM), University of Cambridge, Madingley Road, Cambridge, CB3
OHE, UK

ABSTRACT

There exists a long-standing controversy concerning the nature of the dominant point defect
mechanism in rutile TiO$_2$. Previous classical shell model calculations by Catlow *et al* [1] find a
strong preference for Schottky as opposed to Frenkel-type defects, lending support for oxygen
vacancy rather than titanium interstitial compensation in reduced rutile. However, reviews
of experimental studies [2], show that many conflicting conclusions have been reached. *Ab
initio* total-energy calculations have been performed on a parallel computer to help resolve
this controversy. First results indicate a Schottky formation energy (of the bound Schottky
trio) consistent with the Mott-Littleton values of Catlow *et al* [1]. A first attempt is made at
calculating the heat of reduction through determination of the formation energy of a neutral
oxygen atom vacancy. As a result some interesting insight is gained into the redox chemistry
of reduced rutile.

INTRODUCTION

The material TiO$_2$ affords a range of useful applications ranging from utilisation of its
unusual oxidation properties to application as an electronic sensor. It is also important as
a white pigment and as a catalyst. Rutile, one of the most studied phases of TiO$_2$, exists
in non-stoichiometric form TiO$_{2-x}$ and has an extremely complex defect structure. For low
x, oxygen vacancies dominate, but for greater oxygen reductions (higher x), complex defect
structures form. These are shown by electron microscopy [3] to be extended defects, known
as crystallographic shear planes whose formation is described in terms of planes of oxygen
vacancies, with an associated crystallographic shear. The elimination of vacancies restores
coherence to the non-metal sub-lattice. However within the region of the planar defect,
the structure of the metal sub-lattice changes, as corner sharing between TiO$_6$ octahedra is
replaced by edge sharing. The defect is therefore a shear plane characterised by a planar
fault on the metal sublattice, in which the mode of linking of TiO$_6$ octahedra differs from
the perfect lattice and which can be described in terms of the elimination of point defects.

There are many unanswered questions, in particular why does rutile accommodate non-
stoichiometry by point defect elimination as opposed to aggregation like most grossly non-
stoichiometric compounds. There is major uncertainty as to the nature of point defects in
TiO$_{2-x}$ and their relation to the shear plane structures. It is unknown even over which
composition ranges point defects are the dominant defect, and in these regions, whether
anion vacancies or cation interstitials are the major point defect. These problems are some
of the most intriguing in solid-state chemistry. With present resources it is too ambitious

Mat. Res. Soc. Symp. Proc. Vol. 453 © 1997 Materials Research Society

to calculate shear plane structures *ab initio*, as large unit cells containing regularly spaced defects and spin polarisation are needed, so we have concentrated on attempting to resolve the controversy concerning the dominant point defect mechanism in rutile. This is the first fully *ab initio* attempt at calculating defect energetics in TiO_2, specifically the Schottky and Frenkel formation energies and the cation and anion vacancy formation energies.

TECHNIQUES

In this study we use the plane wave pseudopotential and density functional methods as embodied in the CETEP code [4]. This allows us to carry out full structural relaxation of periodic systems to the equilibrium ground state, and as this is a very demanding task for transition metal oxides the work makes use of parallel computation techniques to achieve this. System sizes ranged from 12-120 atoms and the calculations were run on the 512 node Cray T3D at Edinburgh, U.K. The calculations give important insight into the electronic structure of the defects, perhaps leading to improvements in the empirical modelling of such materials, through revealing important physics which might be incorporated in the empirical models. Previously, treatments of the energetics of defective TiO_2 has relied on empirical interaction models such as the shell model [1]. While this empirical approach has achieved some success, there has long been a strong need for the more fundamental approach provided by *ab initio* methods. For example, oxides are generally difficult to model empirically because of the large deformability of the O^{2-} ion, which is not a stable species in free space.

In the pseudopotential method, only valence electrons are represented explicitly in the calculations, the core-valence interaction being represented by non-local pseudopotentials generated by first principles calculations on isolated atoms. The calculations are performed using periodically repeating geometry, the occupied orbitals being expanded in a plane wave basis. The expansion includes all plane waves whose kinetic energy $E_k = \hbar^2 k^2/2m$ (k is the wavevector, m the electron mass) is less than the cut-off energy E_{cut}, which is chosen to ensure convergence with respect to the basis set.

In this work, the self-consistent ground state of the system was determined using an all-bands conjugate gradient technique to minimise the total energy of the system with respect to the plane wave coefficients. Equilibrium ionic positions were determined by a BFGS method. First principles norm-conserving pseudopotentials in Kleinman-Bylander representation [5] were used, generated by the optimisation scheme of Lee *et al* [6] in order to reduce the required value of the plane-wave cut-off E_{cut}.

Primarily we will obtain important results on excess electron states in TiO_2, through neutral O and Ti atom vacancy calculations, requiring a spin-polarised treatment. Without spin the electronic bands have Pauli degeneracy. When an oxide material is reduced, there are excess electrons which may occupy a variety of states such as F-centres in MgO. In a transition metal oxide, the situation is more complicated as the metal ion may exist in a range of oxidation states. In TiO_2, removal of O implies reduction of Ti^{4+} to Ti^{3+} which needs spin to be included for a correct treatment. Both LDA and GGA calculations [4] are carried out.

PREVIOUS STUDIES

Previous classical Mott-Littleton calculations [1], found that Schottky defects were energetically preferred to Frenkel defects. This leads to the conclusion that vacancy disorder will predominate in the pure crystal and in those regions of the nonstoichiometric phase where

formation energies	Mott-Littleton	exptl[9]
Schottky trio	5.21 eV	$\leq 6.6 eV$
Frenkel pair	11.95 eV	$\leq 5.0 eV$

Table 1: Point defect formation energies in rutile.

point defects are the dominant defect species (low x). So the conclusion from empirical calculations of point defect energies is that anion vacancies are the favoured species on reduction of rutile. This follows directly from the relative magnitudes of the calculated Schottky and Frenkel energies shown in Table 1. According to Catlow it is energetically more favourable by about 5eV to create 2 anion vacancies rather than 1 cation interstitial Frenkel pair in TiO_{2-x}. However, analyses of experimental transport data have favoured the alternative model of cation interstitial formation. Further, recent space charge measurements [9], also favour Frenkel defects. So there is still considerable controversy concerning the predominant point defects in TiO_{2-x}.

RESULTS

From a 120 atom cell (2x2x5 conventional cells) we have removed as a linear neutral unit a Ti atom and 2 O atoms and relaxed the structure to equilibrium. This is the coulombically bound Schottky trio $V_{Ti} + 2V_O$. The converged formation energy E_s is calculated as 4.66 eV, comparing well with the Mott-Littleton values and space charge measurements. The Schottky formation energy may also be calculated as the sum of the energies needed to extract a Ti atom and an O atom (x2) to form isolated relaxed Ti and O atom vacancies, minus the cohesive energy per TiO_2 unit. As E_s is obtained as the small difference between large energies, the same plane wave cutoff and sampling vectors must be used in all the calculations. The isolated defect calculations were carried out in a 12 atom cell, with defects in neighbouring cells being separated by $\sim 5\overset{\circ}{A}$.

In the case of the oxygen vacancy, we find the excess electrons occupy Ti d-states localised on the Ti ions near the vacancy, see Fig. 1. If a neutral Ti atom is removed from the bulk, there is the possibility of holes appearing on the oxygens near the vacancy, i.e., O,O^- being present, so spin polarisation is needed. There are three possible spin states which need to be considered, because if $2O^{2-}$ becomes $2O$ then it is possible that 4 up,2 down electrons on each atom is a low state, i.e.$S_z=+2$, where S_z is the total projected spin. But $S_z=0$ may be a low or lower energy state as one atom may be spin up, the other spin down. Also $S_z=+1$ cannot be ruled out as the system may not behave like separate atoms but the holes may lie over a number of atoms, making the +2 case less likely. To calculate the energy correctly we need to consider the correct spin state, and so electronic convergence has been carried out on all 3 cases to see if one is much lower than all the others. Preliminary results show that they are quite close in energy, decreasing in 0.1 eV steps from spin 0 to 2. Note that full ionic relaxation is yet to be carried out on the Ti vacancy, see Table 2.

The isolated Schottky energy is much higher than the bound trio as we would expect as they are completely different configurations, the energy difference being the binding energy of the trio. In the isolated case there are holes and excess electrons around the isolated defects, not possible in the bound trio.

Ti extraction energy (LDA, $S_z=+2$)	+110.523 eV
O extraction energy (LDA, $S_z=+1$)	+429.236 eV
E_{coh} for TiO_2 unit (LDA, no spin)	-951.421 eV
Schottky energy E_s (unrelaxed ions)	+17.57 eV

Table 2: Isolated defect energies in rutile.

The Heat of Reduction

The aim is to determine the energy required for the following reaction (Kröger-Vink notation):

$$O_o + 2\text{Ti}_{Ti}^{4+} \Longrightarrow \ddot{V}_o + 2\text{Ti}_{Ti}^{3+} + 1/2\ O_2\ (g)$$

Now $\Delta E_{heat} = E_O{}^{ext} + E_f(1/2O_2(g))$,

where $E_O{}^{ext}$ is the O extraction energy and $E_f(1/2O_2(g))$ is the formation energy of the oxygen molecule. We can calculate this in the same level of approximation through an *ab initio* calculation of the free O atom energy using the same O pseudopotential, and knowing the experimental oxygen molecule dissociation energy, $E_{diss}(O_2(g))$.

$$E_f(1/2O_2(g)) = E_O{}^{free} - 1/2E_{diss}(O_2(g))$$

Due to the large core radius of the soft oxygen pseudopotential used in the calculations, it was not possible to calculate E_f by obtaining the free oxygen molecule energy. Since $1/2E_{diss}(O_2(g)) = 2.6$ eV and $E_O{}^{free}$ is -423.223 eV, we have that $E_f(1/2O_2(g)) = $ -425.823 eV. $E_O{}^{ext} = +430.662$ eV and so we calculate ΔE_{heat} to be + 4.8 eV (corresponding to 9.6 eV per oxygen molecule). Now we do not expect this to be negative, but it is perhaps a little too high given that TiO$_2$ is an anion deficient non-stoichiometric oxide. There is a possible error of about 1eV in this first attempt at a calculated value due to system size effects and using the experimental value for the dissociation energy. Also, the O vacancy calculation is unrelaxed.

The Ti Frenkel Pair energy

Rutile has open channels along the c-direction $[0,1/2,z]$, with the position $[0,1/2,1/2]$ being the centre of an octahedron of oxygen atoms. This is said by some authors [7] to be the most stable site for the location of the Ti interstitial. In fact there are equilibrium positions in the planes z=(n/4)c (n integer), where for odd n we have B sites, n even A sites. Bondlength arguments indicate that B sites might be the most stable equilibrium positions. The argument is simply that they afford a more suitable environment for the Ti interstitial, the bondlengths being more bulk-like. At a B site the intersitial has 4 nearest neighbours at 1.82 Å, at an A site only 2 at 2.23 Å. In the bulk each Ti has a coordination of 6. Huntington and Sullivan [8] favour the site $[0,1/2,1/4]$. It might be the case that a B site is in fact a potential maximum, the true minimum being displaced along say [110] in the basal plane. To test this we plan to carry out 3 constrained relaxations in the basal plane (i.e., fix z and allow relaxation only in the basal plane) for the 3 sites $[0,1/2,1/2]$,

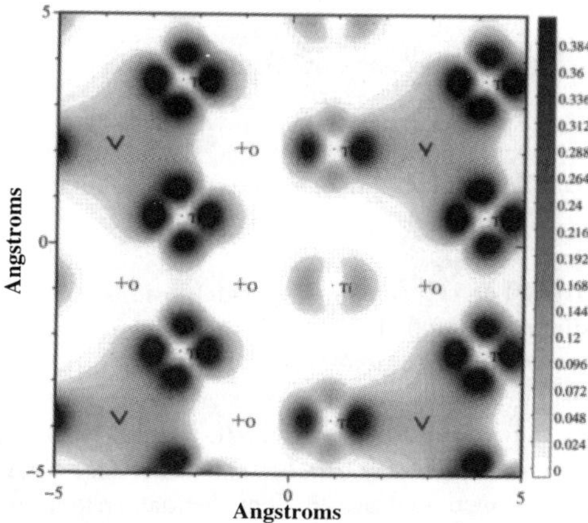

Figure 1: Spin density in (110) plane for periodic 12 atom cell with O vacancy (V), showing localisation of electrons on Ti d-states

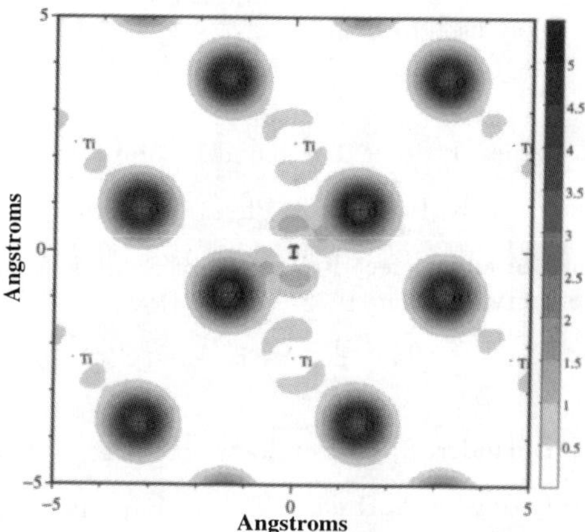

Figure 2: Valence electron density in (001) plane through Frenkel interstitial at $z=1/2$ (I), showing significant charge redistribution

site	E_f
[0,1/2,1/2]	17.52 eV
[0,1/2,1/4]	15.72 eV
[0,1/2,3/8]	16.74 eV

Table 3: Frenkel pair energies (no ionic relaxation)

[0,1/2,1/4] and [0,1/2,3/8]. The cell size in these calculations is 72 atoms, and stoichiometry is maintained by introducing a Frenkel pair with vacancy and interstitial separated by 5 Å. There is therefore no extra charge (4 valence electrons) being introduced due to an interstitial only. So far only electronic relaxation has been carried out, but the first results shown in Table 3 (converged electronic ground state) indicate that the [0,1/2,1/4] site has the lowest formation energy, consistent with [8]. There is seen to be significant charge redistribution around the interstitial, as shown in Fig. 2. The formation energies will be corrected by Classical Mott-Littleton calculations at infinite dilution.

CONCLUSIONS

We have performed the first set of fully self-consistent first principles calculations on the nature of the dominant point defect mechanism in TiO_2. Preliminary results indicate the importance of considering spin polarisation and full ionic relaxation in determining the relative energies of the various isolated and bound point defect types.

ACKNOWLEDGMENTS

This work was supported by the EPSRC under grants GR/K03975 and GR/K41649. The authors are grateful to Professor Richard Catlow at the Royal Institution and Professor Y-M Chiang of M.I.T. for many useful discussions.

REFERENCES

1. C.R.A. Catlow and R. James, Proc. R. Soc. Lond. A **384** (1982).

2. J. Sasaki, N.L. Peterson, and K. Hoshino, J. Phys. Chem. Solids **46** [11] (1985).

3. J.S. Anderson, in "Surface and Defect Properties of Solids", Specialist Periodical Reports, The Chemical Society, London, 1972, Vol. 1., Dover.

4. M.C. Payne, M.P. Teter, D.C. Allan, T.A. Arias, and J.D. Joannopoulos, Rev. Mod. Phys. **64** (1992) 1045.

5. L. Kleinman and D.M. Bylander, Phys. Rev. Lett. **48**, 1425 (1982).

6. M-H. Lee, J.S. Lin, M.C. Payne, V. Heine, V. Milman, and S. Crampin, "Optimised Pseudopotentials with kinetic energy filter tuning", submitted to Phys. Rev B.

7. P.F. Chester, J Appl. Phys **9**, 95 (1963).

8. H.B. Huntington and G.A. Sullivan, Phys. Rev. Lett. **14**, 6 (1965).

9. J.A.S. Ikeda and Y-M. Chiang, J. Am. Ceram. Soc. **76**, 2437 (1993).

ATOMIC-SCALE SIMULATIONS OF STRUCTURAL
PROPERTIES OF CERAMICS

D. J. KEFFER*, F. H. STREITZ**, AND J. W. MINTMIRE*
*Code 6179, Naval Research Laboratory, Washington, DC 20375
**Physics Department, Auburn University, Auburn, AL 36849

ABSTRACT

We have recently developed a novel computational method for molecular dynamics simulations of metal oxide ceramics. This approach explicitly includes variable charge transfer between anions and cations. This method has been used to model the structural properties of bulk and surface alumina and aluminum systems including tensile failure of the bulk systems, as well as to model the rupture under tensile stress of an interface between a (0001) face of α-alumina and a (111) face of aluminum. We have applied this method to perform atomic-scale simulations of nanoindentation of ceramic and model rigid tips onto metal and ceramic substrates.

INTRODUCTION

Friction, wear, adhesion, and lubrication are closely related processes in materials that are of fundamental importance to our understanding of a vast range of technological problems. The mechanical basis of all of these derive from the interaction of two surfaces at an interface, and a fundamental understanding of the dynamics and energetics of the interface is an important component of our understanding of these processes. Progress in understanding the interfacial interactions has historically been limited by the inability to examine what is going on at the atomic scale. As pointed out by several recent reviews,[1-3] experimental techniques have been developed in the last few decades that allow resolutions at or near atomic-scale dimensions, with the most common methods being the surface force apparatus (SFA),[4] the scanning-tunneling microscope (STM),[5] and other proximal probe techniques related to the STM, the atomic-force[6] and friction-force[7] microscopes (AFM and FFM). Theoretical techniques have also evolved to treat problems in this new area of atomic-scale tribology, "nanotribology,"[2,3,8,9] in this case by extending earlier simulation techniques to larger domains both in time and spatial extent so that realistic simulations can be made of the interfacial interactions contributing to adhesion, friction, and wear.

As part of an overall goal in modeling metal-oxide and other ceramic interfaces, we have begun preliminary simulations of nanometer-scale indentations of model tips on ceramic and metal surfaces. Herein we discuss our simulations of the nanoindentation of a tip modeled on diamond into a thin alumina film. These first simulations have been carried out to verify the utility and physical reasonableness of the application of our approach to nanoindentation, and to provide us with guidance for more complex future simulations on this topic. Our initial results are promising and suggestive for several avenues of continuing research in this area.

Mat. Res. Soc. Symp. Proc. Vol. 453 © 1997 Materials Research Society

APPROACH

The work described herein is part of a larger effort to simulate and understand the properties of metal-oxide surfaces and interfaces. One initial goal in this project was to simulate the atomic-scale dynamics and energetics of technologically important metal/metal-oxide interfaces. Modeling of ceramic surfaces and interfaces is difficult because of the complexity of the bonding at the interface. Metal atoms in the bulk metal will form metallic bonds, while those in the metal-oxide will bond ionically. We have recently developed a novel computational method which explicitly includes variable charge transfer between anions and cations in a molecular dynamics simulation for metal oxides.[10] This method has been used to model the structural properties of bulk and surface alumina and aluminum systems including tensile failure of the bulk systems,[11] as well as to model the rupture under tensile stress of an interface between a (0001) face of α-alumina and a (111) face of aluminum. We briefly outline the approach below.[12]

One of the dominant interactions in the metal-oxides is the Coulombic interaction between anions and cations. Earlier models using empirical potentials included such effects by incorporating fixed atomic charges and polarizability functions into the energetics of metal-oxides. Because our ultimate goal in this work is to develop relatively simple empirical potentials for studying the adhesion of metal-oxides with metal substrates, any new approach must include an ability to calculate the local atomic charge (or equivalently, the valence) based on the local environment of each atom. The long-range nature of the Coulomb interaction will lead to this "local" environment typically being relatively large-scale, of the dimensions of the screening length in the metal oxides.

Conceptually then, what is needed is a description of the total electrostatic energy of an array of atoms as a function of atomic charges (valences) and position. We define an atomic energy term E_i in terms of the local charge q_i on atom i

$$E_i(q_i) = E_i(0) + \chi_i^0 q_i + \frac{1}{2}J_i^0 q_i^2 \tag{1}$$

where χ_i^0 and J^0 correspond to local atomic properties traditionally denoted as the electronegativity[13,14] and hardness.[15]

The electrostatic energy, E_{es}, of a set of interacting atoms with total atomic charges q_i is then given by the sum of the atomic energies E_i, and the electrostatic interaction energies between all pairs of atoms,

$$E_{es} = \sum_i E_i(q_i) + \frac{1}{2}\sum_{i \neq j} V_{ij}(\mathbf{R}_{ij}; q_i, q_j) \tag{2}$$

where $\mathbf{R}_{ij}$ denotes the vector $\mathbf{R}_i - \mathbf{R}_j$. The Coulomb pair interaction $V_{ij}(\mathbf{R}_{ij}; q_i, q_j)$ is given by the classic electrostatic interaction between two charge distributions $\rho_i(\mathbf{r}_1; q_i)$ and $\rho_j(\mathbf{r}_2; q_j)$, where $\rho_i(\mathbf{r}; q_i)$ is the charge distribution about atom i (including the nuclear point charge) for total charge q_i and r_{12} is the distance between two atoms. The simplest model for $\rho_i(\mathbf{r}; q_i)$ is as a point charge of charge q_i; this leads to $V_{ij}(\mathbf{R}_{ij}; q_i, q_j) = q_i q_j / R_{ij}$. Rappé and Goddard[16] have suggested the use of spherically symmetric exponential functions to generate a $\rho_i(\mathbf{r}; q_i)$ linear in q_i. We then choose the values of q_i as those that minimize E_{es} subject to the constraint that the sum of the q_i be constant.

The electrostatic energy is just one component of the total energy of a metal-oxide (or metal) system. Indeed, the electrostatic interaction between cations and anions will be strictly attractive at any internuclear separation, so that in the short-range limit a repulsive potential will be needed to maintain physically reasonable internuclear separations in any empirical potential constructed using the above described electrostatic potential. What is needed for a complete potential is a description of the non-Coulombic part of the interatomic interactions. This remaining interaction could be modeled using any standard empirical potential, such as a sum of pair potentials, an embedded atom method (EAM) approach,[17,18] or a many-body potential such as that developed by Abell and others for covalent systems.[19-21] We have chosen to merge our above described electrostatic model with an EAM potential.[10]

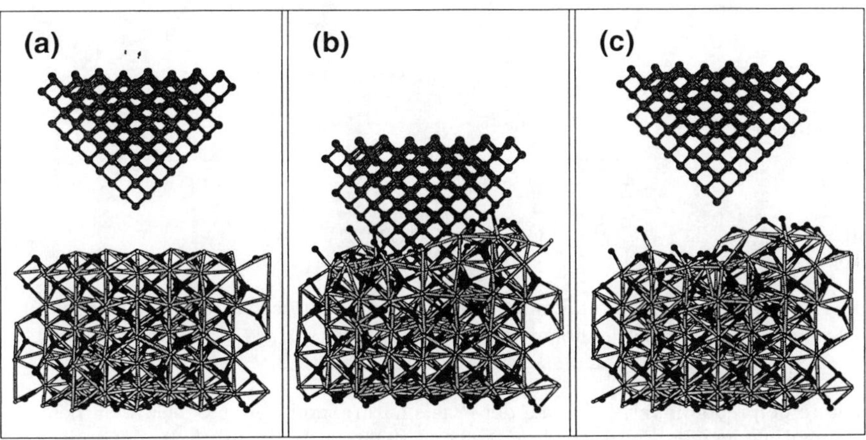

Figure 1. Depiction of "diamond" tip nanoindentation of (0001) surface of α-alumina at (a) 3 Å separation of tip and surface, (b) indented into the surface to a depth of 4 Å, and (c) withdrawn from the surface to a distance of 3 Å.

RESULTS

Our initial nanoindentation simulations have been carried out for a rigid tip with a diamond lattice structure. As depicted in Figure 1a, the tip is composed of 155 atoms arranged in the tetrahedrally-bonded structure of diamond, with carbon-carbon bond distances of 1.55 Å. The tip was constructed by aligning the tip axis along a (100) direction of diamond, and cleaving four (111) faces of diamond to form a four-sided pyramid. A tip structure originally including 203 atoms was reduced to 155 atoms by removing atoms so that the tip would fit into the two-dimensionally periodic supercell imposed for the alumina thin film. In these initial simulations, we are holding the carbon-carbon bond distances and bond angles fixed, corresponding to a completely rigid tip. We plan to relax this constraint in future work. Interactions between the tip atoms and alumina atoms were

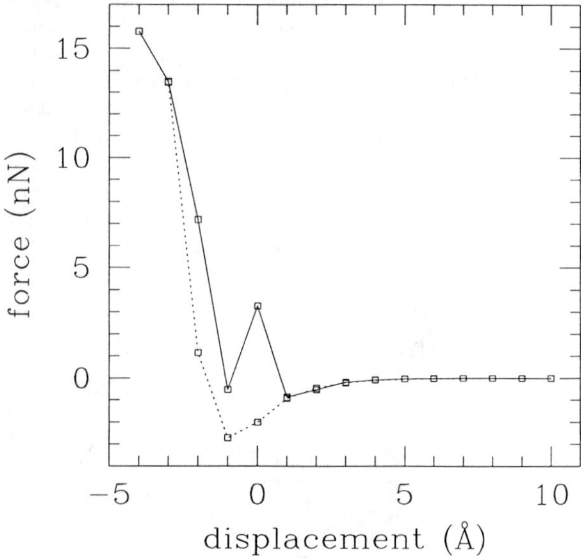

Figure 2. Loading (solid line) and unloading (dashed line) force versus displacement curve for "diamond" tip nanoindentation into (0001) surface of α-alumina.

assumed to be predominantly of a van der Waals nature, and were modeled with Lennard-Jones potentials having a well depth of 0.008 aJ (0.008 × 10^{-18} J or 0.05 eV), and values of σ of 2.29 Å for the C-O interaction and 1.28 Å for the C-Al interaction.

The alumina substrate was modeled with a 3 × 3 supercell of a (0001) slab of α-alumina with 3 layers of the conventional Al_4O_6 crystalline unit cell, for a total of 270 atoms in the alumina slab periodic repeat unit cell and with a thickness of 13 Å. The alumina supercell had hexagonal symmetry in the plane of two-dimensional periodicity, with a supercell lattice translation of 14.24 Å. A 1 × 1 reconstruction was performed using a direct minimization technique on an alumina slab constructed using the bulk structural constants of α-alumina, with the exterior surface a plane of aluminum atoms. The principal result of the reconstruction is to reduce substantially the interplanar distance between surface aluminum atoms and the next interior plane of oxygen anions.

Molecular dynamics simulations were carried out for the system by integrating Newton's equations of motion with a Gear fifth-order predictor-corrector[22] using a 2 fs time step. In addition to the diamond tip being held rigid, an exterior plane of aluminum atoms in the alumina slab was also held rigid. A temperature of 300 K was maintained using a Berendsen thermostat.[23] The simulation was started with a tip to surface separation of 10 Å. The system was evolved for 2 ps to allow equilibration, and then a time average was taken over the subsequent 2 ps of the potential and kinetic energies, as well as forces on the rigid atoms. The tip was then moved toward the alumina slab in 1 Å steps at a rate of 0.25 Å per ps, with a total displacement of 14 Å corresponding to a maximum indentation of 4 Å. The 2 ps equilibration and 2 ps averaging periods were carried out for each 1 Å step. Unloading was carried out with an analogous procedure.

Figure 2 depicts our simulation results for force versus displacement of the diamond tip. On loading the force is seen to be weakly attractive with a maximum magnitude of -1 nN just before contact. At contact the force becomes markedly repulsive with a magnitude of 3 nN, then becomes weakly attractive at the next step before becoming uniformly repulsive at greater indentations. This feature appears to be a result of the tendency of the (0001) surface of α-alumina to reconstruct preferentially at higher temperatures to several larger scale structures, including a $(\sqrt{3} \times \sqrt{3})$ rotated 30°, $(3\sqrt{3} \times 3\sqrt{3})$ rotated 30°, and a $(\sqrt{31} \times \sqrt{31})$ rotated $\pm \tan^{-1}(\sqrt{3}/11)$.[24,25] All three of these structures are in registry with the 1×1 structure. The nanoindentation process appears sufficient to create a lower energy reconstructed surface apparently related to the above reconstructions. On unloading this peak in the force curve at contact is absent, consistent with an irreversible reconstruction of the surface. We are currently carrying out simulations examining the energetics of the possible reconstructions for the (0001) α-alumina surface.

In summary, we have carried out preliminary simulations of nanoindentation of a diamond tip into an α-alumina substrate. The force and energy curves versus displacement are physically reasonable, and our initial results on a 1×1 reconstructed surface are suggestive that reconstruction energetics could be observable with AFM-based nanoindentation experiments. We are currently expanding on these simulations in several areas. First, we are examining the possible reconstructed surfaces to improve our description of the substrate. Second, we plan to implement a non-rigid diamond tip with the effective elastic properties of diamond in the near future. Finally, we are developing a version of our molecular dynamics code on a parallel platform to allow substantially larger simulation sizes on the order of 5000–10000 atoms.

ACKNOWLEDGMENTS

This work was supported by the U.S. Office of Naval Research (ONR), both through the U.S. Naval Research Laboratory and directly from the Chemistry and Materials Divisions of ONR. DJK acknowledges support through the NRC-NRL Postdoctoral Associateship Program. Computational support was provided under a Department of Defense High Performance Computing Modernization Program grant.

REFERENCES

1. I. L. Singer, and H. M. Pollock, *Fundamentals of Friction*, Kluwer, Dordrecht, 1992.

2. B. Bhushan, J. N. Israelachvili, and U. Landman, *Nature* **374**, 607 (1995).

3. I. Singer, *J. Vac. Sci. Technol A* **12**, 2605 (1994).

4. D. Tabor and R. H. S. Winterton, *Proc. Roy. Soc. London A* **312**, 435 (1969); J. N. Israelachvili, D. Tabor, *Nature* **241**, 148 (1973); *Wear* **24**, 386 (1973).

5. G. Binnig, H. Rohrer, Ch. Gerber, and E. Weibel, *Phys. Rev. Lett.* **49**, 57 (1982); *Phys. Rev. Lett.* **50**, 120 (1983).

6. G. Binnig, C. F. Quate, and Ch. Gerber, *Phys. Rev. Lett.* **56**, 930 (1986).

7. C. M. Mate, G. M. McClelland, R. Erlandsson, and S. Chiang, *Phys. Rev. Lett.* **59**, 1942 (1987); G. Meyer and N. M. Amer, *Appl. Phys. Lett.* **56**, 2100 (1990).

8. H. Yoshizawa, Y.-L. Chen, and J. Israelachvili, *J. Phys. Chem.* **97**, 4128 (1993).

9. J. A. Harrison and D. W. Brenner, in *The Handbook of Micro/Nanotribology*, edited by B. Bhushan, CRC Press, Boca Raton, 1995, pp. 397–439.

10. F. H. Streitz and J. W. Mintmire, *J. Adhes. Sci. Technol.* **8**, 853 (1994); *Phys. Rev. B* **50**, 11996 (1994).

11. F. H. Streitz and J. W. Mintmire, *Langmuir* **12**, 4605 (1996).

12. F. H. Streitz and J. W. Mintmire, *Composite Interfaces* **2**, 473 (1994).

13. R. P. Iczkowsky and J. L. Margrave, *J. Am. Chem. Soc.* **83**, 3547, (1961).

14. R. G. Parr, R. A. Donnelly, M. Levy, and W. E. J. Palke, *J. Chem. Phys.* **68**, 3801 (1978).

15. R. G. Parr and R. G. Pearson, *J. Am. Chem. Soc.* **105**, 7512 (1983).

16. A. K. Rappé and W. A. Goddard, *J. Phys. Chem.* **95**, 3358 (1991).

17. M. S. Daw and M. I. Baskes, *Phys. Rev. Lett.* **50**, 1285 (1983); *Phys. Rev. B* **29**, 6443 (1984); S. M. Foiles, M. I. Baskes, and M. S. Daw, *Phys. Rev. B* **33**, 7983, (1986).

18. M. W. Finnis and J. E. Sinclair, *Phil. Mag. A* **50**, 45 (1984).

19. G. C. Abell, *Phys. Rev. B* **31**, 6184 (1985).

20. J. Tersoff, *Phys. Rev. Lett.* **61**, 2879 (1988); *Phys. Rev. Lett.* **56**, 632 (1986).

21. D. W. Brenner, *Phys. Rev. B* **42**, 9458 (1990).

22. C. W. Gear, *Numerical Initial Value Problems in Ordinary Differential Equations*, Prentice-Hall, Englewood Cliffs, NJ, 1971.

23. H. J. C. Berendsen, J. P. M. Postman, W. F. Van Gusteren, A. DiNola, and J. R. Haak, *J. Chem. Phys.* **81**, 3684 (1984).

24. J. M. Charig, *Appl. Phys. Lett.* **10**, 139 (1967).

25. C. C. Chang, *J. Appl. Phys.* **39**, 5570 (1968).

A THEORETICAL AND EXPERIMENTAL STUDY OF THE CHEMICAL BONDING IN AgGaS$_2$

Alessandra Continenza[1], Teresa M. de Pascale[2], Franco Meloni[2],
Marina Serra[2], Ali Shaukat[3], Hans Burzlaff[4] and Roland Spengler[4]
1 INFM-Physics Dept., University of L'Aquila, Italy
2 INFM-Physics Dept., University of Cagliari, Italy, franco@sparc10.unica.it
3 Physics Dept., Punjab University of Lahore, Pakistan
4 Inst. of Physics, University of Erlangen, Germany

ABSTRACT

AgGaS$_2$ is a technologically important semiconductor for its large birefringence coefficient. In this work we compare the theoretical *ab-initio* all-electron FLAPW results with very refined experimental data obtained with accurate X-ray analysis. In particular we focus our attention to the electronic distribution along the significative bonding directions connecting the three different atoms. Furthermore, the charge density contours around Ag provide a clear evidence of the contribution of its *d* orbitals to the chemical bond.

INTRODUCTION

Ternary ABC$_2$ semiconductors crystallise in the chalcopyrite structure and belong to a wide family of materials, which includes A^{I}B^{III}C$_2^{VI}$ and A^{II}B^{IV}C$_2^{V}$ type compounds. They form a group of crystals with diverse optical, electrical and structural properties [1]. For example, CuInSe$_2$ has recently emerged as a very promising material for photovoltaic solar-energy applications [2], while the narrow gap of some pnictide compounds (A=Mg, Zn, Cd; B= Si, Ge, Sn; and C= P, As, Sb) make them suitable as infrared detectors. The larger degree of freedom with respect to that of their binary analogs provides a valid reason for interest in their possible device applications due to the wide variety of optical gaps offered by this class of materials. The study of the relationship between a chalcopyrite and the corresponding binary analog has involved many speculative investigations on the role of the cationic sequence in the non cubic crystallographic phases [3]. In particular it has been demonstrated that the ternary band structures are not a simply direct folded arrangement of the binary correspondent compound [4]. Many effects like band gap anomaly, defined as the difference between the binary and the ternary gap, strongly depend on structural parameters like tetragonal distortions. This effect displaces the anions from their zinc-blende original sites [5]. As a result, the anion C adopts a relaxed configuration that must be taken into account in a reliable comparison between theory and experiment. As a general result we may state that the structural parameters influence systematically the electronic properties of chalcopyrites.

Mat. Res. Soc. Symp. Proc. Vol. 453 © 1997 Materials Research Society

In this work we focus our attention on the $AgGaS_2$ compound and study its electronic properties both theoretically and experimentally. We have taken into account this particular crystal for its interesting photovoltaic properties characterised by a wide energy gap, Eg=2.73eV [6].
A complete analysis requires reliable physical indicators able to compare well defined experimental measures with theoretical results. In this case we analyse the plots of the electronic valence charge density distributions obtained with X-ray analysis and computed in the all-electron FLAPW scheme.

THEORY AND EXPERIMENT

The space group of the ternary ABC_2 chalcopyrite structure is D_{2d}^{21}. The unit cell contains two formula units (eight atoms) and the atomic positions are defined in terms of the atomic parameters (a,a,c), a=10.8476 a.u. The crystal shows a tetragonal distortion $\eta=c/2a=0.8996$.

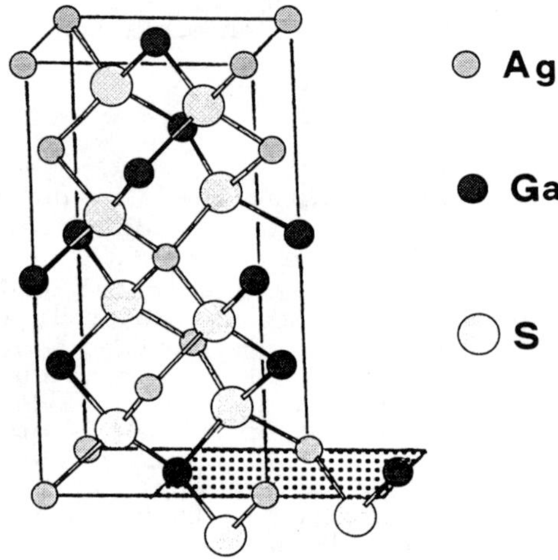

Figure 1. Chalcopyrite crystallographic unit cell. The basal (001) plane contains the cations of Figure 2.

The anion sublattice relaxation makes the internal parameter u, which is 1/4 in the ideal zinc-blende structure, equal to 0.288. With this choice of parameters we use the all-electron full-potential linearized augmented plane wave (FLAPW) method and treat the core states fully relativistically and updated at each iteration [7]. The effect of spin-orbit coupling on the energy bands is considered as a perturbation on the semirelativistic calculation. In the present calculation, the outermost d states of Ag and Ga are considered as part of the valence bands, since they are quite shallow in energy. Inside the muffin-tin spheres the wave functions are expanded in spherical harmonics up to $l_{max}=8$, while $l_{max}=6$ is used for the charge-density and potential expansions. Integrations over the irreducible wedge of the Brillouin Zone were performed using four special k points [8].

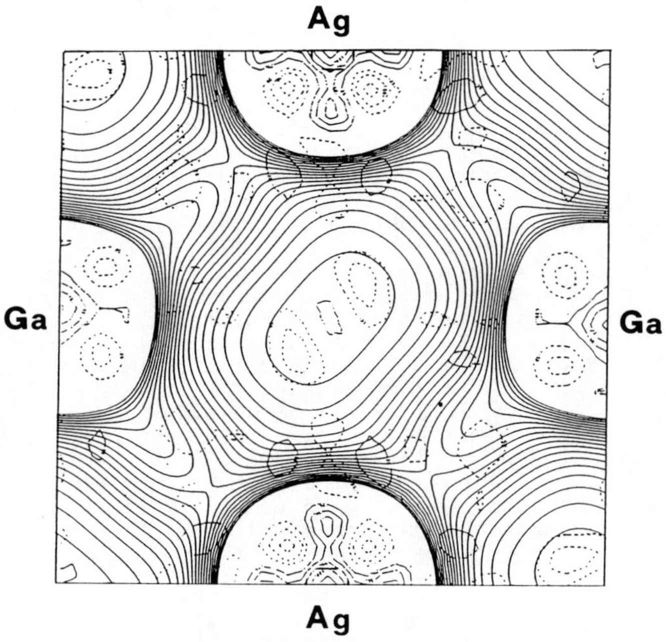

Figure 2. Valence electronic charge density distribution calculated (solid lines) and measured at 25 °K (broken lines) on the shaded plane of Fig. 1.

The experimental charge density distribution has been measured at various temperatures. Single crystal X-ray diffraction data has been collected at 293, 105 and 25°K for various samples. Intensity data were measured on a modified Huber four-circle goniometer with asymmetric Eulerian cradle, equipped with Mo-tube, graphite monochromator, conventional scintillation counter and a two-stage Displex cryostat with closed He circulation. Cell dimensions were determined from measured angles for 18 reflections ($23°<\vartheta<28°$). The crystal size and shape was measured within an accuracy of 1μm on a two-circle goniometer with mounted microscope and CCD-camera. The absorption corrections were evaluated analytically [9]. To avoid umweganregung and Renninger-effects every reflection was determined in three different Psi-settings. The standard reflections were monitored every 100 reflections. The data was corrected for effects like absorption, extinction, multiple scattering before using difference-fourier synthesis to determine the charge densities in different planes. It has been observed that as the temperature is lowered, from 293°K to

25°K, accumulation of charge takes place around the Ag atom, while some
charge depletion is noted on the neighbourhood of the Ga atom.
The charge accumulation has been estimated to be around 0.2 e at
this temperature.

RESULTS

Figure 1 reports the chalcopyrite $AgGaS_2$ in its fully relaxed
configuration. It is possible to notice the anionic displacements
due to the internal distortion. The comparison between
experimental and theoretical charge distribution has been
performed on the shaded part of the basal (xy0) plane. This plane
contains two Ag and two Ga atoms, while the bonds feel the
presence of the off-plane Sulphur atoms. The electronic valence
charge profile is plotted in Figure 2. In this picture the
contour plots obtained by the theory (solid lines) are almost
coincident with the experimental results (broken lines) obtained
at 25°K. It must be noticed that the theoretical calculations
have been performed at 0°K. The general shape, with the obvious
differences due to different methods, indicates how sensitive the
charge distribution is to the presence of the two different
chemical species. By comparing the computed charge densities in
the ideal (u=.250) and in the real (u=.288) configurations we
notice that the variation interests mainly the interstitial
zones. In the atomic spherical regions the difference is quite
negligible (<1%). This fact indicates that the charge
redistribution accounts for the crystallographic rearrangement
and it is not due to an electronegativity difference effect.

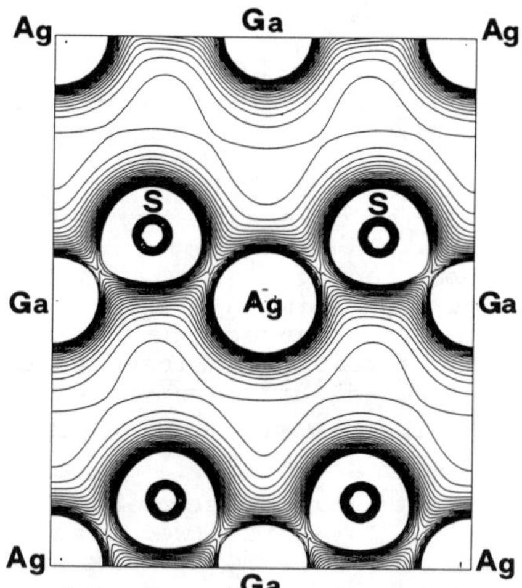

Figure 3. Valence FLAPW electronic charge density computed along the (110)
bonding plane.

The observed charge displacement from Ga toward Ag is also due to the d-states contribution, well evidenced by the experiment.
In order to gain more information on the strength of the chemical bonds we show in Figure 3 the FLAPW charge in the (110) plane cutting the Ag-S-Ga atoms. The asymmetry around S is evident also if the electronegativities of the two cations (1.48 for Ag and 1.46 for Ga) do not contribute to the strength of the chemical bond. The asymmetry with respect to the ideal binary zinc-blende is due to the internal distortions only and in a less extent to the chemical difference.

This report gives a direct comparison between theoretically calculated and experimentally measured charge density plots along the characteristic bonding planes in the chalcopyrite arrangement. Total energy first principle studies, combined with X-ray refined diffraction results seem to be the right tools for a comprehensive analysis of general physical properties in this class of semiconductors of which $AgGaS_2$ is a prototypical example. The analysis may be extended to other ternaries and the same frame may be adapted to gain information about complex crystallographic structures like heterojunctions and superconductors.

REFERENCES

1. For a recent review: Proc. 10th International Conference on Ternary and Multinary Compounds, Cryst. Res. Technol. **31** (1996)
2. H.L. Hwang, Ref.1, pag. 405
3. A, Continenza, S. Massidda, A.J. Freeman, T.M. de Pascale, F. Meloni and M. Serra, Phys. Rev. **B46**, 10070 (1992)
4. S. Massidda, A. Continenza, A.J. Freeman, T.M. de Pascale, F. Meloni and M. Serra, Phys. Rev. **B41**, 12079 (1990)
5. J.E.Jaffe and A. Zunger, Phys. Rev. **B29**, 1882 (1984)
6. L. Artus and Y. Bertrand, J. Phys. C: Solid State Phys. **20**, 1365 (1987); L.M. Suslikov, Yu. A. Khazitarkhanov, Z.P. Gad'mashi, D.S. Kovach and V.Yu. Slivka, Sov. Phys. Solid State **36**, 372 (1990)
7. H.J.F. Jansen and A.J. Freeman, Phys. Rev. **B30**, 561 (1984)
8. M. Gomm, Cryst. Comp. **6**, 1 (1993)
9. H.J. Monkhorst and J.P. Pack, Phys. Rev. **B13**, 5188 (1976)

HYDROGEN VIBRATION IN CUBIC DIHYDRIDES MH_2 ($M=Ti$, Zr), AND LOCALIZATION IN CUBIC LAVES PHASES $ZrM_2H_{1/2}$ ($M=V$, Cr, Fe, Co)

C. ELSÄSSER*, S. SCHWEIZER**, M. FÄHNLE**
* Institut für Werkstoffwissenschaft, Max-Planck-Institut für Metallforschung, Seestrasse 92,
D-70174 Stuttgart, Germany; ** Institut für Physik, Max-Planck-Institut für Metallforschung,
Heisenbergstrasse 1, D-70569 Stuttgart, Germany.

ABSTRACT

A study of cubic dihydrides TiH_2 and ZrH_2 and of H in cubic Laves-phase compounds by means of ab-initio total-energy calculations in the local density-functional approximation is presented. First, for optic vibrational modes in the two dihydrides the calculated local potentials and excitation energies are compared to experimental results obtained by inelastic neutron scattering. Second, the relative stabilities of H on three types of interstitial sites in the cubic Laves-phase compounds, denoted by b, e and g, are investigated. Theoretically, a preference of g sites is found for $M=V$, Cr, and of e sites for $M=Fe$, Co. These results are discussed with respect to experimental observations and to empirical criteria for H accomodation in transition-metal compounds.

INTRODUCTION

The quantitative determination of occupied interstitial sites and excitation spectra with corresponding potentials for vibrations of light particles like hydrogen isotopes absorbed in metal lattices has been a topic of research for many years. (For reviews on metal-hydrogen systems we refer to the books [1], [2] and [3].) Experimentally, most informations were obtained via inelastic neutron scattering (INS) techniques, in which the excitation energies for transitions from the vibrational ground state (zero-point motion) to excited states can be measured directly (see, e.g., ch. 3 in [2], vol. II). The shapes of the local potential wells were usually determined from the measured excitation energies under the assumption that the vibrations are only weakly anharmonic and, hence, can be treated via first-order perturbation theory (see, e.g., ch. 4 in [3]).

The quantitative theoretical prediction of the vibrational potentials and the resulting quantum states for H isotopes in metals without the assumption of weak anharmonicity is now possible by means of ab-initio total-energy calculations based on the local-density-functional theory [4]. For instance, Ho et al. [5] have calculated the vibrational spectra of H in body-centered cubic (bcc) Nb with excitation energies in very good agreement with INS data. Similar calculations for H states in face-centered cubic (fcc) Pd were done by Elsässer et al. [6]. They found that the vibrations can be strongly anharmonic and anisotropic, and they demonstrated the limitations of the common practice of perturbation theory to obtain informations about the potential shapes from INS spectra. Their prediction of a noticeable threefold splitting of the second excited state because of strong anharmonicity and anisotropy, for which there was no evidence in the INS data available then, has been confirmed in a more recent INS experiment [7].

Based on these two paradigmatic metal hydrides, NbH and PdH, a systematic investigation of vibrational quantum states of H isotopes in transition metals (hexagonal close-packed (hcp) Zr and Ti; bcc Nb, Cr and $\alpha - Fe$; fcc Pd, Ni and $\gamma - Fe$) and intermetallic compounds ($FeTi$ and $NiTi$) [8] using an ab-initio mixed-basis pseudopotential method [9] has led to a quantitative microscopic insight into the local interstitial H environments through the transition-metal series. In the present contribution, after a short outline of the computational method, results of ab-initio calculations for vibrational states of protons in the cubic dihydrides ZrH_2 and TiH_2 are presented as one example. As a second example the occupation of intersitital sites by protons in the cubic Laves-phase compounds $ZrM_2H_{1/2}$, with $M=V$, Cr, Fe, Co, is discussed [10].

Mat. Res. Soc. Symp. Proc. Vol. 453 © 1997 Materials Research Society

COMPUTATIONAL METHOD

The calculation of quantum-mechanical states of a transition-metal–hydrogen system can be decoupled into three steps because of large mass differences between the three constituents, i.e. light electrons, H nuclei and heavy metal ions (Born-Oppenheimer approximation).

In the first step, total energies are calculated with the metal ions and the H nuclei placed and fixed at specific positions in the unit cell of a crystal. The electronic structure of the (valence) electrons is calculated using the local density-functional approximation [4]. The interactions between the metal ions and the electrons are incorporated by non-local norm-conserving ionic pseudopotentials [11]. A local pseudopotential is used for the interaction of the electrons with the H nuclei [6]. Perfect translational symmetry of the crystals is assumed by using Born-von-Kármán boundary conditions, and the Bloch electron states are represented by a mixed basis set [9] containing a limited but for the considered properties sufficiently large number of plane waves and a few atom-centered functions constructed from atomic pseudowavefunctions and strictly confined to muffin-tin spheres. Integrals over the electron Brillouin zone are calculated by sampling and Gaussian broadening [12]. In the second step, quantum states of the H nuclei in the electron gas and the rigid lattice of metal ions are calculated. With a choice of crystal unit cells in the first step, which describe frozen displacements of the H and metal sublattices relative to each other according to normal modes of the crystal, this problem reduces to solutions of a single-particle Schrödinger equation. The "particles" move in Born-Oppenheimer or "adiabatic" potentials, which are mapped out by the total energies from the first step [5]. These vibrations belong to optic modes at the center of the phonon Brillouin zone (Γ point). For the third step, at the Γ point acoustic vibrations of the crystal have zero energy (rigid-body translations) and thus require no further computation.

It is essential to map out accurately the adiabatic potentials by ab-initio total-energy calculations for sufficiently many different H positions in the unit cell (e.g. 51 for ZrH_2 and 162 for $ZrFe_2H_{1/2}$).

HYDROGEN IN CUBIC DIHYDRIDES

By loading with hydrogen the fcc transition metals Pd and Ni can form stoichiometric monohydrides, PdH and NiH, where all octahedral interstitial sites of the fcc metal lattices are filled with H atoms (cf. the $NaCl$ structure). The hcp transition metals Ti, Zr and Hf, on the other hand, can be converted to stoichiometric dihydrides, TiH_2, ZrH_2 and HfH_2, with fcc lattices, in which all tetrahedral interstitial sites are occupied by H atoms (cf. the CaF_2 structure).

Experimentally, Ikeda et al. [13] measured excitation spectra of H vibrations in cubic TiH_2 and $ZrH_{1.41}$ as well as in tetragonally distorted $ZrH_{1.93}$ by means of inelastic neutron scattering (INS) over an energy range of 1 eV with a relative energy resolution of about 2 % . These INS spectra are quantitatively very similar and exhibit up to five distinct lines. The lines are spaced almost equidistantly in energy, like for a particle in a harmonic potential well, and the spaces become slightly smaller towards higher energies because of a weak anharmonicity.

Theoretically, in the present work [8] the vibrational H states were calculated for the normal modes at the Γ point of the phonon Brillouin zone of fcc dihydrides. Besides the rigid-body translations of the crystal, the three acoustic Γ modes, there are six optic modes because of the three atoms per formula unit. In one set of three optic modes the two H sublattices move in phase relative to the metal sublattice. The corresponding adiabatic potential has the symmetry of a simple cubic (sc) lattice with half of the lattice constant of the fcc metal lattice (called "sc" modes in the following). In the other set of three optic modes the two H sublattices move in antiphase relative to the metal sublattice at rest, leading to an adiabatic potential with fcc symmetry and the same lattice constant as the metal sublattice (called "fcc" modes).

The adiabatic potentials, i.e. ab-initio total energies as function of H sublattice displacements δ from the tetrahedral positions of the metal sublattice along high-symmetry directions in the cubic crystal according to the two optic modes in ZrH_2 (whose calculated lattice constant is a=4.79Å) are displayed in figure 1. (The corresponding curves for TiH_2, with a=4.38Å, are quantitatively very similar.) For small relative displacements δ/a the adiabatic potential curves of both optic modes nearly coincide and form local potential wells with almost isotropic and harmonic shapes. Anharmonicity and anisotropy become obvious for energies above 0.3 eV.

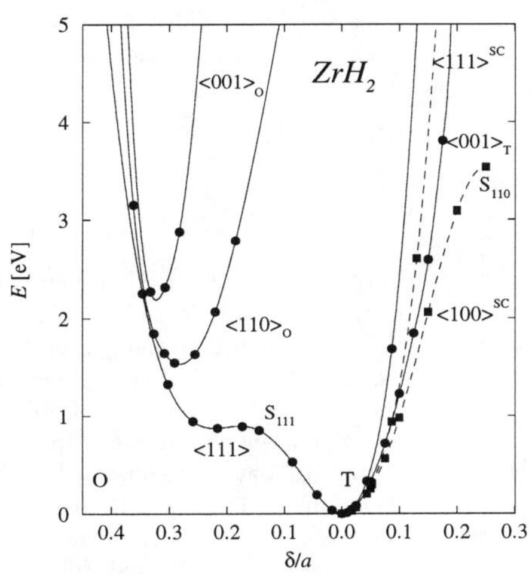

Figure 1: Adiabatic potentials for both types of optic Γ modes of ZrH_2: Energy change E as function of relative displacements δ/a (a=4.79Å is the equilibrium lattice constant). For δ=0 both H sublattices are located at tetrahedral sites T of the fcc Zr lattice. The ab-initio data are marked by black symbols and connected by cubic spline curves: circles and solid lines for the fcc modes, squares and broken lines for the sc modes. $< \ldots >$ denote directions in a cubic crystal. The subscripts T and O mark directions passing through tetrahedral or octahedral sites, respectively. The superscript sc indicates the two directions of the sc modes. Left at $\delta=\sqrt{3}/4$ both H sublattices come together at an octahedral site O. S_{111} and S_{110} mark saddle points in the adiabatic potentials.

Vibrational H states were calculated by solving a single-particle Schrödinger equation for H isotopes moving in the two adiabatic potentials. The resulting excitation energies for protons in ZrH_2 are displayed in figure 2. (Again the results for TiH_2 are almost the same.) For better comparison with the INS spectra, densities of states (DOS) were formed from the discrete Γ-point energy levels ε_ν by broadening with Gaussians, whose width Δ_ν was chosen to be $\Delta_\nu/\varepsilon_\nu$=2% to mimic the energy resolution of an INS spectrometer [13]. The positions of the peak maxima can be compared to those in the measured INS spectra. (The calculated peak heights and experimental intensities, however, should not be compared directly because transition matrix elements between different states were not included in the calculated DOS.) In the INS experiment both types of optic modes are excited, and in general a vibration is a superposition of the two modes. Consequently the INS spectra can be imagined as being formed from a superposition of both DOS. The sharpness of the INS peaks may be attributed to the sc modes, and some fine structures to the fcc modes (except for a tetragonal splitting in the case of $ZrH_{1.93}$ [13]).

In table 1 the peak-maximum positions of the sc modes and of the INS spectra are compared for ZrH_2 and TiH_2. The quantitative agreement between experiment and theory is very good for the lowest excitation. The almost equidistant spacing due to the nearly harmonic potential is also well reproduced. The influence of the weak anharmonicity, however, leads to slightly smaller intervals between the higher excitation lines in the INS spectra, whereas in the DOS the intervals are increasing, but by a smaller amount. This can partly be attributed to the numerical representation of the adiabatic potential and the solution of the Schrödinger equation in a plane-wave basis. Part of the deviation may also be due to the fact that in the INS experiment the intensity decreases strongly with energy, and the peak maxima become less sharp. Within these considerations, however, it can be stated that both ab-initio theory and INS experiment yield the same microscopic picture: The lowest six excited H vibrational states originate from almost harmonic and isotropic local potential wells in the cubic dihydrides. This behaviour is different from the case of cubic monohydrides like PdH and NiH, where strong anharmonicities and anisotropies are noticeable for the lowest two excitation states [6,8].

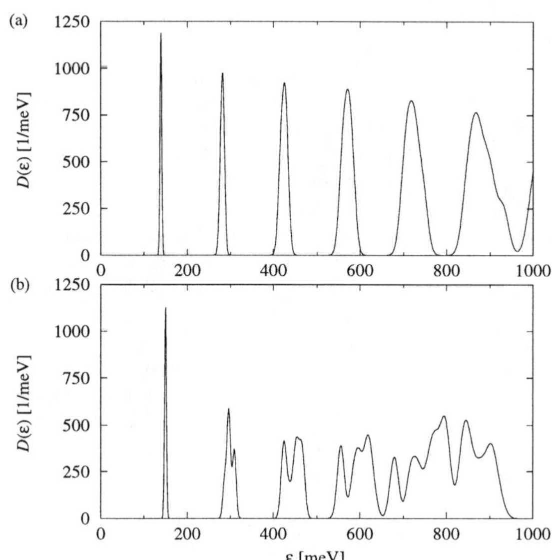

Figure 2: Densities of states of vibrations for protons in ZrH_2, formed by broadening the calculated discrete energy levels ε_ν with Gaussians of relative width $\Delta_\nu/\varepsilon_\nu=2\%$: (a) sc modes; (b) fcc modes.

Table 1: Comparison of the energies of the line maxima, as multiples of the first excitation energy, in the calculated DOS of sc-modes and in the measured INS spectra of ZrH_2 and TiH_2 (energies in meV; the experimental data were obtained from equation (2) in [13], with parameters for the harmonic frequency and the anharmonicity quoted there).

line	ZrH_2 sc modes	$ZrH_{1.41}$ INS [13]	TiH_2 sc modes	TiH_2 INS [13]
1	140	140	148	143
2	281≈ 2·141	274≈ 2·137	298≈ 2·149	281≈ 2·141
3	426≈ 3·142	400≈ 3·133	450≈ 3·150	414≈ 3·138
4	571≈ 4·143	519≈ 4·130	604≈ 4·151	542≈ 4·136
5	718≈ 5·144	632≈ 5·126	759≈ 5·152	666≈ 5·133

HYDROGEN IN CUBIC LAVES PHASES

Cubic Laves phases are binary intermetallic compounds AB_2 with a fcc Bravais lattice and an atomic basis of two species, where atoms of species A are located on a sublattice with diamond structure, i.e. the regular fcc-lattice sites and one of the two tetrahedral-site sublattices. Each four atoms of species B form regular tetrahedra centered at the other tetrahedral-site sublattice. Within this structure there are three types of interstitial sites, which can be occupied by H isotopes (see, e.g., ch. 4 in [2], vol. I): The b sites are formed by B_4 tetrahedra, the e sites by AB_3 and the g sites by A_2B_2. Per formula unit there are one b, four e and twelve g sites. All these sites constitute a complex network for hydrogen diffusion.

In this work [10] cubic Laves phases with $A=Zr$ and a series $B=V$, Cr, Fe, Co are investigated. The compounds ZrV_2 and $ZrCr_2$ are able to accomodate considerable amounts of H, e.g. up to ZrV_2H_6, and rather detailed experimental informations are available for these model systems of H storage materials. For instance, H occupies the g sites [14], whereas the b sites can be excluded because of its much smaller local volume. Also no evidence for e-site occupations was found. Empirical rules were formulated to rationalize these observations (see, e.g. [15,16]). The "chemical" criterion states that sites surrounded by as much Zr as possible are preferred because pure Zr forms more stable hydrides than pure V, Cr, Fe or Co. According to the "geometrical" criterion the interstitial site with the largest local volume is preferred. These rules as well as empirical heat-of-solution calculations [17] lead to the general expectation that g sites should always be preferred in ZrM_2 Laves phases.

The compounds $ZrFe_2$ and $ZrCo_2$ absorb less amounts of H, e.g. $ZrFe_2H_{1/2}$, and hence are not very relevant materials for H storage. Scientifically, however, studies of these materials are important because neutron diffraction experiments on hydrogenated $ZrFe_2$ by Rudnev et al. [18] gave evidence that e instead of g sites are occupied by H. This result, which is in contradiction to all experience and empirical rules for Laves-phase hydrides ZrM_2H_x, is commonly considered suspect [17] and thus deserves further clarification.

The topic of our work is an investigation of energy differences between the three sites for H in the series of Laves-phase compounds via ab-initio total-energy calculations for the formula unit $ZrM_2H_{1/2}$, with one H atom per fcc unit cell. In the calculations structural relaxations were included in several steps, starting from the H atom located at the geometrical center of each one of the three sites in the ideal lattice of ZrM_2, and ending with the unit-cell volume and with all atomic positions fully relaxed. Zero-point energies of H at the three sites in $ZrCr_2$ and $ZrFe_2$ were obtained by calculating vibrational H states in adiabatic potentials of optic Γ modes of the H sublattice in the same manner as for the dihydrides in the previous section.

The first result, in accordance with all previous experience, is that b-site occupations can be ruled out because of considerably higher (potential-minimum + zero-point) energies of proton states centered at b sites. For the other two sites, it turned out that in volume-relaxed unit cells the g site is energetically preferred over the e site, as expected, by 0.34 eV and 0.31 eV in ZrV_2 and $ZrCr_2$, respectively. In $ZrFe_2$ and $ZrCo_2$, however, the e sites were found to be energetically lower than the g sites by 0.15 eV. Despite all further effects of relaxation and zero-point motion for $ZrFe_2H_{1/2}$, the g-e energy difference remained at 0.11 eV in favour of the e site, giving theoretical support to the experimental observation of Rudnev et al. [18].

To understand this behaviour a detailed analysis of the atomic and electronic structures was deemed necessary [10]. In the ZrM_2 series with M from V to Co the local interstitial-site volumes shrink, but by H addition the crystal cohesion becomes stronger. For $M=V$, Cr the metal atoms Zr and M have similar valence-electron structures. Hence the type of neighboring atoms to an interstitial site is not so important. The decisive feature here is the local volume, and this makes the g sites most preferable. For $M=Fe$, Co the Zr and M atoms have more different electron configurations. First, Fe and Co are more electronegative than Zr. This leads to greater delocalization of Zr-centered electron states, and the Zr atoms loose some attraction to H. Second, because of the higher number of valence electrons, antibonding d orbitals are additionally filled in $ZrFe_2$ and $ZrCo_2$. Some depletion of these antibonding states of M-M pairs by the addition of H makes the strengthening of the crystal cohesion plausible. A larger number of neighboring M-M pairs from which antibonding electrons can be taken to screen the H atom, seems to be more important than the smaller local volume. This is further supported by the result that the energetic separation in the adiabatic potential between the b site with four surrounding M atoms and the g site with two M atoms is reduced from 1.25 eV in $ZrCr_2H_{1/2}$ to 0.58 eV in $ZrFe_2H_{1/2}$. But the b site is so small that here electronic effects are not yet sufficient to overcome the volume constraint.

Altogether, our ab-initio study shows that a change of the most stable interstitial site for H in cubic Laves-phase compounds may happen from the g site in $ZrCr_2$ to the e site in $ZrFe_2$. The results of total-energy calculations can be rationalized by changes in the electronic structure. They support the experimental observation of Rudnev et al. [18] for H in $ZrFe_2$ and predict the same behaviour for $ZrCo_2$.

SUMMARY

In the present contribution the capability of ab-initio total-energy techniques for studies of microscopic structural and vibrational properties of metal–hydrogen compounds has been illustrated by two examples. In the first example vibrational quantum states of protons in transition-metal dihydrides were calculated, giving excitation energies, which agree well with INS spectra, and detailed information about the shape of the underlying local vibrational potentials. In the second example significant differences in the interstitial-site preference of H in cubic Laves-phase compounds were found, which may help to reconsider the ranges of validity of commonly used empirical rules.

ACKNOWLEDGEMENTS

The authors gratefully acknowledge valuable discussions with K. M. Ho and C. T. Chan, enabled through the NATO collaborative research grant No. 910439.

REFERENCES

[1] Hydrogen in Metals, edited by G. Alefeld and J. Völkl, volumes I+II, Springer, Berlin, 1978.
[2] Hydrogen in Intermetallic Compounds, edited by L. Schlapbach, volumes I+II, Springer, Berlin, 1988 and 1992.
[3] Y. Fukai, The Metal-Hydrogen System, Springer, Berlin, 1993.
[4] P. Hohenberg and W. Kohn, Phys. Rev. **136**, B864 (1964); W. Kohn and L. J. Sham, Phys. Rev. **140**, A1133 (1965).
[5] K.-M. Ho, H.-J. Tao and X.-Y. Zhu, Phys. Rev. Lett. **53**, 1586 (1984); H.-J. Tao, K.-M. Ho and X.-Y. Zhu, Phys. Rev. B **34**, 8394 (1986).
[6] C. Elsässer, K. M. Ho, C. T. Chan and M. Fähnle, Phys. Rev. B **44**, 10377 (1991); C. Elsässer, K. M. Ho, C. T. Chan and M. Fähnle, J. Phys.: Condens. Matter **4**, 5207 (1992).
[7] A. I. Kolesnikov, I. Natkaniec, V. E. Antonov, I. T. Belash, V. K. Fedotov, J. Krawczyk, J. Mayer and E. G. Ponyatovsky, Physica B **174**, 257 (1991).
[8] C. Elsässer, Ab-initio-Elektronentheorie für Übergangsmetall-Wasserstoff-Verbindungen, Habilitationsschrift, Universität Stuttgart, 1994 (in german).
[9] S. G. Louie, K.-M. Ho and M. L. Cohen, Phys. Rev. B **19**, 1774 (1979); C. Elsässer, N. Takeuchi, K. M. Ho, C. T. Chan, P. Braun and M. Fähnle, J. Phys.: Condens. Matter **2**, 4371 (1990).
[10] S. Schweizer, Ab-initio-Berechnung der Eigenschaften von Wasserstoff in kubischen Laves-Phasen, Dissertation, Universität Stuttgart, 1995 (in german).
[11] D. R. Hamann, M. Schlüter and C. Chiang, Phys. Rev. Lett. **43**, 1494 (1979); D. Vanderbilt, Phys. Rev. B **32**, 8412 (1985).
[12] C.-L. Fu and K.-M. Ho, Phys. Rev. B **28**, 5480 (1983).
[13] S. Ikeda and N. Watanabe, Physica **120B**, 131 (1983); S. Ikeda and N. Watanabe, J. Phys. Soc. Japan **56**, 565 (1987).
[14] J. Didisheim, K. Yvon, P. Fischer and D. Shaltiel, J. Less-Common Met. **73**, 355 (1980); D. Fruchart, A. Rouault, C. Shoemaker and D. Shoemaker, J. Less-Common Met. **73**, 363 (1980).
[15] D. G. Westlake, J. Less-Common Met. **91**, 1 (1983).
[16] D. G. Ivey, D. O. Northwood, J. Mater. Sci. **18**, 321 (1983).
[17] I. Jacob and D. Shaltiel, J. Less-Common Met. **65**, 117 (1979).
[18] I. I. Rudnev, V. N. Bykov and V. I. Shcherbak, Fiz. Metal. Metalloved. **34**, 856 (1972).

MECHANISMS OF SHEAR-INDUCED SOLID-STATE REACTIONS

J. J. GILMAN
Department of Materials Science and Engineering, University of California at Los Angeles,
Los Angeles, California 90095

ABSTRACT

Both experiments and band-structure calculations indicate that shear strains lead to closure of energy band gaps in periodic solids, and closure of the LUMO-HOMO gaps in molecules (this is also expected for non-metallic glasses). It allows athermal and ultrafast mechano-chemical reactions to occur. Simple models account for the phenomena. For periodic solids, shear shortens one axis and extends another. In contrast, hydrostatic compression causes both axes to contract. Thus the former shifts the band gaps oppositely for the two directions, causing a reduction in the minimum gap. For molecules, shear tends to increase the bonding orbitals while decreasing the antibonding orbitals, thereby reducing the LUMO-HOMO gaps. The models are consistent with band structure models for diamond.

Shear-induced gap reductions account for the athermal chemical reactions observed by Bridgman, Enikolopyan, Boldyrev, and others. They account for the ultra-fast reactions that occur in explosives.

INTRODUCTION

The structures of molecules and solids are stabilized by gaps in the energy spectra of their bonding electrons. These gaps separate their lowest energy states from their first excited states [1]. In molecules the gaps are called the chemical hardness [2], in non-metallic solids they are the band gaps, and in metals the work functions. Spontaneous chemical reactions tend to increase the gaps (Table I).

Table I - Increase in chemical hardness values (eV) for some reactions [2].

Reaction	Reactants	Product
$Na + Cl = NaCl$	0.8	4.8
$Li + F = LiF$	1.0	5.4
$C + O = CO$	4.9	7.9
$H + OH = H_2O$	5.7	9.5
$CH_3 + Br = CH_3Br$	3.2	5.8
$Ni + CO = NiCO$	3.3	3.6

In addition to being thermally or photolytically stimulated, chemical reactions in solids can be mechanically stimulated by shear-strain energy. This is more effective than pure compression, or extension, because it breaks the symmetry of the solid. The Inverse behavior in which chemical reactions cause shape changes is also important; particularly for biochemical phenomena such as muscular contraction.

Shear strain causes spheres to become ellipsoids, cubes to become tetragonal or hexagonal, and so on. As will be shown here, this decreases the chemical hardness (the

Mat. Res. Soc. Symp. Proc. Vol. 453 © 1997 Materials Research Society

activation energy), and does it more effectively than does isotropic compression.

Some of the phenomena in which mechanochemistry plays a central role are: phase transitions, friction and wear, detonations, pyrotechnics, toys (cap-guns), solid-state synthesis, mechanical alloying, machining, biochemical phenomena, impact printing, catalyst activation, dissolution kinetics, polymer degradation, and others.

Bridgman [3] reported as long ago as 1935 that he could cause a variety of chemical changes by applying a combination of axial compression and torsion to specimens held between two anvils at room temperature. Under about 50 kbar of compression (+ torsion), he could cross-link polymers; decompose compounds such as PbO and $KMnO_4$; and drive substitution reactions such as, $2Al + Fe_2O_3 \rightarrow Al_2O_3 + 2Fe$, and $Mg + SiO_2 \rightarrow MgO + Si$. Since his specimens were thin wafers, and his anvils were good thermal conductors, he concluded that the reactions were athermal.

Bridgman's experiments were extended by others. Boldyrev [4] has emphasized that the reaction products of mechanochemistry may be distinct from those produced by thermochemistry for the same reactants. Enikolopyan [5] extended the work to various inorganic compounds such as hydrated cupric sulfate; and to the polymerization of various organic monomers (those containing -C=C-, -C=N-, $C \equiv C$, $C \equiv C \equiv C$, heterocyclic, and aromatic groups). Hundreds of reactions have been investigated, and are reviewed by Zharov [6]. Studies of the polymerization of acrylamide have demonstrated that polymerization occurs principally during the shear deformation, not afterwards.

Until recently, there was a great deal of phenomenological knowledge in this field, but no satisfactory mechanisms. The effects were usually rationalized by assuming that strain energy can assist thermal energy. Sometimes this approach is valid, but at low temperatures, the strain energy density is often large compared with the thermal energy density. In many cases, the applied strain becomes highly concentrated by the motion of dislocations [7]; so the important strain is tthe local one.

MECHANISMS

<u>Destabilization of Molecules</u>

When covalent bonds are bent (sheared), their energies of their highest-occupied-molecular-orbitals (HOMO) are raised, while the energies of their lowest-unoccupied-molecular-orbitals (LUMO) are lowered. Thus the gap between these levels is decreased. Since this gap determines a bond's stability, bending reduces the stability, thereby leading to a chenical reaction [8]. Consider the most simple case; the hypothetical H_3^- molecule which is related to the azide ion; N_3^-. A Walsh diagram for it is shown in Figure 1. The diagram correlates the molecular orbitals for the linear and bent versions.

Occupation of the levels is indicated by the small arrows, ↑↓ which represent spin up and spin down electrons. There are three protons and four electrons. The phases of the electronic wave functions are indicated by open circles for $\phi = 0$, and shaded circles for $\phi = \pi$. Thus like pairs with in-phase wave functions form bonds, while unlike pairs form anti-bonds.

For the linear molecule on the left, three orbitals are indicated, ϕ_b (bonding), ϕ_n (non-bonding), and ϕ_a (anti-bonding). ϕ_n is the highest occupied orbital (HOMO).

For the bent molecule on the right, the lowest orbital, ϕ_b is stabilized by increased bonding between atoms a and c. The increase in the a-c interaction lowers the energy of the unoccupied ϕ_A orbital (LUMO), The a-c interaction also increases for the non-bonding ϕ_N orbital (HOMO) so its energy increases.

Since work is done on the molecule to bend it, the total energy rises, but the LUMO-HOMO gap decreases, so the internal stability decreases, and electrons in the HOMO orbital become delocalized as more of ϕ_A becomes mixed with ϕ_N.

If a molecule is bent in its ground state, then straightening it decreases its stability.

Metallization of Solids via Shear

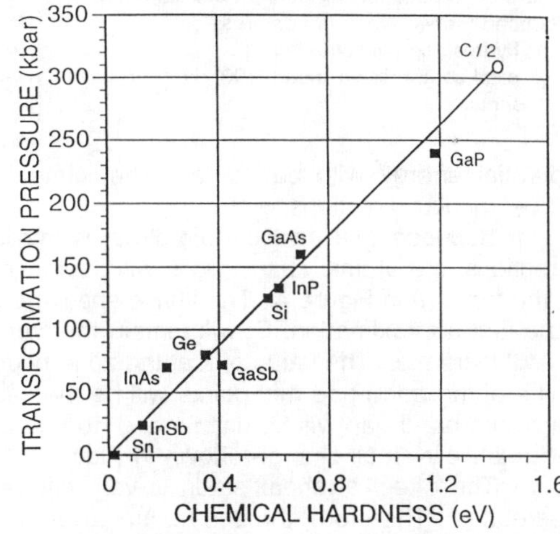

Figure 1 - Walsh diagram for H_3^- showing the effect of bending (shear) on the LUMO-HOMO gap. See the text for more description.

In periodic solids (crystals), the equivalents of LUMO-HOMO gaps are band gaps which are also equal to twice the chemical hardness, η. Bonding orbitals are similar to valence bands, and anti-bonding orbitals are similar to conduction bands. The larger the gap, the more stable the crystal structure [1] as indicated by Figure 2 where the critical pressure needed to cause a change in crystal structure is plotted against the chemical hardness for various crystals.

There are two mechanical paths for converting insulators into metals. One is to reduce the distances between the atoms so the overlaps of their wave functions increase. This is the Herzfeld-Mott approach [9]. The other is to change the bond angles through shear so the symmetry of the solid is changed. thereby reducing the minimum band gap [10]. As mentioned above , shear is more effective than changing the volume.

Calculations [11] have shown this for diamond by comparing the effect of compression along the 4-fold axis of diamond (combination of shear and dilatation) with the effect of isotropic compression (Figure 3).

The amount of compression

Figure 2 - Correlation between lowest critical pressures for phase transformations in various covalent crystals and chemical hardnesses (= 1/2 energy band gap). Note that both scales for the estimated value for diamond must be multiplied by two.

needed to make diamond itself metallic mains controversial. However, the structural transitions of Figure 2 are accompanied by metallization. In the cases of: Si, Ge, Sn, InSb, GaSb, AlSb, InAs, and GaP the structure changes from diamond (or zinc-blende) with the tetrahedral angle of 109.5° to the β-tin structure with bond angles: 149.5 and 94°. The bond lengths change at most by a few percent, and the coordination numbers remain four (contrary to some authors). The materials become metallic.

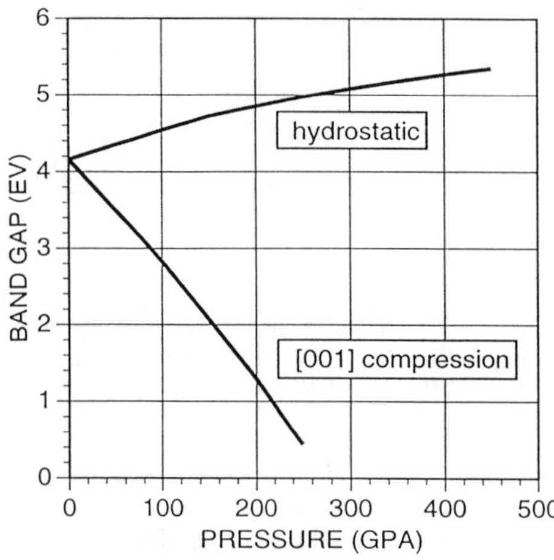

Figure 3 - Calculated variation of the band gap of diamond for two types of compression:
   1. hydrostatic (without shear).
   2. axial compression along [100] direction (with shear).

A simple two-dimensional model of the effect of shear on electronic structure provides some insight [12]. The model is intended to suggest that the effect is general. It is not intended to be quantitative. The essential feature is that while isotropic compression shortens both principal coordinates, shear shortens just one, and lengthens the other one.

The nearly free-electron approximation is assumed (the argument is similar for the LCAO approximation). Figure 4 shows a sketch of energy vs. wave number. Since simple shear is two-dimensional only two coordinates need to be considered. The lattice spacing of a square array is designated a, so the wave vector at the zone boundary is $2\pi/a$, and the energy of the electronic states is $E = \hbar^2 k^2/2m$ where $2\pi\hbar$ = Planck's constant, m = electron effective mass, and $2\pi k$ = electron wavelength, λ. On the left, the energy states are shown with an energy gap at $k = \pi/a$. At the mid-point of the gap the energy level is $E_o = (h/a)^2/8m$. The gap is given by: $E_g = 2|V|$ where V is the periodic potential energy with period, a. The same E vs. k dependence occurs in the two perpendicular directions.

Suppose that an isotropic strain is applied to the square lattice. Then, in both directions, the atomic spacings, a, will decrease and the E vs. k plot will shift as shown at the far right in Figure 4. The lattice spacings will become $a(1-\epsilon)$ where ϵ is the strain. To a first approximation, E_g will remain the same, but the position of the gap relative to E_o will increase. The value of k at the zone boundary will become: $\pi/a(1-\epsilon)$. Since the shifts of the band-gap mid-points will be the same in the two orthogonal directions, the minimum band gap will be unchanged (to a first approximation). This is consistent with the relatively small change shown in Figure 3 for the hydrostatic case.

The effect of shear strain is very different. One axis increases, and the other decreases. Therefore, the zone boundaries will become: $k = \pi/a(1 \pm \epsilon)$, and the band-gap mid-points shift in opposite directions. Thus the minimum gap becomes, E_g^* as indicated in Figure 4. Note that the effect is relatively large because of the parabolic dependence of E on k. Also note that the closure of this simple model refers to an indirect gap; consistent with numerical calculations [13].

When the strain becomes large enough to close the gap, the bonding electrons can move relatively freely so a transformation, or reaction, can proceed athermally. In inter-mediate cases, the electronic process may be assisted by phonons.

REACTIVITY AND REACTION RATES

Through theoretical studies, Fukui [14] has shown that the electron density (density of states) at the HOMO level of a molecule determines reactivity. This density is not isotropic, but varies with position. Fukui's rule applies in most cases, except when a high density position is in a concavity so there is steric blockage.

Fukui's idea can be taken a step further by deriving electrostatic potentials from electron density distributions [15]. From the gradients of these, forces can be obtained. Knowing the force distri-butions (including the signs), the most probable reaction geometries can be predicted, and from these the reaction products.

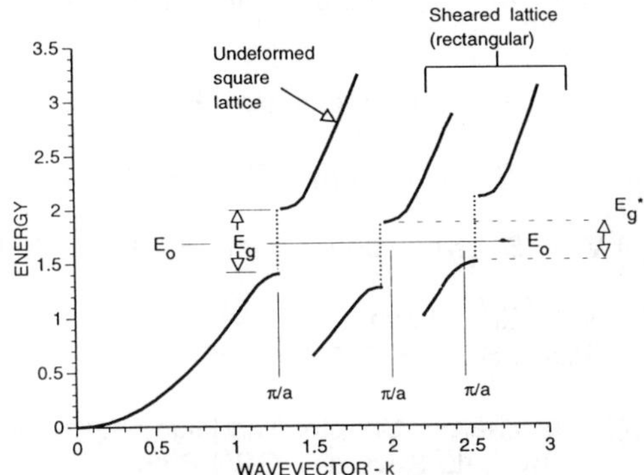

Figure 4 - Showing that shear reduces the mini-mum band gap in the nearly-free-electron approx-imation, while "hydrostatic" compression leaves it nearly unaffected. See text for explanation.

To understand reaction rates at least two regimes need to be considered: thermal and athermal. For the former, the standard theory of thermal activation is adequate. For the latter this is not applicable, and it is likely that a further division is needed between ordinary, and ultra-fast, reactions. Ordinary athermal reactions can be described in terms of Zener tunneling of electrons from HOMO energy states to LUMO states at constant energy [16]. Letting the energy barrier to the reaction be $Q(\gamma)$, a function of the shear strain, γ, the probability, p, of the tunneling process is:
$$\exp\left[-(Q(\gamma)/G\gamma b^3)\right]$$
where G = shear modulus, and b^3.is the bond volume. The attempt frequency, ν, is approximately: $\qquad h/(4\pi m_e b^2)$
which is given by Heisenberg's Principle with h = Planck's constant and m_e = the elec-tron's effective mass. This yields about $\nu \approx 10^{15}$ /sec. The overall rate, r = νp is driven by γ. It has a threshold at $\gamma = \gamma^* = Q/2Gb^3$, and since Q declines with increasing γ, the rate is very sensitive to changes in γ.

In the ultra-fast case, shear strain closes the gap so the material becomes a dense plasma, and the usual considerations of reaction kinetics do not apply.

SUMMARY

Numerous solid-state chemical reactions can be induced by mechanical driving potentials with or without the assistance of temperature. For covalently bonded solids, strained in compression, these reactions occur because the strain causes finite amounts of bond-bending which reduces, or closes, the LUMO-HOMO gaps of the reactants, thereby allowing the bonding electrons of the reactants to rearrange. Shear strain also destabilizes molecules.

REFERENCES

[1] J. K. Burdett, <u>Chemical Bonding in Solids</u>,, Oxford University Press, New York (1995).

[2] R. G. Pearson, Acc. Chem. Res., **26,** 250 (1996).

[3] P. W. Bridgman, Phys. Rev., **48**, 825 (1935).

[4] V. V. Boldyrev, Jour. Chim. Physique, **83**, 821 (1986).

[5] N. S. Enikolopyan, V. B. Vol'eva, A. A. Khzardzhyan and V. V. Ershov, Dokl. Akad. Nauk SSSR, **292**, 1165 (1987).

[6] A. A. Zharov, Chapter 7 in <u>High Pressure Chemistry and Physics of Polymers</u>, Ed. by A. L. Kovarskii, CRC Press, Boca Raton, FL (1994).

[7] R. W. Armstrong, C. S. Coffey and W. L. Elban, p.469 in <u>Advances in Chemical Reaction Dynamics</u>, Ed. by Rentzepis and Capellos, D. Reidel, New York (1986).

[8] J. J. Gilman, in <u>Metal-Insulator Transitions Revisited</u>, P. P. Edwards and C. N. R. Rao, Editors, p. 269, Taylor & Francis Ltd., London (1995).

[9] ibid., P. P Edwards, T. V. Ramakrishnan, and C. N. R. Rao, p. xv.

[10] M. P. Surh, S. G. Louie, and M. L. Cohen, Phys. Rev. B, **45**, 8239 (1992).

[11] J. J. Gilman, Phil. Mag. B, **67**, 207 (1993).

[12] J. J. Gilman, Czech. J. Phys., **45**, 913 (1995).

[13] C. Weigel, R. P. Messmer and J. W. Corbett, Sol. State Comm., **13**, 723 (1973).

[14] K. Fukui, Science, **218**, 747 (1982).

[15] P. Politzer and K. C. Daiker, in <u>The Force Concept in Chemistry</u>, Ed. by B. M. Deb, p.294, Van Nostrand Reinhold Company, New York (1981).

[16] J. J. Gilman and H. S. Tong, J. Appl. Phys., **42**, 3479 (1971).

EFFECT OF CHEMISTRY AND STRUCTURE ON VOLTAGE OF Li INTERCALATION OXIDES

M. K. AYDINOL, G. CEDER, A. F. KOHAN
Massachusetts Institute of Technology, Department of Materials Science and Engineering, Cambridge, MA 02139

ABSTRACT

With the aid of today's powerful computers, quantum mechanical methods are becoming accurate enough to determine the properties of materials before going into laborious and expensive synthesis processes. These "computer experiments" do not require any input other than the atomic numbers of the constituent species. In addition, they allow full control over the "experimental" parameters. In this way the effect of any parameter on the materials properties can be investigated, which is usually not possible in traditional experiments.

We used the *ab initio* pseudopotential method to study the effects of chemistry and structure on the voltage obtained from intercalating Li ions into transition metal oxides. The effect of transition metal and anion chemistry was systematically studied by changing M in $LiMO_2$, (M=Ti, V, Mn, Co, Ni, Cu and Zn) and X in $LiCoX_2$ (X=Se, S and O), keeping the structure fixed at α-$NaFeO_2$ layered structure. In addition, the effect of structure was studied by performing computer experiments on $LiCoO_2$ and $LiMnO_2$ in α-$NaFeO_2$, $LiScO_2$ and Al_2MgO_4 (spinel) structures.

The computed voltages are in good agreement with experimental values. We found that, the electronic band structure and the extent of charge transfer between Li and O ions is correlated to the output voltage. We will explain how the output voltage is affected by the chemistry of the transition metal.

INTRODUCTION

Research for the development of rechargeable lithium batteries has intensified during the last decade, due to the large demand for reliable high energy capacity batteries for electronic devices. The efficient development of these batteries depends critically on our understanding of the relation between chemistry and structure and the properties of the constituent parts of a battery. A battery is composed of an anode, which is generally metallic lithium or a solution of Li in carbon, an ionically conductive electrolyte and a cathodic host. When a battery is discharged, Li ions at the anode travel through the electrolyte and intercalate into the cathodic host structure. This internal flow of Li ions, requires an external flow of electrons between the electrodes, which produces the electrical energy.

In this work we focus on the lithium-metal-oxide compounds which constitute the active component of the cathode. Many different compounds have been investigated as a cathode material for rechargeable Li batteries. Although initially attention was on metal sulfides (e.g. TiS_2, MoS_2 [1]), the focus has shifted to lithium transition metal oxides ($LiMO_2$ [2-5]) since they are more oxidizing towards lithium, resulting in a higher intercalation voltage. Most of the $LiMO_2$ compounds adapt an ordered rocksalt structure [6], such as layered α-$NaFeO_2$, $LiScO_2$ and spinel structures.

To predict the output voltage and to understand the correlation between voltage and electronic structure, we combined a basic thermodynamical model for the battery discharge reaction with the *ab initio* pseudopotential [7] method. The latter is a highly accurate quantum mechanical method to compute the total energy of compounds. The effect of spin polarization

233

Mat. Res. Soc. Symp. Proc. Vol. 453 © 1997 Materials Research Society

has been tested by performing spin polarized and non-spin polarized LMTO-ASA calculations on Co and Ni compounds. We obtained almost identical results, meaning that the effect of spin polarization is negligible for these oxides.

METHODOLOGY

The lithium chemical potential difference ($\Delta\mu_{Li}$) between the anode (high chemical potential side) and the cathode (low chemical potential side) drives the intercalation reaction,

$$Li + MO_2 \rightarrow LiMO_2 \tag{1}$$

The average cell voltage obtained from this reaction is related to the change in the Gibbs free energy ΔG_r,

$$V(x) = \frac{-\Delta G_r}{F} \tag{2}$$

where F is the Faraday constant. Assuming that entropic and volume effects are negligible, ΔG_r ($\approx \Delta E_r + P\Delta V - T\Delta S_r$) can be approximated by only the change in the internal energy (ΔE_r) at $0°K$. To find out the ΔE_r, we only need to compute the internal free energies of the reactant (Li and MO_2) and the product ($LiMO_2$) phases in equation (1),

$$\Delta E_r = E^\circ_{LiMO_2} - E^\circ_{Li} - E^\circ_{MO_2} \tag{3}$$

In this way we can calculate the *average* intercalation voltage. In general, the intercalation voltage will vary as x changes in Li_xMO_2. Obtaining this composition dependent voltage $V(x)$, requires information of the changes in the structure due to changes in the lithium content. This generally involves computing the energy of large supercells, which is tedious with an accurate quantum mechanical method. For this reason we restrict ourselves to the computation of the average intercalation voltage between the cathode limits MO_2 and $LiMO_2$.

The volume as well as the symmetry allowable internal parameters of the structures are relaxed in our total energy calculations. α-$NaFeO_2$ ($R\bar{3}m$) is hexagonal and has two lattice constants, a and c, and one internal coordinate, z. $LiScO_2$ ($I4_1/amd$) also has three degrees of freedom, but is tetragonal. The spinel ($Fd\bar{3}m$) structure is cubic and has only two degrees of freedom.

Ground state energies of periodic crystals can be computed well by *ab initio* methods [8]. Among several methods, we have chosen the pseudopotential method, as its accuracy on oxides has been demonstrated well. In the pseudopotential method the Schrödinger equation is solved within the Density Functional Theory for the valance electrons only. When the solid is formed, the effect of core electrons is represented by a pseudopotential. We used an energy cutoff of 900eV for the expansion of plane waves and 27 k-points in the irreducible Brillouin zone. The details of the calculations will be presented in a forthcoming paper.

RESULTS

The result of calculated lattice parameters are listed in Table I for the MO_2 and $LiMO_2$ compounds in the α-$NaFeO_2$ structure. The calculated crystal volumes are slightly smaller than the experimental values, as expected from LDA calculations. Table II gives the calculated average intercalation voltages for all the compounds in various structures.

Table I. Calculated crystallographic parameters for $LiMO_2$ compounds in the α-$NaFeO_2$ structure. Experimental values are in parenthesis whenever available.

	a($\unicode{x00C5}$)	c($\unicode{x00C5}$)	z	V($\unicode{x00C5}^3$)
TiO_2	3.01	12.8	0.255	33.48
$LiTiO_2$	2.98	14.1	0.25	36.15
VO_2	2.88	12.4	0.248	29.69
$LiVO_2$	2.88 (2.84)[6]	14.2 (14.7)[6]	0.251	34.00 (34.23)
MnO_2	2.90	12.2	0.258	29.62
$LiMnO_2$	2.92	13.5	0.256	33.23
CoO_2	2.88	12.26	0.259	29.36
$LiCoO_2$	2.93 (2.82)[9]	13.2 (14.04)[9]	0.260 (0.264)[9]	32.71 (32.23)
NiO_2	2.87	11.72	0.255	27.87
$LiNiO_2$	2.99 (2.88)[9]	12.85 (14.19)[9]	0.258 (0.259)[9]	33.16 (33.98)
CuO_2	2.98	11.2	0.253	28.71
$LiCuO_2$	3.05	12.7	0.256	34.10
ZnO_2	3.06	11.2	0.248	30.27
$LiZnO_2$	3.10	12.84	0.253	35.62

Table 2. Computed average intercalation voltage, in volts, for $LiMX_2$ compounds. Spinel[a] is the voltage obtained from the intercalation reaction $Li + 2MnO_2 \rightarrow LiMn_2O_4$, and spinel[b] is obtained from the $Li + LiMn_2O_4 \rightarrow 2LiMnO_2$ reaction. The available experimental values are in parenthesis.

Structure	Ti-O	V-O	Mn-O	Co-Se	Co-S	Co-O	Ni-O	Cu-O	Zn-O
α-$NaFeO_2$	2.36	3.03 (3.1)[5]	3.35	1.68	2.36	3.97 (4.1)[10]	3.14 (3.8)[11]	3.88	4.63
$LiScO_2$	-	-	-	-	-	3.48	-	-	-
Spinel[a]	-	-	4.12 (4.2)[12]	-	-	-	-	-	-
Spinel[b]	-	-	2.76 (2.9)[12]	-	-	-	-	-	-

Whenever the experimental data is available, the agreement between calculated and measured experimental voltages is quite good as seen from Table II. It can be observed that the calculated average voltage for oxides increases while going to the right in the 3d-transition metal series. This trend is only interrupted at Ni. The effect of changing anion is quite clear, heavier anions such as sulphur and selenium yield lower average voltage. Structural differences also causes a change in the calculated average voltage. The difference in voltage is about 0.5V in both going from the α-$NaFeO_2$ to the $LiScO_2$ structure for $LiCoO_2$, and going from the α-$NaFeO_2$ to the spinel structure for $LiMnO_2$. The major difference between these structures is only in the cation ordering, while the anion network remains the same.

In Figure 1, we show a schematic bar chart of the calculated electronic band structure of $LiMO_2$ compounds in the α-$NaFeO_2$ structure. In octahedrally coordinated M_d-O_p compounds, the d-bands of the metal split into two higher (e_g) and three lower (t_{2g}) bands with the formation of a band gap Δ_o. The lower lying bands are generally called O_p bands since they have mainly oxygen character. For the early transition metals the t_{2g} band is well separated from the O_p band, whereas for the later transition metals they intermix considerably. Going from early to late in the 3d transition metal series, the level of metal-d character bands, e_g and t_{2g}, drops due to the higher nuclear charge, which we believe, is the cause of getting higher average voltages late in the series. In addition, in case of Ni the intercalated electron has to be accommodated in the high energy e_g band overcoming the band gap, since the Fermi level is at the top of the t_{2g} band in NiO_2. This requires to spent extra energy which causes a reduction in the calculated ΔE_r yielding a lower average voltage. The discrepancy between the calculated and the measured voltages in $LiNiO_2$ is because of this effect, since $LiNiO_2$ in the α-$NaFeO_2$ structure is unstable with respect to Jahn-Teller distortion [13].

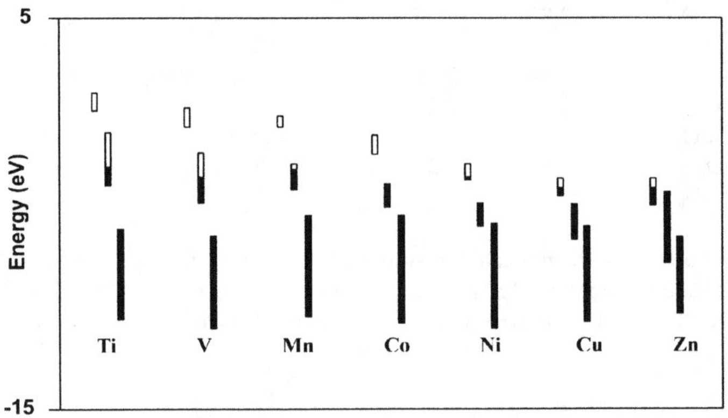

Figure 1. Schematic electronic band structure of $LiMO_2$ compounds in α-$NaFeO_2$ structure. In each elemental column the three bars, from top to bottom, correspond to e_g, t_{2g} and O_p bands. The gap Δ_o is located in between e_g and t_{2g}. The level of electron filling (Fermi level) is indicated by filling of the bars.

Table III. Electron charge transfer values to X and M upon intercalation of Li into α-$NaFeO_2$ MX_2 structures. The radii of spheres for charge integration are 0.98A and 1.15A for M and X respectively. Note that there are two anions per unit cell.

	Ti-O	V-O	Mn-O	Co-Se	Co-S	Co-O	Ni-O	Cu-O	Zn-O
X	0.216	0.245	0.254	0.136	0.152	0.25	0.255	0.261	0.273
M	0.262	0.253	0.289	0.083	0.135	0.276	0.126	0.102	0.067

Figure 2.a shows the valance charge density in $LiCoO_2$ in a plane parallel to c-axis in the α-$NaFeO_2$ structure. The layered nature of the structure can be clearly inferred. The lack of electron density at Li sites indicates that lithium is fully ionized, i.e., electrons are transferred to the O-M-O complexes upon intercalation. In Figure 2.b, we plot the difference in the charge densities between the parent CoO_2 and the fully lithiated $LiCoO_2$. This difference can be attributed to the intercalation of electrons coming from lithium. Clearly, a considerable

proportion of the intercalating electron is transferred to the oxygen ions. This observation contradicts the common assumption that intercalated electron reduces M^{4+} to M^{+3}. Quantitatively, in Table III, we calculated the fraction of the electron transfer to the oxygen and metal ions for all compounds in the α-NaFeO$_2$ structure. Almost half of the electron from lithium is transferred to the oxygen ions (since there are two for each metal). The transfer to oxygen is found to increase as M goes to the right in the series from Ti to Zn, correlating well with the increase in the intercalation voltage.

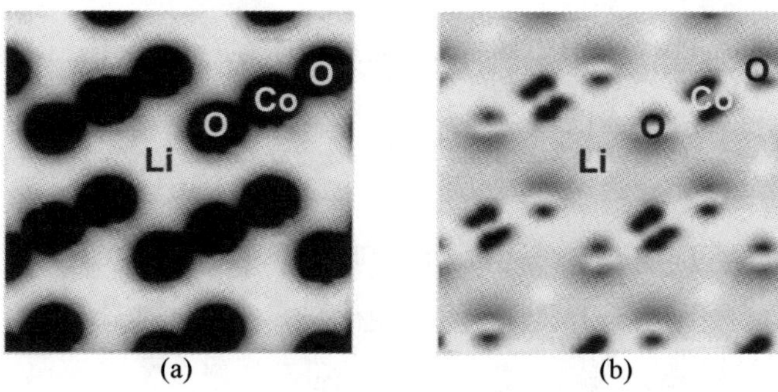

Figure 2. Valence charge density plots in (a) LiCoO$_2$ and (b) the difference between CoO$_2$ and LiCoO$_2$ in the α-NaFeO$_2$ structure. The plane shown is [110] plane in hexagonal settings and it is parallel to the c-axis. Darker indicates higher electron density. Note the layered nature of the structure.

CONCLUSIONS

The good agreement between the calculated and the experimentally measured average intercalation voltages, shows that ab initio methods can be used to predict the average voltages of intercalation compounds reasonably well. This opens up the way for such methods in the design of novel battery materials. By the wealth of information they provide, *ab initio* methods can also contribute considerably to the insight into the relation between structure and properties of these materials.

We investigated the effect of chemistry and structure on the voltage of Li intercalation oxides. The choice of metal, anion and structure can considerably affect the output voltage obtained from an intercalation compound. In determining the output voltage, two factors seems to be important, the level of the bands that is to be filled and the amount of electron transfer to anion during intercalation. The deeper the levels and the more the charge transfer to anion, the higher the average voltage.

ACKNOWLEDGEMENTS

Support from Furukawa Electric and the Idaho National Engineering Laboratory are gratefully acknowledged. Supercomputing calculations were performed at the Pittsburgh Supercomputing Center of the National Science Foundation.

REFERENCES

1. M. S. Whittingham, Prog. Solid St. Chem. **12**, p. 41-99 (1978).

2. K. Mizushima, P. C. Jones, P. J. Wiseman and J. B. Goodenough, Mat. Res. Bull. **15**, p. 783-789 (1980).

3. M. M. Thackeray, W. I. F. David, P. G. Bruce and J. B. Goodenough, Mat. Res. Bull. **18**, p. 461-472 (1983).

4. M. G. S. R. Thomas, W. I. F. David, J. B. Goodenough and P. Groves, Mat. Res. Bul. **20**, p.1137-1146 (1985).

5. L. A. De Picciotto, M. M. Thackeray and G. Pistoia, Solid State Ionics **28-30**, p. 1364-1370 (1988).

6. T. A. Hewston and B. L. Chamberland, J. Phys. Chem. Solids **48**, p. 97-108 (1987).

7. M. C. Payne, M. P. Teter, D. C. Allan, T. A. Arias and J. D. Joannopoulos, Rev. Mod. Phys. **64**, p. 1045-1097 (1992).

8. A. F. Kohan and G. Ceder, Computational Materials Science, in press (1997).

9. T. Ohzuku, A. Ueda, M. Nagayama, Y. Iwakoshi and H. Komori, Electrochimica Acta **38**, p. 1159-1167 (1993).

10. T. Ohzuku and A. Ueda, J. Electrochem. Soc. **141**, p. 2972-2977 (1994).

11. T. Ohzuku, A. Ueda and M. Nagayama, J. Electrochem. Soc. **140**, p. 1862-1870 (1993).

12. J. M. Tarascon, E. Wang, F. K. Shokoohi, W. R. McKinnon and S. Colson, J. Electrochem. Soc. **138**, p. 2859-2864 (1991).

13. A. Rougier, C. Delmas and A. V. Chadwick, Solid State Comm. **94**, p. 123-127 (1995).

Lattice Dynamical Prediction of the Elastic Constants of Diamond

Robert R. Reeber
Department of Materials Science and Engineering, North Carolina State University
Raleigh, NC 27607

ABSTRACT

The thermophysical properties of diamond, a metastable material at room temperature, are difficult to measure at high temperatures. These properties are of interest for testing equation of state and interatomic potential models. Here we utilize a geometrical lattice transformation, one dimensional lattice dynamical theory, and the principle of corresponding states to calculate the elastic constants of diamond over an extended temperature range.

INTRODUCTION

Diamond oxidizes in air above 700 C. The rate of oxidation is severe above 1200 C. Elastic constants for diamond have been measured over a limited temperature range from 4.2 to 300K[1-2]. Because diamond is the hardest material currently known, the temperature dependence of its elastic constants and bulk modulus are of theoretical interest and are important parameters for modeling its behavior. The high pressure/temperature mechanical properties are also of practical importance to mineral physicists who utilize diamond pressure cells for high pressure/temperature experiments.

Thermal expansion directly relates to the strength of a material as it measures the displacement between atoms in a solid as a function of increasing thermal stress. The high temperature/pressure thermal expansion of diamond and other Group IV semiconductors have been calculated [3-4] and are in excellent agreement with available experimental data. The semi-empirical method provides an analytical expression for predicting values in regions difficult to access experimentally .

First principle theories for calculating high temperature elastic constants have many approximations and assumptions and as indicated earlier [3-4] also have difficulty in reproducing thermal expansion measurements over extended temperature ranges. Calculations of high temperature elastic constants [5] based upon the central force model of Keating [6-7] have been questioned on theoretical grounds[8].

CALCULATION METHODS

This work utilizes a lattice dynamical method and the principle of corresponding states to predict the high temperature elastic constants of diamond. Three dimensional lattice dynamical calculations are simplified to one dimension with a geometrical transformation introduced earlier [9]. Equation 1, the definition of an elastic constant, is approximated at the unit cell of a crystal by infinitesimal quantities.

Mat. Res. Soc. Symp. Proc. Vol. 453 © 1997 Materials Research Society

$$C_{ij} \equiv \frac{dF}{dA} \left[\frac{dl}{l} \right]^{-1} \approx \frac{\Delta F}{\Delta A} \left[\frac{\Delta l}{l} \right]^{-1} \qquad (1)$$

Rearranging Eq.(1) yields the linear force constant K_L for a one-dimensional lattice in terms of l, the distance between mass points, Δ, the unit area collapsed on the mass point, and the directional elastic constant where hkl are Miller indices of the crystallographic plane perpendicular to the one dimensional lattice.

$$K_L \approx \frac{\Delta F}{\Delta l} \equiv C_{\langle hkl \rangle} \frac{\Delta A}{l} \qquad (2)$$

Extension of the method to the three principal crystallographic directions allows one to directly relate one-dimensional force constants for each direction to the appropriate elastic constants of the crystal. Table 1 contains geometrical quantities of the diamond lattice required for this transformation.

Table 1. Parameters for a One-Dimensional Diamond Lattice with a Basis

<hkl>	L	ΔA	Mass (atoms)
100	a	a^2	2
110	$a/4\sqrt{2}$	$a^2\sqrt{2}$	4
111	$a/6\sqrt{3}$	$a^2/2\sqrt{2}$	2

Table 2 lists the relationships between directional elastic constants and crystal elastic constants as tabulated by Blackman [10].

Table 2. One-Dimensional Lattice Elastic Constants

<hkl>	$\Delta F/\Delta L$ ($L/\Delta A$)	No. Lattice Modes
$C_{\langle 100 \rangle}$	C_{11}	1 LA
$C_{\langle 110 \rangle}$	$.5\,(C_{11}+C_{12}+2C_{44})$	1 LA
$C_{\langle 111 \rangle}$	$1/3\,(C_{11}+2C_{12}+4C_{44})$	1 LA
$C_{\langle 100 \rangle}$	$1/3\,C_{44}$	2 TA
$C_{\langle 110 \rangle}$	C_{44}	1 TA
$C_{\langle 110 \rangle}$	$1/2\,(C_{11}-C_{12})$	1 TA
$C_{\langle 111 \rangle}$	$1/3\,(C_{11}+C_{12}+C_{44})$	2 TA

For a one dimensional lattice with a basis, $\theta_{<hkl>}$ is the directional characteristic temperature, h and k are Planck and Boltzmann's constants, M is the atomic mass and υ is the appropriate vibration frequency in the crystallographic direction <hkl>.

$$\theta_{<hkl>} \equiv \frac{h\upsilon_{<hkl>}}{k} \equiv \frac{h}{k\pi}\left[\frac{K_{L<hkl>}}{2M}\right]^{\frac{1}{2}} \tag{3}$$

after substituting from Eq. (2) for K_L and rearranging one obtains the general expression for directional elastic constants:

$$C_{<hkl>} \equiv Constant\ M_i\ \frac{\theta^2_{<hkl>}\ l}{\Delta A} \tag{4}$$

From Eq. (4) and utilizing the principle of corresponding states for isostructural materials with similar bonding one obtains:

$$C^D_{<hkl>} \equiv C^G_{<hkl>}\ \frac{M_D\ a_G\ \theta^2_{<hkl>D}}{M_G\ a_D\ \theta^2_{<hkl>G}} \tag{5}$$

where subscripts D and G represent diamond and germanium respectively and a is the lattice constant. In the calculations the corresponding temperature for diamond is calculated by multiplying the germanium data temperature by the ratio of 1 atmosphere melting points as determined from a plot of Group IV semiconductor melting points versus the average of their nine 1D lattice theta temperatures[11]. Characteristic parameters are given in Table 3.

Table 3. Characteristic Parameters for Group IV Semiconductors in Degrees Kelvin

	θ - Elastic Constants	θ - Specific Heat	θ - Thermal Expansion	θ - Lattice (avg.)	Melting Temp. (1atm.)
Ge	368	362 -376	366	368	1210
Si	641	645 - 647.8	644	640	1685
C_{dia}	2233	2219 - 2246	2138	2127	4235(est.)

ELASTIC CONSTANTS

Directional elastic constants for Ge and Si were calculated with elastic constant data digitized from the literature[12-13] and the relationships given in Table 2. These values when combined in Eq. (5) with group IV lattice parameters from recent work[3-4] and atomic masses provide diamond directional elastic constants at equivalent temperatures. Elastic constants of diamond result from algebraic solution of the relationships in Table 2.

RESULTS

Table 4 lists calculated values for diamond elastic constants and adiabatic bulk modulus as a function of temperature.

Table 4 Elastic Moduli of Diamond as a Function of Temperature x 10^{11} dynes/cm^2

Diamond Cij calculated from Germanium					Diamond Cij calculated from Silicon				
T (K)	C_{11}	C_{12}	C_{44}	Ks	T (K)	C_{11}	C_{12}	C_{44}	Ks
25.1	108.01	12.58	57.80	44.39	6.0	108.05	12.55	57.86	44.39
50.3	108.01	12.58	57.80	44.39	80.4	108.03	12.56	57.86	44.38
100.5	108.01	12.58	57.80	44.39	125.7	108.04	12.55	57.84	44.39
201.1	107.98	12.55	57.80	44.36	252.6	107.96	12.51	57.82	44.32
301.6	107.88	12.49	57.77	44.29	334.5	107.82	12.47	57.79	44.29
402.1	107.71	12.42	57.71	44.19	377.0	107.74	12.44	57.74	44.21
502.7	107.48	12.35	57.63	44.06	458.9	107.55	12.38	57.67	44.10
603.2	107.21	12.28	57.52	43.92	544.1	107.32	12.31	57.59	43.98
703.7	106.91	12.21	57.39	43.78	629.6	107.07	12.25	57.45	43.86
804.3	106.58	12.14	57.25	43.62	710.5	106.84	12.20	57.36	43.75
904.8	106.23	12.07	57.09	43.46	783.2	106.63	12.19	57.26	43.67
1005.3	105.88	12.01	56.93	43.30	Diamond (experimental) McSkimin in [12]				
1206.4	105.13	11.87	56.57	42.96	72.9	108.75	12.50	57.89	44.58
1407.4	104.35	11.74	56.20	42.61	100.3	108.69	12.50	57.89	44.56
1608.5	103.55	11.61	55.80	42.26	149.3	108.55	12.50	57.88	44.52
1809.6	102.74	11.48	55.40	41.90	199.1	108.36	12.49	57.88	44.45
					250.2	108.08	12.47	57.84	44.34
					300.7	107.65	12.44	57.81	44.18
					325.1	107.40	12.42	57.79	44.08

DISCUSSION

Elastic constants for diamond calculated from germanium are within 0.05% of those calculated from silicon. Both sets of data are within 0.75% of existing experimental values. Unfortunately higher temperature experimental values are not available for comparison with predicted elastic constants.

CONCLUSIONS

Elastic constants were calculated for diamond over an extended temperature range. The lattice dynamical model when combined with the principle of corresponding states and experimental data from germanium and silicon gives results in excellent agreement with the limited experimental data available. The method provides additional support for Einstein's comments in 1911 on Lindemann's theory of melting that, "this seems to indicate that the Law of Corresponding States for simple solids and liquids holds true within a remarkably good approximation." It is anticipated that such high temperature elastic constant predictions will provide a useful guide for testing equations of state and interatomic potential models for diamond and other covalently bonded materials.

ACKNOWLEDGEMENTS

The author appreciated the support of the Army Research Office through Grant No. DAAL003-89-D-001.

REFERENCES

1. H.J. McSkimin and W.L. Bond, *Phys. Rev.* **105**, 116 (1957).
2. M.H. Grimsditch and A.K. Ramdas, *Phys. Rev.* **B11**, 3139 (1975).
3. R. R. Reeber and K. Wang, *J. Elect. Mat.* **25**, (1) 63 (1996).
4. R. R. Reeber and K. Wang, *Mat. Phys. and Chem.* **46**, (2-3) 259 (1996).
5. A. I. Gubanov and Davydov S.Yu, *Sov. Phys. Sol. St.* **14**, 372 (1972).
6. P.N. Keating, *Phys. Rev.* **145**, 637 (1966).
7. P.N. Keating, *Phys. Rev.* **149**, 674 (1966).
8. J.W. Martin, *J. Phys. C. Sol. St. Phys.* **8**, 2869 (1975).
9. R.R. Reeber and D. MacLachlan, *Canad. J. Phys.*, 2287 (1970).
10. M. Blackman, Handbuch der Physik **VII**, Part 1, Editor S. Flügge, Springer Verlag, Berlin, (1956), p343.
11. R.R. Reeber, *phys. stat sol.* **(a) 26,** 253 (1974).
12 R.F.S. Hearmon, Landolt-Bornstein New Series, Group III, **11**, Editors K.H. Hellwege and A.M. Hellwege, Springer-Verlag, Berlin, (1979), p111, 116.
13 J.J. Hall, *Phys. Rev.* **161**, (3) 756 (1967).

Part V

Optical Properties

Eu^{2+} LUMINESCENCE COLOR: A STRUCTURE-PROPERTY RELATIONSHIP

Douglas A. Keszler and Anthony Diaz
Department of Chemistry, Oregon State University, Gilbert Hall 153, Corvallis, OR   97331-4003, keszlerd@ccmail.orst.edu

ABSTRACT

Structural and spectroscopic characteristics of a series of Eu^{2+}-doped Sr and Ba borates have been studied to establish simple correlations among emission color, excitation energies, Stokes shifts, and O atom environments.

INTRODUCTION

The luminescence of Eu^{2+}-doped solids has been studied for several decades [1]. Emission has been observed from both intraconfigurational $4f^7 \rightarrow 4f^7$ transitions and interconfigurational $4f^65d \rightarrow 4f^7$ transitions. Observation of the intraconfigurational transitions $^6P_{7/2} \rightarrow {}^8S_{7/2}$ [2] and $^6I_{7/2} \rightarrow {}^8S_{7/2}$ [3] was first reported in 1970 and 1994, respectively. These sharp line emission features are present only in those weak-field hosts where the Eu^{2+} $4f^65d$ level occurs at a very high energy. In most oxides, the $4f^65d$ configuration is the lowest excited state, and emission results from this level. The 5d orbitals strongly interact with the nearest-neighbor anions in the host, leading to broad-band emission of varying colors. In oxides, this emission generally occurs in the UV or blue portions of the spectrum, although examples of green luminescence have been known for some time [4]. Recently, however, *red* emission has been observed from two Eu^{2+}-doped borates, $Ba_2LiB_5O_{10}$ and $Ba_2Mg(BO_3)_2$ [5, 6]. These findings have prompted renewed interest in examining the fundamental relationships between Eu^{2+} emission color and the characteristics of the host [7, 8]. And this is the topic of the present contribution.

Aside from this fundamental interest, a high quantum efficiency of the $4f^65d$ emission, coupled with high brightness and suitable chromaticity, can produce useful materials. Eu^{2+}:SrB$_4$O$_7$ and Eu^{2+}:BaMgAl$_{10}$O$_{17}$ are common lamp phosphors, and Eu^{2+}:BaMgAl$_{10}$O$_{17}$ is also employed as the blue component in plasma display panels. The material Eu^{2+}:BaFBr is widely utilized as an X-ray storage phosphor [9], and the sulfides Eu^{2+}:SrGa$_2$S$_4$ and Eu^{2+}:Sr$_{1-x}$Ca$_x$S are being examined as the green and red components, respectively, in electroluminescent and field-emission displays. The high absorption and emission cross sections of the $4f^7 \leftrightarrow 4f^65d$ transition have prompted interest for use in solid-state lasers, but the utility of the transition in this application is hampered by high excited-state absorption cross sections [10]. The models presented below should provide useful guidelines for the synthesis and development of new materials for these and future applications.

EXPERIMENT

All powders were prepared by standard high-temperature methods [6] from high-purity reagents ($\geq$99.9%.) Each sample was nominally doped with 2 mol% Eu^{2+} by heating in an atmosphere of 25 Ar: 1 H$_2$. Excitation and emission spectra were obtained on a computer-

Mat. Res. Soc. Symp. Proc. Vol. 453 © 1997 Materials Research Society

controlled right-angle spectrometer. Excitation light from an Oriel 300-W Xe lamp was passed through a 50-cm water filter, focused initially onto the slits of a Cary Model 15 prism monochromator and then onto the sample. Luminescence was collected at a near right angle to excitation, dispersed through an Oriel 22500 1/8-m monochromator, and detected with a Hamamatsu R636 photomultiplier tube. The signal was collected and amplified with a Keithley model 486 picoammeter; then converted to a digital signal for computer acquisition. Spectrometer control and data acquisition were achieved with computer programs written in this laboratory. The excitaton spectra were corrected by using sodium salicylate as a quantum counter, and the emission spectra were corrected with a tungsten lamp that was calibrated to NIST-traceable sources at Eppley Laboratories, Inc.

RESULTS AND DISCUSSION

The emission color of a Eu^{2+}- doped material is determined by the position of the lowest energy excitation (absorption) band and the magnitude of the Stokes shift. Traditionally, these features have been correlated to the structural and electronic features of the dopant site. For example, the substitution of luminescent ions onto small dopant sites generally produces much different excitation energies and Stokes shifts in comparison to those for substitution on larger sites [11]. Excitation energies are affected by ligand field and nephelauxetic effects [12], while relaxation of the anion environment in the excited state contributes to the magnitude of the Stokes shift. The size of the Stokes shift has also been correlated to dopant site asymmetry. For example, in the compounds $Bi_4Ge_3O_{12}$ (BGO) and $Bi:LaPO_4$, the Bi^{3+} ion occupies an off-center position in the ground state; for BGO, Bi-O distances are 2.16 and 2.60 Å [13]. Each material exhibits an unusually large Stokes shift that has been associated with an extensive reorganization of the Bi-centered coordination polyhedron in the excited state to give a more equal distribution of Bi-O bond lengths [14].

To make comparisons of properties across different families of materials, additional host features must be examined. For example, the degree of covalency affects excitation energies. Also, hosts rich in small electronegative cations are commonly classified as rigid or stiff, while those rich in large electropositive cations are usually classified as soft. The rigid materials tend to afford small Stokes shifts, whereas soft materials tend to produce large Stokes shifts. Considering each of these simple models and the observation of red emission in the Eu^{2+}-doped borates $Ba_2LiB_5O_{10}$ and $Ba_2Mg(BO_3)_2$, we set out to identify the specific host-centered structural and electronic features that could be correlated to the generation of these long emission wavelengths. During the course of our studies, a new model was proposed to account for the unusually long wavelength emission on the basis of a preferential orientation of the 5d orbitals and their electrostatic interactions with the positive field of neighboring cations [8]. We will also consider this model.

To examine the relative importance of each of these features, we have examined the spectroscopic and structural characteristics of fourteen different Eu^{2+}-doped borates (Table 1). The emission of each of these materials at room temperature is dominated by a broad-band transition from the $4f^65d$ level. They represent a unique collection of hosts - each one, with the exception of $Ba_2Ca(BO_3)_2$, contains a single crystallographic site for Eu^{2+} substitution. As such, excitation and emission properties can be readily correlated to specific structural features. The list includes compounds that are rich in heavy, electropositive cations as well as several polyborates that are rich in the small, electronegative B atoms.

Table I

Spectroscopic and Structural Data for Eu^{2+} in Borates *

Emission $\lambda > 500$ nm					
Compound	Excitation** Energy (cm^{-1})	Stokes Shift (cm^{-1})	Emission λ (nm)	Average M-O Distance (Å), M = Sr or Ba	% Angular Deviation (^{o})
$Ba_2Mg(BO_3)_2$	30300	14000	617	2.812 ± 0.123	10.2
$Ba_2LiB_5O_{10}$	27400	11000	612	2.836 ± 0.139	9.9
$BaLiBO_3$	(25000)	(5400)	510	2.833 ± 0.204	7.4
$Ba_2Ca(BO_3)_2$	23300	4300	530	2.841 ± 0.078	6.6
$Sr_2Al_2B_2O_8$	(26500)	(8000)	529	2.675 ± 0.133	11.7
$Sr_3(BO_3)_2$	23300	6000	585	2.601 ± 0.159	7.0
$Sr_2Mg(BO_3)_2$	22200	5600	605	2.657 ± 0.053	6.3
Emission $\lambda < 500$ nm					
$BaBe_2(BO_3)_2$	(27200)	(1700)	392	2.856 ± 0.063	8.5
$BaLiB_9O_{15}$	30300	5300	400	2.878 ± 0.067	14.9
$BaNaB_9O_{15}$	30300	5700	407	2.971 ± 0.087	11.5
BaB_8O_{13}	32200	7200	400	2.879 ± 0.041	7.0
$SrLiB_9O_{15}$	29700	4100	390	2.764 ± 0.059	8.8
SrB_4O_7	33100	5900	367	2.710 ± 0.159	11.1
$SrAl_2B_2O_7$	(31700)	(7600)	415	2.552 ± 0.007	9.3

* Complete references for the data in this table can be found in ref. 7.
** Refers to the lowest energy excitation peak. Values in parentheses are at room temperature, all others are at 4.2 K.

As seen from the data, maximum emission extends from the UV at 367 nm in SrB_4O_7 to the red at 617 nm in $Ba_2Mg(BO_3)_2$. The materials have been sorted into two groups - those emitting at short wavelengths, $\lambda < 500$ nm, in the blue and UV, and those emitting at long wavelengths, $\lambda > 500$ nm, in the green and red. It should be readily apparent that this sorting can not be correlated to any simple average characteristic of the host, such as mean electronegativity or polarizability of the cations. These averages differ dramatically for the B-rich host $Ba_2LiB_5O_{10}$ and the B-poor host $Ba_2Mg(BO_3)_2$; yet they exhibit similar emission wavelengths. Consideration of these two materials alone points to a more local, structural or electronic effect controlling the luminescence properties.

We begin with some very simple considerations and the size of the dopant site. Ba-O distances average 2.85 Å, and Sr-O distances average 2.66 Å. Emission wavelengths in the Ba hosts extend from 392 to 617 nm, and the Sr hosts exhibit a similar range of 367 to 605 nm. From examination of the distances in the Ba or Sr hosts alone, it is also difficult to establish any statistically significant correlation between emission color and the size of a site. Similarly, there is no simple correlation between the size of the site and the excitation energy. Overall, however, the larger Stokes shifts are associated with the Ba rather than the Sr hosts. Interest in examining the Eu^{2+} luminescence of the red-emitting materials $Ba_2LiB_5O_{10}$ and $Ba_2Mg(BO_3)_2$ was originally prompted by the highly asymmetric coordination environments of the Ba atoms in these hosts [5, 6]. The BaO_9 polyhedron in $Ba_2Mg(BO_3)_2$, for example, exhibits six short Ba-O distances, 2.73 Å, and three long lengths, 2.96 Å (Figure 1); a similar asymmetry is observed for the BaO_8 polyhedron in $Ba_2LiB_5O_{10}$ [15]. As noted above such asymmetric sites are proposed to give rise to large Stokes shifts and long wavelength emission. A measure of dopant site asymmetry is provided in Table I by the standard deviations for the average M-O distances. While it is difficult to establish a direct correlation between site asymmetry and any specific spectroscopic property, those hosts containing highly asymmetric sites comprise a large proportion of the long wavelength emitters. Notable exceptions to this correlation are the hosts SrB_4O_7, $Sr_2Mg(BO_3)_2$, and $BaLiBO_3$. The polyborate SrB_4O_7 exhibits the highest asymmetry among the Sr hosts, yet it has an intermediate Stokes shift and the shortest wavelength emission. The simple orthoborate $Sr_2Mg(BO_3)_2$ contains a symmetrical site, but the longest wavelength emission among the Sr hosts. The material $BaLiBO_3$ contains a highly asymmetrical Ba site, but a relatively short wavelength emission at 510 nm.

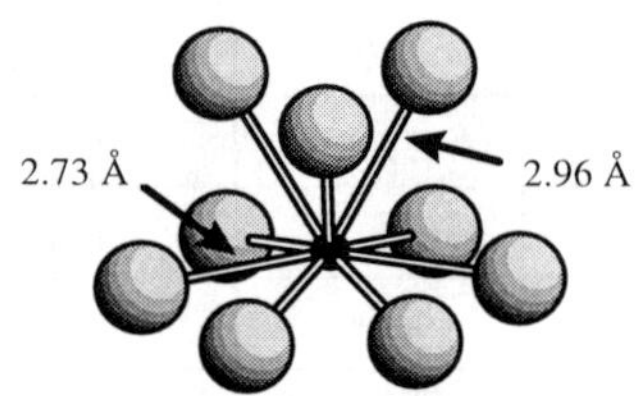

Figure 1. BaO_9 polyhedron in $Ba_2Mg(BO_3)_2$

By concentrating on the coordination environments of the dopant site, we have not been able to establish simple relationships between structure and Eu^{2+} spectral features. In a previous report [7], however, we established a more precise connection between the structure of a host and the magnitude of the Stokes shift leading to Eu^{2+} emission. Specifically, we found direct correlations among the coordination environments of the *O atoms*, emission color, and the magnitude of the Stokes shift. The two groups of materials listed in Table I can be sorted by considering the O-atom environments - those emitting with $\lambda > 500$ nm all contain O atoms bound by at least three Sr or Ba atoms and those emitting with $\lambda < 500$ nm all contain O atoms bound by two or fewer Sr or Ba atoms. For the long wavelength emitters, additional correlations can be established by considering the final column labeled % Angular Deviation. The values in this column represent the deviations of the O-centered polyhedra from idealized geometries. The O environments in the host $Ba_2LiB_5O_{10}$ are distorted tetrahedra, and the environments in $Ba_2Mg(BO_3)_2$ are distorted trigonal bipyramids. From Table I, it is seen that excitation energies and Stokes shifts scale directly with the degree of distortion about the O atoms for those materials emitting with $\lambda > 500$ nm. In our previous work [7], we correlated these distortions to orbital overlap and displacements of the O atoms leading to the observed Stokes shifts.

This same model can now be extended to the excitation energies. As the environments about the O atoms become more regular (smaller % Angular Deviation), orbital overlap increases and excitation energies decrease. The energies of the $4f^65d$ levels decrease because of the relative strength of the nephelauxetic effect with respect to the antibonding interactions associated with the one-electron model. A more complete description of these interactions has been presented elsewhere [16]. This description is consistent with the trend of decreasing $4f^7$-$4f^65d$ energy separation (E) in the series: E(free ion) > E (fluoride) > E (oxide) > E (sulfide). As noted above, intraconfigurational $4f^7 \rightarrow 4f^7$ transitions are commonly observed in fluorides where the energy of the $4f^65d$ level is very high. Such intraconfigurational transitions are rarely observed in oxides, because of the lower energy of the $4f^65d$ level. These Eu^{2+}-doped oxides generally have $4f^7 \rightarrow 4f^65d$ excitation bands in the UV portion of the spectrum, yielding white powders. In the case of sulfides, the $4f^65d$ level can fall to sufficiently low energies that they absorb in the visible, e.g., Eu^{2+}:$Sr_{1-x}Ca_xS$ powders have a deep-red body color. In general, as the polarizability of the host and the Eu^{2+}-anion interactions increase, there is a decrease in the energy separation of the $4f^7$ and $4f^65d$ levels.

There is no simple relationship among % Angular Deviation, excitation energies, and Stokes shifts for those materials emitting with $\lambda < 500$ nm. This is to be expected since excitation energies and Stokes shifts should be dominated by the nature of the cations associated with each O atom. For those materials emitting with $\lambda > 500$ nm, the O coordination environments are dominated by their interactions with the large, electropositive Sr and Ba cations, so structural and spectroscopic properties can readily be correlated.

Consideration of the coordination environments of the O atoms provides a very simple model for anticipating the spectral characteristics of a Eu^{2+}-doped oxide. Poort and co-workers have proposed an alternative model [8] to account for long wavelength Eu^{2+} emission on the basis of the interactions between the electron density in the high lying 5d orbitals and the electrostatic field of the neighboring cations. In this model, certain d orbitals are expected to interact strongly with the second-coordination sphere of cations. As the strength of the field associated with these cations increases, the energy of an electron in one or more of these d orbitals decreases, leading to long wavelength emission. Of course, this model is related to our emphasis on O coordination environments, since the electrostatic field of the nearest neighbor Sr or Ba cations should increase as their coordination numbers to the O atoms increase. The field should also increase as Ba$\cdots$Ba and Sr$\cdots$Sr distances decrease. We have examined some of these distances, but it is difficult to identify any simple trends. Short Ba$\cdots$Ba distances in $Ba_2LiB_5O_{10}$ range from 4.196 to 4.414 Å, while similar distances in $BaLiBO_3$ range from 4.020 to 4.215 Å. Even though Ba$\cdots$Ba distances are shorter in $BaLiBO_3$, the emission, relative to that of $Ba_2LiB_5O_{10}$, is shifted to a shorter wavelength of nearly 100 nm. The d-level stabilization is also expected to be greatest for any orbital that is nonbonding with respect to the ligands. Unfortunately, no such nonbonding levels are present in the materials emitting at long wavelengths.

Among oxides, borate hosts are unique in providing the widest range of Eu^{2+} emission wavelengths, extending from the UV intraconfigurational $4f^7 \rightarrow 4f^7$ transitions in $BaLiB_9O_{15}$ [17] to the red $4f^65d \rightarrow 4f^7$ transition of $Ba_2Mg(BO_3)_2$. We and others have examined the luminescence properties of Eu^{2+} in a number of related oxoanion hosts such as silicates and phosphates. To date, even in those hosts containing O atoms bound by three Ba or Sr atoms, no peak emission at wavelengths longer than those in the green have been observed. Among main-group oxoanion hosts, we speculate that red emission may be unique to borates. The high basicity of the BO_3^{3-} group relative to groups such as SiO_4^{4-}, PO_4^{3-}, and SO_4^{2-} enhances the nephelauxetic effect, lowering the energy of the excited state that leads to long wavelength emission.

SUMMARY

Excitation energies, Stokes shifts, and colors of Eu^{2+}-doped Ba and Sr borates have been correlated to the coordination environments of the O atoms in the hosts. Borates containing O atoms bound by at least three Ba or Sr atoms will exhibit green or red emission. Among these hosts, excitation energies and Stokes shifts scale with the degree of angular distortion about the O atoms. These features are proposed to arise from changes in electron-electron repulsion associated with orbital overlap between the Eu^{2+} ion and its neighboring O atoms.

ACKNOWLEDGMENTS

This work was supported by the National Science Foundation (DMR-9221372) and by ARPA through the Phosphor Technology Center of Excellence.

REFERENCES

1. For example, Jenkins and McKeag described the luminescence of Eu^{2+} in silicates in 1950: H. G. Jenkins and A. H. McKeag, J. Electrochem. Soc. **97**, p. 415 (1950).
2. R. A. Hewes and M. F. Hoffmann, J. Luminescence **3**, p. 261 (1970).
3. A. Ellens, A. Meijerink, and G. Blasse, J. Luminescence **60&61**, p. 70 (1994).
4. G. Blasse, W. L. Wanmaker, J. W. ter Vrugt, and A. Bril, Philips Res. Repts. **23**, p.189 (1968).
5. G. J. Dirksen and G. Blasse, J. Solid State Chem. **92**, p. 591 (1991).
6. A. Akella and D. A. Keszler, Mater. Res. Bull. **30**, p. 105 (1995).
7. A. Diaz and D. A. Keszler, Mater. Res. Bull **31**, p. 147 (1996).
8. S. H. M. Poort, W. P. Blokpoel, and G. Blasse, Chem. Mater. **7**, p. 1547 (1995).
9. M. Sonoda, M. Takano, J. Migahara, Y. Shibahara, Radiology **148**, p. 833 (1983).
10. J. K. Lawson and S. A. Payne in <u>OSA Proceedings on Advanced Solid-State Lasers</u>, <u>Vol. 13</u>, edited by L. L. Chase and A. A. Pinto (Opt. Soc. Amer., Washington, DC 1992), p. 330.
11. G. Blasse and B. C. Grabmaier, <u>Luminescent Materials</u>, Springer-Verlag, New York, 1994.
12. C. K. Jørgensen, <u>Modern Aspects of Ligand Field Theory</u>, North-Holland Publishing Co., Amsterdam, 1971; C. K. Jørgensen, <u>Absorption Spectra and Chemical Bonding in Complexes</u>, Pergamon Press, Oxford, 1962.
13. A. Durif and M. T. Averbuch-Pouchot, Compt. Rend. Hebd. Ser. C, Sci. Chem. **295**, p. 555 (1982).
14. G. Blasse, Prog. Solid State Chem. **18**, p. 79 (1988).
15. R. W. Smith and D. A. Keszler, Mater. Res. Bull. **24**, p. 151 (1989).
16. A. Diaz and D. A. Keszler, results submitted for publication.
17. A. Diaz, Ph.D. dissertation, Oregon State University, 1996.

OPTICAL PROPERTIES OF LANTHANIDE -CONTAINING HALIDE-MODIFIED ZINC TELLURITE GLASSES

D.L. SIDEBOTTOM*, MICHAEL A. HRUSCHKA*, B.G. POTTER**, RICHARD K. BROW**, JAMES J. HUDGENS***

*University of New Mexico, AML Suite 100, 1001 University Blvd. SE, Albuquerque, New Mexico, 87106.
**Sandia National Labs, AML Suite 100, 1001 University Blvd. SE, Albuquerque, New Mexico, 87106.
***presently at Texas Instruments, Plano, Texas.

ABSTRACT

As part of an ongoing investigation to characterize the properties and structure of zinc halide-tellurium oxide glasses, we report preliminary measurements of the optical properties of several Nd- and Er-doped tellurites. Measurements include florescence lifetimes and estimates of the theoretical radiative lifetimes (as obtained by traditional Judd-Ofelt analysis of optical absorption spectra) as well as phonon sideband studies sensitive to vibrational characteristics near the rare earth ion. The response of these optical features to the substitution of alternative halides is examined .

INTRODUCTION

A variety of commercially important photonic devices have inherited their special functionality from the unique optical behavior of rare-earth ions doped into materials[1]. Common examples include lasers and optical amplifiers (e.g., Nd:YAG) as well as phosphor materials (e.g., Eu:Y_2O_3). The optical properties of these materials result from electronic transitions occurring within the partially filled 4f energy shell of the lanthanide series[2,3]. Although normally forbidden by Laporte's selection rule ($\Delta l = \pm 1$), these intrashell, "forced" electric dipole transitions are allowed by virtue of an overlap of the 4f energy levels with nearby 5d levels. This overlap leads to admixing of electron configurations of opposite parity resulting in several optical transitions within the 4f level which are of practical importance for device applications.

Since these transitions are the result of subtle mixing of opposite parity states with pure 4f electron states, they are generally much weaker than normal Laporte-allowed dipole transitions[2]. Given the sensitivity of the forced dipole transition probability to the orbital overlap of 4f and 5d levels, the process of excited state decay is sensitive to certain aspects of the ion local environment. The local crystal field symmetry produces a Stark splitting of the individual 4f levels and thus contributes to the intrinsic 4f-4f transition linewidth and the radiative lifetime, τ_R. The local vibrational density of states of the host lattice can, in addition, provide an alternative, non-radiative pathway for excited state decay, characterized by a rate Γ_{NR}. This non-radiative process can be particularly detrimental to the performance of amplifying devices where a population inversion is required.

253

Mat. Res. Soc. Symp. Proc. Vol. 453 © 1997 Materials Research Society

It is with regard to non-radiative losses that glasses appear to offer several advantages for hosting rare earth ions[4,5]. Unlike many of their crystalline counterparts, glasses possess a wide range of compositions with correspondingly diverse structural features. In principle, this allows one to adjust host properties in a continuous manner, with the potential to select desired optical and mechanical behaviors. Particularly important is the enhanced control this provides over the bulk vibrational characteristics of the host matrix. Not surprisingly this feature has led to several recent examinations of network forming glasses[6,7] in terms of their suitability as rare-earth doped devices. Tellurite glasses[5] appear to possess a suitable compromise of needed properties: (1) low maximum phonon energy (≤ 800 cm^{-1}) which is a prominent factor determining the degree of non-radiative losses, (2) high refractive index and low UV edge, required in many optical component applications, and (3) low glass transition temperatures. These properties can be modified by the substitution of alkali or alkali-earth oxides which convert bridging Te-O-Te bonds to non-bridging oxygens to depolymerize the network.

In this paper, we report measurements of the optical properties of Zn-tellurite glasses doped with Nd^{3+} and Er^{3+}. The novelty of our study is the substitution of zinc halides (ZnCl$_2$, ZnF$_2$) for ZnO which significantly extends the glass forming range of the zinc tellurite system. Our experimental probes include Raman scattering, optical absorption, and fluorescence lifetime techniques designed to evaluate variations in the radiative and non-radiative contributions to the total decay rate upon halide substitution. We also report some intriguing preliminary phonon sideband results which provide clues about the local vibrational environment of the rare earth ion.

EXPERIMENT

Samples of ZnO-TeO$_2$ glass were prepared by melting ZnO and TeO$_2$ powers in a gold crucible for approximately 1 hour at about 800 °C. Halide-substituted samples prepared using ZnCl$_2$ and ZnF$_2$ powders required lower melt temperatures (≈ 650°C) but, owing to a tendency for these halides to react with gold, were melted in silica crucibles. Final compositions were determined ($\pm 10\%$ relative) by X-ray photoelectron spectroscopy. The glassforming range was determined by visual inspection and X-ray powder diffraction analysis. The glass samples were generally transparent at visible wavelengths with a UV cutoff exhibited at around 380 nm. Glass transition and crystallization temperatures were determined using differential thermal analysis[8]. The xZnO-(1-x)TeO$_2$ glassforming range[9] was determined to be $0.2 \leq x \leq 0.35$, and could be increased to $x \approx 0.6$ by halide substitution. Doped glasses were obtained by addition of the desired rare earth oxide to the melt. For optical investigations, the doping levels were varied in an isostructural manner by adding optically inactive lutetium together with the optically active ion so as to maintain 1 mol% total rare earth.

Optical characterization included UV-VIS-NIR absorption, Raman scattering, spectral and temporal photoluminescence, and photoluminescence excitation (phonon sideband) measurements. Absorption measurements were performed in transmission mode at wavelengths from 200 nm to 3000 nm (Perkin-Elmer Lambda 900). Raman scattering was performed with an Ar$^+$ laser (514.5 nm, approx. 200mW) and a Spex 1877 Triplemate spectrometer with a typical spectral resolution of about 2 cm^{-1}. Fluorescence spectra were collected using an Ar$^+$ laser (488 nm) focussed to an approximately 100 μm incident spot size and a 0.275 m grating spectrometer (ARC) with a 2 nm resolution. The emitted light intensity was detected by either a cooled photomutiplier tube (Hamamatsu R2658) or a LN$_2$-cooled, Ge detector (North Coast). For measurements of the fluorescence lifetime the excitation intensity was modulated by a mechanical chopper to provide a temporal resolution of approximately 10 µs. Signal averaging of the emission decay was achieved using a box car integrator (SRS). Phonon sideband measurements were performed using a tunable source comprised of a 400 W Xe lamp and 0.15m monochromator (ARC) with a spectral resolution of 1nm. The emitted intensity was measured using the 0.275 m spectrometer described previously.

The total measured fluorescent decay rate of the excited state of a rare earth ion can be expressed as the sum of three independent processes[10]: radiative (Γ_R), non-radiative (Γ_{NR}) and interion energy transfer (Γ_Q), such that $\Gamma_M = \Gamma_R + \Gamma_{NR} + \Gamma_Q$. Interion energy transfer, or "quenching", is a non-radiative decay mode in which the excited ion's energy migrates from ion to ion, via multipolar electrostatic interactions, to trap sites in the material where the energy is lost non-radiatively[11]. While this process plays a dominant role at high ion concentrations, Γ_Q becomes vanishingly small at concentrations below some critical threshold.

Determination of the intrinsic radiative decay rate, Γ_R, was accomplished through a Judd-Ofelt[3,12] analysis of the absorption spectra. The Judd-Ofelt theory involves a parameterization of the rare earth absorption intensities and provides estimates of the radiative decay rate of all excited states as well as predictions for oscillator strengths and branching ratios. Briefly, the experimentally determined oscillator strengths, P_{if}, of observed transitions (integrated area under corresponding absorption peaks) can be expressed under certain simplifying conditions by a linear sum of standard transition matrix elements, U_{if}, of the form

$$P_{if} = C \sum_{q=2,4,6} \Omega_q \left| U_{if}^{(q)} \right|^2 .$$ The Ω_q (q = 2, 4, 6) are the so-called Judd-Ofelt (JO) parameters whose

values are directly influenced by the host matrix and provide a qualitative measure of the asymmetry of the local crystal field near the rare earth ion[3].

Measurements of the total fluorescence rate at low ion concentration ($\Gamma_Q = 0$) together with Γ_R determined by Judd-Ofelt analysis of the absorption, allow for determination of the non-radiative decay rate Γ_{NR} ($\Gamma_{NR} = \Gamma_M - \Gamma_R$). This non-radiative decay occurs through the generation of lattice vibrations (optical phonons) which carry away the excited state energy. The probability for this process to occur is determined by the strength of the electron-phonon coupling and number of phonons required to bridge the energy gap, E_g, to the next lowest available electronic energy level. A phenomenological theory of multiphonon decay has been developed[13,14], which predicts an energy gap dependence of the form

$$\Gamma_{NR} = Ae^{-\alpha E_g}, \tag{1}$$

where α and A are phenomenological parameters which indicate the sensitivity of Γ_{NR} to E_g, and the value of Γ_{NR} in the limit of vanishing E_g, respectively. The parameters α and A are specific to a given host and can be compared with those obtained from other glass matrices to evaluate the relative contribution of multiphonon decay in the deexcitation of 4f energy levels.

RESULTS

Fluorescence lifetime measurements were performed on $25ZnO-74TeO_2-xNd_2O_3-(1-x)Lu_2O_3$ glass for $0.01 \leq x \leq 1$. Luminescence decay from the $^4F_{3/2}$ state was monitored and the observed rate, Γ_M, was reasonably constant for $x \leq 0.3$ but increased significantly for x=1. This concentration dependence indicates a quenching threshold in the vicinity of $x \approx 0.3$, and so by restricting our optical studies to $x \leq 0.1$, we can safely ignore Γ_Q in the determination of Γ_{NR}.

The absorption spectrum of 25ZnO-74TeO$_2$-0.1Nd$_2$O$_3$-0.9Lu$_2$O$_3$ is presented in Fig.(1). Approximately eight identifiable absorption lines are observed and a UV edge occurs near 360nm. Values of the calculated JO parameters (Ω_q) and the Γ_R for the decay from the $^4F_{3/2}$ state are presented in the inset.

Also included in the figure, are values of Ω_q for two halide substituted base tellurite glasses, with approximate molar compositions of 5ZnF$_2$-25ZnO-70TeO$_2$ and 15ZnCl$_2$-15ZnO-70TeO$_2$. These parameters do exhibit consistent trends upon halide substitution with Ω_2 increasing and Ω_4 decreasing as we go from oxide to fluoride to chloride, while Γ_R increases slightly and Ω_6 remains reasonably unaffected. Thus replacement of oxide by halide anions appears to produce significant changes in the local crystal field asymmetry.

Using the values of Γ_R determined from JO analysis, the non-radiative or multiphonon decay rate, $\Gamma_{NR} = \Gamma_M - \Gamma_R$, was determined. We have determined Γ_{NR} for 25ZnO-74TeO$_2$-0.1Re$_2$O$_3$-0.9Lu$_2$O$_3$ (Re = Nd or Er) for the decay of a variety of electronic states. In this manner we are able to evaluate the

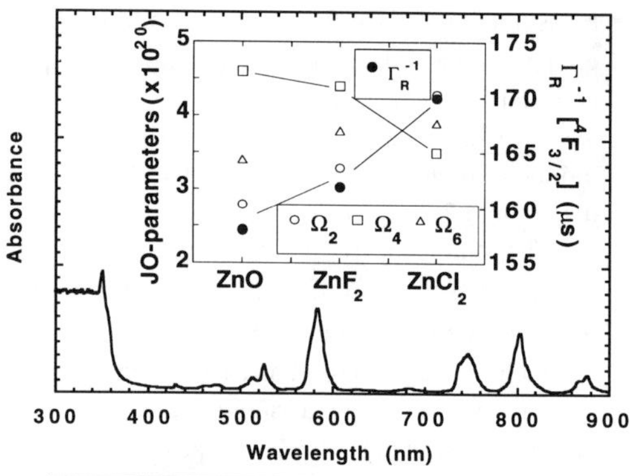

Fig. 1   Absorptivity of 25ZnO-74TeO$_2$-0.1Nd$_2$O$_3$-0.9Lu$_2$O$_3$. Inset shows trends in the JO parameters with anion substitution.

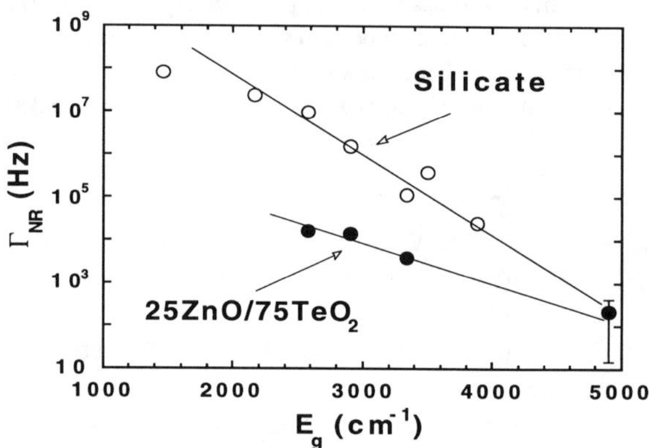

Fig. 2   Non-radiative decay rate vs. energy gap to next nearest level. Closed circles: 25ZnO-74TeO$_2$-0.1Re$_2$O$_3$-0.9Lu$_2$O$_3$ of the present study, Open circles: a silicate glass [2].

dependence of Γ_{NR} upon the energy gap to the next nearest electron energy level. Results of this analysis are provided in Fig. (2). Fits of the phenomenological expression Eq.(1) yield $\alpha = 2.0 \times 10^{-3}$ cm and $A = 3 \times 10^6$ s^{-1}. We have also included in the figure results of a similar analysis performed by Reisfield[2] for a silicate glass. Silicate glasses possess higher maximum phonon energy (≈ 1200 cm^{-1}) than that exhibited by the zinc-tellurite composition (≈ 750 cm^{-1}). As expected the non-radiative decay rate due to multiphonon emission is much lower for the zinc-tellurite glass. What is interesting is the much smaller value for α. This value is about half of that commonly observed in other glassformers (even those with smaller maximum phonon energy[5]). This implies that non-radiative decay in zinc-tellurite is less sensitive to decreasing energy gap and remains low even in the limit of vanishingly small energy steps.

To better understand the vibrational behavior of the lattice in the immediate vicinity of the rare earth ion, we have also performed phonon sideband (PSB) measurements of several zinc tellurite hosts doped with 1 mole % Eu_2O_3. In this experiment[15] the excitation wavelength is varied near the $^5D_2$ absorption band while the 615 nm emission resulting from the $^5D_0$ to $^7F_0$ transition is monitored. Maximum emission occurs when the incident wavelength is coincident (at 465 nm) with direct absorption into the $^5D_2$ level. Our results are shown in Fig. (3). In addition to the peak at 451 nm an additional shoulder is evident at 458 nm. These bands correspond to

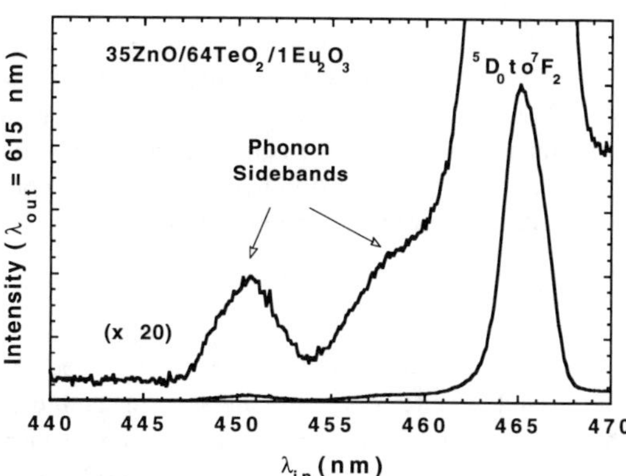

Fig. 3   Phonon sideband spectrum for 35ZnO-64TeO₂-1Eu₂O₃.

an absorption process wherein excess energy from the incident photon is transferred into the lattice in the form of phonons. These phonons can only be generated in the immediate vicinity of the rare earth ion, and so these spectra reflect only the local vibrational environment.

To show how this local vibrational behavior compares with vibrational features obtained from Raman scattering (averaged over all lattice sites), we replot in Fig. (4) our PSB spectra now as a function of energy shift from the pure electronic absorption band (21500 cm^{-1}) together with the Raman spectrum collected from 35ZnO-65TeO$_2$ glass as well as that obtained from amorphous TeO$_2$. The Raman peak at 670 cm^{-1} has been previously assigned[9] to the symmetric Te-O stretch in a TeO$_4$ bipyramidal tetrahedron and that at 740 cm^{-1} to the symmetric Te-O stretch in a deformed TeO$_4$ bipyramidal tetrahedron whose axial bond length is much larger. This deformed unit is referred to as TeO$_{3+1}$. The peak at 420 cm^{-1} is the Te-O-Te bending (deformation) mode. Comparison with our PSB spectra indicates that much

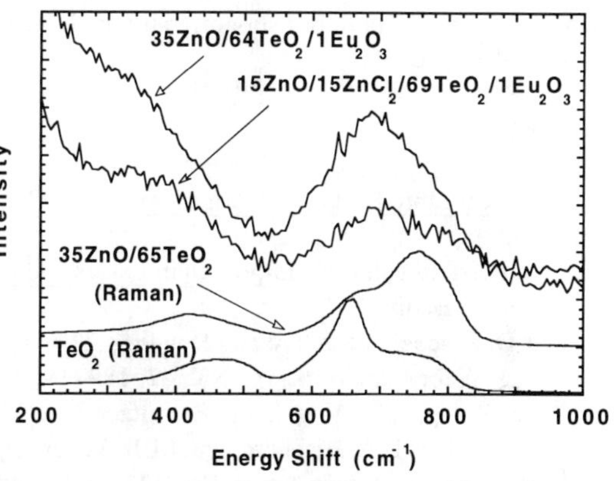

Fig. 4   Phonon sideband intensity vs. energy shift from the 465nm transition for 25ZnO-74TeO$_2$-1Eu$_2$O$_3$ and 15ZnCl$_2$-15ZnO-69TeO$_2$-1Eu$_2$O$_3$. Raman intensity of 35ZnO-65TeO$_2$ and TeO$_2$ for comparison.

of these global vibrational characteristics are reproduced in the vicinity of the rare earth ion. However, it is apparent that the relative ratio of TeO$_{3+1}$/TeO$_4$ is reduced around the rare earth site since the PSB spectra more closely resemble the Raman spectrum from the more polymerized TeO$_2$ glass. While these results are preliminary, we propose this indicates that the rare earth is preferentially incorporated into the lattice near the TeO_4 units and most likely nearest the two lone pair electrons that constitute the bipyramidal tetrahedral TeO_4 unit.

Included in Fig.(4) are recent measurements for a chlorine-substituted glass ($15ZnCl_2$-$15ZnO$-$69TeO_2$-$1Eu_2O_3$). While also preliminary, we see evidence of significant changes in the local vibrational features. In particular we observe that the intensity of the bands (relative to the peak electronic transition intensity) have generally decreased implying a corresponding decrease has occurred in the electron-phonon coupling strength[15] upon substitution of chlorine.

CONCLUSIONS

We have examined a series of rare earth-doped zinc-tellurite glasses in an effort to better understand the interplay of dopant and host and its impact on electron decay dynamics. We find that zinc tellurite glasses possess a low phonon energy and a significantly reduced sensitivity (smaller α) of the non-radiative transition probability to the size of energy gap being encountered. We also have shown that halide substitution in these glasses does influence the crystal field symmetry (as determined by Ω_q) and vibrational behavior. Measurements of phonon sideband spectra reveal that the local vibrational environment at the rare earth ion is different from that inferred from Raman spectroscopy. These results are intriguing as they provide clues to the lattice structure around which the rare earth ion prefers to take residence. Tentative evidence suggests the rare earth is coordinated with TeO_4 bipyramidal tetrahedra.

Acknowledgments: Funding for this work was provided by a grant from the National Science Foundation (contract # DMR-940309). B.G.P. and R.K.B. also acknowledge the resources of Sandia National Laboratory supported by the US Department of Energy under contract DE-ACO4-94AL850000.

REFERENCES

1. Optical Fibre Lasers and Amplifiers, ed. by P.W. France (CRC Press, Boca Raton, FL 1991).
2. R. Reisfeld and C.K. Jorgensen in Lasers and Excited States of Rare Earths, (Springer-Verlag, Berlin 1977).
3. R.D. Peacock, Structure and Bonding **22**, 83 (1975).
4. M.J. Weber, J. Non-Cryst. Sol. **42**, 189 (1980); **47**, 117 (1982).
5. J.S. Wang, E.M. Vogel, and E. Snitzer, Opt. Mats. **3**, 187 (1994).
6. C. Brecher, L.A. Riseberg, and M.J. Weber, Phys. Rev. **B18**, 5799 (1978).
7. M.J. Weber, J. Non-Cryst. Sol. **123**, 208 (1990).
8. M.A. Hruscka, D.L. Sidebottom, R.K. Brow, and B.G. Potter, to be published.
9. H. Burger, etal., J. Non-Cryst. Sol. **151**, 134 (1992).
10. R. Reisfeld, Structure and Bonding **22**, 123 (1975).
11. W.M. Yen, Mod. Probs. in Condens. Matt. Sci. **21**, 185 (1987).
12. B.R. Judd, Phys. Rev. **127**, 750 (1962); G.S. Ofelt, J. Chem. Phys. **37**, 511 (1962).
13. T. Miyakawa and D.L. Dexter, Phys. Rev. **B1**, 2961 (1970).
14. L.A. Riseberg, and H.W. Moos, Phys. Rev. **174**, 429 (1968).
15. S. Tanabe, K. Hirao, and N. Soga, J. Non-Cryst. Sol. **142**, 148 (1992).

THE EFFECTS OF DOMAIN STRUCTURE ON THE ELECTRO-OPTIC RESPONSE OF POTASSIUM NIOBATE THIN FILMS

M. J. NYSTROM *, B. W. WESSELS **
Department of Materials Science and Engineering
Northwestern University, Evanston, IL 60208
* m-nystrom@nwu.edu
** b-wessels@nwu.edu

ABSTRACT

Epitaxial potassium niobate films have been deposited by metalorganic chemical vapor deposition (MOCVD). X-ray diffraction of the films deposited on (100) spinel substrates shows (110) and (001) reflections, indicating a multidomain structure. The electro-optic properties of the thin films were measured using a transmission technique. Effective electro-optic coefficients as large as 470 pm/V were observed. The observed electro-optic response was dominated by domain reorientation effects.

INTRODUCTION

Recently electro-optic switches with modulation frequencies of 100 GHz have been fabricated in bulk ferroelectric crystals.[1] Thin film structures should result in modulators with even higher speeds.[2] If practical thin film ferroelectric devices are to be attained, highly oriented films must be deposited with near bulk optical properties. Electro-optic response of numerous ferroelectric thin films has been reported, but it is often lower than that of a single crystalline material. While the effect of domain structure on the nonlinear optical properties of ferroelectric has been investigated,[3] to date, the mechanisms responsible for the observed electro-optic response have not been described.

We have previously reported the epitaxial deposition of $KNbO_3$ thin films by a metalorganic chemical vapor deposition (MOCVD) process[4] with effective second order nonlinear optical coefficients as high as 12 pm/V, (compared to the bulk values of $d_{32} = 40$ pm/V and $d_{33} = 61$ pm/V at 532nm[5]). In this study the electro-optic response of these thin films are reported and examined in terms of the structure. In addition, a model is reported which describes the observed electro-optic response.

EXPERIMENTAL

Potassium niobate thin films were prepared on $MgAl_2O_4$ substrates by a low pressure metalorganic chemical vapor deposition process as previously described.[4,6] The as-deposited film are orthorhombic and exhibit a mixed domain structure. Figure 1 shows the x-ray diffraction scan of a $KNbO_3$ film, indicating the presence of domain orientations which produce both (110) and (001) x-ray reflections. Analysis of the integrated intensity of the x-ray diffraction peaks for the $KNbO_3$ thin films indicated a volume fraction of 0.70 for the domains associated with the (110) x-ray reflections.

The electro-optic response of the $KNbO_3$ thin films was determined by measuring the phase retardation. Two electrodes, separated by a gap of 200 μm, were evaporated on the surface of the thin film. The apparatus is shown in Figure 2. The electro-optic response signal S generated by the film is given as:

Mat. Res. Soc. Symp. Proc. Vol. 453 © 1997 Materials Research Society

$$S = \Gamma\cos(2\theta)\sin(2\rho) - \Gamma\sin(2\theta)\sin(2\rho)\cos(2\rho) + \cos(2\rho) \qquad (1)$$

where Γ is the phase retardation of the film, and θ and ρ are the angles between the modulation axis of the photoelastic modulator (PEM) and the fast axis of the film and 1/4 wave plate, respectively. Voltages of up to ±500 volts in the plane of the film were applied. A detailed description of the operation of the apparatus is given elsewhere.[6]

To examine contributions by persistent domain orientation changes to the observed electro-optic response, a pulsed voltage was applied to the thin film samples. The waveform used in the measurement is shown in Figure 3. The typical voltage pulses were approximately 3 seconds in duration and were separated by a similar period of zero applied voltage. The signal was measured during (curve b) and between (curve a) the applied pulses and results of each measurement is plotted in Figure 4.

DISCUSSION

As shown in Figure 4b, hysteresis is observed in the electro-optic response of the multidomain $KNbO_3$ thin films. This hysteresis indicates that changes in the domain structure upon application of a bias contribute to the change in birefringence in the film. For the alignment of the thin film crystal axes with the optical modulation axis and the applied electric field utilized in these measurements, six unique domain orientations exist which exhibit different birefringence and electro-optic response. Any changes in domain orientation with different birefringence will result in a change in signal. To separate the signal produced by domain orientation changes from the signal produced by the linear electro-optic effect, measurements both without and with an electric field were recorded and are displayed in Figure 4a and b, respectively.

Since birefringence changes due to the electro-optic effect are only observed while an electric field is applied, the signal changes observed in Figure 4a must result from persistent changes in domain structure. To determine the transient changes in electro-optic response, the signal measured after removing the electric field is subtracted from the signal measured with an applied electric field. This transient electro-optic response is shown in Figure 4c. The V-shaped appearance of the curve suggests the linear response is

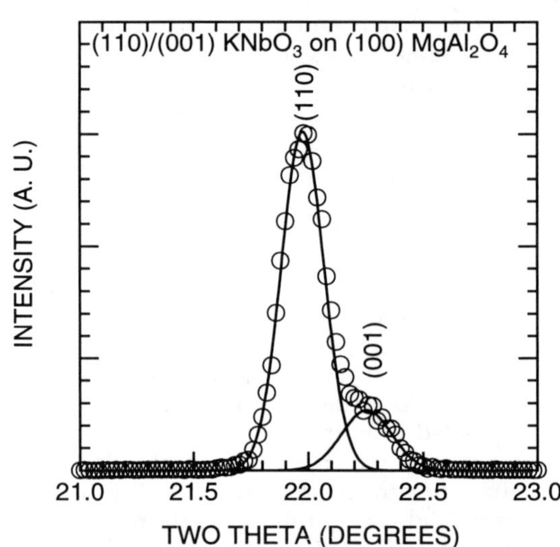

Figure 1 X-ray diffraction scan showing the presence of domains with both (001) and (110) x-ray reflections.

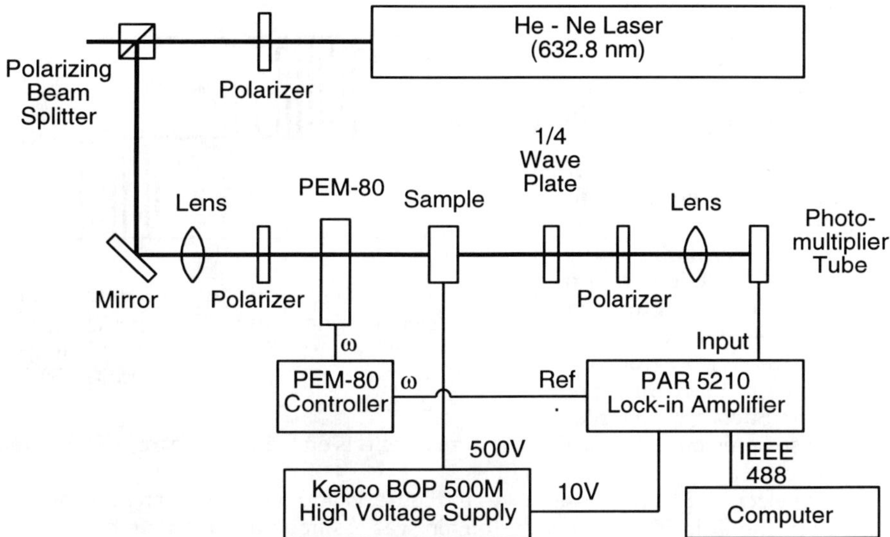

Figure 2   The experimental setup for the measurement of the electro-optic response in thin film samples.

due to the linear electro-optic effect and domain orientation reversal. A reversal of the domain orientations responsible for the linear electro-optic effect due to changing polarity of the applied bias results in a change in the sign of the slope of the linear response.

A model was devised to describe the domain changes and electro-optic properties of the $KNbO_3$ films. For this model, the films were assumed to be composed of a single layer of domains with constant cross-sectional area through the thickness of the film allowing the volume fraction of each domain to be represented by an area fraction. The domains which compose the films were assumed to have bulk birefringence and electro-optic properties. The signal produced by the entire film resulted from the sum of the signal produced by each domain multiplied by the area fraction of the domain. Changes in domain orientation are included in the model by allowing the area fraction of each domain to be a function of electric field. Since only six unique domain orientations exist (I - VI), this summation can be reduced to six terms and is given as:

$$S = \sum_{i=I}^{VI} A_i S_i \qquad (2)$$

where S is the total signal produced by the $KNbO_3$ thin film and A_i and S_i are the area fraction and signal produced by each unique domain type composing film. The area fraction for each domain type is a function of the last applied field and is assumed to persist after the field is removed. Area fraction functions were constructed such that only domain changes which resulted in an increased alignment of the domain polarization with the applied electric field were considered. The signal for each domain is calculated to be that expected for a single domain crystal having the same thickness and orientation.

In a ferroelectric film, a linear electro-optic response will only be observed if a net polarization exists for the film. Since the six unique domain types for the KNbO$_3$ films are composed of three pairs of domains with equal and opposite linear electro-optic induced refractive index changes, the deviation from a centrosymetric state in the film can be represented as the difference in area fraction for each domain type in these three domain pairs. Such a representation was used previously to calculate the domain effects on the second harmonic generation observed in LiTaO$_3$ and KNbO$_3$ thin films.[3,7]

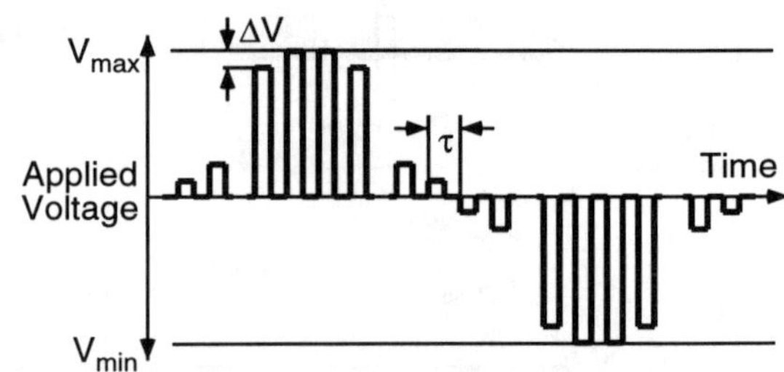

Figure 3   The voltage profile applied to the thin film electro-optic samples. Voltages (V$_{max}$,V$_{min}$) of ±500V were utilized. The typical voltage step (ΔV) was 25V. The pulse duration (τ) was 3 seconds.

The electro-optic response curves in Figure 4a and b were fit using this model, as shown. For the fit in Figure 4a each domain was assumed to have bulk birefringence. Area fraction changes of between 2-4% were necessary to reproduce the observed signal in KNbO$_3$ thin films. The magnitude of these changes compares well with the KNbO$_3$ thin film domain structure observed using second harmonic generation by Gopalan et al.[3] To calculate the fit to Figure 4b, the signal changes due to the linear electro-optic effect were added to the model. However, to reproduce the curve in Figure 4b, area fraction changes of 20-30% were necessary. Based upon the results of the fit in Figure 4a, the magnitude of the area fraction change is not a realistic value for an applied electric field at room temperature.

The probable cause of the large transient electro-optic response is rapid changes in domain structure which are faster than the measurement. To examine this effect, the signal

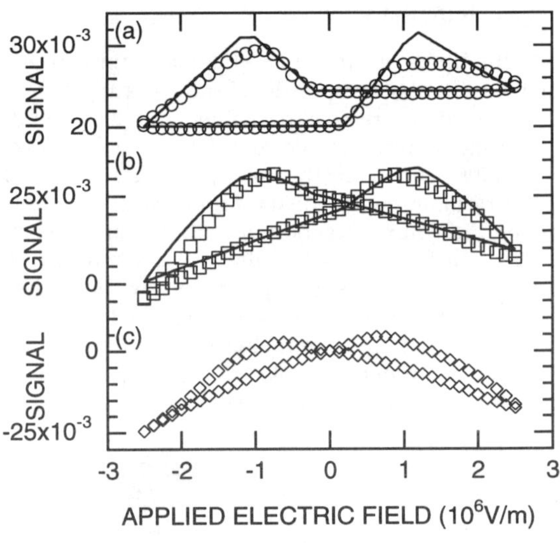

Figure 4   The results of the electro-optic measurement for a KNbO$_3$ thin film. (a) The signal measured with no applied voltage. (b) The signal measured during the applied voltage pulse. (c) The difference of (b) and (a). The calculated fit for curves (a) and (b_ are shown as solid lines superimposed on the data.

decay of the electro-optic response was measured. The signal from the electro-optic apparatus decayed with a range of time constants after the electric field was removed. Time constants between 20 and 4000 seconds were measured. The observed decay rates support rapid domain orientation changes contributing to the transient electro-optic response.

While domain reorientation effects may contribute to the transient electro-optic response at low (<MHz) frequencies, applications such as high frequency modulators will not benefit from the additional electro-optic response. Based upon the observed persistent domain area fraction changes, the linear electro-optic effect expected from the multidomain films will be only a fraction of the bulk electro-optic effect. To increase the response in these films, the domain area fraction difference for complimentary domain pairs must be increased. Potential methods of achieving greater linear electro-optic effects include poling of the as-deposited films and operation of the thin film electro-optic modulators with a dc bias.

To examine the transient change in electro-optic signal for a variety of film micro-structures, an effective linear electro-optic coefficient was calculated for the $KNbO_3$ thin films. Using the average value of the absolute slope of the linear portion of the curve in Figure 4c, the effective linear electro-optic coefficient r_{eff} was calculated using the relationship:

$$r_{eff} = \lambda m / \pi n^3 W \tag{3}$$

where λ is the wavelength of light used (632.8 nm), n is the thin film refractive index, W is the film thickness, and m is the average slope of the curve in Figure 4c. This number represents the contributions to the transient electro-optic response from both the linear electro-optic effect and rapid changes in domain area fraction for the film. For the $KNbO_3$ thin films, r_{eff} values as large as 470 pm/V were observed, but values of approximately 100 pm/V were typical. The effective linear coefficient was found to be closely related to the grain size and film thickness. For the $KNbO_3$ films, the measured effective linear electro-optic coefficient decreased with deceasing film thickness and grain size. These effects are due to constraint to domain reorientation by the film interfaces and low angle grain boundaries. As the film thickness or grain size decreases, the proximity of any portion of the thin film to one of these boundaries will increase.

CONCLUSION

The electro-optic response of multidomain $KNbO_3$ thin films was characterized in this study. Changes in thin film domain structure dominated both the persistent and transient changes in electro-optic signal. Domain area fraction changes as large as 4% were observed for the films. The time response of the transient electro-optic signal revealed contributions from rapid domain motion. Modification of the as-deposited domain structure is necessary if larger linear electro-optic effects are to be obtained in ferroelectric thin films.

ACKNOWLEDGMENT

This work was supported by the Department of Energy (Grant No. FG02 85ER 45209) and by the Office of Naval Research through the National Defense Science and

Engineering Graduate Fellowship program (M. J. N.).

REFERENCES

1 O. Mitomi, K. Noguchi, and H. Miyazawa, 8th Annual Lasers and Electro Optics Meeting (LEOS), San Francisco, 1995.

2 D. K. Fork, F. Armani-Lepplingard, J. J. Kingston *et al.,Proceedings of the Materials Research Society*, Vol. 392, pp. 189 (1995).

3 V. Gopalan and R. Raj, Appl. Phys. Lett. **68** (10), 1323 (1996).

4 M. J. Nystrom, B. W. Wessels, D. B. Studebaker *et al.*, Applied Physics Letters **67**, 365 (1995).

5 Y. Uematsu, Japanese Journal of Applied Physics **13** (9), 1362 (1974).

6 M. J. Nystrom, Ph. D. Thesis, Northwestern University, 1996.

7 H. Xie, W.-Y. Hsu, and R. Raj, J. Appl. Phys. **77** (7), 3420 (1995).

NEW NONLINEAR OPTICAL CRYSTALS OF KTiOPO$_4$ (KTP) FAMILY

V.I. CHANI, K. SHIMAMURA, S. ENDO, T. FUKUDA
Institute for Materials Research, Tohoku University, 2-1-1 Katahira, Aoba-ku, Sendai, 980 Japan

ABSTRACT

Phase formation and crystal growth conditions for different combinations of cations corresponding to chemical formula of KTP were studied. Preparation of new Ti-less arsenates which are isostructural with KTP is reported. In the single crystals grown Ti^{4+} cations were completely substituted with M^{2+} + Nb^{5+} (M = Mg, Zn) or M^{3+} + Nb^{5+} (M = Al, Cr, Ga, Fe, In) cation couples. The crystals were prepared by spontaneous crystallization using flux method. The structure analysis was made by X-ray powder diffraction technique. The crystals composition was measured by electron probe microanalysis (EPMA). Second harmonic generation (SHG) effect was studied by the powder technique with a pulsed YAG:Nd laser (1064 nm). SHG signal of the crystals was found to be slightly greater than that of KTiOAsO$_4$ (KTA).

INTRODUCTION

Crystals with structure of KTP [1-3] are well known for their nonlinear optical properties. The structure of the KTP crystal is constructed by three-dimensional framework of TiO$_6$ octahedras and PO$_4$ tetrahedras and contains spacious channels along c-axis [2]. The cations of monovalent metals are located in these channels. The KTP structure is typified by following chemical formula {K$^+$}[Ti^{4+}]O(P^{5+})O$_4$, where the curly brackets indicate nine- or eight-fold oxygen coordination, the square brackets correspond to six-fold octahedral coordination, and the parentheses correspond to four-fold tetrahedral coordination. Since KTP melts incongruently, the flux method is widely used for the growth of crystals with this structure.

It is known that it is possible to substitute completely K$^+$ cation with Tl$^+$, Rb$^+$ and Cs$^+$. However, all these substitutions result decreasing of stability of KTP structure [1]. At first sight substitution of K$^+$ with Na$^+$ is necessary to increase the stability of the structure. However according to results of [3] the segregation coefficients of K$^+$ and Na$^+$ are related as follows:

$$K_{9,8}(\text{Na}^+) < K_{9,8}(\text{K}^+). \tag{1}$$

Therefore, it can be concluded that incorporation of K$^+$ into the structure is much more preferable than that of Na$^+$. Thus it was anticipated that placement of monovalent K$^+$ into {}-positions correspond to maximal stability of KTP structure.

It is also known that it is possible to substitute all P^{5+} cations in ()-positions with As^{5+} [1]. Placement of As^{5+} instead of P^{5+} was found increases stability of the structure and probability of KTP phase formation [1,4].

There are not evident proofs of corresponding relations between various cations in []-positions. However mostly KTP phase formation was observed in the compounds containing Ti^{4+} in these crystallographic sites.

Nb-doped KTP crystals was observed to be stable. K$_{1-x}$Ti$_{1-x}$Nb$_x$OPO$_4$ single crystals with x ~ 0.25 have been obtained without any charge compensator because Nb^{5+} strongly favors the cis-Ti(1) sites [5,6]. Partial charge compensation of Nb^{5+} in K$_{1-x}$Ti$_{1-x}$Nb$_x$OPO$_4$ crystals by Ga^{3+} [7], Fe^{3+} [5], and Mg^{2+} [8] is also known. Preparation of K[Nb$_{1/2}$Ga$_{1/2}$]OPO$_4$, K[Nb$_{1/2}$Fe$_{1/2}$]OPO$_4$, K[Nb$_{1/2}$Mn$_{1/2}$]OPO$_4$, K[Nb$_{2/3}$Mg$_{1/3}$]OPO$_4$, and K[Nb$_{1/2}$Ga$_{1/2}$]OAsO$_4$ compounds by solid state reaction technique was reported also [9]. However there are not reports related with growth of Ti-less K[Nb$_{1/2}$M$_{1/2}$]OAsO$_4$ (M = Al, Cr, Ga, Fe, In) and K[Nb$_{2/3}$M$_{1/3}$]OAsO$_4$ (M = Mg, Zn) single crystals with KTP structure. It is the aim of this research to grow these crystals by flux method.

Mat. Res. Soc. Symp. Proc. Vol. 453 © 1997 Materials Research Society

EXPERIMENT

The crystals reported were grown by spontaneous crystallization in stoichiometric melts and self-fluxes. Melt compositions are given in Table 1.

Table 1. Composition and melting point (T_m) of the mixtures used for the crystal growth, structure and color of $K[Nb_{1/2}M_{1/2}]OAsO_4$ and $K[Nb_{2/3}M_{1/3}]OAsO_4$ crystals grown

M	#	Melt Composition	T_m, °C	Structure	Color
Al	65-4	$K[Nb_{1/2}Al_{1/2}]OAsO_4$	< 1100	KTP	Colorless
	75-2	$K_{0.91}Nb_{0.64}Al_{0.49}OAs_{0.91}O_4$	< 1150	KTP	Colorless
Cr	62-2	$K[Nb_{1/2}Cr_{1/2}]OAsO_4$	> 1100	KTP	Dark-Green
	63-2	$K[Nb_{1/2}Cr_{1/2}]OAsO_4{:}KAsO_3=1{:}1$	> 1100	KTP	Dark-Green
Ga	52-3	$K[Nb_{1/2}Ga_{1/2}]OAsO_4$	< 1150	KTP	Colorless
	52-2	$K[Nb_{1/2}Ga_{1/2}]OAsO_4{:}KAsO_3=2{:}1$	< 1150	KTP	Colorless
Fe	59-2	$K[Nb_{1/2}Fe_{1/2}]OAsO_4$	< 1100	KTP	Red-Brown
	61-2	$K[Nb_{1/2}Fe_{1/2}]OAsO_4{:}KAsO_3=4{:}3$	< 1100	KTP	Red-Brown
In	64-3	$K[Nb_{1/2}In_{1/2}]OAsO_4$	> 1100	KTP	Colorless
	65-3	$K[Nb_{1/2}In_{1/2}]OAsO_4{:}KAsO_3=2{:}1$	> 1100	KTP	Colorless
	66-3	$K[Nb_{1/2}In_{1/2}]OAsO_4{:}KAsO_3=1{:}1$	< 1100	KTP	Colorless
Yb	66-1	$K[Nb_{1/2}Yb_{1/2}]OAsO_4$	≈ 1100	not KTP	Colorless
Mg	63-1	$K[Mg_{1/3}Nb_{2/3}]OAsO_4$	> 1100	KTP	Colorless
	65-1	$K[Mg_{1/3}Nb_{2/3}]OAsO_4 : KAsO_3 = 3{:}1$	< 1100	KTP	Colorless
Zn	65-2	$K[Zn_{1/3}Nb_{2/3}]OAsO_4$	< 1100	KTP	Colorless
	66-2	$K[Zn_{1/3}Nb_{2/3}]OAsO_4 : KAsO_3 = 3{:}1$	< 1100	KTP	Colorless

Starting mixtures were prepared using desired quantities of salts and oxides, KH_2AsO_4, Nb_2O_5, M_2O_3 (M = Al, Cr, Ga, Fe, In, and Yb), and MO (M = Mg and Zn) which were carefully weighed and mixed. The mixtures were then heated about 10 h at 1100-1150°C and melted in platinum crucibles in a muffle furnace. After that the melts were slowly cooled to 800°C at a rate of about 3°C/h. The crystals were recovered by etching away the solidified melt in hot water. Fig.1 shows the view of the $K[Nb,Ga]OAsO_4$ crystals as-grown from the melts 52-2. View of $K[Nb,Fe]OAsO_4$ crystals grown from 61-2 melt is given in Fig.2.

The structure identification and the lattice parameters measurements of the crystals obtained were made by X-ray powder diffraction technique with CuK_α radiation (1.5418 Å). Results of measurements are given in Table 2.

Second harmonic generation (SHG) of the crystals was studied by the powder technique with a pulsed YAG:Nd laser ($\lambda = 1.064$ µm) similar to [6,8]. Powders were prepared using sieves. Particle size was in the range of 125 - 149 µm. The SHG signal of sample prepared from non-substituted KTA powder was used as standard.

Fig.1 View of $K[Nb,Ga]OAsO_4$ as grown crystals (scale in mm)

Fig.2 View of $K[Nb,Fe]OAsO_4$ as grown crystals (scale in mm)

Table 2. Composition, lattice parameters a, b, c, volume of unit cell V, and results of SHG examination

#	Crystal Composition	a, Å	b, Å	c, Å	V, Å^3	SHG
65-4	$K_{0.76}Nb_{0.64}Al_{0.49}OAs_{0.91}O_4$	10.657	13.039	6.650	924.20	1.0
63-2	polycrystal (not studied)	10.868	13.197	6.626	950.33	-
52-2	$K_{0.52}Nb_{0.53}Ga_{0.51}OAs_{1.06}O_4$	10.868	13.238	6.664	958.73	> 0*
61-2	$K_{0.99}Nb_{0.54}Fe_{0.46}OAs_{0.99}O_4$	10.904	13.286	6.630	960.56	-
66-3	not studied	11.312	13.589	6.820	1048.36	1.1
65-1	$K_{0.72}Mg_{0.32}Nb_{0.71}OAs_{1.02}O_4$	11.036	13.467	6.655	989.0	1.2
66-2	not studied	11.033	13.460	6.712	996.8	1.2
72-4	$KTiOAsO_4$ (SHG standard)	10.888	13.336	6.661	967.19	1.0

* SHG data marked by "> 0" are based on visual observation of green light emission

The composition of the crystals grown was studied by electron probe microanalysis (EPMA). Transparent crystals with well developed habit were selected for these measurements. All cation/cation ratios were measured directly by EPMA. Calculations of the composition were made assuming that the crystals did not have oxygen vacancies. Thus, cation/oxygen ratios were determined assuming that formula units of all compounds contain five oxygen atoms. Concentration of cations in atoms per KTP formula unit were determined from charge balance. EPMA data are presented in Table 2. It was anticipated that in the main possible site occupancies were related with sizes of cations [10] and cation positions (coordination numbers) respectively.

RESULTS AND DISCUSSIONS

The structure of $K[Nb_{1/2}M_{1/2}]OAsO_4$ (M = Al, Cr, Ga, Fe, In) and $K[Nb_{2/3}M_{1/3}]OAsO_4$ (M = Mg, Zn) crystals was similar to that of non-substituted KTP. KTP structure formation was not observed in the mixtures containing $K[Nb_{1/2}Yb_{1/2}]OAsO_4$ compound because of large size of Yb^{3+} cation [10].

In the main the SHG signals of all colorless crystals prepared were found to be close to that of non-substituted KTA as given in Table 2. The Nb-doped KTP crystals is known [6] to be SHG active. According to [5] the SHG responce of such crystals was 10-20% greater than that from KTP. Thus, our results are in good agreement with data reported in [5,6].

$K[Nb_{1/2}M_{1/2}]OAsO_4$ (M = Al, Cr, Ga, Fe, In) crystals

$K[Nb,Al]OAsO_4$ crystals were grown from the $K[Nb_{1/2}Al_{1/2}]OAsO_4$ stoichiometric melt (65-4). K^+ and As^{5+} cations were partially rejected by the crystals during growth as it follows from comparison of melt and crystal composition. Thus, the flux was formed by remain mixture of K_2O and As_2O_5. The segregation coefficients K(C) were calculated by $K(C) = [C]_{crystal}/[C]_{melt}$ formula, where the square brackets indicate concentrations of C-cations in the crystal and melt respectively. It was found that $K(K^+) = 0.81$, $K(Nb^{5+}) = 1.37$, $K(Al^{3+}) = 1.05$, and $K(As^{5+}) = 0.98$. $K(Al^{3+})$ and $K(As^{5+})$ were close to unity. However, $K(K^+)$ and $K(Nb^{5+})$ values observed display us that in spite of cation vacancies formation stability of Nb^{5+} cations in KTP structure is much greater than that of K^+. Deficit of As^{5+} and excess of Nb^{5+} and/or Al^{3+} observed in the crystals grown (Tables 1 and 2) show that part of As^{5+} in tetrahedral sites was substituted with Al^{3+} and perhaps Nb^{5+}. Four-fold coordination is well known for Al^{3+} in different oxide structures and seems to be possible for Nb^{5+}. The possibility of Nb^{5+} incorporation into tetrahedral sites of KTP structure was already mentioned in Ref. [6]. KTP phase formation was also observed in the melt 75-2 (Table 1) with the composition close to the composition of the crystals grown from the melt 65-4. $K[Nb,Al]OAsO_4$ crystals were SHG active.

$K[Nb,Cr]OAsO_4$ compound was prepared from 62-3 and 63-3 mixtures. These mixtures was found have high melting point ($T_m > 1100°C$). Application of this material in non-linear optics is not evidence because of color. Therefore detailed study of this system was not carried out. Perhaps K[Nb,Cr]OAsO$_4$ compound is most stable one within the members of K[Nb$_{1/2}$M$_{1/2}$]OAsO$_4$ subfamily because of greatest melting point observed. Even in case of high flux content (50% of KAsO$_3$ in melt 63-2) complete melting of K[Nb,Cr]OAsO$_4$ containing mixtures was not observed.

K[Nb,Ga]OAsO$_4$ crystals with the KTP structure were grown from both stoichiometric and KAsO$_3$ flux containing melts at T$_m$ < 1150°C (Fig. 1). The crystals were colorless and SHG active. Very high deficit of K$^+$ cations was detected in the crystals grown. Three crystals were selected for the composition analysis. Results of EPMA measurements for all these crystals were very similar.

K[Nb,Fe]OAsO$_4$ crystals were grown from 59-2 and 61-2 melts. Low melting point (T$_m$ < 1100°C) was observed for these mixtures. Similar to K[Nb,Cr]OAsO$_4$ detailed study of this system was not carried out because of color. View of the crystal is given in Fig. 2. Crystal composition was close to the stoichiometric composition of KTP. Segregation coefficient of Nb^{5+} was found to be K(Nb^{5+}) > 1 similar to Al- and Ga-containing crystals.

K[Nb,In]OAsO$_4$ single crystals were grown only from KAsO$_3$ rich mixture 66-3. KTP phase formation was also observed in stoichiometric K[Nb,In]OAsO$_4$ mixture. However, melting of this mixture was not achieved at 1100°C. Size of the crystals grown from 66-3 melt was less than 1 mm. The crystals were colorless and SHG active (Table 2). It is assumed that In^{3+} is the largest cation which can form KTP structure in subfamily of K[Nb$_{1/2}$M$_{1/2}$]OAsO$_4$ compounds.

K[Nb$_{2/3}$M$_{1/3}$]OAsO$_4$ (M = Mg, Zn) crystals

KTP phase formation was observed in K[Mg$_{1/3}$Nb$_{2/3}$]OAsO$_4$ and K[Zn$_{1/3}$Nb$_{2/3}$]OAsO$_4$ stoichiometric mixtures. However melting point of these compounds was found to be greater than 1100°C. Therefore addition of KAsO$_3$ flux was necessary to achieve complete melting at this temperature. The crystals grown from 65-1 and 66-2 melts were also isostructural with KTP. The typical size of the K[Mg$_{1/3}$Nb$_{2/3}$]OAsO$_4$ crystals was about 1-4 mm, but the size of the K[Zn$_{1/3}$Nb$_{2/3}$]OAsO$_4$ crystals was about 1 mm only.

The attempt to prepare K[Mg$_{1/3}$Nb$_{2/3}$]OPO$_4$ crystals with the KTP structure was reported in [8]. Instead, it was found that the crystals grown were not isostructural with KTP. As it was noted above the probability of KTP phase formation can be increased by substitution of P^{5+} with As^{5+}. Thus our experiments prove this statement.

Similar to K[Nb$_{1/2}$M$_{1/2}$]OAsO$_4$ compounds deficit of K$^+$ and excess of Nb^{5+} was observed in K[Mg$_{1/3}$Nb$_{2/3}$]OAsO$_4$ crystals grown as it follows from Table 2. Thus for both subfamilies studied same tendency was observed independently on charge of M cations. SHG response from K[Mg$_{1/3}$Nb$_{2/3}$]OAsO$_4$ and K[Zn$_{1/3}$Nb$_{2/3}$]OAsO$_4$ crystals was detected to be slightly greater than that from non-substituted KTA powder.

Lattice parameters

Chemical formulas of the members of two subfamilies described above are different because of difference of charge of M cations. Therefore for unification of the results of measurements of the orthorhombic lattice parameters a, b, and c for the both above subfamilies using of the mean cation radii R(6) was necessary. According to chemical formulas of the compounds reported the R(6) values were calculated as follows:

$$R(6) = \frac{R(Nb^{5+}) + R(M^{3+})}{2} \quad \text{and} \quad R(6) = \frac{2R(Nb^{5+}) + R(M^{2+})}{3}, \quad (2,3)$$

where R(Nb^{5+}), R(M^{3+}), and R(M^{2+}) are Shannon 'IR' cation radii calculated for coordination number CN = 6 [10]. Figure 3 plots the a and b lattice parameters versus R(6). It was found that a and b parameters depend on R(6) almost linearly.

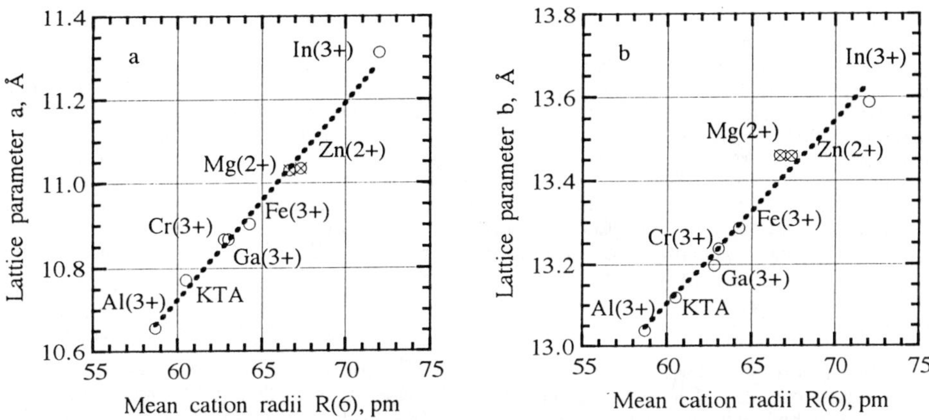

Fig.3 Lattice parameters a (a) and b (b) versus mean radii of cations occupying octahedral sites of KTP structure calculated for coordination number CN = 6 [10].

In general corresponding dependence of c parameter (Fig. 4) was not linear. An explanation for this phenomenon appeared to be that some of the small M^{3+} cations were actually entering tetrahedral positions along with the As^{5+} cations. This explanation is in good agreement with composition data obtained by EPMA for $K[Nb,Al]OAsO_4$ crystals grown from the melt 65-4 (Table 2). However, for the crystals containing large cations (Mg^{2+}, Zn^{2+}, and In^{3+}) the dependence was almost linear.

KTP phase formation in $KAsO_3$ - Nb_2O_5 - Ga_2O_3 system
____

$K[Nb,Ga]OAsO_4$ system is a best representative of the $K[Nb_{1/2}M_{1/2}]OAsO_4$ subfamily. Therefore KTP phase formation was studied in the vicinity of $K[Nb_{1/2}Ga_{1/2}]OAsO_4$ stoichiometric composition (Fig.5).

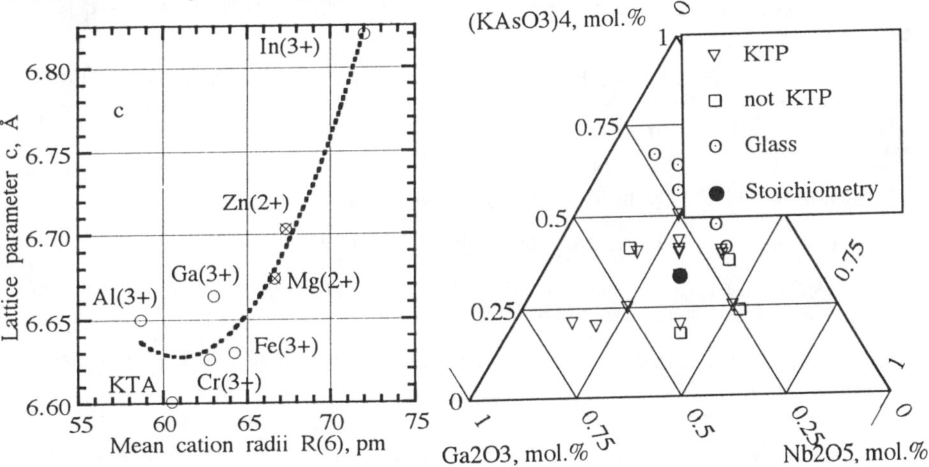

Fig.4 Lattice parameter c versus mean cation radii R(6)

Fig.5 KTP phase formation in $(KAsO_3)_4$ - Nb_2O_5 - Ga_2O_3 system

KTP prime crystallization field was observed to be very broad and include stoichiometric point of $K[Nb_{0.5}Ga_{0.5}]OAsO_4$. It was expectedly, because structure of basic material (KTA) is relatively stable [1]. Moreover, sizes of both Nb^{5+} and Ga^{3+}, cations calculated for octahedral coordination [10] are similar to that of Ti^{4+}. Therefore, it was assumed that stability of both Nb^{5+} and Ga^{3+} in octahedral sites is also high. Phase diagram of K_2O-TiO_2-P_2O_5 system give TiO_2 prime phase formation at the melt corresponding to stoichiometric KTP [11]. Similar result for the KTA follows from [12]. In the $K(Nb_{0.5}Ga_{0.5})OAsO_4$ stoichiometric melt formation of TiO_2 was avoided by substitution of Ti^{4+} with Nb^{5+} and Ga^{3+}. Thus, it was found that melting and further solidification of $K[Nb_{0.5}Ga_{0.5}]OAsO_4$ stoichiometric mixture result KTP single phase formation. This conclusion was confirmed by melting and solidification of the crystals grown from the 52-2 melt (Table 20).

CONCLUSIONS

Preparation procedures and preliminary study of properties of single crystals of new compounds of KTP family were reported. We have shown that Ti^{4+} cations in KTA crystals can be completely substituted with Nb^{5+}-Me^{3+} cation pairs (M = Al, Cr, Ga, Fe, In) and with Nb^{5+}-Me^{2+} cation pairs (M = Mg, Zn) without reducing SHG response. Lattice parameters values in general were found to increase with increasing of M^{3+} and M^{2+} cation radius as it was expected. The results of composition measurements indicate that Ti-less crystals with structure of KTP have tendency to create potassium vacancies. This result is in good agreement with that of reported for KTP crystals substituted with Nb^{5+} only.

REFERENCES

1. L.K. Cheng, E.M. McCarron III, J.C. Calabrese, J.D. Bierlein and A.A. Ballman, J. Cryst. Growth 132, p. 280 (1993).

2. V.I. Voronkova, V.K. Yanovskii, D.Y. Lee, N.I. Sorokina, I.A. Verin, N.G. Furmanova and V.I. Simonov, Crystallography Reports 39, p. 374 (1994).

3. G.M. Loiacono, D.N. Loiacono and R.A. Stolzenberger, J. Cryst. Growth 144, p. 223 (1994).

4. V.I. Chani, K. Shimamura, S.Endo and T. Fukuda, J. Cryst. Growth (1996) (in press).

5. L.T. Cheng, L.K. Cheng, R.L. Harlow and J.D. Bierlein, Appl. Phys. Lett. 64, p. 155 (1994).

6. P.A. Thomas and B.E. Watts, Solid State Comm. 73, p. 97 (1990).

7. G. Blasse and L.H. Brihner, Mat. Res. Bull. 24, p. 1,099 (1989).

8. E.M. McCarron III, J.C. Calabrese, T.E. Gier, L.K. Cheng, C.M. Foris and J.D. Bierlein, J. Solid State Chem. 102, p. 354 (1993).

9. G.D. Stucky, M.L.F. Phillips and T.E. Gier, Chemistry of Materials 1, p. 492 (1989).

10. R.D. Shannon, Acta Cryst. A 32, p. 751 (1976).

11. K. Iliev, P. Peshev, V. Nikolov and I. Koseva, J. Cryst. Growth 100, p. 107 (1990)

12. L.K. Cheng, L.T. Cheng, J. Galperin, P.A. Morris Hotsenpiller and J.D. Bierlin, J. Cryst. Growth 137, p. 107 (1994).

OPTICAL SECOND HARMONIC GENERATION IN TRANSPARENT TELLURITE GLASS-CERAMICS CONTAINING BaTiO$_3$

KATSUHISA TANAKA, HISAKO KURODA, KAZUYUKI HIRAO, NAOHIRO SOGA
Division of Material Chemistry, Faculty of Engineering, Kyoto University, Sakyo-ku, Kyoto 606-01, Japan

ABSTRACT

Optical second harmonic generation has been observed in transparent tellurite glass-ceramics of BaO-TiO$_2$-TeO$_2$ system. BaTiO$_3$ and unidentified crystalline phase precipitate from 15BaO·15TiO$_2$·70TeO$_2$ glass. The temperature of maximum rate of nucleation for these crystalline phases is estimated to be 380°C. The BaTiO$_3$ crystal precipitated is a cubic phase with lattice parameter of 0.4045 nm, which is slightly larger than the value reported in the JCPDS cards. The optical second harmonic intensity of the present transparent tellurite glass-ceramics is comparable to the intensity of poled tellurite glasses although the BaTiO$_3$ phase precipitated is cubic and the crystal face is randomly oriented in the glass-ceramics. For instance, the second harmonic intensity of the present transparent tellurite glass-ceramics is about twice as large as that of 20WO$_3$·80TeO$_2$ glass poled at 280°C under the external electric field of 3 kV, the second-order nonlinear coefficient, d_{33}, of which is 0.10 pm/V.

INTRODUCTION

Transparent glass-ceramics exhibit interesting optical properties which can be applied to optoelectronics devices. Recently, Komatsu *et al.*[1,2] have reported preparation of transparent tellurite glass-ceramics containing dielectric crystals such as LiNbO$_3$ and BaTiO$_3$. Their idea is that the scattering of incident light caused by the crystalline particles precipitated in the glass-ceramics is reduced because the refractive index of tellurite glass matrix is close to those of LiNbO$_3$ and BaTiO$_3$ crystals, leading to high transparency to the visible light. Since ferroelectric phases of LiNbO$_3$ and BaTiO$_3$ possess large nonlinear susceptibility with second-order, it is of interest to examine second-order nonlinear optical properties of transparent tellurite glass-ceramics containing these dielectric crystals. Thus far, optical second harmonic generation has been reported for transparent silicate and borate glass-ceramics containing crystalline phases which exhibit second-order nonlinear optical properties such as β-BaB$_2$O$_4$ and LiNbO$_3$[3-5]. As for the tellurite glass-ceramics containing BaTiO$_3$, the optical second harmonic intensity has not been measured. In addition, the crystal structure of BaTiO$_3$ phase has not been identified. In the present investigation, an attempt was made to identify the crystal structure of BaTiO$_3$, and to observe optical second harmonic generation in the transparent tellurite glass-ceramics.

EXPERIMENTAL PROCEDURE

Glass was prepared using reagent-grade BaCO$_3$, TiO$_2$ and TeO$_2$ as starting materials. The raw materials were weighed to make 15BaO·15TiO$_2$·70TeO$_2$ (in mol%) composition and mixed thoroughly. The mixture was melted in a platinum crucible at 1000 °C for 40 min in air. The melt was poured onto a stainless steel plate and cooled in air. The as-

271

prepared glass specimens were heat-treated at several temperatures for 1 h in air, and subjected to differential scanning calorimetry (DSC) to evaluate the temperature of maximum rate of nucleation. Then, the glass specimens were heat-treated at the nucleation temperature thus determined, and reheated at elevated temperatures for the crystal growth. X-ray diffraction analysis with CuK_α radiation was carried out to ascertain that the as-prepared specimen was amorphous and to identify crystalline phases in the heat-treated specimens if precipitated.

For measurements of optical properties, annealed glass was cut into a plate of about 7 mm x 7 mm x 1 mm, and both sides were polished with CeO_2 powders. The annealing of the as-prepared glass was carried out for 30 min at around glass transition temperature determined by means of differential thermal analysis (DTA). The resultant glass specimens were heat-treated at elevated temperatures for the nucleation and crystal growth. Optical absorption spectra were measured at room temperature using a spectrophotometer with a Xe lamp as a light source. The wavelength was varied from 800 to 400 nm. Maker fringe method[6] was utilized for measurements of optical second harmonic generation. The optical second harmonic intensity was determined at room temperature using a pulsed Nd:YAG laser with a pulse width of about 10 ns. The fundamental wave with the wavelength of 1064 nm was used as an incident light. The situation of polarization was p-excitation and p-detection. The angle of incidence was changed from -70 ° to 70 °. Details of the measurements of optical second harmonic intensity were described elsewhere[7,8].

RESULTS AND DISCUSSION

Figure 1 shows DSC curves of as-prepared $15BaO \cdot 15TiO_2 \cdot 70TeO_2$ glass and specimens heat-treated at several temperatures. The exothermic peak observed for all the specimens is

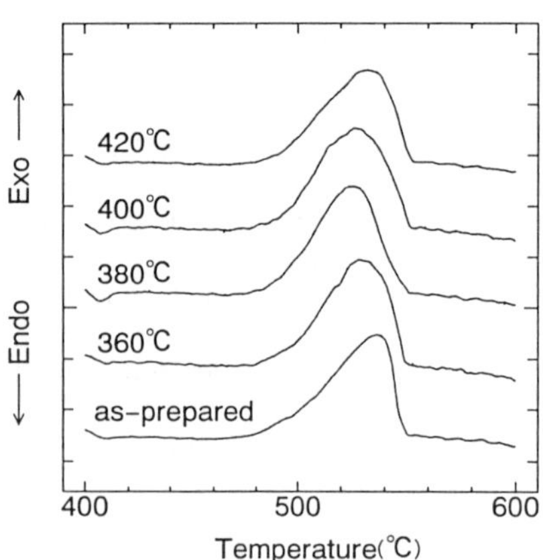

Fig.1 DSC curves of as-prepared glass and specimens heat-treated at several temperatures indicated in the figure.

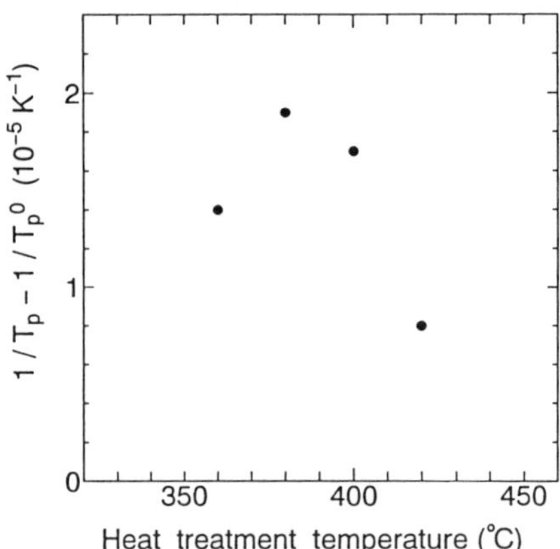

Fig.2 Heat treatment temperature dependence of $1/T_p$-$1/T_p^0$, where T_p and T_p^0 denote the crystallization temperatures of heat-treated specimens and as-prepared glass, respectively.

attributed to crystallization.   The crystalline phases precipitated in this temperature range
are BaTiO$_3$ and unidentified phase as mentioned below.   According to Zhou and
Yamane[9], the logarithm of rate of nucleation at a constant temperature is proportional to
$1/T_p$-$1/T_p^0$, where T_p^0 is the crystallization temperature of as-prepared glass and T_p is the
crystallization temperature of glass heat-treated at a constant temperature.   Hence, the
dependence of $1/T_p$-$1/T_p^0$ on the heat treatment temperature reflects the temperature
dependence of rate of nucleation.   Figure 2 shows the variation of $1/T_p$-$1/T_p^0$ with heat
treatment temperature   for the present 15BaO·15TiO$_2$·70TeO$_2$ glass.   A maximum of
$1/T_p$-$1/T_p^0$ is observed at around 380°C.   In other words, the temperature of maximum rate
of nucleation for the BaTiO$_3$ and unidentified phase is estimated to be 380°C.

X-ray diffraction pattern of specimen heat-treated at 380°C for 0.5 h and reheated at
430°C for 3 h is shown along with the diffraction pattern of as-prepared glass in Fig.3.
Only halo pattern is observed for the as-prepared glass.   The X-ray diffraction diagram of

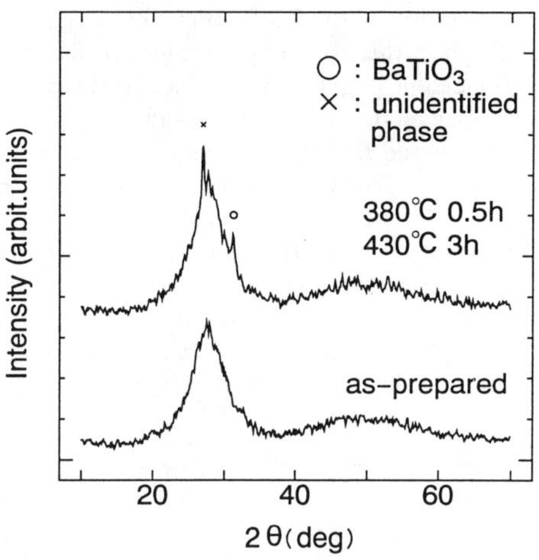

Fig.3 X-ray diffraction patterns of as-prepared glass (lower) and specimen heat-treated at
380°C for 0.5 h and reheated at 430°C for 3 h (upper).

Table 1 Lattice parameters of BaTiO$_3$ precipitated in the
glass-ceramics, and cubic and tetragonal BaTiO$_3$
phases cited from JCPDS cards

Specimen	Lattice parameter	
	a (nm)	c (nm)
BaTiO$_3$ in glass-ceramics	0.4045	
cubic BaTiO$_3$	0.4031	
tetragonal BaTiO$_3$	0.3994	0.4038

specimen heat-treated only at 380°C for 0.5 h also manifested halo pattern.   On the other hand, diffraction lines ascribable to crystalline phases appear in the diffraction pattern of the specimen heat-treated at 380°C for 0.5 h and reheated at 430°C for 3 h.   The diffraction line located at about $2\theta = 31.2°$ is assigned to $BaTiO_3$ and the line observed at around $2\theta = 27.1°$ corresponds to unidentified crystalline phase[2]. The lattice parameter of $BaTiO_3$ precipitated in the glass-ceramics evaluated from the position of X-ray diffraction line is shown in Table 1 along with those of cubic and tetragonal $BaTiO_3$ cited from the JCPDS cards.   The lattice parameter of the $BaTiO_3$ in the present glass-ceramics is rather close to the lattice parameter of cubic phase than that of tetragonal phase.

Optical absorption spectra of specimen heat-treated at 380°C for 0.5 h and specimen heat-treated at 380°C for 0.5 h and reheated at 430°C for 3 h are shown in Fig.4.   The intense absorption at around 400 nm, which is observed for both specimens, is attributed to the absorption edge of tellurite glass.   There are no distinct absorption bands in the wavelength region longer than 450 nm.   The absorbance of the specimen heat-treated at 380°C for 0.5 h and reheated at 430°C for 3 h is somewhat larger than that of the specimen heat-treated at 380°C for 0.5 h in the wavelength region of about 450 to 800 nm.   This is because the scattering of incident light takes place due to the crystalline particles precipitated in the former specimen.   From the results of X-ray diffraction and optical absorption measurements, it is found that the specimen heat-treated at 380°C for 0.5 h and reheated at 430°C for 3 h contains $BaTiO_3$ crystalline phase and is transparent to the visible light.

In Fig.5 is shown the variation of optical second harmonic intensity with angle of incidence for the transparent glass-ceramics.   The optical second harmonic generation is undoubtedly observed in the present transparent tellurite glass-ceramics.   The scattering of the data points presumably arises from the inhomogeniety in microstructure of the specimen. The second harmonic intensity takes a maximum when the angle of incidence is 0°, and

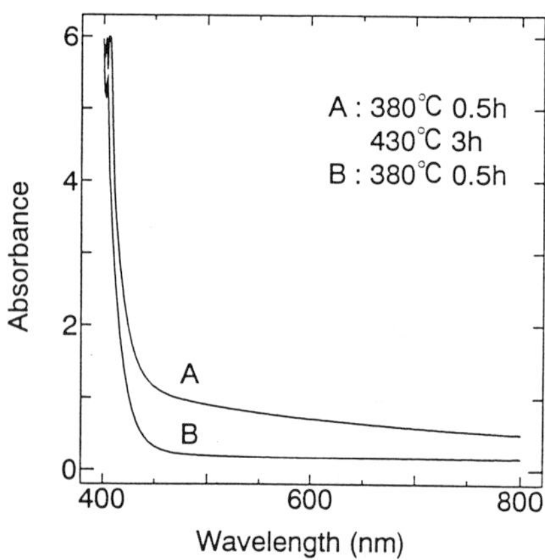

Fig.4 Optical absorption spectra of specimen heat-treated at 380°C for 0.5 h
(lower) and specimen heat-treated
at 380°C for 0.5 h and reheated at
430°C for 3 h (upper).

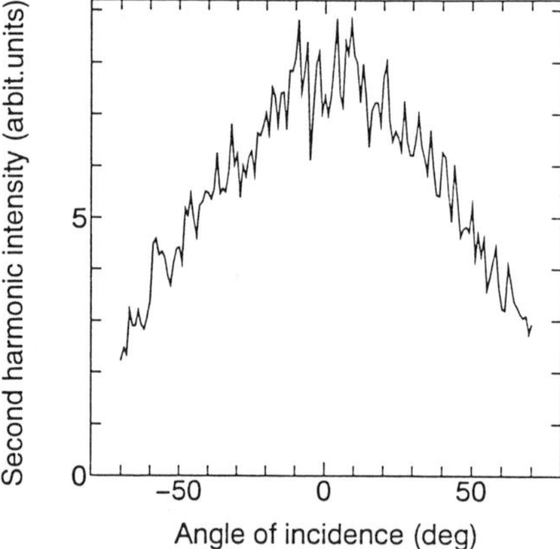

Fig.5 Variation of optical second harmonic
intensity with angle of incidence for
the present transparent tellurite
glass-ceramics.

decreases with an increase in the angle of incidence. This is presumably because the reflectance increases with an increase in the angle of incidence. In other words, the fraction of incident light which passes through the specimen is reduced when the angle of incidence is large. The optical second harmonic intensity of the present transparent tellurite glass-ceramics is comparable to the intensity of poled tellurite glasses reported previously[7,8,10,11]. For instance, the second harmonic intensity shown in Fig.5 is about twice as large as that of $20WO_3 \cdot 80TeO_2$ glass poled at 280°C under an external electric field of 3 kV[10]. It should be noted that the optical second harmonic generation is observed although the present transparent glass-ceramics contains cubic $BaTiO_3$ crystalline phases and the crystal face of the $BaTiO_3$ particles is randomly oriented in the glass-ceramics. A similar phenomenon has been observed by Kim *et al.*[12] recently. They succeeded in observing optical second harmonic generation in transparent tellurite glass-ceramics containing cubic crystalline phases in $K_2O\text{-}Nb_2O_5\text{-}TeO_2$ system. It was speculated that the crystalline phase manifests a small distortion from the cubic structure and exhibits the optical second harmonic generation like $TlNbWO_6$ and $RbNbWO_6$ crystals do, although the cubic crystalline phase was not identified. We infer that a similar phenomenon takes place in the cubic $BaTiO_3$ phase precipitated in the present glass-ceramics. Namely, a small distortion in the cubic phase causes the optical second harmonic generation, although the distortion could not be detected by the present X-ray diffraction measurements. Nonstoichiometry and/or oxygen vacancies in the $BaTiO_3$ crystals precipitated are possible origins for the distortion. However, a possibility that the unidentified phase contributes to the optical second harmonic generation cannot be ruled out at the present time. Furthermore, there exists another possibility that the optical second harmonic wave generated from the interface between crystal particles and glass matrix. It is known that the surface and interface can be origins for the optical second harmonic generation as demonstrated for many kinds of metals, semiconductors, and insulators[13-17]. At this moment, it is unclear which is the most predominant factor to determine the optical second harmonic intensity of the present transparent tellurite glass-ceramics.

CONCLUSIONS

Transparent tellurite glass-ceramics containing $BaTiO_3$ crystalline phase was prepared by heating the $15BaO \cdot 15TiO_2 \cdot 70TeO_2$ glass at 380°C for 0.5 h for nucleation and reheating it at 430°C for 3 h for crystal growth. The X-ray diffraction analysis indicates that the $BaTiO_3$ is a cubic phase. The optical second harmonic generation was clearly observed in the present transparent tellurite glass-ceramics although the crystalline phase precipitated is cubic and the crystal face is randomly oriented in this material. The optical second harmonic intensity is comparable to the intensity of poled tellurite glasses.

ACKNOWLEDGEMENTS

The authors would like to thank Professor T. Kokubo, Faculty of Engineering, Kyoto University, for X-ray diffraction measurements. The authors are also indebted to Professor T. Komatsu, Nagaoka University of Technology, for suggestions on many aspects of the present work. This work was financially supported by a Grant-in-Aid for Encouragement of Young Scientists.

REFERENCES

1. T.Komatsu, H.Tawarayama, H.Mohri and K.Matusita, J. Non-Cryst. Solids **135**, p.105 (1991).
2. T.Komatsu, H.Tawarayama and K.Matusita, J. Ceram. Soc. Jpn. **101**, p.48 (1993).
3. Y.H.Kao, Y.Hu, H.Zheng, J.D.Mackenzie, K.Perry, G.Bourhill and J.W.Perry, J. Non-Cryst. Solids **167**, p.247 (1994).
4. Y.Ding, A.Osaka and Y.Miura, J. Am. Ceram. Soc. **77**, p.749 (1994).
5. Y.Ding, A.Osaka, Y.Miura, H.Toratani and Y.Matsuoka, J. Appl. Phys. **77**, p.2,208 (1995).
6. P.D.Maker, R.W.Terhune, M.Nisenoff and C.M.Savage, Phys. Rev. Lett. **8**, p.21 (1962).
7. K.Tanaka, K.Kashima, K.Hirao, N.Soga, A.Mito and H.Nasu, Jpn. J. Appl. Phys. **32**, p.L843 (1993).
8. K.Tanaka, A.Narazaki, K.Hirao and N.Soga, J. Appl. Phys. **79**, p.3,798 (1996).
9. X.Zhou and M.Yamane, J. Ceram. Soc. Jpn. **96**, p.152 (1988).
10. K.Tanaka, K.Kashima, K.Kajihara, K.Hirao, N.Soga, A.Mito and H.Nasu in Doped Fiber Devices and Systems (Proc. SPIE 2289, San Diego, California 1994), pp.167-176.
11. K.Tanaka, K.Kashima, K.Hirao, N.Soga, A.Mito and H.Nasu, J. Non-Cryst. Solids **185**, p.123 (1995).
12. H.G.Kim, T.Komatsu, K.Shioya, K.Matusita, K.Tanaka and K.Hirao, J. Non-Cryst. Solids (1996), in press.
13. Z.Chen, D.Cui, Y.Zhou. H.Lu, G.Yang and S.Gu, Appl. Phys. Lett. **51**, p.1,301 (1987).
14. S.R.Armstrong, R.D.Hoare, M.E.Pemble, I.M.Povey, A.Stafford and A.G.Taylor, J. Cryst. Growth **120**, p.94 (1992).
15. J.G.Mihaychuk, J.Bloch, T.Liu and H.M.van Driel, Opt. Lett. **20**, p.2,063 (1995).
16. I.V.Kravetsky, L.L.Kulyuk, A.V.Micu, V.I.Tsytsanu and I.S.Vieru, J. Non-Cryst. Solids **187**, p.227 (1995).
17. A.Burgel, W.Kleemann and U.Bianchi, Phys. Rev. B **53**, p.5,222 (1996).

APPLICATION OF PHOTOSTIMULABLE PHOSPHOR, IMAGING PLATE :BaFBr:Eu^{2+}, TO IMAGE STORING FOR UV-VUV RADIATIONS

Masahito KATTO[*], Shunshiro OHNISHI[*], Yutaka KURIOKA[*], Yasuo TAKIGAWA[**], Kou KUROSAWA[***]

[*]Department of Electrical Engineering, KINKI University, Higashi-Osaka, Osaka 577, Japan, mkatto@ex.ee.kindai.ac.jp

[**]Department of Solid State Electronics, Osaka Electro-Communication University, Neyagawa, Osaka 572, Japan

[***]Department of Electrical Engineering, Miyazaki University, Miyazaki 889-21, Japan

ABSTRACT

An imaging plate containing photostimulable phosphor material, BaFBr:Eu^{2+}, was found to be applicable to store images for nanosecond pulses as well as quasi-continuous radiations in 4 - 35 eV photon energy range. The sensitivity, which was defined as photostimulated luminescence intensity for a given incident photon number increased with the photon energy and had a peak at 6.2 eV. The intensity of the luminescence was confirmed to increase in proportion to incident photon numbers of the radiation. The readout system was improved and then pattern images were successfully restored. The resolution of imaging plate was found to be lower than 100 μm with our system, which depends on the intensity of a He-Ne laser for restoring images.

INTRODUCTION

Recently much attention has been paid to photostimulable phosphor materials such as BaFBr:Eu^{2+}, because they enable to store two-dimensional images of photon numbers via only an electronic process which is different from a chemical process in ordinary photographic films [1 - 8]. Incident photons excite electron-hole pairs. The holes are bound at Eu^{2+} ions to form Eu^{3+} ions. The electrons are excited into the conduction band (Fig. 1) [1, 4]. Some of the electrons are captured by halogen vacancies to form F-centers. The electrons captured by the F-centers are released with red stimulus light and then come into the conduction band again. The electrons recombine with Eu^{3+} ions to decay to the ground state through the excited state. While decaying, the electrons emit blue photostimulated luminescence (PSL) [1 - 5, 9 - 11]. The PSL intensity is proportional to the F-center number, and thus we can estimate the incident photon number by measuring the PSL intensity.

The small crystalline photostimulable material, BaFBr:Eu^{2+}, fixed on a flexible plastic sheet, has been developed as an image storing medium mainly for medical X-ray diagnosis and is commercially available as Imaging Plate (IP). Now, they are used in wide fields of diffraction crystallography, auto radiography and others with the attractive properties of wide dynamic range, good linear response, high sensitivity, and easy data acquisition with computers [2 - 8]. We proposed the IP as a storing media for the radiation from ultraviolet (UV) to vacuum and extreme ultraviolet (VUV, XUV) [11, 12] in order to diagnose the output beam from the excimer

Mat. Res. Soc. Symp. Proc. Vol. 453 © 1997 Materials Research Society

lasers, because few detectors for the two-dimensional diagnostics are available in this spectral range. We constructed a readout system controlled by personal computer and successfully measured the intensity distributions of nanosecond lasers. Beam profiles of a VUV pulsed laser were successfully observed and all the process was performed in the vacuum chamber [12, 13]. We also found that the PSL sensitivity for pulsed lasers was different from the continuous wave.

In this paper, we summarize the PSL response of the IP and also report a spatial resolution with our readout system. From the results, we discuss issues of the IP as storing media for the UV - VUV radiations.

EXPERIMENT

We used two types of imaging plates, both of which are available from Fuji Photo Film Co. Ltd. ST type of plates, which have been prepared mainly for medical X-ray diagnosis, have a BaFBr: Eu^{2+} layer of 150 μm thickness. They are normally covered with a thin polymer film for protection of the layer against damages such as stain and scratch, but we used the plates without the film, because it absorbs strongly ultraviolet and vacuum ultraviolet photons. TR type of plates have been developed for detection of low energy β-ray and covered with nothing. The plates have a BaFBr: Eu^{2+} layer as thin as 50 μm and contain dyes to prevent scattering of the red stimulus light. The latter makes us expect an improvement of spatial resolution.

In order to measure PSL intensity, we have designed and constructed a two-dimensional readout system with an imaging plate [11]. The system is shown in Fig. 2. The one shot of laser beam exposes an imaging plate fixed on a two-dimensional stage which is driven by two stepping motors. A He-Ne laser beam for photostimulation is focused by a objective lens on the imaging plate to 10 μm in diameter. The blue PSL with a photon energy of 3.2 eV was collected with the lens and reflected separately by a dichroic mirror to a photomultiplier tube (R1635, Hamamatsu Photonics), whose outputs were converted to 12 bit digital signals with an analog-to-digital converter and then acquired with a personal computer (PC-9801, NEC). The PSL intensity data were taken with the sampling time of 30 μsec for 50 msec equal to the He-Ne laser irradiation time. The time integrated PSL intensity is proportional to the incident photon number. In order to prevent the reflected and scattered He-Ne laser beam from entering the photomultiplier tube, colored glass filters (C-40C, Toshiba), were placed in front of the tube.

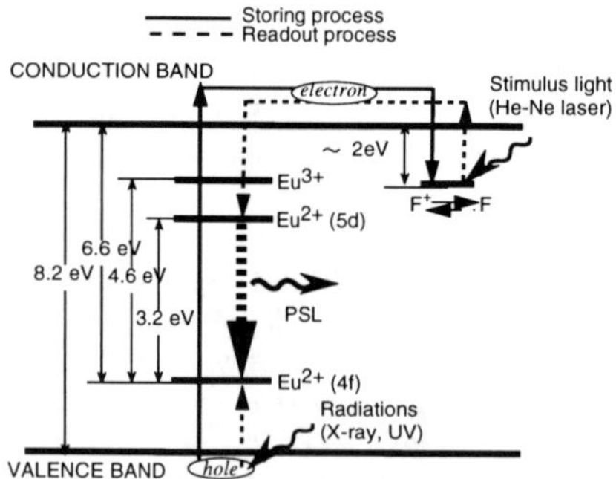

Fig. 1   Energy diagram of BaFBr: Eu^{2+} [1, 4].

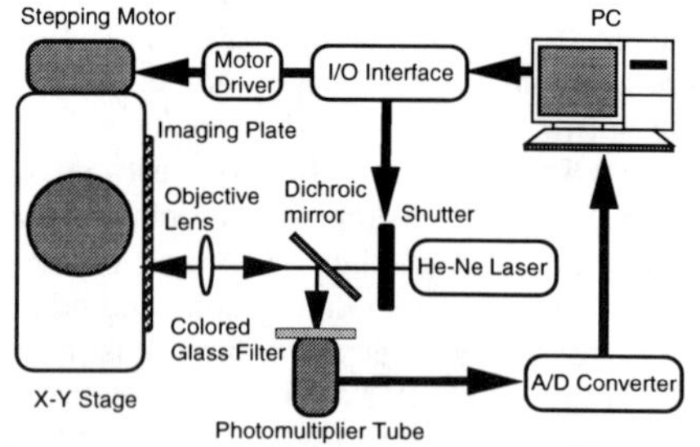

Fig. 2 Two-dimensional readout system.

After taking the PSL intensity from a point on the imaging plate, we moved it to next and measured the PSL intensity. After measuring the intensity from all points in the area, we illuminated the plate with white light for a few minutes to erase the image.

KrF (5.0 eV/20 nsec) and ArF (6.4 eV/20 nsec) and Ar_2 (9.8 eV/ 5 nsec) excimer lasers were used as incident sources. We changed the incident photon numbers with changing the output energy per shot. We also used synchrotron radiation (SR) from a storage ring (UVSOR) installed at the Institute for Molecular Science. The SR has 500 psec pulses with repetition rate of 90 MHz and it was monochromatized with a 1 m Seya-Namioka type monochromator equipped in the BL-1B beam line. The photon energy of the beam was varied from 4 eV to 35 eV. The photon number was changed by changing the exposure time to the radiation and was calibrated by the SR intensity.

RESULTS AND DISCUSSION

PSL Sensitivity Function

In order to estimate PSL sensitivity as a function of incident photon energy over a wide energy range, we exposed the imaging plate to various energy photons from SR for 10 sec. PSL sensitivity is defined as PSL intensity normalized by the incident photon number. In Fig. 3 is shown the relative PSL intensity as a function of incident photon energy for a ST plate. The intensity increases exponentially with the incident photon energy except for a peak at 6.2 eV. In the whole energy range, the sensitivity of a TR plate is 35 % lower than that of a ST plate. On the basis of the energy diagram in Fig. 1, an electron transition probability from the valence band and ground state of Eu^{2+} ions to the conduction band should have large value for the photon energy above 6.6 and 8.2 eV. The increment of the probability results in high sensitivity. The fluorescence spectra indicated the relatively broad ground and excited levels of Eu^{2+} [1,4]. Broad band levels induced the sensitivity below the photon energy of 6.6 eV. We can roughly say that photons with energy lower than the band gap energy of 8.2 eV can penetrate the imaging plate more deeply beneath the surface than those with energy higher than the band gap energy. On the other hand, an electron transition probability mentioned above has small value. The total number of electrons excited in the conduction band is proportional to the product of the transition probability and the penetration depth. Conclusively, this is a reason why the relative PSL intensity had a maximum value around 6.2 eV. The IP was found to have a sensitivity for photons whose energy were above 4 eV. Below 5 eV, the PSL was so low, it is necessary to improve the collection efficiency of the PSL.

Photon Number Dependence

In Figs. 4, is shown the PSL intensity measured as a function of incident photon numbers from SR when the photon energy was fixed at (a) 5.0, (b) 6.4, and (c) 9.8 eV, respectively. In Fig. 5, is also shown the PSL intensities for the excimer lasers. The PSL intensity increased linearly in proportion to the photon number from nanosecond pulsed lasers as

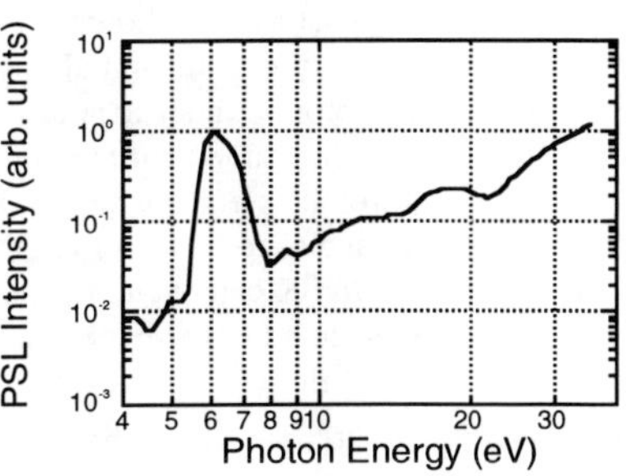

Fig. 3 Relative PSL intensity plotted as a function of photon energy.

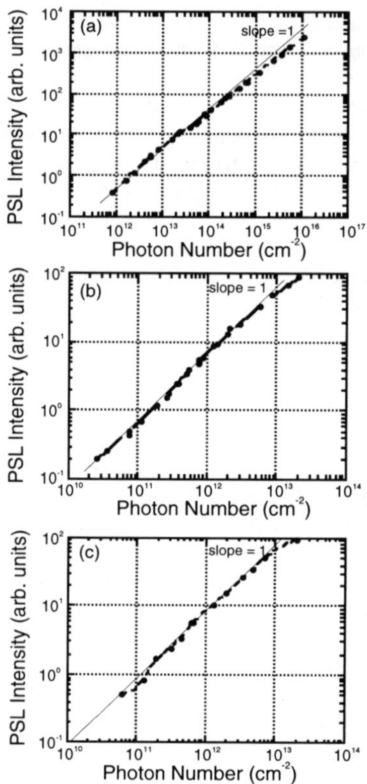

Fig. 4   PSL intensity from ST plate plotted as a function of incident photon numbers of SR with photon energy of (a) 5.0 eV, (b) 6.4 eV, and (c) 9.8 eV.

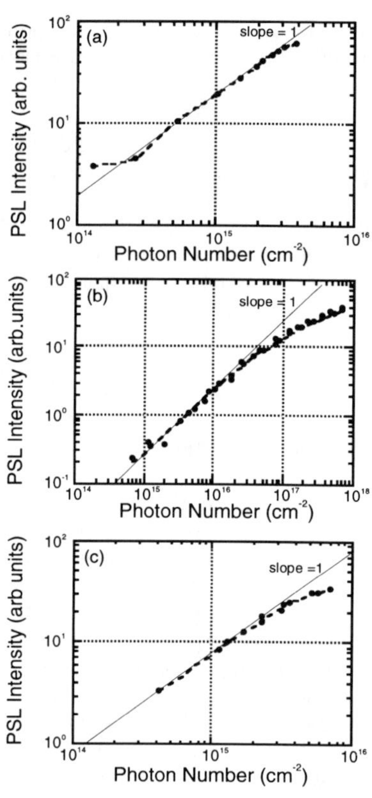

Fig. 5   PSL intensity from ST plate plotted as a function of incident photon numbers of excimer lasers with photon energy of (a) 5.0 eV, (b) 6.4 eV, and (c) 9.8 eV.

well as SR with high repetition rate, quasi-CW radiation. The saturation was observed for both radiations and the saturation photon number depended on the photon energy. These results were obtained with a ST type plate and almost the same results were obtained with a TR type plate. The saturation photon number for pulsed lasers were larger than those for SR by 10^4. The PSL intensity was considered to saturate with the total number of electron-hole pairs or halogen vacancies. Among the electrons that were excited by the incident photons, only a part of them met halogen vacancies to be captured. Remainder decayed back to the ground state.The quasi-CW radiation from SR can excite the decayed electrons again. If the total number of halide vacancies was assumed to be much lower than the excited electron-hole pairs, the PSL saturation occurred by filling the vacancies. On the other hand, a large number of UV photons were

irradiated for 5 - 20 nsec much shorter than the decay time of 0.8 μsec from the excited to ground state. The electron-hole pairs never meet photon and could not be excited again. The saturation took place by the number of electron-hole pairs for the nanosecond pulsed lasers.

We confirmed the sensitivity and linearity for both nanosecond pulsed lasers and quasi-

CW radiations. But, the saturation occurred at different photon numbers for the nanosecond radiations and the quasi-CW radiations. Using a detector for pulsed radiations, such as picosecond and femtosecond pulsed radiations, the linear response range must be confirmed for each radiation.

<u>Spatial Resolution</u>

We stored and restored the two-dimensional images of ArF excimer laser through a resolution test target, which has a patterned aluminum film deposited on a SiO_2 plate. In Fig. 6, we show images restored by (a) high and (b) low intensity He-Ne laser beams with TR plate. In Fig. 6(c), we show an image from a ST plate restored by the low intensity laser. Scan distance was set as 100 μm and the images were displayed in 32 scales. With TR plate, image were clearly observed in contrast to that with ST plate, as we expected. The He-Ne laser has output of (a) 4 mW, and (b)(c) 0.1 mW within a diameter of 10 μm, respectively. With a high intense readout, degradation of the images was observed. With a low intensity HeNe laser, patterns above the scan distance of 100 μm were successfully restored. It concluded that the resolution of our system should be lower than 100 μm and the resolution was largely influenced by the He-Ne laser scattering. The TR plate included the dye to absorb the He-Ne laser. The scattered length of the He-Ne laser in the TR plate should be shorter than in ST plate. It resulted in improvement of the resolution and decrement of the sensitivity.

We pointed out here that a UV photon should also has a large scattering cross section and it may cause the degradation of the images. It must be clarified that the resolution of the IP was limited by whether scattering of the incident UV radiation or He-Ne laser for readout.

CONCLUSIONS

Imaging plates containing a photostimulable phosphor material, $BaFBr:Eu^{2+}$, can store the images of incident photon numbers in the spectral range from 4 eV to 35 eV. The IP was confirmed to show a linear response to the both nanosecond pulsed and quasi-CW radiations. We also pointed out that the PSL saturation limit need to be estimated using a detector for pulsed radiation because the limit has dependence on the pulse width of the incident radiations. We

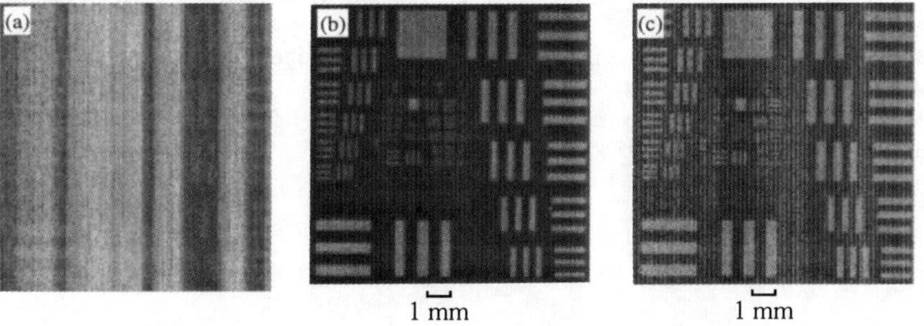

Fig. 6 Two-dimensional intensity distribution images of ArF excimer laser
outputs through a test target with TR plate restored with (a) high and (b) low
intense He-Ne laser . In (c), shown the images with ST plate.

successfully measured the intensity distributions and the resolution of the IP was found to be influenced by the He-Ne laser intensity for readout. The IP was concluded to be a useful storage medium from the fact that two-dimensional images, holograms and spectra can be recorded on it in all spectral range of ultraviolet, vacuum ultraviolet, extreme ultraviolet and X-ray radiations.

ACKNOWLEDGEMENTS

We gratefully acknowledge Profs. K. Mima, S. Nakai, and T. Yamanaka of the Institute of Laser Engineering at Osaka University for their continuous encouragement and support to this work. We would like to thank Dr. J. Miyahara of Fuji Photo Film Co. Ltd. for supplying imaging plates and also useful information and suggestions about the imaging plates. We are indebted to all stuff members of the UVSOR facility of Institute for Molecular Science for measurements with a synchrotron radiation.

REFERENCES

1. K. Takahashi, K. Kohda, J. Miyahara, Y. Kanemitsu, and S, Shionoya, J. Lumi. **31 & 32**, 266 (1984).
2. J. Miyahara, K. Takahashi, Y. Amemiya, N. Kamiya, and Y. Satow, Nucl. Instrum. & Methods **A246**, 572 (1986).
3. Y .Amemiya, and J. Miyahara, Nature **336**, 89 (1988).
4. Y. Iwabuchi, N. Mori, K. Takahashi, T. Matsuda, and S. Shionoya, Jpn. J. Appl. Phys. **33**, 178 (1994).
5. C. Mori and A. Matsumura, Nucl. Instrum. & Methods **A312**, 39 (1992).
6. T. Bücherl, C. Rausch, and H. von Seggern, Nucl. Instrum. & Methods **A333**, 502 (1993).
7. C. Mori, A.Matsumura, T. Suzuki, h. Miyahara, T. Aoyama, and K. Nishizawa, Nucl. Instrum. & Methods **A339**, 278 (1994).
8. N. Niimura, Y. Karasawa, I. Tanaka, J. Miyahara, K. Takahashi, H. Saito, S. Koizumi, and M. Hidaka, Nucl. Instrum. & Methods **A349**, 521 (1994).
9. K. Takahashi, J. Miyahara, and Y. Shibahara, J. Electrochem. Soc. **132**, 1492 (1985).
10. Y. Iwabuchi, C. Umemoto, K. Takahashi, and S. Shionoya, J. Lumi. **48 & 49**, 481 (1991).
11. M. Katto, K. Kurosawa, W. Sasaki, Y. Takigawa, and M. Okuda, Jpn. J. Appl. Phys. **A30**, 2806 (1991).
12. M. Katto, R. Matsumoto, K. Kurosawa, W. Sasaki, Y. Takigawa, and M. Okuda, Rev. Sci. Instrum. **64**, 319 (1993).
13. M. Katto, M. Okuda, W. Sasaki, K. Kurosawa, and Y. Kato, IEEE J. Selected Topics in Quantum Electron. **1** , 924 (1995).

CRYSTAL GROWTH OF $Ca_3(Li,Nb,Ga)_5O_{12}$ GARNETS FROM MELT

V.I. CHANI*, Y.M. Yu**, K. SHIMAMURA*, Y. SAIKI*, T. FUKUDA*
*Institute for Materials Research, Tohoku Univ., 2-1-1 Katahira, Aoba-ku, Sendai, 980 Japan
**Korea Research Institute of Chem. Technology, 100 Jang-dong, Yuseoung-ku, Taejon, Korea

ABSTRACT

Single crystals of $Ca_3(Li,Nb,Ga)_5O_{12}$ garnets have been grown from stoichiometric melts by micro-pulling down and Czochralski methods using Pt crucibles. It was found that the mixture of oxides with atomic ratio $Ca : Li : Nb : Ga = 3 : 0.275 : 1.775 : 2.95$ correspond to the garnet composition which melts congruently at about 1450°C. Solid state reaction data of the compounds related with this material are also reported. Lattice parameter of all $Ca_3(Li,Nb,Ga)_5O_{12}$ crystals grown was about 1.254 nm. Transparent and bubble-free crystals of $Ca_3Li_xNb_{(1.5+x)}Ga_{(3.5-2x)}O_{12}$ ($x = 0.25$ and 0.275) were grown by Czochralski technique in air. An optical transmission spectrum of the crystals was studied. No absorption was detected in 400-1200 nm wavelength range.

INTRODUCTION

The Bi-containing iron garnet films are important for application in magneto-optic devices due to high Faraday rotation [1]. The magnitude of Faraday rotation for these materials is proportional to the concentration of large Bi^{3+} cation. Thus, the incorporation of large Bi^{3+} cation into the garnet structure demands preparation of non-magnetic garnet substrates with large lattice parameters ($a > 12.5$ Å).

There are a few garnets which have been grown by Czochralski (CZ) method as transparent single crystals suitable for devices [2-7]. $Ca_3Nb_{1.6875}Ga_{0.1875}Ga_3O_{12}$ (CNGG) crystal [3] seem to be very useful for this purpose because of large lattice parameter (12.51 Å) and low absorption. However, quality of the films grown by liquid phase epitaxy (LPE) on these substrates is low [8]. These results suggested to us that it might be possible to prepare new garnets which have following properties: congruent melting, low melting point ($T_m < 1500$ °C), lattice parameter greater than that of CNGG, and low absorption in 400-1200 nm wavelength range.

In this paper we describe our attempts which were made to obtain new garnets by substitutions in congruently melting ($Ca_3Nb_{1.6875}Ga_{0.1875}Ga_3O_{12}$) and stoichiometric ($Ca_3Nb_{1.5}Ga_{3.5}O_{12}$) CNGG crystals. Solid state reaction technique, single crystal fiber growth by micro-pulling down (μ-PD) and bulk crystal growth by CZ method were used to select best composition and prove congruent melting of the crystals reported.

EXPERIMENTAL AND RESULTS

Solid state reaction

As a first step dependence of garnet structure stability on different isovalent substitutions in octahedral sites was studied by solid state reaction technique. Desired quantities of $CaCO_3$, Nb_2O_5, Ga_2O_3, In_2O_3, Yb_2O_3, Y_2O_3, and Gd_2O_3 of at least 99.99% purity were weighed and mixed by grinding with an agate mortar and pestle. The mixtures were then pressed to pellets and heated 3 h at 1350 °C or for 10 min at 1460 °C. Phase purity and lattice parameters of all the pellets were studied using the X-ray powder diffractometer with $CuK_{\alpha 1}$ (30 kV, 15 mA). The garnet phase purity and concentration of second phase were estimated using relative intensity ratio $I_{(420)}/I_{(2nd)}$, where $I_{(420)}$ was intensity of the greatest peak corresponding to the garnet phase (420), and $I_{(2nd)}$ was intensity of the largest peak corresponding to any non-garnet phase. It was assumed that this ratio is proportional to the amount of the garnet phase in the material.

Mat. Res. Soc. Symp. Proc. Vol. 453 © 1997 Materials Research Society

Following substitution reactions:

$$(Ga^{3+})_6 \rightarrow (Ln^{3+})_6, \; (Ln = Yb, \; Y, \; and \; Gd) \tag{1}$$
$$(Ga^{3+})_6 \rightarrow (In^{3+})_6 \tag{2}$$

were selected to increase lattice parameter of CNGG by isovalent substitution of Ga^{3+} in six-fold sites (the indexes correspond to octahedral sites with coordination number CN = 6). Single garnet phase formation was not observed in all the mixtures prepared. It was assumed that it follows from incorporation of rare-earth cations into dodecahedral sites, because these cations can enter eight-fold positions easily. In these compounds a part of dodecahedral positions was occupied by rare-earth elements. Therefore, excess of Ca^{2+} cations was rejected by garnet structure and form the second phase.

Substitution of Ga^{3+} with In^{3+} was found results almost pure garnet phase formation. $Ca_3In_{0.5}Nb_{1.5}Ga_3O_{12}$ garnet obtained according to substitution (2) seems to be one of the best among the garnets studied due to large lattice parameter (12.58 Å) and low concentration of second phase. However it is impossible to grow this crystal from Pt crucible because of high melting point. Dependence of coefficient of garnet phase purity, $I_{(420)}/I_{(2nd)}$ on size of cations substituted for Ga^{3+} is given in Fig. 1. Thus, it was observed that the higher the size of Ln cations was, the higher amount of the second phase was.

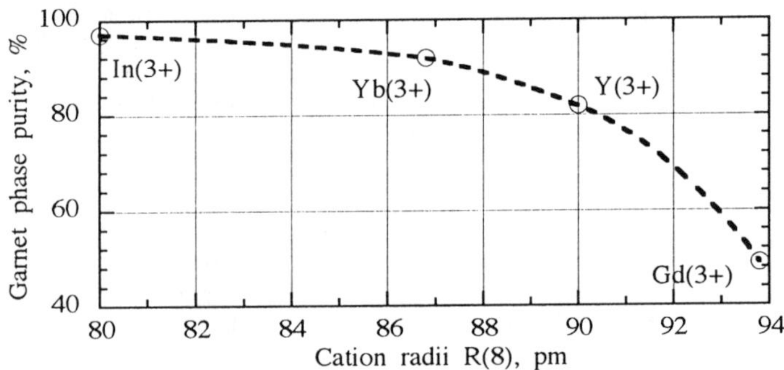

Fig. 1 Dependence of coefficient of garnet phase purity $I_{(420)}/I_{(2nd)}$ on size of C^{3+} cations in $Ca_3Nb_{1.6875}[C^{3+}]_{0.1875}Ga_3O_{12}$ compounds prepared by solid state reaction technique.

The substitution of Ca^{2+} cations in dodecahedral sites (CN = 8) with Sr^{2+} according to following reaction:

$$(Ca^{2+})_8 \rightarrow (Sr^{2+})_8 \tag{3}$$

did not result single garnet phase formation also. Total amount of second phase was found to be very high. It is due to the large size of Sr^{2+} cation.

Different aliovalent substitutions were studied also. Greatest garnet phase purity was observed in the mixtures which were prepared according following substitution reaction:

$$2(Ga^{3+})_6 \rightarrow (Li^+)_6 + (Nb^{5+})_6. \tag{4}$$

The garnet phase purity in the $Ca_3Li_xNb_{(1.5+x)}Ga_{(3.5-2x)}O_{12}$ mixtures versus x is given in Fig. 2. Maximal concentration of the garnet phase was observed correspond to x = 0.2 ~ 0.3. High phase purity and low melting point ($T_m \sim 1450\,°C$) of this compound allow us to continue study of this material by μ-PD and CZ methods using Pt crucibles. Details of garnet crystal growth from the melts are given in following subsections.

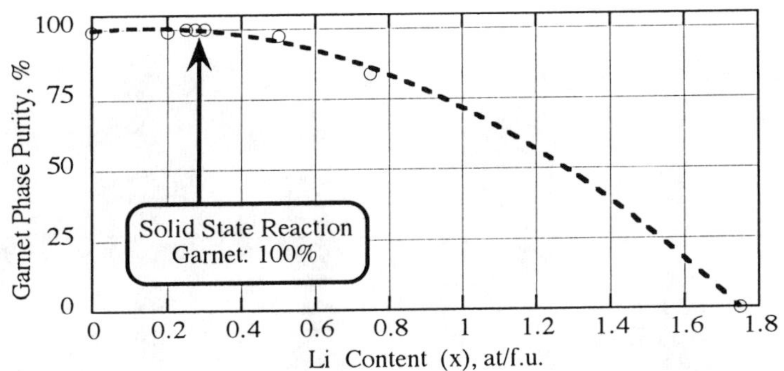

Fig.2. Dependence of coefficient of garnet phase purity $I_{(420)}/I_{(2nd)}$ on Li^+ content (x) in $Ca_3Li_xNb_{(1.5+x)}Ga_{(3.5-2x)}O_{12}$ compounds prepared by solid state reaction technique

<u>Growth of single crystal fibers</u>

Fiber crystals of $Ca_3Li_xNb_{(1.5+x)}Ga_{(3.5-2x)}O_{12}$ (CLNGG), where x=0.0-0.5, were grown by μ-PD [9] method using Pt/Rh crucible in air. The crucibles were made with relatively large diameter of capillary channel (0.8 mm) and short length (1 mm). Pulling down rate was in the range of 3 ~ 18 mm/hr. The pre-calcinated or pre-melted polycrystals prepared by solid state reaction were used as starting materials. Parameters of μ-PD crystals grown are given in Table I.

Table I. Melt composition, maximal growth (pulling) rate f (mm/h), density of cracks D, total crystallization fraction C (wt. %), average lattice parameter of grown fiber crystal a ($\mathring{A}$) and relative intensity ratio $I_{(2nd)}/I_{(420)}$ of second phase for the raw materials (RM) and solidified remaining melts (SM) after μ-PD crystal growth

Melt Composition	f	D	C	a	RM	SM
$Ca_3Li_{0.00}Nb_{1.50}Ga_{3.5}O_{12}$	3	Very high	21	-	0.011	0.193
$Ca_3Li_{0.25}Nb_{1.75}Ga_3O_{12}$	12	High	49	12.541	0.014	0.012
$Ca_3Li_{0.2625}Nb_{1.7625}Ga_{2.975}O_{12}$	12	Low	37	12.542	0.005	-
$Ca_3Li_{0.27}Nb_{1.77}Ga_{2.96}O_{12}$	12	Low	51	12.541	0.000	0.005
$Ca_3Li_{0.275}Nb_{1.775}Ga_{2.95}O_{12}$	15	Very low	63	12.540	0.000	0.005
$Ca_3Li_{0.2875}Nb_{1.7875}Ga_{2.925}O_{12}$	12	High	27	12.541	0.000	0.132
$Ca_3Li_{0.30}Nb_{1.80}Ga_{2.9}O_{12}$	12	Low	16	12.540	0.000	-
$Ca_3Li_{0.50}Nb_{2.0}Ga_{2.5}O_{12}$	6	Low	8	12.542	0.022	0.037

We could not obtain transparent single crystals from the melts which correspond to $I_{(2nd)}/I_{(420)} > 0.03$ according to Fig. 2. Transparent and crack-free $Ca_3Li_{0.30}Nb_{1.80}Ga_{2.9}O_{12}$ single crystal was grown up to 15 mm in length and up to 0.8 mm in diameter. The grown crystal was formed in a single phase of garnet, but the solidified melt included the garnet and non-garnet phases. For more detailed investigation of congruency, fiber crystals with the various compositions were grown. Fig. 3 shows CLNGG fiber crystals grown by the μ-PD method. The crystals were transparent and well developed in crystal form. Bubbles and inclusions were not detected under the microscopic observation. Considering the hexagonal cross-section of the crystal, we assumed that the direction of crystal growth was close to <111>. The non-garnet phases were not detected in the crystals. They were detected only in the starting raw materials and in the solidified melt remaining in the crucible. In the sintered raw materials, when the concentration of Li^+ ion was between 0.27 and 0.3, non-garnet phases were not detected also.

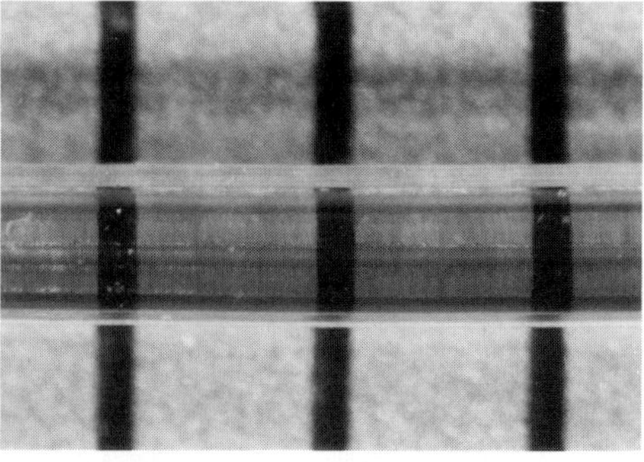

Fig.3 Single fiber crystals grown from $Ca_3Li_{0.25}Nb_{1.75}Ga_3O_{12}$ (left) and $Ca_3Li_{0.275}Nb_{1.775}Ga_{2.95}O_{12}$ (right) melts

In the solidified remain melts, the amount of second phase was minimal in the melt corresponding to $Ca_3Li_{0.275}Nb_{1.775}Ga_{2.95}O_{12}$. The composition of $Ca_3Li_{0.275}Nb_{1.775}Ga_{2.95}O_{12}$ appeared to have minimum content of non-garnet phases in the raw material, grown fiber crystal and solidified melt. It is necessary to note that in all experiments the amount of non-garnet phases in the solidified melt was much higher than that of raw materials, which is a result of segregation phenomena in the melts.

It was impossible to use charged melt completely because of partial evaporation of the melt and the structure of the Pt/Rh crucible. Maximum amount of crystallized melt was slightly larger than 60% of the total melt for the composition of $Ca_3Li_{0.275}Nb_{1.775}Ga_{2.95}O_{12}$. When the concentration of rejected impurities reaches a certain critical value, because of flux nature of non-garnet phases, the diameter of fiber crystals rapidly decreases and then crystals were separated from the melt automatically. In this experiment it appeared that the total crystallization fraction depends on the composition. Thus, maximum solidification fraction achieved corresponds to congruently melting composition.

Maximal growth rate which results in transparent single crystal growth depends on composition also. It follows from Table I. Fiber crystals grown from $Ca_3Nb_{1.5}Ga_{3.5}O_{12}$ and $Ca_3Li_{0.5}Nb_2Ga_{2.5}O_{12}$ stoichiometric melts were not transparent in spite of low growth rate (3-6 mm/hr). In case of $Ca_3Li_{0.275}Nb_{1.775}Ga_{2.95}O_{12}$ and its vicinity, transparent and high quality were grown at 12-15 mm/h pulling rate (Fig. 3).

Thus, it was shown that (1) growth rate resulting in high quality transparent single crystal growth, (2) garnet phase purity of raw material and remaining melt, and (3) total solidification fraction are maximum at the composition of $Ca_3Li_{0.275}Nb_{1.775}Ga_{2.95}O_{12}$.

To estimate uniformity of the crystals grown the variation of lattice parameter as a function of the solidification fraction was studied. In spite of the fact that lattice parameters of raw materials were widely distributed from 12.534 Å to 12.546 Å, the parameters of grown crystals were almost same for all compounds (12.539 Å < a < 12.542 Å). Therefore it was expected that the lattice parameter at congruent melting composition is 12.541 ± 0.001 A.

Thus, we conclude that within the compounds studied $Ca_3Li_{0.275}Nb_{1.775}Ga_{2.95}O_{12}$ is closest to congruently melting composition. This compound was selected as basic one for the CZ experiments, which will be discussed below.

Growth of single crystals by Czochralski method

The crystals of CLNGG were grown by CZ technique from the $Ca_3Li_{0.25}Nb_{1.75}Ga_3O_{12}$ and $Ca_3Li_{0.275}Nb_{1.775}Ga_{2.95}O_{12}$ melts using the RF power supply of 30 kW. The starting materials were of 99.99% purity as described before. Two kinds of the cylindrical Pt crucibles were used ; Ø 40 x 40 mm and Ø 50 x 50 mm, respectively. The crystals grown at the pulling rate of 1 mm/hr are shown in Fig. 4.

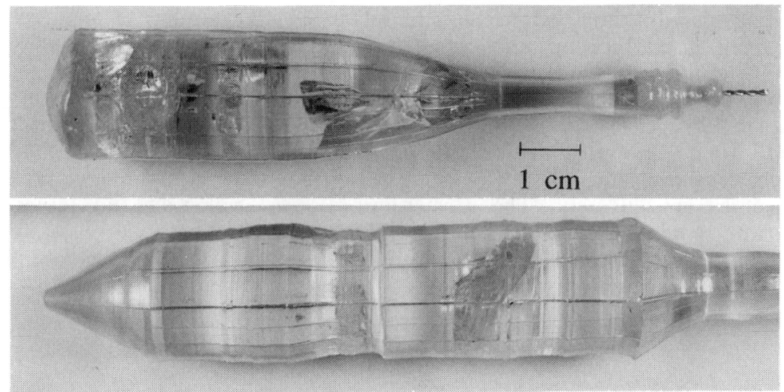

Fig. 4 Single crystals of garnet grown by Czochralski technique from the $Ca_3Li_{0.25}Nb_{1.75}Ga_3O_{12}$ (above) and $Ca_3Li_{0.275}Nb_{1.775}Ga_{2.95}O_{12}$ (below) melts

The neck of crystals was grown at pulling rate of 3 mm/hr. The rotation rate of 20 rpm and air atmosphere with flow rate of 0.6 l/min were used. As a first step polycrystalline material was crystallized on the Pt wire by multiple dipping into the melt. After that one crystallographic direction was developed by necking process. Melting point of $Ca_3Li_{0.275}Nb_{1.775}Ga_{2.95}O_{12}$ was measured by heating and melting of CNGG crystals in resistive (SiC) furnace. 63 and 36 wt.% of the starting melts were crystallized into single crystals of $Ca_3Li_{0.25}Nb_{1.75}Ga_3O_{12}$ and $Ca_3Li_{0.275}Nb_{1.775}Ga_{2.95}O_{12}$ compositions, respectively.

In Fig. 5, an optical transmission spectrum of CLNGG crystal is shown. The thickness of CLNGG plate was 1 mm. No optical absorption was detected in 400-1200 nm wavelength range. The spectra of both crystals (Fig. 4) were very similar. Both crystals were transparent and slightly yellow. Bubbles and inclusions were not detected. The crystals had some cracks. It was found that the density of cracks of crystals grown from $Ca_3Li_{0.275}Nb_{1.775}Ga_{2.95}O_{12}$ melt was lower than that from $Ca_3Li_{0.25}Nb_{1.75}Ga_3O_{12}$. The crystals were sliced and then the lattice parameter was measured as a function of solidification fraction.

The variation of lattice parameter was similar that of described for μ-PD grown crystals.

Phase composition of solidified melts remained after crystal growth was studied also. It was found that $0.05 < I_{(2nd)}/I_{(420)} < 0.06$ in case of $Ca_3Li_{0.25}Nb_{1.75}Ga_3O_{12}$, and $I_{(2nd)}/I_{(420)} = 0$ in case of $Ca_3Li_{0.275}Nb_{1.775}Ga_{2.95}O_{12}$, which corresponds to pure garnet phase.

After the phase composition was studied, both of the remaining melts were heated to temperature 1460 °C (just above the melting point), held for 1 h and then cooled down to 1445°C in 3 h. In the case of $Ca_3Li_{0.25}Nb_{1.75}Ga_3O_{12}$ melt, many spontaneous garnet crystals with size of about 1 mm were found on the surface of the melt as given in Fig.6.

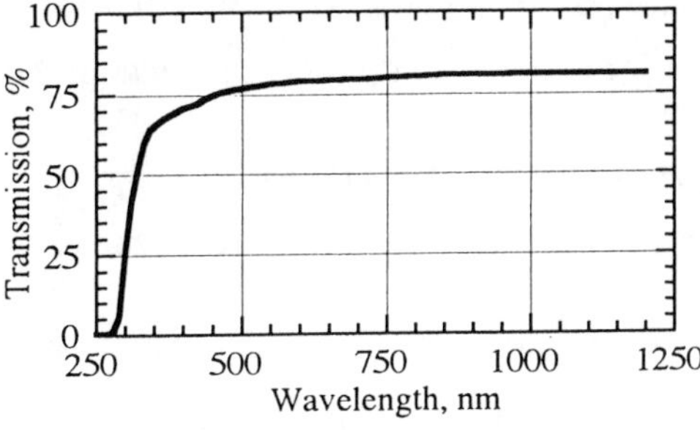

Fig. 5. Transmission spectrum of CLNGG

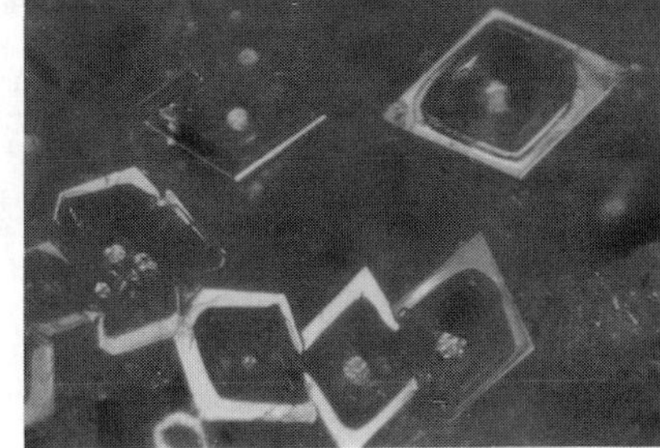

Fig.6. View of recrystallized remaining melt

The crystals had mainly rhombododecahedral facets <110>, similar to that observed in ref. [10]. Judged from good faceting, the crystals were grown at relatively low growth rate as in the flux technique, with second phase compound acting as a flux. In $Ca_3Li_{0.275}Nb_{1.775}Ga_{2.95}O_{12}$ remaining melt, this phenomenon was not observed. It was assumed that in this case solidification rate was much higher in comparison with the previous melt. It is known that for the congruently melting composition liquidus and solidus temperatures are same: $T_l = T_s$, but for all the other compositions (except eutectic point) these values are different. The greater the $(T_l - T_s)$ value is, the lower solidification rate at the same cooling rate is. Therefore we conclude that $Ca_3Li_{0.275}Nb_{1.775}Ga_{2.95}O_{12}$ composition is closest to the congruent one. After this recrystallization process phase compositions of the solids were studied again, but the results were same as before.

CONCLUSIONS

It was found that in $Ca_3Li_xNb_{(1.5+x)}Ga_{(3.5-2x)}O_{12}$ system the synthesized materials are mostly in the garnet phase in the range of $0 < x < 0.75$. Pure garnet phase formation was observed in the vicinity of stoichiometric composition of $Ca_3Li_{0.25}Nb_{1.75}Ga_3O_{12}$.

Fiber crystals of CLNGG were grown by μ-PD method using melts with $0 < x < 0.50$. Maximal growth rate (15 mm/h), highest solidification fraction (63 wt.%), lowest density of cracks and highest garnet phase purity of raw material and remain melt were observed for the crystals grown from $Ca_3Li_{0.275}Nb_{1.775}Ga_{2.95}O_{12}$ melt. Transparent and bubble-free single crystals of garnet were grown from $Ca_3Li_{0.25}Nb_{1.75}Ga_3O_{12}$ and $Ca_3Li_{0.275}Nb_{1.775}Ga_{2.95}O_{12}$ melts by CZ method also. Lattice parameters of all the crystals grown by different methods and from different melts were in the range of $12.539 \text{ Å} < a < 12.542 \text{ Å}$. It was found that $Ca_3Li_{0.275}Nb_{1.775}Ga_{2.95}O_{12}$ crystal melts congruently at $1450°C$ in air atmosphere and can be grown using Pt crucible. No optical absorption of CLNGG was detected in the 400-1200 nm wavelength range.

The crystal of CLNGG seems to be promissing substrate material for the liquid phase epitaxy of Bi-containing iron garnet films for magneto-optical applications.

REFERENCES

1.   P.Hansen and J.P.Krumme, Thin Solid Films **114**, p.69 (1984).

2.   S.A.Markgraf, S.Kimura, T.Sawada and M.Göbbers, J.Cryst.Growth **135**, p. 253 (1994).

3.   K.Shimamura, M.Timoshechkin, T.Sasaki, K.Hoshikawa and T.Fukuda, J.Cryst.Growth **128**, p. 1,021 (1993).

4.   D.Mateika, E.Völkel and J.Haisma, J.Cryst.Growth **102**, p. 994 (1990).

5.   M.Kestigian, W.R.Bekebrede and A.B.Smith, J.Cryst.Growth **42**, p. 343 (1977).

6.   Y.Miyazawa, H.Toshima, S.Hanita and N.Kodama, J.Cryst.Growth **99**, p. 854 (1990).

7.   M.Kawata, H.Toshima, Y.Miyazawa and S.Morita, J.Cryst.Growth **128**, p. 1,011 (1993).

8.   N.V.Vasiljeva, V.I.Chani, V.V.Randoshkin and N.A.Es'kov, Proc.SPIE, **1126**, p. 99 (1989).

9.   D.H.Yoon, I.Yonegawa, T.Fukuda and N.Ohnishi, J.Cryst.Growth **142**, p. 339 (1994).

10.   V.I.Chani and A.M.Balbashov, J.Cryst.Growth **66**, p. 616 (1984).

Part VI
Electronic and Magnetic Properties

ELECTRON TRANSPORT PHENOMENA IN $NiS_{2-x}Se_x$ SINGLE CRYSTALS

X. YAO*[a], S. EHRLICH[b], G. LIEDL[b], T. HOGAN**[c], C. KANNEWURF[c], AND J.M. HONIG[a]

*Beckman Institute, University of Illinois, 405 N. Mathews Avenue, Urbana, IL 61801
**University of Houston, Physics Department, Houston TX 77204
a. Department of Chemistry, Purdue University, West Lafayette, IN 47907, b. School of Materials Engineering, Purdue University, West Lafayette, IN 47907, c. Department of Electrical Engineering and Computer Science, Northwestern University, Evanston, IL 60208.

ABSTRACT

Structural studies, electrical resistivity, and Seebeck coefficient measurements are reported in the range 4.2 - 300 K for single crystals of $NiS_{2-x}Se_x$ ($0 \leq x \leq 0.71$) grown from a Te melt. Over the entire temperature and composition ranges there are no large scale structural changes concomitant to a variety of magnetic ordering phenomena, and to a changeover from insulating to metallic characteristics as x increases. Thus, the evolution in transport characteristics with x can be studied without interference from the lattice; moreover, the electron count is unaffected by substitution of Se for S. The existence of anomalous peaks in resistivity as a function of temperature is attributed to significant electron correlation phenomena which allow the entropy of charge carrier to play a dominant role. The complex temperature dependence of the Seebeck coefficient is attributed to the participation of both electrons and holes in charge transport.

INTRODUCTION

The $NiS_{2-x}Se_x$ system lends itself well to systematic studies of electron correlation effects without the complications generally attendant to changes in lattice structure or substitutional disorder. Pioneering studies of electrical properties [1-6], carried out over the last three decades, have shown that nickel disulfide is a very good insulator, whereas nickel diselenide is metallic. These two compounds can be mixed in all proportions; the intermediate $NiS_{2-x}Se_x$ obeys Vegard's law [7]. With increasing x in the range $0.35 \leq x \leq 0.55$ one encounters correlation effects associated with the gradual shift toward the metallic state. What is almost unique to this system is that the changes in electronic properties can be achieved by isoelectronic compositional adjustment of the anionic sublattice while keeping the cationic component fixed. Thus, the valency of nickel is kept invariant as long as the anion to cation ratio remains at 2:1. Furthermore, there are no large scale shifts in crystal structure either as a function of temperature or of composition, though the possibility of very small changes in atomic configurations cannot at this point be excluded.

Mat. Res. Soc. Symp. Proc. Vol. 453 © 1997 Materials Research Society

By way of background information attention is directed to the phase diagram shown in Fig. 1; this represents an attempt to summarize somewhat divergent reports in the literature [8-12]. The various phases are identified in the caption to this diagram. Phase boundaries have been sketched in with some fairing and interpolation. In particular, the horizontal lines map out uncertainties in the placement of the boundary separating the insulating and metallic phases. The dashed line indicates uncertainties in the range over which parasitic ferromagnetism (the WFI or WFM regime) extends. One should note the variety of magnetically ordered and disordered phases that exist.

Portions of the research cited below have been summarized in other publications [13-15].

EXPERIMENTAL

Ni, S, and Se in elemental form were mixed in proper stoichiometric proportions and heated at 730 °C for up to eight days in an evacuated silica tube. The resultant material was usually dissolved in Te flux at temperatures between 500 and 670 °C and subjected to a modest temperature gradient for two weeks, according to the procedure detailed elsewhere [13]. The Te was removed with aqua regia. Single crystals of NiS_2 approximately 5 mm on an edge were harvested; these tended to become smaller as x was increased toward 0.71. Electron microprobe techniques showed that the S/Se ratio was uniform across the samples and that the cation/anion ratio was close to 2. No incorporation of Te was encountered above the detection limit of roughly 0.1 wt. %. Any small occlusion does not present a serious problem since the flux is isoelectronic with S and Se. Four-probe resistivity

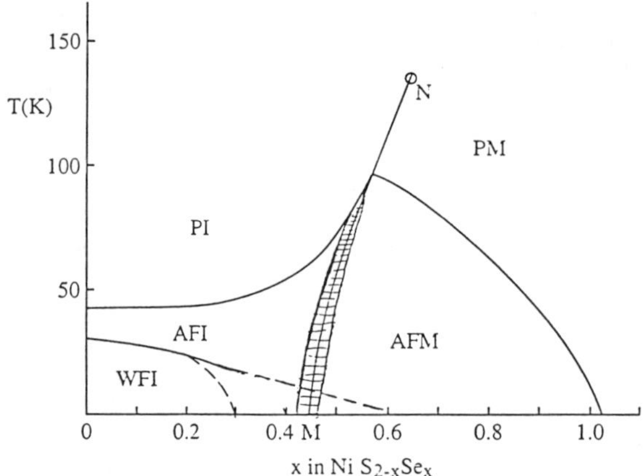

Fig 1. Proposed Composite Phase Diagram for the $NiS_{2-x}Se_x$ System in the Range $0 \leq x \leq 1$, Based on Published Literature and Present Results. PI = paramagnetic insulator, AFI = antiferromagnetic insulator, WFI = weakly ferromagnetic insulator(canted antiferromagnet), AFM = antiferromagnetically ordered metal, PM = paramagnetic metal. Hatched area indicates regions where various investigators reported different findings.

measurements were carried out on specimens of sufficient size; the van der Pauw technique [16] was employed in the remaining cases. For simplicity we set the adjustable ratio of resistances, f, normally lying in the range $0.8 \le f \le 1$, equal to unity; this choice does not affect the conclusions cited below. Seebeck coefficient measurements were carried out by use of a slow ac technique [17,18]. Corrections were applied for the Seebeck voltage of the gold leads. Below 30 K the measurements became unreliable because of interference by the large phonon drag effects in Au at low temperatures. All electrical measurements were executed under computer control.

EXPERIMENTAL RESULTS

X ray diffraction patterns for alloys with $x = 0.55$ are shown in Fig. 2. Careful examination of the diffractograms shows no extraneous phases; all peaks could be indexed on the pyrite structure. Also, there is no discernible difference in the patterns taken at 300 K and at very low temperatures. Aside from a small shift along the 2θ axis there is no difference between the diffractograms for samples with $x = 0.51$ and $x = 0.55$. A more detailed study of peak shapes as a function of temperature T and composition x is planned for the future; however, it is clear that no large-scale changes were encountered.

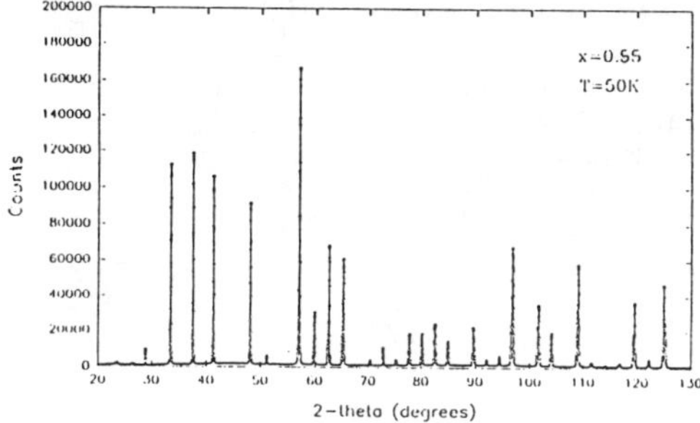

Fig 2. X ray Diffractograms for $NiS_{2-x}Se_x$ for $x = 0.55$ and at T=50 and 298 K.

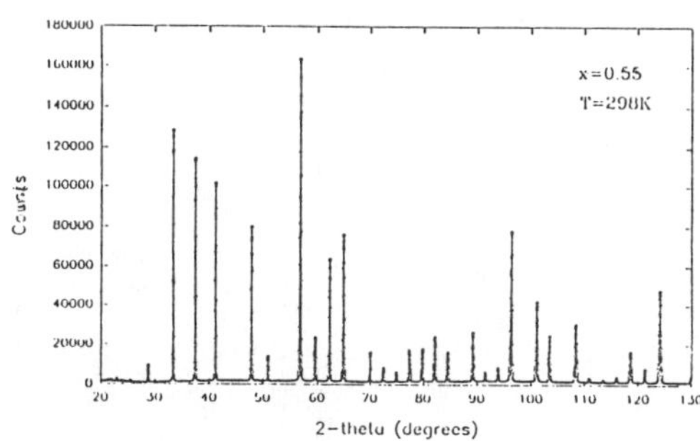

The resistivity (ρ) data can be grouped into several distinct categories;

(a) $0 \le x \le 0.24$: As shown in Fig. 3, the resistivities are extremely high at low temperatures where the WFI phase prevails. The drop of ρ with increasing T is attenuated in the temperature range from

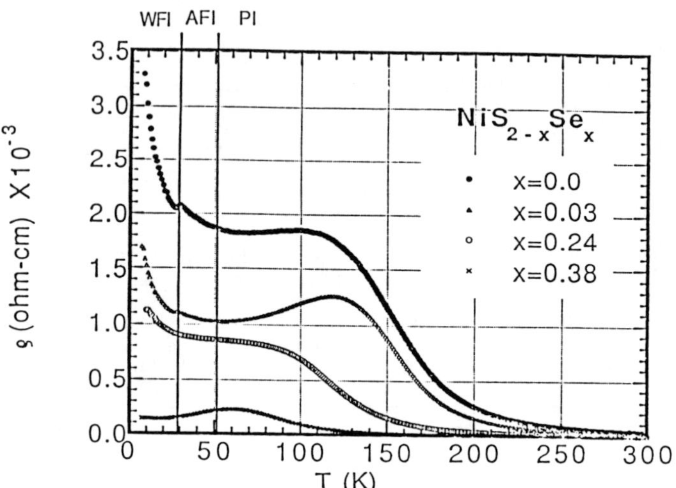

Fig 3(a). Plots of Resistivity ρ vs. Temperature T for Single Crystals of NiS$_{2-x}$Se$_x$ in the Composition Range $0 \leq x \leq 0.24$.   Data taken using the van der   Pauw Technique.   The regions of the various phases   of Fig. 1 are indicated at the top of the diagram.

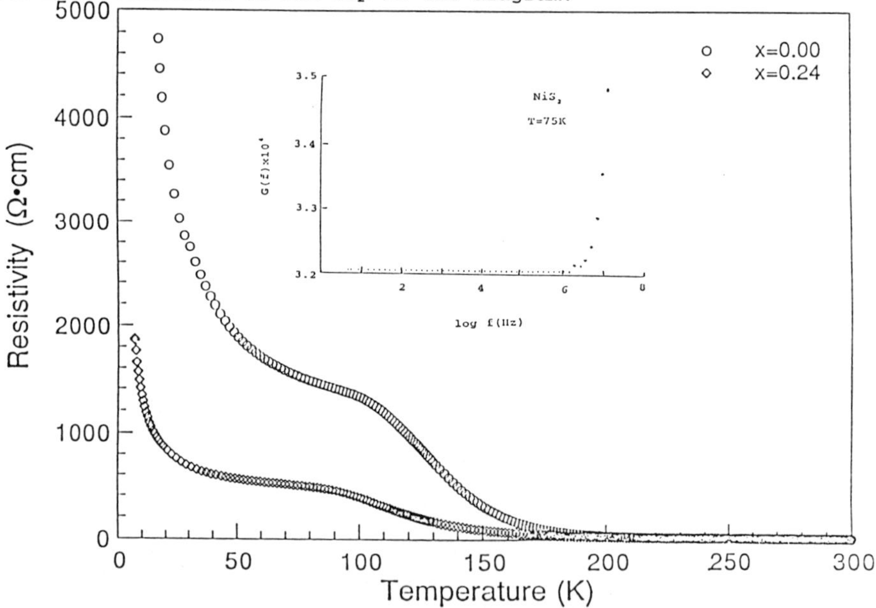

Fig 3(b). Four-probe Resistivity Measurements of Single Crystals of NiS$_{2-x}$Se$_x$ with x = 0, 0.24.   Insert shows frequency dependence of A   C conductivity for NiS$_2$.

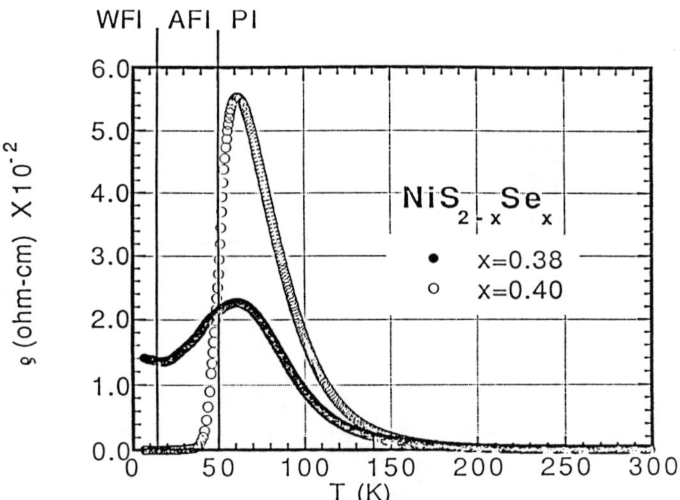

Fig 4. ρ vs. T for Single Crystal NiS$_{2-x}$Se$_x$ for x = 0.38, 0.40. The regions of the various phases of Fig. 1 are indicated at the top of the diagram.

~30 to ~120 K; ρ flattens out or passes through a gentle maximum before diminishing again with rising T. For T < 30 K or T > 140 K the Arrhenius plots are linear, with activation energies of 300 $\pm$ 20 meV and 105 $\pm$ 10 meV respectively, independent of x. The data taken by the van der Pauw method and by the four-probe technique (Figs. 3a and 3b) are in reasonable accord. AC resistivity measurements on NiS$_2$ for T = 75 K are shown as an inset to Fig. 3b. It is seen that ρ remains essentially independent of applied frequency up to approximately 1 MHz, beyond which some dispersion sets in; this effect appears exaggerated in the inset because of the very large scale adopted for the ordinate. The value of ρ = 3.2 X 10^{3}ohm-cm is in rough accord with the value 2 X 10^3 ohm-cm read off from Fig. 3. The temperature dependence of the AC resistivity over the whole range of applied frequencies matches that of the DC resistivities.

(b) 0.38 $\leq$ x $\leq$ 0.51. The data in Figs. 4 to 6 show that ρ is nearly independent of T at lowest temperatures, then rises steeply with T in a range coincident with the passage of the system from the WFI through the AFI to the PI phase. After passing through a sharp maximum the resistivities drop again toward the metallic range at elevated temperatures. For T > 80 K plots of log ρ vs 1/T are roughly linear, with conductivity activation energies still in the range 105 $\pm$ 10 meV, independent of x. The resistivities in the cryogenic temperature region diminish from 10^2 ohm-cm for x = 0.38 to 10^{-3} ohm-cm for x = 0.55. Thus, there is an increasing degree of metallicity with increasing replacement of S by Se, in accord with expectation. As a check on reproducibility and cycling we show in the inset of Fig. 5 a set of resistivity measurements on a second alloy specimen with nominal composition x = 0.51. The general trend in the ρ vs T dependence is quite similar to

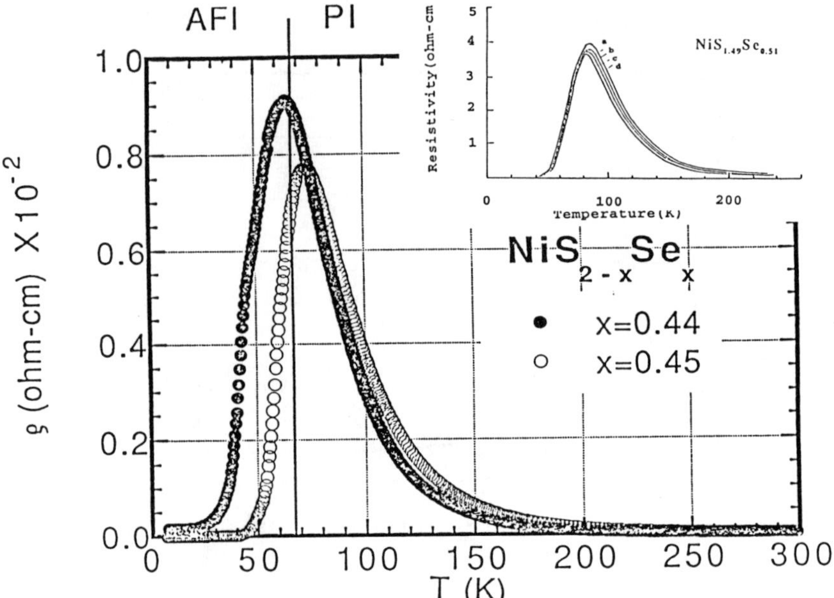

Fig 5. ρ vs. T for Single Crystal NiS$_{2-x}$Se$_x$ for x = 0.44, 0.45. The regions of the various phases of Fig. 1 are indicated at the top of the diagram. Insert shows Reproducibility of van der Pauw Resistivity Measurements in NiS$_{1.49}$Se$_{0.51}$. Curve (a) 1st Run, T increasing; Curve (b) 1st Run, T decreasing; Curve (c) 2nd Run, T increasing; Curve (d) 2nd Run, T decreasing.

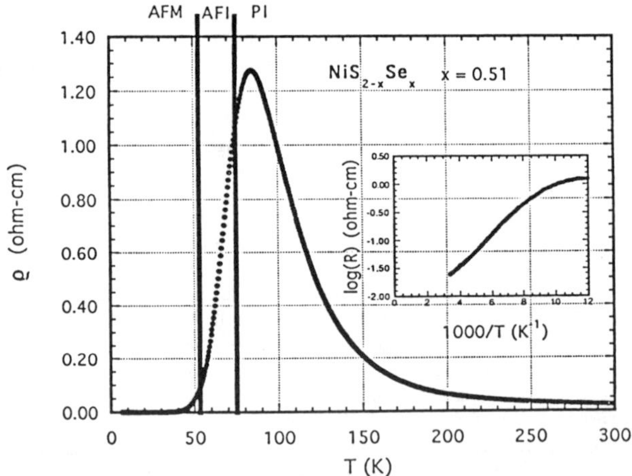

Fig 6. ρ vs. T for Single Crystal NiS$_{1.49}$Se$_{0.51}$. Insert shows plot of log ρ vs. 1/T. The regions of the various phases of Fig. 1 are indicated at the top of the diagram.

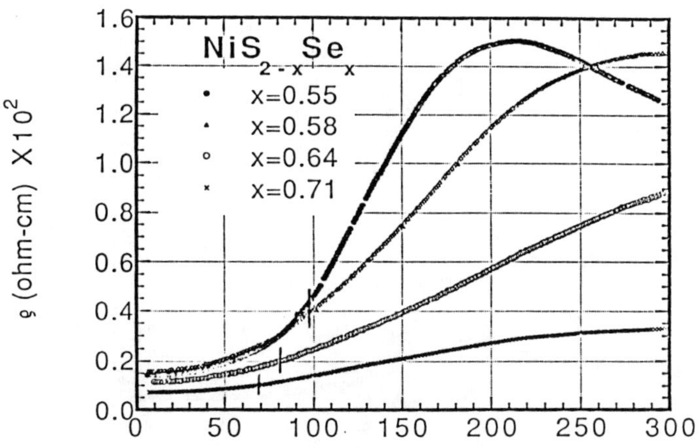

Fig 7. ρ vs. T for Single Crystal NiS$_{2-x}$Se$_x$ in the Composition Range 0.55 ≤ x ≤ 0.71. The vertical bars indicate the temperature of the AFM-PM transition shown in Fig. 1.

that shown in Fig. 6, except that the peak value of ρ now ranges from 3.5 to 3.8 ohm-cm, as compared to the value of 1.3 ohm-cm read off from Fig. 6. This discrepancy is readily explained by noting the very steep decline in peak resistivities as x increases from 0.4 to 0.55. One should also note the degree of reproducibility as the sample is thermally cycled. Subsequent annealing of this sample in vacuum at 750 °C for five days led to a severe degradation in electrical properties: thermal cycling during these latter measurements resulted in irreproducible behavior and in a marked increase of ρ at 300 K relative to the unannealed sample.

(c) 0.55 ≤ x ≤ 0.71: Here ρ falls into the range consistent with correlated metals. The data are shown in Fig. 7. For sample with x = 0.55 and 0.58 ρ still passes through a shallow maximum which by now has moved to temperatures beyond 200 K. Specimens with x = 0.64 and 0.71 exhibit metallic-like characteristics; in this composition range ρ rises monotonically with T.

Our findings are qualitatively consistent with published results, but in the insulating regimes the present resistivities exceed those in the literature by two orders of magnitude, presumably reflecting the higher degree of purity of samples prepared by the flux gradient technique employed in the present study.

Seebeck coefficient (α) measurements are displayed in Figs. 8 and 9 for the ranges 0 ≤ x ≤ 0.51 and 0 ≤ x ≤ 0.71 respectively. We distinguish three regimes:
(a) 0 ≤ x ≤ 0.24: The thermopower is positive, passes through a pronounced maximum, and approaches zero as T → 0.

(b) 0.38 ≤ x ≤ 0.44: In this composition range α is negative both

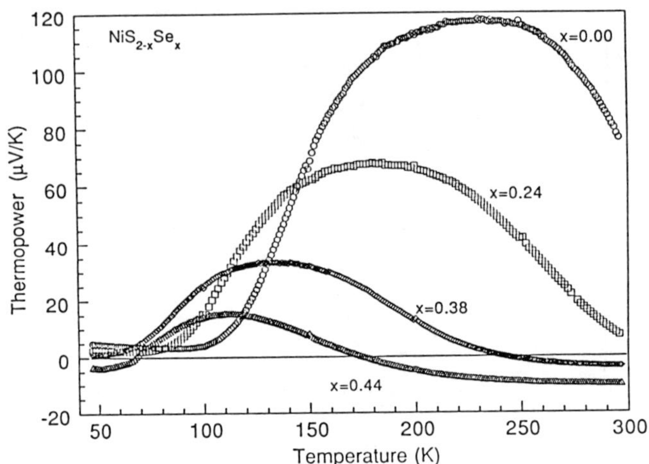

Fig 8. Thermoelectric Properties for $NiS_{2-x}Se_x$ in the Composition Range $0 \leq x \leq 0.44$.

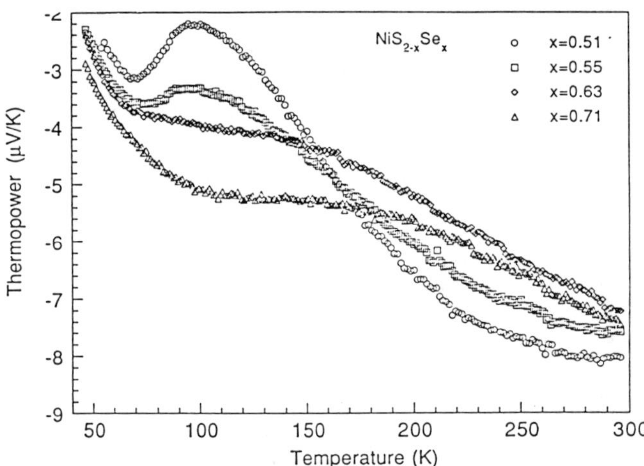

Fig 9. Thermoelectric Properties for $NiS_{2-x}Se_x$ in the Composition Range $0.51 \leq x \leq 0.71$.

at low and at high temperature and remains positive in the intermediate T range.

(c) $0.51 \leq x \leq 0.71$: Here α is negative and remains numerically small; its magnitude increases roughly linearly with T. Note that the scatter in the data arises from the enlarged scale of Fig. 9.

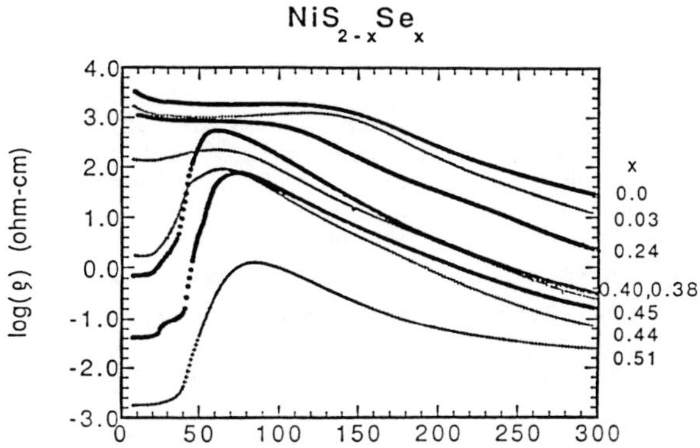

Fig 10. Summary of Experimental Resistivity Measurements as Plots log ρ vs. T for the Composition Range $0 \leq x \leq 0.51$.

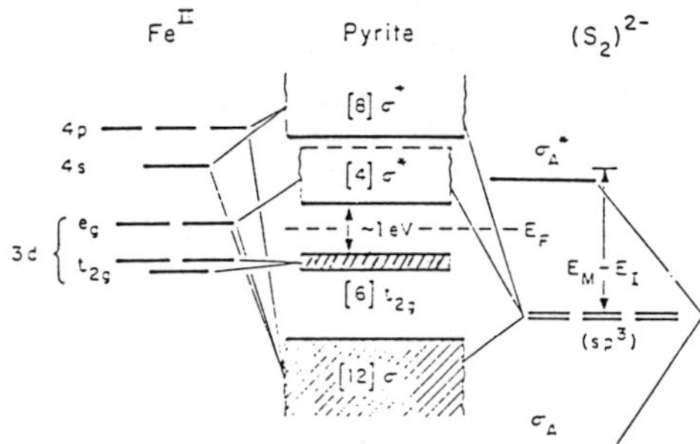

Fig 11. Band Structure Diagram for Pyrite, after Goodenough, Ref. 19; See Text.

A summary of the electrical resistivities is provided as plots of log ρ vs. T for the composition $0 \leq x \leq 0.51$ in Fig. 10. The plots indicate very clearly the trend toward increasing metallicity with increasing Se substitution, as well as the enormous range of resistivities encountered in the cryogenic temperature range.

DISCUSSION

The above information can be rationalized in terms of the energy band diagram proposed by Goodenough [19] for pyrite structures, as depicted in Fig. 11. A similar diagram is likely to apply to the NiS$_{2-x}$Se$_x$ system, except that the bands are expected to be wider and any gaps, narrower than in FeS$_2$. It is seen that the Fermi level is close to a band of primarily t$_{2g}$ character, separated by a gap from a wider band of primarily σ* parentage, derived from sigma-type bonding between cationic and nearest neighbor anionic constituents. In NiS$_{2-x}$Se$_x$ the bands gradually widen and the gaps are gradually narrowed with increasing substitution of Se, because of the greater radial extension of its p - type wave functions compared to those of S. The increasing cation-anion wave function overlap ultimately results in a metallic -

like configuration in which electron near the Fermi level populate mixed band states.

We now attempt to rationalize the observed electrical characteristics. The ac conductivity data displayed in Fig. 3(b) are consistent with delocalized electronic states in NiS_2 and therefore, in all alloys with $x > 0$. The resistivity changes with temperature for $0 \leq x \leq 0.24$ can be interpreted as arising from a semiconductor with a bandgap of approximately 600 and 220 meV for the WFI and PI phases respectively. One should note what happens in the limit as $T \to 0$ K: whereas ρ increases without limit α becomes small, contrary to the rule of thumb that the variation of α should parallel that of ρ. This forces the conclusion that the system must be characterized by a complex Fermi surface which features both hole and electron pockets. From purely thermodynamic arguments one may show [20] that the Seebeck coefficient is then given by

$$\alpha = \left[\alpha_p \sigma_p - |\alpha_n| \alpha_n\right]/\sigma,$$

where σ_n and σ_p are the partial conductivities of the electrons and holes, and α_n and α_p are the corresponding partial Seebeck coefficients; $\sigma = 1/\rho$ is the total conductivity. Clearly, a near cancellation from the positive and negative contributions would then be resposible for the small α values in the low temperature range. This argument is supported by a study of plots of α/ρ vs T (shown in Figs. 12a and 12b) which are close to zero for $0 \leq x \leq 0.44$ and for $T < 120$ K. This indicates that the material is essentially intrinsic in this concentration and temperature range.

The near constancy in ρ between 50 and 120 K apparently is a precursor to the effect discussed below. Beyond 120 K the Fermi surface is evidently shifted in such a manner as to render the hole contribution more dominant; the Seebeck coefficient rises correspondingly. This trend is ultimately cut off by the increasing thermal promotion of carriers across the band gap. As the overall conductivity of the sample rises with temperature the increasing degree of metallicity diminishes the values of α_n and α_p, thereby causing α to drop with rising T. This also explains why the α vs T curves are successively reduced in

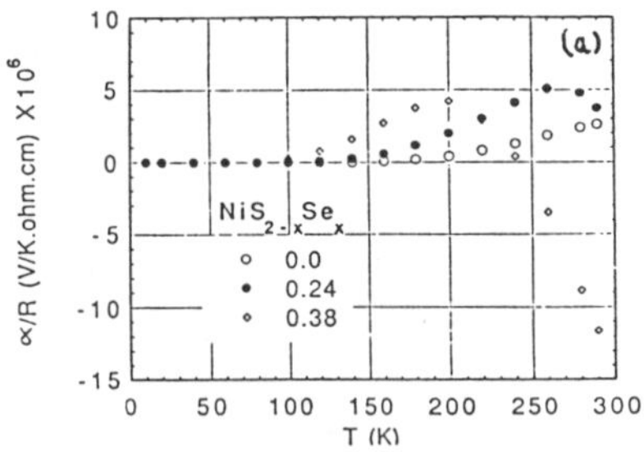

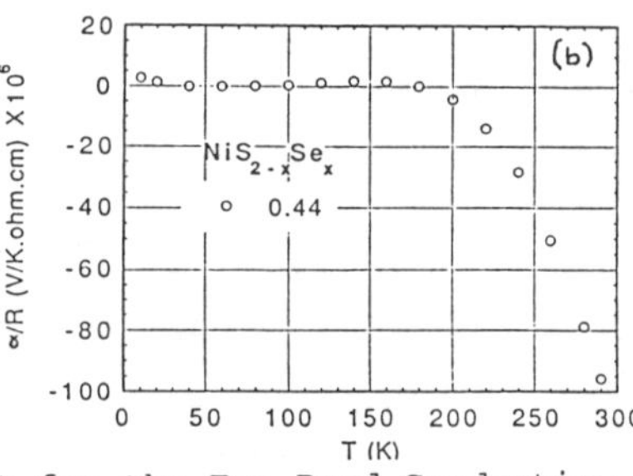

Fig 12. Plots of α/ρ vs. T as a Test for the Two-Band Conduction Model for $x = 0$, 0.24, 0.38 and $x = 0.44$ Respectively.

magnitude as the Se content of the alloys is increased.

In the range $0.40 \leq x \leq 0.51$ the most striking characteristic is the enormous peaks in ρ close to 70 - 80 K, while the variation of α with T either passes through a very shallow maximum ($x = 0.44$) or is quite unremarkable. In fact, for alloys with $x > 0.51$, α remains negative, numerically small, and varies roughly linearly with T. This indicates that the latter alloys have become quasimetallic. The very large low-temperature resistivities shown in Figs. 4-6 for compositions $x = 0.44, 0.45, 0.51$ indicate that we are dealing with a highly correlated metallic regime. For such systems the stabilization of a more insulating state with rising temperatures between roughly 30 and 80 K reflects the opening of a band gap, as a result of qualitative changes in the nature of the electronic states. For these highly correlated systems there is a near compensation of the negative band energies of the itinerant charge carries and their positive electronic repulsion energies; this matter is discussed elsewhere in detail [21-24]. With a net energy close to zero it is the entropy of the charge carriers that then increasingly dominates the electronic free energy as T rises. As is well established, the entropy of the carriers constituting the correlated Fermi liquid in these alloys is smaller when the carriers are delocalized than when they are more nearly localized. One should recall that for an occupancy of one localized electron per site the entropy per carrier is $k_B \ln 2$ since an electron can be accommodated in either the spin-up or spin-down configuration. The increasing localization process is ultimately interrupted because at sufficiently elevated temperatures the entropy of the delocalized carriers does ultimately exceed its more localized counterpart.

The remarkably high values in ρ of 0.1 to 1 ohm-cm for these quasimetallic alloys (See Fig. 10) show that alloys in the composition range $0.4 \leq x \leq 0.5$ are highly correlated metals. These findings may validate theoretical predictions of a very sharp spike in the density of states at the Fermi level. This case arises close to the crossover point between the highly correlated metal and the onset of the semiconducting regime. The theoretical justification is provided in recent studies of the Hubbard model for a lattice of infinite dimensions [25-32].

Lastly, for the regime with $x > 0.55$ one deals with alloys of increasing metallicity which exhibit either vestiges of the ρ peak at higher temperatures or monotonic changes of ρ with T. The near cancellation of kinetic and potential energies no longer holds when electron interactions are screened out. The Seebeck coefficients then also display the characteristics expected of a normal metal.

It should be emphasized again that the rich variety of different electrical properties in the $NiS_{2-x}Se_x$ system is encountered without any significant changes in the lattice structure, and thus is a direct consequence of alterations in the degree of electron correlations achieved by varying the S/Se ratio, while leaving the Ni cation sublattice intact.

ACKNOWLEDGMENTS

The authors thank Patricia Metcalf for valuable advice during the preparation of the single crystals used in this study. The research was supported by Grant DMR 92-22986 at Purdue University. TPH was supported on NSF Grant DMR 92-22245 at Northwestern University. The thermoelectric studies were carried out at the MRL Central Facilities supported by Grant DMR 91-20521 at Northwestern University. Fruitful discussions with Prof. J. Spałek are gratefully acknowledged.

REFERENCES

1. R.J. Bouchard, J.L. Gillson, and H.S. Jarrett, Mater. Res. Bull. 8, 489 (1973).
2. G. Krill, M.F. Lapierre, F. Gautier, C. Robert, G. Czjzek, J. Fink, and H. Schmidt, J. Phys. C: Solid State 9, 761 (1976).
3. M. Kamada, N. Mōri, and T. Mitsui, J. Phys. Soc. C: Solid State Phys. 10, L643 (1977).
4. P. Kwizera, M.S. Dresselhaus, and D. Adler, Phys. Rev. B21, 2328 (1980).
5. T. Miyadai, M. Saitoh, and Y. Tazuke, J. Magn. Magn. Mater. 104-107, 1953 (1992).
6. S. Sudo, J. Magn. Magn. Mater. 114, 57 (1992).
7. D.D. Klemm, N. Jahr. Miner. Monat. 1962, 32.
8. H.S. Jarrett, R.J. Bouchard, J.L. Gillson, G.A. Jones, S.M. Marcus, and J.F. Weiher, Mater. Res. Bull. 8, 877 (1973).
9. F. Gautier, G. Krill, M.F. Lapierre, P. Panissod, C. Robert, G. Czjzek, J. Fink, and H. Schmidt, Phys. Lett. 53A, 31 (1975).
10. G. Czjzek, J. Fink, H. Schmidt, G. Krill, M.F. Lapierre, P. Pannisod, F. Gautier, and C. Robert, J. Magn. Magn. Mater. 3, 58 (1976).
11. S. Ogawa, J. Appl. Phys. 50, 2308 (1979).
12. H. Takano and A. Okiji, J. Phys. Soc. Jpn. 50, 3835 (1981).
13. X. Yao and J.M. Honig, Mater. Res. Bull. 29, 709 (1994).
14. X. Yao, S.N. Ehrlich, G. Liedl, and J.M. Honig, The Metal-Nonmetal Transition Revisited edited by P.P. Edwards and C.N.R. Rao, (Taylor & Francis, London, 1995) p. 127 ff.
15. X. Yao, J.M. Honig, T. Hogan, C. Kannewurf, and J. Spałek, Phys. Rev. B, in the press.
16. L. J. van der Pauw, Philips Res. Repts. 13, 1 (1958).
17. P.M. Chaikin and J.F. Kwak, Rev. Sci. Instr. 46, 28 (1975).
18. H.O. Marcy, T.J. Marks, and C.R. Kannewurf, IEEE Trans. Instr. Meas. 32, 756 (1990); this program also made use of the SB100 thermopower control system provided through the courtesy of MMR Technologies Inc.; Mountain View, CA.
19. J.B. Goodenough, J. Solid State Chem. 3, 26 (1971).
20. T.C. Harman and J.M. Honig, Thermoelectric and Thermomagnetic Effects and Applications, (McGraw-Hill, New York, 1967), Chap. 3.
21. J. Spałek, J.M. Honig, M. Acquarone, and A. Datta, J. Magn. Magn. Mater. 54-57, 1047 (1986).
22. J. Spałek, A. Datta, and J.M. Honig, Phys. Rev. Lett. 59, 778 (1987).
23. J. Spałek, J. Solid State Chem. 38, 70 (1990).

24. A. Datta, J.M. Honig, and J. Spałek, Phys. Rev. $\underline{B44}$, 8459 (1991).
25. W. Metzner and D. Vollhardt, Phys. Rev. Lett. $\underline{62}$, 324 (1989).
26. E. Müller-Hartmann, Z. Phys. $\underline{B74}$, 507 (1989).
27. U. Brandt and Ch. Mielsch, Z. Phys. $\underline{75}$, 365 (1989); $\underline{79}$, 295 (1990).
28. A. Georges and G. Kotliar, Phys. Rev. $\underline{B45}$, 6479 (1992).
29. M. Jarrell, Phys. Rev. Lett. $\underline{69}$, 168 (1992).
30. A. Georges and W. Krauth, Phys. Rev. $\underline{B48}$, 7167 (1993).
31. Th. Pruschke, D.L. Cox, and M. Jarrell, Phys. Rev. $\underline{B47}$, 3553 (1993).
32. X.Y. Zhang, M.J. Rozenberg, and G. Kotliar, Phys. Rev. Lett. $\underline{70}$, 1666 (1993).

Ln$_3$Cu$_4$P$_4$O$_2$: A New Lanthanide Transition Metal Pnictide Oxide Structure Type

R. J. CAVA*, H. W. ZANDBERGEN**, J. J. KRAJEWSKI, T. SIEGRIST, H. Y. HWANG, A. P. RAMIREZ, AND B. BATLOGG
*Department of Chemistry and Princeton Materials Institute, Princeton University, Princeton, NJ 08540, Bell Laboratories, Lucent Technologies Murray Hill, NJ 07974,**National Center for High Resolution Electron Microscopy, Technical University,
Delft, Holland

ABSTRACT

A new type of lanthanide transition metal pnictide-oxide, Ln$_3$Cu$_4$P$_4$O$_2$ (for Ln = La, Ce, and Nd), is reported. The crystal structure is made from Ln$_2$O$_2$ layers of the Pb$_2$O$_2$ type stacked with layers of Cu$_2$P$_2$ tetrahedra. The existence of a homologous series of compounds with similar chemistry connecting the BiCuOSe and ThCr$_2$Si$_2$ structure types is indicated. The resistivity of La$_3$Cu$_4$P$_4$O$_2$ is metallic, with no superconductivity observed above 2K. The Ce and Nd compounds have magnetic susceptibilities characteristic of the trivalent Lanthanides. The Ce compound orders magnetically at approximately 0.8 K.

INTRODUCTION

Layered transition metal oxides continue to be the source of interesting and surprising physical phenomena. The same holds true for chemically layered intermetallic compounds. There is a very small class of layered materials which represent a combination of these distinct structural chemistries: the transition metal pnictide or chalcogenide oxides (see, for instance, refs. 1 and 2). These materials combine commonly found layered metal-oxide structural components and layered metal-pnictide or metal-chalcogenide structural components in a surprising manner. Pnictogen, chalcogen and oxygen all act as anions: e.g. no phosphate or selenate is formed. We began an exploration of copper oxide based pnictide-oxides in the hope of finding a cuprate analog to Ba$_2$Mn$_3$P$_2$O$_2$, in which square planes of both Mn$_2$P$_2$ and MnO$_2$ are found (3). Although a lanthanide copper pnictide oxide does exist, its crystal structure does not contain CuO$_2$ sheets, but rather Cu$_2$P$_2$ sheets only. These are stacked with Ln$_2$O$_2$ sheets with the Pb$_2$O$_2$ structure, as are for example found in Nd$_2$CuO$_4$ (4). The synthesis, crystal structure, relation to known structure types, and some basic physical properties of these compounds are briefly described here.

SYNTHESIS

The starting materials were: fresh lanthanide metal shavings, lump red phosphorus, and Cu$_2$O (source of O and Cu). Small amounts of excess P were added to the initial mixtures, which were of the form Ln$_3$Cu$_4$P$_{4+x}$O$_2$; x=0.05, 0.10, and 0.15 (approximately 1-3% excess). Best results were consistently found for the lanthanides La, Ce and Nd for x = 0.05 and 0.10, where polycrystalline samples of greater than 95% purity (by powder X-ray diffraction) could be obtained. Attempts to synthesize the phase with the smaller rare earths Y, Gd and Ho were not

Mat. Res. Soc. Symp. Proc. Vol. 453 ©1997 Materials Research Society

successful by the same method, suggesting that the phase is stable only for the larger rare earths. The powders, in 1 gram quantities, were placed in dense Al_2O_3 crucibles and sealed in evacuated quartz tubes. The heating schedule was as follows: 450°C, 550°C, 650°C, 1 day soaks; 700°C, 750°C, 800°C, 3 day soaks; grind; 850°C, 900°C, 1 day soaks; grind; 950°C, 1000°C, 1 day soaks; grind; 1050°C, 2 day soak; grind; 1050°C, 8 day soak.

STRUCTURAL CHARACTERIZATION

The new compound was discovered in an electron microscope equipped with elemental microanalysis in one of a set of exploratory syntheses. The crystallographic unit cell was determined by electron diffraction to be I centered tetragonal with approximate dimensions 4.05 x 26.8 angstroms. A more complete electron microscopy study was then performed with a Philips CM30ST electron microscope with a field emission gun and Link EDX equipment operated at 300kV. High resolution images for through focus exit wave reconstruction were recorded using a Tietz software package and a 1024 x 1024 pixel Photometrix CCD camera having a dynamic range of 12 bits. The reconstructions were done using 15-20 images (with focus increments of 5.2 nm) with a software program developed by Coene et al (5). HREM on crystals in [100] and [110] orientation clearly revealed layers of Lanthanide and copper atoms in the structure, and on consideration of alternative models for the light atom positions, a stacking sequence of La layers, La_2O_2 layers and Cu_2P_2 layers was found to be consistent with the experimental exit waves indicating an overall stoichiometry $La_3Cu_4P_4O_2$.

A small single crystal (0.05 x 0.05 x 0.07mm) of $La_3Cu_4P_4O_2$ was obtained from grain growth in a polycrystalline pellet of that stoichiometry heated at 1100°C. Data were collected on an Enraf-Nonius CAD04 Diffractometer with Mo Kα radiation. To correct for absorption, psi-scan followed by spherical absorption correction was employed. The structure was refined with the NRCVAX suite of programs. The refined crystal structure atomic coordinates are presented in table 1. The coordinates are consistent with the structure observed in the HREM study. The powder diffraction patterns of $Ce_3Cu_4P_4O_2$ and $Nd_3Cu_4P_4O_2$ were employed to obtain crystallographic cell parameters of a = 3.985(1), c = 26.573(9) and a = 3.964(1), c = 26.551(5) angstroms, respectively, for those compounds. Unlike the case for the quaternary layered boro-carbides (7), for the present materials (in the La case a = 4.033(1), c = 26.765(8)), all cell parameters decrease with decreasing rare earth size.

Table 1. $La_3Cu_4P_4O_2$ atomic parameters x,y,z and B_{iso} (Mean of the Principal Axes of the Thermal Ellipsoid). Standard deviations refer to the last digit printed.

	x	y	z	B_{iso}
LA1	0	0	0	0.43(8)
LA2	1/2	1/2	0.20156(10)	0.90(7)
CU	1/2	0	0.09406(12)	0.87(7)
P1	0	0	0.1405(4)	0.32(19)
P2	1/2	1/2	0.0417(5)	0.55(20)
O	1/2	0	1/4	3.3(9)

The crystal structure of $La_3Cu_4P_4O_2$ is shown in figure 1. The figure shows the manner in which the $La-O_2-La$ layers with the Pb_2O_2 type structure are stacked with the $P-Cu_2-P-La-P-Cu_2-P$ layers with the $ThCr_2Si_2$ type structure to build up the full crystal structure. The Cu_2P_2 layers have copper in tetrahedral coordination with phosphorous. The coppers form a square planar net. The oxide and pnictide layers are joined by bonding between the La in the La_2O_2 layer and the outer P in the Cu_2P_2 layer. This results in an interesting square antiprismatic coordination for that La atom: to all O in a small square on one side and all P in a larger square on the opposing side, in a PbFC1 type arrangement. The compatibility of the sizes of the different layers is due to the fact that $d(Cu - P) \approx d(La - O) \sim d(La - P)/\sqrt{2}$ allowing the layers to stack without excessive strain. We note that the P-P bondlength across the La layer in the $LaCu_2P_2$ sandwich, 2.23Å, is rather short. It is close to that found in elemental phosphorus (~2.25Å) and shorter than that found for P-P bonds in Ni-P (~2.4Å) (8). This indicates that the P in the current compound cannot be thought of as being the P^{3-} anion, but rather acts more like $(P_2)^{4-}$.

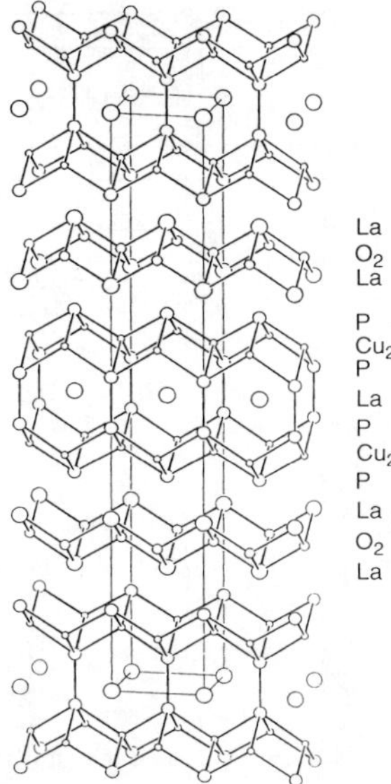

Fig 1. The crystal structure of $La_3Cu_4P_4O_2$. Outline shows the crystallographic unit cell. The layers are identified on the side of figure: the layered nature of the crystal structure is emphasized via the interatomic bonds shown.

COMPARISON TO OTHER STRUCTURES

Consideration of the crystal structure of $La_3Cu_4P_4O_2$ in the context of other known compounds (9-13) suggests the existence of a homologous series between those of the relatively rare BiCuOSe structure type and those of the ubiquitous $ThCr_2Si_2$ intermetallic structure type. The homologous series is illustrated schematically in figure 2. where the known compounds with related chemistry ThCuPO (9) and $LaNi_2P_2$ (14) are selected to represent the end-members, and $Th_3Ni_3P_3O$ (9) and $La_3Cu_4P_4O_2$ as the first few members of the series. The series is generated by the stacking of "$Th_2Cu_2P_2O_2$" type layers with "$LaNi_2P_2$" type layers, as shown in the figure. The formula for the series is therefore:

$$(A_2O_2T_2P_2)_m \cdot (AT_2P_2)_n$$

Members of the series have been observed for m = 1, and n = 0,1,2 and ∞, as shown in the figure. To our knowledge, no examples are known for m $\neq$ 1 or n larger than 2 and less than ∞. This relationship between the relatively rare pnictide oxides and the common $ThCr_2Si_2$ structure type suggests that the potential exists for the discovery of many new materials with related chemistry and structure.

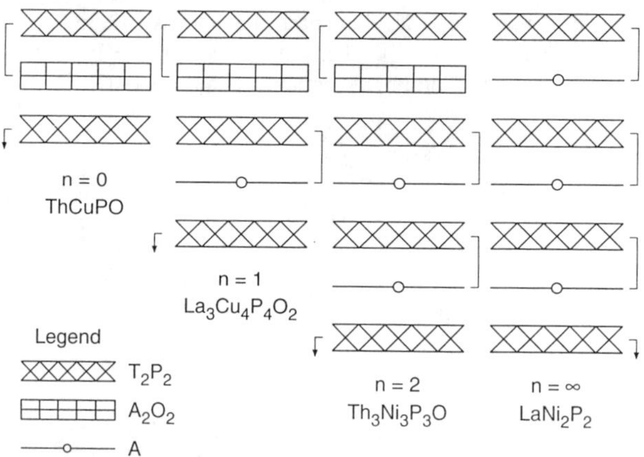

Fig 2. Schematic comparison of the crystal structures of known members of the new homologous series $(A_2O_2T_2X_2)_m(AT_2X_2)_n$ for m = 1 and n as shown.

PHYSICAL PROPERTIES

The electrical resistivity of a $La_3Cu_4P_4O_2$ polycrystalline pellet was measured between 300 and 4.2K in a standard 4-probe geometry. The behavior of ρ (T) is shown in figure 6. The room temperature resistivity is 250μ ohm cm, and the resistivity at 4K 30μ ohm cm, showing this material to be a good metal even in polycrystalline form where there may be a significant contribution from grain boundary resistivity. The inset to figure 3 shows the variation of magnetic susceptibility between 300 and 4K measured in the field of 5000 Oe (SQUID magnetometer). The susceptibility is basically diamagnetic and temperature independent, dominated by core diamagnetism and only a weak Pauli paramagnetism from the conduction electrons. A very small contribution from uncompensated Cu^{2+} spins is observed at low temperature, possibly from impurity phases, at the level of 0.4%. The magnetic susceptibility of $Nd_3Cu_4P_4O_2$ shows Curie-Weiss behavior at high temperature with a deviation at low temperature. A fit to the high temperature data indicates a moment of 3.68μ_B/Nd, comparing favorably to the expected moment of 3.62μ_B/Nd. The Curie-Weiss θ from the fit is 23.0K, indicating antiferromagnetic correlations between the Nd spins. However, no magnetic ordering is observed above 4.2K indicating at least a moderate degree of frustration of the Nd spins.

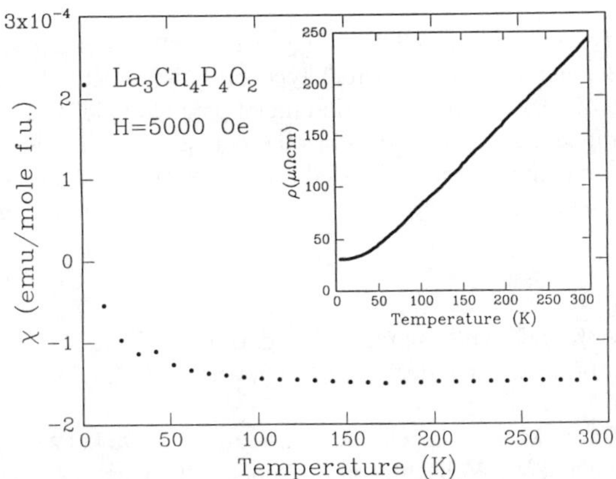

Fig 3. Magnetic susceptibility of $La_3Cu_4P_4O_2$ measured in a 5kOe field. Inset: Temperature dependent resistivity of a polycrystalline sample of $La_3Cu_4P_4O_2$.

The magnetic susceptibility of $Ce_3Cu_4P_4O_2$ shows curvature for the entire temperature range between 4 and 300K. No signature of magnetic ordering is observed. The Ce moment obtained from the high temperature data is $2.33\mu_B$/Ce (with a θ of 39.2K), slightly reduced from the expected value of $2.54\mu_B$/Ce. The temperature dependence of $1/\chi$ vs. T deviates from Curie-Weiss behavior significantly at low temperatures due to crystal field effects, which depopulate the excited 4f states with decreasing temperature and decrease the apparent moment. The specific heat at low temperatures (fig. 4) shows the presence of magnetic ordering at approximately 0.8 K.

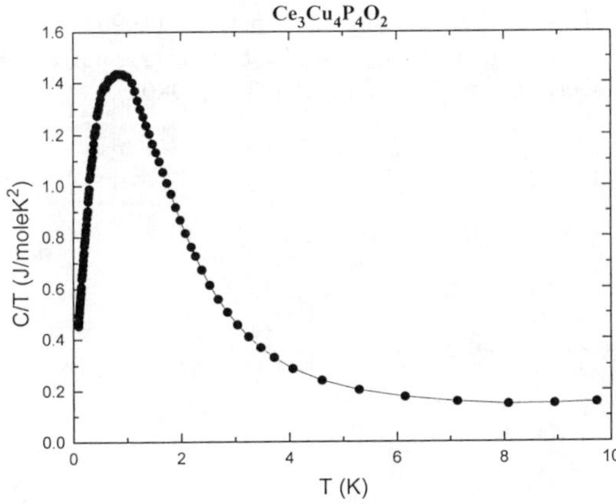

Fig 4. Specific heat of $Ce_3Cu_4P_4O_2$ at low temperatures showing magnetic ordering at approximately 0.8 K.

CONCLUSION

The synthesis, crystal structure and basic physical properties of a few members of a new layered transition metal pnictide oxide structure type of stoichiometry $Ln_3Cu_4P_4O_2$ have been reported. Comparison with other known transition metal pnictide-oxides suggests the possibility that this so far rather limited class of materials has the potential for discovery of a considerably larger number of members with a variety of related crystal structures. More detailed study of the physical properties of some of these compounds would be of significant interest.

REFERENCES

1. S. L. Brock and S. M. Kauzlarich, Comments Inorg. Chem. *17*, p.213 (1995).
2. Y. Park, D. C. DeGroot, J. L. Schindler, C. R. Kannewurf and M. G. Kanatzidis, Chem. Mater. *5*, p.8 (1993).
3. Ned T. Stetson and Susan M. Kauzlarich, Inorg. Chem. *30*, p.3969 (1991).
4. H. Muller-Buschbaum and W. Wollschlager, Z. Anorg. Allg. Chem. *414*, p.76 (1975).
5. W. J. M. Coene, A. Thust, M. op de Beek, and D. Van Dyck, *Ultramicroscopy*, to be published.
6. D. Tang, J. Jansen, H. W. Zandbergen, M. op de Beek, and D. Van Dyck, *Ultramicroscopy*, to be published.
7. T. Siegrist, R. J. Cava, J. J. Krajewski and W. F. Peck, Jr., J. Alloys and Compounds *216*, p.135 (1994).
8. A. F. Wells, *Structural Inorganic Chemistry*, Clarendon Press, Oxford, 1984, pp. 838-875.
9. D. Kaczorowski, J. H. Albering, H. Noel and W. Jeitschko, J. Alloys and Compounds *216*, p.117 (1994).
10. P. S. Berdonosov, A. M. Kusainova, L. N. Kholodkovskaya, V. A. Dolgikh, L. G. Akselrud and B. A. Popovkin, J. Sol. St. Chem. *118*, p.74 (1995).
11. B. I. Zimmer, W. Jeitschko, J. H. Albering, R. Glarum, and M. Reehuis, J. Alloys and Compounds *229*, p.238 (1995).
12. J. H. Albering and W. Jeitschko, Z. Naturforsch *51b*, p.257 (1996).
13. D. Kaczorowski, J. H. Albering, H. Noel, J. Sol. St. Chem. *112*, p.228 (1994).
14. W. Jeitschko and B. Jaberg, J. Sol. St. Chem. *35*, p.312 (1980).

LAYERED CUPRATES

P. A .SALVADOR *, K. OTZSCHI [§], H. ZHANG *, J. R. MAWDSLEY *, K. B. GREENWOOD [§],
B. M. DABROWSKI [†], L. D. MARKS *, T. O. MASON *, K. R. POEPPELMEIER [§‡]
* Materials Science and Engineering Department , Northwestern University, 2225 N. Campus Dr.,
Evanston, IL 60208-3108
§ Department of Chemistry, 2145 Sheridan Rd., Northwestern University, Evanston IL 60208
† Department of Physics, Northern Illinois University, DeKalb, IL 60115
‡ Author to whom correspondence should be addressed

ABSTRACT

Layered copper-oxide superconductors exhibit the highest critical transition temperatures of any materials. Yet all of the known double perovskites $A'A''B'B''O_6$ containing copper have a random or rock salt distribution of the B cations with the exception of the unique layered arrangement found in La_2CuSnO_6. Only the layered arrangement contains the CuO_2^{2-} planes which are necessary for high-temperature superconductivity. The occurrence of layered or two dimensional structures increases markedly when vacancies are introduced on the oxygen sublattice, as evidenced in $Ln_2AE_mCu_2Ti_mO_{5+3m}$ (Ln = lanthanide, Y: AE = Ba, Ca: $2 \leq m \leq 4$). Similarities among oxygen-deficient structures, especially those with two-dimensional solid-state features, are discussed. Combined conductivity and thermopower analysis are presented to elucidate their unique internal chemistry, defect structure, and conduction parameters. In particular, data for $La_{2-x}Sr_xCuSnO_6$ are presented and related to the crystal chemistry of the copper-oxygen layer. These data are compared with $La_2Ba_2Cu_2Sn_2O_{11}$ and $La_2Ba_2Cu_2Ti_2O_{11}$ to illustrate the significance of oxygen vacancies on the properties of the copper oxygen planes. New layered cuprates are discussed including the mixed A-site stoichiometries $Ln'Ln''AE_mCu_2Ti_mO_{5+3m}$ (Ln = lanthanide, Y: AE = Ba, Ca: $2 \leq m \leq 4$) which contain the smaller lanthanide (Ln'') ordered between the closely spaced, facing sheets of Cu-O square pyramids.

INTRODUCTION

While the field of solid state chemistry is vast and its parts can often seem unrelated, even amongst specific types of inorganic materials, common underlying principles can be applied to all materials [1]. A variety of chemical factors, which are only partially understood at present, govern the structure-property relationships in materials [1]. We have been investigating the relationships between the inner architecture, internal chemistry, and physical properties of layered copper oxides, and how these relationships can be used to rationally design materials [2-6]. While these concepts are applied specifically to potential high-T_c superconductors, similar problems are encountered in all fields of solid state materials design.

The structure-property relationships in layered cuprates have received great attention over the last decade, owing to the extremely high critical transition temperatures for superconductivity observed in layered cuprates. Despite this intense effort, no general theory for the mechanism of superconductivity has been universally accepted. The observation that two-dimensional sheets of stoichiometry CuO_2^{2-} occur in all high-T_c cuprates, has resulted in a focus on materials having this structural feature, i.e. layered cuprates. A variety of structural families have been discovered which contain these sheets and support superconductivity, however, rational design of new compounds is only in its early stages. Moreover, several structural families exist which contain CuO_2^{2-} sheets and have not exhibited superconductivity. Materials which have the prerequisite structural elements for high temperature superconductivity, yet are not superconductors, are exemplary systems for the development of rational materials design.

This communication reports some of the relationships between the internal chemistry, inner architecture, and macroscopic properties in 'pure'-perovskite layered cuprates. Structural and physical properties for several of these materials are reviewed to highlight the basis for the aforementioned relationships. In addition, high temperature electrical properties are presented for $La_{2-x}Sr_xCuSnO_6$, where oxidation of the CuO_2^{2-} sheets is observed to be quite different than in other layered cuprates. The ability to rationally design specific properties and chemical features is discussed in relation to oxygen-deficient perovskite materials. Structural features of new members

Mat. Res. Soc. Symp. Proc. Vol. 453 © 1997 Materials Research Society

of the $Ln'Ln''AE_mCu_2Ti_mO_{5+3m}$ (Ln = lanthanide, Y: AE = Ba, Ca: $2 \leq m \leq 4$) family are presented alongside a review of previously known members.

EXPERIMENTAL

The syntheses of the materials in the series $La_{2-x}Sr_xCuSnO_6$ (x = 0.0, 0.05, 0.10, 0.15, and 0.20) were in accord with previous reports [7, 8]. Single-phase materials were obtained for x = 0.05, 0.10, and 0.15, having structural parameters in agreement with previous reports. Kinetic difficulties (long reaction times and small, < 5%, amounts of impurities) were encountered in the synthesis of the unsubstituted (x = 0.0) material, rendering it unsuitable for electrical property characterization. In-situ high temperature electrical characterization of the phase-pure materials is described elsewhere[4, 5, 9]. Synthesis and structural characterization of the $Ln'Ln''AE_mCu_2Ti_mO_{5+3m}$ (Ln = lanthanide, Y: AE = Ba, Ca: $2 \leq m \leq 4$) family followed procedures described elsewhere [6].

RESULTS AND DISCUSSION
La_2CuSnO_6

The term, layered cuprates, is closely associated with those materials which exhibit the highest critical temperatures for superconductivity. To define layering in cuprates, which exhibit appropriate structural features for high-temperature superconductivity, we must adopt an electronic perspective. Ferrodistortive order of Jahn-Teller distortions of the d^9 Cu^{2+} cations results in a plane of copper-oxygen interactions involving the $d_{x^2-y^2}$ copper orbitals and the oxygen p orbitals, with an ideal geometric configuration involving 180° Cu-O-Cu interactions (in two dimensions). The d_{z^2} copper orbitals are aligned perpendicular to this plane, and stabilize both particular structural motifs and carrier concentrations. Two-dimensional copper-oxygen bond networks of this type can be considered as derivatives of a perovskite, or oxygen-deficient perovskite, structural element. Intergrowths of the perovskite structural element and a variety of other motifs result in the three dimensional solid state structure. In this communication we will restrict our attention to materials having only perovskite and oxygen-deficient perovskite structural elements. Several reviews of materials with other structural elements exist [10-14], as well as a review of two-dimensional features of transition metal oxides and chalcogenides [1], to which the reader is recommended for a broader introduction to these concepts.

Materials with the stoichiometry A_2CuMO_6 adopting perovskite based structures have been reviewed with respect to geometric configuration of the B-cations (Cu, M = transition/post-transition metal), and the chemical factors which govern these configurations [15]. Three B-cations sublattices are observed, as illustrated in Figure 1(a-c), for these stoichiometric double perovskite compounds. Figure 1(a) is the basic perovskite unit, having two distinct cations randomly distributed over a single B-site. Figure 1(b) is termed the rock-salt B-cation sublattice, owing to the alteration of distinct B-cations along any of the three crystallographic axis, similar to the cation-anion alteration in the rock-salt crystal structure. Figure 1(c) is the layered arrangement, where the B-cations segregate into distinct (100) planes of the perovskite structure. Only the last structure has the two dimensional CuO_2^{2-} sheets necessary for superconductivity. Moreover, to date, the only material that has been observed to adopt this structure is La_2CuSnO_6 [7, 8, 15].

While only the structure depicted in Figure 1(c) has the appropriate two dimensional characteristics to exhibit superconductivity, the rock-salt ordered structure also has two-dimensional features. The rock-salt structure can be viewed as an ordering of the two B-cations into alternate (111) planes, and can be considered layered with respect to the alteration of the B-cations. Thus two-dimensional features are present in the rock salt B-cation structure, and are perpendicular to the close-packed planes. It is important to stress the type of layering with respect to specific properties. The electronic properties of copper-oxide perovskites are derived predominantly from the Cu-O-Cu interactions, and therefore, an 'electronically' layered copper-oxide perovskite has the two dimensional structural order illustrated in Figure 1(c).

Despite having the necessary structural requirements for superconductivity, La_2CuSnO_6 has not been observed to superconduct. Aliovalent substitution of strontium for lanthanum was successful, e.g. $La_{2-x}Sr_xCuSnO_6$, (over the range of $0.0 \leq x \leq 0.175$ [8]), and led to p-type semiconductivity. The copper-oxygen environment is similar to that in the superconductors $La_{2-x}AE_xCuO_4$ (AE = alkaline earth cation, $T_c \approx 42$ K[16]), having a single layer of corner sharing

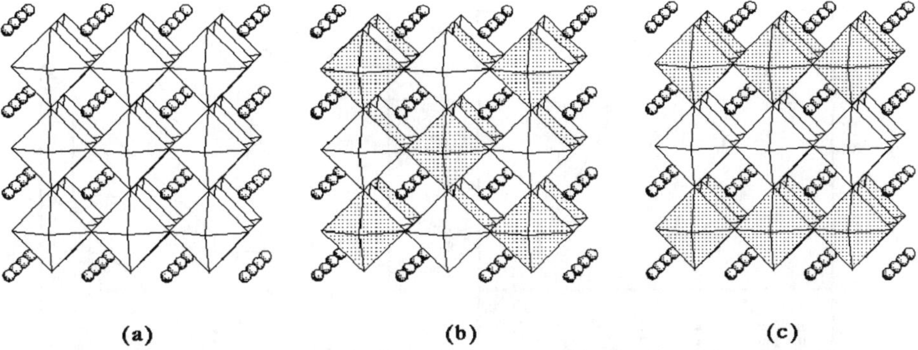

(a)   **(b)**   **(c)**

Figure 1. (a) Random, (b) Rock Salt, and (c) Layered arrangement of the B-cations (BO_6 octahedra) in oxygen stoichiometric double perovskites. Spheres represent A-cations, while light and dark octahedra in (b) and (c) denote Cu and M centered polyhedra , respectively. Adapted from Ref. [15].

copper-oxygen octahedra. However, the in-plane bond lengths are nearly 0.1 Å larger in La_2CuSnO_6 ($\approx$ 1.99 Å[7, 8]) than in La_2CuO_4 ($\approx$ 1.89 Å[17, 18]). The apical Cu-O bond lengths are about 0.05 Å shorter in La_2CuSnO_6 ($\approx$ 2.37 - 2.40 Å) than in La_2CuO_4 ($\approx$ 2.43 Å). The substitution range in $La_{2-x}Sr_xCuSnO_6$ spans from the parent antiferromagnetic insulator (x $\approx$ 0.0) to the expected region of superconductivity (x > 0.15) [8]. Theoretical investigations into the absence of superconductivity in $La_{2-x}Sr_xCuSnO_6$ concluded that the electronic structure near the Fermi level is quite different than that of other layered cuprate superconductors [8, 19]. The long bond lengths and structural distortion result in the d_z^2 orbital having a prominent role in the hole-states near the top of the valence band, and a correspondingly three-dimensional electronic structure [8, 19].

High-temperature electrical property data have demonstrated that the transport properties of layered cuprates are all quite similar, both for superconducting and non-superconducting materials having two-dimensional structural and electronic features [4, 5, 20, 21]. The $La_{2-x}Sr_xCuSnO_6$ system, on the other hand, can be expected to behave quite differently than the previous materials, based on prior investigations [8, 19]. Figure 2(a) illustrates the high temperature electrical data as a plot of the thermopower versus the natural logarithm of the conductivity (as first described by Jonker [22]) for both $La_{2-x}Sr_xCuSnO_6$ and $La_{2-x}Ba_xCuO_{4\pm\delta}$ [23] at T $\approx$ 760 °C. Data for the copper-tin system collected at other temperatures (650 °C- 800°C) overlap the data illustrated here, and for clarity have only been included for the x = 0.05 compound. Each group of data points for a specific doping level (and temperature) corresponds to values resulting from variations in the oxygen partial pressure (log pO_2 = -5, -4, -3, -2, -1, 0). The usefulness of this analysis technique for interpreting the transport properties of the parent semiconducting and undoped layered cuprates has been described elsewhere[4, 5, 20, 21, 24].

As shown in Figure 2(a), both cuprates follow a linear dependence (slope -k/e) indicating p-type extrinsic behavior. The conductivity data for $La_{1.85}Sr_{0.15}CuSnO_6$ material is slightly shifted to lower conductivity values than expected, probably owing to microstructural effects. However, the thermopower is less affected by these microstructural problems and is a better experimental method for probing the relative carrier concentrations across a series of polycrystalline materials. The electrical properties display a negligible dependence on oxygen partial pressure or temperature. The intercept of this linear behavior with the zero thermopower axis for $La_{2-x}Sr_xCuSnO_6$, seen in Figure 2(a), is similar to that for $La_{2-x}Ba_xCuO_{4\pm\delta}$ and other layered cuprates [4, 5, 20, 21, 24]. However, a dramatic difference between these two systems is seen in the absolute magnitude of the thermopower at any particular substitution level. The copper-tin system has much higher thermopower values (larger negative reduced thermopower, see below) than observed in the superconducting cuprate. The values of the thermopower observed in the copper-tin materials, even at the maximum substitution level are too high to expect superconductivity [4].

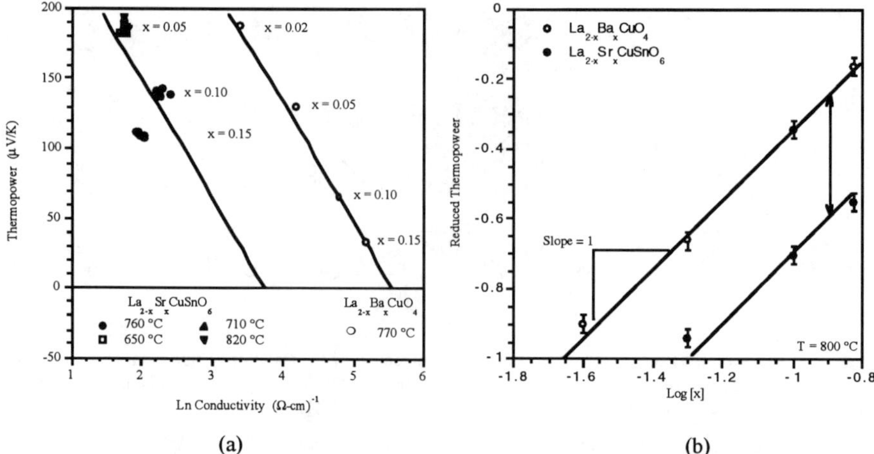

(a)   (b)

Figure 2. (a) Plot of the thermopower versus the natural logarithm of the conductivity for $La_{2-x}Sr_xCuSnO_6$ and $La_{2-x}Ba_xCuO_{4\pm\delta}$ (from Refs. [20, 23]). (b) Reduced thermopower versus the logarithm of the dopant concentration for $La_{2-x}Sr_xCuSnO_6$ and $La_{2-x}Ba_xCuO_{4\pm\delta}$. The vertical arrow in (b) illustrates the dramatic difference in the transport parameters for these materials (see text).

To address the origin of this difference, we convert the thermopower to a parameter which directly corresponds to the underlying transport parameters. A useful conversion of the thermopower value for an extrinsic semiconductor, to a form called the reduced thermopower, Q_R, is given by:

$$Q_R = -(\frac{e}{2.303k})Q = \log p - \log Ne^A \tag{1}$$

where Q is the thermopower, e is the unit of electronic charge, k is Boltzmann's constant, p is the hole concentration, N is the density of states for holes, and A is the transport constant. Therefore, if the hole concentration is scaling with the dopant concentration, then a linear relationship between the reduced thermopower and the logarithm of the dopant level is expected. A plot of these values is given in Figure 2(b), for both $La_{2-x}Sr_xCuSnO_6$ and $La_{2-x}Ba_xCuO_{4\pm\delta}$ [23]. It is apparent that both materials follow this linear relationship, which demonstrates that the hole concentrations are fixed in both materials by the doping concentrations (according to the relationship $[AE_{La}'] = p$, $AE =$ Sr, Ba [20, 23]) over this substitution range. Stoichiometric oxygen contents were observed for all compositions in the copper-tin system, in agreement with an electronic compensation mechanism for the substituted strontium cations [8].

The above argument precludes differences in the overall hole concentration as the origin of the differences in the reduced thermopower values. This implies that the values in the second term on the right hand side of Eq. 1 are fundamentally different in the two systems. The transport coefficient appears to be similar between the two systems, as demonstrated by the temperature-independent Jonker behavior of both extrinsic semiconducting materials [25]. However, this does not rule out the possibility of different transport constants in these materials, and further work is warranted. However, differences in the density of states between these two materials is in agreement with the discrepancy between the thermopower values, prior crystal chemical results [8] and band structure calculations [19] on these materials. The absolute difference between the two lines in Figure 2(b) is 0.35 logarithmic units, corresponding to a density of states which is ≈ 2.2 times larger in La_2CuSnO_6 than in the superconducting systems. This is consistent with speculation that the doped orbital is the d_z^2 orbital, which has two times the density of states (in the simplest approximation) than the corresponding lower Hubbard band derived from the $d_{x^2-y^2}$ orbitals.

314

Crystal chemical investigations of $La_{2-x}Sr_xCuSnO_6$, with which the samples used for electrical characterization agree well, demonstrated that variations in the lattice parameters with strontium substitution are consistent with hole doping the d_z^2 orbital [8]. Contractions along the unit cell direction perpendicular to the copper-oxygen planes, in the direction of the d_z^2 orbital, were observed as the strontium concentration was increased. Furthermore, the persistence of antiferromagnetic order in the highly doped samples (suggesting that the $d_{x^2-y^2}$ states are relatively unperturbed by substitutions) and theoretical band calculations support this view of an increased density of states for holes owing to a prominent role of the d_z^2 orbital in the electronic structure of this material [8, 19].

Therefore, the extremely short bond length in La_2CuO_4 (1.89 Å) can be considered as an important structural element for retaining a favorable electronic structure of the copper oxygen sheets. Furthermore, the copper-tin material reminds us that the structural occurrence of two-dimensional copper oxygen networks is only a necessary prerequisite for superconductivity, and that other features of the inner architecture/internal chemistry are equally as important in determining the ultimate properties. The large tin-octahedra leave the copper-oxygen octahedra in tension, however n-type doping is not favored owing to the existence of coordinating ligands of copper on the d_z^2 axis, raising this energy to be in competition with the layered $d_{x^2-y^2}$ orbital which has its energy lowered owing to the increased bond lengths compared to known superconductors. That is, the decreased Jahn-Teller distortion reduces the $(d_{x^2-y^2}) - d_z^2$ energy difference.

While $La_{2-x}Sr_xCuSnO_6$ has two-dimensional features, the competition between the outer orbitals of copper leads to an electronic structure which does not favor superconductivity. Moreover, we can consider the dearth of materials which adopt this structure to be closely related to the subtle features of the electronic structure which play a role in its stabilization. Any method to obtain superconductivity in the layered La_2CuSnO_6 compound, via either chemical or synthetic procedures (i.e. high-pressure), should involve stabilizing the d_z^2 orbital (i.e., lowering its energy away from the Fermi level) and force doping (p- or n-type) in the $d_{x^2-y^2}$ orbital. A simple chemical approach to achieve this goal is to introduce oxygen vacancies in the coordination sphere of copper. Although this results in somewhat different structural motifs, we can begin to design materials where two-dimensional structural and electronic features are favored.

$Ln'Ln''AE_mCu_2Ti_mO_{5+3m}$ (Ln = lanthanide, Y; AE = alkaline earth, m = integer)

Introduction of oxygen vacancies in the coordination sphere of copper via aliovalent substitutions of the A-site cations leads us from the stoichiometric double perovskite La_2CuSnO_6 to the oxygen-deficient quadruple perovskite $La_2Ba_2Cu_2Sn_2O_{11}$ [26]. Reduction of the A-site total valence leads to hole doping, as discussed above, for low levels of doping in $La_{2-x}Sr_xCuSnO_6$ ($x \le 0.175$), but at the $x = 1.0$ level a concerted order of cations and anion vacancies results in the structure depicted in Figure 3(c) [26]. The oxygen vacancies align themselves in the d_z^2 orbital direction and are located between planes of two copper-oxygen square pyramids, leading to an electronically layered material. This structural motif is similar to that observed in oxygen-deficient perovskite-based superconductors, i.e. $YBa_2Cu_3O_{7-x}$ [27, 28] and $Y_{1-x}Ca_xSr_2Cu_2GaO_7$[29], as well as intergrowth superconductors, i.e., $Bi_{2.1}Sr_{1.9}CaCu_2O_{8\pm\delta}$ [30], $La_2CaCu_2O_6$ [31], $YBa_2Cu_4O_8$ [32]. Focusing on the first two materials in this list, which belong to the 'pure'-perovskite family of superconductors, it is clear that oxygen vacancies play an important structure-directing role in perovskite layered cuprates, as reviewed elsewhere[10]. It is clear that the occurrence of two-dimensional structures increases markedly upon the introduction of vacancies on the oxygen sublattice.

The quadruple perovskite $La_2Ba_2Cu_2Sn_2O_{11}$ belongs to a family of layered cuprates which have the general formula $(A_2Cu_2O_5)(ABO_{3-x})_m$, which highlights their pure-perovskite stoichiometry. Figure 3 illustrates the structures of the materials which have varying m values and $x = 0$, to which the $La_2Ba_2Cu_2Sn_2O_{11}$ material adopts the $m = 2$ structure. The $2 \le m \le 4$ structures are further discussed below, while it should be pointed out that $YBaCuFeO_5$ [33, 34]

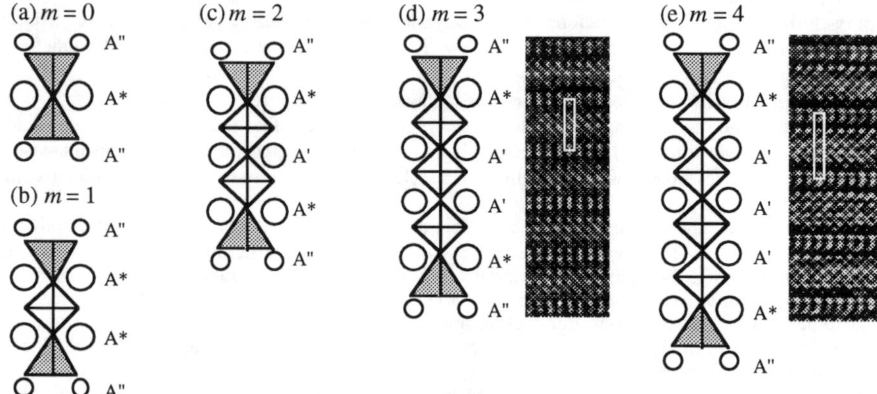

Figure 3. The structures of the c-axis-aligned perovskite series $Ln_2AE_mCu_2Ti_mO_{5+3m}$; (a) $m = 0$, (b) $m = 1$, (c) $m = 2$: $Ln_2Ba_2Cu_2Ti_2O_{11}$ (quadruple perovskite), (d) $m = 3$:$Ln_2Ba_2CaCu_2Ti_3O_{14}$ (quintuple perovskite) and (e) $m = 4$: $Ln_2Ba_2Ca_2Cu_2Ti_4O_{17}$ (sextuple perovskite). Shaded triangles stands for CuO_5 pyramids and open squares with cross are TiO_6 octahedra. The positions for A-cations are shown by open circles. Largest ions sitting between CuO_5 pyramids and TiO_6 octahedra, which are usually Ba^{2+}, are marked by A*. Y^{3+} or smaller lanthanide ions which tend to reside between two CuO_5 pyramids are marked by A'', and other ions which are located in the TiO_6-blocking layer are marked by A'. High resolution electron microscopy images of $Eu_2Ba_2CaCu_2Ti_3O_{14}$ ($m = 3$) and $Dy_2Ba_2Ca_2Cu_2Ti_4O_{17}$ ($m = 4$) taken along the [100] directions are also shown next to the each figure. Unit cells are outlined by white squares.

adopts the $m = 0$ structure, Figure 3(a), albeit with 1/2 of the copper replaced by iron, and the $LnAE_2Cu_2BO_8$ (B = Ta, Nb, Ru) [35-38] materials adopt the $m = 1$ structure.

Since the discovery of the quadruple perovskite structure for $La_2Ba_2Cu_2Sn_2O_{11}$ [26], several chemical approaches to modify this structure have been attempted. Substitution of the smaller Ti^{4+} ion for Sn^{4+} [39] has been achieved, which reduces the structural distortion/modulation in the BO_6-CuO_5 polyhedral skeleton. The substitution of smaller lanthanides ions (Nd^{3+} - Tb^{3+}) for La^{3+} [2, 39-42] stabilizes the titanium-substituted materials and decreases the in-plane Cu-O distance, and thereby achieves better Cu_{3d}-O_{2p} overlap essential to p-type high temperature superconductivity. Finally, selective substitution of the much smaller Y^{3+} ion [3], which tends to reside between the two CuO_2 planes, both stabilizes long-range order in specific materials and prevents oxygen intercalation (in the plane of the A'' site in Figure 3) which is considered to be detrimental to superconductivity [3, 4].

The similarity of the quadruple perovskite structure to the high temperature superconductor $YBa_2Cu_3O_7$ is a major reason why materials adopting this structure have attracted attention. In fact, the band structure of the quadruple perovskite structure has recently been shown to be quite similar to other cuprate superconductors [43]. Thus, the oxygen vacancies remove the contribution of the d_z^2 orbital from the Fermi level (which are important in La_2CuSnO_6) and the 'layered' copper $d_{x^2-y^2}$ - oxygen p manifold governs the electronic properties. The major difference between the quadruple perovskites and the 123 phases is in the structure of the blocking/conditioning layer. Whereas the former contains insulating TiO_6 double layers, which are rather "rigid", expand the CuO_2^{2-} network, and are intolerant of oxygen nonstoichiometry, the latter has the "chain-copper" layer which not only controls the hole concentration and T_c, but also compresses the CuO_2^{2-} network and leads to various structural transitions [12]. Investigation into the structural and internal chemical properties of the quadruple perovskites, and the related compounds (including $YBa_2Cu_3O_7$) is crucial for both understanding high-T_c superconductivity and comprehending the nature of "layering" in cuprates as well as transition metal oxides in general.

Recently we have reported the sextuple perovskite $Ln_2Ca_2Ba_2Cu_2Ti_4O_{17}$, whose structure is depicted in Figure 3(e), which forms with $Ln_2 = Pr_2$ - Dy_2 [6]. The high resolution electron microscopy image of $Dy_2Ca_2Ba_2Cu_2Ti_4O_{17}$ (Figure 3e) taken along the [100] directions clearly shows the c-axis array of four TiO_6 octahedra separated by two CuO_5 pyramids, consistent with the given chemical formula [6]. This is the $m = 4$ member in the new homologous series $Ln_2AE_mCu_2Ti_mO_{5+3m}$ [6]. In this structure, the largest A-cation Ba^{2+} is located between CuO_5 pyramids and TiO_6 octahedra (positions indicated by A*), assisting the layering of the structure in cooperation with the c-direction alignment of the Jahn-Teller axis (d_z^2 orbital) of the $Cu^{2+}O_5$ pyramids. The combination of the smaller lanthanide Dy^{3+} (Tb^{3+} is the smallest ion which can hold the quadruple perovskite structure, $m = 2$ [42]) and the smaller Ca^{2+} decreases of lattice constant to $a = 3.8442$Å for $Dy_2Ca_2Ba_2Cu_2Ti_4O_{17}$ in comparison to $a = 3.8769$Å for $Tb_2Ba_2Cu_2Ti_2O_{11}$. This makes in-plane Cu-O bond distances shorter and more amenable to hole-doped superconductivity.

The $m = 3$ member (the quintuple perovskite) illustrated in Figure 3(d), which has the formula $Ln_2AE_3Cu_2Ti_3O_{14}$, was first reported by Zhu et al. [44] for $Sm_2CaBa_2Cu_2Ti_3O_{14}$. Independent investigations in our lab (XRD and HR-TEM) and by Weller et al. (XRD and neutron diffraction)[45] on isovalent lanthanide substitutions in the $m = 3$ structure have led to the extension of the phase field for this structure. The high resolution electron microscopy image of $Eu_2CaBa_2Cu_2Ti_3O_{14}$, taken along the [100] directions, is shown in Figure 3(d). As was seen for the $m = 4$ material, Figure 3(e), the HR-TEM image clearly illustrates the c-axis array of cation alteration, in this case three TiO_6 octahedra separated by two CuO_5 pyramids, consistent with the chemical formula. Lanthanide substitutions lead to structural distortions depending on the size of the lanthanide in $Ln_2CaBa_2Cu_2Ti_3O_{14}$ ($Ln_2 = La_2$ - Dy_2), as discussed below.

The left part of Figure 4 shows the changes of the XRD profile of the $m = 3$ phase as the lanthanide cation is varied. $La_2CaBa_2Cu_2Ti_3O_{14}$ (bottom) shows only a few peaks owing to its cubic-perovskite-like structure. Actually, all four peaks in the figure can be indexed with a simple cubic cell; 100, 110, 111 and 200, respectively. However, large peak-widths and deviations of the peak-ratio from that of an ideal cubic perovskite (the 111 peak is smaller than expected) indicate that short range ordering is probable, as in the quadruple perovskite $La_2Ba_2Cu_2Ti_2O_{11}$ [5, 41]. $Nd_2CaBa_2Cu_2Ti_3O_{14}$ exhibits a similar pattern, but the peaks are sharper. Upon close inspection, the 200 peak reveals a shoulder on its left side, which is due to the emergence of the 0010 peak, and is indicative of the c-axis-ordering/layering associated with the $m = 3$ structure. The XRD patterns of both $Eu_2CaBa_2Cu_2Ti_3O_{14}$ and $Dy_2CaBa_2Cu_2Ti_3O_{14}$ show a clear splitting of the 0010 and 200 reflections, and all peaks can be indexed with the space group $P4/mmm$. A weak but clear peak is due to the 106 reflection, which does not belong to a cubic perovskite, around 36° for these compounds. This weak superstructure peak is another indication of the $m = 3$ layering, in addition to the 0010/200 peak-splitting. Weller et al. also observe similar structural trends [45]. For the largest lanthanide ion ($Ln_2 = La_2$), the structure is a cubic perovskite with A/B-cation disordering, while for $Ln_2 = Nd_2$, partial disordering on both the A- and B-cation sites occurs. With the middle-size lanthanide ions ($Ln_2 = Sm_2$ - Gd_2), however, the structure becomes tetragonal as a result of complete ordering of B-cations (Cu^{2+}/Ti^{4+}) and partial ordering of A-cations, in agreement with our results. With the smallest ions ($Ln_2 = Tb_2$ and Dy_2), the structure distorts slightly to an orthorhombic cell [45], but the extra peaks which result from the orthorhombic distortion are not detected by XRD. Moreover, Weller et al. have observed tilting of the TiO_6 octahedra in all cases [45].

Substitution of 1/2 of the lanthanide ions by the smaller yttrium cation increases the propensity for long-range order, as observed in the $m = 2$ materials, $Ln'Ln''Ba_2Cu_2Ti_2O_{11}$ [3]. The right part of Figure 4 shows XRD profiles for $LnYCaBa_2Cu_2Ti_3O_{14}$ (Ln = La, Nd, Eu, Dy). All the XRD patterns, including the Ln = La, show both a clear splitting of the 0010 and 200 peaks and the small superstructure related 106 peak. No evidence of other phases are seen in these diffraction patterns. Preliminary HR-TEM (not shown) observations on $LaYCaBa_2Cu_2Ti_3O_{14}$ reveal a phase-pure, well-ordered quintuple perovskite structure. Therefore, selective substitution of the lanthanide cations, based on the occurrence of distinct crystallographic sites (the A'' and A' sites in Figure 3)

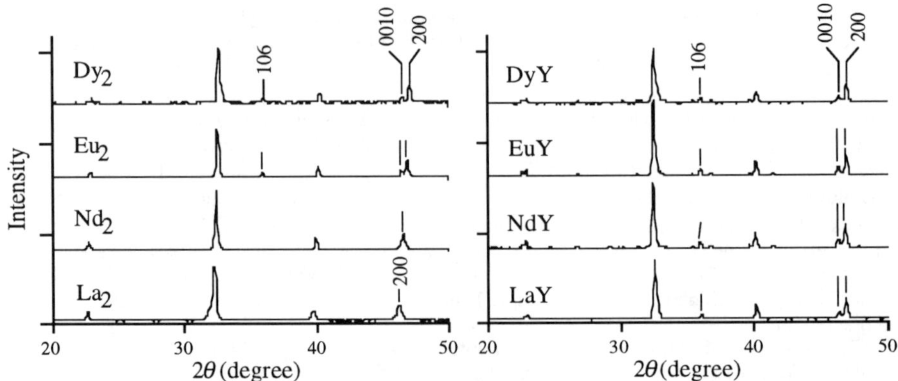

Figure 4. XRD (CuK$_\alpha$) profiles of the quintuple perovskite :Ln$_2$Ba$_2$CaCu$_2$Ti$_3$O$_{14}$ (m = 3). The left part shows ones with single lanthanide ion ; Ln$_2$ = La$_2$, Nd$_2$, Eu$_2$ and Dy$_2$, and the right part shows ones containing Y ; Ln$_2$ = LaY, NdY, EuY and DyY. The peaks which indicate the c-axis ordering, indexed by 0010 and 200, are marked in the figure (around 47°). The small peak around 36°, which is indexable as the 106, and also appears as a result of ordering, is also marked.

leads to an increase in the long-range order of these materials, as in the quadruple perovskites [3]. In the quadruple perovskites [3], yttrium substitution is only observed to be stable for Ln' = La, while in the m = 3 compounds, yttrium can be substituted in compounds having Ln' = La - Dy. Moreover, the stability of the m = 3 compounds having the lanthanide combinations of Dy$_2$ or DyY implies that the small rare earth dysprosium occupies a 12-coordinate site. This has been experimentally confirmed by Weller et. al. with EXAFS investigations on Dy$_2$CaBa$_2$Cu$_2$Ti$_3$O$_{14}$ [45]. Whereas the quadruple perovskite (m = 2) structure is not stable with dysprosium as the 12-coordinate (Ln') cation [3], the addition of CaTiO$_3$ units leads to an extension of the small rare-earth solubilities on both the Ln' and Ln'' sites.

Figure 5 shows the change of the lattice constant a with varying lanthanide ions and/or increasing m. Certainly, the lattice constant a within a fixed m series decreases with decreasing lanthanide-ion radius. And it is also clearly shown that a decreases with increasing m. The general formula Ln$_2$AE$_m$Cu$_2$Ti$_m$O$_{5+3m}$ can also be written as Ln$_2$Ba$_2$Cu$_2$Ti$_2$O$_{11}$(CaTiO$_3$)$_n$, where n = m - 2. This formula shows that the quintuple (n = 1) and sextuple (n = 2) perovskites are related by the addition of one or two CaTiO$_3$ units, respectively, to the quadruple perovskite (although some A-site mixing occurs). The average unit cell parameter of CaTiO$_3$, reduced to a cubic cell (the true CaTiO$_3$ crystallizes in a orthorhombic cell), is 3.8254Å, and is much shorter than that of the quadruple perovskite. Therefore, the addition of the smaller CaTiO$_3$ units results in a contraction of the cell. Conversely, the n = ∞ limit of this system is simply CaTiO$_3$, and by adding the larger Ba^{2+} and lanthanide ions, the cell expands. A competition exists between the large BaTiO$_3$-like regions and the small CaTiO$_3$ regions in the stabilization of the layered structures. BaTiO$_3$ has an average unit cell parameter of 4.0023Å (the true BaTiO$_3$ crystallizes in a tetragonal cell), and the largest quadruple perovskites, La$_2$Ba$_2$Cu$_2$Sn$_2$O$_{11}$ and La$_2$Ba$_2$Cu$_2$Ti$_2$O$_{11}$, which have no CaTiO$_3$ like regions have the lattice parameters a = 3.9868Å and 3.9348Å, respectively [26, 46]. A possible reason why the La$_2$ phase for m = 4 does not exist is that the Ca concentration (two per unit cell) may be too high to retain the site-mixed structure with the larger cations.

Figure 5 also shows the in-plane lattice constants, a, of the yttrium-substituted compounds. The contraction of the cell from the quadruple (m = 2) LaYBa$_2$Cu$_2$Ti$_2$O$_{11}$ to the quintuple (m = 3) LaYCaBa$_2$Cu$_2$Ti$_3$O$_{14}$ is steeper than that in the non-Y system. Moreover, the a lattice parameter of LaYCaBa$_2$Cu$_2$Ti$_3$O$_{14}$ (3.8675Å) is even shorter than that of the smallest quadruple perovskite Tb$_2$Ba$_2$Cu$_2$Ti$_2$O$_{11}$ (3.8769Å [42]), despite that the former contains the largest rare earth. Thus, this compound has great potential as a hole-doped superconductor. The lattice constant of the compound with smaller lanthanide, $DyYCaBa_2Cu_2Ti_3O_{14}$, remains almost the same as that of non-Y equivalent, $Dy_2CaBa_2Cu_2Ti_3O_{14}$, because the ionic radius of Dy (8 C.N. = 1.027Å [47]) is comparable with that of Y (8 C.N. = 1.019Å [47]). Thus, changes resulting from the Y substitution are more dramatic in those compounds with larger lanthanide ions.

There are several characteristics of the inner architecture and internal chemistry across the 'pure'-perovskite homologous series illustrated in Figure 3. The inter-plane Cu-Cu distance, or the size of the blocking/conditioning layer, can be changed by the number of the BO_6 octahedra in this layer. In the general formula for the titanate compounds, $Ln'Ln''AE_mCu_2Ti_mO_{5+3m}$, m corresponds to the number of the TiO_6 octahedra. In all structures illustrated in Figure 3, the large Ba^{2+} cations are located between the CuO_5 pyramids and BO_6 octahedra, stabilizing planes of the Jahn-Teller distorted $Cu^{2+}O_5$

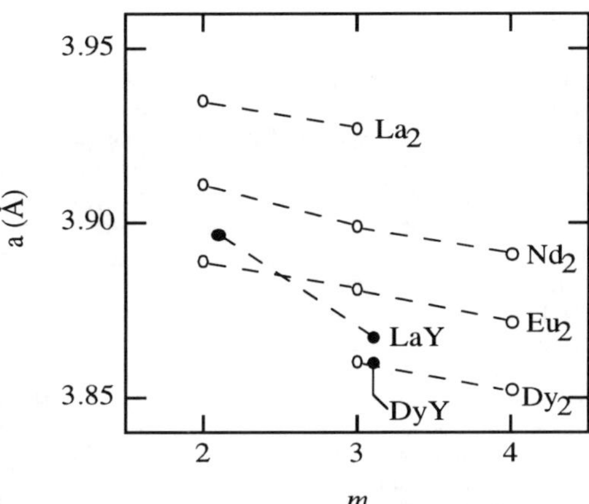

Figure 5. Changes of the lattice constant a with increasing number of TiO_6 octahedra in the blocking layer (m in the formula $Ln_2AE_mCu_2Ti_mO_{5+3m}$). The open circles are for ones with single lanthanide ion; Ln_2 = La_2, Nd_2, Eu_2 and Dy_2, and the closed circles are for ones containing Y ; Ln_2 = LaY and DyY. The lines between circles are only the guide for eyes.

pyramids. This mutual association is the essential force for layering in this series. The smallest A-cation (i.e. Y^{3+} for $LaYBa_2Cu_2Ti_2O_{11}$ [3], Dy^{3+} for $NdDyBa_2Cu_2Ti_2O_{11}$ [3]) tends to reside between two CuO_5 pyramids on the A'' site, while the others are disordered over the A' positions.

Another characteristic feature of this series is a great tolerance for lanthanide substitutions. As already mentioned, the m = 2, 3 and 4 phases form for Ln_2 = La_2 - Tb_2, La_2 - Dy_2 and Pr_2 - Dy_2, respectively. This is a rather common feature in other 'pure'-perovskite-related oxide superconducting systems, such as $LnBa_2Cu_3O_7$ [12] and $LnSr_2Cu_2GaO_7$ [29], which are related to the m = 1 phases but have oxygen-deficient blocking/conditioning layers, i.e., x = 2. The latter two compounds both form with all the rare earth ions. Structural elasticity of the blocking/conditioning layer polyhedra, smaller in-plane lattice parameters, and the single A'' site for lanthanide occupation contribute to this tolerance for substitution. The rigidness of the BO_6 octahedra in the new series (Figure 3) and relatively large in-plane bond lengths are considered to be the reasons why smaller rare earth cations do not support the m = 2 - 4 structures. In fact, the m = 1 compounds, which are related to the $LnBa_2Cu_3O_7$ and $LnSr_2Cu_2GaO_7$ compounds but have oxygen-stoichiometric perovskite blocking/conditioning layers, do not show tolerance to lanthanide variations; $LnBa_2Cu_2TaO_8$ forms only with Ln = La [35] and $LnSr_2Cu_2BO_8$ with Ln = Pr - Gd, B = Nb, Ta, Ru [37, 38] .

Materials with the stuffed blocking/conditioning layer motif, i.e., materials adopting the structures depicted in Figure 3, have not to date been observed to support superconductivity. A similar behavior was observed for La_2CuSnO_6, however its behavior can now be understood on the basis of both theoretical studies of the band structure and the experimental transport investigations described previously. The absence of superconductivity in the materials depicted in Figure 3, however, remains somewhat anomalous. In fact band structure calculations for $LaBa_2Cu_2NbO_8$ [48] and $LaYBa_2Cu_2Ti_2O_{11}$ [43] demonstrate that these compounds do have similar band structures to known high-T_c superconducting cuprates. We have been investigating the relationships between the transport properties and the inner architecture/internal chemistry to understand the reason behind the absence of superconductivity and to design these materials to support superconductivity.

The high temperature electrical properties of $La_2Ba_2Cu_2Sn_2O_{11}$, $Eu_2Ba_2Cu_2Ti_2O_{11}$, and $YBa_2Cu_3O_{7-x}$ (at T ≈ 650 and 750 °C) are plotted in Figure 6 [5, 20]. At each isotherm , the data points correspond to various oxygen partial pressures. Large portions of the Jonker curves are traced out as a function of the oxygen partial pressure for each material. $YBa_2Cu_3O_{7-x}$ traces out the largest portion owing to the large range of x values accessible in this material. The oxygen variations are associated with the oxygen content in the blocking/conditioning layer and are located in the coordination sphere of the 'chain'-site copper, exclusively. In both $La_2Ba_2Cu_2Sn_2O_{11}$ and $Eu_2Ba_2Cu_2Ti_2O_{11}$, smaller portions of the curve are traversed as the oxygen partial pressure is varied, and more importantly, the excess oxygen intercalates between the copper square pyramids (in the plane of the Ln'' cation) and is deleterious to superconductivity [5, 43].

Interestingly, the curves are related to the in-plane Cu-O bond length. The shortest bond lengths are for $YBa_2Cu_3O_{7-x}$ ($\approx$ 1.94 Å) with experimental values far down the extrinsic, p-type leg, near the high conductivity intercept. This region is associated with materials that have appropriate carrier concentrations for superconductivity, as seen for $La_{2-x}Ba_xCuO_4$ in Figure 2(a). The titanate quadruple perovskite, with a bond length near 1.96 Å [3], exists on the p-type extrinsic leg for high oxygen partial pressures, with a transition into the intrinsic carrier region (the back-side of the curve) as the oxygen pressure is reduced or the temperature is increased. The tin-based material has the longest bond lengths, $\approx$ 2.00 Å, and exists entirely on the intrinsic region of the curve, and displays a p-to-n transition as a function oxygen partial

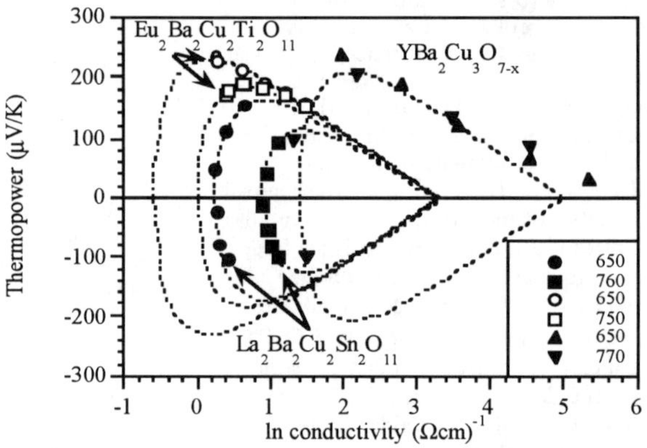

Figure 6. Thermopower plotted as a function of the logarithm of the conductivity for $La_2Ba_2Cu_2Ti_2O_{11}$, $Eu_2Ba_2Cu_2Ti_2O_{11}$, and $YBa_2Cu_3O_{7-x}$ at T $\approx$ 650 and 750 °C. Points for each isotherm correspond to variations in the oxygen partial pressure (10^{-5}, 10^{-4}, 10^{-3}, 10^{-2}, 10^{-1}, 10^0 atm) increasing clockwise around the curves. Dashed lines are best fit lines to two band conduction model. From Refs. [5, 20]

pressure. This intrinsic behavior suggests that the equilibrium bond-length, where neither electron or hole formation is favored structurally, is about 2.00 Å. Decreases in the bond distances lead to a structural driving force for hole formation, owing to the compressive stress on the antibonding copper-oxygen bond network [13]. Thus materials having smaller bond lengths favor increased carrier concentrations (oxidation). The empirical upper limit of the in-plane bond distance in cuprate superconductors is closely related to the ability of the copper oxygen sheets to support an appropriate amount of charge carriers. Recently, we demonstrated both metallic behavior (bond lengths of $\approx$ 1.95 Å) and chemical control over the structural location of oxygen defects for the m = 2 compounds, $Ln'Ln''Ba_2Cu_2Ti_2O_{11}$, by selective substitution of cations over the crystallographically distinct sites [4].

Of the structures shown in Figure 3 ($1 \leq m \leq 4$), only the new m = 3 an 4 phases have in-plane bond lengths (or lattice parameters as shown in Figure 5) below that of observed superconductors. However, long-range order associated with the structural integrity of the copper-oxygen sheets and the exclusion of excess oxygen in the plane between the copper oxygen sheets will also need to be chemically controlled if superconductivity is to be observed. Figure 4 demonstrates that the preference of yttrium to occupy the Ln'' site results in the stabilization of long range order, even in disordered parent structures. Exclusion of oxygen in this plane has recently been demonstrated in the m = 2 materials when yttrium occupied this site [4]. Therefore, the Ln'YCaBa$_2$Cu$_2$Ti$_3$O$_{14}$ and Ln'YCa$_2$Ba$_2$Cu$_2$Ti$_4$O$_{17}$ materials have a high potential for supporting superconductivity. Doping studies are currently being pursued.

CONCLUSIONS

Layered copper oxides derive their electronic properties from the two-dimensional copper-oxygen orbital interactions. Out-of-plane ligand interactions cannot be ignored, however, as demonstrated in the sole oxygen-stoichiometric 'pure' perovskite, La$_2$CuSnO$_6$. The out of plane copper-oxygen orbital energy is nearer to the Fermi energy than that of the in-plane orbitals, leading to significant differences in the crystal chemistry and physical properties upon hole-doping. In the short bond-length material, La$_2$CuO$_4$, the in-plane orbital interactions are at the highest occupied energies and lead to superconductivity. Introduction of oxygen vacancies produces a similar outcome, as the removal of a ligand from the antibonding interaction lowers the energy of the out-of-plane states. A series of materials can be formed with a single plane of ordered oxygen vacancies between double square pyramids of copper, and a varying number of oxygen-stoichiometric perovskite blocks interleaved, according to the formula, $(Ln''$BaCu$_2$O$_5)(ABO_3)_m$, and as illustrated in Figure 3. The oxygen-stoichiometric perovskite blocks keep the copper oxygen sheets at what appear to be a near equilibrium distance for this motif of ≈ 2.00 Å in La$_2$Ba$_2$Cu$_2$Sn$_2$O$_{11}$, as observed by the intrinsic nature of the electrical properties. The blocking/conditioning layers decrease in size as the value of m is increased when $B =$ Ti. This should lead to a structural driving force for hole incorporation. Long range order is stabilized by site-selective substitution in the titanates, and materials with the small yttrium cation between the copper-oxygen sheets have been attained for the $m = 3$, $B =$ Ti phases. These materials have the necessary structural characteristics required for superconductivity, including in-plane bond lengths smaller than those of YBa$_2$Cu$_3$O$_{7-x}$, and therefore are ideal prospective superconductors.

ACKNOWLEDGMENTS

This work was supported by the National Science Foundation (Award No. DMR-91-20000) through the Science and Technology Center for Superconductivity and the Summer Undergraduate Research Program (JRM). This work also made use of MRL Central Facilities supported by the National Science Foundation, at the Materials Research Center of Northwestern University (Award No. DMR-9120521). One of the authors (KO) is partly supported by a Research Fellowship of the Japan Society for the Promotion of the Science for Young Scientists.

REFERENCES

1. P. A. Salvador, T. O. Mason, M. E. Hagerman and K. R. Poeppelmeier, in <u>Chemistry of Advanced Materials: A New Discipline</u>, edited by L.V. Interrante and M. Hampden-Smith (VCH, In Press).
2. K. B. Greenwood, M. T. Anderson, K. R. Poeppelmeier, D. L. Novikov, A. J. Freeman, B. Dabrowski, S. A. Gramsch and J. K. Burdett, Physica C **235-240**, p. 349 (1994).
3. K. B. Greenwood, G. M. Sarjeant, K. R. Poeppelmeier, P. A. Salvador, T. O. Mason, B. Dabrowski, K. Rogacki and Z. Chen, Chem. Mater. **7**, p. 1355 (1995).
4. P. A. Salvador, T. O. Mason, K. Otzschi, K. B. Greenwood, K. R. Poeppelmeier and B. Dabrowski, J. Am. Chem. Soc. (Submitted).
5. P. A. Salvador, L. Shen, T. O. Mason, K. B. Greenwood and K. R. Poeppelmeier, J. Solid State Chem. **119**, p. 80 (1995).
6. K. D. Otzschi, K. R. Poeppelmeier, P. A. Salvador, T. O. Mason, H. Zhang and L. D. Marks, J. Am. Chem. Soc. **118**, p. 8951 (1996).
7. M. T. Anderson and K. R. Poeppelmeier, Chem. Mater. **3**, p. 476 (1991).
8. M. T. Anderson, K. R. Poeppelmeier, S. A. Gramsch and J. K. Burdett, J. Solid State Chem. **102**, p. 164 (1993).
9. A. Trestman-Matts, S. E. Dorris and T. O. Mason, J. Am. Ceram. Soc. **66**, p. 589 (1983).
10. M. T. Anderson, J. T. Vaughey and K. R. Poeppelmeier, Chem. Mater. **5**, p. 151 (1993).
11. B. Raveau, C. Michel and M. Hervieu, J. Solid State Chem. **88**, p. 140 (1990).
12. B. Raveau, C. Michel, M. Hervieu and D. Groult, <u>Crystal Chemistry of High-T$_c$ Superconducting Copper Oxides</u> (Springer-Verlag, Berlin, 1991).

13. J. B. Goodenough and A. Manthiram, J. Solid State Chem. **88**, p. 115 (1990).
14. M. T. Anderson and K. R. Poeppelmeier, Appl. Supercon. **1**, p. 493 (1993).
15. M. T. Anderson, K. B. Greenwood, G. A. Taylor and K. R. Poeppelmeier, Prog. Solid St. Chem. **22**, p. 197 (1993).
16. N. Yamada and M. Ido, Physica C **203**, p. 240 (1992).
17. J. M. Longo and P. M. Raccah, J. Solid Stae Chem. **6**, p. 526 (1973).
18. J. D. Jorgensen, H.-B. Schüttler, D. G. Hinks, D. W. Capone, H. K. Zhang and M. B. Brodsky, Phys. Rev. Lett. **58**, p. 1024 (1987).
19. D. L. Novikov, A. J. Freeman, K. R. Poeppelmeier and V. P. Zhukov, Physica C **252**, p. 7 (1995).
20. M.-Y. Su, C. E. Elsbernd and T. O. Mason, J. Am. Ceram. Soc. **73**, p. 415 (1990).
21. G. W. Tomlins, N.-L. Jeon, T. O. Mason, D. A. Groenke, J. T. Vaughey and K. R. Poeppelmeier, J. Solid State Chem. **109**, p. 338 (1994).
22. G. H. Jonker, Philips Res. Repts **23**, p. 131 (1968).
23. M.-Y. Su, K. Sujata and T. O. Mason, in <u>Ceramics Superconductors II</u>, edited by M.F. Yan (The American Ceramic Society, Westerville, Ohio, 1988), p. 99.
24. M.-Y. Su, C. E. Elsbernd and T. O. Mason, Physica C **160**, p. 114 (1989).
25. B.-S. Hong and T. O. Mason, in <u>Superconductivity and Ceramic Superconductors II</u>, edited by K.M. Nair, U. Balachandran, Y.-M. Chiang and A.S. Bhalla (American Ceramic Society, Westerville OH, 1991), p. 95.
26. M. T. Anderson, K. R. Poeppelmeier, J. P. Zhang, H.-J. Fan and L. D. Marks, Chem. Mater. **4**, p. 1305 (1992).
27. M. K. Wu, J. R. Ashburn, C. J. Torng, P. H. Hor, R. L. Meng, L. Gao, Z. J. Huang, Y. Q. Wang and C. W. Chu, Phys. Rev. Lett. **58**, p. 908 (1987).
28. M. A. Beno, L. Soderholm, I. D. W. Capone, D. G. Hinks, J. D. Jorgensen, J. D. Grace, I. K. Schuller, C. U. Segre and K. Zhang, Appl. Phys. Lett. **51**, p. 57 (1987).
29. J. T. Vaughey, J. P. Thiel, E. F. Hasty, D. A. Groenke, C. L. Stern, K. R. Poeppelmeier, B. Dabrowski, D. G. Hinks and A. W. Mitchell, Chem. Mater. **3**, p. 935 (1991).
30. H. Maeda, Y. Tanaka, M. Fukutomi and T. Asano, Jp. J. Appl. Phys. **27**, p. L209 (1988).
31. R. J. Cava, B. Batlogg, R. B. v. Dover, J. J. Krajewski, J. V. Waszczak, R. M. Fleming, W. F. P. Jr, L. W. R. Jr, P. Marsh, A. C. W. P. James and L. F. Schneemeyer, Nature **345**, p. 602 (1990).
32. P. Fischer, J. Karpinski, E. Kaldis, E. Jilek and S. Rusiecki, Solid State Comm. **69**, p. 531 (1989).
33. L. Er-Rakho, C. Michel, P. Lacorre and B. Raveau, J. Solid State Chem. **73**, p. 531 (1988).
34. J. T. Vaughey and K. R. Poeppelmeier, in <u>Chemistry of Electronic Ceramic Materials</u>, edited by P.K. Davies and R.S. Roth (National Institute of Standards and Technology, Washington D.C., 1991), p. 419.
35. N. Murayama, E. Sudo, K. Kani, A. Tsuzuki, S. Kawakami, M. Awano and Y. Torii, Jpn. J. Appl. Phys. **27**, p. L1623 (1988).
36. C. Greaves and P. R. Slater, Physica C **161**, p. 245 (1989).
37. M. Vybornov, W. Perthold, H. Michor, T. Holubar, G. Hilscher, P. Rogl, P. Fischer and M. Divis, Phys. Rev. B **52**, p. 1389 (1995).
38. B. Hellebrand, X. Z. Wang and P. L. Steger, J. Solid State Chem. **110**, p. 32 (1994).
39. A. Gormezano and M. T. Weller, J. Mater. Chem. **3**, p. 771 (1993).
40. A. Gormezano and M. T. Weller, J. Mater. Chem. **3**, p. 979 (1993).
41. M. R. Palacín, A. Fuertes, N. Casañ-Pastor and P. Gómez-Romero, Adv. Mater. **6**, p. 54 (1994).
42. M. R. Palacín, F. Krumeich, M. T. Caldés and P. Gómez-Romero, J. Solid State Chem. **117**, p. 213 (1995).
43. D. L. Novikov, A. J. Freeman and K. R. Poeppelmeier, Phys. Rev. B **53**, p. 9448 (1996).
44. W. J. Zhu, Y. Z. Huang, T. S. Ning and Z. X. Zhao, Mat. Res. Bull. **30**, p. 243 (1995).
45. M. J. Pack, A. Gormezano and M. T. Weller, Personal Communication.
46. P. Gómez-Romero, M. R. Palacín and J. Rodriguez-Carvajal, Chem. Mater. **6**, p. 2118 (1994).
47. R. D. Shannon, Acta Cryst. **A32**, p. 751 (1976).
48. L. F. Mattheiss, Phys. Rev. B **45**, p. 2442 (1992).

NONSTOICHIOMETRY AND DOPING OF ZINC OXIDE

ARTHUR W. SLEIGHT *, RUIPING WANG **
*Department of Chemistry, Oregon State University, Corvallis, OR 97331-4003,
sleighta@ccmail.orst.edu
**Applied Materials, Applied Materials, 3100 Bowers Ave/Mail Stop 1125, Sunnyvale, CA
95054

ABSTRACT

Contrary to the commonly accepted model, we conclude that the zinc interstitials in $Zn_{1+x}O$
are not ionized. Instead, the electron pair associated with the interstitial zinc is trapped in a
metal–metal bond. The increase in conductivity on initial reduction of zinc oxide is associated
with impurities such as silicon which are compensated by extra oxygen unless removed by
reduction. Zinc oxide powders and thin films were prepared with the dopants B, Al, Ga, In, and
Ge. By using high temperatures and highly reducing conditions, conductivities 1000 times higher
than previously reported for powders were obtained for some of our powders.

INTRODUCTION

Zinc oxide is an inexpensive, nontoxic material with numerous applications. Despite
hundreds of publications on this material, it remains poorly understood. Applications for zinc
oxide include catalysts, UV blocking in cosmetics, phosphors, fungicides, varistors, and surface
acoustic wave guide devices. Our primary interest is in zinc oxide as a transparent conductor.
Doping of zinc oxide with fluorine or higher valent cations can give high electrical conductivity
with very low optical absorption in the visible region of the spectrum.

We have prepared nonstoichiometric and doped zinc oxide as powders and as thin films.
Characterization has included electrical resistivity as a function of temperature, Hall data at 77
and 298 K, chemical titrations, optical absorption, EPR, NMR, X-ray diffraction, electron
microscopy, electron microprobe chemical analyses, perturbed angular correlation, and iron
Mössbauer of Fe doped material. This paper summarizes both our published [1–5] and
previously unpublished results.

NONSTOICHIOMETRY IN ZINC OXIDE

Zinc oxide which has been annealed in air is a white insulator. When zinc oxide is annealed
under reducing conditions, such as in the presence of zinc metal vapors, it becomes electrically
conducting and red in color. There is a widespread belief that the changes in conductivity and
color are directly related, but the relationship has never been established. Reduced zinc oxide is
frequently referred to as a textbook example of a metal oxide with excess metal present as
interstitials [6]. It is generally believed that these interstitial zinc atoms are singly or doubly
ionized, giving rise to significant electrical conductivity. We agree that the excess zinc in $Zn_{1+x}O$
is likely present at interstitial sites of the zinc oxide lattice, but we believe that these interstitials
are not ionized.

Mat. Res. Soc. Symp. Proc. Vol. 453 © 1997 Materials Research Society

Figure 1 shows how the conductivity and color of zinc oxide change as it becomes reduced. Despite many publications on the nonstoichiometry of zinc oxide, the information in Figure 1 has never before been presented. This curve is partly schematic because some of the details depend highly on the purity of the zinc oxide. There is a consensus that the upper limit of x in $Zn_{1+x}O$ is about 0.0003. At that composition, zinc oxide appears red or yellow depending on particle size. For x equal to 0.0000, zinc oxide is a white insulator.

As x begins to increase from 0.0000, there is a very rapid increase in electrical conductivity. During this dramatic rise in conductivity, zinc oxide does not become red in color. Instead it adopts the expected colors on introduction of conduction electrons. The color change is first white to blue and then to gray. Only after the conductivity rise is complete does further reduction result in a red color. The green intermediate color is simply due to a combination of red and blue absorption.

We believe that the initial rise in conductivity of zinc oxide on reduction is not related to nonstoichiometry of pure zinc oxide. Instead, it is directly related to impurities in zinc oxide. It is a sad commentary on the vast literature on zinc oxide that few papers mention the purity of their material. Electrons in the conduction band of zinc oxide have high mobility. Small concentrations of such electrons dramatically enhance conductivity; something which we are more familiar with in classical semiconductors such as silicon.

A common impurity in zinc oxide is silicon. For the conducting situation, we may assume doping as $Zn_{1-x}Si_xO$ where two electrons may be donated to the conduction band for every silicon atom substituting for zinc at a normal zinc lattice site. However, when $Zn_{1-x}Si_xO$ is annealed in air, it oxidizes and becomes $Zn_{1-x}Si_xO_{1+x}$. The conduction electrons have been withdrawn, and $Zn_{1-x}Si_xO_{1+x}$ is a white insulator. The effect of the Si doping is lost due to compensation by extra oxygen. However, when zinc oxide is heated under reducing conditions, such as vacuum, the extra oxygen is withdrawn and the conduction electrons reappear due to silicon impurity. Other common impurities, such as aluminum, have an analogous effect. Further reduction of zinc oxide, after the extra oxygen is removed, leads ultimately to the formation of $Zn_{1+x}O$ and the red color.

Most workers have concluded that the defect mechanism for nonstiochiometry in zinc-rich zinc oxide is zinc interstitials. Although there is no firm proof of this model, as opposed to the oxygen vacancy model, we tend to accept it. When zinc metal is dissolved in zinc chloride and

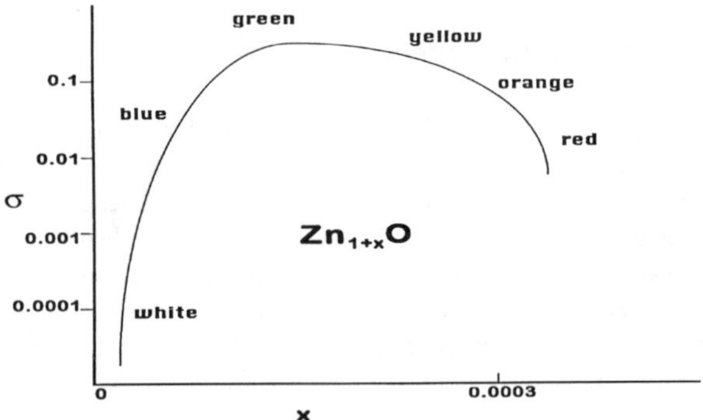

Figure 1. Electrical conductivity and color of zinc oxide on reduction.

quenched, the resulting glass is red [7]. Raman studies on this glass suggest the presence of $(Zn_2)^{2+}$ dimers. These are analogous to the better known $(Hg_2)^{2+}$ and $(Cd_2)^{2+}$ dimers. The zinc oxide lattice can be described as an array of hexagonally close packed oxygens with zinc atoms occupying one-half of the tetrahedra interstitial sites in an ordered manner. An interstitial zinc atom might then occupy one of the tetrahedral sites normally vacant. This would give a Zn–Zn distance of 1.5 Å before lattice relaxation. This distance is much too short, even if a metal–metal bond formed between the lattice Zn and the interstitial Zn. Placing the interstitial Zn in an octahedral site leads to Zn–Zn distances of 2.28 Å. Such a distance is reasonable for clusters, considering that Hg–Hg distances in various $(Hg_2)^{2+}$ clusters range from 2.4 to 2.9 Å.

Our model of zinc oxide is shown schematically in Figure 2. The levels just below the bottom of the conduction band are due to impurities such as Si and Al. These states are empty if zinc oxide is annealed in air. They are occupied if zinc oxide is slightly reduced. These states are so close to the conduction band that electrons in them are easily promoted to the conduction band, resulting in significant conductivity. Low energy excitations of these electrons give rise to absorption in the infrared which tails into the visible, giving a blue or gray color. Impurities at the 10^{-5} level could easily account for the highest conductivity shown in Figure 1. The state just above the valence band (Figure 2) is created during severe reduction of zinc oxide and is due to interstitial zinc. The electrons associated with this zinc interstitial are not ionized. Instead, they are in a deep trap such as a $(Zn_2)^{2+}$ cluster. The red color may be explained as due to an excitation from this trap to the conduction band.

Additional evidence for our model for zinc oxide might come from studying highly pure $Zn_{1+x}O$ as a function of x. In this case, we predict reduction would produce the red color with very little increase in conductivity. However, preparing $Zn_{1+x}O$ samples with high x and high purity is an extreme challenge. Zinc metal and zinc oxide can be obtained in sufficient purity for such studies. However, red zinc oxide has only been prepared by heating zinc oxide under highly reducing conditions at high temperatures (1000 to 1200°C). At such temperatures, zinc oxide is highly volatile, significantly dissociated, and easily doped by its container. Silica containers are convenient but give significant contamination as $Zn_{1-x}Si_xO$. Likewise, alumina containers cannot be used without risking $Zn_{1-x}Al_xO$ doping. Metals containers are also problematic because of reaction with the zinc metal vapors.

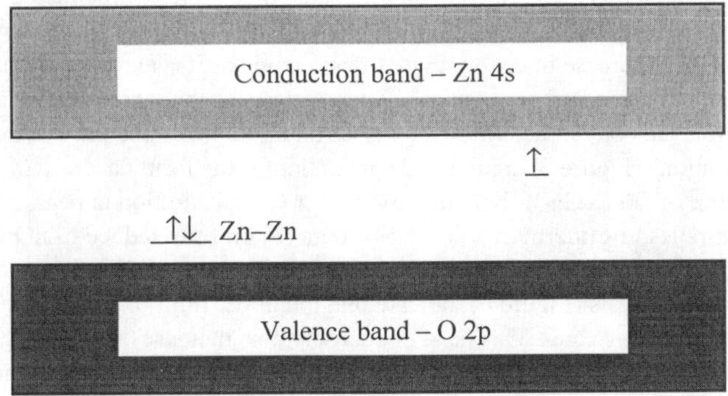

Figure 2. Schematic band structure of zinc oxide. Upper states in gap are due to impurities such as Si and Al; the lower state is due to a Zn interstitial.

DOPED ZINC OXIDE

Doping zinc oxide with fluorine ($ZnO_{1-x}F_x$) or higher valent cations can give good n-type electrical conductivity. Attempts to dope zinc oxide p-type have failed. Such attempts (e.g., Li doping) compensate for the impurities in zinc oxide which can given n-type conductivity. Thus, attempts to produce p-type conductivity in zinc oxide actually make it a better insulator.

Powders

Our interest in doped zinc oxide has been primarily for powders that would be added to polymers, paper, etc. to impart sufficient conductivity to avoid build-up of static electricity. Previous to our work, conductivities achieved in zinc oxide were disappointing in that the conductivities were much lower than in doped tin oxide powders and zinc oxide films. In fact, the room temperature conductivities of doped zinc oxide powders were so low that the only conductivity values reported were for temperatures well above room temperature, for example, 0.1 ohm^{-1} cm^{-1} for ZnO:Al at 400°C. We have now achieved conductivities in doped zinc oxide powders as high as 300 ohm^{-1} cm^{-1} at room temperature [1].

High conductivity in zinc oxide powders was obtained through synthesis at high temperatures and highly reducing conditions. The temperature range used for producing highly conducting materials was 1000 to 1200°C. The room temperature conductivity of the powders normally increased with synthesis temperature over that temperature range. Temperatures higher than 1200°C are not practical due to the volatility of zinc oxide. The electrical resistivity of doped zinc oxide powders rises dramatically if synthesis temperatures less than 1000°C are employed. We used the most reducing conditions possible, i.e., the oxide was generally in equilibrium during synthesis with metal vapors of either zinc or the dopant.

Powders were prepared in sealed evacuated silica tubes. The reactants were generally high purity zinc oxide and high purity dopants as elements. The dopant-to-zinc ratio was increased in a series of experiments for each dopant until no further change in the doped zinc oxide was detected. This method worked well for the dopants B, Ga, In, and Ge. This method was not successful when the dopant was Al; thus, another method was used. A nitrate solution of zinc and aluminum was evaporated to dryness and then heated at 500°C for 10 hours. This zinc-aluminum oxide powder was then sealed in an evacuated tube either with zinc metal powder or zirconium metal foil. Both reducing agents gave products with essentially the same properties.

Figure 3 shows the increase in conductivity of zinc oxide as it is annealed with various quantities of elemental Ga. Changes in unit-cell parameters for the same samples are shown in Figure 4. The actual Ga content determined by electron microprobe and the electron doping level determined by titrations (Figure 5) are in good agreement for the formula $Zn_{1-x}Ga_xO$ assuming complete ionization of Ga to Ga^{3+}. No other model for Ga substitution appears plausible for our highly reduced samples. Furthermore, ^{69}Ga NMR data [8] on fully reduced ZnO:Ga samples confirm the formula $Zn_{1-x}Ga_xO$ with Ga fully ionized. Both X-ray diffraction results and electron microprobe analysis (Figure 6) indicate that the upper limit of Ga substitution in our most reducing conditions is about 3% Ga. Conductivity also increases to 3% Ga level, an effect which is more apparent if the conductivity is plotted on a linear scale instead of the log scale used in Figure 3. When ZnO:Ga samples are prepared under more oxidizing conditions, higher Ga

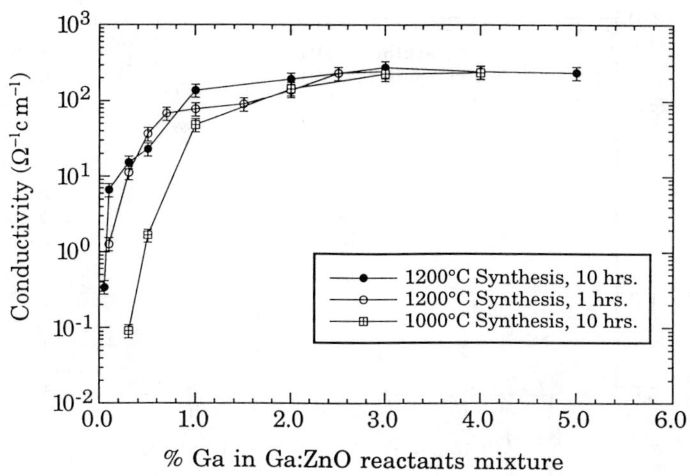

Figure 3. The conductivity of Ga-doped ZnO samples at two different synthesis temperatures.

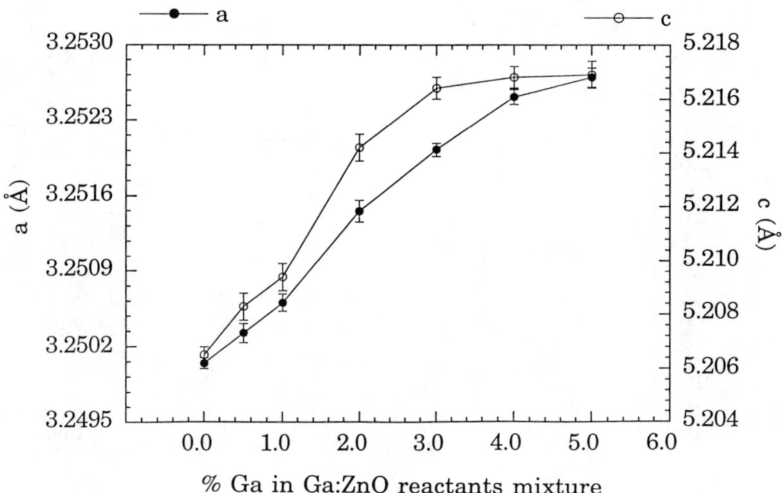

Figure 4. Cell dimensions of Ga-doped ZnO as a function of the percent Ga in the reactant mixture.

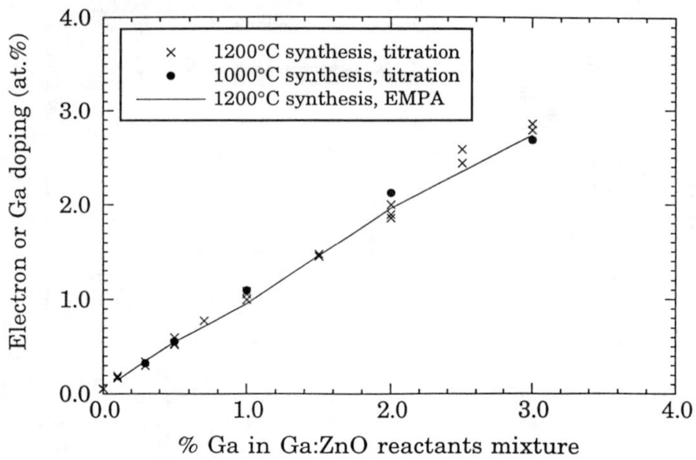

Figure 5. The apparent electron doping levels of Ga-doped ZnO obtained from iodimetric titration and the Ga-doping concentration analyzed by EMPA.

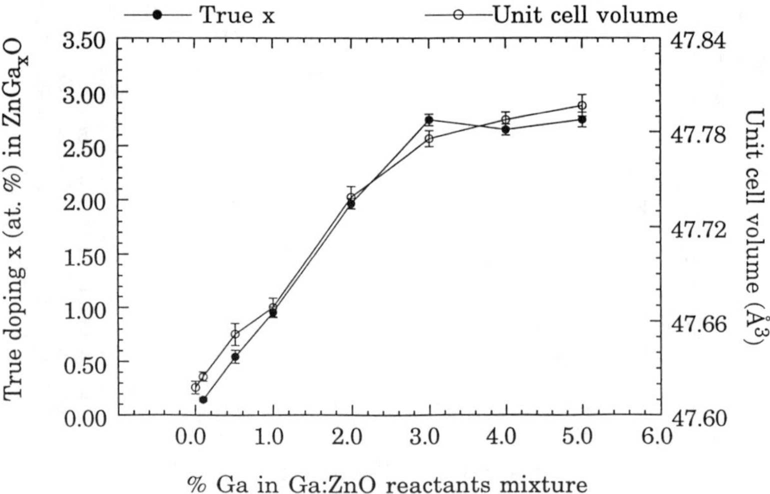

Figure 6. The actual doping level and the unit cell volume of Ga-doped ZnO obtained by increasing the percent Ga in the synthesis.

Table I. Dopant concentration (x) and electrical properties at 25°C of doped zinc oxide.

Dopant	Powders			Thin Films				
	x	x′	σ	σ	x′	x	mobility	carriers
B	1.0	1.0	20	–	–	–	–	–
Al	2.5	2.0	90	1600	2.6	8.0	22	3.4
Ga	3.0	1.0	300	1800	4.0	10.0	20	5.5
In	0.5	0.5	40	1100	4.2	8.0	12	5.7
Ge	0.3	0.3	80	1100	3.5	10.0	15	4.8

x is the highest percentage of dopant achieved under highly reducing conditions.

x′ is the dopant concentration for which the conductivity is the highest.

σ is the conductivity in $ohm^{-1}\ cm^{-1}$, mobility units cm^2/Vs, and carrier units $\times\ 10^{20}/cm^3$.

doping levels are obtained, but conductivities are lower [1]. Similar results were obtained with the dopant B, Al, In, and Ge (Table I). We are not aware of any previous studies of Ge doping of zinc oxide, but Ge is, in fact, a very effective dopant even though the doping level is very low in powders.

<u>Thin Films</u>

Doped zinc oxide thin films were prepared by sputtering [3]. The electrical resistivity and doping concentrations are compared to values we obtained in powders (Table I). Also, optical data and Hall data, considered unreliable on powders, were obtained on the thin films (Table I). Optical transmittance throughout the visible range was always higher than 85%. The conductivity values obtained in the thin films were always higher by a factor greater than ten. Several factors may be important for this difference. One is that the compacted powders were not fully dense, and no correction was made for this porosity. The other factor is that higher dopant concentrations could be achieved in thin films. The conductivities of doped zinc oxide thin films reported in Table I and similar to those prepared by other methods such as MOCVD [9].

DISCUSSION

Our synthesis procedure gives very high conductivities for doped zinc oxide powders. Furthermore, this conductivity is stable in dry or moist air for prolonged periods at temperatures of 400°C or below. A potential disadvantage of our synthesis method is the large particles produced due to the high temperature employed. However, fine particle zinc oxide is produced commercially by burning zinc metal vapors in air. Thus it seems likely that highly conducting zinc oxide could be prepared as fine powder if the dopant is added as metal vapor and oxygen partial pressure is kept low during the oxidation at high temperatures.

ACKNOWLEDGMENTS

This work was supported in part by E. I. DuPont de Nemours and Company.

REFERENCES

1. R. Wang, A.W. Sleight and D. Cleary, Chem. Mater. **8**, 433 (1996).
2. R. Wang, A.W. Sleight, R. Platzer and J.A. Gardner, J. Solid State Chem. **122**, 166 (1996).
3. R. Wang, L.L.H. King and A.W. Sleight, J. Mater. Res. **11**, 1659 (1996).
4. R. Wang, A.W. Sleight and M.A. Subramanian, J. Solid State Chem. **125**, 224 (1996).
5. H. Sawada, R. Wang and A.W. Sleight, J. Solid State Chem. **122**, 148 (1996).
6. W.D. Kingery, H.K. Bowen and D.R. Uhlmann, *Introduction to Ceramics*, 2nd ed. (Wiley, New York, 1975), p. 159 & 891.
7. D.H. Kerridge and S.A. Tariq, J. Chem Soc. (A), 1122 (1967).
8. N. Roberts, R. Wang, A.W. Sleight and W.W. Warren, unpublished manuscript.
9. J. Hu and R.G. Gordon, Mat. Res. Soc. Symp. Proc. **242**, 743 (1992) and references therein.

CRYSTAL CHEMISTRY AND ELECTRONIC PROPERTIES OF THE N = 2 RUDDLESDEN-POPPER MANGANATES: UNCONVENTIONAL CMR MATERIALS

P.D. BATTLE[*], S.J. BLUNDELL[**], D.E. COX[***], M.A. GREEN[*], J.E. MILLBURN[*], P.G. RADAELLI[****], M.J. ROSSEINSKY[*], J. SINGLETON[**], L.E. SPRING[*], J.F. VENTE[*]

[*] Inorganic Chemistry Laboratory, University of Oxford, South Parks Road, Oxford, OX1 3QR, U.K.
[**] Department of Physics, Clarendon Laboratory, University of Oxford, Parks Road , Oxford OX1 3PU, U.K.
[***] Department of Physics, Brookhaven National Laboratory, Upton, NY 11973, U.S.A.
[****] Institut Laue-Langevin, P.O. Box 156, 38042 Grenoble Cedex 9, France

ABSTRACT

The crystallography and electronic properties of the $Ln_{2-x}Sr_{1+x}Mn_2O_7$ manganese oxides adopting the n = 2 Ruddlesden-Popper (RP) structure are discussed, focusing on the structural phase diagrams and electronic properties in the vicinity of the Mn +3.5 oxidation state and in particular the ease of synthesis of single phases of these materials.

INTRODUCTION

The observation that the metal-insulator transition at the Curie temperature (T_c) of the perovskites $La_{1-x}A_xMnO_3$ could be shifted by the application of magnetic fields at temperatures just above T_c to produce a massive decrease in the electrical resistance[1,2] has become known as colossal magnetoresistance (CMR). Problems of temperature range, saturation field and device noise remain in the way of applications, but recent advances have been made in all of these directions.[3-5] It has become accepted that CMR and field-induced ferromagnetism are closely interrelated, leading to the use of the double exchange mechanism[6] to interpret the experimental observations. Recently, the n = 2 member of the RP homologous series, of which the perovskite is the n = ∞ end member, has become the focus of attention.[7-10] We have demonstrated CMR without bulk ferromagnetism in ceramic samples of $Sr_{2-x}Nd_{1+x}Mn_2O_7$.[9] In the course of this study, we have been forced, through our use of high resolution powder diffraction characterisation, to address detailed and sometimes difficult issues of phase separation and phase purity, which turn out to be sensitive functions of the lanthanide cation in question and the manganese oxidation state. In this paper, we report on the evolution of the phase diagram at the $Sr_2LnMn_2O_7$ composition with lanthanide and Sr/Ln ratio for La, Tb and Nd. We emphasise the need for the application of very high resolution powder diffraction techniques and profile refinement to confirm the synthesis of single phases. We also discuss cation ordering in the single and biphasic samples, and briefly address the evolution of magnetic and magnetotransport behaviour across the series.

The n = 2 member of $(La,Sr)_{n+1}Mn_nO_{3n+1}$ series had been investigated by several groups before the current interest in manganese oxides for CMR applications[11-13]. Ferromagnetic behaviour had been identified in the La,Sr n = 2 phases with a saturation moment per Mn which appeared to vary strongly as a function of La/Sr ratio. CMR at the ferromagnetic Curie temperature of 126K was found in single crystal $La_{1.8}Sr_{1.2}Mn_2O_7$[7] - the saturation moment of 3 m_B with the moment direction in the xy plane has been confirmed by powder neutron diffraction[14]. A particularly interesting feature of this material from the point of view of applications is the low magnetic fields (less than 1T) required to produce magnetoresistance. The recent observation of low-field CMR in cation substituted perovskites[4] with a manganese oxidation state close to +3.5, and an ordered array of the +III and +IV oxidation states in zero field, makes the behaviour of the n = 2 RP phases in this oxidation state range of interest.

Mat. Res. Soc. Symp. Proc. Vol. 453 © 1997 Materials Research Society

EXPERIMENT

The samples discussed in the paper are prepared by subsolidus reaction of the constituent oxides and $SrCO_3$ in alumina crucibles in air or flowing nitrogen at temperatures between 1350 °C and 1500 °C: precise details of the synthesis conditions are given in [15]. Magnetisation experiments were carried out with an MPMS SQUID magnetometer over the temperature range $5 \leq T/K \leq 300$ and $0.01 \leq B/T \leq 5$. High resolution X-ray powder diffraction data were recorded at room temperature with the Bragg-Brentano geometry diffractometer at station 2.3 at the Synchrotron Radiation Source, Daresbury Laboratory and at the X7A beamline at the NSLS, Brookhaven National Laboratory. Neutron powder diffraction data were recorded on the D1B and D2B neutron diffractometers at the Institut Laue-Langevin, Grenoble over the temperature range $1.5 \leq T/K \leq 300$.

RESULTS

$\underline{Sr_2LaMn_2O_7}$

High resolution powder X-ray and neutron diffraction measurements show that this composition phase separates into two n = 2 Ruddlesden-Popper phases with lattice parameters of $a = 3.8748(1)$Å, $c = 19.9984(8)$Å for the majority phase (~80%) and $a = 3.8753(3)$Å, $c = 20.072(1)$Å for the minority phase (~20%). SQUID magnetisation measurements had suggested the presence of both antiferromagnetic and ferromagnetic ordering in this sample[8], consistent with earlier measurements on materials of this composition.[11,12] Variable temperature powder neutron diffraction measurements show that this is attributable to the onset of ferromagnetic order at 124K in the minority phase, whose properties are very similar to those of $Sr_{1.8}La_{1.2}Mn_2O_7$,[7,10,14] while the majority phase is a previously uncharacterised antiferromagnet with a Neel temperature of 220K.[16] Rietveld refinement yields phase fractions of the two phases which, in combination with the assumption that the ferromagnetic phase is $Sr_{1.8}La_{1.2}Mn_2O_7$, leads to the conclusion that the new antiferromagnetic phase has a composition of $Sr_{2.04}La_{0.96}Mn_2O_7$, corresponding to a manganese oxidation state of +3.52. This has implications for a comparison of the homologous phases in the Sr/La system as a function of perovskite block thickness - n= ∞ is metallic at all temperatures and ferromagnetic at the +3.5 oxidation state, whereas both n = 1 and n = 2 are antiferromagnets, although it appears that the precise composition is unstable in the n = 2 case. This may have consequences for the existence of commensurate charge ordering in the n = 2 phase, which has not been detected in the current experiments.

$\underline{Sr_{2-x}Nd_{1+x}Mn_2O_7}$

High resolution diffraction experiments indicate that phase separation is also an important problem to be addressed in this system. The ease of single phase synthesis increases with x, and depends on the reaction temperature and time. Representative data collected on station 2.3 at Daresbury are shown in Figure 1. The {0 0 10} reflection is especially sensitive to the existence of multiple phases. The example of $Sr_{1.6}Nd_{1.4}Mn_2O_7$ shows clearly the issues involved. Figure 1(a) shows data from a sample with a {0 0 10} reflection which can be fitted with a single pseudo-Voigt function with an FWHM of 0.075(2)° (only slightly broadened over the instrumental resolution of 0.05° in this angular range at a wavelength of 1.39937Å). We judge that this is a single phase, and the magnetic data shown in Figure 2(a) show only a low temperature spin-glass freezing transition. The data in Figure 1(b) are from a second, larger sample with the same nominal composition and similar thermal treatment. The { 0 0 10} reflection now has an FWHM of 0.127(2)°, which we judge to be representative of a biphasic sample. The magnetic data shown in Figure 2(b) are a strong indication that our conclusion that { 0 0 10 } widths much in excess of 0.08° identify inhomogeneities in the sample which are significant from the point of view of controlling the electronic properties. These data clearly show the existence of a high temperature feature in the magnetisation similar to that found for samples with slightly lower values of x (≤ 0.3),[9] together with the lower temperature freezing

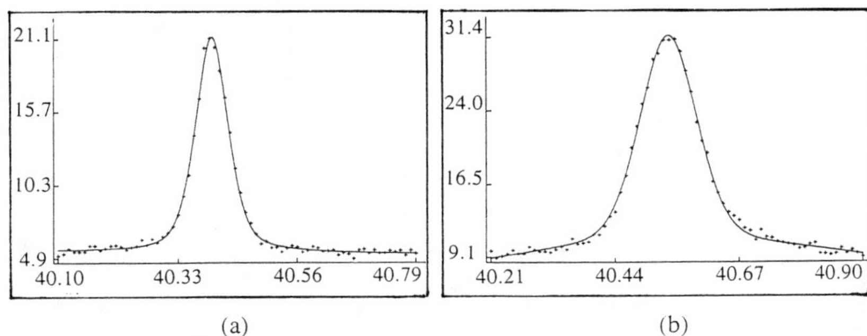

Figure 1: Synchrotron X-ray powder diffraction measurement of the {0 0 10} reflection of (a) single phase $Sr_{1.6}Nd_{1.4}Mn_2O_7$ (b) biphasic $Sr_{1.6}Nd_{1.4}Mn_2O_7$. Fits are to a single pseudo-Voigt function with widths quoted in the text.

transition attributable to a second phase. Samples of $Sr_2NdMn_2O_7$ prepared with different thermal treatments illustrate the problems involved in preparing single phases and identifying them with certainty. Samples prepared either in air at 1350 °C or with a Bi_2O_3 flux at 1500 °C[10] have FWHM of 0.182(4)°, while extended annealing at 1400 °C, 1450 °C and 1500 °C give reflections with widths of 0.131(2)°, 0.131(3)° and 0.113(2)° still in excess of our single phase limiting width. This is consistent with the complex magnetism, suggestive of biphasic behaviour,[9] and in agreement with two-phase Rietveld refinements of x = 0.0 and x = 0.1 samples.[17] The x = 0.2 composition (prepared according to a recently published protocol[10]) has a {0 0 10} width of 0.103(3)°, which is again broader than our single phase criterion. These effects are hard to detect using a conventional laboratory diffractometer.

$Sr_2LnMn_2O_7$

At the 2:1 composition (which is phase-separated under all the subsolidus synthesis conditions we have investigated for Ln = Nd, La) changing the size of the lanthanide cation produces a transition to truly single phase behaviour beyond $Sr_2EuMn_2O_7$. The precise details of the synthetic procedure remain important in reproducibly preparing a single phase. In Figure 3, we show the {0 0 10} reflections of single phase samples of $Sr_2YMn_2O_7$ and $Sr_2HoMn_2O_7$ with FWHM of 0.0681(6)° and 0.072(2)° respectively. The electronic properties of both samples suggest that they are electronically homogeneous, with spin-glass like freezing below 20K signalled by a divergence between field cooled (FC) and zero field cooled (ZFC) magnetisations. Rietveld refinement of laboratory X-ray data indicates that the smaller 9 co-ordinate A cation site in the rock-salt layer of the structure is preferentially occupied by Ln^{3+} to an increasing extent as the size of the lanthanide cation decreases into this single phase region. In $Sr_2DyMn_2O_7$ this site is occupied by 94% of the Dy^{3+} cations while the larger 12 co-ordinate rock-salt layer site is almost exclusively occupied by Sr^{2+} cations. The site ordered single phases have a magnetic response which is quite different to that of the biphasic, larger lanthanide 2:1 compositions discussed earlier. In particular, there is no evidence for high temperature ordering or freezing of the manganese spins, and the magnetisation is non-hysteretic and obeys the Curie law well above 50K. Below approximately 30K, a divergence between the field-cooled (FC) and zero-field cooled (ZFC) magnetisations indicates a spin-glass like freezing of the manganese moments. The lower freezing temperature compared with $Sr_2NdMn_2O_7$ implies an increasing frustration of antiferromagnetic interactions as the size of the lanthanide cation decreases. This is particularly important as the low temperature spin-glass phase of $Sr_2YMn_2O_7$ does not display CMR at fields of up to 14T in the 5-300K temperature

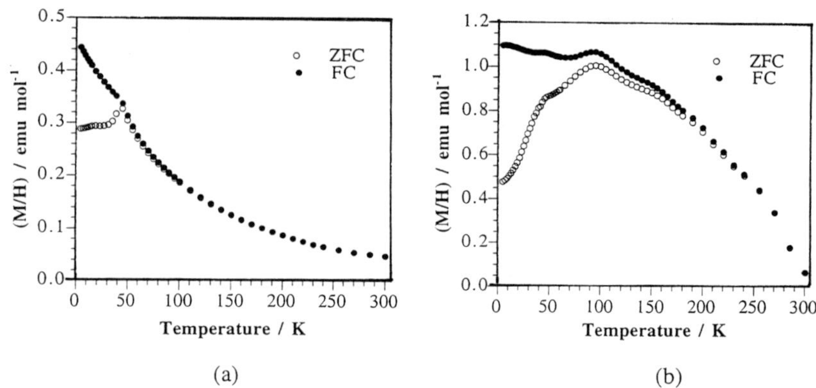

(a)   (b)

Figure 2: Magnetisation measurements on the (a) mono and (b) biphasic samples of $Sr_{1.6}Nd_{1.4}Mn_2O_7$ corresponding to the X-ray measurements in Figure 1.

range, consistent with the absence of metamagnetic features in M(H) isotherms measured for $5 \leq T/K \leq 300$.

CONCLUSIONS

We begin with a discussion of the implications of the phase separation at the $Sr_2LaMn_2O_7$ composition - in contrast to the $Sr_2NdMn_2O_7$ case, the cell volume differences between the two phases are large and different syntheses of this composition by different groups [11,12,16] appear to give consistent results. In particular, reaction conditions which lead to single phase $Sr_{1.8}La_{1.2}Mn_2O_7$ yield for $Sr_2LaMn_2O_7$ a material whose electronic and diffraction properties can only be understood in terms of a separation into two distinct phases.

The minority ferromagnetic component has already been well characterised from the electronic and diffraction point of view[7,14], but the new finding here is the existence of an antiferromagnetic phase close to the +3.5 oxidation state, where commensurate charge ordering is found in the n =1 system $La_{0.5}Sr_{1.5}MnO_4$ [18,19] but the n = ∞ $La_{0.5}Sr_{0.5}MnO_3$ remains metallic and ferromagnetic.[20] The reduction in the bandwidth of the e_g carriers reduces the significance of the double exchange interaction sufficiently to suppress ferromagnetic order and leave the antiferromagnetic interactions dominant. The question of possible charge ordering in the n = 2 phase, where the carrier concentration is now no longer at the commensurate 1:1 +III/+IV value, is one that needs to be addressed by variable temperature electron diffraction measurements.

The behaviour of the Nd system is in sharp contrast to that of the La system and also that of other CMR systems: compositions x = 0.0, 0.1 and 0.2 in the $Sr_{2-x}Nd_{1+x}Mn_2O_7$ display CMR below 150K in fields of between 2T and 14T without any sign of bulk ferromagnetism.[9,10] Magnetisation measurements show a sharp rise in magnetisation just below 300K: this corresponds to the onset of hysteresis in fields of less than 0.05T but not magnetic long range order (no magnetic Bragg peaks appear in the neutron diffraction pattern) or CMR. A further local maximum in M occurs at lower temperature and corresponds to the onset of long-range antiferromagnetic order. The Neel temperature decreases as the e_g electron concentration increases (from 137K at x = 0.0 to 100K at x = 0.1).[17] The complete absence of even field-induced ferromagnetism in the field range over which CMR is observed suggests that this is a new type of CMR effect, which may in principle arise in either the spin-glass or the antiferromagnetic phase or at the interfaces between them. Further studies should concentrate on the preparation of both truly single phase powder samples, with close to resolution-limited reflection widths, and highly perfect single crystals to establish the possible

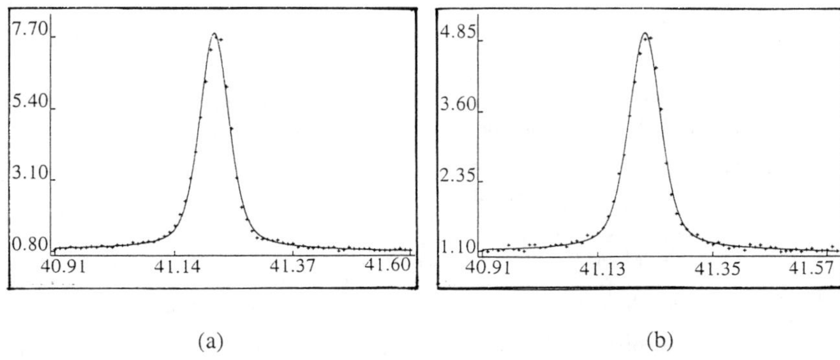

(a)   (b)

Figure 3: Synchrotron X-ray powder diffraction measurement of the {0 0 10} reflection of single phase $Sr_2LnMn_2O_7$ samples. (a) $Sr_2YMn_2O_7$ (b) $Sr_2HoMn_2O_7$. Fits are to a single pseudo-Voigt function with widths quoted in the text.

role of grain boundaries[3] in the unusual observation of CMR without ferromagnetism. Increased Mn(III) concentration into the single phase region for $x \geq 0.4$ reduces the Neel temperature due to increasing frustration of the antiferromagnetic interactions, and lead to the suppression of high temperature features (Neel ordering or spin-glass like freezing) in the magnetisation measurements. The magnetotransport properties of these single phases are currently unknown and promise to be particularly important for the interpretation of the behaviour of the earlier members of the series. The $Sr_{2-x}Nd_{1+x}Mn_2O_7$ series illustrates the difficulty of preparing single phase monophasic samples of the n = 2 RP phases and the utility of high resolution powder diffraction techniques in assessing the outcomes of syntheses and the electronic property measurements on the samples in question.

ACKNOWLEDGEMENTS

We thank the Donors of the Petroleum Research Foundation, administered by the American Chemical Society, and the UK EPSRC for their support of this work. Work at Brookhaven is supported under contract number DE-AC02_76CH00016, Division of Materials Sciences, U.S. Department of Energy.

REFERENCES

1.	R. von Helmolt; J. Wecker; B. Holzapfel; L. Schultz; K. Samwer, Physical Review Letters **71**, 2331 (1993).
2.	S. Jin; T. H. Tiefel; M. McCormack; R. A. Fastnacht, Science **264**, 413 (1994).
3.	H. Y. Hwang; S. W. Cheong; N. P. Ong; B. Batlogg, Physical Review Letters **77**, 2041 (1996).
4.	H. Kuwahara; Y. Tomioka; Y. Moritomo; A. Asamitsu; M. Kasai; R. Kumai; Y. Tokura, Science **272**, 80 (1996).
5.	T. Kimura; Y. Tomioka; H. Kuwahara; A. Asamitsu; M. Tamura; Y. Tokura, preprint (1996).
6.	P. G. de Gennes, Physical Review **118**, 141 (1960).
7.	Y. Moritomo; A. Asamitsu; H. Kuwahara; Y. Tokura, Nature **380**, 141 (1996).
8.	P. D. Battle; M. A. Green; N. S. Laskey; J. E. Millburn; M. J. Rosseinsky; S. P. Sullivan; J. F. Vente, Chem. Comm. 767 (1996).
9.	P. D. Battle; S. J. Blundell; M. A. Green; W. Hayes; M. Honold; A. K. Klehe; N. S. Laskey; J. E. Millburn; L. Murphy; M. J. Rosseinsky; N. A. Samarin; J. Singleton; N. E. Sluchanko; S. P. Sullivan; J. F. Vente, Journal of Physics Condensed Matter **8**, L427 (1996).

10.   R. Seshadri; C. Martin; A. Maignan; M. Hervieu; B. Raveau; C. N. R. Rao, Journal of Materials Chemistry **6**, 1585 (1996).

11.   J. B. MacChesney; J. F. Potter; R. C. Sherwood, Journal of Applied Physics **40**, 1243 (1969).

12.   R. A. M. Ram; P. Ganguly; C. N. R. Rao, Journal of Solid State Chemistry **70**, 82 (1987).

13.   M. Lamire; A. Daoudi, Journal of Solid State Chemistry **55**, 327 (1984).

14.   J. F. Mitchell; D. N. Argyriou; J. D. Jorgensen; D. G. Hinks; C. P. Potter; S. D. Bader, preprint (1996).

15.   P. D. Battle; M. A. Green; N. S. Laskey; J. E. Millburn; L. E. Murphy; M. J. Rosseinsky; S. P. Sullivan; J. F. Vente, Chemistry of Materials in press.

16.   P. D. Battle; D. E. Cox; M. A. Green; J.E. Millburn, L.E. Spring, P. G. Radaelli; M. J. Rosseinsky; Chemistry of Materials, submitted for publication (1996).

17.   P. D. Battle; M. A. Green; N. S. Laskey; J. E. Millburn; P. G. Radaelli; M. J. Rosseinsky; S. P. Sullivan; J. F. Vente, Physical Review B **54**, 15967 (1996).

18.   W. Bao; C. H. Chen; S. A. Carter; S. W. Cheong, Solid State Communications **98**, 55 (1996).

19.   B. J. Sternlieb; J. P. Hill; U. C. Wildgruber; G. M. Luke; B. Nachumi; Y. Moritomo; Y. Tokura, Physical Review Letters **76**, 2169 (1996).

20.   A. Urushibara; Y. Moritomo; T. Arima; A. Asamitsu; G. Kido; Y. Tokura, Physical Review B **51**, 14103 (1995).

Critical Transport and Magnetization of $La_{0.67}Ca_{0.33}MnO_3$

G. Jeffrey Snyder[1], R. Hiskes[2], S. DiCarolis[2], M. R. Beasley[1], and T. H. Geballe[1]

[1]*Department of Applied Physics, Stanford University, Stanford, CA 94305-4090*
[2]*Hewlett-Packard, Palo Alto, California 94303-0867*

Abstract

The critical magnetic properties of bulk $La_{0.67}Ca_{0.33}MnO_3$ and magnetoresistance of a thin film with the same T_C were measured and critical exponents determined. The magnetization data can be scaled with $\beta = 0.3$ and $\gamma = 0.9$, except above T_C which appears to be affected by a region where χ_3 ($M = \chi H + \chi_3 H^3$) is positive. An M^2 dependence of the magnetoconducitvity is observed above T_C. Below T_C, however, such a correspondence between the critical magnetic and transport behavior is not found.

Introduction

The mixed Mn^{3+}/Mn^{4+} perovskite manganites are being actively studied because they show very large negative magnetoresistance at the ferromagnetic-metal to paramagnetic-semiconductor transition [1], and are being considered for use in the magnetic recording industry [2]. In homogeneous materials, the magnetoresistance maximizes near the ferromagnetic Curie temperature T_C. The conductivity behaves much like the magnetization: above T_C the conductivity and magnetization are low, while below T_C the conductivity and magnetization rapidly increase.

Several authors have shown a correlation between the resistivity and magnetization by plotting one as a function of the other [3-5]. These plots show that the resistivity is roughly proportional to the square of the magnetization M^2 for small M. For larger M however, the resistance decreases more slowly in relationship to M^2. This relationship may instead be well described by an M^2 dependence of the conductivity in a slightly more complicated circuit [6]. In order to determine if the conductivity, σ can be represented by a term proportional to M^2, the temperature dependencies and magnitudes of σ and M need to be examined.

Experiment

A well annealed $La_{0.67}Ca_{0.33}MnO_3$ thin film on $LaAlO_3$ grown by MOCVD was used for transport measurements as described in [1, 7]. Repeated resistance R *vs.* H curves were measured and fit with three parameters (ρ_∞, σ_0, and σ_{H^2} or σ_H) to $\rho = \rho_\infty + 1/(\sigma_0 + \sigma_{H^2}H^2)$ for $T > T_C$ and $\rho = \rho_\infty + 1/(\sigma_0 + \sigma_H|H|)$ for $T < T_C$. The data fit well to these forms except for a few degrees near T_C (where a combination of the two forms is better). The polycrystalline pellets and a float zone crystal [1] used for magnetization measurements have very similar magnetic properties.

Mat. Res. Soc. Symp. Proc. Vol. 453 © 1997 Materials Research Society

Results

The square of the magnetization M^2, plotted *vs.* H/M (Figure 1) facilitates understanding the critical behavior of $La_{0.67}Ca_{0.33}MnO_3$. The isotherms below T_C should be approximately linear and intersect $H/M = 0$ at M_0. From these $M_0(T)$ the critical exponent $\beta \approx 0.3$ and $T_C = 262.2$ K can be estimated by fitting $M_0(T) \propto (1 - T/T_C)^\beta$ in the critical region. Similarly, $T > T_C$ isotherms intersect $M^2 = 0$ at $1/\chi(T, H = 0) = 1/\chi_0 \propto (1 - T/T_C)^\gamma$ and can be used to estimate $\gamma \approx 0.7$.

The faster than linear increase in M with H, as seen in Figure 1, is not characteristic of a ferromagnet, but may be seen in an unsaturated ferromagnet or spin glass. This is unlikely however, since no hysteresis is found. Alternatively, the paramagnetic coupling of the moments may change from canted-antiferromagnetic to ferromagnetic in a field (like a metamagnet). Most of the $M(H, T)$ data obeys the scaling hypothesis with $\beta \approx 0.27$ and $\gamma \approx 0.90$. The regions of negative slope on the M^2 vs. H/M plot do not obey the scaling relation, indicating that this is not part of the ferromagnetism, which complicates the estimation of γ.

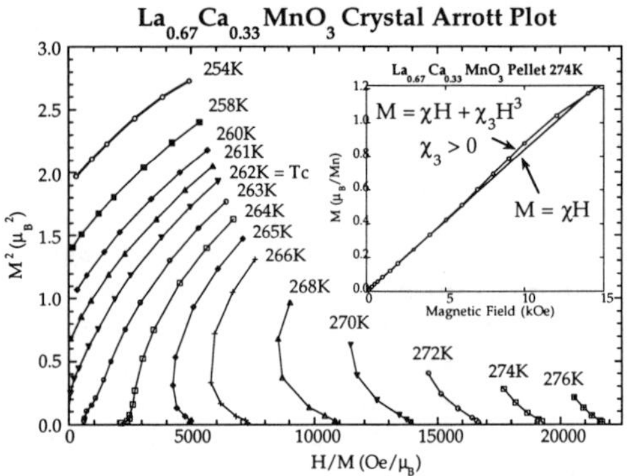

Figure 1. M^2 vs. H/M plot for $La_{0.67}Ca_{0.33}MnO_3$ float zone crystal. A mean field ferromagnet has linear isotherms with a positive slope. The negative slope for $T > T_C$ indicates a faster than linear increase in M (inset) due to a highly unusual positive non-linear susceptibility χ_3.

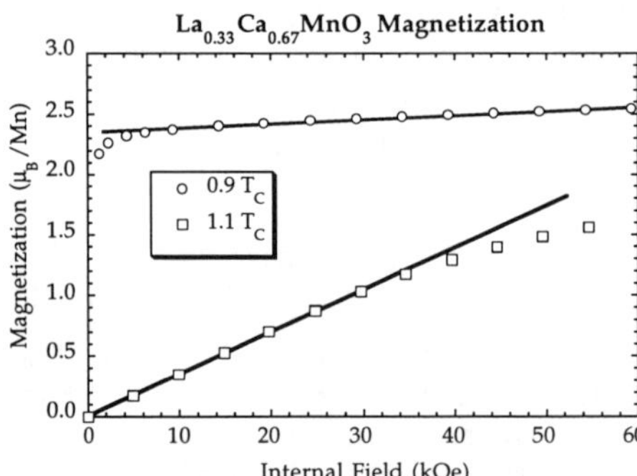

Figure 2. Magnetization in a magnetic field for a $La_{0.67}Ca_{0.33}MnO_3$ polycrystalline pellet at 0.9 and 1.1 T_C. The solid lines indicate the linear regions in each case.

To illustrate the need for a magnetoconductivity expression for the magnetoresistance, we consider first the two temperatures 0.9 T_C and 1.1 T_C. The magnetization of a polycrystalline pellet at these two temperatures is shown in Figure 2. At $T = 0.9$ T_C, $M \approx M_0 + \chi H$ except in fields less than a few kOe when the material is not yet magnetically saturated. At $T = 1.1\,T_C$, $M \approx \chi H$ is a good approximation particularly for $H < 40$ kOe. Since the resistance in $La_{0.67}Ca_{0.33}MnO_3$ near T_C decreases rapidly in small magnetic fields and then begins to saturate in higher fields, one might attribute this to an M^2 dependence on the resistivity and the saturation of M in large fields. For comparison, the magnetoresistance at $0.9\,T_C$ is shown along with $-M^2$ in Figure 3.

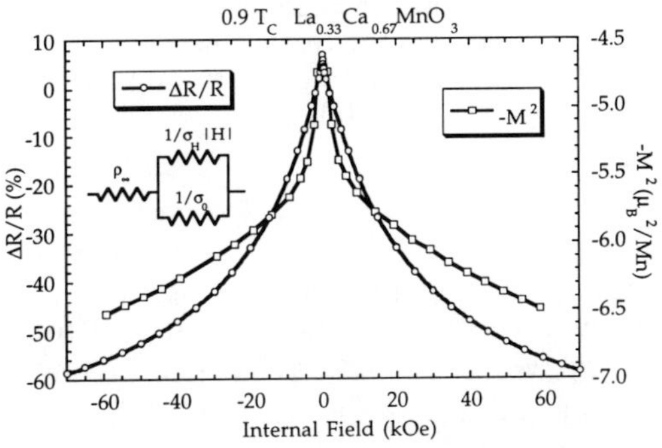

Figure 3. Magnetoresistance of La$_{0.67}$Ca$_{0.33}$MnO$_3$ film compared with $-M^2$ of a pellet, both at 0.9 T_C. The solid line for the magnetoresistance data show the fit using the indicated equivalent circuit.

The saturation of $-M^2$ is clearly unrelated to the saturation of $\Delta R/R$. Similarly, Figure 4 shows the comparison of $-M^2$ and $\Delta R/R$ at $1.1\,T_C$. Again, the saturation of these two quantities is visibly different. The saturation of the magnetoresistance is better described by the fit to the magnetoconductivity [6] circuit mentioned above, which assumes a linear M vs. H. If the magnetoconductivity is indeed proportional to M^2 as suggested above, then the nonlinearity of M vs. H (Figure 2) will somewhat alter the fitting parameters (ρ_∞, σ_0, and σ_{H^2} or σ_H).

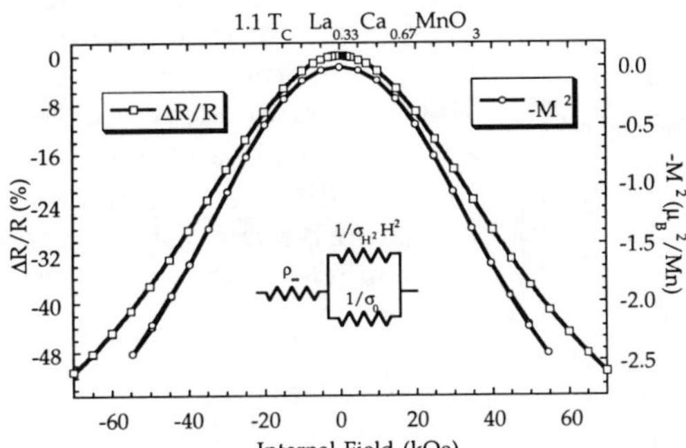

Figure 4. Magnetoresistance of La$_{0.67}$Ca$_{0.33}$MnO$_3$ film compared with $-M^2$ of a pellet, both at 1.1 T_C. The solid line for the magnetoresistance data show the fit using the indicated equivalent circuit.

The magnetoresistance data at various temperatures above T_C were fit to the expression $\rho = \rho_\infty + 1/(\sigma_0 + \sigma_{H^2}H^2)$. The resulting parameters σ_0 and ρ_∞ are shown in Figure 5. σ_0 remains relatively constant, while ρ_∞ decreases uniformly. This can be expected since above T_C La$_{0.67}$Ca$_{0.33}$MnO$_3$ is an insulator with $d\rho/dT < 0$.

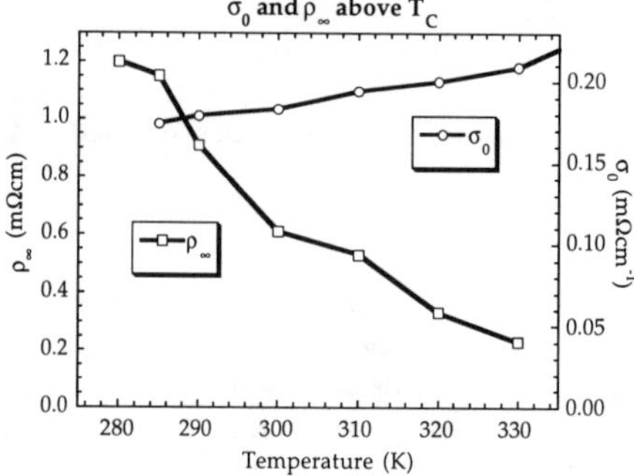

Figure 5. Fitting parameters σ_0 and ρ_∞ for $T > T_C$ in a $La_{0.67}Ca_{0.33}MnO_3$ film. The temperature dependence of these two parameters reflect the insulating behavior of the material.

The parameter σ_{H^2}, which determines the magnetoresistance, is highly temperature dependent (Figure 6), becoming very large at T_C. The data can be fit with a critical exponent of -2.0 ± 0.1: $\sigma_{H^2}(T) = \sigma_{H^2}(2T_C)\,(T/T_C - 1)^{-2}$, where $T_C = 262$ K ± 1 K and $\sigma_{H^2}(2T_C) \approx 8 \times 10^{-13}$ Oe^{-2} mΩcm^{-1}. This result supports an M^2 dependence of the magnetoconductance (or magnetoresistance) since $M^2 \propto H^2(T/T_C - 1)^{-2\gamma}$ above T_C, where, as experimentally determined above, $\gamma \approx 0.9$, or in the Curie-Weiss theory, $\gamma = 1$. Specifically we have found $\sigma(H,T) = \sigma_0 + \sigma_{H^2}(T)H^2 \approx \sigma_0 + \sigma_{M^2}M^2$. Using the Curie Weiss susceptibility from [1], $M = \chi H = \approx 5 \times 10^{-4}$ (emu/ G cm^3) $(T/T_C - 1)^{-1} H$, gives $\sigma_{M^2} = 3 \times 10^{-6}$ Oe^{-2} mΩcm^{-1}.

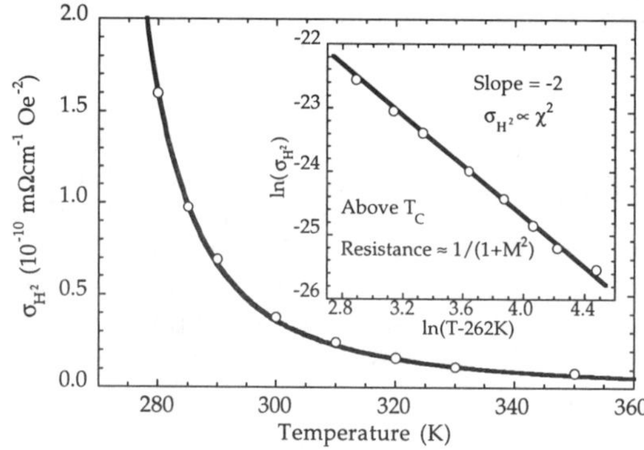

Figure 6. Fitting parameter σ_{H^2} as a function of temperature in a $La_{0.67}Ca_{0.33}MnO_3$ film for $T > T_C$. The temperature dependence of σ_{H^2} and the square of the susceptibility are the same, indicating a relationship between the magnetoconductance and M^2.

Below T_C, ρ_∞ can be interpreted as due to the non-magnetic scattering processes, which are unaffected by a magnetic field. As shown in Figure 7, the minimum resistivity in a infinite field, ρ_∞, follows the $A + BT^2$ fit to the low-temperature intrinsic resistivity [1]. The parameter σ_0, which can be related to the resistivity in zero field $\rho(H = 0) = \rho_\infty + 1/\sigma_0$, describes the field-dependent, spin-disorder scattering processes present in zero applied field, $H = 0$. Unlike ρ_∞, $1/\sigma_0$ diverges as the Curie temperature is approached. A critical exponent of about 1.8 (Figure 7) is found: $\sigma_0(T) \approx \sigma_0(0)\,(1 - T/T_C)^{1.8}$, where $\sigma_0(0) \approx 2 \times 10^{-3}$ mΩcm^{-1}.

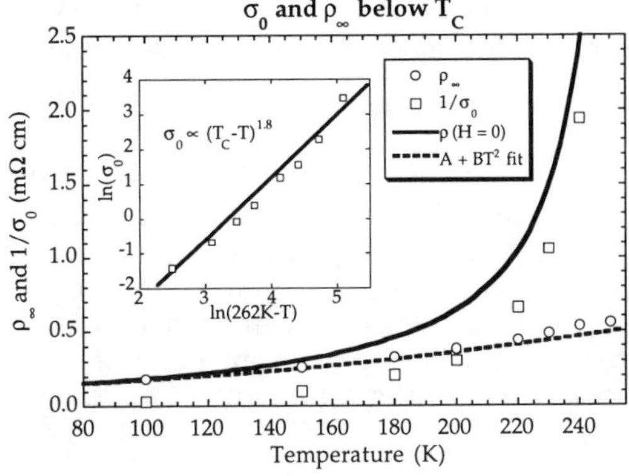

Figure 7. Fitting parameters σ_0 and ρ_∞ for $T < T_C$ in a La$_{0.67}$Ca$_{0.33}$MnO$_3$ film. ρ_∞ is governed by the A + BT^2 terms in the resistivity while σ_0 behaves critically as $(T_C - T)^{1.8}$. The zero field resistivity $\rho(H = 0) = \rho_\infty + 1/\sigma_0$ is shown for comparison.

The parameter σ_H is more difficult to interpret than the other parameters. Although the magnetoresistance, $\Delta R/HR$ proportional to σ_H/σ_0^2, increases as T_C is approached, σ_H decreases. Thus the large magnetoresistance found below T_C is due tot he divergance of $1/\sigma_0$ and not σ_H. The parameter σ_H does not appear to vanish as T approaches T_C, making it difficult to fit to a $(1 - T/T_C)^n$ power law. Furthermore, we have found different temperature dependencies for different samples with $n \approx 1 \pm 1$. The data in Figure 8 are from the same sample used for Figure 5-Figure 7. Assuming $T_C = 262K$ (determined from the critical properties of σ_{H^2} and σ_0) the critical exponent for σ_H is about 0.7.

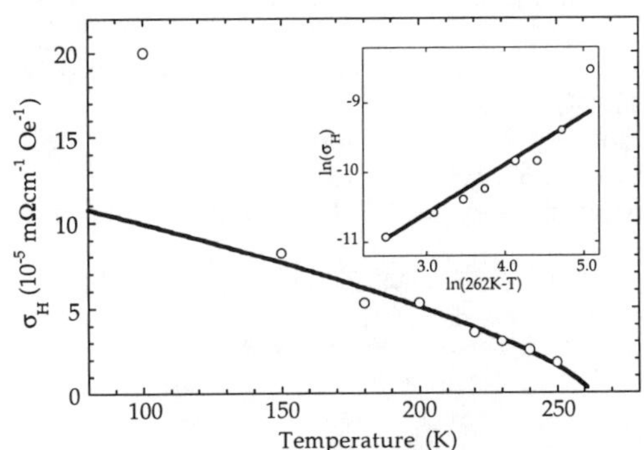

Figure 8. Fitting parameter σ_H as a function of temperature in a La$_{0.67}$Ca$_{0.33}$MnO$_3$ film for $T < T_C$. The solid line shows the best fit to the data using a critical exponent of 0.7.

Below T_C, the simple relationship between M^2 and the magnetoconductance found above T_C no longer holds. As seen in Figure 2, the magnetization below T_C can be approximated with $M(H) = M_0 + \chi H$, and therefore $M^2 \approx M_0^2 + 2\chi M_0 H$. Assuming $\sigma(H,T) \approx \sigma_{M^2} M(H,T)^2$ as suggested above, then below T_C the following relations should hold: $\sigma_0 = \sigma_{M^2} M_0^2$ and $\sigma_H = 2\chi\sigma_{M^2} M_0$. The temperature dependence of M_0 and χ below T_C should be governed by the critical exponents β and γ, via $M_0 \propto$

$(1 - T/T_C)^\beta$ and $\chi \propto (1 - T/T_C)^{-\gamma}$. Thus one would expect $\sigma_0 \propto (1 - T/T_C)^{2\beta}$ and $\sigma_H \propto (1 - T/T_C)^{\beta-\gamma}$ where experimentally $\beta = 0.3$ and $\gamma \approx 0.9$ which are near typical values $\beta \approx 1/3$ and $\gamma \approx 1.2$. Experimentally, however we find $\sigma_0 \propto (1 - T/T_C)^{1.8}$ and $\sigma_H \propto (1 - T/T_C)^{0.7}$, which would require quite unusual critical exponents $\beta \approx 0.9$ and $\gamma \approx 0.2$.

Conclusion

In conclusion, we find that the magnetic properties of $La_{0.67}Ca_{0.33}MnO_3$ in the critical region near T_C can be approximately described by the scaling hypothesis with $\beta = 0.3$ and $\gamma = 0.9$, except for a region above T_C which appears to be affected by a positive non-linear susceptibility. The magnetoconductance above T_C may be accurately described by a simple M^2 dependence of the conductivity. However, below T_C this correlation appears to break down. The experimental critical exponents for the magnetoconductance below T_C differ significantly from those expected from the magnetic properties.

Acknowledgments

The work at Stanford was supported in part by the Air Force Office of Scientific Research, the Stanford Center for Materials Research under the NSF-MRL program and the Hertz foundation.

References

[1]G. J. Snyder, R. Hiskes, S. DiCarolis, M. R. Beasley, and T. H. Geballe, Phys. Rev. B **53**, 14434 (1996).

[2]J. A. Brug, T. C. Anthony, and J. H. Nickel, MRS Bulletin **21(9)**, 23 (1996).

[3]J. O'Donnell, M. Onellion, M. S. Rzchowski, J. N. Eckstein, and I. Bozovic, Phys. Rev. B (1996).

[4]Y. Tokura, A. Urushibara, Y. Moritomo, T. Arima, A. Asamitsu, G. Kido, and N. Furukawa, J. Phys. Soc. Jpn. **63**, 3931 (1994).

[5]M. F. Hundley, M. Hawley, R. H. Heffner, Q. X. Jia, J. J. Neumeier, J. Tesmer, J. D. Thompson, and X. D. Wu, Appl. Phys. Lett. **67**, 860 (1995).

[6]G. J. Snyder, R. Hiskes, S. DiCarolis, M. R. Beasley, and T. H. Geballe, Appl. Phys. Lett. (in press).

[7]D. C. Worledge, G. J. Snyder, R. Hiskes, S. DiCarolis, M. R. Beasley, and T. H. Geballe, J. Appl. Phys. (in press).

LAYERED CMR MANGANITES: STRUCTURE, PROPERTIES, AND UNCONVENTIONAL MAGNETISM

J.F. MITCHELL, D.N. ARGYRIOU, C.D. POTTER, J.D. JORGENSEN, D.G. HINKS, AND S.D. BADER

MATERIALS SCIENCE DIVISION, ARGONNE NATIONAL LABORATORY, 9700 S. CASS AVE., ARGONNE, IL 60439

ABSTRACT

Neutron powder diffraction studies of the layered compounds $R_{1.2}Sr_{1.8}Mn_2O_7$, (R = La,Pr, Nd), $RSr_2Mn_2O_7$ (R = Pr,Nd), and $La_{1.4}Sr_{1.6}Mn_2O_7$ show that the degree of distortion of the MnO_6 octahedra do not correlate with the appearance of a metal-insulator (MI) transition in these compounds. Instead, the in-plane Mn-O bond length appears to be a better indicator of the electronic behavior. Detailed bulk magnetization studies on single crystal $La_{1.2}Sr_{1.8}Mn_2O_7$ show that there are three magnetic regimes as a function of temperature: paramagnetic insulator, short-range ordered (SRO) ferrromanget, and long-range ordered (LRO) ferromagnet. Scaling analysis indicates that a 2D finite-size XY model is an appropriate description of the magnetic state in the SRO regime.

INTRODUCTION

Current research in the mixed valent manganite perovskite materials has expanded to include the search for materials with other structure types exhibiting colossal magnetoresistance (CMR). Shimakawa, et al.[1] have reported CMR in a pyrochlore ($Tl_2Mn_2O_7$) synthesized at high pressure, and Cheong, et al.[2] have studied In substitution in this sytem. Moritomo, et al.[3] have published a single crystal study of layered $La_{1.2}Sr_{1.8}Mn_2O_7$, the n=2 member of the Ruddlesden-Popper series $(La,Sr)_{n+1}Mn_nO_{3n+1}$. As shown in Fig. 1, this compound consists of double layers of MnO_6 octahedra separated by $(La,Sr)_2O_2$ layers. This material has coincident ferromagnetic (FM) and metal-insulator (MI) transitions at $T_C \sim 120$ K. Moritomo, et al. report a ~20,000 % MR (129 K, H=7 T) in this layered material and speculate on the role of reduced dimensionality in the electronic transport. This material is also significant due to its pronounced CMR in low field, ~200% at 129 K in 0.3 T.

We have synthesized a number of two-layer manganites ($R_{1.2}Sr_{1.8}Mn_2O_7$, R = La, Pr, Nd; $RSr_2Mn_2O_7$, R = Pr, Nd; and $La_{1.4}Sr_{1.6}Mn_2O_7$) to study the effect of doping and/or ionic size on the structural, magnetic, and electronic properties of this class of materials. Seshadri, et al.[4] have published a study of the $R_{1.2}Sr_{1.8}Mn_2O_7$ (R=La,Pr,Nd) series using x-ray powder diffraction and conclude that the smaller ionic size of the Pr and/or Nd results in an enhanced Jahn-Teller distortion of the MnO_6 octahedra in the double layers of $Pr_{1.2}Sr_{1.8}Mn_2O_7$ and $Nd_{1.2}Sr_{1.8}Mn_2O_7$. This increased distortion was presumed to suppress an MI transition observed in the $La_{1.2}Sr_{1.8}Mn_2O_7$ analog that coincides with the onset of long range FM order (LRO). This last compound was reported to have a much smaller distortion.

In contrast to the findings of Seshadri, et al., our neutron powder diffraction studies[5] of these compounds show no correlation between the octahedral distortion and the suppression of the MI transition. Rather, our data suggest that perhaps the in-plane Mn-O bond is the metrical parameter sensitive to the electronic transition. We also demonstrate in the $La_{1.2}Sr_{1.8}Mn_2O_7$ compound the existence of an unusual magnetic state extending from T_C to near room temperature.[6] Detailed magnetization measurements reveal that a short-range ordered (SRO) ferromagnetic (FM) state persists in this temperature regime. Furthermore, despite the absence of a long-range ordered state in the Pr and Nd analogs, these compounds also show evidence of this SRO state.

Mat. Res. Soc. Symp. Proc. Vol. 453 © 1997 Materials Research Society

Fig. 1. Structure of $(La,Sr)_3Mn_2O_7$

EXPERIMENTAL

Results reported below are for a single-crystal specimen of $La_{1.2}Sr_{1.8}Mn_2O_7$ grown by the traveling-solvent floating-zone technique and polycrystalline samples of the remaining compounds. Details of the preparation of single crystal specimens of the $La_{1.2}Sr_{1.8}Mn_2O_7$ have been published elsewhere.[5] The remaining compounds were synthesized in polycrystalline form by routine solid state reaction of La_2O_3 (Johnson-Matthey, dried at 900 ºC in flowing O2), Pr_6O_{11} (Johnson-Matthey, as received), Nd_2O_3 (Johnson-Matthey, dried at 900 ºC in flowing O2), $SrCO_3$ (Johnson-Matthey, as-received), and MnO_2 (Johnson-Matthey, as received). Stoichiometric quantities of the starting materials were mixed and fired in three steps: (1) 900 ºC for 12 hr in dynamic vacuum; (2) 1250 ºC in flowing oxygen; and finally, (3) 1400 ºC in flowing oxygen (with the exception of $La_{1.4}Sr_{1.6}Mn_2O_7$ which was fired in 2% O2/balance Ar). Materials were reground between each firing cycle. For the final firing, pellets were cold-pressed at 20 kpsi. In each case, the final products were single-phase as judged by powder x-ray diffraction. Neutron powder neutron diffraction (NPD) was carried out at the SEPD at Argonne National Laboratory's Intense Pulsed Neutron Source (IPNS). NPD measurements were made either in the Earth's field or in an applied field of 0.6 T generated by permanent magnets.[7] Diffraction measurements were made between 20 K and 300 K on a Displex refrigerator and between 300 K and 500 K in a radiation furnace. Magnetic and transport measurements were made in a Quantum Design PPMS system equipped with a 9 T superconducting solenoid. The trapped flux in the solenoid was monitored and field values adjusted accordingly. Magnetization measurements reported here on single crystal $La_{1.2}Sr_{1.8}Mn_2O_7$ utilize a field applied along the easy axis of the crystal, which is in the a-b plane.

RESULTS AND DISCUSSION

Fig. 2 shows the temperature-dependent resistance of the series $R_{1.2}Sr_{1.8}Mn_2O_7$ (R=La,Pr,Nd) normalized to the resistance at 300 K. In agreement with Seshadri, et al.,[4] the Pr and Nd compounds are insulating throughout the temperature regime measured, while the La compound has an MI transition at $T_C \sim 120$ K, with the resistance dropping by a factor of 100 through the transition. Although the data are not shown, the compound $La_{1.4}Sr_{1.6}Mn_2O_7$ also has an MI transition near 120 K, while $NdSr_2Mn_2O_7$ and $PrSr_2Mn_2O_7$ are insulators with no transition observed between 20 K and 350 K.

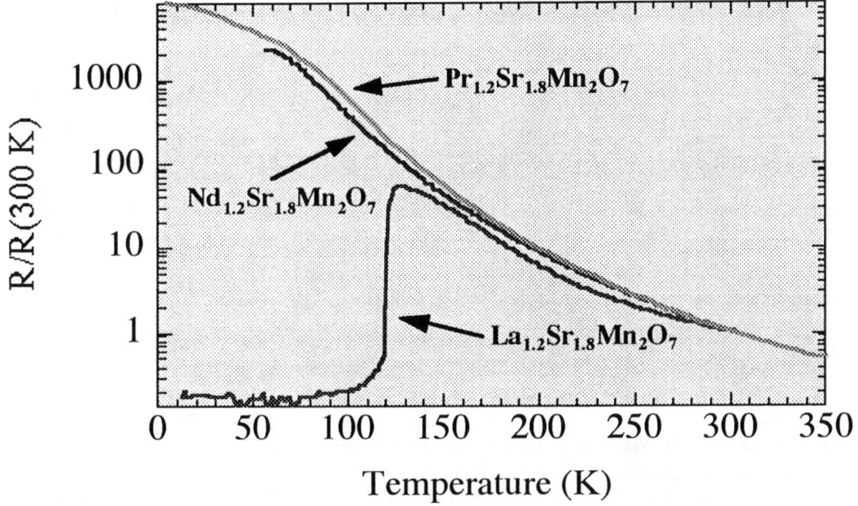

Fig. 2. Normalized R vs. T for the series $R_{1.2}Sr_{1.8}Mn_2O_7$ (R=La,Pr,Nd)

Results of room temperature neutron powder diffraction on the above series of compounds are presented in Table I, which focuses on the Mn-O bond lengths in the double layer of MnO_6 octahedra. The first two samples in Table I—$La_{1.2}Sr_{1.8}Mn_2O_7$ and $La_{1.4}Sr_{1.6}Mn_2O_7$—exhibit an MI transition; the remaining samples do not. We note that there are some differences in the bond lengths reported here compared to those in reference 4, but neutron powder diffraction is expected to more precisely locate the oxygen atoms than the x-ray diffraction. We note first that the bond lengths in like-doped Pr and Nd compounds are quite similar; this is consistent with the similar ionic radii of these two trivalent ions. Increasing the Sr content in any member of these series results in a decrease of the apical Mn-O bonds—consistent with an oxidation of the Mn center— while the in-plane Mn-O bond remains essentially unchanged. In the last column of Table I, we present the octahedral distortion defined by Seshadri, et al.[4] as the percent elongation of the apical (Mn-O(1) and Mn-O(2)) bond lengths compared to the in-plane (Mn-O(3)) bond length. If we look only at the $R_{1.2}Sr_{1.8}Mn_2O_7$ series, it would seem that this parameter separates those compounds with an MI transition from those without, as the $La_{1.2}Sr_{1.8}Mn_2O_7$ distortion (1.57%) is substantially less than that in the Pr or Nd analogs (3.07% and 3.30%, respectively). However, when the entire range of compounds is considered, this rule is seen to no longer hold true. Consider that $La_{1.4}Sr_{1.6}Mn_2O_7$ (with an MI transition) has a distortion of 2.74%, while $Pr_{1.0}Sr_{2.0}Mn_2O_7$ (an insulator at all T) is distorted by only 1.98%, and $Nd_{1.0}Sr_{2.0}Mn_2O_7$ by 2.31%. This fact, combined with the observation that the Mn-O(2) bond length in both insulators and compounds exhibiting an MI transition can be quite similar leads us to believe that the room temperature value of these parameters does not adequately predict the electronic behavior at low temperature. On the other hand, the Mn-O(3) in-plane bond is considerably longer (1.937 Å) in the MI compounds than in the insulators (~1.925 Å). This bond length may be a more important parameter characterizing the electronic behavior of these materials. A larger sample population will be required to judge whether this rule will hold true in general for other compounds in this family.

Table I. Mn-O bond lengths in $(R,Sr)_3Mn_2O_7$ compounds at room temperature

	Mn-O(1) (Å)	Mn-O(2) (Å)	Mn-O(3) (Å)	Distortion (%)
$La_{1.2}Sr_{1.8}Mn_2O_7$	1.942	1.994	1.937	1.57
$La_{1.4}Sr_{1.6}Mn_2O_7$	1.953	2.030	1.937	2.74
$Pr_{1.2}Sr_{1.8}Mn_2O_7$	1.937	2.035	1.925	3.07
$Pr_{1.0}Sr_{2.0}Mn_2O_7$	1.927	2.003	1.926	1.98
$Nd_{1.2}Sr_{1.8}Mn_2O_7$	1.935	2.038	1.921	3.30
$Nd_{1.0}Sr_{2.0}Mn_2O_7$	1.927	2.010	1.923	2.31

Additional evidence that a large apical Mn-O bond length distortion is inconsistent with a metallic state is found in the temperature dependence of this parameter for the $La_{1.2}Sr_{1.8}Mn_2O_7$ sample. As can be seen in Fig. 3, the in-plane Mn-O(3) bond is essentially temperature independent, showing only a small anomaly at $T_C \sim 120$ K. Likewise, the Mn-O(1) apical bond shared between the layers (not shown) does not change with temperature.[5] However, the Mn-O(2) apical bond *expands* by ~1% as the compound becomes metallic below T_C. This results in an increased octahedral distortion in the metallic state. This result is particularly interesting when compared to similar studies in the perovskite CMR family reported by Caignaert, et al.[8] These researchers showed that the octahedral distortion coordinate in $Pr_{0.7}Ca_{0.2}Sr_{0.1}MnO_3$ became substantially reduced in the metallic regime. They attributed this reduced distortion to the delocalization of the Mn^{3+} e_g electron, rendering the Mn sites Jahn-Teller inactive. In the layered compunds this does not seem to be the case; a large octahedral distortion is not inconsistent with an MI transition.

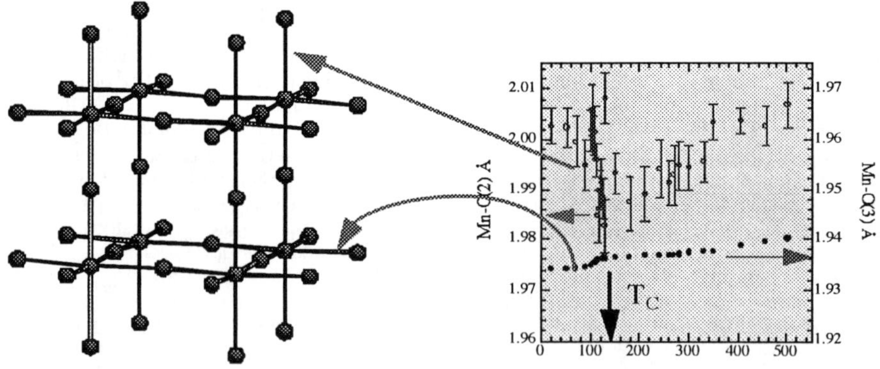

Fig. 3. Temperature dependence of Mn-O bonds of single crystal $La_{1.2}Sr_{1.8}Mn_2O_7$.

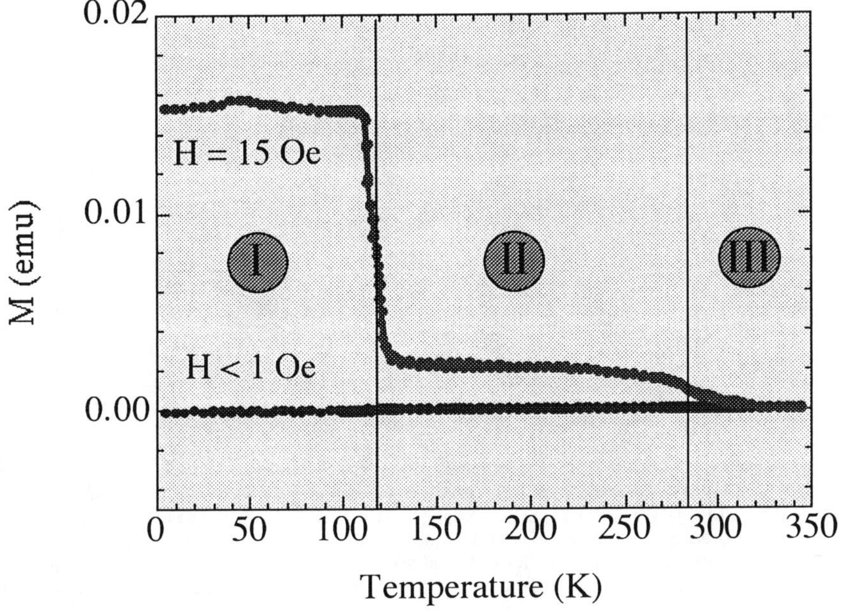

Fig. 4. Magnetization of single crystal $La_{1.2}Sr_{1.8}Mn_2O_7$

We now present the results of magnetization measurements on single-crystal specimen of $La_{1.2}Sr_{1.8}Mn_2O_7$. Similar features are found in the $La_{1.4}Sr_{1.6}Mn_2O_7$,which shows a MI transition. The insulating compounds have a somewhat different magnetic behavior that will be discussed below. What is immediately apparent from Fig. 4 is that the magnetic behavior of $La_{1.2}Sr_{1.8}Mn_2O_7$ falls into three regimes as a function of temperature. At low temperature (region I), in the metallic phase, a long-range, field-induced FM state is observed with Curie temperature $T_C \sim 120$ K. Neutron diffraction detects magnetic scattering in this regime.[5] For $T > 287$ K (region II), the material is paramagnetic and insulating. Interestingly, for $T_C < T < 287$ K (region II), a weak FM state (with hysteresis) is observed. Neutron diffraction detects no magnetic scattering in this temperature regime,[5] indicating that the FM correlations are short range. For $H = 15$ Oe, M scales as $(1-T/T^*)^\beta$, with $T^* = 287$ K and $\beta = 0.25\pm0.02$. This critical exponent is within experimental error of that predicted for the finite-size 2D XY model ($\beta = 3\pi^2/128 = 0.231$), suggesting that indeed the magnetic state in region II is of a short range nature.

Figure 5 shows the phase diagram of $La_{1.2}Sr_{1.8}Mn_2O_7$ in the H-T plane. This diagram delineates three regions in this phase plane. At higher temperatures a fourth region, the paramagetic insulating region, would appear on the phase diagram. Also shown on this diagram as dark circles are the results of *in situ* neutron powder diffraction measurements of $La_{1.2}Sr_{1.8}Mn_2O_7$ in the Earth's field or in a field of 0.6 T. These points are the Curie temperatures in the respective fields. The H = 0.6 T point lies on the line determined by magnetization measurements, while the Earth's field measurement lies on an extrapolation of the line determined by these measurements. Together they yield a consistent picture of the magnetic behavior of this material from both bulk magnetization and structural measurements.

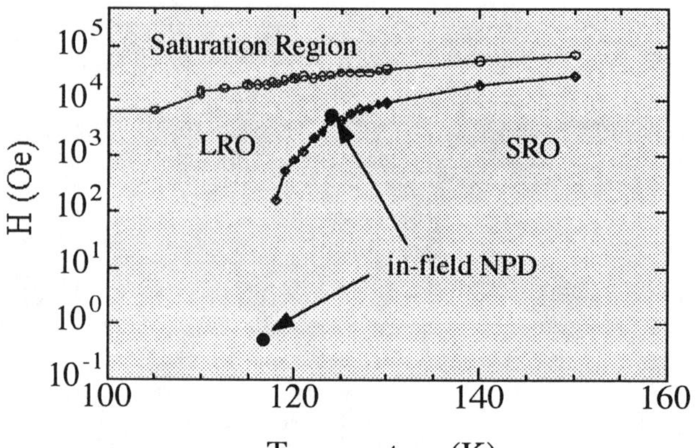

Temperature (K)

Fig. 5. Phase Diagram of $La_{1.2}Sr_{1.8}Mn_2O_7$ in the H-T plane.

The insulating compounds $R_{1.2}Sr_{1.8}Mn_2O_7$ and $RSr_2Mn_2O_7$ (R = Pr,Nd) do not develop the LRO FM state, but there is evidence of a transition at higher temperature (near 300 K for each compound). The similarity to the $La_{1.2}Sr_{1.8}Mn_2O_7$ and $La_{1.4}Sr_{1.8}Mn_2O_7$ materials suggests that these insulating compounds have an SRO state that never fully develops into the LRO FM state associated with the MI transition. Further detailed study of these compounds will be required to better understand the similarities and differences in the magnetic ordered states of the insulating and metallic compounds.

ACKNOWLEDGMENTS

This work was supported by the U.S. Department of Energy, Basic Energy Sciences-Materials Sciences, under Contract W-31-109-ENG-38 (JFM, CDP, DGH, JDJ, SDB) and by the NSF Office of Science and Technology Centers under Contract No. DMR 91-20000 (DNA)

REFERENCES

1. Y. Shimakawa, Y. Kubo and T. Manako, Naure **379**, 53 91996).

2. S-W. Cheong, H.Y. Hwang, B. Batlogg and L.W. Rupp, Jr., Solid State Comm. **98**, 163 (1996).

3. Y. Moritomo, A. Asamitsu, H. Kuwahara and Y. Tokura, Nature **380**, 141 (1996).

4. R. Seshadri, C. Martin, A. Maignan, M. Hervieu, B. Raveau and C.N. Ramachandra Rao, J. Mater. Chem. **6**, 1585 (1996).

5. J.F. Mitchell, D.N. Argyriou, J.D. Jorgensen, D.G. Hinks, C.D. Potter and S.D. Bader Phys. Rev. B., in press.

6. C.D. Potter, M. Swiatek, D.N. Argyriou, J.F. Mitchell, D.G. Hinks, J.D. Jorgensen and S.D. Bader, submitted to Phys. Rev. Lett.

7. D.N. Argyriou, C.D. Potter, J.F. Mitchell, J.D. Jorgensen, R. Kleb and S.D. Bader, submitted to Phys. Rev. Lett.

8. V. Caignaert, E. Suard, A. Maignan, C. Simon and B. Raveau, J. Magn. Magn. Mater. **153**, L260 (1996).

The Narrow-Band System $La_{1-x}Sr_xTiO_3$

Charles C. Hays, John B. Goodenough,* and John T. Markert
Department of Physics, *Center for Materials Science & Engineering, ETC 9.102
University of Texas at Austin, Austin, TX 78712

Abstract

Magnetic, transport, and structural data on polycrystalline samples of the system $La_{1-x}Sr_xTiO_3$, $0 \leq x \leq 0.8$, show an abrupt, but apparently smooth, transition from p-type polaronic conduction in a canted-spin antiferromagnetic system for $x \leq 0.04$ to metallic, mass-enhanced Pauli paramagnetic behavior for $x \geq 0.1$ with evidence for a spin glass and a discontinuous drop in T_N in the transition range. We interpret these results as a manifestation of the coexistence of two distinguishable electronic domains within a single crystallographic phase as a result of cooperative displacements of the oxygen atoms.

Introduction

Unusual physical properties have been found in oxides with the perovskite structure where there is a cross-over from localized to intinerant electronic behavior. In this paper we report transport, magnetic, and structural data on the perovskite system $La_{1-x}Sr_xTiO_3$, which undergoes a transition from a Mott insulator at $x = 0$ to a Pauli paramagnetic metal for $x \geq 0.08$. The parent compound $LaTiO_3$ has a Goldschmidt tolerance factor $t < 1$ that is accommodated by a cooperative rotation of the corner-shared TiO_6 octahedra about the cubic [110] axes to reduce the symmetry from cubic to orthorhombic. It has a type-G antiferromagnetic order below a Néel temperature T_N, but with a weak ferromagnetic component induced by an antisymmetric exchange term having a Dzialoshinskii vector $\mathbf{D}_{ij}$ parallel to the orthorhombic "b" axis. The magnetic-ordering temperature T_N has been reported [1] to increase with hydrostatic pressure P,

which indicates the applicability of superexchange perturbation theory [2]; but the π^* conduction band is only 1/6-filled, which would stabilize a half-band ferromagnetism for narrower bands where intraatomic-exchange interactions are stronger than the interatomic interactions. Indeed, the narrower isoelectronic perovskite $YTiO_3$ is a ferromagnetic insulator [3]. Of particular interest for our investigation were: (1) whether the transition is smooth or abrupt or involves a segregation of hole-rich and hole-poor domains, (2) how the long-range magnetic-ordering temperature T_N changes with x, (3) how the thermoelectric power changes on crossing the critical composition range for the transition, and (4) whether changes in the kinetic energy of the electron system are manifest in deviations from Végard's law for the cell volume.

Experimental

The $La_{1-x}Sr_xTiO_3$ specimens were prepared by arc melting stoichiometric quantities of La_2O_3, Ti_2O_3, Ti, and $SrTiO_3$. The specimens were annealed for 48 hours at 850 °C. Powder x-ray diffraction results, obtained with a Philips diffractometer and Cu K_α radiation, showed the specimens to be single phase. The peak positions were calibrated against Si as an internal standard, and the lattice parameters were obtained with a least-squares fitting procedure. The oxygen contents of the specimens were obtained with a Perkin-Elmer TGA-7 thermogravimetric

Mat. Res. Soc. Symp. Proc. Vol. 453 © 1997 Materials Research Society

analyzer (TGA) from the weight gain recorded upon heating in air to 1000 °C to oxidize the titanium to Ti^{4+}.

Magnetic susceptibility, $\chi = M/H$, measurements for the $La_{1-x}Sr_xTiO_3$ specimens were obtained from 10 to 300 K with a Quantum Design MPMS SQUID magnetometer.

The four-point-probe method was used to obtain the temperature dependence of the resistance of the specimens over the range 80 to 310 K in a homemade device controlled by a Lakeshore temperature controller. The thermoelectric-power data were recorded over the temperature range 80 to 310 K in a homemade device described elsewhere [4].

Results

Fig. 1 shows the variation with x of the room-temperature cell volume per formula unit for the system $La_{1-x}Sr_xTiO_3$, which forms a solid solution for $0 \leq x \leq 1.0$. A linear extrapolation of V from the range $0.3 \leq x \leq 0.7$ to $x < 0.3$ shows an increasing deviation from Végard's law to larger volumes as x decreases, but there is no evidence of a discontinuous change at a first-order transition in the range $0 < x < 0.7$. A first-order contraction of the volume occurs on increasing x through the orthorhombic to cubic transition between $x = 0.7$ and $x = 0.75$.

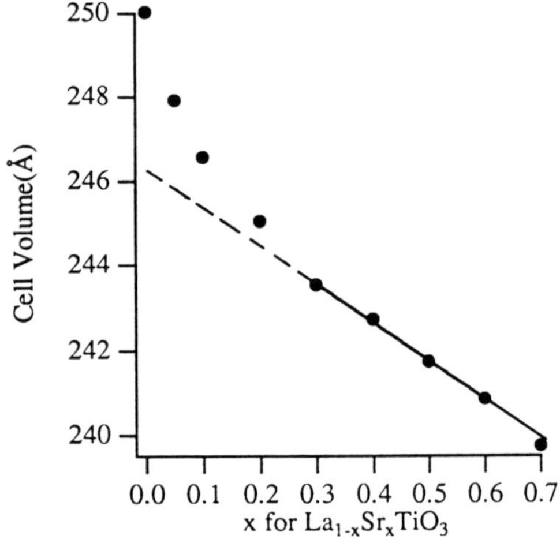

Fig. 1. Variation with x of orthorhombic unit-cell
volume for $La_{1-x}Sr_xTiO_3$, $0 \leq x \leq 0.7$.

Fig. 2(a) shows the temperature dependence of the magnetic susceptibility, $\chi(T)$, of $LaTiO_3$ taken on heating in a field of 1 kOe after cooling in zero field (ZFC) and in 1 kOe (FC). Long-range order at a $T_N \approx 145$ K with a weak, canted-spin ferromagnetism agrees with the published data [5]. A Curie-Weiss temperature dependence of χ^{-1} is found at higher temperatures, but with changes of slope setting in near 280 and 170 K that signal the onset of weak ferromagnetism in domains of short-range magnetic order. Fig. 2(b) shows similar $\chi(T)$ data for $x = 0.05$. These curves are characteristic of spin glass behavior, and indeed other magnetic data support this inference. The spin-glass characteristics increase with x in the range $0.04 \leq x \leq 0.075$.

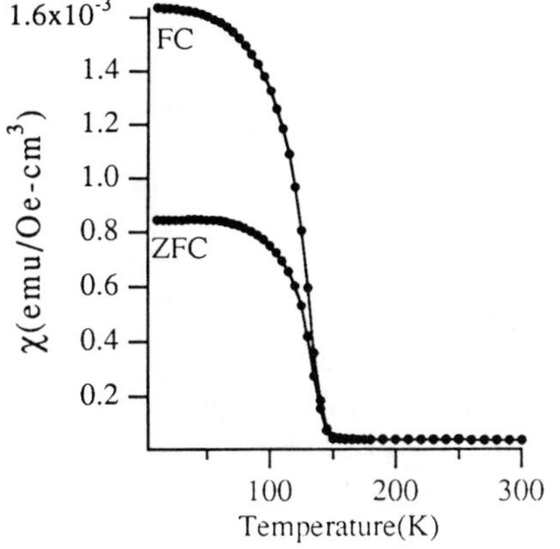

Fig. 2(a). Magnetic susceptibilty (H = 1 kOe) vs. temperature for LaTiO$_3$.

Fig. 2(b). Magnetic susceptibilty (H = 1 kOe) vs. temperature for La$_{0.95}$Sr$_{0.05}$TiO$_3$.

Fig. 3 shows the variation of T$_N$ with x; it falls abruptly at the transition to spin glass behavior , but is nearly 100 K for x = 0.075; by x = 0.1, long-range magnetic order no longer exists and the specimens exhibit a mass-enhanced Pauli paramagnetism.

Fig. 4 shows the temperature variation of the thermoelectric power, S(T), for different values of x. A dramatic change from p-type polaronic behavior to n-type metallic behavior occurs in the narrow composition range $0.044 < x < 0.05$. The increase with x in the magnitude of S(300K) is compatible with a metallic $S \approx k_B T / \varepsilon_F$ dependence as ε_F decreases on oxidation and the curvature of the $\varepsilon(\mathbf{k})$ dispersion curve also increases on approaching the bottom of the conduction band.

Discussion

Fig. 5 shows schematic energy vs. density of states diagrams for two possible strongly correlated electronic systems: one applies to antiferromagnetic LaTiO$_3$ and the other to isoelectronic, ferromagnetic YTiO$_3$. The π^* bands shown refer to the antibonding bands of Ti: 3d$_t$ - O: 2p$_\pi$ parentage where d$_t$ refers to the xy, yz $\pm$ i zx degenerate orbitals in a cubic crystalline field that π bond with the coordinating oxygen atoms. Where these orbitals are degenerate, the bands are 1/6-filled in LaTiO$_3$ and YTiO$_3$. We can expect retention of antiferromagnetic order, as in LaTiO$_3$, where the interatomic Ti-O-Ti bonding competes with the intraatomic exchange coupling responsible for Hund' s highest multiplicity rule in the free atom. For narrower π^* bands, the intraatomic exchange interaction should give rise to ferromagnetic order as is found in YTiO$_3$. In the system La$_{1-x}$Sr$_x$TiO$_3$, oxidation of the TiO$_3$ array of Fig. 5(b)

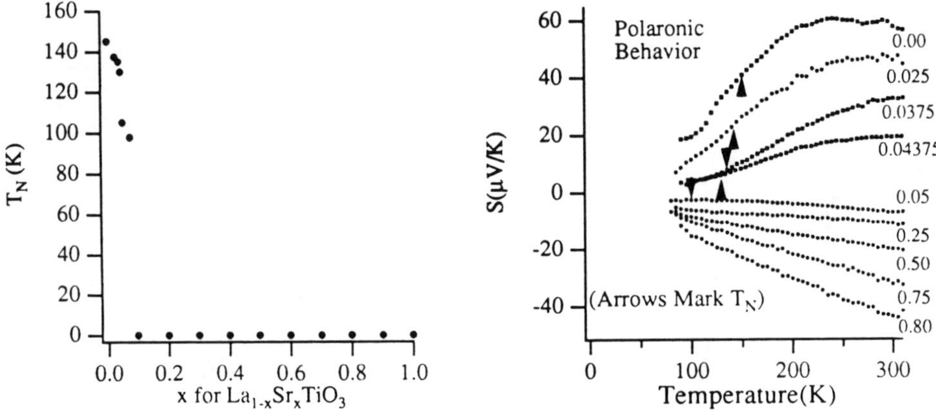

Fig. 3. Variation with x of the Néel temperature T_N for La$_{1-x}$Sr$_x$TiO$_3$.

Fig. 4. Thermoelectric power vs. temperature for La$_{1-x}$Sr$_x$TiO$_3$.

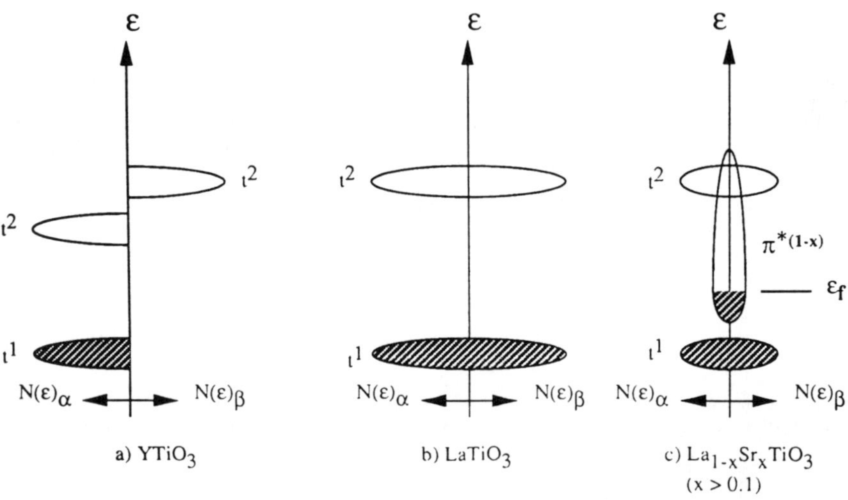

Fig. 5. Schematic energy diagrams for (a) YTiO$_3$, (b) LaTiO$_3$, and (c) metallic La$_{1-x}$Sr$_x$TiO$_3$, for x > 0.1.

is occurring; the different size and electronegativity of Sr^{2+} vs. La^{3+} also perturbs the electron potentials of the π^* bands, which can be expected to introduce Anderson localized states at the edges of the band.

The S(T) data, Fig. 4, as well as our resistance data indicate retention of Mott-Hubbard splitting of the $Ti^{4+/3+}$ and $Ti^{3+/2+}$ redox couples as shown in Fig. 5(b) for $0 \leq x \leq 0.044$; metallic behavior for $x \geq 0.05$ indicates an abrupt transition from the strongly correlated regime of Fig. 5(b) to the more weakly correlated regime of Fig. 5(c) where the π^* bands are 1/6-filled.

The magnetic data indicate that there is spin-glass behavior in the range $0.04 \leq x \leq 0.075$, which implies a segregation into hole-rich and hole-poor domains in the transition range with a percolation threshold for the hole-rich domains at $x = 0.05$. The Néel temperature T_N drops sharply with the onset of spin-glass behavior, and vanishes abruptly with the disappearance of the hole-poor domains.

A segregation into hole-rich and hole-poor domains in the transition region implies a first-order phase change between strongly correlated and more weakly correlated electronic systems, which would be consistent with the evidence from photoemission data for the coexistence of Mott-Hubbard and coherent states [6]. A first-order transition between these electronic behaviors can be expected from the Virial theorem, which states that a discontinuous change in the kinetic energy of an electronic system must be compensated by a discontinuous change in the potential energy. For antibonding electrons, the equilibrium Ti-O-Ti bond length for more localized electrons would be greater than that for itinerant electrons. Moreover, the perovskite structure can accommodate the coexistence of domains with different Ti-O bond lengths by cooperative displacements of the oxygen atoms from a Ti atom on one side toward the Ti atom on the other. In the copper-oxide superconductors, there is mounting evidence for such a segregation into hole-rich and magnetic stripes within the CuO_2 sheets. The deviation from Végard's law in Fig. 1 is consistent with an expansion of the mean volume with an increase in the volume of the hole-poor, strongly correlated domains.

Conclusion

The system $La_{1-x}Sr_xTiO_3$ exhibits a smooth transition from p-type polaronic conduction in a canted-spin antiferromagnet to n-type metallic behavior with mass-enhanced Pauli paramagnetism across the range $0.04 < x < 0.1$. However, the magnetic, structural and transport properties indicate that the transition may be first-order, the smooth change in properties resulting from the coexistence of two electronic phases that are accommodated within a single crystallographic phase by cooperative oxygen displacements that may be static or dynamic.

References

1. Y. Tokura, et al, Phys. Rev. B **48**, 9677 (1993).

2. J. B. Goodenough, Prog. Solid State Chem. **5**, 145 (1972).

3. J. P. Goral, J. E. Greedan, and D. A. MacLean J. Solid State Chem. **43**, 244 (1982).

4. J. B. Goodenough, J.-S. Zhou, and J. Chan, Phys. Rev. B **47**, 5275 (1993).

5. Y. Tokura, et al, Phys. Rev. B **48**, 9677 (1993).

6. I. H. Inoue, et al, Phys. Rev. Lett. **74**, 2539 (1995).

LITHIUM DOPING IN La$_2$CuO$_4$ AND La$_2$NiO$_4$

J.L. SARRAO, M.E. TORELLI, and Z. FISK
National High Magnetic Field Laboratory, Florida State University
1800 E. Paul Dirac Dr., Tallahassee, FL 32306

ABSTRACT

We have studied the effect of hole doping via Li substitution in La$_2$CuO$_4$ and La$_2$NiO$_4$. The response to hole doping of the Ni (S=1) sublattice is distinct from that of the Cu (S=1/2) sublattice. These effects are seen most clearly for La$_2$Cu$_{0.5}$Li$_{0.5}$O$_4$ and La$_2$Ni$_{0.5}$Li$_{0.5}$O$_4$, in which the Li ions form an ordered superlattice. It appears that Ni is in a low-spin d^7 configuration whereas Cu is in a d^9-ligand "Zhang-Rice" configuration. Despite this difference, the response of both La$_2$CuO$_4$ and La$_2$NiO$_4$ to hole doping is surprisingly independent of whether Sr (which substitutes out of plane for La) or Li (which substitutes in the plane for Cu/Ni) is used as the dopant.

INTRODUCTION

The evolution of the properties of La$_2$CuO$_4$ with hole doping into the CuO$_2$-layers via alkali metal (e.g. Sr) substitution for the out-of-plane La ion has been extensively studied in recent years. However, many aspects of this problem remain poorly understood, including the loss of Cu's magnetic moment, the evolution from semiconducting to metallic to superconducting transport behavior, and the coupled nature of the (micro-)structural and transport/magnetic properties.

In-plane hole doping is also possible via Li-substitution, and Li is a nearly perfect lattice match for Cu [1]. The properties of the series of compounds La$_2$Cu$_{1-x}$Li$_x$O$_4$ are somewhat surprising. The materials are metallic for no x, indicating that the introduced holes are bound, presumably in the vicinity of the Li [2]. However, many of the other properties of La$_2$Cu$_{1-x}$Li$_x$O$_4$ are remarkably similar to those found for La$_{2-x}$Sr$_x$CuO$_4$ for the same x. The in-plane plaquette size, as determined by lattice constant measurements, depends only on x [1], the magnetism of La$_2$CuO$_4$ is suppressed at essentially the same rate for Sr or Li substitution [3], and even the rate of suppression of T$_c$ in optimally-doped La$_{1.85}$Sr$_{0.15}$CuO$_4$ by added Li is relatively weak, being similar to that due to excess Sr doping and being several times smaller than that found for Zn substitutions [1].

Li can be substituted up to the limiting stoichiometry La$_2$Cu$_{0.5}$Li$_{0.5}$O$_4$, which is an atomically ordered, diamagnetic, and insulating material [4]. This material contains formally Cu^{+3}, and the diamagnetism is therefore unexpected. Equally surprising is that an NQR study of Cu in this material finds a magnetic relaxation mechanism with activation energy 1500 K [5]. In view of the diamagnetism of the material, it appears that the holes brought into the lattice by the Li couple into singlets with the Cu^{+2} spins, reminiscent of the Zhang-Rice singlet picture [6], but in this case on an ordered lattice with apparently immobile holes.

Many studies have examined hole-doping in La$_2$NiO$_4$ in an attempt to observe high-T$_c$ superconductivity in this isostructural material. Here the Ni ions form a spin-1 lattice rather than the spin-1/2 lattice present in the cuprates. In La$_{2-x}$Sr$_x$NiO$_4$, Sr substitutions are possible to beyond x=1.5 [7]. These materials, however, do not superconduct for any x and only become metallic for x>1.1. Furthermore, it appears that the physics starts to change beyond x=0.5, where charge ordering has been reported [8].

355

Mat. Res. Soc. Symp. Proc. Vol. 453 © 1997 Materials Research Society

There is also, although the data are much less extensive, a strong similarity between Sr and Li doping in La_2NiO_4, with many physical properties being principally determined by the hole count. It is possible to dope with Li to x=0.5, creating formally Ni^{+3}, and in this case it appears that Ni is in fact in the d^7-configuration [9]. Here, we discuss these two families of compounds, $La_2Cu_{1-x}Li_xO_4$ and $La_2Ni_{1-x}Li_xO_4$, $0 \leq x \leq 0.5$, and compare them with their Sr-doped analogues, hoping to shed further light on their similarities and differences and thereby better understand these materials.

EXPERIMENTAL DETAILS

All of the data reported here were taken with polycrystalline samples. The sintered specimens were synthesized using standard ceramic preparation techniques. The appropriate stoichiometric amounts of La_2O_3, $SrCO_3$, CuO, NiO, and Li_2CO_3 were mixed and fired at 900° C over a multi-day period with several intermediate regrindings. The rather low sintering temperature was dictated by the volatility of Li_2O; however, no appreciable Li losses were observed in our specimens. x-ray powder diffraction revealed the samples to be essentially single phase, and the lattice constants were refined from the observed reflections using Si as an internal standard. Magnetic susceptibility measurements were made using a Quantum Design SQUID magnetometer.

RESULTS

$La_2Cu_{1-x}Li_xO_4$

The effect of Li-substitution on the in-plane lattice constant of pure and Sr-doped La_2CuO_4 is shown in Figure 1. The data for the in-plane lattice parameters collapse onto one plot versus hole count: the effect of Sr or Li addition is essentially identical structurally. This is true for both the concentration dependence of the orthorhombic-tetragonal transition temperature, as well as the variation of tetragonal a_0 lattice parameter at room temperature [1]. It appears that the introduced hole has a particular size independent of its origin or whether or not the hole is mobile. Perhaps the way to think about the contraction of the in-plane lattice parameter with hole addition is as a relaxation about the hole in the highly negatively charged CuO_2 background. Li substitution can be made all the way to $La_2Cu_{0.5}Li_{0.5}O_4$, which forms an ordered superstructure of alternating Cu/Li positions in the planes. Rietveld refinement of neutron powder diffraction data and electron diffraction studies show the in-plane ordering to be at least 80% [10].

The behavior of the c-axis lattice parameter with Sr and Li substitution is distinct and may explain the different solubility limits for the two dopants. The maximum Sr concentration that can be accommodated is near 0.35 and c appears to saturate near $La_{1.75}Sr_{0.25}CuO_4$ [11]. On the other hand, the variation of c with Li concentration is small and essentially linear up to and including the point at which the Cu and Li form an ordered superlattice [1,4]. Thus, it would appear that the structural effects in the out-of-plane direction limit solubility rather than an in-plane hole-concentration effect.

Sr and Li also have nearly identical effects on the magnetic phase diagram of La_2CuO_4: the long range magnetic order of the Cu-spins is lost by x=0.03 [3] (see inset, Figure 2). This behavior contrasts strongly with the effect of in-plane substitution of Zn or Mg. For both of these

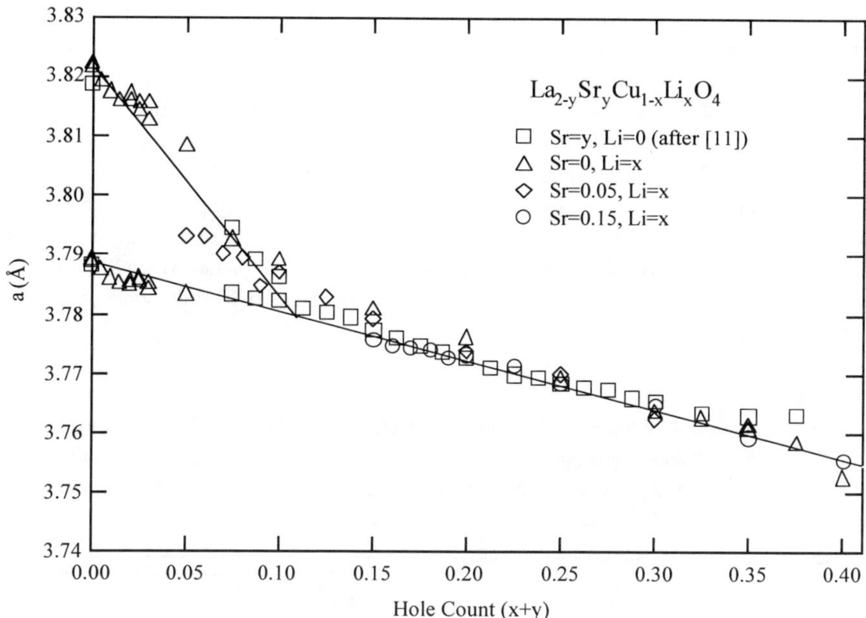

Figure 1. In-plane lattice constant, a_0, of La$_{2-y}$Sr$_y$Cu$_{1-x}$Li$_x$O$_4$ at room temperature. a_0 is a function only of the net hole count (i.e., x+y) and not the individual Sr and Li concentrations. Data for y=0 are after Radaelli et al. [11].

cases, the 3-D magnetic ordering temperature is depressed at nearly but less than the percolation rate, with roughly 30% substitution finally suppressing T_N completely [12]. Muon spin rotation experiments also suggest that microscopically the magnetism in La$_2$Cu$_{1-x}$Li$_x$O$_4$ is closer to that of Sr-doped La$_2$CuO$_4$ than that of Zn-doped La$_2$CuO$_4$ [13].

The evolution of the temperature dependent magnetic susceptibility with Li-substitution is shown in Figure 2. The susceptibility evolves with increasing Li concentration from Curie-Weiss-like to, aside from a small and sample-dependent extrinsic Curie tail, temperature-independent diamagnetism for La$_2$Cu$_{0.5}$Li$_{0.5}$O$_4$. The magnitude of the diamagnetism at room temperature (-60 x 10^{-6} emu/mol) is consistent with the expected core diamagnetism and van Vleck paramagnetism, so that, despite the presence of a Cu^{+2} (S=1/2) sub-lattice, no spin susceptibility is observed. Although the Cu is formally trivalent here, the stability of the material suggests that the diamagnetism is unlikely to be a property of Cu^{+3}. Furthermore, Yoshinari, et al. [5] have reported, using Cu NQR relaxation measurements, the observation of a magnetic excitation with an activation energy near 1500 K. These data are more easily understood by assuming that Cu possesses a spin (i.e. is Cu^{+2}) that is compensated in some sort of singlet rather than by assuming a trivalent Cu configuration.

The evolution of the magnetic susceptibility of $La_{2-y}Sr_yCuO_4$ for intermediate and high Sr concentrations (i.e. $y \geq 0.05$) is qualitatively different from that of $La_2Cu_{1-x}Li_xO_4$ [14]. The susceptibility in pure and Sr-doped La_2CuO_4 is characterized by a broad peak as a function of temperature which moves to lower temperature as the Sr concentration is increased. By normalizing the data for various y with respect to a scaling temperature T_{max}, proportional to the intralayer Cu-Cu coupling constant J, the susceptibilities can be made to overlap and are in good agreement with calculation [14]. Thus, the dominant effect of the Sr-produced holes is to reduce J and thereby shift the peak in susceptibility to lower temperature. No such peak is observed in Li-doped La_2CuO_4 . Although we do not have a quantitative understanding of these differences, it seems probable that the differences are due to the defects in the CuO_2 planes which Li introduces - the effect of which may be to mask the above-mentioned peak - and to the finite screening length of the bound Li holes.

$La_2Ni_{1-x}Li_xO_4$

Although not as striking as in the case of La_2CuO_4, the change in lattice constants of La_2NiO_4 due to Sr or Li doping is similar. In both cases for doping concentrations up to 0.5, c increases slowly while a decreases rapidly (at a rate 3-4 times faster than the expansion of c).

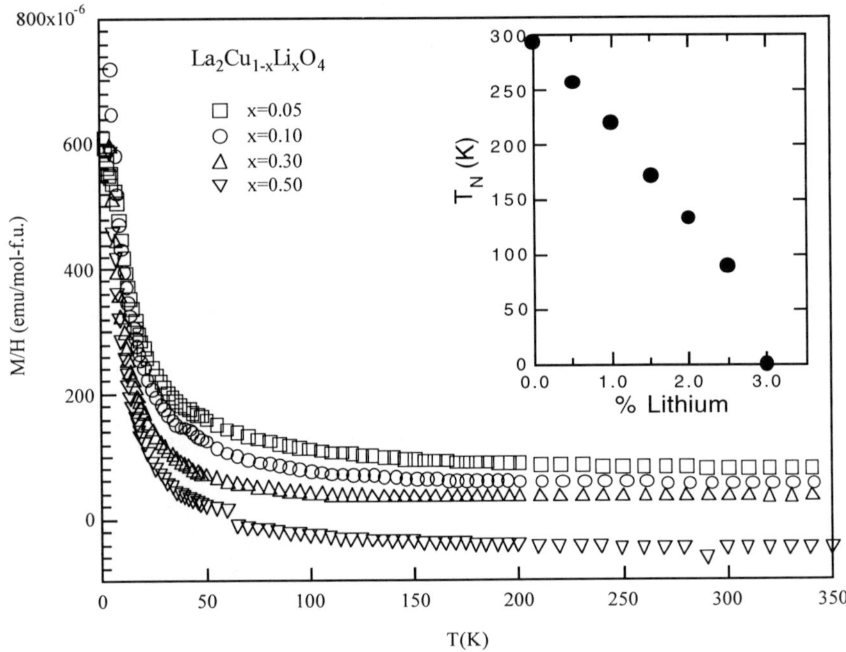

Figure 2. Magnetic susceptibility as a function of temperature for $La_2Cu_{1-x}Li_xO_4$ for various values of x. The inset shows the suppression of the T_{Neel} with Li substitution.

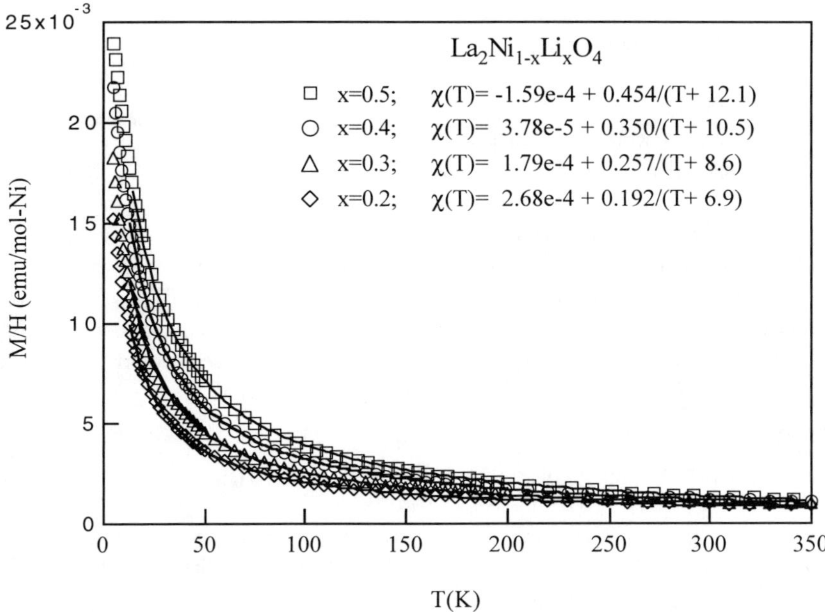

Figure 3. Magnetic susceptibility as a function of temperature for $La_2Ni_{1-x}Li_xO_4$ for various values of x. The solid lines are fits to the data using a Curie-Weiss form. The extracted effective moments per Ni are as follows: x=0.2, μ_{eff}= 1.24 μ_B; x=0.3, μ_{eff} = 1.43 μ_B; x=0.4, μ_{eff} = 1.67 μ_B; x=0.5, μ_{eff} = 1.91 μ_B. See text for details.

Thus, the Ni-O octahedra become more distorted as a function of hole doping (as determined by the increase in c/a), suggesting the influence of strong crystal fields. It has not proved possible to produce samples with Li concentrations greater than 0.5. For Sr concentrations greater than 0.5 a qualitative change occurs: c drops rapidly while a is essentially independent of doping up to the maximum doping level of 1.5, perhaps suggesting a change in the valence configuration of the Ni ions [7]. These effects may also be associated with the observation of charge ordering reported near $La_{1.5}Sr_{0.5}NiO_4$ [8].

The evolution of the temperature-dependent magnetic susceptibility of $La_2Ni_{1-x}Li_xO_4$ is shown in Figure 3. In striking contrast to the data of Figure 2, in the nickelate case the susceptibility per Ni actually increases with Li concentration. The solid lines are fits of the data to $\chi(T)=\chi_0+C/(T+\theta)$. The fitted parameters are given in the figure. The effective moment per Ni increases from 1.24 μ_B for Li=0.2 to 1.91 μ_B for Li=0.5. This is close to the value expected for a g=2 spin-1/2 lattice, μ_{eff} =1.73 μ_B, and the difference may be due to an enhanced g factor associated with crystal field effects [15].

Thus, it would appear that rather than forming in a d^8-ligand configuration (analogous to the d^9-ligand configuration suggested by $La_2Cu_{0.5}Li_{0.5}O_4$) $La_2Ni_{0.5}Li_{0.5}O_4$ possesses a low-spin d^7 configuration. In support of this argument, electron paramagnetic resonance measurements find a spin-1/2 resonance with strong g-shift, consistent with crystal field effects dominating Hund's Rule coupling [15], as would be required in the d^7 configuration.

CONCLUSIONS

Although more work is needed, especially in the case of the doped nickelates, it seems clear that a detailed understanding of the similarities and differences between Sr and Li doping in La_2CuO_4 and La_2NiO_4 will provide valuable insight into the relevant physics of these systems. In particular, the extent to which doped holes reside on the Ni ions whereas they form ligand states in the Cu-O plaquettes may be at the origin of why Sr-doped La_2CuO_4 conducts so well whereas Sr-doped La_2NiO_4 is a rather poor conductor. On the other hand if $La_2Cu_{0.5}Li_{0.5}O_4$ does possess a low-spin Cu^{+3} configuration analogous to that in $La_2Ni_{0.5}Li_{0.5}O_4$, it remains to be understood how the system evolves from a ligand-like configuration, which seems to be appropriate at low doping, to an atomic orbital picture, which might explain the fully-doped data.

REFERENCES

1. J.L. Sarrao, D.P. Young, Z. Fisk, E.G. Moshopoulou, J.D. Thompson, B.C. Chakoumakos, and S.E. Nagler, Phys. Rev. B **54** (1996) 12014.

2. M.A. Kastner, R.J. Birgeneau, C.Y. Chen, Y.M. Chiang, D.R. Gabbe, H.P. Jenssen, T. Junk, C.J. Peters, P.J. Picone, T. Thio, T.R. Thurston, and H.L. Tuller, Phys. Rev. B **37** (1988) 1111.

3. A.I. Rykov, H. Yasouka, and Y. Ueda, Physica C **247** (1995) 327.

4. J.P. Attfield and G. Ferey, J. Solid State Chem. 80 (1989) 112.

5. Y. Yoshinari , P.C. Hammel, J.A. Martindale, E. Moshopoulou, J.D. Thompson, J.L. Sarrao, and Z. Fisk, Phys. Rev. Lett. **72** (1996) 2069.

6. F.C. Zhang and T.M. Rice, Phys. Rev. B **37** (1988) 3759.

7. R.J. Cava, B. Batlogg, T.T. Palstra, J.J. Krajewski, W.F. Peck, Jr., A.P. Ramirez, and L.W. Rupp, Jr., Phys. Rev. B **43** (1991) 1229.

8. S.-W. Cheong, C.H. Chen, H.Y. Hwang, B. Batlogg, L.W. Rupp, Jr., A.S. Cooper, Physica B **199&200** (1994) 659.

9. G. Demazeau, J.L. Marty, M. Pouchard, T. Rojo, J.M. Dance, and P. Hagenmuller, Mat. Res. Bull. **16** (1981) 47

10. E. Moshopoulou, G.H. Kwei, J.D. Thompson, J.L. Sarrao, Z. Fisk, B.C. Chakoumakos, and S.E. Nagler, unpublished.

11. P.G. Radaelli, D.G. Hinks, A.W. Mitchell, B.A. Hunter, J.L. Wagner, B. Dabrowski, K.G. Vandervoort, H.K. Viswanathan, and J.D. Jorgensen, Phys. Rev. B **49** (1994) 4163.

12. S.-W. Cheong, A.S. Cooper, L.W. Rupp, Jr., B. Batlogg, J.D. Thompson, and Z. Fisk, Phys. Rev. B **44** (1991) 9739.

13. L.P. Le, R.H. Heffner, D.E. MacLaughlin, K. Kojima, G.M. Luke, B. Nachumi, Y.J. Uemura, J.L. Sarrao, and Z. Fisk, Phys. Rev. B **54** (1996) 9538.

14. D.C. Johnston, Phys. Rev. Lett. **62** (1989) 957.

15. G. Demazeau, B. Buffat, M. Pouchard, and P. Hagenmuller, J. Solid State Chem. **54** (1984) 389.

SYNTHESIS OF $A_{1-x}A'_xNbO_3$ ($A=Na^+$, K^+; $A'=La^{3+}$, Sr^{2+})
AND ITS ELECTRICAL PROPERTIES

W. Sugimoto, T. Tahara, Y. Sugahara, and K. Kuroda
Department of Applied Chemistry, Waseda University, Ohkubo 3, Shinjuku-ku, Tokyo 169,
Japan

ABSTRACT

Polycrystalline samples of $A_{1-x}A'_xNbO_3$ ($A=Na^+$, K^+; $A'=La^{3+}$, Sr^{2+}) were prepared by solid-state reactions and the relationship between the structure and electrical properties was examined. Single-phase products were obtained for $Na_{1-x}La_xNbO_3$ ($x \leq 0.1$), $K_{1-x}La_xNbO_3$ ($x \leq 0.15$), $Na_{1-x}Sr_xNbO_3$ ($x \leq 0.3$), and $K_{1-x}Sr_xNbO_3$ ($x \leq 0.5$). The structures of the sodium-containing compounds and potassium-containing compounds were indexed according to a pseudocubic and cubic symmetry, respectively. The resistivity measurements showed semiconductive behavior for the lanthanum-containing samples. The strontium-containing samples showed semiconducting properties for $x \leq 0.2$, while a transition to metallic conduction at low temperature was observed for $x > 0.2$.

INTRODUCTION

Niobates are known to show metallic, semiconducting, and even superconducting properties, when the niobium ion is in a low valence state. Layered M_xNbO_2 ($M=Li$, Na)[1-3] and the defect NaCl type NbO^4 have been well studied mainly concerning their superconducting properties. High electrical conductivity has also been reported for perovskite niobates ($ANbO_3$). Metallic conduction in $ANbO_3$ has been reported for binary systems where divalent ions occupy the A site, such as A-site deficient M_xNbO_3 ($M=Sr$,[5-10] Ba,[5, 6, 11-14] and Eu^{15-21}) and stoichiometric $BaNbO_3$.[22] Various electrical properties have been reported for ternary systems where the A site is occupied by monovalent and divalent ions. For example, semiconducting properties with a linear $\log\rho - T$ relationship for $Na_{1-x}Sr_xNbO_3$ ($0.2 \leq x \leq 0.6$)[23, 24] and a very weak temperature dependence with a change in the $d\rho/dT$ sign at low temperature for $K_{0.5}Sr_{0.5}NbO_3$[22] have been reported. Also, a metal-semiconductor transition with change in the composition has been reported for $K_{1-x}Ba_xNbO_3$ ($0.2 \leq x \leq 0.5$).[22] The latter two compounds were synthesized through high pressure techniques. The electrical properties of these ternary niobates suggest that high conductivity can be obtained by sufficient doping of low valence niobium through substitution in $ANbO_3$ ($A=M^+$) with La^{3+}. The substitution of a trivalent ion for a monovalent ion can alter the carrier concentration two times more compared to a divalent ion substitution; however, no such compounds have been reported so far. Here we report the synthesis of ternary $A_{1-x}A'_xNbO_3$ niobates where $A=Na^+$ or K^+ and $A'=La^{3+}$. Substitution with Sr^{2+} was also conducted for comparison. Emphasis was placed on the structural variation with substitution of ions with different valency. The relationship between the structure and electrical properties is also discussed.

EXPERIMENT

Polycrystalline samples of $Na_{1-x}La_xNbO_3$ (NLN) and $K_{1-x}La_xNbO_3$ (KLN) were prepared by solid-state reactions of $ANbO_3$ ($A=Na$, K), La_2O_3, NbO_2, and Nb. $ANbO_3$ was

361

Mat. Res. Soc. Symp. Proc. Vol. 453 © 1997 Materials Research Society

prepared by mixing appropriate amounts of A_2CO_3 (A=Na, K) and Nb_2O_5 in an ambient atmosphere at 1000 °C for 1 h. NbO_2 was prepared by firing a mixture of Nb_2O_5 and Nb in an argon atmosphere for 1 h at 1200 °C. The starting mixtures were thoroughly ground in acetone and dried before firing. Pelletized NLN was fired at 1200 °C for 10 h under an argon atmosphere. Dried powder of KLN was prefired at 1100 °C for 0.5 h under an argon atmosphere. The powder was then reground, pelletized, and fired for 5 h under the same conditions for the prefiring process. To minimize oxidation and reaction with the alumina boat, each pellet was surrounded by powder with the same composition and set in an alumina tube which had been evacuated to $\sim$13 Pa before argon purging. Samples of $Na_{1-x}Sr_xNbO_3$ (NSN) and $K_{1-x}Sr_xNbO_3$ (KSN) were synthesized in a similar manner to that of NLN.

X-ray powder diffraction (XRD) patterns were obtained with CuKα radiation (Shimadzu XD-3A). Unit-cell parameters were obtained by precise measurements with FeKβ radiation (Rigaku Rad II-A) using Si as the internal standard. Resistivity was measured by the standard dc four-probe method in the range of 5 to 280 K.

RESULTS AND DISCUSSION

The end members $NaNbO_3$ and $KNbO_3$ were white-colored. The color of the samples changed to light blue with substitution and darkened with further substitution. XRD measurements revealed the structures of $NaNbO_3$ and $KNbO_3$ to be orthorhombic, consistent with previous reports.[25, 26] Single phases for La substitution were obtained only near the end members $NaNbO_3$ and $KNbO_3$ and the solubility limits under the present synthetic conditions were x =0.10 for NLN and x =0.15 for KLN. Higher solubility limits were obtained for Sr substitution, x =0.3 for NSN and x =0.5 for KSN. Traces of peaks due to $LaNbO_4$ (for NLN and KLN) or unidentified peaks (for NSN and KSN) were detected for samples beyond the solubility limit. The XRD patterns for $Na_{0.90}La_{0.10}NbO_3$ and $K_{0.85}La_{0.15}NbO_3$ are shown in Fig. 1. An increase in the symmetry was observed with La or Sr substitution.

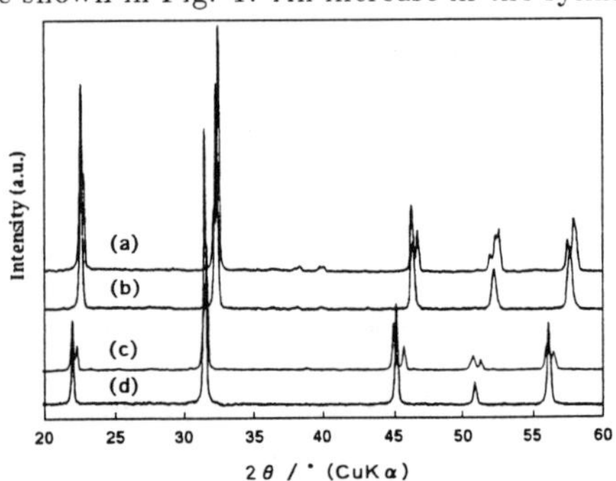

Figure 1. XRD patterns of (a)$NaNbO_3$, (b)$Na_{0.90}La_{0.10}NbO_3$, (c)$KNbO_3$, and (d)$K_{0.85}La_{0.15}NbO_3$.

Figure 2. (200) XRD peaks (FeKβ) of (a)$Na_{0.9}La_{0.1}NbO_3$ and (b)$K_{0.9}La_{0.1}NbO_3$.

Precise measurements of the (200) XRD (FeKβ) peaks for NLN and KLN with $x = 0.1$ are shown in Fig. 2. A relatively sharp (200) peak was observed for KLN (Fig. 2(b)). Similar peak sharpness was obtained for KLN ($x = 0.05$ and $x = 0.15$) and KSN ($0.2 \leq x \leq 0.5$), thus these samples were indexed according to a cubic symmetry. On the otherhand, relatively broader peaks were observed for KSN ($x = 0.1$), NSN, and NLN (Fig. 2(a)). For these compounds, indexing according to a tetragonal symmetry was difficult due to poor resolution of the XRD patterns. Thus a pseudocubic symmetry was employed for the calculation of the unit-cell parameter. It must be noted that previous studies on NSN ($0.15 \leq x \leq 0.6$) revealed it has a tetragonal symmetry.[23] The unit-cell parameters of the single-phase products are shown in Fig. 3. The structures of the K-containing compounds did not vary with the x value, whereas a linear increase in the unit-cell parameter was observed with increase in x values for the Na-containing compounds. The resemblance with La and Sr substitution can be explained by the difference in the mean-ionic radius of the A and B sites. The ionic radii[27] are listed in Table I. For NSN, the mean-ionic radius of both the A and B sites increases. On the contrary, the mean-ionic radius of the A site decreases slightly for NLN, but the mean-ionic radius of the B site increases twice as much compared to NSN (since more Nb^{4+} will be generated with the same amount of substitution). Hence, similar unit-cell parameters are obtained when the substitution amount is equal, increasing with the increase in the x value. Likewise for KLN and KSN, the mean-ionic radius of the A site decreases more for KLN than that for KSN, but the mean-ionic radius of the B site is larger compared to KSN. The mean-ionic radius of the A site decreases much more compared to the Na-containing compounds, since K^+ is considerably larger. This compensates for the increase in the mean-ionic radius of the B site, thus the unit-cell parameter varies slightly.

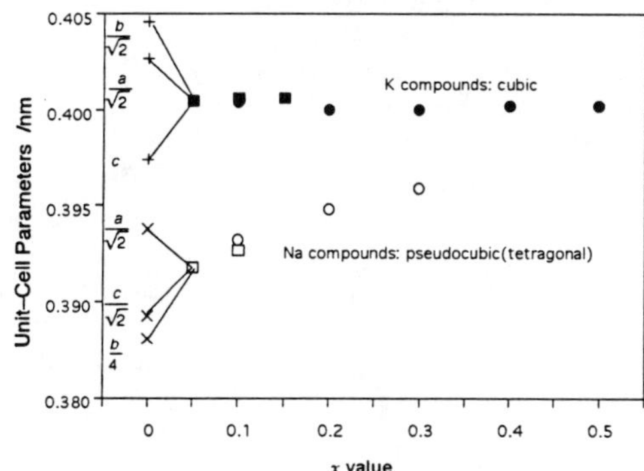

Figure 3. The unit-cell parameters as a function of composition. $\square$: $Na_{1-x}La_xNbO_3$, $\blacksquare$: $K_{1-x}La_xNbO_3$, $\bigcirc$: $Na_{1-x}Sr_xNbO_3$, and $\bullet$: $K_{1-x}Sr_xNbO_3$.

Table I. The ionic radii of various ions.[27]

ion	coordination number	ionic radius /nm
Na^+	12	0.139
K^+	12	0.164
Sr^{2+}	12	0.144
La^{3+}	12	0.136
Nb^{5+}	6	0.064
Nb^{4+}	6	0.068

The logarithm of resistivity for the La- and Sr-containing compounds is plotted against temperature in Figs. 4 and 5. A decrease in $d\rho/dT$ is observed with increase in the x value for all compounds. The conductivity behavior for NSN and the $x = 0.5$ sample of KSN are consistent with previous studies.[22, 24]

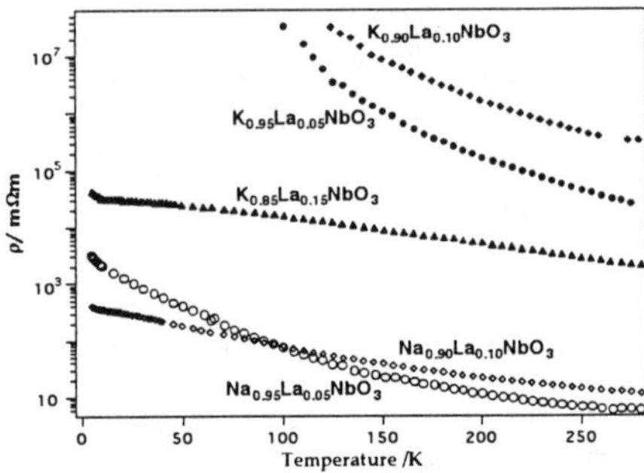

Figure 4. The temperature dependence of normalized resistivity of the La-containing compounds.

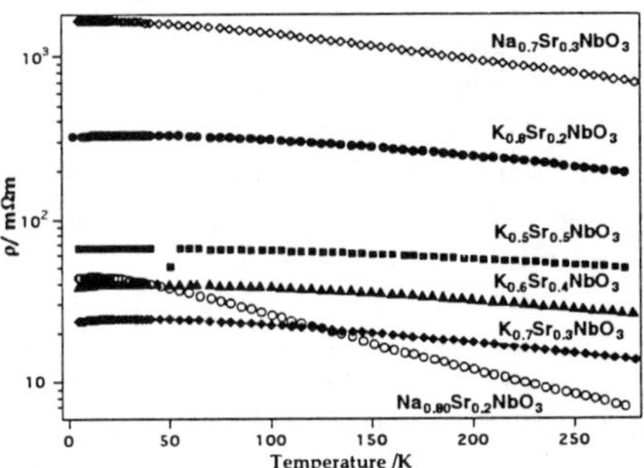

Figure 5. The temperature dependence of normalized resistivity of the Sr-containing compounds.

The resistivity normalized to the resistivity at 280 K is plotted against temperature for the compounds with $Nb^{4.8+}$ in Fig. 6. The La-containing compounds exhibited semiconducting properties throughout the measured temperature region whereas, a change of sign in $d\rho/dT$ is observed for the Sr-containing compounds (a change of sign in $d\rho/dT$ was also observed for $K_{0.5}Sr_{0.5}NbO_3$[22] and Sr_xNbO_3[7, 10]). Also, the temperature dependence of resistivity is greater (larger $d\rho/dT$) for the La-containing compounds than the Sr-containing compounds. This is interesting since KLN ($x = 0.1$) and KSN ($x = 0.2$) have similar unit-cell parameters. Furthermore, the unit-cell parameter of NLN ($x = 0.1$) is smaller than NSN ($x = 0.2$), but $d\rho/dT$ for NLN is larger than NSN. Since no relation is evident between the observed resistivity behavior and the unit-cell parameter, factors besides the structure of the compounds must also influence the resistivity behavior for the studied compounds. The synthesis of compounds with higher x values may help to clear the anomalous behavior of the resistivity.

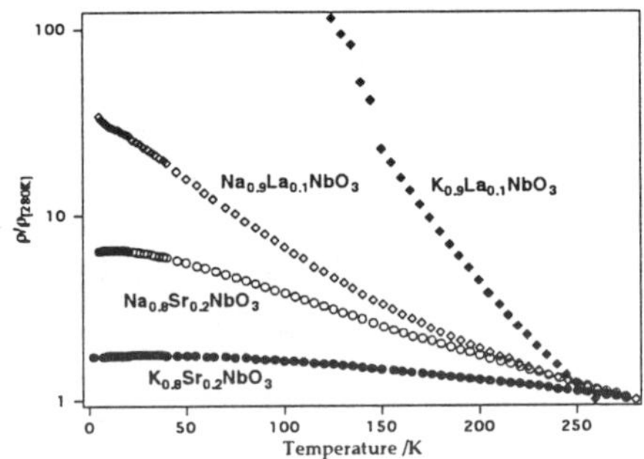

Figure 6. The temperature dependence of normalized resistivity for the compounds with niobium valence of 4.8+.

CONCLUSION

We have demonstrated that both $NaNbO_3$ and $KNbO_3$ can be doped with carriers by substitution of trivalent ions for monovalent ions. Solid-solutions of $Na_{1-x}La_xNbO_3$, $K_{1-x}La_xNbO_3$, $Na_{1-x}Sr_xNbO_3$ and $K_{1-x}Sr_xNbO_3$ were synthesized through solid-state reactions. The symmetry of the structure was strongly dependent upon the end members (NaNbO$_3$ and KNbO$_3$): Na$_{1-x}$La$_x$NbO$_3$ and Na$_{1-x}$Sr$_x$NbO$_3$ were pseudocubic, while K$_{1-x}$La$_x$NbO$_3$ and K$_{1-x}$Sr$_x$NbO$_3$ were cubic. In contrast with the unit-cell parameters, the resistivity behavior was mainly influenced by the substituting ion. Na$_{1-x}$La$_x$NbO$_3$ and K$_{1-x}$La$_x$NbO$_3$ showed semiconducting behavior throughout the measured temperature region, while Na$_{1-x}$Sr$_x$NbO$_3$ and K$_{1-x}$Sr$_x$NbO$_3$ exhibited weaker temperature dependence with a slight decrease in resistivity at low temperature. Through the substitution of trivalent and divalent ions, it has been revealed that the electrical properties in A$_{1-x}$A'$_x$NbO$_3$ are dominated by the substituting ion (A') and the structures of these compounds have a weak influence.

REFERENCE

1. N. Kumada, S. Muramatu, F. Muto, and N. Kinomura, J. Solid State Chem. **73**, 33 (1988).

2. M.J. Geselbracht, T.J. Richardson, and A.M. Stacy, Nature **345**, 324 (1990).

3. M.A. Rzeznik, M.J. Geselbracht, M.S. Thompson, and A.M. Stacy, Angew. Chem. Int. Ed. Engl. **32**, 254 (1993).

4. V.W. Meissner, H. Franz, and H. Westerhoff, Ann. Phys. **17**, 593 (1933).

5. D. Ridgley and R. Ward, J. Am. Ceram. Soc. **77**, 6132 (1955).

6. B. Hessen, S.A. Sunshine, T. Siegrist, and R. Jimenez, Mat. Res. Bull. **26**, 85 (1991).

7. K. Isawa, J. Sugiyama, K. Matsuura, A. Nozaki, and H. Yamauchi, Phys. Rev. B **47**, 2849 (1993).

8. K. Isawa, J. Sugiyama, and H. Yamauchi, Phys. Rev. B **47**, 11426 (1993).

9. K. Isawa, R. Itti, J. Sugiyama, K. Koshizuka, and H. Yamauchi, Phys. Rev. B **49**, 3534 (1994).

10. J. Sugiyama, K. Isawa, and H. Yamauchi, Solid State Commun. **95**, 181 (1995).

11. E.I. Krylov, Dokl. Akad. Nauk. S.S.S.R. **104**, 246 (1955).

12. R.R. Kreiser and R. Ward, J. Solid State Chem. **1**, 368 (1970).

13. G. Svensson and P-E. Werner, Mat. Res. Bull. **25**, 9 (1990).

14. M.T. Casais, J.A. Alonso, I. Rasines, and M.A. Hidalgo, Mat. Res. Bull. **30**, 201 (1995).

15. G.J. McCarthy and J.E. Greedan, Inorg. Chem. **14**, 772 (1975).

16. J.P. Fayolle, F. Studer, G. Desgurdin, and B. Raveau, J. Solid State Chem. **13**, 57 (1975).

17. F. Studer, G. Allais, and B. Raveau, J. Phys. Chem. Solids **41**, 1187 (1980).

18. K. Ishikawa, G. Adachi. M. Tanida, and J. Shiokawa. Bull. Chem. Soc. Jpn. **54**, 159 (1981).

19. K. Ishikawa, G. Adachi, M. Hasegawa, K. Sato, and J. Shiokawa, J. Electrochem. Soc. **128**, 1374 (1981).

20. K. Ishikawa, G. Adachi, and J. Shiokawa, Bull. Chem. Soc. Jpn. **55**, 3317 (1982).

21. K. Ishikawa, G. Adachi, and J. Shiokawa, Mat. Res. Bull. **18**, 653 (1983).

22. E.M. Kopnin, S.Y. Istomin, O.G. D'yachenko, E.V. Antipov, P. Bordet, J.J. Capponi, C. Chaillout, M. Marezio, S.de Brion, and B. Souletie, Mat. Res. Bull. **30**, 1379 (1995).

23. B. Ellis, J.P. Doumerc, M. Pouchard, and P. Hagenmuller, Mat. Res. Bull. **19**, 1237 (1984).

24. B. Ellis, J.P. Doumerc, P. Dordor, M. Pouchard, and P. Hagenmuller, Solid State Commun. **51**, 913 (1984).

25. A.C. Sakowski-Cowley, K. Lukaszewicz, and H.D. Megaw, Acta Cryst. **18**, 851 (1965).

26. A.W. Hewat, J. Phys. C **6**, 2559 (1973).

27. R.D. Shannon, Acta Crystallogr. **A32**, 751 (1976).

THE Ir$_3$Ge$_7$ (D8$_f$) STRUCTURE: AN ELECTRON PHASE RELATED TO γ-BRASS

D. SWENSON, Department of Metallurgical and Materials Engineering
Michigan Technological University, 1400 Townsend Drive, Houghton, MI 49931

ABSTRACT

The crystallographic similarity between the Ir$_3$Ge$_7$ (D8$_f$) and γ-brass (Cu$_5$Zn$_8$, D8$_2$) structures is discussed. By considering all phases known to possess the Ir$_3$Ge$_7$ structure, it is shown that phases of this structure type obey essentially the same electron concentration rules as does γ-brass, possessing on average 22 atoms/formula unit. Based on crystal-chemical considerations, it is further suggested that the Ir$_3$Ge$_7$ structure may be considered to be a structural derivative of γ-brass, in which 12 atoms have dropped out of the γ-brass lattice and the remaining atoms have undergone a certain amount of rearrangement in order to accommodate such a large concentration of vacancies. Finally, as further support of this idea, the phase NiGa$_4$ is presented as an example of another γ-brass derived structure which contains crystallographic features of both γ-brass and the Ir$_3$Ge$_7$ structure.

INTRODUCTION

It is well-known that certain metallic phases tend to be stable at rather specific electron concentrations; as such, these phases are referred to as electron phases. The correlation between electron concentration and alloy phase stability was first noted by Hume-Rothery[1] when studying the constitution of several Noble Metal-Group B binary systems. It is quite remarkable that he inferred the existence of this relationship prior to the determination of the crystal structures of many of these electron phases, relying instead on the commonalities of topography of the various binary systems.

The quintessential electron phase is γ-brass, or Cu$_5$Zn$_8$ (also denoted the D8$_2$ structure in Strukturbericht notation). Gamma-brass possesses an unusual, large, body-centered cubic unit cell containing 52 atoms, or four formula units[2]. Its structure is most often described as being a superlattice of 27 unit cells of β-brass (which has the CsCl or B2 structure), with two atoms removed and the remaining atoms shifted a certain amount from their ideal positions. The γ-brass structure is ubiquitous among Noble Metal-Group B Metal binary systems. Additionally, many examples of the γ-brass structure have been found in Platinum Metal-Group IIB binary systems. (For a compilation of all known phases crystallizing in this structure, the reader is referred to Reference 3.) Curiously, however, there is no known example of a γ-brass phase in binary systems comprising a Platinum Group Metal and a Group IIIB or IVB element. As stoichiometries such as Ni$_6$Al$_7$ would satisfy the electron concentration requirements of the phase, there is no apparent reason for the absence of the γ-brass structure from these binary systems.

In many of the Platinum Metal-Group IIIB and IVB systems, however, another rather unusual body-centered cubic structure type is found. This is the Ir$_3$Ge$_7$ (D8$_f$) structure, which contains 40 atoms per unit cell[4]. To date, 30 binary and ternary phases have been found that crystallize in the Ir$_3$Ge$_7$ structure[2,5-8]. Interestingly, there is no known instance of a binary system containing both a phase with the γ-brass structure and a phase with the Ir$_3$Ge$_7$ structure. This fact strongly suggests that the phases may be related, and that certain criteria exist that determine which of the two structure types will appear in a given system.

Recently, during an investigation of phase equilibria in the In-Rh-As system, the present author and coworkers[8] discovered a ternary phase with the somewhat unexpected composition Rh$_3$In$_5$As$_2$. This phase was subsequently found to possess the Ir$_3$Ge$_7$ structure. This has led to a general interest in the nature of the Ir$_3$Ge$_7$ structure, and in particular to speculation as to whether the unusual stoichiometry of Rh$_3$In$_5$As$_2$ can be rationalized in terms of electron concentration rules, in direct analogy to γ-brass.

In the present paper, crystallographic data are presented for Rh$_3$In$_5$As$_2$, along with a conventional representation of the Ir$_3$Ge$_7$ structure. Subsequently, it is shown that the Ir$_3$Ge$_7$ structure obeys the same electron concentration rules as γ-brass, possessing on average 22 electrons per formula unit, as compared with 21 electrons per formula unit for γ-brass. An

Mat. Res. Soc. Symp. Proc. Vol. 453 © 1997 Materials Research Society

alternative crystal-chemical comparison of the two phases is then presented that highlights the close relationship between the two structure types. Ultimately, it is suggested that the Ir_3Ge_7 structure may be considered to be a derivative of extremely defective γ-brass, in which 12 atoms have fallen out of the lattice in order to obtain the requisite electron concentration. As further support of this contention, $NiGa_4$ is presented as an example of a hybrid structure that displays characteristics of both the γ-brass and Ir_3Ge_7 structure types. Moreover, $NiGa_4$ obeys the same electron concentration rules as do γ-brass and Ir_3Ge_7. Based on the present study, it seems likely that several distinct structure types similar to both the γ-brass and Ir_3Ge_7 structures remain to be discovered.

THE Ir_3Ge_7 STRUCTURE

The Ir_3Ge_7 structure type was first identified by Nial[4] for the phases Ru_3Sn_7 and Ir_3Sn_7. However, his structure determinations were published posthumously by his colleagues, and no crystal-chemical analysis of the structure was undertaken. At a later date, Schubert[9] mentioned in passing that the structure might be related to that of γ-brass through vacancy formation and atomic rearrangement.

Pearson[10] made the first attempt to classify the Ir_3Ge_7 structure. His representation of the Ir_3Ge_7 structure is depicted in Fig. 1. Pearson has considered the structure to be built up of layers of squares of two sizes, quadrilaterals and triangles located at $z = 0$ and 1/2, with additional squares of atoms located at $z \sim$ 1/6, 1/3, 2/3 and 5/6 and rotated 45° with respect to the squares at $z = 0$ and 1/2. A few additional atoms located at $z = 1/4$ and 3/4 lie in the centers of diamonds formed from back-to back triangles in the square-quadrilateral-triangle nets.

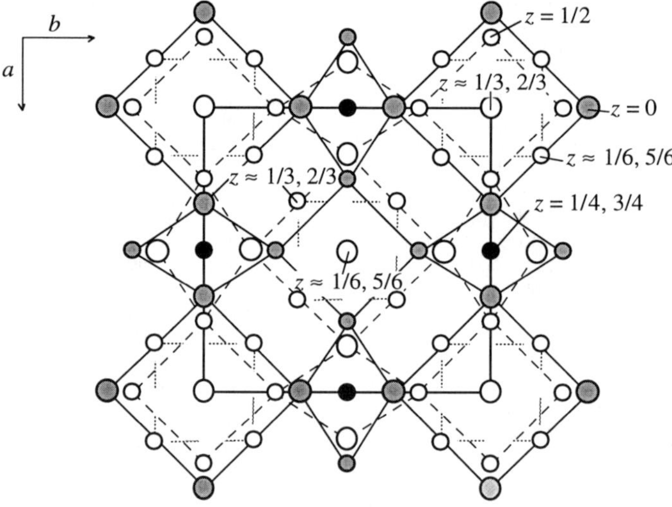

Fig. 1. The crystal structure of Ir_3Ge_7 according to Pearson[10], in projection along an a axis. Large circles represent Ir atoms, while small circles denote Ge. The bold square represents the boundaries of unit cell of the structure.

As was mentioned previously, $Rh_3In_5As_2$ has been found to possess the Ir_3Ge_7 structure. Table 1 gives crystallographic data for the phase, as determined by the present investigator. Single phase $Rh_3In_5As_2$ was prepared from elements of at least 99.99% purity. Sample preparation has been described elsewhere[8] and is not repeated here. X-ray analysis was performed using a Scintag PAD V automated diffractometer, employing Cu K_α radiation and a scanning rate of 0.5° 2l/min. Separate scans were taken for lattice parameter refinement and peak intensity

Table I. Crystallographic Data for the Phase $Rh_3In_5As_2$					

Space Group: *Im$\bar{3}$m* (No. 229)
Lattice Parameter: $a = 0.91653(6)$ nm
Prototype: *Ir$_3$Ge$_7$*
Pearson Symbol: *cI40*
Strukturbericht Designation: *D8$_f$*
Formula Units/Unit Cell: $Z = 4$
Powder Diffraction Analysis, 10°-90° 2θ, *Cu K$_\alpha$* Radiation, 32 lines, $R = 0.075$

Atom	Site	x/a	y/a	z/a	Occupancy
In1	12*d*	1/4	0	1/2	1.0
Rh	12*e*	0.338(1)	0	0	1.0
M1	16*f*	0.163(1)	x	x	0.5In + 0.5As

determination, employing narrower and wider slit sizes, respectively. Si (NIST Standard 640b, $a = 0.543\ 094$ nm) was used as an internal standard for lattice parameter refinement. Refinement was performed utilizing a commercial nonlinear least squares computer program that was included in the diffractometer software package. As the Ir$_3$Ge$_7$ structure contains only two adjustable atomic positions, the x-y-z coordinates of individual atomic sites were determined through trial and error. No attempt was made to determine the temperature coefficients either for the structure as a whole or for individual atomic sites. Various occupancies were tested for the As atoms, and no attempted structural arrangement yielded a residual error that was nearly as small as that obtained by assuming a random occupancy of 50% In and 50% As on the 16*f* site.

γ-BRASS AND Ir$_3$Ge$_7$: ELECTRON PHASES

Hume-Rothery[1] developed his electron concentration rules by assigning a valence of zero to the Platinum Metals, 1 to the Noble Metals, 2 to the Group IIB Metals, 3 to the Group IIIB Metals, 4 to the Group IVB Metals and 5 to the Group VB Metals. This scheme is equivalent to counting only the *s* and *p* electrons of the outermost shell of each atom, and assuming that atoms with no *s* or *p* electrons in their outermost shells contribute no electrons to the phase. He suggested a value of 1.6 electrons per atom for γ-brass. Later, Westgren and Phragmén[11] determined the crystal structure of γ-brass, and rationalized this electron concentration as being equivalent to the ratio 21/13 (1.615), which corresponds to the idealized stoichiometry Cu$_5$Zn$_8$.

When the number of *s* and *p* electrons per atom are determined for phases crystallizing in the Ir$_3$Ge$_7$ structure, a narrow range of electron concentrations is also obtained. In Fig. 2, the *s* and *p* electrons per formula unit (*i.e.*, per 10 atoms) are plotted for the 30 phases known to exhibit the Ir$_3$Ge$_7$ structure. (For reasons that will become clear, it is more appropriate to plot electrons per formula unit than electrons per atom.) It may be seen that there are on average approximately 22 electrons per formula unit, which is nearly the same as that found in γ-brass on an electron per formula unit basis. It is noteworthy that many of the binary phases with electron concentrations that are not equal to 22, *e.g.*, Ni$_3$In$_7$ and Ir$_3$Ge$_7$, may be alloyed with other elements so that their phase fields extend to this value. Only 4 of the 30 phases listed have electron concentrations in excess of 23 electrons per formula unit, excluding those that may be alloyed to achieve lower electron concentrations. Of these, the composition of Rh$_3$Sn$_6$In is known only approximately[12], and hence its electron concentration may in fact be lower than it is given here. The remaining exceptions comprise Group VB and VIB Metals: Nb$_3$Te$_5$Sb$_2$, Tc$_3$As$_7$ and Re$_3$As$_7$. They may represent a special case, owing to the presence of semimetals and nonmetals, but nevertheless it is noteworthy that no known attempts have been made alloy these phases with other elements in order to reduce their electron concentrations.

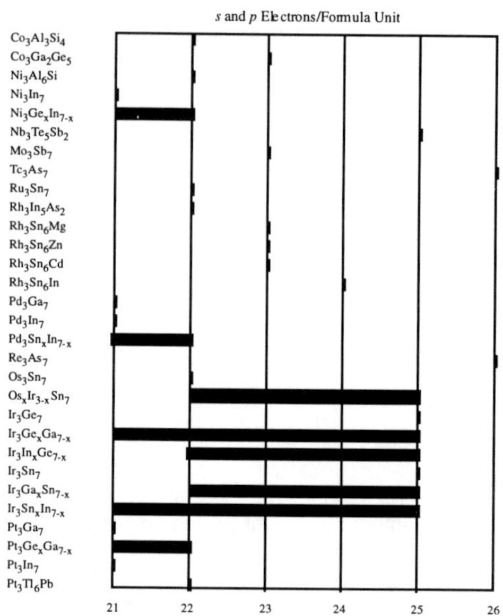

Fig. 2. *Electron concentrations of phases with the Ir₃Ge₇ structure.*

It should also be pointed out that in order to make the electron concentration per formula unit of Ir₃Ge₇ consistent with that of γ-brass, it was necessary to assign different valences to the transition elements than those used by Hume-Rothery[1]. Transition elements with unfilled *d* orbitals were assumed to absorb *s* and *p* electrons to an extent such that their *d* orbitals were filled. In other words, a Group VA Metal, such as Nb, was assumed to have a valence of -5, whereas a Group IXA Metal, such as Co, was assigned a valence of -1. However, it should be noted that Hume-Rothery *et al.*[13] have stated that owing to the wide ranges of homogeneity of the γ-brasses, the transition elements could be assigned negative valences without invalidating the electron concentration rules that have been established for the phase.

THE STRUCTURE OF Ir₃Ge₇ REVISITED: "γ-BRASS WITH VACANCIES"

It is clear from the preceding discussion that the Ir₃Ge₇ structure follows electron concentration rules that are analogous to those of γ-brass. That being the case, it would be surprising if strong structural similarities did not exist between the two structural types. However, the structural similarities between the two phases are not readily apparent if one considers only the conventional crystallographic descriptions of the two phases, which have been described earlier. Bradley and Jones[14] have developed an alternative description of the γ-brass structure that is much more useful for comparing these two structure types. Their interpretation of the γ-brass structure is depicted in Fig. 3(a). Gamma brass is considered to consist of near-icosahedra comprising 26 atoms stacked together in a body-centered arrangement. Each near-icosahedron is made up of four crystallographically distinct polyhedra: an inner tetrahedron (IT) containing of 4 atoms, an outer tetrahedron (OT) comprising 4 atoms, an octahedron (OH) consisting of 6 atoms, and an outer cubo-octahedron (CO) of 12 atoms.

Using Bradley and Jones' model of γ-brass, a similar interpretation has been made of the Ir₃Ge₇ structure, as depicted in Fig. 3(b). By comparing Figs. 3(a) and 3(b), it may be seen that the structure of Ir₃Ge₇ strongly resembles that of γ-brass. Both structures contain an OH. In the case of Ir₃Ge₇, the IT and OT found in γ-brass have merged into a single cube (CU) made up of 8 atoms. The greatest difference between the structures is found in the cubo-octahedra (CO). In γ-

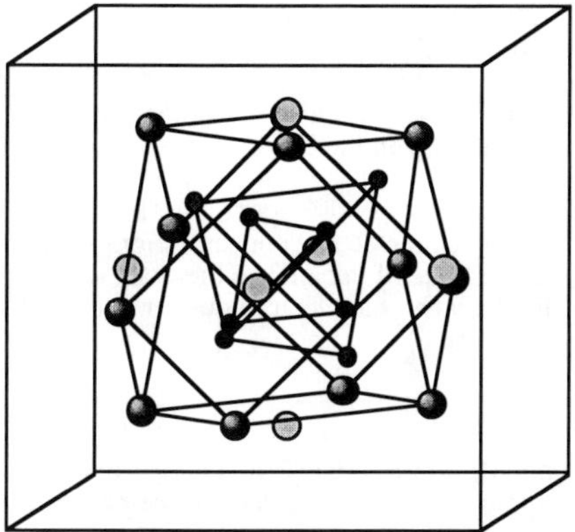

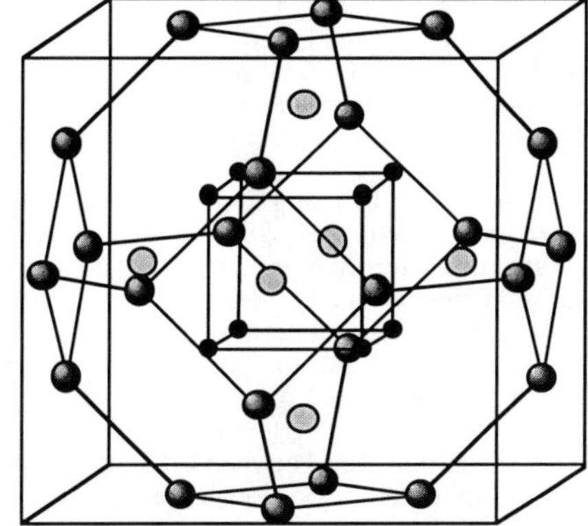

- ● Inner Tetrahedron (IT)
- ● Outer Tetrahedron (OT)
- ◉ Octahedron (OH)
- ● Cubo-Octahedron (CO)

- ● Cube (CU)
- ○ Octahedron (OH)
- ● Cubo-Octahedron (CO)

Fig. 3. The crystal-chemical environment about a lattice point in two electron phases. (a) γ-brass. (b) Ir₃Ge₇. In γ-brass, OT is occupied by Cu, and all other polyhedra are occupied by Zn. In Ir₃Ge₇, OH is occupied by Ir, and the remaining polyhedra are occupied by Ge.

brass, each near-icosahedron has its own CO. However, in the case of Ir$_3$Ge$_7$, adjacent CO's share faces, forming a continuous network of atoms throughout the crystal. Because the faces of the CO's are shared in the Ir$_3$Ge$_7$ structure, each unit cell of Ir$_3$Ge$_7$ contains only one full CO, while each unit cell of γ-brass contains two full CO's. As one CO comprises 12 atoms, this conveniently accounts for the fact that γ-brass contains 52 atoms/unit cell, whereas Ir$_3$Ge$_7$ contains 40 atoms/unit cell.

Ir$_3$Ge$_7$ therefore seems to represent an extreme case of atoms dropping out of a lattice to maintain a required electron concentration. Because 12 atoms, or nearly 25% of the atoms present, must be eliminated from the γ-brass unit cell to achieve the appropriate electron concentration, a rearrangement of the CO atoms occurs, leading to the formation of the Ir$_3$Ge$_7$ structure. This structural rearrangement is not surprising, as it does not seem plausible that such a high concentration of vacancies could be sustained by the γ-brass crystal lattice.

It is interesting to note a remark made by Hellner and Laves[15], who discovered several cubic Platinum Metal-Group IIIB phases that would later be shown to possess the Ir$_3$Ge$_7$ structure. They stated that while the structures of the phases could not be determined, the phases themselves seemed to be like "γ-brass with vacancies." The present investigation suggests that their observation was rather astute.

HYBRID STRUCTURES BETWEEN γ-BRASS AND Ir$_3$Ge$_7$

If the Ir$_3$Ge$_7$ structure may be considered to be a deformed derivative of the γ-brass lattice, it is natural to wonder if other permutations of the γ-brass structure exist, with varying amounts of atoms removed from the γ-brass lattice in order to maintain the desired electron concentration of 21-22 atoms per formula unit. NiGa$_4$ appears to be one such phase. The Ni-Ga system might be expected to include a phase crystallizing in the Ir$_3$Ge$_7$ structure. However, instead, it contains the phase NiGa$_4$, which according to Liang and Xie[16] exhibits a structure ideally containing 40 atoms per unit cell, like that of Ir$_3$Ge$_7$, but at the same time retaining some characteristics of the γ-brass structure. For example, in NiGa$_4$ the IT and OT are preserved from the γ-brass structure, one being occupied by Ni and the other by Ga. However, while consisting of different types of atoms, they are positioned such that together they nearly form a CU, as is found in Ir$_3$Ge$_7$. Additionally, the CO's of neighboring lattice points share faces, again in analogy to the Ir$_3$Ge$_7$ structure. Therefore, the environment about a lattice point in NiGa$_4$ strongly resembles Fig. 3(b), but with occupation of the CU such that nearest neighbor atoms within the cube are of different types.

According to Liang and Xie, the Ga sites of NiGa$_4$ are only about 90% occupied. Therefore, there are approximately 29 Ga atoms per unit cell, which leads to an electron concentration of about 22 atoms per formula unit, assuming that regardless of the presence of vacancies, there are 4 formula units per unit cell. Again, good agreement is found between the electron concentration of NiGa$_4$ and that characteristic of phases with the γ-brass structure.

CONCLUSIONS

The similarities between the Ir$_3$Ge$_7$ (D8$_f$) structure and the γ-brass (Cu$_5$Zn$_8$, D8$_2$) structure were highlighted. Additionally, experimental crystallographic data were obtained for the phase Rh$_3$In$_5$As$_2$. By considering all phases known to crystallize in the Ir$_3$Ge$_7$ structure, it was demonstrated that this phase obeys essentially the same electron concentration rules as does γ-brass, possessing on average 22 atoms per formula unit. A crystal-chemical analysis indicated that Ir$_3$Ge$_7$ may represent an extreme case of atoms dropping out of a lattice to maintain a required electron concentration. It therefore may be considered to be a γ-brass structural derivative, in which 12 atoms have dropped out of the lattice and the remainder of the lattice undergoes a certain amount of rearrangement in order to accommodate such a large number of vacancies. Finally, as further support of this idea, the phase NiGa$_4$ was presented as an example of another structure derived from γ-brass which contains features of both γ-brass and the Ir$_3$Ge$_7$ structure. Additionally, NiGa$_4$ was shown to obey the same electron concentration rules as both Ir$_3$Ge$_7$ and γ-brass. Based on the present study, it seems likely that several distinct structure types similar to both the γ-brass and Ir$_3$Ge$_7$ structures remain to be discovered. In this regard, it would be of great interest to study a ternary system comprising a binary γ-brass structured phase and a binary Ir$_3$Ge$_7$ structured phase, both to further test the electron concentration rules for Ir$_3$Ge$_7$ established in the present study and to determine whether new γ-brass derivative structures become stable in the ternary region of the system.

REFERENCES

1. W. Hume-Rothery, J. Inst. Met. **35**, 295 (1926).
2. A. F. Westgren and G. Phragmén, Trans. Faraday Soc. **25**, 379 (1929).
3. P. Villars and L. D. Calvert, Pearson's Handbook of Crystallographic Data for Intermetalllic Phases, 2nd Edition, ASM International, Materials Park, OH, 1991.
4. O. Nial, Svensk Kem. Tidskr. **59**, 172 (1947).
5. T. Chattopadhyay, K. Khalaff and K. Schubert, Metall **28**, 1160 (1974).
6. I. Kosovinc, M. El-Boragy and K. Schubert, Metall **26**, 917 (1977).
7. M. El-Boragy, T. Rajasekharan and K. Schubert, Z. Metallkd. **73**, 193 (1982).
8. D. Swenson, Sutopo and Y. A. Chang, J. Alloys Comp. **216**, 67 (1994).
9. K. Schubert, Kristallstrukturen Zweikomponentiger Phasen, Springer-Verlag, Berlin, 1964, p. 318.
10. W. B. Pearson, The Crystal Chemistry and Physics of Metals and Alloys, Wiley-Interscience, New York, 1972, pp. 724-725.
11. A. F. Westgren and G. Phragmén, Trans. Faraday Soc. **25**, 379 (1929).
12. A. S. Cooper, Mater. Res. Bull. **15**, 799 (1980).
13. W. Hume-Rothery, J. O. Betterton and J. Reynolds, J. Inst. Met. **80**, 609 (1951-1952).
14. A. J. Bradley and P. Jones, J. Inst. Met. **51**, 131 (1933).
15. E. Hellner and F. Laves, Z. Naturforsch. A **2**, 177 (1947).
16. J.-K Liang and S.-S. Xie, Scientia Sinica A **26**, 1305 (1983).

ON THE BRASS- AND SILVER-COLORED FORMS OF PTGA$_2$

B. MOROSIN*, D. SWENSON**
* Sandia National Laboratories, Org. 1153, MS 1421, Albuquerque, NM 87185-1421
** Materials Science Program, University of Wisconsin, 1509 University Avenue, Madison,
WI 53706; currently at Department of Metallurgical and Materials Engineering, Michigan
Technological University, 1400 Townsend Drive, Houghton, MI 49931-1295

ABSTRACT

PtGa$_2$ has previously been reported to exist as both a brass-colored, fluorite-structured phase
and as a silver-colored alloy of unknown crystal structure. The crystal structure of the silver-
colored form is reported here, along with a discussion of the stoichiometric factors responsible for
the polymorphy of this phase. The silver-colored form belongs to the space group *I4$_1$/acd*, with
the lattice parameters $a = 8.5544(4)$ and $c = 21.574(17)$ Å, and $Z = 32$. The structure consists of
layers of Pt and Ga stacked along the c axis, in which two crystallographically different Pt and Ga
atoms have similar environments involving eight and nine nearest neighbors, respectively. This
structural arrangement may be the prototype for a family of ternary platinum metal-Group
B-based phases. The present investigation also necessitates changes in the currently accepted
Ga-Pt phase diagram. Stoichiometric PtGa$_2$ undergoes a structural transformation upon cooling
from a brass-colored fluorite structure to this silver-colored, tetragonal structure. However, the
transformation is inhibited if the composition of the fluorite-structured phase is Ga-poor.

INTRODUCTION

Many of the intermetallic compounds possessing the fluorite (CaF$_2$, Fm$\bar{3}$ m, *C1*) structure
were first synthesized in the 1930's by Zintl et al.[1], who noted that the phases exhibited a
striking variety of colors, ranging from violet (AuAl$_2$) to copper (PtIn$_2$) to brass (PtAl$_2$, PtGa$_2$).
Such unusual characteristics prompted much additional work to characterize both their physical
properties and phase equilibria with, in the case of PtGa$_2$, continued confusion existing as to its
phase stability at low to moderate temperatures [2].

Zintl et al.[1] noted that PtGa$_2$ possessed the fluorite structure only at high temperatures; upon
slowly cooling brass-colored PtGa$_2$ from 200 °C to room temperature, it suddenly turned "white"
at about 150 °C. X-ray diffraction patterns of this slowly cooled alloy were described as being
complicated. The changes in color and X-ray diffraction pattern were found to be reversible upon
heating the sample above 150 °C, and moreover it was demonstrated that the fluorite structure
could be retained at room temperature by rapidly quenching a powdered sample from 200 °C.
This behavior suggests a polymorphic transformation of PtGa$_2$ near 150 °C. However, several
other investigators concluded that PtGa$_2$ decomposed eutectoidally at 153 °C into Pt$_3$Ga$_7$ and
Pt$_2$Ga$_3$[3].

Several recent investigators have not found PtGa$_2$ to show any tendency toward a structural
transformation or decomposition at low temperatures. Most notable among these is Baughman,
who grew large boules of PtGa$_2$ using the Bridgman method and reported no difficulties in
retaining them in their brass-colored states at room temperature [4]. At the time of their
fabrication, one of the present investigators (B.M.) conducted various unpublished single crystal
diffraction experiments on sections of these boules without noting any changes in sample color or
X-ray diffraction pattern upon cooling from room temperature to approximately 10 K. Moreover,
large portions of these boules are in the possession of the present investigators, and although over
20 years have passed since their fabrication, they have retained their fluorite crystal structure and
brass color to this day. Similar long-term, low temperature stability has been observed in the thin
films of PtGa$_2$ grown by other researchers [2].

Mat. Res. Soc. Symp. Proc. Vol. 453 © 1997 Materials Research Society

Recently, as part of an investigation of phase equilibria in the Ga-In-Pt-As system, one of the present authors (D. S.) prepared alloys with the composition $PtGa_2$ [2]. Upon quenching in iced water from 875 °C, these alloys changed from brass-colored to silver-colored. An X-ray diffraction pattern of pulverized alloy was found to be very complicated. However, the pattern did not correspond to a combination of Pt_3Ga_7 and Pt_2Ga_3, as is predicted by the Ga-Pt phase diagram [5], nor did it correspond to that of any of the known platinum gallides. Moreover, pieces of alloy were observed to change reversibly from silver to brass in color when placed on and removed from a 300 °C hot plate. This behavior corresponds rather closely to Zintl et al.'s original observations suggesting a polymorphic transformation of the fluorite phase of $PtGa_2$ upon cooling. Therefore, the present single crystal X-ray diffraction structural determination of this silver-colored form of $PtGa_2$ was undertaken. Some additional experiments performed on both the new alloys and on the brass-colored crystals of $PtGa_2$ grown by Baughman are also reported. While all questions regarding the low temperature stability of fluorite-structured $PtGa_2$ are not answered, these new data provide insight as to how many of the seemingly disparate observations described above may be made consistent with each other.

EXPERIMENTAL DETAILS

New samples were fabricated from elemental Pt powder of 99.95 at.% purity and Ga ingots of 99.999 at.% purity (both from Johnson Matthey Chemicals, Limited). The elements were weighed in the appropriate ratios for the composition $PtGa_2$, sealed in an evacuated (10^{-2} Torr) quartz ampoule, and annealed at 1100 °C for three hours. The samples were then homogenized at 875 °C for two months, after which time they were quenched in iced water.

These new samples turned from brass-colored to silver-colored during quenching. X-ray diffraction patterns of pulverized new sample corresponding primarily to the fluorite structure, along with many weaker peaks of another unidentified phase. After much experimentation, the intensities of the fluorite diffraction peaks relative to those of the unidentifiable peaks were determined to be functions of annealing temperature, quenching rate, and heat treatment subsequent to sample pulverization. In particular, it was determined that if the sample were annealed at low temperatures after pulverization, the intensities of the unidentifiable diffraction peaks became very great, whereas those peaks corresponding to the fluorite structure virtually disappeared. Additionally, if a pulverized sample were rapidly quenched from about 300 °C or higher, its diffraction pattern corresponded to that the fluorite structure.

In the present investigation, chips of the old Baughman samples were heated to about 200 °C and cooled at various rates to nearly 77 K both under an optical microscope and in an X-ray diffraction precession instrument without any color change or change in diffraction pattern being observed. Additionally, powdered specimens of the old samples were annealed at 300 °C for three days, furnace-cooled to room temperature, and then analyzed by X-ray diffraction. No differences in X-ray diffraction patterns were observed between the annealed powders and freshly pulverized powders of the old samples.

Both the new sample and the old Baughman sample were compositionally analyzed using electron probe microanalysis, employing elemental Pt and a commercially produced GaAs wafer as standards for wavelength-dispersive spectroscopy. A significant difference in composition was found between the old brass-colored and new silver-colored samples. The new sample contained 33.1(2) at.% Pt, whereas the old sample comprised 34.3(1) at.% Pt, based on the average of five different measurements for each sample. Thus, while the new sample is very close in composition to the ideal stoichiometry of $PtGa_2$, the old sample is substantially Ga-poor, corresponding more closely to the formula $PtGa_{2-x}$, where $x \sim 0.085$.

Tiny fragments of the silver-colored samples were examined by precession photography, and those shown to be single rather than twinned crystals were examined further in order to characterize their crystal structure. On several such fragments, Mo K_α radiation intensity data sets (to 3330 reflections) were collected. These extensive data sets were used to determine whether any weak maxima might be present which violated the body centered lattice or the extinctions expected for the space group $I4_1/acd$. Averaged cell dimensions from four smaller fragments are $a = 8.5544(4)$ and $c = 21.574(17)$ Å. Because of the high absorption cross section for Pt, as well as the irregular shapes of the X-ray specimens, empirical radiation absorption corrections were necessary. Data from the thinner, more needle-like fragments were eventually combined (579 symmetry-inequivalent hkl reflections) and employed in the final structural determination. The crystal structure was determined by Patterson and successive Fourier methods, coupled with least-squares refinements. Final atomic positional and displacement (thermal) parameters ($R = 0.0392$, $wR = 0.0434$) are given in Table 1. The final difference Fourier synthesis was carefully examined for any residual electron density peaks, particularly at interstitial sites which should be vacant, for any indication of the presence of impurity atoms.

Table 1. Atomic Position Coordinates and Displacement Parameters of Silver-Colored $PtGa_2$

Space Group: $I4_1/acd$ (No. 142)
Origin Choice 2 (Origin at $\bar{1}$ at 0, -1/4, 1/8 from $\bar{4}$)

Atom	Site	x/a	y/b	z/c	U_{eqv}
Ptl	16 d	0	1/4	0.01858(3)	100(3)
Pt2	16 f	0.19080(4)	x+ 1/4	1/8	64(2)
Gal	32 g	0.27711(12)	0.20392(12)	0.06168(5)	98(5)
Ga2	32 g	-0.04891(12)	0.52520(12)	0.06053(6)	145(6)

U_{eqv} is calculated from the anisotropic displacement parameters using the definition $U_{eqv} = 1/3\,\mathrm{TrU}$ and are in units 1000 $Å^2$. Numbers in parentheses following refined parameters represent one standard deviation in the last digit(s).

RESULTS AND DISCUSSION

The Crystal Structure of Silver-Colored $PtGa_2$

The structure of the silver-colored form of $PtGa_2$ may be described as alternating layers of Pt and Ga atoms stacked along the c axis of the tetragonal unit cell. Two crystallographically different Pt layers, Ptl at $z \sim n/4$ and Pt2 at $z = (2n + 1)/8$, are interwoven by layers containing both Gal and Ga2 at $z \sim (2n + 1)/16$ (where $n = 0, 1, 2, \ldots$). This complex structure may best be visualized by examining a basic unit consisting of the first four layers of atoms of the unit cell. The complete crystal structure may be obtained by applying symmetry operations to this unit of four layers, as will be discussed below. These layers are depicted on the right-hand side of Fig. 1, in a projection along the c-direction of the unit cell. Ptl (large, white circles) form an essentially square, but slightly buckled array centered about $z = 0$. Ptl containing a "+" sign are located at $z=0.019$, whereas Ptl containing a "-" sign are located at $z = -0.019$. Owing to this buckling, the arrangement of Ptl may best be visualized as a face-centered square net in which the face-centered atoms are slightly displaced along c with respect to their four nearest neighbors. Pt2, depicted as large shaded circles, form an array of near squares and rhombi at $z = 1/8$, the squares being rotated about 10° relative to the a axes. Gal and Ga2 (small, black circles) are virtually coplanar, and form nets of near squares, near rectangles and trapezoids at $z \sim 1/16$ and $z \sim 3/16$. In the Ga net at $z \sim 3/16$, the positions of individual Gal and Ga2 are virtually transposed with respect to their positions in the net located at $z \sim 1/16$. On the scale of Fig. 1, the Ga nets at $z \sim 1/16$ and $z \sim 3/16$ are coincident in a projection along the c-direction; the two nets therefore appear to be

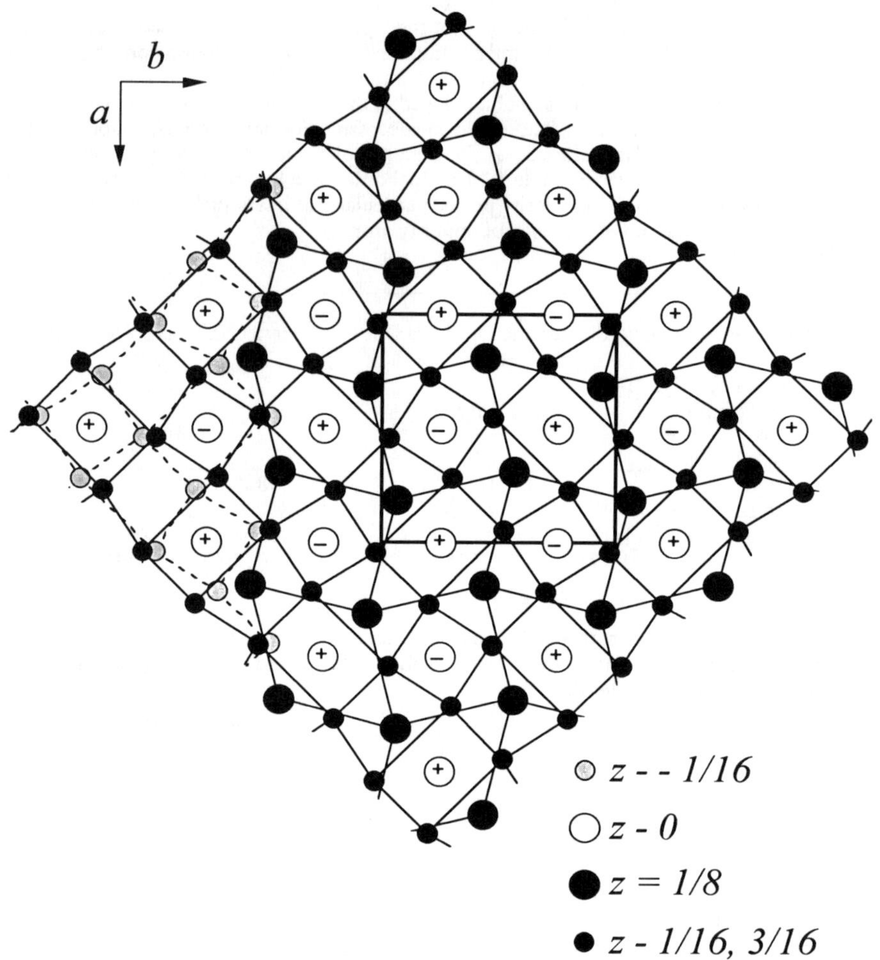

Fig. 1. A partial projection of the structure of silver-colored PtGa$_2$ as viewed along the c axis. The boundaries of the unit cell are denoted by the bold square near the center of the figure. Refer to the text for further details pertaining to the figure.

superimposed in the figure. However, in actuality the individual Ga possess slightly different x-y coordinates from one layer to the next.

In order to show the nearest neighbor environment of Ptl, a small portion of the Ga net at $z \sim$ -1/16 (lightly shaded, small circles connected by dashed lines) has been included in the far left-hand region of Fig. 1. In addition, for the sake of clarity, the net of Pt2 has been removed from this region of the figure.

The full unit cell of silver-colored PtGa$_2$ consists of 16 layers of atoms: 4 of Ptl, 4 of Pt2 and 8 of Gal + Ga2. The remaining 12 layers of atoms may be generated either by 1.) translating the four atomic layers described above by (1/2, 0, 1/4) and then applying the body-centering translations (1/2, 1/2, 1/2) to each of these sets of layers, or 2.) successively applying the 4_1 symmetry operation about the screw axis at (1/4, 0). (The entire unit cell may be produced making three transparent foils of Fig. 1 arranged atop the original figure according to these symmetry operations.)

The idealized near-neighbor environment about both Ptl and Pt2 consist of eight Ga which form the corners of trapezoidal prisms. The edges of these prisms are shared by neighboring Pt. The trapezoidal prisms surrounding Ptl may be seen in the far left-hand region of Fig. 1. Their tops and bases are perpendicular to the c axis, and are formed by the near squares and near rectangles, respectively, found within the Ga nets. The tops and bases are slightly twisted with respect to each other, with the remaining faces folded slightly along one of the diagonals. The trapezoidal prisms about Ptl at $z = 0.019$ are upside-down (base facing up), whereas the prisms about Ptl at $z = -0.019$ are rightside-up (top facing up). The trapezoidal prisms surrounding Pt2 lie on their sides (i.e., with their tops and bases essentially perpendicular to the c direction), as may be seen in the right-hand side of Fig. 1. They comprise a larger, near rectangular, base and a smaller, near rectangular, top, which alternate about a vacant interstitial site at (0, 1/4, 1/8), situated about $1/8c$ above a Ptl. Two sides of the prisms are formed by the trapezoids found within the Ga nets.

Both Ga have similar environments as well, consisting of four Pt and five Ga. The neighboring Ga form a distorted pentagonal pyramid, while the neighboring Pt form a very distorted tetrahedron (one angle of 78.7° opposite the tip of the pyramid; other angles grouped 114.7 +2°).

The nearest neighbor Ga-Pt distances range from 2.55 to 2.65 Å, whereas the Ga-Ga distances range from 2.70 to 2.78 Å. The Ga-Pt contacts are, on average, slightly smaller for Ptl than Pt2, while Gal has a slightly larger spread of both Pt and Ga contacts than does Ga2. Similar Pt-Ga (2.54 to 2.70 Å) and Ga-Ga (2.61 to 2.93 Å) distances have recently been observed in $Ce_2Pt_6Ga_{15}$ [6]. The longest Ga-Ga separations observed in this latter phase are influenced by the large Ce-Ga separations (3.13 to 3.37 Å) found in that structure. Although the Ga-Pt distances found in silver-colored $PtGa_2$ are close to those observed in its fluorite-structured modification (2.56 Å), the Ga-Ga distances are significantly shorter than those present in the cubic polymorph (2.96 Å).

The Phase Stability of PtGa$_2$

Our results suggest that the accepted Ga-Pt phase diagram [5] is incorrect in the compositional vicinity of $PtGa_2$. First, rather than decomposing eutectoidally into Pt_3Ga_7 and Pt_2Ga_3, $PtGa_2$ undergoes a transformation near 150 °C from the brass-colored fluorite structure to the silver-colored tetragonal structure. The experimental data of previous researchers may have been misinterpreted in this regard (i.e., the complicated diffraction pattern of tetragonal $PtGa_2$, decomposed microstructure of $PtGa_2$ and the thermal arrests at 153 °C are due to a polymorphic transformation rather than eutectoidal decomposition [2]).

Second, the compositional analysis of the sample grown by Baughman indicates that fluorite-structured $PtGa_2$ exists over a range of homogeneity and is not a line compound as depicted in the assessed Ga-Pt phase diagram. Its chemical formula is more appropriately expressed as $PtGa_{2-x}$, where at the very least $0 < x < 0.085$. However, because Baughman's boules were grown at high temperatures and subsequently quenched, it is unclear whether fluorite-structured $PtGa_2$ exhibits a range of homogeneity at low temperatures under equilibrium conditions.

Why are Ga-poor samples resistant to transformation while stoichiometric samples transform readily? We believe that the effect is due to kinetic limitations induced by the low temperature of the structural transformation. If the silver-colored tetragonal form of $PtGa_2$ were a line compound, then Pt_2Ga_3 would have to precipitate from Ga-poor, fluorite-structured $PtGa_2$ before it could transform to the tetragonal polymorph. However, such precipitation would necessarily entail long-range diffusion, a process that would be exceedingly slow at 153 °C, the assumed equilibrium temperature of transformation. If precipitation of Pt_2Ga_3 were precluded, it is plausible that nonstoichiometric $PtGa_2$ could remain indefinitely in the fluorite-structured modification.

A striking parallel may exist between $PtGa_2$ and ZrO_2. ZrO_2 possesses the fluorite structure at high temperatures, transforming to a tetragonal and then monoclinic modification upon cooling. It is not possible to retain ZrO_2 in its fluorite-structured modification at room temperature. However, doping ZrO_2 with an aliovalent cation, such as Ca^{2+} or Y^{3+}, leads to the development of anion vacancies and a cation-rich phase. Such defect- or impurity-stabilized ZrO_2 may easily be retained at room temperature by quenching.

CONCLUDING REMARKS

Recently the crystal structure of Ga_8Ir_4B was determined [7]. Ga_8Ir_4B is essentially isostructural with silver-colored $PtGa_2$, except that in Ga_8Ir_4B, the B atoms occupy an atomic site (0, 1/4, 1/8) which is vacant in $PtGa_2$. The existence of an isostructural, ternary boride raises the question of whether the silver-colored form of $PtGa_2$ might be stabilized by impurities, such as O or C, on these or similar vacant atomic sites. The Fourier maps generated in the present investigation were carefully examined without discovering any electron densities on such atomic sites. Specifically with respect to the (0, 1/4, 1/8) sites, the electron density is zero.

The high-temperature fluorite to the low-temperature tetragonal transformation of $PtGa_2$ appears to be driven by the substantial decrease in atomic volume afforded by such a change. If the atomic volumes of the intermediate phases in the Ga-Pt system are plotted as a function of composition, one notes that the atomic volume of fluorite-structured $PtGa_2$ is anomalously large relative to the atomic volumes of the remaining phases from elemental Ga to elemental Pt; however, tetragonal $PtGa_2$ lies in its anticipated position. Therefore, upon transforming from the cubic to the tetragonal structure, $PtGa_2$ achieves an atomic volume that is consistent with those of the other platinum gallides.

ACKNOWLEDGMENTS

This work was supported at the University of Wisconsin by the DOE under Grant No. DE-FG02-86ER452754 and at Sandia by the DOE under Contract No. DE-AC04-76DP00789.

REFERENCES

1. E. Zintl, A. Harder and W. Haucke, Z. Phys. Chem. B35 (1937) 354.

2. For a historical perspective on these intermetallics, seeone of the authors' Ph. D. Thesis (D. Swenson, University of Wisconsin-Madison, 1994) and in particular with regard to $PtGa_2$ and the accepted Ga-Pt phase diagram, refer to D. Swenson and B. Morosin, J. Alloys Comp., to appear.

3. E. Hellner and F. Laves, Z. Naturforsch. A2 (1947) 177; K. C. Jain and S. Bhan, Trans. Indian Inst. Metals 21 (1968) 41; P. Guex and P. Feschotte, J. Less-Common Metals 46 (1976) 101.

4. R. J. Baughman, Mater. Res. Bull. 7 (1972) 505.

5. T. B. Massalski (Ed.), *Binary Alloy Phase Diagrams*, 2nd Ed., ASM International, Materials Park, OH (1990).

6. G.H. Kwei, A.C. Lawson, A.C. Larson, B. Morosin, E. M. Larson and P.C. Canfield, Acta Cryst. B52 (1996) 580.

7. W. Klunter and W. Jung, Z. Anorg. Chem. 621 (1995) 197.

SYNTHESIS, STRUCTURES AND MAGNETIC PROPERTIES OF A FAMILY OF ONE-DIMENSIONAL OXIDES

H.-C. ZUR LOYE*, P. NÚÑEZ**, M. A. RZEZNIK***
*Department of Chemistry and Biochemistry, The University of South Carolina, Columbia, SC 29208, hanno@psc.sc.edu **Departamento de Química Inorgánica, Universidad de La Laguna, E-38200 La Laguna,Tenerife (Spain). ***Department of Chemistry, MIT, Cambridge, MA 02139.

ABSTRACT

The one-dimensional compounds Sr_3MgPtO_6, Sr_3MgIrO_6, Sr_3MgRhO_6, Sr_3GdRhO_6, have been synthesized and structurally characterized by Rietveld refinement of powder X-ray diffraction data. All four compounds are isostructural with the rhombohedral K_4CdCl_6-type structure. The structure consists of infinite one-dimensional chains of alternating face-shared MO_6 octahedra (M = Pt, Ir, Rh) and $M'O_6$ (M' = Gd, Mg) trigonal prisms. The strontium cations are located in a distorted square antiprismatic environment. Magnetic susceptibility data show that both Sr_3MgIrO_6 and Sr_3MgRhO_6 obey the Curie-Weiss law with $\theta = -6(1)$ K, and $\theta = -15(3)$K, respectively. Sr_3GdRhO_6 obeys the Curie law with $\mu_{eff} = 7.80$ B.M, consistent with an oxidation state of +3 for both rhodium and gadolinium.

INTRODUCTION

The investigation of one-dimensional materials has been of interest to chemists and physicists because of their unique electronic and magnetic properties. The strong directionality of low-dimensional structures can produce highly anisotropic physical properties, since interactions between electrons, such as magnetic coupling, can be strongly dependent on the crystallographic directions along which they occur [1, 2]. We recently reported the synthesis and characterization of several members of this family of materials including Sr_3ZnPtO_6 [3], Sr_3ZnIrO_6 [4] and $Sr_3CuPt_{0.5}Ir_{0.5}O_6$ [5], which exhibit magnetic behavior ranging from 1-D Ising to random quantum spin chain paramagnetism for Sr_3ZnIrO_6 and $Sr_3CuPt_{0.5}Ir_{0.5}O_6$, respectively. These one-dimensional compounds, as well as those described in this paper, are isostructural with Sr_4PtO_6 [6,7], which has the K_4CdCl_6 structure type [8]. The structure contains 1-D chains consisting of alternating face-sharing octahedra $[MO_6]$ (M = Pt, Ir, Rh) and trigonal prisms $[M'O_6]$ (M' = Mg, Gd). Each chain is surrounded by six parallel neighboring chains that are separated by the strontium cations, which are in a distorted square antiprismatic coordination. Two papers reporting isostructural compounds, Sr_3CaRhO_6 [9] and Sr_3CaIrO_6 [10], have recently been published, and the relationship between the Sr_4PtO_6 structure-type and that of the hexagonal perovskites was recently discussed [11].

Mat. Res. Soc. Symp. Proc. Vol. 453 © 1997 Materials Research Society

EXPERIMENTAL

Polycrystalline samples of $Sr_3MM'O_6$ (M= Pt, Ir, Rh) (M' = Gd, Mg) were prepared by firing stoichiometric quantities of strontium carbonate (Alfa, 99.99%), magnesium acetate tetrahydrate (Alfa, 99.99%), gadolinium acetate tetrahydrate (Alfa, 99.99%) and the corresponding powdered metal (Engelhard, 99.95%) in a platinum crucible overnight at 850 °C to decompose the carbonate and acetate. Subsequently, the powders were ground and heated to 1050 °C in alumina crucibles. Sr_3MgIrO_6 is almost completely formed after one day at 1050 °C, although the reaction was allowed to proceed for a total of four days, with intermittent grinding. A final heating step at 1200 °C for 12 hours was added to improve the crystallinity of the sample. Sr_3MgPtO_6 was heated for 2 weeks at 1050 °C and 4 days at 1150 °C, with intermittent grinding. The same heating schedule was applied to Sr_3MgRhO_6; however, a significant amount of MgO impurity was observed by powder X-ray diffraction in the final product. For this reason the citrate method was used to synthesize the rhodium compound. Due to the insolubility of rhodium metal in acid, stoichiometric quantities of $SrCO_3$, $MgAc_2.4H_2O$ and Rh metal grains were heated at 1150 °C for two days. The resultant oxide mixture was dissolved in nitric acid; and citric acid and ethylene glycol were added in a molar ratio of 1:2:2 (total metal:CA:EG). The pH was adjusted to 6.5 using ammonium hydroxide and the solution was concentrated by heating to about 80 °C. Further heating to 200 °C resulted in the formation of a black foam, which auto-ignites to form a fine powder. This powder was ground and heated at 800 °C for several hours, followed by a heat treatment at 1000 °C for four days with one intermittent grinding. Long heat treatment of these phases at high temperature leads to the precipitation of magnesium oxide out of the material. A careful analysis of the powder X-ray diffraction data, described below, revealed that a very small MgO impurity is present in all three compounds. Sr_3MgPtO_6 is brownish-yellow in color; Sr_3MgIrO_6 and Sr_3MgRhO_6 are black and Sr_3GdRhO_6 is olive green.

CRYSTAL STRUCTURE DETERMINATION

The X-ray powder diffraction data were collected on a Siemens D5000 powder X-ray diffractometer using a Bragg-Brentano geometry with Cu K_α radiation. The step-scan covered the angular range 15-115° 2θ in steps of 0.03° 2θ. Three step-scans were collected and summed. Structure refinements of Sr_3GdRhO_6, Sr_3MgRhO_6, Sr_3MgIrO_6, Sr_3MgPtO_6, Figures 1-4, respectively, were carried out in the space group $R\bar{3}c$ (no. 167), using the structure of Sr_4PtO_6 [6, 7], as the starting model. Based on our experience with the structurally related compounds Sr_3MgMO_6 (M= Pt, Ir, Rh) [5], Sr_3NiPtO_6 [12] and Sr_3ZnPtO_6 [3], the rhodium, platinum and iridium atoms were placed in the octahedral site $6b$, and the magnesium and the gadolinium atom in the trigonal prismatic site $6a$ and the strontium atom in the $18e$ site. Structure refinements were performed using the Rietveld method [13] implemented in the computer program Fullprof [14]. The profile of the diffraction peaks were described by a pseudo-Voigt function varying two half-width parameters, U and W, and the parameter eta, which take into account the

Rietveld Refinement of Sr_3GdRhO_6

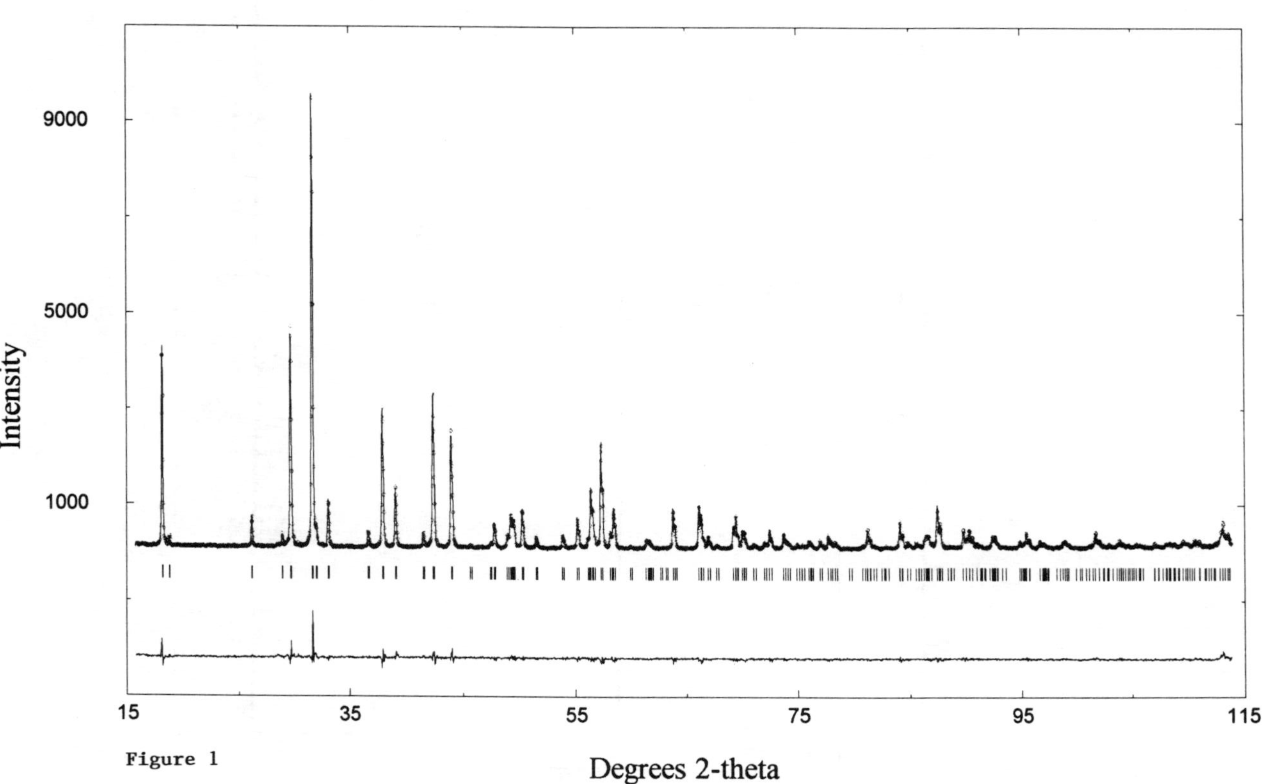

Figure 1

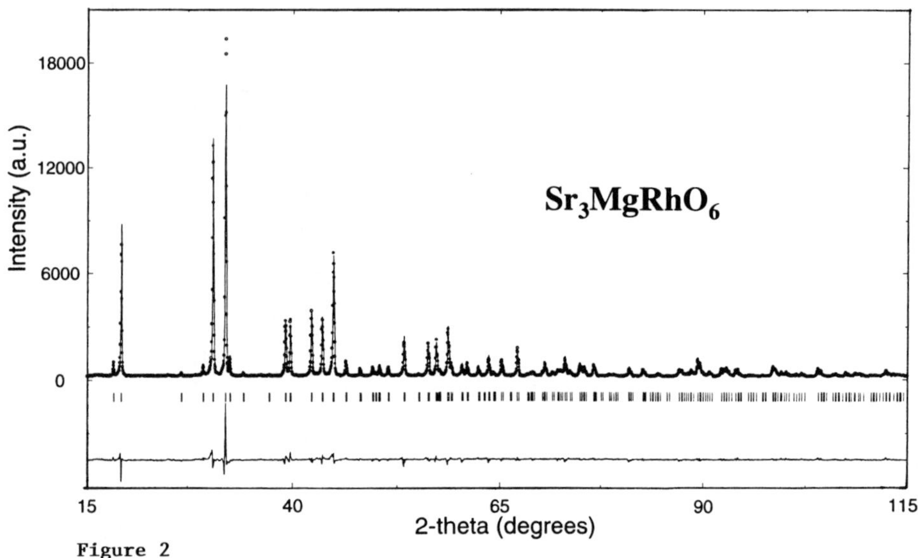

Figure 2

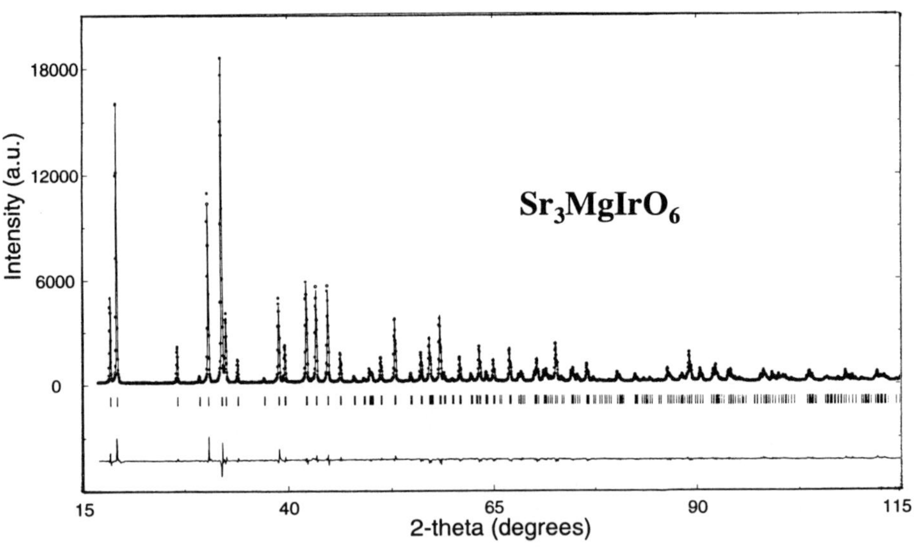

Figure 3

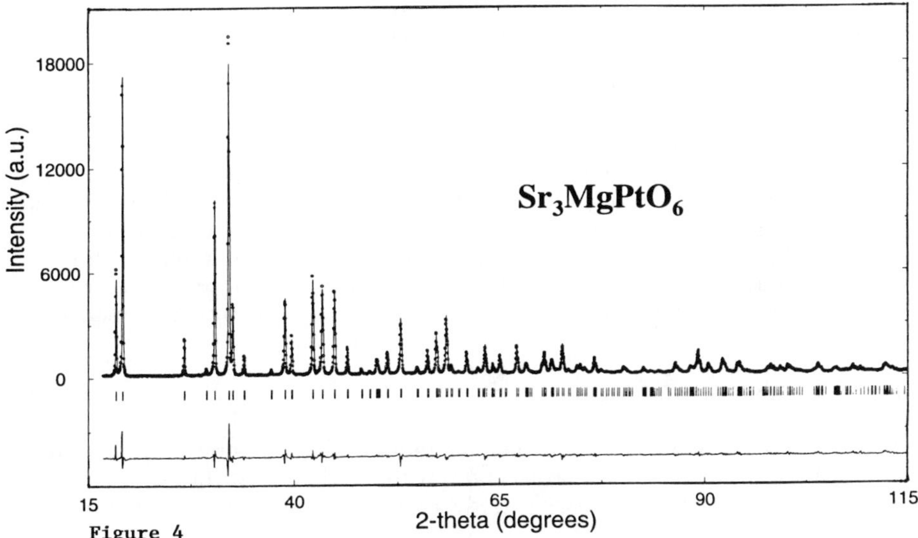

Figure 4

TABLE I

Formula	Sr$_3$MgPtO$_6$	Sr$_3$MgIrO$_6$	Sr$_3$MgRhO$_6$	Sr$_3$GdRhO$_6$
Space Group	R$\bar{3}$c	R$\bar{3}$c	R$\bar{3}$c	R$\bar{3}$c
a(Å)	9.6481(5)	9.6661(3)	9.6526(6)	9.7840(5)
c(Å)	11.1169(7)	11.1028(4)	11.0184(7)	11.4196(7)
no. of reflections	312	312	312	323
no. of refined parameters	22	22	22	22
no. of profile parameters	8	8	8	8
R$_B$ [b]	4.38	4.67	5.69	4.28
R$_p$ [b]	7.61	6.52	7.13	6.30
R$_{wp}$ [b]	9.88	8.48	9.21	8.07

[a] The e.s.d.'s have been corrected by a calculated factor {2.55 (Sr$_3$MgPtO$_6$), 2.46 (Sr$_3$MgIrO$_6$), 2.58 (Sr$_3$MgRhO$_6$) and 1.73 (Sr$_3$GdRhO$_6$)} based on the method described by Bérar.

Gaussian-Lorentzian contribution. Refinements of the peak asymmetry was allowed [18]. The background was described by a polynomial function with six refineable coefficients. One parameter was used for the refinement of the zero point offset and another for the pattern scale factor. Table I.

MAGNETIC MEASUREMENTS

Magnetic susceptibilities of Sr_3MgIrO_6, Sr_3MgRhO_6 and Sr_3GdRhO_6 were measured using a Quantum Design SQUID magnetometer in applied fields of 0.1 T, 0.5 T and 4 T in the temperature range 5 - 305 K. The samples were zero field cooled and the magnetization was measured on heating. No significant differences were observed in the susceptibility vs. temperature data collected in different applied fields. A correction was made for the clear gel capsules used as sample containers. No correction was made for the diamagnetic contribution of the non-paramagnetic species .

RESULTS AND DISCUSSION

Analysis of powder X-ray diffraction data indicates that the series of compounds, Sr_3MgMO_6 (M= Pt, Ir, Rh) is isostructural with the rhombohedral K_4CdCl_6 structure type, Figure 5, as was expected based on the fairly large number of

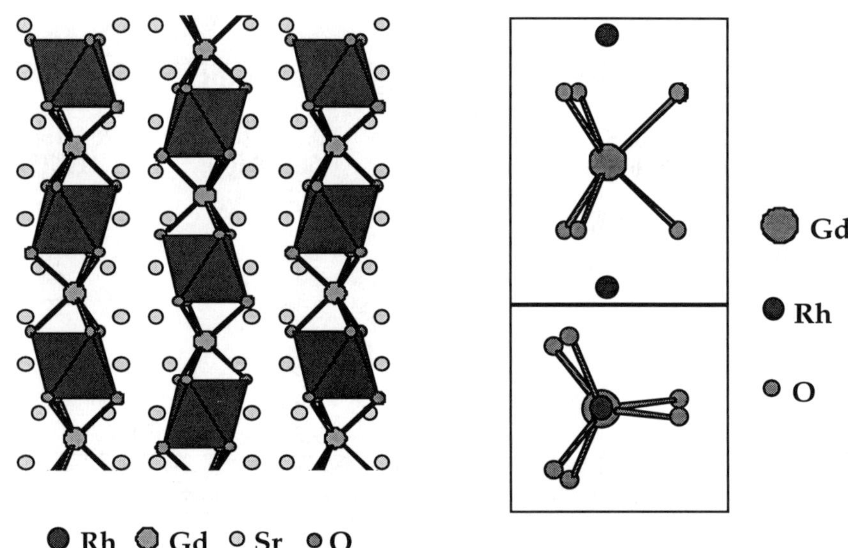

Figure 5. Structure of Sr_3GdRhO_6

compounds that have been synthesized with this structure type. Magnetic measurements show that both Sr_3MgIrO_6 and Sr_3MgRhO_6 order antiferromagnetically at ~13 and ~14 K, respectively. Sr_3GdRhO_6 follows the Curie law and exhibits no long range magnetic order. The magnetic moment of μ_{eff} = 7.80 B.M, is consistent with an oxidation state of +3 for both rhodium and gadolinium. This structure type is rather versatile and has led to the observation of a variety of interesting magnetic behavior. We succeeded in introducing a rather small cation (Mg^{+2}) into a large trigonal prismatic size. In addition, a rare-earth--rhodium combination was synthesized and, especially for this system, it is likely that additional isostructural materials can be made.

ACKNOWLEDGMENT

This work was supported primarily by the MRSEC Program of the National Science Foundation under award number DMR:94-00334.

REFERENCES

1)   L.J. de Jongh and A.R.Miedema, Adv.Phys.,**23**, 1(1974).
2)   Several authors in "Organic and Inorganic Low Dimensional Crystalline Materials" NATO-ARW, P.Delhaes and M.Drillon Editors, Plenum Press, N.Y. (1987).
3)   C.Lampe-Önnerud and H.C. zur Loye, Inorg.Chem., **35**, p. 2155 (1996).
4)   C.Lampe-Önnerud, M.Sigrist and H.C. zur Loye, J.Solid StateChem., in print.
5)   T.N.Nguyen,P.A.Lee and H.C. zur Loye, Science,**271**, p. 489 (1996).
6)   J.J. Randall and L. Katz, Acta Crystallogr., **2,** 519 (1959).
7)   I.Ben-Dor, J.T.Suss and S.Cohen, J.Crystal Growth, **64**, p. 395 (1983).
8)   G.Bergerhoff and O.Schmitz-Dumont, Z.Anorg.Chem., **284**, p. 10 (1956).
9)   J.F.Vente, J.K.Lear and P.Battle, J.Mater.Chem. **5**, p. 1785 (1995)
10)   N.Segal,J.F.Vente, T.S.Bush and P.Battle, J.Mater.Chem. **6**, p. 395 (1996)
11)   J. Darriet and M. Subramanian, J. Mater.Chem., **5**, p. 543 (1995).
12)   T.N.Nguyen,D.M.Giaquinta;H.C. zur Loye, Chem.Mat., **6**, p. 1642 (1994).
13)   H.M.Rietveld, J.Applied Crystallogr.,**2**, p. 65 (1969).
14)   J.Rodriguez-Carvajal, FULLPROF, ver.3.1c, (1996), LLB-CE,Saclay(France).

LOW TEMPERATURE MAGNETIC PROPERTIES OF SOME NEW HIGH-PRESSURE PEROVSKITE PHASES OF IRON II: OBSERVATION OF NOVEL SLOW PARAMAGNETIC RELAXATION IN $CaFe_2Ti_2O_6$ AND $CaFe_3Ti_4O_{12}$

W. M. REIFF*, K. LEINENWEBER**, J. PARISE***
*Department of Chemistry, Northeastern University, Boston, MA 02115.
** Department of Chemistry, Arizona State University, Tempe, AZ 85287-1604
***Center for High Pressure Research, Dept. of Earth and Space Science, University at Stony Brook, Stony Brook, NY 11794-2100

ABSTRACT

Results of iron-57 Mössbauer spectroscopy and ac and dc susceptibility study of the double perovskite phases, $CaFeTi_2O_6$ (A) and $CaFe_3Ti_4O_{12}$ (B) are presented. Both phases exhibit spectral broadening effects with decreasing temperature and incipient magnetic hyperfine splitting attributable to novel slow paramagnetic relaxation at the tetrahedral and square planar sites of (A) and the square planar sites of (B). The temperature dependence of magnetic moment for (A) corresponds to classical single ion zero field splitting, while that of (B) indicates antiferromagnetic exchange interactions. There is no evidence of cooperative long range magnetic order in these materials.

INTRODUCTION

The high pressure syntheses and structures of the title high-spin ferrous phases have recently been reported. (A) contains Fe^{2+} in tetrahedral and square planar coordination [1] while (B) only in square planar [2]. Herein we report the low temperature nuclear gamma resonance (Mössbauer effect) spectra and magnetic susceptibility behavior of (A) and (B).

Slow paramagnetic relaxation leading to resolved nuclear Zeeman splitting is not uncommon for ^6A ground state high-spin ferric, owing to zero field splitting of its ground spin sextet (Figure (1)) and the fact that $L \approx 0$ for this ground state, thus largely eliminating rapid spin-lattice relaxation as a line-width narrowing effect. One is then left with slow spin-spin relaxation as the primary Mössbauer spectrum line width broadening effect where spin-spin relaxation times become progressivley longer with increasing magnetic dilution and/or progressive population of slowly relaxing Kramers doublets with decreasing temperature namely, $M_S = \pm 3/2$ and $\pm 5/2$ in Figure (1).

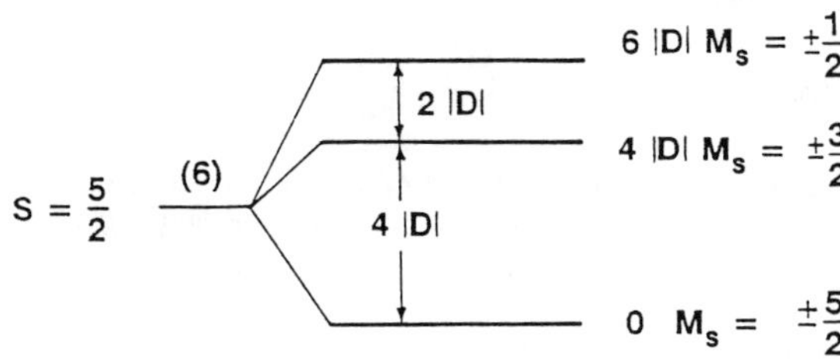

FIGURE (1)   Negative zero field splitting, D, of the spin sextet ground state of high spin FeIII.

Mat. Res. Soc. Symp. Proc. Vol. 453 © 1997 Materials Research Society

Such slow paramagnetic relaxation is truly rare for high-spin ferrous. Among the few examples that come to mind are the cases of $ZnCO_3$ and $MgCO_3$ doped with Fe^{2+} [3,4], unground, large crystals of $[Fe(pyridine-n-oxide)_6] (ClO_4)_2$ [5] and the neutral tetrahedral ferrous complex $[Fe(2,9-Di-CH_3-phen) (NCS)_2]$ [6]. Simple grinding of the foregoing N-oxide material introduces a non-axial ligand field component that results in the complete disappearance of the resolved magnetic hyperfine splitting of its 4.2K Mossbauer spectrum, i.e., the mechanical distortion leads to rapid paramagnetic relaxation. The present perovskites are discussed in these contexts.

EXPERIMENTAL

AC susceptibility studies were conducted at 1 Oe and 125 Hz and 2.51 Oe and 500 Hz respectively, using a Lake Shore Model 7000 susceptometer while DC susceptibility measurements were determined over the range 0.5 to 1.9 T using a conventional Faraday balance. Iron-57 Mössbauer spectra were measured using a standard constant acceleration spectrometer (calibrated versus the center of the hyperfine split spectrum of magnetically ordered a-iron at ambient temperature) in association with an exchange gas type liquid helium cryostat.

The sample of $CaFeTi_2O_6$ used for these measurements was synthesized in a single (approx. 30 mg.) high pressure run in the Sumitomo USSA-2000 multianvil at Stony Brook [7], using an 8 mm truncation edge length (TEL). The original purpose of this run was to produce single crystals for structural measurements [1]. A starting material consisting of $Ca(OH)_2+2TiO_2+1/3(Fe+Fe_2O_3)$ was welded into a gold capsule, and treated at a pressure of 14.5 GPa and a temperature of 1200 $^\circ$C for 3 hours. The product upon opening the capsule was a loose aggregate of black crystals, with a thin whitish rind immediately adjacent to the capsule wall. X-ray diffraction indicated that the sample was primarily $CaFeTi_2O_6$, with a small amount of unidentified impurity possibly due to iron loss to the gold capsule.

The $CaFe_3Ti_4O_{12}$ sample was produced in 3 runs (approx. 90 mg) in a multianvil apparatus at Arizona State University, specially for the present magnetic measurements. In each of these runs, a mixture of $CaTiO_3+Fe_2O_3+Fe+4TiO_2$ was treated at a pressure of 8.5 GPa and a temperature of 1200 $^\circ$C. A slight excess of $CaTiO_3$ was also added to the starting material in order to ensure that any impurity present due to compositional error would be $CaTiO_3$, which admits only about 3 mole % Fe^{2+} [1], and not $FeTiO_3$. The x-ray patterns of the resulting samples showed majority conversion to $CaFe_3Ti_4O_{12}$, with a small amount (<5%) of $CaTiO_3$ perovskite. TEM work on samples such as this made in iron capsules [8] indicates that some metallic Fe may be present due to the reducing environment. In any event, impurities for any of the samples studied herein were at too low of a level to be detectable in the x-ray patterns or Mössbauer spectra.

RESULTS AND DISCUSSION

μ vs T for (A) suggests depopulation of excited spin-orbit states or, in a simple spin only approximation the M_S states for a zero field split quintet ground state of high spin FeII for the case of zero field splitting, D<0 (Figure (2) and a concomitant decrease in magnetic moment.

This behavior is shown in Figure (3) (Faraday balance data) wherein one sees a rise in moment just below ~20K and then a sharp drop at lower temperatures.

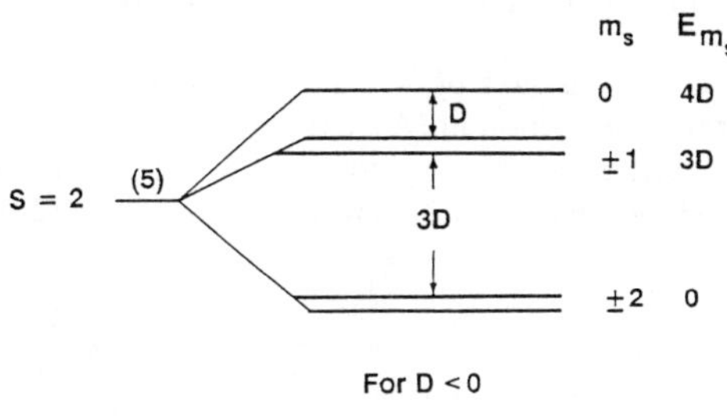

FIGURE (2)   Negative zero field splitting of the spin quintet ground state or high spin FeII.

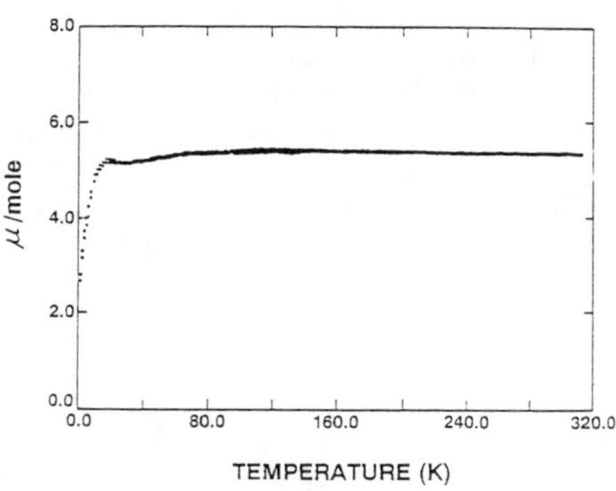

FIGURE (3)   Magnetic moment (μ) per mole versus temperature for Ca Fe Ti$_2$ O$_6$

The results are as expected for (i) significant local magnetic anisotropy, (ii) partial alignment of the sample polycrystals by the applied field (in this case ranging from 0.5 to 1.9 T) and (iii) the case of D<0 and $g_{\parallel} > g_{\perp}$ [9]. The moment decrease observed for (B), Figure (4), is more consistent with anti-ferromagnetic exchange coupling. In particular, the theoretical spin-only moment per mole for three non-interacting S=2 ions is 8.49 Bohr magnetons, whereas μ for (B) at ambient temperature is 7 Bohr magnetons and decreases to 4.6 B.M. at 78 K. However, the susceptibility data for (A) and (B) give no indication of the onset of cooperative long range order to as low as 4.2K. This observation is further confirmed by the <u>absence</u> of <u>sudden,</u> fully resolved nuclear Zeeman splitting of the Mössbauer spectra that usually accompanies long range ordering processes.

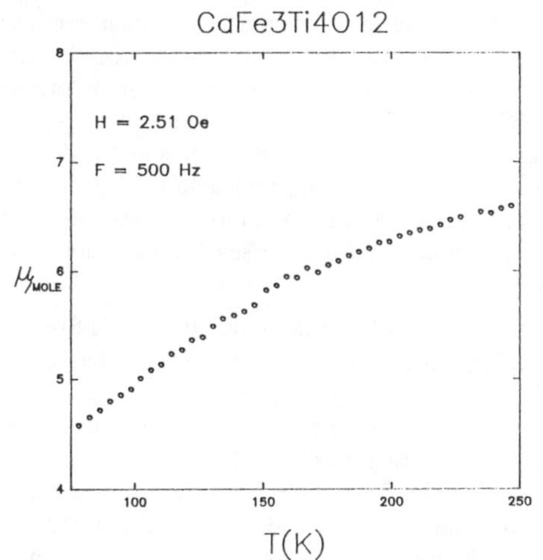

FIGURE (4)

Magnetic moment (μ) per mole versus temperature for Ca Fe$_3$ Ti$_4$ O$_{12}$

On the other hand, and quite to our surprise, the zero field Mössbauer spectra of these phases exhibit a gradual, progressive line width broadening and incipient magnetic hyperfine splitting on decreasing the temperature from ambient to 4.2K and approach to long spin-lattice and spin-spin relaxation times. In addition, as now discussed in detail, the paramagnetic relaxation rates differ for the two ferrous sites of (A).

The local coordination environments of the ferrous sites of (A) and (B) are illustrated in Figure (5), i.e., both tetrahedral and square planar coordination in (A) and only the latter in (B) [1,2]. The isomer shifts δ (relative to iron metal) and quadrupole splittings (ΔE) for the square planar sites of (A) and (B) are very similar (δ=1.04 mm/sec (A), 1.02 mm/sec (B) while ΔE=2.94 mm/sec) (A) and 2.85 mm/sec (B)).

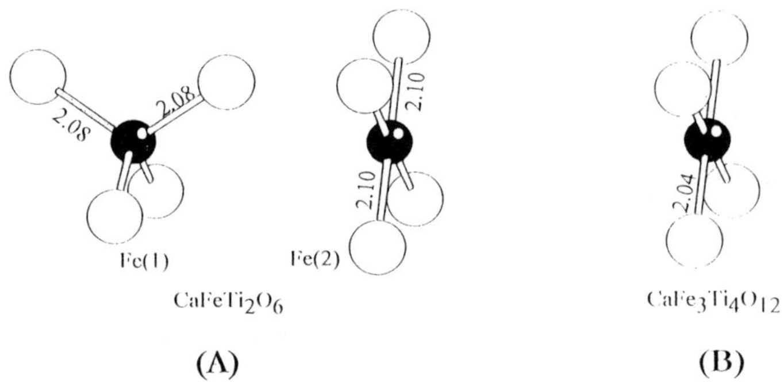

FIGURE (5)   Local coordination environments in Ca Fe Ti$_2$ O$_6$ (A) and Ca Fe$_3$ Ti$_4$ O$_{12}$ (B).

These isomer shift values are fairly typical of the higher end of the range observed for Fe^{2+} in tetrahedral (oxygen) coordination or the lower end of the range for Fe^{2+} in octahedral coordination [10]. The large (temperature independent) quadrupole splittings for the square planar sites of (A) and (B) are consistent with a large electric field gradient resulting from the absence of axial oxygen ligation at these sites. In fact, the sole example of well characterized [11] (in a Mössbauer spectroscopy context) high spin Fe^{2+} in rigorous square planar coordination and with which one might wish to make comparisons to the present phases is that of the mineral Gillespite, $BaFeSi_4O_{10}$, for which δ (relative to iron metal) = 0.75 mm/sec and ΔE=0.51 mm/sec. The latter isomer shift is anomalously low and perhaps related to its very short (1.95Å) Fe-O bond lengths. Its quadrupole splitting is also unusually low but explicable [12] in terms of an opposing lattice contribution to the electric field gradient tensor at the ferrous sites. In any event, all of the materials are exceptional in their own right and further useful comparisons to other ferrous phases are not appropriate at this time.

The quadrupole doublet of the tetrahedral sites or (A) corresponds to the inner pair of resonances at the top of the Figure. This splitting (ΔE=0.57 mm/sec, δ=1.02 mm/sec) is considerably smaller than that of the square planar sites and much more temperature dependent over take decreasing temperature range 293 K to 77 K. Such small, highly temperature dependent quadrupole splittings are characteristic [10] of relatively undistorted tetrahedrally coordinated high-spin ferrous accompanied by a small (~100 to 300 cm^{-1}) Jahn-Teller splitting effect of their ^5E ground state. It is clear that these resonances are already significantly broadened at ambient temperature and essentially disappear on cooling to 4.2 K. At the same time, below 45K the narrow line-width

Sample Mössbauer spectra for (A) are shown in Figure (6).

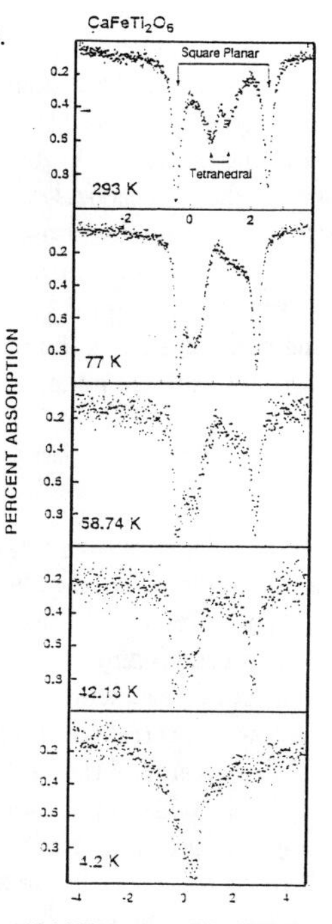

FIGURE (6)

Sample Mössbauer spectra
for Ca Fe Ti$_2$ O$_6$

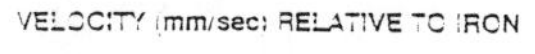

FIGURE (7)

Sample Mössbauer spectra
for Ca Fe$_3$ Ti$_4$ O$_{12}$

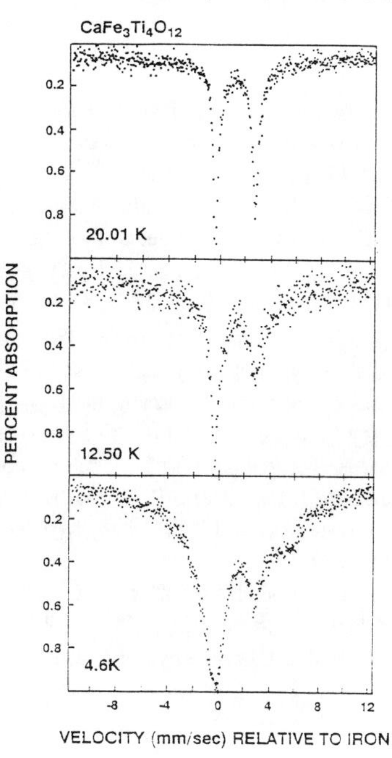

resonances of the square planar sites begin to broaden and similarly decrease in intensity. This gradual behavior, with <u>no</u> coincidental anomaly in the temperature dependence of the magnetic susceptibility, is typical of classical slow paramagnetic relaxation wherein a non-zero time averaged internal hyperfine field grows at the Mössbauer nucleus owing to slow relaxation among electron spin Zeeman states of Figure (2). From the spectra of Figure (6) it is apparent that the relaxation time at the tetrahedral sites is longer or alternatirely the <u>rate</u> is smaller than at the square planar sites. Generally similar behavior is observed for perovskite (B) (Figure 7) except that the situation is less complicated owing to the presence of only one type of Fe^{2+}. The larger velocity sweep spectra of Figure (7) and similar spectra for (A) give no evidence of narrow line width, pure Zeeman transitions indicating that even at 4.2 K, we are not at the limit of infinitely long relaxation times in these materials.

CONCLUSION

Wignall [13] has shown that long spin-spin relaxation times and related Mössbauer spectra broadening effects require Fe-Fe separations $\gtrsim 7$ Å. Thus, in view of the fact that the smallest Fe-Fe separations [1,2] in (A) and (B) are only ~5.3Å and 3.7Å respectively, the observed line width broadening is <u>not</u> predominantly slow spin-spin in nature. This leaves spin-lattice broadening combined with a large, negative D value, i.e., $D\sim-10cm^{-1}$ at the square planar sites and $-30cm^{-1}$ at the tetrahedral sites. With reference to Figure (2), these <u>estimates</u> give an overall zero field splitting of the S=2 spin manifold of 40 and 120 cm^{-1} respectively, and are reasonably consistent with the observed onset of line broadening in the Mössbauer spectra. Finally, in passing we note that asymmetric line width broadening for Gillespite has likewise been [12] attributed to slow spin-lattice relaxation. However, to our knowledge, this material has not been studied below 80 K and clearly deserves additional study in the present context.

REFERENCES

1. K. Leinenweber and J.B. Parise, J. Solid State Chem. **114**, pp. 277-281 (1995).
2. K. Leinenweber, J. Linton, A. Navrotsky, Y. Fei and J.B. Parise, Phys. and Chem. Minerals **22**, pp. 251-258 (1995).
3. D.C. Price, C.E. Johnson and I. Maartense, J. Phys. Chem. **10**, p. 4843 (1977).
4. K.K.P. Srivastava, PhD Thesis, Australian National University (1976).
5. B.D. Howes, D.C. Price and D.J. Mackey, Colloquium C2, Supplement Number 3, Journal de Physique **40**, p. 286-289 (1979).
6. W.M. Reiff, L. Nicolini and B.W. Dockum, Colloquium C2, Supplement Number 3, Journal de Physique **40**, p. 230-232 (1979).
7. R.C. Liebermann and Y. Wang in <u>High-Pressure Research: Application to Earth and Planetary Sciences</u>, edited by Y. Syono and M.H. Manghnani, TERRAPUB, 1992.
8. N. Yao, A. Navrotsky and K. Leinenweber, J. Solid State Chem. **123**, pp. 73-82 (1996).
9. M. Gerloch, J. Lewis and R.C. Slade, J. Chem. Soc. A, p. 1422 (1969).
10. N.N. Greenwood and T.C. Gibb, <u>Mössbauer Spectroscopy</u>, Chapman and Hall, Ltd., London, 1971.
11. M.G. Clark, G.M. Bancroft and A.J. Stone, J. Chem. Phys. **47**, p. 4250 (1967).
12. M.G. Clark, J. Chem. Phys. **48**, p. 3246 (1968).
13. J.W. Wignall, J.Chem. Phys. **44**, p. 2462 (1966).

EFFECTS OF NON-STOICHIOMETRY
ON ANISOTROPIC MAGNETIC PROPERTIES IN $CuFeO_{2+\delta}$ SINGLE CRYSTALS

M. HASEGAWA*[&]**, T. R. ZHAO**, M. I. BARTASHEVICH**, H. TAKEI***, T. GOTO**
* Department of Chemistry, Pennsylvania State University, University Park, PA 16802,
masa@chem.psu.edu
** Institute for Solid State Physics, University of Tokyo, Roppongi, Minato-ku, Tokyo 106,
Japan, masa@takeilab.issp.u-tokyo.ac.jp
*** Department of Earth and Space Science, Osaka University, Toyonaka, Osaka, 560, Japan

ABSTRACT

Effects of non-stoichiometry on anisotropic magnetic states in $CuFeO_2$ with a so-called delafosite structure have been investigated using $CuFeO_{2+\delta}$ (δ=-0.061, -0.026, 0.014, 0.085) single crystals. Each single crystal is grown using the floating zone method in each controlled atmosphere and found to have good quality. All crystals show two anisotropic successive antiferromagnetic transitions in the temperature dependence of magnetic susceptibilities. The higher antiferromagnetic transition temperature is not affected by the non-stoichiometry. On the other hand, the lower one shows distinct effects of the non-stoichiometry, depending on the kind of oxygen defects. Oxygen deficiency leads to an increase in T_{N2}, while excess oxygen leads to a decrease. It is also found that all samples show clear anisotropic step anomalies in the magnetization curve up to 38 T and that the anomalies and hystereses depend on the non-stoichiometry.

INTRODUCTION

Magnetic materials with the two-dimensional crystal structure are of interest to investigate the correlations in the spins and electrons. With the exception of the two-dimensional tetragonal materials such as high temperature oxide superconductors, two-dimensional triangular structure materials are the next important targets. The magnetic and electric states of these materials are usually quite affected by non-stoichiometry and impurity. The family of so-called delafosite structure oxides ABO_2 is one of the important materials in this type.

Cuprous ferrite $CuFeO_2$ with a so-called delafosite structure is one of these typical magnetic materials. $CuFeO_2$ has a crystal structure with the space symmetry group R3m and lattice parameters of a_h = 3.03 Å and c_h= 17.09 Å in hexagonal description [1], as shown in Figure 1. The structure consists of hexagonal layers of Cu, Fe and O with Cu at (0, 0, 0), Fe at (1/2, 1/2, 1/2) and two O at (1/9, 1/9, 1/9) and (-1/9, -1/9, -1/9), which stack in a sequence of

Mat. Res. Soc. Symp. Proc. Vol. 453 © 1997 Materials Research Society

A-B-C [A(Cu), A(O), B(Fe), C(O), C(Cu), C(O), A(Fe), etc.] to form a layered triangular lattice antiferromagnet with the anisotropic successive transitions at T_{N1} = 16 K (paramagnetic - antiferromagnetic I) and T_{N2} = 11 K (antiferromagnetic I - antiferromagnetic II)[2].

Ajiro et al. investigated impurity effects on the stability of the ground-state spin configuration in the triangular lattice antiferromagnetic $CuFeO_2$ [3]. They reported that distinct magnetization anomalies disappeared in only small amounts of Al or Cr impurities substituting from Fe in $CuFeO_2$. Bunko and Koffyberg studied the opto-electronic properties of $CuFeO_2$ doped with MgO or SnO_2 [4]. They identified that $CuFeO_2$ made either p- or n-type semiconductors by suitable doping.

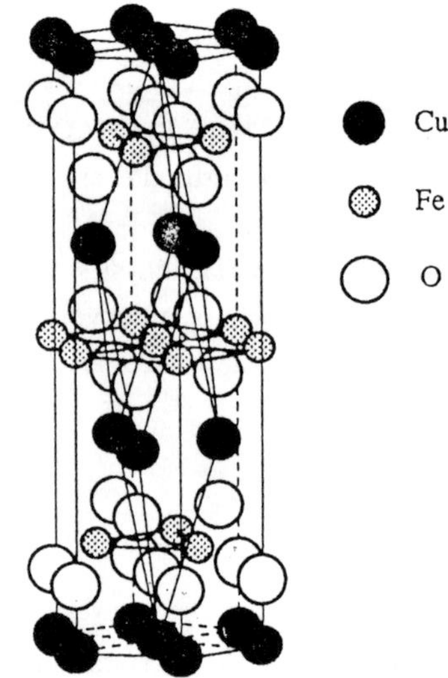

Figure 1 Crystal structure of $CuFeO_2$.

In this study effects of non-stoichiometry on anisotropic magnetic states in $CuFeO_2$ with a so-called delafosite structure have been investigated using $CuFeO_{2+\delta}$ (δ=-0.061, -0.026, 0.014, 0.085) single crystals. Since this is a typical antiferromagnetic material with spin frustrations, the magnetic states are expected to be affected by the non-stoichiometry. All the samples are single crystals grown using the floating zone method. They are also well-characterized and found to have high quality. This paper will report and discuss, particularly, effects on the magnetic transition temperature T_{N2} and anisotropic magnetizations.

EXPERIMENTAL

All of the samples are as-grown single crystals grown in an infrared radiation furnace under CO_2, Ar, Ar+0.5%O_2, Ar+2%O_2 atmosphere, respectively, by means of the travelling Cu_2O-solvent floating-zone method. Compositions of the single crystals were investigated by a scanning electron microscope with an electron probe microanalyzer (SEM-EPMA) using metal copper and iron as standards. The densities were also measured by a so-called Archimedean method. The formula of each sample was determined from these measurements as $CuFeO_{1.939}$ (δ = -0.061), $CuFeO_{1.974}$(δ = -0.026), $CuFeO_{2.070}$ (δ = 0.070) and $CuFeO_{2.085}$ (δ = 0.085). The crystal structures were also investigated by an X-ray powder diffraction method. From these results, the single crystals were found to have good qualities. The details of the crystal growth and characterizations have been described elsewhere [5-9].

Each sample to measure anisotropic magnetic properties was cut from each single crystal. The Laue back-reflection pattern was used to establish the orientation of the single crystals. Magnetic susceptibilities were measured in the magnetic field of 1 kOe by a superconducting quantum interference device (SQUID) magnetometer (Quantum Design MPMS) between 5 and 300 K. The high-field magnetization was measured by an induction method using pulse field up to 38 T at 4.2 K.

RESULTS AND DISCUSSION

The magnetic susceptibilities in both directions for all of the samples increase with decreasing temperature following the Curie-Weiss law [10]. Then, they suddenly decrease anisotropically at T_{N1} and more at T_{N2}, as for example shown in Figure 2. This is the temperature dependences of magnetic susceptibilities parallel and perpendicular to the c-axis for the $CuFeO_{1.974}$ below 25 K [10]. These drops correspond to the successive antiferromagnetic transitions [2]. All of the samples showed almost the same temperature dependence and anisotropy of the magnetic susceptibilities.

Effects on the temperature dependence of magnetic susceptibilities parallel to the c-axis below about 25 K are shown in Figure 3. The T_{N1} does not change with the non-stoichiometry. On the other hand, the T_{N2} systematically depends on the oxygen content and changes almost linearly, as shown in Figure 4. It should be noted that

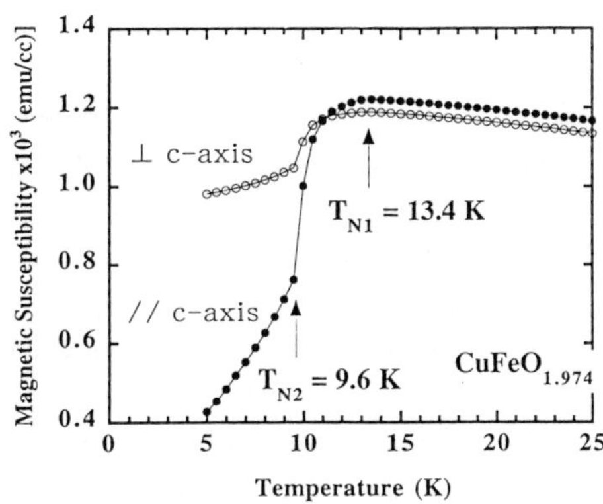

Figure 2 Temperature dependences of magnetic susceptibilities parallel and perpendicular to the c-axis below 25 K.

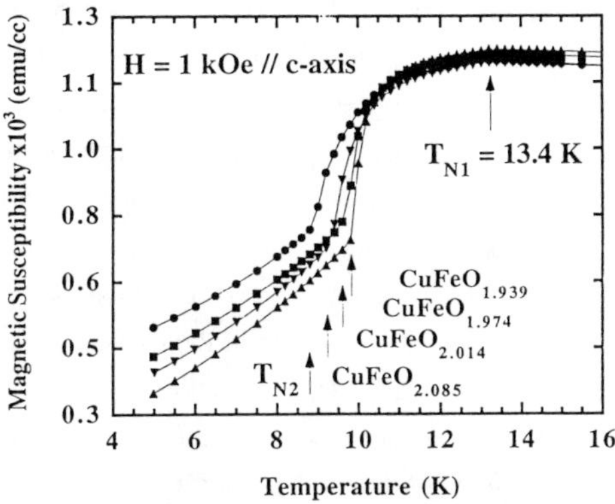

Figure 3 Effects of non-stoichiometry on temperature dependence of magnetic susceptibilities parallel to the c-axis.

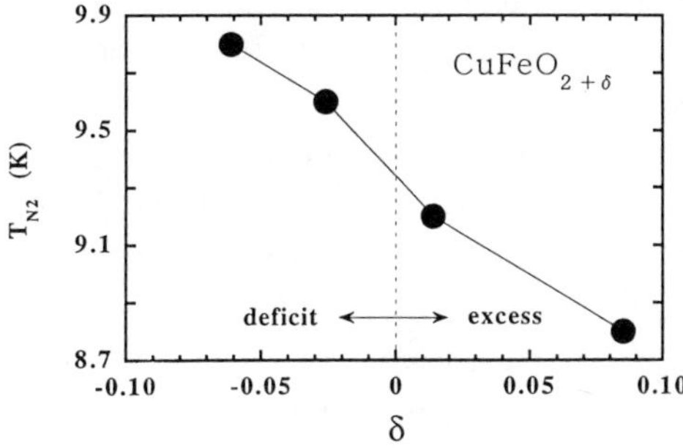

Figure 4 Oxygen defect dependence of T_{N1}.

the effects of the oxygen defects on the low temperature antiferromagnetic phase is definitely different in accordance with the defect type of either deficiency or excess of oxygen. The effects of the oxygen non-stoichiometry on T_{N2} apparently depend on the type of oxygen defect. In the case of oxygen deficit, the T_{N2} moves towards high temperature with decreasing oxygen content and the transition tends to be sharply. On the other hand, in the case of oxygen excess, the T_{N2} shifts towards the low temperature region. The low temperature antiferromagnetic phase becomes instable and the transition tends to be loose.

The effects with the advance of valence state transformation of copper or iron ions are explained in the following considerations. In the case of the oxygen deficit, the valence state of copper is kept monovalent, while iron is in the mixed valence state, i.e., the trivalent state with extremely few of the divalent state. As $CuFeO_{2+\delta}$ is in a layered antiferromagnetic structure with the frustrated triangular spin structure, as shown in Figure 1, the slight change of the valence state in iron ions results in an insignificant change of the magnetic interaction. Besides, this occurs only within the iron layers, which means that the two-dimensional magnetic properties in $CuFeO_{2+\delta}$ are preserved. Accordingly, the two different valence states of iron ions relieve the spin frustration and lead to stabilize the low-temperature antiferromagnetic phase. On the other hand, in the case of the oxygen excess, the valence state of iron is kept trivalent, while, the valence state in copper appears in the mixed state, i.e., the monovalent state with extremely few of the divalent state. As the divalent copper ions posses magnetic moment, the perfect non-magnetic copper layers below and above the iron layers in the $CuFeO_{2+\delta}$ structure disappear gradually. In addition, the c-axis in the $CuFeO_{2+\delta}$ structure shrinks with increasing oxygen [9]. As a result, the completeness of the two-dimensional magnetic property is lost, and this results in the instability of the low temperature antiferromagnetic phase, whose phase transition at T_{N2} shifts towards the lower temperature and the transition becomes broad.

Effects of the non-stoichiometry on the magnetization curve are shown in Figure 5. Distinct anisotropic step anomalies with hystereses, which are typical characteristic of the Ising-like system, are observed clearly for all of the samples, even if the oxygen defect concentration increases. The step magnetic fields, behavior and numbers estimated from dM(H)/dH, are found to grossly depend on the oxygen defect concentration.

It is interesting that these results are quite different from impurities effects on the magnetization curve reported by Y. Ajiro et al. [3]. They investigated effects of Al or Cr impurities substituting Fe in $CuFeO_2$ using powder samples. In their case the step anomalies almost disappeared rapidly by the impurities only less than 5 %. In addition, the step magnetic fields, behavior and numbers were reported to be almost independent of the concentration and type (Cr : magnetic or Al : non-magnetic) of the impurities. Therefore, they concluded that the reason for the independency was that the impurities did not change electronic states and exchange interaction but only lead to geometrical disturbance of the subtle balance in a frustrated situation. On the other hand, the non-stoichiometry in this study is considered to affect the electronic states and exchange interaction, as described in the previous section. Therefore, this difference may lead to the differences between the substituting impurities effects on the magnetization curve and the non-stoichiometry ones.

It should be also noted that the hysteresis behaviors in the magnetization curves depend on the non-stoichiometry. Figure 6 shows effects of the non-stoichiometry on the integrated magnetization hysteresis in magnetization curve. The hysteresis values are normalized by the value of $CuFeO_{1.974}$. The effects on the magnetization hysteresis perpendicular to the c-axis are found to be much larger than those parallel to the c-axis. It is also found that the effects depend on the oxygen defect type and are anisotropic. In the case of the oxygen deficit, the magnetization hysteresis increases with deficit content in both of directions. The effect on the hysteresis is much effective in the magnetization parallel to the c-axis than in perpendicular to the c-axis. On the other hand, in the case of the excess oxygen the effect on the hysteresis shows

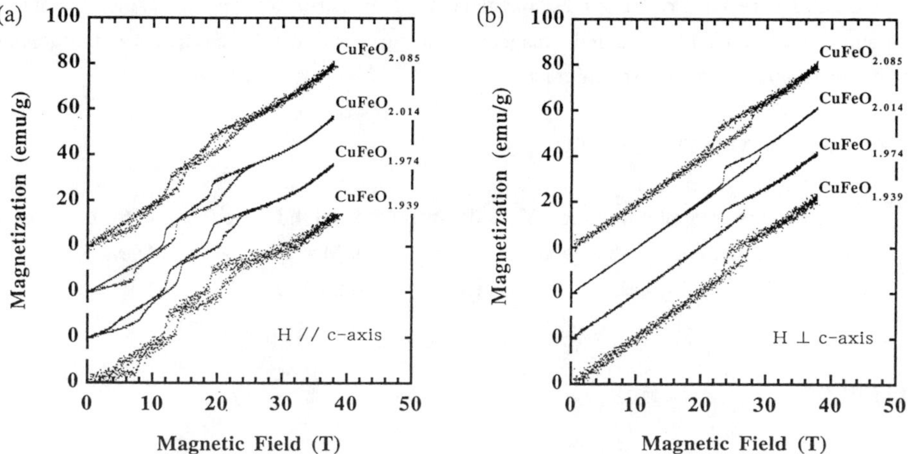

Figure 5 Effects of non-stoichiometry on magnetization curve.
(a) parallel to the c-axis, (b) perpendicular to the c-axis.

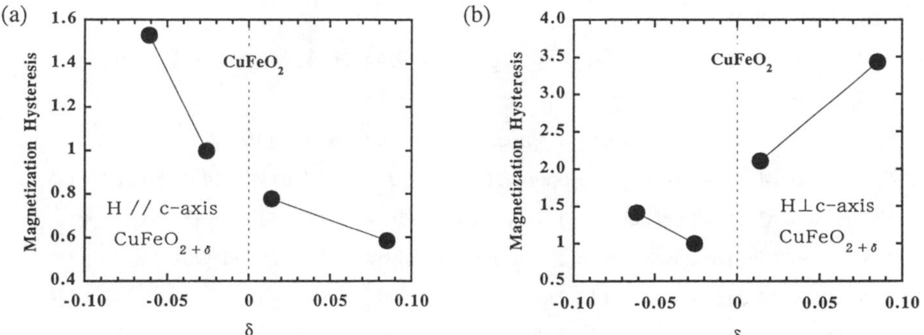

Figure 6 Oxygen defect dependence of normalized hysteresis area in magnetization curve.
(a) parallel to the c-axis, (b) perpendicular to the c-axis.

opposite tendency between the magnetization parallel to the c-axis and that perpendicular to the c-axis. That is, the magnetization hysteresis parallel to the c-axis decreases with the excess oxygen content, while that perpendicular to the c-axis increases. In addition, the effects in the latter direction are more effective than the former direction.

CONCLUSIONS

The effects of non-stoichiometry on anisotropic magnetic states in $CuFeO_2$ with a so-called delafosite structure have been investigated using $CuFeO_{2+\delta}$ (δ=-0.061, -0.026, 0.014, 0.085) single crystals. The higher antiferromagnetic transition temperature T_{N1} is not affected by the non-stoichiometry. On the other hand, the lower one T_{N2} shows distinct effects of the non-stoichiometry, depending on the kind of oxygen defects. It is also found that all samples show clear anisotropic step anomalies in the magnetization curve and that the anomalies and hysteresis behaviors depend on the non-stoichiometry.

ACKNOWLEDGMENTS

The authors are grateful to Prof. Y. Ueda, Institute for Solid State Physics, for his support to the SQUID magnetometer. They would also like to thank Mrs. F. Sakai and Mr. M. Koike, Institute for Solid State Physics, for all their helps throughout this work.

REFERENCES

1. W. Soller and A. J. Thompson, Phys. Rev., **47**, 644 (1935).

2. M. Mekata, N. Yaguchi, T. Takagi, S. Mitsuda and H. Yoshizawa, J. Mag. Mag. Mater., **104-107**, 823 (1992).

3. Y. Ajiro, K. hanasaki, T. Asano, T. Takagi, M. Mekata, H. A. Katori and T. Goto, J. Phys. Soc. Jpn., **64**, 3643 (1995).

4. F. A. Benko and F. P. Koffyberg, J. Phys. Chem. Solid, **48**, 431 (1987).

5. T. R. Zhao, M. Hasegawa, M. Koike and H. Takei, J. Crstal Growth, **148**, 189 (1995).

6. T. R. Zhao, M. Hasegawa and H. Takei, J. Crstal Growth, **154**, 322 (1995).

7. T. R. Zhao, M. Hasegawa and H. Takei, J. Crstal Growth, **148**, 189 (1996).

8. T. R. Zhao, M. Hasegawa and H. Takei, J. Mater. Sci., in press.

9. T. R. Zhao, M. Hasegawa and H. Takei, J. Solid State Chem., in press.

10. T. R. Zhao, M. Hasegawa and H. Takei, J. Mag. Mag. Mater., in press.

LAYERED 3-D FERROMAGNETS AND ANTIFERROMAGNETS, $M^{2+}(ReO_4)_2$ (M = Mn, Fe, Co, Ni, Cu) : IMPORTANCE OF DIPOLAR INTERACTIONS

C. C. TORARDI[a], W. M. REIFF[b], B. C. DODRILL[c], and T. VOGT[d]

[a] DuPont Company, CR&D, Experimental Station, Wilmington, DE 19880-0356
[b] Northeastern University, Department of Chemistry, Boston, MA 02115
[c] Lake Shore Cryotronics, Inc., 64 E. Walnut Street, Westerville, OH 43081
[d] Brookhaven National Laboratory, Department of Physics, Upton, NY 11973

ABSTRACT

Powder samples of the anhydrous compounds $M^{2+}(ReO_4)_2$ (M = Mn, Fe, Co, Ni, and Cu) were prepared. All have trigonal symmetry, $P\bar{3}m1$, except M = Cu which is Jahn-Teller distorted to monoclinic symmetry, I2/m. Single crystals of the Fe and Cu compounds were grown by vapor transport and their layered structures determined. Ferromagnetism, deduced from ac susceptibility and dc magnetization studies, is exhibited by the Fe (T_{Curie} = 8.5 K), Co (T_{Curie} = 4.7 K), and Ni (T_{Curie} = 12.5 K) phases. They show 2-D ferromagnetic (intralayer) interactions as well as a 3-D ferromagnetic (interlayer) ground state with both of these effects obvious in the heat capacity of $Ni(ReO_4)_2$. Low-temperature powder neutron diffraction confirms the ferromagnetism of $Ni(ReO_4)_2$ below 12 K, and also reveals a structural phase transition occurring between room temperature and 77 K. The origin of the 3-D ferromagnetic ground states of the Fe, Co, and Ni perrhenates is the result of dipolar interlayer interactions, as opposed to superexchange between the layers. Susceptibility results of the Cu perrhenate indicate intralayer ferromagnetism and any long-range 3-D order occurring at $T \le 1.5$ K. The Mn compound is antiferromagnetic.

INTRODUCTION

Previously, we have reported the synthesis, crystal structure, and magnetism of $M(ReO_4)_2$ for M = Fe, Co, Ni.[1,2] These layered compounds are isostructural, crystallizing with trigonal symmetry, space group $P\bar{3}m1$, with the $\bar{3}$ axis perpendicular to the layers. An individual layer consists of a covalently-bonded network of corner-sharing MO_6 octahedra and ReO_4 tetrahedra with each MO_6 oxygen atom bonded to a separate Re atom. Alternatively, the layer can be viewed as a sandwich composed of atomic planes with sequence O^{ap}-Re-O-M-O-Re-O^{ap}. Layers stack so that the unshared, apical perrhenate O-atoms of one layer interpenetrate those of an adjacent layer. The layers are held together solely by van der Waals interactions between the interleaving oxygen atoms. Preliminary magnetic susceptibility measurements indicated that these compounds are ferromagnetic and should exhibit spontaneous magnetization in zero field. However, the possibility of metamagnetism could not be discounted. In this paper, we present the results of a.c. susceptibility studies, d.c. magnetization measurements, and low temperature neutron diffraction experiments that conclusively demonstrate that these compounds have genuine 3-D ferromagnetic ground states.

We have also synthesized the M = Mn and Cu analogs and present here the distorted structure of the copper phase, as well as preliminary magnetic studies on both compounds.

399

EXPERIMENT

M(ReO$_4$)$_2$ compounds were prepared in single crystal or powder form as summarized by the three chemical reactions below. M = Fe was synthesized by procedures A and B, all other samples were prepared by procedure C. In addition, crystals of anhydrous Cu(ReO$_4$)$_2$ were grown from premade powder by TeCl$_4$ vapor transport. Lattice constants for the M(ReO$_4$)$_2$ compounds are given in Table I.

$$\text{Fe}_2\text{O}_3 + \text{Fe} + 3\ \text{Re}_2\text{O}_7 \xrightarrow[\text{sealed glass tube}]{\text{TeCl}_4,\ 450°\text{C},} 3\ \text{Fe(ReO}_4)_2 \quad \text{(A)}$$
plate crystals

$$2\ \text{FeCl}_3 + 2\ \text{Fe}_2\text{O}_3 + 6\ \text{ReO}_3 \xrightarrow[\text{sealed glass tube}]{450°\text{C},} 3\ \text{Fe(ReO}_4)_2 + 3\ \text{FeCl}_2 \quad \text{(B)}$$
plate crystals

$$\text{MCO}_3 + \text{Re}_2\text{O}_7\ \text{(aq)} \xrightarrow[\text{dry 150-300°C}]{\text{stir, filter, evap,}} \text{M(ReO}_4)_2 + \text{CO}_2 \quad \text{(C)}$$
powder

TABLE I. Lattice Constants[a] for M(ReO$_4$)$_2$ Compounds (M = Mn, Fe, Co, Ni, Cu)

M	a (Å)	b (Å)	c (Å)	β (°)	V(Å^3)
Mn	5.869(1)		6.076(2)		181.2
Fe	5.767(1)		6.129(1)		176.5
Co	5.729(1)		6.120(8)		174.0
Ni	5.669(1)		6.144(8)		171.0
Cu	10.178(3)	5.570(1)	12.005(4)	92.71(1)	170.0 x 4 = 679.8

[a] Mn, Fe, Co, and Ni are trigonal, P$\bar{3}$m1. Cu is monoclinic, I2/m.

A summary of a.c. susceptibility data (showing strong imaginary components $\chi''_m \neq 0$ near T_{Curie}), and d.c. magnetic measurements (observation of classical hysteresis, remenance decay with T increasing towards T_{Curie}, and magnetization in very low applied fields with T decreasing towards T_{Curie}) is given in Table II for the M = Fe, Ni, and Co perrhenates. Neutron powder diffraction data for Ni(ReO$_4$)$_2$ were obtained at the Brookhaven National Laboratory High Flux Beam Reactor.

TABLE II. Magnetic Data for Ferromagnetic $M(ReO_4)_2$ (M = Fe, Co, Ni)

	T_{Curie} (K)	M_{sat} (emu/mole)	Coercivity, H_c (Oe)	M_{rem} (emu/mole)	S (calc/expected)
Fe[a]	8.5	20,500	≈ 0	≈ 0	1.84 / 2
Ni[a]	12.5	10,600	$\approx 2,600$	$\approx 3,200$	1 / 1
Co[b]	4.7	12,300	≈ 85	$\approx 2,250$	1.1 / 1.5

[a] for T = 4.2 K.
[b] for T = 3.1 K, except H_c at 2.4 K.

RESULTS

The trigonal structure of $M(ReO_4)_2$ for M = Mn, Fe, Co, and Ni is shown in Figure 1. Monoclinic $Cu(ReO_4)_2$ (Fig. 2) has a similar structure, with individual layers built from corner-sharing CuO_6 distorted octahedra and ReO_4 tetrahedra. The symmetry of the MO_6 unit is lowered from D_{3d} (M = Mn, Fe, Co, Ni) to C_s (M = Cu) because of the expected Jahn-Teller distortion. The CuO_6 units are axially elongated with four equatorial Cu-O bond lengths at 1.95 Å (Cu-O3 and Cu-O4) and two apical bonds at 2.30 Å (Cu-O6) and 2.34 Å (Cu-O5) (Fig. 2).

The susceptibility and magnetization studies (information summarized in Table II) offer proof to the existence of three-dimensional ferromagnetic ground states in the M = Fe, Co, and Ni compounds[2]. The ferromagnetism of $Ni(ReO_4)_2$ was also confirmed by a low-temperature neutron diffraction study of the polycrystalline powder (Figure 3). Below RT, but above 77 K, the trigonal structure transforms to a lower symmetry phase which appears monoclinic. Below T_{Curie}, the structure remains monoclinic, but the 100 reflection (on the basis of the trigonal cell) clearly increases in intensity due to the ferromagnetic order. This indicates the spins are perpendicular to the $Ni(ReO_4)_2$ layers, similar to that found for the layered $Ni(OH)_2$.[3] In addition, the heat capacity of $Ni(ReO_4)_2$ clearly shows both the 2-D (intralayer) and 3-D (interlayer) ferromagnetic ordering.

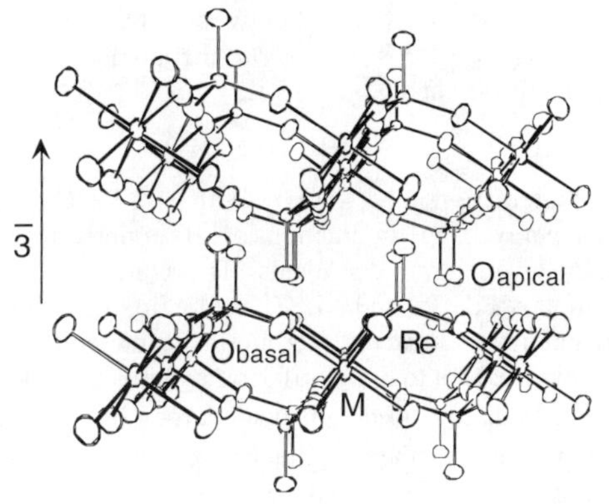

FIGURE 1.
Trigonal layered structure of $M(ReO_4)_2$ for M = Mn, Fe, Co, and Ni viewed perpendicular to the c axis and showing two stacked layers. Each layer is composed of corner-sharing MO_6 octahedra and ReO_4 tetrahedra.

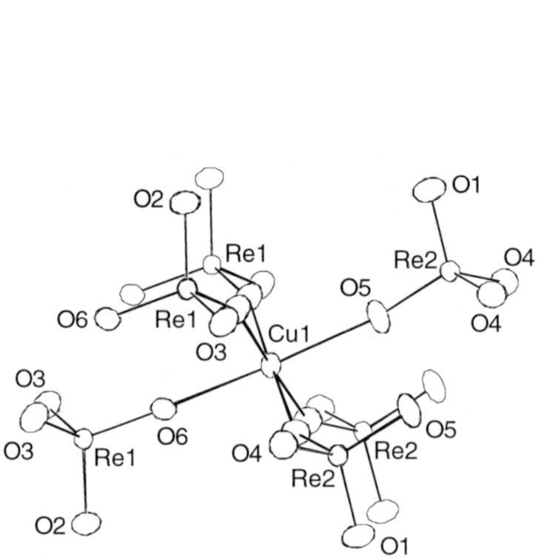

FIGURE 2.
Monoclinic structure of $Cu(ReO_4)_2$ showing the Jahn-Teller distorted (axially elongated) CuO_6 octahedron sharing corners with ReO_4 tetrahedra.

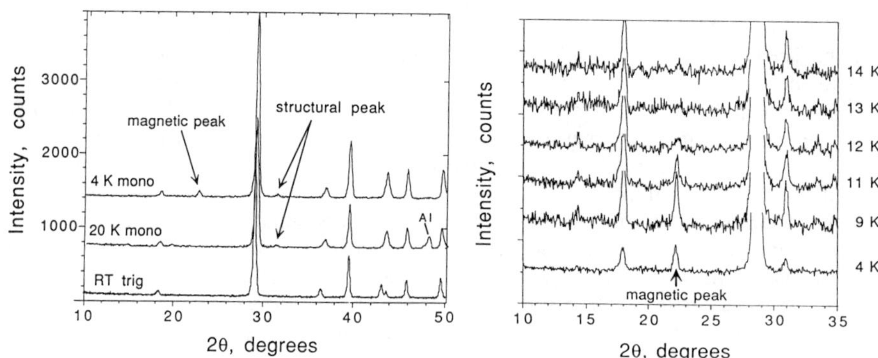

FIGURE 3.

Neutron powder diffraction patterns ($\lambda = 1.8857$ Å) of $Ni(ReO_4)_2$ as a function of temperature showing, *left*: the trigonal to monoclinic structural change that occurs below room temperature, and the magnetic diffraction peak occurring below $T_{Curie} \sim 13$ K, and, *right*: the intensity of the ferromagnetic peak decreasing to zero at $T_{Curie} \sim 13$ K.

The origin of the interlayer ferromagnetic interaction in $M(ReO_4)_2$, with M = Fe, Co, and Ni, is likely *dipolar* in nature. Strong interlayer superexchange interactions are discounted for these compounds because the bonding between the layers is van der Waals. This is in contradistinction to the situation in $M(OH)_2$ with M = Fe, Co, and Ni. These hydroxides, like the present perrhenates, show intralayer ferromagnetism. However, the interlayer interactions, via hydrogen-bonded superexchange-coupled pathways, lead to 3-D antiferromagnetic ground states and metamagnetism.[3] When the interlayer spacing in $Co(OH)_2$ is made larger, by exchanging some OH^- with larger organic or inorganic anions, the coupling between the layers is weakened and ferromagnetism results.[4] This is entirely analogous to the behavior of the Fe, Co, and Ni perrhenates.

Low-temperature susceptibility measurements for $Cu(ReO_4)_2$ show a steady increase in χ_M with decreasing T. The magnetic moment above ~ 30 K is about 1.9 μ_B (close to the expected 1.73 μ_B for S = 1/2 Cu^{2+}). However, as the temperature decreases below ~ 30 K, the moment increases rapidly and reaches a value of 3.2 μ_B at 1.5 K, substantially above the spin-only expectation. This suggests that ferromagnetic interactions exist at least within the layers. However, further studies are needed to determine the nature of the three-dimensionally ordered ground state.

Susceptibility measurements on $Mn(ReO_4)_2$ indicate both intralayer and interlayer antiferromagnetic behavior with $T_N = 3.1$ K.

CONCLUSIONS

The $M(ReO_4)_2$ compounds, with M = Fe, Co, Ni, and possibly Cu, are transparent and insulating solid-state materials that are true ferromagnets at low temperature. The origin of the interlayer ferromagnetic interaction is likely dipolar in nature. We are continuing our work to gain further insights into this interesting class of compounds.

ACKNOWLEDGMENTS

The authors thank J. C. Calabrese, W. J. Marshall, and C. M. Foris for single-crystal and powder x-ray data, and J. Paul Attfield and Marc Drillon for helpful discussions.

REFERENCES

1. C. C. Torardi, J. C. Calabrese, K. Lázar, and W. M. Reiff, Inorg. paper no. 155, 187th National A.C.S. Meeting, St. Louis, MO, 1984.

2. W. M. Reiff, B. C. Dodrill, and C. C. Torardi, Mol. Cryst. Liq. Cryst. **274**, 137-143 (1995).

3. T. Takada, T. Bando, M. Kiyama, H. Myamoto, and T. Sato, J. Phys. Soc. Japan **21**, 2726 (1966).

4. V. Laget, S. Rouba, P. Rabu, C. Hornick, and M. Drillon, J. Magnetism Magnetic Mater. **154**, L7 (1996).

CONSEQUENCES OF FLUORINE ATOM BRIDGING AND INCREASING METAL ATOM SPIN DENSITY FOR THE LOW TEMPERATURE MAGNETISM OF THE SERIES: $Na_3[FeF_6]$, $[NH_4]_3[FeF_6]$, $[Co(NH_3)_6][FeF_6]$ AND $[NH_4CoFeF_6]$

W.M. REIFF *, M.J. KWEICIEN*, M.A. SUBRAMANIAN**
*Department of Chemistry, Northeastern University, Boston, MA 02115
**E.I. Du Pont, CR&D, Experimental Station, P.O. Box 80328, Wilmington, DE 19880-0328

ABSTRACT

Aspects of the low temperature magnetism of some distorted cryolite-like compounds of high-spin iron III, namely $Na_3[FeF_6]$, $[NH_4]_3[FeF_6]$ are presented. The behavior of these systems is compared to iron-57 Mössbauer spectroscopy and ac-susceptibility results for the related $[Co(NH_3)_6][FeF_6]$ and the modified pyrochlore $[NH_4CoFeF_6]$ obtained via thermolysis of $[Co(NH_3)_6][FeF_6]$ under nitrogen.

INTRODUCTION

The pyrochlore ($A_2M_2X_6X'$, X and X' = O, F, S, OH) structure may be viewed as an interpenetrating network of M_2X_6 octahedra and anti-cristobalite type, A_2X' [1]. A majority of these materials crystallize with cubic symmetry and belong to the space group $Fd3m$ (no. 227), Z = 8. In $AMM'X_6$ modified pyrochlores, the A_2X' sublattice can be replaced by larger monovalent cations such as Rb^+ or Cs^+. These ions are then located at the former site of the seventh anion, X' (8b sites). In $AM^{2+}M'^{3+}F_6$ pyrochlores where the M and M' are transition metals, the distribution of M and M' cations in octahedral sites greatly influences their magnetic properties. A random (disordered) distribution of M^{2+} and M'^{3+} leads to a ferrimagnetic or spin-glass behavior, as observed in $CsMnFeF_6$ [2]. On the other hand, an <u>ordered arrangement</u> (as shown in $NH_4Fe^{2+}Fe^{3+}F_6$) results in an antiferromagnetic ordering [3]. Herein we extend these studies to the disordered modified pyrochlore $NH_4Co^{2+}Fe^{3+}F_6$, its precursor $[Co(NH_3)_6][FeF_6]$ and the related distorted cryolite (Na_3AlF_6) like variants, Na_3FeF_6 and $(NH_4)_3FeF_6$.

EXPERIMENT

Polycrystalline samples of $NH_4CoM^{3+}F_6$ (M = V, Cr or Fe) were prepared by heating hexaamminecobalt hexafluorometallate, $[Co(NH_3)_6][MF_6]$ at 280°C for 3 hours in nitrogen. Thus for M = Fe:

$$2\,[Co(NH_3)_6][FeF_6] \rightarrow 2NH_4Co^{2+}Fe^{3+}F_6\,(s) + N_2\,(g) + 8NH_3\,(g) + 2H_2\,(g) \qquad (1)$$

$$\text{(A)} \qquad\qquad\qquad \text{(B)}$$

The hexaamine molecular solids were synthesized from $[Co(NH_3)_6]Cl_3$ and MCl_3 by the precipitation method using 48% HF as described in the literature [4]. X-ray powder diffraction data were obtained with a Scintag (PAD IV) diffractometer using $CuK\alpha$ radiation. Cell dimensions were refined by least squares. AC susceptibility studies were conducted at 1 Oe and 125 Hz using a Lake Shore Model 7000 susceptometer.

Mat. Res. Soc. Symp. Proc. Vol. 453 © 1997 Materials Research Society

Single crystals of NH$_4$CoCrF$_6$ were grown by hydroflurothermal route and the details are given elsewhere [5].

X-ray diffraction pattern of the NH$_4$CoMF$_6$ (M = V, Cr or Fe) synthesized from hexaammine complex precursor showed the formation of a cubic phase with a lattice parameter of ~10Å. Single crystal x-ray structure refinement of NH$_4$CoCrF$_6$ as well as powder x-ray structure refinements of NH$_4$CoFeF$_6$ and NH$_4$CoVF$_6$ showed these phases are isostructural to that of the modified pyrochlore, RbNiCrF$_6$ [6], where the Co^{2+} and M^{3+} are statistically distributed in the octahedral sites (16c sites in space group $Fd3m$) forming a modified (Co,M)F$_3$ pyrochlore-type network, in which the NH$_4^+$ ions occupy the 8b sites [X′ sites in the A$_2$M$_2$X$_6$X′ pyrochlore with the A sites (16d) being vacant]. The Co/M^{3+}-F-Co/M^{3+} angle is close to ~138°.

RESULTS AND DISCUSSION

On the basis of iron-57 Mössbauer spectroscopy, we find that the distorted cryolite (Na$_3$[AlF$_6$]) variants, Na$_3$[FeF$_6$] and [NH$_4$]$_3$[FeF$_6$] (structure shown in Figure (1)) are unordered, rapidly relaxing paramagnets to as low as 0.29K. Sample data for the related Na$_3$FeF$_6$ at 0.52K are presented in Figure (2) where it is seen that this material is a rapidly relaxing paramagnet at 0.52K in zero field. The results were obtained using a sorption pumped helium-3 cryostat that has been previously described [7].

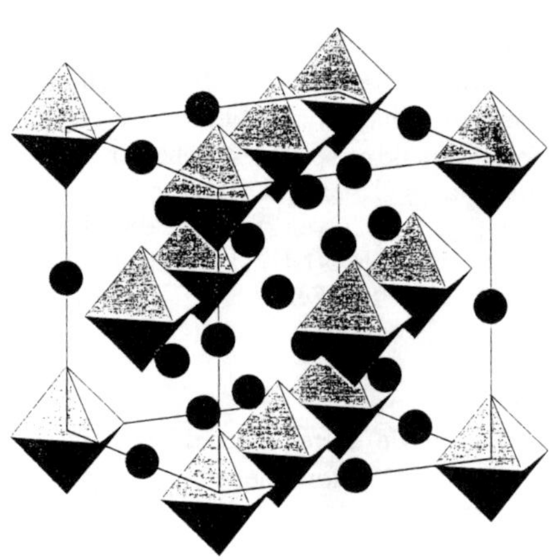

Figure 1.   Basic distorted cryolite structure of (NH$_4$)$_3$FeF$_6$.

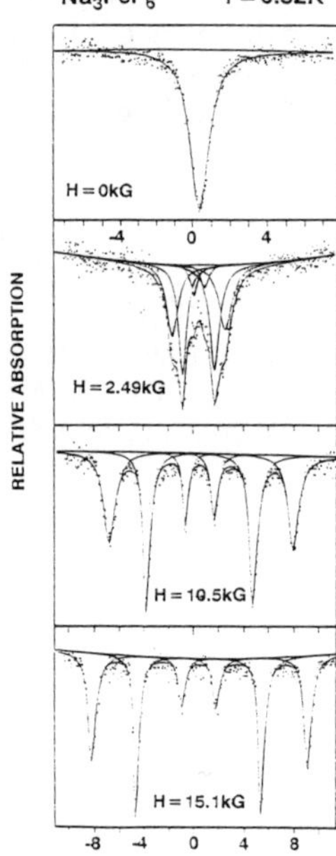

Figure 2.

Some zero and transverse applied field Mössbauer spectra for a polycrystalline sample of Na$_3$FeF$_6$ at 0.52K.

However, it is seen that relatively small (transverse) applied magnetic fields (~1T) are able to induce slow paramagnetic relaxation, fully resolved magnetic hyperfine splitting, and a very large (~50T) effective internal field for this material. Similar field dependence behavior is observed for $(NH_4)_3FeF_6$ i.e. rapidly relaxing paramagnetism in H=O. Essentially complete polarization of the nuclear and electron spin moments to the direction of the applied field (perpendicular to the direction of gamma ray propigation in our experimental set-up) is evidenced by the strong intensification of the ΔM_I=O transitions (transitions 2 and 5 of the spectra) relative to the $\Delta M_I = \pm$ 1 transitions.

The simple replacement of the cation in these systems by the complex low-spin Co III species, $[Co[NH_3)_6^{3+}]$ to give the cubic salt $[Co(NH_3)_6]FeF_6]$ (A) causes no change in magnetic behavior (i.e. simple parmagnetism for T≥0.33 K) even though hydrogen bonded magnetic exchange pathways are now possible. Very weak exchange interactions are anticipated in that these compounds contain "structurally isolated" S=5/2 FeF_6^{3-} anion polyhedra and intervening <u>diamagnetic cations</u> as shown in Figures (1) and (3). Thermolysis of (A) under nitrogen yields the cobaltous species $[NH_4CoFeF_6]$, (B), where the spin-singlet Co III ion is now replaced by NH_4^+ and spin quartet CoII in a disordered metal atom (based on Mössbauer spectra, vide infra) fluorine bridged modified pyrochlore structure (Figure (4)).

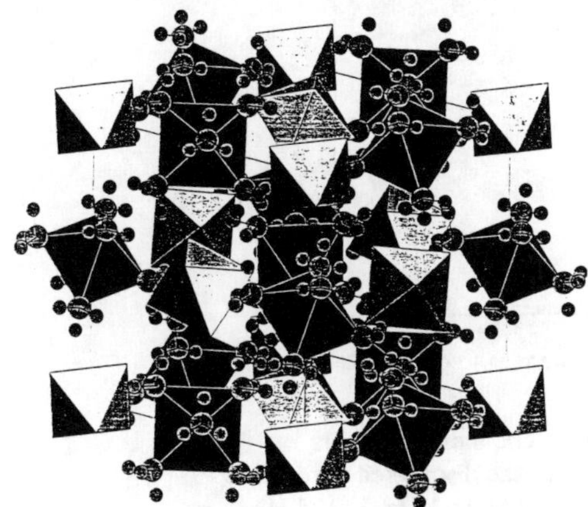

Figure 3.

Structure of $[M(NH_3)_6][M'FeF_6]$. $[M(NH_3)_6]^{3+}$ units are shown as filled octahedra whereas $M'F_6^{3-}$ units are shown as partially shaded octahedra. These units are joined by weak H · · · · F hydrogen bonds.

Figure 4.

Structure of NH_4CoMF_6. Co and M atoms are statistically distributed in the octahedral sites and NH_4^+ cations (filled circles) are located above and below the hexagonal puckered rings formed by $(Co,M)F_6$ octahedral network.

The resulting S=3/2, S=5/2 magnet is a member of a larger family of compounds that can exhibit any of ferrimagnetism, magnetic frustration, spin-glass behavior or antiferromagnetic order depending on the specific metal species and the degree of order of the cation distribution. In any event, the structural and magnetic condensation that result on going from (A) → (B) has a dramatic effect as $T_{(critical)}$ for (B) is ~11K via Mössbauer spectra and AC susceptibility measurements. That is, one has simultaneously introduced extensive fluorine atom bridging leading to Co^{II}-F-Co^{II}, Co^{II}-F-Fe^{III} and Fe^{III}-F-Fe^{III} super exchange pathways and increased the metal-atom spin densitiy (S=0 Co^{III} → S=3/2 Co^{II}) in a highly bridged network structure. Specifically, one sees (Figure (5)) a sharp rise in the magnetic moment (μ) of (B) near T_c ~11K riding along on a more general overall decreasing μ clearly suggesting ferrimagnetism.

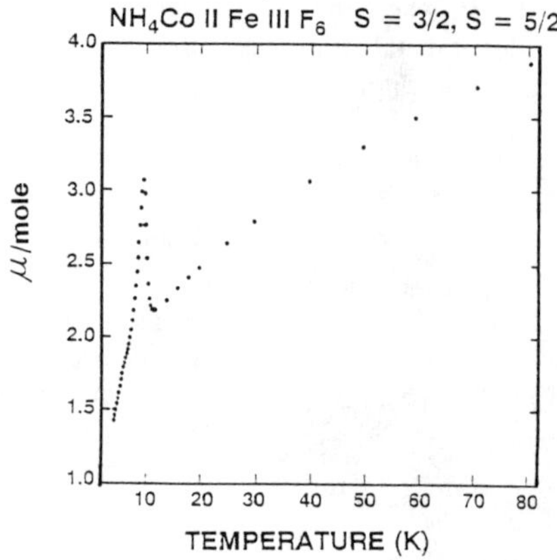

Figure 5.   Effective magnetic moment versus temperature for the disordered pyrochlore NH$_4$CoFeF$_6$, H$_0$(AC) =1 Oe, F=125 Hz.

These are AC susceptibility data taken at $H_{o(AC)}$ = 1 Oe and a frequency of 125 Hz. A more detailed field and especially frequency dependence study of (B) in the context of possible spin glass behavior is planned. Finally, paramagnetic cation disorder leading to spin glass behavior for (B) is clearly suggested via its highly broadened Mössbauer spectrum (Figure (6)) at 1.38K where it is magnetically saturated. That is, one has a distribution of internal hyperfine fields at the Fe III sites of (B) corresponding to the aforementioned variation of M-F-M' exchange pathways. This strongly contrasts with the narrow line width Mössbauer spectrum observed for the cation ordered pyrochlore NH$_4$Fe^{2+}Fe^{3+}F$_6$ [8].

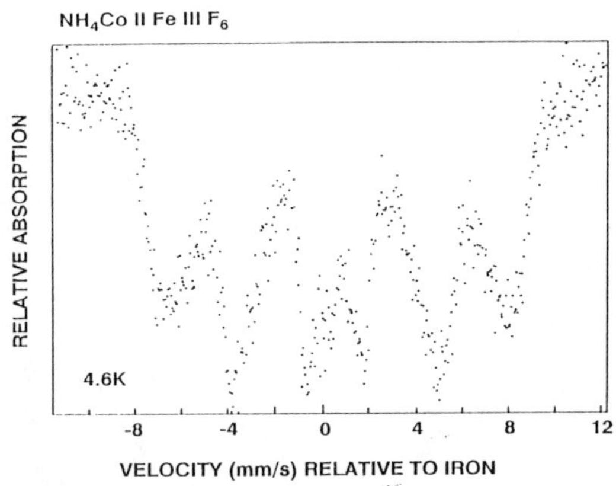

Figure 6.   Zero field Mössbauer spectrum
of NH4CoFeF6 at 1.38K.

CONCLUSION

The distorted cryolite variants $Na_3[FeF_6]$, $(NH_4)_3[FeF_6]$ as well as $[Co(NH_3)_6][FeF_6]$ (A) all contain magnetically "isolated" FeF_6^{3-} polyhedra with intervening diamagnetic cations. These structural features lead to paramagnetism for $T \geq 0.33K$. The condensation resulting from replacement of all terminal fluoride anions in (A) by all bridging in NH_4CoFeF_6 (B) results in a highly super-exchanged coupled fluorine-bridged network and a substantially enhanced critical temperature.

REFERENCES

1.   M.A. Subramanian, G. Aravamudan and G.V. Subba Rao, Prog. Solid State Chem. **15**, p. 55 (1983).
2.   J. Villain, Z. Phys. **B33**, p.31 (1979).
3.   G. Ferey, M. LeBlanc and R. dePape, J. Solid State Chem. **40**, p. 1 (1981).
4.   K. Wieghardt and H. Siebert, J. Mol. Structure 7, p. 305 (1971).
5.   M.A. Subramanian, W.H. Marshall and R.H. Harlow, Mater. Res. Bull. **31**, p. 585 (1996).
6.   D. Babel, Z. Anorg. Allg. Chem. **387**, p.161(1972).
7.   W.M. Reiff, Hyperfine Interactions **40**, p. 195 (1988).
8.   N.N. Greenwood, F. Menil and A. Tressaud, J. Solid State Chem. **5**, p. 402 (1972).

EFFECT OF HYDROSTATIC PRESSURE ($\leq$2GPa) ON THE TRANSPORT PROPERTIES IN PEROVSKITE-TYPE OXIDES

*M.ITOH, *J-D.YU, *K.NISHI, *T.KATSUMATA, *Y.INAGUMA, **T.HUANG
*Materials and Structures Laboratory, Tokyo Institute of Technology, 4259 Nagatsuta, Midori, Yokohama 226, Japan, *itoh1@rlem.titech.ac.jp*
**Department of physics, Shanghai University, Shanghai 201800, China

ABSTRACT

A clamp-type hydrostatic pressure device for the measurement of properties of perovskite-related compounds under pressure was established in our laboratory and applied to the measurements of resistivity, AC susceptibility, Hall effect and ionic conductivity. In usual perovskite-type oxides, loading of a pressure of 1 GPa (10kbar) corresponds to a contraction of interionic distance as large as 0.2%. This small value is not enough to evaluate the whole scheme of phase transition behavior for perovskite-type oxides but very effective for the observation of changes in interactions between electrons and ions. Some problems concerned with this method have been pointed out.

INTRODUCTION

At low temperature, high pressure device is a useful tool to investigate the phenomena of solid state physics. Under high pressure some intrinsic behaviors of the material can be changed: Fermi surface, the overlap of bands, interaction between electrons and ions, and Coulomb and repulsion potentials in oxides[1]. In the case of single crystal, probably, the variations of electronic and magnetic behavior along different directions are not the same under high pressure if its structure is anisotropic. From the variation of resistivity, susceptibility, capacitance and so on, the mechanical information of the material can be recognized.

By now, the high hydrostatic pressure device used at cryogenic temperature has two main designs. One is the anvil cell that can produce the pressure from 10 GPa to 30 GPa or higher pressure depending on the material strength of the anvil. The other is the piston cell and normally the maximum pressure is 2 GPa. From their mechanism they have their own advantages. The anvil cell can produce much higher pressure and the piston cell has relatively large sample chamber that can hold two or more samples. The latter is much important in the study of anisotropy of the property because it is necessary to measure the anisotropic behavior under the same ambient pressure.

In order to research the properties of perovskite-related compounds under high pressure at low temperature, a self-clamping piston cylinder high pressure device was assembled in our laboratory. This device was applied to the measurements of resistivity, magnetic resistance, Hall coefficient and AC susceptibility of samples under pressure. Compressibilities of $Bi_{2.2}Sr_{1.8}Ca_1Cu_2O_y$ and $IBi_{2.2}Sr_{1.8}Ca_1Cu_2O_y$ were determined in order to evaluate the change in the carrier concentration caused by the volume contraction under pressure.

EXPERIMENT

Design of High Pressure Device

The schematic drawing of the device, which is based on refs. [2] and [3] is given in Fig. 1. The device mainly includes: (A) press bar, (B) shaft, (C) supporting ring, (D) holder, (E) plug support, (F) locking nut, (G) piston, (H) cylinder, (I) sample room and (J) backing plate. All the parts of the device are made of copper-beryllium(Cu-Be), except for the piston and plug support, which are made of tungsten carbide(WC), and the press bar, sample room, which are made of hardened steel, Teflon, respectively. In the center of the plug support, a small hole with 0.5 mm diameter is made to let the lead wires pass through. The dimension of the Teflon bucket are 6 mm in outer diameter, 4 mm in internal diameter and 20 mm in length, which are large enough to put in two samples and the manometer. In the Teflon bucket, pressure medium of organic liquid, Idemitsu Kosan Daphnee 7373, is filled and sealed by the electrode plug. The antiextrusion ring and the electrode plug are made of unhardened Cu-Be. In the center of the electrode plug, a hole of 0.3 mm diameter is drilled and 10 coated wires of 0.06 mm diameter are through the hole and are sealed by epoxy resin.

Mat. Res. Soc. Symp. Proc. Vol. 453 © 1997 Materials Research Society

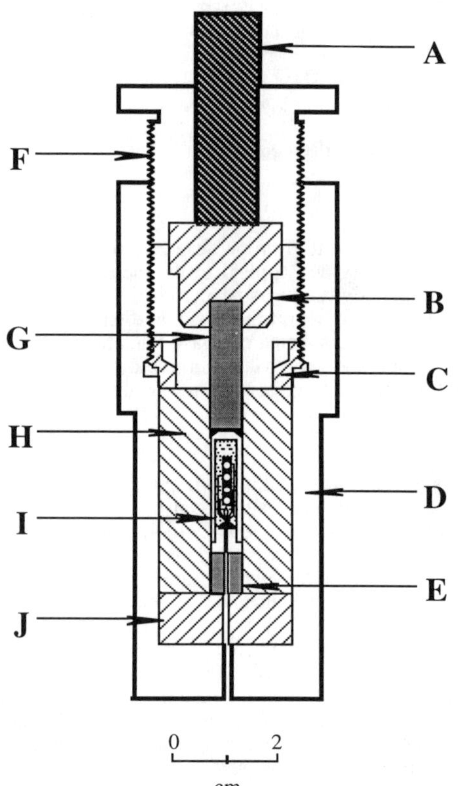

$$\begin{array}{cc} 0 & 2 \end{array}$$

cm

Fig.1 Cross section of high pressure device.:(A) press bar, (B) shaft, (C) supporting ring, (D) holder, (E) plug support, (F) locking nut, (G) piston, (H) cylinder, (I) sample room, (J) backing plate.

Temperature Measurement

Because the pressure medium of organic liquids has a small conductivity (nearly a few of mW/cm•K[4]), it is better to use the thermocouple and install it in the pressure cell for the accurate temperature measurement. Temperature measurement was carried out by using Chromel-Alumel thermocouple with a diameter 0.1 mm. The influence of the pressure on the thermopower of the thermocouple was checked at 77 K and room temperature under pressures of 0 GPa and 1.4 GPa. The result indicated that the temperature shift due to the pressure is negligible compared to the accuracy of the thermocouple.

Pressure Measurement

Because the contraction percentages of materials in high pressure device are different, usually, the pressure in the cell changes with temperature so the pressure loss of device has to be identified in experiments. In our case, manganin wire resister was chosen as a manometer because it is sensitive and easiness of measurement. At room temperature the resistivity of manganin wire is linear up to 3 GPa and this linearity is believed to be independent of temperature. Thus, the curves of manganin wire resistivity under pressure can be thought to raise the curve of atmospheric pressure parallelly. The change of relative manganin wire resistance with pressure has the value of $(\Delta R / R) / P = 2.5 \times 10^{-2} (GPa)^{-1}$ below 3 GPa and this value is thought to be a constant for the temperature range 4.2 to 300 K[5]. Figure 2 shows the temperature dependence of pressure loss in the temperature range 4.2 to 273 K. This figure shows that the pressure loss from room temperature to 77 K is 0.3 GPa and that from 77 K to 4.2 K is less than 0.1 GPa. Then the total pressure loss is within 0.4 GPa in our device.

Sample Preparation

Single crystalline samples, $La_{0.85}Sr_{0.15}MnO_3$ and $Bi_{2.2}Sr_{1.8}Ca_1Cu_2O_y(Bi2212)$, were prepared by Floating Zone(FZ) and Traveling Solvent Floating Zone(TSFZ) methods, respectively. The chemical composition of these samples were determined using Induced Coupled Plasma(ICP) emission spectroscopy. Valence of manganese ions were determined by the redox titration. Iodine was intercalated into single crystalline $Bi_{2.2}Sr_{1.8}Ca_1Cu_2O_y$ and its composition was determined as $IBi_{2.2}Sr_{1.8}Ca_1Cu_2O_y(IBi2212)$.

Measurement of Compressibility

Compressibility of Bi2212 and IBi2212 were determined using diamond anvil up to 10 GPa.

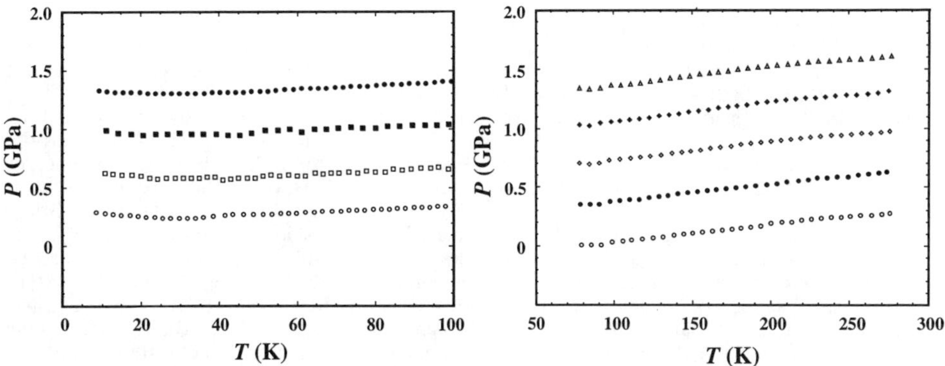

Fig.2. Measured pressure loss in the temperature ranges below 100 K(left) and above 77 K(right)

X-ray source were a tube with Mo target or synchrotron radiation in photon factory in Tsukuba.

Measurement of Resistivity

Electrical resistivity of the sample was measured in the pressure device by the conventional four probe method using the current 1 mA. Magnetic resistance was measured under a magnetic field of 1 tesla.

MD calculation

Compressibility of lithium conducting material $La_{0.6}Li_{0.2}TiO_3$ was obtained by using the MD calculation program MXDORTO at room temperature.

RESULTS

Problems Involved in the Device

Through our high pressure researches during last several years[6-11], we became aware of the following problems in the high pressure device. Most serious problem is the pressure loss at low temperatures. As explained in the previous section, this is 0.4 GPa from room temperature down to 4 K. Since this pressure loss is reversible against the temperature change, this is caused simply by the difference in the thermal expansion coefficients of the materials used in the device. This unavoidable problem forces us to recalculate the data of temperature dependence of properties when we draw the pressure phase diagram. Due to this reason, the pressure inside the cell is always monitored by manometer. Second problem is a ferromagnetism of the WC piston and plug. For sintering of tungsten carbide powder, metallic cobalt, which shows ferromagnetism, is most often used as a binder. This prevents the precise measurement of magnetic susceptibility and sometimes causes an accident under high magnetic field. Quite recently, nonmagnetic WC sintered body became available in a commercial base, so such a material should be used in the devise for high magnetic field. Third problem is the tensile strength of Cu-Be alloy, 0.95 GPa for annealed one. This gives a limit of high pressure which allows a reversible generation of pressure. Above this pressure, the Cu-Be cylinder should experience a plastic deformation. Due to this reason, the cylinder must be replaced after each experimental run.

<u>Compressibility Measurements</u>

When we measure the physical properties of materials under high pressure, we must consider the contribution of volume contraction. In order to evaluate the contribution of this term, we have carried out the measurement of compressibility of Bi2212 and IBi2212 single crystals, which have large anisotropies both in structure and chemical bonds and are considered to have large compressibilities compared to other perovskite-related compounds. Figure 3 shows the normalized pressure dependence of lattice constants a/a_0, c/c_0, and V/V_0 for Bi2212 and IBi2212. It is easy to estimate that the compressibility of IBi2212 is larger than that for Bi2212 because the lattice of IBi2212 is expanded by the intercalation of iodine into Bi_2O_2 layers. Bulk moduli of these compounds were obtained by the least square regression analysis for the data using Brich-Murnagham equation of state.   In Fig.4, we compare the compressibility determined in this study with those in the literature[12-15]. The closed circles show the results determined by the molecular dynamics(MD) calculation[17]. It is interesting that compressibility for $SrTiO_3$, ReO_3(<0.4 GPa), $BaBiO_3$, and $La_{0.6}Li_{0.2}TiO_3$ show almost same tendency in the range <2 GPa, i.e., a linear dependence with almost same slope appears in this pressure range. Bi2212, IBi2212, ReO_3 with high pressure form, and $PbTiO_3$ lay on other lines with larger slopes. Table 1 shows the bulk moduli for the materials given in Fig.4. These results suggest that effect of volume contraction can

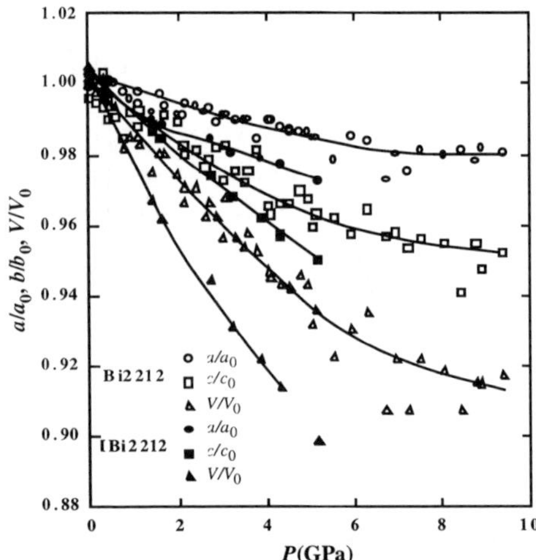

Fig.3 Pressure dependence of normalized a, c, and V for Bi2212 and IBi2212.

Fig.4. Comparison of compressibilities of some perovskite-related compounds.

be ignored in the calculation of carrier density for electronic conducting materials determined by the measurement of Hall coefficient under pressure.

Table 1 List of bulk module for perovskite-related materials.

Material	B_0(GPa)	Ref.	Material	B_0(GPa)	Ref.
$SrTiO_3$	176	12	$BaBiO_3$	157	15
$PbTiO_3$	81.56	13	$La_{0.6}Li_{0.2}TiO_3$	165	This work
ReO_3 (<0.4 GPa)	192	14	Bi2212	68.6	This work
ReO_3(>2G Pa)	33.3	14	IBi2212	41.4	This work

<u>Resistivity Measurement Under High Pressure</u>

Using the bulk modulus for $SrTiO_3$, we can estimate per cent linear compressibility of perovskite-type compound. Since the compressibility curve can be linear in the low pressure region, linear compressibility of 0.2 %/GPa is obtained from the bulk modulus. For $SrTiO_3$, we get a contraction of 0.0039 A/GPa for Ti-O distance. Figure 5 shows the pressure dependence of resistivity of as-grown and annealed single crystalline $La_{0.85}Sr_{0.15}MnO_3$. The annealing of single crystal was carried out at 1073 K for 200 h in an oxygen gas flow. This annealing changes the crystal from orthorhombic to rhombohedral symmetry with no change in the oxygen content. When magnetic field of 1 tesla is applied to the sample, the negative magnetic resistance appears as reported in the previous works. When applying hydostatic pressure, the resistance also decreases as well as the effect of magnetic field. These results can be explained by the strengthening of double exchange interactions among Mn^{3+} and Mn^{4+}. Magnetic field assist t_{2g} spins to align each other, and application of isostatic pressure results in the shortening the Mn-O bond by 0.004 A/GPa. This shortening strengthens the ferromagnetic interaction and decreases the resistivity. We have measured the AC susceptibility for the same sample under high pressure using the same clamp cell. The results indicated that Tc increases and secondary phase, which is considered to have a rhombohedral symmetry, appears with increasing pressure. As a consequence, the orthorhombic phase completely disappears at higher pressures. Pressure dependence of Tc for orthorhombic and rhombohedral phases were 20 K/GPa and 15 K/GPa, respectively, below 1.5 GPa. The value for rhombohedral phase is same for the polycrystalline $La_xA_yMn_zO_3$ system[10]. The difference in the pressure dependence are due to the difference in the compressibilities of the two phases.

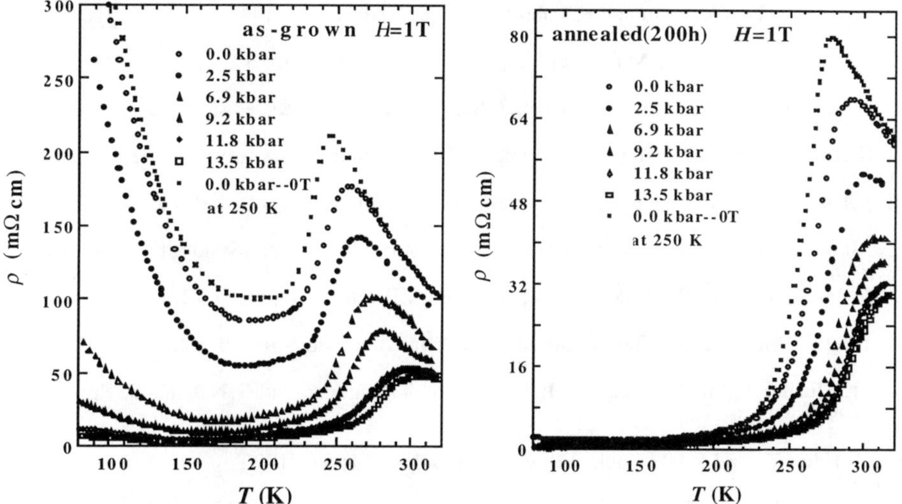

Figure 5. Pressure dependence of resistivities under various pressures for as-grown(left) and annealed(right) $La_{0.85}Sr_{0.15}MnO_3$.

CONCLUSIONS

A clamp-type hydrostatic pressure device facilitates the measurement of transport properties under pressure, <2GPa. The compressibilities of some perovskite-related compounds including Bi2212 and IBi2212 indicate that a change in the carrier density under high pressure less than 2 GPa can not be the principal factor in the change of properties. For example, the change in Tc of Bi2212 and IBi2212 [6,7,10] with pressure is considered to be mainly correlated to the change in the chemical bond with pressure. Although the pressure of 2 GPa is too low to evaluate the whole scheme of the properties of perovskite-type compounds under pressure, we can understand the tendency of the pressure-induced changes such as in resistivity, phase transition temperature, and magnetic phase transition. Loading of a pressure of 1 GPa (10 kbar) on typical perovskite-type oxides causes a contraction of interionic distance as large as 0.2%. Even such a small value changes the overlap of bands in electronic conducting materials and repulsion potential in ionic conducting materials[18]. Using this high pressure device, measurements of transport properties under magnetic field such as Hall effect, magnetic resistance and Seebeck coefficient are easy and may give a fruitful result in the evaluation of electronic structure of oxides.

ACKNOWLEDGMENTS

The authors wish to express their thanks to professor Sugiura of Yokohama City University and professor Sasaki of Tokyo Institute of Technology for their help in the compressibility measurement. Also thanks are due to professor Kawamura for his permission to use the computer simulation program MXDORTO. Part of this work was supported by Grant-in-Aid for Scientific Research from Ministry of Education, Science and Culture.

REFERENCES

1. C.W.Chu and J.A.Woolam, High Pressure and Low Temperature Physics, Plenum, New York, (1978).
2. J.D.Thompson, Rev.Sci.Instrm. **55**, 231 (1991).

3. H.Fujiwara, H.Kadomatsu and K.Tohma, Rev.Sci.Instrum. **51**, 1345 (1980).

4. P.M.Chaikin, C.Weyl, G.Malfait and D.Jerome, Rev.Sci.Instrum. **52**, 1397 (1981).

5. A.Jayaraman, A.R.Hutson, J.H.Mcfee, A.S.Coriell and R.G.Maines, Rev.Sci.Instrum. **38**, 44 (1967).
6. T-K.Huang, M.Itoh, J-D.Yu, Y. Inaguma, and T.Nakamura, Phys.Rev.B **48**, 7712(1993).

7. T-K.Huang, M.Itoh, J-D.Yu, Y. Inaguma, and T.Nakamura, Phys.Rev.B **49**, 9885(1994).

8. M.Shikano, T-K.Huang, Y.Inaguma, M.Itoh, and T.Nakamura, Solid State Commun. **90**, 115(1994).
9. Y.Inaguma, J-D.Yu, Y-J.Shan, and T.Nakamura, J.Electrochem.Soc. **142**, L8(1995).

10. M.Itoh, T.Shimura, J-D.Yu, T.Hayashi, and Y.Inaguma, Phys.Rev.B **52**, 12522(1995).

11. T-K.Huang, L.Yu, Y.Zhang, M.Itoh, J-D.Yu, Y.Inaguma, and T.Nakamura, Physica C, in press(1996).

12. L.R.Edwards and R.W.Lynch, J.Phys.Chem.Solids **31**, 573(1970).

13. S.A.Kabalkina,and L.F.Vereshchagin, Soviet Phys., Doklady, **1**, 310(1962).

14. J.E.Schirber and B.Morosin, Phys.Rev.Lett. **42**, 1485(1979).

15. H.Sugiura and T.Yamadaya, Solid State Commun. **49**, 499(1984).

16. T.Katsumata, Y.Inaguma, M.Itoh, and K.Kawamura, unpublished work.

17. Y.Inaguma, L.Chen, M.Itoh, T.Nakamura, T.Uchida, M.Ikuta, and M.Wakihara, Solid State Commun. **86**,689(1993).

18. Y.Inaguma, Y.Matsui, J-D.Yu, Y-J.Shan, T.Nakamura, and M.Itoh, J.Phys.Chem.Solids, accepted.

Part VII
Dielectrics and Ferroelectrics

DOMAINLIKE ORGANIZATIONS IN FERROELECTRICS CONTAINING QUENCHED RANDOMNESS

Dwight Viehland, M.Y. Kim, Z. Xu, and Jie-Fang Li
Department of Materials Science and Engineering, and the Materials Research Laboratory,
University of Illinois, Urbana, IL 61801

ABSTRACT

Transmission electron microscopy studies of the $(1-x)Pb(Mg_{1/3}Nb_{2/3})O_3-(x)PbTiO_3$ (PMN-PT $(1-x)/x$) crystalline solution have been performed for $x=0.1$, 0.2, 0.35, and 0.60. Bright-field imaging has revealed a common sequence of domain-like states with increasing x. Normal micron-sized ferroelectric domains were found for $x>0.40$. Tweed-like structures were found for $x\sim0.35$. These tweedlike structures are similar to those previously reported in pre-martensitic states. Paraelectric clusters were found for $x<0.30$. The paraelectric cluster state was characterized by the lack of self-assembly amongst embryos and the presence of relaxor behavior in the macroscopic response characteristics. The composition PMN-PT 65/35 was then modified with La for a detailed study of the transition between the tweed-like precursor and paraelectric cluster states with increasing impurity content.

INTRODUCTION

Transmission electron microscopy (TEM) studies have previously revealed the presence of polar nanodomains in both mixed A-site cation (La-modified lead zirconate titanate (PLZT))[1] and mixed B-site cation (lead magnesium niobate (PMN))[2] relaxor ferroelectrics. Coarsening of polar nanodomains with decreasing temperature has been observed in rhombohedral-structured La-modified $Pb(Zr_{0.70}Ti_{0.30})O_3$ relaxors with 8.2 at.% La (PLZT 8.2/70/30).[3] Decreasing temperature resulted in enhanced tendencies towards long-range ordering (LRO), however the system did not transform the mixture of polar nanodomains and the paraelectric matrix into a completely LRO ferroelectric. Application of an electric field has been found necessary to transform the relaxor to a LRO ferroelectric in rhombohedral-structured PLZT.[4-6] This transition is often referred to in the literature as the field induced micro- to macro-domain transition.[4] The purpose of this investigation was to systematically study the changes in the domain structure of PMN by alloying the base composition with $PbTiO_3$ (PMN-PT). Modification of PMN with PT is known to decrease the relaxor-like characteristics.[7] It was anticipated that a common sequence of domain-like transitions might be observed with increasing compositional and/or structural disorder between normal micron-sized ferroelectric domains typical of a LRO ferroelectric state and polar nanodomains typical of a relaxor state.

EXPERIMENT

Various compositions in the $(1-x)Pb(Mg_{1/3}Nb_{2/3})O_3-(x)PbTiO_3$ (PMN-PT $(1-x)/x$) crystalline solution were made by a mixed oxide method. The compositions chosen for study, included, $x=0.10$, 0.20, 0.35, and 0.60. All chemicals used had purities greater than 99.9%. Processing conditions can be found in previous publications[8]. La-modified (y) PMN-PT 65/35 (PLMN-PT y/65/35) compostions were also fabricated. The compositions chosen for study included y=0.00, 0.01, 0.03, 0.05, and 0.07. Thin sections were prepared for TEM studies by ultrasonic drilling 3-mm diameter discs which were mechanically polished to ~100 μm. The central portions of the discs were further ground by a dimpler to $\sim10\mu$m, and argon ion-milled to perforation on a cold stage. Specimens were coated with carbon before examination. The TEM studies were carried out on a Phillips EM-420 microscope at 120 kV. The dielectric response was measured using a Hewlett Packard 4284A LCR meter which can cover a frequency range from 20 to 10^6 Hz. The samples were placed in a Delta Design 9023 test chamber, which can be operated between -200 and 250°C.

Mat. Res. Soc. Symp. Proc. Vol. 453 © 1997 Materials Research Society

RESULTS

Figures 1(a)-(d) illustrate the room temperature bright-field images for PMN-PT 90/10, 80/20, 65/35 and 40/60, respectively. Various types of domain structures can readily be seen in these micrographs. For PMN-PT 90/10 (Figure 1(a)), polar nanodomains are clearly evident. The average size of these polar nanodomains was ~50 Å. With increase of the Ti-content, the scale of the polar nanodomains increased. However, the morphology of these domains remained irregular. For PMN-PT 80/20, polar nanodomains remained obvious in the bright-field images. For this composition, the size increased to ~100-150 Å. Correspondingly, relaxorlike characteristic were observed in the dielectric response for both PMN-PT 90/10 and 80/20, as shown in Figure 2. The degree of relaxor characteristics increased with decreasing Ti-content. For the composition PMN-PT 65/35 (Figure 1(c)), tweed-like structures were observed. Cross-hatched patterns are clearly evident. The tweeds are ~2000 Å in length and ~200 Å wide. The corresponding dielectric response for 65/35 shown in Figure 2 exhibited only weak relaxor characteristics. On increasing the Ti-content to 60 at.% (Figure 1(d)), normal micron-sized domains were found to exist. The texture of the tweedlike structures for 65/35 was found to be sensitive to changes in the e-beam intensity. Figure 1(e) illustrates that the cross-hatched texture in the tweeds was changed to an elongation along a particular variant by increasing the beam intensity.

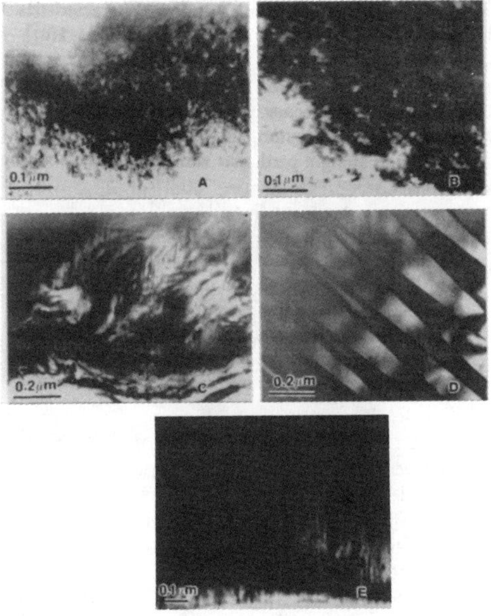

Figure 1.   Bright field images for various PMN-PT compositions. (a) 90/10, (b) 80/20, (c) 65/35, (d) 40/60, and (e) 65/35 under strong beam illumination.

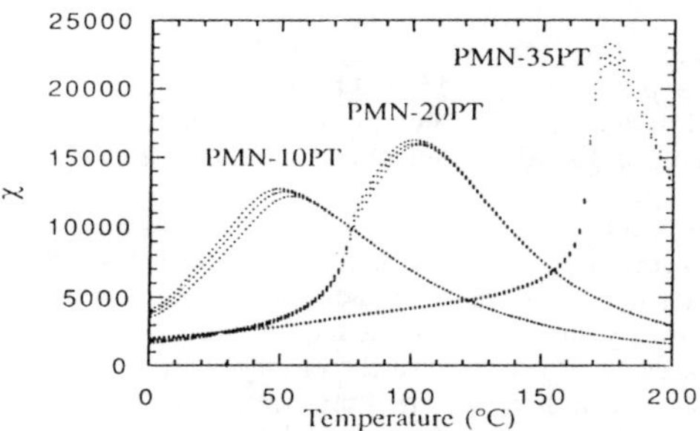

Figure 2.   Dielectric constant as a function of temperature and frequency for various PMN-PT compositions.

Much effort has focused on studying a variety of precursor or premonitory phenomena that occur prior to phase transitions in martensitic transformations.[9,10] Tweedlike striations identical to that shown in Figure 1(c) have been observed by TEM at temperatures significantly above the average transformation temperature in FePd and NiAl.[11-15] These tweedlike structures are characterized by a crosshatching in the bright-field images, are orientated along particular crystallographic directions, and are hundreds of angstroms in length. Upon approaching T_c on cooling, the tweedlike patterns become increasingly pronounced and begin to order into normal micron-sized twin structures. The origin of the tweedlike patterns is believed to be due to a high density of embryos above T_c.

Mesoscopic tweedlike texture has more recently been found in $YBa_2Cu_3O_{7-\delta}$ and $YBa_2(Cu_{1-x}Fe_x)_3O_{7-\delta}$.[16,17] In these materials, metastable tweedlike patterns were found after cooling through a tetragonal to orthorhombic transformation. In the current study, metastable tweedlike structures were found in PMN-PT 65/35 at least 100°C below the average transformation temperature, as determined by dielectric measurements. For example, tweedlike precursor structures were observed in PMN-PT 65/35 at room temperature, however the dielectric maximum was found to occur near 170°C. In this metastable tweed state, no evidence of normal micron-sized ferroelectric domains was observed. The changes in the domain-like states (between normal ferroelectric and tweedlike) for PMN-PT 65/35 strongly resembles the pattern of domain-like states previously reported in $YBa_2(Cu_{1-x}Fe_x)_3O_{7-\delta}$ and $YBa_2Cu_3O_{7-\delta}$ with increasing δ. In both types of systems, metastable tweed-like structures evolved from normal micron-sized domain structures with increasing compositional and/or structural disorder in the low temperature state.

Bratkovsky, Salje, and Heine[18] have recently simulated the ordering of metastable tweedlike structures into thin microdomains and the subsequent coarsening of the thin microdomains into a more traditional domain structure. Thermodynamic fluctuations were shown to control the length and width of the metastable tweeds. Texturing and coarsening were controlled by long-range elastic strains, which resulted in an ordering of the embryos. Some of the characteristic features of the sequence of domain-like states observed in PMN by varying the PT-content might be accounted for by the simulations of Bratkovsky, Salje, and Heine. However, the inability of the tweed-like and polar nanodomain states to order and coarsen with temperature and time for a particular composition would need additional considerations to be explained. Semenovskaya and Khachaturyan[19,20] have shown in $YBa_2Cu_3O_{7-\delta}$ and $YBa_2(Cu_{1-x}Fe_x)_3O_{7-\delta}$ for $0.5<\delta<0.8$ that tweedlike structures cannot order and coarsen into microtwins, rather a metastable or "glassy" mesophase persists in the low temperature phase. Ordering and coarsening was only possible in the stoichiometric range $\delta<0.5$. Kartha, Castan, Krumhansal, and Sethna[21] have also shown that long-time metastable tweed (or "glassy" mesophase) may exist below the transformation temperature in a martensitic phase. They modelled tweed theoretically using a nonlinear, nonlocal elastic free energy coupled to quenched local compositional inhomogeneities. It would appear that compositional and/or structural disorder in ferroelectrics (i.e., La in PZT) gives rise to long-time metastable or "glassy" mesophases.

Long-time metastable tweed-like structures in ferroelectrics may develop due to an inability of cation displacements to completely order, due to compositional or structural nonuniformities. In the metastable tweedlike state, long-range interactions may exist between embryos. These long-range interactions can obviously suppress to some extent the spatially inhomogeneity in the order parameter, resulting in texturing or ordering of the embryos. However, on increasing the degree of compositional and/or structural disorder, the long-range interactions between embryos would be suppressed, resulting in decreased texturing. Above a critical range, it might be more appropriate to consider the embryos as large paraelectric clusters rather than microdomains or tweeds, as boundaries would not exist between local regions of the order parameter.

It is unclear how the paraelectric cluster (or polar nanodomain) state evolves from the tweed-like precursor state with increasing compositional and/or structural disorder. To thoroughly investigate this evolution, a series of PMN-PT 65/35 compositions were made with varying amounts of La-content. Figures 3(a)-(d) show bright-field images for PMN-PT 65/35 with La-contents of 0, 1, 5, and 7 at.% (PLMN-PT x/65/35), respectively. Tweed-like precursor structures are clearly evident in Figure 3(a). For this composition, a significant degree of regularity or periodicity was found between tweeds. Upon doping with 1 at.% La (Figure 3(b)), tweed-like precursor structures were again observed. However, the tweed-like structures were more irregular in shape and orientation than for the undoped specimen. With increase of the La-content to 5 at.%, several significant changes were observed in the bright-field images. First, polar nanodomains were found to coexist with the tweed-like precursor structures. And secondly, the tweed-like structures found were smaller and more irregular in shape than for the lower La-content compositions. For the composition 7/65/35, only polar nanodomains were observed. With increase in La-content, the degree of ordering in the tweed-like precursor structures is clearly decreased. Upon reaching a critical La-content range, polar nanodomains began to evolve from irregularities in the tweed-like precursors.

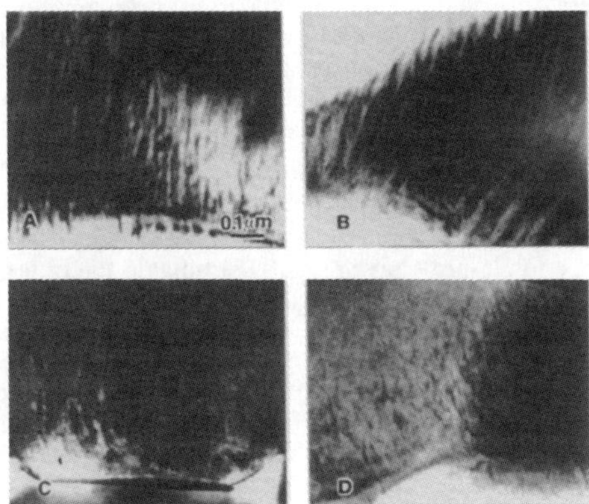

Figure 3.   Bright field images for various PLMN-PT x/65/35 compositions. (a) 0/65/35, (b) 1/65/35, (c) 5/65/35, and (d) 7/65/35.

Schmidt[22] has previously tried to explain the relaxor ferroelectric behavior using a model described as cubically-stabilized perovskite. He speculated the possible resemblance of the relaxor and pre-martensitic states. We believe that there is some validity to this concept, however it might be more correct to consider the relaxor as a continuously degenerate long-time metastable paraelectric cluster state which evolves from the tweed-like precursor state with increasing disorder. This paraelectric cluster state is conceptually similar to that originally conjectured by Cross.[23] It should be mentioned that relaxor ferroelectric behavior is only found in the metastable paraelectric cluster state, significant dielectric dispersion is not observed in the metastable mesoscopic tweedlike state, rather the dielectric response is characterized by a strong broadening of the phase transformational characteristics. It would appear that the inability of the paraelectric clusters to self-organize into tweed-like patterns results in the relaxor ferroelectric state. In this continuously degenerate metastable paraelectric cluster state, low frequency polarization fluctuations occur, somewhat similar to that observed in classic dipolar materials. On cooling below a characteristic temperature, the polarization fluctuations undergo a Vogel-Fulcher-like slowing down, exhibiting nonergodic-like behavior similar to dipolar and structural glasses.[24,25]

CONCLUSIONS

The important conclusions of the present study can be summarized as follows:

(1) The paraelectric-ferroelectric transformation is destroyed by quenched randomness in the PMN-PT system. Rather, a pretransitory state is stabilized which is characterized by the presence of tweedlike precursors for PMN-PT 65/35. The tweedlike precursor state is observed at temperatures significantly below that of the average transformation characteristics, clearly indicating strong metastability.

(2) With increasing disorder in the tweedlike precursor state (i.e, increasing La-content or decreasing PT-content), the degree of self-organization between embryos was decreased. Above a certain point, a paraelectric cluster state develops which is characterized by the lack of self organization between polar droplets. This continuously degenerate metastable paraelectric cluster state is the relaxor ferroelectric.

ACKNOWLEDGEMENTS

The research was supported by ONR N00014-95-1-0805 and by NUWC contract No. N66604-95-C-1536. The use of facilities in the Center for Microanalysis of Materials in the Materials Research Laboratory at the University of Illinois is gratefully acknowledged.

REFERENCES

1. C. A. Randall, D. J. Barber, and R. W. Whatmore, J. of Microscopy 45, 275 (1987).
2. A. Hilton, C. Randall, D. Barber and T. Shrout, Ferroelectrics 93, 379 (1989).
3. C. A. Randall, D.J . Barber, R. W. Whatmore, and P. Groves, Ferroelectrics 76, 311 (1987).
4. E. T. Keve and A. D Annis, Ferroelectrics 5, 77 (1973).
5. X. Yao, Z. L. Chen, and L. E. Cross, J. Appl. Phys. 54, 3399(1983).
6. S. M. Fan, J. W. He, and X. Yao, Ferroelectrics 77, 181 (1988).
7. Namchul Kim, PhD Dissertation, The Pennsylvania State University (1994).
8. S. Swartz, T. Shrout, Mat. Res. Bull. 17, 1245 (1982).
9. S. Moss, J. Phys. (Paris), Colloq. 38, C7-440 (1977).
10. Y. Morii and M. Iizumi, J. Phys. Soc. Jpn. 54, 2948 (1985).
11. I.M. Robertson and C.M. Wayman, Phil. Mag. A 48, 421, 443, 629 (1983).
12. L.E. Tanner, Phil Mag. 14, 111 (1966).

13. L.E. Tanner, A. Pelton, and R. Gronsky, J. de Phys. (Paris) 43, Colloq. C4, C4-169 (1982).
14. M. Mori, Y. Yamada, and G. Shirane, Solid State Commun. 17, 127 (1975).
15. S. Shapiro, J. Larese, Y. Noda, S. Moss and L.E. Tanner, Phys. Rev. Lett. 57, 3199 (1986).
16. A. Moodenbaugh, C. Yang, Y. Zhu, R. Sabatini, D. Fischer, Y. Xu, and M. Suenaga, Phys. Rev. B 44, 6991 (1991).
17. Y. Zhu, M. Suenaga, and A. Moodenbaugh, Phil. Mag. 62, 51 (1990).
18. A. Bratkovsky, E.K.H. Salje and V. Heine, Phase Transitions 52, 77 (1994).
19. S. Semenovskaya and A.G. Khachaturyan, Phys. Rev. B 47, 12182 (1993).
20. S. Semenovskaya and A.G. Khachaturyan, Physica D 66, 205 (1993).
21. S. Kartha, T. Castan, J.A. Krumhansl, and J. Sethna, Phys. Rev. Lett. 67, 3630 (1991).
22. G. Schmidt, Phase Transitions, 20, 127 (1990).
23. L. E. Cross, Ferroelectrics, 76, 249 (1987).
24. D. Viehland, M. Wuttig, and L.E. Cross, Ferroelectrics 120, 71 (1991).
25. D. Viehland, S. Jang, L.E. Cross, and Manfred Wuttig, Phil. Mag. B 64, 335 (1991).

PHASE EQUILIBRIA AND DIELECTRIC PROPERTIES OF RARE EARTH CONTAINING BaTiO₃ CERAMICS

D. KOLAR

Jožef Stefan Institute, University of Ljubljana, 1000 Ljubljana, Slovenia, Drago.Kolar@ijs.si

ABSTRACT

Addition of rare earth oxides affects several properties of barium titanate ceramics such as dielectric constant, Curie temperature and electrical conductivity. Electrical properties depend on the composition, firing conditions and mode of incorporation of rare earth ions into perovskite or perovskite-like lattice.

Incorporation of rare earth ions into the $BaTiO_3$ lattice was studied by quantitative wavelength dispersive spectroscopy in combination with scanning electron microscopy and X-ray powder diffraction. Broad solubility regions between rare earth titanates and barium titanate were detected, influencing the properties.

In present review, phase equilibria and properties of rare earth (La,Nd,Ce) containing $BaTiO_3$ ceramics are compared with particular emphasis on preparation conditions.

INTRODUCTION

Ceramics based on the ternary $BaO\text{-}TiO_2\text{-}R_2O_3$ (R = La - Gd) system are extensively used in manufacturing electronic ceramic components. $BaTiO_3$ slightly doped with rare earth oxides (up to a concentration of around 0.3-0.5 at.%) is semiconductive with a large PTCR effect and is used for preparation of a wide variety of switching, heating and regulating devices. With larger amounts of rare earths, $BaTiO_3$ exhibits high electrical resistivity and is used for manufacture of highly temperature stable ceramic capacitors ("NPO" type). Raise in conductivity is interpreted as a consequence of a change in the mode of donor compensation from electron to cation vacancies [1,2]. Ceramics with composition in the vicinity of $BaO:R_2O_3:TiO_2 = 1:1:4$ are used in the manufacture of high permittivity microwave resonators.

The electrical properties of $BaO\text{-}TiO_2\text{-}R_2O_3$ ceramics strongly depend on their crystal structure, grain size and phase composition. Therefore, for understanding their properties, phase equilibrium data are of great importance. However, the reported data are limited, inconclusive and frequently contradictory.

It is the aim of the present paper to review current knowledge of the phase equilibrium data of two rare-earth containing ternary oxide systems, $BaO\text{-}TiO_2\text{-}La_2O_3$ and $BaO\text{-}TiO_2\text{-}CeO_2/Ce_2O_3$, and to correlate the structural characteristics with electrical properties.

It is well known that 3-valent rare earth ions, such as La^{3+} or Nd^{3+} can be incorporated into the $BaTiO_3$ perovskite lattice ("ABO_3") on "A" sites. Cerium with a $4f^1 5d^1 6s^2$ electronic configuration is distinguished in the lanthanide series by its stability in both Ce^{4+} and Ce^{3+} states. From a comparison of the effective ionic radii of the ions involved [3] (Table I) it is

Mat. Res. Soc. Symp. Proc. Vol. 453 © 1997 Materials Research Society

expected that Ce^{4+} should enter the $BaTiO_3$ lattice at octahedrally co-ordinated B sites, while Ce^{3+} should enter the dodecahedrally co-ordinated A sites and thereby replace Ba ions.

TABLE I

Effective ionic radii of Ba^{2+}, Ti^{4+}, Ce^{3+}, and Ce^{3+} in different co-ordination states[1]

Ion	Effective ionic radius (nm) CN 6	Effective ionic radius (nm) CN 12
Ba^{2+}	-	0.161
Ti^{4+}	0.0605	-
Ce^{4+}	0.087	0.114
Ce^{3+}	0.101	0.134
La^{3+}	-	0.136

The stability of Ce(+3) and (+4) valent states depends on the oxygen partial pressure, temperature and crystal lattice environment. Low oxygen partial pressure and high temperatures stabilize the Ce^{3+} valent state.

The content of the paper is based on literature data and recent extensive research in the author's laboratory. The main experimental techniques used in determination of phase compositions were quantitative wavelength dispersive spectroscopy - electron probe microanalysis (WDS - EPMA) in combination with X-ray powder diffractometry (XRPD), scanning electron microscopy (SEM) and thermogravimetric analysis (TGA). A detailed description of the methods used has been published in [4,5].

Cerium dioxide when heated and cooled in air remains in the +4 valent state unless rapidly quenched. When calcined with BaO in air it forms $BaCeO_3$ with the perovskite structure[6]. $BaCeO_3$ melts incongruently at 1480°C with formation of solid CeO_2 and a liquid[7]. CeO_2 was also observed to be stable when heated in air with TiO_2. It forms the compound $CeTi_2O_6$ with the brannerite type ($ThTi_2O_6$) of structure, which melts incongruently at approximately 1400°C[8]. However, on melting CeO_2-TiO_2 mixtures in air Maister et al.[9] registered formation of Ce^{3+} compounds: $Ce_2O_3 \cdot TiO_2$, $Ce_2O_3 \cdot 2TiO_2$ and $2Ce_2O_3 \cdot 9TiO_2$. Compounds with the same molar ratio are known to exist in other systems of 3 valent rare earth oxides with TiO_2. The stoichiometry of TiO_2 on melting CeO_2-TiO_2 mixtures did not change appreciably.

On annealing at lower temperatures (1300°C) in air, compounds of Ce^{3+} with TiO_2 reoxidized and decomposed to CeO_2, TiO_2 and $CeO_2 \cdot 2TiO_2$. However, small amounts of impurities may stabilize the Ce^{3+} titanates.

BaO-TiO$_2$-CeO$_2$ Phase Equilibria

Equilibrium in the ternary BaO-TiO_2-CeO_2 system was described by Guha and Kolar[10]. Extensive mutual solid solubility was observed between $BaTiO_3$ and $BaCeO_3$. At 1200°C, the solid solubility of $BaCeO_3$ in $BaTiO_3$ is 20 mol.% and that of $BaTiO_3$ in $BaCeO_3$ 22 mol.%. Existence of the $CeTi_2O_6$ compound was confirmed.

Measurement of the dielectric properties of $Ba(Ti,Ce)O_3$ solid solution ceramics revealed a moderate shift of the maximum dielectric constant. 10 mol.% $BaCeO_3$ lowered the Curie point from 128°C to ~100°C[11] and caused significant broadening of the dielectric maximum vs. T curve. The ceramic was electrically highly insulating. However, it was also noted that the shift of Curie temperature was considerably more pronounced in $BaTiO_3$-$BaCeO_3$ ceramics sintered with excess TiO_2.

Other experimental observations were also not in accord with phase equilibrium data. In particular, $BaTiO_3$-CeO_2 compositions with low CeO_2 content (02-05 at.%) show the PTCR effect, characteristic of donor doped $BaTiO_3$ with +3 valent lanthanides on Ba sites, and an accompanying "grain growth anomaly" effect.[12,13,14]. It is obvious that the stability of Ce(+4) at high temperatures in the presence of TiO_2 is much lower than in the presence of BaO.

BaO-TiO$_2$-La$_2$O$_3$ Phase Equilibria

Literature data on phase relations in the BaO-Re_2O_3-TiO_2 systems are fragmentary and frequently inconsistent.

Data on La_2O_3 and Nd_2O_3 systems are the most abundant. Both rare-earth oxides are closely related and form compounds with BaO and TiO_2 with the same compositions and structures.

The latest data on the defect structure and phase relations of highly lanthanum doped barium titanate give the following picture: In the binary La_2O_3-TiO_2 system, Mc Chesney and Sauer[15] described 3 rare earth titanates with compositions La_2O_3:TiO_2 = 1:1 (La_2TiO_5, "LT"), 1:2 ($La_2Ti_2O_7$, "LT$_2$") and 2:9 ($La_4Ti_9O_{24}$, "L$_2$T$_9$"). Later on, Fedorov et al.[16] identified a fourth compound with La_2O_3:TiO_2 ratio 2:3($La_4Ti_3O_{12}$, "L$_2$T$_3$"). It was described as a perovskite-like, 12-layer hexagonal structure. It was reported to be stable in air up to 1450°C, whereupon it decomposes into 1:2 and 1:1 compounds. The identical binary compounds were also found to occur in the Nd_2O_3-TiO_2 system.

The BaO-TiO_2 binary system is well known [17-20]. In the TiO_2-rich portion of the system, the stable compounds are $Ba_6Ti_{17}O_{40}$ ("B$_6$Ti$_7$"), $Ba_4Ti_{13}O_{30}$ ("B$_4$T$_{13}$"), $BaTi_4O_9$ ("BT$_4$") and $Ba_2Ti_9O_{20}$ ("B$_2$T$_9$"). The first three compounds decompose to liquid + solid phases at 1350°, 1365°, and 1446°C, respectively. In the BaO - rich part of the system, the binary compound is Ba_2TiO_4 ("B$_2$T").

In the BaO-La_2O_3 system, highly refractory $BaLa_2O_4$ ("BL") has been described [21].

Several ternary phases occur in the BaO-TiO_2-La_2O_3 system and similar ones in the BaO-Nd_2O_3-TiO_2 system. Kolar et al. [22] reported the ternary compounds $BaNd_2Ti_3O_{10}$ ("BNT$_3$") and $BaNd_2Ti_5O_{14}$ ("BNT$_5$"). Razgon et al. [23] investigated tie-lines connecting $BaTiO_3$ with $La_2Ti_2O_7$ and La_2TiO_5. Three ternary compounds were described, $BaLa_2Ti_2O_8$ (BLT$_2$), $BaLa_2Ti_3O_{10}$ (BLT$_3$) and $BaLa_2Ti_4O_{12}$ (BLT$_4$). BLT$_4$ has the same structure as "BNT$_5$" reported by Kolar et al. [22]. At present, it is generally agreed that the composition "BNT$_5$" is not a single phase [24-27]. Varfolomeev et al. [24] studied the quasi-binary $BaTiO_3$-$La_{2/3}TiO_3$ system. They found a broad solid solubility region, which can be described by the formula $Ba_{6-x}La_{8+2/3x}Ti_{18}O_{54}$ with x = 0 - 3. Jonker and Havinga [28] examined the tie-lines pointing from $BaTiO_3$ to LT$_3$, LT$_2$ and L$_2$T$_3$ phases. They found a ternary compound with a BaO:La_2O_3:TiO_2 ratio of 1:2:4 ($BaLa_4Ti_4O_{15}$, "BL$_2$T$_4$") and solid solubility along the $BaTiO_3$-$La_4Ti_3O_{12}$ tie-line. Saltykova et al. [29] published a quasi-binary phase diagram of the $BaTiO_3$-La_2T_3 system. Besides the BL$_2$T$_4$ compound, two new phases were determined with BaO:La_2O_3:TiO_2 ratios of 2:2:5 ($Ba_2La_4Ti_5O_{18}$, "B$_2$L$_2$T$_5$") and 4:2:7 ($Ba_4La_4Ti_7O_{24}$, "B$_4$L$_2$T$_7$"). The last compound was reported to be stable only above approximately 1450°C. The BLT$_2$ compound could not be confirmed and it was proposed that the correct composition is B$_2$L$_2$T$_5$.

Our examination of sintered microstructures of the $BaTiO_3$ rich part of the $BaO-TiO_2-La_2O_3$ system by electron probe wavelength dispersive spectroscopic (WDS) microanalysis resulted in the phase diagram shown in Fig. 1 [30]. Important conclusions were that incorporation of La^{3+} ions on Ba sites in $BaTiO_3$ is compensated by $V_{Ti}^{''''}$, i.e. solid solution extends along the $BaTiO_3$-L_2T_3 tie-line and not along the $BaTiO_3$-LT_3 tie-line. The maximum solid solubility at 1400°C extends to the composition $Ba_{0.70}La_{0.30}Ti_{0.925}(V_{Ti}^{''''})_{0.075}O_3$. This finding explains the good sinterability of La-doped $BaTiO_3$ ceramics, when La_2O_3 is added with TiO_2 in a molar ratio of 1:3, and much worse sinterability of ceramics when La_2O_3 is added with TiO_2 in a molar ratio of 2:3. Other conclusions were that the limiting $(Ba,La)TiO_3$ solid solution is compatible with La-doped Ba_2TiO_4, La_2O_3, and $BaLa_2Ti_3O_{10}$. The solid solubility limit of $(Ba,La)TiO_3$ is slightly lower in equilibrium with $BaLa_2Ti_4O_{12}$ ("114" phase) and $BaTi_4O_9$.

Phase Equilibria In The System $BaO-TiO_2-Ce_2O_3/CeO_2$ [31]

Thermogravimetric analysis of $BaTiO_3-CeO_2-TiO_2$ mixtures clearly demonstrated that excess TiO_2 decreases the thermal solubility of CeO_2 [4,32]. Ce^{3+} thus formed is incorporated with excess Ti^{4+} in $BaTiO_3$.

We may use the $BaTiO_3$-rich part of the $BaO-TiO_2-La_2O_3$ system as a template to interpret the results of quantitative microstructural analysis of $BaO-TiO_2$ ceramics containing ceria in (+4) and (+3) valent states.

In interpreting the results, the following assumptions were made:

1. Ce could be incorporated exclusively as Ce^{4+} at Ti sites or as Ce^{3+} at Ba sites. The assumption is based on the size of the ionic radii listed in Table I, and on the electrical properties of the sintered ceramics. Ce^{3+} in small amounts at Ba sites causes semiconductivity, Ce^{4+} at Ti sites causes high resistivity.
2. Ce^{3+} donor charge could be compensated by reduction of Ti^{4+} to Ti^{3+} (electron compensation), by creation of ionized vacancies at Ti^{4+} sites ($V_{Ti}^{''''}$), or by a combination of both mechanisms
3. Ce^{4+} is more easily reduced than Ti^{4+} [33]. Therefore, Ce^{4+} cannot coexist with Ti^{3+} in solid solution.

18 $BaO-TiO_2-CeO_2/Ce_2O_3$ compositions were prepared by the solid state method and sintered at 1400°C in air or $Ar-8\%H_2$. Microstructures of sintered pellets were subjected to quantitative microanalysis. Considering the assumptions listed above, the analytical results gave compositions shown in Table II.
The results confirm the mechanism of Ce incorporation into the $BaTiO_3$ lattice already proposed by Hennings et al [32] and Makovec et al. [4]. When sintered in air, Ce may be incorporated in $BaTiO_3$ as Ce^{4+} at Ti sites, as well as Ce^{3+} at Ba sites. In the presence of excess Ba, cerium is incorporated preferentially at Ti sites. The solubility strongly depends on temperature and at 1400°C extends up to the composition $BaTi_{0.65}Ce_{0.35}O_3$. In the presence of excess TiO_2, cerium is reduced even in air and enters the $BaTiO_3$ lattice as Ce^{3+} at Ba sites. The solid solution may be expressed by the formula $Ba_{1-x}Ce_x^{3+}Ti_{1-x/4}^{4+}(V_{Ti}^{''''})_{x/4}O_3$ where x reaches ~ 0.08 at 1400°C. When CeO_2 is added to $BaTiO_3$ without either a Ba or Ti excess, cerium may be incorporated

into the $BaTiO_3$ lattice simultaneously at Ba and Ti sites in the ratio 4 : 3 as expressed by formula

$$Ba_{1-4x/7}Ce^{3+}_{4x/7}\ Ti^{4+}_{1-4x/7}\ Ce^{4+}_{3x/7}\ (V''''_{Ti})_{x/7}O_3$$

In a reducing atmosphere (for example 92 % Ar - 8 % H_2), Ce is incorporated into the $BaTiO_3$ lattice exclusively as Ce^{3+} independent of the starting composition. When added to $BaTiO_3$ with a high excess of TiO_2 ($TiO_2 : CeO_2 > 3 : 4$), $Ba_{1-x}Ce^{3+}_x\ Ti^{4+}_{1-x}\ Ti^{3+}_x\ O_3$ is formed. If the concentration of Ti is lower than required by this formula, then a Ti-deficient solid solution is formed. The solid solubility limit of $Ba_{1-x}\ Ce^{3+}_x\ Ti^{4+}_{1-x}\ Ti^{3+}_x\ O_3$ solid solutions is at $x \approx 0.52$ at $1400^{\circ}C$. The solid solubility limit in the Ti-deficient solid solution

$$Ba_{1-x}\ Ce^{3+}_x\ Ti^{4+}_{1-x/4-2\delta}\ Ti^{3+}_{2\delta}\ (V''''_{Ti})_{x/4}O_{3-\delta}(V_o)_\delta$$

is lower, $x \sim 0.28$ at $1400^{\circ}C$.

TABLE II

Formulas of Ce-BaTiO$_3$ solid solutions, calculated from results of
wavelength dispersive spectroscopy-electron probe microanalysis

A. Samples fired in air at $1400^{\circ}C$

Starting composition (mol.%)			Calculated Formula
BaO	CeO$_2$	TiO$_2$	
51.00	25.00	24.00	$Ba_{1.01}Ti_{0.64}Ce_{0.35}O_3$
47.00	6.00	47.00	$Ba_{0.98}Ce^{3+}_{0.02}\ Ti_{0.986}Ce^{4+}_{0.008}\ (V''''_{Ti})_{0.005}O_3$
			$Ba_{0.98}Ce^{3+}_{0.03}\ Ti_{0.99}(V''''_{Ti})_{0.01}O_3$
41.25	10.00	48.75	$Ba_{0.92}Ce^{3+}_{0.08}\ Ti_{0.98}(V''''_{Ti})_{0.02}O_3$
42.50	6.00	51.50	$Ba_{0.91}Ce^{3+}_{0.09}\ Ti_{0.98}(V''''_{Ti})_{0.02}O_3$

B. Samples fired in 92 % Ar/8 % H_2 at $1400^{\circ}C$

Starting composition (mol.%)			Calculated Formula
BaO	CeO$_2$	TiO$_2$	
50.00	15.00	35.00	$Ba_{0.72}\ Ce^{3+}_{0.28}\ Ti^{4+}_{0.925}\ (V''''_{Ti})_{0.07}O_3$
38.00	15.00	47.00	$Ba_{0.73}Ce^{3+}_{0.27}\ Ti^{4+}_{0.94}\ (V''''_{Ti})_{0.07}O_3$
36.88	15.00	48.12	$Ba_{0.72}\ Ce^{3+}_{0.29}\ Ti^{4+}_{0.92}\ (V''''_{Ti})_{0.07}O_3$
23.75	30.00	46.25	$Ba_{0.58}\ Ce^{3+}_{0.42}\ Ti^{4+}_{0.71}\ Ti^{3+}_{0.27}\ (V''''_{Ti})_{0.02}O_3$
20.00	30.00	50.00	$Ba_{0.45}Ce^{3+}_{0.55}\ Ti^{4+}_{0.51}\ Ti^{3+}_{0.45}\ O_3$
31.25	15.00	53.75	$Ba_{0.65}Ce^{3+}_{0.35}\ Ti^{4+}_{0.66}\ Ti^{3+}_{0.34}\ O_3$
21.43	22.86	55.71	$Ba_{0.49}Ce^{3+}_{0.51}\ Ti^{4+}_{0.46}\ Ti^{3+}_{0.53}\ O_3$

Dielectric properties

As already mentioned, BaTiO$_3$ slightly doped with rare earth oxides (0.3 - 0.5 mol.%) shows semiconductive behaviour. With larger amounts of rare earths BaTiO$_3$ becomes insulator showing a strong dependence of permittivity versus composition of ceramics (Fig. 1). 3-valent rare-earths ions on Ba sites affect strong lowering of BaTiO$_3$ Curie temperature (1 mol. % of La^{3+} shifts T$_c$ for app. 30°). Since excess TiO$_2$ induces reduction of Ce^{4+} to Ce^{3+} the sensitivity of T$_c$ shift of CeO$_2$ doped BaTiO$_3$ on excess TiO$_2$ [11] is easily understandable.

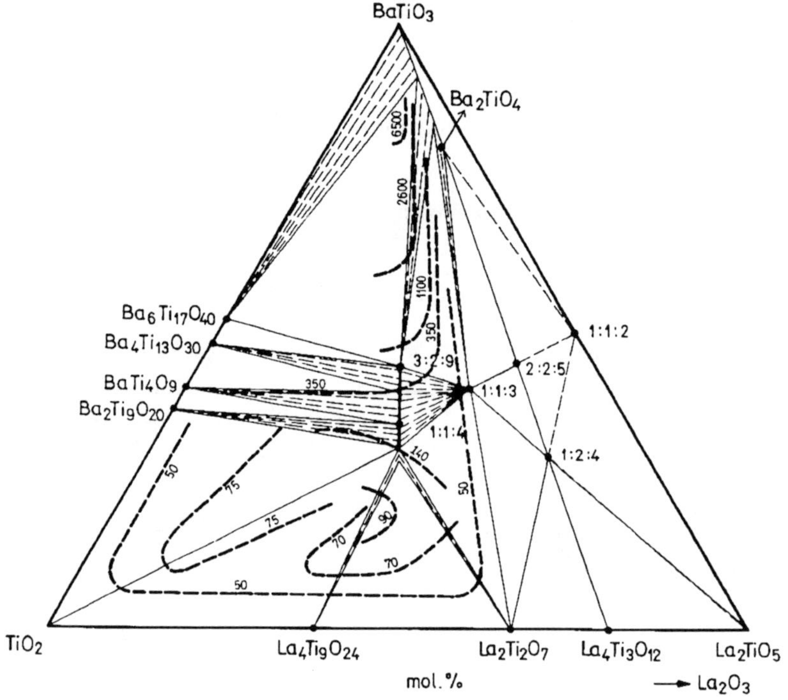

Fig. 1: Isodielectric curves in the system BaTiO$_3$-TiO$_2$-La$_2$TiO$_5$

Recently it was determined [32,34] that incorporation of Ce as Ce^{3+} - donor at Ba sites lowers T$_c$ for app. 22°/mol. % Ce^{3+} and increases the dielectric constant in a maximum (Fig. 2).

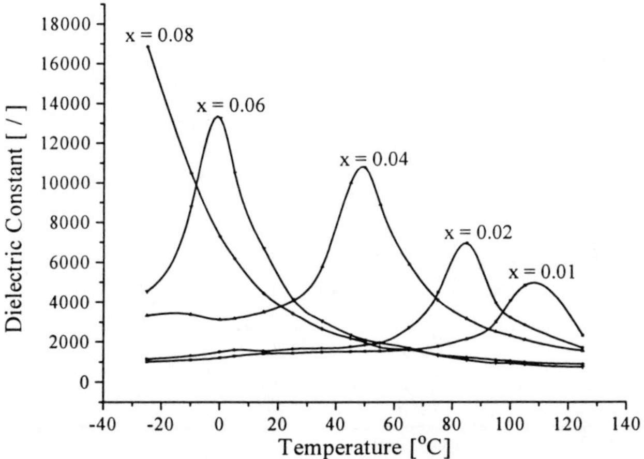

Fig. 2: Dielectric constant vs. temperature for $Ba_{1-x}Ce^{3+}_x Ti^{4+}_{1-x/4}$ ($V_{Ti}^{''''}$)$_{x/4}O_3$ solid solutions

When Ce^{4+} enters in the $BaTiO_3$ by replacing Ti^{4+} the shift of Curie temperatre is much smaller, namely app. $6°$/mol. % Ce^{4+} (Fig. 3), and a decrease of dielectric constant in maximum is observed.

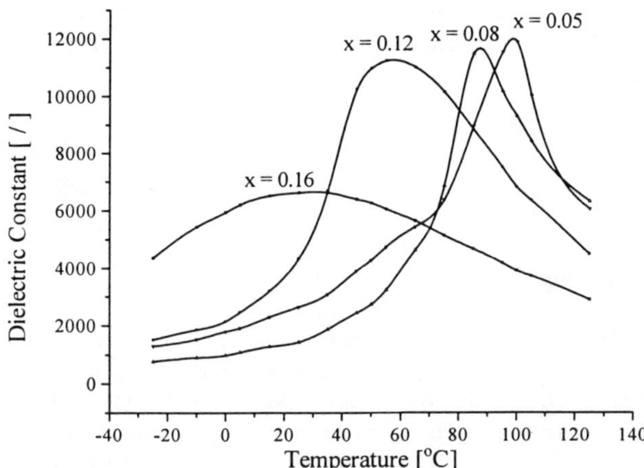

Fig. 3: Dielectric constant vs. temperature for $BaTi_{1-x}Ce_xO_3$ solid solutions

Especially important part of the $BaTiO_3$-R_2O_3-TiO_2 system is in the vicinity of 114 solid solutions. Such ceramics are nowadays used for high-frequency applications (higher than 0.5 GHz) due to the extremely low losses (high Q-values), a small temperature coefficient of resonant frequency (τ_f) and permittivity (k') higher than 80.

La 114 isotypical solid solutions are formed with rare earth oxides from La to Gd. On decrease in the rare earth ionic radius, the dielectric properties change, as shown in Table III.

TABLE III

Permittivities and temperature coefficients
(k'/τ_K) of rare earth family 114 compositions (at 1 MHz).

Composition	References				
	(35)	(36)	(37)	(36)	(38)
$BaLa_2Ti_4O_{12}$	100/-155*	110/-740			
$BaCe_2Ti_4O_{12}$		90/-410			
$BaPr_2Ti_4O_{12}$		91/-315			
$BaNd_2Ti_4O_{12}$	85/-160	85/-120	92/-113		
$Ba(Nd,Sm)_2Ti_4O_{12}$				83/-(6-10)	
$BaSm_2Ti_4O_{12}$	81/+30	80/+40			
$BaEu_2Ti_4O_{12}$	76/+60	78/+100			
$BaGd_2Ti_4O_{12}$					68/+147

* ppm/K

It is noticeable that permittivity decreases and temperature coefficient of capacitance (τ_k) increases from 114-La to 114-Gd solid solutions. The decrease in permittivity, according to the Clausius-Mosotti equation:

$$k' = \frac{3Vm + 8\pi\alpha_D}{3Vm - 4\pi\alpha_D} \qquad (1)$$

Vm = molar volume
α_D = net dielectric polarizability

can be ascribed to the decrease of the net dielectric polarizability α_D as a consequence of decreased ionic polarization from La to Gd [39]. The same phenomena occur at microwave frequencies, but the decrease of permittivity is accompanied by a decrease of temperature coefficient of resonant frequency (τ_f), even though there is no valid explanation for the reverse value of τ_f from Nd to the Sm analogue of 114 solid solution. (La, Nd based are positive, Sm, Eu, Gd are negative). However this phenomenon is used for τ_f-compensation in Nd-based commercial microwave ceramics.

Formation of multiphase ceramics in such systems can strongly effect the resulting microwave properties. It is typical that in the Nd- and La-114 ceramics a second phase (Bapolytitanates or $R_4Ti_9O_{24}$) generally lowers the net permittivity. On the other hand, negative effect of multiphase ceramic formation in some systems (La and Nd - based) can turn into a positive one in other systems. The beneficial influence of TiO_2 addition on tuning τ_f to 0 in Sm-114 based ceramics has been reported [40,41]. A small addition of TiO_2 turns a monophase ceramic into a multiphase one, also containing, besides 114-type solid solution, TiO_2 and $Ba_2Ti_9O_{20}$ phase. The slightly negative τ_f of Sm-based 114 solid solution (τ_f = -8 ppm/K) is compensated by the very positive τ_f of TiO_2, while the net permittivity remains nearly unchanged.

TABLE IV

Dielectric properties of various compounds in $BaTiO_3$-R_2TiO_5-TiO_2 systems

Phase	k'	τ_f (ppm/K)	$Q \cdot f_r$	f_r (GHz)	lit.
$Ba_2Ti_9O_{20}$	40	+4	8000	4.0	42
$BaTi_4O_9$	38	+30	8000	4.0	42
$Nd_4Ti_9O_{24}$	37.5	+65	3000	8.0	43
$Nd_2Ti_2O_7$	36	-118	1800	9.1	43
TiO_2	14.1	+427	14.600	3.0	44
$La_4Ti_9O_{24}$	37	+15	3060	8.0	43
$La_2Ti_2O_7$	47	-10	1090	7.8	43
114-Nd, ss	85	+88	6000	3.0	45
114-Sm, ss	80	-8	9000	3.0	45

SUMMARY

Present knowledge of phase equilibria in the ternary systems BaO-TiO_2-R_2O_3 (R = La^{3+}, Nd^{3+} and Ce^{3+}/Ce^{4+}) has been reviewed and correlated with the electrical properties of ceramics based on the ternary systems. Solid solutions and ternary compounds in the described systems exhibit a wide variety of properties, such as semiconductivity, high apparent dielectric constant, thermally stable dielectric constant and a valuable combination of dielectric properties at microwave frequences. Ability of Ce to exist in (+3) and (+4) valence state make it possible to be incorporated in $BaTiO_3$ structure on Ba sites and/or Ti sites, respectively and the electrical properties vary accordingly. Since the electrical properties are strongly influenced by the microstructure of ceramics, the phase equilibrium data are of valuable and indispensable help to understand and design the processing parameters needed to optimize the desired electrical properties of ceramics in the discussed systems.

ACKNOWLEDGEMENTS

This publication is based on work sponsored by the U.S.-Slovene Science and Technology Joint Fund in cooperation with the Ministry of Science and Technology of Slovenia under Project Number NIST 95-351.

REFERENCES

1. G.H. Jonker, Solid State Electronics **7**, p. 895-903 (1964)
2. H.M. Chen, M.P. Harmer and D.M. Smyth, J. Am. Ceram. Soc. **69**, p. 507-510 (1986)
3. D.S. Shannon, Acta Crystallogr. **A 32**, p. 751-67 (1976).
4. D. Makovec, Z. Samardžija, and D. Kolar, pp 961-966 in Third Euro-Ceramics, Vol. 1, Processing of ceramics (P. Duran and J. F. Fernandez, eds.) Faenza Editrice Iberica, Castellon de la Plana, 1993
5. Z. Samardžija, M. Čeh, D. Makovec and D. Kolar, Microchim. Acta [Suppl.] **13**, p. 517-523 (1996)
6. A. Hoffman, Z. Physik. Chem **28B**, p. 65 (1935).
7. J.P. Guha and D. Kolar, J. Mat. Sci. **6**, p. 1174-1177 (1971).
8. R.S. Roth, T. Negas, H.S. Parker, D.B. Minor and C. Jones, Mat. Res. Bull. **12**, p. 1173-1182. (1977).
9. J.M. Maister, A.V. Shevchenko and L.M. Lopato, Neorgan. Mat **18** (9), 1589-1590 (1982) (in Russian).
10. J.P. Guha and D. Kolar, J. Am. Cer. Soc. **56** (1), p. 5-6 (1973).
11. D.Kolar, J.P. Guha and M. Buh, Ceramic and dielectric properties of selected compositions in the BaO-TiO_2-CeO_2 system, pp 152-8 in Proceedings of the Conference on Electrical, Magnetic and Optical Ceramics, London, Dec. 1972 Brit. Ceram. Soc.
12. O. Saburi, J. Phys.Soc.Japan **14** (9), 1159-74 (1959).
13. Z. Ali Nemati, M. Tabib-Azaz and M. R. De Guire, Br. Ceram. Trans. **92** (3), p. 109-113 (1993).
14. L.A. Xue, Y. Chen and R.J. Brook, J. Mat. Sci. Letters **7** , p. 1163-1165 (1988).
15. J. B. Chesney and H. A. Sauer, J. Am. Cer. Soc. **45** (9), p. 416-422 (1962).
16. N. F. Fedorov, O. V. Melnikova, V. A. Saltykova and M. Chistiakova, Zh. Neorg. Himii **24** No. 5, p. 166-1170 (1979).
17. K.W. Kirby, and B. A. Wechsler, J. Am. Ceram. Soc. **74** (8), p. 1841-47 (1991).
18. D. E. Rase and R. Roy, J. Am. Cer. Soc. **38** (3), p. 102-13 (1955).
19. T. Negas, R. S. Roth, H. S. Parker and D. Minor, J. Solid State Chem. **9**, p. 297-307 (1974).
20. H. M. O'Bryan, Jr. and J. Thomson, Jr. J. Am. Ceram. Soc. **57** (12), p. 522-26 (1974).
21. L.M. Lopato, Ceramurgia Intern. **2** (1), p. 18-32 (1976).
22. D. Kolar, S. Gaberšček, B. Volavšek, H. S. Parker, and R. S. Roth, J. Solid State Chem. **38**, p. 158-64 (1981).
23. R. S. Razgon, A. M. Gens, M. B. Varfolomeev, S. S. Korovin, and V. S. Kostomarov, Russ. J. Inorg. Chem. **25** (8), p. 1274-75 (1980).
24. M. B. Varfolomeev, A. C. Mironov, V. S. Kostomarov, L. A. Golubčova, and T. A. Zolotova, (in Russ.), Zh. Neorg. Khim. **33** (4), p. 1070 (1988).
25. T. Jaakola, A. Uusimaki, R. Rautioaho, and S. Leppa-Vuori, J. Am. Ceram. Soc. **69** (10), p. C-234 (1986).
26. J. Takahashi and T. Ikegami, J. Am. Ceram. Soc. **74** (8), p. 1873-79 (1991).
27. D. Kolar, S. Gaberšček, and D. Suvorov, Structural and Dielectric Properties of Perovskite-like Rare Earth Titanates; pp. 229-34 in Third EuroCeramics, Vol. 2 Edited by P. Duran and J. F. Fernandez, Faenza Editrice Iberica, Castellón de la Plana, Spain,1993.
28. G. H. Jonker and E. E. Havinga, Mater. Res. Bull. **17**, p. 345-50 (1982).
29. V. A. Saltikova, O. V. Meljnikova, N. V. Leonova, and F. N. Fedorov, (in Russ.), Zh. Neorg. Khim. **30** (1), p. 190-93 (1985)

30. D. Makovec, Z. Samardžija, U. Delalut and D. Kolar, J. Am. Ceram. Soc. **78** (8), p. 2193-97 (1995).
31. D. Makovec, Z. Samardžija and D. Kolar, J. Solid State Chem. **123**, p. 30-38 (1996).
32. D.F.K. Hennings, B. Schreinemacher and H. Schreinemacher, J. Eur. Ceram. Soc. **13**, p. 81-88 (1994).
33. A. J. Leonov, M. I. Piryutko and E. K. Keller, Bull. Akad. Nauk USSR, Div. Chem. Sci. (Engl. Transl.) 5, p. 756 (1966).
34. D.Makovec and D. Kolar, Electroceramics V, edited by J.L. Baptista, Vol. 1, (1996). p. 557-60.
35. A. M. Gens, M. B. Varfolomeev, V. S. Kostomarov and S. S. Korovin, Zh. Neorgan. Khim. **26** (4), p. 896-898 (1981).
36. L. P. Mudrolubova, K. E. Lisker, B. A. Rotenberg, T. F. Limar and A. N. Borsch, Ser. Radiodetali & Radiokomponenti **1** (46), p. 3-8 (1982).
37. D. Kolar, Z. Stadler, S. Gabršček, D. Suvorov, Ber. Dtsch. Keram. Ges. **55** (7). p. 346-348 (1978).
38. M. Valant, D. Suvorov and D. Kolar, Jpn. J. Appl. Phys. **35**, p. 144-150 (1996).
39. R. D. Shannon, J. Appl. Phys. **73**, p. 348- 366 (1983).
40. T. Negas, G. Yeager, S. Bell, R. Amren, p. 21-38 in Chemistry of Electronic Ceramic Materials, (Eds. P.K. Davies and R.S. Roth), NIST, SP 804, 21-38 (1991)
41. S. Nishigaki, M. Kato, S. Yano and R. Kamimura, Am. Ceram. Soc. Bull. **66** (9), p. 1405-1410 (1987).
42. H. M. O'Bryan, Jr., J. Thomson and J. K. Plourde, Ber. Dt. Keram. Ges. **55** (7), p. 348-351 (1978).
43. J. Takahashi, K. Kageyama and M. Kodaira, Jpn. J. Appl. Phys. **32** (9B), p. 4327-4331 (1993).
44. K. Wakino, K. Minai and H. Tamura, J. Am. Ceram. Soc. **67** (4), p. 278-281 (1984).
45. T. Negas and P. K. Davies, Ceramic Transactions., Vol. 53, p. 179-196 (1995).

CORRELATION BETWEEN STRUCTURAL PARAMETERS, POLARISABILITIES AND DIELECTRIC PROPERTIES IN NON-FERROELECTRIC COMPLEX PEROVSKITES

E.L.COLLA, N. SETTER
Laboratoire de Céramique, EPFL, MX-D Ecublens, 1015 Lausanne, Switzerland,
ecolla@mxsg1.epfl.ch.

ABSTRACT

A combination of different empirical methods which allow the prediction of the dielectric properties of non-ferroelectric complex perovskites has been proposed and its present limits verified. The long term purpose is to engineer new microwave materials with increased relative permittivity ε_0 and low thermal coefficient of the resonant frequency τ_f.

The estimation of the dielectric properties has been accomplished by means of structural predictions based on the tolerance factor and a microscopic model. Furthermore, the optical permittivity based on the Gladstone-Dale relation and the total polarisability according to the molecular/ion additivity rule were also estimated. Ceramics of new compositions were then fabricated and analysed to verify the dielectric properties predicted by the empirical methods. Results show good agreement with the expected structural features and optical permittivities, and the method is sufficiently reliable to estimate the end values, or at least to estimate trends of the dielectric properties induced by partial substitution of cations on site A. The importance of the electronic polarisability on the end value of τ_ε has been demonstrated.

INTRODUCTION

Non-ferroelectric complex perovskite (CP) ceramics, due to their high permittivity, low dielectric loss and low temperature coefficient of the resonance frequency, are used in microwave resonators [1,2]. In previous work [3,4] it was reported that the temperature dependence of their dielectric properties was mainly determined by structural changes. Equation 1 shows the different components of the temperature coefficient of the permittivity derived using the Clausius-Mossotti (C.M.) formula [3,5].

$$\tau_\varepsilon = \frac{1}{\varepsilon}\left(\frac{\partial \varepsilon}{\partial T}\right) = \frac{(\varepsilon-1)\cdot(\varepsilon+2)}{3\cdot\varepsilon}\cdot\frac{1}{\alpha_D}\cdot\left\{-\frac{\alpha_D}{V}\left(\frac{\partial V}{\partial T}\right)_P + \left(\frac{\partial \alpha_D}{\partial V}\right)_T\cdot\left(\frac{\partial V}{\partial T}\right)_P + \left(\frac{\partial \alpha_D}{\partial T}\right)_V\right\} =$$

$$= \frac{(\varepsilon-1)\cdot(\varepsilon+2)}{3\cdot\varepsilon}\cdot\frac{1}{\alpha_D}\cdot\left\{\text{Dilution} + \text{Displacement} + \text{DirectDependence}\right\} \tag{1}$$

In general the "dilution" and the "direct dependence" terms have negative values. The only positive contribution arises from the ionic polarisability ("displacement" term). Assuming that the thermal expansion is structure independent to a first approximation, only the two last terms of equation 2 are expected to show a structure dependence.

$$\tau_\varepsilon = \frac{1}{\varepsilon}\left(\frac{\partial \varepsilon}{\partial T}\right) = \frac{(\varepsilon-1)\cdot(\varepsilon+2)}{3\cdot\varepsilon}\cdot\frac{1}{\alpha_D}\cdot$$

$$\cdot\left\{-\left[\frac{\alpha_e}{V}+\frac{\alpha_i}{V}\right]\cdot\left(\frac{\partial V}{\partial T}\right)_P + \left[\left(\frac{\partial \alpha_i^A}{\partial V}\right)_T + \left(\frac{\partial \alpha_i^B}{\partial V}\right)\right]_T\cdot\left(\frac{\partial V}{\partial T}\right)_P + \left[\left(\frac{\partial \alpha_i^A}{\partial T}\right)_V + \left(\frac{\partial \alpha_i^B}{\partial T}\right)_V\right]\right\} \tag{2}$$

Since the "direct dependence" term (the last one in square brackets of equation 2) is normally much smaller than the ionic term, one can suggest that the most important coupling between structure and permittivity is given by the second term which consists in the sum of the volume dependence of the ionic polarisabilities at the A and B-sites.

437

Mat. Res. Soc. Symp. Proc. Vol. 453 © 1997 Materials Research Society
">

In reference [6] a microscopic mechanism responsible for this dependence was proposed. It consists in the different local thermal expansion at the A and B-sites in the low temperature phases combined with a nearly constant global thermal expansion over the three phases (Orthorhombic, Tetragonal, Cubic). Due to the decrease of the tilting angle of the oxygen octahedra with increasing temperature, their relative volume increases slower than the volume of the A-sites. This model can estimate $\varepsilon_0(T)$ by combining the C.M. formula with the prediction of the "relative" evolution of $\alpha_D(T)$ in the different structural phases. However, the initial values for α_D or ε_0 still need to be determined experimentally. The first attempt to obtain such initial values using basic properties of simple oxides and cations was presented in [7]. In the present work, new experimental results on Sr substitution by Pb in the systems $SrZn_{1/3}Nb_{2/3}O_3$ and $SrMg_{1/3}Ta_{2/3}O_3$ are reported and discussed (hereafter named SZN and SMT, respectively).

PROPOSED EMPIRICAL METHODS

The possibility of estimating α_D and V_m (see equation 3) from some basic data would allow a complete calculation of $\varepsilon_0(T)$. In a previous work we reviewed the empirical methods that can help to estimate the total polarisability ($\alpha_D = \alpha_e + \alpha_i$) and the unit cell volume V_m. These methods are briefly summarised hereafter to facilitate the understanding of what follows.

In order to relate the polarisability α_D (a microscopic quantity) to the dielectric permittivity ε_0 (a macroscopic quantity), the C.M. was applied:

$$\varepsilon_0 = \frac{3V_m + 8\pi\alpha_D}{3V_m - 4\pi\alpha_D} \tag{3}$$

To allow a simple estimation of V_m as a function of composition, the following formula was used:

$$a^* = \frac{(\overline{R}_A + R_o)}{\sqrt{2}} + \overline{R}_B + R_o \tag{4}$$

where R_A, R_B and R_O are the radii of the A- and B-site cations and the oxygen ion, respectively and a* is the pseudo-cubic lattice parameter.
The tolerance factor t, used to determine the structure of the perovskite (cubic or tetragonal or orthorhombic) is defined by the formula [8]

$$t = \frac{(R_A + R_o)}{\sqrt{2} \cdot (R_B + R_o)} \tag{5}$$

The additivity rule of ion polarisabilities states that the molecular polarisabilities α_D of a complex substance can be broken up into the molecular polarisabilities of simpler substances by [9]:

$$\alpha_D(M_2 M' X_4) = 2\alpha_D(MX) + \alpha_D(M' X_2) \tag{6}$$

where M are the cations and X is either O^{2-} or F^{1-}. Furthermore, it is possible to break up the molecular polarisabilities of complex substances into ions according to

$$\alpha_D(M_2 M' X_4) = 2\alpha(M^{2+}) + \alpha(M'^{4+}) + 4\alpha(X^{2-}) \tag{7}$$

In 1864 Gladstone & Dale [10] formulated a simple rule to predict the permittivity at optical frequencies ε_∞ for oxides. The relationship is best explained by the equation

$$K = \frac{n-1}{\delta} = \frac{k_1 p_1}{100} + \frac{k_2 p_2}{100} + \dots \frac{k_n p_n}{100} \tag{8}$$

where K is the specific refractive energy of a substance, n its refractive index, δ its density, k_i the specific refractive energies of its different components and p_i the weight percentages of the components. Recently Mandarino [11] undertook a complete recalculation of a large number of k_i. It is therefore possible to calculate reliable $\varepsilon_\infty = n^2$ values and their trend for our complex oxides.

The first objective is to verify the applicability of these empirical rules and the available data on the ion and simple oxide properties.

RESULTS AND DISCUSSION

The chosen starting compounds are SZN and SMT. New compositions were obtained by replacing Sr by Pb at the A-site of the perovskite, with concentrations well below the critical concentration at which the compounds become ferroelectric. In SZN 10% and 30% of Sr were substituted and 30% in SMT. The essential ionic data of Sr^{2+} are compared with those of Pb^{2+} in the Table 1 in view of using the additivity rule of ion polarisability. In the same table, the parameters of the related oxides used in the Gladstone-Dale formula are also presented.

	R_A	α_D		α_D	k_i	M_W	$k_i \cdot M_W$
Sr^{2+}	1.54 Å	4.25 Å^3	SrO	6.48 Å^3	0.145	103.6 g	15.0
Pb^{2+}	1.63 Å	6.58 Å^3	PbO	8.50 Å^3	0.133	223.2 g	29.6
Zn^{2+}	0.89 Å	2.09 Å^3	ZnO	4.13 Å^3	0.158	81.37 g	12.8
Mg^{2+}	0.86 Å	1.33 Å^3	MgO	3.33 Å^3	0.200	40.30 g	8
Nb^{5+}	0.78 Å	3.98 Å^3	Nb$_2$O$_5$	19.5 Å^3	0.268	265.81 g	66.7
Ta^{5+}	0.78 Å	4.73 Å^3	Ta$_2$O$_5$	19.72 Å^3	0.151	441.89 g	29.6

Table 1: Ion and oxide data used in the calculations. (M_W = molecular weight)

Pb has a larger ionic radius, a larger polarisability and the oxide has a larger refractive index. Therefore the partial substitution of Sr by Pb is expected to provoke a change of the unit cell dimension, of the tolerance factor, of the polarisabilities and of the index of refraction.

	t	α_D [Å^3]	ε_∞	a* [Å]	α_e [Å^3]	ε_0	ε_0*
SZN	0.953	13.60	4.473	4.002	8.21	25	41
10% Pb	0.956	13.83	4.611	4.009	8.40	28	49
30% Pb	0.963	14.30	4.890	4.022	8.76	37	78
SMT	0.956	13.84	4.171	3.993	7.82	29	21
30% Pb	0.957	14.54	4.582	4.012	8.40	47	30

Table 2: Calculated parameters for the starting compounds Sr(Zn$_{1/3}$Nb$_{2/3}$)O$_3$ and Sr(Mg$_{1/3}$Ta$_{2/3}$)O$_3$ and their derivatives.

Table 2 shows the calculated new features. The equations of the previous section were used to determine t, α_D, ε_∞ and a* and the C.M. formula was used to calculate α_e, ε_0 and ε_0*.

The increase of the tolerance factor should be confirmed by a decrease of the critical temperature of the O-octahedra tilting phase transitions by nearly 70°C and 200°C respectively. Figure 1 shows the normalised permittivity curves of the measured systems. The small anomalies correspond to the structural phase transitions (at high temperature Cubic-Tetragonal and at lower temperature Tetragonal-Orthorhombic). The shifts of the transition temperatures to lower values coincide with the calculated ones. The phase with the lowest symmetry shows in general the larger gradient whereas the cubic phase for all the presented systems has a negative slope [12]. However, the addition of Pb provokes a general "rotation" of the $\varepsilon_0(T)$ towards negative gradients. We suggest that this effect is related to an increase of the electronic polarisability.

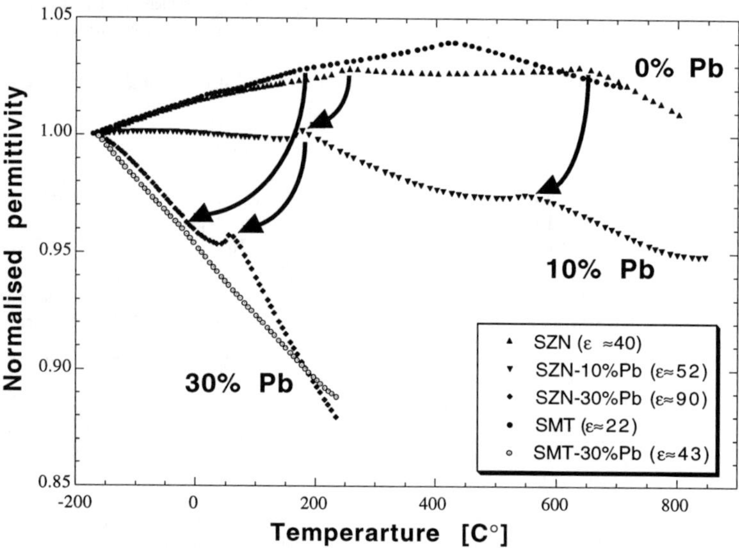

Figure 1: Normalised permittivities as a function of temperature for the pure systems (SZN and SMT) and their derivatives.

Figure 2a, b and c show the measured permittivity ε_0 and the temperature coefficient of capacitance τ_c for the system SZN-PbSZN as a function of temperature. ε_0 increases with increasing Pb-content according to the trend predicted in table 2. However, the calculated permittivity of the starting compound (SZN) is much smaller than the expected one (25 and 41, respectively). In the case of SMT (Figure 2e) the situation is exactly the opposite (calculated 29, measured 21). This discrepancy should arise from the polarisability values used for the Nb^{5+} and Ta^{5+} ions. Table 1 shows that the ion polarisability of Ta^{5+} is much larger than that of Nb^{5+} although in general Nb-based perovskites show higher permittivity than Ta-based ones. It is therefore pertinent to ask how reliable are the polarisability data of these two cations as obtained for other oxides [9] when applied to perovskite structures.

In table 2 the column ε_0^* has been calculated using different values of α_D for both cations: α_D =4.95 and 4.00 for Nb^{5+} and Ta^{5+}, respectively. These values were adjusted to fit the permittivity of the pure compounds. The calculated increase of ε_0^* by adding Pb is nevertheless still too small. For the system SZN the measured values are $\varepsilon_0=51$ and $\varepsilon_0=92$ and the calculated values are $\varepsilon_0^*=49$ and $\varepsilon_0^*=77$ for 10% and 30% Pb, respectively.

As pointed out in [7], the calculation of large ε_0 with the C.M. formula can be affected by a lack of accuracy. An error of 1% in α_D, for ε_0 values above 30, can lead to an error of 30-50% in ε_0. However, the SMT system shows the same type of deviation (with 30% Pb: calculated 30, measured 44). This could be due to the indirect effect of increasing volume with Pb-addition. In other words, in our calculation we didn't take into account the fact that by "expanding" the size

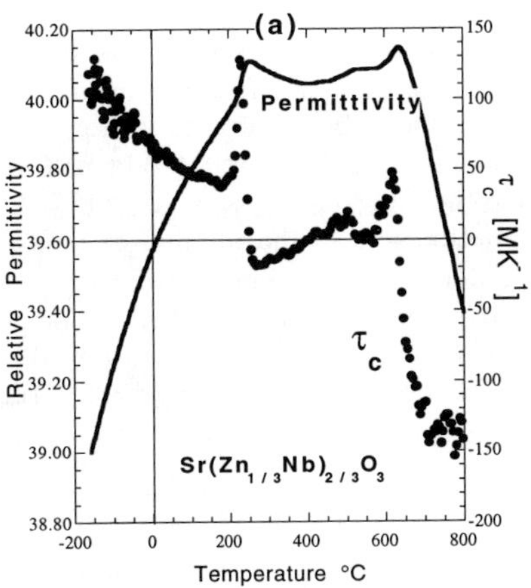

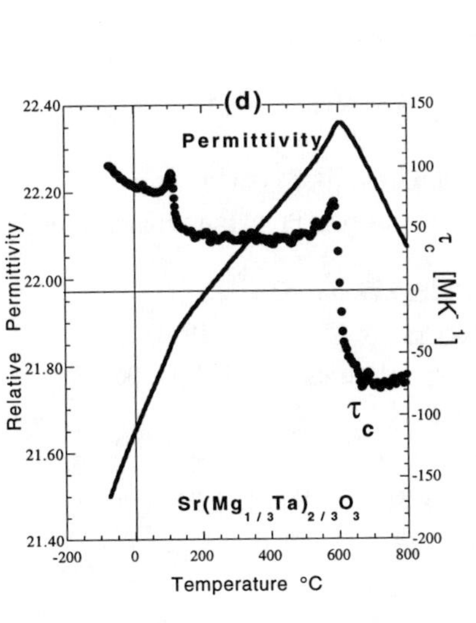

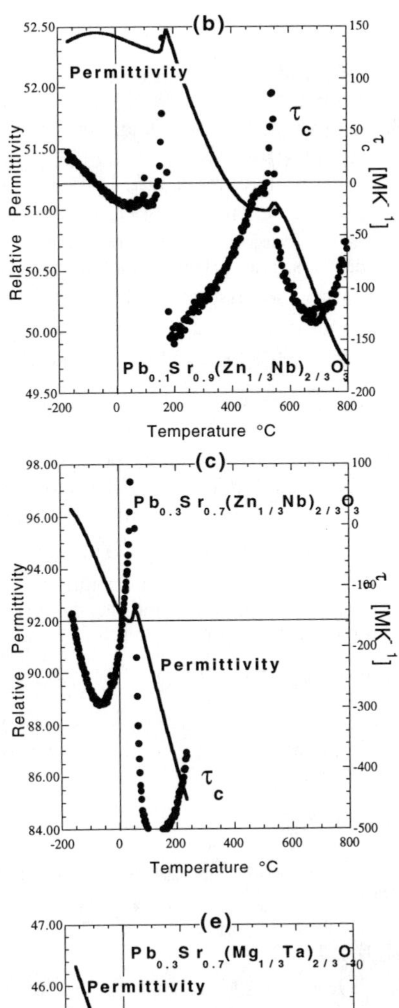

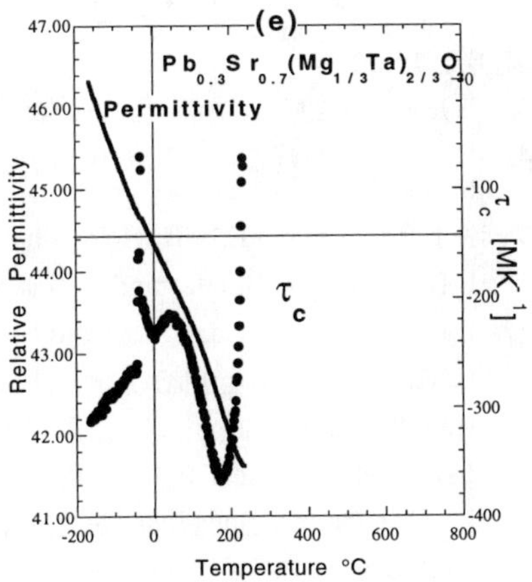

Figure 2: ε_0 (solid lines) and $\tau_c = \tau_\varepsilon + \alpha_L$ (filled circles) as a function of T of SZN (a) of its new derived compounds (b and c) and of SMT (d) and its Pb modified system (e).

of the structure, the ionic contribution is not only diluted but also increased according to the elongation of the dipoles. This should at least partially explain the observed discrepancy.

The behaviour of τ_C with increasing Pb-concentration corresponds to the previously suggested mechanism related to the electronic polarisability. In the orthorhombic phase for the system SZN, τ_C varies from +100ppm for the pure compound, to about 0ppm for 10% Pb and finally to -250ppm for 30% Pb. The values of α_e/α_i for the same sequence are: 1.36 -> 1.38 -> 1.42. We explain this feature in terms of polarisability dilution due to the thermal expansion. The temperature dependence of α_e is dominated by the dilution effect, which lowers its value with increasing temperature. For α_i, where $\alpha_D=\alpha_e+\alpha_i$, the "negative" dilution effect is compensated by the "positive" effect of increasing dipole length which can lead to a global positive value. The larger the relative contribution of α_e to α_D, the more negative should be τ_C. The same behaviour is observed in the SMT system. τ_C changes from 100ppm down to -280ppm with the addition of 30% Pb and the fraction α_e/α_i increases from 1.41 to 1.48.

CONCLUSIONS

The sign of τ_ε in complex perovskites is strongly influenced by the structure. It has been possible to explain this dependence by means of different local thermal expansions at the A and B sites. τ_ε is also determined by the ratio $\mu= \alpha_e/\alpha_i$ where $\alpha_D = \alpha_e + \alpha_i$. The larger μ is, the more negative τ_ε becomes. The applicability of the additivity rule, of the Gladstone-Dale relationship and of the CM-formula to predict α_D , ε_∞ and ε_0 in CP is acceptable. However, new tables of ion polarisabilities calculated by only using data for non-ferroelectric perovskites would certainly improve the reliability.

ACKNOWLEDGEMENTS

This work was supported by the Swiss National Science Foundation. Additional support from the Foundation for Promotion of Material Science and Technology of Japan is gratefully acknowledged.

REFERENCES

1.   W. Wersing, Electronic Ceramics, (Elsevier Applied Science, London and New York, 1991), 67.

2.   K. Wakino and H. Tamura, Ceramic Transactions, 8, 305 (1990). K. Wakino, Ferroelectrics, 91, 69, (1989).

3.   E.L. Colla, I.M. Reaney and N. Setter, J. Appl. Phys., 74, 3414 (1993).

4.   O. Steiner, E.L. Colla, I.M. Reaney and N. Setter, Conf. Proc. European Ceram. Soc., Vol. 2, 223 (1993).

5.   A.J. Bosman and E.E. Havinga, Phys. Rev., 129, 1593 (1963).

6.   E.L. Colla, I.M. Reaney and N. Setter, Ferroelectrics, 154, 1173, (1994).

7.   E.L. Colla, N. David, C. Rau and N. Setter, Ferroelectrics, 184, 151, (1996).

8.   H. A. Megaw, Proc. Phys. Soc., 58, 133, (1946).

9.   R.D. Shannon, J. Appl. Phys., 73, 348, (1993).

10.   J.H. Gladstone and T.P. Dale, Philos.Trans.Roy.Soc.London, 153, 317 (1864).

11.   J.A. Mandarino, Canadian Mineralogist, part I, 14, 498, (1976).

12.   I.M. Reaney, E.L. Colla and N. Setter, J.J.A.P., 33, 3984 (1994).

FORMATION AND CHARACTERIZATION OF "1:1" ORDERED PHASES IN $AM^{4+}O_3$ - $A(B^{2+}_{1/3}Ta_{2/3})O_3$ (A= Ba, Sr; B^{2+} = Mg, Zn; M^{4+} = Ti, Sn, Zr, Ce) PEROVSKITES

L. CHAI, M. A. AKBAS, and P. K. DAVIES
Department of Materials Science & Engineering, University of Pennsylvania, Philadelphia, PA
19104-6272, lchai(or akbas, davies)@lrsm.upenn.edu

ABSTRACT

Previous research has demonstrated that the formation of nano-sized "1:1" cation ordered domains in Zr^{4+} substituted $Ba(Zn_{1/3}Ta_{2/3})O_3$ may be responsible for the anomalies in the dielectric loss properties of these perovskite ceramics. To explore the structure and crystal chemistry of these phases, the effect of a series of M^{4+} substitutions (Zr^{4+}, Ce^{4+}, Sn^{4+}, and Ti^{4+}) on the B-site cation ordering in $Ba(Zn_{1/3}Ta_{2/3})O_3$, $Ba(Mg_{1/3}Ta_{2/3})O_3$, and $Sr(Mg_{1/3}Ta_{2/3})O_3$ has been examined. In all cases the substitution of low levels (~ 5 mole %) of "over-sized" tetravalent cations (e.g. Zr^{4+}, Ce^{4+}) promotes a transformation to a 1:1 ordered phase, while the introduction of "under-sized" cations (e.g. Ti^{4+}) stabilizes a disordered perovskite. Using Rietveld methods the structures of the 1:1 ordered $A(\beta^I_{1/2}\beta^{II}_{1/2})O_3$ phases were found to conform to a "random layer" model in which the β^{II} positions are occupied by Ta, and the β^I sites by a random distribution of the Mg(Zn), M^{4+} and residual Ta cations.

INTRODUCTION

Mixed metal "1:2" tantalate perovskites with the general formula $A^{2+}(B^{2+}_{1/3}Ta_{2/3})O_3$ (A= Ba, Sr; B^{2+} = Mg, Zn) constitute an important class of dielectric materials. In particular their very low dielectric losses in the microwave region have led to extensive applications as frequency filters in wireless communication systems [1,2 and references therein]. In their stable form the B-site cations in the $A^{2+}(M^{2+}_{1/3}Ta_{2/3})O_3$ phases order onto individual (111) planes of the perovskite subcell [3-5]. The long-range ordering and associated correlated displacements of the oxygen anions between the cation planes yields a trigonal supercell with a $\{M^{2+}TaTa\}$ layer repeat, see Figure 1a. For $Ba(Zn_{1/3}Ta_{2/3})O_3$ (BZT) and $Ba(Mg_{1/3}Ta_{2/3})O_3$ (BMT) the formation of a defect free, fully ordered cation arrangement has been shown to be critical in optimizing the dielectric response [6,7]. However, the long high temperature anneal times that are required to achieve perfect order have led to the development of alternate more effective methods for stabilizing a low loss state with a short sintering time. Tamura et al. [8] first showed this can be achieved through the substitution of small concentrations of $BaZrO_3$ into BZT. Recent HRTEM investigations of the effect of this substitution on the cation ordering [9] have revealed that the low levels of Zr (~5 mole %) promote a transformation to a cubic "1:1" ordered structure $(Ba\{\beta^I_{1/2}\beta^{II}_{1/2}]O_3)$ with a doubled perovskite repeat, see Figure 1b. In that study it was proposed that the 1:1 phase could be described by a "random layer" model in which the β^{II} sites were occupied exclusively by Ta and the β^I positions by a random mixture of the Zn, Zr, and remaining Ta cations. For this model the compositions of the 1:1 ordered phases in the (1-x)BZT - (x)BZ solid solution can be represented by $Ba\{(Zn_{2-z/3}Ta_{1-2z/3}Zr_z)_{1/2}(Ta)_{1/2}\}O_3$ where z = 2x. It is apparent from this formulation that the random layer structure model could operate up to substitution levels as high as 25 mole% BZ, i.e. z = 0.5. In this paper we summarize the results of our investigations of the phase stabilities of the entire BZT-BZ system, examine how changes in the chemistry of the 1:2 perovskite end-member and tetravalent substituent affect the

443

Mat. Res. Soc. Symp. Proc. Vol. 453 © 1997 Materials Research Society

formation of the 1:1 phase, and report on the refinements of these phases using the Rietveld method.

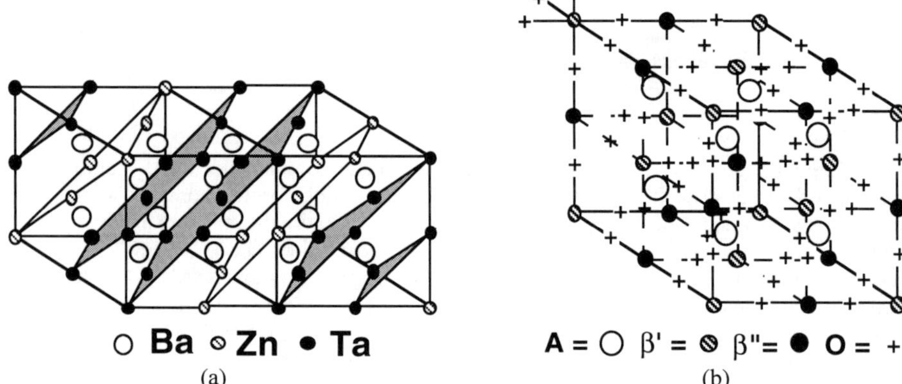

Figure 1. Schematic illustrations of (a) 1:2 cation ordering in $Ba(Zn_{1/3}Ta_{2/3})O_3$. Oxygen omitted for clarity. (b) the double cubic perovksite cell of the 1:1 $A\{\beta^I_{1/2}\beta^{II}_{1/2}]O_3$ ordered structure.

EXPERIMENTAL

All samples were prepared by conventional solid state reaction methods using oxide and/or carbonate precursors. For BZT-BZ and BMT-BZ mixing was done by hand in an agate mortar; the other systems were homogenized by mechanical milling. In all cases the initial mixture was heated at 1100 °C in air to remove CO_2. After another milling the powders were pressed into pellets and calcined at high temperature. The final sintering temperature for the Mg^{2+} containing samples was 1600 to 1650 °, while a lower temperature (1425 - 1500 °C) was used for the Zn^{2+} systems to inhibit volatilization. Final equilibrium was gauged by the lack of any change in the x-ray patterns after a repeat firing. Typically 2 or 3 firings were required to reach this state. Phase identification was accomplished using powder x-ray diffraction and the unit cell parameters were refined by the least squares method using silicon as an internal standard. Synchrotron x-ray diffraction data were collected on beamline X7A at the National Synchrotron Light Source at Brookhaven National Laboratory. A Ge(111) crystal was used as a monochrometer for the incident beam. Constant wavelength powder neutron data were also collected for selected samples using the H1A diffractometer at Brookhaven National Lab. Data were collected at room temperature using a neutron of wavelength of 1.8857 Å. Both GSAS and RIETAN programs [10,11] were used for the Rietveld refinements. Samples in each system were also examined using analytical TEM.

RESULTS

<u>$(1-x)Ba(Zn_{1/3}Ta_{2/3})O_3$ - $(x)BaZrO_3$ system</u>

Powder x-ray diffraction patterns collected from samples across the entire BZT-BZ system are shown in Figure 2. The strong ordering peaks associated with the 1:2 ordered structure of pure BZT weaken and eventually disappear when the concentration of BZ reaches 4 mole %. Examination of this sample using TEM revealed that it was comprised of small domains of a cubic 1:1 ordered structure with a doubled perovskite repeat. As the substitution level is raised to 7.5 mole % the superstructure peaks from the 1:1 ordered structure become visible in the powder x-ray patterns. These new ordering reflections, which can all be indexed using a cubic (Fm$\bar{3}$m) cell with a = $2a_{sub\text{-}cell}$, become stronger and sharper with increasing substitution and reach their maximum intensity at 25% BZ. For 35% substitution these reflections are very weak and broad; they are completely absent in samples with 50% BZ which could be indexed in terms of a single phase disordered (Pm$\bar{3}$m) perovskite.

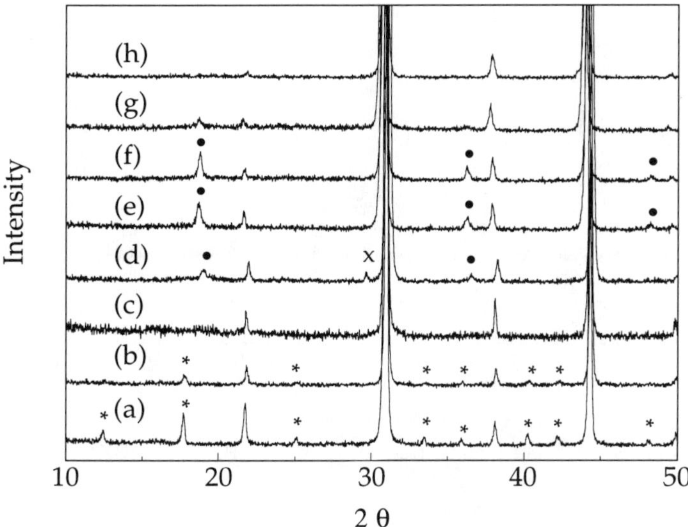

Figure 2. X-ray diffraction patterns for (1-x)BZT - xBZ: (a) 1%, (b) 2%, (c) 4%, (d) 7.5%, (e) 15%, (f) 25%, (g) 35%, (h) 50% BZ. Superstructure peaks from 1:2 phase indicated by *; peaks from cubic (Fm$\bar{3}$m) 1:1 ordered supercell by dots. A minor impurity in 7.5% BZ is labeled by x.

The boundary between the region of stability of the 1:2 and 1:1 ordered structures lies close to 4 mole % BZ, and evidence for a very narrow 2-phase region was found in electron diffraction patterns collected from samples with 3% BZ. The transformation to a fully disordered perovskite occurs at substitution levels beyond 25 mol %, at this stage it is unclear whether this is a continuous or first order transition. As expected the unit cell volumes increase with increasing BZ, though small discontinuities were apparent at each of the phase boundary regions.

The range of homogeneity of the 1:1 ordered structure is consistent with that predicted by the random layer structure model. As stated earlier, for this model the β^{II} positions in the $(Ba\{\beta^{I}_{1/2}\beta^{II}_{1/2}]O_3)$ structure are assumed to be occupied by Ta and the β^{I} sites contain a random distribution of Zn, Zr and Ta with the stoichiometry, $(Zn_{2-z/3}Ta_{1-2z/3}Zr_z)$. To test this model 1:1 ordered samples with 15% (z = 0.3) and 25% (z = 0.5) BZ were examined using synchrotron x-rays and refined using the Rietveld method. In both cases all of the peaks in the synchrotron patterns, Figure 3, could be indexed within the Fm$\bar{3}$m space group of the doubled perovskite cell. Because the supercell reflections were consistently broader than the sub-cell reflections, two sets of profile parameters were used for the refinements. Final reliability factors of $R_{wp}=$ 0.11, $R_F(sub)=$ 0.018, $R_F(super)=$ 0.023 and $R_{wp}=$ 0.088, $R_F(sub)=$ 0.012, $R_F(super)=$ 0.020 were obtained for the 15 and 25% samples respectively. The final refined occupancies are listed in Table 1 and compared to those predicted by the random layer model. Complete details of the final atomic positions will appear elsewhere.

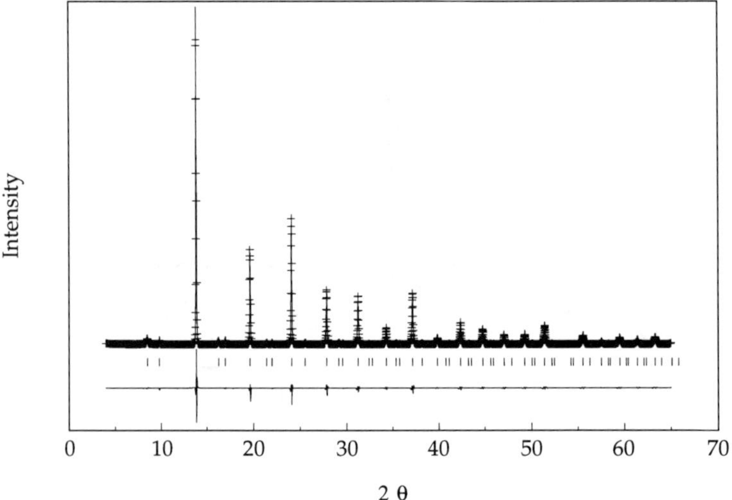

Figure 3. Experimental, calculated, and difference profile for BZT - 15%BZ.

Table 1. Refined Occupancies for $(1-x)Ba(Zn_{1/3}Ta_{2/3})O_3 - (x)BaZrO_3$

Atom	x = 0.15		x = 0.25	
	Refined	"Random layer"	Refined	"Random layer"
Ta (β^I)	0.250	0.133	0.162	0.0
Zn (β^I)	0.457	0.567	0.346	0.5
Zr (β^I)	0.294	0.300	0.492	0.5
Ta (β^{II})	0.884	1.0	0.838	1.0
Zn (β^{II})	0.110	0.0	0.154	0.0
Zr (β^{II})	0.006	0.0	0.008	0.0

The refined occupancies are in good agreement with those predicted by the random layer model. While the small differences in the predicted and experimental occupancies suggest that the structure does show some degree of anti-site disorder, it should be noted that both samples are comprised of ~300Å ordered domains. Therefore the refined occupancies also contain contributions from the locally disordered anti-phase boundaries that separate the different translational variants of the 1:1 structure.

$(1-x)Ba(Mg_{1/3}Ta_{2/3})O_3$ - $(x)BaZrO_3$, - $(x)BaCeO_3$, - $(x)BaTiO_3$ systems

The phase stabilities in BMT-BZ were essentially identical to those reported above for BZT-BZ. A single phase 1:1 ordered structure was formed between 10 to ~ 25 mole % BZ. Structure refinements on compositions in the 1:1 region were conducted using both synchrotron x-rays and neutrons. All the refinements yielded excellent agreement with the random layer model. At the microstructural level these samples were again comprised of approximately 300Å ordered domains separated by APB's.

Ordered 1:1 phases are also formed when Ce^{4+} is introduced onto the B-sites. The solubility of $BaCeO_3$ in the 1:2 structure of BMT is very limited and a two-phase mixture of the 1:2 and 1:1 structures was observed between ~ 1 and 7 mol % $BaCeO_3$. A single phase 1:1 structure is stable in samples with 7 - 25 mole% Ce^{4+}. Higher levels of $BaCeO_3$ produce a two-phase mixture of the 1:1 structure and a disordered $BaCeO_3$ solid solution that contains ~ 25 mole % BMT. Preliminary structure refinements have again confirmed that the random layer model for the 1:1 structure gives good agreement with the experimental profiles.

While the larger substituents such as Ce^{4+} (0.87Å) and Zr^{4+} (0.72Å) lead to the stabilization of a 1:1 random layer structure, studies of the BMT- $BaTiO_3$ system revealed that this phase does not form when smaller tetravalent cations such as Ti^{4+} (0.605 Å) are introduced onto the octahedral sites. In this case a 1:2 ordered solid solution of BMT was formed for compositions up to 10 mole % $BaTiO_3$ and higher levels of substitution induced a direct transformation to a fully disordered perovskite.

Sr-based 1:2 perovskites: $Sr(Mg_{1/3}Ta_{2/3})O_3$ - $SrZrO_3$

Preliminary studies indicate that the phase stabilities in solid solutions of $Sr(Mg_{1/3}Ta_{2/3})O_3$ with $SrZrO_3$ are very similar to the corresponding Ba system. However, the shorter Sr-O bonds and reduced tolerance factor appear to lead to small deviations of the 1:1 ordered phases from cubic symmetry. Work is currently in progress to refine this structure.

DISCUSSION AND CONCLUSIONS

Although the 1:1 ordered $A\{\beta^I_{1/2}\beta^{II}_{1/2}]O_3$ structure is commonly observed in mixed-metal perovskites with a 1:1 B-site chemistry [3] (where both the β^I and β^{II} positions are occupied by a single type of cation), it is rarely the superstructure of choice for the 1:2 $A(B^I_{1/3}B^{II}_{2/3})O_3$ systems - exceptions being lead-based relaxor ferroelectrics (e.g. PMN, Pb $(Mg_{1/3}Nb^{5+}_{2/3})O_3$) and hexavalent $A(M^{6+}_{1/3}M^{3+}_{2/3})O_3$ phases. From our studies of $AM^{4+}O_3$ substitutions into Ba and Sr-based $A(B^{2+}_{1/3}Ta_{2/3})O_3$ perovskites, it is evident that small concentrations (≤ 4 mole %) of tetravalent substituents such as Zr and Ce are effective in stabilizing a 1:1 ordered random layer structure in a much broader range of systems . The transformation in the ordering at such low concentrations of M^{4+} implies that these substitutions are effective in frustrating long-range interactions in the 1:2 structure, such as the concerted anion displacements that accompany the ordering of the B^{2+} and Ta^{5+} ions. This conclusion is supported by the observed decrease in the stability of the 1:2 structure with increasing mismatch in the Ta-O and M^{4+}-O bond lengths. The transformation to the 1:1 random layer structure, where one site is occupied by Ta and the other by a random distribution of the B^{2+}, M^{4+}, and Ta cations, appears to be an energetic compromise in which the long-range ordering of Ta in one site is preserved at the expense of the others. From an enthalpic viewpoint the stability of the 1:1 random layer structure is clearly less favorable than that of a pristine un-substituted 1:2 ordered phase, but more so than a completely disordered perovskite or perhaps a partially substituted 1:2 structure. However, this is offset by the configurational entropy associated with the mixing of the cations on the random layer position. The magnitude of the entropy is very sensitive to the concentration of M^{4+}, and large increases can occur at very low levels of $AM^{4+}O_3$ (e.g. a 3 mole% substitution produces a 27% increase in ΔS_{config}). For the BZT-BZ system the transitions in the cation ordering lie at the same compositions that had previously been shown to exhibit anomalies change the dielectric loss. The correlations between the loss properties and cation ordering in the new systems reported in this paper are currently under investigation.

ACKNOWLEDGMENTS

We thank P. Woodward and F. Izumi for helpful discussions and J.B. Parise, Q. Zhu, D. Cox, W. Dmowski, and T. Egami for assistance in collecting synchrotron and neutron data. This work was supported by the National Science Foundation (DMR94-21184) and also made used of MRSEC Shared Experimental Facilities supported by the National Science foundation under award number DMR96-32598. The facilities at the National Synchrotron Light Sources, Brookhaven National Laboratory are supported by the U.S. Department of Energy, Division of Materials Science and Chemical Science.

REFERENCES

1. T. Negas, G. Yeager, S. Bell and R. Amren, in <u>Chemistry of Electronic Ceramic Materials</u>, edited by P.K. Davies and R.S. Roth, (NIST Special Publication 804, 1991), p. 21-37.
2. P. K. Davies, in <u>Materials and Processing for Wireless Communication</u>, edited by T. Negas and H. Ling, (Am. Ceram. Soc., Westerville, OH, 1994), p. 137-152.
3. F. Galasso, <u>Perovskites and High Tc Superconductors</u>, Gordon and Breach Science, New York, 1990.
4. A. J. Jacobson, B. M. Collins, and B.E.F. Fender, Acta Cryst., B32, 1083 (1976).
5. H. Vincent, Ch. Perrier, Ph. Theririer, and M. Labeyrei, Mat. Res. Bull. 28, 951 (1993).
6. S. Kawashima, M. Nishida, I. Ueda, and H. Ouchi, J. Am. Ceram. Soc, 66, 421 (1983).
7. K. Matsumoto, T. Hiuga, K. Takada, and H. Ichimura, in <u>Proceedings of the 6th IEEE International Symposium on Application of Ferroelectrics</u>, (Institute of Electrical and Electronic Engineers, New York, 1986), p. 118-121.
8. H. Tamura, T Konoike, Y. Sakabe, and K. Wakino, J. Am. Ceram. Soc, 67, C-59 (1984).
9. P. K. Davies, J. Tong, and T. Negas, J. Am. Ceram. Soc., (1996, submitted).
10. A.C. Larson and R.B. Von Dreele, Los Alamos Laboratory Report No. LAUR-86-748.
11. F. Izumi, in <u>The Rietveld Method</u>, edited by R.A. Young, Oxford University Press, Oxford, 1993, p. 111.

MODELING PHASE STABILITY IN $A(B_{1/3}B'_{2/3})O_3$ PEROVSKITES

R. McCORMACK AND BENJAMIN P. BURTON
Materials Science and Engineering Laboratory, Ceramics Division
National Institute of Standards and Technology, Gaithersburg, MD 20899

ABSTRACT

Order-disorder phenomena on the simple-cubic B-site sublattice in $A(B_{1/3}B'_{2/3})O_3$ perovskites is examined. A simple cubic groundstate analysis in the cube approximation reveals that this approximation is inadequate for $A(B_{1/3}B'_{2/3})O_3$ perovskites, because it cannot predict a common experimentally-observed [111] superstructure. A partial vertex enumeration technique is used to demonstrate that a 12 interaction subset of the cube + linear triplet approximation is sufficient. First-principles calculations were performed for $Ba(Zn_{1/3}Ta_{2/3})O_3$ (BZT) and $Pb(Mg_{1/3}Nb_{2/3})O_3$ (PMN), two technologically interesting materials, in an effort to construct generalized Ising model Hamiltonians to enable simulations of these materials. Both ionic model (SSCAD) and pseudopotential (PP) calculations were done, enumerating the relative energies of a series of B-site superstructures. Structural hierarchies are reasonble and predict the [111] groundstate for both BZT and PMN, despite the fact that it is not observed in the latter. The pseudopotential results also indicate the possibility of metal-insulator transitions in BZT as a function of B-site configuration.

INTRODUCTION

It has been seen experimentally that the state of B-site cation order in mixed $A(B',B'')O_3$ perovskites has a strong effect on their electronic properties. For example, the degree of order in $Ba(Zn_{1/3}Ta_{2/3})O_3$ (BZT) is correlated with dielectric loss at microwave frequencies.[1] The state of order has also been shown to impact the dielectric dispersion in the relaxor ferroelectric $Pb(Mg_{1/3}Nb_{2/3})O_3$ (PMN), a compound that is widely used in multilayer capacitors.[2] BZT and PMN possess analogous stoichiometries and B-site cation valences, hence one might expect them to possess similar groundstates and electronic properties. This does not appear to be the case. The low-temperature ordered structure (groundstate) of BZT has been determined experimentally [3] as a simple 1:2 stacking modulation of Zn and Ta atoms along the [111] direction of the simple cubic B-site sublattice. The structure of PMN at low-temperatures appears to be more complicated, with small regions that possess local ratios of Mg:Nb of 1:1, incommensurate with the global ratio of 1:2.[5, 6, 7] This implies a competition between a 1:1 ordering tendency and the large electrostatic energy created by regions of charge imbalance. To understand order-disorder behavior in BZT and PMN, it is desirable to study the stability of different compounds in these systems at T = 0 K and also to simulate order-disorder at finite temperatures. To this end, we determine a sufficient model to perform such simulations and perform first-principles calculations of phase stability in these materials.

ISING MODELS IN $A(B_{1/3}B'_{2/3})O_3$ PEROVSKITES

A central issue is to determine what constitutes a sufficient theoretical model to perform a simulation of thermodynamics in $A(B_{1/3}B'_{2/3})O_3$ perovskites. The typical methodology for treating substitutional order-disorder phenomena in alloy materials is to model the alloy as a collection of interacting objects on a lattice (a generalized Ising model). In the present case, we are interested in disorder on the simple cubic B-site sublattice of the perovskite structure. Compounds in the present study of the form $A(B_{1/3}B'_{2/3})O_3$ correspond to B-site structures of composition BB'_2; if we adopt the notation A = B' and B = B, then the mixed B-site compounds correspond to binary compounds of composition A_2B.

Different configurations on the simple cubic lattice are assigned using pseudospin variables (occupation operators) $\sigma_i = \pm 1$ if site i is occupied by a B' or B" atom, respectively. The configuration on the N-site lattice is then specified with the vector $\sigma = (\sigma_1, ..., \sigma_N)$. In a classic paper by Sanchez, Ducastelle, and Gratias [8], it was shown that any function that depends on the configuration σ (e.g., the energy) can be written as a linear, orthonormal cluster expansion using a complete set of *cluster functions*:

$$E(\sigma) = \sum_{\alpha} m_{\alpha} V_{\alpha} \langle \phi_{\alpha}(\sigma) \rangle \tag{1}$$

where V_{α} are the so-called *effective cluster interactions* (ECI's), $<\phi_{\alpha}(\sigma)>$ are lattice-averaged values of the cluster functions that depend on the configuration, and m_{α} are the number of clusters of type α per lattice site. The sum in Eq. [1] runs over all symmetry-distinct clusters α of sites in the lattice. With all 2^N terms, the expansion is *exact*; in practice however, the expansion is truncated at a level that is acceptable for the problem at hand. One specifies a range (distance) over which atomic interactions are deemed important, and then uses all clusters of sites within this range (the *interaction clusters* α_I).

Within such a cluster expansion framework, there are two separate, yet interrelated, questions: (1) What are the ground states (configurations with minimum energy at T = 0 K) that are possible for the given interaction clusters? and (2) What is the behavior of such a system at finite temperature? The first problem is the so-called ground-state (GS) problem, which has a rich history.[9] The GS problem can be solved by enumerating the vertices of the so-called *configuration polyhedron* (CP) in the space of lattice-averaged cluster functions (LACF).[9] A more detailed discussion of the GS problem and configuration polyhedra can be found in Ref. [9]

The ground-state problem can be solved generally for all possible ground states for a given set of interaction clusters (independent of the values of the effective cluster interactions), or it can be solved to find the specific ground-state structures for a given set of ECI's. The first analysis involves enumerating all of the vertices of the CP, a problem that rapidly becomes intractable as the number of interaction clusters (range of interaction) increases.[10] A more straightforward problem is to determine the groundstates for a set of cluster interactions with fixed numerical values; this can be solved by linear-programming techniques.[10] For the purposes of the present study, an even more specific *partial vertex enumeration* (PVE) is possible [14]. This allows determination of a sufficient level of approximation to yield a specified structure as a vertex of the CP. In $Ba(Zn_{1/3}Ta_{2/3})O_3$, the experimentally observed groundstate is a [111] stacking modulation of composition A_2B. Any simulation of disorder in these materials must allow this structure as a groundstate. Hence, the present problem reduces to finding an approximation in which the A_2B [111] structure is a vertex of the CP. A more detailed descritpion of the PVE method can be found in Ref. [15].

Application to Complex Perovskites

To date, only three studies of groundstates on the simple cubic lattice have been performed. Katsura and Narita [11] and Kaburagi and Kanamori [12] both performed vertex enumerations assuming pair interactions from nearest- through third-nearest-neighbor (NN-3NN) distance Lipkin [13] performed a partial analysis, but no third neighbor interactions were included and it was assumed that all three body effective cluster interactions were equal. The study of Kaburagi and Kanamori [12] is the most complete; it revealed a total of 19 ordered groundstates possible within the range of 3NN pair interactions.

Kaburagi and Kanamori found no structures of composition A_2B, which implies that for NN-3NN pair interactions, any A_2B structure will disproportionate into an appropriate mixture of the groundstates that bound A_2B in composition space. It turns out that the groundstate problem can be solved exactly if one uses all cluster interactions within the simple cube (NN-3NN *and* all many-body clusters within the unit cube). Within this *cube* approximation, there are a total of 22 cluster interactions and 22 symmetry-distinct configurations of atoms on the cube, which implies that the CP in this case is a *simplex*: each vertex of the polyhedron corresponds to a structure that possesses cubes of only one configuration.[9] The three new groundstates at the level of the cube approximation (all with composition AB) must be stabilized by many-body interactions; again no A_2B groundstates exist in this approximation.

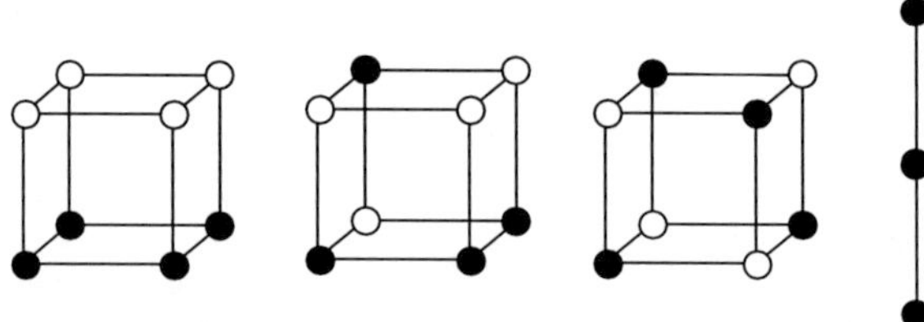

Figure 1: An A_2B [111] groundstate can be generated using the minimum interaction clusters indicated by shaded sites and all of their subclusters (a total of 12 interactions). the sites

The cube approximation is insufficient to treat ordering into an A_2B [111] superlattice, hence to find a sufficient approximation, we successively add clusters to the cube approximation and perform the PVE analysis for each group of new interaction clusters. This procedure demonstrates that with the cube and linear triplet (LT) as interaction clusters, the [111] superstructure is a vertex. Extensive analysis of subsets of this cube + LT approximation reveals the minimal set of clusters required to generate this A_2B vertex. A minimum of 12 interactions (the interaction clusters shown in Fig. 1 and their subclusters) is required to give a [111] A_2B GS. These results cast doubt on previous simulations (cf. Bursill et al. [16]) by demonstrating that they utilize insufficient Ising models, and therefore cannot capture the essential physics of the problem.

STUDIES OF PHASE STABILITY IN $Ba(Zn_{1/3}Ta_{2/3})O_3$ and $Pb(Mg_{1/3}Nb_{2/3})O_3$

The type of Ising model required to do simulations of BZT and PMN was discussed in the previous section. All that remains is to calculate the effective cluster interactions to perform a first-principles simulation. Several methods exist to calculate ECI's [17, 18, 19, 20, 21], and all essentially revolve around the use of total energy calculations for a series of different configurations on the lattice. Such total energy calculations also help to elaborate the electronic structure of relevant materials, the importance of structural relaxations, and whether the materials should behave as metals or insulators.

Total energy calculations were performed for a series of superstructures on the simple cubic B-site sublattice (Fig. 2). Four of these ([111], [110], [001] and tP6) represent simple stacking modulations along high-symmetry directions; the remaining three are compact or potentially stable structures. Two methods were used to perform total energy calculations, the first based on simplified ionic models for the total energy, the second using the more accurate pseudopotential method.

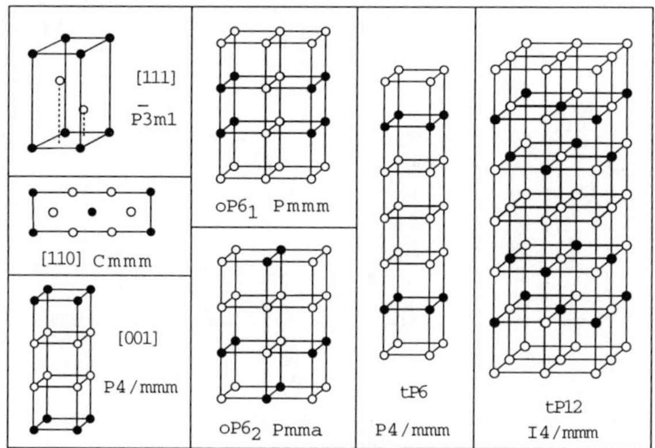

Figure 2: Simple cubic superstructures used in total energy calculations. For each structure, a label is given along with the space group. In the actual first-principles calculations, all of the A sites and Oxygen atoms (not shown) were of course included.

Ionic Model Approach

Perovskites are expected to be predominantly ionic, hence one might expect that electronic structure techniques tailored to ionic systems would be useful (i.e., modified Gordon-Kim [22] approaches). One method that has received attention is the self-consistent atomic deformation (SCAD) model [23, 24]. In the present formulation of the SCAD, the atomic densities are assumed to be spherically-symmetric (SSCAD), although it is possible to relax this approximation to the non-spherically-symmetric case.[24]

SSCAD calculations were performed in BZT and PMN for all structures in Figure 2. All structural degrees of freedom (cell internal and cell external) were optimized. An interatomic interaction cutoff of 0.79 nm (15 $a.u.$) was used and total energies were converged to within 1.313 kJ/mol (0.5 $mHartree/atom$). Nominal ionic valences (closed shells) were assumed for all atoms; charge transfer was not allowed in the calculations. SSCAD equilibrium volumes, lattice constants, and energies relative to the [111] superstructure are given in Table 1. Equilibrium volumes for all compounds are quite similar. In addition, the volume of the [111] superstructure is fairly close to the experimental value for the ordered BZT structure. [3] Comparison is made in PMN between the volume of the [111] structure and the experimental low-temperature (average) structure, even though the this structure is indeterminate; agreement is also quite good. There is substantial structural relaxation along the direction of the stacking modulation in the [111], [110], [001] and tP6 structures; for all structures, internal relaxations were important and strongly coupled to cell external relaxations.

Table 1: Structural properties for a group of simple cubic superstructures (shown in Figure 2). Lattice constants (a, b, c) are in nm; Volume (Ω_0) in cm^3/mol (experimental volumes are given in parentheses where available). E_{rel} corresponds to the energy relative to the [111] superstructure in kJ/mol ($mHartree/atom$) for the SSCAD and pseudopotential calculations. For the pseudopotential results, N_k is the number of **k**-points used and $N(E_F)$ indicates whether the material should be a metal (M, $\mathcal{N}(E_F) > 0$) or Insulator (I, $\mathcal{N}(E_F) = 0$). A – indicates that the calculation was not performed.

Structure	a	b	c	Ω_0	E_{rel} (SSCAD)	E_{rel} (Pseudo)	N_k	$N(E_F)$
BZT:								
[111]	0.58	0.58	0.72	8.50 (8.26)	0.0 (0.0)	0.0	8	I
[110]	1.82	0.57	0.41	8.50	3.2 (1.2)	8.1 (3.1)	20	I
[001]	0.41	0.41	1.30	8.64	19.7 (7.5)	11.6 (4.4)	12	I
oP6$_1$	0.85	0.40	1.25	8.56	9.2 (3.5)	10.0 (3.8)	12	I
oP6$_2$	0.83	0.40	1.28	8.56	5.0 (1.9)	4.2 (1.6)	12	I
tP6	0.41	0.41	2.71	8.95	54.6 (20.8)	15.8 (6.0)	20	M
tP12	0.59	0.59	2.49	8.55	10.2 (3.9)	–	–	–
PMN:								
[111]	0.56	0.56	0.69	7.50 (8.00)	0.0 (0.0)	–	–	–
[110]	1.72	0.55	0.39	7.50	6.8 (2.6)	–	–	–
[001]	0.38	0.38	1.30	7.66	19.7 (7.5)	–	–	–
oP6$_1$	0.82	0.38	1.19	7.54	6.3 (2.4)	–	–	–
oP6$_2$	0.81	0.38	1.23	7.65	2.9 (1.1)	–	–	–
tP6	0.40	0.40	2.32	7.47	78.0 (29.7)	–	–	–
tP12	0.56	0.56	2.40	7.51	9.2 (3.5)	–	–	–

Examination of the total energy relative to that of the [111] superstructure reveals that this is indeed the lowest energy structure of those considered for both BZT and PMN. For both BZT and PMN, the highest energy structures considered are the [001] superlattice and tP6 (a longer-period modulation along [001]). This is quite reasonable given that these layered structures possess a larger interplanar spacing (between successive layers along the modulation direction) than those of the other stacking modulations. In addition, the tP6 structure separates successive B' cation layers more completely than the other structures, creating large regions of local charge-imbalance. This indicates a strong correlation between the Madelung energy and the total energy in these compounds, as one might expect for an ionic system.

Pseudopotential Method

The ionic model calculations are simplified in their treatment of the total energy. Such simplifications are necessary due to the large unit cells for which the calculations are performed. In an effort to benchmark these ionic calculations, a limited set of pseudopotential (PP) calculations for BZT only were undertaken (a complete discussion of the pseudopotential method can be found in Ref. [25]). The number of crystallographic degrees of freedom in the structures in Figure 2 is quite large, making fully-relaxed pseudopotential calculations quite time-consuming. Hence, PP calculations were done at the optimized geometries predicted in the SSCAD calculations for the superstructures in Figure 2.

Optimized norm-conserving pseudopotentials built using the method of Sadigh et al. [26] were used. A plane-wave cutoff of 25 Hartree (50 Ry = 680 eV) was used. Brillouin zone integration was done using Gaussian broadening; the number of **k**-points used for each structure was $8 \leq N_k \leq 20$. Convergence with respect to **k**-point sampling was tested for the [111] and [001] superstructures. The plane-wave cutoff was found to be suitable in atomic calculations,[26] hence it was deemed reasonable to use this cutoff for the solid. All of the pseudopotential total energies are contained in Table II, as well as the number of **k**-points used for each calculation and whether the PP calculation predicts a finite density-of-states at the Fermi level.

Discussion

A comparison of the SSCAD and pseudopotential predictions for structural energies relative to the [111] superstructure is given in Figure 3. It should be emphasized that the pseudopotential calculations are for crystal geometries *constrained* to the equilibrium SSCAD values. Hence, the pseudopotential predictions act more as a benchmark of SSCAD's ability to reproduce structural energy differences for a fixed geometry. With these caveats, several conclusions can be drawn from these calculations.

SSCAD appears to be qualitatively reasonable in terms of predicting the structural energy hierarchy, especially for the low-energy excitations. Quantitative differences for low-energy structures are small, again indicating that SSCAD is probably sufficient to treat these energetics. The overestimation of E(tP6)-E([111]) is problematic and points out that the SSCAD is probably not very useful in treating higher-energy excitations properly. The fact that tP6 is a metal in BZT indicates that the fundamental nature of bonding in BZT superstructures is a function of the configuration, i.e., some configurations are metals and some are insulators. Given that the SSCAD method is inherently ionic, this implies that it could be inadequate for the construction of a universally valid cluster expansion in the BZT system (and by extension the PMN system and all $A(B_{1/3}B'_{2/3})O_3$ Perovskites).

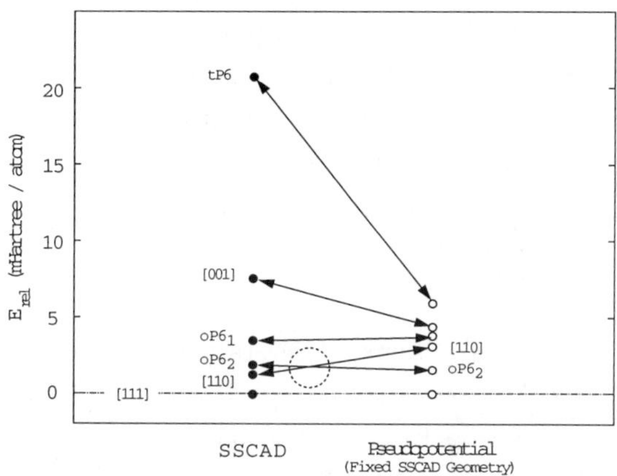

Figure 3: A comparison of SSCAD and pseudopotential calculations for several BZT superstructures. For each structure, the energy relative the the [111] superstructure is given. Pseudopotential results are for a geometry that is *fixed* to that of the equilibrium SSCAD result.

The configuration dependence of the electronic structure of BZT (i.e., the metal-insulator transitions that may occur in response to cation order-disorder phenomena) may have implications for the electronic properties of these materials. As the temperature increases, disorder on the B-sites may give the bonding more metallic character, and thereby alter the dielectric response. Such behavior may also be correlated with the observations of 1:1 ordering in PMN; to the authors' knowledge, no measurements of the transport properties of these materials have been made as a function of temperature. It should be noted that metallic behavior is being predicted for bulk *superstructures* with a given configuration. States of partial order will effectively contain a mixture of superstructures, with short length-scale regions corresponding to a given structure. Based on this length scale inhomogeneity, it is diffcult to make any comments about the *bulk* transport properties in these partially-ordered materials. However, the configuration-dependence of the transport properties could be a window into explaining the unusual physics in these mixed B-site systems.

ACKNOWLEDGEMENTS

RPM wishes to acknowledge G. Garbulsky and G. Ceder at MIT for assistance in the simple cubic groundstate problem, and for help related to partial vertex enumeration. V. Ozolins provided the pseudopotential codes and patient instruction about their use. RPM was supported by a National Research Council Postdoctoral Fellowship. This work supported by the Office of Naval Research under ONR contract No. N00014-94-F0017.

References

[1] S. Kawashima, M. Nishida, I. Ueda, and H. Ouchi, J. Am. Ceram. Soc., **66**, 421 (1983)

[2] S.G. Porter, Ferroelectrics, **33**, 193 (1981)

[3] A. J. Jacobson, B.M. Collins and B. E. F. Fender, Acta Cryst. **B32**, 1083 (1976)

[4] D. A. Sagala and S. Nambu, J. Phys. Soc. Jpn. **61**, 1791 (1992)

[5] H.D. Rosenfeld and T. Egami, Ferroelectrics **158**, 351 (1994)

[6] D. Viehland and J.-F. Li, J. Appl. Phys. **74**, 4121 (1993)

[7] N. de Mathan, E. Husson, G. Calvarin, J.R. Gavarri, A. Hewatt, and A. Morell, J. Phys.: Cond. Matter **3**, 8159 (1991)

[8] J.M. Sanchez, F. Ducastelle, and D. Gratias, Physica **128A**, 334 (1984).

[9] F. Ducastelle, "Order and Phase Stability in Alloys" (Kluwer-Dordrecht, Amsterdam, 1990)

[10] G. Ceder, G.D. Garbulsky, D. Avis, and K. Fukuda, Phys. Rev. B, **49**, 1 (1994).

[11] S. Katsura and A. Narita, Progr. Theor. Phys., **50**, 1426 (1973).

[12] M. Kaburagi and J. Kanamori, Progr. Theor. Phys., **54**, 30 (1975).

[13] M.D. Lipkin, Physica A, **150**, 18 (1988).

[14] G. Ceder (private communication)

[15] R. McCormack and B.P. Burton, Computational Materials Science (in press)

[16] L.A. Bursill, H. Qian, J.L. Peng, and X.D. Fan, Physica B, **216** 1 (1995).

[17] J.W.D. Connolly and A.R. Williams, Phys. Rev. B **27**, 5169 (1983).

[18] F. Ducastelle and F. Gautier, J. Phys. F **6**, 2039 (1976).

[19] A. Gonis, X.-G. Zhang, A.J. Freeman, P. Turchi, G.M. Stocks, and D.M. Nicholson, Phys. Rev B **36**, 4630 (1987).

[20] H. Dreyssé, A. Berera, L.T. Wille, and D. de Fontaine, Phys. Rev. B **39**, 2442 (1989).

[21] G.D. Garbulsky and G. Ceder, Phys. Rev. B **51**, 67 (1995).

[22] R.G. Gordon, and Y.S. Kim, J. Chem. Phys. **56**, 3122 (1972)

[23] L.L. Boyer and M.J. Mehl, Ferroelectrics **150**, 13 (1993)

[24] L.L. Boyer, H.T. Stokes, and M.J. Mehl, Ferroelectrics (in press)

[25] M.C. Payne, M.P. Teter, D.C. Allan, T.A. Arias, and J.D. Joannopoulos, Rev. Mod. Phys., **64**, 1045 (1992).

[26] B. Sadigh and V. Ozolins, Phys. Rev. B (submitted).

INVESTIGATION OF POLARIZATION MECHANISM OF RELAXOR FERROELECTRICS

Z. –Y. CHENG[+*], R. S. KATIYAR[+], YAO XI[*]

[+] Department of Physics, P. O. Box 23343, University of Puerto Rico, San Juan, PR 00931-3343, USA

[*] Electronic Materials Research Laboratory (EMRL) , Xi'an Jiaotong University, Xi'an 710049, CHINA

ABSTRACT

Addition to thermally activated flips of polar regions in relaxor ferroelectrics, a new polarization mechanism, which originates from the vibrations (breathing) of surface of polar regions, is introduced to explain the dielectric behavior of relaxor ferroelectrics. This new mechanism plays an important role in the dielectric behavior of such materials at low temperature. Based on the above assumption and general dielectric theory, a formula is given to characterize the temperature dependence of the dielectric constant. The correctness of the formula is verified by using it to fit the experimental results of the two typical relaxors. The fitted results show that the method is of high precision and that the temperature of the dielectric constant maximum is decided by the two polarization behavior. It also indicates that the new polarization is a resonance polarization.

INTRODUCTION

Relaxor ferroelectrics (RFE) are an important kind of functional materials with high dielectric constant, high electrostriction coefficient, switchable pyroelectric and piezoelectric properties [1]. Although the materials of RFE have been used in many commercial devices, and many models have been proposed to explain the dielectric relaxation mechanism in these materials, such as the inhomogeneous micro-region model [2], the micro-macro domain transition model [3,4], the superparaelectric model [1,5], the dipolar glass models [6,7], the order-disorder model [8,9,10], the local electric field model [11], etc., the polarization mechanism in the materials is still not clear. However, it is widely accepted that the dielectric behavior of the materials is mainly dependent on the dynamical behavior of the polar regions in the materials [12].

Most of the above proposed models correlate the dielectric relaxation behavior in RFE with the orientation of polar regions under external electric field. Based on these models, it is difficult to explain the dielectric behavior of the quenched samples, in which the dielectric loss is very large [13], and the dielectric behavior of the sample under DC bias or large amplitude of the applied AC field [13,14]. All of the difficulties occur at the temperatures lower than the temperature (T_m) of the dielectric constant maximum. Meanwhile, the estimated dielectric constant by various models in the low temperature is always smaller than the experimental results, and the peak of measured dielectric constant is always broader than that of the estimated dielectric constant. It is widely known that the polar regions in RFE appear at the temperatures much higher than T_m and that the volume fraction of polar regions in the materials increases with decreasing temperature. Thus, it is reasonably assumed that the surface or interface of the polar regions in the materials has an important contribution to the dielectric behavior of the materials in the low temperature range [2,13,14]. Anyway, it is not clear so far how the surface of polar regions contributes to the dielectric behavior of the material.

455

Mat. Res. Soc. Symp. Proc. Vol. 453 © 1997 Materials Research Society

In this paper, we propose that the RFE materials have two polarization mechanisms, one corresponds to thermally activated flips of the polar regions and the other corresponds to moving of the interface between the polar region and non-polar substrate. Based on it, a practical formula is proposed to fit the temperature dependence of the dielectric constant for RFE. The temperature dependence of the dielectric constant of the two typical RFE, namely PMN-PT and PZN, was measured. The results are used to verify the formula. The correctness of the formula is confirmed by the measured results.

DIELECTRIC MEASUREMENTS

Two solid solution ceramics were used in the experiments, one (PMN-PT) is 10 mol% $PbTiO_3$ in $Pb(Mg_{1/3}Nb_{2/3})O_3$ and another (PZN12) is $0.87(Pb_{1-x}La_{x/2}K)_{x/2}(Zn_{1/3}Nb_{2/3})O_3-0.08PbTiO_3-0.05BaTiO_3$ with x=0.12. Both the samples, which were in disk forms with the thickness of about 1mm, are of pure perovskite-phase as identified by x-ray diffraction. The samples were prepared by the conventional ceramic processing. The gold electrodes were deposited on the sample surfaces by the DC sputtering.

The temperature dependence of both the capacitance and loss tangent were measured with a cooling rate of about 0.8°C per minute using an HP4274A LCR meter with the high resolution function. The oscillating level of the signal was about 0.5V/mm to ensure the measured results are the small-signal dielectric properties. Before the measurement, the samples had been thermally treated at the temperature much higher than T_m to remove the effect of electric history. The four test leads, which were connected directly to the electrodes of the sample, were used to increase the measuring accuracy. Before each measurement, a standard calibration was performed to remove any stray capacitance, lead and contact resistance. The dielectric constant was calculated from the measured capacitance, with the geometric parameters of the sample and using a parallel-plate capacitance model. The temperature was measured using an HP3455A multimeter via a 100Ω platinum resistance thermometer. To increase the measuring accuracy of the temperature, a 4-wire configuration was used to measure the resistance of the 100Ω platinum resistance thermometer. The temperature controller, HP3455A and HP4274A were interfaced to a PC computer.

Some measured results are shown in Fig.1. It demonstrates the typical dielectric behavior of RFE. Referring to Fig. 1, we find the following most essential features of the temperature dependence of the dielectric constant:

1). For temperatures much higher than T_m, the temperature dependence of the dielectric constant is an exponential function [7,13], that is, the dielectric constant (ε_s) is dependent on the temperature (T) by the following relationship:

$$\varepsilon_s(T) = \exp(\alpha - \beta T) \tag{1}$$

where α (>0) and β (>0) are constants, which are independent of the frequency, T is the absolute temperature.

2). For temperatures much lower than T_m, the temperature dependence of the dielectric constant is nearly an exponential function [13], which can be expressed as follows:

$$\varepsilon_L(\omega,T) = \exp(\alpha' + \beta' T^{1+\delta}) \tag{2}$$

where $\alpha'\,(>0)$, $\beta'\,(>0)$ and $\delta\,(\geq 0)$ are constants, ω is the frequency.

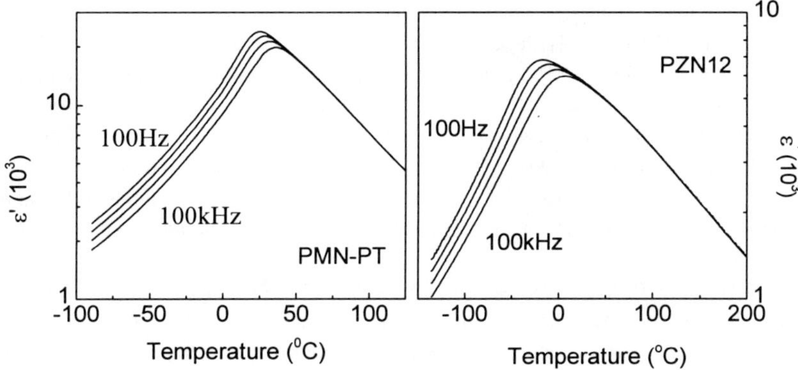

Fig. 1 Dielectric constant as a function of temperature at some measured frequencies (0.1, 1, 10, 1100 kHz) for PMN-PT and PZN12.

It is also found that the value of δ is much smaller than the unit and the value of β' is independent of the frequency.

THEORETICAL ANALYSIS

For an RFE, as discussed in the introduction, there should be two polarization mechanisms contributing to the dielectric behavior of the material. One corresponds to the thermally activated flips of the polar regions, and another corresponds to the surface moving of the polar regions. When the temperature is much higher than T_m, the dielectric behavior of RFE can be easily explained using relaxation polarization, thus we can assume that the dielectric behavior of RFE at temperatures much higher than T_m is mainly dependent upon the thermally activated flips of the polar regions. The lower the temperature is, the more the contribution of the surface moving of polar regions to the dielectric behavior of RFE is. Thus, it is reasonable to assume that the dielectric behavior of RFE at low temperatures is mainly dependent on the surface moving of polar regions as the thermal activity of the flips of polar region reduces. That is, the total dielectric constant (ε_T) of RFE's response can be written as:

$$\varepsilon_T(\omega,T) = \varepsilon_1(\omega,T) + \varepsilon_2(\omega,T) \tag{3}$$

where ε_1 and ε_2 are the dielectric response of the flips and the surface moving, respectively, of the polar regions.

Based on the general dielectric theory [15], the dielectric response of both ε_1 and ε_2 can be simply written as:

$$\varepsilon(\omega,T) = \frac{A(\omega,T)}{1+B(\omega,T)} \qquad (4)$$

where $A(\omega, T)$ and $B(\omega,T)$ are dependent on the polarization mechanism.

For ε_1, $A(\omega,T)$, which is only dependent on the temperature, is the static dielectric constant; $B(\omega,T)$, which is dependent on both the temperature and frequency, is very small when the temperature is very high and increases with decreasing temperature or with increasing frequency. Based on their physical meanings, $A(\omega,T)$ is related to both the concentration and dipole moment of the dipoles, while $B(\omega,T)$ is dependent on an interaction among the dipoles and dynamical behavior of the dipoles. Thus $B(\omega,T)$ is also related to both the concentration and dipole moment of the polar regions and the interaction between polar regions and its neighboring non-polar substrate.

For ε_2, assuming $A(\omega,T)$ represents the low-temperature behavior of the dielectric constant contributed by the polarization mechanism, then $B(\omega,T)$ will be very small in the low temperature range and it increases with increasing temperature. Thus, both $A(\omega,T)$ and $B(\omega,T)$ are dependent on the temperature, the concentration of the polar regions, the volume fraction of the surface of the polar regions, the damping constant and the amplitude of the surface moving, and so on.

Therefore, both $A(\omega,T)$ and $B(\omega,T)$ in ε_2 has the relation to $B(\omega,T)$ in ε_1, because the surface area is related to the volume and shape of the polar region, and both the damping force and amplitude of the surface moving are related to the concentration and dipole moment and neighborhood of the polar regions.

Based on the above discussion, we can assume that $A(\omega,T)$ in ε_1 can be expressed with Eq. (1), while $A(\omega,T)$ in ε_2 can be expressed with Eq. (2). $B(\omega,T)$ in both ε_1 and ε_2 is related to $A(\omega,T)$ in both ε_1 and ε_2. Based on the feature of $B(\omega,T)$, we can assume the function of ε_1 and ε_2 as follows:

$$\varepsilon_1 = \frac{\varepsilon_S(T)}{1+C\left(\varepsilon_S(T)\big/\varepsilon_L(\omega,T)\right)^m} \qquad (5a)$$

$$\varepsilon_2 = \frac{\varepsilon_L(\omega,T)}{1+D\left(\varepsilon_L(\omega,T)\big/\varepsilon_S(T)\right)^n} \qquad (5b)$$

where C, D, m, and n are constants.

Using Eqs. (1), (2), (3) and (5) we fitted the experimental results of both PMN-PT and PZN12, to a high precision. Two examples are shown in Fig. 2. It is clearly shown that the dielectric constant around T_m is decided by the two polarization mechanism. Compared with the fitted results of the temperature dependence of the dielectric constant for RFE using other method [16,17,18], our results are of very high precision.

It is also very interesting to compare the fitted ε_1 and ε_2 under different frequencies, which is shown in Fig. 3.

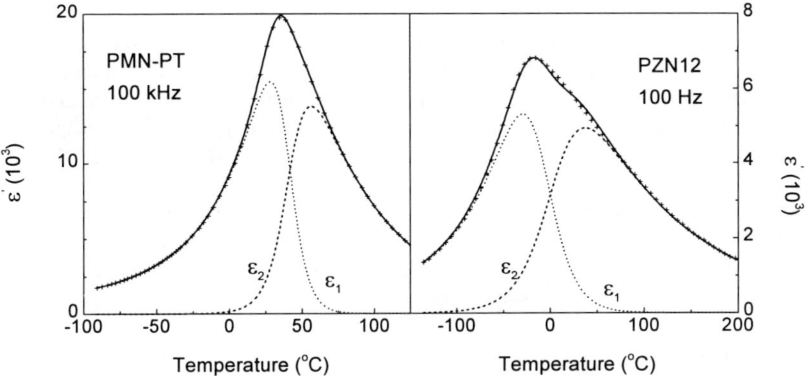

Fig. 2 Temperature dependence of the dielectric constant for PMN-PT at 100 kHz
and for PZN12 at 100 Hz. Measured results (+) and fitted results (solid line)
with Eq. (5)

From Fig. 3, it is clearly shown that ε_1 is a typical results of relaxation polarization and ε_2 is a results of resonance polarization, because the dielectric constant (ε_2) increases with increasing frequency in the high temperature range. It also indicates that the damping constant of the resonance polarization increases with decreasing temperature.

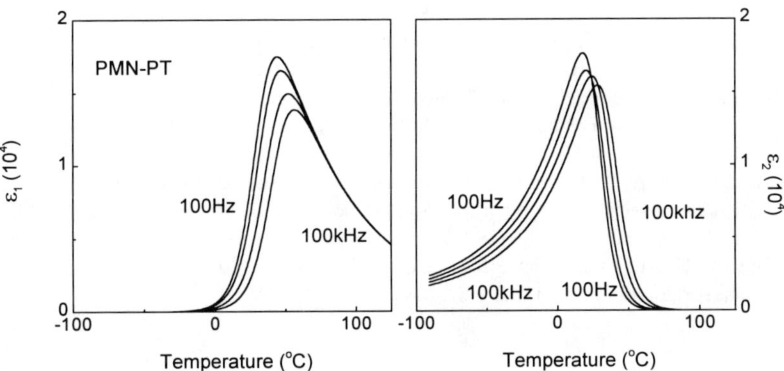

Fig. 3 Temperature dependence of fitted ε_1 and ε_2 under under different frequencies
of 0.1, 1, 10, 100 kHz for PMN-PT.

CONCLUSION

A new polarization mechanism, which corresponds to the moving of the surface of polar regions in the materials, is introduced to fit the temperature dependence of the dielectric constant for RFE. The experimental relations, Eqs (1) and (2), between the temperature and dielectric constant at the temperatures much higher and much lower than T_m are found respectively for RFE. The formula, Eq. (5), is given to express the temperature dependence of the dielectric constant.

The measured temperature dependence of the dielectric constant of PMN-PT and PZN12 is fitted with the above formula. The fitted results is of a high precision and show that the new polarization mechanism is a resonance polarization. It also shows that the dielectric behavior of RFE at temperatures around T_m is decided by the two polarization mechanism.

ACKNOWLEDGMENT

This work was partly supported by the National Natural Science Foundation of China under the contract No. 59232042 and EPSCOR-NSF grant NSF-OSR-9452893 of USA.

REFERENCES

1. L.E. Cross, Ferroelectrics, **76**, 241 (1987).
2. G. A. Smolensky, J. Phys. Soc. Jpn., **28**, Suppl. 26 (1970)
3. X. Yao, Z. Chen, L. E. Cross, J. Appl. Phys., **54**, 3399, (1983).
4. Z. -Y. Cheng, L. Y. Zhang, X. Yao, IEEE Trans. Elect. Insul., **27**, 773 (1992).
5. A. J. Bell, J. Phys.: Condens. Matter, **5**, 8773 (1993).
6. D. Viehland, S. J. Jang, L. E. Cross, M. Wutting, J. Appl. Phys., **68**, 2916 (1990).
7. Z. -Y. Cheng, L. Y. Zhang, X. Yao, J. Appl. Phys., **79**, 8615 (1996).
8. N. Setter and L. E. Cross, J. Mater. Sci., **15**, 2478 (1982).
9. C. A. Randall, A. S. Bhalla, J. Appl. Phys. Japn., **29**, 327 (1990).
10. X. W. Zhang, Q. Wang, B. L. Gu, J. Am. Ceram. Soc., **74**, 2846 (1991).
11. W. kleemann, Int. J. Mod. Phys. B, **7**, 2469 (1993).
12. G. Burns, F. H. Dacol, Ferroelectrics, **104**, 25 (1990).
13. Z. -Y. Cheng, Ph.D. Thesis, Xian Jiaotong University (China), 1995.
14. A. E. Glazounov, A. K. Tagantsev, and A. J. Bell, Phys. Rev. B, **53**, 11281, (1996).
15. A. K. Jonscher, <u>Dielectric Relaxation in Solid</u> (Chelsea Dielectric Press, London, 1983).
16. P. Groves, Ferroelectrics, **76**, 81 (1987).
17. H. Wang, Y. H. Xu, Acta Phys. Sin. (in Chinese), **5**, 605 (1986).
18. J. Kuwata, K. Uchino, S. Nomura, Ferroelectrics, **7**, 151 (1979)

STRUCTURAL MODELING BASED ON THE OXYGEN SUBLATTICE FEATURES IN THE PEROVSKITE FAMILY

L.E. DEPERO *, L. SANGALETTI *, F.PARMIGIANI**, E.GIULOTTO***,
L.ROLLANDI***
* Istituto Nazionale di Fisica per la Materia e Dipartimento di Chimica e Fisica per i Materiali,
Università di Brescia, Via Branze, 38 - 25123 Brescia, ITALY
**INFM & Politecnico di Milano, Dipt. di Fisica - Milano, ITALY
***INF & Univ.di Pavia, Dipt. di Fisica A.Volta -Pavia, ITALY

ABSTRACT

A new approach based on the study of distortion in the AO_3 layer of ABO_3 perovskites is proposed with the aim to discuss their ferroelectric properties. Rhombohedral distortions exist for either A-O average distances shorter than 2.7 A (space group Nr.161) or A-O distances larger than 3 Å (space group Nr.160). In the case of small distortions tetragonal or orthorhombic symmetries are preferred. Moreover, a relationship between the dimension of the B cation and the distortion in this type of structure has been established. Preliminary results show interesting correlations between distortions in the AO_3 plane and the critical temperature of ferroelectric materials.

INTRODUCTION

Structural studies of perovskites with the general formula ABX_3 are critical in view of their relevant physical properties such as ferroelectricity, antiferroelectricity, piezoelectricity and optical activity. All these properties largely depend on the dipole moments caused by ionic displacements.

The perovskite structure consists of a three dimensional network of BX_6 octahedra, with A ions forming AX_{12} cuboctahedra to fill the spaces among octahedra. The common approaches for describing the analogies and differences among the structures of this family are based on the study of the coordination polyhedra of the B cation.

Indeed, it was stated that the framework of BO_6 octahedra plays an essential role in the phenomenon of ferroelectricity of these compounds. Matthias [1] advanced an empirical rule for the occurrence of ferroelectricity in perovskite-type compounds, namely that the B ion must have the electronic configuration of a noble gas, a rule that is obeyed by such ions as Ti, Zr, Nb, and Ta. It also appears that the framework of BO_6 octahedra is essential to the ferroelectric activity of other non-perovskite oxides, such as $Cd_2Nb_2O_7$ and $PbNb_2O_6$. In this frame, a comparison of the $Cd_2Nb_2O_7$ structure with the perovskite shows that the main difference lies in the relative orientation of the oxygen octahedra. In $Cd_2Nb_2O_7$, the arrangement is such that the O-Nb-O chains lie on zig-zag lines approximately along the [110] direction. The Cd ions and one of seven oxygens occupies open spaces in the structural framework [2].

The geometrical relationships governing the structures of rhombohedral perovskites have been study in the past [3,4], and recently it was shown that the ratio of the volumes of the A- and B-cation polyhedra can be directly correlated with octahedra tilt angle [5]. Orthorhombic and tetragonal perovskites present a wider range of polyhedra distortions and tilting possibilities. A classification of octahedral tilting in terms of ten alternative tilts systems was proposed by Glazer [6, 7]. Recently, a new parametrization have been defined for a quantitative description of octahedral tilting [8].

New ferroelectric inorganic materials have been predicted [9, 10, 11, 12] on the basis of the relationship proposed by Abrahams, Kurts, and Jamieson [13], in which the T_c is a

Mat. Res. Soc. Symp. Proc. Vol. 453 © 1997 Materials Research Society

function of the displacement along the polar direction of the metal ion which forms the strongest bonds in the ferroelectric structure.

In all these approaches, attention is paid to the cation coordination and the properties of the material are related to these distortions.

Recently a model for the superstructure diffraction peaks measured in $PbMg_{1/3}Nb_{2/3}O_3$ (PMN) single crystal [14] has been proposed [15], based on the structural modulations in the oxygen sublattices as well as the relationships among space groups in distorted perovskites.
In the present study, a general approach based on the distortion in the AO_3 planes is proposed and analogies and differences of many crystal types are shown. Moreover, preliminary results indicating an interesting correlations between distortions in the AO_3 planes and the T_c of ferroelectric perovskites are reported.

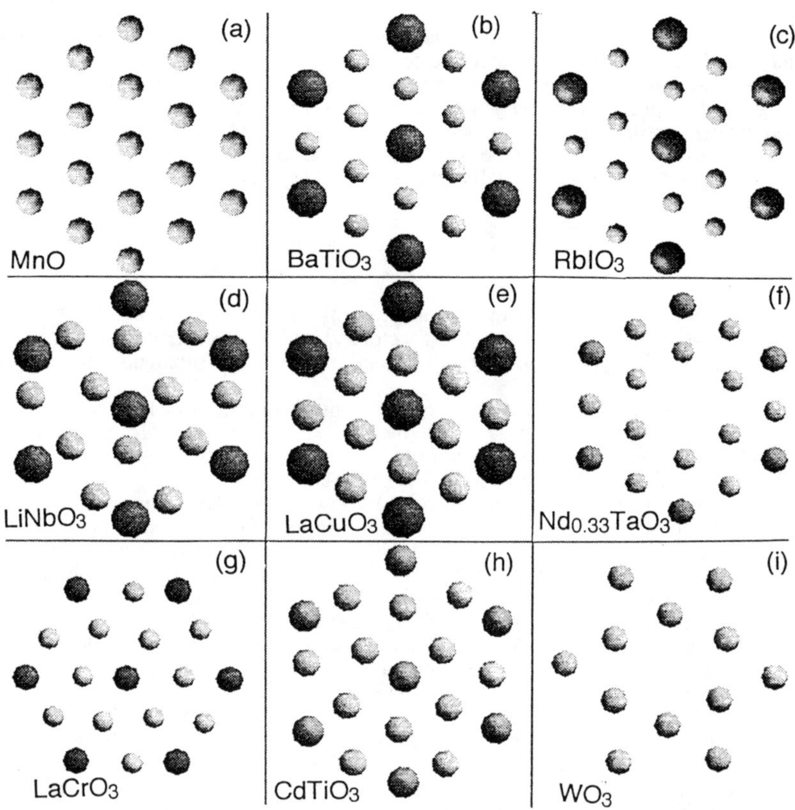

FIGURE 1: Examples of AO plane in some structure type. Oxygen anions: small balls; A cation: large balls.
a) MnO, space group Nr. 225 ICSD Nr.30520; b) $BaTiO_3$, space group Nr. 221, ICSD Nr.67518; c) $RbIO_3$, space group Nr.160, ICSD Nr.10283; d) $LiNbO_3$, space group Nr.161, ICSD Nr. 28296; e) $LaCuO_3$, space group Nr.167, ICSD Nr.73554 ; f) $Nd_{0.33}TaO_3$ space group Nr.99, ICSD Nr.62319 ; g) $LaCrO_3$ space group Nr.62, ICSD Nr. 100185; h) $CdTiO_3$ space group 33, ICSD Nr.62151;   i) WO_3, space group 14, ICSD Nr.16080.

RESULTS AND DISCUSSION

Many monoxides have the NaCl-type of structure (space group Fm3m, Nr.225), in which a perfect hexagonal layer of oxygens can be identified parallel to the [111] direction (Figure 1-a). A similar hexagonal plane is found in the ideal perovskite structure (space group Pm3m, Nr.221), in which A-O and O-O distance are exactly identical for symmetry. In Figure 1-b the structure of $BaTiO_3$ is shown as an example. For all the structures related to perovskite, similar pseudo-hexagonal planes can be identified with possible both in the plane and out of plane distortions. In Figure 1-c the AO_3 layer for $RbIO_3$ structure is shown (space group R3m, Nr.160). In this case, all the A-O distance are equal to 3.23 Å, while the O-O distances are 2.76 and 3.64 Å. This distortion seems to be determined by the large size of the A cation. On the contrary, in the rhombohedral distorted perovskites with $LiNbO_3$ type of structure (space group R3c, Nr.161) two different A-O distances are present, one short (1.92Å) and the second large (3.24Å) (Figure 1-d), indicating that the A cation is too small to coordinate all the six oxygens in the plane. The same kind of distortion is present in structures belonging to space group Nr. 167, as $LaCuO_3$ (Figure 1-e). The similarities in AO_3 planes of structure belonging to space group 161 and 167 can be originated by the possibilities of having in these space groups a tilting of the B coordination octahedra [5].

In the tetragonal and orthorhombic distortions fo the perovskite type of structure (Figure 1-e, -f, -h) a more complex situation is found. In Figure 1-i the extreme case of the V_AO_3 layer in monoclinic WO_3 structure is shown, where V_A indicates the vacancy in the A sites.

The large number of inorganic structures determined with high accuracy available nowadays on inorganic crystal structure database makes possible to establish theoretical generalizations based on statistical studies and to develop empirical models.

To verify the relationship among A radii, distortions, and space groups, ABO_3 compounds have been selected in the ICSD database [16] with R≤0.05 and no disorder in cation sites. Structures refined at non standard pressure or temperature have been discarded.

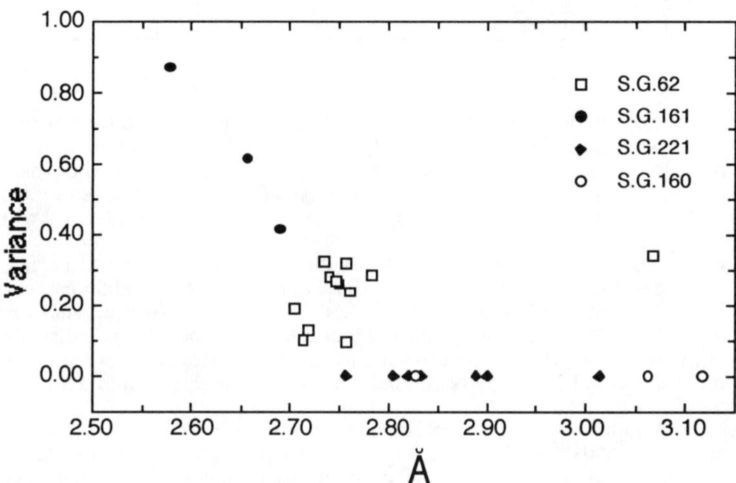

FIGURE 2: Average A-O distance <d_{A-O}> in the perovskite type of structure against the variance of d_A

In Figure 2 the average A-O distance in the AO_3 plane for structure belonging to the most common space groups is shown against the variance of the average distance. The correlation between the distortion in the oxygen plane and the average A-O distance is confirmed. Indeed, for very large A-O distance, structures in space group Nr.160 have been found, for A-O distance in the range of 2.8-3.0 Å space group 221 is preferred, for A-O distance shorter than 2.7 Å a rhombohedral space group Nr.161 is detected, while in the intermediate range of 2.7-2.8 tetragonally distorted structures are found. A correlation between the distortion and the average distance in the AO_3 plane is also evident: structures with short average A-O distances have high values of the variance.

A search has been done in ICSD for structures belonging to space group 221, with a calculated R factor, and with a A-O distance smaller than 2.80 Å. The results are reported in Table I. Only in the case of $SrTiO_3$, $(Sr_3Ti_2TeO_9)_{0.333}$, and PMN the compounds are given as stoichiometric, while in all the other cases the A-O planes contain vacancies in the anionic or cationic sites.

TABLE I

Compounds in ICSD belonging to 221 space group with A-O <2.80Å

Formula	ICSD Nr.	A-O distance (in Å)	R value
$PbMg_{1/3}Nb_{2/3}O_3$	72226	2.63	0.037
$SrVO2.93$	71449	2.71	0.017
$Sr(Co_{0.81}Fe_{0.19})O_{2.78}$	79022	2.72	0.084
$Sr(Fe_{0.5}Ti_{0.5}O_{2.877})$	72040	2.75	0.014
$SrTiO3$	201256	2.76	0.027
$Pb_{0.56}La_{0.30}TiO_3$	9442	2.77	0.032
$(Ba_{0.5}La_{0.5})FeO_{2.995}$	68873	2.78	0.029
$(Sr_3Fe_2TeO_9)_{0.333}$	9408	2.79	0.025

In the above discussion it has been shown that the A radius plays a fundamental role in determining the perovskite distortions. Indeed, in the case of metal monoxide belonging to space group Fm3m, the general A-O distance is in the range of 2.90-3.00 Å and when the A-O distance in the perovskite structures is in the same range of values, an ideal hexagonal layer is obtained. In the case of vacancies, ideal hexagonal AO_3 planes can be still obtained with shorter A-O distance.

In Figure 3, the x oxygen content of ABO_x (with x<3) of Table I is plotted against the A-O distance. Three of the four structures can be linearly fitted. The minimum calculated A-O distance for a ABO_3 compound (x=3) is 2.785 Å. In the majority of ABO_3 structures belonging to space group 221 the R factor is not given and in many compounds the A-O distances is less than 2.8 Å. In view of the presence of heavy atoms in all these structures, we suggest that the oxygen content could not be properly refined and many structures could actually be sub-stoichiometric.

When the A radius is short, its coordination in the A-O plane goes from six to three oxygens, causing a rhombohedral distortion in space group Nr.161. For intermediate A cation radius tetragonal distortions are present. When large A cation exists, another rhombohedral distortion in space group Nr.160 is determined, with six long identical A-O distances.

The variance of the average A-O distance in the AO_3 plane has been taken as an index of distortion. In Table II ferroelectric compounds are reported with the R factor and the critical temperature. In the case of WO_3 a baricenter of the oxygens is considered as the A site. In Figure 4 the variance in logarithmic scale against the critical temperature in logarithmic scale is plotted. Two behaviors can be recognized, as indicated by open and black circles. For low symmetry structures, belonging to space groups Nr.14, 26, 33, 38, the critical temperature increases when the variance of the A-O distances in the AO_3 plane decreases, while for high symmetry structures, belonging to space groups Nr. 99 and 161, the critical temperature increases with the distortion. A tentative explanation can be given by considering that when a field is applied, a ferroelectric behavior of the material requires iso-oriented dipole moments. An highly symmetric dipole moment distribution could make this process easier along some direction with respect to a low symmetry situation.

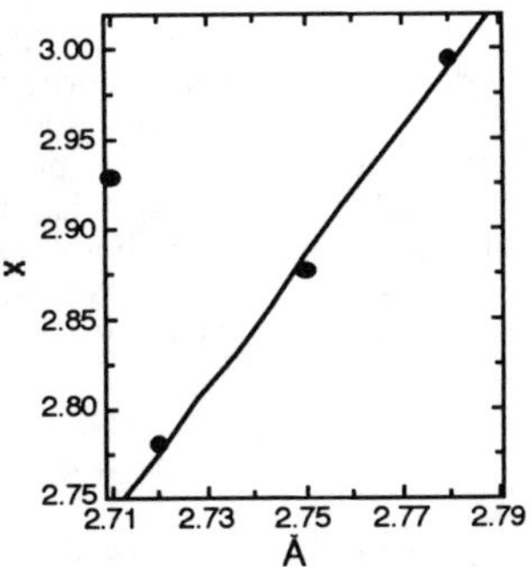

Figure 3 the x oxygen content of ABO_x (with x<3) is plotted against the A-O distance

TABLE II

ICSD number, space group, R factor, variance of the A-O distances in the AO_3 plane, and critical temperature for selected ferroelectric compounds.

Compounds	ICSD Nr.	S.G.	R value	variance	Tc
WO_3	16080	14	0.048	0.0302	223 [2]
$NaNbO_3$	39624	26	0.028	0.1020	123 [8]
$CdTiO_3$	62151	33	0.018	0.1933	55 [2]
$KNbO_3$	9533	38	0.021	0.0022	708 [2]
$Ba_{0.88}Ca_{0.12}TiO_3$	62667	99	0.012	0.0038	595 [11]
$Nd_{0.33}TaO_3$	62319	99	0.025	0.0042	550 [11]
$BaTiO_3$	67520	99	0.013	0.0015	393 [2]
$Pb_{0.64}La_{0.206}Ti_{0.949}O_{2.846}$	68975	99	0.041	0.0255	670 [11]
$Pb_{0.634}La_{0.209}Ti_{0.948}O_{2.844}$	68976	99	0.041	0.0306	700 [11]
$PbTiO_3$	61168	99	0.025	0.1012	763 [2]
$LiNbO_3$	28296	161	0.040	0.5231	1483[8]

CONCLUSIONS

A new approach to the structural modeling of ABO_3 perovskites based on the study of the AO_3 layer is proposed. This approach provides a framework for relating chemical composition with physical properties.

An important result is that perovskites belonging to space group Nr. 221 have A-O distance larger than 2.80 Å. It is proposed that, when average A-O distances shorter than 2.80

Å are found in the structures, vacancies may be present in the AO_3 layer. For example, in the case of disordered structures as PMN (ICSD Nr.72226, R=0.037), an average A-O distance in the pseudo-hexagonal plane shorter than 2.80 Å can be considered as an indication of vacancies in the oxygen or A cation sites. Indeed, a recent work shows that oxygen vacancies stabilize the PMN perovskite with respect to the pyrochlore structure [17].

It was already suggested that the most promising way of obtaining ferroelectric properties is through a perovskite of rhombohedral R3c symmetry [8]. In the present study it is shown that in this space group the largest distortion in the AO_3 hexagonal plane is found and this fact can be correlated with the critical temperature.

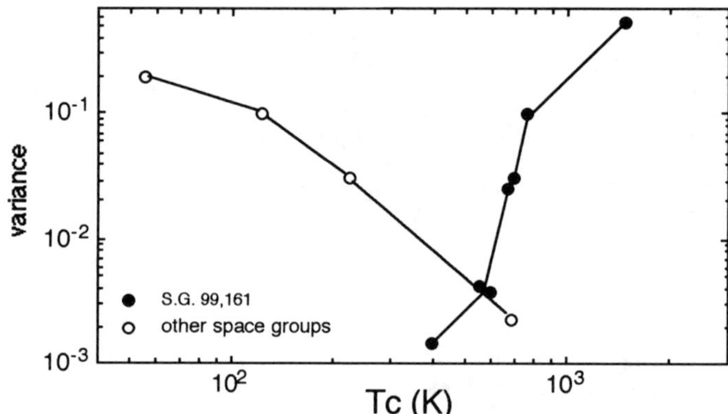

FIGURE 4: Variance of the A-O distance in the AO_3 planes against the critical temperature of selected ferroelectric compounds. Both variables are in a logarithmic scale. The open circles indicates the structure with low symmetry space group, while the black circles indicates high symmetry structures.

REFERENCES

1.   B.T.Matthias, Science, **113**, p.591 (1951)
2.   F.Jona, G.Shirane, <u>Ferroelectric Crystals</u>, Dover Publications, Inc., NewYork, 1993, pp.108-260
3.   C.Michel, J.M.Moreau , W.J.James, Acta Cryst. **B27**, p.501 (1971)
4.   H.D.Megaw, C.N.W.Darlington, Acta Cryst. **A31**, p.161(1995)
5.   N.W.Thomas, A.Beitollahi, Acta Cryst. **B50**, p.549 (1994)
6.   A.M.Glazer, Acta Cryst., **B28**, p.3384 (1972)
7.   A.M.Glazer, Acta Cryst., **A31**, p.756 (1975)
8.   N.W.Thomas, Acta Cryst. ,**B52**, p.16 (1996)
9.   S.C.Abrahams, Acta Cryst., **B44**, p.585 (1988)
10.   S.C.Abrahams, Acta Cryst.,**B45**, p.228 (1989)
11.   S.C.Abrahams, Acta Cryst., **B52**, p.790 (1996)
12.   S.C.Abrahams, K.Mirsky, R.M Nielson, Acta Cryst., **B52**, p.806 (1996)
13.   S.C.Abrahams,S.K.Kurts, P.B.Jamieson, Phys.Rev.,**172**, p.5651 (1968)
14.   Q.M.Zhang, H.You, M.L.Mulvihill, S.J.Jang, Solid State Comm., **97**, p.693 (1996)
15.   L.Depero, L.Sangaletti, submitted for pubblication
16.   Inorganic Crystal Structure Database (ICSD) by Gmelin Institut and FIZ Karlsruhe, Release 1996-1
17.   G. R.Fox, S.B.Krupanidhi, J. of Mat.Res., **9**, p.699 (1994).

CONTROLLED TEMPERATURE COEFFICIENT OF DIELECTRIC CONSTANT THROUGH PHASE TRANSFORMATION IN $(Ba, Sr)(Mg_{1/3}Ta_{2/3})O_3$

T. Nagai*, M. Sugiyama**, M. Sando[+], and K. Niihara[++]

*Synergy Ceramics Lab., Fine Ceramics Research Association

1-1 Hirate-cho, Kita-ku, Nagoya 462, Japan, tnagai@nirin.go.jp

**Advanced Materials & Technology Research Lab., Nippon Steel Corporation

20-1 Shintomi, Futtsu, Chiba 293, Japan

[+]National Industrial Research Institute of Nagoya, AIST / MITI

1-1 Hirate-cho, Kita-ku, Nagoya 462, Japan

[++]The Institute of Scientific and Industrial Research, Osaka University

Mihogaoka 8-1, Ibaraki, Osaka 567, Japan

ABSTRACT

A structural phase transformation from hexagonal to monoclinic has been found in an ordered perovskite compound $(Ba, Sr)(Mg_{1/3}Ta_{2/3})O_3$ (BSMT). The transformation temperature of BSMT increases and the dielectric constant also changes from 24 to 27 with an increase in Sr concentration. The temperature coefficient of the dielectric constant is also changed with the variation of Sr content. In order to apply BSMT for practical use of dielectric resonators these changes in the dielectric properties, which is dominated by the phase transformation, must be controlled. In order to clarify the relation between the dielectric constant change and the structural phase transformation, phonon mode of BSMT has been studied using Raman spectroscopy. Change in the Raman spectrum of BSMT is classified into four steps. The first and the fourth steps correspond to the phase transformation. The change in the phonon mode detected by Raman spectroscopy is discussed in connection with the structural change.

INTRODUCTION

Complex perovskite compound $Ba(Mg_{1/3}Ta_{2/3})O_3$ (BMT) is known to have the ordered hexagonal structure, that is $Ba(Sr_{1/3}Ta_{2/3})O_3$-type, and an excellent dielectric property with its high quality factor, $Q \geqq 35000$ at 10GHz. Because of the ultra low dielectric loss property, BMT is expected as a promising material for a dielectric resonator operated at microwave frequency region. In the real applications, however, the temperature coefficient of the resonant frequency (TC_f) must be controlled around 0 ppm/℃. TC_f of BMT is 4 ppm/℃, and it is related to the temperature coefficient of the dielectric constant (TC_ϵ). For the dielectric property control, $(Ba_{1-x}Sr_x)(Mg_{1/3}Ta_{2/3})O_3$ system (BSMT), in which Ba ion in the A-site of perovskite is substituted by Sr ion, has been studied. In the system, TC_ϵ is changed between 100 to -200 ppm / ℃ with Sr concentration [1].

Mat. Res. Soc. Symp. Proc. Vol. 453 © 1997 Materials Research Society

Recently, the new structural phase transformation has been found in BSMT. It is also found that the change in TC $_\varepsilon$ with Sr concentration is strongly correlated to the phase transformation [2, 3]. Furthermore, it has been reported that the phase transformation happens commonly in the (Ba, Sr) (B'B'')$_3$-typed complex perovskite compound [4 - 6].

The phase transformation originates from tilting of oxygen octahedra. Theoretical study revealed that BSMT changes its structure from Ba(Sr$_{1/3}$Ta$_{2/3}$)O$_3$-type hexagonal to monoclinic [7] due to the phase transformation. Similar structural phase transformation upon tilting of oxygen octahedra has been well studied in the perovskite compounds, such as SrTiO$_3$. The soft-mode of the phase transformation is at the Brillouin zone boundary of the mother phase.

To clarify the relationship between the phase transformation and the dielectric constant, the polar phonon mode of BSMT was studied using infrared-reflection spectroscopy [8]. Kramers-Kronig analysis revealed an appearance of a new polar mode in the monoclinic phase. Frequency shift of the polar mode is correlated to the positive TC $_\varepsilon$ of the transformed phase. But infrared-active modes of BSMT is rather broad, and it does not give enough information to discuss the phonon change due to the phase transformation.

In the present study, change in the phonon mode of BSMT has been studied using Raman spectroscopy. It enables us to observe directly the change in the phonon mode at Γ point. Raman spectroscopy turns out to be sensitive for phonon change.

EXPERIMENT

Ceramic pellets of (Ba$_{1-x}$Sr$_x$)(Mg$_{1/3}$Ta$_{2/3}$)O$_3$ were prepared by a conventional mixed oxide technique. Reagent-grade BaCO$_3$, SrCO$_3$, MgO and Ta$_2$O$_5$ were weighed and mixed in a nylon mill for 16 hours with ZrO$_2$ balls. Ethyl alcohol was used as a medium.   The dried powder was calcined at 1250 ℃ for 4 hours in air and then ground.   The powder was formed into pellets by a cold isostatic pressing and sintered at 1600 ℃ for 64 hours in air.   Seven kinds of compositions were prepared as samples. The compositions are denoted by Sr content, x=0 (BMT), 0.6, 0.65, 0.7, 0.8, 0.9 and 1.0 (SMT).   The density of samples was greater than 96% of the theoretical one for each composition.

The Raman spectra for well-polished samples were measured using a JASCO-NR1100.   The 514.5 nm line of an Ar-ion laser operating at 400 mW was used.   A measurement was done on 90° geometry, and an instrumental resolution was 1 cm^{-1}.   In order to measure a change in the Raman spectrum with temperature, a sample stage equipped with a heater unit was used.   The temperature of the sample was measured by a thermocouple attached to the sample.

RESULTS

A series of Raman spectra of (Ba$_{1-x}$Sr$_x$)(Mg$_{1/3}$Ta$_{2/3}$)O$_3$ compounds measured at room temperature

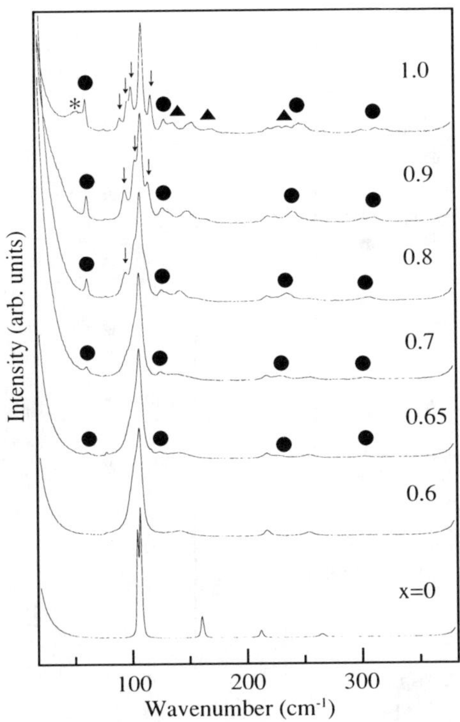

Fig. 1. Change in Raman spectrum of $(Ba_{1-x}Sr_x)(Mg_{1/3}Ta_{2/3})O_3$ with a variation of Sr content.

are shown in Fig. 1. Raman spectrum of BMT, x=0, exhibits five peaks in the frequency range measured. Two peaks at 105 and 107 cm^{-1} are considered as a main-peak. With the increase in Sr content in the range between x=0 and x=0.6, only a frequency shift of the Raman peaks is observed. In $(Ba_{0.4}Sr_{0.6})(Mg_{1/3}Ta_{2/3})O_3$, the main-peak degenerates into a single broad peak at around 107 cm^{-1}. In the Raman spectra of $(Ba_{0.35}Sr_{0.65})(Mg_{1/3}Ta_{2/3})O_3$ and $(Ba_{0.3}Sr_{0.7})(Mg_{1/3}Ta_{2/3})O_3$, on the other hand, four new peaks are observed at room temperature. They are marked by solid circles. With increasing Sr content to x=0.8, 0.9 and 1.0, the main-peak splits into doublet, quartet and quintet, respectively. Additional peaks in the main-peak are shown by arrows. In the range of Sr content between x=0.9 and 1.0, not only the main-peak but also other Raman peaks exhibit a splitting. These additional peaks are denoted by solid triangles and an asterisk.

Raman spectra of $(Ba_{0.35}Sr_{0.65})(Mg_{1/3}Ta_{2/3})O_3$ measured at a temperature range between 25 and 100 °C are shown in Fig. 2. The intensity of the four additional peaks gradually decreases with increasing temperature. They disappear at a certain temperature between 80 and 100 °C. The four peaks do not exhibit distinct shift in the temperature change.

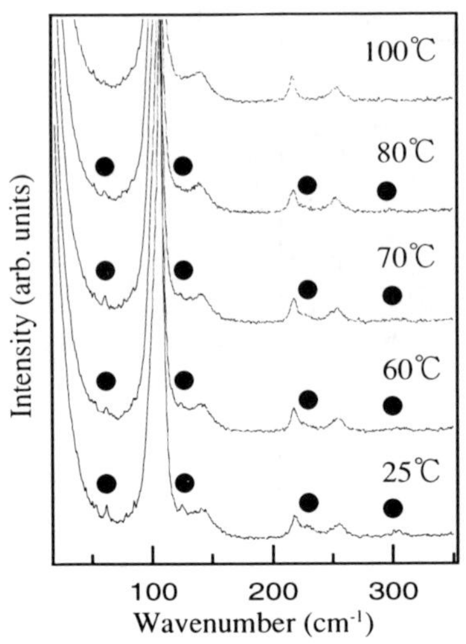

Fig. 2. Change in the Raman spectrum of $(Ba_{0.35}Sr_{0.65})(Mg_{1/3}Ta_{2/3})O_3$ with temperature.

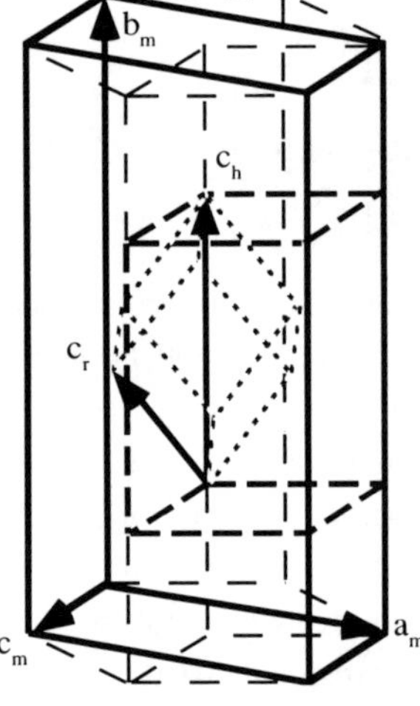

Fig. 3. Change in the Raman spectrum of $Sr(Mg_{1/3}Ta_{2/3})O_3$ with an increase in temperature.

Fig. 4. Unit cell for complex perovskite compound with 1:2 ordering of B-site ions and one antiphase titlting axis of oygen octahedra.

Raman spectra of $Sr(Mg_{1/3}Ta_{2/3})O_3$ measured between 25 and 700℃ is shown in Fig. 3.   With increasing temperature between 25 and 100℃, the two lowest-frequency peaks of the main-peak approach each other gradually, and they merge into a single peak. The Raman peaks shown by solid triangles decrease in intensity and disappear at around 90 ℃. It should be noted that the broad peak denoted by an asterisk exhibits a characteristic frequency shift different from other peaks; the

Raman peak decreases in frequency with an increase in temperature, and it finally merges into the Rayleigh wing at around 100℃. Between 100 and 500℃, the number of the peaks in the main-peak is reduced further; from four to two, and one. At 600 ℃, four peaks marked by solid circles still remain. They disappear between 600 and 700℃.

DISCUSSIONS

The change in the Raman spectrum of BSMT with temperature or Sr concentration is classified into four steps. With cooling, four additional peaks appear first; this is the first step. Other steps; second, third and fourth step are characterized by the split of the main-peak into 2, 4 and 5 peaks, respectively.

Judging from the temperature change, the first step of the Raman change is correlated with the anomaly of the dielectric constant [3], and the structure changes from hexagonal (space group D_{3d}^{3}) to monoclinic (space group C_{2h}^{6}). The structural phase transformation is due to an antiphase tilting of oxygen octahedra. Due to the shift of oxygen ions, the unit cell is doubled along the c-axis of the hexagonal one, as shown in Fig. 4 [7]. The doubling of the unit cell results in the folding of the Brillouin zone. Due to the folding, a phonon mode standing at the Brillouin zone boundary in the hexagonal phase may appear at the Γ-point in the monoclinic phase. Since the four additional peaks do not exhibit distinct frequency shift with temperature, as shown in Fig. 2, it is considered that; the additional Raman peaks are not the soft mode, and they arise from the zone boundary modes of the hexagonal phase due to the unit cell doubling.

Normal modes at the Γ-point of the monoclinic phase are:

$$\Gamma = 40A_g + 46A_u + 44B_g + 50B_u ,$$

where A_g and B_g are Raman active modes and A_u and B_u are infrared active modes. It should be noted that the oxygen ions on $(00 \cdot 1)_h$ plane do not have Raman active modes in the $Ba(Sr_{1/3}Ta_{2/3})O_3$-type structure, while they have Raman active modes, $8A_g + 10B_g$, in the monoclinic structure. Thus the four additional Raman peaks denoted by solid circles are correlated with the vibration of the oxygen ions.

At the fourth step of the Raman spectrum change, the Raman peak marked by an asterisk in Fig.1 exhibits softening with increasing temperature. The shift of the peak agrees with one of Cochran's soft-mode. Thus it is estimated that the fourth step of the Raman change is caused by the phase transformation due to another tilting of oxygen octahedra. This structural change has not been detected by electron and X-ray diffraction analyses.

On the other hand, there is not hard proof that second and third step of the Raman change correspond to other phase boundaries, and it may be apparent because of the change in inter ionic force upon an increase in antiphase tilting angle of the oxygen octahedra.

In the previous study, change in the polar modes in the transformed phase has been reported and discussed in connection with the dielectric constant [8]. In the present study, Raman-active mode is related to the structural change; unit cell doubling and so on. Thus, it is suggested that the change in the polar mode and the dielectric constant due to the phase transformation should be the result of the unit cell change. Further analysis of the phonon mode and the structure is in progress.

CONCLUSIONS

A change in the phonon mode of $(Ba_{1-x}Sr_x)(Mg_{1/3}Ta_{2/3})O_3$ compound due to the structural phase transformation is studied using Raman spectroscopy. The results are summarized as follows.
(1) The observed change in Raman modes of $(Ba_{1-x}Sr_x)(Mg_{1/3}Ta_{2/3})O_3$ with a variation of temperature or Sr concentration is categorized into four steps.
(2) The first step takes place at the highest temperature of the four steps. It corresponds to the structural phase transformation from hexagonal to monoclinic. The change in the Raman modes is attributed to the Brillouin zone folding due to the structural phase transformation.
(3) The fourth step of the Raman spectrum change, a peak shows a behavior of softening with increasing temperature. The existence of the soft mode suggests the possibility of further phase transformation in $(Ba_{1-x}Sr_x)(Mg_{1/3}Ta_{2/3})O_3$ compound.

ACKNOWLEDGMENTS

Work promoted by AIST, MITI, Japan as part of the Synergy Ceramics Project under the Industrial Science and Technology Frontier (ISTF) Program. Under this program, part of the work has been supported by NEDO. T. Nagai, M. Sando and K. Niihara are members of the Joint Research Consortium of Synergy Ceramics.

REFERENCES

1. T. Nagai, T. Inuzuka and M. Sugiyama, Jpn. J. Appl. Phys. 31, p. 3132 (1992).
2. M. Sugiyama and T. Nagai, Jpn. J. Appl. Phys. 32, p. 4360 (1993).
3. T. Nagai and M. Sugiyama in Advanced Materials '93 I /B, edited by M. Homma et al. (Trans. Mat. Res. Soc. Jpn. 14B, Tokyo, Japan 1993), p. 1735.
4. E. L. Colla, I. M. Reaney and N. Setter, Ferroelectrics 133, p. 217 (1992).
5. E. L. Colla, I. M. Reaney and N. Setter: J. Appl. Phys. 74, p. 3414 (1993).
6. O. Steiner, E. L. Colla, I. M. Reaney and N. Setter in Third EURO-CERAMICS, edited by P. Duran and J. F. Fernandez, (Madrid, Spain 1993) v. 2, p. 223.
7. T. Nagai, M. Sugiyama, M. Sando and K. Niihara, Jpn. J. Appl. Phys. [in contribution].
8. T. Nagai, M. Sugiyama, M. Sando and K. Niihara, Jpn. J. Appl. Phys. 35, p. 5163 (1996).

INFLUENCE OF Li ADDITION ON THE SINTERING AND DIELECTRIC PROPERTIES OF PNN-PMW-PT RELAXOR FERROELECTRICS

RAGHU NATARAJAN*, JOSEPH P. DOUGHERTY
Center for Dielectric Studies, Materials Research Laboratory, The Pennsylvania State University, University Park, PA 16802.
* Permanent address: C-MET, Dept. of Electronics, Thrissur, INDIA- 680 771.

ABSTRACT

Compositions in the PNN-PMW-PT system can be tailored to give high dielectric constant for MLC applications. Low temperature firing of MLC is always advantageous in using the less expensive electrode systems. Controlled addition of $LiNO_3$ reduces the sintering temperature, increases the fired density and influences the microstructure of the ceramics. It also raises the peak dielectric constant and shifts the Curie temperature. The possible role of Li^+ in modifying the characteristics of the ceramics is discussed.

INTRODUCTION

High K dielectric ceramic materials find a variety of applications such as decoupling capacitors and energy storage capacitors in power planes [1]. There has always been an increased interest in low-temperature sintering of dielectric materials without compromising the performance [2-4]. The driving force behind achieving high dielectric constant with low firing ceramics are two major electronic industrial applications, namely multilayer capacitors and integrated ceramics [1,5-9]. The major benefit of using a low-firing capacitor in MLC applications is the reduction in cost of utilizing the much cheaper Ag-Pd electrodes [1]. The future trend in the integrated ceramics is to achieve a Low Temperature Co-fired Ceramic (LTCC) in which active and passive components are co-fired to obtain 3–dimensional integration. This application also demands a low-firing high K passive component that matches the sintering characteristics of the LTCC dielectric substrate [5-12].

Most common capacitor materials are based on barium titanate (BT) which sinters at a high temperature (T > 1300°C). Attempts to decrease the sintering temperature of BT with glass or flux addition usually leads to reduction in K-value [13]. The distinctive advantages of relaxor ferroelectrics [1] over barium titanate based are relatively low-sintering temperature (900–1200°C), high K values, and better matching of thermal properties with active electronic materials. Traditionally, the sintering temperatures of most ferroelectric ceramic materials have been lowered by either adding dopants or sintering aids to create liquid phase sintering [14]. Considerable literature exists in the low temperature sintering of different Pb-based relaxors [15-18]. Li-salt has been found to be highly effective in decreasing the sintering temperature of both barium titanate and relaxor materials [19-26]. Li addition to Pb-based relaxors not only helps sintering in an air atmosphere, it also improves the dielectric properties in certain systems [25-26]. Many studies exist on the effect of Li-salts on the sintering of perovskite materials, yet the underlying role played by Li and its effect on the perovskite chemical inhomogeneity is not well understood. Whether or not Li enters into the structure is still unresolved and numerous contradictions exist in the literature.

$Pb(Ni_{1/3}Nb_{2/3})O_3$–$Pb(Mg_{1/2}W_{1/2})O_3$–$PbTiO_3$ (PNN-PMW-PT) has been studied extensively for MLC applications [27-28]. The antiferroelectric component of PMW moderates or weakens the ferroelectricity of PNN and PT. Its distinctive features are high dielectric constant, high specific resistivity and its sintering temperature and Curie temperature can be tailored by varying the composition [4,28]. For the present study $(PNN)_{0.3}$–$(PMW)_{0.29}$–$PT_{0.41}$ has been chosen because its T_c is near room temperature. This paper describes the influence of Li on the sintering and dielectric properties of this relaxor material.

Mat. Res. Soc. Symp. Proc. Vol. 453 © 1997 Materials Research Society

EXPERIMENT

Reagent grade PbO, NiO, Nb_2O_5, MgO, WO_3, and TiO_2 were vibratory milled in ethyl alcohol for 24 hours using zirconia media. The resulting slurry was dried at 100°C for 24 hours. The dry powders were calcined at 800°C for 6 hours. Different mol% $LiNO_3$ solutions were added to the calcined powder which are pressed into pellets and then sintered at 850–1000°C for 1 hour. A heating rate of 10°C/min. and different cooling rates were employed. Densities of the samples were measured by the Archimedes' method and electrical properties were measured after sputtering Au with a Ag overcoat. Dielectric measurements were made during cooling cycle from 125°C to -55°C at a rate of 2°C per minute. Capacitance and loss were measured at different frequencies using HP4274A and 4275A LCR meters. Phase analysis of the calcined and sintered materials were performed using Cu-K_α radiation with a Scintag X-ray diffractometer. Microstructural analysis of the fractured surface of ceramics was performed using an ISI DS 130 SEM.

RESULTS

X-ray diffraction analysis of the PNN-PMW-PT powder calcined at 800°C for 6 hours revealed the presence of $\leq 4\%$ pyrochlore. Without the addition of $LiNO_3$, samples sintered at 900°C for 1 hour showed 100% perovskite structure. Careful examination of a slow scan (0.5°/min) of the (200) peak for the sample sintered at 900°C for 1 hour revealed it to be a double peak (Fig. 1.a). But when the same sample is sintered at 1000°C for 1 hour, it becomes a single peak (Fig. 1.b). With $LiNO_3$ addition in the range of 1-4 mole%, all the samples exhibit 100% perovskite structure when sintered at 850°C for 1 hour. Unlike the sample without $LiNO_3$ addition, the (200) peak remains as a single peak, even with 1% $LiNO_3$ after sintering at 850°C for 1 hour (Fig. 1.c and d).

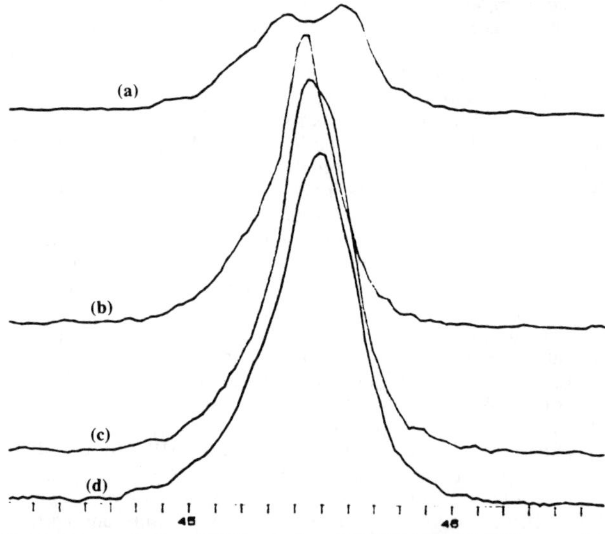

Fig 1. XRD pattern of the (200) peak of PNN-PMW-PT sample sintered at (a) 900°C-1h. and (b) 1000°C-1h. (c) and (d) are for 1 and 2 mole% $LiNO_3$ added samples sintered at 850°C–1h respectively.

Table 1 presents the density of samples sintered with different mole% of $LiNO_3$ at different sintering conditions. Without $LiNO_3$ addition, samples sinter only at 1000°C for 1 hour. But in the presence of even 1% $LiNO_3$, the sintering temperature could be brought down to 850°C. Densities $\geq 95\%$ could be achieved with 2 mole% $LiNO_3$ when sintered at 900°C for 1 hour.

Table I.

The density and dielectric properties of PNN-PMW-PT ceramics with different mol% $LiNO_3$ addition.

% of $LiNO_3$	Sintering Condition	Relative Density	T_{max} °C	K_{max}
0%	850–1 h.	86.2%	—	—
	900–1 h.	88.3%	—	—
	1000–1 h.	94.5%	20	5,000
1%	850–1 h.	92.8%	31	7,300
	900–1 h.	93.85%	32	8,120
2%	850–1 h.	94.1%	44	10,050
	900–1 h.	95.5%	44	12,100
4%	850–1 h.	94.4%	56	9,200
	900–1 h	95.93%	58	9,290

Fig. 2 reveals the microstructure of the fractured samples with different mol% Li after firing at 900°C for 1 hour. All the specimens exhibit predominantly intergranular fracture. The sample sintered without $LiNO_3$ is porous (Fig. 2.a), in agreement with the density data presented in Table 1, and the grain size is 0.6 μm, which is nearly the same as the average particle size of the starting material. Without the addition of $LiNO_3$ the material must be heated at 1000°C for 1 hour to sinter. Microstructural evaluation of this ceramic revealed a grain size of 3 μm. With the addition of $LiNO_3$, all the ceramics attained well faceted uniform grains, as revealed in Fig. 2.b and c. With 1% $LiNO_3$, the average grain size is ~ 2.2 μm and with 2 and 4 mol%, ~ 3.5 μm. The microstructure and properties of Li added ceramics remained unchanged with sintering temperature variation between 850 and 950°C. Similarly different cooling rate from quenching to 1.5°/min also had little effect on the dielectric properties.

The dielectric constant and loss of these ceramics measured at 1 KHz are shown in Fig. 3A&B. For the ceramic sintered at 900°C for 1 hour, without $LiNO_3$, the sample displays two distinct peaks. When the same sample is sintered at 1000°C it shows a single K_{max} at ~ 20°C. With the addition of $LiNO_3$ there is a definite shift in T_c and increase in K value as revealed in Table I. With just 2 mol% of $LiNO_3$, K_{max} optimizes to ~ 12,000. The dielectric constant as a function of temperature at different frequencies for the 2 mole% $LiNO_3$ added sample is shown in Fig. 4. It shows the typical characteristics of a relaxor ferroelectric with diffused phase transition, and the diffuseness decreases with increasing lithium.

Fig 2. SEM micrographs of PNN-PMW-PT ceramics sintered at 900°C with different mole% of LiNO$_3$. (a) 0%, (b) 1%, (c) 2%, and (d) 4%.

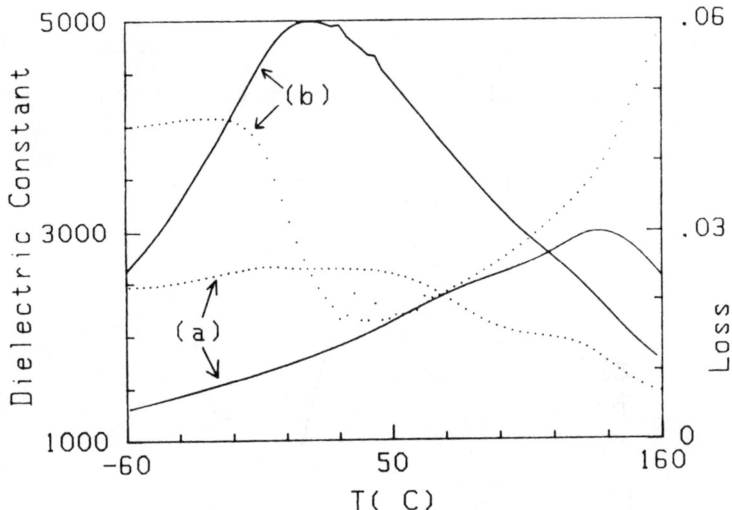

Fig. 3A. Dielectric constant and loss of PNN-PMW-PT ceramics without Li addition sintered at (a) 900°C-1h & (b) 1000°C-1h.

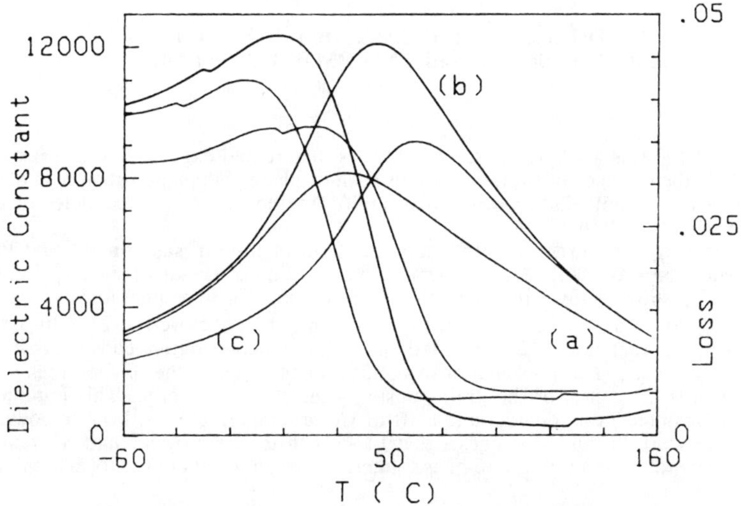

Fig. 3B. Dielectric constant and loss of PNN-PMW-PT with different mole % of Li addition at 1 KHz. (a) 1%, (b) 2% and (c) 4%.

DISCUSSION

From the observation of a double peak for the (200) line in the XRD and two peaks in K vs. T curve for the sample without $LiNO_3$ sintered at 900°C–1 h., it is obvious that at 900°C, the sample exists as a mixture of two perovskite phases. Only when fired at 1000°C for 1 hour

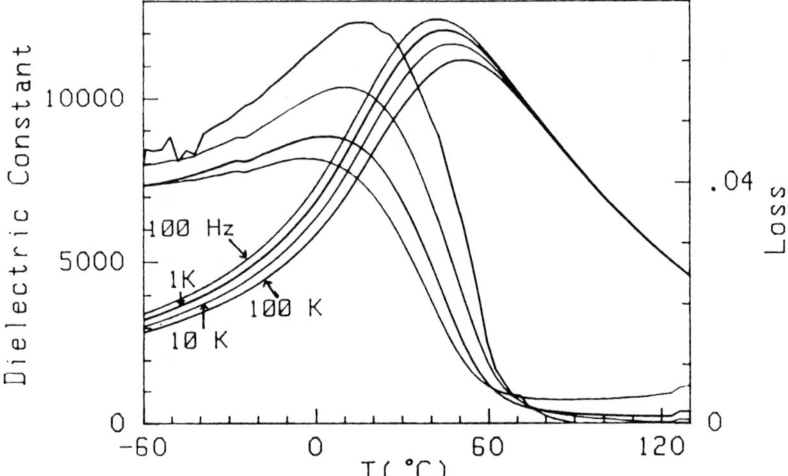

Fig. 4 Diffused phase transition behaviour of 2 mole% LiNO3 added PNN-PMW-PT ceramic.

PNN-PMW-PT forms a single phase perovskite structure and exhibits a single peak in K-T curve. But in the presence of Lithium it forms a single phase perovskite structure at 850°C in 1 hour. Moreover lithium addition brings in a significant improvement of the dielectric constant and a considerable shift in T_c.

Considerable literature [19-26] on the effect of lithium salts on BT and Pb-based relaxor materials is available. It is observed that addition of Li-salts (Li_2CO_3, $LiNO_3$, and LiCl) to BT results in lowering both the sintering and Curie temperature. The authors suggested based on the Curie shift that Li^+ enters into the structure [20-21]. In the case of $SrTiO_3$ (ST), Laurent et al. [22] suggested that Lithium incorporation leads to A-site, B-site, and anion vacancies. The presence of oxygen-vacancies caused the oxygen ions to diffuse rapidly towards the center of the grain causing grain inhomogeneity. This inhomogenous distribution resulted in a considerable shift of Curie temperature. All the hypotheses and experimental work discussed suggests that Li salts in the case of BT and ST reduces the sintering temperature through liquid phase sintering, which results in the formation of a solid solution.

In the case of relaxor materials, Fu and Chen observed lowering of the sintering and Curie temperatures in the PFW-PFN-PT systems with Bi_2O_3/Li_2O addition [24]. The authors suggest the possibility of Bi^{3+} and Li^+ incorporation into either the A or B-site. In the case of PFN-PFT system, it is observed that both the K-value and sinterability increase with lithium addition [25]. The Curie temperature decreased with Li content. It is suggested that the decrease in T_c is because of the partial substitution of Li ion for Nb and Ta. In PMN-PZN and PMN-PT systems also $LiNO_3$ is found to decrease the sintering temperature with modified uniform large grain microstructure and high dielectric constant [26]. Based on the very minor shift in T_c, diffuseness coefficient and lack of decrease in resistivity with Li content, it is suggested that Li remains as an impurity in the grain boundary as isolated features or at triple points.

The role of $LiNO_3$ in lowering the sintering temperature of perovskites may be understood from the fact that $LiNO_3$ decomposes, as given by [26],

$$LiNO_3 \xrightarrow{350°C} LiNO_2 + 1/2\ O_2 \qquad\qquad (1)$$

$$LiNO_2 + LiNO_3 \xrightarrow{635°C} Li_2O + 2\ NO_2 \qquad\qquad (2)$$

The second reaction is vigorous. This reaction results in the formation of a melt (liquid phase) at ~ 635°C and results in the enhanced sinterability of the ceramic. Interestingly, both in BT and Pb-based relaxor systems so far only a decrease in Curie temperature with Li content has been observed. In the present PNN-PMW-PT system we observe an increase in Curie temperature with Li content.

From a simple consideration of the ionic size of Li^+ (0.76 Å) in a six coordinated site one can expect it to substitute for Mg^{2+} (0.72 Å) and for Ni^{2+} (0.69 Å). Moreover, if Li^+ substitutes for a $2+$ site, the resistivity is expected to decrease for a p-type material [29]. From Table II one can observe that there is a slight but definite decrease in resistivity with Li^+ content. Moreover, from the analysis of the slow scanned (200) X-ray peak, we observe a shift in the lattice parameter with Li addition (Table II). From the above observations, it is clear that in the case of PNN-PMW-PT system Li addition reduces the sintering temperature through liquid phase sintering and results in the formation of a solid solution.

Table II.

Resistivity and lattice parameter of $LiNO_3$ added PNN-PMW-PT Ceramics.

% of $LiNO_3$	Resistivity x 10^{12} (ohm-cm)	Lattice Parameter (Å)
0	4.0	3.9898
1	3.0	3.9870
2	1.31	3.9858
4	0.52	3.9844

CONCLUSIONS

$LiNO_3$ addition to PNN-PMW-PT system results in:

(i)   reduced sintering temperature because of liquid phase sintering.

(ii) formation of single-phase perovskite at lower temperatures.
(iii) modified microstructure leading to uniform, well-faceted grains of increased grain size.
(iv) considerable increase in K-value and shift in T_c to higher temperatures.
(v) substitution of Li into the lattice structure.

ACKNOWLEDGEMENT

One of the authors, Raghu Natarajan, wishes to thank the Ministry of Science and Technology, Government of India for financial support.

REFERENCES

1. T.R. Shrout and J.P. Dougherty, Ceramic Transactions, Edited by H.C. Long and M.F. Yan, **8**, pp. 3-19 (1990).

2. L. Li, in First Pacific Rim International Conference on Advanced Materials and Processing (PRICH), p. 899 (1993).

3. G. Zhilun, Y. Liu and H. Sun, Proceedings, Second International Conference on Properties and Applications of Dielectric Materials, IEEE, **1**, p. 125 (1988).

4. M. Yonezawa, Ceramic Bulletin, **62**, p. 1375 (1983).

5. S. J. Stein, R.L. Wahlers, P.W. Bless, D. H. Dychala and C. Y. D. Huang, Proceedings of the 8th International Microelectronic Conference, p. 291 (1994).

6. Y. Fujioka, T. Fujisaki, I. Sakaguchi, Y. Yamakuchi, S. Shimo, K.Onitsuka and N. Fujikawa
Proceedings of the seventh International Microelectronic Conference, p. 355 (1992).

7. L. Drozdyk, Proceedings of the International Symposium on Hybrid Microelectronics, p. 209 (1993).

8. M. Megherhi, J.P. Dougherty, G.O. Dayton, and R.E. Newnham, 7th International Symposium on Application of Ferroelectricity, p. 31 (1990).

9. J.H. Adair, D.A. Anderson, G.O. Dayton, and T.R. Shrout, Journal of Materials Education, **9**, p.71 (1987).

10. Y. Shimida, K. Utsumi, T. Ikeda, and S. Nagarako, Proceedings of International Microelectronics Conference, Tokyo, Japan, p. 227 (1984).

11. K.A. Vorotilov, Microelectronic Engineering, **29**, p.41 (1995).

12. K. Utsumi, Ceramic Bulletin, **70**, p. 1050 (1991).

13. J.M. Herbert, <u>Ceramic dielectrics and capacitors</u>, Electrocomponent Science Monographs, **6**, Gordon and Breach, New York, (1985).

14. K.R. Choudhary and E.C. Subbarao, Ferroelectrics, **37**, p. 689 (1981).

15. S.J. Jang, W.A. Schulze, and J.V. Biggers, Am. Ceram. Soc. Bull., **62**, p. 216 (1983).

16. M.H. Megherhi, "The Effect of Additives, Microstructure, and Macrostructure on the Fracture Behaviour of Lead Magnesium Niobate-based Ceramic Actuators," M.S. Thesis, The Pennsylvania State University (1988).

17. J.P. Guha, D.J. Houg and H.V. Anderson, J. Am. Cer. Soc., **71**, C152 (1988).

18. V. Srikanth and E.C. Subbarao, Journal of Mat. Research, **6**, p. 1308 (1991).

19. R.B. Amin, H.V. Anderson and C.E. Hodgekins, U.S. Patent # 4,082,906 (1978).

20. J.M. Haussonne, G. Desgardin, P. Bajolet, and B. Raveau, J. of Am. Cer. Soc., **66,** p. 801 (1983).

21. J.M. Haussonne, O. Regreng, J. Lostec, G. Desgardin, M. Halmi, and B. Raveau, Mat. Sci. Monograph, **38B**, p. 1515 (1987).

22. M.J. Laurent, G. Desgardin, B. Raveau, J.M. Haussonne, and J. Lostec, J. Mater. Sci., 23, 4481 (1988).

23. G. Desgardin, H. Bali, and B. Raveau, Mater. Chem. & Phys., **8**, p. 469 (1983).

24. S.L. Fu, and G.F. Chen, The International Journal for Hybrid Microelectronics, **10**, p. 1 (1987).

25. M. Halmi, G. Desgardin and B. Raveau, Advanced Ceramic Materials, **3**, p. 32 (1988).

26. M. Megherhi, "Interaction Studies of Lead Magnesium Niobate-based Capacitor Materials with Integrated Ceramic Packaging," Ph.D. Thesis, The Pennsylvania State University (1991).

27. A. Ochi, T. Mori, M. Furuya, Proceedings of the Eighth International Symposium on Applications of Ferroelectrics, IEEE, p. 66 (1992).

28. M. Yonezawa, K. Utsumi, A. Ochi, and T. Mori, Proceedings of the 7th International Symposium on Application of Ferroelectrics, IEEE, p. 159 (1990).

29. B. Jaffe, W.R. Cook, Jr., and H. Jaffe, "Piezoelectric Ceramics," Academic Press, New York, p. 59 (1971).

CHARACTERIZATION OF CATION-ORDERING INDUCED STRUCTURES IN Ba(Mg$_{1/3}$Nb$_{2/3}$)O$_3$-BaZrO$_3$ MICROWAVE CERAMICS.

M.A. AKBAS and P.K. DAVIES
Dept. of Materials Science & Engineering, University of Pennsylvania, 3231 Walnut St., Philadelphia, PA 19104-6272

ABSTRACT

Compositions in the Ba(Mg$_{1/3}$Nb$_{2/3}$)O$_3$-BaZrO$_3$ (BMN-BZ) pseudo-binary system were prepared to evaluate the effect of BaZrO$_3$ on the B-site cation ordering. While pure BMN adopts a 1:2 ordered structure, space group $p\bar{3}m1$, 5 to 15 mole% additions of BZ stabilize a cubic (Fm$\bar{3}$m) 1:1 ordered phase with a doubled perovskite repeat. At higher levels of substitution (>25 mole% BZ), the B-site cations are disordered. Analytical TEM studies revealed that the 1:1 ordering is limited to nanometer-sized domains and the microstructure was found to strongly depend on the thermal history of the ceramics. For example, a cooling rate of 300°C/hr from a sintering temperature of 1640°C produced a partially ordered microstructure comprised of 4-5nm 1:1 domains dispersed in a disordered matrix. A completely ordered structure with 30-40 nm domains could be achieved when the cooling rate was reduced to 10°C/hr. The structure of the ordered 1:1 phases is interpreted using a "random layer" model in which one site is occupied exclusively by Nb, and the second contains a random distribution of all the remaining cations. Similar thermal annealing experiments were also conducted in an attempt to grow the 1:1 ordered domains in PMN-type (PbMg$_{1/3}$Nb$_{2/3}$O$_3$) relaxor ferroelectric ceramics. Through experiments on PbZrO$_3$ (PZ)-doped PMT (PbMg$_{1/3}$Ta$_{2/3}$O$_3$), the first observation of thermally induced domain growth in this class of relaxor ceramics are presented. These results cannot be explained by the widely accepted "space charge" model for relaxor ferroelectrics, but instead favor a charge-balanced random layer model for the 1:1 cation order.

1. INTRODUCTION

Low-loss dielectric ceramics such as Ba(Zn$_{1/3}$Ta$_{2/3}$)O$_3$ (BZT) and Ba(Mg$_{1/3}$Ta$_{2/3}$)O$_3$ (BMT) are widely used in commercial wireless communication systems. The B-site cations in these "1:2" perovskites are stoichiometrically ordered in a hexagonal unit cell, space group $p\bar{3}m1$. Previous research has shown that microwave loss properties of BZT and BMT are intimately related to the degree of cation order and the losses can be reduced by maximizing the cation order through extended annealing at high temperature [1]. Tamura et al. [2] reported that the dielectric loss of BZT can also be enhanced by doping with 4 mole% BaZrO$_3$ (BZ). Surprisingly, the B-site cations in the BZT-BZ ceramics were reported to be disordered which seemed to contradict the observations of the beneficial effect of cation order on dielectric loss in the pure, undoped systems. These apparent contradictions have been resolved through recent investigations conducted in our laboratory, which have shown that BZ-doped BZT is locally ordered, and the low losses arise from the stabilization of 1:2 ordering induced domain boundaries by the partial segregation of Zr [3]. Those studies also revealed that when the level of substitution of BZ in BZT is increased to 3-5 mole%, the symmetry of the cation ordering is

Mat. Res. Soc. Symp. Proc. Vol. 453 © 1997 Materials Research Society

changed and the system adopts a cubic "1:1" ordered cell with a doubled perovskite repeat [3]. In this paper we present data for the corresponding niobate system, $Ba(Mg_{1/3}Nb_{2/3})O_3$-$BaZrO_3$ (BMN-BZ), focusing in particular on the microstructural features of the 1:1 ordered phases.

Pb-based 1:2 perovskites such as $Pb(Mg_{1/3}Nb_{2/3})O_3$ (PMN) have also attracted widespread interest as a result of their relaxor ferroelectric properties. Although the chemical ratio of the B-site cations in PMN and other related relaxors is 1:2, several investigators have shown they adopt a structure comprised of nano-sized 1:1 ordered domains surrounded by a disordered matrix [4, 5]. The ordering in these systems has been interpreted in terms of a "space charge" model in which the chemically ordered nano-domains are claimed to contain a 1:1 ratio of the divalent and pentavalent cations, with the resultant charge imbalance being compensated by an equal and oppositely charged, pentavalent-rich disordered matrix [4, 5]. Support for this "two-phase" model has come primarily from the apparent lack of coarsening of the ordered domains upon annealing [5]. However, as yet there has been no unambiguous experimental evidence to support or contradict the space charge model. In this paper, we present the first direct evidence that the space charge model for the 1:2 relaxors is incorrect.

2. EXPERIMENTAL METHODS

Ceramics in the BMN-BZ and PMT-PZ systems were prepared by conventional mixed oxide methods using high purity (>99.9%) oxides and/or carbonates. Details of the ceramic processing can be found in our previous publications [6, 7]. Phase analysis was conducted using X-ray diffraction (XRD) and the microstructures of the ceramics were characterized using TEM. Thin foils for the TEM investigations were prepared by mechanically grinding the ceramic pellets to a thickness of 20-30μ and polishing both sides with 5, 1, 0.25μ alumina powder. Copper hole grids were then mounted and final thinning was done by argon ion milling followed by a carbon coating to prevent charging.

3. RESULTS AND DISCUSSION.

Stability of Ordered Phases in BMN-BZ: Electron Diffraction Studies

Figure 1 shows the [110] zone axis electron diffraction pattern of the (1-x)BMN-(x)BZ ceramics as a function of x. Pure BMN exhibits 1:2 cation order which is characterized by the superlattice reflections located at $(h \pm 1/3, k \pm 1/3, l \pm 1/3)$ (figure 1a). The substitution of BZ stabilizes a cubic (Fm$\bar{3}$m) 1:1 ordered structure which is identified by the additional reflections at $(h \pm 1/2, k \pm 1/2, l \pm 1/2)$ (figure 1b). A narrow mixed phase 1:2 + 1:1 region was observed for 0<x<0.05, while "pure" 1:1 ordering is stabilized between 0.05<x<0.25 (figure 1.b). Further addition of BZ reduces the cationic correlations and ultimately (x>0.25) the structure assumes a disordered cation distribution. Details of the energetic driving forces for this transformation have been previously discussed by Davies et. al [3]. While the phase stability regions in BMN-BZ are quite similar to those we have reported for the corresponding tantalates, the nano-structural features of the niobate system were found to be especially sensitive to thermal history.

Nanostructural Features of 1:1 Ordered Niobates.

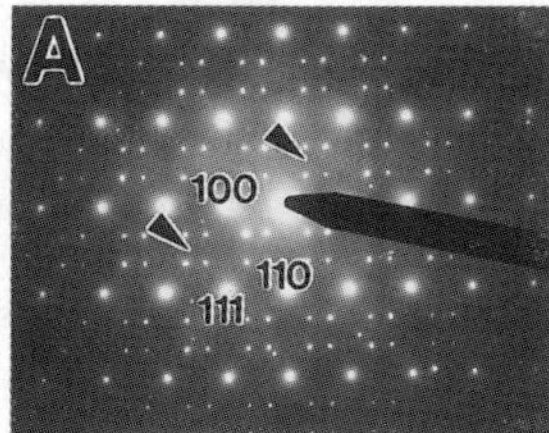
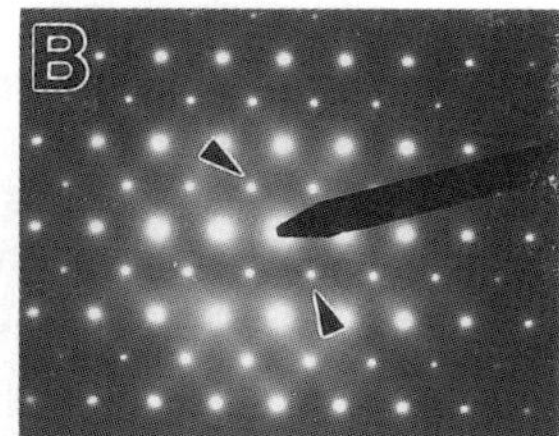

Figure 1. Electron diffraction patterns collected along [110] of the perovskite sub-cell for a) pure b) 10 mole% BZ doped BMN ceramics. Superlattice reflections are marked.

The dark-field TEM micrograph (collected using the (3/2, 3/2, 3/2 supercell reflection) in figure 2a shows a typical microstructure of the 1:1 ordered BZ-BMN phases after normal sintering at 1640°C. The grains contain 4-5 nm 1:1 ordered domains surrounded by a disordered perovskite matrix; approximately 50% of the sample is ordered. This type of microstructure is analogous those observed in 1:2 Pb-based relaxor compositions such as PMN and PMT [5, 6]. For the relaxors the formation of this phase-separated 1:1 order has been interpreted using the "space charge" model, where it is suggested that the ordered domains contain a 1:1 distribution of Mg and Nb(Ta) [4, 5]. If this model is correct any growth of the nano-sized 1:1 domains would

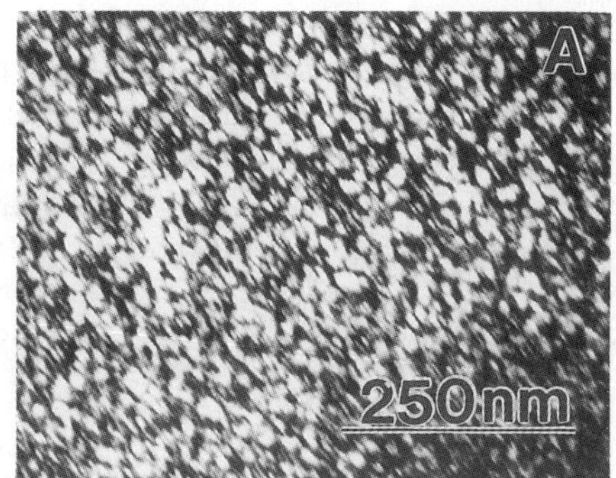

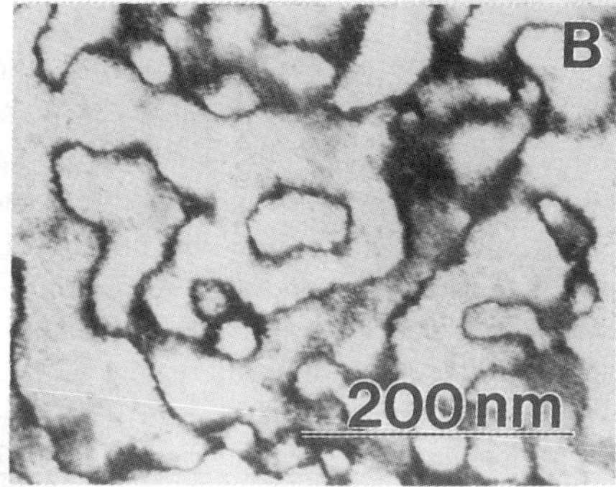

Figure 2. DF image collected using the (3/2, 3/2, 3/2) reflection for BMN-10 mole%BZ samples, showing final domain structure after a) furnace cooling (300°C/hr) b) slow cooling (10°C/hr).

be severely inhibited by the associated space charge. The primary support for the model has come from the apparent absence of any domain coarsening with extended thermal annealing [5].

Growth of the 1:1 Ordered Domains in BMN-BZ.

To promote the growth of the ordered domains it is clearly necessary to utilize conditions that lie in the region of thermodynamic stability of the 1:1 phases and yet allow significant cation diffusion. For the BMN-BZ systems the range of temperature that satisfied both conditions was found to be very narrow. In figure 2 we show dark field images of BMN-10%BZ (a) after normal sintering, where the samples were cooled from 1640°C to 1000°C at 300°C/hr, and (b) after cooling from 1640°C at 10°C/hr. While the normal sintering procedure yields a phase separated mixture comprised of 4-5nm ordered domains, the slow-cooled samples are fully ordered with a domain size that exceeds 30nm. The large changes in the degree of order and domain size were also evident in the x-ray patterns collected from samples cooled at different rates (figure 3).

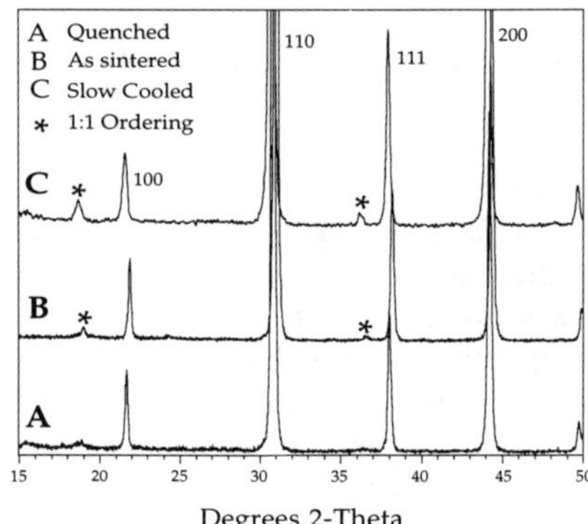

Figure 3. XRD pattern of BMN-10 mole%BZ ceramics, a) quenched b) furnace cooled (300°C/hr) c) slow cooled (10°C/hr).

From these results it is clear that a space charge model cannot be accepted as a valid explanation for the 1:1 order in BMN-BZ ceramics. Instead the experimental data favors a "random layer" model for the ordered $Ba(\beta^I_{1/2}\beta^{II}_{1/2})O_3$ structure in which the β^{II} positions are exclusively are occupied by Nb^{5+} and the β^I sites contain a random distribution of Mg^{2+}, Zr^{4+} and the remaining Nb^{5+} cations. Within this model the structures of the 1:1 ordered (1-x)BMN-(x)BZ solid solutions can be represented as, $Ba[(Nb)_{1/2} (Mg_{(2-z)/3}Ta_{(1-2z)/3}Zr_z)_{1/2}]O_3$, where z=2x. For a random layer model the 1:1 ordered domains are charged balanced and the overall 1:2 cation stoichiometry is preserved in all regions of the samples.

<u>High Temperature Phase Equilibria in BMN-BZ.</u>

To examine the phase stabilities at elevated temperatures, a series of BMN-BZ compositions were air quenched from temperatures as high as 1640°C and examined using XRD. Figure 3 shows the patterns collected from samples of 10 mole% BZ-BMN that were sintered, slow cooled, and quenched from 1640°C. The absence of any evidence for cation order in the quenched samples (figure 3a) indicates that the system undergoes an order-disorder transformation at high temperature; for this sample the transformation temperature was determined to lie between 1550 and 1600°C. The transformation temperature was found to decrease as the concentration of BZ was increased beyond this level.

Recognizing the existence of this transformation at temperatures below those used in sintering, the variation of the microstructures of 1:1 BMN-BZ with thermal history can be readily explained with the aid of the schematic phase diagram in figure 4. Because the sintering temperature lies in the stability field of a disordered perovskite structure, the transformation to the 1:1 ordered state and therefore the growth of the ordered domains can only occur during the cooling cycle. Rapid cooling can completely preserve the meta-stable disordered phase, the

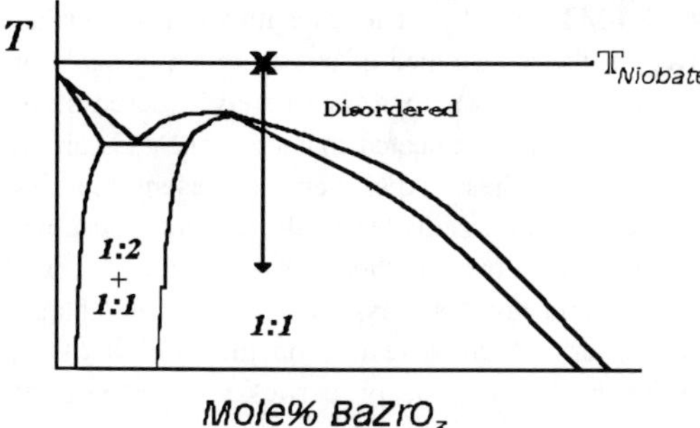

Figure 4. A schematic phase diagram for the BMN-BZ system.

slower cooling employed in the standard sintering procedure results in nucleation and partial growth of the 1:1 ordered phase and a metastable two-phase assemblage, while slow cooling induces complete order and a large domain size.

Domain Growth in $Pb(\beta^{I}_{1/3}\beta^{II}_{2/3})O_3$ Type Relaxors

In light of the results obtained on the BMN-BZ system, we have recently extended our experiments to re-examine the structures and properties of Pb-based 1:2 perovskite relaxor

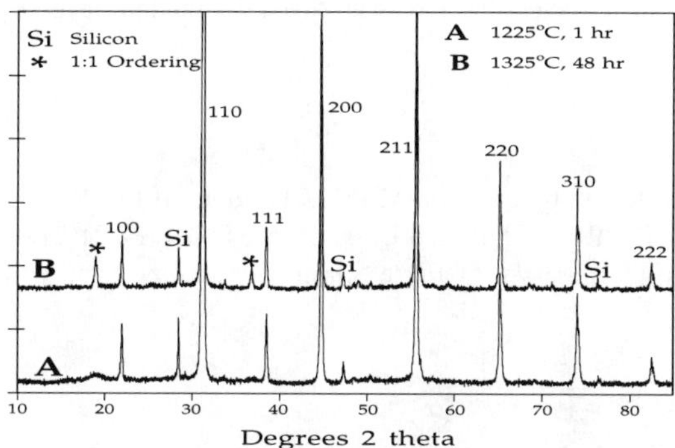

Figure 5. XRD pattern of PMT ceramics doped with 5 mole% PZ, after firing at a) 1225, I hour and b) 1325, 48 hours.

ceramics. Our initial experiments have been conducted on a tantalate member of the relaxor family, $Pb(Mg_{1/3}Ta_{2/3})O_3$ (PMT) [6]. Ceramics of pure PMT were found to exhibit a large increase in the degree of 1:1 cation order after extended annealing at 1325°C; while some domain growth was also observed, the transformation was quite slow. When relaxor compositions that contained small substitutional levels of $PbZrO_3$ (PZ) were subjected to similar high temperature treatments, very dramatic changes in the degree of order and domain in size were observed. For example, figure 5 shows the XRD patterns collected from (a) as sintered (1225°C, 1 hr), and (b) thermally annealed (1325°C, 48 hr) samples of PMT-5%PZ. The as sintered samples are essentially completely disordered to x-rays; however, the appearance of strong ordering peaks in the annealed sample were found to arise from a transformation to a completely ordered state. This interpretation was confirmed by analytical TEM which indicated that the annealed sample was comprised of ordered domains that exceed 50 nm. These experiments represent the first report of domain coarsening in the 1:2 perovskite relaxors and indicate that the currently accepted space charge models for the cation ordering are incorrect. Instead these results for PMT, and experiments that are being conducted on other related relaxor systems, seem to favor a "charge-balanced" random layer model for the 1:1 cation order. Detailed results on the growth of the ordered domains in 1:2 relaxors and their effect on the ferroelectric properties will be reported elsewhere in the near future.

4. CONCLUSIONS

The substitution of Zr into BMN stabilizes an ordered 1:1 perovskite structure. The transformation of the 1:1 ordered phase to a fully disordered perovskite just below the sintering temperature permits the tailoring of the nanostructure using different cooling rates. The growth of the 1:1 ordered domains with extended annealing supports a random layer model for the cation ordering. Preliminary experiments carried out on PMT-based relaxor ferroelectrics indicate that the 1:1 ordered domains in these systems also respond to extended thermal treatment. The order of magnitude changes in domain size cannot be explained using the currently accepted space charge model for relaxor ferroelectrics and again favor the random layer model as a more likely alternative.

ACKNOWLEDGMENTS

This work was supported in part by the MRSEC Program of the National Science Foundation under Award number DMR96-32598 and by grant no, DMR 94-21-21184 (Ceramic Division). We also acknowledge the support of the electron microscopy facility by the NSF, MRSEC program.

REFERENCES

1) S. Kawashima, M. Nishiada, I. Ueda and H. Ouchi, J. Am. Ceram. Soc., 66[6], 421 (1983).
2) H. Tamura, T. Konoike, Y. Sakabe and K. Wakino, Commun. Am. Ceram. Soc., 67[4], c-59 (1989).
3) P.K. Davies, J. Tong and T. Negas, J. Am. Ceram. Soc. (submitted 1996).
4) M.P. Harmer, J. Chen, P. Peng, H.M. Chen and D.M. Symth, Ferroelectrics, 97, 263 (1989).
5) A.D. Hilton, D.J. Barber, A. Randall and T.R. Shrout, J. Mater. Sci., 25, 3461 (1990).
6) M.A. Akbas and P.K. Davies, J. Mater. Res. (submitted 1996).
7) M.A. Akbas and P.K. Davies, to be submitted to the J. Am. Ceram. Soc. (1996).

HIGH-RESOLUTION ELECTRON MICROSCOPY OF MAGNETIC DIELECTRIC OXIDES IN THE BaO:TiO$_2$Fe$_2$O$_3$ SYSTEM

L.A. BENDERSKY, T.A. VANDERAH AND R.S. ROTH
Materials Science and Engineering Laboratory, National Institute of Standards and Technology, Gaithersburg Md 20899, leoben@nist.gov

ABSTRACT

In a recent work on phase equilibria relations in the BaO:Fe$_2$O$_3$:TiO$_2$ system the existence of sixteen ternary compounds was confirmed. Most of the new found compounds apparently have a new structural type. In this work we employed electron diffraction and high-resolution microscopy in attempt to understand the structures of the new compounds. Since the present work was done on the compounds prepared by solid-state reaction, it also provides a correlation with a single-crystal x-ray work on some of the similar compounds prepared by crystallization from a melt.

INTRODUCTION

Ceramic magnetic oxides are essential components in a wide variety of electronic applications. Materials in current use include various garnets, spinels, and the BaFe$_{12}$O$_{19}$-type family of hexaferrites. Improved new materials with higher dielectric constants (to enhance miniaturization) and high saturation magnetization values are looked for in different systems, particularly in the BaO:Fe$_2$O$_3$:TiO$_2$ system. In a recent work by Vanderah et al [1] on phase equilibria in the BaO:Fe$_2$O$_3$:TiO$_2$ system at 1250-1270°C the existence of sixteen ternary compounds has been reported. Most of the new found compounds apparently have a new structure type and require complete structural determination. In this paper we report the preliminary results on high-resolution transmission electron microscopy (HREM) and electron diffraction (ED) on some of these compounds. The objective of this work was to understand structural features of the new compounds as well as the presence and nature of structural defects. In addition, it is important to establish the correlation between single-crystal x-ray diffraction and HREM results for the following reasons: (a) the HREM work is performed on the compounds prepared by solid-state reaction, as compared to the compounds for x-ray prepared by crystallization from a melt; (2) ED is much more sensitive then x-ray to the subtlety of structural ordering.

The studied compounds have designations and compositions given in Table 1 according to reference [1]. Unit cell and space group of the compounds are also given in Table 1 according to x-ray measurement [1]. Most of the new compounds appears to have a structure consisting of close-packed (cp) Ba/O or O layers, reminiscent of the known BaFe$_{12}$O$_{19}$ (hexaferrite) [2] and BaTiO$_3$ (either cubic or hexagonal) [3]. Interplanar spacing between the cp layers is about 0.235 nm, and therefore the number of the close-packed (cp) layers n was estimated as d(001)/0.235 (see Table 1). (001) corresponds to the largest dimension of the unit cell, usually c parameter. The studied phases were reported to be in an apparent thermodynamic equilibrium with each other.

EXPERIMENTAL

Polycrystalline samples of the studied phases were prepared by solid-state reaction of BaCO$_3$, Fe$_2$O$_3$, and TiO$_2$ in molar ratios given in Table 1. Before each heating mixtures were ground, pelletized, and placed on sacrificial powder of the same composition in alumina combustion boats.

Mat. Res. Soc. Symp. Proc. Vol. 453 © 1997 Materials Research Society

A preliminary overnight calcine at 1000 °C was following by three one-week heatings at 1265 °C; the attainment of equilibrium conditions was presumed when the detailed X-ray powder diffraction patterns exhibited no further changes.

Samples for electron microscopy were prepared by crushing the samples with an agate mortar and pestle. Crushed fragments were mixed with acetone and deposited onto a copper grid covered with holey carbon film. TEM specimens were examined for high-resolution imaging in a JEM 3010UHR microscope having -420 nm Scherzer defocus, and for high-tilt electron diffraction in a Philips 430 microscope. High-resolution images were recorded with a CCD camera and analyzed using Gatans' *Digital Micrograph* software package.

TABLE I

Compounds Label	Composition	Unit cell parameters, nm	Suggested space group	# of cp layers
SS	*"1:0:1"* - *Ba(Fe,Ti)O$_3$*	*a = 0.572,* *c = 1.396*	P6$_3$/mmc	6
E	*"4:2:3"* - *Ba$_4$Fe$_4$Ti$_3$O$_{16}$*	*a = (0.57618)* 0.99* *c = 2.3738*	?* P3	10
F	*"1:6:3"* - *Ba$_3$Fe$_2$Ti$_4$O$_{14}$*	*a = 1.0024,* *b = 1.7378,* *c = (3.5373)* 2.46* *β = (90.76°)* 106.7°*	C2/m	(15)* 10
I	*"8:3:16"* *Ba$_8$Fe$_6$Ti$_{16}$O$_{49}$*	*a = 0.99917,* *c = 4.2252*	P6$_3$/mcm	18
K	*"8:5:8"* - *Ba$_8$Fe$_{10}$Ti$_8$O$_{39}$*	*a = 0.9977,* *c = 6.141*	Pseudotrigonal	26
M	*"8:3:6"* - *Ba$_8$Fe$_{10}$Ti$_8$O$_{39}$*	*a = 0.57538,* *c = 6.1482*	Trigonal	26

()* - this values were later corrected to the value shown on the right

RESULTS

<u>Electron diffraction.</u>

Series of selected area electron diffraction (SAD) patterns were obtained for different phase by tilting a TEM specimen around the axis normal to the cp planes (most dense row of reflections). Fig. 1 shows two major SAD patterns of the series, A and B type. The patterns are rotated to each other about 30°. SAD patterns of the known Ba(Fe,Ti)O$_3$ hexagonal phase are given for the reference. The outlined cell in the pattern A ([1$\overline{1}$00] zone axis) of the Ba(Fe,Ti)O$_3$ represents the fundamental distances of the cp layers - spacing between Ba/O layers, about 0.23 nm, and a second-neighbors distance between O and Ba atoms in the layer, about 0.49 nm. From these SAD patterns we identified the following features common for all ternary phases studied here:
1. The tilting experiment suggests the presence of either true or pseudo hexagonal/rhombohedral lattice for all studied phases. Observation of a 6-fold symmetric ([0001]-type) SAD pattern is an

additional support for that.

2. The lattice parameter values were found to be in accord with x-ray results for all studied phases, with the exception of the M phase. All ternary phases have additional (as compared to the $Ba(Fe,Ti)O_3$ hexagonal phase) rows of intensity at 1/3 and 2/3 positions of the A-type SAD patterns (indicated by arrows in Fig.1,a between 0000 and 1120 spots of $Ba(Fe,Ti)O_3$). Failure to detect the 1/3-2/3 intensity by x-ray for the M phase is attributed to its weakness because the rows are diffused. These additional rows indicate tripling of the idealized hexagonal/rhombohedral unit cell by changing **a** parameter to $\sqrt{3}a_{Ba(Fe,Ti)O3} \approx 0.99$ nm).

3. Diffuseness of the 1/3-2/3-type rows of intensity for the K and M phases does not effect other rows which have sharp, well separated Bragg reflections (both for A and B-type patterns). Such behavior can be understood as a structural disorder which is effected only by the structural features related to the tripling of a unit cell.

More specifically about different phases the following can be said:

F and I phases usually have a well ordered structure, with only occasionally seen intergrowth-type defects. The reciprocal lattice of the F phase can be viewed as rhombohedral ($a_r \approx 0.59$ nm; $c_r \approx 3.5$ nm), with ordering which reduces its symmetry to monoclinic ($c_m \approx 2/3 c_r \approx 2.4$ nm). Structures of both

E phase has a well ordered hexagonal Bravais lattice. The possible space group of the phase was suggested based on the absence of reflections, symmetry of convergent beam electron diffraction and high-resolution images [3]. The only space group that can be considered for the E phase is trigonal P3.

M and K phases appear to have similar SAD patterns with 1/3-2/3 rows of diffuse intensity suggesting a structure of 26 cp layers. The difference comes in the extinctions (the K phase has weak odd 0001 reflections which are absent for the M phase in A-type pattern) and the degree of diffuseness of the 1/3-2/3 rows. For the M phase the intensity of the 1/3-2/3 rows is continuous, and therefore it is impossible to assign a definite Bravais lattice based only on the SAD observation. As for the K phase, there is an apparent condensation of the intensity into separate spots. Positions of these spots suggest that for the K phase may have either rhombohedral or monoclinic (pseudo-rhombohedral) lattices in its final, ordered state are possible.

<u>High-resolution imaging.</u>

HREM images of the E, M and K phases taken at A and B-type orientations are shown in Fig. 2. HREM images of two other phases, F and I, structures of which were recently obtained by a single crystal x-ray work [2], are shown in Fig. 3. The images of the E, M and K phases suggest that these structures can be seen as built out of two types of structural blocks (designated in the figure as P (perovskite-type) and H). At B-type orientation the P blocks have a typical zigzag-like appearance. Such contrast was readily observed for a wide range of defocus-thickness conditions. It is suggested here that the P blocks have perovskite-type structures, similar to $BaTiO_3$ but with different stacking sequences of the cp layers. HREM image simulation of known perovskite-type structures (e.g., cubic and hexagonal $BaTiO_3$) shows that zigzag-like sequence of white dots (spaced about 0.5 nm) represents the sequences of cp layers ([ccc] for cubic and [chcchc] for hexagonal) in wide range of imaging conditions. With such an assumption the following sequences of cp layers in the P blocks can be suggested: M phase - [hcchcchcch]; for E phase - [hcchcch]; for K phase - [hcchcch]

At A-type orientation the H block can be easily identified by its tripled periodicity in the direction normal to **c** direction and compared to a single periodicity of the P block. These tripled periodicity reveals the observed on the SAD patterns 1/3-2/3 rows of intensity (Fig. 1). From the HREM it is clear that tripling of a unit cell is localized in the block H and not throughout the whole unit cell for these three phases.

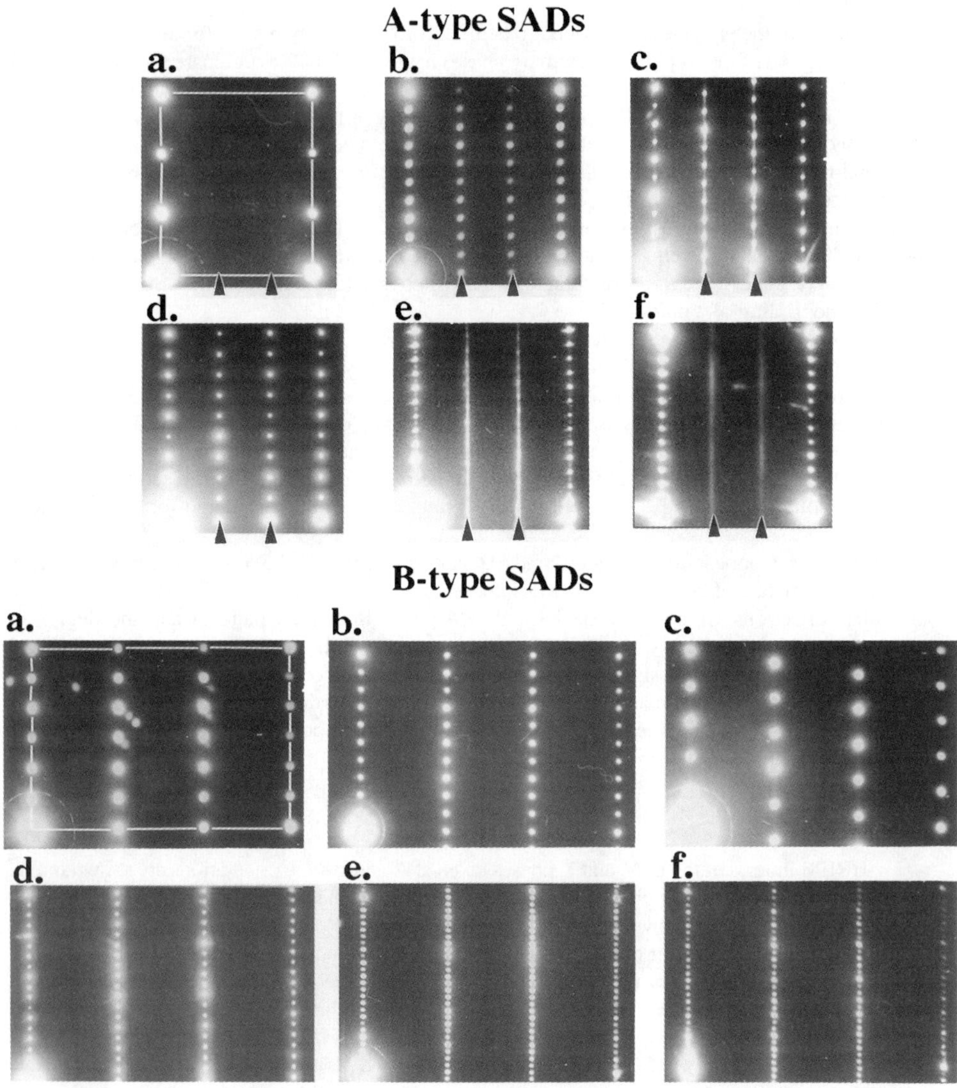

Fig. 1. SAD patterns of A- and B-type obtained by rotating a thin crystal of (a) hexagonal $Ba(Fe,Ti)O_3$, (b) E - $Ba_4Fe_4Ti_3O_{16}$, (c) F- $Ba_3Fe_2Ti_4O_{14}$, (d) I - $Ba_8Fe_6Ti_{16}O_{49}$, (e) K - $Ba_8Fe_{10}Ti_8O_{39}$, (f) M - $Ba_8Fe_{10}Ti_8O_{39}$ phases around a 0001 row of reflections. The angle of rotation between A and B patterns is about 30 deg. The patterns for the hexagonal $Ba(Fe,Ti)O_3$ phase are indexed as <1100> for A and <2110> for B, while for hexagonal I phase ($Ba_8Fe_6Ti_{16}O_{49}$) -<2110> for A and <1100> for B. Arrows in the A-type patterns point on location of 1/3, 2/3 intensities.

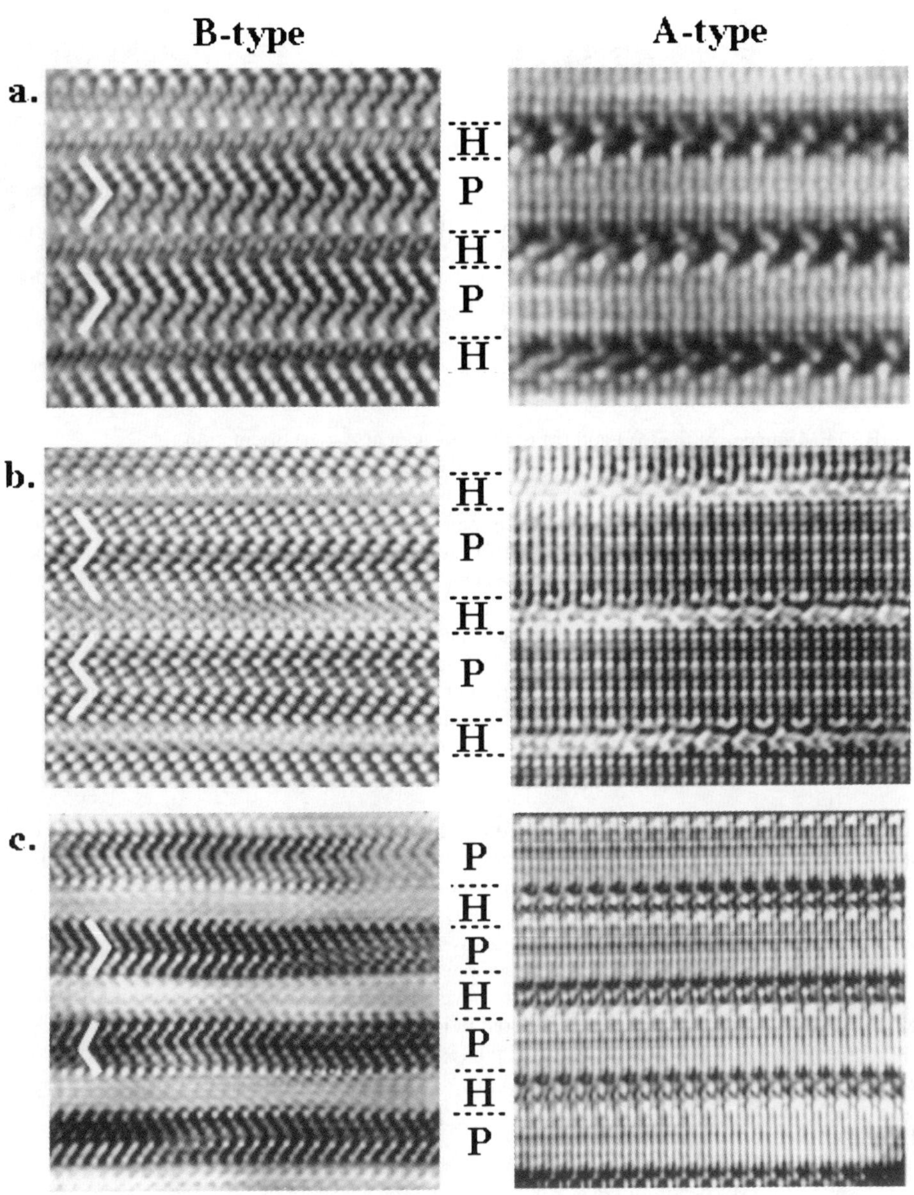

Fig. 2. HRTEM images of the (a) E phase, (b) M and (c) K phases taken at A-type and B-type orientations.

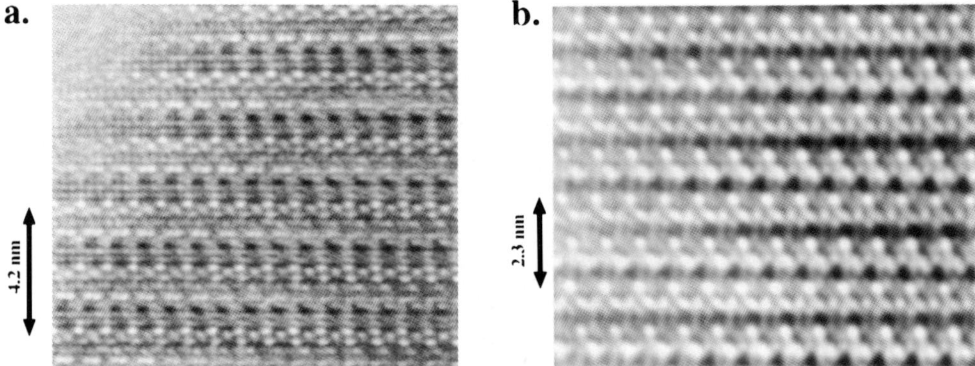

Fig. 3. HRTEM images of the (a) I and (b) F phases taken at A-type orientation.

The situation is different for the F and I phases where the tripled periodicity spreads throughout a whole unit cell (Fig. 3). Analysis of the F and I phase structures derived from x-ray [4] suggests that both ordering between Ba and O atom in cp layer and ordering of cations are responsible for the tripling. The tripled Ba/O layers are distributed uniformly thought the cell in these structures.

Concerning the structural disorder of the M and K phases, the following can be learned from HREM images. When translations of the H blocks (with respect to each other) in (0001) plane were analyzed, it was found that the spacial correlation of the translations was poor. That results in the one-dimensional disorder and observation of 1/3 and 2/3 diffuse rods.

CONCLUSIONS

Five different compounds of the $BaO:Fe_2O_3:TiO_2$ system, the F, I , M, E and K phases, were studied by electron diffraction and HREM. The following preliminary conclusions were obtained:
1. There is a good correspondence between phases obtained by solid-state reaction and by crystallization from a melt. The only exceptions are E and M phases where the tripling of a unit cell was not detected by x-ray.
2. The E, M and K phases apparently belong to a new class of structures built out of two type of structural blocks. One block, P, is suggested to have a structure of polytypic perovskite. Another block, H, of yet known structure is triple ordered in its basal plane with respect to the P block.
3. The observed 1-D disorder of the M and E phases was interpreted as a poor spatial correlation between P blocks.

REFERENCES

1. T.A. Vanderah, J.M.Loezos and R.S. Roth, J. Solid State Chem., **121**, 38 (1996)
2. W.D.Townes, J.H.Fang and A.J.Perrotta, Z. Kristallog., **125**, 437 (1967)
3. J.Akimoto, Y.Gotoh and Y.Oosawa, Acta Cryst., **C50**, 160 (1994)
4. Theo Siegrist, unpublished research, AT@T Bell Lab (1996)
5. L.A.Bendersky, T.A. Vanderah and R.S. Roth, J. Solid State Chem., **125**, 281 (1996)

EFFECT OF DIVALENT DOPANTS ON PROPERTIES OF
$Ba_{3.75}Nd_{9.5}Ti_{18}O_{54}$ MICROWAVE DIELECTRIC RESONATORS

R. UBIC[*], I.M. REANEY[*], W.E. LEE[*], J. SAMUELS[†], E. EVANGELINOS[†]
[*]Department of Engineering Materials, The University of Sheffield
 Hadfield Bldg, Mappin Street, Sheffield S1 3JD (United Kingdom)
[†]Morgan Matroc Ltd. Unilator Division, Vauxhall Industrial Estate, Ruabon,
 Wrexham, Clwyd LL14 6HY (United Kingdom)

ABSTRACT

Pellets in the ternary solid-solution $Ba_{6-3x}Nd_{8+2x}Ti_{18}O_{54}$ (x=0.75) have been processed to near theoretical density by mixed-oxide synthesis, and their dielectric properties determined to be $\varepsilon_r \approx 89$, $Q_f > 10,000$, and $\tau_f \approx +60ppm/°C$ at microwave frequencies. Similar ceramics with Ca, Sr, and Pb partially substituted for Ba were also processed and analysed. Both Ca and Sr were found to increase ε_r to 100-105 but significantly lower Q_f and inflate τ_f. While Pb substitution also lowered Q_f, it had no significant effect on ε_r and beneficially suppressed τ_f to values from -16 to +8ppm/°C, tunable with Pb content, strongly suggesting that these materials may be useful for commercial applications.

INTRODUCTION

Microwave resonators are used most commonly in communications equipment like mobile telephones and satellites. For such applications a high dielectric constant (ε_r) ceramic can reduce the circuit size, and low $\tan\partial$ (high Q) means high selectivity. In addition, they play a crucial role in compensating for frequency drift because of their tunable temperature coefficients of resonant frequency (τ_f). Recently a suitable phase has been identified in the system $BaO \cdot Nd_2O_3 \cdot TiO_2$.

Much attention has been given to doping ternary compositions to effect substitution on either the rare-earth ion sites or the barium sites to maximise dielectric properties. In general it has been found[1] that as the ionic radius of the lanthanide decreases (atomic number increases), τ_f decreases. The most common dopant on the rare-earth sites, apart from other rare-earths,[2,3] has been bismuth.[2,4-6] In all cases, adding Bi_2O_3 to ternary compositions increased ε_r but lowered Q and depressed τ_f. The most frequently added divalent cations have been strontium (usually added to Sm-based compositions) and lead (usually added to Nd-based compositions), to substitute on barium sites. Nishigaki et al.[7] and Sun et al.[1] reported that strontium doping in Sm-based 1:1:5 (oxide ratio) compositions increased ε_r while lowering Q and raising τ_f, and Wu et al.[8] used varying processing parameters to tune the properties of Sr-doped material. Wakino et al.[9] reported that addition of PbO to 1:1:5 Nd-based powders increased both ε_r and Q at 3GHz.

The present work focuses on the x=0.75 structure. This composition was chosen for the reason cited in reference [10] (*i.e.* that Q increases with increasing x) and x = 0.75 is thought to be at or near the upper solubility limit for the Nd-system at the processing temperatures used. The aim of this research was to establish the influence of divalent dopants, specifically Ca, Sr, and Pb, on the dielectric properties of these ceramics and relate these changes to the phase evolution and microstructure of sintered resonators.

EXPERIMENT

Synthesis of various ternary compositions was conducted via the conventional mixed oxide route. The compositions tested are shown in Table I. To minimise Pb loss, a two-stage pre-reaction process was necessary, and calcination and sintering were conducted in closed Pt-lined crucibles. Batches not containing Pb were processed in the same way to maintain self-consistency of the results.

Mat. Res. Soc. Symp. Proc. Vol. 453 © 1997 Materials Research Society

Table I
Compositions (in wt%) of powder batches

Batch	$BaCO_3$	Nd_2O_3	TiO_2	$CaCO_3$	$SrCO_3$	PbO
A (un-doped)	19.6	42.3	38.1	0	0	0
B ($^1/_2$-Ca)	10.3	44.5	40.0	5.2	0	0
C ($^1/_2$-Sr)	10.0	43.4	39.0	0	7.5	0
D ($^1/_2$-Pb)	9.7	41.8	37.6	0	0	10.9
E ($^1/_4$-Pb)	14.6	42.1	37.8	0	0	5.5
F ($^1/_8$-Pb)	17.1	42.2	38.0	0	0	2.7

First, a 1:1 stoichiometric mix of TiO_2 and $BaCO_3$ (and dopant-carbonates, if appropriate) was milled for four hours, dried in an oven ($\approx80°C$), ground, sieved to under 500µm, and reacted in air at 1000°C for two hours, forming $BaTiO_3$ (and dopant-titanate, if present). This product was then added to the appropriate amount of Nd_2O_3 and the remaining TiO_2 and milled, dried, ground, and sieved as before. The mix was then reacted at 650°C for two hours to drive off the water of hydration from the $Nd(OH)_3$ which inevitably forms during wet milling. If the batch was to be Pb-doped, the PbO was added at this point and ground into the powder dry. The whole batch was then sealed in a Pt crucible and calcined at 1150°C for four hours.

After calcination, the powder batches were re-milled with organic binders (1wt% PVA and 1wt% PEG1500) and spray dried to a final granule size of 50-100µm. Powders were then uniaxially pressed (360MPa) into green rods 8mm in diameter and ≈1.5cm long. These rods then underwent a binder burn-out at 700°C (0-hold) prior to being sintered in closed Pt-lined crucibles at temperatures from 1300-1400°C for four hours. After sintering, rods were reduced in diameter to 6mm and sliced into pellets ≈2mm thick. These pellets finally were fired at 700°C (0-hold) again to remove traces of epoxy used in the wafering process.

To determine the reaction mechanism leading to the ternary phase and the optimal temperature for calcination, DTA (1600DTA, T.A. Instruments, USA) and TGA (TG760 Series, Stanton Redcroft, UK) were performed in air, in conjunction with XRD (PW 1730/10, Philips, Holland), on un-pre-reacted powders of various compositions. The particle size of these powders was typically <10µm. Density measurements of the as-sintered rods were performed using Archimedes' principle in deionised water.

Thin TEM samples were made from finished pellets, and some pellets were polished and etched for observation in the SEM (Camscan Mk. II, Cambridge Scanning, England). TEM observations were also made (Jeol models 200CX and 3010, Japan). High-resolution electron diffraction images and diffraction patterns were made using the EMS software developed by Stadelmann (V3.3, P.A. Stadelmann, I2M-EPFL, CH-1015 Lausanne, Switzerland).

Measurements of Q and τ_f were made at Morgan Matroc Ltd. Unilator Division on a vector network analyser (Wiltron model 37247A, USA) and scalar network analyser (Wiltron model 562, USA), respectively. Values of ε_r were calculated from the resonant frequency and dimensions of the sample and cavity using the appropriate system of transcendental equations published by Itoh and Rudokas[11] with software developed at the University of Sheffield.

RESULTS

<u>Ceramic Processing</u>

Thermal analysis for all compositions tested gave similar results, shown graphically in figure 1 for a non-pre-reacted, un-doped composition. At 400°C the $Nd(OH)_3$ (which inevitably results from the mixing of h-Nd_2O_3 in water) has lost two thirds of its water of hydration and has probably become NdOOH, though there is considerable uncertainty in the x-ray pattern for this phase. Above 550°C the NdOOH dries further to become c-Nd_2O_3, a much less hydratable polymorph of the oxide. From ~800°C-1000°C the $BaCO_3$ reacts completely with the available TiO_2 to form $BaTiO_3$ (there is a large excess of TiO_2 in this case). At 810°C there is a sharp endotherm where c-Nd_2O_3 reverts back to h-Nd_2O_3 via a first-order polymorphic transformation. The subsequent formation of the ternary phase is slow and is not apparent at these heating rates.

The densification behaviour of test rods is partly summarised in figure 2, which clearly shows that, as expected, Pb-doped rods attain the highest density until 1400°C, where density falls sharply. This drop off in density is apparent in both the $^1/_2$-Pb and $^1/_4$-Pb compositions .

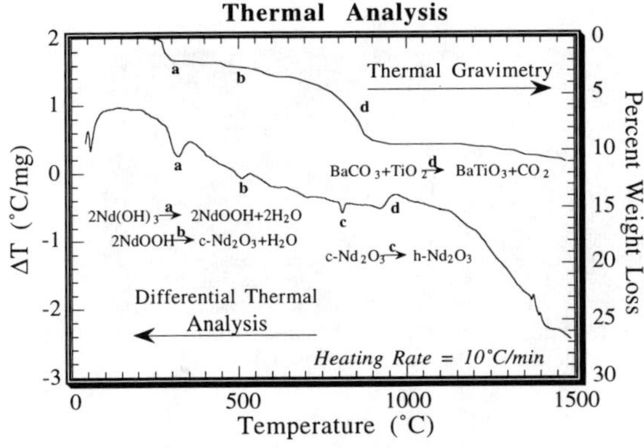

Figure 1: Thermal analysis of the ternary composition showing the reaction mechanism on heating

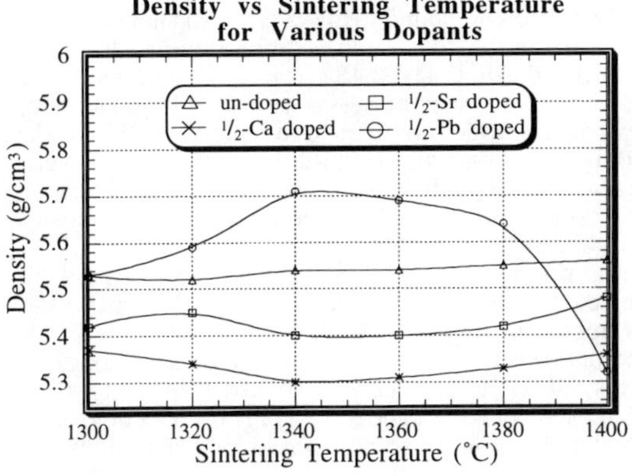

Figure 2: Densification behaviour for various dopants

Phase Evolution

Most thermally etched pellets showed clear evidence of a two-phase microstructure with a markedly different grain size. More abundant at low temperatures, the large grains were shown by EDS and electron diffraction to most likely be the perovskite phase $NdTiO_3$. It is suggested that this phase ($\approx 30\mu m$ grains) forms along with the solid-solution phase ($\approx 1\mu m$ grains) because of the large proportion of Nd and Ti in the stoichiometry. It persists in the microstructure until the divalent species have sufficient time and energy to diffuse into it, thus forming the solid-solution phase. The micrograph in figure 3a shows a Pb-containing pellet sintered at 1380°C, showing many such large grains. Figure 3b shows a pellet of identical composition sintered at 1400°C, showing a complete absence of this phase. Figure 3c shows the same pellet as in figure 3a after being re-sintered at 1400°C, again showing the absence of any large-grain phase.

This transformation has also been observed by TEM[12] where the ternary phase can be seen growing into a grain of $NdTiO_3$, transforming it as the boundary advances. High-resolution images, which show zig-zagged Ba sites between perovskite columns continuous across the interface, also suggest this mechanism. It should be noted here that whether this perovskite phase is the same as the large grains observed by SEM has yet to be confirmed. Nevertheless, the large grains only completely disappear in the Pb-doped pellets . For the other compositions tested some $NdTiO_3$ persists even at 1400°C, indicating that it is easiest for Pb, being the least refractory of the divalent species tested, to diffuse through the material and transform it to the desired phase.

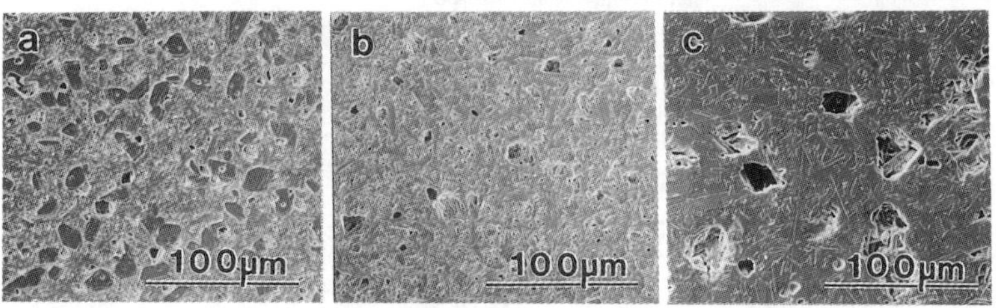

Figure 3: Batch D. The pellets in (a) and (b) were sintered at 1380°C and 1400°C, respectively; and (c) shows the same pellet as in (a) after re-sintering at 1400°C.

Crystal Chemistry

The atomic positions for HREM and XRD simulations were obtained from the work of Matveeva et al.[13] who refined the analogous Pr-based compound. However, it was noted during the present analysis that there exists a small error in the 4c oxygen position reported by these authors. This has been corrected, substituting 0.104,0.410,0.487 for the listed value of 0.104,0.485,0.487. In addition, extra reflections corresponding to cell-doubling along c, neglected by Matveeva et al.,[13] have been verified by electron diffraction patterns in this work. Calculated XRD traces (based on the atomic positions of the un-doubled unit cell) were compared to experimental patterns and reveal a reasonable match. The crystal structure of $Ba_{3.75}Nd_{9.5}Ti_{18}O_{54}$ is thus confirmed to be orthorhombic, with lattice parameters calculated from the indexed XRD traces: a_o=12.252Å, b_o=22.316Å, c_o=7.660Å.

The perovskite phase identified in the microstructure by SEM was also analysed by TEM and found to be heavily tilted.[14] Selected area diffraction patterns (SADPs) from the [001], [110], and [111] pseudocubic zones appear in figure 4. Both the [001] and the [110] diffraction patterns show $^1/_2\{h00\}$ type superlattice reflections, indicating the presence of anti-parallel A-site (Nd^{+3}) cation displacements. The [110] SADP also shows $^1/_2\{hkl\}$ type reflections, indicating anti-phase rotation of oxygen octahedra about the [111] pseudocubic axis. Also present in the [110] pattern are very weak reflections corresponding to $^1/_2\{hk0\}$ which are attributed to double diffraction.

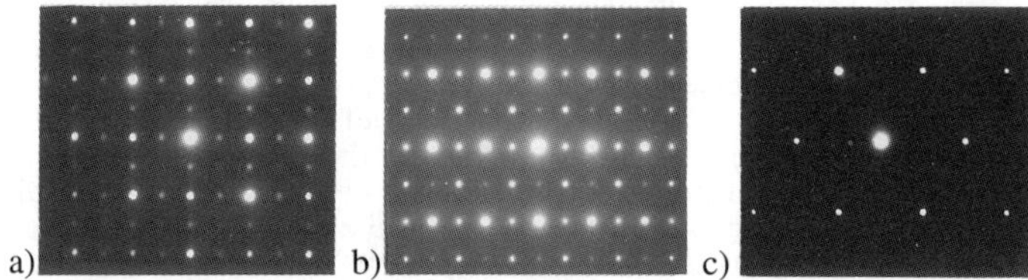

Figure 4: (a) [001], (b) [110], (c) [111] SADPs of the perovskite phase

Microwave measurements

The effect of the various dopants on the dielectric properties is summarised in Table II. Essentially, all dopants tested lower Q; but while Ca and Sr increase ε_r, Pb leaves it virtually unaffected, and may even lower it slightly. Also, only Pb was effective in suppressing τ_f. In fact, Pb addition was found to be critical in tuning τ_f, as illustrated by figure 5. The jump in τ_f observed at 1400°C correlates with the disappearance of the $NdTiO_3$ phase.

Table II
Summary of Dielectric Properties

Batch	A	B (Ca)	C (Sr)	D (Pb)	E (Pb)	F (Pb)
ε_r	88-90	100-103	98-100	80-85	75-80	85-88
Q_f	>10,000	≈900	≈3500	≈4000	≈5400	≈6050
τ_f (ppm/°C)	+60	+150	+130	-16	-10	+8

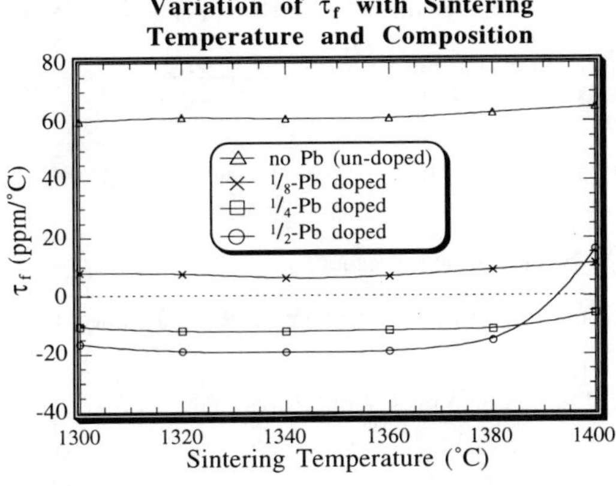

Figure 5: Variations in temperature coefficient of resonant frequency for various levels of Pb doping

CONCLUSIONS

Both Ca and Sr doping increase ε_r to >100, but significantly lower Q and increase τ_f to unusable levels. While Pb doping has the disadvantage of lowering Q as well, it does not significantly affect ε_r; and the high ε_r and Q_f of the lead-doped material, combined with its tunable τ_f, make this phase extremely attractive for microwave devices.

ACKNOWLEDGEMENTS

This work has been graciously supported by the Council of Vice-Chancellors and Principals of British Universities in the form of an Overseas Research Scholarship, by a grant from the Unilator Division of Morgan Matroc Ltd., Wrexham, and by the Department of Engineering Materials at the University of Sheffield.

REFERENCES

1   J.-S. Sun, C.-C. Wei, and L. Wu, J. Mater. Sci. **27** [21], p. 5818-22 (1992).

2   P. Laffez, G. Desgardin, and B. Raveau, J. Mater. Sci. **30** [1], p. 267-273 (1995).

3   H. Ohsato, H. Kato, M. Mizuta, S. Nishigaki, and T. Okuda, Jap. J. Appl. Phys., Part 1 **34** [9B], p. 5413-5417 (1995).

4   J.M. Durand and J.P. Boilot, J. Mater. Sci. Lett. **6**, p. 134-36 (1987).

5   M. Valant, D. Suvorov, and D. Kolar, in <u>Proceedings of the Fourth International Conference on Electronic Ceramics and Applications</u> **1**, (European Ceramic Society, 1994), p. 69-72.

6   D. Kolar, S. Gaberšček, Z. Stadler, and D. Suvorov, Ferroelectrics **27,** p. 269-72 (1980).

7   S. Nishigaki, H. Kato, S. Yano, and R. Kamimura, Bull. Am. Ceram. Soc. **66** [9], p. 1405-10 (1987).

8   J.-M. Wu and M.-C. Chang, J. Am. Ceram. Soc. **73** [6], p. 1599-1605 (1990).

9   K. Wakino, K. Minai, and H. Tamura, J. Am. Ceram. Soc. **67** [4], p. 278-281 (1984).

10   T. Negas and P.K. Davies in <u>Materials and Processes for Wireless Communications</u>, edited by. T. Negas and H. Ling, (Ceramic Transactions 53, Am. Ceram. Soc., Westerville, Ohio USA 1995), p. 179-96.

11   T. Itoh and R. Rudokas, IEEE Trans. on Microwave Theory and Tech **MTT-25** [1], p. 52-59 (1977).

12   R. Ubic, I.M. Reaney, and W.E. Lee, to be published.

13   R.G. Matveeva, M.B. Varfolomeev, and L.S. Il'yushenko, Russ. J. Inorg. Chem. **29** [1], p. 17-19 (1984).

14   I.M. Reaney, E.L. Colla, and N. Setter, Jap. J. Appl. Phys., Part 1 **33** [7], p. 3984-3990 (1994).

FABRICATION AND DIELECTRIC PROPERTIES OF PHASE-PURE Ba$_2$Ti$_9$O$_{20}$ MICROWAVE RESONATORS

WEN-YI LIN, ROSARIO A. GERHARDT, AND ROBERT F. SPEYER
School of Materials Science and Engineering, Georgia Institute of Technology
Atlanta, Georgia 30332
WESLEY S. HACKENBERGER AND THOMAS R. SHROUT
Materials Research Laboratory, The Pennsylvania State University
University Park, Pennsylvania 16802

ABSTRACT

The thermal processing schedules necessary to form phase pure Ba$_2$Ti$_9$O$_{20}$, with and without substitutional solid solution were investigated. Undoped compositions formed the compound most easily under rapid (500°C/min) heating rates, where diffusional agglomeration of the TiO$_2$ batch constituent was minimized. The compound formed most easily with minor (0.82%) substitutions of SnO$_2$ and ZrO$_2$ for TiO$_2$. The solubility of Sn in Ba$_2$Ti$_9$O$_{20}$ was higher than that of Zr. Extended heat treatment (16 vs. 6 h at 1390°C) resulted in volitalization of grain boundary liquid phase, leading to a more porous and slightly degraded resonators. The effects of dopant concentrations and soak periods at 1390°C on dielectric constants and temperature-dependent quality factors are reported.

INTRODUCTION

Efforts toward further miniaturization of microwave circuitry have stimulated the development of smaller, highly stable filters and compatible oscillators. Dielectric resonators with large permittivities and quality factors (Q) are a means for miniaturizing these functions. Two candidate resonator materials are BaTi$_4$O$_9$ and Ba$_2$Ti$_9$O$_{20}$; they both have high dielectric constants and quality factors (39 and $\sim$10000, respectively[1, 2]). However, Ba$_2$Ti$_9$O$_{20}$ has a lower temperature coefficient (drift of resonance frequency with temperature) of 2 ppm/°C than BaTi$_4$O$_9$ (14 ppm/°C). This makes Ba$_2$Ti$_9$O$_{20}$ a more favorable choice. The difficulty in fabricating undoped monophase Ba$_2$Ti$_9$O$_{20}$ has daunted its development for years[2]. There is disagreement about the ability to form phase-pure Ba$_2$Ti$_9$O$_{20}$[3]-[6]. Jonker et. al.[3] reported that the formation of Ba$_2$Ti$_9$O$_{20}$ could not be achieved without stabilizing dopants such as tin and zirconium oxides . Several different formation mechanisms of Ba$_2$Ti$_9$O$_{20}$ based on doped systems have been proposed[4, 5, 6]. Although such dopants have been shown to stabilize Ba$_2$Ti$_9$O$_{20}$, it is suspected that undoped phase pure Ba$_2$Ti$_9$O$_{20}$ would have a higher Q. The current work focuses on the rapid thermal processing of undoped Ba$_2$Ti$_9$O$_{20}$, as well as formation of this compound via ZrO$_2$ and SnO$_2$ batch substitutions, and their corresponding microwave properties.

EXPERIMENTAL PROCEDURE

BaTiO$_3$ (electronic grade) of 0.6 μm average particle size, and TiO$_2$ (anatase, electronic grade) of 0.6 μm average particle size, were batched to form Ba$_2$Ti$_9$O$_{20}$ (22.22 mol% BaTiO$_3$, 77.78 mol% TiO$_2$). The powder was mixed with distilled water and an ammonia-based dispersing agent (Darvan) in plastic bottles with ZrO$_2$ grinding media overnight. The slurry was then dried for 48 h at 80°C. The particle size distribution of the as-milled powder was examined (Microtrac II, Leeds & Northrup Co.); 98% of the particles were

501

Mat. Res. Soc. Symp. Proc. Vol. 453 © 1997 Materials Research Society

less than 6 μm. The powder was uniaxially pressed into pellets and soaked at 1250°C for 2 h, after heating at rates of 5°C/min to 500°C/min, using conventional and infrared[7] radiant heating furnaces. Varying soaking times were applied to determine the progress of formation of $Ba_2Ti_9O_{20}$. Some pellets were exposed to second heat treatments at 1390°C for 8, 16, and 24 h to foster the formation of $Ba_2Ti_9O_{20}$.

To form doped $Ba_2Ti_9O_{20}$, raw materials of $BaCO_3$ (electronic grade) and TiO_2 (anatase, electronic grade) were used. For these compositions, TiO_2 was substituted with 0.82, 1.64, and 2.46 mol% ZrO_2 or SnO_2. The powders were mixed as previously indicated. The dried slurries were then calcined/pre-reacted at 1200°C for 4 h. Pre-reaction was also initially attempted at 1150°C for 4 h. The powders were then re-milled in deionized water. Fifteen gram powder samples were uniaxially pressed into 22.5 mm diameter pellets and sintered at 1390°C for 6 and 16 h in flowing oxygen.

The phases formed after heat-treatments, interrupted after various soak times, were analyzed using X-ray diffraction (XRD, Model 12054, Philips Electronic Instruments Co.) with a step size of 0.015° and a sampling time of 1 or 3 s. Diffraction patterns were taken from pellet cross sections. The relative amounts of $Ba_2Ti_9O_{20}$ and $BaTi_4O_9$ were calculated from ratios of peak intensities[8]. Specimen microstructure was examined using scanning electron microscopy (SEM, Model S-800, Hitachi, Ltd.) with secondary and back scattered detectors. The micrographs were digitally scanned into bit-map computer files (30 pixels/cm) and greyscale ranges were differentiated to quantify percentages of each phase[8].

The sintered doped-composition pellets were cut perpendicular to the cylindrical axis to form parallel faces and their side walls were ground to specific dimensions. The microwave properties of the $TE_{01\delta}$ mode[9] of these pellets were measured at 3 GHz (8510C network analyzer, Hewlett Packard). The unloaded Q of a specimen within the microwave cavity of brass coated with silver was measured from the reflection coefficient and peak width. The Kobayashi method[10] was applied to measure the dielectric constants.

RESULTS

Unreacted TiO_2 in heat-treated undoped specimens was apparent in SEM microstructures (Figure 1) as black (low atomic mass) regions in secondary electron images. These TiO_2 grains were distributed along with porosity (secondary electron images), as well as $Ba_2Ti_9O_{20}$ and $BaTi_4O_9$ (back scattered electron images). The slower heating rate resulted in a greater percentage of unreacted TiO_2. Agglomerations of TiO_2 were enhanced in size by the slower heating rates; the large agglomerate size after heating at 5°C/min was ~80 μm, as compared to ~10 μm after heating at 500°C/min. These micrographs typify the trend of decreasing TiO_2 agglomerate size with increasing heating rate.

For the undoped specimens soaked at 1250°C for 2 h, increased heating rate from 5 to 500°C/min fostered an increased fraction of $Ba_2Ti_9O_{20}$ (Figure 2), from 12 to 64 vol%. The scatter in the data is clearly greater with analysis using relative XRD peak heights as compared to the use of image analysis. Figure 3 shows that after rapid thermal processing, reaction of residual $BaTi_4O_9$ with TiO_2 formed 98 vol% $Ba_2Ti_9O_{20}$ after 8 h at 1390°C. However, specimens heated at 5°C/min to 1250°C required 24 h at 1390°C to achieve 87 vol% of $Ba_2Ti_9O_{20}$.

Pre-reacting the 0.82 mol% ZrO_2-substituted composition (henceforth referred to as 0.82%Zr) at 1200°C and holding for 4 h led to the formation of $Ba_2Ti_9O_{20}$ and $BaTi_4O_9$; minute amounts of $BaCO_3$ and TiO_2 were also identified. These phases were identified in all SnO_2- and ZrO_2-substituted samples, with $BaCO_3$ and TiO_2 remaining as minor phases.

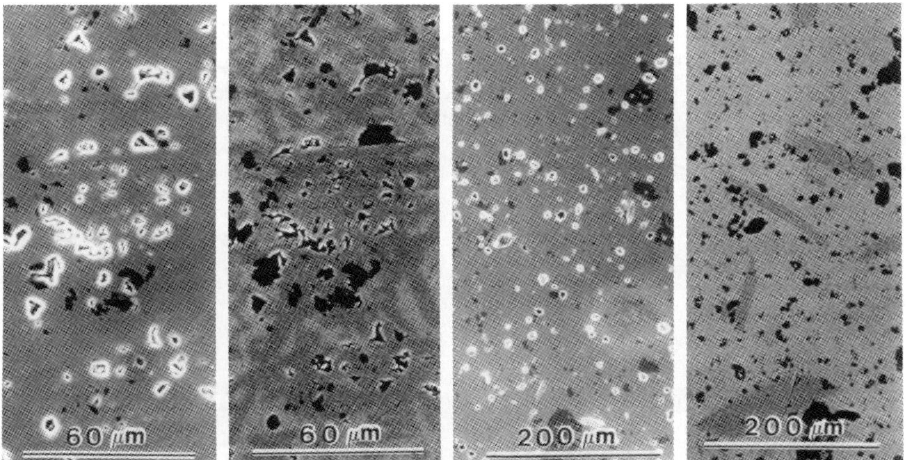

Figure 1: Polished section of a specimen heated at 500°C/min to 1250°C and soaked for 2 h: (a) Secondary electron image (left), where small TiO_2 grains appear as black regions distributed in a (grey) indistinguishable mixture of $Ba_2Ti_9O_{20}$ and $BaTi_4O_9$. The white regions are pores where the white rings are from charging effects. (b) Back scattered image (center left) of the same microstructure showing needle-like $Ba_2Ti_9O_{20}$ (dark gray) dispersed in the $BaTi_4O_9$ (light grey) matrix. Polished section of a specimen heated at 5°C/min to 1250°C and soaked for 2 h: (c) Secondary electron image (center right). (d) Back scattered electron image(right).

Heat treatment of the 0.82%Zr composition resulted in significantly greater stabilization of $Ba_2Ti_9O_{20}$, than when the same concentration of SnO_2 was used (Figure 4). The relative percentages of $Ba_2Ti_9O_{20}$ after this heat treatment decreased with increasing concentration of SnO_2. This was markedly the case for the ZrO_2-substituted samples, with the exception of the highest concentration (2.46%Zr).

After pre-reacting and then sintering at 1390°C for 6 and 16 h, monophase $Ba_2Ti_9O_{20}$ was observed in all specimens doped with tin and zirconium ions; an exception was the 2.46%Zr specimens, which were composed of $Ba_2Ti_9O_{20}$, $BaTi_4O_9$, TiO_2, and ZrO_2. Based on relative XRD intensities, the longer soak period for this latter composition decreased the relative proportion of $Ba_2Ti_9O_{20}$. The porosity increased (in size and frequency) with sintering time for 0.82%Zr. Unreacted titania was present in the 2.46%Zr specimens, which also contained higher porosity than the 0.82%Zr heat-treated for the same time period.

Figure 5 shows that the dimensional density of the sintered pellets increased with increasing SnO_2 content, while the dielectric constant decreased, for both soak periods. Longer soak period, e.g. 16 h vs. 6 h, fostered decreases in densities and dielectric constants. The dielectric constants followed the trend of densities, going from the 0.82% to 1.64%Zr compositions. An exception was the 2.46%Zr specimen which exhibited the highest dielectric constant after soaking for 16 h.

The quality factors of specimens sintered for 6 h increased from 11909 to 12919 as SnO_2 content was increased from 0.82 to 1.64%, and remained largely unchanged for the 2.46%Sn pellets. A similar behavior was indicated for samples soaked for 16 h. For these samples, the soak period had no significant influence on the Q. Increasing the Zr doping level from

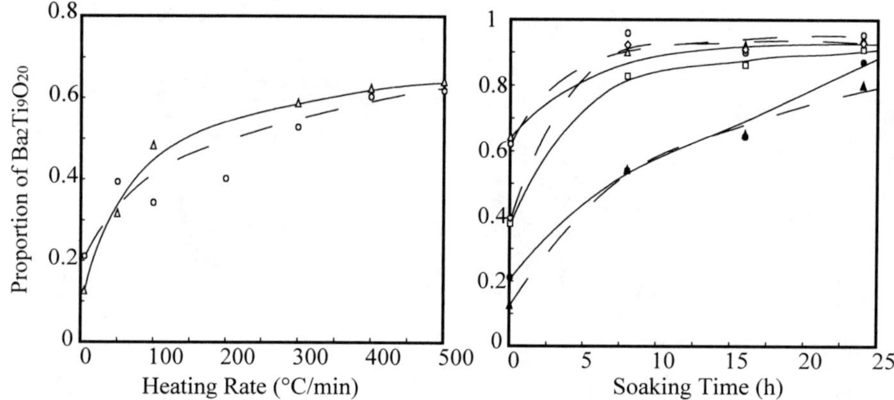

Figure 2: (left) Proportion of $Ba_2Ti_9O_{20}$ to the sum of $BaTi_4O_9$ and $Ba_2Ti_9O_{20}$ measured from XRD relative peak intensities (circles) and SEM image analyses (triangles).

Figure 3: (right) Relative amounts of $Ba_2Ti_9O_{20}$ for samples heated at 1250°C for 2 h, followed by various soaking periods at 1390°C. Triangle, square, and filled triangle are specimens heated at 500, 50, and 5°C/min, respectively, via image analysis. Circle, diamond, and filled circle are specimens heated at 500, 50, and 5°C/min, respectively, via relative XRD peak intensities.

0.82 to 1.64% improved the Q to a maximum (13895). The Q then degraded precipitously in the 2.64%Zr specimens, especially for the long soak period. Figure 6 shows that the Q of all Sn-doped and the 0.82%Zr specimens, sintered for 6 h, decreased by about 14% as temperature increased from 25 to 117°C. The other pellets with ZrO_2 substitutions exhibited a greater diminution in Q with increasing temperature.

DISCUSSION

Under rapid heating to 1250°C, the average grain size of TiO_2 was the same as the starting particle size. However, the TiO_2 agglomerates under slow (e.g. 5°C/min) heating were an order of magnitude larger. Such agglomeration was motivated by reduction of interfacial area by shrinking of smaller (larger surface to volume ratio) TiO_2 grains, while larger ones grew, both via lattice diffusion. Corresponding to the lower heating rates, significantly more time was allotted for such diffusional processes. After slow heating, the larger TiO_2 agglomerations were (comparatively) more stable, as well as less distributed in the microstructure (resulting in greater average diffusion distances between reactants). Hence the high temperature formation of $Ba_2Ti_9O_{20}$ by $2BaTi_4O_9 + TiO_2 \longrightarrow Ba_2Ti_9O_{20}$ occurred more sluggishly. The needle-like $Ba_2Ti_9O_{20}$ phase which formed after slow heating was less frequent, but of significantly larger size (Figure 1). This resulted from the extended time for reaction local to TiO_2 agglomerates, and the plentiful supply of TiO_2 reactants, creating large $Ba_2Ti_9O_{20}$ grains. However, the less frequent presence of TiO_2 grains resulted in less frequent formation of $Ba_2Ti_9O_{20}$ grains.

An attempt at pre-reaction of doped batches at 1150°C showed that mostly $BaTi_4O_9$ formed, implying that $BaTi_4O_9$ formed first, which then further reacted with TiO_2 over

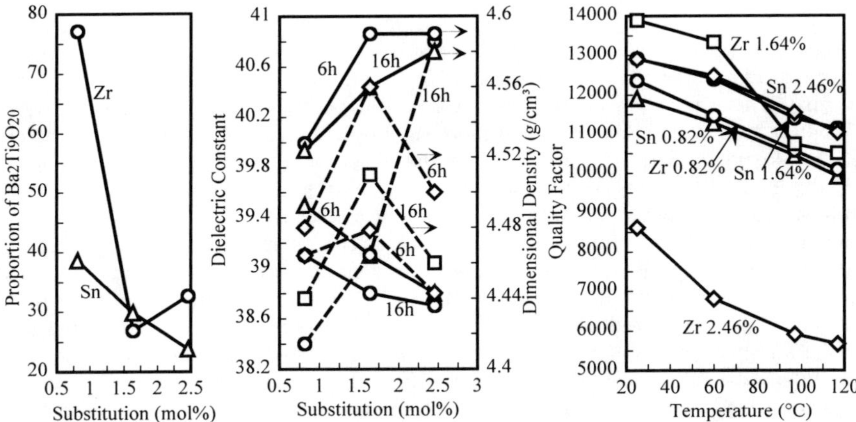

Figure 4: (left) Relative proportion of $Ba_2Ti_9O_{20}$ as a function of ZrO_2 or SnO_2 substitution. Values were determined via ratio of most intense XRD peak height of $Ba_2Ti_9O_{20}$ to the sum of most intense peak heights of $Ba_2Ti_9O_{20}$ and $BaTi_4O_9$.

Figure 5: (center) The effects of SnO_2 (solid lines) and ZrO_2 (dashed lines) additions and sintering times on dimensional densities and dielectric constants in the microwave frequency range (3 GHz).

Figure 6: (right) Influence of tin and zirconium oxide substitutions on Q as a function of temperature. Specimens were sintered at 1390°C for 6 h.

time to form $Ba_2Ti_9O_{20}$. It is postulated that $BaTi_4O_9$ has a higher solubility for large cation replacement of Ti^{4+} than does $Ba_2Ti_9O_{20}$ at 1250°C. If this is the case, higher doping concentrations would resist conversion to $Ba_2Ti_9O_{20}$ since some of the soluble dopant would have to be rejected. Since formation of $Ba_2Ti_9O_{20}$ with no doping requires significantly more time (and a soaking temperature of 1390°C), the dilation of the $BaTi_4O_9$ unit cell via small concentrations of dopants appears to facilitate reaction with TiO_2 to most rapidly form $Ba_2Ti_9O_{20}$ at 1390°C.

Sintering at 1390°C for 6 and 16 h resulted in formation of monophase $Ba_2Ti_9O_{20}$ for all substituted compositions, except the 2.46%Zr batch which still contained unreacted ZrO_2. This indicates that the solubility of Sn^{4+} in $Ba_2Ti_9O_{20}$ was higher than that of Zr^{4+}. This can be attributed to the smaller ionic size of the tin ion, compared to zirconium, for substitution into Ti^{4+} positions with less distortion.

All specimens sintered for 6 h had either equal or higher densities than those after 16 h. This suggests volitalization of interior regions over time. Higher porosity was apparent in the microstructures soaked for 16 h. In addition, thermogravimetric analysis was used, which showed a 0.044 wt%/h during an identical thermal schedule. Above the incongruent melting temperature of $BaTi_4O_9$, TiO_2 is in equilibrium with liquid phase. Dopant additions likely locally lowered the liquid-forming temperature below 1390°C. Such grain boundary liquid phase then volitalized over time, leaving increased porosity.

The increasing dimensional density of specimens with increasing SnO_2 content corresponds to the substitution of titanium ions with heavier tin ions, where the mass of the unit cell was increased more than its volume was dilated. The same explanation may be used in the Zr-doped specimens. An exception was the over-doped 2.46%Zr case, where lower density second phases lowered the overall dimensional density. The dielectric constant decreased with increasing soak period (for a given composition), with the exception of 2.46%Zr. The porosity formed with the 16 h soak introduced low dielectric constant vapor gaps in the material, decreasing the overall dielectric constant. For the 2.46%Zr sample, after 16 h soaking at 1390°C, a more significant percentage of TiO_2 was detected by XRD (as compared to after 6 h). Since the dielectric constant of TiO_2 ($\sim$100[2]) is appreciably higher than $Ba_2Ti_9O_{20}$, the excess TiO_2 content in the 2.46%Zr sample soaked for 16 h was responsible for the high measured dielectric constant.

Soaking time showed a minimal effect on the Q. Dielectric loss is known to be only minimally influenced by levels of porosity, since pores contain vapor which do not contribute to energy loss unless moisture is trapped inside the pores[11]. However, Q is strongly dependent on the condensed phases present in the microstructures. The degradation of Q in 2.46%Zr specimens was a consequence of dielectric losses via anharmonicities associated with boundaries between multiple phases[12]. When Q was measured with increasing temperature, thermal agitation disturbed the harmonic resonance of the oscillators (electron clouds and ions), leading to greater losses. The effects of dopants on temperature coefficients of resonance frequency were reported elsewhere[13].

CONCLUSION AND FUTURE WORK

The feasibility of fabricating phase-pure undoped $Ba_2Ti_9O_{20}$ was clearly proven in this work. Introduction of Zr and Sn solid solution ions facilitates formation of the compound. The trends in microwave properties with dopant concentration, soak period, and measurement temperature are reported herein. Microwave property measurements of the undoped compound is presently being undertaken.

References

[1] T. Negas, G. Yeager, S. Bell, N. Coats, and I. Minis, Am. Ceram. Soc. Bull. **72**, 80 (1993).

[2] A.J. Moulson and J.M. Herbert, *Electroceramics: Materials, Properties, and Applications,* (Chapman and Hall, New York, 1990).

[3] G.H. Jonker and W. Kwestroo, J. Am. Ceram. Soc. **41**, 390 (1958).

[4] J.-M. Wu and H.-W. Wang, J. Am. Ceram. Soc. **71**, 869 (1988).

[5] H.M. O'Bryan, J. Thompson, and J.K. Plourde, J. Am. Ceram. Soc. **57**, 450 (1974).

[6] J. Yu, H. Zhao, J. Wang and F. Xia, J. Am. Ceram. Soc. **77**, 1052 (1994).

[7] W. Hackenberger and R.F. Speyer, Rev. Sci. Instrum. **65**, 701 (1994).

[8] W.-Y. Lin, and R.F. Speyer, submitted for publication, *Journal of Materials Research,* November, 1996.

[9] J.K. Plourde, and D.F. Linn H.M. O'Bryan, J. Thompson, J. Am. Ceram. Soc. **58**, 418 (1975).

[10] Y. Kobayashi and M. Katoh, IEEE Trans. on Microwave Theory Tech. **MTT-33**, 586 (1985).

[11] H. Tamura, Am. Ceram. Soc. Bull. **73**, 92 (1994).

[12] B.D. Silverman, Phys. Rev. **125**, 1921 (1962).

[13] W.-Y. Lin, R.F. Speyer, W.S. Hackenberger, and T.R. Shrout, submitted for publication, *Journal of American Ceramic Society,* December, 1996.

EFFECT OF DEFECTS ON DIELECTRIC PROPERTIES IN KTiOPO$_4$, KTiOAsO$_4$, RbTiOAsO$_4$ AND CsTiOAsO$_4$ SINGLE CRYSTALS

A. R. Guo*, Z. –Y. Cheng*, R. S. Katiyar*, Ruyan Guo** and A. S. Bhalla**
*   Department of Physics, University of Puerto Rico, P. O. Box 23343, San Juan, PR 00931-3343, USA
** Materials Research Laboratory, The Pennsylvania State University, University Park, PA 16802-4801, USA

ABSTRACT

Dielectric measurements were carried out in single crystals of KTiOPO$_4$, KTiOAsO$_4$, RbTiOAsO$_4$ and CsTiOAsO$_4$. All of the materials exhibit a clear dielectric relaxation process in the low temperature range and a conductance mechanism in the high temperature range. The dielectric relaxation process can be well described by the Debye dielectric model with an activation energies of   0.8 eV, 0.5 eV and 0.4 eV respectively. The relaxation process is associated with the deviation of the alkali ions from its ideal lattice positions. The high temperature conductance is associated with the motion of the alkali ions from one lattice site to another. Therefore, both the low temperature relaxation process and the high temperature conductance originate from different features of defect behavior of alkali ions in the cage structure of these materials.

INTRODUCTION

Materials with the general formula M$^+$TiOX^{5+}O$_4$, where M=[K, Rb, Cs, Tl] and X=[P, As], have been of immense interest to the applied and fundamental research in recent years, because most of these materials possess rather large nonlinear and electro-optic coefficients, large birefringence and low dielectric constants [1-14]. Research on these materials has mainly been focused on the fabrication of the materials, the determination of the structure, analysis of the phase transition, measurements of the nonlinear optical and electro-optic properties, *etc* [1-12]. Some of these materials have been used in many commercial devices. However, the research on the dielectric behavior of these materials is relatively limited due to the fact that it is known to have high ionic conductivity at higher temperatures. Moreover, there are conflicting reports whether the materials are ferroelectrics and also what is the mechanism for the origin of polarization in these materials [13-14].

The high ionic conductivity of the materials not only limits some applications of these materials, but also makes the measurements on ferroelectric properties and polarization behavior difficult because the phase transition temperature of these materials is generally very high (600°C ~ 950°C) [1]. Thus, the high ionic conductance limits the research on the polarization mechanism of the materials. However, the ability to characterize, predict, and control the ionic conductivity of the materials is of tremendous technological and scientific importance. In order to do these, the studies on both the dielectric properties and micro-mechanism are necessary and important. Although the conductivity is known to be mainly contributed by the movement of M$^+$ ions in the above materials [7,12], to our knowledge, there is no prior systematic dielectric studies on these materials in a wide temperature range.

Mat. Res. Soc. Symp. Proc. Vol. 453 © 1997 Materials Research Society

In this paper, we report the experimental results of dielectric properties of KTiOPO$_4$ (KTP), KTiOAsO$_4$ (KTA), RbTiOAsO$_4$ (RTA), and CsTiOAsO$_4$ (CTA) single crystals in the temperature range from −170°C to 200°C. A dielectric relaxation process is observed in all of the materials at low temperatures for the first time. The high ionic conductance appears in the high temperature region. The micro-mechanism of both the relaxation process and high temperature conductance are discussed with the different defect features of alkali ions in the materials.

EXPERIMENT

Single crystals of KTP, KTA, RTA, and CTA were grown using a tungstate flux by a process which is very similar to that previously described for growing KTP [17]. The crystals were oriented by x-ray and were cut into rectangular shapes having (100), (010) and (001) faces. The gold electrodes were deposited by vacuum sputtering on two (001) surfaces. Thus, the dielectric properties of the materials, the capacitor and the loss of the samples, along <001> direction are measured with an HP 4275A LCR meter. The dielectric constants are calculated from the measured capacitance, with the geometric parameters of the samples and using a parallel-plate capacitance model.

The experimental procedure and the parameters used in the measurement are the following: the sample is at first heated to a temperature higher than 100°C from room temperature, then cooled down to −170°C. After that, the sample is heated again to 200°C from −170°C. Both the heating and the cooling rate is 4°C per minute. The dielectric properties were measured on-line during both heating processes and one cooling process. The oscillating level of the measuring signal was about 1 V/mm. The chosen frequencies are 10, 20, 100, 200 kHz, 1 and 2 MHz. In order to increase the measuring accuracy, the four test leads were used and a standard calibration process was performed before each measurement to remove any stray capacitance, lead and contact resistance. All systems were controlled and the measured results were recorded using a PC computer.

RESULTS AND DISCUSSION

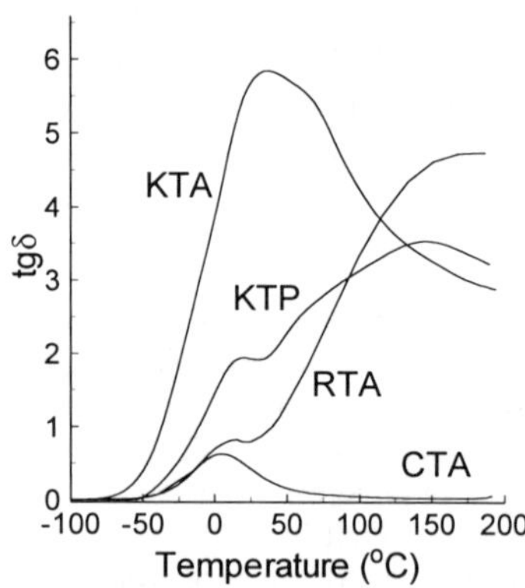

Fig. 1. Temperature dependence of dielectric loss (tgδ) for all the samples (KTP, KTA, RTA, CTA) at 10kHz.

The temperature dependence of both the dielectric constant and loss for all of the samples at the same frequency are shown in Fig. 1. A dielectric loss maximum appears in its temperature dependence. To confirm this results, the temperature dependence of the dielectric constant and loss during heating and cooling process are compared. It is found that both experimental approaches have the same results. This indicates that the materials have a dielectric relaxation process, which is the common behavior of all materials, and that the relaxation behavior is the intrinsic characteristics of the materials, because the experimental processes do not influence the experimental results. Therefore, the results obtained during heating from −170°C to 200°C are used in the following discussion.

Because the phase transition temperatures of the above materials are very high, besides the contributions by the relaxation process, the dielectric constant contributed by the polarization of electrons and ions should be nearly independent of the temperature and frequency in the studied temperature range. The intrinsic dielectric loss must be very small beside the contribution of the conduction and the relaxation process to the dielectric loss. Based on the general dielectric theory, the dielectric relaxation strength for the relaxation process are nearly independent of the temperature in limited temperature range. Thus, we can use the temperature dependence of both the dielectric constant and dielectric absorption at a frequency to study the relaxation process by the Cole-Cole plot, which will make relaxation behavior more clear and will give us additional information about the process. The Cole-Cole curves of the materials at the same frequency are given in Fig. 2. The dielectric relaxation process in the materials can clearly be seen from the Cole-Cole plot in Fig. 2. Because all of the materials have the same structure, we can assume that they have the same relaxation mechanism and relaxation characteristics. Thus, we can use one of them to study the relaxation process. Among the four studied materials, the conductance of CTA is the smallest. Thus, it is the best candidate to study the relaxation process. Based on the results of CTA, we find that the relaxation process can be well described by the following Cole-Cole formula as shown in Fig. 3.

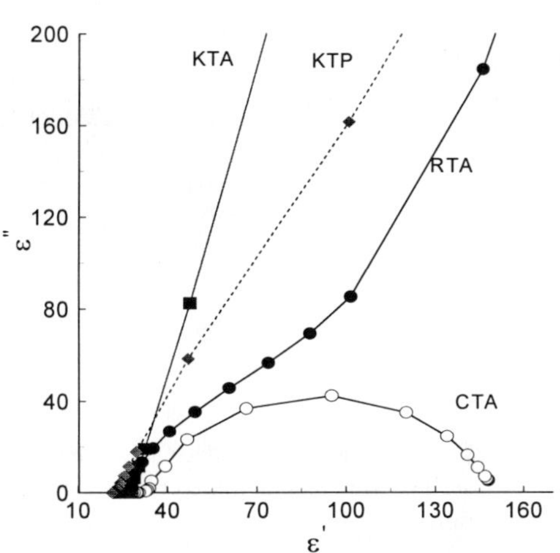

Fig. 2. The Cole-Cole plot of the relaxation process in the materials.

$$\varepsilon^{*} = \varepsilon^{'} - i\varepsilon^{''} = \varepsilon_{\infty} + \frac{\varepsilon_{S} - \varepsilon_{\infty}}{1 + \left(i\omega\tau_{o}\right)^{h}} \tag{1}$$

where ω is the angular frequency, τ_0 the center relaxation time, h ($0<h\leq1$) a constant, ε_S and ε_∞ the static and high frequency dielectric constants respectively. The best-fitted parameters for CTA are: $h=0.83$, $\varepsilon_S=149$ and $\varepsilon_\infty=35$.

Based on Eq. (1) and above discussion, we can conclude that the loss maximum at a frequency (ω) appears at the temperature (T_m) when:

$$\omega\tau_0(T_m) \approx 1 \tag{2}$$

Thus, we can use the experimental frequency dependence of T_m to study the temperature dependence of τ_0. We find that the temperature dependence of τ_0 can be well described by the Debye relation as follows:

$$\tau_0 = \tau_\infty \exp\left(\frac{T_0}{T}\right) \tag{3}$$

where τ_∞ is the inverse of the attempt frequency of the dipole, T_0 the equivalent temperature of the activation energy.

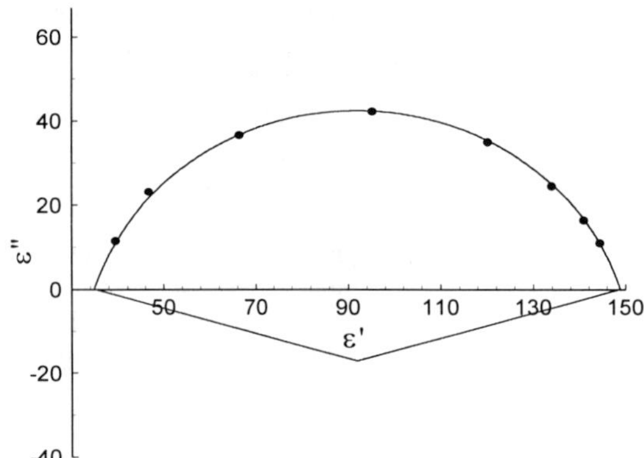

Fig. 3. The Cole-Cole plot for CTA at 10 khz. The best-fitted parameters are: $\varepsilon_\infty=35$, $\varepsilon_s=149$, h=0.83.

Fitting the experimental results with Eqs. (2) and (3), we get the activation energy of the relaxation process, which is shown in Table I. All of the above results indicate that the relaxation process originates from a kind of the dipole relaxation.

TABLE I. Best-fitted parameters of Eqs. (3) and (5) for all the materials

	KTP	KTA	RTA	CTA
Polarizability of M^+ ion ($10^{-40}Fm^2$)	0.922	0.922	1.56	2.69
T_0 (Relaxation) (10^3 K)	N/A	9.10	5.83	4.63
Radius of M^+ ion (10^{-10}m)	1.33	1.33	1.48	1.69
T_A (Conduction) (10^3 K)	4.05	3.93	4.66	N/A

When the temperature is much higher than T_m, the dielectric absorption contributed by the relaxation process can be ignored, so the temperature dependence of the dielectric absorption at high temperatures can be used to study the conductance behavior of materials. The conductance (σ) of the materials is:

$$\sigma = \omega \varepsilon_0 \varepsilon'' \tag{4}$$

Based on the general conductance theory, of the materials is dependent on the temperature by the following relation:

$$\sigma(T) = \sigma_0 \exp\left(-\frac{T_A}{T}\right) \tag{5}$$

where σ_0 is a constant, T_A the activation energy of the conductance.

Fitting the experimental results, the relationship between $ln\varepsilon''$ and $1/T$, in the high temperature range with Eq. (5), we get the best-fitted parameters, the activation energy (T_A), which are shown in Table I.

It is known that the micro-mechanism of the high temperature conduction is due to the moving of alkali ions (K^+, Rb^+, Cs^+) in the KTP family materials [7,12]. That is, the high temperature conductance originates from the defect of alkali ions in the materials. In the structure of these materials, the alkali ions have more than one site to occupy [18]. So, the alkali ions are easy to move from one site to the other, as a result these produce the high temperature conduction. Thus, the activation energy of the conductance are related to the radius of the alkali ions and the size of the lattice. The bigger the alkali ion, the higher the activation energy. For the largest Cs^+, the activation energy is so high that the conductance is too small to measure for temperatures lower than 200°C.

In each lattice position of alkali ions, there is a big space which is much larger than the volume of the alkali ion [18]. Thus, the equilibrium position of the alkali ions is not at the ideal lattice site. Therefore, an alkali ion in the materials is similar to a dipole, which is something like the Li^+ ion in the solid solution of $(K_{1-x}Li_x)NbO_3$ [19]. Thus, the dipoles originate from the defects of alkali ions in the materials. That is, besides moving from one site to another, the alkali ions also vibrate in each lattice position among the equilibrium positions. The vibration of alkali ions among the equilibrium positions produces a dielectric relaxation process. Based on the above discussion, the activation energy of the relaxation process should decrease with increasing polarizability of the alkali ions. In the materials studied, the alkali ions have two kind of positions to occupy and the vibration behavior of the alkali ions are influenced by the near neighbor ions. Thus, the relaxation time of the vibration of alkali ions must have a distribution, but it is not very broad. This is the mechanism of the dipole relaxation process observed experimentally in the low temperature range.

CONCLUSION

In the electrical property measurements of the KTP family materials, beside the contribution from the high-temperature conduction, these materials exhibit a clear dielectric relaxation process in the temperature range below 200°C. The relaxation process is a dipolar relaxation. The temperature dependence of the relaxation time follows the Debye model. The relaxation process can be well described by the Cole-Cole formula. The activation energy of the relaxation process in this family of materials increases with increasing the polarizability of the alkali ion, while the activation energy of the high temperature conduction increases with increasing the radius of the alkali ion. All of these properties were explained on the basis of the open crystallographic structure of the KTP family materials with the defect feature of the alkali ions, that is, the deviation of alkali ions from its ideal position. The position and freedom of movements of alkali ion, from its ideal lattice position in the structure, determines both the conduction and relaxation behavior of the materials.

ACKNOWLEDGMENTS

We thank Mr. R. Tao and Dr. J. F. Meng for helpful discussions. This work is supported by the DOE Grant DE-FG02-94ER75674, and NSF-OSR-9452893.

REFERENCE

1. P. A. Thomas, A. M. Glazer, B. E. Watts, Acta Cryst. **B46**, 333-343 (1990).
2. Z. Y. Ou, S. F. Peereira, E. S. Polzik and H. J. Kimble, Opt. Lett. **17**, 640 (1992).
3. L. K. Cheng, L. T. Cheng, J. D. Bierlein, J. Parise, Appl. Phys. Lett. **64**, 1321 (1994).
4. G. Marnier, B. Boulanger and B. Menaert, J. Phys.: Condens. Matter, **1**, 5509 (1989).
5. P. A. Thomas, S. C. Mayo, B. E. Watts, Acta Cryst. **B48**, 401-407 (1992).
6. G. M. Loiacono, D. N. Loiacono, R. A. Stolzenberger, J. Cryst. Growth, **131**, 323-330 (1993).
7. L. T. Cheng, L. K. Cheng, J. D. Bierlein, F. C. Zumsteg, Appl. Phys. Lett. **63**, 2618-20 (1993).
8. H. P. Li and X. Wang, Ferroelectrics, **17**, 91 (1994).
9. C. -S. Tu, A. R. Guo, R. W. Tao, R. S. Katiyar, R. Guo, A. S. Bhalla, J. Appl. Phys. **76**, 3235-40 (1996).
10. L. K. Cheng, L. -T. Cheng, J. D. Bierlein, F. C. Zumsteg, and A. A. Ballman, Appl. Phys. Lett, **62**, 346 (1993).
11. A. R. Guo, C. -S. Tu, R. Tao, R. S. Katiyar, R. Guo and A. S. Bhalla, Ferroelectric letter, **21**, 71 (1996).
12. G. M. Loiacono, D. N. Loiacono, J. J. Zola, and R. A. Stolzenberger, T. McGee and R. G. Norwood, Appl. Phys. Lett. **61**, 895 (1992).
13. V. D. Kugel, G. Rosenman, N. Angert, E. Yaschin, and M. Roth, J. Appl. Phys. **76**, 4823 (1994).
14. Y. V. Shaldin and R. Poprawski, Ferroelectrics, **106**, 399 (1990).
15. L. K. Cheng and J. D. Bierlein, Ferroelectrics, **142**, 209 (1993).
16. I. Tordjman, R. Masse and J. C. Guitel, Z. Kristallogr. **139**, 103-115 (1974).
17. A. A. Ballman, H. Brown, D. H. Olson, and C. E. Rico, J. Cryst. Growth, **75**, 390 (1986).
18. I. Tordjman, R. Masse and J. C. Guitel, Z. Kristallogr. Bd, **139**, 12, (1974).
19. U. T. Hochli, K. Knorr, A. Loidl, Adv. Phys., **39**, 405 (1990).
20. V. Kalesinskas, N. Pavlova, I. Rez, and J. Grigas. Liet. Fiz. Rink. **22**, 87 (1982).

SYNTHESIS OF FERROELECTRIC STRONTIUM BISMUTH TANTALATE FILMS FROM METAL ALKOXIDE PRECURSORS

WEI-WEI ZHUANG,* LUMEI LIU,* NAIJUAN WU,** ZHIDONG HAO,** DAVID M. HOFFMAN,* ALEX IGNATIEV,** ALLAN J. JACOBSON* AND SCOTT S. PERRY*
*Department of Chemistry, University of Houston, Houston, TX 77204
**Texas Center for Superconductivity and Space Vacuum Epitaxy Center, University of Houston, Houston, TX 77204

ABSTRACT

Ferroelectric $SrBi_2Ta_2O_9$ (SBT) films were prepared by the spin coating technique on platinum, quartz and $YBa_2Cu_3O_{7-x}$/$LaAlO_3$ substrates from a methoxyethanol solution of bismuth isopropoxide $(Bi(OCH(CH_3)_2)_3)$ and strontium tantalum isopropoxide $(SrTa_2(OCH(CH_3)_2)_{12})$. X-Ray diffraction studies showed some crystallization occurred after annealing the films under oxygen flow at 600 °C and excellent crystallinity was achieved after annealing at 750 °C for 0.5 h. Electron microprobe analysis gave a composition close to that expected for SBT, and atomic force microscopy gave a root mean square surface roughness of 101 Å. An hysteresis measurement (1 kHz) gave remnant polarization $(2P_r)$, saturation polarization (P_s) and coercive field (E_c) values of 14.5 $\mu C/cm^2$, 14.5 $\mu C/cm^2$ and 59 kV/cm, respectively.

INTRODUCTION

There is strong current interest in developing solid state non-volatile ferroelectric random access memory (FRAM) technology. FRAMs have substantial advantages over normal nonvolatile memories including faster write speeds, the ability to operate at low voltages and higher resistance to radiation damage [1]. Ferroelectric fatigue, imprint, retention and the integration of FRAMs into silicon-based transistor technology are major problems in achieving wide-spread FRAM use [2]. The recently discovered ferroelectric material $SrBi_2Ta_2O_9$ (SBT) has shown significant improvements in fatigue, imprint and retention when integrated with Pt electrodes [3]. The current spray/spin-on synthetic methods for SBT film growth, however, require annealing temperatures that are too high (>800 °C) for standard semiconductor processing technologies [4].

The purpose of this work was to develop new precursors for spin-coated SBT films [5] and to examine integration of SBT with conducting oxide electrodes, which could further enhance ferroelectric properties, including SBT's low polarization, as demonstrated for the $PbMO_3$ (M = Ti, Zr) class of ferroelectrics [6]. Previously, spin coated SBT films have been prepared on Pt by using 2-ethylhexanoate complexes and/or 2-ethylhexanoic acid in the precursor solution preparation [7,8]. Herein we report the preparation of high quality SBT films at 750 °C on Pt, quartz and $YBa_2Cu_3O_{7-x}$ substrates from a methoxyethanol solution of metal alkoxide precursors.

EXPERIMENTAL

Bismuth isopropoxide $(Bi(OCH(CH_3)_2)_3)$ and strontium tantalum isopropoxide $(SrTa_2(OCH(CH_3)_2)_{12})$ were used to prepare the precursor solution. Both complexes were synthesized by following established literature procedures [9,10]. The strontium-tantalum

Mat. Res. Soc. Symp. Proc. Vol. 453 © 1997 Materials Research Society

compound was purified by crystallizing from a cold saturated toluene solution and the bismuth compound was precipitated from the hexane reaction solution and then washed with hexane.

Manipulations of the precursor solution were carried out inside a nitrogen-filled glovebox or by using standard Schlenk techniques. Bismuth isopropoxide (0.80 g, 2.1 mmol) and $SrTa_2(OCH(CH_3)_2)_{12}$ (1.2 g, 1.0 mmol) were added to dry, degassed methoxyethanol (10 mL), giving an approximately 0.1 M solution of "$SrBi_2Ta_2(OR)_{18}$." The mixture was heated under nitrogen in a closed flask at 90 °C for 3 hours. After cooling to room temperature, the clear pale yellow solution was filtered through a 0.2 μm filter and then stored under nitrogen until it was needed.

Depositions were carried out on polished platinum, quartz and $YBa_2Cu_3O_{7-x}$ (YBCO) substrates. The platinum substrates (1 mm x 1 cm^2) were polished by using 1 μm Metsup 70 (Metallurgical Supply Co., Inc., Friendswood, TX). They and the quartz substrates were prepared for deposition by soaking in Chromerge®/concentrated H_2SO_4, exhaustively rinsing with de-ionized water, washing with acetone and finally heating in air at 500 °C for 1 h. The substrates were then stored in an oven (200 °C) until they were needed. The YBCO substrates, which were used because of their metallic conductivity at room temperature and their chemical and structural compatibility with SBT, were prepared by pulsed laser deposition of YBCO (2500 Å) on $LaAlO_3$ substrates.

RESULTS AND DISCUSSION

Depositions were carried out by using the spin coating method. A flow diagram of the deposition procedure is given in Figure 1. The same procedure was used for all three types of substrates except that the coatings on YBCO were heated at 450 °C for 2 h under O_2 flow after annealing the films to reactivate the substrate conductivity.

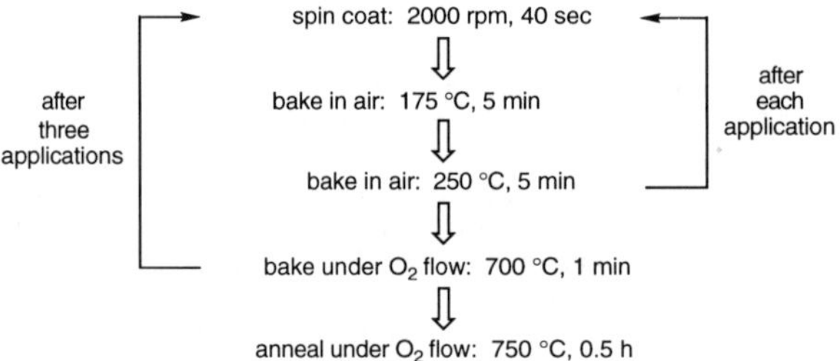

Figure 1. Spin coating procedure used to deposit SBT on platinum, quartz and YBCO substrates.

Electron microprobe analysis gave the results in Table I for a film prepared from six sequential solution applications. The composition data indicates the films are Sr rich, and the film thickness measurement suggests that each application from the precursor solution added approximately 700 Å to the film. X-ray diffraction studies for films deposited on platinum and quartz substrates indicate that the onset of crystallization is ≈600 °C and excellent crystallinity is achieved on both substrates by annealing at 750 °C for 0.5 h (Figure 2)

Table I. Composition of a film deposited on quartz from six applications of 0.1 M precursor solution.[a]

Sr	Bi	Ta	thickness (Å)
1.09	1.97	1.98	4190

[a] The error in composition is ±2% and in thickness ±30 Å.

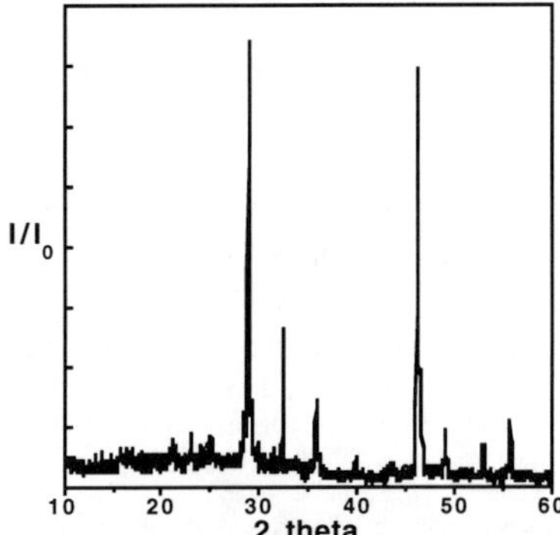
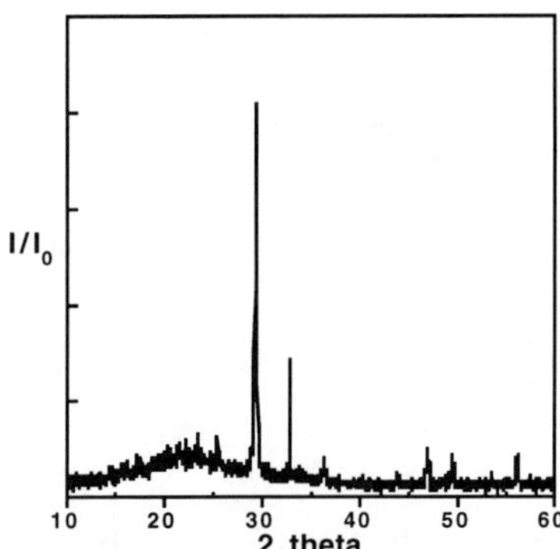

Figure 2. X-ray powder pattern for SBT films approximately 6300-Å thick on platinum (left) and quartz (right) deposited by using the procedure outlined in Fig. 1.

The surface morphology of a film approximately 8400-Å thick on YBCO was examined by atomic force microscopy (AFM). The topographic images (Figure 3) show the film is continuous and pinhole-free with an rms surface roughness of 101 Å. The surface roughness is calculated as the rms deviation in the surface height for each pixel of the image from the height of a best fit plane.

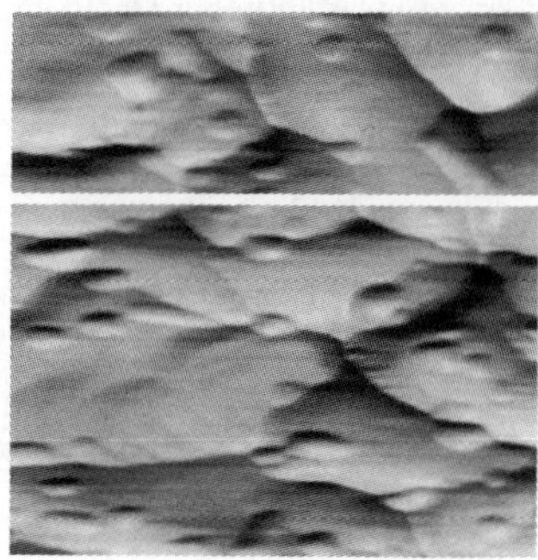
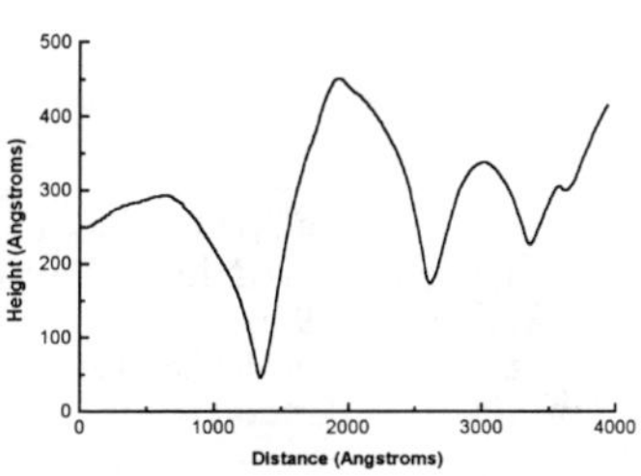

Figure 3. 4000 x 4000 Å topographic AFM image (left) and a plot of the surface topography for the line in the upper part of the image (right).

The ferroelectric properties of an SBT film approximately 6700-Å thick on Pt were examined (Figure 4). The Pt substrate was the bottom electrode in the measurements and sputter-deposited gold dots with areas of approximately 2.6×10^{-4} cm^2 were the top electrodes. A hysteresis (1 kHz) measurement gave remnant polarization ($2P_r$), saturation polarization (P_s) and coercive field (E_c) values of 14.5 μC/cm^2, 14.5 μC/cm^2 and 59 kV/cm, respectively. A fatigue measurement at 1 MHz and ±5 V (i.e., at less than the saturation polarization) showed there was no significant degradation after 10^{10} cycles.

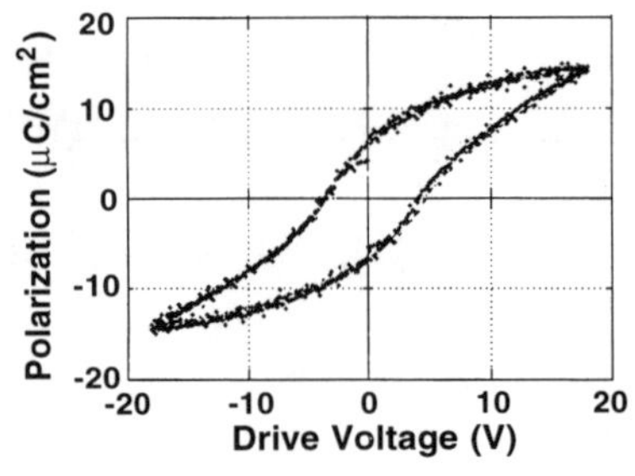
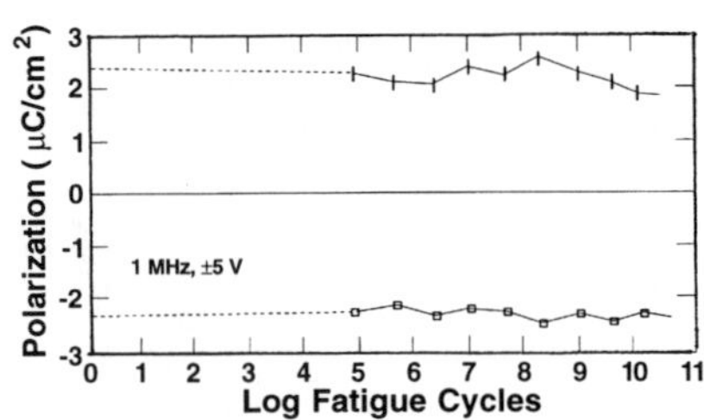

Figure 4. Hysteresis loop (left) and fatigue (right) measurements for an SBT film on Pt.

CONCLUSIONS

High quality $SrBi_2Ta_2O_9$ films were prepared by spin coating from a mixture of bismuth isopropoxide and strontium tantalum isopropoxide in methoxyethanol at a processing temperature of 750 °C. Current work is focused on lowering the annealing temperature and isolating putative Sr-Bi-Ta alkoxide clusters from the precursor solution.

ACKNOWLEDGEMENTS

We thank the Texas Center for Superconductivity, Robert A. Welch Foundation, NASA and the Texas Advanced Technology Program for funding. This work was also supported in part by the MRSEC program of the National Science Foundation under award number DMR-9632667.

REFERENCES

1. Y. Miyasaka, "Material and Process Technologies for High Density Ferroelectric Non-Volatile Memories Developed at NEC," 8th International Symposium on Integrated Ferroelectrics, (Tempe, AZ, March 1996).
2. R. E. Jones, S. B. Desu, MRS Bulletin 21, 55 (1996).
3. C. A. Araujo, J. D. Cuchiaro, L. D. McMillan, M. C. Scott, J. F. Scott, Nature, 374, 627 (1995).
4. O. Auciello, R. Ramesh, MRS Bulletin 21, 31 (1996).
5. B. A. Tuttle, R. W. Schwartz, MRS Bulletin 21, 49 (1996).
6. H. Lin, N. J. Wu, K. Xie, Y. Y. Li, A. Ignatiev, Integrated Ferroelectrics, 5, 197 (1994).
7. H. Watanabe, T. Mihara, H. Yoshimori, C. A. Paz de Araujo, Jpn. J. Appl. Phys. 34, 5240 (1995).
8. K. Amanuma, T. Hase, Y. Miyasaka, Appl. Phys. Lett. 66, 221 (1995).
9. M. A. Matchett, M. Y. Chiang, W. E. Buhro, Inorg. Chem. 29, 358 (1990).
10. S. Govil, P. N. Kapoor, R. C. Mehrotra, J. Inorg. Nucl. Chem. 38, 172 (1976).

Part VIII

Solid-State Ionics

NON-STOICHIOMETRY AND STRUCTURE OF $SrCe_{1-x}Yb_xO_3$ PEROVSKITE-TYPE OXIDES

IGOR KOSACKI , MARK SHUMSKY AND HARLAN U.ANDERSON
Department of Ceramic Engineering , University of Missouri-Rolla , Rolla , MO 65401 , USA

ABSTRACT

The structural and electrical properties of $SrCe_{1-x}Yb_xO_3$ ceramics have been studied as a function of temperature and Yb-concentration using x-ray diffraction and impedance techniques. The influence of Yb-dopants on electrical transport and structural disorder has been studied. A correlation between the structural properties, electrical conductivity is observed and discussed. These measurements allow us to determine the mechanism of charge carrier compensation and also the concentration and mobility of the electrical species.

INTRODUCTION

The $SrCeO_3$-based oxides are materials where the type of electrical conductivity is controlled by impurity level and ambient atmosphere. As was first reported by Iwahara [1] , these materials when doped with trivalent cations, such as - Yb, Y and Sc exhibit protonic conductivity in hydrogen atmospheres. The study of the electrical transport of acceptor doped $SrCeO_3$ in atmospheres with different hydrogen partial pressures showed a $P_{H2}^{+1/4}$ behavior which is attributed to electronic conductivity [2,3] , while in oxygen atmosphere this material exhibits hole, ionic and electronic conductivity at high, intermediate and low oxygen partial pressure respectively [4,5]. These relationships of the electrical conductivity in different atmospheres makes this material very attractive for use in semiconductor-type gas sensors. For example a hydrogen sensor based on a $SrCe_{1-x}Yb_xO_3$ thin film has been recently proposed [6,7].

The correlation of the stoichiometry of the oxide with impurity concentration, ambient atmosphere and microstructure is very important from the applications point of view. The substitution of the trivalent elements-Yb, Y and Sc for Ce^{4+} ions plays an important role in the electrical transport since this substitution creates negatively charged point defects - $[Me_{Ce}']$ which are believed to be compensated by oxygen vacancies with one vacancy for every two Me^{3+} ions ($[Me_{Ce}']=2[V_o^{..}]$) [2,4,8,9,]. The charge compensation in acceptor doped $SrCeO_3$ can be also occur by electron holes which has been observed at high P_{O2} atmospheres by many authors [4,10,11].

The generation of oxygen vacancies in $SrCe_{1-x}Yb_xO_3$ leads to the structural disorder which has been observed in Raman scattering [12]. Analyze of strain can also give additional information on structural disorder. In the present work we studied the relationship between the electrical conductivity and the structure of $SrCe_{1-x}Yb_xO_3$ as a function of Yb-concentration and temperature. The influence of the structural disorder connected with the generation of oxygen vacancies on the lattice parameters , strain and the electrical conductivity was studied by x-ray diffraction and impedance spectroscopy. These measurements allowed us to determine the concentration and mobility of the electrical species.

EXPERIMENTAL

$SrCe_{1-x}Yb_xO_3$ ceramic samples with x= 0.00 ; 0.02 ; 0.05 and 0.20 were prepared from oxide powders of $SrCO_3$, CeO_2 and Yb_2O_3 using a procedure reported by Iwahara [13]. The

Mat. Res. Soc. Symp. Proc. Vol. 453 © 1997 Materials Research Society

structure of the specimens was analyzed by x-ray diffraction patterns using a XDS-2000 SCINTAG spectrometer. Specimens were analyzed with Si powder as an internal standard at room temperature using $CuK\alpha_1$ radiation λ=1.54060A. This experimental procedure allowed the determination of the d-spacing to an accuracy of 0.0005A. The x-ray patterns were analyzed with the Rietveld program to obtain the distances between ions. Initial estimates of the atom positions were taken from data of Knight and Bananos [14]. The analyze of the diffraction line profiles was performed using the Shadow program which yielded estimates of both the lattice parameters and strain [15]. From scanning electron microscopy (SEM) measurements, the grain size of all of the $SrCe_{1-x}Yb_xO_3$ specimens was found to be about 5μm, so we assumed that any broadening of x-ray lines was due to the strain. In this model, the broadening of x-ray line, (β_ε) is related to the strain, $\varepsilon=\Delta d/d$ by the equation - $\beta_\varepsilon = 4\varepsilon \tan\Theta$, [16] and than determined from the slope of the function $\beta_\varepsilon \cos\Theta$ vs. $4\sin\Theta$, where Θ is the diffraction angle.

The electrical conductivity of $SrCe_{1-x}Yb_xO_3$ was studied using a Solartron 1260 Frequency Analyzer by the determination of the impedance spectra in the frequency range of 10^{-3}Hz to 10^7Hz in the temperatures range of 200-1000°C in air. The bulk equilibrium conductivity values were obtained from the real axis intercept of the high frequency semicircle obtained in the impedance spectra. Platinum electrodes were used in all electrical measurements.

RESULTS

Figure 1 shows x-ray diffraction patterns obtained for $SrCe_{1-x}Yb_xO_3$. These spectra compare well with the standards indicating a single perovskite phases. However, with increasing Yb-concentration , we observed a change in the crystal symmetry from orthorhombic presented for x=0.00 , 0.02 and 0.05 to tetragonal for x=0.20. This observation agrees with the structural

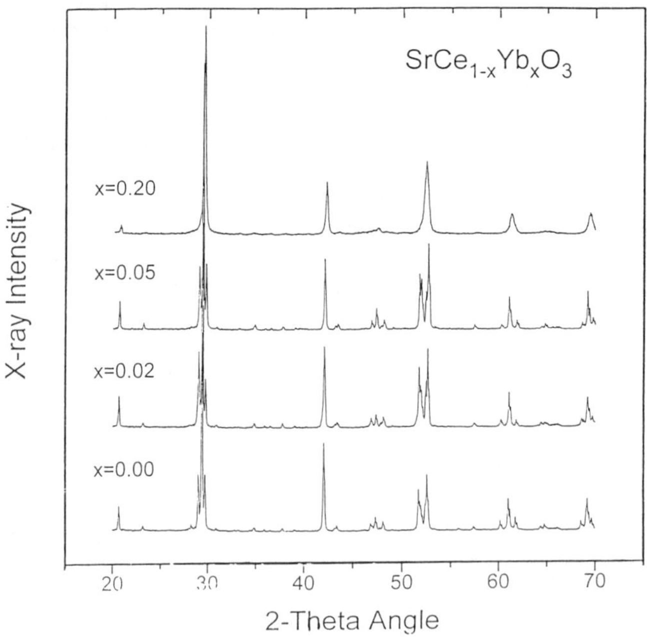

Fig.1
X-ray diffraction patterns of $SrCe_{1-x}Yb_xO_3$.

data reported for $BaCe_{1-x}Gd_xO_3$, where crystal the lattice approached the tetragonal symmetry at 20% Gd [17]. Fig.2 shows the lattice parameters as the function of Yb-concentration calculated from the experimentally determined d-spacing. As can be seen the lattice parameters a and c decreased with increasing Yb level while the parameter b increased until it become similar to c at about 20% Yb indicating tetragonal symmetry. The influence of Yb-doping on the elemental cell volume (V=abc) is also shown in Fig.2. These data plus Rietvield analyze of the XRD spectra were used to determine the oxygen vacancy concentration and the spacing between ions as the function of Yb-doping. A summary of the calculated parameters obtained from this analysis is presented in Table1.

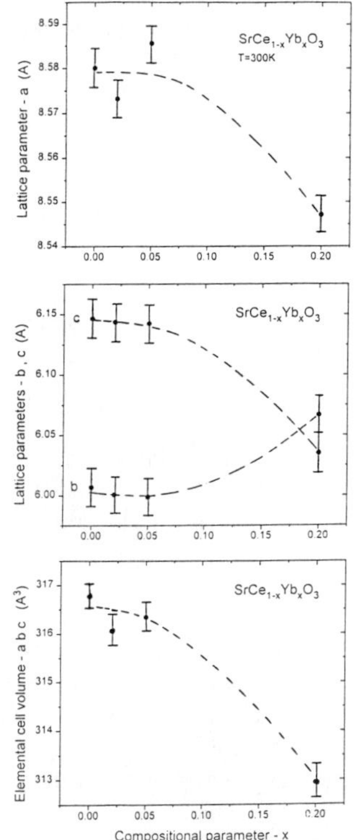

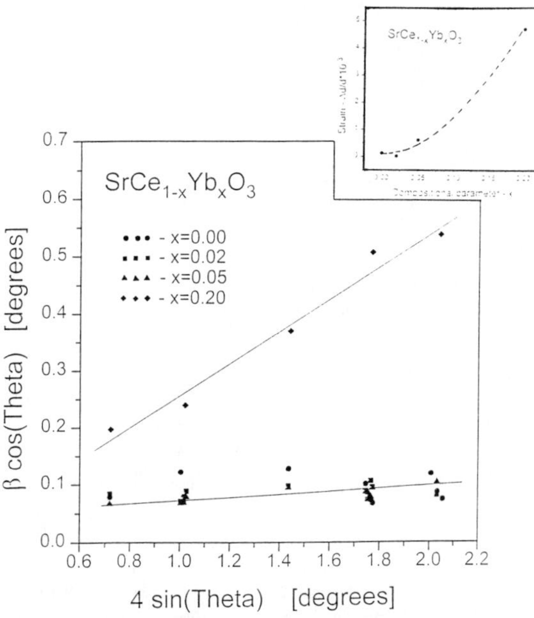

Fig.2 (left) Compositional dependence of the lattice parameters and elemental cell volume for $SrCe_{1-x}Yb_xO_3$.
Fig.3 (upper) The dependence of the strain in $SrCe_{1-x}Yb_xO_3$ on the Yb concentration.

The influence of Yb-concentration on the strain in $SrCe_{1-x}Yb_xO_3$ is shown in Fig.3. As can be seen only small changes in strain are observed for levels up to 5%Yb while a larger increase is observed for higher Yb-concentration. Comparing the ionic radii of Yb^{3+} and Ce^{4+} , 0.998A to 0.940A respectively we conclude that the difference in ionic radius is to small to induce such a large effect and observed structural disorder is due to oxygen vacancies which compensate some of the acceptors, $Yb^{3+} => [Yb_{Ce}'-1/2V_o^{..}]^x$. A comparison of the Figs.2 and 3 supports this conclusion in the fact that the elemental cell volume and strain appear to be inversely related.

The presence of Yb ions in the $SrCeO_3$ lattice also change the electrical properties which is manifested in a rise in conductivity and a decrease of activation energy with increasing Yb-concentration (Fig.4).

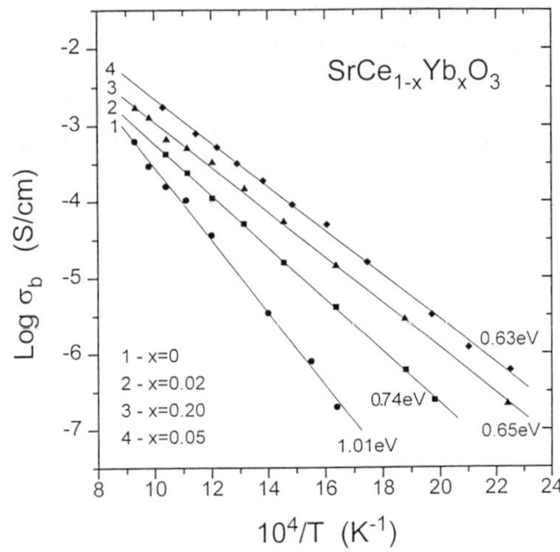

Fig.4
Temperature dependence of the electrical conductivity for SrCe$_{1-x}$Yb$_x$O$_3$.

The dependence of electrical conductivity of Sr$_{1-x}$Ce$_x$O$_3$ has been analyzed using the expression:

$$\sigma T = A \exp(-E_{act}/kT) \qquad (1)$$

from which the activation energy of conductivity, E_{act} and preexponential factor, A were determined. These values for SrCe$_{1-x}$Yb$_x$O$_3$ are presented in Table1. The preexponential A is related to the mobility factor and in the model of hopping diffusion is defined as

$$A = B\,\lambda^2 q^2 \nu N / 6\,V\,k \qquad (2)$$

where B is the symmetrical factor, λ is the jump distance, taken as the oxygen-oxygen separation, q is the charge of the hopping ion, ν is the frequency factor taken as the thermal vibration frequency, V is elemental cell volume and N is the charge carrier concentration [18,19]. Since all of the quantities except N are known, the charge carrier concentration in SrCe$_{1-x}$Yb$_x$O$_3$ can be determined. In our calculations, we used the ion distances and elemental cell volumes from our x-ray study (Table 1). For ν we taken value of $4*10^{13}$s^{-1} (hν=kT) which is similar to that used for KTaO$_3$:Fe oxides, 10^{14}s^{-1}[18]. The results of this estimation of the charge carrier concentration in SrCe$_{1-x}$Yb$_x$O$_3$ oxides as the function of Yb are presented in Fig.5 and compared

Table 1 Structural and electrical parameters for SrCe$_{1-x}$Yb$_x$O$_3$ as a function of Yb-doping level.

[Yb]	a (A)	b(A)	c(A)	λ_{O-O} (A)	A (S/cm K)	E$_{act}$ (eV)	N (cm^{-3})	[V$_O^{\cdot\cdot}$](cm^{-3})
x=0.00	8.5801	6.0067	6.1465	3.1223	$3.5*10^4$	1.01	$3.7*10^{20}$	------
x=0.02	8.5732	6.0007	6.1436	3.1200	$3*10^3$	0.74	$3.2*10^{19}$	$1.3*10^{20}$
x=0.05	8.5855	5.9987	6.1423	3.1154	$3.36*10^3$	0.63	$3.6*10^{19}$	$3.2*10^{20}$
x=0.20	8.5471	6.0668	6.0350	3.0681	$2.06*10^3$	0.65	$2.3*10^{19}$	$1.3*10^{21}$

with the concentration of oxygen determined from x-ray measurements. As can bee seen the charge carrier concentration is much lower than oxygen vacancy concentration and is independent of Yb-concentration which strongly suggests that only a very small fraction of the Yb-dopants is compensated by oxygen vacancies. Similar results have been observed by Nowick for Fe-doped $KTaO_3$ [18] and for $SrCeO_3$:Yb with lower concentration of Yb [19].

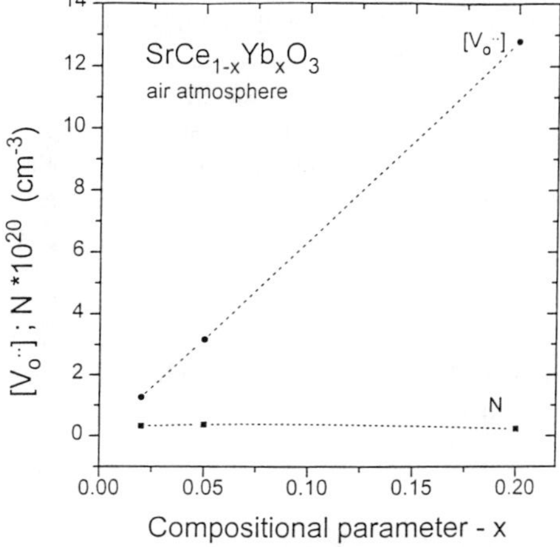

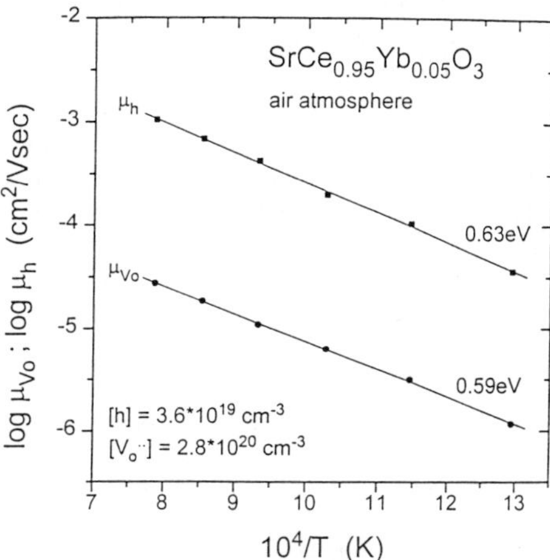

Fig.5 Compositional dependence of charge carriers, (N) and oxygen vacancies, $(V_o^{..})$ concentration for $SrCe_{1-x}Yb_xO_3$

Fig.6 Temperature dependence of electron hole and oxygen vacancy mobility for $SrCe_{1-x}Yb_xO_3$

These observations suggests that at high oxygen activity the charge carrier compensation has a mixed character and consists of both holes and oxygen vacancies, $[Yb_{Ce}'] = 2[V_o^{..}] + p$. Therefore, we conclude that the electrical conductivity at higher oxygen activities is defined by electron holes and oxygen ion transport, and that the observed behavior of the conductivity with the Yb-concentration is probably connected with the changes in the energetic barrier for hopping since the charge carrier concentration is independent with acceptor level. In support of these observations, the results of electrical conductivity measurements for both $SrCe_{0.95}Yb_{0.05}O_3$ and $BaCeO_3$:Nb as the function of oxygen activity show the hole type conductivity dominates at high P_{O2} [4,5].

Based on the results of electrical conductivity measurements of $SrCe_{0.95}Yb_{0.05}O_3$ as the function of oxygen activity [4] and the calculated concentration of electron holes and oxygen vacancies (Fig.5), we estimated the mobilities for the electrical species. The result of these estimations are shown in Fig.6. These values are in good correlation with data obtained for others oxides [20,21].

CONCLUSIONS

The electrical and structural properties of $SrCe_{1-x}Yb_xO_3$ has been investigated as a function of Yb-concentration. The effect of Yb doping is observed by the broadening of the x-ray lines and a dependence the elemental cell volume. The change in crystal symmetry from orthorhombic to tetragonal is observed for x=0.20. For this Yb-concentration a large increase in the strain has also been observed. These effects are connected with structural disorder induced by introduction of oxygen vacancies which compensated the Yb-dopants.

The electrical conductivity also is dependent on the Yb concentration and at high oxygen activity has a mixed, electron hole and oxygen character. The combination of the x-ray and electrical conductivity data allowed an estimation of the concentration of charge carriers.These values are about one order of magnitude lower than the acceptor concentration. The charge carrier concentration appears to be independent of Yb content and is related to holes indicating the mixed hole and oxygen vacancy character of charge compensation. Based on these results, the temperature behavior of oxygen vacancies and electron holes mobilities has been estimated. These results indicate that the mobilities of the charge carriers rather than the carrier concentration control the electrical conductivity. Thus at high oxygen activity, the more mobile holes control the conductivity eventhough the concentration of the oxygen vacancies is at least one order of magnitude larger.

ACKNOWLEDGMENTS

Support for this program from the New Energy Development Organization (NEDO) International Joint Research Grant in Protonics is gratefully acknowledged.

REFERENCES

1. H.Iwahara , T.Esaka , H.Uchida and N.Maeda , Solid State Ionics ¾,359(1981).
2. I.Kosacki , J.G.Becht , R.van Landschoot and J.Schoonman , Solid State Ionics59,287(1993).
3. U.Reichel , R.R.Arnos and W.Schilling , Solid State Ionics 86/88,639(1996).
4. I.Kosacki , H.L.Tuller , Solid State Ionics 80,223(1995).
5. N.Bonanos , Solid State Ionics 53/56,967(1992).
6. I.Kosacki and H.U.Anderson, Appl.Phys.Letters, 69(1996) , in print.
7. I.Kosacki and H.U.Anderson, Solid State Ionics, 1997 , in print.
8. H.L.Tuller , in Ceramic Materials for Electronics , ed.R.Buchman (Marcel Dekker , New York , 1986) pp.425-473.
9. H.L.Tuller , Solid State Ionics 52,135(1992).
10. H.Iwahara , Solid State Ionics 55,99(1992).
11. H.Iwahara , in Ionic and Mixed Conducting Ceramics , eds.T.A.Ramanarayanan , W.L.Worrell and H.L.Tuller (The Electrochem.Soc. Pennington , NY , 1994) pp1-8.
12. I.Kosacki , J.Schoonman and M.Balkanski , Solid State Ionics 57,345(1992).
13. H.Iwahara , H.Ushida and S.Tanaka , Solid State Ionics 9/10,1021(1983).
14. K.S.Knight and N.Bananos , Materials Research Bulletin 30,347(1995).
15. Shadow Program , Materials Data Inc. Livermore CA , 1996.
16. S.A.Howard , J.K.Yau and H.U.Anderson , J.Appl.Phys. 65,1492(1989).
17. N.Taniguchi , K.Hatoh , J.Niikura , T.Gamo and H.Iwahara , Solid State Ionics 53/56,998(1992).
18. T.Scherban and A.S.Nowick , Solid State Ionics 53/56,1004(1992).
19. T.Scherban and A.S.Nowick , Solid State Ionics 35(1989)189.
20. Y.Larring and T.Norby , Solid State Ionics 70/71,305(1994).
21. J.-H.Park and R.N.Blumenthal, J.Electrochem.Soc.136,2868(1989).

La$_2$Zr$_2$O$_7$ FORMATION BETWEEN YTTRIA-STABILIZED ZIRCONIA AND La$_{0.85}$Sr$_{0.15}$MnO$_3$ at 1373 K

A. MITTERDORFER and L. J. GAUCKLER

ETH Zurich, Department of Materials, Chair of Nonmetallic Materials,
Swiss Federal Institute of Technology, CH–8092, Zurich, Switzerland
gauckler@nonmet.mat.ethz.ch

ABSTRACT

Oxygen ion conducting yttria-stabilized zirconia and perovskite-type Sr-doped LaMnO$_3$ have been widely used as solid electrolyte and cathode materials in solid oxide fuel cells. The electrochemical properties of the cathode depend largely on the nanostructure and the phase composition of the interface between cathode and electrolyte. Interfaces between single crystals of 9.5mol% Y$_2$O$_3$-stabilized ZrO$_2$ and porous La$_{0.85}$Sr$_{0.15}$Mn$_y$O$_3$ ($y = 0.95...1.10$) were investigated. Atomic force microscopy was used for interface studies of YSZ single crystals after removal of sintered perovskite cathodes. High resolution transmission electron microscopy, electron diffraction and electrochemical impedance spectroscopy were employed for interface characterization. Pyrochlore-type lanthanum zirconate formed at the interface during sintering at 1100°C. Nucleation, growth kinetics, and morphology were largely depending on cathode stoichiometry. Lanthanum zirconate formation was retarded in case of A-site deficient perovskite.

INTRODUCTION

Increasing interest in operating solid oxide fuel cells (SOFC) at temperatures considerably lower than 1000°C has raised the need for a better understanding of the factors limiting the performance of cathodes consisting of Sr-doped lanthanum manganites (LSM). The interaction between cathode and electrolyte material which occurs during high temperature fabrication as well as under operating conditions is of particular interest. Pyrochlore-type La$_2$Zr$_2$O$_7$ and perovskite-type SrZrO$_3$ have been found to form at interfaces between yttria-stabilized zirconia (YSZ) and Sr- or Ca-doped LaMnO$_3$ at high temperatures in air [1]. The amount of La$_2$Zr$_2$O$_7$ formed has been found to decrease with increasing Sr or Ca substitution for La, and to increase with longer reaction periods or higher firing temperatures [2]. Yamamoto et al. [3] showed that lanthanum zirconate (LZO) formation during sintering is depressed with A-site deficient (La,Sr)MnO$_3$ (LSM$_y$). For non-stoichiometric material an incubation period for zirconate formation was observed [4].

The purpose of the present work was to investigate the influence of the manganese stoichiometry in LSM$_y$ upon nucleation and growth of LZO at low sintering temperatures of 1100°C in air. The study was performed with porous cathodes on YSZ single crystals. Atomic force microscopy (AFM) was used for topographic investigations of YSZ surfaces after the cathodes were removed. Transmission electron microscopy (TEM) was employed for phase analysis. Distinct differences in nuclea-

Mat. Res. Soc. Symp. Proc. Vol. 453 © 1997 Materials Research Society

tion and growth of LZO were found depending on the stoichiometry of LSM$_y$. Surface diffusion of cations is predominant and enhanced in case of A-site deficient LSM$_y$.

EXPERIMENTAL

Reaction couples of 9.5 mol%-Y$_2$O$_3$-stabilized ZrO$_2$ single crystals and screen printed LSM$_y$ cathodes were prepared. Single crystals (25×25×0.5 mm, (100) orientation, polished surfaces) were obtained from Zirmat Corp., N. Billerica, MA, USA; La$_{0.85}$Sr$_{0.15}$Mn$_y$O$_3$ powders (calcined at 950 °C for two hours, ball milled) were obtained from SSC Inc., Woodinville, WA, USA. The powder compositions were confirmed with ICP-AAS. Screen printing pastes with manganese excess were prepared by adding Mn(NO$_3$)$_2$ (dissolved in ethanol) to the calcined powders. The compositions of the cathode powders were: y = 0.95, 0.98, 1.02, 1.05, 1.10. Cathodes of 1 cm^2 and 15 µm thickness were screen printed on solid electrolytes and fired at 1100°C in air. Sintering periods ranged from 15 minutes to 10 hours. A three-electrode four-terminal configuration was employed for electrochemical measurements. Pt gauze and Pt wires were used to contact the cathodes. Pt paste was used as counter electrode. Further details about sample preparation and measuring techniques are given elsewhere [5].

Topographic investigations of YSZ surfaces were conducted with AFM in contact mode. For these measurements, we first removed the cathode layers from the YSZ single crystals with concentrated hydrochloric acid. YSZ and LZO are stable against hydrochloric acid whereas LSM$_y$ dissolves rapidly. TEM was employed for phase analysis of the YSZ surfaces. The crystals were prepared for TEM investigation by dividing, polishing, and Ar-ion beam thinning. Plane-view samples were cut into disks of 3 mm diameter using an ultrasonic disc cutter and then dimpled from the substrate side to a final thickness of 20 nm. The final thinning to electron transparency was done by Ar-ion beam thinning.

RESULTS

<u>Manganese Deficient (y = 0.95) Cathodes on YSZ</u>

Fig. 1a shows the microstructure of a cathode/electrolyte interface after firing at 1100°C for 2 h in air. In Fig. 1b the topography of the YSZ surface is shown after removing the cathode layer. The complete surface of the YSZ single crystals is covered with polycrystals of a foreign phase. XPS, HRTEM, and Selected Area Electron Diffraction (SAED) showed these grains to consist of La$_2$Zr$_2$O$_7$ (LZO). Note that the YSZ single crystals were completely flat before heat treatment and that the average height of the LZO grains shown in Fig. 1b is ~20 nm.

Formation of LZO starts near the contact points of the LSM$_{0.95}$ grains and the YSZ surface (cf. Fig. 1a) and quickly spreads over the entire YSZ surface during sintering. A further growth of the layer is controlled by bulk diffusion of La^{3+} and Zr^{4+} through the growing layer. We observed layer growth kinetics following a square root dependence on time. The LZO layer grows in direction of the cathode as well as into the abutting YSZ. The reason for the low activation energy for LZO nucleation can be found in the close lattice match of the (100) surfaces of YSZ and LZO. In addi-

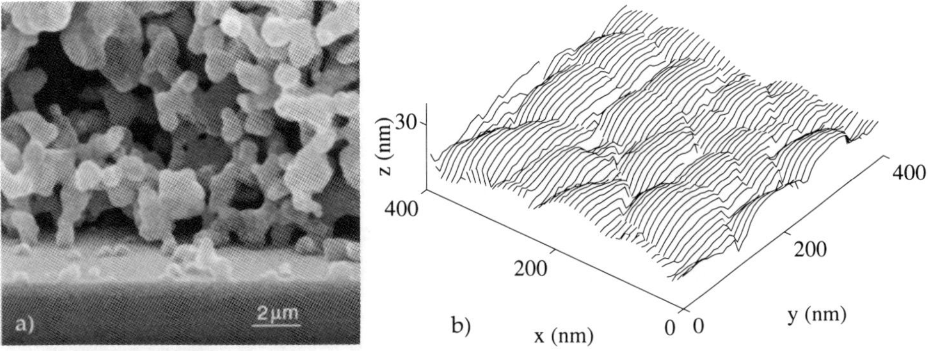

Fig. 1. Microstructure of a $La_{0.85}Sr_{0.15}Mn_{0.95}O_3$ cathode and YSZ surface after sintering.—(a) Shows a SEM of the cathode/solid electrolyte interface. Note the connectivity of the perovskite grains. There is no indication of a foreign phase at the interface between the cathode grains and the YSZ surface. (b) AFM of the YSZ surface after removing the cathode. Note that the YSZ surface is covered with extremely small and thin grains (~20 nm) of $La_2Zr_2O_7$.

tion, excess La_2O_3 from the Mn-deficient $LSM_{0.95}$ is chemically active and reacts immediately with YSZ. The process is facilitated by the use of porous cathodes which allow for surface diffusion of the participating elements and oxygen exchange with the surrounding atmosphere.

Slight Manganese Deficient ($y = 0.98$) Cathodes on YSZ

$LSM_{0.98}$ cathodes were sintered on YSZ single crystals for 15 min to 10 h at 1100°C in air and were subsequently removed. In Fig. 2a the topography of a YSZ surface is shown. Cube-shaped islands of LZO formed epitaxially on top of the YSZ substrate. The cubes grow in height and diameter with increasing sintering periods. The length of the boundary line between the YSZ surface and the new phase was estimated from AFM data to be ~4 $\mu m/\mu m^2$. Surface coverages of LZO ranged from ~50% to ~90%, depending on the sintering period.

Fig. 2b shows a cross-section TEM of a single LZO island. HRTEM showed that the LZO islands are covered with a few monolayers of cubic zirconia. This indicates that island growth is dominated by surface diffusion of Zr^{4+} along the LZO surface towards the interface between LSM_y and LZO. Surface diffusion also explains the rapid growth in z direction of the LZO islands which can be readily seen by comparing the z axis of the AFM in Fig. 1a and Fig. 2a. With increasing sintering period the islands grow in diameter and height until a dense layer of LZO forms. Further growth of the LZO layer is then limited by bulk diffusion of La^{3+} and Zr^{4+}.

The influence of the LZO islands on the electrochemical properties of $LSM_{0.98}$ cathodes on YSZ were determined with electrochemical impedance spectroscopy. Screen-printed cathodes were sintered and the impedance was monitored during heating, sintering at 1100°C, and cooling. A strong increase in the polarization resist-

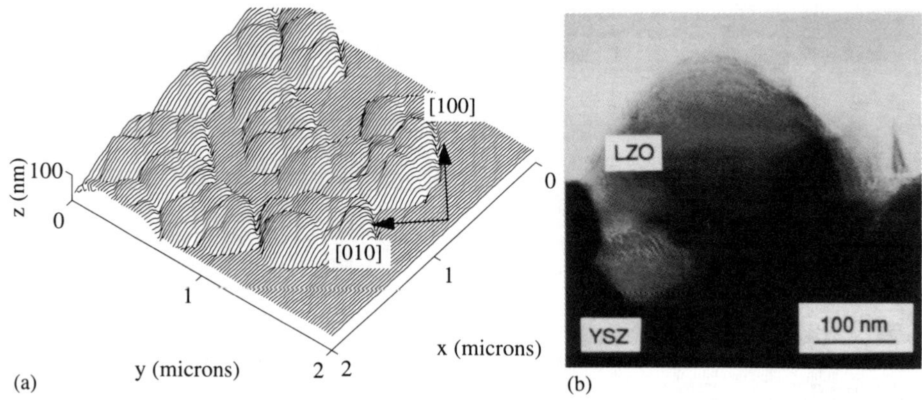

Fig. 2. La$_2$Zr$_2$O$_7$ islands on top of a YSZ single crystal surface.—(a) AFM of YSZ surface which was sintered with a porous layer of La$_{0.85}$Sr$_{0.15}$Mn$_{0.98}$O$_3$ for 15 min at 1100°C in air. Note that island growth is extremely fast and that an orientationship between LZO islands and YSZ substrate can be clearly seen. (b) Cross-section TEM of LZO island. LZO islands grow epitaxially on (001) YSZ surface ({001}$_{LZO}$ | | {001}$_{YSZ}$ and <100>$_{LZO}$ | | <100>$_{YSZ}$). A dislocation network can be seen in the YSZ facing the LZO island.

ance (up to factor of ten) as well as a change in the rate limiting factors of the electrochemical reaction was observed due to the LZO islands. Details were presented elsewhere [6].

Manganese Excess ($y \geq 1.02$) Cathodes on YSZ

Porous LSM$_{1.02}$ cathodes were sintered with YSZ single crystals for 15 minutes to 10 hours at 1100°C in air and were subsequently removed. The surface topographies of YSZ single crystals showed rings of new material which formed as necks around the LSM grains during sintering (Fig. 3). TEM showed these necks to be mainly YSZ with a high density of lattice defects. SAED from plane-view samples indicated the presence of small amounts of LZO. HRTEM of cross-section samples revealed that the tips of the necks consist of LZO (see inset of Fig. 3).

LZO formation was generally retarded in case of A-site deficient LSM$_y$. The incubation time, i.e. the time before LZO can be detected, increased with increasing manganese content of the LSM$_y$. Samples with $y = 1.02$ showed significant amounts (~10% with regards to the total volume of the necks) of LZO after 2 h of sintering at 1100°C, whereas samples with $y = 1.10$ did not show any LZO after this time. However, with increasing sintering periods all samples showed the presence of LZO. However, after the LZO nuclei appeared, their further growth was much slower than in the case of manganese deficient LSM$_y$. Upon further growth shape of the LZO islands changed into cubes, as in the case of LSM$_{0.98}$.

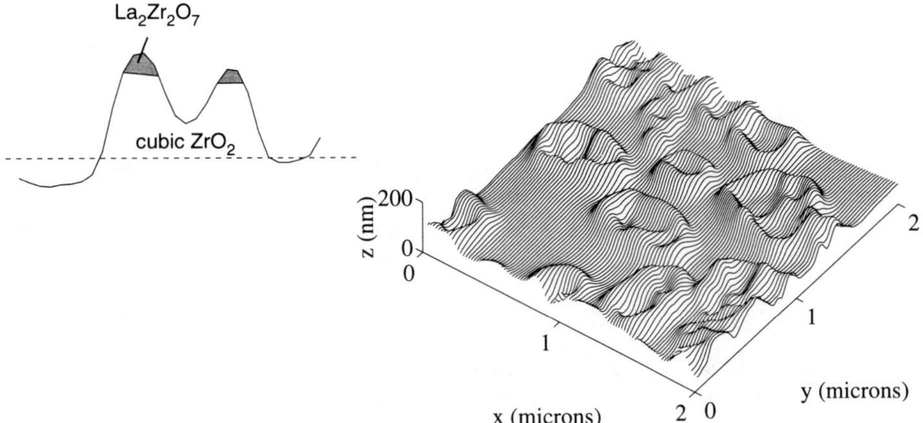

Fig. 3. AFM of YSZ surface with ring-shaped cubic zirconia islands (necks). The La$_{0.85}$Sr$_{0.15}$Mn$_{1.02}$O$_3$ layer was sintered for 1 h at 1100°C in air and subsequently removed with hydrochloric acid. Obviously, the necks formed around the contact point between the LSM$_{1.02}$ grains and the YSZ surface. The inset shows a schematic cross-section of a single neck (results from HRTEM). The tips of the necks consist of LZO. There is a sharp boundary between the cubic zirconia and the LZO embryo. - - - denotes the original YSZ single crystal surface level.

CONCLUSIONS

Nucleation and growth of lanthanum zirconate on YSZ single crystal surfaces at 1100°C has been examined. Dense layers of lanthanum zirconate form when YSZ surfaces are in contact with porous La$_{0.85}$Sr$_{0.15}$Mn$_{0.95}$O$_3$ layers at 1100°C already after a few minutes. Layer growth is limited by bulk diffusion of cations through the layer. This is supported by the observed growth kinetics of the layer. The electrochemical properties severely degrade due to the presence of the dense layer of badly conducting (electronically and ionically) lanthanum zirconate. Cube-shaped lanthanum zirconate islands form in between YSZ surfaces and La$_{0.85}$Sr$_{0.15}$Mn$_{0.98}$O$_3$ grains. Island growth is limited by surface diffusion of Zr^{4+} along the zirconate islands. Although no dense layer forms, the electrochemical properties of this cathode/electrolyte interface is strongly affected. The triple phase boundary, i.e. the region where the electronation and incorporation of adsorbed oxygen intermediates is assumed to take place, is replaced by a low conducting inter-layer. Apparently, the electronation reaction rate decreases due to the change in the electronic properties of the material at the triple phase boundary. The formation of La$_2$Zr$_2$O$_7$ at the cathode/electrolyte interface can be delayed and retarded by excess manganese in the La$_{0.85}$Sr$_{0.15}$Mn$_y$O$_3$. The chemical activity of the lanthanum is obviously reduced by excess manganese.

The combined application of AFM and TEM has given a detailed picture of the nucleation and growth of lanthanum zirconate at the interface between Sr-doped LaMnO$_3$ and yttria-stabilized zirconia. The additional employment of electrochemical impedance spectroscopy has helped to further understand the influence of small amounts of lanthanum zirconate at the triple phase boundary upon the electrochemical properties.

ACKNOWLEDGEMENTS

This work was funded in part by the Swiss Federal Department of Transportation, Communications, and Energy. The authors would like to thank M. Cantoni who performed the TEM measurements, M. Gödickemeier for helpful discussions, and W. Tobler for experimental assistance (all our laboratory).

REFERENCES

[1]   S. K. Lau and S. C. Singhal in <u>Proc. Corrosion 85, Boston,</u> (1985), p. 79–80.

[2]   J. A. M. van Roosmalen and E. H. P. Cordfunke. Solid State Ionics **52**, p. 303, (1992).

[3]   O. Yamamoto, Y. Takeda, R. Kanno, and T. Kojima in <u>Proceedings of the First International Symposium on Solid Oxide Fuel Cells</u>, edited by S. C. Singhal (The Electrochemical Society, Inc., Pennington, NJ, 1989), p. 240–246

[4]   H. Taimatsu, K. Wada, H. Kaneko, and H. Yamamura. J. Am. Ceram. Soc. **75**, p. 401 (1992).

[5]   A. Mitterdorfer and L. J. Gauckler, J. Am. Ceram. Soc., to be submitted.

[6]   A. Mitterdorfer and L. J. Gauckler in <u>Proc. on the 17th Risø Int. Symp. on Mat. Sci.,</u> edited by F. W. Poulsen, N. Bonanos, S. Lindenroth, M. Mogensen, and B. Zachau-Christiansen (Risø National Laboratory, Roskilde, Denmark, 1996), p. 357–362.

ELECTRICAL CONDUCTIVITY AND NONSTOICHIOMETRY IN $Pr_{0.545}Ce_{0.455}O_{2-x}$

O. PORAT[a] , H.L. TULLER[a], M. SHELEF[b] and E.M. LOGOTHETIS[b]

[a] Crystal Physics and Electroceramics Laboratory, Department of Materials Science and Engineering, MIT, Cambridge MA 02139

[b] Ford Research Laboratory, P.O. Box 2053, Dearborn MI 48121

ABSTRACT

The oxygen nonstoichiometry of $Pr_{0.545}Ce_{0.455}O_{2-x}$ was studied by solid state coulometric titration and was found to be extensive ($0 \leq y \leq 0.28$) at temperatures of 400-600 °C. The P_{O_2} - x curves showed evidence of the existence of a number of single and two phase regions. Electrical conductivity measurements, performed under similar conditions, exhibited a p-n transition for temperature of 500 °C and below. The activated p-type conductivity was modeled in terms of the small polaron hopping mechanism. The p-n transition was correlated with the onset of phase separation.

INTRODUCTION

Solid solutions of $Pr_yCe_{1-y}O_{2-x}$ (PCO) are of interest due to their unusually high oxygen storage capacity [1, 2] and levels of mixed ionic-electronic conductivity (MIEC) [3, 4].

In oxidation-reduction studies of mixed Ce/Pr oxides performed by Logan and Shelef [2] using temperature programmed desorption (TPD), reduction (TPR) and oxidation (TPO), the solid solution $Pr_{0.55}Ce_{0.45}O_{2-x}$, was observed to lose oxygen at temperatures as low as 400 °C. Pd additions were found to further promote oxygen loss and lower the onset of reduction down to the 200-300 °C range.

In the search for improved cathodes for solid oxide fuel cells (SOFC) exhibiting MIEC, Takasu et al [3] examined the electrical properties of $Pr_yCe_{1-y}O_{2-x}$ as a function of temperature in air. They reported a 2,000 fold increase in total conductivity for $Pr_{0.4}Ce_{0.6}O_{2-x}$ over that of CeO_{2-y} at 600 °C and an oxygen ion transference number of 0.11. Their X-ray studies indicated a single phase fluorite structure for $0 < x < 0.8$. More recently, Nauer et al. [4] examined the electrical properties of the same system with y=0.2, 0.4, and 0.5. They report conductivities exceeding 0.1S/cm at 800 °C, which they attribute to be largely electronic. Their X-ray analysis disagrees with that of Takasu et al [3] in that they find a second fluorite phase coexisting with the first for $0.3 \leq x \leq 0.6$.

The large oxygen storage capacity reported for PCO is a reflection of the large deviations from stoichiometry, x, attainable in this system, even at relatively reduced temperatures. This is expected, given the variable valent natures of both Ce (+3, +4) and Pr (+3, +4). To date only limited TGA data are reported for a solid solution with x ~ 0.8 [1]. The relationship between reported electrical properties and nonstoichiometry, while essential for interpreting the defect and transport mechanisms, remain unknown.

In this study, we present the first extensive measurements of nonstoichiometry and electrical conductivity reported for a PCO solid solution. We selected the composition $Pr_{0.545}Ce_{0.455}O_{2-y}$ for study, since it showed rapid redox kinetics and high oxygen storage capacity in the study of Logan and Shelef [2] and was close in composition to the best conductors reported by Takasu et. al. [3] and Nauer et al. [4].

EXPERIMENTAL

Nitrates of cerium and praseodymium were coprecipitated out of solution using oxalic acid.

Mat. Res. Soc. Symp. Proc. Vol. 453 © 1997 Materials Research Society

Precipitates were calcinated at 650 °C for 6h and at 750 °C for 2h. Palladium nitrate was introduced into the calcined samples by wet impregnation and ultrasonic mixing with a resultant Pd level after calcining of 0.24 mol of PdO. Further details may be found in ref. 2. Specimens were prepared for electrical measurements by sintering isostatically pressed pellets (40 kB) at 1200 °C for 16 h.

Due to the highly conductive nature of these materials, the conventional 4 probe DC method was used to measure the electrical conductivity. Nonstoichiometry was investigated by titrating precise levels of oxygen into and out of the sample chamber with a stabilized zirconia cell. Subsequent to equilibration, the zirconia cell, in the open circuit potentiometric mode, was used to monitor the P_{O_2} in equilibrium with the newly established level of nonstoichiometry.   A more detailed description of this cell and its operation may be found in [5].

X-ray powder diffraction studies were performed, using a Rigaku 300 diffractometer, on specimens annealed at 450 °C to achieve a series of values of x and subsequently quenched to room temperature.

RESULTS

The deviations from stoichiometry of the PCO solid solutions were measured as a function of P_{O_2} at 400 °C with and without Pd. The curves were nearly identical showing that the small addition of Pd has little or no effect on the oxygen equilibrium.

While the equilibrium results were independent of Pd, the re-equilibration kinetics were found to be enhanced with Pd by at least a factor of 3, in general agreement with the observations of Logan and Shelef [4]. Consequently, the remaining results are reported for PCO:Pd. Fig. 1 shows the log Po2 versus x results for isotherms ranging from 400-600 °C at 50 °C intervals.

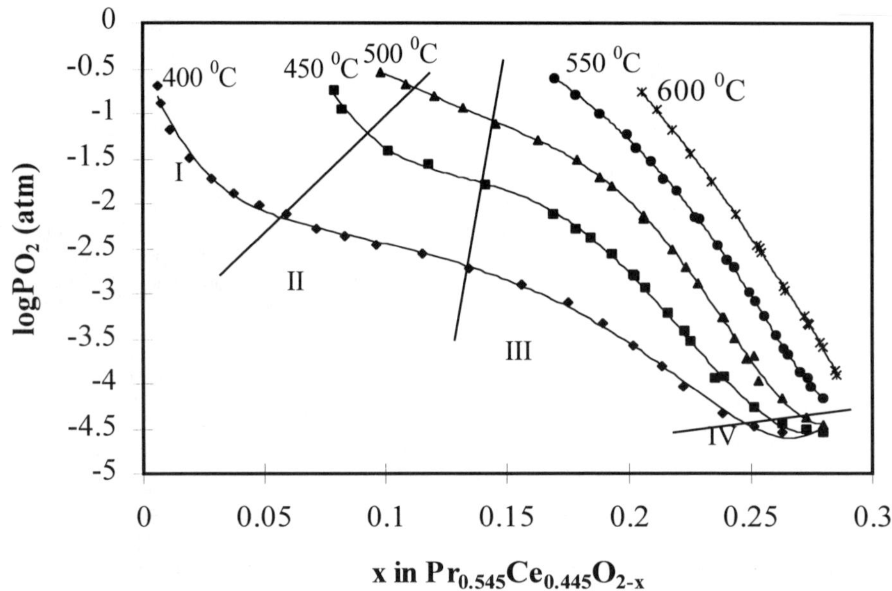

Fig. 1: Nonstoichiometry, x, in Pr$_{0.545}$Ce$_{0.445}$O$_{2-x}$ as a function of P_{O_2}, at different temperatures

Four regimes, characterized by different slopes, can be isolated. In regions I and III, P_{O_2} changes relatively rapidly with variations in x while in II and IV the P_{O_2} changes are slow. The significance of these observations are discussed below. X-ray diffraction patterns for specimens annealed at 450 °C in 0.21, 10^{-3} and 10^{-5} atm respectively and subsequently quenched, are shown in Fig. 2. While all spectra correspond to the fluorite structure, note the shift in peak positions between 0.21 and 10^{-3} atm.

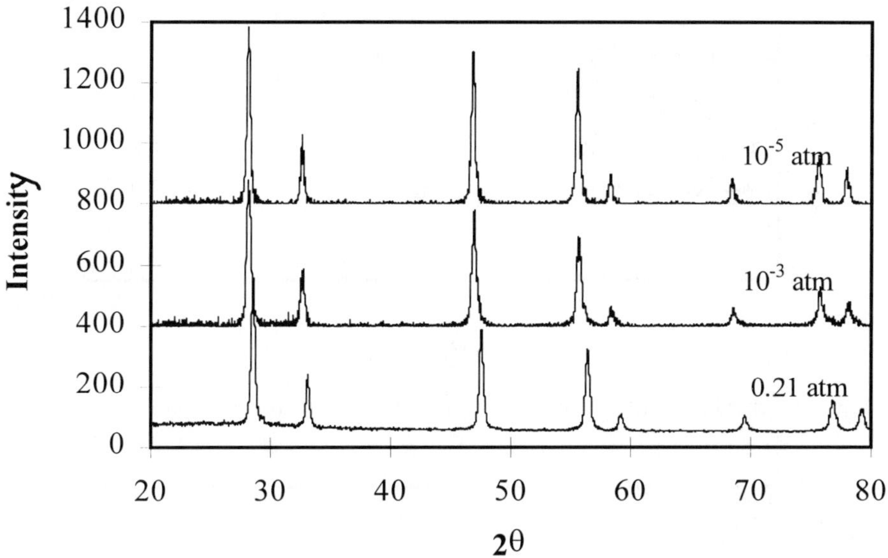

Fig. 2: X-ray diffraction patterns of samples annealed at 450 °C and different atmospheres, and quenched to room temperature.

The P_{O_2} -dependence of the electrical conductivity is shown in Fig. 3. For T=550 and 600 °C, a relatively montonic rise in σ with increasing P_{O_2} is observed, characteristic of p-type semiconduction. For 500 °C and below, a maximum in σ is obtained at intermediate values of P_{O_2} with the peak shifting to higher P_{O_2} with increasing T. The n-type region at high P_{O_2} and lower T follows an approximate $P_{O_2}^{-1/4}$ dependence. By combining the data in Figs. 2 and 3, it becomes possible to examine the dependence of σ on x, explicitly as in Fig. 4.

DISCUSSION

As expected from earlier studies, the range of nonstoichiometry can be large as evidenced in Fig. 1. At 400 °C, $\Delta x \sim 0.25$ between 10^{-5} and 0.21 atm. Since Ce in CeO_2 only begins to reduce significantly from Ce^{4+} to Ce^{3+} at much higher temperature and /or much lower P_{O_2}'s [6], it is reasonable to assume that the observed changes in x are largely due to reduction of Pr^{4+} to Pr^{3+}. This is in agreement with the conclusions of Gortsman et al. [7] and Logan and Shelef [2].

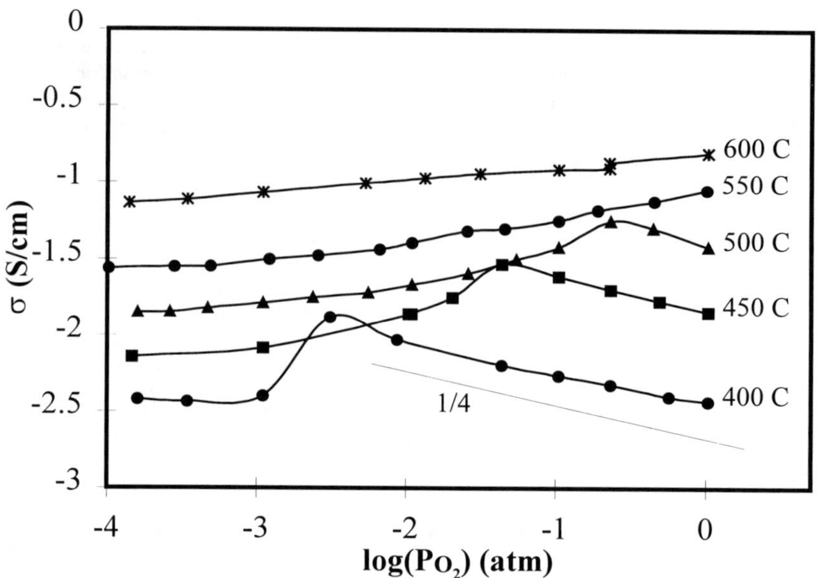

Fig. 3: Electrical conductivity of $Pr_{0.545}Ce_{0.445}O_{2-x}$ vs. P_{O_2} at different temperatures

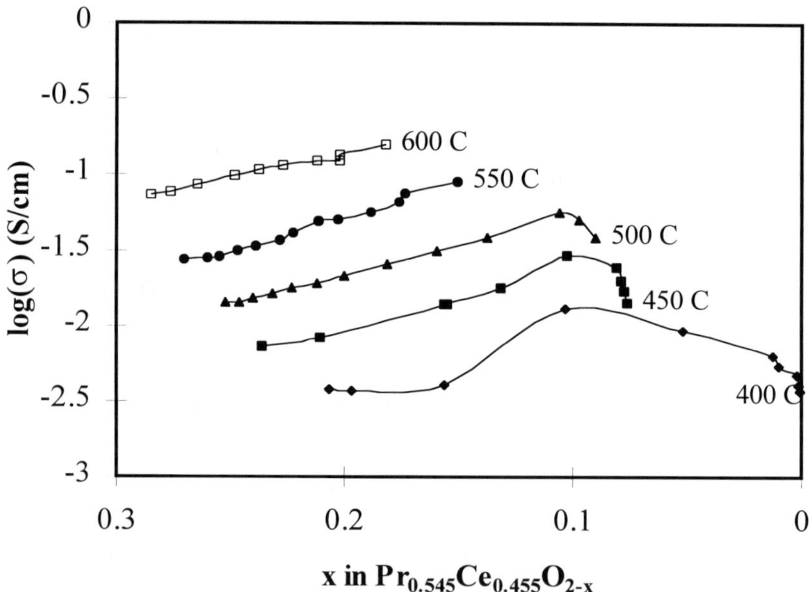

Fig. 4: Electrical conductivity of $Pr_{0.545}Ce_{0.445}O_{2-x}$ vs. **x** at different temperatures

The shape of the curves in Fig. 1 strongly suggests that phase transitions accompany the changes in composition. An invariant P_{O_2} accompanying changes in x points to a two phase region. Region II appears to reflect incipient if not actual phase separation between a near stoichiometric phase represented in region I and highly oxygen deficient phase represented in region III. Indeed, a number of oxygen deficient intermediate ordered structures are known to exist in the CeO_{2-x} and PrO_{2-x} systems characterized by long range ordering of oxygen Vacancies [8]. The apparent shrinkage of region II with increasing temperature is consistent with the disappearance of the miscibility gap with increasing entropic contributions. Although there is limited experimental data, a second-phase separation region, region IV, appears to initiate at large x>0.25 and $P_{O_2} < 3 \cdot 10^{-5}$ atm.

In support of the interpretation of the existence of different phases in regions I and III, we refer to our X-ray diffraction results shown in Fig. 2. Indeed, the spectrum obtained at 450 $^{\circ}$C and 0.21 atm (region I) differs from that obtained at 10^{-3} atm (region III). While both correspond to the fluorite structure, there is a shift in lattice parameter. This observation is now consistent with both the findings of Nauer et al. [4], who reported two co-existing fluorite phases and Takasu et al [3], who found only one phase. Obviously, depending on the thermal history, one or the other, or both, fluorite-like phases would exist. Since oxygen sublattice ordering is difficult to detect by X-ray diffraction, a proper description of the various phases formed under reduction awaits appropriate neutron diffraction studies.

By comparing Figs. 1 and 4, it is clear that most of the conductivity data, which is p-type in nature (see Fig. 3) fall within region III. Since only Pr is expected to be multivalent under these condition (see above), conduction via small polaron hopping between Pr^{3+} and Pr^{4+} states is the likely conduction mechanism. The activated nature of conduction is evident from Fig. 4 where one observes increasing σ with increasing T for fixed x with activation energy of ~ 0.9 eV.

Assuming an ideal small polaron model, one expects a p-n transition to occur under the condition of a half-filled band, i.e., $[Pr^{3+}] = [Pr^{4+}]$. For our solid solution (y=0.545), this condition is predicted to occur for x=0.14. While the p-n transition falls close to this value at x~0.11, it also happens to fall within region II of Fig. 1. The n-type contribution may therefore be due to the n-type character of the phase in region I. We have previously observed similar p-n transitions induced by phase transitions in manganese ferrites [9].

ACKNOWLEDGMENT:

Central X-ray facilities of the MIT Center of Materials Science and Engineering, under the support of the National Science Foundation, award No. 9400334-DMR, were utilized in this study.

REFERENCES

[1] J.A. Jones and G.D. Blue, J. Spacecraft **25**, 202 (1988).
[2] A.D. Logan and M. Shelef, J. Mat. Res. **9**, 468 (1994).
[3] Y. Takasu, T. Sugino, and Y. Matsuda, J. Appl. Electrochem. **14**, 79 (1984).
[4] M. Nauer, C. Ftikos, and B.C.H. Steele, J. European Ceram. Soc. **14**, 493 (1994).
[5] O. Porat and H.L. Tuller, J. Am. Ceram. Soc., in press.
[6] H.L. Tuller and A.S. Nowick, J. Electrochem. Soc. **126**, 210 (1979)
[7] K.G. Gortsman, N.F. Kartenko, B.T. Melekh, S.V. Nikitin, I.A. Smirnov, Y.N. Filin, N.V. Shorenkova, A.T. Shunaev, B.Y. Kehl'mer, and F.M. Ovsynnikov, Fiz. Tverd. Tela [Sov. Phys. Solid State] **32**, 1868 (1990).
[8] O. T. Sorensen, in Nonstoichiometric Oxides, edited by O.T. Sorensen (Academic Press, New York, 1981), pp.15-20.
[9] H.I. Yoo and H.L. Tuller, J. Mater. Res. **3**, 552 (1988)

ELECTRICAL CONDUCTIVITY AND STRUCTURE OF NANOCRYSTALLINE CeO₂ THIN FILMS

IGOR KOSACKI AND HARLAN U.ANDERSON
Department of Ceramic Engineering , University of Missouri-Rolla , Rolla , MO 65401 , USA

ABSTRACT

The results of structural and electrical measurements of nanocrystalline CeO_2 thin films are presented. A correlation between the electrical conductivity and microstructure has been observed and discussed. The electrical properties of nanocrystalline CeO_2 thin films are attributed to a dominant role of grain boundary phase.

INTRODUCTION

The dependence of the electrical conductivity of ionic conductors on impurity concentration and ambient atmosphere is important for many applications. These parameters play an important role in controlling the material stoichiometry and type of electrical conductivity. For instance , high ionic conductivity is achieved in a number of oxides - ZrO_2 [1] , CeO_2 [2] , $SrCeO_3$ [3] , $Gd_2Ti_2O_7$ [4] by acceptor doping due to generation the oxygen vacancies. In this case ionic conductivity is P_{O2} independent [1,3,4]. The influence the ambient atmosphere on electrical properties is observed at low P_{O2} and is related to extrinsic-intrinsic transition where electron conductivity dominant acceptor impurity level.

The electrical conductivity of ionic conductors can also be modified by their microstructure, when the grain size decreases the microstructure can reach a point where the properties become dominated by interfacial area , grain boundary and impurity segregation effects [5,6]. These effects are observed when microstructure enters the nanometer grain size range. For nanocrystalline CeO_2 a different electrical conductivity behavior related to the microstructure has been recently reported [7]. Since , the electrical properties of nanocrystalline materials are controlled by interfacial effects , we can expect the occurrence of much faster ion diffusion and chemical reaction rates as a function of gas atmosphere than for polycrystalline specimens. These observations makes it possible to construct semiconductor-type gas sensors using nanocrystalline materials. For example a hydrogen sensor based on a $SrCe_{0.95}Yb_{0.05}O_3$ nanocrystalline thin film has been recently proposed [8].

In the present work , we examine the relationship between the microstructure and electrical properties of nanocrystalline CeO_2 thin films. The results of electrical conductivity measurements of these films are presented and compared with the measurements performed on bulk nano- and polycrystalline specimens.

EXPERIMENTAL

We have prepared dense films of CeO_2 with a thickness of about $1\mu m$ using a polymer precursor process [9]. This process allows the preparation of the oxides from polymeric precursors which can be converted to the oxide film by spin-coating , drying and crystallization at temperatures in the range 350-900°C. Thin films have been obtained on both single crystal Si and polycrystalline Al_2O_3 substrates. X-ray diffraction , scanning electron (SEM) and atomic

Mat. Res. Soc. Symp. Proc. Vol. 453 © 1997 Materials Research Society

force microscopy (AMF) techniques have been used to determine the structure , thickness and grain size obtained films.

The electrical conductivity was studied using a Solartron 1260 Frequency Response Analyzer. The impedance spectra were recorded for frequencies in the range of 10^{-3} Hz to 10^{7} Hz over the temperature of 300-900°C in ambient atmospheres using air and a mixture containing $0.9N_2+0.1H_2$ and $0.9CO+0.1CO_2$. The low-frequency conductivity was also determine from 2-probe dc measurements which were performed using a Solartron 1286 Electrochemical Interface. In all electrical measurements , platinum electrodes have been used.

RESULTS

Figure 1 shows x-ray diffraction patterns for CeO_2 thin films obtained at the different crystallization temperatures. Analyze of the shape of x-ray lines has been used to determine the grain size for each specimen. As can be seen , the microstructure of the CeO_2 films is strongly dependent upon annealing temperature with a grain size of 5 , 9 and 30 nm for 350 , 600 and 900°C respectively. These data are in excellent correlation with results obtained from AFM studies. Fig.2 shows x-ray spectra obtained for CeO_2 films on both Si and Al_2O_3 substrates. The x-ray patterns from the films compares well with the standards indicating the successful preparation the nanostructured CeO_2 films with the fluorite structure. The influence of the substrate on the microstructure of CeO_2 films was observed by differences in grain size. For both Si and Al_2O_3 substrates the average grain size is in the nanocrystalline range but after annealing at 900°C that obtained on Al_2O_3 is slightly larger , 36nm as compared to 30nm on Si.

The existence of such small grain size suggests the possibility of observing unique defect structures and physical properties due to the extended interfacial area in these films. As has been recently observed for bulk CeO_2 and $SrCe_{0.95}Yb_{0.05}O_3$ thin films the electrical conductivity through interfacial and grain boundary effects has been enhanced [7,8].

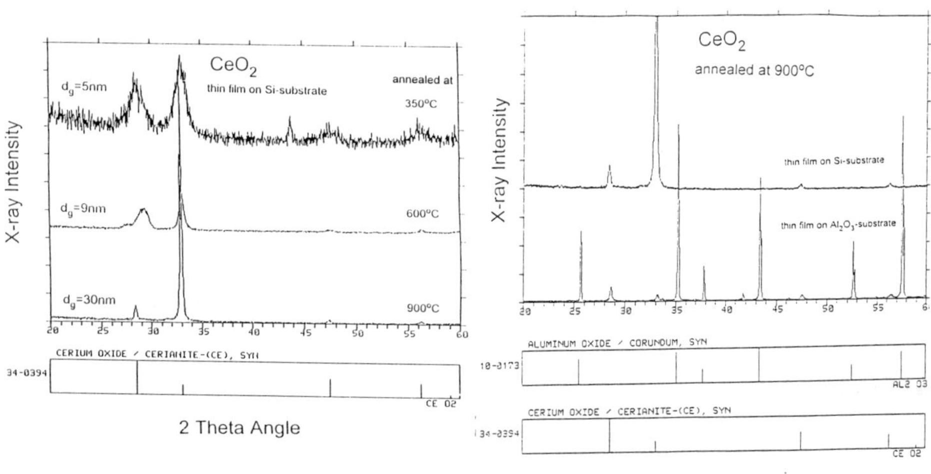

Fig.1 X-ray diffraction patterns of CeO_2 thin films deposited on Si substrates and crystallized at different temperatures.

Fig.2 X-ray diffraction patterns of CeO_2 thin films deposited on Si and Al_2O_3 substrates.

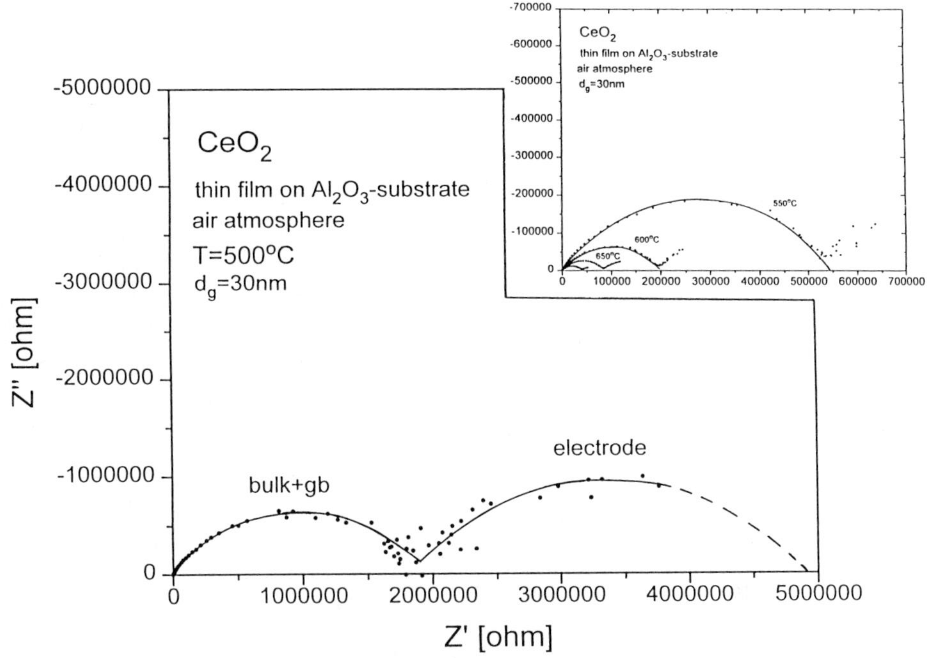

Fig.3 Impedance spectra of CeO_2 thin films at different temperatures in an air atmosphere.

The electrical conductivity of the CeO_2 thin films has been studied by impedance spectroscopy. Representative impedance spectra for these films are presented in Fig.3. Distinctly different spectrum were obtained for the nanocrystalline thin films and for the polycrystalline specimens with the grain size in the micrometer range. The spectra of polycrystalline ceramics usually exhibit three semicircles which are attributed to bulk , grain boundary and electrode resistance [10] , while the nanocrystalline materials showed only two semicircles. This probably is connected with the increasing volume of grain boundary phase when grain size decreases. The result of this will be significant higher the grain boundary capacitance which dominates the bulk contribution. This observation correlated well with the data for nanocrystalline CeO_2 and also for nanocrystalline $SrCe_{0.95}Yb_{0.05}O_3$ thin films specimens [7,8]. The low-frequency semicircle in the impedance spectra of CeO_2 films (Fig.3) is probably due to the electrode/film interface and can be identify by 2 and 4-probe dc measurements [11]. The high-frequency semicircle is related to the thin film resistance and is the superposition of bulk and grain boundary contributions.

Fig.4 presents the conductivity behavior related to the high-frequency semicircle for bulk nano- and polycrystalline CeO_2 [12] and also for nanocrystalline thin films specimens. As can be seen , the conductivity of the polycrystalline material is about 2 orders of magnitude lower when compared the high-frequency conductivity for the nanocrystalline material with a quite different activation energy for the conductivity. Note , the same activation energy is observed for both

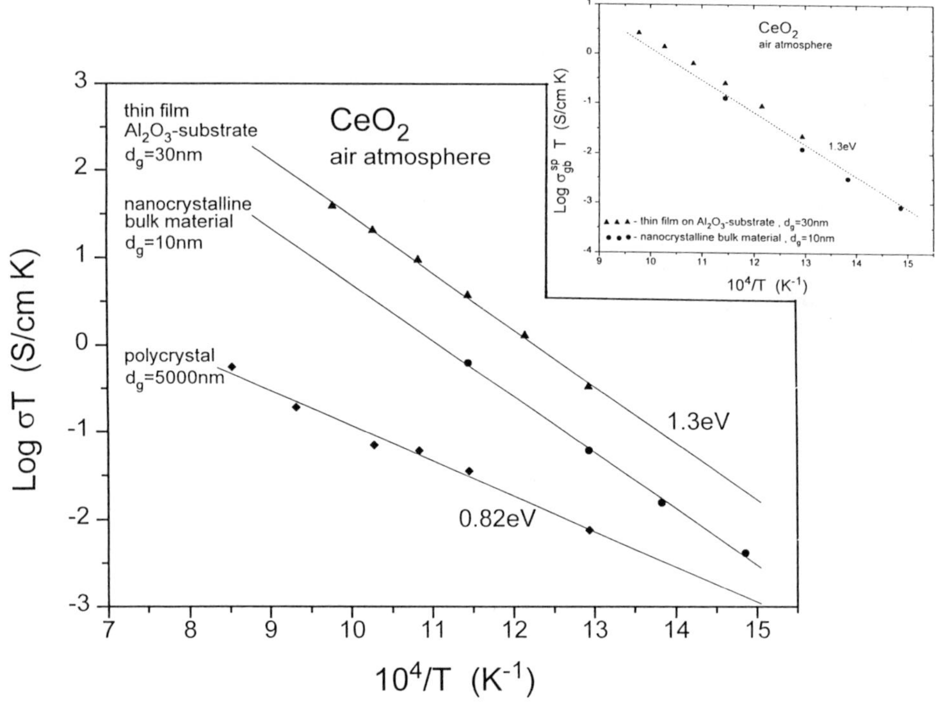

Fig.4 Comparison of the high-frequency electrical conductivity determined for bulk poly- and nanocrystalline CeO_2 [12] and nanocrystalline thin films.
The insert is the temperature dependence of the high-frequency conductivity of nanocrystalline bulk and thin films CeO_2 normalized to the single grain boundary.

bulk and thin film nanostructured CeO_2 which suggested that in both cases the high-frequency conductivity is controlled by the same mechnism. This observation and analysis of the impedance spectra strongly supports the hypothesis that the high-frequency semicircle for the CeO_2 thin film can't be attributed to the bulk resistance and is connected with grain boundary effects.

In order to show that the electrical properties of nanocrystalline CeO_2 thin films are determined by the grain boundary, the high-frequency conductivity of the thin film and bulk nanocrystalline CeO_2 was normalized by their grain size by assuming the grain boundary thickness to be 2 nm [5] . This conductivity is related to the single grain boundary and is a good parameter to compare the conductivity of materials with different grain size[5,13]. An excellent correlation between the conductivity behavior for bulk and thin film nanocrystalline specimens is presented in the insertion of Fig.4 strongly suggesting that the electrical properties of nanocrystalline CeO_2 are defined by grain boundary effects and are directly related to material microstructure.

Since the electrical transport in nanocrystalline CeO_2 thin film is controlled by interfacial effects , we can expect much faster ion diffusion and chemical reaction rates to occur as a function of gas atmosphere then is the case for bulk polycrystalline specimens. The interaction between the oxide film and gas atmosphere can lead to generation of ionic defects , electrons or holes at the interfacial areas which causes a change the magnitude and type of conductivity. The kinetics of the electrical conductivity changes due to the gas atmosphere allow us to examine the conductivity mechanism in nanocrystalline CeO_2 thin films. If the kinetics are bulk diffusion controlled , regarding to a classical diffusion model , the slope of the function of log conductivity versus log time equals ½. Such a value has been observed for bulk polycrystalline $Gd_2Ti_2O_7$ specimens [14].

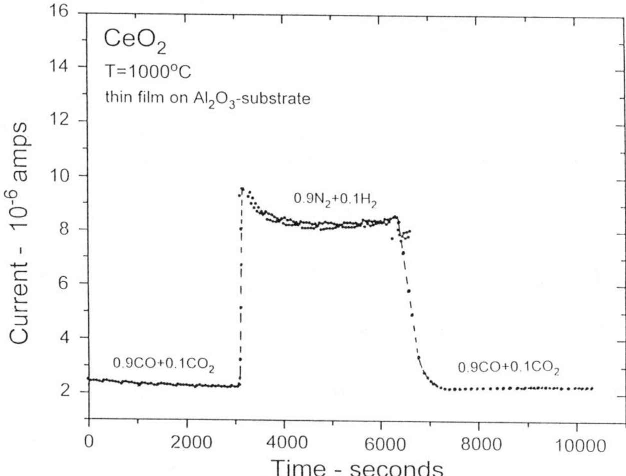

Fig.5   Conductance of CeO_2 thin films as a function of time after switching gases between 0.9CO+0.1CO_2 and 0.9N_2+0.1H_2 atmospheres.

Fig.5 presents the time dependent conductance of CeO_2 thin film when atmosphere is changed from $0.9CO+0.1CO_2$ to $0.9N_2+0.1H_2$. In this measurement a typical slope of ca. 200 and 120 for reduction and oxidation reactions has been determined. This observation suggests that interfacial phenomena prevail in nanostructured materials giving us an opportunity to use these films in electrochemical devices , especially gas sensors.

CONCLUSIONS

The electrical properties of CeO_2 thin films have been investigated as a function of temperature , gas atmosphere and correlated with microstructure. Dense CeO_2 thin films (~1μm) with grain size in the nanometer range were obtained on Si and Al_2O_3 substrates using a polymer precursor spin coating technique. The grain size in obtained films is controlled by annealing temperature.

The electrical transport in nanocrystalline CeO_2 thin films is determined by grain boundary conductivity and is related to their microstructure and ambient atmosphere. The observed rate of change of the electrical conductivity due to changes in the gas composition is about two orders of magnitude faster compared with polycrystalline materials. These observation confirm the unique grain boundary phenomena observed in nanostructured ion conducting materials. The presented electrical and structural properties of nanocrystalline CeO_2 thin films are a good example of material in which microstructure plays important role for controlling its stoichiometry.

ACKNOWLEDGMENTS

Support for this program from the New Energy Development Organization (NEDO) International Joint Research Grant in Protonics is gratefully acknowledged.

REFERENCES

1. Advances in Ceramics, Vol.3, Science and Technology of Zirconia, eds. A.H.Heuer and L.W.Hobbs, American Ceramic Society, Columbus, OH 1981.

2. R.N.Blumental, F.S.Brunger and J.E.Garnier, J.Electrochem. Soc.120,1230(1973).

3. I.Kosacki , H.L.Tuller , Solid State Ionics 80, 223 (1995).

4. I.Kosacki and H.L.Tuller, unpublished.

5. M.Aoki, Y-M.Chiang, I.Kosacki, L.J-R.Lee, H.L.Tuller and Y.Liu, J.Am.Ceram.Soc.79, 1169 (1996).

6. Nanophase Materials: Synthesis-Properties-Applications , eds. G.C.Hadjipanayis and R.W.Siegel , Kluwer Academic Publ. , The Netherlands , 1994.

7. Y-M.Chiang, E.B.Lavik, I.Kosacki, H.L.Tuller and J.Y.Ying, Appl. Phys. Letters 69, 185 (1996).

8. I.Kosacki and H.U.Anderson, Appl.Phys.Letters, 69(1996) in print.

9. H.U.Anderson, U.S.Patent #5,494,700,Feb.1996.

10. I.Kosacki, J.G.Becht, R.van Landschoot and J.Schoonman, Solid State Ionics 59, 287 (1993).

11. I.Kosacki and H.U.Anderson, Solid State Ionics, 1996 , in print.

12. I.Kosacki, Y.-M.Chiang and H.L.Tuller, unpublished.

13. G.M.Christie and F.P.F. van Berkel, Solid State Ionics 83,17(1996).

14. I.Kosacki and H.L.Tuller, Sensors and Actuators B, 24/25, 370 (1995).

SYNTHESIS OF TERBIA DOPED YTTRIA STABILIZED ZIRCONIA THIN FILMS BY USING THE ELECTROSTATIC SPRAY DEPOSITION (ESD) TECHNIQUE

N.H.J. STELZER, A. GOOSSENS, J. SCHOONMAN

Laboratory for Applied Inorganic Chemistry, Delft University of Technology, Julianalaan 136, 2628 BL Delft, The Netherlands, n.h.j.@stm.tudelft.nl

ABSTRACT

Terbia doped yttria stabilized zirconia (Tb-YSZ) thin films have been deposited on various substrates by applying the electrostatic spray deposition (ESD) technique. Yttrium- and zirconium-acetylacetonate as well as terbium-acetate hydrate were used as precursors. Butyl carbitol-ethanol mixed alcohol's with ratios of 50:50 vol% and 80:20 vol% were used as solvents for precursor solutions with total metal concentrations between 0.09 M and 0.009 M. Tb-YSZ thin films were deposited at temperatures between 520 K and 730 K. Surface morphology, chemistry, and crystal structure were studied using scanning electron microscopy (SEM), energy dispersive X-ray (EDX), and X-ray diffraction (XRD). The Tb-YSZ layers consist of a fully stabilized cubic phases. The surface morphology of deposited Tb-YSZ layers can be controlled by changing different deposition parameters. The adhesion of Tb-YSZ to the substrate can be improved by a high temperature treatment.

INTRODUCTION

Terbia-doped yttria-stabilized zirconia (Tb-YSZ) exhibits oxygen ion and p-type conductivity at high oxygen partial pressures and elevated temperatures (e.g. $P_{O_2} > 0.21$ atm at 1073 K) [1]. The oxygen ion conductivity in Tb-YSZ is due to the high concentration of oxygen ion vacancies in the lattice which have significant mobilities at temperatures higher than 873 K. The p-type conductivity in terbia-doped zirconia-based cubic fluorite structured solid solutions such as $(ZrO_2)_{1-x-y}(Tb_2O_{3.5})_x(Y_2O_3)_y$ is believed to be due to hopping of mobile electronic charge carriers [2]. Due to this mixed conduction, Tb-YSZ may have potential electrode and catalytic applications in high-performance solid oxide fuel cells (SOFCs) [3] and in electrocatalysis. Therefore, the preparation of porous as well as dense Tb-YSZ thin films is of high interest.

Recently, dense Tb-YSZ thin films have been prepared by electrochemical vapour deposition (EVD) [4]. This technique, often used to obtain dense and gas tight thin films, has the disadvantage of high processing costs and is rather unsuitable for mass production. In this paper, we report the use of the electrostatic spray deposition (ESD) technique [5] to prepare Tb-YSZ thin films. ESD is a non-vacuum thin film method based on electrostatic atomization of liquid precursors. It has been applied for preparing thin film metal oxides for application in SOFCs and lithium-ion batteries, such as yttria stabilized zirconia (YSZ) and $BaCeO_3$, and $LiMn_2O_4$ and $LiCoO_2$, respectively [6-9]. This technique has the advantage of requiring only a simple and cheap set-up compared to conventional gas phase deposition techniques like Electrochemical Vapour Deposition (EVD). Due to a well defined trajectory of spray droplets towards a grounded substrate by the electric field, a high deposition efficiency and reliability for controlled layer compositions and morphologies can be obtained. In this respect, it is similar to the so-called

Mat. Res. Soc. Symp. Proc. Vol. 453 © 1997 Materials Research Society

corona spray pyrolysis technique [10]. It is a promising route to the synthesis of dense as well as porous thin films at relatively low deposition temperatures and, thereby, lower materials processing and fabrication costs. There are different chemical and physical processes involved which influence the morphology of the deposited layers.

Here we present the results of Tb-YSZ thin films prepared by ESD. Butyl carbitol / ethanol mixed solutions containing terbium-, yttrium- and zirconium metal organic precursors are used. The control of the morphology formed under certain deposition parameters is investigated.

EXPERIMENTAL ASPECTS

An ESD set-up with a vertical configuration, as shown in Fig. 1, was used for the deposition of thin Tb-YSZ layers. A resistively heated substrate holder was used to control the substrate temperature. Nickel, stainless steel, and YSZ ($Zr_{0.84}Y_{0.16}O_{1.92}$, ECN) disks (12 or 20 mm in diameter) are used as substrates. The metal disks had a thickness of 1 mm, while the YSZ substrates had a thickness of 0.13 mm. A thin stainless steel disk was used to obtain a good contact between the thin YSZ substrates and the heated substrate holder. The nickel substrates having a porosity of 75% were sintered at 1173 K for 2 h in an Ar/H_2 atmosphere. The nickel disks were polished on one side in order to decrease their surface roughness. A positive high voltage of +10 kV was applied to the nozzle. The nozzle-to-substrate distance was 30 mm. The feed rate of the precursor solution was set to 0.5 ml/hour using a syringe pump.

Zirconium-(IV)-acetyl acetonate ($Zr(O_2C_5H_7)_4$, Alfa), yttrium-(III)-acetyl acetonate ($Y(O_2C_5H_7)_3$, 99.9%, Alfa) and terbium-(III)-acetate hydrate ($Tb(CH_3COO)_3 \cdot H_2O$, 99.9%, Aldrich) have been used as precursors. Butyl carbitol (99.9%)-ethanol (96%) mixed alcohols with ratios of 50:50 vol% and 80:20 vol% were used as solvents for precursor solutions. The Zr:Y:Tb ratio was set to 0.71:0.14:0.15 in the solutions with total metal concentrations of 0.09 M, 0.05 M, and 0.009 M.

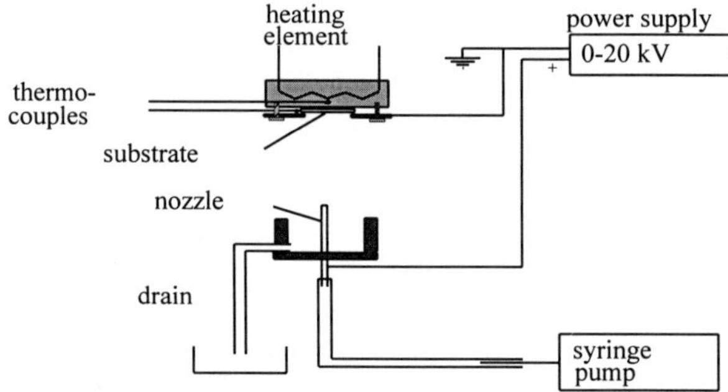

Fig. 1 Schematic view of a vertical ESD set-up.

The Tb-YSZ layers deposited on nickel and YSZ substrates were heat-treated at 1373 K for 3 to 5 hours in air. The heating and cooling rates were set at 120 K/hour. The morphology and composition of the layers were analyzed using a scanning electron microscope (SEM) (JEOL JSM-35 model) equipped with an energy dispersive X-ray microanalyzer (EDX) (Link ISIS, Oxford Instruments Ltd.). Both methods were applied before and after the heat-treatment. For

EDX analysis a cobalt probe was used for calibration. The crystal structure of the deposited layers was determined by X-ray diffraction (XRD) (Phillips PW 1840).

RESULTS AND DISCUSSION

During our investigations of the ESD of Tb-YSZ thin layers, five types of layer morphologies are observed. Type I is a relatively dense but cracked layer, type II is a dense layer, type III consists of a dense bottom layer with some particles incorporated or some lamellar particles on top of the dense layer, type IV is a porous layer, and type V is a very porous layer of fractal-like agglomerates. These five types of morphologies are formed under certain deposition conditions, as will be described below.

XRD analysis on non-annealed Tb-YSZ thin films reveals only a slight and broad (111) diffraction peak of cubic phase YSZ, while the peaks with lower intensity can not be resolved from the background. This indicates a predominantly amorphous structure. These results may be due to the low deposition temperatures (773 K) preventing the formation of the fluorite phase, which is in agreement with the phase diagram of YSZ, the cubic fluorite phase being formed at temperatures above 973 K [11]. After annealing at 1373 K for 3 h in air the XRD pattern of Tb-YSZ thin films can be attributed to the diffraction peaks of the cubic fluorite structure of YSZ and cubic phase nickel oxide (NiO). No monoclinic and tetragonal phases can be observed which indicates that the cubic ZrO_2 phase is fully stabilized by the dopants Y_2O_3 and $Tb_2O_{3.5}$.

Elemental analysis by EDX for a Tb-YSZ layer deposited at 603 K on top of a nickel substrate after annealing at 1373 K for 3 h in air confirms that the overall composition of the layer (Zr:Y:Tb ratio) is the same as that of the precursor solution within experimental error. This is in agreement with earlier investigations on various oxide compounds [8,9] and shows another advantage of the ESD technique compared to other deposition techniques.

<u>Effect of Annealing</u>

As mentioned above XRD analysis indicates that electrostatic spray deposited Tb-YSZ films by using metal organic precursors are amorphous and form the cubic fluorite phase after annealing at 1373 K in air. For annealing times of 3 to 5 hours at 1373 K in air the morphology of porous Tb-YSZ films changes slightly to a less porous layer and the adherence to the substrate increases. After annealing at 1373 K for 15 h in air a typ IV layer morphology changes to a dense bottom layer with small particles on it, belonging to a type III morphology. This result implies that a sintering of the small particles occurs which also increases the adherence to the YSZ substrate. However, no change in the morphology can be observed for dense Tb-YSZ thin films after annealing. Thermal cycling from 293 K to 1373 K for 3 h in air with a heating and cooling rate of 300 K/hour is repeated 5 times. No cracks or change in the morphology can be observed. The deposited Tb-YSZ layer shows very good adherence to the YSZ substrate.

<u>Effect of Substrate</u>

In Fig. 3, two layers deposited at 603 K on different substrates are shown. The layer deposited on a nickel substrate [Fig. 3(a)] is quite porous and consists of large agglomerates of lamellar particles, belonging to a type IV morphology. In comparison, the layer deposited on stainless steel [Fig. 3(b)] consists of a fractal-like morphology, belonging to type V. This implies, that for the same deposition parameters the incoming droplets on the stainless steel substrate are completely dried, where as on the nickel substrate the arriving droplets at the surface are still wet. This result can be explained by the different densities of the substrates. Though, both substrates have the same thickness of 1 mm, the porosity of the used nickel substrate is much higher than that of the stainless steel substrate. Therefore, the thermal conductivity, which is similar for dense nickel and stainless steel, decreases with increasing porosity which lowers the heat necessary for the evaporation of the solvent near or at the surface of the nickel substrate. To verify this hypothesis, a dense nickel substrate was used to deposit Tb-YSZ at 603 K using the same deposition parameters and solvent composition. The obtained layer shows the same fractal-like structure as shown for the dense stainless steel substrate in Fig. 3(b). This result shows that the porosity of a substrate affects the morphology of a deposited layer.

Investigations on the effect of the size of the substrate, by using YSZ substrates of 12 and 20 mm in diameter, showed that the morphology of a Tb-YSZ layer is not affected by changing the surface area of the substrate.

Fig. 3   Surface morphologies of layers deposited at 603 K for 2 h on different substrates: (a) porous nickel; (b) stainless steel. A 50 vol% butyl carbitol + 50 vol% ethanol mixture as solvent for a 0.05 $mol \cdot dm^{-3}$ precursor solution was used.

<u>Effect of Solvent</u>

By changing the solvent composition, the physical properties such as boiling point, solubility of the precursor solutes, spreading behaviour on the substrate of incoming droplets, and the electrical conductivity of the solution will change. Therefore, the morphology of a deposited layer may also be modified. As shown in Fig. 4 the morphology changes quite radically from a fractal-like type V to a dense type II layer by using a mixture of 80 vol% butyl carbitol + 20 vol% ethanol instead of a mixture of 50 vol% butyl carbitol + 50 vol% ethanol as the solvent.

This extreme change in the layer morphology can be explained by the higher boiling point of the 80:20 alcohol mixture, considering that due to the used ethanol the water content in this mixture decreases. Using high-boiling point solvents causes the evaporation rate to decrease and the spreading of the still wet droplets on the substrate will be slower than using butyl carbitol-ethanol mixtures with higher ethanol content and accordingly lower boiling point. Therefore, a relatively dense layer can be obtained.

Preliminary results showed that with the 50 vol% butyl carbitol + 50 vol% ethanol mixture no dense layers could be obtained on stainless steel and YSZ substrates. A very narrow temperature range, i.e. around 10 K, was found, in which the morphology of a deposited layer changes from type I to type IV on stainless steel, respectively, type I to type V on YSZ substrates.

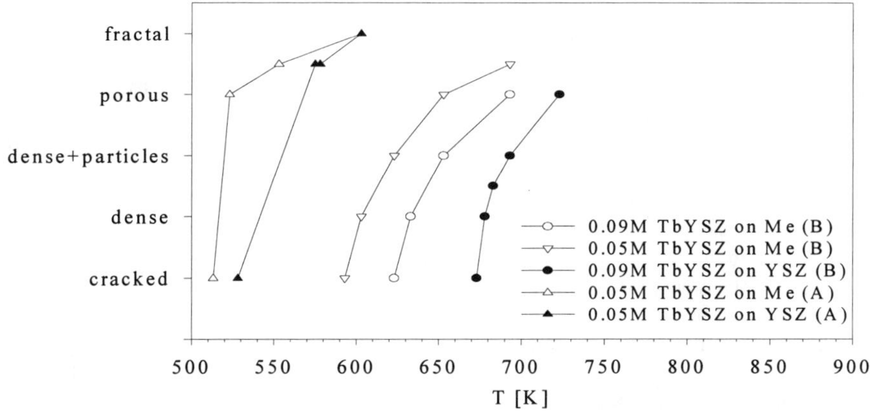

Fig. 4   Surface morphologies of Tb-YSZ layers deposited on stainless steel and YSZ using precursor solutions with different solvent compositions: (A) 50 vol% butyl carbitol + 50 vol% ethanol solution; (B) 80 vol% butyl carbitol + 20 vol% ethanol solution.

Effect of Deposition Temperature

The morphology of layers deposited at different temperatures on YSZ substrates are shown in Fig. 5. It can be seen that the Tb-YSZ film deposited at 633 K [Fig. 5(a)] for 1 h is cracked and tends to spill-off from the YSZ substrate. This morphology, (type I), is always found to be present if the deposition temperature is too low. Increasing the deposition temperature up to 673 K [Fig. 5(b)] a dense Tb-YSZ layer can be obtained, belonging to the type II morphology. Further increase in deposition temperature to 693 K leads to minor agglomeration on top of a dense bottom layer and the deposited film [Fig. 5(c)] shifts to a type III morphology. At a deposition temperature of 723 K [Fig. 5(d)], the deposited layer shows a porous type IV morphology. With increasing deposition temperature the morphology of the deposited layer for solutions with a total metal concentration of 0.09 $mol \cdot dm^{-3}$ in 80 vol% butyl carbitol + 20 vol% ethanol mixed alcohol changes from type I to type V.

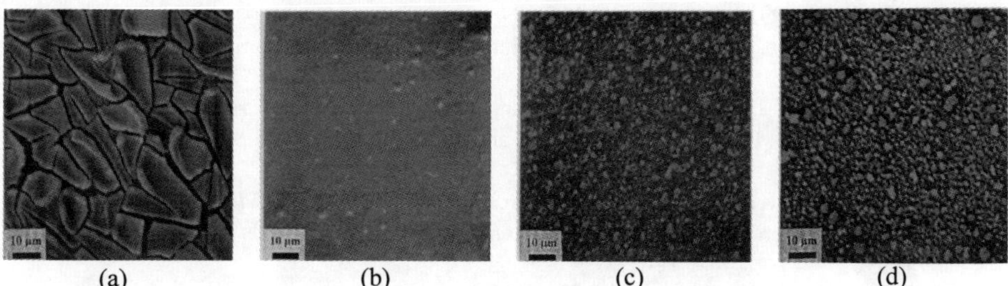

   (a)   (b)   (c)   (d)

Fig. 5   Surface morphologies of layers deposited on YSZ at different temperatures for 1 h: (a) 633 K; (b) 673 K; (c) 693 K; (d) 723 K. An 80 vol% butyl carbitol + 20 vol% ethanol mixture as solvent for a 0.09 $mol \cdot dm^{-3}$ precursor solution was used.

CONCLUSIONS

The investigations on electrostatic spray deposited Tb-YSZ thin films show that the morphologies of these layers can be controlled, and are mainly determined by the deposition temperature, the solvent and the substrate.

By increasing the deposition temperature, the morphology of a layer deposited on a certain substrate can shift from a dense to a highly porous structure.

The boiling point of the solvent determines the spreading behaviour of the solvent droplets on the substrate. By using a solvent with a high boiling point the morphology of a layer becomes more dense due to the slower evaporation and precipitation rate of the droplet. At the same time a higher deposition temperature is required in order to obtain a certain evaporation rate. Otherwise, a cracked layer will be obtained.

The substrate can influence the morphology of a deposited layer due to its density, surface roughness and surface tension. These properties may change during deposition due to the different properties of the deposited layer.

Therefore, in order to obtain the desired morphology of a Tb-YSZ layer the above specified parameters have to be controlled in detail during deposition.

It is shown that the ESD technique is very promising to deposit thin Tb-YSZ films. However, with regard to a possible use in SOFCs further investigations on the influence of the morphology on the electrical properties of Tb-YSZ layers deposited by using the ESD technique is necessary. This work is presently in progress in our laboratory.

ACKNOWLEDGEMENTS

N.H.J. Stelzer is grateful to the Delft University of Technology for a research fellowship. The authors are indebted to Dr. F.P.F. van Berkel of the Netherlands Energy Research Foundation (ECN) for providing the YSZ substrates.

REFERENCES

1. C.Z. Cao, J. Appl. Electrochem., **24**, 1222 (1994).
2. M.P. van Dijk, K.J. de Vries, and A.J. Burggraaf, Solid State Ionics, **16**, 211 (1985).
3. P. Han and W.L. Worrell, J. Electrochem. Soc., **142**, 4235 (1995).
4. G.-Z. Cao, J.Meijerink, H. W. Brinkman, K.J. de Vries, and A.J. Burggraaf, J. Mater. Chem., **3**, 773 (1993).
5. A.A. van Zomeren, E.M. Kelder, J.C.M. Marijnissen, and J. Schoonman, J. Aerosol Sci., **25**, 1229 (1994).
6. E.M. Kelder, O.C.J. Nijs, and J. Schoonman, Solid State Ionics, **68**, 5 (1994).
7. C.H. Chen, A.A.J. Buysman, E.M. Kelder, and J. Schoonman, Solid State Ionics, **80**, 1 (1995).
8. C.H. Chen, E.M. Kelder, and J. Schoonman, J. Mater. Sci., **31**, 5437 (1996).
9. C.H. Chen, K. Nord-Varhaug, and J. Schoonman, J. Mater. Synthesis and Processing, **4**, (3), 189 (1996).
10. W. Siefert, Thin Solid Films, **120**, 267 (1984).
11. V.S. Stubican, R.S. Hink, and M. Henry, Prog. Solid State Chem., **18**, 259 (1988).

DEFECTS AND ION TRANSPORT PROPERTIES OF YTTRIUM DOPED ZIRCONIA

P.J. CHABA, P.E. NGOEPE
Department of Physics, University of the North, Sovenga 0727, South Africa.

ABSTRACT

A comparison of calculated and experimental temperature variation of elastic constants were used to predict types of oxygen–vacancy dopant clusters in yttria stabilised cubic zirconia, which serves as an electrolyte in solid oxide fuelcells. Such clusters were incorporated in supercells set up for molecular dynamics studies, where oxygen transport properties were investigated at concentrations of 9.4 and 24 mol % of yttrium oxide and up to 1600K.

INTRODUCTION

Solid oxide fuel–cells that use yttrium–stabilized cubic zirconia, $ZrO_2(xY_2O_3)$, as electrolyte are being studied extensively owing to their outstanding ionic conductivity and mechanical properties [1]. Oxygen vacancies are created for charge compensation as the concentration of yttrium is increased. EXAFS [2] and previous computer modelling [3] studies have shown that vacancies are located at next–nearest neighbour sites to yttrium. In a recent molecular dynamics study [4] it was shown that ionic conductivity decreases with an increasing number of nearest Y^{3+}–Y^{3+} neighbour pairs. However, the choice of dopant oxygen vacancy clusters was random.

Previous comparison of experimental and calculated elastic constants have assisted in clarifying types of defects in LaF_3 [5]. In the current study temperature variation of elastic constants, calculated from energy minimisation methods will be compared with Brillouin scattering results [6] in order to predict the nature of defect complexes that occur in cubic zirconia. Such clusters will be embedded in supercells used for molecular dynamics studies where diffusion coefficients and ion migration mechanisms are determined at different yttrium concentrations and temperatures.

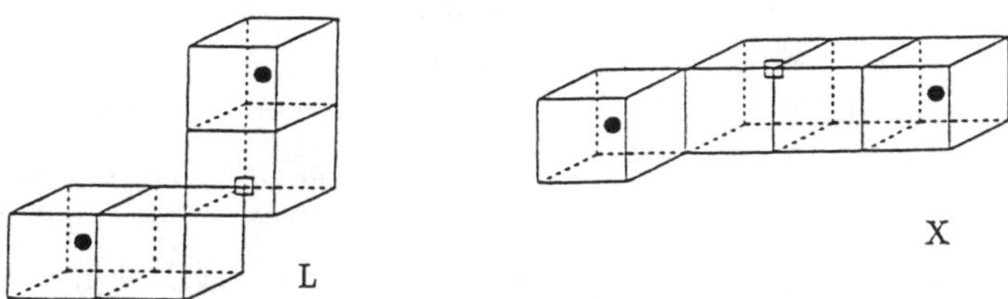

Figure 1   Different orientations of oxygen–vacancy yttrium clusters, namely, L and X shaped. Solid circles are Y positions.

COMPUTATIONAL PROCEDURE

Distorted fluorite structured $ZrO_2(xY_2O_3)$ supercells containing approximately 96 ions were set up, where x=9.4 and 24 mol%. The supercells included oxygen–vacancy yttrium clusters, with orientations shown in figure 1.

Mat. Res. Soc. Symp. Proc. Vol. 453 © 1997 Materials Research Society

Table 1

Interionic potentials for $ZrO_2(xY_2O_3)$

	$A(eV)$	$\rho(A)$	$C(eVA^{-6})$
$Zr^{4+}-O^{2-}$	985.87	0.376	0
$Y^{3+}-O^{2-}$	1345.10	0.349	0
$O^{2-}-O^{2-}$	22764.30	0.149	27.89

The energy minimisation procedure was applied on resulting structures at constant volume, to determine equilibrium configurations. Associated thermodynamic quantities were calculated. In all cases, the computer code THBREL [5] was used. Temperature changes of elastic constants were determined by lattice expansion, within the quasi–harmonic approximation. Interionic potentials employed in our calculations are given in table 1.

Molecular dynamics technique involves solution of Newton's second law of motion. Using suitable algorithms, trajectories and velocities of ions are determined. Diffusion coefficients, D_i, may be obtained from Einstein like relationship

$$<r_i^2> = 6D_it + B$$

where $<r_i^2>$ is the mean square displacement and B is a constant. FUNGUS [7] program was used in the present study and details of the calculation are given in table 2:

Table 2

Details of molecular dynamics calculations

Simulation box	8 unit cells of $ZrO_2(xY_2O_3)$
Timestep	5 fs
Startup	Initial structure: distorted fluorite
	1000 steps for equilibration
	7000 steps (35ps) for a run
Temperatures	$300 - 1600K$
Rigid ion potentials were used	

RESULTS

Energy Minimisation Calculations

Table 3 shows a comparison of calculated temperature variation of elastic constants related to various cluster models with experimental values. The L shaped model yields results that are consistent with experiments as compared to other configurations.

Table 3

Calculated and experimental slopes of temperature variation of elastic constants of $ZrO_2(xY_2O_3)$, at different values of x.

a) $ZrO_2(9.4\ mol\%\ Y_2O_3)$

Configuration	$\partial C_{11}/\partial T$	$\partial C_{12}/\partial T$ (GPa/K)	$\partial C_{44}/\partial T$
X	0.030	0.010	0.015
L	0.051	0.021	0.015
XL	0.036	0.015	0.012
Random	0.125	0.010	0.033
Experimental [6]	0.050	0.020	0.012

b) $ZrO_2(24\ mol\%\ Y_2O_3)$

Configuration	$\partial C_{11}/\partial T$	$\partial C_{12}/\partial T$ (GPa/K)	$\partial C_{44}/\partial T$
X	0.037	0.025	0.013
L	0.055	0.032	0.024
Random	0.036	0.015	0.019
Experimental [6]	0.055	0.036	0.020

<u>Molecular Dynamics Calculations.</u>

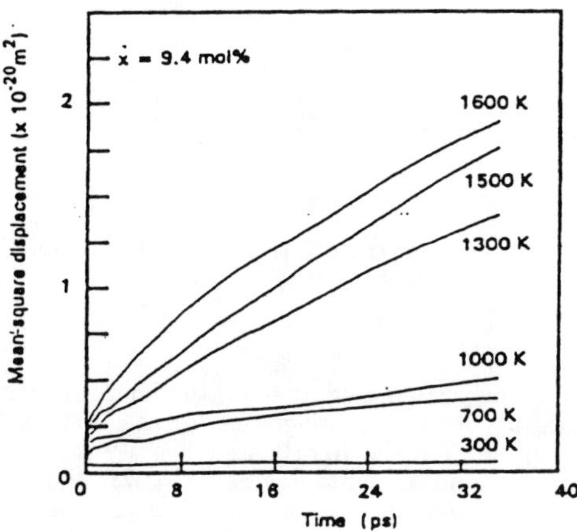

Figure 2a   Change of mean square displacement of oxygen with time of $ZrO_2(xY_2O_3)$ at different temperatures for x=9.4

551

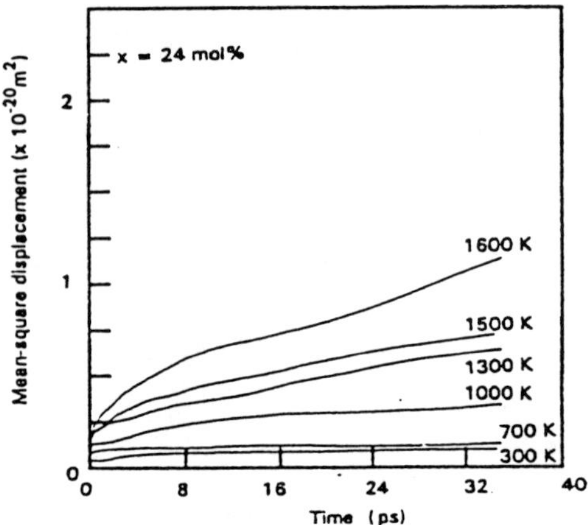

Figure 2b   Change of mean square displacement of oxygen with time for $ZrO_2(xY_2O_3)$ at different temperatures for x=24 mol%

It is noted that no oxygen diffuses at 300K in figure 2. However, diffusion increases with temperature above 700K and is suppressed by higher yttrium content.

In figure 3 it is shown that oxygen migrates predominantly by a hopping vacancy mechanism at higher temperatures.

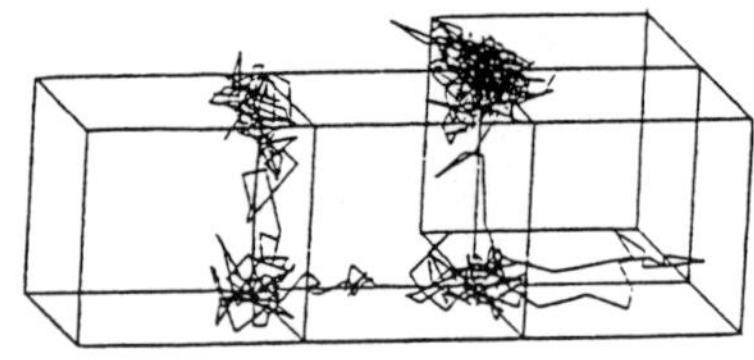

Figure 3   Oxygen ion trajectories of $ZrO_2(9.4$ mol% $Y_2O_3)$ at 1600K for a period of 35ps.

DISCUSSIONS

A comparison of our calculated temperature dependences of elastic constants and Brillouin scattering results [6] shows preference for the L shaped oxygen—vacancy yttrium clusters. This trend applies for the lightly and heavily doped cubic zirconia. A random distribution of vacancies and dopants yields $\partial C_{ij}/\partial T$ that deviate substantially from experimental values. It therefore suggests a low probability for occurence of this configuration. Furthermore, vacancies are located at next—nearest neighbour sites to yttrium, particularly at low concentrations, in agreement with previous EXAFS [2] and static simulation studies [3].

Molecular dynamics studies, based on preferred L shaped yttrium—vacancy clusters indicate that oxygen diffusion is negligible at ambient temperatures and becomes significant above 700K. Furthermore, an increase in yttrium concentration tends to suppress diffusion, in agreement with previous ionic conductivity studies [1].

The current study, has also indicated the nature of ion migration mechanism found at high temperatures. Whereas, interstitialcy mechanism is dominant in rare—earth doped alkaline—earth fluorides [7], the vacancy mechanism is prevalent in cubic zirconia. Similarly, a difference in behaviour of bulk moduli of fluorites and cubic zirconia at high temperatures was noted [6]. These features suggest that the nature of disorder responsible for the fast—ion phase in these systems is different.

REFERENCES

1.   Y. Suzuki, T. Takahashi, and N. Nagae, Solid State Ionics, **314** p. 483 (1981).
2.   C.R.A. Catlow, A.V. Chadwick, G.N. Greaves and L.M. Moroney, J. Am. Cer. Soc. **69**, p. 272 (1986).
3.   A. Dwivedi, A.N. Cormack, J. Solid State Chem. **79**, p. 218 (1989).
4.   Li Xiaoyun and B. Hafskjold, J. Phys. Condens. Matter 7, p. 1225, (1995).
5.   J.D. Comins, P.E. Ngoepe and C.R.A. Catlow, J. Chem. Soc. Faraday Trans. **86**, p. 1183 (1990).
6.   J. Botha, J. Chiang, J.D. Comins, P.M. Mjwara and P.E. Ngoepe, J. Appl. Phys. **73**, p. 7287 (1993).
7.   Fungus
8.   P.E. Ngoepe and C.R.A. Catlow, Radiation Effects and Defects in Solids **119**, p. 399 (1991).

Grain boundary conductivity and microstructure study of 4% Y₂O₃ doped CeO₂ thin films

Chunyan Tian and Siu-Wai Chan

Dept. of Chemical Engineering, Materials Science and Mining Engineering,
Columbia University, New York, NY 10027.

Abstract

The thin films of 4% Y_2O_3 doped CeO_2 have been deposited on different substrates of Pd film/(001) LaAlO₃, Pd film/r-cut sapphire, and Pd film/Quartz using an e-beam deposition technique. The microstructures and electrical properties of the films were investigated by means of x-ray diffraction, transmission electron microscopy, and ac impedance spectroscopy. Both textured and polycrystalline films were produced on different substrates. A brick layer model was adopted to correlate the microstructure and electrical property of the films. Only the grain boundary arc was observed in the film complex impedance plots. The conductivities of the films were similar to the conductivity of 6% Y_2O_3 doped CeO_2 bulk grain boundary because of lower preexponential factor, although the activation energies were smaller than that of bulk grain boundary. The resistive grain boundaries were found to dominate the conductivities of the films.

Introduction

Yttria stabilized zirconia (YSZ) is a traditional solid electrolyte material used in high temperature solid oxide fuel cells (SOFC) because of its relatively high ionic conductivity[1,2]. However, YSZ based devices has to be operated at 1000°C or higher to achieve an applicable power density. Fluorite structure cerium oxide with trivalent dopants are a type of materials that exhibit higher oxygen ion conductivity over YSZ for the same dopant concentration[3]. By adopting CeO_2 based electrolytes would allow SOFC to operate in intermediate temperatures (500-700°C)[4]. Presently, thick film ceramic technology can provide reliable material down to 10-15μm which can lower the operating temperature to 450°C for CeO_2 based electrolytes.[5]. Therefore, it is highly desirable to produce thinner layers of these materials to lower working temperature further by reducing the ohmic loss.

On the other hand, from a scientific point of view, the thin film technology may present many advantages. First of all, the experimental conditions of vacuum-deposition techniques, such as electron-beam, can be well defined to yield reproducible deposits. Secondly, by varying the experimental parameters (source, oxygen partial pressure, substrate temperature, etc.), deposits of various composition or structure can be obtained[6]. In this work, we have deposited 4% Y_2O_3 doped CeO_2 thin films on various substrates using an e-beam deposition technique and investigated the relation between the microstructure and the electrical properties.

555

Experimental

The substrates were prepared in three different ways. For quartz substrates, they were cleaned by successive treatments in acetone, hydrofluoric acid, and distilled water. Some of the r-cut sapphire and (001) LaAlO$_3$ substrates were prepared by chemical etching[7], the others were first chemically etched as above, followed by annealing at 700°C in O$_2$ for an hour. The annealing process is very crucial for obtaining the textured films.

The film deposition parameters can be found from the previous study[8]. During and after deposition, oxygen was introduced to fully oxidize the films. The microstructure of the as-deposited films were studied using x-ray diffraction and transmission electron microscopy (TEM). The TEM samples of the films on r-cut sapphire and LaAlO$_3$ substrates were prepared by mechanically polishing, dimpler thinning, and ion milling to electron transparency[7]. For films on quartz substrates, the substrates were chemically etched off using hydrofluoric acid.

The electrical property measurements were performed from room temperature to 350°C and the frequency ranging from 0.1Hz to 10MHz using a Solartron 1260 Impedance/Gain-phase analyzer. A sandwich structure of the measurement sample was obtained by utilizing Pd as base electrode and Ag as top electrodes[8].

Results and Discussion

The imbedded diffraction pattern and the TEM bright field image from Fig. 1 show the polycrystalline nature of the as-deposited film on (001) LaAlO$_3$. The average grain size is about 500Å. Gaps were observed along grain boundaries. They were probably from incomplete coalescence during film deposition process or from ion milling during TEM sample preparation. The polycrystalline films were also produced on quartz substrates (Fig. 2), with a average grain size of 250Å, and some (110) textured clusters. The (111) textured films were obtained on r-cut sapphire substrates as shown by x-ray diffraction pattern (Fig.3).

Figure 4 shows the complex impedance plot of 4% Y$_2$O$_3$ doped CeO$_2$ film/Pd film/(001) LaAlO$_3$ measured at different temperatures in ambient atmosphere. The real part of the impedance (Z') corresponds to the conduction of the film, and the imaginary part Z" corresponds to the polarization of the film. The conductivity (σ_{ac}) of the film can be calculated from Z'. Figure 5 is the Arrhenius plot showing the variation of $\sigma_{ac}T$ as a function of 1/T. From the slope of the plot, one can obtain the value of the activation energies. For comparison, the conductivities of 4% Y$_2$O$_3$ doped CeO$_2$ bulk grain lattice[3] and 6% Y$_2$O$_3$ doped CeO$_2$ bulk grain boundary[9] were both included in Fig. 5. It should be noted that the activation energies of the films were smaller than that of bulk grain boundaries, but the preexponential factor was much smaller than that of bulk grain lattice and grain boundary. As the results, the conductivities of the films were all approximate to the conductivity of the bulk grain boundary, and much lower than that of bulk grain lattice.

Ideally, two arcs should be observed in the complex impedance plots. One is due to grain lattice, while another one is due to grain boundary. However, through all the measurement temperatures, only one arc was observed. There are three senarios that could result in one arc. The first senario is that the grain boundaries are highly conducting, with only the lattice arc observed, which implies that the film conductivities should be approximately equal to that of the bulk lattice, but that was not the case according to Fig. 5. Second senario is that the grain size is comparable to the thickness of the films. Therefore, the overall

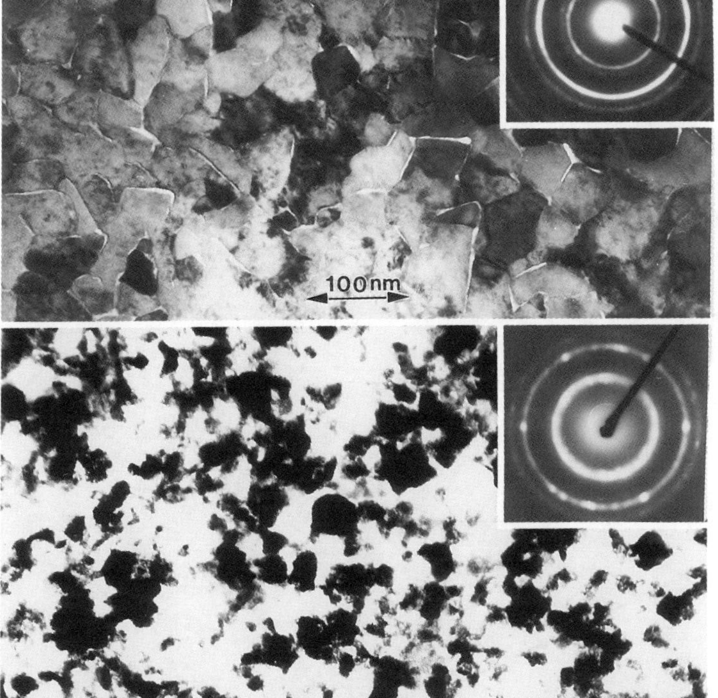

Fig. 1
TEM micrographs showing the microstructure of 4% Y_2O_3 doped CeO_2 film deposited on Pd/(001)LaAlO$_3$ at 700°C.

Fig. 2
TEM micrographs of 4% Y_2O_3 doped CeO_2 film on Pd film/quartz grown at 700°C.

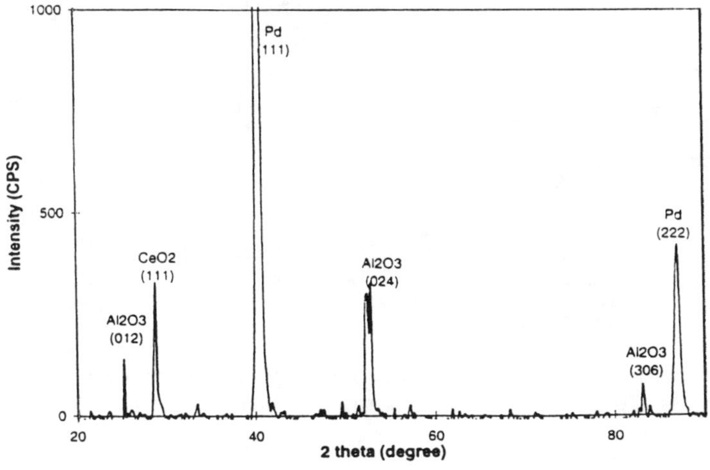

Fig. 3
X-ray diffraction pattern showing the textured alignment of 4% Y_2O_3 doped CeO_2 film on Pd/r-cut sapphire.

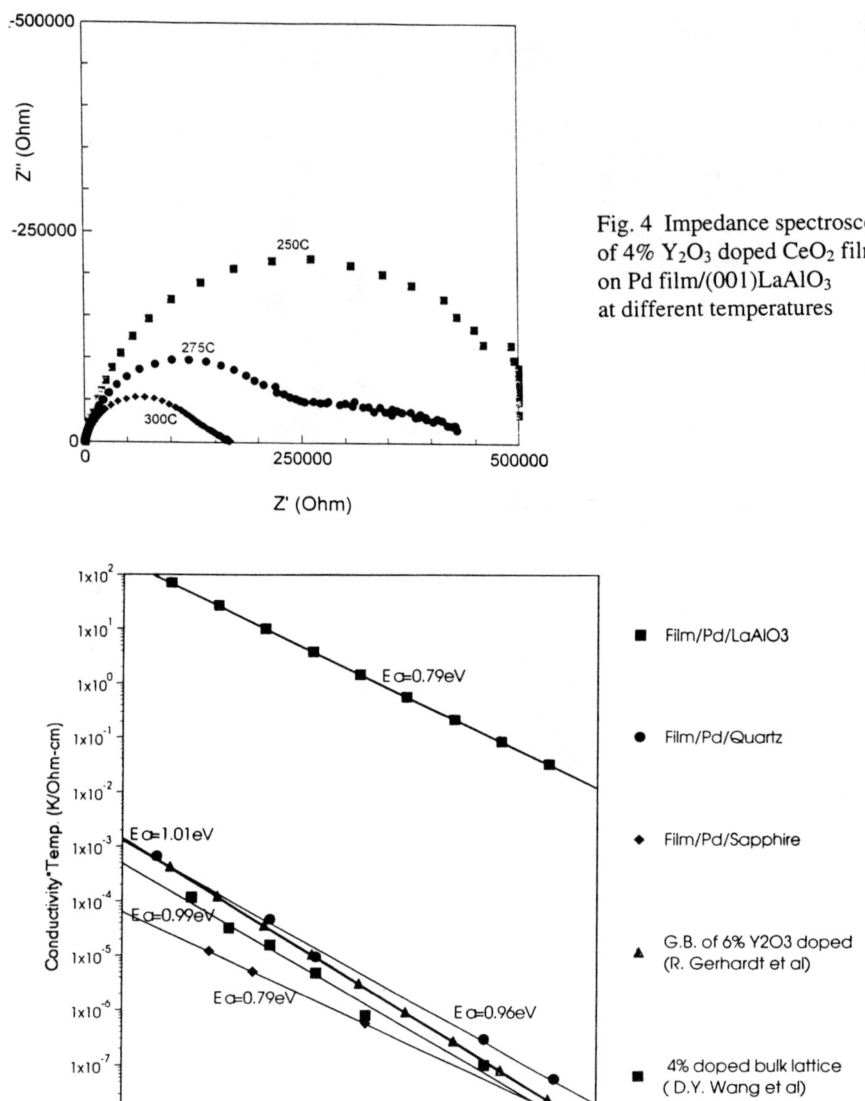

Fig. 4 Impedance spectroscopy of 4% Y_2O_3 doped CeO_2 film on Pd film/(001)$LaAlO_3$ at different temperatures

Fig. 5 Plot of the conductivities of 4% Y_2O_3 doped CeO_2 film on different substrates as a function of temperatures. For comparison, the conductivities of bulk lattice and grain boundary are included.

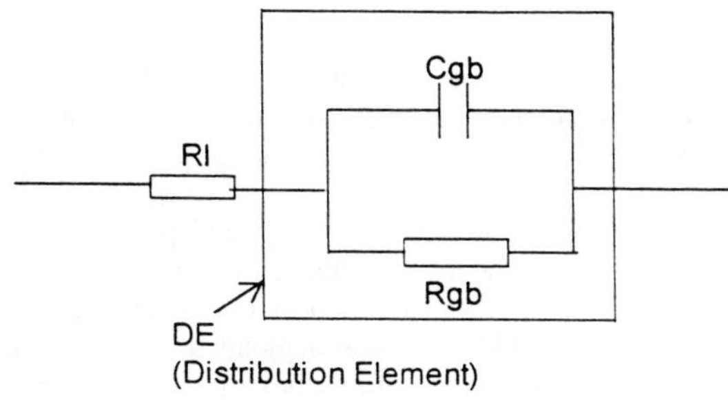

Fig. 6
A simplified equivalent circuit based on the brick layer model.

$$Zgb = \frac{Rgb}{1+(jwRgbCgb)^n}, \quad n: 0{\sim}1$$

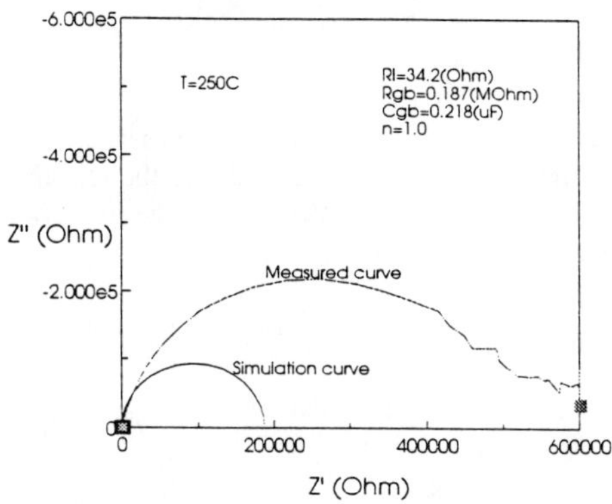

Fig. 7
Impedance spectroscopy of simulated plots at 250°C using the equivalent circuit in Fig. 6, R_l, R_{gb}, and C_{gb} were calculated using the bulk lattice and grain boundary conductivities and film geometry.

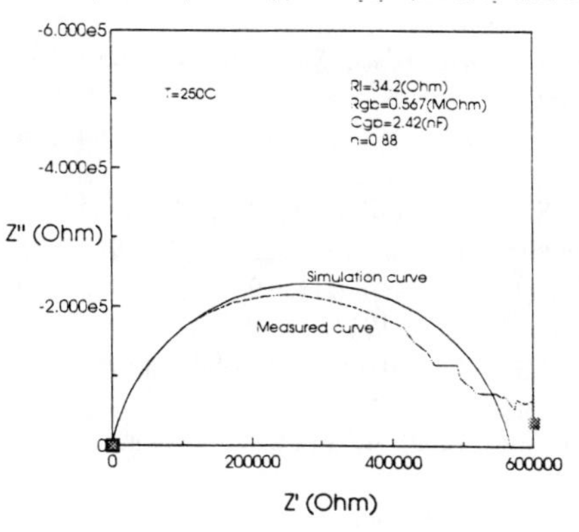

impedance from grain lattice and grain boundary results in one arc represented by a parallel equivalent circuit. However, it was not the case either, since the average grain sizes of 500Å (Fig. 1) and 250Å (Fig. 2) are both much smaller than the film thickness of 0.5μ. The third senario is that grain lattice arc is too small to be perceptible, so that the measured arc is from grain boundaries, which means the conductivities of the films would be that of grain boundary, that was the case as shown in Fig. 5.

Here, the brick layer model which illustrates how the conduction and polarization of both the grain lattice and grain boundary contribute to the overall impedance[10] can be adopted to correlate the microstructure and the electrical property of the films. The series equivalent circuit representing the brick layer model of the electrolyte was simplified and the grain boundaries were treated as a time distribution element because of its inhomogeneity[11] (Fig. 6). The measured impedance data was then fitted with a simulation using the conductivity of bulk grain lattice[3] as well as of the bulk grain boundary[9], and the geometry of the film as the simulation parameters. The simulated and measured curves were shown in figure 7 for different parameters. It can be seen that the measured resistance is about three times of the resistance of grain boundary calculated using the conductivity and permittivity of bulk lattice and grain boundary, while the measured capacitance was about 10^{-2} times of the calculated value. This further proved that the measured arc is from grain boundary.

Summary

We have shown that 4% Y_2O_3 doped CeO_2 films with different microstructure can be successfully prepared using e-beam deposition technique. It was found that the resistive grain boundaries dominate the conductivities of the films. It maybe attributed to the grain boundary impurity segregation or to the ion-blocking at internal interfaces.

Acknowledgments

This work was supported by National Science Foundation under grant DMR-93-50464.

References

1. E. C. Subbarao and H. S. Maiti, *Solid State Ionics*, **11**, 317 (1984).
2. B. C. H. Steele, *Solid State Ionics*, **12**, 391 (1984).
3. D. Y. Wang, D. S. Park, J. Griffith and A. S. Nowick, *Solid State Ionics*, **2**, 95 (1981).
4. N. Q. Minh, *J. Am. Ceram. Soc.*, **76**, 563 (1993).
5. B. C. H. Steele, *Proc. of 1st European Solid Oxide Fuel Cell Forum* (Oct. 1994, Lucerne, Switzerland) Vol. **1**, p.375, Ed. U. Bossel.
6. M. Croset, J. P. Schnell, G. Velasco and J. Siejka, *J. Appl. Phy.*, **48**, 775 (1977).
7. C. Tian, Y. Du and S.-W. Chan, to be published in *J. Vac. Sci. Tech. A*, Jan./Feb. 1997.
8. C. Tian and S.-W. Chan, *Mat. Res. Soc. Symp. Proc.*, Vol. **411**, 1996.
9. R. Gerhardt and A. S. Nowick, *J. Am. Ceram. Soc.*, **69** [9], 641 (1986).
10. J. R. Macdonald, in <u>Impedance Spectroscopy: Emphasizing Solid Materials and Systems</u>, New York: Wiley, c1987.
11. C. K. Chiang, J. R. Bethin, A. L. Dragoo, A. D. Franklin, and K. F. Young, *J. Electrochem. Soc.*, **129** [9], 2113 (1982).

MECHANOCHEMICAL PREPARATION OF CERAMIC METAL OXIDE ELECTROLYTES WITH ENHANCED CONDUCTIVITY

R. VITLOV-AUDINO, F. J. LINCOLN
Special Research Centre for Advanced Mineral and Materials Processing, Department of Chemistry, University of Western Australia, Nedlands, WA, 6907, Australia.

ABSTRACT

In this study, mechanochemical milling (also known as mechanical alloying), has been used as an alternative means of synthesis of ceramic metal oxide electrolytes at room temperature and compared to the conventional calcination methods. The oxide electrolytes prepared by mechanical milling, were, the lanthanide-doped *fcc* Bismuth Oxides and Cerium Oxides, both of which are oxygen deficient. Conductivities for some of these milled oxides, measured by the four probe technique, were found to be enhanced when compared to those for conventionally prepared materials.

INTRODUCTION

The lanthanide-doped Bismuth Oxides and Cerium Oxides possess the cubic fluorite-type structure. Pure Bismuth Oxide is monoclinic at room temperature (α-Bi_2O_3), however above 730° C it transforms to the fluorite-type phase δ-Bi_2O_3, the best ionic conductor known with a conductivity of ~1 Ohm^{-1} cm^{-1}, which is about two orders of magnitude higher than the stabilised zirconias. This form is characterised by a disordered, partially occupied oxygen sublattice [1] which can be stabilised at room temperature by doping with various lanthanide oxides [2]. Cerium Oxide is an electronic conductor in its pure form, however doping it with various lanthanide oxides introduces a stoichiometric oxygen deficiency, which enhances its ionic conductivity [3]. In both these systems it is this oxygen deficiency which is responsible for the ionic conductivity.

Conventionally, these oxides have been synthesised by either calcination or coprecipitation techniques. Calcination is the more common method, involving the direct heating of reactant powders at temperatures between 800 and 1500° C, sometimes for long periods, since these oxides are characterised by high melting temperatures. This method has several disadvantages, reaction with the containment vessel can occur, additional heat treatments may be necessary to ensure complete solid state reaction, and reaction times may also be long. Coprecipitation generates finer particles, which have greater surface area, hence reactivity, making the coprecipitated oxide more sinterable, however the technique involves strict control of a number of variables, which can lead to irreproducibility.

Mechanochemical alloying (hereafter MA) has been successfully used for the synthesis of intermetallic compounds, composites and oxides [4-6]. The nature of the high energy milling process is such that the particles are being fractured continuously, exposing fresh atomic surfaces, while the reactant powders are in intimate contact throughout the reaction. Reaction times are therefore generally reduced compared to thermal methods, because of the enhanced solid state diffusion processes occurring within the system , due to its dynamic nature compared to calcination, where diffusion occurs more slowly [7]. Recently, MA has been used successfully in synthesising CeO_2-ZrO_2 solid solutions by direct milling of CeO_2 and ZrO_2 powders [8]. From these considerations, MA appears to be an attractive alternative to conventional synthesis and in this project it was used to prepare solid solutions in the above systems.

Mat. Res. Soc. Symp. Proc. Vol. 453 ©1997 Materials Research Society

EXPERIMENT

Samaria and Yttria-doped Bismuth Oxides and Samaria and Bismia-doped Cerium Oxides were prepared by MA. Stoichiometric quantities of the starting oxides, Sm_2O_3 (Cerac, 99.9%), CeO_2 (Cerac, 99.9%), Bi_2O_3 (Sigma, 99.9%) and Y_2O_3 (Fluka, 99.98%), were placed in a hardened steel reaction vial (radius, 2.2 cm, length, 7.5 cm) with hardened steel balls (radius, 0.6 cm) using a powder to ball mass ratio of 1:13. The vial was then placed in a high energy vibratory ball mill (Spex-8000 Shaker Mill). High energy milling was performed for 6-8 hours depending on the system, in both air and argon atmospheres. The resulting oxides were characterised by Powder X-Ray Diffraction (PXRD), using Cu K_α radiation and a Siemens D5000 diffractometer with a graphite diffracted beam monochromator, and Scanning Electron Microscopy (SEM), coupled with Energy Dispersive Spectrometry (EDS).

The electrical conductivity measurements were performed in a high temperature furnace (HPS-501E, Morris Research Co.) equipped with a four point probe to measure sample resistances [9]. All measurements were taken in the "pulse drive" mode to offset thermal effects. This mode toggles the source current between pulsed and DC, which cancels the thermal emfs in the test lead connections and the unknown resistance. Sample bars (10 mm x 2 mm x 2 mm) were prepared by pressing sample powders into pellets (10 mm x 2 mm), which were then sintered at temperatures between 900 and $1000°$ C for 4 hours, before being cut into bars. Electrical connections were made by looping four separate platinum wires (0.25 mm diam.) around the sample bar, with electrical contact between the sample and platinum wires, made at four points in a collinear array with "silver paint". The platinum wires were then connected to the four platinum probe wires. Measurements were taken at $50°$ increments, in the range 500 to $900°$ C in air, with an equilibration time of 30 minutes at each temperature.

RESULTS

Direct reaction between Sm_2O_3, Y_2O_3 and Bi_2O_3, CeO_2 and Bi_2O_3, and CeO_2 and Sm_2O_3 was achieved by MA for the compositions and milling conditions shown in Table I. Each preparation, whether conducted in air or argon, resulted in the formation of a single phase, cubic material with the fluorite-type structure.

Table I. Starting compositions and structures of oxides before and after MA.

Starting Composition (mol %)	Milling Time (hours)	Structure-type before milling	Structure-type after milling	Lattice Parameter (Å), this study	Lattice Parameter (Å), lit. values [ref]
40 Sm_2O_3 60 Bi_2O_3	6	Monoclinic Monoclinic	Fluorite	5.5150	5.518 [10]
20 CeO_2 80 Sm_2O_3	8	Fluorite Monoclinic	Fluorite	5.4386	5.4312 [11]
30 Bi_2O_3 70 CeO_2	8	Monoclinic Fluorite	Fluorite	5.4209	not previously reported
40 Y_2O_3 60 Bi_2O_3	8	C-type Monoclinic	Fluorite	5.4691	5.475 [12]
20 Y_2O_3 20 Sm_2O_3 60 Bi_2O_3	8	C-type Monoclinic Monoclinic	Fluorite	5.4892	not previously reported

Initially, all the starting oxides were individually milled on their own, in both argon and air, to determine whether any milling-induced phase transformations occurred. PXRD patterns of these milled materials, exhibited no structural changes, only peak broadening, due to a reduction in particle size. The particles, observed with SEM, were much reduced in size and often agglomerated into sub-micron clusters, as shown in Figure 1 for the phase $(Sm_2O_3)_{0.4}(Bi_2O_3)_{0.6}$, which was typical for all as-milled phases. Lattice parameters, calculated for all the as-milled phases shown in Table I, were found to be in good agreement with literature values.

Figure 1. SEM micrograph of as-milled $(Sm_2O_3)_{0.4}(Bi_2O_3)_{0.6}$.

<u>Doped Bi$_2$O$_3$ phases.</u>

Initially, the Sm_2O_3-Bi_2O_3 system was investigated to find out whether the cubic δ-Bi_2O_3 type phase, stable between 730 and 825° C, could be stabilised at room temperature, by MA Sm_2O_3 with α-Bi_2O_3 (the monoclinic, room temperature form). As shown in Table I, a starting mixture of 40 mol % Sm_2O_3 and 60 mol% Bi_2O_3 produced a cubic fluorite phase after milling for six hours; the same phase was produced by conventional means by heating at 1000° C for 24 hours. Figure 2 shows the observed PXRD pattern for the cubic, fluorite-type phase produced by MA. The broadened PXRD reflections indicate that some particle size reduction has occurred, as it did in all milling experiments.

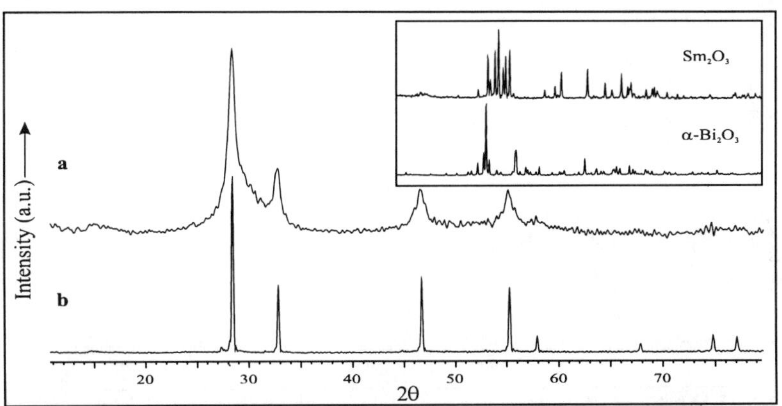

Figure 2. PXRD patterns of the stabilised cubic Bi_2O_3 phase $(Sm_2O_3)_{0.4}(Bi_2O_3)_{0.6}$ made by (**a**) milling and (**b**) calcination of the starting oxides α-Bi_2O_3 and Sm_2O_3 (INSET).

There are literature reports of other rare earth oxides being used as dopants [2, 10] to stabilise the cubic δ-Bi_2O_3 phase, and in this study, cubic fluorite phases were produced when Y_2O_3, and a mixture of Y_2O_3 and Sm_2O_3, were milled with α-Bi_2O_3 (Figure 3).

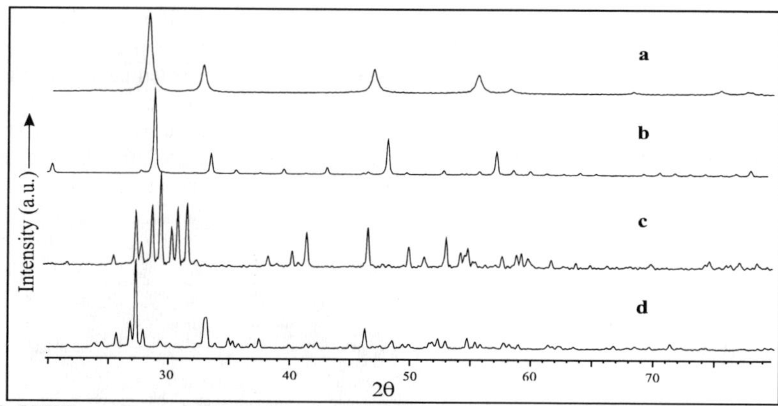

Figure 3. PXRD patterns of the stabilised cubic Bi_2O_3 solid solution phase
(a) $(Sm_2O_3)_{0.2}(Y_2O_3)_{0.2}(Bi_2O_3)_{0.6}$, made by milling the starting oxides **(b)** C-type Y_2O_3,
(c) monoclinic Sm_2O_3 and **(d)** monoclinic α-Bi_2O_3.

<u>Doped CeO$_2$ phases.</u>

When cubic CeO_2 is doped with rare earth oxides, oxygen vacancies are created to compensate for charge imbalance (Sm^{3+} for Ce^{4+}). The solid solution $(CeO_2)_{0.8}(Sm_2O_3)_{0.2}$ has been reported in the literature as having the highest ionic conductivity of all the Ceria-based oxide ion conductors [11, 13], and, as such, provided a model system to investigate preparation by MA. Figure 4 shows the PXRD patterns of the starting oxides and the as-milled product, which had the expected cubic, fluorite-type structure.

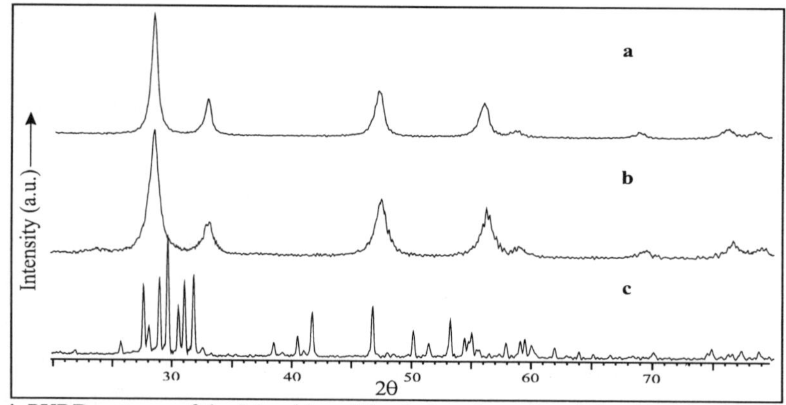

Figure 4. PXRD patterns of the **(a)** cubic solid solution phase $(Sm_2O_3)_{0.2}(CeO_2)_{0.8}$ formed by milling the starting oxides **(b)** cubic CeO_2 and **(c)** monoclinic Sm_2O_3.

In order to investigate whether another good Ceria-based conductor could be produced, 30 mol% Bi_2O_3, as dopant, was milled with 70 mol% CeO_2. This system is not reported in the literature, and attempts to make a solid solution with this stoichiometry, by calcination, proved unsuccessful. Figure 5 shows PXRD patterns of the starting materials, and the as-milled cubic solid solution, produced after a relatively short milling time.

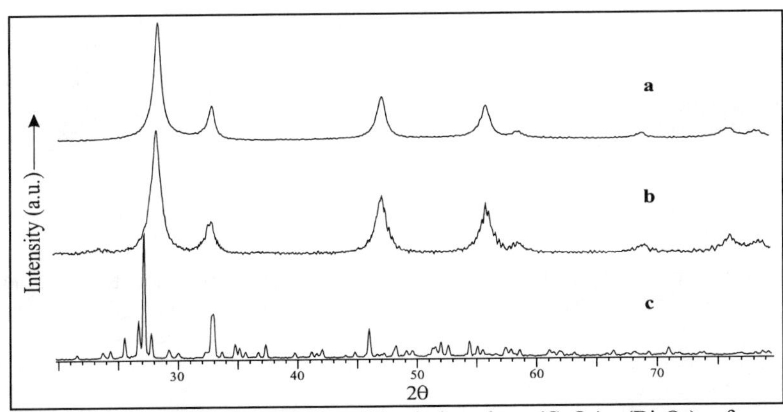

Figure 5. PXRD patterns of the (**a**) cubic solid solution phase $(CeO_2)_{0.7}(Bi_2O_3)_{0.3}$ formed by milling the starting oxides (**b**) cubic CeO_2 and (**c**) monoclinic α-Bi_2O_3.

<u>Electrical Conductivity Measurements.</u>

The electrical conductivities of the Sm_2O_3-doped Bi_2O_3 and the Sm_2O_3-doped CeO_2 system, considered in this study, are well documented [3, 10-12], and both have been found to be ionically conducting over the temperature ranges in which the milled oxides in this study were monitored. The Arrhenius plot for the milled Sm_2O_3 doped Bi_2O_3, shown in Figure 6, exhibits the expected linear relationship between temperature and conductivity, but there is a discontinuity, at ~700°C, where the conductivity increases disproportionately. More recent studies, using high temperature X-Ray diffractometry, have shown that these room temperature "stabilised" oxides do decompose at temperatures between 650 and 750°C, when annealed in this temperature range for 24 hours [14, 15]. These workers have shown that there is a transition from the cubic, fluorite-type phase to a rhombohedral phase, which is more highly conducting. This phase reverts back to the cubic, fluorite-type phase as the temperature increases beyond 750° C, indicated by the change in the gradient of the Arrhenius plot. Conductivities for the as-milled oxide, before and after the discontinuity, are slightly higher than the literature values for samples prepared conventionally [10, 15], as shown in Figure 6.

In the case of the Sm_2O_3-doped CeO_2 phase, $(Sm_2O_3)_{0.2}(CeO_2)_{0.8}$ (Figure 7), conductivities were found to be about the same as those for calcined samples [11, 13]. The CeO_2-doped Bi_2O_3 solid solution, prepared by MA, was found to be highly conducting, when compared with other doped CeO_2 oxides (Figure 7), for example, the Sm_2O_3-doped CeO_2 discussed above, and exhibits conductivities of the order expected in doped Bi_2O_3 systems. The Arrhenius plot for this milled sample also exhibited the linear relationship expected of a single phase oxide solid solution over the temperature range monitored, but there are no reports of this system in the literature.

In this preliminary study only DC measurements were obtained, thus representing only bulk conductivities of the samples. Grain boundary contributions for MA materials as opposed to calcined materials may be different, and therefore a comparison of AC conductivities between materials made by the two methods may prove insightful.

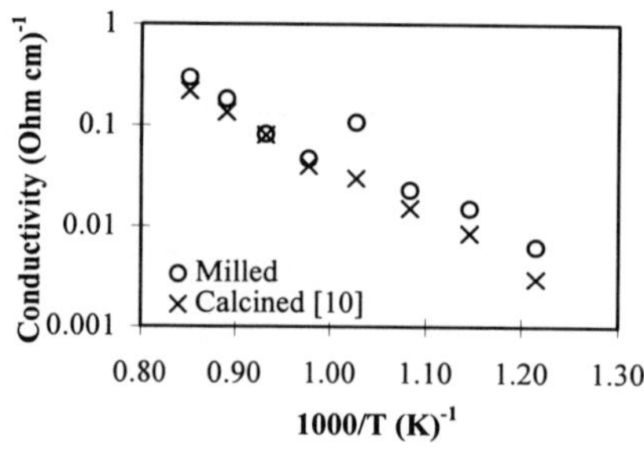

Figure 6.Arrhenius plots for the milled and calcined $(Sm_2O_3)_{0.4}(Bi_2O_3)_{0.6}$ phase.

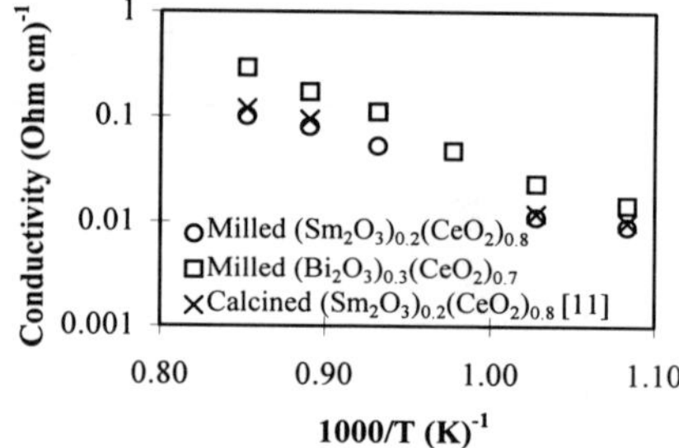

Figure 7. Arrhenius plots for the milled and calcined $(Sm_2O_3)_{0.2}(CeO_2)_{0.8}$, and the milled $(Bi_2O_3)_{0.3}(CeO_2)_{0.7}$ phase.

CONCLUSIONS

It has been established that mechanochemical milling is a suitable technique for the preparation of ceramic metal oxide electrolytes, with an increased reactivity between solid phases and, in some cases, an enhanced ionic conductivity. Further work is in progress.

ACKNOWLEDGEMENTS

The Materials Institute of WA is thanked for partial funding of the High Temperature furnace, and RVA thanks the Special Research Centre for a postgraduate scholarship.

REFERENCES

1. H. A. Harwig and A. G. Gerards, J. Solid State Chem. **26**, 265 (1978).
2. T. Takahashi and H. Iwahara, Mat. Res. Bull. **13**, 1447 (1978).
3. T. H. Etsell and S. N. Flengas, Chem. Rev. **70**, 339 (1970).
4. C. C. Koch, in Mechanical milling and Alloying in Materials Science and Technology. A Comprehensive Treatment, **15**, ed R. W. Cahn, (VCH, Weinheim, 1991), p. 195.
5. C. C. Koch and J. D. Whittenberger, Intermetallics, **4**, 339 (1996).
6. D. Michel, E. Gaffet and P. Berthet, NanoStructured Materials, **6**, 667 (1995).
7. B. J. M. Aikin, T. H. Courtney and D. R. Maurice, Mat. Sci. Eng. **A147**, 229 (1991).
8. C. da Leitenburg, A. Trovarelli, F. Zamar, S. Maschio, G. Dolcetti and J. Llorca, J. Chem. Soc. Chem. Commun. 2181 (1995).
9. H. H. Wieder in Laboratory Notes on Electrical and Galvanomagnetic Measurements in Materials Science Monographs 2, (Elsevier Scientific Publishing Co. 1979), p. 6.
10. H. Iwahara, T. Esaka and T. Sato, J. Solid State Chem. **39**, 173 (1981).
11. H. Yahiro, Y. Eguchi, K. Eguchi and H. Arai, J. Applied Electrochem. **18**, 527 (1988).
12. T. Takahashi, H. Iwahara and T. Arao, J. Applied Electrochem. 5, 187 (1975).
13. T. Inoue, T. Setoguchi, K. Eguchi and H. Arai, Solid State Ionics, **35**, 285 (1989).
14. A. Watanabe, Solid State Ionics, **40/41**, 889 (1990).
15. P. Conflant, C. Follet-Houttemane and M. Drache, J. Mater. Chem. **1**, 649 (1991).

RIETVELD X-RAY POWDER PROFILE ANALYSIS AND ELECTRICAL CONDUCTIVITY OF FAST-ION CONDUCTING $Gd_2(Ti_{1-y}Sn_y)_2O_7$ SOLID SOLUTIONS

KEVIN EBERMAN, PER ÖNNERUD, TAE-HWAN YU, HARRY L. TULLER, AND BERNHARDT J. WUENSCH
Department of Materials Science and Engineering, Crystal Physics and Electroceramics Laboratory, Massachusetts Institute of Technology, Cambridge, MA 02139.

ABSTRACT

The $A^{3+}_2B^{4+}_2O_7$ pyrochlores have fast-ion conduction properties that make them attractive candidates for applications in fuel cells. Rietveld powder-profile analysis of x-ray diffraction data has been used to determine structural parameters for $Gd_2(Ti_{1-y}Sn_y)_2O_7$ solid solutions (y = 0.20, 0.40, 0.60, and 0.80) for correlation with composition-dependence of the specimens' electrical conductivity. In accord with Vegard's law, the lattice constants increase linearly with y as the larger Sn^{4+} ion replaces Ti^{4+}. The structures in the system remain remarkably ordered in view of the fact that it had been suggested that increasing the average radius of the ions that occupy the B^{4+} site relative to A^{3+} drives the system toward a disordered defect-fluorite state. A small but significant increase in mixing of the occupancy of the cation sites was found with increasing y to a maximum of $[Gd_B] = 0.05$. We could detect no significant disorder in the anion array. The activation energy and pre-exponential term for oxygen ion conduction were found to increase linearly and exponentially with y, respectively, such that at a given temperature, the ionic conductivity changes by less than an order of magnitude across the system. The behavior is in marked contrast to other systems where the substitution of a larger cation in the B site increases conductivity by up to three orders of magnitude as a consequence of substantial disorder that is introduced in both the cation and anion arrays.

INTRODUCTION

The pyrochlore structure type, $A^{3+}_2B^{4+}_2O_7$, space group Fd3m, is a superstructure with a lattice constant twice that of a parent fluorite-like array of ions. The larger A^{3+} and smaller B^{4+} cations order along alternating <110> rows in positions 16c 000 and 16d ½½½, respectively, of the space group. The oxygen ions O(1) and O(2) occupy positions 48f $x\frac{1}{8}\frac{1}{8}$ (x = $\frac{3}{8}$ for an undistorted cubic array) and 8a $\frac{1}{8}\frac{1}{8}\frac{1}{8}$, respectively, accounting for 7 of the 8 anion positions that would be occupied in a fluorite structure type. An oxygen ion, O(3), placed in 8b $\frac{3}{8}\frac{3}{8}\frac{3}{8}$ would complete the anion arrangement found in fluorite.

The pyrochlore structure disorders, upon heating, to a non-stoichiometric fluorite structure, $(A,B)O_{1.75}$. Progressive disorder also accompanies increasing amounts of substitution of a larger cation in solid solution in the B site. It is generally believed that the difference between the ionic radii of the species occupying the A and B sites serves as the main driving force for formation of the ordered pyrochlore superstructure. Since the cation and anion arrays are both driven to a state of complete disorder as the fluorite structure is assumed, an interesting basic question arises: is the progress of disorder in the two arrays coupled, or does it proceed independently? There is evidence in one system [1] that the latter is the case.

We have found, in solid solution series such as $Gd_2(Ti_{1-y}Zr_y)_2O_7$ [2,3] and $Y_2(Ti_{1-y}Zr_y)_2O_7$ [3], that ionic conductivity increased by several orders of magnitude with increasing y. Values were attained that make the materials of interest for possible application as the electrolyte in a fuel cell. A further attractive feature is that appropriate doping can result in mixed conductors in which electronic conductivity predominates. It might thus be possible to employ thermally- and chemically-compatible materials with the same pyrochlore structure type for both electrodes and electrolyte of a fuel-cell system [4].

Mat. Res. Soc. Symp. Proc. Vol. 453 © 1997 Materials Research Society

Detailed structural studies have been performed with Rietveld analysis of neutron diffraction data obtained for solid solutions in the system $Y_2(Ti_{1-y}Zr_y)_2O_7$ [1] to permit correlation of the state of structural disorder. Conductivity measurements have recently been extended to three additional systems: $Gd_2(Ti_{1-y}Sn_y)_2O_7$, $Gd_2(Zr_{1-y}Sn_y)_2O_7$, and $Y_2(Ti_{1-y}Sn_y)_2O_7$ [5]. These systems were selected for study because the ionic radius of Sn^{4+} is intermediate to those of Ti^{4+} and Zr^{4+}, and the experiments were designed to determine whether the state of structural disorder and accompanying electrical behavior were indeed determined by the relative ionic radii of the cations residing in the A and B sites. Rather different results were determined for the dependence of conductivity on Sn content. The present work presents a preliminary determination of the composition dependence of the distribution of ions in $Gd_2(Ti_{1-y}Sn_y)_2O_7$ in an attempt to shed light on this behavior.

EXPERIMENTAL

Preparation of Samples

Powders of $Gd_2(Ti_{1-y}Sn_y)_2O_7$ with y = 0.2, 0.4, 0.6, and 0.8 were prepared by the Pechini liquid-mix process [6]. Metal-organic compounds of the cations were dissolved in citric acid and mixed in appropriate proportions with ethylene glycol. The solution was then heated at 150°-200°C to form a polyester in which the homogeneity of the liquid solution is preserved. Excess ethylene glycol and water were removed by further heating, resulting in the formation of a hard resin. This resin, in turn, was charred at 400°C to oxidize the organic material, and next calcined at 800°-900° to provide pyrochlore.

Conductivity Measurements

Pyrochlore powder was pressed into disks at 4000 psi followed by isostatic pressing at 40 Kpsi. The resulting disks were sintered at 1570°C to 93-95% theoretical density. The final grain size was ca. 3 μm [5]. Bars measuring 2 x 2 x 8 mm were cut from the sintered disks and sputter-coated with Pt. Wires of Pt were then attached for 4-probe measurements of electrical conductivity as a function of temperature (750°-1050°C) and oxygen partial pressure ($1-10^{-20}$ atm) using complex impedance techniques. Further details are reported elsewhere [5,7].

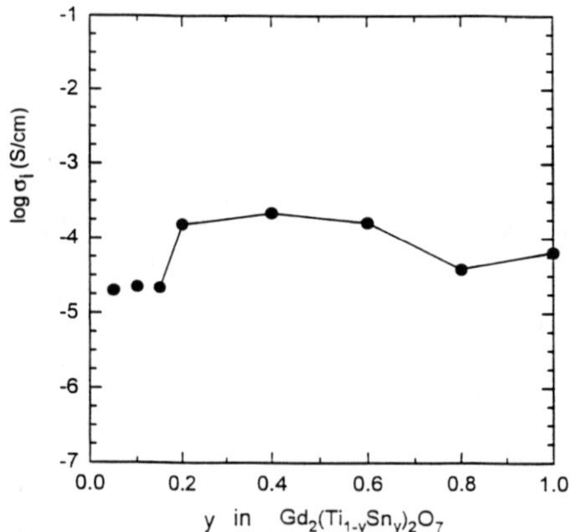

Figure 1. Ionic portion of the electrical conductivity of $Gd_2(Ti_{1-y}Sn_y)_2O_7$ at 1000°C as a function of the atomic fraction, y, of Sn substituted for Ti. For y < 0.2 the conductivity is determined by oxygen vacancy concentrations that are determined by accidental impurities.

X-ray Analysis

The powder samples employed for Rietveld analysis were annealed for 24 hours at 1400°C. The resulting powder was ground in an agate mortar and sifted to provide a particle size of less than 75 μm, then mounted on a glass slide. Data were collected with a Rigaku RU300 x-ray powder diffractometer equipped with a graphite diffracted-beam monochromator. Copper Kα radiation ($\lambda_{CuK\alpha_1}$ = 1.54051, $\lambda_{CuK\alpha_2}$ = 1.54433 Å) provided by an 18 kW rotating-anode generator was employed. Data were recorded in steps of 0.03° 2θ between 10°-150° 2θ.

REFINEMENT OF THE STRUCTURES

One must proceed with care in refinement of the structural parameters of pyrochlore solid solutions. Information on the perturbations of the fluorite structure that result in the formation of the pyrochlore superstructure is contained solely in the set of weak superstructure diffraction maxima. Such intensities contain contributions from (a) the difference in average scattering power of the species occupying the A and B sites, (b) scattering by the fractional oxygen occupancy of the normally-vacant 8b site, and (c) the magnitude of the displacement of the O(1) atom in position 48f x $^1/_8$ $^1/_8$ from its ideal location at x = $^3/_8$. All three contributions approach zero as the structure approaches the highly-disordered states that are of greatest interest. The data available for analysis thus become few and acquire large standard deviations.

Two other complications are associated with analysis of the pyrochlore superstructure. The anion and cation arrays both experience disorder. The occupancy of no site in the structure is known at the outset. This creates serious problems in establishing a correct scale factor for analysis of the superstructure intensities. A procedure for systematically treating this problem through initial refinements with special classes of reflections has been described in earlier work [1]. Secondly, it is not possible for any solid solution to determine, using a single set of diffraction data, the distribution of three or more chemical species over two crystallographically independent sites. All that can be determined, even, in a refinement with occupancies constrained to the known chemical composition, is a measure of the average scattering density in each site and *not* a unique distribution of species that might provide that density [1,8].

Neutron powder diffraction had proved advantageous in analysis of the $Y_2(Ti_{1-y}Zr_y)_2O_7$ system. Oxygen is a stronger scatterer relative to the cations than is the case with x-rays. Moreover, the scattering length of Ti is negative for thermal neutrons, providing unusually large

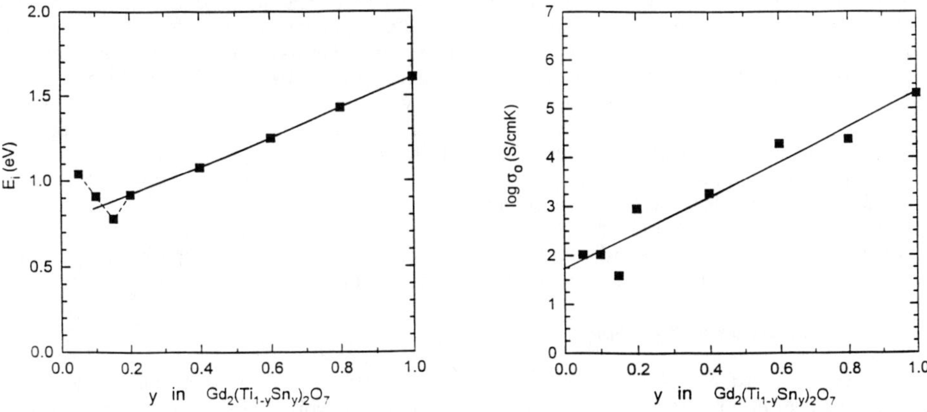

Figure 2. (a) Variation of the activation energy for ionic conduction in $Gd_2(Ti_{1-y}Sn_y)_2O_7$ as a function of the atomic fraction, y, of Sn substituted for Ti. (b) Variation with y of the pre-exponential term, σ_0, for oxygen ion conduction.

contrast between scattering from the atoms that occupy the cation sites [1,8]. The contributions to the superstructure intensities for all three of the perturbations listed above are much larger for neutrons than for x-rays and it was possible to analyze highly disordered states with precision — especially the behavior of the oxygen ions.

Pyrochlore phases that contain Gd present a special challenge. This species possesses the largest absorption cross-section for neutrons of any element in the periodic table. The present analysis was accordingly undertaken with x-rays (in spite of the fact that Gd also has the largest mass absorption coefficient for Cu $K\alpha$ radiation of any element in the periodic table!). In doing so we were aware that this would decrease sensitivity to the oxygen distribution and, because of weaker superstructure intensities, limit the extent of disorder that could be successfully analyzed.

Refinements were performed with the GSAS system of programs [9]. A total of 15 structural parameters were, in principle, necessary to describe each structure: a scale factor, the lattice constant, the x coordinate for O(1), the scattering power of one cation site, two independent site occupancies for O(1) and O(2), and nine anisotropic temperature factor coefficients. Some 19 additional instrumental parameters were required to describe the peak shapes, background, and the zero point in 2θ for the profile. Final weighted residuals for the refinements ranged from 5.8 to 9.8 %, comparing favorably with corresponding expected residuals of 6.5 to 10.6 % based on counting statistics.

RESULTS

Ionic Conductivity

Measurement of the total bulk electrical conductivity, σ, as a function of oxygen partial pressure, P_{O_2}, between 1 and 10^{-20} to 10^{-11} atm (depending upon the temperature) [5] showed a P_{O_2}-independent plateau at intermediate partial pressures that was attributed to predominance of ionic conduction arising from oxygen ion migration. At high P_{O_2} a p-type electronic conduction was observed with log σ proportional to $P_{O_2}^{-\frac{1}{4}}$. At low P_{O_2} a transition to n-type electronic behavior with $P_{O_2}^{+\frac{1}{4}}$ dependence was observed for specimens with compositions that had $y \leq 0.8$. Specimens of endmember $Gd_2Sn_2O_7$ decomposed before an atmosphere sufficient to produce the transition from ionic behavior could be reached. Details of the interpretation of the electronic components of the conductivity will be published elsewhere [7].

The ionic portion of the conductivity was extracted from measurements of the total conductivity as a function of P_{O_2} by fitting the data with a relation that was derived from a model for the defect structure. The ionic conductivity at 1000°C is shown as a function of composition in Fig. 1. The scatter in data for $y < 0.2$ arises from the fact that oxygen vacancy concentrations are controlled by accidental impurities; results ranging between 10^{-4} and 10^{-6} S/cm have been found for samples with $y = 0$, $Gd_2Ti_2O_7$ [2,10]. The variation of σ with the composition of the solid solutions may be seen to be very slight in contrast to the changes of several orders of magnitude that have been observed for systems such as $Gd_2(Ti_{1-y}Zr_y)_2O_7$, and $Y_2(Ti_{1-y}Zr_y)_2O_7$ [2,3]. The ionic conductivity found for endmember $Gd_2Sn_2O_7$ at 1000°C is only 10^{-2} that of $Gd_2Zr_2O_7$ [2]. The latter phase has a structure that is known to be significantly disordered. The measurements of ionic conduction with temperature provide an activation energy for O^{-2} conduction that varies linearly with composition, as shown in Fig. 2a. The logarithm of the pre-exponential term, σ_o, was found to increase linearly with y, Fig. 2b.

Variation of Structure with Composition

The lattice constants of the $Gd_2(Ti_{1-y}Sn_y)_2O_7$ solid solutions increased linearly with the amount of substitution of the larger Sn^{4+} ion in accord with Vegard's law as may be seen in Fig. 3a. Values for the lattice constants of the $Gd_2Ti_2O_7$ and $Gd_2Sn_2O_7$ endmembers are taken from the literature [11,12] and agree well with the present data for the solid solutions. Somewhat surprisingly, no difficulty was experienced in obtaining sufficient superstructure intensity to permit the refinement. The reason is that the structures of the solid solutions remained almost fully ordered for the complete range of compositions. Refinement of the occupancy of the normally-vacant O(3) site provided values that were statistically insignificant compared with their standard deviations. These were, in spite of the relative insensitivity of the analysis to the behavior of the oxygen ions, on the order 0.03 oxygen ion per O(3) site. Within limits of this order of magnitude, therefore, the oxygen ion array may be considered to remain fully ordered and was held as such in the final cycles of refinement.

The deviation of the x coordinate of the O(1) ion from the ideal value of $^3/_8$ for an undistorted cubic fluorite-like array corresponds to a relaxation of this ion in the direction of the O(3) site. The experimental values for this parameter, Fig. 3b, show no systematic variation, remain essentially constant at ca. 0.426, and are close to the coordinates for the endmembers [11,12]. This supports the conclusion that there is little disorder in the anion array. For $Y_2(Ti_{1-y}Zr_y)_2O_7$, in which the O(3) site fills with a linear dependence on y between y =0.3 and complete disorder

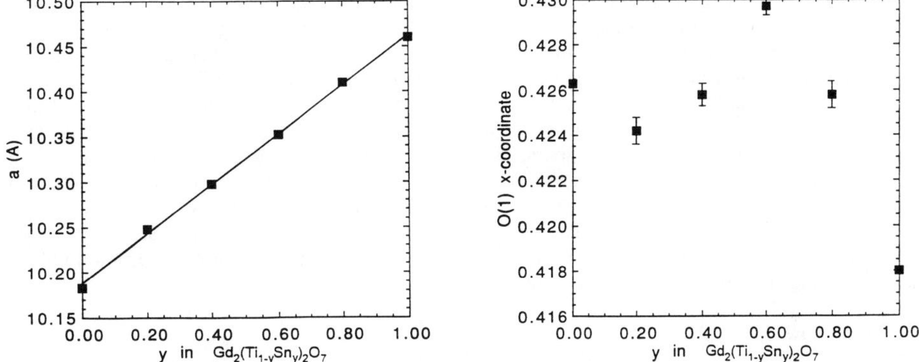

Figure 3. (a) The lattice constant of $Gd_2(Ti_{1-y}Sn_y)_2O_7$ as a function of composition. Data for the endmembers, y = 0 and 1, are taken from the literature [11,12]. (b) Variation of the coordinate x for O(1) in 48f $x^1/_8{}^1/_8$. The coordinate is essentially constant and represents marked relaxation toward the unoccupied O(3) position. Disorder in the oxygen array would have resulted in the filling of O(3), accompanied by return to an undistorted location at x = $^3/_8$.

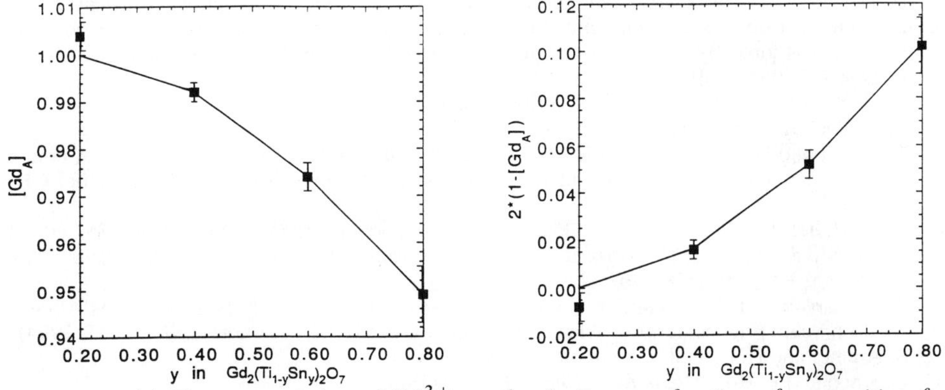

Figure 4. (a) The concentration of Gd^{3+} on the A site as a function of composition for $Gd_2(Ti_{1-y}Sn_y)_2O_7$ solid solutions under the assumption that $Ti^{4+}{}_{1-y}Sn^{4+}{}_y$ exchanges with each Gd^{3+}. (b) The same result for mixing of species between cation sites in terms of a disorder parameter that ranges 0 to 1 for fully ordered and disordered cation distributions, respectively.

at y = 0.9, the x coordinate for O(1) decreases continuously and rapidly to 0.375 with a quadratic dependence on y.

The refinement revealed a very slight mixing of the species that would occupy the A and B sites as y increased across the system. As mentioned above, a set of single diffraction data cannot be used to specify a unique distribution for three cation species over two independent sites. For present purposes, therefore, we assumed that Ti^{4+} and Sn^{4+} will interchange with Gd^{3+} in the A site in the same proportion in which they are present in the solid solution: that is, that $Ti^{4+}_{1-y}Sn^{4+}_y$ exchanges with Gd^{3+}. Under this assumption the refinements show that the fractional occupancy of the A site by Gd^{3+} decreased monotonically with increasing y by a small amount, Fig. 4a. The same result is presented in a different fashion in Fig. 4b in terms of a disorder parameter defined such that it assumes a value zero for an ordered distribution of cations and unity for fully disordered occupancies.

The results of the structural analyses provide an explanation for why the variation of ionic conductivity with y, Fig. 1, remains essentially flat upon substitution of the larger Sn^{4+} cation rather than increasing several orders of magnitude. The most interesting conclusion, however, upon comparison of the behavior of the several pyrochlore systems that we have examined, is that the state of disorder does *not* depend solely on the difference in average radii of the species that occupy the A and B sites. The Sn-O bond in pyrochlores has been shown by infrared absorption spectroscopy [13] to be more rigid and have greater covalent character than Ti-O bonds in the pyrochlore system. It appears, therefore, that the state of disorder depends at least in part on bonding characteristics between the cations and their oxygen neighbors and not solely on differences in ionic radii. The Sn^{4+} ion seems to display marked preference for the octahedral environment of the B site. We have observed very similar behavior in solid solutions in the $Y_2(Ti_{1-y}Sn_y)_2O_7$ system for which the absence of an ion with large absorption permitted a precise analysis of neutron powder diffraction data [14].

REFERENCES

1. C. Heremans, B. J. Wuensch, J. K. Stalick and E. Prince, *J. Solid State Chem.* **117**,108 (1995).
2. P. K. Moon and H. L. Tuller, *Solid State Ionics* **28-30**, 470 (1988).
3. P. K. Moon and H. L. Tuller in *Solid State Ionics*, edited by G. Nazri, R. A. Huggins and D. F. Shriver (*Mat. Res. Soc. Symp. Proc.* **135**, Pittsburgh, PA, 1989) p.149-163.
4. H. L. Tuller in *High Temperature Electrochemistry: Ceramics and Metals*, edited by F. W. Poulsen, N. Bonanos, S. Linderoth, M. Morgensen and B. Zachau-Christiansen (Proc. 17th Risø International Symposium on Materials Science, Risø National Laboratory, Roskilde, Denmark, 1996) p.139-153.
5. T.-H. Yu, Ph.D. thesis, Massachusetts Institute of Technology, 1996.
6. M. P. Pechini, U.S. Patent No. 3,330,697 (1967).
7. T.-H. Yu and H. L. Tuller in *Role of Ceramics in Advanced Electrochemical Sysems*, edited by P. N. Kumta, G. S. Rohrer and U. Balachandran (*Ceram. Trans.* **65**, Westerville, OH, 1996) p.3-11.
8. S. M. Haile, B. J. Wuensch and E. Prince in *Neutron Scattering for Materials Science*, edited by S. M. Shapiro, S. C. Moss and J. D. Jorgensen (*Mat. Res. Soc. Symp. Proc.* **166**, Pittsburgh, PA, 1990) p.81-86.
9. A. C. Larson and R. B. Von Dreele, *GSAS General Structure Analysis System*, Los Alamos Neutron Scattering Center, Los Alamos National Laboratory, Los Alamos, NM (1985, 1994).
10. S. Kramer and H. L. Tuller, *Solid State Ionics* **82**, 15 (1995).
11. O. Knop and F. Brisse, *Canad. J. Chem.* **46**, 859 (1968).
12. O. Knop and F. Brisse, *Canad. J. Chem.* **47**, 971 (1969).
13. M. T. Vandenborre, E. Husson, J. P. Chartry and D. Michel, *J. Raman Spectroscopy* **14**, 63 (1983).
14. P. Önnerud, K. Eberman, B. J. Wuensch and J. K. Stalick, to be published.

ELECTROCHEMICAL CHARACTERIZATION OF MIXED CONDUCTING $Ba(Ce_{1-(x+y)}Pr_yGd_x)O_{3-x/2}$ CATHODES

R. Mukundan, P. K. Davies and W. L. Worrell
Department of Materials Science, University of Pennsylvania, Philadelphia, PA 19104-6272

ABSTRACT

The high protonic conductivities of barium cerium gadolinium perovskites ($Ba(Ce_{1-x}Gd_x)O_{3-x/2}$, $0 < x < 0.2$) has led to their incorporation as electrolytes in fuel cells operating at 873-1073K. In an effort to develop new mixed-conducting electrodes that are compatible with the cerate electrolyte, we have examined the electrochemical properties of Pr-doped barium gadolinium cerate. The Pr doping increases the electronic contribution to the total conductivity and for $Ba(Ce_{0.5}Pr_{0.4}Gd_{0.1})O_{2.95}$, a conductivity exceeding ≈ 0.1 S/cm at 1173K in dry air has been obtained. EMF measurements indicate the presence of both protonic and oxygen ionic contributions to the conductivity in the Pr substituted samples. The cathodic overpotentials of $Ba(Ce_{1-(x+y)}Pr_yGd_x)O_{3-x/2}$ mixed conductors on a $Ba(Ce_{0.8}Gd_{0.2})O_{2.9}$ electrolyte have been measured. $Ba(Pr_{0.8}Gd_{0.2})O_{2.9}$ has the best cathodic performance and could be a viable alternative to Pt cathode for a barium cerate based fuel cell.

INTRODUCTION

A new class of protonic conducting oxides having the perovskite structure was reported by Iwahara and co-workers in 1981[1]. These $Sr(Ce_{1-x}M_x)O_{3-x/2}$ based perovskites were the first group of protonic conductors that could withstand thermal cycling and the high temperatures encountered in fuel cell applications. Since that time several cerate-based systems have been studied, and the gadolinium-substituted barium cerates have been found to have one of the highest conductivities in the 973-1273K range[2,3]. The substitution of trivalent gadolinium for tetravalent cerium in $Ba(Ce_{1-x}Gd_x)O_{3-x/2}$, creates oxygen ion vacancies (Equation 1)[4] and at high temperatures under reducing conditions this is the majority defect in the absence of moisture or hydrogen.

$$BaO + \tfrac{1}{2}Gd_2O_3 \xrightarrow{BaCeO_3} Gd'_{Ce} + \tfrac{1}{2}V_O^{\cdot\cdot} + \tfrac{5}{2}O_O^x + Ba_{Ba}^x \qquad (1)$$

Under oxidizing atmospheres these oxides can incorporate holes[4] according to equation 2.

$$V_O^{\cdot\cdot} + \tfrac{1}{2}O_2 \rightarrow O_O^x + 2h^{\cdot} \qquad (2)$$

While in the presence of moisture or hydrogen they can incorporate protons[4] according to equation 3.

$$H_2O + V_O^{\cdot\cdot} \rightarrow 2H^{\cdot} + O_O^x \qquad (3)$$

Under fuel cell conditions the ionic conduction in these cerates arises primarily from the transport of protons at 873K and from the movement of oxygen ions at 1273K[2]. The high concentration and mobility of protons accommodated by the structure at lower temperatures results in a very high protonic conductivity; for example, the protonic conductivity (4×10^{-2} S/cm) of $Ba(Ce_{0.8}Gd_{0.2})O_{2.9}$ at 873K under fuel cell conditions is comparable to the oxygen ionic conductivity of yttria stabilized zirconia at 1073K. As a result $(Ba(Ce_{1-x}Gd_x)O_{3-x/2}, x = 0.1 - 0.2)$ has received considerable attention for application as an electrolyte in fuel cells operating in this temperature regime[2,5]. A short circuit current density of 800 mA/cm^2 has been reported for a fuel cell operating at 1073K and using $Ba(Ce_{0.8}Gd_{0.2})O_{2.9}$ electrolyte and platinum electrodes[2]; this is one of the best reported cell performance for a perovskite-electrolyte system.

To improve the performance and reduce the overall cost of cerate-based protonic fuel cells, there is considerable interest in developing new mixed-conducting oxides to replace the platinum electrodes. Manganese-based perovskite cathodes have been tested in $Ba(Ce_{1-x}Nd_x)O_{3-x/2}$ fuel cells, and $La_{0.6}Ba_{0.4}MnO_3$ was found to have the lowest polarization resistance (0.15 Ω/cm^2 at 1273K)[6]. However, the polarization resistance of these pure electronic conducting systems was significantly higher than platinum (≈ 0.07 Ω/cm^2 at 1273K)[6].

In our studies we have attempted to develop new mixed-conducting perovskites that are chemically compatible with the barium cerate-based electrolytes. In particular we are investigating schemes for the chemical modification of barium cerate to produce oxides with a high electronic and ionic (protonic and oxide ion) conductivity. A series of dopants that can adopt a mixed-valence state, and therefore impart a higher electronic contribution to the total conductivity of $Ba(Ce_{1-x}Gd_x)O_{3-x/2}$, have been investigated. In this paper we report our results on the conductivity, transference number, and cathodic performance of cerate solid solutions containing Pr.

EXPERIMENTAL

Single phase perovskite samples of $Ba(Ce_{1-(x+y)}Pr_yGd_x)O_{3-x/2}$ (x = 0.1, 0.2 ; y = 0, 0.2, 0.4, 0.6, 0.8) were prepared using standard solid state methods by heating appropriate stoichiometric mixtures of $BaCO_3$, CeO_2, Gd_2O_3, and Pr_6O_{11} at 1673K for 32 hours with several intermediate grinding steps. The formation of a single phase orthorhombic perovskite was confirmed using powder x-ray diffraction. Dense ceramic samples (typically $\geq$ 95% theoretical) were obtained by isostatically pressing the powders into pellets at 620 MPa for 5 minutes, and sintering at 1923K for 10 hours.

Both 4-probe DC and 2-probe AC measurements were used to measure the total conductivity of the ceramics. The conductivity measurements were made under dry ($P_{H2O} = 2.3 \times 10^2$ Pa) and wet ($P_{H2O} = 2.5 \times 10^3$ Pa) atmospheres in oxygen, air, and ultra pure argon ($P_{O2} \approx 1$ Pa). The ionic transference number of these mixed conducting oxides was determined using an EMF technique. The EMF measurements were carried out under various wet and dry atmospheres to determine both the protonic and oxygen ionic transference numbers ($10^5 < P_{O2} < 1$ Pa : $2.3 \times 10^2 < P_{H2O} < 2.5 \times 10^3$ Pa). The details of the experimental setup used for the conductivity and EMF measurements are given elsewhere[7].

The cathodic overpotential of these mixed conductors on a $Ba(Ce_{0.8}Gd_{0.2})O_{2.9}$ electrolyte was measured using a 3-electrode current interruption technique[8]. The current was applied using a Solartron 1286 electrochemical interface and interrupted using a fast ($<1\mu$s) switch. A Tektronix TDS 320 oscilloscope was used to measure the voltage decay. $Ba(Ce_{1-(x+y)}Pr_yGd_x)O_{3-x/2}$ powders were mixed with ethylene glycol, brush painted on the electrolyte and fired at 1723K for 3 hours to form the working electrode (cathode). Pt gauze applied with Pt paste was used as the current collector, counter (anode) and reference electrodes. The measurements were made under fuel cell conditions with dry air as the oxidant and H_2/3% H_2O as the fuel.

RESULTS AND DISCUSSION

Two sets of samples $(Ba(Ce_{0.9-y}Pr_yGd_{0.1})O_{2.95}$ (y = 0. 0.2, 0.4) and $Ba(Ce_{0.8-y}Pr_yGd_{0.2})O_{2.9}$ (y = 0, 0.4, 0.6, 0.8) were prepared for analysis. Both of the undoped (y = 0) end-members $(Ba(Ce_{0.9}Gd_{0.1})O_{2.95}$ and $Ba(Ce_{0.9}Gd_{0.2})O_{2.9})$ have received considerable attention as protonic conducting electrolytes. The conductivity of $Ba(Ce_{0.9}Gd_{0.1})O_{2.95}$ has been studied in detail under various conditions and its ionic and electronic transference numbers are available in the literature[2,3]. Compounds in the $Ba(Ce_{0.9-y}Pr_yGd_{0.1})O_{2.95}$ series were synthesized to compare our data for the undoped sample with that available in the literature and to illustrate the effect of Pr doping on the electronic and ionic conductivities of the barium cerates. $Ba(Ce_{0.9}Gd_{0.2})O_{2.9}$ on the other hand, has a higher ionic conductivity than the 10 mole% Gd doped sample and has been studied primarily under fuel cell conditions as an electrolyte[3]. The compounds in the $Ba(Ce_{0.8-y}Pr_yGd_{0.2})O_{2.9}$ series were studied to identify materials with the highest electronic and ionic conductivities to be used as cathodes on a $Ba(Ce_{0.9}Gd_{0.2})O_{2.9}$ electrolyte.

The X-ray diffraction patterns of all the samples studied could be fit to an orthorhombic perovskite structure[9]. The decreasing lattice volume of the perovskite unit cell with increasing Pr content suggests that Pr is primarily present in the 4+ state $(r_{Ce4+} = 0.87Å, r_{Pr4+} = 0.85Å$, and $r_{Pr3+} = 0.99Å)$. EPR measurements on $BaPrO_3$ reported in the literature[10] also confirm the presence of tetravalent Pr ions in these perovskites. However the Pr doped samples are easier to reduce than the undoped samples indicating that the Pr^{3+} cation is more stable than the Ce^{3+} cation. For example, there is no significant reduction in $Ba(Ce_{0.9}Gd_{0.1})O_{2.95}$ at 973K in $5\%H_2/H_2O$ $(P_{O2} = 10^{-18}$ Pa), whereas $\approx 70\%$ of the Pr cations can be reduced to the 3+ state in $Ba(Ce_{0.5}Pr_{0.4}Gd_{0.1})O_{3-\delta}$. The samples in the as prepared state were black in color and changed to brown upon reduction, indicating a p-type electronic conductivity.

$\underline{Ba(Ce_{0.9-y}Pr_yGd_{0.1})O_{2.95}}$

The total conductivity (four-probe DC) of $Ba(Ce_{0.9-y}Pr_yGd_{0.1})O_{2.95}$, y = 0.2 and 0.4 under dry oxygen and dry air is shown in Fig. 1. The conductivity of undoped $Ba(Ce_{0.9}Gd_{0.1})O_{2.95}$ in air is

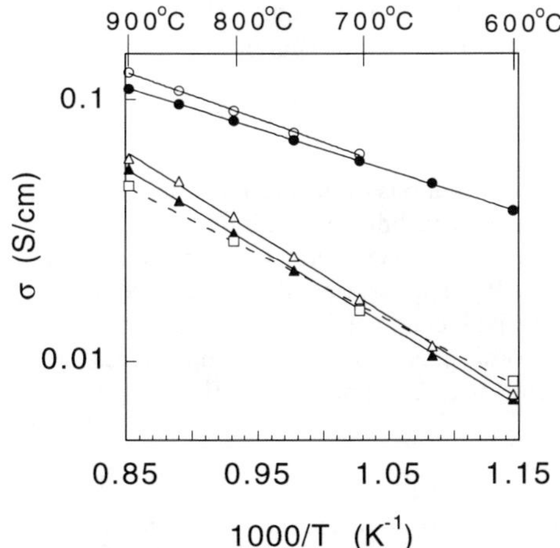

Fig 1. Four probe DC conductivity of $Ba(Ce_{0.9-y}Pr_yGd_{0.1})O_{2.95}$, (y = 0, 0.2, 0.4) in dry air and dry oxygen

also plotted for comparison. The total conductivity is significantly enhanced by the substitution of Pr for Ce. The total conductivity (0.1 S/cm) of $Ba(Ce_{0.5}Pr_{0.4}Gd_{0.1})O_{2.95}$ at 1173K in dry air is more than double the conductivity (4.7 x 10^{-2} S/cm) of undoped Ba(Ce$_{0.9}$Gd$_{0.1}$)O$_{2.95}$ under the same condition. The total conductivity also increases with increasing P$_{O2}$, indicating a p-type electronic contribution to the conductivity. This indicates that the electronic conductivity of the barium cerate increases with increasing concentration of Pr^{4+} dopant. However, the conductivity of the 20 mole% Pr doped sample was not significantly larger than that of the undoped sample, suggesting a decrease in ionic conductivity with increased Pr substitution for Ce.

EMF measurements confirm the presence of up to 10% ionic (protonic and oxygen ionic) conductivity in the Pr doped samples. The protonic transference number (t$_{H+}$) of Ba(Ce$_{0.5}$Pr$_{0.4}$Gd$_{0.1}$)O$_{2.95}$ in a wet argon / dry argon gradient was $\approx$ 0.1 at 1093K. At this temperature the total conductivity in argon is 4 x 10^{-2} S/cm, giving a protonic conductivity of $\approx$ 4 x 10^{-3} S/cm. The oxygen ionic contribution to the conductivity estimated from the measured EMF in a dry air / dry argon chemical potential gradient was $\approx$ 3 x 10^{-3} S/cm. In comparison, the protonic, and oxygen ionic conductivity of Ba(Ce$_{0.9}$Gd$_{0.1}$)O$_{2.95}$ are 2.5 x 10^{-2} S/cm in wet air and 1.4x10^{-2} S/cm in dry argon[3] respectively at 1073K. Thus, the Pr substitution leads to a decrease in the ionic conductivity in addition to the increase in the electronic conductivity.

The apparent activation energy for the total conductivity in Ba(Ce$_{0.5}$Pr$_{0.4}$Gd$_{0.1}$)O$_{2.95}$ is dependent upon oxygen partial pressure (Fig. 1), and also upon temperature (Fig. 2). The

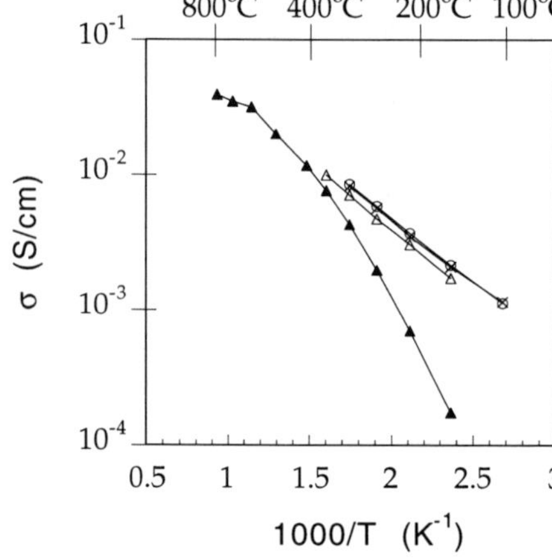

Fig. 2. Bulk Conductivity in oxygen, air and argon; and total conductivity in dry argon; of Ba(Ce$_{0.5}$Pr$_{0.4}$Gd$_{0.1}$)O$_{2.95}$

variations in the activation energy were due to the contributions of the grain boundaries in the polycrystalline samples. Using two-probe AC impedance techniques, the bulk and grain boundary contributions to the resistance were separated at low temperatures. The activation energy for the bulk conductivity was independent of P$_{O2}$ and corresponded to 0.23 eV (Fig. 2). At high temperatures and reducing atmospheres, the bulk conductivity dominates, giving a low total activation energy. However, under oxidizing atmospheres and at low temperatures the grain boundary conductivity predominates and results in a higher apparent activation energy.

Ba(Ce$_{0.8-y}$Pr$_y$Gd$_{0.2}$)O$_{2.9}$

The total conductivity of Ba(Ce$_{0.8-y}$Pr$_y$Gd$_{0.2}$)O$_{2..9}$, y = 0.4, 0.6, and 0.8 under dry oxygen is shown in Fig. 3. The estimated ionic conductivity of Ba(Ce$_{0.8}$Gd$_{0.2}$)O$_{2.9}$ under fuel cell condition[2] is also shown for comparison. The increasing total conductivity with increasing Pr

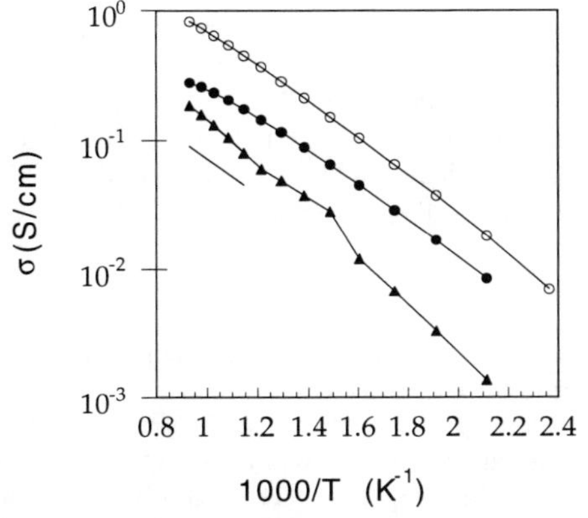

Fig. 3. Four probe DC conductivity of $Ba(Ce_{0.8-y}Pr_yGd_{0.2})O_{2.9}$ in dry oxygen. The estimated[2] conductivity of the undoped (y=0) sample is shown for comparison

dopant is consistent with the results obtained for compounds in the $Ba(Ce_{0.9-y}Pr_yGd_{0.1})O_{2.95}$ series. The total conductivity of $Ba(Pr_{0.8}Gd_{0.2})O_{2.9}$ in air at 1073K is 0.75 S/cm compared to a reported[2] conductivity of $5x10^{-2}$ S/cm for $Ba(Ce_{0.8}Gd_{0.2})O_{2.9}$ under wet air. Thus the complete substitution of Pr for Ce increases the total conductivity by more than one order of magnitude.

EMF measurements confirmed that the increase in conductivity was primarily due to an increase in the electronic conductivity. The protonic conductivity of $Ba(Pr_{0.8}Gd_{0.2})O_{2.9}$ in a wet argon / dry argon gradient at 1073K was $\approx 3 \times 10^{-2}$ S/cm, and the oxygen ionic conductivity in a dry oxygen / dry argon gradient at 1073K was $\approx 2 \times 10^{-2}$ S/cm. In comparison, the ionic conductivity reported for $Ba(Ce_{0.8}Gd_{0.2})O_{2.9}$ under fuel cell conditions[2] at 1073K is ≈ 0.1 S/cm. The compounds in the $Ba(Ce_{0.8-y}Pr_yGd_{0.2})O_{2.9}$ series have greater ionic conductivity than the compounds in the $Ba(Ce_{0.9-y}Pr_yGd_{0.1})O_{2.95}$ series due to the increased Gd^{3+} dopant content (eqn. 1,3). Moreover, both series of compounds indicate a significant increase in the electronic contribution to the conductivity with increasing Pr^{4+}.

Cathodic Overpotentials

The IR free cathodic overpotentials in air for the Pr doped samples on a $Ba(Ce_{0.8}Gd_{0.2})O_{2.9}$ electrolyte are shown in Figure 4. The performance of Pt and $La_{0.6}Ba_{0.4}MnO_3$ (the best available

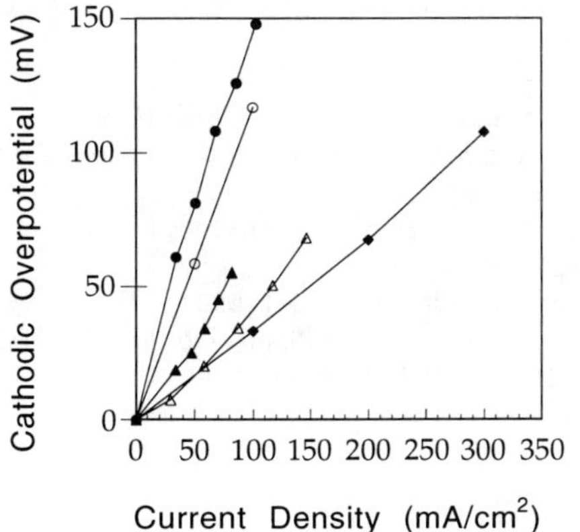

Fig. 4. Cathodic overpotentials at 1073K in dry air on a $Ba(Ce_{0.8}Gd_{0.2})O_{2.9}$ electrolyte.

oxide electrode) on a $Ba(Ce_{0.9}Nd_{0.1})O_{2.95}$ electrolyte are also given for comparison[6]. The results indicate that the cathodic performance improves with both increasing ionic (increasing Gd content) and increasing electronic (increasing Pr content) conductivity. The best performance at 1073K was obtained for the $Ba(Pr_{0.8}Gd_{0.2})O_{2.9}$ cathode which had an overpotentialof 70mV at 150 mA/cm^2. This is the best performance reported for an oxide cathode on a $Ba(Ce_{0.8}Gd_{0.2})O_{2.9}$ electrolyte, and is comparable (up to $\approx$ 100 mA/cm^2) to that of sputtered Pt.

CONCLUSION

Mixed-conduction has been introduced into $Ba(Ce_{1-x}Gd_x)O_{3-x/2}$ through the substitution of Ce by Pr. The Pr dopant leads to a significant increase in the electronic conductivity accompanied by a slight decrease in ionic conductivity. The electronic conductivity in $Ba(Ce_{0.5}Pr_{0.4}Gd_{0.1})O_{2.95}$ was found to be grain boundary limited at low temperatures under oxidizing atmospheres. These mixed conductors have been tested as cathodes in a $Ba(Ce_{0.8}Gd_{0.2})O_{2.9}$ electrolyte-fuel cell. The cathodic overpotentials of the Pr doped barium cerates improves with increasing electronic and ionic conductivity of the cathode material, with $Ba(Pr_{0.8}Gd_{0.2})O_{2.9}$ having the best performance at 1073K. The morphologies of these mixed conducting cathodes needs to be optimized in order to further improve their performance at high current densities. These optimized mixed conducting cathodes could be viable alternatives to Pt for a $Ba(Ce_{0.8}Gd_{0.2})O_{2.9}$ electrolyte-fuel cell.

ACKNOWLEDGMENTS

The authors gratefully acknowledge the financial support of the EPRI and the GRI. This research was conducted in the LRSM at the University of Pennsylvania, which is supported by the National Science Foundation. The authors also thank W. Romanov and R. Chao for their help in sample preparation and H. Ping for useful discussions and help in assembling the equipment.

REFERENCES

1. H. Iwahara, T. Esaka, H. Uchida and N. Maeda, Solid State Ionics, **3/4**, pp. 359-363 (1981)
2. N. Taniguchi, K. Hatoh, J. Niikura, T. Gamo and H. Iwahara, Solid State Ionics, **53-56**, pp. 998-1003 (1992)
3. N. Bonanos, B. Ellis, K.S. Knight, and M.N. Mahmood, Solid State Ionics, **35**, pp. 179-188 (1989)
4. H. Iwahara, Solid State Ionics, **28-30**, pp. 573-578 (1988)
5. N. Bonanos, B. Ellis, and M.N. Mahmood, Solid State Ionics, **44**, pp. 305-311 (1991)
6. H. Iwahara, and H. Uchida, Subrea C: Science and technology for energy conversion, Energy conversion and utilization with high efficiency, pp. 83-88 (1990)
7. R. Mukundan, P. K. Davies, and W. L. Worrell, Role of Ceramics in Advanced Electrochemical Systems, Cer. Trans., **65**, pp. 13-21 (1996)
8. D. Y. Wang, and A. S. Nowick, J. Electrochem. Soc., **126**, pp. 1155-1165 (1979)
9. A. J. Jacobson, B. C. Tofield, and B. E. F. Fender, Acta. Cryst., **B28**, pp. 956-961 (1972)
10. Y. Hinatsu, Journal of Solid State Chemistry, **102**, pp. 362-367 (1993)

DEFECT STRUCTURE OF THE MIXED-CONDUCTING Sr-Fe-Co-O SYSTEM

B. MA, U. BALACHANDRAN, C.-C. CHAO, and J.-H. PARK
Energy Technology Division, Argonne National Laboratory, Argonne, Illinois 60439

ABSTRACT

Electrical conductivity of the mixed-conducting Sr-Fe-Co-O system was investigated at elevated temperatures and various oxygen partial pressures (pO_2). The system exhibits not only high combined electrical and oxygen ionic conductivities but also structural stability in both oxidizing and reducing environments. The conductivity of $SrFeCo_{0.5}O_x$ increases with increasing temperature and increasing pO_2, within our experimental pO_2 range ($1 \geq pO_2 \geq 10^{-18}$ atm). p-type conduction behavior was observed. The activation energy of which increases with decreasing pO_2. A model of the defect chemistry in the Sr-Fe-Co-O system is proposed. The pO_2-dependent conducting behavior can be understood by considering the trivalent-to-divalent transition of the transition metal ions in the system.

INTRODUCTION

Mixed-conducting ceramic oxides have potential uses in solid-oxide fuel cells, batteries, sensors, oxygen-permeable membranes and other electrochemical devices. The Sr-Fe-Co-O system exhibits not only high combined electrical and oxygen ionic conductivities but it is also structurally stable [1,2] in both oxidizing and reducing environments. Dense ceramic membranes made of this material can be used to separate oxygen from air without the need for external electrical circuitry. The oxygen permeation flux rate through dense membranes made of this material can be considered commercially feasible. and use of this material would greatly improve the economics of fuel production [3-8].

When used as a ceramic membrane in gas separation, $SrFeCo_{0.5}O_x$ is exposed to large oxygen chemical potential gradients. Oxygen transport from high to low oxygen partial pressure (pO_2) in the gas separation reactor will occur if the oxygen partial pressure difference, ΔpO_2, can develop substantial oxygen flux. Therefore, an understanding of the oxygen permeation process, which is closely related to the electrical transport properties, is important to the search for better oxygen-permeable membrane materials.

In this paper, we discuss the temperature- and pO_2-dependent conductivity of $SrFeCo_{0.5}O_x$. Activation energy, E_a, is obtained by fitting the conductivity data to the equation $\sigma \cdot T = A \exp(-E_a / kT)$. A defect chemistry model of this system is proposed and examined with the conductivity data.

EXPERIMENTAL

$SrFeCo_{0.5}O_x$ powder was made by the solid-state reaction method. Appropriate amounts of $SrCO_3$, $Co(NO_3)_2 \cdot 6H_2O$, and Fe_2O_3 were mixed and then ground in isopropanol with zirconia medium for 15 h. After drying, the mixture was calcined in air at 850°C for 16 h, with intermittent grinding. After the final calcination, the powder was ground with an agate mortar and pestle to an average particle size of ≈ 7 μm. The resulting powder was pressed with a 1.2 GPa load into pellets 21.5 mm in diameter and ≈ 5 mm thick. The pellets were covered by powder of the same composition to eliminate contamination and then sintered at ≈ 1200°C for 5 h. Subsequently, the pellets were cut into thin bars with a diamond saw for conductivity measurement.

Mat. Res. Soc. Symp. Proc. Vol. 453 © 1997 Materials Research Society

X-ray powder diffraction results obtained for the $SrFeCo_{0.5}O_x$ show that the sintered sample contained a single phase [9]. The theoretical density of $SrFeCo_{0.5}O_x$ was measured on the powder with an AccuPyc 1330 pycnometer and confirmed by X-ray powder diffraction results. Bulk density of the sample used in our experiments was $\approx 95\%$ of theoretical value.

Conductivity of the specimen was measured by the DC four-probe method. Four platinum wires, of 0.2 mm in diameter, were wound around the specimen to serve as current and voltage leads. A Keithley 263 current source was used to supply the current, and the voltage was measured with a Keithley 196 system digital multimeter. Conductivity of the specimen was calculated by

$$\sigma = \frac{I \cdot d_{vv}}{V \cdot S} \tag{1}$$

where I is the applied current (in A), V is the voltage across the two voltage leads (in V), d_{vv} and S are the separation of the voltage probes (in cm) and the cross-sectional area of the specimen (in cm^2), respectively. Activation energy was calculated by fitting the temperature-dependent conductivity data to the equation

$$\sigma = \frac{A}{T} \exp\left(\frac{E_a}{kT}\right) \tag{2}$$

where A is a constant, E_a is the activation energy, k is the Boltzmann constant, and T is the absolute temperature. To measure the conductivity in various pO_2 environments, premixed gas cylinders were used.

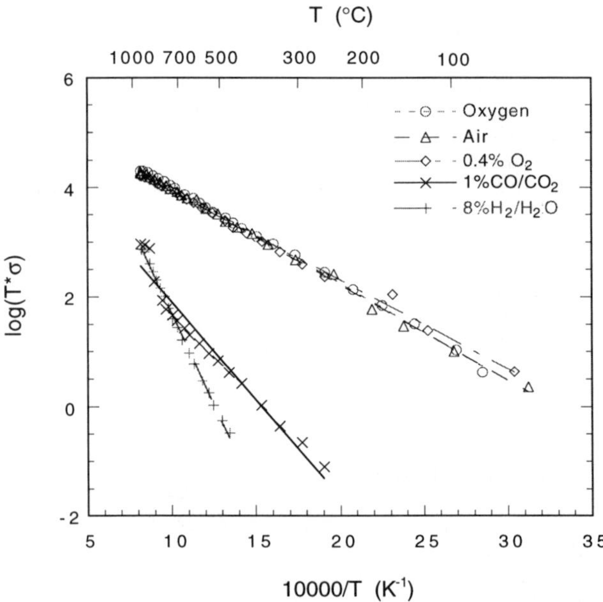

Fig. 1. Log(T·σ) of $SrFeCo_{0.5}O_x$ as a function of reciprocal temperature for various pO_2 levels.

RESULTS AND DISCUSSION

The conductivity of $SrFeCo_{0.5}O_x$ increases with increasing temperature and pO_2. Figure 1, a plot of $\log(T\cdot\sigma)$ as a function of reciprocal temperature in various oxygen environments, shows that $\log(T\cdot\sigma)$ has good linear dependence on reciprocal temperature. According to Eq. 2, the activation energy of $SrFeCo_{0.5}O_x$ can be calculated from the slopes of $\log(T\cdot\sigma)$ vs. $10000/T$ curves. The activation energy of $SrFeCo_{0.5}O_x$ thus obtained is plotted in Fig. 2 as a function of pO_2. The activation energy increases with decreasing pO_2 in the low-pO_2 range ($< 10^{-3}$ atm), whereas in the high-pO_2 range ($> 10^{-3}$ atm) activation energy is independent of pO_2 and has lower value (≈ 0.35 eV) compare to those of other $Sr(Fe,Co)O_x$ systems [10,11].

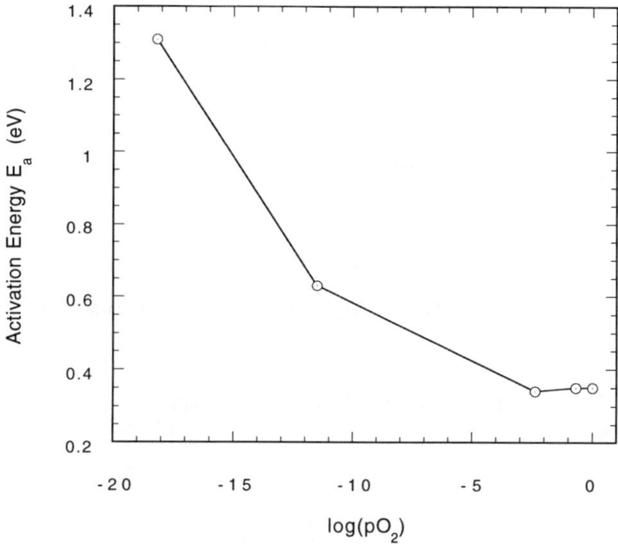

Fig. 2. Activation energy of $SrFeCo_{0.5}O_x$ as a function of pO_2.

The plot of $\log(\sigma)$ vs. $\log(pO_2)$ of $SrFeCo_{0.5}O_x$ at 950°C is shown in Fig. 3. This result is consistent with that we obtained earlier with a gas-tight electro-chemical cell [12]. In the intermediate-pO_2 range ($10^{-12} < pO_2 < 10^{-3}$ atm), the slope of $\log(\sigma)$ vs. $\log(pO_2)$ is $\approx 1/6$. This can be explained with the defect chemistry model described below. The interaction between transition metal ions and oxygen in the surrounding atmosphere can be represented as

$$2Fe_{Fe}^{x} + \frac{1}{2}O_2 \leftrightarrow 2Fe_{Fe}^{\bullet} + O_i'' \tag{3}$$

The electroneutrality equation for the intermediate range can be written as

$$2[O_i''] = p \tag{4}$$

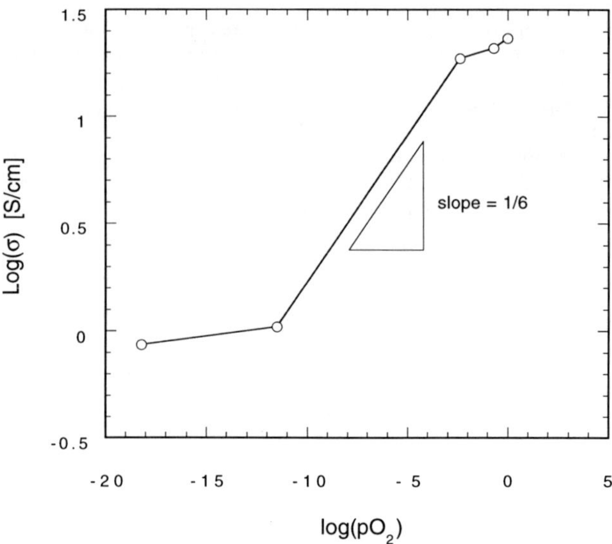

Fig. 3. Conductivity of SrFeCo$_{0.5}$O$_x$ as a function of pO$_2$ at 950°C.

where p = $\left[Fe^{\bullet}_{Fe} \right]$ is the concentration of holes. Thus, the mass-balance relationship of the transition metal ion is

$$\left[Fe^{\bullet}_{Fe} \right] + \left[Fe^{\times}_{Fe} \right] = 1 \tag{5}$$

Here, we use Fe to represent both Fe and Co transition metal ions. Equation 3 leads to the mass reaction relationship

$$K = \frac{\left[Fe^{\bullet}_{Fe} \right]^2 \cdot \left[O''_i \right]}{\left[Fe^{\times}_{Fe} \right]^2 \cdot \left(pO_2 \right)^{1/2}} \tag{6}$$

where K is the temperature-dependent reaction constant. If we let x = $\left[Fe^{\bullet}_{Fe} \right]$, and use Eqs. 4 and 5, we obtain

$$\left[O''_i \right] = \frac{x}{2} \tag{7}$$

and

$$\left[Fe^{\times}_{Fe} \right] = 1 - x \tag{8}$$

Substituting Eqs. 7 and 8 into Eq. 6, we obtain

$$K = \frac{x^3}{2(1-x)^2 \cdot (pO_2)^{1/2}}$$ (9)

Under the condition $x \ll 1$, i.e., the concentration of trivalent transition metal ions is much lower than the concentration of divalent transition metal ions, Eq. 9 gives us

$$p = x \propto (pO_2)^{1/6}$$ (10)

and

$$[O_i''] = \frac{x}{2} \propto (pO_2)^{1/6}$$ (11)

Because the conductivity is proportional to the concentrations of charge carriers, it leads to

$$\sigma \propto (pO_2)^{1/6}$$ (12)

The 1/6 dependence of the slope in Fig. 3 indicates that the trivalent-to-divalent transition of the transition metal ions plays an important role in the interaction between the $SrFeCo_{0.5}O_x$ sample and oxygen in the surrounding atmosphere.

CONCLUSIONS

$SrFeCo_{0.5}O_x$ exhibits not only high electronic and oxygen ionic conductivities but it is also structurally stable in both oxidizing and reducing atmospheres. It is a technologically important material for use in high-temperature electrochemical applications and holds particular promise as a dense ceramic membrane for separation of oxygen from air. The conductivity of $SrFeCo_{0.5}O_x$ increases with increasing temperature and pO_2. It is a p-type conductor and its $\log(T \cdot s)$ vs. $1/T$ curve exhibits good linear dependence. The activation energy of $SrFeCo_{0.5}O_x$, which decreases with increasing pO_2, is ≈ 0.35 eV in air. The pO_2-dependent conductivity behavior can be understood by considering the divalent-to-trivalent transition of the transition metal ions in the sample.

ACKNOWLEDGMENTS

Work at Argonne National Laboratory is supported by U.S. Department of Energy, Pittsburgh Energy Technology Center, under Contract W-31-109-Eng-38.

REFERENCES

1. U. Balachandran, J. T. Dusek, S. M. Sweeney, R. B. Poeppel, R. L. Mieville, P. S. Maiya, M. S. Kleefisch, S. Pei, T. P. Kobylinski, C. A. Udovich, and A. C. Bose, *Am. Ceram. Soc. Bull.*, **74**, 71 (1995).

2. B. Ma, J.-H. Park, C. U. Segre, and U. Balachandran, *Mater. Res. Soc. Symp. Proc.*, **393**, 49 (1995).

3. Y. Teraoka, H. M. Zhang, S. Furukawa, and N. Yamozoe, *Chem. Lett.*, 1985, 1743.

4. Y. Teraoka, T. Nobunaga, and N. Yamazoe, *Chem. Lett.*, <u>1988</u>, 503.

5. U. Balachandran, S. L. Morissette, J. J. Picciolo, J. T. Dusek, R. B. Poeppel, S. Pei, M. S. Kleefisch, R. L. Mieville, T. P. Kobylinski, and C. A. Udovich, *Proc. Int. Gas Research Conf.*, edited by H. A. Thompson, (Government Institutes, Inc., Rockville, MD, 1992), p. 565-573.

6. T. J. Mazanec, T. L. Cable, and J. G. Frye, Jr., *Solid State Ionics*, **111**, 53 (1992).

7. A. C. Bose, J. G. Stigel, and R. D. Srivastava, "Gas to Liquids Research Program of the U.S. Department of Energy: Programmatic Overview," paper presented at the *Symp. on Alternative Routes for the Production of Fuels*, Am Chem. Soc. National Meeting, Washington, DC, Aug. 21-26, 1994.

8. U. Balachandran, S. L. Morissette, J. T. Dusek, R. L. Mieville, R. B. Poeppel, M. S. Kleefisch, S. Pei, T. P. Kobylinski, and C. A. Udovich, *Proc. Coal Liquefaction and Gas Conversion Contractors Review Conf., edited* by S. Rogers et al., (U.S. Dept. of Energy, Pittsburgh Energy Technology Center, Pittsburgh, PA, 1993), Vol. 1, pp. 138-160.

9. B. Ma, U. Balachandran, J.-H. Park, and C. U. Segre, *Solid State Ionics*, **83**, 65 (1996).

10. T. Ishigaki, S. Yamauchi, K. Kishio, J. Mizusaki, and K. Fueki, *J. Solid State Chem.*, **73**, 179 (1988).

11. K. Nisancioglu and T. M. Gür, *Solid State Ionics*, **72**, 199 (1994).

12. B. Ma, U. Balachandran, J.-H. Park, and C. U. Segre, *J. Electrochem. Soc.*, **143**, 1736 (1996).

PHASE STABILITY AND PROPERTIES OF SUPERIONIC $PbSnF_4$ AS A FUNCTION OF THE METHOD OF PREPARATION

Raimondo CALANDRINO, Anthony COLLIN, Georges DENES, M. Cecilia MADAMBA and Juanita M. PARRIS
Laboratory of Solid State Chemistry and Mössbauer spectroscopy, Laboratories for Inorganic Materials, Department of Chemistry and Biochemistry, Concordia University, Montreal, Quebec, H3G 1M8, Canada, gdenes@vax2.concordia.ca

ABSTRACT

Superionic $PbSnF_4$ can be prepared using a variety of different methods. It crystallizes in various different unit-cells. All structures are closely related to that of the fluorite type structure, with various degrees of order/disorder and different types of superstructures. Reaction of lead(II) nitrate and stannous fluoride in water gives highly stressed and highly oriented tetragonal α-$PbSnF_4$, however, in HF/H_2O, it gives orthorhombic o-$PbSnF_4$, which is also stressed but less. Reactions of solid α-PbF_2 with an aqueous solution of SnF_2, in excess SnF_2, also gives stressed α-$PbSnF_4$. Solid state reactions of SnF_2 and PbF_2 give unstressed and much less oriented α-$PbSnF_4$ at 250°C, whereas tetragonal β-$PbSnF_4$ is obtained above 270°C and is quenchable to ambient temperature. At 390°C, cubic γ-$PbSnF_4$ is obtained, however, this is not quenchable. Ball milling of o-$PbSnF_4$ gives stressed α-$PbSnF_4$, and ball milling of α- and β-$PbSnF_4$ gives microcrystalline γ-$PbSnF_4$. Annealing the latter gives unstressed and non-oriented α-$PbSnF_4$. Stirring a slurry of microcrystalline γ-$PbSnF_4$ in water gives α-$PbSnF_4$ except the one originating from o-$PbSnF_4$, which gives back o-$PbSnF_4$. The texture of these phases, and therefore the macroscopic properties related to it, can be drastically modified by modifying the method of preparation.

INTRODUCTION

In 1967, Donaldson and Senior were the first to prepare $PbSnF_4$ [1]. Further studies of $PbSnF_4$ were carried out by Denes et al. in 1975, and isotypic $SrSnF_4$ and $BaSnF_4$ were also prepared and characterized [2]. Two high temperature phases were identified in 1979, i.e. quenchable β-$PbSnF_4$ and non-quenchable γ-$PbSnF_4$ [3]. More recently, we solved the question of lower symmetry in some preparations from aqueous solutions, and we showed that the presence of HF in the aqueous solution of SnF_2 creates an orthorhombic distortion [4]. The high performance fluoride ion conductivity of $PbSnF_4$ was established in 1978, and it has remained the highest up to date [5, 6]. We showed recently that the exceptionally high fluoride ion conductivity of $MSnF_4$ (M= Sr, Ba, Pb) cannot be explained based on the easy formation of Frenkel defects the way they exist in the fluorite type structure [7]. The high fluoride ion conductivity of $PbSnF4$ was applied recently to the fabrication of an amperometric oxygen sensor with fast response at ambient temperature [8].

In this work, we have discovered a new method for preparing $PbSnF_4$ from aqueous solution, and also how to have some control over the texture of the material. In

Mat. Res. Soc. Symp. Proc. Vol. 453 © 1997 Materials Research Society

addition, we have discovered a method for preparing microcrystalline γ-PbSnF$_4$ and stabilizing it at ambient conditions, when crystalline γ-PbSnF$_4$ is stable only at high temperature. Some of the properties of PbSnF$_4$ depend on the phase present, and more particularly, of its texture.

EXPERIMENTAL PROCEDURES

The following reactants were used in the synthesis: SnF$_2$ (99%) from Ozark Mahoning, Pb(NO$_3$)$_2$ (99%) from Sharpe, PbF$_2$ (99%) from Alfa, HF in 40% aqueous solution from Mallinckrodt, and doubly distilled water. Elemental analysis for Pb and Sn was carried out by atomic absorption spectrometry (AAS) on a Perkin-Elmer 503 absorption spectrometer. Fluorine analysis was carried out by use of a specific fluoride ion electrode from Orion. X-ray powder diffraction was performed on a Philips X-ray powder diffractometer that has been automated by Sietronics. The Ni-filtered K$_\alpha$ peak of copper was used [λ(K$_{\alpha 1}$= 1.54051Å). Diffraction patterns were run at the angular velocity of 1°(2θ)/min with a step size of 0.02°(2θ). KCl was used as an internal standard for line position, line intensity and linewidth. Bulk density measurements were carried out by displacement in CCl$_4$ (Archimedean method) by using a special home made apparatus which allows wetting the degassed sample with CCl$_4$ while still under vacuum. DTA was performed on an Omnitherm DTA/TGA instrument equipped with a STA1500 module and ATII hardware and software. Ball milling was carried out on a Wig L Bug Amalgamator from Crescent Dental MFG, using 100mg of sample and two steel balls.

RESULTS AND DISCUSSION

<u>Preparation of the Various Phases of PbSnF$_4$</u>

When a 1.7M aqueous solution of Pb(NO$_3$)$_2$ is added to a freshly prepared 1.5M solution of SnF$_2$ in the molar ratio Pb/Sn= 0.2, an immediate precipitate of α-PbSnF$_4$(aq$_1$) is obtained (fig. 1a). The labels **aq$_1$** and **aq$_2$** (see below) refer to two different methods of preparation of α-PbSnF$_4$ from aqueous solutions. If the SnF$_2$ solution contains some HF, o-PbSnF$_4$ precipitates. The magnitude of the orthorhombic splitting is a function of the volume ratio VR of the SnF$_2$ solution defined as the volume of 40% HF over the total volume of the solution [4]. Above VR$\cong$ 0.005, the orthorhomic splitting reaches a constant maximum value. Figure 1c shows o-PbSnF$_4$ obtained for VR= 0.05. We discovered just recently a second method of preparing α-PbSnF$_4$ from aqueous solutions: stirring solid α-PbF$_2$ in an aqueous solution of SnF$_2$ in the ratio Pb/Sn= 0.5 for 2 hours at ambient temperature gives α-PbSnF$_4$(aq$_2$) (fig. 2a). It should be pointed out that if the Pb/Sn ratio is higher than 0.5, another tin(II)/lead(II) fluoride material is obtained. In addition, we found that β-PbF$_2$ does not react at all with a solution of SnF$_2$ under the same conditions.

PbSnF$_4$ can also be prepared by the dry method, i.e. the direct reaction of PbF$_2$ and SnF$_2$. This was carried out in sealed copper tubes under nitrogen, according to a procedure we have designed [9]. If the reaction is carried out at 250°C, α-PbSnF$_4$(ssr) is obtained (fig. 1b), where **ssr** refers to the direct solid state reaction under dry

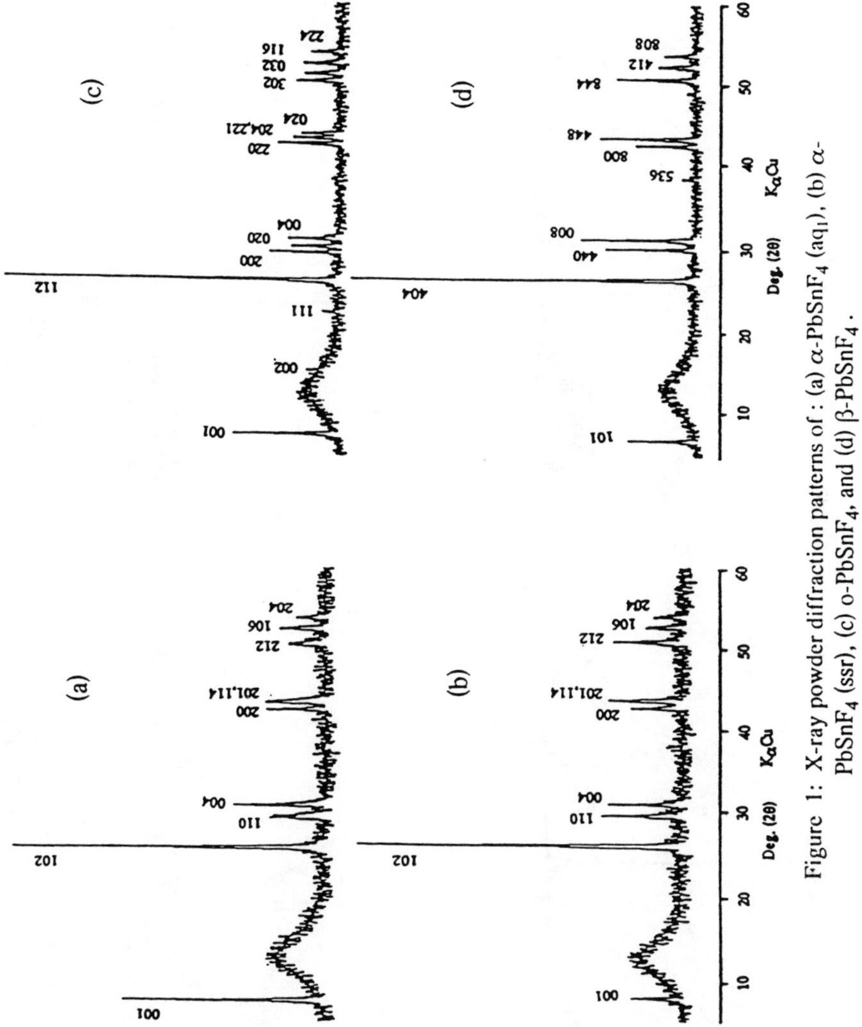

Figure 1: X-ray powder diffraction patterns of : (a) α-PbSnF$_4$ (aq$_1$), (b) α-PbSnF$_4$ (ssr), (c) o-PbSnF$_4$, and (d) β-PbSnF$_4$.

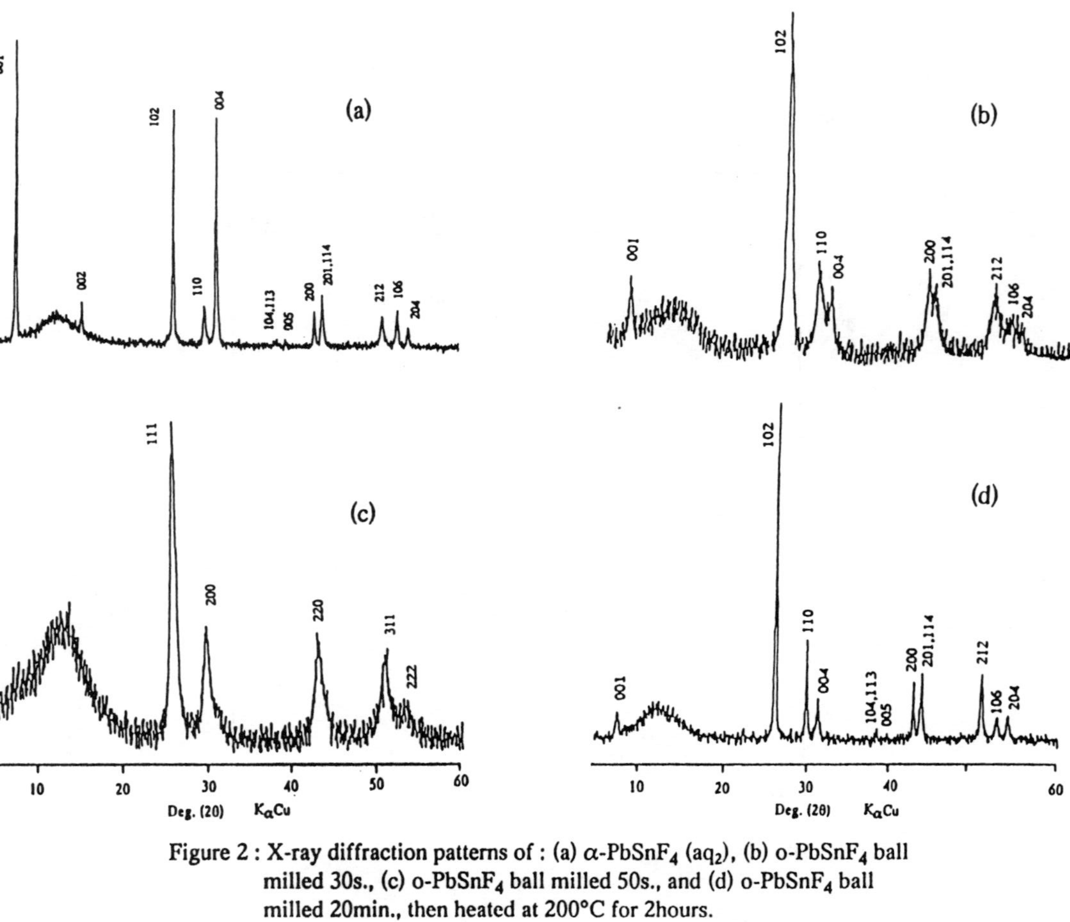

Figure 2 : X-ray diffraction patterns of : (a) α-PbSnF$_4$ (aq$_2$), (b) o-PbSnF$_4$ ball milled 30s., (c) o-PbSnF$_4$ ball milled 50s., and (d) o-PbSnF$_4$ ball milled 20min., then heated at 200°C for 2hours.

conditions. If the same reaction is carried out above 290°C, β-PbSnF$_4$ is obtained (figure 1d). Between 270 and 280°C, a mixture of α- and β-PbSnF$_4$ is observed. When β-PbSnF$_4$ is heated further, it changes to γ-PbSnF$_4$ at 390°C [3]. However, since the β --> γ transition is fully reversible without measurable hysteresis, γ-PbSnF$_4$ exists only in-situ for T>390°C. Furthermore, between 200 and 270°C, metastable β-PbSnF$_4$ changes slowly to α-PbSnF$_4$, whereas the α --> β transition starts at ca. 270°C.

Additional changes can be obtained by use of mechanical energy. PbSnF$_4$ was ball milled for various periods of time, and the ball milled materials were examined by X-ray diffraction, then they were submitted to other tests. After 30s ball milling, o-PbSnF$_4$ gives α-PbSnF$_4$ (fig. 2b). α-PbSnF$_4$(aq$_1$), o-PbSnF$_4$ and α-PbSnF$_4$(ssr) give a diffraction pattern similar to that of β-PbF$_2$, with broadened lines and a similar unit-cell parameter, after 50s ball milling (fig. 2c). The same is obtained with β-PbSnF$_4$, however, this requires 5min., i.e. six times as long as for the other phases. The reasons for this are not clear. Thus, all the milled phases look identical to one another, i.e. the broadened diffraction pattern of a F lattice that could be microcrystalline β-PbF$_2$. If PbSnF$_4$ decomposed to microcrystalline β-PbF$_2$, then the SnF$_2$ produced would have to be amorphous. Alternatively, the milled phases could be microcrystalline γ-PbSnF$_4$, stabilized for the first time at ambient conditions, possibly by the lattice defects created upon milling. The tests performed in order to identify the nature of the milled phases included annealing the milled phases, stirring them in water, mill 1:1 mixtures of PbF$_2$ and SnF$_2$ then anneal them, and stir PbF$_2$ in an aqueous solution of PbF$_2$ for Pb/Sn= 1. The results, which cannot be detailed here, prove unambiguously that the milled phases are microcrystalline γ-PbSnF$_4$. In addition, they keep some memory of their origin, which is evidenced as follows:

(i) After stirring one hour in water, all give α-PbSnF$_4$, except the sample originating from o-PbSnF$_4$, which gives back o-PbSnF$_4$;

(ii) Although their diffraction patterns look identical, their unit-cell parameters are different: 5.968Å (α-aq$_1$), 5.967 Å (α-ssr), 5.979 Å (o-) and 5.937 Å (β-), compared to 5.924 Å for β-PbF2. The average particle diameter was calculated from the linewidth; it varies from 117 Å to 134 Å and does not show any particular trend.

<u>Texture and Properties</u>

Figures 1 and 2 show very impressive changes in the relative intensity of the Bragg peaks of PbSnF$_4$, even when the same phase is concerned. For example, for α-PbSnF$_4$, the (00l) peaks are much stronger for α-aq$_1$ and α-aq$_2$ than for α-ssr. This is due to the layered structure, which gives the crystallites the shape of very thin plates [10]. From solution, they grow quite large but remain very thin and it results in a high degree of preferred orientation, which can be enhanced by careful filtration on a büchner funnel, to give nearly perfect preferred orientation along the *c* axis of the unit-cell, whereas random orientational disorder is present in the (*a*,*c*) plane. On the other hand, α-PbSnF$_4$(ssr) and o-PbSnF$_4$ have much weaker (00l) peaks. This is due to the fact that the crystallites, although they have the same shape with a similar thickness, have a much smaller surface (we observed this by SEM), and therefore they do not stack parallel to one another as efficiently. There is also less preferred orientation in β-PbSnF$_4$. This is also probably due to the method of preparation (β-PbSnF$_4$ cannot be prepared from aqueous solutions), and also to its structure, which is not known, but is believed to have less efficient cleavage planes. α-PbSnF$_4$ can also be obtained from microcrystalline γ-PbSnF$_4$ by annealing at 200°C for 2 hours (fig. 2d), or by stirring in water (except the one originating from o-PbSnF$_4$). Figure 2 shows that this gives α-PbSnF$_4$ with much less preferred orientation that α-PbSnF$_4$ from any other source.

The properties that are related to the crystal structure or the texture show substantial variations with the method of preparation. For example, the conductivity should be very anisotropic in the layered structures. Further, the peaks that result from the splitting of (200) of β-PbF$_2$, i.e. (110) & (004) for α-PbSnF$_4$, (200) & (020) & (004) for o-PbSnF$_4$, and (400) & (008) for β-PbSnF$_4$, show considerable variation in linewidth. The two main reasons for line broadening are microcrystallinity and non-uniform stress. Microcrystallinity would have to occur perpendicularly to the sheets, i.e. along c, and therefore the (00l) peaks should be broadened. (004) is indeed moderately broadened in α-PbSnF$_4$, more particularly when obtained by heating microcrystalline γ-PbSnF$_4$, while (008) of β-PbSnF$_4$ is narrow, which seems to confirm that cleavage is not as efficient in β-PbSnF$_4$. However, the broadening of (110) in α-PbSnF$_4$ cannot be due to microcrystallinity. It can be attributed to non-uniform strain in the (a,b) plane due to ferroic behavior. This is greatly enhanced in o-PbSnF$_4$ when VR is very small, and nearly disappears at very high VR at full orthorhombic splitting.

ACKNOWLEDGMENTS

The Natural Science and Engineering Research Council of Canada and Concordia University are acknowledged for financial support.

REFERENCES

1. J.D. Donaldson and B.J. Senior, J. Chem. Soc. A, 1821 (1967).
2. G. Denes, J. Pannetier and J. Lucas, C. R. Acad. Sc. Paris, C **280**, 831 (1975).
3. J. Pannetier, G. Denes and J. Lucas, Mater. Res. Bull. **14**, 627 (1979).
4. G. Denes, M.C. Madamba and J.M. Parris, in <u>Solid State Ionics IV</u>, edited by G.A. Nazri, J.M. Tarascon and M. Schreiber (Mater. Res. Symp. Soc. Proc. **369**, Pittsburgh, PA 1995), p. 463-468.
5. J.M. Réau, C. Lucat, J. Portier, P. Hagenmuller, L. Cot and S. Vilminot, Mater. Res. Bull. **13**, 877 (1978).
6. G. Denes, M.C. Madamba, G. Milova and M. Perfiliev in <u>Electrically Based Microstructural Characterization</u>, edited by R.A. Gerhardt, S.R. Taylor and E.J. Garboczi (Mater. Res. Soc. Proc. **411**, Pittsburgh, PA 1996), p. 151-156.
7. G. Denes, in <u>Solid State Ionics IV</u>, edited by G.A. Nazri, J.M. Tarascon and M. Schreiber (Mater. Res. Soc. Proc. **369**, Pittsburgh, PA 1995), p. 295-300.
8. A. Wakagi and J. Kuwano, J. Mater. Chem. **41**, 973 (1994).
9. G. Denes, J. Solid State Chem. **77**, 54 (1988).
10. G. Denes, Y.H. Yu, T. Tyliszczak and A.P. Hitchcock, J. Solid State Chem. **91**, 1 (1991).

CHEMICAL DIFFUSION IN MIXED CONDUCTORS WITH COMPARABLE IONIC AND ELECTRONIC TRANSPORT NUMBERS: RESULTS FOR $Ag_{1.92}Te$

W. PREIS, W. SITTE

Institute of Physical and Theoretical Chemistry, Graz University of Technology, Rechbauerstraße 12, A-8010 Graz, Austria

ABSTRACT

Galvanostatic polarization of a mixed conductor located between an ionically blocking electrode and an electronically blocking electrode in an asymmetric electrochemical cell is treated in detail. Evaluation formulae for the determination of the chemical diffusion coefficient of mixed conductors with comparable ionic and electronic transport numbers are introduced. They allow the determination of the chemical diffusion coefficient of $Ag_{1.92}Te$ as a function of composition at 160°C from galvanostatic polarization and depolarization experiments on the asymmetric cell $Ag \mid AgI \mid Ag_{1.92}Te \mid Pt$. The chemical diffusion coefficient shows composition dependent values between 0.002 and 0.004 cm^2s^{-1}. The electronic transport numbers are obtained independently from four-point van der Pauw measurements with typical values around 0.8-0.9.

INTRODUCTION

A number of new materials have been introduced which show ionic and electronic conductivities of the same order of magnitude [1,2]. Such substances are regarded as promising electrode materials in SOFC′s. Using mixed conducting electrodes the electrochemical transfer reaction will occur across the entire surface area of the electrode-gas interface instead of three-phase regions (gas / electrode / electrolyte) in the case of materials with electronic conductivities being higher by orders of magnitude than the ionic conductivities [e.g. $(La,Sr)MnO_3$]. Therefore, lower polarization losses can be expected when mixed conducting electrodes like e.g. yttria-stabilized zirconia-terbia are employed in SOFC′s. In addition, mixed conductors with comparable ionic and electronic conductivities may be suitable as oxygen selective membranes, since no external short-circuit is required during the operation.

If the electronic transport number of the mixed-conductivity compound deviates considerably from unity, some modifications concerning the theoretical description of chemical diffusion processes in the mixed conductor located between blocking electrodes in an asymmetric electrochemical cell are necessary [3,4]. Asymmetric cells containing blocking electrodes are usually used for coulometric titrations to alter the composition of the mixed conductor with high stoichiometric resolution [5] and for measurements of the chemical diffusion coefficient of the mixed conductor by means of dc-polarization experiments or impedance spectroscopy [6-9]. The solution of Fick′s second law of diffusion for galvanostatic polarization of mixed conductors with electronic conductivities being higher by orders of magnitude than the ionic conductivities (i.e. the electronic transport number is equal to one) is well developed [7,8]. Recently, the general case of arbitrary electronic transport numbers has been investigated by the present authors and exact solutions of Fick′s second law for galvanostatic polarization processes in asymmetric cells have been proposed [3,4]. Here, the evaluation formulae for the determination of the chemical diffusion coefficient will be summarized briefly and experimental results for the mixed-conductivity compound $Ag_{1.92}Te$ will be presented. Our theoretical findings have been applied to $Ag_{1.92}Te$, since this compound is a typical example of a mixed conductor with comparable ionic and electronic conductivities [10,11] and experiments can be performed at moderate temperatures.

Mat. Res. Soc. Symp. Proc. Vol. 453 © 1997 Materials Research Society

THEORY

Chemical diffusion processes in mixed conductors are characterized by a simultaneous flow of ionic and electronic charge carriers through the solid. Neglecting any internal reactions between the ionic and electronic defects the Onsager cross-coefficients in the flux equations can be disregarded [11-14], leading to

$$J_i = -\tilde{D}\,\frac{\partial c}{\partial x} + \frac{t_i}{zF}\,j \tag{1a}$$

and

$$J_e = -z\tilde{D}\,\frac{\partial c}{\partial x} - \frac{t_e}{F}\,j \tag{1b}$$

with J_i, J_e, j, $\tilde{D}$, t_i, t_e, z, F and c denoting the fluxes of the ionic and electronic components, the total electric current density, the chemical diffusion coefficient, the ionic and electronic transport number, the charge number of the ionic component, the Faraday constant and the concentration of the mobile species, respectively. The symbol c can also be interpreted in terms of the concentration of stoichiometric excess of metal atoms in the case of binary metal chalcogenides. The first term of the right hand side of eqs. (1a) or (1b) represents the driving force caused by the concentration gradient of the stoichiometric excess of metal atoms, whereas the second term describes the driving force due to the electric field applied to the mixed conductor. Using the equation of continuity for the ionic or electronic fluxes one obtains Fick´s second law of diffusion

$$\frac{\partial c}{\partial t} = \tilde{D}\,\frac{\partial^2 c}{\partial x^2} \tag{2}$$

which is an exact description of the chemical diffusion process when spatial variations of the chemical diffusion coefficient can be neglected due to small concentration gradients (i.e. small current densities are applied in order to satisfy the linear response limit).

Locating the mixed conductor (e.g. $Ag_{1.92}Te$) between an electronically blocking electrode $(Ag|AgI)$ and an ionically blocking electrode (Pt) results in the asymmetric electrochemical cell

$$Ag \mid AgI \mid Ag_{1.92}Te \mid Pt \,. \tag{I}$$

The voltage response U of cell (I) is given by [3,4]

$$U = -\frac{jL}{\sigma} - \frac{t_e}{zF}\,\frac{RT\vartheta}{c_0}\,\Delta c(0) - \frac{t_i}{zF}\,\frac{RT\vartheta}{c_0}\,\Delta c(L) \,, \tag{3}$$

where L, σ, R, T, ϑ, $\Delta c(0)$ and $\Delta c(L)$ are the length of the mixed conductor, the total conductivity, the gas constant, the absolute temperature, the thermodynamic factor ($\vartheta = d\ln a/d\ln c$; a is the activity of the neutral mobile species) and the deviation of the diffusant concentration (stoichiometric excess of metal atoms) from its initial value c_0 at the electronically blocking electrode / mixed conductor interface and the mixed conductor / ionically blocking electrode interface, respectively. The first term of the right-hand side of eq. (3) describes the ohmic voltage drop and the remaining two terms are due to the chemical diffusion process occurring inside the mixed conductor when a constant direct current is fed through cell (I). All interface resistances as well as the ohmic voltage drop owing to the bulk resistance of the solid electrolyte are disregarded in eq. (3). As the flux of the electrons vanishes at the electronically blocking electrode and the flux of the ions is equal to zero at the ionically blocking electrode, eqs. (1a) and (1b) yield the boundary conditions for galvanostatic polarization of the mixed conductor

$$\frac{\partial c}{\partial x} = -\frac{t_e j}{zF\tilde{D}} \tag{4a}$$

at the AgI | $Ag_{1.92}Te$ interface and

$$\frac{\partial c}{\partial x} = \frac{t_i j}{zF\tilde{D}} \tag{4b}$$

at the $Ag_{1.92}Te$ | Pt interface. In eqs. (4a) and (4b) any complications owing to space charge regions at the interfaces have been disregarded. In the case of comparable ionic and electronic conductivities the ionic transport number considerably deviates from zero and, therefore, chemical diffusion processes are stimulated at both electrodes when a direct current step is applied to cell (I), see eqs. (4a) and (4b). For the depolarization process the boundary condition at both electrodes of cell (I) reads

$$\frac{\partial c}{\partial x} = 0 \ , \tag{5}$$

as the total current density j is equal to zero when the current source is switched off.

Inserting the mathematical solution of the partial differential equation system into eq. (3) yields the exact expression for the voltage response of cell (I) for galvanostatic polarization of a mixed conductor with comparable ionic and electronic transport numbers. The exact mathematical procedure can be found elsewhere [3], here only the main results will be summarized briefly.

According to the short-time approximation of the solution for the polarization voltage a straight line (slope: k) results in the regime of short polarization times satisfying $t < 2L^2 / (3\tilde{D}\pi^2)$, when the voltage response is plotted versus the square root of time (see Fig. 1a). If the polarization voltage is plotted as a function of time, a straight line (slope: k′) will occur at sufficiently long polarization times ($t \gg L^2 / 4\tilde{D}$), see Fig. 1b. Finally, the chemical diffusion coefficient of the mixed conductor is obtained from the ratio of the slopes k and k′

$$\tilde{D} = \left[2\frac{k'L}{k\sqrt{\pi}} (t_e^2 + t_i^2) \right]^2 \tag{6}$$

The difference ΔU between the voltage response at the end of the preceding polarization process and the voltage response of the depolarization process is usually used for the evaluation of the depolarization experiments. A straight line (slope: k) is obtained at very short times in a ΔU versus $t^{1/2}$ plot, as can be seen from Fig. 1c. When the difference $kt^{1/2} - \Delta U$ is plotted as a function of t, a straight line (slope: k′) will result at sufficiently short times (see Fig. 1d). Again, the chemical diffusion coefficient is determined by calculating the ratio between the slopes of these two straight lines as given in eq. (6).

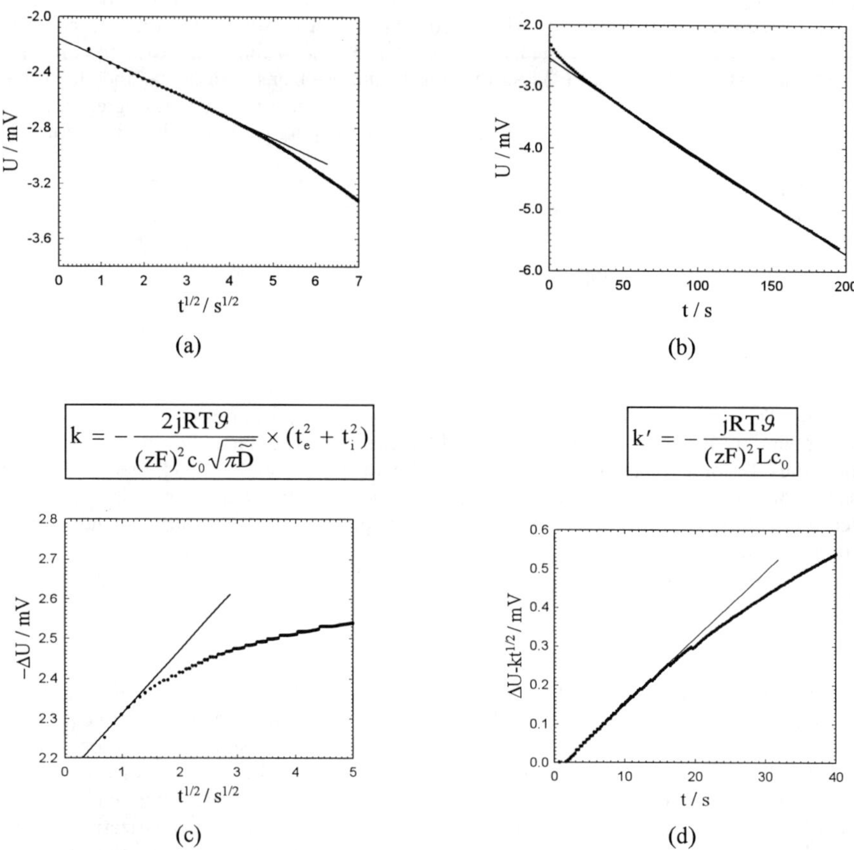

Fig. 1 (a) Polarization voltage versus square root of time. (b) Polarization voltage versus time. (c) Voltage response ΔU for the depolarization process as a function of square root of time. (d) Difference $kt^{1/2}$ - ΔU versus time. (emf of the cell Ag | AgI | Ag$_{1.92}$Te | Pt: 174 mV; polarization current: 0.5 mA; temperature: 160°C)

EXPERIMENT

Polycrystalline samples of the mixed-conductivity compound Ag$_{1.92}$Te were prepared from commercially available Ag$_2$Te (99.99 %, Johnson Matthey) using the solid-state coulometric titration technique. As the composition of the specimen can be varied by passing a direct current through cell (I), this asymmetric electrochemical cell was used for the preparation of Ag$_{1.92}$Te. At an emf of 150.6 mV corresponding to the desired composition Ag$_{1.920}$Te the coulometric titration process was stopped and the material was ground and pressed into a disc-shaped polycrystalline sample for the chemical diffusion coefficient measurements.

Galvanostatic polarization as well as depolarization experiments were performed on cell (I) which consisted of a combination of a silver tablet (silver source or sink), a silver iodide tablet (solid electrolyte) and a disc-shaped powder compact of $Ag_{1.92}Te$ (sample). The electronic electrode was a platinum plate and a silver wire wound around the silver iodide tablet served as a reference electrode. All compartments were held together by light spring action and the experiments were performed in a sealed quartz apparatus under a constant helium flow. The sample dimensions were 6 mm diameter, 6.090 mm length and 1.3623 g weight. The composition of the specimen was changed in situ with a high stoichiometric resolution by coulometric titrations.

The voltage response of cell (I) was measured between the platinum plate and the reference electrode (silver wire) using a high-impedance multimeter (Keithley 199). The constant direct current for the polarization experiments and the coulometric titrations was supplied by a precision current source (Knick J152). The constant temperature of 160°C was controlled by a precision temperature controller (Eurotherm 818) employing chromel-alumel thermocouples.

RESULTS

We measured the chemical diffusion coefficient of the mixed-conductivity compound Ag_yTe ($1.912 < y < 1.920$) as a function of composition at 160°C. The experimental data of the galvanostatic experiments were evaluated by using the theoretical considerations developed above. According to eq. (6) the electronic transport numbers have to be measured independently in order to determine the chemical diffusion coefficient. Applying four-point van der Pauw dc-measurements [15-18] the ionic and electronic conductivities of Ag_yTe were obtained within its thermodynamic stability range at 160°C. The ionic and electronic transport numbers are available from the partial conductivities and Fig. 2a shows the variation of the electronic transport number of $Ag_{1.92}Te$ as a function of composition.

The variation of the chemical diffusion coefficient of $Ag_{1.92}Te$ at 160°C with the silver content y in Ag_yTe is shown in Fig. 2b. The chemical diffusion coefficient of the specimen decreases slightly with decreasing silver content with values between 0.002 and 0.004 cm^2s^{-1}. Polarization as well as depolarization experiments yield the same results within the limits of error.

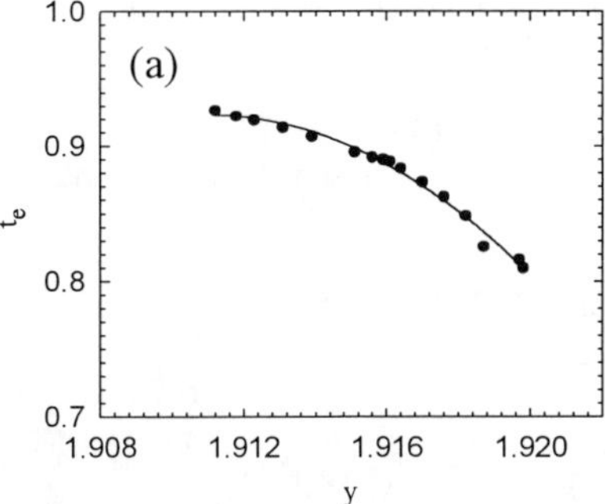
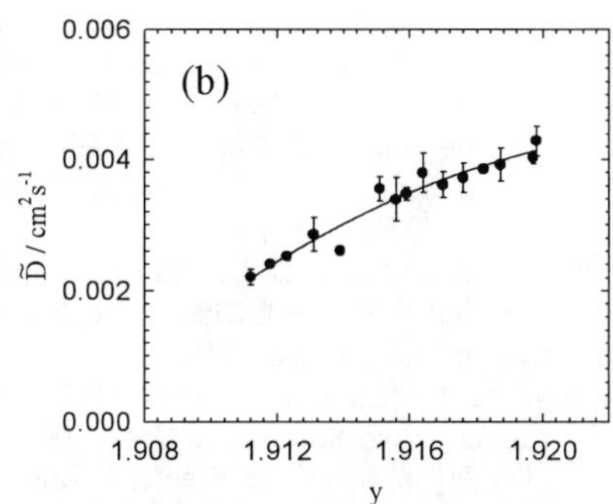

Fig. 2 (a) Electronic transport number as a function of y in Ag_yTe at 160°C. The solid line is a least square fit as a guide to the eye. (b) Chemical diffusion coefficient as a function of y in Ag_yTe at 160°C. The filled circles correspond to mean values of the chemical diffusion coefficient obtained from polarization as well as depolarization experiments on cell (I). The error bars refer to the standard variation of the experimental data and the solid line is a least square fit as a guide to the eye.

CONCLUSIONS

Evaluation formulae for the determination of the chemical diffusion coefficient of a mixed conductor with comparable ionic and electronic transport numbers employing galvanostatic polarization as well as depolarization experiments on an asymmetric electrochemical cell have been introduced. The formulae have been applied to obtain the chemical diffusion coefficient of $Ag_{1.92}Te$ as a function of composition at 160°C from galvanostatic polarization / depolarization experiments. The electronic transport numbers have been obtained independently from measurements of the partial conductivities using a van der Pauw technique. As the electronic transport numbers deviate from unity throughout the thermodynamic stability range of $Ag_{1.92}Te$ at 160°C, our evaluation formulae have to be used for the determination of the chemical diffusion coefficient. The case of comparable ionic and electronic transport numbers has been taken into account by introducing the correction factor ($t_e^2 + t_i^2$). Evaluation of the experimental data by means of the method proposed in this work avoids the determination of the thermodynamic factor, as this parameter does not occur in the final evaluation formula [eq. (6)].

ACKNOWLEDGMENT

We are indebted to the Austrian Science Foundation (FWF) for supporting this work under the project No. F915.

REFERENCES

1. P. Hahn and W. L. Worrell, in Ionic and Mixed Conducting Ceramics, edited by T. A. Ramanarayanan, W. L. Worrell and H. L. Tuller (The Electrochemical Society, Pennington, NJ, 1994), pp. 317-324.
2. M. A. Spears and H. L. Tuller in Solid State Ionics IV, edited by G. A. Nazri, J. M. Tarascon and M. Schreiber (Mater. Res. Soc. Proc. **369**, Pittsburgh, PA, 1995) pp. 271-288; I. Kosacki and H. L. Tuller, ibid., pp. 703-708.
3. W. Preis and W. Sitte, J. Chem. Soc. Faraday Trans. **92**, 1197 (1996)
4. W. Preis and W. Sitte, Solid State Ionics **86-88**, 779 (1996)
5. C. Wagner, J. Chem. Phys. **21**, 1819 (1953)
6. W. Weppner and R. A. Huggins, J. Electrochem. Soc. **124**, 1569 (1977)
7. C. J. Wen, C. Ho, B. A. Boukamp, I. D. Raistrick, W. Weppner and R. A. Huggins, Intern. Met. Rev. **5**, 253 (1981)
8. A. Honders and G. H. J. Broers, Solid State Ionics **15**, 173 (1985)
9. W. Preis and W. Sitte, J. Chem. Soc. Faraday Trans. **91**, 2127 (1995)
10. S. Miyatani, J. Phys. Soc. Japan **13**, 341 (1958)
11. I. Yokota, J. Phys. Soc. Japan **16**, 2213 (1961)
12. C. Wagner, Prog. Solid State Chem. **10**, 3 (1975)
13. G. J. Dudley and B. C. H. Steele, J. Solid State Chem. **31**, 233 (1980)
14. J. Maier, Z. Phys. Chem. (NF) **140**, 191 (1984)
15. L. J. van der Pauw, Philips Res. Rep. **13**, 1 (1958)
16. I. Riess and D. S. Tannhauser, Solid State Ionics 7, 307 (1984)
17. D. Grientschnig and W. Sitte, Z. Phys. Chem. (NF) **168**, 143 (1990)
18. W. Preis and W. Sitte, Solid State Ionics **76**, 5 (1995)

A QUANTITATIVE TUNNELING/DESORPTION MODEL FOR THE EXCHANGE CURRENT AT THE POROUS ELECTRODE/BETA"-ALUMINA/ALKALI METAL GAS THREE PHASE ZONE AT 700-1300K

R. M. Williams, M. A. Ryan, C. Saipetch, H. G. LeDuc
Jet Propulsion Laboratory, California Institute of Technology, Pasadena, CA 91109
roger.m.williams@jpl.nasa.gov

ABSTRACT

The exchange current observed at porous metal electrodes on sodium or potassium beta"-alumina solid electrolytes in alkali metal vapor is quantitatively modeled with a multi-step process with good agreement with experimental results. No empirically adjusted parameters were used, although some physical parameters have poor precision. Steps include: (1) diffusion of Na^+ ions to the reaction site; (2) stretching of the Na^+ ionic bond with the β"-alumina surface to reach a configuration suitable for accepting an electron to form a surface bound Na^0 atom; (3) electron tunneling from the Mo electrode to the ions; and (4) desorption of Na^0 atoms weakly bound at the reaction site on the β"-alumina surface; (5) electron tunneling between Na^0 or Na^+ on the defect block and (6) Na^0 and/or Na^+ mobility on the spinel block surface may extend the reaction area substantially.

The rate is increasingly dominated by the region close to the three-phase boundary as temperature increases and the rate near the three-phase boundary increases fastest, because desorption has a higher energy than reorganization. At high temperatures, surface diffusion of Na+ ions from the defect block edges to the spinel block edges is responsible for an increase in the total effective reaction zone area near the three-phase boundary.

INTRODUCTION

It is generally accepted that heterogeneous electron exchange electrochemical reactions exhibit high rates due to electron transfer via tunneling, but quantitative modeling of real reactions is difficult because of to the complexity of the ancillary processes to electron transfer, especially when liquid electrolytes are involved. Gurney proposed that tunneling must play an important role in electrochemical electron transfer reactions.[1] This conclusion was extended and more thoroughly treated with the work of Gerischer, and is generally presented in electrochemistry texts, but fundamental models showing quantitative agreement with experiment are lacking.[2] Some of the problems inherent to solution electrochemistry are eliminated in developing a quantitative model of heterogeneous electron transfer reactions when solid electrolyte/solid electrode/low pressure gas electrochemical systems are investigated, since the mechanics of mass and charge transfer are somewhat simpler. However, other problems arise because an adequate definition of electrolyte and electrode morphology is necessary.

Experimental exchange currents for the reduction/oxidation of alkali metal ions and atoms at porous metal electrodes on alkali metal β"-alumina solid electrolytes (BASE) in low pressure alkali metal vapor have been evaluated at temperatures from 700K to 1250K. [3-5] Measurements have been made for a variety of electrodes including Mo, Mo/Na$_2$MoO$_4$, WRh alloys, WPt alloys, and TiN on sodium β"-alumina in sodium vapor, as well as for Mo electrodes on potassium β"-alumina in potassium vapor.[3-6] At typical pressures of alkali metal vapor encountered in these experiments, metal electrodes have an adsorbed less-than-monolayer coverage of the alkali metal.[7] Extensive measurements have been made of the work functions of Mo and W, with alkali metal sub-monolayer films. We use the work function for Na on W, 2.5 eV, since the value for Na on Mo is less well characterized.[8-10] All experimental data in this paper have previously been reported.

Measurements have been carried out in power producing alkali metal thermal to electric

Mat. Res. Soc. Symp. Proc. Vol. 453 ©1997 Materials Research Society

converter (AMTEC) cells where the anode is liquid alkali metal in contact with BASE, which were first described in detail by Weber.[11] The impedance of the liquid metal electrode/BASE interface at high temperatures is negligible.[5,12] As a result the impedance of AMTEC cells contains significant contributions from BASE and lead ohmic resistances and small inductances from the lead configurations, and a major Faradaic impedance only from the porous electrode/BASE/vapor interphase region on the low pressure side of the AMTEC cell. The exchange current has been determined from these cells by deconvoluting the contributions of kinetics and transport from the mixed control Faradaic component of the impedance measured as a function of cell potential.[5] We have reported exchange currents for most of the electrodes described above either referenced to the saturated vapor pressure of the alkali metal, or using an empirical parameter, which relates the exchange current at any alkali metal vapor pressure to the collision rate of alkali atoms in the vapor with the electrolyte surface.[3,5] Measurements performed on oxide free thin (0.5μm) Mo and (1.0-1.5μm) W/Rh and W/Pt electrodes have shown fairly consistent exchange currents which are nearly proportional to collision rates of Na gas at the interface over several hundred K.[3] The experimental exchange current of the porous Mo electrode/ Na^0 vapor/ $Na\beta$"-alumina system varies by a factor of about 100 over this temperature range.[3] The electrochemical transfer coefficient, α, has been determined to be close to 0.5 based on high quality impedance data on mature electrodes at high temperatures, $T > 1100K$, and no systematic deviation from this value is observed on cooling, although values show significant scatter. This argues for a simple, symmetric reaction. The near proportionality of the exchange current to the alkali metal gas collision frequency, with the empirical parameter independent of temperature and pressure, suggested that the exchange current could be modeled fairly simply.[3] While differences exist among the results from various electrode materials, the exchange currents of oxide-free metallic electrodes show a similar temperature dependence and a relatively narrow distribution at any temperature. There is a decrease in the exchange current for Mo electrodes at high temperature as grain-growth occurs, reducing the length of the three-phase boundary between electrode, electrolyte, and vapor.[13] This suggests that the reaction zone is confined to a region quite close to the three-phase boundary, within about 0.25μm, and that Na^+ diffusion onto [001] grain surfaces is also less than 0.25 μm, a radius of grains or pores in Mo electrodes at which further grain growth reduced exchange currents.

Three morphological parameters which directly affect the reaction rate have been characterized by scanning electron microscopy (SEM) and surface decoration experiments: (a) the length of the three-phase boundary, about 3-$6x10^6$m/m^2; (b) the fraction of the BASE surface composed of electrochemically active [hk0] crystallite faces, about 5-30%, and (c) the contact angle between electrode grains and BASE, estimated at 45°.[3] The reaction rate increases linearly with (a) and (b) and decreases with (c).

The crystal structure of alkali β"-alumina is well known, with quasi-two dimensional defect blocks containing conducting ions isolated by spinel blocks.[14,15] A surface structure and sodium ion population for the defect block is assumed which is similar to the known structure in the crystal interior. We make the assumption that reactive surface grains are oriented with [001] or [hk0] faces exposed, because grain fracture with any component along the basal plane [001] results in predominately basal plane surfaces.

The rate of the Na^+ diffusion to the reaction site can be estimated from the conductivity of Na β"-alumina single crystals at high temperatures. The thickness of the defect block (>0.216 nm) is somewhat greater than the diameter of the Na^+ ion, because some ions are above and some are below the plane of the center of the defect block, and the BASE defect block structure readily accommodates the larger K^+ ion (0.266 nm) with slight c axis expansion. The spinel block is about 0.87 nm thick, and migration of Na^+ by less than 0.5 nm onto spinel block surfaces could multiply the rate by 5 if surface migration over this distance is competitive with the desorption rate of the alkali metal from the defect block.

The vapor pressure of Na and K have been characterized over a large temperature range. [16,17] The desorption rate of Na^0 is equal to the collision rate of alkali metal vapor at equilibrium with the liquid. The desorption of Na^+ deposited on sodium β-alumina of composition $Na_2O \cdot 8(Al_2O_3)$, cut along the a-c plane, shows a lowest energy weak peak at 1.68eV and a second lowest, most prominent peak at 2.18eV dominating at high rates of desorption.[18]

Wetting characteristics of liquid sodium on BASE provide information about the surface energy between liquid sodium and BASE. Liquid sodium does not wet BASE at low temperatures, but does wet it at temperatures above about 673 K, resulting in a low resistance contact.[12] This moderate wetting suggests that the interfacial free energy between liquid sodium and BASE is probably positive, but smaller than the surface free energy of sodium, at least at higher temperatures. A molecular orbital calculation of the binding of a single sodium atom to an AlO_3^{2-} cluster at the surface of an ionic solid has been carried out and results indicate that if appropriate charge compensation occurs, Na^0 will bind to a β''-alumina surface, but that because β''-alumina has no low-lying unoccupied electronic levels, the sodium 3s electron will remain associated with the sodium atom.[19] We therefore use an approximation to the binding energy of a sodium atom at the surface of the bulk, with its number of nearest neighbors reduced from about 12 to about 9, and binding energy to 2/3, recognizing that this value is uncertain, especially at low pressures and coverages in which the adsorbed layer does not have the characteristics of bulk liquid sodium.

We define the exchange current to consist of the equal and opposite redox processes which occur at reaction sites which can contribute to a continuous dc current when the cell is perturbed from the stationary state: $Na^0_{(g)} \rightleftharpoons Na^+_{(\beta''\text{-alumina})} + e^-_{(metal)}$. Then the reaction zone will be the region where the incoming and outgoing sodium atom fluxes become different when a potential is applied across the cell. If there is no effective transport mode away from a reaction site, the contribution of the exchange current at that reaction site to the dc exchange current is negligible. Local transport of ions or atoms at the reaction site may have important effects on the exchange current; if, for example, Na^+ ions can diffuse rapidly on the spinel block surface between defect block edges, the effective reaction area at the three-phase boundary would be quadrupled. A high frequency, larger exchange current which does not require escape of the reaction product might also be defined; chemical exchange with the gas occurs on the whole BASE surface.

This paper discusses the microscopic mechanisms which account for the observed electrode kinetics of the alkali metal coated porous metal electrode/β''-alumina/alkali metal gas three-phase region. Some details of standard calculations have been omitted for conciseness, and will be presented in a later publication. The model has no empirically adjusted parameters. This model is constructed using physical parameters which are known or can be estimated (sometimes with poor precision) and includes only credible physical processes. The discussion of reaction steps generally will describe the cathodic reaction, an electron tunnelling from a Mo electrode to a Na^+ ion, with eventual desorption of a sodium atom, but will apply equally well to the reverse, anodic reaction. At zero volts, all of the reverse steps are equally likely. The Na^+/Na^0 electrochemical couple over the BASE surface is close to equilibrium with the chemical potential of the electrode; the model is close to adiabatic. K may be substituted for Na, and W or other metal electrodes may take the place of Mo, with change in the work function, but the model treats Na at 0.5 μm thick Mo electrodes, as only in this case are the morphological parameters and exchange current dependence on temperature experimentally well characterized.

The model describes the initial reaction as a four step sequence: (1) diffusion of Na^+ ions to the reaction site; (2) stretching of the Na^+ ionic bond with the β''-alumina surface to reach a configuration suitable for accepting an electron to form a surface bound Na^0 atom; (3) electron

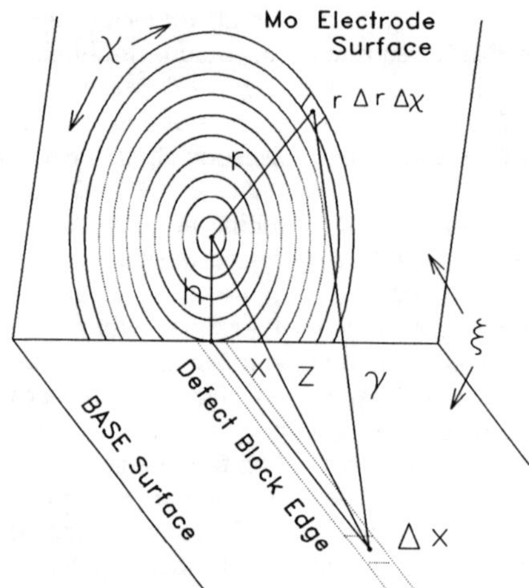

Fig. 1. Cylindrical coordinate system used to compute tunneling current to an element of BASE surface. Tunneling contributions over the Mo surface within 3 nm of x are summed. Symbols illustrate definitions in the text.

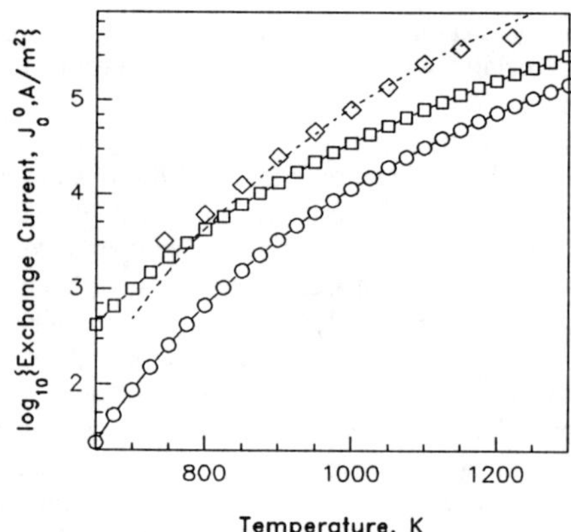

Fig. 2. Diamonds are average experimental exchange currents for Mo electrodes. Dashed line is a semi—empirical fit to dependence of the exchange current on collision frequency. $J_0^0 = B \, P(Na)/T^{0.5}$, where B = 138. Circles are calculated from the basic model; the squares include electron hopping along the defect block.

tunneling from the Mo electrode to the ions; and (4) desorption of Na^0 atoms weakly bound at the reaction site on the β''-alumina surface. Steps (1) and (2) are always faster than step (4) at the temperatures under consideration, and neither has any dependence on position. The tunnelling step (3) is very fast close to the three phase boundary line, but is slow at distances of > 1.0 nm from the three phase line, and hence defines the primary reaction area. Additional processes including (5) electron tunneling between Na^0 or Na^+ on the defect block and (6) Na^0 and/or Na^+ mobility on the spinel block surface may extend the reaction area substantially.

MODEL DEVELOPMENT AND RESULTS

For high performance 0.5μm thick Mo electrodes, following operation in AMTEC cells or sodium or potassium vapor exposure cells at temperatures from 1150-1200K for times >25hrs, a network of interconnected, fairly uniform Mo grains is formed.(see Fig. 1 of Ref. [3]) The cylindrical coordinate system shown in **Fig. 1** is used to be as consistent as possible with the observed morphology. We adopt a microscopic model in which one defect block, 0.22 nm in width, extends away from the three-phase boundary. We arbitrarily choose a perpendicular orientation of the defect block from the three phase line in **Fig. 1**, but all possible orientations exist. If only the defect block edges are reactive, the reactive length of the three phase boundary is in the range of 1.5x105 m/m2, but this value could be in error by as much as a factor of 5. Then 6.7x1014 defect block edges/m2 oriented perpendicularly to the three-phase line would be required to account for all the defect block area of BASE surface coated with porous electrode. The calculations and description are referenced to one isolated defect block in order to facilitate description of the microscopic process. In the cylindrical coordinate system, z is the distance from an area element of a defect block along a line normal to the electrode surface, x is the projection of this normal on the BASE surface, r is a radial coordinate on the electrode surface away from the normal's intersection, and χ is the angular coordinate about r. The contact angle is ξ. For acute ξ, the normal to the electrode surface is above the three-phase boundary by a distance, h, such that h completes the right triangle with x and z. For values of $r > h$, only that part of the ring of width Δr and angle χ above the three-phase boundary contributes to the tunneling current.

Na^+ Diffusion Rate into Surface Sites: The rate of sodium ion diffusion into defect block surface sites is assumed to be equal to the rate of diffusion into sites within the bulk crystal of the electrolyte, except that attempts occur from fewer neighboring sites. The activation energy, $E = 2.628 \times 10^{-20}$ J/ion, is an average of several reported measurements; the pre-exponential for conductivity is $\sigma_0 = 3 \times 10^4$ K/(ohm-cm).[15] We derive parameters for high temperature ion diffusion using equations from Sato which relate conductivity, diffusion, and attempt frequencies assuming nearest jumps of 0.324 nm between pairs of Beevers-Ross sites or between mid-oxygen sites.[20] The attempt frequency for diffusion is $\omega_0 = 7.11 \times 10^{13}$, which allows the jump frequency vs. temperature to be calculated from the Boltzmann distribution.

Na Rearrangement Probability: Although bond length changes accompanying electron transfer are very rapid compared with the residence time prior to desorption, the amount of time the participating atoms are in an appropriate configuration for electron transfer limits the probability of the transfer. The rearrangement which is most obvious and is easiest to calculate is a stretch of the Na^+ bond to oxide ions on the BASE surface, and its probability is used in both the electrode to Na^+, and Na^0 to Na^+ tunneling calculations. The equilibrium distances of Na^+ ion and the Na^0 atom to the β''-alumina surface may be estimated from atomic and ionic radii. We can also construct what should be fairly reasonable simple potential energy vs. distance diagrams for the Na^+ and Na^0 on the β'' surface, because the attractive term in the ion's case is a ion-ion interaction, and in the atom's case is an induced dipole-ion interaction. The Na^+-O^{2-} potential well is first constructed as a sum of an r^{-1} attractive term and an r^{-12} Lennard-Jones repulsive term, constraining the binding energy to be 2.18eV or 3.49×10^{-19} J/atom, and the bond length to 0.242 nm.[15] The binding energy of the Na^0 atom to the β''-alumina surface is estimated to be equal to 2/3 that of the atom in bulk sodium: 1.62×10^{-19} J/atom, and the bond distance is about 0.331 nm. The Na^+-O^{2-} potential well is then approximated to a harmonic oscillator which will give the same potential difference when a bond stretch of 0.089 nm to the Na^0-O^{2-} equilibrium distance occurs. The probability that the Na^+ ion can accept an electron is taken to be 1/2 if the ion is in one of about 30 vibrationally excited stated of the sodium ion on BASE which result in stretches to within $k_B T$ of the Na^0-O^{2-} equilibrium distance; these make contributions to the overall tunnelling probability, and their occupancy is determined from a Boltzman distribution.

Tunneling Probability and Frequency: Expressions for the density of states, $D(\epsilon)$, and electron velocity, $v(\epsilon)$, are based on the free-electron model. Transition metals are not considered well described by the free electron model, but W and Mo are among the most conductive transition metals, and their low temperature electronic contributions to the heat capacity, 1.3 and 2.0 mJ/(mol-K^2), respectively, agree well with calculated values of 1.112 and 1.099 mJ/(mol-K^2).[21] Calculations of the Mo band structure give a similar density of states at the Fermi energy, $D(\epsilon_f)$, and the band structure shows no sharp features near ϵ_f.[22] The tunneling attempt frequency, $X_e(\epsilon)$, is taken as the collision rate at the electrode surface of electrons within $\pm k_B T$ of ϵ_f, calculated from $v(\epsilon)$, $D(\epsilon)$, and the Fermi-Dirac distribution function, $F(\epsilon)$, where ϵ is electron energy, and $k_B = 1.38 \times 10^{-23}$ J/K is Boltzmann's constant.

The probability, $P_t \approx 16\epsilon_e(V-\epsilon_e)V^{-2} \exp[-4\pi\gamma(2m_e(V-\epsilon_e))^{0.5}/h]$, for electron tunnelling to Na^+ at distance γ from the electrode, is given by the equation for transmission through a rectangular barrier of height above the Fermi level equal to the electrode's work function, $\Phi \approx$

2.5 eV $= 4.00 \times 10^{-19}$ J, and a total barrier energy of $\Phi + \epsilon f \approx 3.46 \times 10^{-18}$ J, where the electron mass, $m_e = 9.11 \times 10^{-31}$ kg; and Planck's constant: $h = 6.626 \times 10^{-34}$ J-s.[23] The product of the tunneling transmission probability, $P_t(\chi, r, \epsilon, x)$, with the probability that the Na^+ is in an appropriate vibrational state, $P(Na^{+*})$, and with the tunneling attempt frequency, $W_e(\epsilon)$, for electrons near the Fermi level, gives the tunneling rate X_t to an electrolyte surface area element at distance x from the electrode, when integrated over energy, ϵ, and the cylindrical coordinates, r and χ: $X_t(\epsilon, r, \chi) = W_e(\epsilon, r, \chi) \, P(Na^{+*}(x)) \, P_t(\epsilon, r, \chi, x)$.

Na Desorption and Reaction Rate: The collision rate, $X_{Na} = P/[2*\sqrt{(m_{Na}k_BT)}]$, for sodium atoms with the surface come from the kinetic theory of gases, since gas pressure is due to momentum exchange when gas molecules strike a surface. At equilibrium, the sticking coefficient is taken as 1.0; for each colliding and adsorbed atom there is a desorbed atom.

The integral which gives the primary reaction rate (ignoring here electron tunneling from Na^0 to Na^+, and Na^+ surface diffusion) is equal to the tunnelling rate from the electrode followed by Na^0 desorption. It is calculated by evaluation at area elements over the two cylindrical spatial coordinates and the energy coordinate. The triple integral over two spatial coordinates, χ and r, in conical coordinates, and over energy, ϵ, determines the primary exchange current: $J_0^0(x,T) = {}_0\int^{2\pi} {}_0\int^\infty {}_{\epsilon F-kBT}\int^{\epsilon F+kBT} (\, 1/KT + 1/X_e(\epsilon_f))^{-1} \, d\epsilon, dr, d\chi$. The total rate integral apparently cannot be solved in closed form. The integral is readily evaluated numerically. The reaction rate contributions are obtained as a function of distance, x, and are summed over all distances and all the defect blocks, with results shown in **Fig. 2**.

Tunneling along the Defect Block: Electron migration away from the three phase region occurs by a random walk, and both the rearrangement of sodium ions prior to tunneling from sodium atoms and the desorption of sodium atoms are activated processes. We refer to this process as electron hopping, between Na^0 and Na^+ sites at a distance of about 0.324 nm. The rate of desorption is about 3000 times faster at 1300K than at 700K, while the sodium ion stretch required for rearrangement is about 30 times more probable at the higher temperature. The tunneling rate is not very temperature dependent, so the number of hops is the average residence time of the sodium atoms before desorption times the tunneling step frequency, so that about 100 times more hops occur at 700K than at 1300K. Equilibration through the gas phase and the entire BASE and electrode surfaces will tend to occur, so net Na^0 transport proceeds via a random walk, with $U(x,t)$ giving the probability that the electron has travelled to point x in time t: $U(x,t) = 1/(2\pi t)^{0.5} \exp[-x^2/2t]$. Hopping cannot occur efficiently between adjacent defect blocks separated by 1.1nm (center to center) as only about 2 sites are close enough to the adjacent defect block site to be within the range for low probability tunneling. Hopping brings the calculation to about 50% lower than experiment at 700K and a factor of three lower at 1250K, as shown in **Fig. 2**.

Surface Diffusion of Na^+: At 1200K the calculated rate including hopping is only 1/3 of the observed rate, while at 750K the calculated rate is about 50% too low. The greater high temperature discrepancy suggests that an additional activated transport process such as surface diffusion of sodium ions onto the spinel block surfaces between the defect blocks may contribute to the rate. Rapid surface diffusion of either Na^0 and Na^+ on the spinel block surface will expand the reaction zone, and is physically reasonable, but detailed calculation would involve extensive assumptions about reaction parameters, including an attempt frequency which is orientation dependent, as well as activation energies for both jumping from the defect block to the spinel block surface as well as surface diffusion on the spinel block. Since e^- hopping betwen Na^0 and Na^+ on the spinel block is expected to be as fast as on the defect block, the reaction zone should fill the region between all active defect block regions. An average activation energy for Na^+ diffusion onto and on the spinel block of 1.5×10^{-19} J/atom would double the reaction area at 750K, but would quadruple it at 1250K, well within the uncertainty of the model's parameters.

DISCUSSION

Comparison of the model results and the experimental data lead to a fairly complete picture of the reaction mechanism and remarkably good agreement with experiment as shown in **Fig. 2**. Initial tunneling from the metal electode is only important to a distance of about 1.0nm, and at closer distances desorption limits the overall rate. Hopping of electons between defect blocks is unimportant, but hopping away from the three phase boundary on each defect block is important, and contributes substantially to the overall rate, especially at lower temperatures. A significant portion of the high temperature rate involves migration of Na^+ or Na^0 onto the spinel block surface between the defect blocks, which has a higher activation energy than hopping.

There are several sources of uncertainty in the model. Morphological parameters including the length of three phase boundary, the contact angle between electrode and solid electrolyte, and the area of electrochemically active solid electrolyte, are not known with great precision but will not effect the temperature dependence of the calculated exchange current. The surface structure of the β"-alumina has been assumed to be similar to the bulk structure, but reorganization may be significant at the temperatures of the measurements. The binding energy of Na^0 to BASE surface was estimated from qualitative arguments and is not known with certainty; experimental measurement or calculation of this quantity would be very useful. Because the tunneling probablility drops very rapidly with distance, and the rate is limited by desorption at shorter distances, the rate is only weakly dependent on the rearrangment probability and attempt frequency. For accurate assessment of the exchange current, it will be necessary to include any dependence of each parameter in the rate equation on alkali metal activity; here these parameters have been approximated as constants. Additional experimental data will be required to determine what changes in the model are justified.

CONCLUSIONS

Given the experimental observation that the activation of the interfacial reaction rate is due mostly to the desorption process, and can be accounted for by a reaction zone near the three-phase interface line, the most likely reaction mechanism is electron tunnelling from the electrode to a Na^+ ion at the surface of the β"-alumina. The rate involves tunnelling through a barrier equal to the work function of the electrode surface (~2.5eV) with a barrier thickness of ≤ 1.0 nm. The alkali metal atom is produced on the BASE surface and it may desorb and diffuse in the gas phase, or it may either transfer its electron to a sodium ion further away from the boundary by hopping, or may move by activated surface diffusion on the BASE surface to increase the reaction area before desorption. Electron transfer by hopping is clearly important to explain the magnitude of the reaction rate. Surface diffusion of either Na^+ or Na^0, especially onto the spinel block surfaces near the three-phase boundary may be a slower, thermally activated process which becomes more competitive with hopping at higher temperature. Either experiment or detailed calculation must determine which surface diffusion is an important process. Na^+ ions require some vibrational activation to achieve a stretched bonding configuration to accommodate electron transfer to form the larger sodium atom. Comparison of calculated and experimental values for exchange currents for the analogous $K_{(gas)}/K$-β" alumina system with Mo or W electrodes may be expected to provide additional insights. The agreement between experiment and model suggest that the model provides realistic insight into the reaction mechanism.

ACKNOWLEDGEMENTS

The research described in this paper was performed by the Jet Propulsion Laboratory, California Institute of Technology, and was supported by the Caltech President's Fund, the National Aeronautics and Space Administration, and the Air Force Weapons Laboratory.

REFERENCES

1. R.W. Gurney, Proc. Royal Soc. A, **134**, 137 (1931)

2. H. Gerischer, Zein. Phys. Chemin N. F., **26**, 223; 325 (1960)

3. R.M. Williams, B. Jeffries-Nakamura, M.L. Underwood, C.P. Bankston, and J.T. Kummer, J. Electrochem. Soc., **137**, 1716 (1990)

4. R.M. Williams, B. Jeffries-Nakamura, M.L. Underwood, B.L. Wheeler, M.E. Loveland, S.J. Kikkert, J.L. Lamb, T. Cole, J.T. Kummer and C.P. Bankston, J. Electrochem. Soc., **136**, 893 (1989)

5. R.M. Williams, M.E. Loveland, B. Jeffries-Nakamura, M.L. Underwood, C.P. Bankston, H. LeDuc, and J.T. Kummer, J. Electrochem. Soc., **137**, 1709 (1990)

6. R.M. Williams, A. Kisor, M. A. Ryan, B. Jeffries-Nakamura, S. Kikkert, and D.E. O'Connor, 29th Intersociety Energy Conversion Engineering Conference Proceedings, **2**, 888 (1994)

7. L. Schmidt and R. Gomer, J. Chem. Phys., **42**, 10 (1965)

8. V.K. Medvedev, A.G. Naumovets, and A.G. Fedorus, Soviet Phys.-Solid State, **12**, 301 (1970)

9. R. Morin, Surface Science, **155**, 187 (1985)

10. Z. Li, R. Lamb, W. Allison, and R. Willis, Surface Science, **211/212**, 931 (1989)

11. N. Weber, Energy Conversion, **14**, 1, (1974)

12. C. Mailhe, S. Visco, and L. DeJonghe, J. Electrochem. Soc., **134**, 1121 (1987)

13. R.M. Williams, B. Jeffries-Nakamura, M.l. Underwood, D. O'Connor, M.A. Ryan, S. Kikkert, and C.P. Bankston, 25th Intersociety Energy Conversion Engineering Conference Proceedings, 2, 413 (1990)

14. Y. F. Y. Yao and J. T. Kummer, J. Inorg. Nucl. Chem., **29**, 2453, (1967)

15. P. T. Moseley, "The Solid Electrolyte" in The Sodium-Sulfur Battery edited by J.L. Sudworth and A. R. Tilley,(Chapman and Hall, New York, 1985) p44

16. R. Ditchburn and J. Gilmore, Rev. Mod. Phys., **13**, 310 (1941)

17. U. Buck and H. Pauly, Z. Phys. Chem., **44**, 345 (1965)

18. M. Knotek, Phys. Rev. B, **14**, 3406 (1976)

19. W. W. Lee, and S. Choi, J. Chem. Phys., **12**, 3884, (1980)

20. H. Sato, "Some Theoretical Aspects of Solid Electrolytes" in Solid Electrolytes edited by S.Geller,(Springer-Verlag, New York, 1977)

21. C. Kittel, Introduction to Solid-State Physics,(4th ed., John Wiley & Sons, New York, NY, 1977) pp. 239-265

22. A.R. Jani, G.S. Tripathi, N.E. Brener, and J. Callaway, Phys. Rev. B, **40** 1593-1602 (1989)

23. L. I. Schiff, Quantum Mechanics, (3rd edition, McGraw-Hill Book Co., New York, 1968) pp 102-104

IONIC DISTRIBUTION IN RARE EARTH ION-DOPED NA$^+$ β''- ALUMINA

YUHU WANG, ALASTAIR N. CORMACK
New York State College of Ceramics, Alfred, NY 14802, wangyh@bengal.alfred.edu

ABSTRACT

Structure of β''-alumina has been studied through atomistic simulation. Distribution of Mg^{2+} in the spinel blocks, Na^+ as well as Nd^{3+} in the conduction planes have been calculated and compared with the structure models derived experimentally. Mg^{2+} ions, as the stabilizer of β''-alumina structure, prefer to occupy the tetrahedral Al(2) sites in the spinel block. Two configurations for Mg^{2+} distribution are proposed. Due to the presence of Mg^{2+}, oxygen ions in or adjacent to the conduction plane, O(5), O(4) and O(3), are slightly displaced from their original sites, leading to the removal of the local symmetry at the mO site. Strongly affected by the Mg^{2+} distribution, Nd^{3+} ions, originally introduced into the BR site in the conduction plane, are found to move towards the distorted mid-oxygen sites, which is in excellent agreement with the previous estimate from spectroscopic study.

INTRODUCTION

β''-alumina is of great interest due to its application as the solid electrolyte material in Na/S batteries. Stoichiometric β''-alumina has a chemical composition of $Na_2MgAl_{10}O_{17}$ but, normally, it appears in a non-stoichiometric form, typically with a composition of $Na_{1.66}Mg_{0.66}Al_{10.33}O_{17}$. The structure of β''-alumina has rhombohedral symmetry, space group, $R\bar{3}m$. There are three spinel-like blocks, which are sandwiched by three conduction planes, in a unit cell. The spinel blocks have close packed oxygen layers with Al and Mg atoms and are structurally similar to the mineral spinel ($MgAl_2O_4$). The interlayer conduction plane has a more open structure, where two Na^+ and one bridging oxygen, O(5) are present. The spinel blocks are held together by the O(5), around which Na^+ ionic migration takes place through two dimensional honeycomb-like pathways. Due to this disordered, liquid like feature, the conventional diffraction techniques have encountered intrinsic difficulties in providing detailed structural information.

In early X-ray studies, Bettman and Peters [1] suggested that Mg^{2+} is located in the vicinity of the conduction plane. Later neutron diffraction studies, however, showed that Mg^{2+} prefers the tetrahedral Al(2) sites in the spinel blocks [2]. How Mg is distributed among the Al(2) sites remains unclear. For the cations in the conduction planes, there are two non-equivalent sites available: the seven- coordinated Beevers-Ross site (BR) and the eight-coordinated mid-oxygen site (mO), as shown in Fig. 1. These two sites are different in that the BR has no inversion symmetry but the mO site has. In spite of extensive studies, the distribution of cations in the conduction plane has been more controversial, mostly because of the liquid-like feature of the conduction plane. It is thought that Na^+ are distributed among the BR sites and a position midway between the BR and mO sites, the 18h site. For other cations, e.g., rare earth dopants, some studies suggested a preference for the mO site but others have argued for the BR site [3, 4, 5].

Since the ideal mO site has a C_{2h} point group symmetry, and is a center of an inversion symmetry, electric-dipole transitions are forbidden and no strong absorption should be observed in the optical spectra, unless the symmetry is broken, for example, by local environmental changes. The experimentally observed strong absorption with Nd^{3+} doped β''-aluminas, which is dependent on the dopant concentration, is difficult to explain if all the Nd^{3+} are located in the ideal mO or BR sites [6]. To be consistent with experimental observation, Alfrey *et al.* proposed a model in which the bridging oxygen O(5) and the Nd^{3+} relaxed slightly off their original sites and thus the inversion symmetry was broken [6]. In this study, we examined the structure of Nd(III) β''-

Mat. Res. Soc. Symp. Proc. Vol. 453 © 1997 Materials Research Society

alumina through detailed static lattice energy calculations, with a focus on the ionic distribution and the local environment change upon the substitution of Na^+ with Nd^{3+}.

SIMULATION METHODS

The simulation methods are the exactly the same as reported previously [7]. They are based on the Born model description of solid, which treats the solid as a collection of point ions with short range forces acting between them. The potential and other parameters used in this studies were taken from the compilation of Lewis and Catlow [8] and were calibrated with consideration of the coordination number within the spinel blocks.

Lattice energy was calculated in the Born model of the solid and then minimized through a second derivative Newton-like procedure, coded into METAPOCS. Calculation of the equilibrium atomic configuration involves adjusting the coordinates until the internal basis strains are totally removed. The defect energy calculation were carried out with the two region treatment. The outer region is treated as a polarizable dielectric continuum, while the atomic coordinates of the distorted inner region are explicitly relaxed using appropriate interatomic potentials. The theory of defect energy calculation has been outlined by Catlow *et al.* [9] and is coded into CASCADE.

Calculations were performed on either the Silicon Graphics Workstations in our laboratory or the IBM SP2 Parallel Supercomputer at the Cornell Theory Center, Cornell University, depending on the lattice structure input models chosen.

SIMULATION RESULTS AND DISCUSSION

<u>Mg distribution in the spinel blocks</u>

The structure derived by Bettman and Peters through single crystal X-ray diffraction on a non-stoichiometric β''-alumina ($Na_{1.66}Mg_{0.66}Al_{10.33}O_{17}$) was used for the initial input model. But our calculation was started from a stoichiometric lattice $Na_2MgAl_{10}O_{17}$. Since all the BR sites are occupied, we only need to consider the distribution of Mg^{2+}, which turned out to be a crucial factor, affecting other ions. There are altogether four non-equivalent Al sites that Mg^{2+} may occupy. Only the lowest energy configuration has Mg ions in each spinel blocks in the single unit cell. Our calculation shows that the three Mg^{2+} prefer to disperse in the tetrahedral Al(2) sites as evenly as possible in reaching the most stable state. Location of Mg^{2+} in any other sites will cause an increase of the lattice energy by about 1~4 eV (Table 1), implying that other Al sites are not favorable for Mg^{2+}. This is in agreement with the neutron diffraction observation.

Table 1 Lattice energy of β''-alumina

Mg position	type of site	single cell lattice	supercell Model I	supercell Model II
Al(1)1,7,13	octahedral	-2596.47		
Al(2)2,4,6	tetrahedral	-2598.98	-7796.94	
Al(3)2,4,6	tetrahedral	-2594.13		
Al(4)1,2,3	octahedral	-2597.67		
Al(2)1,2,3,4,5,6	tetrahedral			-7798.59

However, if we establish a $\sqrt{3}a \times \sqrt{3}a \times c$ triple superlattice structure as shown in Figs. 2 (a, c are the lattice constants of single unit cell), there are two possible, non-equivalent configurations for Mg^{2+} to be evenly distributed in the Al(2) sites. In the following discussion, we refer to Fig. 2a as Model I, and Fig. 2b as Model II. In Fig. 2a, three Mg ions are introduced into each spinel block and occupy either the Al(2) 2, 4, 6 positions, or equivalently, Al(2) 1, 3, 5 positions. In the same plane, all the Mg^{2+} are separated by equal distance (Fig. 2a). While there is no energetic difference between the basic unit cell and the triple supercell for Model I, the configuration of Model II is only available with a triple lattice, as shown in Fig. 2b. Similarly to Model I, three Mg ions are introduced into each spinel block. But they occupy the Al(2) 1, 2, 3, 4, 5, 6 positions instead. In the same plane, Mg ions are no longer separated by equal distances, rather they form three-membered Mg clusters which constitute a 2D triple cell. The calculated lattice energies for Models I and II (also shown in Table 1) indicates that the Model II configuration is slightly more stable.

For the non-stoichiometric β''-alumina, in a single unit cell, it is not possible to allow the two Mg^{2+} ions to be evenly distributed into the three spinel blocks, the one Na vacancy into three conduction planes. A $\sqrt{3}a \times \sqrt{3}a \times c$ triple superlattice (derived either from Model I or II) is necessary. The most stable configuration turned out to be that which takes one Na from each conduction plane and one Mg from each Mg plane in the triple superlattice, for both Models I and II.

The conduction planes for both the stoichiometric and non-stoichiometric β''-alumina are different in that all the Na sites on the stoichiometric conduction plane are equivalent, whilst two kinds of Na^+ site are created on the non-stoichiometric plane: one is adjacent to the V_{Na} and the other is not. The formers tend to relax towards to the V_{Na} and is located close to the 18 h position. Such a Na vacancy distribution is in good agreement with observation of X-ray diffraction by Boilot *et al* [10].

Nd^{3+} Preferred Site

With the structure models we established for the unexchanged parent β''-alumina, we were then able to calculate the position of Nd^{3+} exchanged for Na^+ in the conduction plane. Basically, to introduce one Nd^{3+}, one needs to remove three Na^+ for charge neutrality. But for a more complete analysis, the following substitution processes which include defect interactions are considered:

$$Na_{plane} \rightarrow V_{Na} + Na^+ \qquad\qquad E_1 \qquad (1)$$

$$Nd^{3+} + Na_{plane} \rightarrow Nd_{plane} + Na^+ + V_{Na} \qquad E_2 \qquad (2)$$

$$Nd^{3+} + 2Na_{plane} \rightarrow Nd_{lattice} + 2Na^+ + 2V_{Na} \qquad E_3 \qquad (3)$$

$$Nd^{3+} + 3Na_{plane} \rightarrow Nd_{plane} + 3Na^+ + 3V_{Na} \qquad E_4 \qquad (4)$$

In process (1), a single V_{Na} is formed, whilst in processes (2) or (3), one or two Na ions are replaced by one Nd^{3+}. Note that these configurations are not electrostatically neutral; there is no requirement for this as long as other V_{Na} are simultaneously created, to keep the crystal on a whole charge balanced. While many configurations for V_{Na} are possible, those reaction energy E_i summarized in Table 2 are calculated for the most simple, but basic, configurations, i.e. the V_{Na} are adjacent to each other. For example, the E_3 value is for the process where Na at positions 1 and 2 are replaced by a Nd (Fig. 3); the E_4 is for the process where Na at positions 1, 2, 3 are replaced by a Nd (Fig. 5).

Table 2. Reaction energy E_i (eV) of substitution processes

Reaction energy type	Model I	Model II
E_1	4.484, 4.424	4.447, 4.468
E_2	-29.72	-29.78
E_3	-26.27	-26.29
E_4	-21.88	-21.72

The X-ray derived space group symmetry requires that all the BR site be equivalent [1]. However, because of the distribution of Mg, our calculations on the Na vacancy energy gave two different values for E_1, especially in Model I, implying that the Na sites are not exactly equivalent.

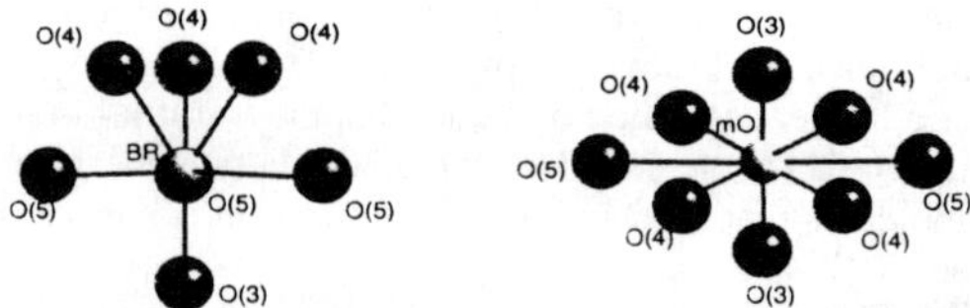

Fig. 1. Beevers-Ross site(BR) and mid-oxygen site (mO) in the conduction plane of β"-alumina

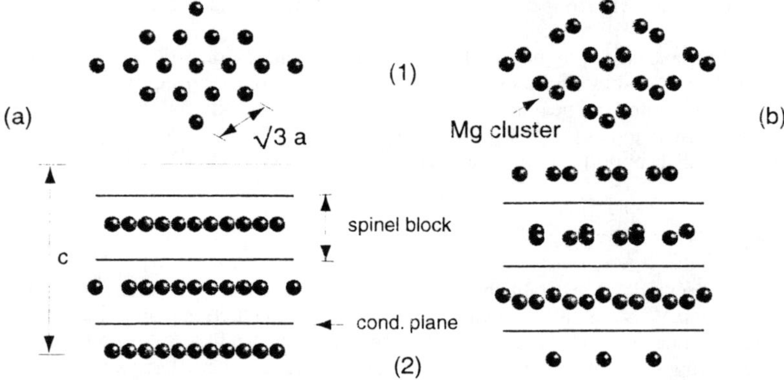

Fig. 2. Configurations of Mg (a) Model I (b) II, planar view (1) and [110] direction view (2)

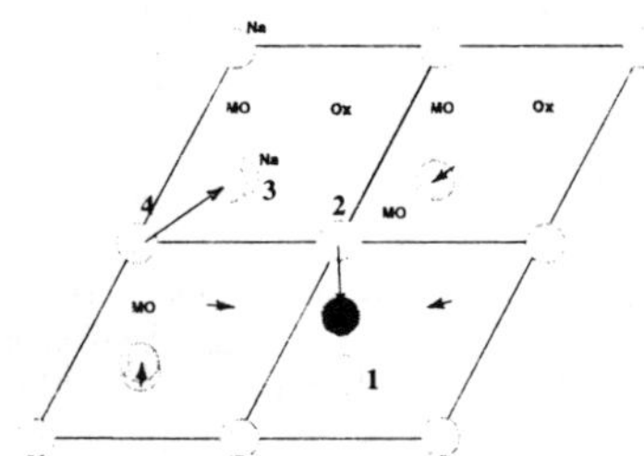

Fig. 3. Subsitution process for Na by Nd in the conduction plane

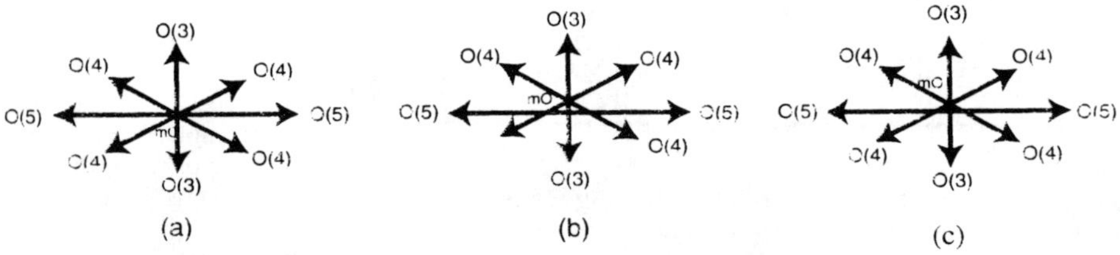

Fig 4. (a) mO site in ideal β"-alumina. (b) Model I and (c) II

Since the difference among E_1 values is less evident for Model II, we can suggest that the Na is more uniformly distributed in the Model II structure. Although no experimental data are available to make comparison, our calculations indicate that the existence of Mg^{2+} actually causes the crystal structure to be distorted, with the oxygen ions shifted from their ideal positions. This can be seen from the magnified mO sites (Fig. 4). Whilst the mO site in the X-ray derived structure is a center of symmetry(Fig. 4a), there is no inversion symmetry at the mO sites in either Model I or II, due to the local displacement of oxygen, especially that of the O(3), from their original sites. Model I (Fig. 4b) shows more significant displacement than Model II (Fig. 4c). Such a displacement is caused by Mg^{2+} and results in the non-equivalent BR positions in the conduction plane.

For the stoichiometric lattice, Table 2 shows that E_3 is lower than (E_1+E_2) by about 1 eV, for either Model I or II, which implies that the defect structure in which two adjacent V_{Na} are associated with one dopant Nd is more stable than the defects V_{Na}, Nd_{Na} formed independently. Process (2) and (3) can be treated as extreme cases of (4) when some V_{Na} are created simultaneously far away from Nd^{3+}. For Process (4), which describes the stoichiometric ion exchange process of Na by Nd, the simplest one is to replace three adjacent Na^+ with one Nd^{3+}, as depicted in Fig. 3. This turned out to be an unstable configuration: the second-nearest Na vacancy, position 3, became occupied by another Na^+, which relaxed from position 4. Further calculations confirmed that the Nd^{3+} tends to be associated with only two nearest Na vacancies, with the third V_{Na} remaining distant from the Nd^{3+}. This means that the Nd^{3+} will not at the center of the triple Na^+ vacancy cluster.

Only in process (2), where one Na is replaced by one Nd^{3+}, does Nd stay at the original Na site, the seven-coordinated BR site, because of the highly symmetric configuration. Otherwise, as in processes (3) and (4), the Nd^{3+} always relaxes off the BR site towards the mO site, as long as there are two adjacent V_{Na} available.

Displacement from the Ideal Sites

The result that the eight-coordinated mO site is preferred by Nd^{3+} is consistent with the fact that the rare earth ions normally tend to reside in high coordination sites in oxides. The actual equilibrium Nd^{3+} site, however, is displaced from ideal mO position, and by considerably different amounts for Models I and II. From above discussion, it is clear that even in the unexchanged parent β"-alumina, the inversion symmetry of the mO site has already been broken, due to the displacement of each surrounding oxygen from their original positions. Therefore, even without further relaxation of nearest oxygen, the ion exchanged Nd^{3+} will not sit at the center of an inversion symmetry. Actually, the introduction of Nd^{3+} causes further displacement of the surrounding oxygen non-uniformly. In Model I, Nd relaxes off the mO site by about 0.3Å but in

Table 3. Interionic distance between Nd and nearest oxygen

	Interatomic Distance (Å)			
	X-ray [4]	EXAFS [5]	Model I	Model II
Nd-O(5)	2.460	2.458	2.487	2.512, 2.506
Nd-O(4)	2.758	2.686	2.785, 2.645	2.665, 2.649, 2.71, 2.744
Nd-O(3)	2.781	2.724	3.104, 2.424	2.907, 2.548

Model II, the displacement is reduced to 0.17 Å. Based on their analysis on the polarizing absorption spectra, Afrey et al. proposed a model in which the displacement of Nd^{3+} is set at about 0.2Å and that of O(5) 0.3~0.5Å to fit the experimental data [6]. Our calculated displacement of Nd^{3+} from Model II is in excellent agreement with their model.

Another feature is the displacement of all the nearest oxygen atoms around the Nd. Displacement of the bridging oxygen, O(5), toward the Nd has already been clarified by single crystal X-ray diffraction and EXAFS studies[4, 5]. Our calculations shows that the Nd...O(5) ionic distance shortens by 0.39Å, which is again in excellent agreement with the value from those experimental studies. Our results also show the displacement of other surrounding oxygens. In either Model, we see significant displacement of one nearest O(3) towards the Nd^{3+}, which contributes most to the removal of the symmetry at the mO site. Comparing between Models I and II, the displacement of O(5) and O(3) are seen to be similar but that of O(4) are different. Although neither of them are uniform, the relaxation of O(4) is more uniform in Model II than Model I, implying again that the strain for Model II structure should be less. Table 3 summarizes our calculated data and those obtained from experiments. The primary difference is that the O...Nd bond length is not uniform in our model, whereas space group imposed symmetry requires this to be the case, experimentally.

CONCLUSIONS

Our atomistic calculations on β"-alumina show that in the unexchanged parent crystal, due to the existence of Mg^{2+} in the spinel blocks, the crystal structure is distorted and the inversion symmetry at mO site has been broken. Ion exchanged Nd^{3+} causes the surrounding oxygen to be further displaced, non-uniformly. Nd^{3+} sitting at the eight-coordinated, distorted mO site, associated with two adjacent Na vacancies is a stable and favorable defect structure. With this structure model, optical properties of Nd^{3+} doped β"-alumina can be well explained.

ACKNOWLEDGMENTS

We are grateful to the U.S. Department of Energy for financial support under grant number DE-FG02-91ER45451. Some of the calculations were performed at Cornell Theory Center.

REFERENCES

1. M. Bettman and C.R. Peters, J. Phys. Chem., **73**[6], p. 1774 (1969).

2. J.O. Thomas and G.C. Farrington, Acta Cryst., **B39**, p.227 (1983).

3. P.Davies, A. Petford and M. O'Keefe, Solid State Ionics, **18-19**, p.624 (1986).

4. W. Carrillo-Cabrera, J.O. Thomas and G.C. Farrington, Solid State Ionics **28-30**, p.317 (1988).

5. F.Rocca, A. Kuzmin, J. Purans and G. Mariotto, Phys. Rev., **B50**, p.6662 (1994).

6. A.J. Alfrey, O.M. Stafsudd, B. Dunn and D.L. Yang, J. Chem. Phys., **88**, 707 (1988).

7. J.G. Park and A.N. Cormack, J. Solid State Chem., **121**, p. 278 (1996).

8. G.V. Lewis and C.R.A. Catlow, J.Phys. C: Solid State Phys., **18**, p.1149 (1985).

9. C.R.A. Catlow, R. James, W. C. Mackrodt, and R.F. Stewart, Phys. Rev. B: Condens. Matter, **25**, 1006(1982).

10. J.P. Boilot, G. Collin, Ph. Colomban and R. Comes, Phys. Rev. **B22**, p.5912 (1980).

CONDUCTION PROPERTIES OF Na+-ION IMPLANTED GLASS-CERAMICS IN THE SYSTEM Na_2O-R_2O_3-P_2O_5-SiO_2 (R=RARE EARTH)

TOSHINORI OKURA, MIKI TANAKA, GIICHI SUDOH
Department of Applied Chemistry, Faculty of Engineering, Kogakuin University,
Nishi-shinjuku 1-24-2, Shinjuku-ku, Tokyo 163-91, Japan

ABSTRACT

The glass-ceramics of the phosphorus containing $Na_5RSi_4O_{12}$-type (R=rare earth) Na^+-superionic conductors (Narpsio-V) in the system Na_2O-Sm_2O_3-P_2O_5-SiO_2 were prepared by crystallization of glasses with the composition $Na_{3+3x-y}Sm_{1-x}P_ySi_{3-y}O_9$. The optimum conditions for crystallization were discussed with reference to the conduction properties and the preparation of uncracked bulky glass-ceramic Narpsio-V. The ionic conductivity of the glass-ceramic $Na_{3.9}Sm_{0.6}P_{0.3}Si_{2.7}O_9$ at 300°C was drastically enhanced from 2.38×10^{-2} S/cm to 7.15×10^{-2} S/cm upon implantation of 200 keV Na^+ ions to a fluence of 1×10^{15} cm^{-2}.

INTRODUCTION

The glass-ceramics of the phosphorus containing $Na_5RSi_4O_{12}$-type (R=rare earth) [1,2] Na^+-superionic conductors (Narpsio-V) [3,4] have been developed by crystallization of glasses with the composition $Na_{3+3x-y}R_{1-x}P_ySi_{3-y}O_9$ [5-7]. These materials have been reported as being comparable to the conventional ceramic Na^+-conductors such as NASICON, β- and β''-Aluminas. The R elements have a significant effect on the crystallization of glasses [6], as well as on the conduction properties [1]. To date, polycrystalline Narpsio-V has been obtained with Sc, Y, Gd or Sm as the R element. These Narpsio-V compounds will be referred to as Nascpsio-V, Naypsio-V, Nagpsio-V and Naspsio-V, respectively, in the following. The size of the R ions has been expected to have a significant effect on the crystallization of the phase. Although the precise structure analysis of the silicophosphate Narpsio-V has not yet been completed, it is currently assumed from the analogy with $Na_5RSi_4O_{12}$ that all the R ions can be octahedrally coordinated with the non-bridging oxide ions of the (SiO_4, PO_4)-tetrahedra of the 12-membered rings. The reported results on the silicate ceramics [1] show that the conductivity of Narpsio-V increases with increasing size of its R ions, giving the order Naspsio-V>Nagpsio-V>Naypsio-

Mat. Res. Soc. Symp. Proc. Vol. 453 © 1997 Materials Research Society

V>Nascpsio-V. It can be expected that Naspsio-V is the most conductive [8]. However, this order was not always true in glass-ceramics [9]. Although most of the Narpsio-V compounds were obtained as uncracked bulky glass-ceramics, Naspsio-V was difficult to prevent from cracking during crystallization. It was found that uncracked Nagpsio-V with larger Gd^{3+} ions was the most conductive; however, Naspsio-V with the largest Sm^{3+} ions was less conductive than Naypsio-V with medium Y^{3+} ions. In the present study, the Naspsio-V ionic conductors were prepared by crystallization of glasses. The optimum conditions for crystallization were discussed with reference to the conduction properties and the preparation of uncracked glass-ceramic Naspsio-V. We consider that a material processing technique to introduce a large amount of mobile Na^+ ions into samples is required to realize superionic conduction in Narpsio-V compounds. Here we report that a large enhancement of electrical conductivity, probably due to Na^+ ions, has been obtained in the glass-ceramic Naspsio-V by ion implantation of Na^+ ions.

EXPERIMENT

Samples were prepared according to the chemical formula mentioned above of $Na_{3+3x-y}Sm_{1-x}P_ySi_{3-y}O_9$. The precursor glasses were made by melting mixtures of reagent-grade powders of anhydrous Na_2CO_3, Sm_2O_3, SiO_2 and $NH_4H_2PO_4$ at 1300°C for 1 h, followed by annealing for several hours at an optimum temperature (ca. 25°C below the glass transition temperature (Tg)) determined by DTA analysis. The temperatures employed for nucleation and

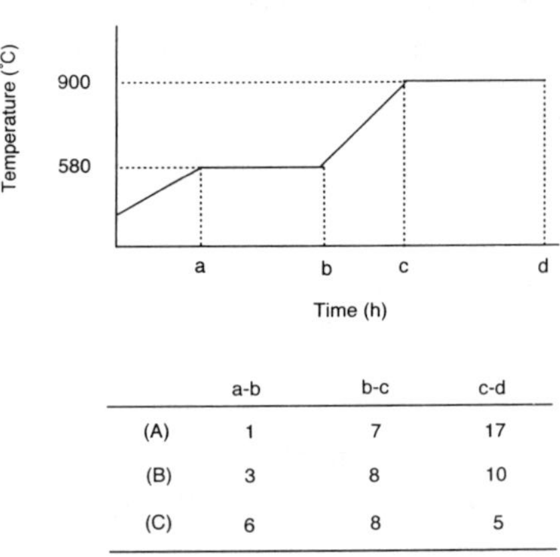

	a-b	b-c	c-d
(A)	1	7	17
(B)	3	8	10
(C)	6	8	5

Figure 1. The program of temperature and time for the production of glass-ceramic Naspsio-V.

crystallization of glass specimens were also determined by the results of DTA analysis. Figure 1 shows the program of temperature and time for the production of glass-ceramic Naspsio-V employed in the present work. Crystalline phases were identified by the X-ray diffraction (XRD) method. Microstructure was investigated with SEM. Substrate Naspsio-V compounds used for ion implantation were the glass-ceramic $Na_{3.9}Sm_{0.6}P_{0.3}Si_{2.7}O_9$. Approximately 5 mm thick glass-ceramic disks were implanted with 200 keV Na^+ ions to fluences of 10^{14} to 10^{15} cm^{-2} at room temperature. The dose rate of each ion was $3\,\mu A/cm^2$. The depth profiles of implanted Na^+ ions was calculated with the LSS theory. Ionic conductivities of sintered and Na^+-ion implanted disks were measured by the ac two-probe method with Au-sputtered blocking electrodes with a LF impedance analyzer. The conductivities of grains and grain boundaries were respectively calculated from the intercepting points of two semicircles on the horizontal in a complex admittance diagram. The temperature dependence of the conductivity was measured in a similar way at several temperatures ranging from room temperature to 350°C.

RESULTS AND DISCUSSION

The Naspsio-V ionic conductors were successfully produced by crystallization of glasses. Although the glass samples heated by the program pattern (A) shown in Fig. 1 broke during crystallization and the glass-ceramic Naspsio-V obtained by the pattern (B) was difficult to prevent from cracking during crystallization, most of the Naspsio-V compounds by the pattern (C) were obtained as uncracked bulky glass-ceramics (the glass samples broke during crystallization when heating time for crystallization was over 5 h). Figure 2 shows the phase-composition diagram of samples crystallized at 900°C by the pattern (C). The crystallization of the single-phase grass-ceramic Naspsio-V was dependent strongly on the concentrations of both |R| and |P| (or x and y in the composition parameters) and the temperature for crystallization of glass specimens. Figure 3 shows SEM photograph of microstructure of specimen with the $Na_{3.9}Sm_{0.6}P_{0.3}Si_{2.7}O_9$ composition heated at 900°C by the pattern (C). The grain size of the

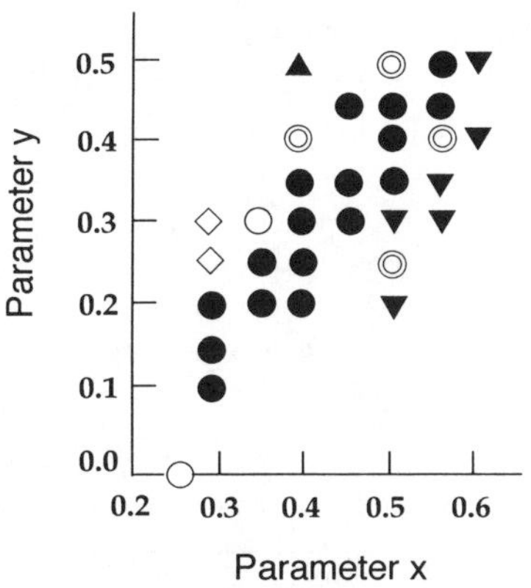

Figure 2. The phase-composition diagram of sample crystallized at 900°C by the pattern (C). The iso-structural compounds with $Na_3SmSi_3O_9$, $Na_5SmSi_4O_{12}$ and $Na_9SmSi_6O_{18}$ containing phosphorusare referred to as N3, N5 and N9, respectively. ●,N5; ▲,N3; ▼,N9; ○,N5+N3; ◎,N5+N9; ◇,N3+N9

Figure 3. SEM photograph of specimen with the $Na_{3.9}Sm_{0.6}P_{0.3}Si_{2.7}O_9$ composition heated at 900°C by the pattern (C).

specimen was about $3\sim5\,\mu$m. The grain growth is promoted with increase of heating temperature and heating time for crystallization. Although grain growth may cause high conductivity, it was difficult to prevent the sample heated for a long time from cracking during crystallization. Studies are underway to produce an uncracked sample. Figure 4

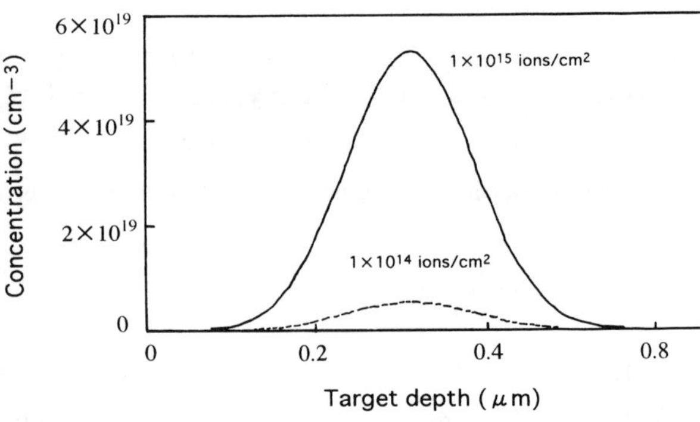

Figure 4. Calculated depth profiles of implanted Na$^+$ ions with the LSS theory.

shows calculated depth profiles of implanted Na+ ions with the LSS theory. The calculated range is $\sim$0.8$\,\mu$m for 200 keV Na+ ions. Figure 5 shows the XRD pattern of the Na+-ion implanted specimen with the $Na_{3.9}Sm_{0.6}P_{0.3}Si_{2.7}O_9$ composition. The single phase of Naspsio-V was

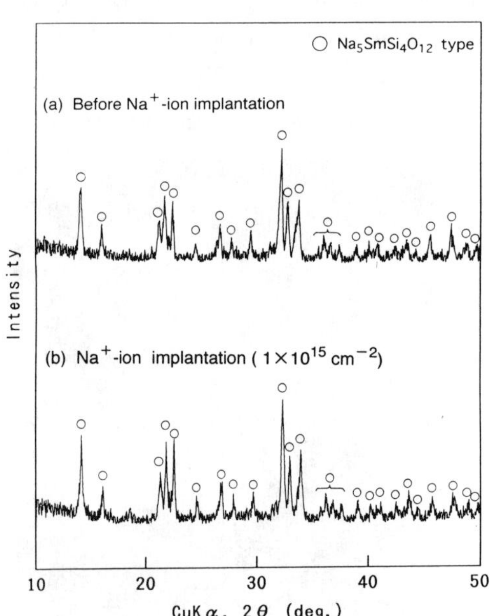

Figure 5. X-ray diffraction patterns of specimens with the $Na_{3.9}Sm_{0.6}P_{0.3}Si_{2.7}O_9$ composition heated at 900 °C by the pattern (C).

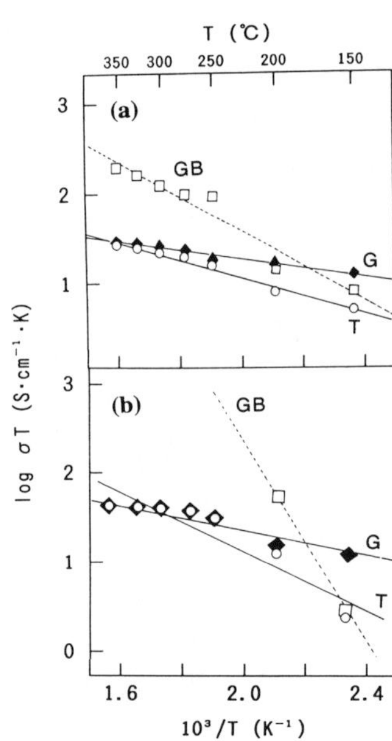

Figure 6. Arrhenius plots of conductivities of grains (G), grain boundaries (GB) and total (T) for the specimen with the $Na_{3.9}Sm_{0.6}P_{0.3}Si_{2.7}O_9$ composition heated at 900℃ by the pattern (C). (a) Before Na$^+$-ion implantation; (b) Na$^+$-ion implantation (1 × 10¹⁵ cm^{-2})

obtained. No cracking was perceived for all the implanted surfaces. Figure 6 shows the temperature dependence Arrhenius plots of conductivity for the Na^+-ion implanted specimen with the $Na_{3.9}Sm_{0.6}P_{0.3}Si_{2.7}O_9$ composition. The activation energy obtained by the conductivity of the grain boundaries was 104.1 kJ/mol, while that by the sum of the conductivities of grains and grain boundaries was 33.3 kJ/mol. Table 1 summarizes the conduction properties of the glass-ceramic Naspsio-V specimens. The sample with the $Na_{3.9}Sm_{0.6}P_{0.3}Si_{2.7}O_9$ composition crystallized at 900°C for 5 h showed the ionic conductivity of 2.38×10^{-2} S/cm at 300°C. It was

Table 1. Conduction properties of the glass-ceramic Naspsio-V specimens.

Mix proportion		Heat-treatment		Dose	σ_{300}	Ea (kJ/mol)		
x	y	Temp.	Time	(ions/cm^2)	($\times 10^{-1}$S/cm)	T	G	GB
0.40	0.30	900	5	Before Implantation	0.238	27.6	17.9	51.4
0.40	0.30	900	5	1×10^{14}	0.512	36.7	16.9	88.8
0.40	0.30	900	5	1×10^{15}	0.715	33.3	16.3	104.1

σ_{300} : conductivity at 300°C   Ea : activation energy

T : total   G : grain   GB : grain boundary

found that Naspsio-V containing the largest Sm^{3+} ions was less conductive than Naypsio-V with medium Y^{3+} ions as the grain sizes of the presented specimens were very small. The ionic conductivity of the glass-ceramic $Na_{3.9}Sm_{0.6}P_{0.3}Si_{2.7}O_9$ at 300°C was drastically enhanced from 2.38×10^{-2} S/cm to 7.15×10^{-2} S/cm upon implantation of 200 keV Na^+ ions to a fluence of 1×10^{15} cm^{-2}.

CONCLUSIONS

The uncracked glass-ceramic Naspsio-V superionic conductors were successfully produced by crystallization of glasses with the composition $Na_{3+3x-y}Sm_{1-x}P_ySi_{3-y}O_9$. The crystallization of the single-phase grass-ceramic Naspsio-V was dependent strongly on the concentrations of both |R| and |P| and the temperature employed for crystallization of glass specimens. The ionic conductivity of the glass-ceramic $Na_{3.9}Sm_{0.6}P_{0.3}Si_{2.7}O_9$ was 2.38×10^{-2} S/cm at 300°C. The Naspsio-V was less conductive than Naypsio-V as the grain sizes of the presented specimens were very small. Although grain growth may cause high conductivity, it was difficult to prevent the sample heated for a long time from cracking during crystallization. Large enhancement of electrical conductivity has been observed in the glass-ceramics by ion implantation of Na^+ ions at 200 keV to a fluence of 1×10^{15} cm^{-2}. Upon implantation ionic conductivity at 300°C was drastically increased from 2.38×10^{-2} S/cm to 7.15×10^{-2} S/cm.

ACKNOWLEDGMENTS

One of the authors (T. Okura) would like to express his thanks to the New Energy Development Organization (NEDO), the Ministry of International Trade and Industry of Japan, and the Showa-hokokai Foundation for their support to part of this research.

REFERENCES

1. R. D. Shannon, B. E. Taylor, T. E. Gier, H. Y. Chen and T. Berzins, Inorg. Chem., **17**, 958 (1978).

2. H. U. Beyeler and T. Himba, Solid State Commun., **27**, 641 (1978).

3. K. Yamashita, T. Nojiri, T. Umegaki and T. Kanazawa, Solid State Ionics, **35**, 299 (1989).

4. K. Yamashita, T. Nojiri, T. Umegaki and T. Kanazawa, Solid State Ionics, **40/41**, 48 (1990).

5. K. Yamashita, S. Ohkura, T. Umegaki and T. Kanazawa, Solid State Ionics, **26**, 279 (1988).

6. K. Yamashita, S. Ohkura, T. Umegaki and T. Kanazawa, J. Ceram. Soc. Jpn., **96**, 967 (1988).

7. K. Yamashita, T. Umegaki, M. Tanaka, T. Kakuta and T. Nojiri, J. Electrochem. Soc., **143**, 2180 (1996).

8. T. Okura, M. Tanaka, H. Kanzawa and G. Sudoh, Solid State Ionics, **86-88**, 511 (1996).

9. K. Yamashita, M. Tanaka, T. Kakuta, M. Matsuda and T. Umegaki, J. Alloys and Compounds, **193**, 283 (1993).

PREPARATION AND ELECTRICAL CONDUCTIVITY MEASUREMENT
OF LiSbO₃ THIN FILMS

K. OZAWA, Y. SAKKA, M. AMANO
National Research Institute for Metals, 1-2-1, Sengen, Tsukuba-shi, Ibaraki, 305 Japan

ABSTRACT

$LiSbO_3$ thin films have been prepared by the sol-gel process with metal alkoxides using a spin-coating method. The (011)-oriented and randomly oriented $LiSbO_3$ thin films are obtained by the precursor film crystallizing under an atmosphere of a flowing mixture of water vapor and oxygen, and an air atmosphere, respectively. The two different atmospheres also affect the crystallization temperature of the films. The electrical conductivity of the (011)-oriented $LiSbO_3$ thin film is approximately one order of magnitude larger than that of the randomly oriented $LiSbO_3$ thin film in the temperature range of 380 to 600°C.

INTRODUCTION

$LiSbO_3$ exhibits lithium ion conduction. However, there are very few papers dealing with the electrical conductivity of $LiSbO_3$, except for one which discussed the lithium ion conductivity of the sintered specimen in the temperature range of 500 to 1000°C [1]. Furthermore, $LiSbO_3$ is important as the starting material for the synthesis of monoclinic antimonic acid, $Sb_2O_5 \cdot nH_2O$. Monoclinic antimonic acid, which shows not only high ion-exchange capacity and selectivity for lithium ion but also high protonic conduction at a room temperature [2], can be synthesized uniquely by Li^+ / H^+ ion exchange reaction with a concentrated nitric acid solution from $LiSbO_3$ [3]. The crystal structure of $LiSbO_3$ is orthorhombic [4], while monoclinic antimonic acid has basically the same crystal structure as $LiSbO_3$. Several methods on the synthesis of $LiSbO_3$ have been reported; $LiSb(OH)_6$ prepared from $LiOH$ and $SbCl_5$ is heated at 900°C [3], and Li_2CO_3 is reacted with Sb_2O_3 or Sb_2O_5 at 900°C [5]. However, it seems difficult for these methods to produce $LiSbO_3$ fine particles or a $LiSbO_3$ film suitable for electrical conductivity measurement, owing to the high temperature-treatment as high as 900°C.

We have already reported that $LiSbO_3$ fine particles and $LiSbO_3$ films can be synthesized by the sol-gel process with lithium n-propoxide, $LiO\text{-}n\text{-}C_3H_7$, and antimony n-propoxide, $Sb(O\text{-}n\text{-}C_3H_7)_3$, [6], [7]. In the present study, $LiSbO_3$ thin films suitable for electrical conductivity measurement are prepared by the sol-gel process, and the morphologies and the electrical conductivity of the films are discussed.

EXPERIMENT

Preparation of Films

Figure 1 illustrates the flow diagram for the preparation of $LiSbO_3$ films. Equal molar amounts of $LiO\text{-}n\text{-}C_3H_7$ and $Sb(O\text{-}n\text{-}C_3H_7)_3$ were dissolved in 1.34 mol of absolute n-propanol to give the 2.0 mol% mixed alkoxide solution. The solution was added with a very small amount of acetylacetone (below 5×10^{-3} mol), then refluxed at 80°C for 24h under a nitrogen gas atmosphere (solution-1). As shown in Fig. 1, two different procedures, namely, Procedure-1 and -2, were employed for the preparation of $LiSbO_3$ thin films in this study. Each procedure is as

Mat. Res. Soc. Symp. Proc. Vol. 453 ©1997 Materials Research Society

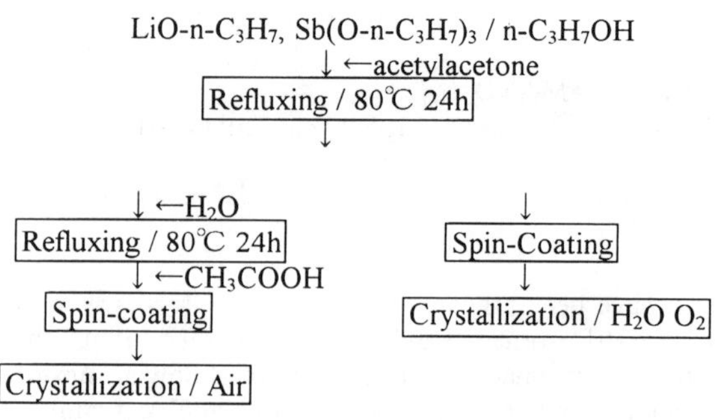

Fig. 1 Flow diagram for the preparation of $LiSbO_3$ thin films.

follows. Procedure-1: Solution-1 was added with H_2O (0.03 mol) dropwise sufficiently slowly so that the solution did not become turbid, then refluxed again at 80℃ for 24h to hydrolyze the alkoxides. After cooled, the solution was added with the mixture of CH_3COOH (0.02 mol) and n-C_3H_7OH (0.134 mol) (solution-2). The precursor films were prepared on a quartz glass substrate using solution-2 by a spin-coating method and subsequently crystallized at a temperature ranging from 300 to 800℃ for 30min under an air atmosphere. Procedure-2: Using Solution-1 the precursor films were prepared on a quartz glass substrate by a spin-coating method under a nitrogen gas atmosphere. The resulting precursor films were crystallized at a temperature ranging from 300 to 800℃ for 30min in a flowing mixture of water vapor and oxygen. For both Procedure-1 and -2 the substrates ($1 \times 1 cm^2$) were rotated at a constant speed of 1×10^4rpm during the spin-coating process, and the spin-coating and crystallization cycle was repeated several times, as required, to have a desired film thickness. Prior to coating, the substrates were cleaned in a mixed solution of acetone and benzene, an aqueous HCl solution and deionized water in an ultrasonic bath, then dried at 140℃. All the resulting $LiSbO_3$ films were characterized by X-ray diffraction at 35kV and 300mA with CuKα radiation and by scanning electron microscopy .

<u>Electrical Conductivity Measurement</u>

Electrical conductivity was measured for two typical $LiSbO_3$ thin films prepared by Procedure-1 and -2. The measurement was carried out using the ac current complex impedance method under an air atmosphere in the temperature range of 380 to 600℃. As an electrode, gold was sputtered on both of the edge of the film and platinum wire was attached to gold sputtered with silver paste, so that the distance of the electrodes was approximately 1.0mm. A complex impedance analysis was examined in the frequency range of 100Hz to 10MHz using an impedance analyzer (HP 4194A, Hewlett-Packard Inc.).

RESULTS AND DICCUSSION

<u>Characterization of Films</u>

Figure 2 shows X-ray diffraction patterns of $LiSbO_3$ thin films prepared by Procedure-1 and -2, in comparison with that of a quartz glass substrate itself. The pattern of the film crystallized at

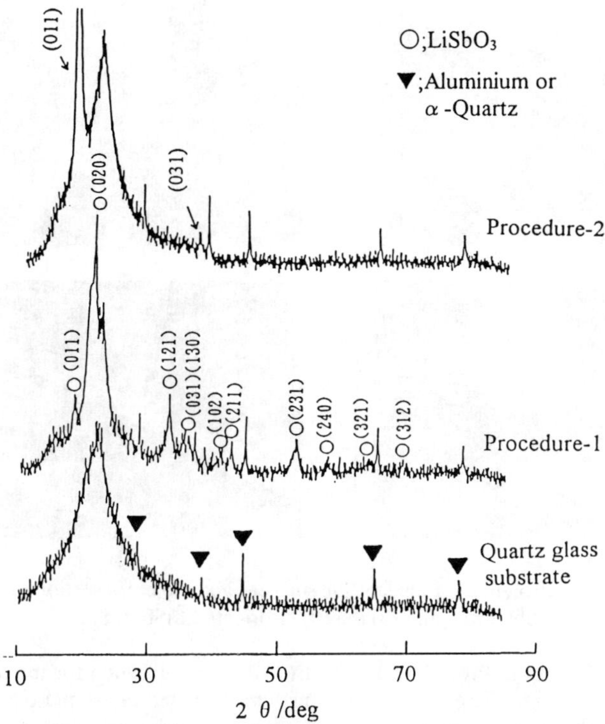

Fig. 2 X-ray diffraction patterns of LiSbO₃ thin films prepared by Procedure-1 and -2, together with the pattern of a quartz glass substrate.

at 650℃ by Procedure-1 is almost the same as that of the powder specimen [6], except for several reflection peaks that also appear in the pattern of the substrate. This suggests that the film is randomly oriented. In contrast, the pattern of the film crystallized at 600℃ by Procedure-2 shows the fairly large reflection peak from LiSbO₃ (011) and the very small reflection peak from LiSbO₃ (031), as indicated in Fig. 2. This means that the film is mainly (011)-oriented. As to the reflection peaks appearing in the patterns of both the films and the substrate, it has been clarified that the peaks at $2\theta = 38°$, $44°$, $65°$ and $78°$ belong to aluminium used as a sample holder and the peak at $2\theta = 28°$ to α-quartz.

Figures 3 shows the scanning electron micrographs of the surface and the cross section of the randomly oriented LiSbO₃ thin film and the (011)-oriented LiSbO₃ thin film. As can be seen from Fig. 3 (a), fine pores with 1μm or smaller in diameter resulting from the combustion or the pyrolysis of the organic residue, are observed to be distributed throughout the surface of the randomly oriented film. On the other hand, in case of the (011)-oriented LiSbO₃ thin film, no pore but crystallite with approximately 1μm in diameter is observed to disperse over the surface, and the cross section appears more uniformly and densely structured than the randomly oriented film, as shown in Fig. 3 (b).

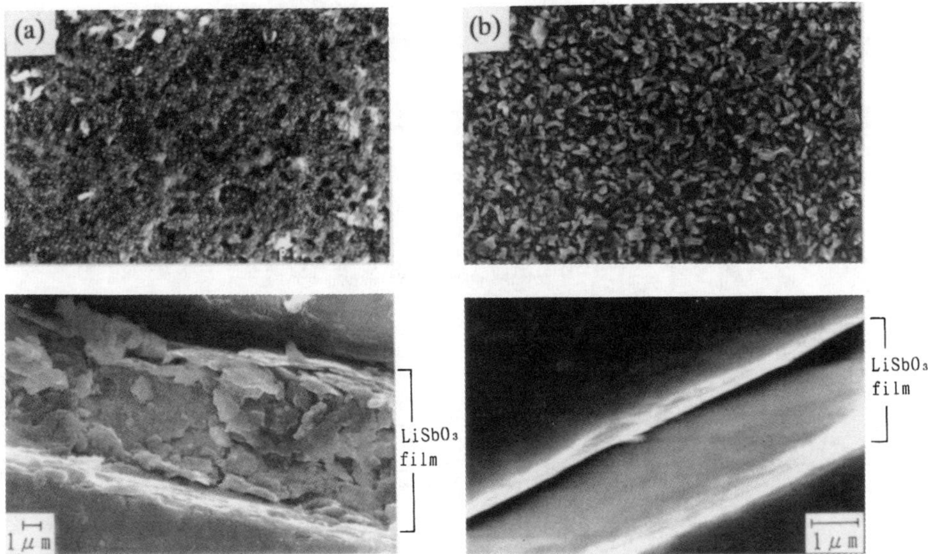

Fig. 3 Scanning electron micrographs of the surfaces and the cross section of (a) the randomly oriented LiSbO$_3$ film and (b) the (011)-oriented LiSbO$_3$ film.

Procedure-2 differs from Procedure-1 in that (1) the crystallization for the precursor films is carried out under an atmosphere of a flowing mixture of water vapor and oxygen, and (2) the hydrolysis of the alkoxides seems to take place simultaneously with the crystallization of the precursor film. Ref. [8] demonstrated that the (012)-, (110)-, and (110)-oriented LiNbO$_3$ films can be prepared by a sol-gel process with metal alkoxides, when the precursor films formed on the sapphire substrates with less lattice mismatch are crystallized in a flowing mixture of water vapor and oxygen. According to the paper, a mixture of water vapor and oxygen during the crystallization of the precursor films is said to exhibit remarkable effects both on the low-temperature crystallization and on the elimination of the free carbon in the films. In the present study, we also examined the crystallization temperature of the precursor films prepared by both Procedures. As a result, it has become apparent that the crystallization of the precursor film prepared by Procedure-1 occur in the temperature range higher than about 400°C, as is the case with the powder specimen [6]. On the other hand, the precursor film prepared by Procedure-2 begins to crystallize at temperature of as low as about 330°C. However, the mechanism for the preparation of the LiSbO$_3$ oriented film is not clear yet, which should be further substantiated.

<u>Electrical Conductivity</u>

The alternating complex impedance method, referred to as the Cole-Cole plot, was employed to determine the electrical conductivity of the LiSbO$_3$ films. Then the equivalent circuit of the polycrystalline ion conductor was applied. According to the equivalent circuit, a Cole-Cole plot is composed of a semicircle in the high frequency region, representing the parallel combination of bulk capacitance and resistance, and a line in the low frequency region, representing electrode impedance [9]. The electrical conductivity of the specimen is calculated by extrapolating the semicircle onto the real impedance axis.

Figure 4 shows typical Cole-Cole plots of the randomly oriented $LiSbO_3$ thin film and the (011)-oriented $LiSbO_3$ thin film, measured at temperatures as indicated in Fig. 4. Each plot is obtained as a relative clear semicircle in the frequency of the present research range, i. e., 100Hz - 10MHz. These suggest that ion conductivity is of single relaxation and the film adheres to the substrate tightly [10].

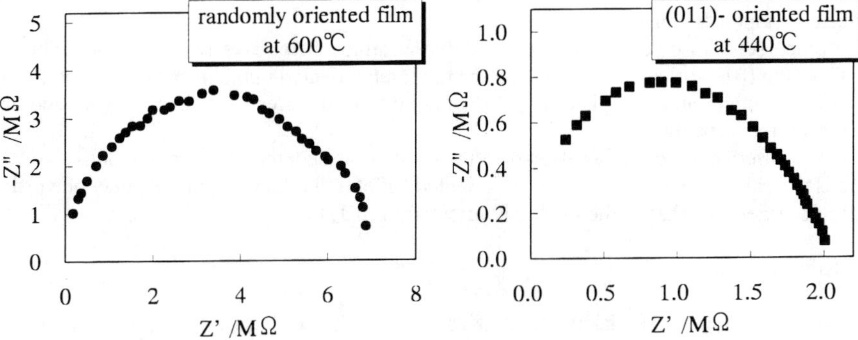

Fig. 4 Cole-Cole plots of the randomly oriented $LiSbO_3$ film and the (011)-oriented $LiSbO_3$ film.

Figure 5 shows the temperature dependence of electrical conductivity of the randomly oriented film and the (011)-oriented film. In Fig. 5 the results of the sintered specimen derived from the solid-phase reaction between Li_2CO_3 and Sb_2O_5 [1], are also cited, where ion transference number of the sintered specimen was nearly unity [11]. The electrical conductivity of the (011)-oriented film shows approximately one order magnitude larger than that of the randomly oriented film. This is probably because lithium diffusion path in the randomly oriented film is more disconnected due to mismatched or detached crystal lattices at a grain boundary, than in the (011)-oriented film. Activation energy of the randomly oriented film calculated from Fig. 5 is 83.9 kJmol^{-1} in the temperature rang of 500 - 600℃, and that of the (011)-oriented film is 83.5kJmol^{-1} and 10.9kJmol^{-1} in the

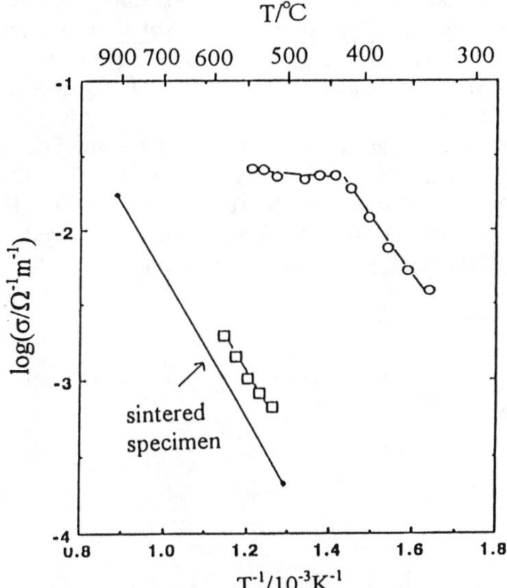

Fig. 5 Temperature dependence of the conductivity of (□) the randomly oriented $LiSbO_3$ film and (○) the (011)-oriented $LiSbO_3$ film, together with the sintered specimen [1].

temperature range of 330 - 420°C and 520 - 580°C, respectively. The former two activation energies being in good agreement with that of the sintered specimen indicates that the lithium ion conduction seems to occur in the same mechanism. However, the mechanism depressing activation energy as low as $10.9 kJmol^{-1}$ is not clear yet.

CONCLUSIONS

$LiSbO_3$ thin films were prepared from metal alkoxides, $LiO-n-C_3H_7$ and $Sb(O-n-C_3H_7)_3$ by two different procedures, and their electrical conductivities were measured. The overall conclusions can be drawn as follows.
1. The precursor film crystallizing under a different atmosphere gives two types of $LiSbO_3$ thin film. Thus, the (011)-oriented thin film and the randomly oriented thin film can be derived from the crystallization under an atmosphere of a flowing mixture of water vapor and oxygen, and an air atmosphere, respectively.
2. The two different atmospheres also affect the crystallization temperature of the films.
3. The electrical conductivity of the (011)-oriented $LiSbO_3$ thin film is approximately one order of magnitude larger than that of the randomly oriented $LiSbO_3$ thin film.

REFERENCES

1. M. Amano, T. Ogata, H. Nakamura, A. Kasahara, and M. Kobayashi, Bull. National Research Institute for Metals, **16**, p.23-29(1995).
2. K. Ohi and M. Abe, Bull. Ceram. Soc. Jpn., **27**, p.401-405(1992).
3. R. Chitrakar and M. Abe, Mater. Res. Bull., **23**, p.1231-40(1988).
4. W. G. Wyckoff, <u>Crystal Structures, Vol. II</u>, Interscience Pub. (1948) Chap. VII,illus. P.81.
5. M. Edstrand and N. Ingri, Acta Chem. Scand., **8**, p.1021-31(1954).
6. K. Ozawa, Y. Sakka and M. Amano, J. Japan Society of Powder and Powder Metallur., **42**, p.603-7(1995).
7. K. Ozawa, Y.Sakka and M. Amano, J. Ceram. Soc. Japan, **104**, p.185-9(1996).
8. S. Hirano and K. Kato, Adv. Ceram. Mater., **3**, p.503-6(1988).
9. N. Miura, Y. Ozawa and N. Yamazoe, Chem. Soc. Jpn., **12**, p.1954-59(1988).
10. Y. Shimizu, H. Arai and T. Seiyama, Denki Kagaku, **53**, p.300-5(1985).
11. H. Nakamura, private communication.

PREDOMINANT FACTORS OF LITHIUM ION CONDUCTIVITY IN PEROVSKITE-TYPE OXIDES

Y. INAGUMA*, T. KATSUMATA, J. YU AND M. ITOH
Materials and Structures Laboratory, Tokyo Institute of Technology, 4259 Nagatsuta, Midori-ku, Yokohama 226, Japan, *inaguma1@rlem.titech.ac.jp

ABSTRACT

The relation between the structure and the lithium ion conductivity in perovskite-type oxides ABO_3 was investigated from the viewpoint of the the arrangement of A-site ions. It was showed that the conductivity in $La_{2/3-x}Li_{3x}\square_{1/3-2x}TiO_3$ is strongly influenced by not only the concentration but also the ordered arrangement of skeletal A-site ions. Further the activation energy for ion conduction was found to be dominated by the covalency of B-O bond in addition to the bottleneck size.

INTRODUCTION

Studies on lithium ion conducting solid electrolytes have been of much interest because of potential applications and the fundamental research of the transport mechanism. Since lithium ion conducting oxides with perovskite structure show the ionic conductivity as high as $10^{-5} \sim 10^{-3}$ S·cm^{-1} at room temperature[1-11] and are stable at high temperature, they are expected as promising materials for various applications. The high conductivity originates from the migration of lithium ion via vacancies among the A-sites in the perovskite oxides ABO_3 (Fig.1).

Usually, the ionic conductivity is given by:

$$\sigma = Zen\mu, \qquad (1)$$

where Ze is the ionic charge, n is the carrier concentration and μ is the mobility of the ion. The mobility mainly depends on the activation energy for ion conduction. In our recent studies[8-10], the influence of carrier concentration on the lithium ion conductivity in perovskite-type oxides was investigated and it was found that the effective carrier concentration is strongly influenced by site percolation [12]and the ratio of lithium to vacancy concentration.

In perovskite-type lithium conducting oxides as shown in Fig.1, the smallest cross-sectional area of the conduction channel(called bottleneck) is located in the space surrounded by two adjacent A-sites and four oxygens(for example, in a cubic lattice with space group $Pm3m$ corresponding to the $3c$-site when the B-site ion occupies the $1a$-site). Lithium ion conductivity in perovskite-type $Ln_{1/2}Li_{1/2}TiO_3(Ln$=La, Pr, Nd and Sm) series was reported to decrease with a decrease in the ionic radius of lanthanide ions and to increase by the substitution of Sr with a larger ionic radius than La for $La_{1/2}Li_{1/2}$ in $La_{1/2}Li_{1/2}TiO_3$[4, 5, 10, 11].

These results indicate that the A-site space contraction decreases the bottleneck size and consequently disturbs the lithium ion migration while the A-site space expansion increases the bottleneck size and promotes the ion migration. It was recognized that the bottleneck size, which effectively determines the value of the repulsion potential between oxygen and lithium ions, is much important factor for the ion conductivity, especially the activation energy[11].

In this study, the structure and the lithium ion conductivity in perovskite-type oxides were elucidated in order to find out other predominant factors of high lithium ion conduction in perovskite-type oxides.

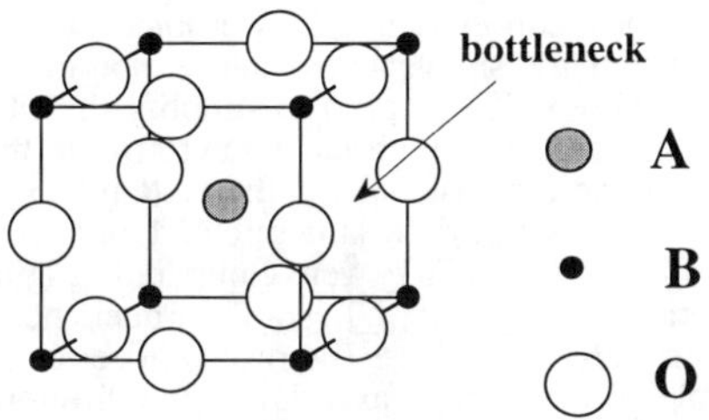

Fig.1. Structure of perovskite-type oxides.

Mat. Res. Soc. Symp. Proc. Vol. 453 ©1997 Materials Research Society

EXPERIMENTAL

Samples were prepared by a conventional solid state reaction. La_2O_3(4N), Pr_5O_{11}(4N), Nd_2O_3(3N), Sm_2O_3(3N), Li_2CO_3(3N), Na_2CO_3(3N), TiO_2(3N), Nb_2O_5(3N) and Ta_2O_5(3N) were used as starting materials. The metal content of lanthanide oxides was determined by the titration using EDTA. A mixture of the starting materials was calcined at 1073 K for 1~4 h and 1373~1423 K for 12h in air with intermediate grindings. The calcined powder was ground, pelletized and fired again at 1423~1623 K in air, and cooled in the furnace or quenched into the air at room temperature or liquid nitrogen. The phase identification and the determination of the lattice parameter were carried out using Mac Science MXP[18] X-ray diffractometer equipped with a graphite monochromator. The structure analysis was performed by Rietveld method using the code RIETAN[13]. Chemical composition of samples except for Nb and Ta oxides was determined by the inductive coupled plasma spectrometry (ICP, SEIKO Model SPS1500 or Shimazu ICPS-5000)[3]. The lithium ion conductivity for the bulk part of the sample pellet with electrodes of Au paste was measured by an ac complex impedance method using Hewlett Packard 4192A impedance analyzer over the frequency range 5Hz to 13MHz and over the temperature range of 238K to 720K. The size of the pellet was about 10mm in diameter and 1.5mm in thickness. Gold electrodes were formed on the pellet surface for the complex impedance method. The plot of the imaginary part of the impedance Z" against the real part Z' exhibited a straight line at low frequency and a distorted semicircle at high frequency. In our earlier work[3], the distorted semicircle was attributed to the bulk crystal part of the sample. The resistivity of the bulk part was determined by the extrapolation of the semicircle in the low frequency side to the Z' axis.

RESULTS AND DISCUSSION

Relation between the arrangement of A-site ions and the ion conduction in perovskite-type oxides

When viewing only the arrangement of A-ions in the cubic perovskite structure, they construct a simple cubic lattice. In the A-site lattice, there are lithium ions, vacancies and skeletal ions, for example, La, Sr and Na with larger ionic radius than lithium ion. Since lithium ions migrate among the A-site lattice, the skeletal ion behaves as an obstacle for the ionic conduction, resulting in the decrease in the essential carrier concentration. The dc ionic conduction is attained when the cluster which made up of lithium ions and vacancies percolates through the system. Therefore the total concentration of the lithium ion and vacancy or skeletal ion will strongly influence the ionic conductivity. In fact, the variation of the lithium ion conductivity at room temperature versus p of cubic perovskite-type pLiTaO$_3$-(1-p)SrTiO$_3$ solid solution could be explained from the view of the site percolation for simple cubic lattice[8]. It was proved that the lithium ion conduction occurs three-dimensionally among the A-sites. Furthermore, in the system that the ratio of lithium to vacancy is constant irrespective of composition yLa$_{0.5}$Na$_{0.5}$TiO$_3$-(1-y)La$_{0.55}$Li$_{0.35}\square_{0.1}$TiO$_3$ solid solutions, the conductivity can be also explained by site percolation for simple cubic lattice. While in the system that the ratio of lithium to vacancy concentration is variable, La$_{2/3-x}$Li$_{3x}\square_{1/3-2x}$TiO$_3$, the ratio in addition the site percolation for simple cubic lattice dominates the ionic conductivity[9]. At that time, the quenched sample was used to simplify the situation because the distribution of La approaches to the random state at high temperature (more than 1300K) and consequently the ionic conduction occurs three-dimensionally in the system La$_{2/3-x}$Li$_{3x}\square_{1/3-2x}$TiO$_3$.

In this study, the structure and ionic conductivity for quenched and furnace cooled La$_{2/3-x}$Li$_{3x}\square_{1/3-2x}$TiO$_3$ were compared. Figure 2 (a) and (b) shows the powder X-ray diffraction patterns for La$_{2/3-x}$Li$_{3x}\square_{1/3-2x}$TiO$_3$ quenched and furnace cooled from the temperature 1623 K, respectively. In these figures, solid circles refer to the reflection lines of supperlattice with double period along c-axis due to the difference of the site occupancy of La ions between La(1) and La(2) sites as shown in Fig.3. This superstructure is same as that seen in La$_{2/3}$TiO$_3$[14]. As shown in Fig.2, the intensity of supperlattice lines for the quenched sample is rather smaller than that of furnace cooled sample and decreases with an increase in the Li content, *i.e.*, a decrease in La content. This result indicates that the difference of the site occupancy of La ions between the quenched and furnace cooled samples increases as the La content increases. Figure 4 shows the logarithmic conductivity at 300K versus lithium concentration $3x$ for quenched and furnace cooled

La$_{2/3-x}$Li$_{3x}$[ ]$_{1/3-2x}$TiO$_3$. Though the tendency among two cases is similar, the values of the ionic conductivity are seen to be significantly different when x is rather smaller. Especially, in La$_{0.62\pm0.01}$-Li$_{0.16\pm0.01}$TiO$_{3.02\pm0.02}$, the ion conductivity of furnace cooled sample is 2.5 times higher than that of quenched sample. According to the Rietveld analysis, the site occupancy of La(1) and La(2) for the furnace cooled sample in La$_{0.62}$Li$_{0.16}$TiO$_3$ are 0.93 and 0.31, respectively. This implies that the lithium ion is easy to migrate two-dimensionally in the plain including La(2). The difference of the ionic conductivity between quenched and furnace cooled sample originates from the difference of La distribution like this.

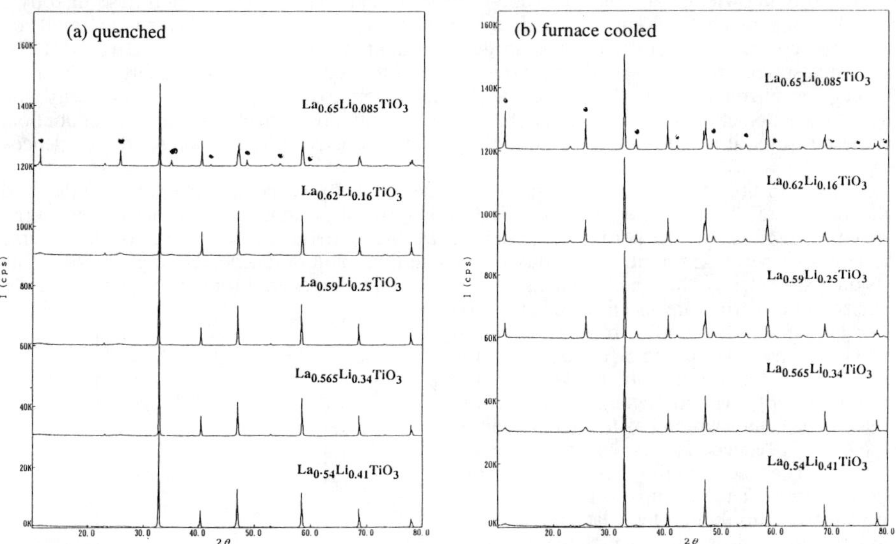

Fig. 2. Powder X-ray diffraction patterns for (a) quenched and (b) furnace cooled La$_{2/3-x}$Li$_{3x}$□$_{1/3-2x}$TiO$_3$, respectively.

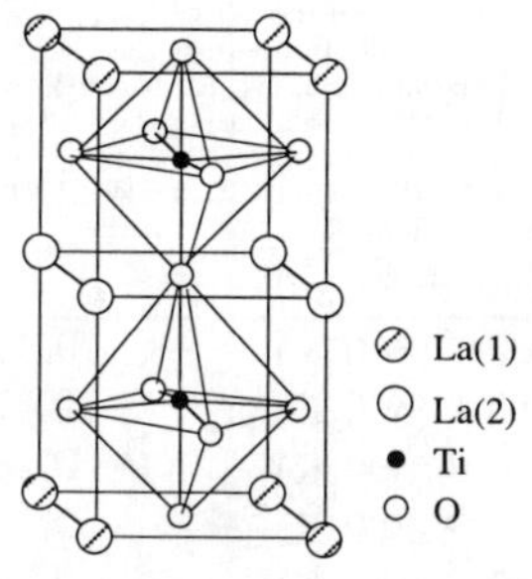

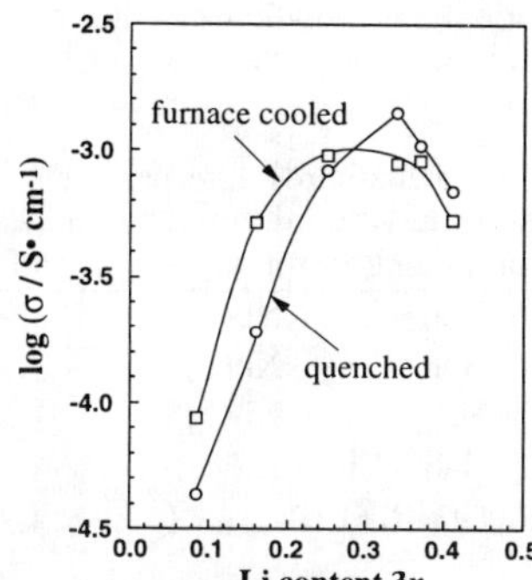

Fig.3. The arrangement of La ions in double perovskites.

Fig.4. Variation of the logarithmic ionic conductivity at 300K versus lithium content 3x for quenched and furnace cooled La$_{2/3-x}$Li$_{3x}$ □$_{1/3-2x}$TiO$_3$.

Predominant factors of the activation energy for ion conduction in perovskite-type oxides

The lithium ion conductivity in $Ln_{1/2}Li_{1/2}TiO_3$(Ln=La, Pr, Nd and Sm) series was found to decrease accompanied by the increase in the activation energy(Fig.5) and activation volume for the ion conduction when the ionic radius of Ln ions was decreased or high pressure was applied[5, 7, 11]. The increase in the activation energy results from the decrease in the bottleneck size by the decrease in the lattice parameter and the tilting of BO_6 octahedra with orthorhombic distortion. Since the bottleneck is estimated to be smaller than the diameter of a Li ion using the ionic radius[15], a dilatation of the bottleneck must occur when a Li ion jumps to a adjacent vacancy. This knowledge and the calculation of the potential energy of a Li ion described by a sum of the long range Coulomb potential energy and short range Born-Mayer-type repulsion potential energy, suggest that the increase in the activation energy in $Ln_{1/2}Li_{1/2}TiO_3$(Ln=La, Pr, Nd and Sm) series strongly depends on the increase in the repulsive potential energy due to the four oxygens surrounding the bottleneck where a Li ion experiences the highest activation state[11]. The calculation also showed that there is an optimum bottleneck size for ion conduction. From these results, it is expected to attain higher ionic conductivity by the control the bottleneck size.

Hence, in this study the various perovskite-type lithium ion conducting were synthesized and the ionic conductivity was measured. Figure 5 shows the variation of the activation energy for ion conduction versus the perovskite parameter a_p, being defined as cube root of the volume of the primitive perovskite unit, of various perovskite-type lithium ion conducting oxides. The perovskite parameter a_p, the ionic conductivity and the activation energy for the novel compounds synthesized in this work are listed in Table I. The data for the other compound is referred to ref.[8] and [11]. The data in Fig.5 divides into two groups. The compounds which B-site ion is composed of only Ti ion belong to one group, while the compounds which B-site ions include the ions, Nb or Ta except for Ti ion belong to the other. The activation energies of some compounds in the latter group are higher than that in the former group though their lattice parameters become larger. This result suggests that the effect of the chemical bond of B-O, *i.e.* covalency must be taken into account. It is thought that Ti-O bond decreases the repulsion potential between oxygen and lithium ion at the bottleneck site preferably than Ta-O or Nb-O bond does.

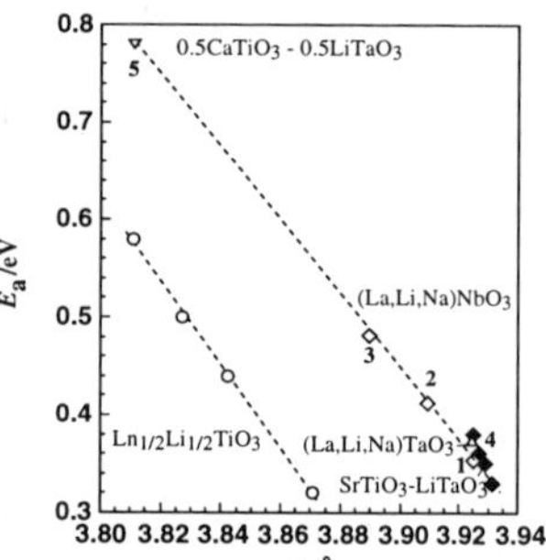

Fig.5. Variation of the activation energy E_a, for ion conduction versus the perovskite parameter a_p, for perovskite-type lithium ion conductors.

Table I. The list of the perovskite parameter a_p, the ionic conductivity and the activation energy for novel perovskite-type lithium ion conducting oxides appeared in Fig.5.

No.	Compound(norminal)	a_p / Å	conductivity(S•cm^{-1})	E_a / eV
1	$La_{0.25}Na_{0.20}Li_{0.05}NbO_3$	3.925	2.1×10^{-5} (300 K)	0.35
2	$La_{0.20}Na_{0.25}Li_{0.15}NbO_3$	3.910	5.3×10^{-6} (300 K)	0.41
3	$La_{0.15}Na_{0.30}Li_{0.25}NbO_3$	3.890	2.9×10^{-7} (300 K)	0.48
4	$La_{0.25}Na_{0.20}Li_{0.05}TaO_3$	3.925	2.2×10^{-6} (300 K)	0.37
5	$0.5CaTiO_3$-$0.5LiTaO_3$	3.811	1.9×10^{-5} (700 K)	0.78

CONCLUSIONS

In this study the influence of the arrangement of A-site ions and chemical effect on the lithium ion conductivity in perovskite-type oxides was demonstrated. In conclusion, it was found that (a) the site percolation considering the ordering of the distribution of A-site ions and (b) the covalency of B-O bond, in addition to (c) the bottleneck size, (d) the ratio of lithium to vacancy concentration and (e) the site percolation for simple cubic lattice are predominant factors of the lithium ion conductivity in perovskite-type oxides.

ACKNOWLEDGMENTS

Authors wish to express their thanks to the Grant-in-Aid for Scientific Research in Priority area (No.260) from the Ministry of Education, Science and Culture.

REFERENCES

1. L. Latie, G. Villeneuve, D. Conte and G. L. Flem, J. Solid State Chem. **51,** 293(1984).

2. A. G. Belous, G. N. Novitsukaya, S. V. Polyanetskaya and Yu. I. Gornikov, Izv. Akad. Nauk SSSR, Neorg. Mater. **23,** 470(1987).

3. Y. Inaguma, L. Chen, M. Itoh, T. Nakamura, T. Uchida, M. Ikuta and M. Wakihara, Solid State Commun. **86,** 689(1993) .

4. Y. Inaguma, L. Chen, M. Itoh, T. Nakamura, Solid State Ionics **70/71**, 196(1994).

5. M. Itoh, Y. Inaguma,W. H. Jung, L. Chen, M. Itoh, T. Nakamura, Solid State Ionics, **70/71**, 203(1994).

6. M. Oguni, Y. Inaguma, M. Itoh and T. Nakamura, Solid State Commun. **91**, 627(1994).

7. Y. Inaguma, Jianding Yu, Y.J. Shan,, M. Itoh and T. Nakamura, J. Electrochem. Soc. **142**, L8(1994).

8. Y. Inaguma, Y. Matsui, Y.J. Shan, M. Itoh and T. Nakamura, Solid State Ionics **79**, 91(1995).

9. Y. Inaguma and M. Itoh, Solid State Ionics **86-88**, 257(1996).

10. T. Katsumata, Y. Matsui, Y. Inaguma and M. Itoh , Solid State Ionics, **86-88** 165(1996).

11. Y. Inaguma, Y. Matsui, J. Yu, Y.J. Shan, T. Nakamura and M. Itoh , J. Phys. Chem. Solids in print(1997).

12. D. Stauffer, Introduction to Percolation Theory, 1st Ed. (Taylor and Francis, London, 1985); D. Stauffer and A. Aharnony, Introduction to Percolation Theory, 2nd Ed. (Taylor and Francis, London, 1992).

13. F. Izumi, J. Crystalogr. Soc. Jpn. **27**, 23(1985) and in The Rietveld Method(R. A. Young, Ed.), Chap. 13, Oxford Univ. Press, Oxford(1993).

14. M. Abe and K. Uchino, Mat. Res. Bull. **9**, 147(1974).

15. R. D. Shannon, Acta Cryst. **A32**, 751(1976).

CHARACTERIZATION OF PROTON CONDUCTING ANTIMONIC ACIDS WITH AMORPHOUS, CUBIC AND MONOCLINIC STRUCTURES

Y. SAKKA,* K. SODEYAMA,** T. UCHIKOSHI, * K. OZAWA* and M. AMANO*
*National Research Institute for Metals, 1-2-1, Sengen, Tsukuba, Ibaraki 305 Japan
**Kagoshima Preferectural Institute of Industrial Technology, Oda, Kagoshima 899-51, Japan

ABSTRACT

Powders and films of amorphous, cubic and monoclinic antimonic acids are prepared. Their water sorption isotherms and the water desorption spectra are measured. The protonic conductivities are very sensitive to water vapor pressure and are discussed with the amount of water. The amorphous and cubic antimonic acids contain larger amounts of water than the monoclinic acid, which can be attributed to the existence of micropores in the structures.

INTRODUCTION

Three types of antimonic acids, $Sb_2O_5 \cdot nH_2O$, with amorphous, cubic and monoclinic structures exist [1,2]. The cubic antimonic acid is known to have a high protonic conductivity at room temperatures [3-6]. It can offer applications in gas sensors, electrochromic devices, and energy storage materials. The protonic conductivity is sensitive to water vapor pressure. However, studies on water sorption-desorption characteristics of the antimonic acids are limited.

In the present paper, water sorption-desorption characteristics of three types of antimonic acids, amorphous (Am-Sb), cubic (C-Sb) and monoclinic (M-Sb) structures, are presented by using the temperature-programmed desorption (TPD) spectra and sorption isotherms of water. The differences in protonic conductivity and water content for the antimonic acids are discussed in relation to the presence of micropores.

EXPERIMENT

C-Sb powder was synthesized by the hydrolysis of $SbCl_5$ in distilled water, which has been produced at Toa-Gosei Co. Ltd., (IXE-300). M-Sb powder was produced by the ion-exchange reaction of $LiSbO_3$ in 10M HNO_3 solution at 303K for 1 month [1,7]. Am-Sb powder was prepared by a direct reaction of $Sb(O-n-C_3H_7)_3$ with H_2O_2 in a mixed solution of H_2O_2, H_2O and $n-C_3H_7OH$ at 273K[8]. It was confirmed by an X-ray diffraction analysis that each powder was single phase. The powders were also characterized using a TG-DTA and FT-IR. The TG-DTA analysis was conducted in air to determine the content of water of each powder.

Water vapor isotherms on the powders were measured at 298K by means of a computer-controlled automatic adsorption apparatus [9]. The powders were evacuated for 10 h at room temperature with a turbo molecular pump. The adsorption and desorption isotherms were determined by measuring pressure change. The nitrogen isotherms at 77K were determined with an automatic adsorption apparatus. The surface areas of adsorbents are calculated using the BET equation where the adsorbed areas of nitrogen molecule and water molecule are assumed to be 0.162 and 0.125 nm^2, respectively. To determine micropore distribution and total micropore volume by the Dubinin-Radushkevich (DR) plots [10,11], the isotherms of methanol and benzene at 298K were also measured.

TPD experiments were also performed to determine the water sorption-desorption characteristics on the Am-Sb, C-Sb and M-Sb powders. After sufficient evacuation (approximately 10^{-8} Torr) at room temperature, the samples were heated at a constant rate of 5K/min. The gases evolved were monitored with a quadruple mass spectrometer. To check the mobility of hydrogen and oxygen in each structure, TPD experiments were also conducted after exposure to D_2O and $H_2^{18}O$ at room temperature [9].

Mat. Res. Soc. Symp. Proc. Vol. 453 © 1997 Materials Research Society

RESULT

Sample characterization

It was found by the thermogravimetric analyses that Am-Sb, C-Sb and M-Sb can be expressed as $Sb_2O_5 \cdot 3.4H_2O$, $Sb_2O_5 \cdot 3.1H_2O$ and $Sb_2O_5 \cdot 1.3H_2O$ under an air atmosphere, respectively. The BET specific surface areas of M-Sb determined by nitrogen and water sorption are 12.0 and $11.8 m^2/g$, and those of C-Sb determined by nitrogen and water sorption are 7.5 and 102 m^2/g, respectively. In case of Am-Sb, the BET specific area by nitrogen was 2.7 m^2/g, but the BET analysis could not be conducted by the water sorption.

Figure 1 shows FTIR spectra of three kinds of antimonic acids. The absorption bands at 2400 cm^{-1} for Am-Sb, 2460 cm^{-1} for C-Sb and 2320 cm^{-1} for M-Sb are assigned to the overtones of the SbOH deformation vibrations [1, 6, 12]. The absorption at 1670 cm^{-1} is the deformation band of molecular H_2O, the intensity difference of which indicates that Am-Sb and C-Sb include a larger amount of H_2O than does M-Sb. The absorption bands found at from 2700 to 3800 cm^{-1} result from the stretching vibration of free or hydration water.

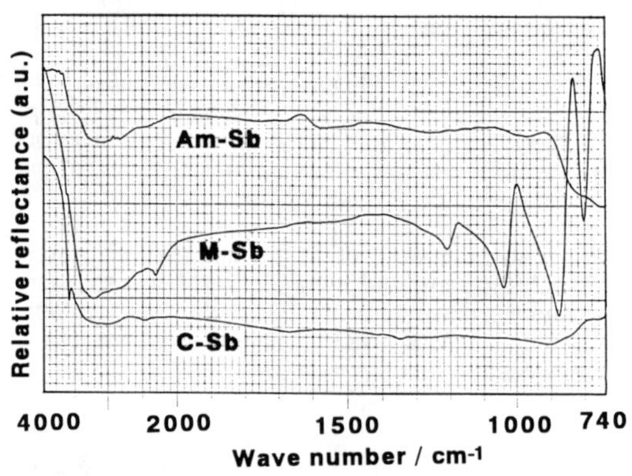

Figure 1 FTIR spectra of Am-Sb, C-Sb and M-Sb

Sorption isotherm

Figure 2 shows the water sorption isotherms at 298K for the samples after evacuation at room temperature, where the vertical line V indicates water adsorbed (mol) per 1mol antimonic acid. Initial n values for $Sb_2O_5 \cdot nH_2O$ of Am-Sb, C-Sb and M-Sb after evacuation at room temperature were 3.1, 2.3 and 1.2, respectively. It is seen that Am-Sb and C-Sb adsorb much more water than M-Sb. The steep increase in water adsorption for Am-Sb and C-Sb at low relative pressures indicates the presence of micropores, which will be discussed later.

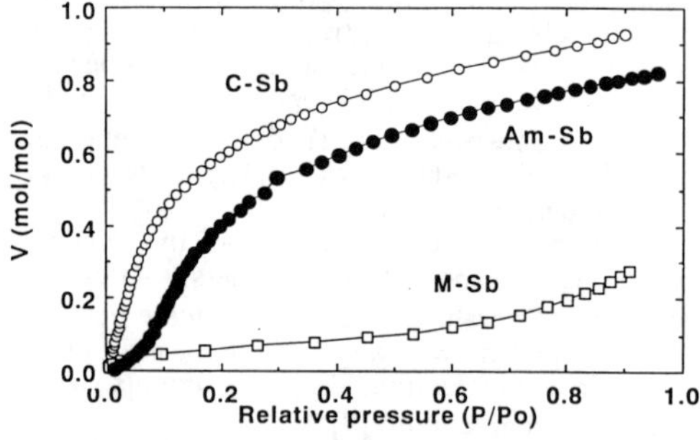

Fig. 2 Adsorption isotherms of H_2O for Am-Sb, C-Sb and M-Sb at 298K.

TPD spectra

The evolved gases from the powders were mainly H_2O, and evolution of oxygen was observed above 560 K for C-Sb and above 660 K for M-Sb. Figure 3 shows TPD spectra of as-prepared Am-Sb, C-Sb and M-Sb powders. Two desorption peaks are observed for C-Sb and M-Sb at around 450 and 530 K, and at around 450 and 630 K, respectively. In case of C-Sb, the intensity of the first peak (450 K) is larger than that of the second peak (530 K). In case of M-Sb, however, the intensity of the second peak (630 K) is larger than that of the first peak (450 K). On the other hand, at least four peaks are observed for Am-Sb at around 450, 600, 670 and 780K. The second peak (600K) might be due to the crystallization of Am-Sb.

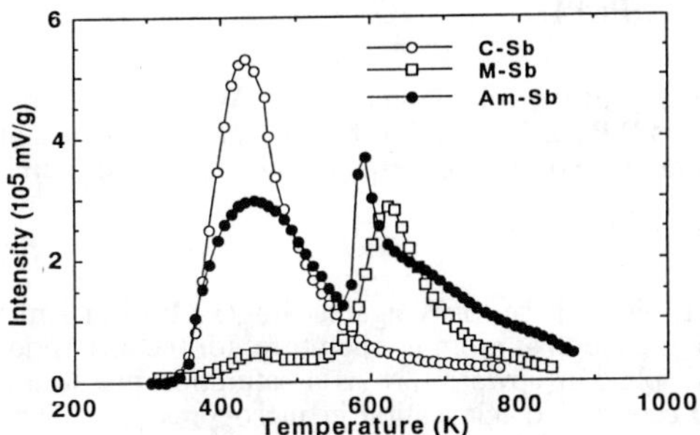

Fig. 3 H_2O desorption spectra of as prepared Am-Sb, C-Sb and M-Sb at heating rate 5K/min.

TPD spectra of C-Sb after exposure to D_2O and $H_2^{18}O$ for 10 h at room temperature are shown in Figure 4. The peak temperatures of H_2O(M/Z=18), HDO(19) and D_2O(20) after exposure to D_2O are nearly the same as those of $H_2^{16}O$(18) and $H_2^{18}O$(20) after exposure to $H_2^{18}O$, but the intensity ratios of the first peaks and the second peaks are different. After exposure to D_2O, the intensity ratios among H_2O(18), HDO(19) and D_2O(20) are nearly equal, while after exposure to $H_2^{18}O$ the intensity ratio of $H_2^{18}O$(20) to $H_2^{16}O$(18) at the second peak (530 K) is small in comparison with that at the first peak (around 450 K) as is seen in Fig.3(b). Similar results were also obtained for M-Sb [9].

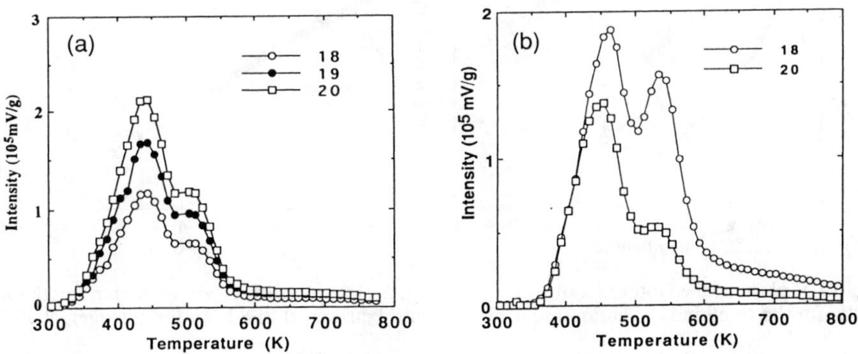

Fig. 4 TPD spectra of the C-Sb after exposure to (a) D_2O and (b) $H_2^{18}O$ for 10 h at room temperature. (a) H_2O(18), HDO(19), D_2O(20). (b) $H_2^{16}O$(18), $H_2^{18}O$(20).

DISCUSSION

The specific surface areas of M-Sb determined by nitrogen and water molecules are nearly equal. However, the specific surface area of C-Sb determined by water molecule is extremely larger than that determined by nitrogen molecule. These results indicate that the micropores, where H_2O molecule can enter but nitrogen cannot, exist in the cubic (pyrochlore) structure of C-Sb[12], but such micropores do not exist in the monoclinic structure of M-Sb [13].

The process of gas adsorption into very fine pores is considered to be the volume filling micropores rather than layer-by-layer adsorption on the pore walls. The degree of filling of the micropores is expressed by the following equation [10,11].

$$W/W_0 = \exp[-B(T/\beta)^2 \log(P_0/P)^2] \qquad (1)$$

$$B = 2.303R^2/k$$

Where W_0 is the total volume of the micropore system, W is the volume that has been filled when the relative pressure is P/P_0, β is a scaling factor (similarity constant), k is characteristic parameter, and R is gas constant. For plotting, eq. (1) may be transformed into

$$\log W = \log W_0 - D\log(P_0/P)^2 , \qquad (2)$$

where $D = B(T/\beta)^2$.

Figure 5 shows the DR plots of the $\log W$ against $[\log(P_0/P)]^2$ of Am-Sb, C-Sb and M-Sb, which have been evacuating enough at room temperature, for the adsorption isotherms of water at 298 K. The DR plots display an upward turn as a saturation pressure is approached due to multilayer adsorption and capillary condensation in mesopores [11]. Therefore the micropore volume is calculated from the straight lines at lower relative pressures. In case of Am-Sb, however, the DR plot falls into three straight lines, which may indicate that different types of micropores exist. The total pore volume of the micropores obtained from the water adsorption is 0.043 cm^3/g for C-Sb and nearly zero for M-Sb. Figure 6 shows the DR plots of C-Sb for the adsorption of water, methanol and benzene at 298 K, and nitrogen at 77 K. The total pore volumes obtained from the methanol, nitrogen and benzene are nearly zero. The micropores under the size of water molecule only exist in C-Sb. Similar situation can be expected to hold for Am-Sb.

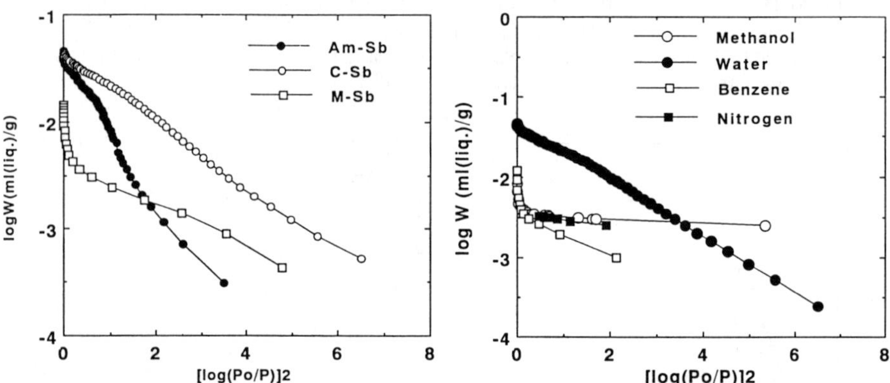

Fig. 5 DR plots of the isotherms of water at 298K for the tree kinds of antimonic acids.

Fig. 6 DR plots of C-Sb of water, methanol and benzene at 298 K and of nitrogen at 77 K.

In the cubic (pyrochlore) structure, pore channel (the micropore) exists [12]. The size of the micropore evaluated from the lattice parameter is 0.247 nm, where the distance between Sb and O atoms is assumed to be 0.195 nm and ionic radius of OH^- is 0.153 nm [5]. The evaluated size is consistent with the size where water molecule can enter but nitrogen, methanol and benzene molecules cannot.

Figure 7 shows the previously reported protonic conductivities of three kinds of antimonic acid films as a function of relative humidity at 296 K [6,8]. The Am-Sb film was synthesized by the spin coating of the product by a direct reaction of $Sb(O-n-C_3H_7)_3$ with H_2O_2 in a mixed solution of H_2O_2, H_2O and $n-C_3H_7OH$ [8]. The films of C-Sb and M-Sb were prepared by spreading slurries of their powders[6]. Therefore the density of the Am-Sb film is higher than those of the C-Sb and M-Sb films. The higher conductivity of Am-Sb than those of C-Sb and M-Sb is partly attributed to the difference in densities of the films. The conductivities of Am-Sb, C-Sb and M-Sb decrease with decreasing water vapor pressure. By a combination of Fig. 2 with Fig. 7, it is possible to show the effect of water contents on the protonic conductivity at 296 K. Figure 8 shows the relation between the protonic conductivities and the water content. Significant decrease in conductivity of M-Sb is seen with decreasing in water content.

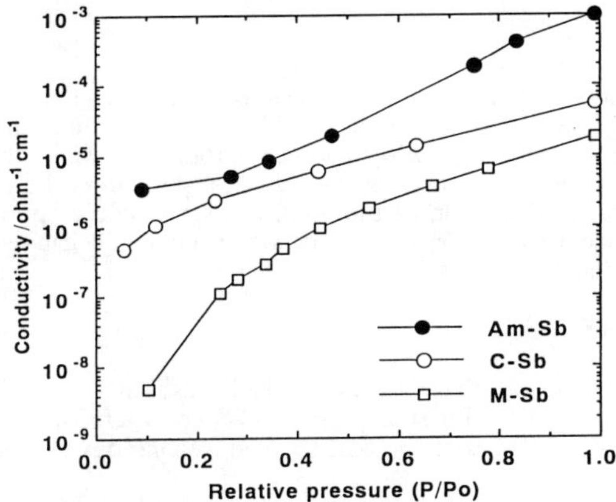

Fig. 7 Relative humidity dependence of the conductivity of Am-Sb, C-Sb and M-Sb at 296 K.

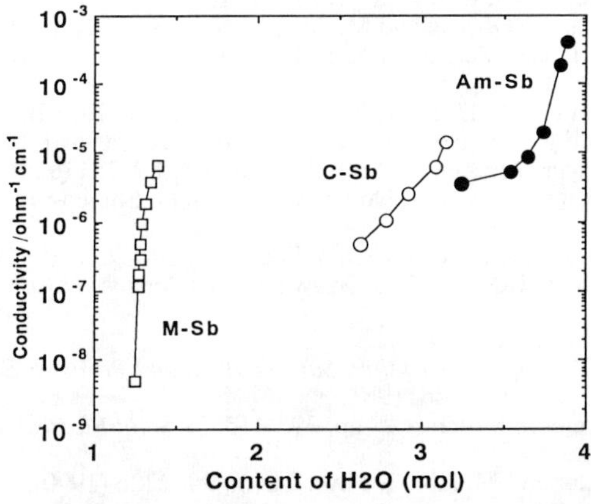

Fig. 8 Relation between conductivity and H_2O contents for Am-Sb, C-Sb and M-Sb at 296 K.

It has been clarified from the TPD desorption measurements that the water desorption of Am-Sb, C-Sb and M-Sb takes place in several steps. The water existing in the micropores (channels of the cubic antimonic acid) is considered to correspond to the adsorbed and first water desorption peak, where proton can migrate with oxygen easily as is seen from Fig. 4.

Similar results were also observed from the change in the IR absorption peaks of hydroxyl groups after exposure to a D_2O-saturated N_2 stream at room temperature [6]. The lack of the micropores of M-Sb may reflect to the small amount of water sorption and the significant decrease in conductivity with decreasing in water vapor pressure.

Figures 4 shows that the mixing of H and D has taken place almost equally for both peaks of water-desorption, whereas the mixing of ^{16}O and ^{18}O is more profound for the first peak than for the second one. This may indicate the followings; (1) The water molecules or their constituents being desorbed at the second peak (II) are less active in the oxygen-exchange reaction at room temperature than those at first peak(I). This suggests that Species II are bonded to the lattice more strongly or are located at the sites less accessible from outside. (2) The ease of H-D mixing indicates that both Species I and II are reactive in the proton exchange reaction, presumably reflecting the proton-conductive nature of the materials.

CONCLUSION

Water adsorption and desorption characteristics are studied by using the water adsorption isotherm and TPD experiments for amorphous, cubic and monoclinic antimonic acids. The amorphous and cubic antimonic acids adsorb larger amounts of water than the monoclinic antimonic acid. The water desorption takes place in several steps from the TPD measurements. The larger amount of water adsorption of the amorphous and cubic antimonic acids can be attributed to the existence of micropores, where water molecule can enter but methanol, nitrogen and benzene molecules cannot.

ACKNOWLEDGMENTS

We acknowledge H. Nakamura, A. Kasahara and M. Kobayashi of NRIM for supplying the M-Sb powder and Toa-Gosei Co. Ltd., for supplying the C-Sb powder. This study was performed through Special Coordination Funds of the Science and Technology Agency of the Japanese Government.

REFERENCES

1. R. Chitrakar and M. Abe, Mat. Res. Bull, **23**, 1231 (1988).
2. Y. Kanzaki, R. Chitrakar, T. Ohsaka and M. Abe, Chem. Soc. Jpn., **1993**, 1299 (1993).
3. W. A. England, M. G. Cross, A. Hamnett, P. J. Wiseman and J. B. Goodenough, Solid State Ionics, **1**, 231 (1980).
4. U. Chowdhry, J. R. Barkley, A. D. English, A. W. Sleight, Mat. Res. Bull, **17**, 917 (1982).
5. N. Miura, Y. Ozawa and N. Yamazoe, Chem. Soc. Jpn., **1988**, 1954 (1988).
6. T. Uchikoshi, Y. Sakka and M. Amano, Solid State Ionics, **89**, 351 (1996).
7. M. Kobayashi, H. Nakamura, A. Kasahara and M. Amano, Shigen-to-Sozai, **111**, 949 (1995).
8. K. Ozawa, Y. Sakka and M. Amano, J. Ceram. Soc. Jpn., **104**, 756 (1996).
9. Y. Sakka, K. Sodeyama, T. Uchikoshi, K. Ozawa and M. Amano, J. Am. Ceram. Soc., **79**, 1677 (1996).
10. M. M. Dubinin, Carbon, **27**, 457 (1989).
11. S. J. Gregg and K. S. W. Sing, "*Adsorption, Surface Area and Porosity, Second Edition*"; pp.195-247, Academic Press,London, 1981.
12. J. B. Goodenough, "Skeleton Structures" pp. 393-415 in *Solid Electrolytes*, Academic Press, New York, 1978.
13. Y. Kanzaki, R. Chikrakar and M. Abe, J. Phys. Chem. **94**, 2206 (1990).

SYNTHESIS, CHARACTERIZATION, AND ELECTROCHEMICAL PERFORMANCES OF SUBSTITUTED LAYERED TRANSITION METAL OXIDES, $LiM_{1-y}M'_yO_2$, (M=Ni and Co, M'= B and Al)

G.A. Nazri, A. Rougier, and K.F. Kia, Physics and Physical Chemistry Department, General Motors Research and Development Center, RCEL, Warren, MI 48090-9055

ABSTRACT

The synthesis, characterization and electrochemical performances of lithiated nickelate and cobaltate doped with Al and B are reported. The synthesis involves solid state reaction between lithium hydroxide, nickel or cobalt oxides and several sources of aluminum and boron. Careful selection of precursors and heat treatment conditions are required to prepare single phase impurity free samples. X-ray diffraction and Rietveld refinement analysis indicate that the layered structure is preserved upon considerable substitution of aluminum and boron. X-ray diffraction line intensities and positions remained in good agreement with the $R\bar{3}m$ space group. The IR spectra of the samples indicate formation of compressed CoO_6 and NiO_6, and elongated LiO_6 octahedra. The IR vibrational mode of the LiO_6 remains in the 200-300 cm^{-1} and the vibrational modes of the MO_6 expand over 400-650 cm^{-1}. Results of long charge-discharge cycling of the samples as cathode materials in lithium cells showed long cycle life. The capacity of the electrodes upon substitution were reduced almost linearly as the concentration of substitution was increased. The solubility limit for the formation of solid solutions upon substitution of Al and B in $LiNiO_2$ and $LiCoO_2$ was found to be around 25%. Specific capacities of the samples were between 120 to 160 mAh/g depending on the amount of substitution.

INTRODUCTION

High oxidation states of transition metals in layered oxides have attracted attention of many research groups for application in high voltage electrochemical cells. The layered structure is preferred for their structural flexibility to adopt the geometry of the guest species by expansion of their interlayer spacing or by the gliding movement of the slabs with respect to each other, which may result in changes in stacking order (1). The lithiated layered transition metal oxides in general have been considered for application in rechargeable lithium batteries, due to their high number of available sites per volume and mass for lithium intercalation, and their high redox potential of the tri- and tetravalent cations. Redox potentials of the first row transition metal oxides are highly correlated with the number of d-electrons when they remain within the same space group, and it is possible to tune the voltage by the choice of transition metals in the metal oxide slabs.

The most studied cathode in this series are the $LiNiO_2$ and $LiCoO_2$ (2-17). These compounds are formed by slabs of NiO_2 and CoO_2 respectively, where the lithium ions reside between the slabs to stabilize the structure and compensate the repulsion of the negatively charged MO_2 layers. The cations are octahedraly coordinated with compressed NiO_6 and CoO_6 octahedra and elongated LiO_6 octahedra. The atomic positions are according to the D^5_{3d} - $R\bar{3}m$ space group. The transition metals are located at the 3a (000) and the lithium resides at 3b (00$^1/_2$) sites. The oxygen anions are coordinated with three lithium and three transition metal ions and occupy the 6c (00z,

Mat. Res. Soc. Symp. Proc. Vol. 453 © 1997 Materials Research Society

$00\bar{z}$) sites. The oxygen parameters z = 0.25 is in good agreement with the observed and calculated x-ray diffraction lines for the LiNiO$_2$.

Structural stability of the LiMO$_2$ in general dependence on the amount of lithium between the MO$_2$ slabs. The LiNiO$_2$, in particular is very sensitive to stoichiometry of lithium per transition metal (2,3). Synthesis with lithium deficient mixture, or when the precursors are not well mixed, yield products with some Ni^{II} ions which are located in the lithium sites. In most cases, nickel ions migrates from the slab of NiO$_2$ and occupy the lithium sites to form a more three dimensional structure with expansion of lattice parameters. The nickel ions between the NiO$_2$ sheets remain in divalent state and stabilize the structure by reducing the interlayer repulsion. However, when the concentration of lithium is adjusted for the partial evaporation of lithium salt at high temperature and the initial components are well mixed, high quality single phase LiNiO$_2$ can be obtained. The nature of lithium and the nickel salts are also important factors for obtaining high quality layered lithiated nickelate compound. The decomposition temperatures of the lithium and nickel sources must be carefully considered, so mobile lithium ions would be available during formation of the NiO$_2$ slabs. In addition, long range ordering in triangular lattice such as LiNiO$_2$ is difficult due to magnetic interactions of Ni^{III} ions. LiCoO$_2$ is less sensitive to synthesis condition due to the higher bond covalency in the CoO$_2$ slabs, and single phase impurity free compound is normally obtained through solid state reaction of lithium salts and cobalt oxides. Formation of layered two dimensional oxide is more favorable for the compounds with lower bond ionicity according to the Mooser-Pearson diagram (18-22) . A strong bond covalency in LiCoO$_2$, with reduced Co-O bond distance, results in stabilization of Co^{III} in low-spin (23) ground state, [d^6 = (t$_{2g}$)^6(e$_g$)^0, S=0], and reduces the electronic conductivity of the compound (band gap = 2.7 eV). Strong hybridization with an enhanced bond covalency in MO$_2$ slabs may promote formation of layer compounds with long range order perpendicular to the slab direction. In addition, synthesis condition may affect the formation of compounds with different stacking orders (1, 24,25).

Although the two dimensional oxides are ideal host structures for ion-electron transfer processes, the amount of lithium which can be topochemically inserted and extracted from the layered LiMO$_2$ compounds (without structural degradation) is limited. It is well established that complete removal of the lithium from lithiated layered oxides causes structural phase transformation. As a result of intercalation or deintercalation beyond the reversible limit, stacking disorder and structural transformation may occur. The reversible intercalation process is restricted to a limited concentration zone corresponding to the limit of elastic deformability of the lattice which is dictated by the nature of the chemical bonds and the capacity of the redox process for electron-ion transport.

EXPERIMENTAL

<u>SYNTHESIS</u>

The LiCo$_{1-y}$B$_y$O$_2$ and LiNi$_{1-y}$B$_y$O$_2$ samples were synthesized by direct reaction of LiOH, LiBO$_2$ or B, and CoO or Co$_3$O$_4$, and NiO respectively. The homogenous mixture of the powders were packed in a shallow ceramic boat and heat treated at 450 °C (melting point of LiOH) for 8 hours under a flow of oxygen. The product was then ground, packed in the ceramic boats and heat treated at 750 °C for 15 hours under oxygen. The last treatment was repeated twice. All powder mixing and grinding were done inside a glove box under an Argon atmosphere to prevent carbonate formation. Loss of lithium compound at high temperature (due to a high volatility of lithium oxide) has been adjusted by adding excess lithium hydroxide (10 %). The final composition of each sample was determined by elemental analysis using atomic absorption (AA) and inductive coupling plasma emission spectroscopy (ICPES).

The same process of synthesis was used to prepared the $LiCo_{1-y}Al_yO_2$ and $LiNi_{1-y}Al_yO_2$ samples from $LiOH$, γ-$LiAlO_2$, and Co_3O_4 or NiO respectively. An attempt to use Al_2O_3 for the Al source leads to a final product not free from impurity.

X-RAY DIFFRACTION

X-ray diffraction analysis of the sample was made using a Siemens D-5000 diffractometer equipped with a diffracted beam graphite single crystal monochromator and Cu-K_α radiation source. Data were recorded at a step width of 0.01 degrees 2θ and a scan rate of 0.02 degree 2θ per minute. Structural refinements are made based on $R\overline{3}m$ space group and calculated intensities are generated using the Philips Rietveld refinement procedure.

VIBRATIONAL SPECTROSCOPY

Infrared spectra were collected using Bruker FT-IR spectrometer model 113 equipped with a MCT mid-IR, DTGS, and liquid helium cooled far-IR blometer detectors. Samples were mixed with KBr or CsI, and pellets were made at 12 tons/cm^2 with good transparency for IR radiation. Data were collected in transmission mode after 100 scans in an argon atmosphere at 10^{-3} torr. IR spectra of some of the samples were also collected using a Nicolet FT-IR spectrometer after 100 scans. Data were plotted and analyzed using GRAMS 386 and Spectra Calc. software (from Galactica Industries Corp.).

Raman spectra of the samples were collected with a Spex 1403 double monochromator using the 514.5 nm laser line from the Spectra-Physics 164 Ar ion laser. For data analysis, Raman spectra files were imported to Spectra Calc. software. Raman spectra were also collected using a Spectra Physics spectrometer, where different excitation lines of an Ar ion laser were used (blue and green lines). Care has been taken to use a low energy laser beam to prevent reduction of cathode materials by the laser during collection of Raman spectra. The laser power was kept below 25 mW.

ELECTROCHEMICAL STUDIES

About 1 gram of active cathode materials was mixed with 10 weight percent conductive carbon black (Shawinagan Black), and about 4 w% EPDM (from a 2% EPDM in cyclohexane solution). A slurry of the mixture was applied to a carbon coated aluminum foil. The coated sample was dried in a vacuum oven antechamber at 100 °C for 10 hours in an argon atmosphere. The coated sample was pressed at 5 tons/10 cm^2 for 5 minutes. A disk with 5 cm^2 apparent surface area was cut form the pressed sample and used as a cathode in an electrochemical cell against a lithium anode. A Maccor battery cycler was used for charge-discharge of the cells, and data were collected using a 386 IBM computer. Cells were charged and discharged at 0.1 mA/cm^2 from 3.0 V to 4.5 V in a 1M $LiPF_6$ in ethylene carbonate - dimethylcarbonate (65/35 mole ratio) electrolyte at ambient temperature.

RESULTS AND DISCUSSIONS

<u>ELEMENTAL ANALYSIS</u>

Elemental analysis of samples indicated loss of lithium when heat treatments were made at high temperature (700 - 800 °C). Usually about 7-10% loss of lithium was observed when the heat treatment was made in a tube furnace under 20 ml/min oxygen flow between 750 to 850 °C for 12 hours. The loss of lithium can be adjusted during the preparation of samples by addition of excess lithium compounds (LiOH, Li_2O_2, or Li_2O). The elemental analysis of samples prepared with 8% excess lithium showed formation of stoichiometric compounds. The elemental analysis of each sample was done three times and average results on all samples were close to the stoichiometry. Adjustment of lithium concentration is an important factor for preparation of good quality impurity free $LiNiO_2$ and its substituted derivatives.

<u>X-RAY DIFFRACTION ANALYSIS</u>

X-ray diffraction of $LiNiO_2$ and $LiCoO_2$ are reported in Fig. 1. The diffraction lines can be indexed according to $R\bar{3}m$ space group. The ideal structure with $R\bar{3}m$ space group is shown in Fig. 2. The indexed diffraction lines correspond to an oxygen packing where lithium resides in octahedral 3b sites, between the MO_2, (M=Ni or Co) slabs. The Ni and Co are located in octahedral 3a sites and oxygen anion are in a cubic close-packing, occupying the 6c sites. The transition metals and lithium ions are occupying the alternating (111) planes. The Rietveld refinement analysis of our x-ray diffraction data show good agreement between the observed and calculated x-ray diffraction line intensities. The structural analysis of the $LiNiO_2$ indicate less than 1% nickel between the NiO_2 layers. Magnetic measurements have been very useful in determination of Ni^{II} content between the NiO_2 slabs (13-16). The lattice parameter for the $LiNiO_2$ were a_h=2.88 and c_h = 14.20 with c/a = 4.93.

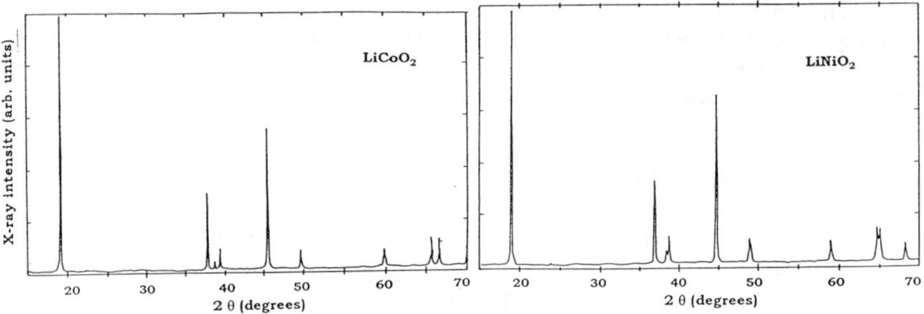

Fig. 1. X-ray diffraction of $LiCoO_2$ and $LiNiO_2$ with $R\bar{3}m$ space group.

X-ray diffraction of aluminum substituted lithium nickelate samples are shown in Fig. 3. The intensity ratios for Al sustituted lithiated nickelate are affected by Al concentration and the hexagonal cell dimension are also modified due to Al substitution. The degree of trigonal distortion and departure from ideal layered structure are observed from changes of (c-a)/a values, which is indicative of the local inhomogenity around the substituted elements. Further information can be obtained form, the intensity ratios of I_{104}/I_{003} and $I_{(006+102)}/I_{101}$. The cation ordering normally decreases the of I_{104}/I_{003} and $I_{(006+102)}/I_{101}$ ratios.

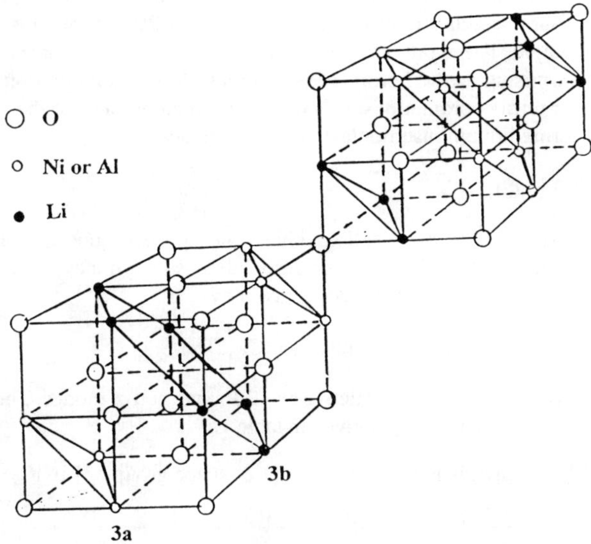

Fig. 2. Ideal structure of Layered $LiMO_2$ with $R\bar{3}m$ space group. Li and Ni are located at alternate (111) planes. Ni ions are located at 3a, and Li at 3b sites. The oxygen anions are in 6c sites.

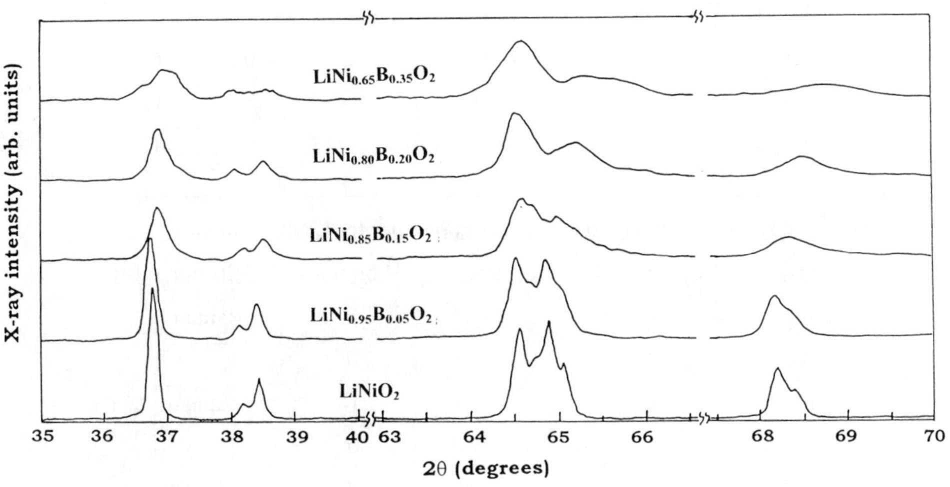

Fig. 3. X-ray diffraction of lithium nickelate, $LiNiO_2$, at various levels of boron substitution.

The x-ray diffraction of Al doped lithium nickelate with Al concentration higher than 25% shows some residual impurity, and it was difficult to prepare single phase impurity free at higher Al concentrations. This may indicate that there is a solubility limit for Al substitution in lithium nickelate, (the solubility limit may be around 25%). Substitution of boron to lithium nickelate and lithium cobaltate also follow similar trend in which higher than 25% born substitution generate impurity phase in the sample. The x-ray diffraction line broadening was enhanced by the addition of boron. This is not surprising as the boron oxide is a glass former. In addition boron tends to bond to oxygen in tri- or tetra coordinations. It is important to emphasis that the B and Al remain in three valency state during lithium insertion/extraction processes.

VIBRATIONAL SPECTROSCOPY

Materials with layered structure and crystallographic $R\bar{3}m$ space group have a corresponding spectroscopic space group of D^5_{3d} (26). The corresponding site symmetry for the D^5_{3d} which contains both the IR and the Raman active modes is given by

$$C_s(6), 2C_2(6), 2C_{2h}(3), C_{3v}, 2D_{3d}.$$

The factor group analysis of D^5_{3d} yields 4 IR active and 2 Raman active modes. The invariance conditions for the $LiNiO_2$ and the $LiCoO_2$ is given in table 1.

Table 1. Invariance condition for D^5_{3d} space group of $LiNiO_2$

D_{3d}	E	$2C_3$	$3C_2$	I	$2S_6$	$3_{\sigma d}$
3 (a) D_{3d}	E	$2C_2$	$3C_2$	I	$2S_6$	$3_{\sigma d}$
Co^{III}, Ni^{III}	3	3	3	3	3	3
3 (b) Li^+	3	3	3	3	3	3
3 (c) C_v	E	$2C_3$				$3_{\sigma d}$
$6O^{2-}$	6	6	0	0	0	6
ω_p	12	12	6	6	6	12
$\chi_p(hex)$	36	0	-6	-18	0	12

Table 2. Normal modes and selection rules for $LiNiO_2$ structure

D_{3d}	Total modes	Acoustic	Vibrations	Selection rules
A_{1g}	1		1	Raman
A_{2g}	0		0	
E_g	1		1	Raman
A_{1u}	0		0	
A_{2u}	3	z	2	IR E$\|$C
E_u	3	(x,y)	2	IR E$\perp$C

Excluding the acoustic modes the active vibrational modes are $A_{1g} + E_g + 2A_{2u} + 2E_u$, where the four $(2A_{2u} + 2E_u)$ vibrational modes are IR active. The A_{1g} and E_g are Raman active modes and they correspond mainly to vibrations of oxygen cages, (oxygen - oxygen vibration in the c direction and in parrallel to the Li and transition metal planes). The two Raman active modes for $LiMO_2$ (M= transition metal) are located at 400 - 650 cm^{-1} region. The position of these bands are less sensitive to the transition metals, but their intensities are very sensitive to the long rang order in MO_2 slabs (27).

The IR spectra of the lithium nickelate doped with boron and aluminum are shown in Fig. 4. There is considerable difference between ionic radii of B^{+3} and Ni^{III} according to Shannon and Prewitt (33) which may enhance the cation ordering in $LiNi_{1-y}B_yO_2$. The pure $LiNiO_2$ phase has two IR peaks between 400 - 600 wavenumbers. These two peaks are characteristic of the NiO_6 vibrations. Substitution of boron and aluminum reduces the resolution of the two peaks. The lower frequency IR peak is more sensitive to the addition of boron. The systematic change in band frequency as a function of aluminum or boron doping is not surprising, as in solid solutions with various concentrations of substitution, a systematic shift has been observed (28).

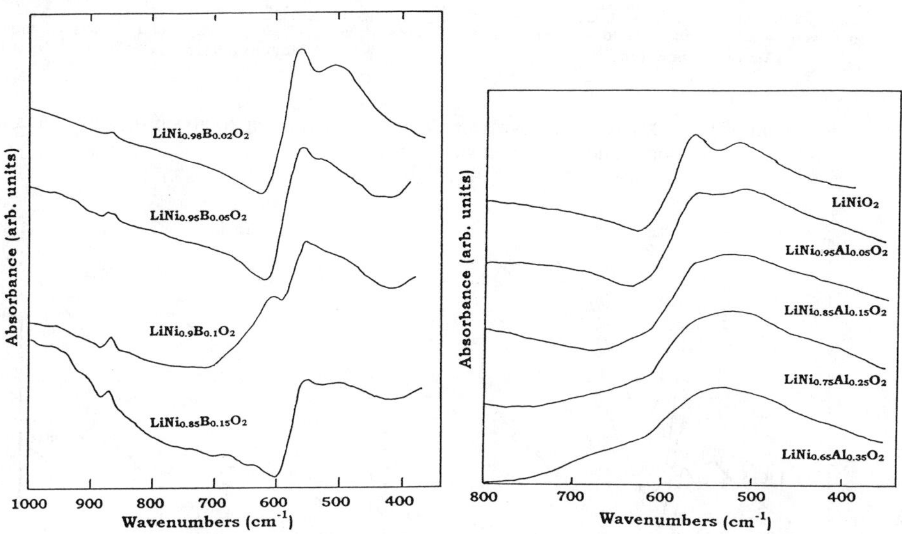

Fig. 4. IR spectra of $LiNiO_2$ at various levels of boron and aluminum substitution.

The isotopic ^6Li-^7Li replacement in $LiMO_2$ with D^5_{3d} space group has proven that the Far infrared peak between 200-300 cm^{-1} is related to an asymmetric stretching vibration of LiO_6 (29-32). The IR vibrational peaks between the 400 -700 cm^{-1} are related to the MO_6 octahedra (M = metal in MO_2 slabs). In order to better understand the vibrational spectra of the layered $LiMO_2$ compounds with $R\bar{3}m$ space group, we consider a structure which is consists of compressed MO_6 and elongated LiO_6 octahedra that yields distinct vibrations in two different frequency regions. The frequency of the LiO_6 octahedra is sensitive to the ionic radius of the M^{III} and ionicity of the bond in MO_6 octahedra. Vibrational spectra of the MO_6 octahedra is also affected by the way in which these octahedra are linked to each other.

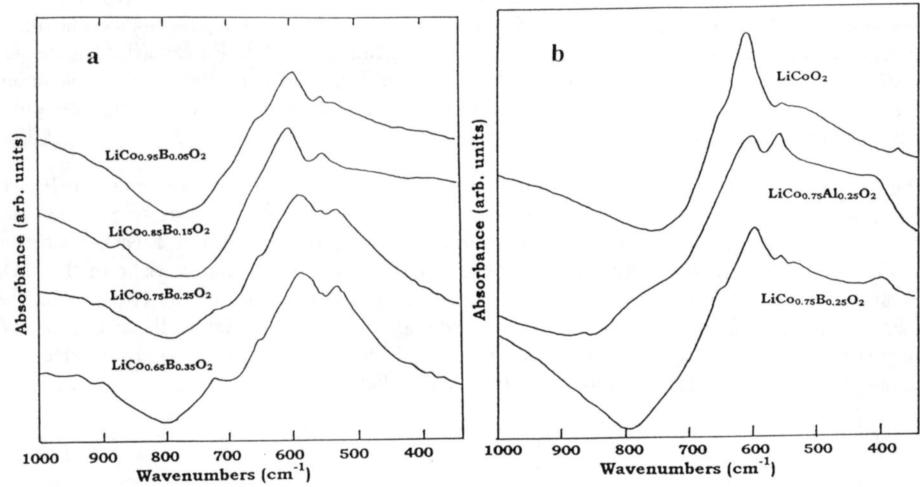

Fig. 5. IR spectra of $LiCoO_2$ at various levels *sp* elements substitution, a) boron substitution, and b) comparison of boron and aluminum substitution.

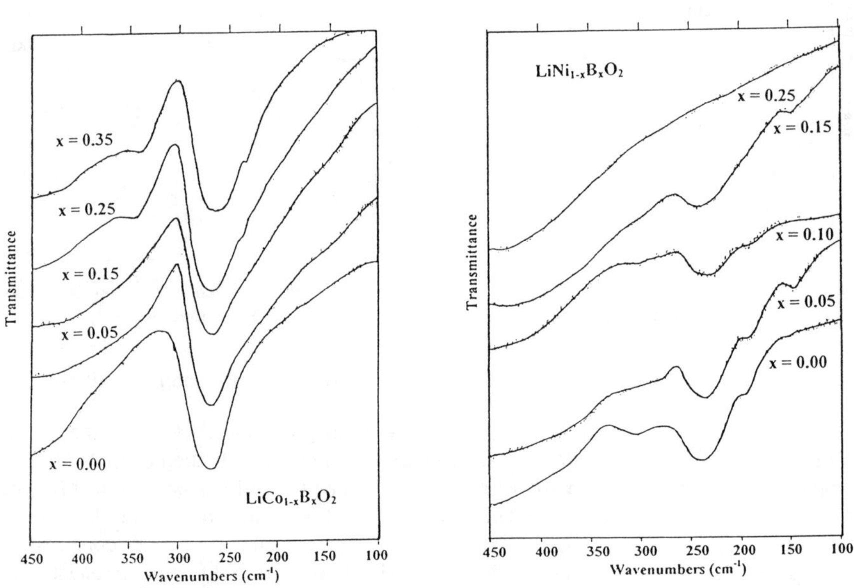

Fig. 6. Far IR spectra of LiO_6 octahedra in $LiNiO_2$ and $LiCoO_2$ with boron substitution.

Fig. 5 shows IR spectra of $LiCoO_2$ with various amounts of B substitution. The broadening of the IR peak can be interpreted as an increase in MO_6 distortion due to the addition of boron in the MO_2 layers. It is well known that boron oxide is used as glass former, and it may reduces the long range order in the $LiNiO_2$. This effect is also seen in x-ray diffraction of lithium nickelate doped with boron. Similar distortion of lattice is also observed when the $LiCoO_2$ is doped with Al (Fig 5b). The IR vibrational modes of the LiO_6 octahedra appear in far infrared region, 200-300 cm^{-1}. Fig. 6. shows the effect of B substitution in lithium nickelate and lithium cobaltate. The band broadening of the peak related to the vibration of LiO_6 is clearly shown upon substitution with the boron, *sp*, element.

ELECTROCHEMICAL STUDIES

The layered transition metal oxides such as $LiCoO_2$ and $LiNiO_2$ are the most attractive cathode materials for the rechargeable lithium batteries, because they have high specific capacity (Ah/cm^3 and Ah/g), high operating cell voltage, and excellent rechargeability. The $LiNiO_2$ is a more attractive choice because of lower cost as compared with that of $LiCoO_2$. However, the formation of low lithium content Li_xNiO_2 (x < 0.2) causes cycle life failure. In addition, the material become highly catalytic toward electrolyte oxidation, and some of the nickel ions may migrate to lithium sites. The formation of pure $LiNiO_2$ is difficult, and residual Ni^{II} (up to 1-2%) exist between the NiO_2 slabs. In fact, the first cycle irreversibility during charge discharge is mainly related to the amount of Ni^{II} between the slabs of NiO_2, which require extra charge for oxidation to higher valency state (34), when electrolyte decomposition is controlled. Through careful synthesis and adjustment of lithium concentration in the material during heat treatment, we obtained $LiNiO_2$ very close to stoichiometry. Fig. 7 shows first charge discharge of the samples at 0.1 mA/cm^2. Total irreversibility of the first cycle is an indicative of the stoichiometric compound with negligible Ni^{II} in the lithium layer.

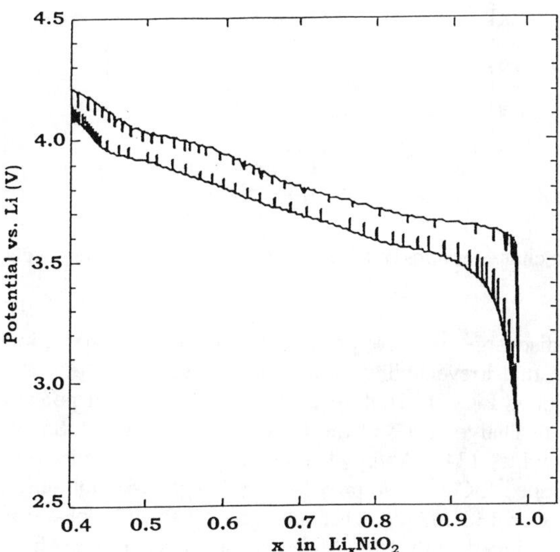

Fig. 7. Charge discharge characteristic of $LiNiO_2$ at 0.1 mA/cm^2 in (1M $LiPF_6$ + 65/35 EC-DMC)

In order to stabilize the structure of LiNiO$_2$ at low lithium content, we have doped the material with sp elements such as B and Al. These dopants do not participate in oxidation reduction processes during charge-discharge of the cell. Fig. 8. shows the charge discharge of LiNiO$_2$ doped with various amounts of Al. The specific capacity of the cell reduces as a function of Al content, as the Al does not participate in the redox process. The capacity changes almost linearly with respect to the Al content up to 25 atomic%. This result may indicate a solubility limit in the formation of a solid solution between LiAlO$_2$ and LiNiO$_2$. It is also observed that the first charge-discharge irreversibility also increases as the amounts of Al in the samples were increased. The extra nickel in the lithium sites may cause the increase in first cycle irreversibility.

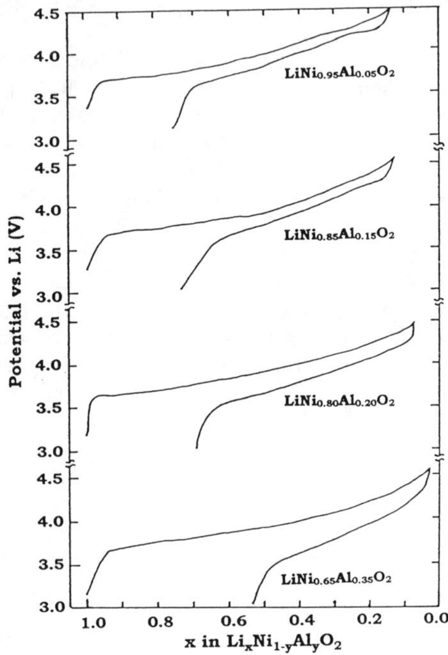

Fig. 8. Charge-discharge characteristics of Lithium nickelate with various levels of Al substitution.

Fig. 9 shows charge-discharge voltage profiles for LiCoO$_2$ with various amounts of Al substitution. It is shown that irreversibility during the first cycle is a less severe problem in the case of Al substituted doped LiCoO$_2$. High reversibility with low over potential polarization was observed during charge-discharge cycles of the LiCoO$_2$ doped with Al. Sample of LiCoO$_2$ with 10% Al substitution provides 134 mAh/g, when charged up to 4.3 volts versus metallic lithium anode (Fig. 9). Boron doped LiCoO$_2$ also provide very low polarization during charge-discharge cycling, with capacity over 130 mAh/g when charged up to 4.3 volts versus lithium anode. Capacity of the LiCoO$_2$ doped with 15% boron remain over 125 mAh/g after 100 charge-discharge cycles. It appears that a less electrolyte decomposition occur using B or Al substituted LiNiO$_2$ and LiCoO$_2$ cathodes in lithium batteries.

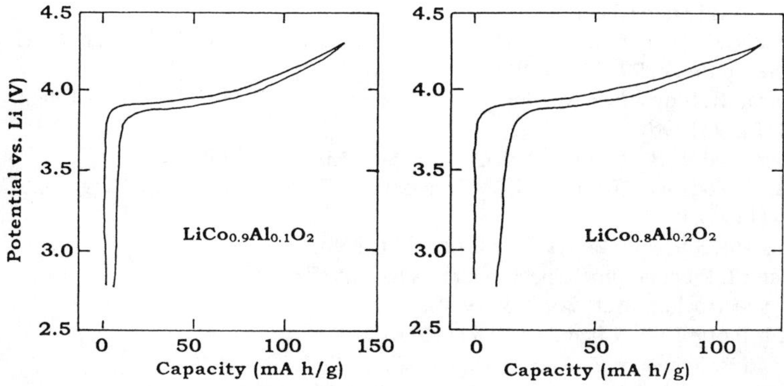

Fig. 9. Charge discharge characteristics of $LiCoO_2$ at various levels of Al substitution.

REFERENCES:

1. C. Delmas, J. Braconnier, P. Hagenmuller, Mat. Res. Bull. **17**, 117, 1982.
2. M. Broussley, F. Perton, P. Biensan, J.M. Bodet, J. Labat, A. Lecerf, C. Delmas, A. Rougier, J.P. Peres, J. Power Sources **54**, 109 (1995).
3. A. Rougier, P. Gravereau and C. Delmas, J. Electrochem. Soc. **143**, 1168 (1996).
4. M.G.S.R. Thomas, P.G. Bruce, J.B. Goodenough, Solid State Ionics **17**, 13 (1985).
5. R.J. Gummow, M.M. Thackeray, W.I.F. David, S. Hull, Mat. Res. Bull. **27**, 327 (1992).
6. E. Antolini and M. Ferretti, J. Solid State Chem. **117**, 1, (1995).
7. J.N. Reimers and J.R. Dahn, J. Electrochem. Soc. **139**, 2091 (1993).
8. J.R. Dahn, U. Von Sacken, C.A. Michal, Solid State Ionics **44**, 87 (1990).
9. T. Ohzuku, U. Ueda, M. Nagayama, Electrochim. Acta **38**, 1159 (1993).
10. T. Ohzuku and A. Ueda, J. Electrochem. Soc., 141, 2972 (1994).
11. T. Nagaura and K. Tozawa, Progress in Batteries & Solar Cells **9**, 209 (1990).
12. T. Nagaura, Progress in Batteries & Materials **10**, 218 (1991).
13. J.B. Goodenough, D.G. Wickham, W.J. Croft, J. Phys. Chem. of Solids **5**, 107 (1958).
14. J.P. Kemp, P.A. Cox, J.W. Hodby, J. Phys. Condens. Matter **2**, 6699 (1990).
15. R. Stoyanova, E. Zhecheva, C. Friebel, J. Phys. Chem. Solids **54**, 9 (1993).
16. H. Taguchi and Y. Takahashi, J. Mater. Sci. **19**, 3347 (1984).
17. G. Dutta, A. Manthiram, G.B. Goodenough, J-C. Grenier, J. Solid State Chem. **96**, 123 (1992), and A. Rougier, C. Delmas, G. Chouteau, J. Phys. and Chem. of Solids **57**, 1101 (1996).
18. E. Mosser and W.B. Pearson, Acta Cryst. **12**, 1012 (1959).
19. J. Rouxel, in <u>Intercalated Layered Materials</u>, edited by F. Levy (D. Reidel, Dordrecht, 1979), p. 201.
20. J. Rouxel and P. Monceau, J. Phys. C. **11**, 4117 (1978).
21. R. Schollhorn, in <u>Physics of Intercalation Compounds</u>, edited by L. Pietronero and E. Tosatti, (Springer Verlag, Berlin/NY, 1981), p. 33.

22. J. Rouxel and M. Tournoux, Solid State Ionics **84**, 141 (1996).

23. M. Oku, J. Solid State Chem. **23**, 177 (1978).

24. V.R. Galakhov, E.Z. Kurmaev, St. Uhlenbrock, M. Neumann, D.G. Kellerman, M. Gorshkov, Solid State Commun. **99**, 221 (1996).

25. E. Zhecheva, R. Stoyanova, M. Gorova, R. Alcantara, J. Morales, J.L. Tirado, Chem. Mater. **8**, 1429 (1996).

26. R.K. Moore and W.B. White, J. Am. Ceramic Soc. **53**, 679 (1970).

27. M. Inaba, Y. Todzuka, H. Yoshida, Y. Grincourt, A. Tasaka, Y. Tomida, Z. Ogumi, Chem. Lett. 889 (1995).

28. I.F. Chang and S.S. Mitra, Phys. Rev. **172**, 924 (1968).

29. P. Tarte and J. Preudhomme, Spectrochim. Acta **26A**, 747 (1970).

30. P. Tarte, Spectrochim. Acta **20**, 238 (1964).

31. P. Tarte, J. Inorg. Nucl. Chem. **29**, 915 (1967).

32. P. Hope and B. Schepers, Z. Anorg. Allgem. Chem. **295**, 233 (1958).

33. R.D. Shannon and C.T. Prewitt, Acta Cryst. B, **25**, 925 (1989).

34. J.P. Peres, C. Delmas, A. Rougier, M. Broussely, F. Perton, P. Biensan, P. Willmann, J. Phys. Chem. of Solids **57**, 1057 (1996).

SYNTHESIS, STRUCTURE, LATTICE DYNAMICS AND ELECTROCHEMISTRY OF LITHIATED MANGANESE SPINEL, LiMn$_2$O$_4$

C. JULIEN*, A. ROUGIER**, G.A. NAZRI**
*Laboratoire de Physique des Solides, CNRS-ERS 113, Universite Pierre et Marie Curie,
4 place Jussieu, 75252 Paris cedex 05, France
**Physics and Physical Chemistry Department, General Motors R & D Center, RCEL,
Warren, MI 48090-9055

ABSTRACT

We report synthesis, crystal structure, lattice dynamics, and electrochemical features of the lithiated manganese oxide spinel prepared through solid state reaction by careful selection of precursors and synthesis conditions. Elemental analysis shows that the material is a lithium-rich spinel phase. X-ray diffraction data and Rietveld refinement indicate formation of a single phase, impurity free, normal spinel of LiMn$_2$O$_4$. Lattice dynamics have been investigated by vibrational spectroscopy and group theoretical analysis has been carried out. Electrochemical performances of the lithiated spinel manganese oxide have been investigated, and the voltage profile of the cathode during lithium intercalation-deintercalation processes, close to equilibrium, has been obtained. The upper 4-volt plateau provides over 130 mA h/g with an excellent cyclability.

INTRODUCTION

The recent interest in developing advanced lithium batteries such as rocking-chair or lithium-ion type cells, has stimulated investigation on high-performance positive electrodes [1]. The spinel manganese oxide LiMn$_2$O$_4$ belongs to the class of 4-volt intercalation hosts and remains potentially attractive because of its low cost and low toxicity. Since the electrochemical characteristics of LiMn$_2$O$_4$ depend on its crystal nature, size and shape, various kinds of preparation methods have been developed to improve the characteristics of the LiMn$_2$O$_4$ spinel [2-5]. Also, it has been pointed out that the rechargeability can be markedly improved and the fading capacity loss upon cycling can be minimized for the lithium-rich material [6].

This work reports on the preparation and characterization of the LiMn$_2$O$_4$ cathodic material. The Rietveld method has been used to refine the structure. Structural properties have been also studied using Raman and infrared spectroscopy. Electrochemical features show our prepared manganese oxide spinel is a promising material for lithium ion batteries.

SYNTHESIS AND STRUCTURAL PROPERTIES

The LiMn$_2$O$_4$ cathode material was prepared by mixing fine powder of LiOH and MnO$_2$ in 1:2 mole ratio with 7 (mole%) excess of LiOH to compensate for the loss of lithium during final process. The mixture was ground inside of a dry box filled with Argon to prevent carbonate formation. The mixture then was packed in a ceramic boat and heat treated at 470°C for four hours. At this temperature the LiOH melts and uniformly coat the MnO$_2$ grains. The product was then ground to fine powder and heat treated at 750°C for 12 hours under flow of oxygen. The final product was cooled to room temperature under continuous flow of oxygen. Loss of lithium due to evaporation at 750°C has been estimated from our previous work by elemental analysis to be about 7 mole%. The final product of the synthesis was kept inside the dry box for characterization.

Mat. Res. Soc. Symp. Proc. Vol. 453 © 1997 Materials Research Society

The final composition of the sample was determined by elemental analysis for Li and Mn using atomic absorption (AA) and inductive coupling plasma emission spectroscopy (ICPES). The reported stoichiometry is the average of the three measurements. The calculated stoichiometry of our sample was $Li_{1.008}Mn_{1.980}O_4$, a lithium-rich manganese oxide spinel.

X-ray diffraction (XRD) analysis of the sample was made using a Siemens D-5000 diffractometer equipped with a diffracted beam graphite single crystal monochromator and Cu-K$_\alpha$ radiation source. Data were recorded at a step width of 0.01 degrees 2θ and a scan rate of 0.02 degree 2θ per minute as shown in Fig. 1. Structural refinements are made based on Fd3m space group and calculated intensities are generated using the Siemens Rietveld refinement procedure. The XRD patterns are summarized in Table I. A good agreement is obtained between observed and calculated values.

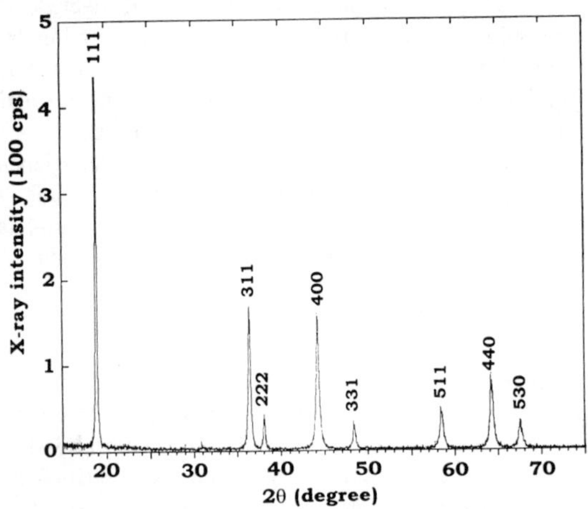

Fig. 1. X-ray diffraction pattern of lithium-rich $LiMn_2O_4$ spinel.

Table I. Structural refinement based on Fd3m space group of $LiMn_2O_4$.

h k l	$D_{calculated}$	$D_{observed}$
1 1 1	4.73078	4.71700
3 1 1	2.47057	2.46900
2 2 2	2.36539	2.36500
4 0 0	2.04849	2.04900
3 3 1	1.87982	1.88100
5 1 1	1.57693	1.57900
4 4 0	1.44850	1.45100
5 3 0	1.40525	1.38800

The TGA measurements have been carried out using a Perkin-Elmer TGA-7 apparatus. Fig. 2 shows the thermogram of a $LiMn_2O_4$ sample heat treated in air up to 900°C at a rate of 5°C/min and cooling down at the same rate. As can be seen in Fig. 2, oxygen loss occurred at about 700°C and re-adsorption of oxygen is observed during cooling the sample.

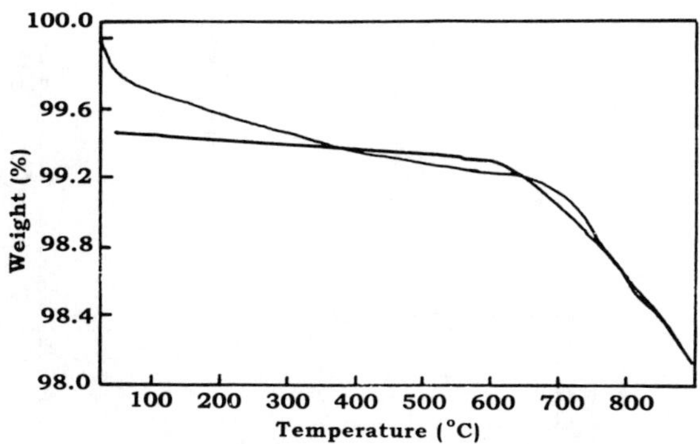

Fig. 2. Thermogravimetric analysis of $LiMn_2O_4$ in air.

Vibrational spectra of $LiMn_2O_4$ sample have been studied by Raman scattering (RS) and Fourier transform infrared (FTIR) absorption using a Jobin-Yvon U1000 spectrometer and a Bruker IFS113v interferometer, respectively. Figs. 3 and 4 show the FTIR and RS spectra of a lithiated $LiMn_2O_4$ sample and an electrochemically delithiated $Li_{0.3}Mn_2O_4$ sample, respectively. Analysis of vibrational spectra of $LiMn_2O_4$ with Fd3m space group yields nine optic modes: 5 modes are Raman active ($A_{1g}+E_g+3F_{2g}$) and 4 modes are infrared active ($4F_{1u}$). It is also convenient to analyse these spectra in terms of localized vibrations, considering the spinel structure built of MnO_6 octahedra and LiO_4 tetrahedra [7].

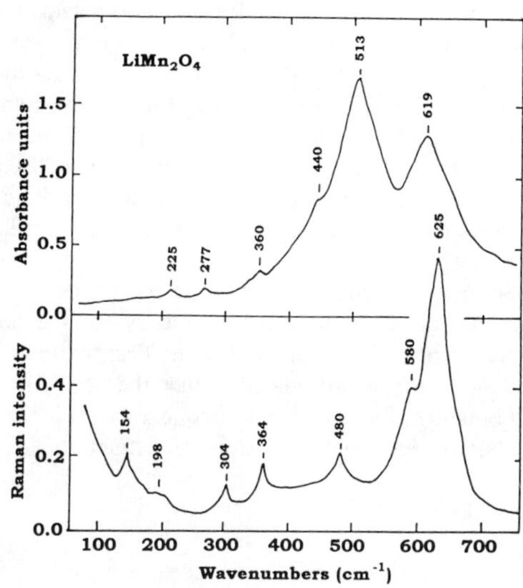

Fig. 3. FTIR and Raman spectra of lithiated $LiMn_2O_4$ spinel recorded at room temperature.

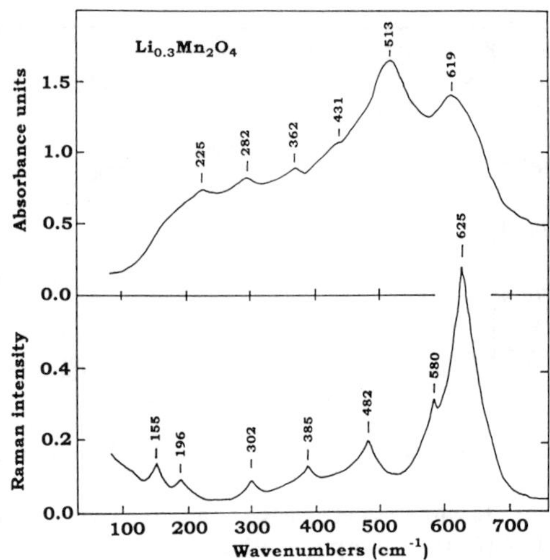

Fig. 4. FTIR and Raman spectra of delithiated $Li_{0.3}Mn_2O_4$ spinel recorded at room temperature.

The Raman band located at 625 cm^{-1} can be viewed as the symmetric Mn-O stretching vibration of MnO_6 groups. The position and halfwidth of this band remain almost unchanged upon delithiation. This band is assigned to the A_{1g} symmetry in the O_h^7 spectroscopic space group. The RS peak at 304 cm^{-1} has the E_g symmetry whereas the peaks located at 154, 364 and 480 cm^{-1} belong to F_{2g} symmetry. The high-frequency bands of the FTIR spectra located at ca. 619 and 513 cm^{-1} are attributed to the symmetric stretching modes of MnO_6 groups whereas bands, which are observed in the low-frequency region at ca. 255, 277 and 370 cm^{-1}, are assigned to the bending modes of $LiMn_2O_4$. Because FTIR spectroscopy is capable of probing directly the near-neighbor environment of the cation, we can consider the local environment of lithium ions in this material. The vibrational frequency of the LiO_4 tetrahedron has been pointed out at 435 cm^{-1} in the spinel $LiFeCr_4O_8$ [8]. It has been also demonstrated that the IR resonant frequencies of alkali metal cations in their equilibrium position into inorganic oxide glasses can be described as cation mass dependent bands [9]. This leads to the frequency at ca. 440 cm^{-1} for oscillation of the Li^+ ion with O^{2-} near-neighbors in $LiMn_2O_4$.

The RS and FTIR spectra of $Li_{1-x}Mn_2O_4$ delithiated phases have the same shape than the $LiMn_2O_4$ spinel. The bands have almost the same frequency for the composition in the range $0 \leq x \leq 0.7$. Only the peak located at ca. 360 cm^{-1} shifts significantly upon lithium extraction. Thus the resonance frequency at 360 cm^{-1} is an indication that the configuration of the F_{2g} mode is predominantly due to Li-O bonding. The increase in frequency of this mode is in good agreement with the structural modification of the spinel phase upon Li extraction.

ELECTROCHEMICAL STUDIES

About 1 gram of active cathode materials was mixed with 10 weight percent conductive carbon black (Shawinagan Black), and about 4 w% EPDM (from a 2% EPDM in cyclohexane solution). A slurry of the mixture was applied to a carbon coated aluminum foil. The coated sample was dried in a vacuum oven antechamber at 100°C for 10 hours in an argon atmosphere. The coated sample was pressed at 5 tons/10 cm^2 for 5 minutes. A disk with 5 cm^2 apparent surface area was cut form the pressed sample and used as a cathode in an electrochemical cell against a lithium anode. A Maccor battery cycler was used for charge-discharge of the cells, and data were collected using a 386 IBM computer. Cells were charged and discharged at 0.2 mA/cm^2 from 3.7 V to 4.5 V in a 1M LiPF$_6$ in ethylene carbonate-dimethylcarbonate (65/35 mole ratio) electrolyte at ambient temperature. Fig. 5. shows the voltage profile of a Li//LiMn$_2$O$_4$ cell during charge-discharge cycle. The results indicate the good charge and discharge performance. Sample shows a two-step voltage profile characteristic of the spinel Li$_{1.008}$Mn$_{1.980}$O$_4$ cathode material.

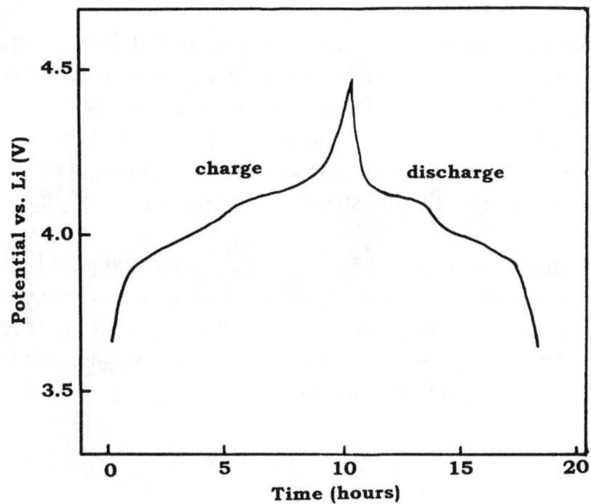

Fig. 5. Voltage profile of LiMn$_2$O$_4$ during charge-discharge cycle at 0.2 mA/cm^2.

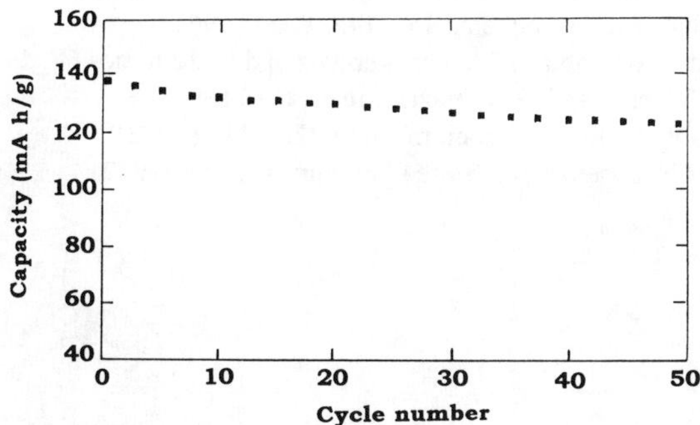

Fig. 6. Discharge capacity of LiMn$_2$O$_4$ at a rate of 0.2 mA/cm^2.

The rechargeability of the LiMn$_2$O$_4$ sample was examined at 0.2 mA/cm^2 current density between 3.5 and 4.5 V. The capacity is plotted as a function of the cycle number in Fig. 6. The upper 4-volt plateau provides over 130 mA h/g with an excellent cyclability. The capacity fading of the cell is very slow. The capacity loss is about 10% of initial capacity and the loss factor is 0.15% per cycle. Considering the low loss factor upon charge-discharge cycling, we can conclude that our sample is able to deliver a rechargeable capacity of more than 100 mA h/g for long cycles.

CONCLUSION

The $LiMn_2O_4$ cathode material has been prepared through solid state reaction by careful selection of precursors and synthesis conditions. Elemental analysis shows that the material is a lithium-rich spinel phase. X-ray diffraction data and Rietveld refinement indicate formation of a single phase, impurity free, normal spinel of $LiMn_2O_4$.

Lattice dynamics have been investigated using Raman and infrared vibrational spectroscopy. Vibrational spectra has been investigated using either a classical group theoretical analysis or a local environment model. The RS and FTIR spectra of $Li_{1-x}Mn_2O_4$ delithiated phases have the same shape than the $LiMn_2O_4$ spinel. The resonance frequency at 360 cm^{-1} is an indication that the configuration of the F_{2g} mode is predominantly due to Li-O bonding. The increasing frequency of this mode is in good agreement with the structural modification of the spinel phase upon Li extraction.

Electrochemical performances of the $LiMn_2O_4$ have been investigated, and the voltage profile of the cathode during lithium intercalation-deintercalation processes, close to equilibrium, has been obtained. The upper 4-volt plateau provides over 130 mA h/g with an excellent cyclability. The capacity loss factor of 0.15% per cycle results that our $Li_{1.008}Mn_{1.980}O_4$ sample is able to deliver a rechargeable capacity of more than 100 mA h/g for long cycles.

REFERENCES

1. C. Julien and G.A. Nazri, <u>Solid State Batteries: Materials Design and Optimization</u>, Kluwer, Boston, 1994.
2. J. M. Tarascon and D. Guyomard, Electrochim. Acta **38**, 1221 (1993).
3. T. Ohzuku, M. Kitagawa and T. Hirai, J. Electrochem. Soc. **137**, 767 (1990).
4. J.N. Reimers and J.R. Dahn, J. Electrochem. Soc. **139**, 2091 (1992).
5. Y. Xia and M. Yoshio, J. Electrochem. Soc. **143**, 825 (1996).
6. R.J. Gummow, A. de Kock and M.M. Thackeray, Solid State Ionics **69**, 59 (1994).
7. C. Julien, C. Perez-Vicente and G.A. Nazri, Ionics **2**, 1 (1996).
8. J. Preudhomme and P. Tarte, Spectrochim. Acta **27A**, 845 (1972).
9. G.J. Exarhos and W.N. Risen, Solid State Commun. **11**, 755 (1972).

NEW MANGANESE OXIDES BY HYDROTHERMAL REACTION OF PERMANGANATES

RONGJI CHEN, PETER ZAVALIJ AND M. STANLEY WHITTINGHAM
Chemistry Department and Materials Research Center, State University of New York at Binghamton, Binghamton, NY 13902-6000, USA

ABSTRACT

The mild hydrothermal decomposition of aqueous permanganate solutions has been found to lead to new layered manganes oxides. In the case of alkali permanganates, $AMnO_4$, layered birnessite-type compounds are formed with the general formula - $A_xMnO_2 \cdot nH_2O$. These compounds have R 3 m rhombohedral structures analogous to the layered disulfides. The water is reversibly lost on heating, and the compounds readily intercalate lithium ions in a reversible manner. In the case of the hydrothermal decomposition of tetramethylammonium permanganate in the presence of nickel, the new compound $NiMnO_3$ is formed that has a novel structure. This compound also has interesting magnetic properties, being a paramagnet above 111K and a ferromagnetic below that temperature.

INTRODUCTION

There has been much interest in the synthesis and characterization of new materials by soft chemistry approaches over the last decade [1]. Such materials might be expected to exhibit new structures and different properties than known materials as they are formed under conditions of kinetic rather than thermodynamic control. One particularly successful synthesis approach [2] is hydrothermal under mild conditions, that is no higher than 200°C. The TMA cation was found to be particularly adept in forming new structures [3, 4], and it is not necessarily retained in the structure formed[5]. Layered manganese oxides have also been formed in our laboratory by the hydrothermal decomposition of aqueous permanganate solutions. In the case of the potassium and sodium permanganates well crystalline birnessite type phases were formed [6, 7]. Thus, we explored the possibility of forming new manganates by the hydrothermal decomposition of solutions containing both tetramethylammonium and permanganate ions.

Manganese oxides, such as $LiMn_2O_4$, are of particular interest as the cathode of lithium batteries [8]. However, only 0.5 lithium can be cycled per manganese atom so that their energy densities are not high. In contrast the SONY lithium ion cell uses the very expensive $LiCoO_2$ cathode. Hence, extensive research is underway to find promising candidates for cathode materials in lithium secondary batteries to replace the expensive cobalt used in the present commercial cells, and a manganese oxide that behaved like the layered $LiCoO_2$ would be a prime candidate for this application because of its high free energy of reaction with lithium and relatively low cost.

This paper is concerned with the formation of manganese oxides under mild hydrothermal conditions, and will discuss in particular the role of the cation in determining the structure formed, and the reactivity of the formed manganese oxides for both redox and intercalation reactions. Nickel, cobalt and manganese oxides are of particular interest as cathode materials in lithium batteries and electrochromic devices, as catalysts, and for their magnetic behavior. The $NiMnO_3$ reported here has a new structure and might be of interest as a battery cathode or as a magnetic material.

Mat. Res. Soc. Symp. Proc. Vol. 453 © 1997 Materials Research Society

EXPERIMENTAL

Potassium and sodium manganese oxides were synthesized by the methods previously described [1,2]. Since LiMnO$_4$ is not commercially available, It took several steps to synthesize lithium manganese oxide. First, HMnO$_4$ was synthesized by passing KMnO$_4$ solution through a column containing proton-charged cationic ion exchanger; then LiOH was used to neutralize HMnO$_4$ until pH about 9, thus LiMnO$_4$ formed, this solution was finally acidized to pH about 5.2; final the resulting solution was transferred to a 125-ml Teflon lined autoclave (Parr bomb), sealed, and reacted hydrothermally for 3.5 days at 160°C. The resulting brown lithium manganese oxide was filtered, and dried at 45°C in air for use. The pH of the solution after reaction was alkaline, pH 12-13. As we should mention here that the direct hydrothermal reaction of HMnO$_4$ at 160°C led to MnO$_2$ instead of desired layered H$_x$MnO$_2$·yH$_2$O.

The nickel manganese oxide was synthesized by reacting hydrothermally for 2 days at 200°C a mixture of 0.001 moles of [N(CH$_3$)$_4$]MnO$_4$, 0.0015 moles Ni(OOCCH$_3$)$_2$•4H$_2$O and 0.001 moles Li$_2$CO$_3$ in 25 ml water. The crystals formed were filtered and washed with 2M HNO$_3$ to remove any Ni(OH)$_2$ impurity. The [N(CH$_3$)$_4$]MnO$_4$ starting material was synthesized by mixing together, in a 1:1 molar ratio, separate aqueous solutions of N(CH$_3$)$_4$Cl and KMnO$_4$; [N(CH$_3$)$_4$]MnO$_4$ precipitated immediately. The reader is warned that the tetramethyl ammonium permanganate is potentially explosive and should not be heated in the dry state, as for example on a TGA.

X-ray powder diffraction was performed using CuKa radiation on a Scintag θ−θ diffractometer. The TGA data was obtained on a Perkin Elmer model TGA 7, the FTIR on a Perkin Elmer 1600 series, the Electron Microprobe on a JEOL8900, and the chemical analysis was obtained on an ARL Spectrospan-7 DCP Atomic Emission Spectrometer. The degree of reduction of the vanadium oxide by lithium was determined by reaction with n-butyl lithium. Electrochemical studies were conducted in lithium cells using LiAsF$_6$ in PC:DME=1:1 as the electrolyte; a MacPile potentiostat was used to cycle the cells. The manganate was hot-pressed, at ≈150°C, with 10 wt% teflon powder and 10 wt% carbon black into a stainless steel Exmet™ grid, and discharged at 0.1 mA/cm^2.

RESULTS AND DISCUSSION

<u>Alkali Manganates</u>

The yield of product based on manganese approached 100%. Consistent with this yield analysis, the basic liquid product contained potassium species (as potassium carbonate due to reaction with CO$_2$ in the air) and no manganese species; there was no indication of the formation of K$_2$MnO$_4$ as noted in the dry thermal decomposition of KMnO$_4$ [29]. The overall reaction is consistent with the equation:

$$\text{AlkMnO}_4 + (1-x+2y)/2\text{H}_2\text{O} \rightarrow \text{Alk}_x\text{MnO}_2\bullet y\text{H}_2\text{O} + (1-x)\text{AlkOH} + (3+x)/4\text{O}_2$$

The X-ray pattern of the unreported Li$_x$MnO$_2$·yH$_2$O, as shown in Fig. 1 shows similar layered characteristics to the reported K$_{0.27}$MnO$_2$·0.54H$_2$O [6] and Na$_{0.35}$MnO$_2$·0.78H$_2$O [7] with repeat distance around 7Å. It loses about 12.5% H$_2$O and the d-spacing contracts to 4.85Å upon TGA heating. Li content was determined by DCP-AES as x=0.47. The contents of this layered oxide is very similar to that of spinel, LiMn$_2$O$_4$. In fact, it is metastable and converts to spinel upon heating over 400°C.

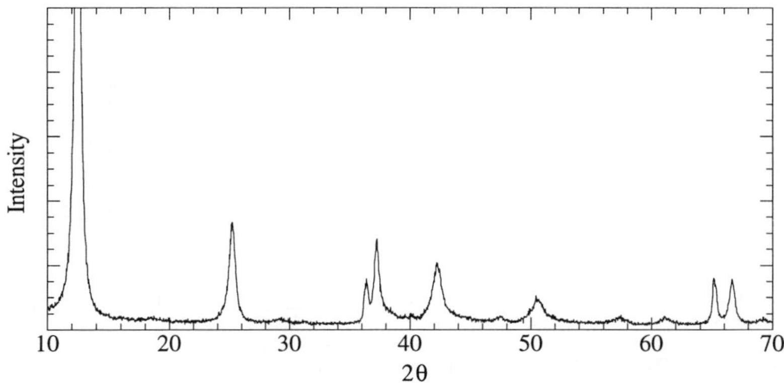

Fig. 1. X-ray diffraction pattern of the layered lithium manganate, $Li_xMnO_2 \cdot nH_2O$.

A typical thermal gravimetric analysis of the manganese oxides phase $Alk_xMnO_2 \cdot yH_2O$ in oxygen at a heating rate of 3°C/min is shown in Fig. 2. The water is totally lost on heating; after cooling to room temperature and exposure to the ambient environment the sample took up water again; however, if the sample was maintained in a dry atmosphere no water was absorbed and no weight loss was observed on a second heating cycle. Although the sodium and potassium samples were totally reversible, the lithium sample was not. This is related to a partial conversion to the spinel phase on drying, which is not reversible. The uptake and discharge of water was readily confirmed by x-ray diffraction studies for all three alkalis. The schematics of these layered manganates are shown in figure 2, anhydrous material in figure 2a, one water layer (birnessite) in figure 2b, and two water layers (buserite) in figure 2c.

The x-ray diffraction pattern of the dried layered phase Li_xMnO_2 is almost identical to that of the spinel $LiMn_2O_4$ so that x-ray diffraction is not a reliable method for determining the structure in this case. Both structures contain oxygen in cubic close-packing, but in the case of the layered form the lithium and maganese ions are ordered giving alternating planes containing all lithium or all manganese. However, the degree of water uptake by the sample gives an indication of the proportion of layer to cubic spinel; as the cubic spinel phase does not swell in water.

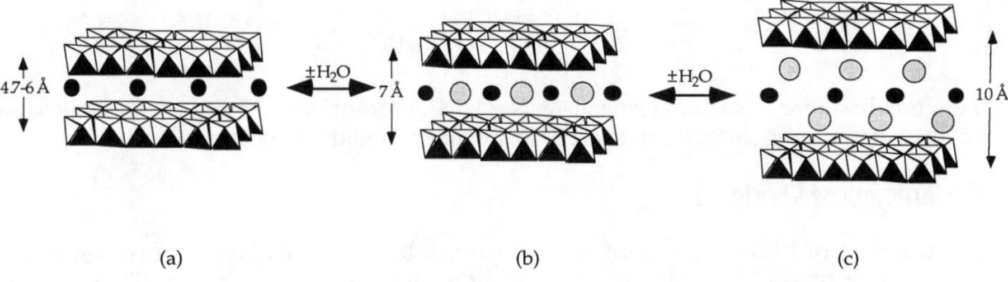

Fig. 2. Schematic of the interaction of the layered manganates with water, ◯ = H_2O; ● = M ion. Left: dry sample; center: one water layer; and right: two water layers [7].

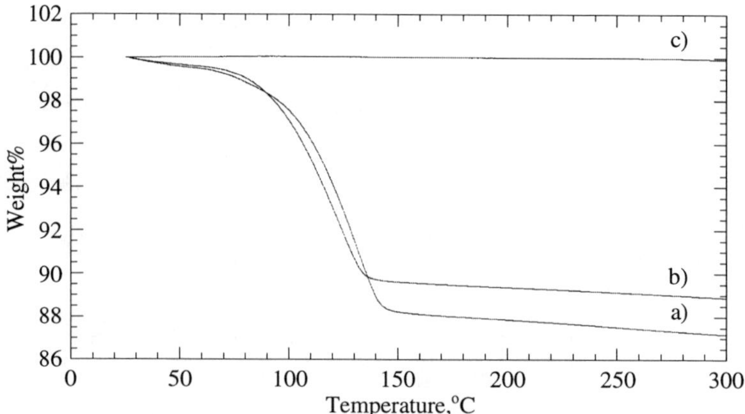

Fig. 3. Thermal analysis of a layered sodium manganate [7].

All the alkali manganates, after drying to remove any water, were evaluated as cathodes in lithium cells. The results for the sodium manganate are shown in figure 4. Good reversibility is observed, with only slight fading as cycling progresses.

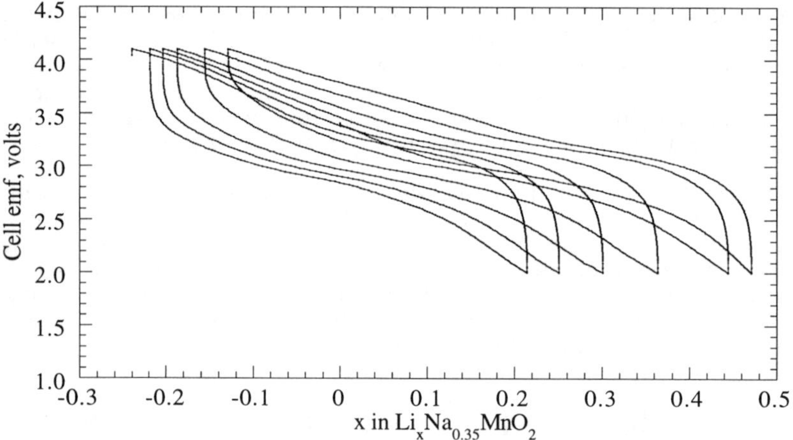

Fig. 4. Cycling behavior of sodium manganate, Na_yMnO_2. Part of the shift to the left is due to an apparent overcharge on each cycle, and part probably due to sodium removal.

Nickel Managanese Oxide [9]

The structure of $NiMnO_3$ is shown in Figure 5. It comprises layers of nickel oxide and manganese oxide. Nickel is octahedrally coordinated by oxygen. These NiO_6 octahedra form sheets by edge and face sharing. The manganese is in square pyramidal sites with the base Mn-O distances of 1.90 and 1.91 Å and the apex distance 2.11 Å. These MnO_5 square pyramids share edges and corners to form the manganese sheet.

These manganese square pyramids also share corners with nickel octahedra to form the 3-dimensional framework with one trigonal prismatic cavity per formula unit. The distance from the center of this cavity to the center of the neighboring oxygens is from 2.1 to 2.3Å, giving a cavity size of radius of about 0.8Å. This is about the same size as a lithium ion. These cavities form a tunnel down the a axis, down which small ions should be able to diffuse. On heating to above 400°C this structure converts into the known ilmenite form [10, 11].

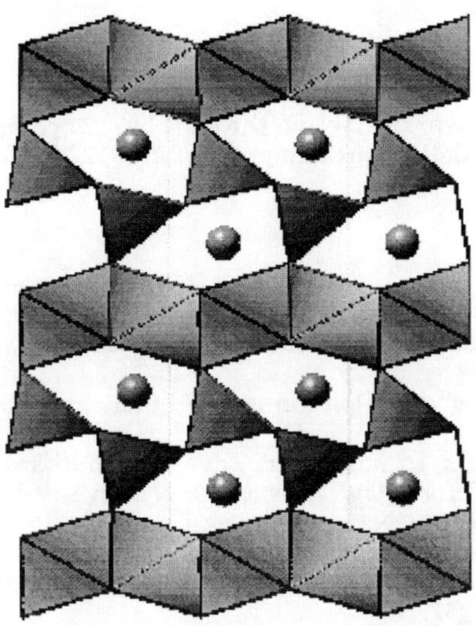

Figure 5. Structure of NiMnO$_3$ showing the probable position of the lithium ion.

Reaction of this nickel manganese oxide with n-butyl lithium showed the uptake of 0.91 Li per formula unit of NiMnO$_3$. The lithium content was further confirmed by DCP analysis. The intercalated compound Li$_{0.91}$NiMnO$_3$ appeared to be stable in air. X-ray diffraction pattern indicates the same structure as NiMnO$_3$ itself. This composition is consistent with a lithium ion entering each of the vacant cavities in the NiMnO$_3$ structure, as shown in figure 4. However, the final determination of the atomic positions within this structure will have to await a neutron diffraction study, which will also help elucidate the magnetic behavior.

This compound was found [9] to be paramagnetic above 111K, unlike the ferrimagnetic ilmenite form of NiMnO$_3$ [10]. The formula Ni^{3+}Mn^{3+}O$_3$ gives a calculated magnetic moment of 4.42 μ_B, in good agreement with the experimental value of 4.34 μ_B.

CONCLUSIONS

The hydrothermal decomposition of permanganate solutions is a clean and rapid synthetic route to layered manganates. The alkali manganates all form a simple layered rhombohedral structure analogous to the well known layered disulfides. They all intercalate lithium readily and reversibly and so are of potential interest as battery cathodes.

A new structural form of $NiMnO_3$ has been formed where the manganese are in planes containing MnO_5 square pyramids and the nickel in planes containing NiO_6 octahedra. It converts into the known ilmenite form of $NiMnO_3$ at around 400°C. The structure contains empty tunnels into which lithium ions can be intercalated. Above 111K it exhibits paramagnetism, consistent with Ni^{3+} and Mn^{3+} ions; below 111K it shows feromagnetism.

ACKNOWLEDGEMENTS

We thank the Department of Energy, through Lawrence Berkeley Laboratory, and the National Science Foundation through grant DMR-9422667 for partial support of this work.

REFERENCES

1. J. Rouxel, M. Tournoux, and R. Brec, ed. *Soft Chemistry Routes to New Materials - Chimie Douce-*. Materials Science Forum, Vol. 152-153. 1994, Trans Tech Publications: Switzerland.
2. M. S. Whittingham, Current Opinion in Solid State and Materials Science, **1** (1996) 227.
3. M. S. Whittingham, J. Li, J. Guo, and P. Zavalij. *Hydrothermal Synthesis of New Oxide Materials using the Tetramethyl Ammonium Ion.* in *Soft Chemistry Routes to New Materials*. **152-153** (1993) 99. Nantes, France: Trans Tech Publications Ltd.
4. Y. J. Li and M. S. Whittingham, Solid State Ionics, **63** (1993) 391.
5. M. S. Whittingham, J. Guo, R. Chen, T. Chirayil, G. Janauer, and P. Zavalij, Solid State Ionics, **75** (1995) 257.
6. R. Chen, P. Zavalij, and M. S. Whittingham, Chem Mater, **8** (1996) 1275.
7. R. Chen, T. Chirayil, and M. S. Whittingham, Proceedings of the 10th International Symposium on Solid State Ionics, Singapore, December 1995. Solid State Ionics, **86-88** (1996) 1.
8. J-M. Tarascon, editor, *Lithium Batteries* in Solid State Ionics, **69** (1994)
9. R. Chen, P. Y. Zavalij, and M. S. Whittingham, Chem. Mater., (1997) in press.
10. M. Pernet, J. C. Joubert, and B. Ferrand, Solid State Communications, **16** (1975) 503.
11. W. H. Cloud, Phys. Rev., **111** (1958) 1046.

A COMPARATIVE STUDY OF LITHIUM AND SODIUM INSERTION IN TWO BLOCK STRUCTURE TYPE PHASES, $W_3Nb_{14}O_{44}$ AND $W_4Nb_{26}O_{77}$

ANTONIO F. FUENTES, A. MARTINEZ DE LA CRUZ AND LETICIA M. TORRES-MARTINEZ

Facultad de Ciencias Químicas, División de Estudios Superiores, Universidad Autónoma de Nuevo León. ApartadoPostal 1864, Monterrey, Nuevo León, México.

ABSTRACT

A comparative study of lithium insertion in $W_4Nb_{26}O_{77}$ and $W_3Nb_{14}O_{44}$, two block structure type phases, has been carried out. This process is, in both cases, reversible for a similar number of lithium ions incorporated to total metal atoms ratio (Li/ΣM), 1.15 in $W_3Nb_{14}O_{44}$ and 1.23 in $W_4Nb_{26}O_{77}$. No significant structural changes are seen by X-ray diffraction upon insertion. Sodium insertion in $W_3Nb_{14}O_{44}$ has been also studied observing a substantial reduction on the number of ions incorporated when compared with lithium (Na/ΣM: 0.03).

INTRODUCTION

Considerable attention has been focused over the years in the Nb_2O_5-WO_3 binary system [1-3]. Those phases containing up to 48% of WO_3, such as $W_3Nb_{14}O_{44}$ and $W_4Nb_{26}O_{77}$, present ordered shear structures made up of corner-sharing MO_6 octahedra forming $n \times m \times \infty$ blocks (where n and m are the length and width of each block in number of octahedra, respectively). Blocks are joined up on the same plane by tetrahedrally coordinated tungsten atoms and, at different planes, along the short axis by edge sharing [4]. This atomic arrangement originates a large number of interconnected channels which can be occupied by additional ions and makes these phases potential host materials for insertion reactions. Additionally, edge-sharing adds stability to the corner-sharing octahedra framework minimizing structural distortion which might take place during insertion. The idealized structures of $W_3Nb_{14}O_{44}$ (n and m = 4) and $W_4Nb_{26}O_{77}$ are presented in Fig.1 [3, 5]. This latter phase contains blocks of two different sizes (3 x 4 x ∞ and 4 x 4 x ∞) which alternate in a regular sequence. Although some previous works on lithium insertion in analogous phases have been already published [6, 7], not much information is available about sodium insertion in this type of oxides. A comparative work on lithium and sodium reactions would be very convenient when studying site occupancy. We have carried out in this work a comparative study of lithium insertion in $W_3Nb_{14}O_{44}$ and $W_4Nb_{26}O_{77}$ and of sodium insertion in $W_3Nb_{14}O_{44}$ in order to draw some conclusions about the importance of ion size in insertion reactions in these materials.

EXPERIMENTAL

Pristine materials were synthesized by solid state reaction. Starting oxides, WO_3 (Aldrich Chem. Co, 99+%) and Nb_2O_5 (Alfa Products, 99.5%), were weighed in the appropriate stoichiometric ratio, mixed and ground in an agate mortar. This mixture was pressed into 10 mm diameter pellets, placed in a platinum crucible and taken to an electrical furnace where temperature was slowly raised to reach reaction conditions: 1200°C/45h for $W_4Nb_{26}O_{77}$ and 1150°C/98h for $W_3Nb_{14}O_{44}$.

Mat. Res. Soc. Symp. Proc. Vol. 453 © 1997 Materials Research Society

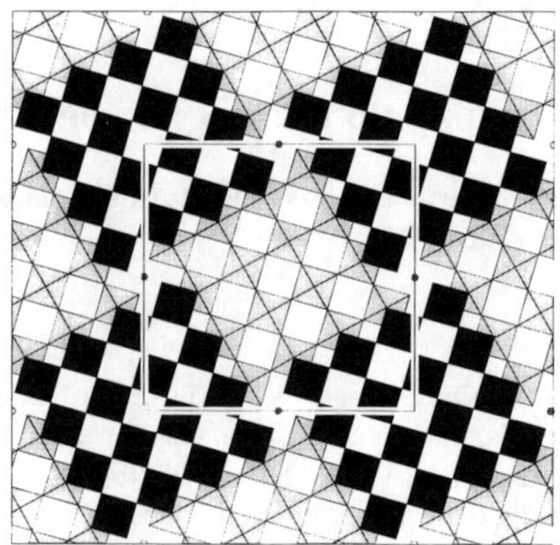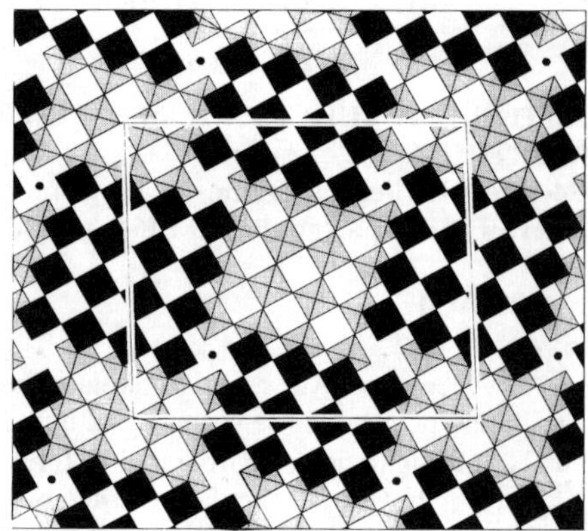

Fig.1: Idealized structures of $W_4Nb_{26}O_{77}$ (right) and $W_3Nb_{14}O_{44}$ (left). Dark and grey squares are octahedra at y = 0 and ½ respectively. Solid and light circles are tungsten atoms at y = ¼ and ¾. Unit cell for each phase is also shown [3, 5]

Phase identification was carried out by X-ray powder diffraction in a Siemens D-5000 diffractometer using Cu-Kα radiation (λ = 1.5418 Å) and by HRTEM in a JEOL 4000 electron microscope. In each case, products obtained through these reactions could be indexed on the basis of the tetragonal and monoclinic cells reported in literature for $W_4Nb_{26}O_{77}$ [3] and $W_3Nb_{14}O_{44}$ [4] respectively. These products were used as active materials when preparing the cells for the electrochemical study.

Electrochemical experiments were carried out in a multichannel potentiostatic-galvanostatic system MacPile II [8] using a Swagelok type cell [9] with metal lithium or sodium acting as negative and reference electrode simultaneously. Cathodes were used as pellets prepared by mixing each niobium-tungsten oxide with carbon black and a binder, EPDT (0.5% Ethylene-propilen-diene terpolymer in ciclohexane), in a 89:10:1 ratio. The electrolytes used were 1 $mol \cdot dm^{-3}$ solutions of $LiClO_4$ in propylene carbonate (PC) and in an 50:50 mixture of ethylene carbonate (EC) and diethoxy ethane (DEE) [10] when inserting lithium and an 1 $mol \cdot dm^{-3}$ solution of $NaClO_4$ in PC when inserting sodium. Cells assembly was carried out in a MBraun argon glove box with continuous purge of oxygen and water vapour. Two types of electrochemical experiments were performed in this study, either in a current- or in a potential-controlled mode. Galvanostatic experiments were carried out at continuous or intermittent discharge rates using a current density of 0.08 mA/cm^2. In galvanostatic intermittent titrations, GITT [11], this current density was applied for 0.5 hours and the system relaxed for 0.5 hours or until voltage changes with time were smaller than 10 mV/h. Potentiodynamic titrations were carried out by a stepwise technique also known as "Step Potential Electrochemical Spectroscopy", SPECS [12]. In this technique, the potential is stepwise increased or decreased with time recording charge increments vs. time at each potential level, thus allowing the insertion reaction kinetics study. Simultaneous observation of reaction kinetics and incremental capacity ($\partial x/\partial V$ vs. x) enables us to study transformations, single phase domains and two phase equilibria which might take place on insertion. Typical experimental conditions were set at 10 mV/h potential steps with a charge recording resolution of 5 μAh.

Some lithiated phases were prepared by chemical reaction of pristine materials with n-butyllithium in n-hexane. Lithium incorporated was quantified by AAS in a Varian SpectrAA5 spectrometer using a λ of 670.8 nm.

RESULTS AND DISCUSSION

Lithium insertion

As it can be seen in Fig.2 where two voltage vs. lithium content plots obtained in GITT experiments are shown, $W_4Nb_{26}O_{77}$ and $W_3Nb_{14}O_{44}$ present a similar behavior on lithium insertion. In both cases there is an initial voltage drop followed by a plateau at around 1.7 volts and a second voltage drop resulting in a comparable number of lithium ions incorporated to total metal atoms ratio (Li/ΣM) : 1.23 in $W_4Nb_{26}O_{77}$ vs. 1.15 in $W_3Nb_{14}O_{44}$. These values are in agreement with previous results observed by Cava et al. (6) in analogous phases. As it can also be appreciated in this graph, lithium insertion seems to be reversible in those working voltage limits used in this study (from 3.2 to 1.2 volts vs. Li^+-Li^0), and nearly all the ions incorporated when discharging the cells, are removed when charging (96.5% in $W_3Nb_{14}O_{44}$, C/217, and 97.3% in $W_4Nb_{26}O_{77}$, C/300). A larger number of lithium ions were incorporated in these phases (Li/ΣM = 1.65 in $W_3Nb_{14}O_{44}$ and 1.53 in $W_4Nb_{26}O_{77}$) in experiments run down to 1 volts (not shown) although the insertion of such numbers of ions could not be considered reversible reactions with 20% of them remaining in the structure after completing a charge-discharge cycle. To further test lithium insertion reversibility between voltage limits mentioned above, a few charge-discharge cycles were completed in galvanostatic mode and at continuous discharge rates, in two cells similar to the ones described previously. Results confirmed that lithium insertion in $W_4Nb_{26}O_{77}$ and $W_3Nb_{14}O_{44}$ is a reversible reaction with no significant capacity losses on insertion. Three cycles completed in a cell with $W_3Nb_{14}O_{44}$ as cathode, are shown in Fig.3. A comparison between SPECS response for $W_4Nb_{26}O_{77}$ and $W_3Nb_{14}O_{44}$ is shown in Fig.4 as charge increments vs. V plot. Lithium insertion and deinsertion reactions in both phases seem to go through analogous intermediates as observed by the similarity between curves. Two steps, labelled as A and B in Fig.4 and slightly shifted when comparing between phases, are present in both reactions at around 1.2 and 1.7 volts respectively. These steps are better marked in oxidative than in reductive current. In order to identify the formation of solid solution regions during insertion, representations of incremental capacity, $\partial x/\partial V$, and V vs. lithium content were plotted for both phases. Plots obtained in each case were equivalent supporting the idea of comparable reduction and oxidation mechanisms on insertion. As an example, Fig.5 shows an $\partial x/\partial V$ and V vs. lithium content plot for $W_4Nb_{26}O_{77}$. As it can be observed in this curve, at least two solid solution regions, labelled as I and II, are formed during lithium insertion in these materials. To study the influence of lithium insertion on the structure of the pristine materials, we proceeded with the chemical synthesis and characterization of some lithiated compositions. Chemical insertion was carried out as described previously while phase identification was performed by X-ray powder diffraction. Powder patterns were indexed on the basis of the unit cells described in literature for the pristine phases: tetragonal for $W_3Nb_{14}O_{44}$ [4] and monoclinic for $W_4Nb_{26}O_{77}$ [3] respectively. Cell parameters were calculated by a least-square refinement programme and compared with those obtained for the parent oxides. Results are shown in Table I and II. As it can be seen, a small expansion of the unit cell along the three directions, around 0.7% in volume between the limiting values, is observed in both cases as the number of lithium ions incorporated increases.

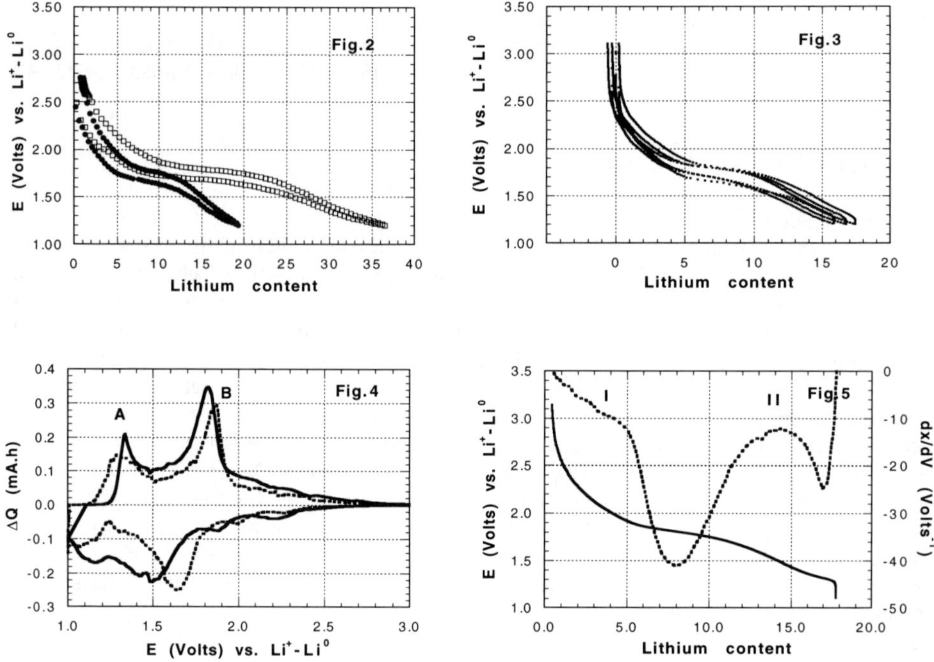

Fig.2.- E (volts) vs. lithium content plots obtained in GITT experiments for $W_3Nb_{14}O_{44}$, C/217, (solid circles) and $W_4Nb_{26}O_{77}$, C/300, (open squares). **Fig.3**.- E (Volts) vs. lithium content plot obtained after three complete charge-discharge cycles at continuous discharge rates, C/72, in a cell using $W_3Nb_{14}O_{44}$ as active material. **Fig.4**.- Total charge increments for each potential level vs. E (Volts) plot obtained in SPECS experiments for $W_4Nb_{26}O_{77}$ (dashed line) and $W_3Nb_{14}O_{44}$ (solid line). **Fig.5**.- E (solid line) and $\partial x/\partial V$ (dashed line) vs. lithium content plot obtained for $W_4Nb_{26}O_{77}$ from SPECS data showing the formation of two solid solution regions on insertion.

Table I.- Cell parameters and cell volumes for $Li_xW_4Nb_{26}O_{77}$

	a (Å)	b (Å)	c (Å)	V (Å³)
$W_4Nb_{26}O_{77}$ (r)	29.74 (1)	3.824 (2)	25.98 (1)	2951.1
$W_4Nb_{26}O_{77}$	29.66 (1)	3.822 (2)	25.95 (2)	2940.7 ± 2.9
$Li_{0.6}W_4Nb_{26}O_{77}$	29.73 (1)	3.821 (2)	25.99 (1)	2950.9 ± 2.6
$Li_{4.0}W_4Nb_{26}O_{77}$	29.73 (2)	3.822 (2)	25.99 (2)	2953.0 ± 3.1
$Li_{32.5}W_4Nb_{26}O_{77}$	29.73 (5)	3.821 (7)	26.10 (5)	2964.3 ± 9.8

(r) Reported values taken from reference (3)

Table II.- Cell parameters and cell volumes for $Li_xW_3Nb_{14}O_{44}$

	a (Å)	b (Å)	c(Å)	V (Å^3)
$W_3Nb_{14}O_{44}$ (r)	21.02	21.02	3.824	1689.6
$W_3Nb_{14}O_{44}$	21.024 (6)	21.024 (6)	3.829 (2)	1692.7 ± 1.0
$Li_3W_3Nb_{14}O_{44}$	21.11 (1)	21.11 (1)	3.826 (3)	1705.6 ± 2.2
$Li_{19.2}W_3Nb_{14}O_{44}$	21.094 (9)	21.094 (9)	3.830 (3)	1704.4 ± 1.8

(r) reported values taken from reference (4)

For $W_3Nb_{14}O_{44}$, the insertion of a small number of lithium ions resulted in a rapid expansion of the unit cell with no further enlargement as the number of ions incorporated increased. A great similarity was observed between the X-ray powder diffraction patterns of lithiated and pristine materials which is in agreement with the good reversibility observed in the electrochemical test cells. A drastic colour change was observed as the number of lithium ions incorporated increased: the original colour of the mixed parent oxides, pale yellow, became successively royal blue, dark blue and black for $W_4Nb_{26}O_{77}$ and grey and black for $W_3Nb_{14}O_{44}$. It is important to point out that the incorporation of a very small amount of lithium in $W_4Nb_{26}O_{77}$ (0.6 Li ions per formula) already produced an important colour change of the parent oxide from yellow to blue. Therefore, when quantifying lithium, the complete extraction of these lithium ions incorporated in each case could be guessed by the colour change of the remaining powder back to pale yellow.

Sodium insertion

Sodium insertion was carried out in a cell already described elsewhere in this work and using $W_3Nb_{14}O_{44}$ as active material. SPECS response of this cell between 3 and 1 volts vs. Na^0/Na^+ is presented in Fig.6 as an incremental capacity ($\partial y/\partial V$) and V vs. sodium content plot. With the experimental conditions set in this test, 10 mV/h, only 0.5 sodium ions can be incorporated in $W_3Nb_{14}O_{44}$ (Na/ΣM = 0.03). Therefore, there is a dramatic decrease in the number of ions incorporated depending on whether lithium or sodium are the guest species. Additionally, sodium insertion does not seem to be a reversible reaction with only 88% of these ions incorporated, being removed after completing a charge-discharge cycle. The formation of two solid solution regions on insertion, labelled in Fig.6 as I and II, can be observed in this plot.

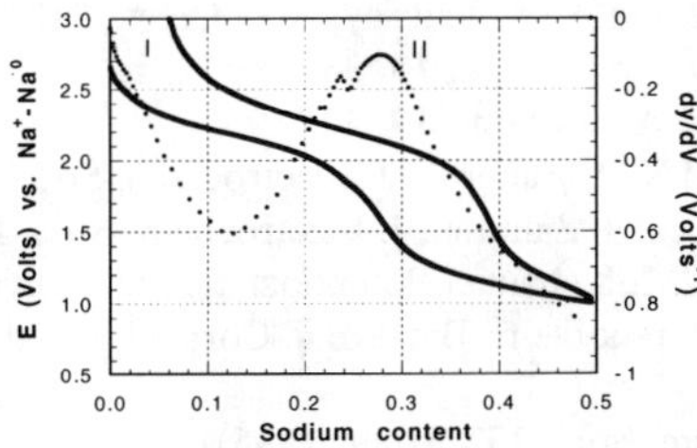

Fig.6.- SPECS response of W3Nb14O44 upon sodium insertion. This plot represents $\partial y/\partial V$ (dotted line) and E (solid line) vs. sodium content. I and II represent solid solution regions.

CONCLUSIONS

A comparative study of lithium and sodium insertion in two block-structure type phases, $W_3Nb_{14}O_{44}$ and $W_4Nb_{26}O_{77}$, has been carried out. Lithium insertion in these phases and between these voltage limits used in this study (3.2 to 1.2 volts vs. Li^+-Li^0), can be considered as a reversible reaction. Thus, both phases can reversibly accomodate a relatively large number of lithium ions (Li/ΣM = 1.15 for $W_3Nb_{14}O_{44}$ and 1.23 for $W_4Nb_{26}O_{77}$) without major structural changes. Only a small expansion of the unit cell could be noticed by X-ray powder diffraction as the number of lithium ions incorporated increases making for a topotactic reaction (0.7% in cell volume related to the pristine materials for these phases with the largest lithium content in each case). Additionally, lithium insertion and deinsertion in $W_3Nb_{14}O_{44}$ and $W_4Nb_{26}O_{77}$ seem to go through comparable intermediates as infered by the similarity between both phases response to SPECS experiments: reduction and oxidation go through two analogous steps. From incremental capacity vs. E (Volts) plots, the formation of at least two solid solution regions is proposed for lithium insertion in $W_3Nb_{14}O_{44}$ and $W_4Nb_{26}O_{77}$. A much smaller number of sodium ions can be accomodated in $W_3Nb_{14}O_{44}$ via an electrochemical method (Na/ΣM: 0.03). This reaction does not seem to be reversible with 12% of these sodium ions incorporated when charging remaining in the structure after completing a reduction-oxidation cycle. Although sodium ionic radius is only 35% larger than that of lithium, there is a dramatic decrease in the number of ions incorporated in $W_3Nb_{14}O_{44}$ depending on whether lithium or sodium, is the metal inserted. This situation can not be only explained on the basis of differences in atomic radii. Different occupational preferences and the possibility that a large structural distortion is taking place on sodium insertion, will contribute to those differences.

We have given a comparative example of lithium and sodium insertion on $W_4Nb_{26}O_{77}$ and $W_3Nb_{14}O_{44}$, two block-type structures showing that for a given charge, size plays an important role in intercalation reactions. However some additional factors have to be taken into account if a solid explanation of differences in the number of ions inserted, is to be given.

ACKNOWLEDGEMENTS

We would like to thank CONACYT for the financial support given to this work (projects 1488-E9207, DO603-9201 and 476200-5-4917E). Our acknowledgement goes also to Dr. M. José-Yacamán for giving us access to the electron microscope at IFUNAM (México D.F.).

REFERENCES

1.- R.S. Roth and A.D. Wadsley, Acta Cryst., **19**, 26 (1965).

2.- R.S. Roth and A.D. Wadsley, Acta Cryst., **19**, 32 (1965).

3.- S. Andersson; W.G. Mumme and A.D. Wadsley, Acta Cryst., **21**, 802 (1966).

4.- R.S. Roth and A.D. Wadsley, Acta Cryst., **19**, 38 (1965).

5.- R.S. Roth and A.D. Wadsley, Acta Cryst., **19**, 42 (1965).

6.- R.J. Cava; D.W. Murphy and S.M. Zahurak, J. Electrochem. Soc., **130**, 2345 (1983).

7.- R.J. Cava; D.W Murphy and S.M. Zahurak, J. Electrochem. Soc., **130**, 243 (1983).

8.- C. Mouget and Y. Chabre, "Multichannel Potentiostatic and Galvanostatic System MacPile", Licensed from CNRS and UJF Grenoble to Bio-Logic Corp., 1 Av. de l'Europe, F-38640, Claix, France.

9.- J.M. Tarascon, J. Electrochem. Soc., **132**, 2089 (1985).

10.- D. Guyomard and J.M. Tarascon, J. Electrochem. Soc., **139**. 937 (1992).

11.- W. Weppner and R.A. Huggins, J. Electrochem. Soc., **124**, 1569 (1977).

12.- Y. Chabre, J. Electrochem. Soc., **138**, 329 (1979).

KINETICS OF ELECTROCHROMIC EFFECT IN AMORPHOUS TUNGSTEN TRIOXIDE FILMS

I.A. KONOVALOV*, V.B. NECHITAJLO**, E. SHUSHAKOVA*
*Toronto, Canada, cl591@freenet.toronto.on.ca
**Institute of Physics, 46 Pr. Nauki, Kiev, Ukraine.

ABSTRACT

Reversible coloration occurs in tungsten trioxide (WO_3) films when double injection of electrons and ions is provided in its volume. Experiments were performed in electrochemical cell with liquid electrolyte, where protons were used for injection in amorphous WO_3 films. Kinetics of electrochromic coloration process was studied in spectral range of 300 nm to 1200 nm. Dependence of optical density on injected charge was found to be different in various spectral regions for coloring and bleaching. Infrared vibrational spectra of electro-chemically colored crystalline and amorphous WO_3 were obtained. Electronic structure features of amorphous WO_3 related to electorchromism are discussed.

INTRODUCTION

Transition metal oxides of the types WO_3, V_2O_5, TiO_2, etc., are able to change their optical characteristics reversibly in the visible range. When this is achieved by electrical (usually electrochemical) treatment then it is referred to as electrochromic (EC) effect. This phenomenon offers a lot of applications not only in the field of display technologies and information recording, but also in energy storage devices because the electrochemical coloration process is equivalent to charging a battery [1,2].

It is well established that EC coloration of oxide films occurs due to a double injection of electrons and cations (such as H^+, Li^+, Na^+) in film volume. The best-known EC materials, WO_3 and MoO_3, possess perovskite-like structure in which the metal oxygen polyhedra are joined at their vertices. This type of structure produces the high mobility for inserted cations, and compounds as tungsten (or molybdenum) bronzes A_xWO_3 (A-metal atom or hydrogen, $0 < x < 1$) can be easily formed. When the hydrogen is used, the reaction leading to coloration of the WO_3 can be written as [3]:

$$\text{(transparent) } WO_3 + xH^+ + xe^- \rightarrow H_xWO_3 \text{ (colored)}.$$

Speed of coloration process is essential for many practical applications. Amorphous tungsten trioxide (a-WO_3) films possess higher rates of EC colouring and bleaching, that's why they became the principal object of applied and fundamental investigations. But this feature (amorphousy) produces additional problems related to the nonequilibrium thermodynamic and nonstoichiometry nature of the films, which are the main barriers towards the development of commercial electrochromic devices.

For most of display applications integrated optical transmittance/reflectance are used. To understand the mechanism of electrochromism spectral measurements are of

Mat. Res. Soc. Symp. Proc. Vol. 453 © 1997 Materials Research Society

a great importance. For the present purpose, we have investigated the kinetics effects on spectral changes of WO_3 coloration.

EXPERIMENT

WO_3 films were grown in vacuum by thermal evaporation of WO_3 crystal powder at the pressure $P=(2-5)\ 10^{-3}$ Pa. Films were deposited at the rate of 10 nm/s on glass substrates coated by transparent conducting SnO_2 layers of thickness 200-300 nm. The thickness of WO_3 films was about 500 nm. The as-deposited WO_3 films were transparent and were later found to be amorphous. Electron microscopy studies of these films showed diffuse diffraction patterns characteristic of amorphous phases.

The EC effect kinetics was studied in electrochemical cell as described in details in [4]. The electrolyte was a 1-M aqueous solution of H_2SO_4, and the electrode voltages were measured relatively to an Ag/AgCl reference electrode. The current, potential and transmitted light intensity were plotted by high-speed recorder with response time better than 20 ms. IR absorption spectra were measured on UR-20 spectrophotometer in the range of 400-4000 cm^{-1}. Optical data were obtained in the spectral range of 300-1200 nm.

RESULTS

Kinetic effect on spectral changes

As known [1,3], optical density, D, of a-WO_3 films linearly depends on the charge, Q, passing trough the cell during coloring or bleaching. Our results confirm this conclusion in the visible range, but for near infrared range this dependence has another behavior (Fig.1). As it can be seen, EC coloration is more strong (e.g., electrochromic efficiency $K(\lambda) = D(\lambda,t)/Q(t)$ is higher [4]) for the near IR range than it is for the visible one. It means that the same D value for IR range can be obtained for essentially lower amount of Q and as a result for shorter time than in the visible range. An interesting behavior is observed for bleaching in IR range (curve 1, Fig.1, b): after applying voltage bleaching polarity some "dead" time exists, after which bleaching begins.

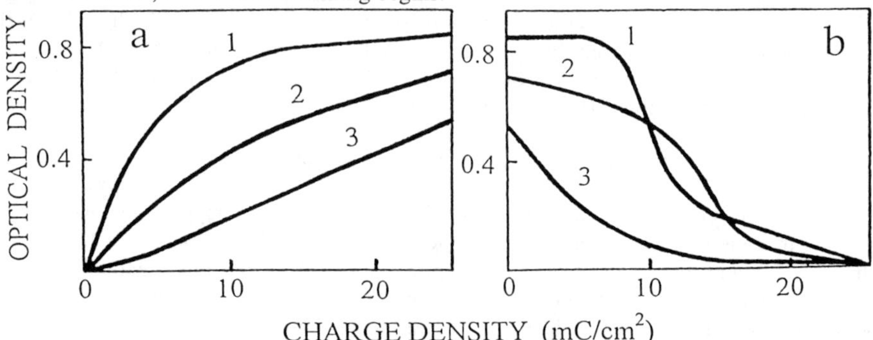

Fig.1. Optical density as a function of charge passing through the cell during coloring (a) and bleaching (b): 1 - λ=1200 nm, 2 - λ=800 nm, 3 - λ=600 nm. Voltage amplitude U=1.5 V.

Unusual behavior of light transmittance in coloring and bleaching was observed in the spectral range of 340 nm to 400 nm. In this interval the fundamental absorption edge for a-WO$_3$ is located [5]. These results are shown in Fig.2. For visible range, the behavior of transmittance during coloring-bleaching cycle has well-known character (curves 1, Fig.2, a, b). But then we can see that curves 3-6 show different degrees of "bleaching," when usual voltage polarity for coloration is applied. The similar temporal behavior is observed when higher voltage is applied to the EC cell but complicated details appear (Fig.2,b). These results will be discussed below.

Structural rearrangement due to coloration

Most of coloration models consider injected proton only to compensate the negative charge of injected electrons. Structural transformations in crystalline H$_X$WO$_3$ were observed using X-ray analysis [6]. These transformations are mainly related to changes in texture and translational symmetry of crystalline samples. Strong lattice distortion due to coloration process was observed in vibrational spectra of crystalline hydrate WO$_3$·1/3H$_2$O films [7].

Results of our spectral-structural experiments in IR range are presented in Fig.3. Comparison of curves 1 and 2 shows the dramatic changes of absorption band in the range of 600-1000 cm^{-1}. This band is related to different stretching modes of W-O vibrations in ...-O-W-O-W... chains [8,9]. The dissapearance of this band may be due to the injected proton which strongly interact with bridge oxygen in W-O-W chains, leading to local lattice distortion. Such distortion is weaker in colored a-WO$_3$ (Fig.3, curves 3,4). This may be related to the fact that W-O bonds in a-WO$_3$ are less stressed than in crystalline WO$_3$.

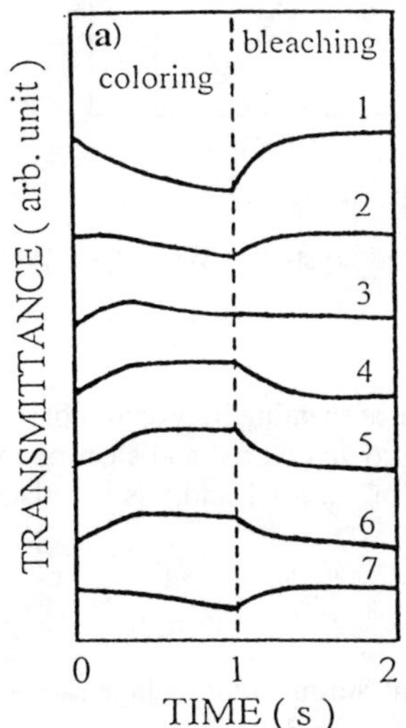

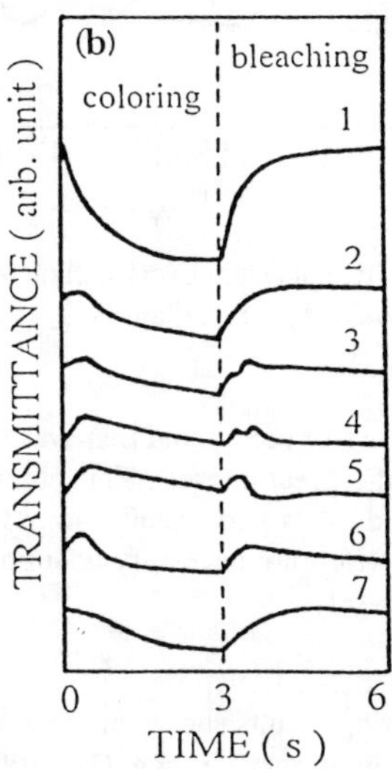

Fig.2. Temporal behavior of the transmittance during coloring-bleaching cycle: a)U= 1.0 V, b) U=1.5 V; 1 - λ=420 nm, 2 - λ=400 nm, 3 - λ=380 nm, 4 - λ= 370nm, 5 - λ=360 nm, 6 - λ=350 nm, 7 - λ=340 nm.

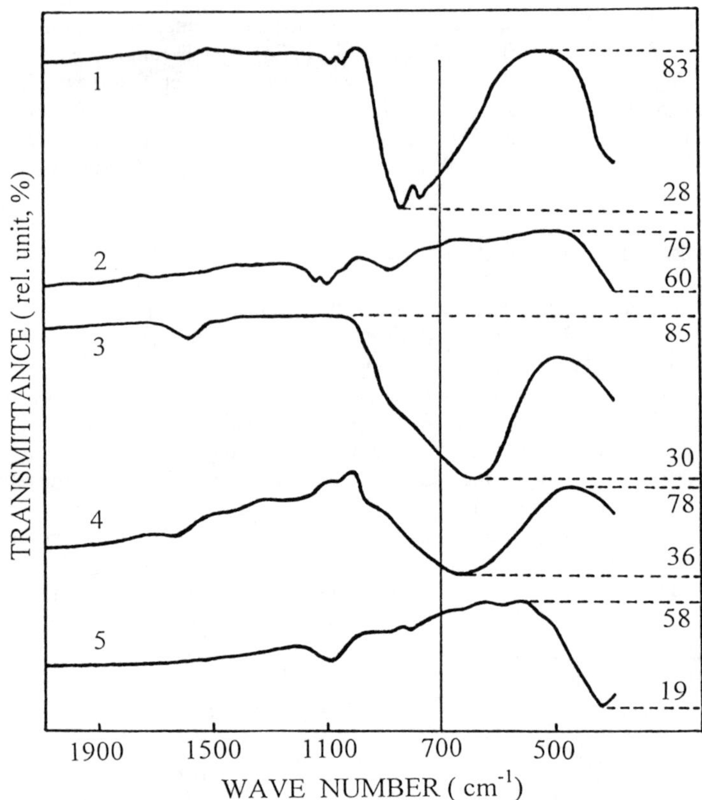

Fig.3. IR spectra: 1- crystalline WO_3, 2 - colored crystalline WO_3, 3 - a-WO_3, 4 - colored a-WO_3, 5 - crystalline WO_{3-x}

IR spectra of compound that was gotten as result of heating in vacuum of crystalline WO_3 may be of a great interest. This compound (dark-brown colored and electroconductive) was WO_{3-x} and it is very similar to WO_2. Similarity of curves 1 and 5 is surprising.   But future investigations are necessary to support this result.

Discussion

Comparing results shown in Fig.1, we can see that during coloring high rate absorption begins in the long wave range (λ=1200 nm), and some time delay exists for bleaching. This result may be explained using electron structure model of a-WO_3 where new electron states are generated in the band gap at the distance around 1 eV under the bottom of conduction band [10]. In this case during coloring the occupation of these states occurs first and optical transitions to empty states at the conduction band become available. During bleaching these states become empty last that should owe time delay observed in our experiments (curve 1, Fig. 1,b).

The above electron structure model is based on X-ray photoelectron spectroscopy experiments [10]. Accordingly to ones, Fermi level increases in proportion to the amount of injected charge: it's shift achieves about 0.34 eV for injected charge density 19.6 mC/cm^2. This fact is consistent with our results related to "bleaching" of the absorption edge (curves 2-6, Fig.2, a, b) during coloration process. Obviously, the Fermi level shift in colored a-WO_3 effectively excludes electrons of valence band from absorption because upper states in conduction band are occupied by injected electrons. This shift can be evaluated from our experiments as an order of 0.4 eV for charge density 0.25 mC/cm^2.

Different models of electrochromism exist up to date. Above mentioned density of states maximum around 1 eV below conduction band is present in almost all models but it's origin is not established till now. We would like to emphasize some theoretical results which may be useful for any model. One of the proposed possible origins of these states is hybridization of tungsten 5d levels with 2p oxygen ones in amorphous WO_3. Such states don't appear in theoretical calculations of WO_3 and Na_xWO_3 [11, 12], where conduction band is formed from only 5d tungsten states and valence band - from pure 2p oxygen ones. It is possible to suppose that these states may appear in theoretical consideration of amorphous WO_3. But cluster calculations, which are widely used for amorphous structures, showed that electron structure of different clusters related to WO_3 possess the same features as crystalline WO_3 [13]. Very interesting result for electrochromism is that addition of extra electron in different clusters does not change the charge of W atom essentially, although this electron is located on W5d level. In this case, displacement of electron density along the chemical bond from W atom towards O atoms occurs, and W atom charge state changes weakly. So, for WO_6^{-6} cluster, addition of one electron leads to W charge increasing only by 0.2 [14].

We suppose that nature of these local states in amorphous WO_3 may be connected with H_2O molecules, which are usually present in a-WO_3 films. Theoretical estimations show that energy levels of physically adsorbed H_2O molecule are located in the 1eV range below the bottom of conduction band [15].

CONCLUSIONS

Strong coloration is observed in near IR range. Kinetics in this range may be explained using electron structure model [10] of amorphous WO_3 where unoccupied states are located around 1 eV below conduction band. Occupation of these states during coloration process leads to Fermi level shift, that achieves 0.4 eV (for charge density 0.25 mC/cm^2) in our experiments.

Protons, injected in WO_3 during coloration process, not only compensate electron charge but strongly distort lattice. This result emphasizes that models and, particularly, theoretical calculations of colored state electronic structure should consider structural rearrangements that occur during coloring.

ACKNOWLEDGMENTS

This work was supported by National Academy of Sciences of Ukraine.

REFERENCES

1. S. Deb, Sol. Ener. Mater. Sol. Cells, **39**, 2 (1995).
2. C. Bechinger, S. Ferrere, A. Zaban, J. Sprague, Nature, **383**, 608 (1996).
3. B. Faughnan, R. Crandall, P. Heyman, RCA Rev., **36**, 197 (1975)
4. I. Konovalov, Yu. Kulyupin, V. Nechitajlo, A. Sukharier, Zhur. Tekh. Fiz., **31**, 552 (1986) [Sov. Phys. Tech. Phys. **31**, 431 (1986)].
5. W. Daumremont-Smith, M. Green, K. Kang, Electrochimia Acta, **22**, 751 (1977).
6. A. Gavrilyuk, B. Prokhvatilov, F. Chudnovskii, Fiz, Tver. Tela, **24**, 982 (1982). [Sov.Phys. Solid State, **24**, 558 (1982)].
7. I. Shiyanovskaya, H. Ratajczak, J. Baran, M. Marchewka, J. Mol. Struc., **348**, 99 (1995).
8. G. Ramans, E. Gabrusenoks, A. Veispals, Phys. St. Sol. A, **74**, K41 (1982).
9. R. Mercier, O. Bonke, C. Bonke, G. Robert, Mat. Res. Bull., **18**, 1, (1983).
10. S. Hashimoto, H. Matsuoka, J. Appl. Phys., **69**, 933 (1991).
11. L. Kopp, B. Harmon, S. Liu, Sol. St. Com., **22**, 677 (1977).
12. D. Bullet, J. Phys. C, **16**, 2197 (1983).
13. A. Gubskii, A. Kovtun, V. Sachenko, Ukr. Fiz. Zh. (USSR), **28**, 441 (1983), (in Russian).
14. R. Vedrinskii, A. Gubskii, S. Prosandeev, V. Sachenko, Izv. Akad. Nauk USSR, **46**, 742 (1982), (in Russian).
15. A. Gubskii, A. Kovtun (private communication).

Part IX

Surfaces and Interfaces

CHEMISTRY AND STRUCTURE OF INTERNAL INTERFACES IN INORGANIC MATERIALS

M. RÜHLE, A. RECNIK* and M. CEH*
Max-Planck-Institut für Metallforschung, Seestr. 92, D-70174 Stuttgart, Germany
*J. Stefan Institute, University of Ljubljana, Jamova 39, 1001 Ljubljana, Slovenia

ABSTRACT

Physical properties of polycrystalline materials are controlled by intergranular as well as intragranular effects. In most cases materials' properties adhere to the bonding character of internal interfaces and plasticity of bulk parts. The nature of interfaces is in general difficult to understand, although they have far reaching consequences to overall properties of processed materials. To properly understand the effects brought by such interfaces one has to correlate particular properties with their local chemistry and structure. The availability of modern instrumentation that can provide a detailed chemical and structural information down to a subnanometer scale has helped greatly in tackling many fundamental problems in materials science. Our systematic studies of several metal/ceramic and ceramic/ceramic interfaces has given us a wealth of information that brings us closer to the answer on the formation mechanisms of the internal interfaces and other structural transformations in materials.

INTRODUCTION

Internal interfaces in inorganic materials control most materials properties [1]. It is well established that already small changes in composition at grain boundaries (segregation) in metals may lead to brittle intergranular fracture [2]. High concentration of defects at internal interfaces may influence the electrical properties of oxide ceramics [3]. It is, therefore, most desirable to know the composition (chemistry) and structure of the relevant internal interfaces. There exist, however, only a few techniques which address the problems. These are either X-ray scattering experiments [4] or transmission electron microscopy (TEM) studies [5].

TEM studies will be reported in this paper. In the first section a short summary of the different advanced TEM techniques will be given. In the second section we will present two examples on atomic level studies of the chemistry and structure of internal interfaces. First we will present studies on the Cu/Al_2O_3 hetero-interface (metal/ceramic interface) and then the study on SrO-doped $CaTiO_3$ ceramics will be reported. The two examples should demonstrate the potentials of advanced TEM techniques.

Within the last decade astonishing advances in the instrumentation in TEM have occurred [6,7]. Besides conventional TEM (CTEM), new techniques are emerging which are based on very specialised instruments [8,9,11]. With those advanced methods atomic structures can be studied and the chemical composition can be determined with very high lateral resolution. Table I presents the different TEM techniques. Each column represents a specialised independent technique in the area of microstructural characterisation. The recent advancements in specialised instrumentation also promote and necessitate further development in the methodology of TEM. All information included in a TEM micrograph

Mat. Res. Soc. Symp. Proc. Vol. 453 © 1997 Materials Research Society

should be evaluated and whenever possible quantitative data on physical parameters should be extracted. In many cases this requires extensive image processing and frequently also computer simulation to retrieve the information content. Whereas in many fields these new quantitative techniques are only at their beginning, in the field of materials science a number of different methods are already well established.

The different TEM techniques (Table I) are shortly described in the following section.

Table I. Summary of the different TEM techniques

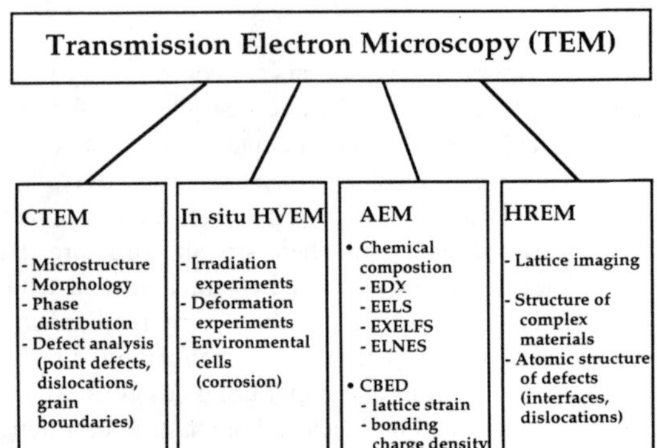

CTEM: Conventional Transmission Electron Microscopy
HVEM: High-Voltage Electron Microscopy
AEM: Analytical Electron Microscopy
HREM: High Resolution Electron Microscopy
EDS: Energy Dispersion X-Ray Spectroscopy
EELS: Electron Energy Loss Spectroscopy
EXELFS: Extended Electron Energy Loss Fine Structure
ELNES: Electron Energy Loss Near Edge Structure
CBED: Convergent Beam Electron Diffraction

CONVENTIONAL TRANSMISSION ELECTRON MICROSCOPY (CTEM)

CTEM plays a central role in microstructural analysis. It is widely developed and heavily used in materials science and inorganic chemistry. CTEM stands for all imaging modes which only use one beam, either the primary beam or one of the diffracted beams for imaging. There have been major advances in the quantitative and qualitative description of the morphology and phase distribution of a materials. By CTEM lattice defects and their strain field can be analysed [10].

Recently, dynamical experiments were done in situ in an microscope giving insight into the processes of deformation and radiation damage of materials. Those experiments were usually done in a high-voltage microscope operating at extremely low pressures and equipped with special specimen holders for heating, cooling and straining [4].

ANALYTICAL ELECTRON MICROSCOPY (AEM)

AEM allows the determination of the chemical composition and gradients of the chemical composition with extremely high spatial resolution. In AEM signals resulting from inelastic scattering processes are used to identify the chemical composition [12,13]. There exist different techniques to form a small probe. The best lateral resolution is momentarily available with a dedicated scanning transmission electron microscope (STEM) working in ultra high vacuum (UHV). The probe size is < 1 nm and the lateral resolution can be better than 1 nm.

A completely different approach is realised in the new energy filtering TEMs (EFTEMs) [14,15]. In these instruments a large area on the specimen is illuminated. Spatially resolved analytical information is obtained by forming images with inelastically scattered electrons which suffered a particular energy-loss. These electrons can be selected by means of the energy filter and the resulting images are projected onto a CCD camera for effective two-dimensional recording. Over the last five years a remarkable improvement took place in AEM due to improved electron detectors [7,16]. The diode arrays in energy-loss spectrometers are absolutely noisefree and all electrons can be detected. Energy dispersive X-ray spectrometers now operate digitally allowing the detection of electrons with very high density.

In an inelastic scattering event energy is transferred from a primary electron to an electron of the specimen. The scattering can therefore be detected by analysing the energy-loss of the primary electron beam, this techniques is known as electron energy-loss (EEL) spectroscopy. Fig. 1 shows a typical EEL spectrum. One possibility is the excitation of an inner shell electron of the material. This does not only lead to an absorption edge in the EEL spectrum which has an energy characteristic for the element but also to the emission of an x-ray when the excited atom recombines by filling the hole in the inner shell with a more outward electron. This signal is used in energy dispersive X-ray spectroscopy (EDS). Here there are element specific peaks which reveal the chemical composition of the specimen under investigation. Both, electron energy-loss spectroscopy (EELS) and EDS, can be used qualitatively and quantitatively to determine the local chemical composition of the material [12,13].

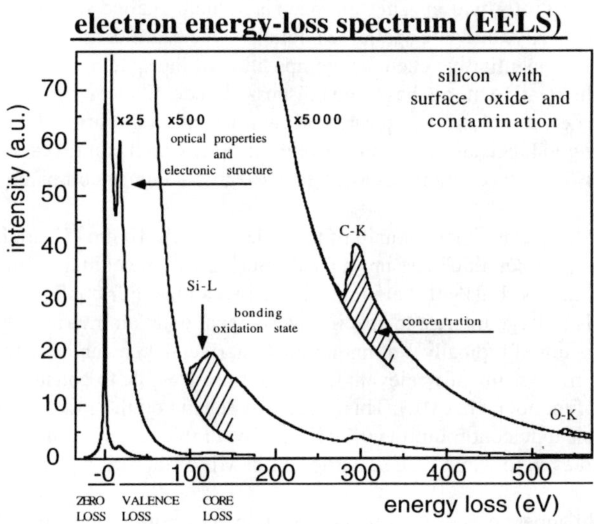

Fig. 1: Typical EEL spectrum outlining the different areas. The example was acquired in Si.

APPLICATIONS OF THE DEDICATED STEM

In the application to interfaces and other crystal defects it is mandatory to have a high spatial resolution which approaches atomic dimensions. This is achieved by focusing the primary beam into a spot. A dedicated STEM with a cold field emission electron gun is capable of forming such a small probe with a high beam current.

In principle it is possible to obtain the analytical signals with the same spatial resolution as the image. Atomic column resolved EEL spectra have already been obtained from some interfaces [9]. The beam current under these condition is extremely low (less than 50 pA) and therefore a detector with optimum detection sensitivity is necessary. Despite the recent improvements in linear photodiode arrays (the noise component is diminished) a slow scan CCD camera provides a signal-to-noise ratio which is one order of magnitude larger at these low signal intensities. Therefore, high quality atomic column resolved spectroscopy will be limited to a few specially equipped microscopes around the world.

Another method which was recently developed is the so called spatial difference method [17]. This method deliberately sacrifices some of the spatial resolution, in order to facilitate the experiment and obtain reasonable signal levels. Typically a spectrum is acquired from a region of about 3x4 nm^2 which contains the interface. The bulk components contributing to this spectrum are then removed by subtracting suitably scaled reference spectra obtained in bulk material away from the interface. This yields the interface sensitive component of the spectrum. The advantage of the method is that beam damage is minimised and that specimen drift can easily be corrected manually.

More recently an extensive software packages and improvements in the hardware are available [18]. This allows to scan the electron beam under computer control and simultaneously acquire EEL spectra. Thereby either one dimensional profiles (often sufficient for interfaces problems) or full two dimensional images are obtained, where each pixel contains the full spectral information. This approach is normally called spectrum imaging [19], which has intrinsically the same spatial resolution as the beam diameter.

For determining the quantitative chemical composition of the specimen, both EELS and EDS can be used. The relative merits have been discussed in detail elsewhere [20]. In the sampled volume the detection limit is typically below 1 at% of an impurity element. This might not seem very good, because there are several methods which can reveal the impurities at the ppm level. However, those methods average over large volumes compared to regions analysed by STEM.

The sampled volumes can be very small, of the order of a few 10 nm^3 (1 nm beam diameter and 30 nm specimen thickness are typical). Such a volume contains only a few hundred to thousand atoms. 1 at% of that yields then a few atoms. For applications in materials science it is of large interest to specify the segregation at an interface. Considering the geometry segregation of typically 1 atom per nm^2 interfacial area can be detected [21]. While this is the thermodynamically relevant unit, it is more feasible to consider the equivalent number of monolayers (ML). This value will depend on the exact definition of one monolayer, but a typical detection limit is 0.1 ML. In favourable cases it can be one order of magnitude better and even in worst case scenarios a full ML of impurity segregation will be detectable.

While EDS might appear easier, it suffers from a number of drawbacks. Exact quantification is much more difficult than for EELS. Beam broadening in the specimen degrades the ultimate spatial resolution. Most important is its limitation to only determine chemical composition. EELS on the other hand contains a wealth of extra information. Since the energy resolution in an EELS spectrum is well below 1 eV, a number of spectral fine structures can be observed. Each absorption edge arising from an inner shell has superimposed on it two types of fine structure. The energy-loss near edge structure (ELNES) dominates from the edge onset up to about 30 eV, whereas the extended energy-loss fine structure (EXELFS) is responsible for intensity modulations from there on up to several hundreds of eV above the edge onset [12].

One can view the excited electron as a wave which is scattered in the crystal at neighbouring atoms. The interference at the location of the excited atom decides the intensity measured in a spectrum. The mean free path for the excited electrons within the solid are short. For EXELFS this means that only paths with single scattering events have to be considered. The measured spectrum can be analysed and yields the radial distribution function. However, this method is not used very often in the electron microscope because the achievable signal-to noise ratio is not sufficient. For ELNES the mean free path becomes longer, so that multiple scattering has to be considered. ELNES is therefore complex and cannot be analysed directly. From this picture it is clear that the ELNES depends on coordination number, bond distances and bond angles. Multiple Scattering calculations for clusters yield the expected ELNES for a given arrangement of atoms. This approach is amenable to interfaces [22], where the more traditional band structure calculations are unfeasible, because the low symmetry requires large supercells, which are in most cases beyond the current scope of computing power. ELNES can also be viewed as transitions from occupied inner shell levels into unoccupied states in the conduction band. In this view the transition rate and therefore the measured spectrum is proportional to the unoccupied density of states (DOS) in the conduction band. Selection rules apply, because under typical experimental conditions only dipole transitions are allowed. K-shell excitation probe the p-unoccupied DOS, whereas L-shells are a sum of s- and d-DOS. Furthermore the local DOS has to be considered because of the high localisation of the initial state.

Changes in the ELNES are observed for different oxidation states (chemical shifts, white line intensities) and coordination [23,24]. The first is a more atomic effect whereas the latter depends on the environment of the probed atom. The ELNES can be used to determine the bonding, electronic structure and real space structure. The interpretation makes use of reference spectra which for some classes of material can be used as fingerprints, i.e. specific structures in the ELNES correspond to specific structural units [24]. A more complex way of interpretation compares ELNES calculated for model structures with experimental data. This yields valuable insight into the correspondence of spectral features and real space structures. The important point about ELNES is that it contains information about the three dimensional atomic arrangement and therefore complements the information obtained from imaging, which is limited to two dimensional projections.

HIGH-RESOLUTION ELECTRON MICROSCOPY (HREM)

HREM allows the determination of the atomistic structure of internal interfaces. Quantitative evaluation yields an exact coordinate of atomic columns including errors bars. It should be emphasized that straightforward interpretation of experimental micrographs is not possible since a shift of the contrast can be caused by interference effects within the specimen and by the spherical aberration of the objective lens. Reconstruction of information from high-resolution images is possible through careful quantification of beheld artefacts. To do this we first have to know the range of experimental imaging conditions, and then we need a good starting model which most closely conveys the observed structural feature (interface).

Even the very best atomic resolution microscope with 1 Å resolution does not allow a naive interpretation of the experimental micrograph. It is therefore essential to follow the scheme shown in Fig. 2. Besides experimental investigations also image simulation had to be done and the experimental micrograph has to be compared to the simulated image. Modification of the model of the defect for a specific configuration leads then to a best agreement between theoretically and experimentally determined micrographs. This involves extensive and time-consuming image processing. The best accuracy for column positions is ± 0.1 Å. The accuracy strongly depends, however, on the scattering factor of the elements. Light elements have larger error bars than heavier elements [27].

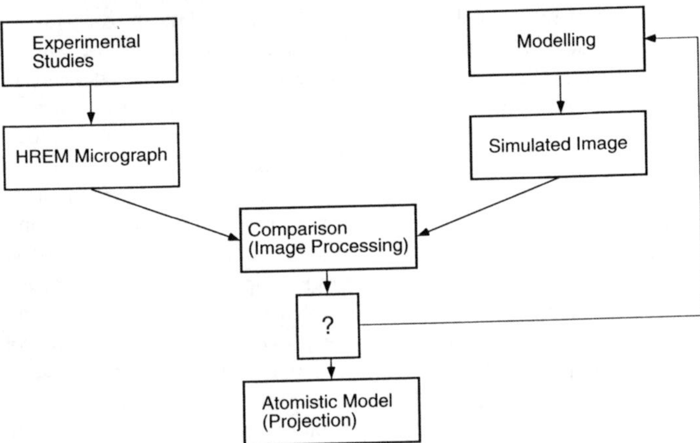

Fig. 2 Schematic of quantitative HREM

THE Cu/Al$_2$O$_3$ INTERFACE

The Cu/Al$_2$O$_3$ interface plays an important role in packaging problems in microelectronics [28]. Furthermore, it serves as a model system for studying the structure and bonding between metals and ceramics. Thin films of Cu were grown by molecular beam epitaxy (MBE) on (0001) α-Al$_2$O$_3$ substrates [29]. The orientation relationship was determined by selected area diffraction and it was observed that the close-packed planes were parallel to each other, i.e. (0001) Al$_2$O$_3$ parallel to (0001) Cu [29].

HRTEM micrographs of the Cu/Al$_2$O$_3$ interface edge-on illustrate that both, the close-packed planes and close-packed directions of Cu and Al$_2$O$_3$ are parallel to one another. However, the interface itself is not coherent. The The Cu/Al$_2$O$_3$ interface is atomically flat and reveals a sharp transition between the α-Al$_2$O$_3$ substrate and the copper overgrowth [29,30].

The epitaxial orientation relationship leads to a mismatch of about 7 % between the correspondent spacings of the adjacent Cu and α-Al$_2$O$_3$ lattices. Only the volume change perpendicular to the basal plane can be determined with high accuracy out of a HRTEM micrograph. Since the (0001) plane of α-Al$_2$O$_3$ consists of an alternating stacking sequence of oxygen and aluminium ions, the terminating layer could be either an oxygen or aluminium layer at the interface. Then, either Cu-O bonds or Cu-Al bonds are expected to establish across the Cu/Al$_2$O$_3$ interface. The type of bonding can be determined by EELS studies of the fine structure of the Al, O, and Cu ionisation edges [23]. The ELNES of an ionisation edge contains the information about local coordination and electronic structure as shown above. The EELS spectra were background subtracted and the different results are represented in Fig. 3. A measured Cu/Al$_2$O$_3$ spectra of the Cu-film (I_{Cu}) and the interface (I_{IF}) are presented. The difference spectrum of the interface (ΔI_{IF}) is shown in Fig. 3. The difference spectrum of the interface reveals an ELNES structure which is clearly different from bulk Cu. At an energy loss of 933eV a L$_3$ white line is observed for interfacial Cu-atoms. White lines are typical for transition metals and their oxides and arise from electron transitions from 2p-states into empty 3d-states. In metallic Cu^0 all 3d states are occupied and no white line exists. Since the white line is present it has to be assumed that Cu is oxidized near the interface. Comparing the interfacial Cu$_{L2,3}$ ELNES (Fig. 3) to reference spectra of Cu$_2$O (nominally Cu^{1+}) and CuO (nominally Cu^{2+}, Fig. 3) indicates that the nominal oxidation state of Cu at the interface is Cu $^{1+}$, since no chemical shift, which is typical for Cu^{2+}, is detected.

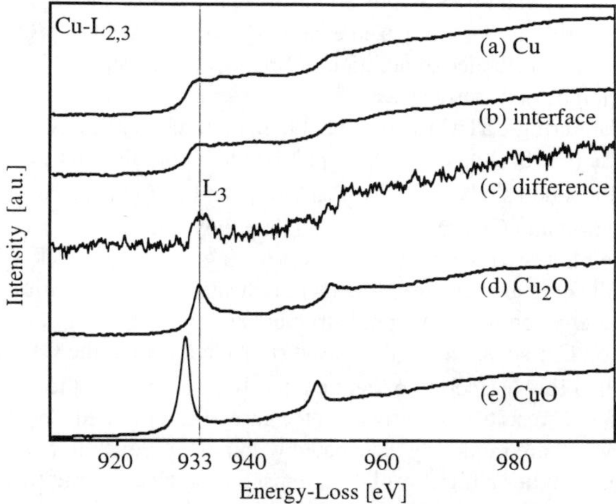

Fig. 3: Cu$_{L2,3}$ ELNES acquired in a region of 10x12nm^2 in (a) the Cu-film and (b) at the Cu/Al$_2$O$_3$ interface. The (c) difference spectrum, which contains the interface specific component shows a white line at 933eV for interfacial Cu atoms due to Cu-O bonds at the interface. Reference spectra of (d) Cu$_2$O and (e) CuO indicate a nominal Cu^{1+} oxidation state for Cu atoms at the Cu/Al$_2$O$_3$ interface

Examination of the O$_K$ interface difference ELNES indicates that oxygen has changed its local environment at the interface. A direct comparison of the difference spectrum with the bulk α-Al$_2$O$_3$ shows a broadening of the main peak at 540 eV and a shift to slightly lower energy-losses. At approximately 532 eV electron transitions cause a shoulder in the difference spectrum which is not present in bulk α-Al$_2$O$_3$. At the same energy-loss a peak is observed in the O_K ELNES of bulk Cu_2O due to transitions into a hybridised Cu-3d-O-2p state. Furthermore, the peaks above 550 eV change their shape compared to those observed in the O_K ELNES of $\alpha-Al_2O_3$.

The interface specific components found in $Cu_{L2,3}$ and O_K edge imply the presence of ionic-covalent Cu-O bonds across the interface. Image simulations of experimental HRTEM micrographs were then carried out under the assumption of Cu-O bonds at the interface as found by our ELNES studies. A reasonable atomistic model of the interfacial region must result in realistic interatomic spacing across the interface. The best agreement between the experimental and simulated interface image was achieved for an average projected bonding distance of 0.2nm $\pm$ 0.03nm between the terminating oxygen layer and copper layer. The individual atomic columns fluctuate by the denoted error bar of $\pm$ 0.03nm around the mean monolayer position. The projected bonding distance of 0.2nm between the terminating oxygen layer and copper layer at the interface appears realistic considering the minimum interatomic spacing in Cu_2O to be 0.18 nm.

INTERNAL INTERFACES IN Sr-DOPED $CaTiO_3$

In another example, we studied the chemistry and structure of planar defects in SrO-doped $CaTiO_3$. It turns out that an excess of SrO leads to the formation of planar defects which are antiphase boundaries as shown by Ceh et al. [32]. For the $CaTiO_3$-SrO materials it is of great interest to study the influence of addition of SrO to $CaTiO_3$ since this material is being used for encapsulation of radioactive waste.

When sintering $CaTiO_3$ with SrO at temperatures as high as 1550 °C, ordered structures of the $(Ca_xSr_{1-x})_{n+1}Ti_nO_{n+1}$ polytypoidic homologous series are formed. Crystallographic studies of $Ca_3Ti_2O_7$, $Ca_4Ti_3O_{10}$ and $Ca_{3.6}Sr_{0.4}Ti_3O_{10}$ phases employing neutron diffraction and CBED have shown that polytypoids consist of coherent intergrowths of n perovskite blocks with single CaO layers having a distorted rock-salt structure [33]. Furthermore, the crystal structure refinement indicated that the addition of Sr had a significant influence on the polytypoid structure and that larger Sr atoms substituted preferentially on Ca_I sites, within the perovskite layers, whilst the Ca_{II} sites remained fully Ca occupied. In a low temperature regime it has been found that the sintering of SrO-doped $CaTiO_3$ at 1350 °C results, similarly as in the case of CaO-doped $CaTiO_3$, in the formation of planar faults which are randomly distributed within the perovskite matrix [34].

For HRTEM studies a high-resolution transmission electron microscope operating at 400 kV was used (JEM 4000 EX). The acquisition and processing of EEL spectra was controlled as described above [18]. The spatial difference method [17] was applied in addition to line scans.

STRUCTURE OF INTERNAL INTERFACES IN Sr-DOPED $CaTiO_3$

The most characteristic structural features that are observed in SrO-doped $CaTiO_3$ materials are shown in bright field HRTEM images in Figs. 4a,b. When $CaTiO_3$ with additions up to 25 mol% of SrO is sintered at lower temperatures (T=1350 °C), isolated planar faults are observed in $CaTiO_3$ grains. Such defects never occur in a stoichiometric $CaTiO_3$. These planar faults and polytypoids in the $CaTiO_3$-SrO were studied by conventional and high-resolution transmission electron microscopy (CTEM and HRTEM). It has been found that planar faults occupy {100} perovskite planes and form square networks. Similarly as do the CaO layers in the polytypoid single faults would have a two-dimensional rock-salt structure embedded in the host perovskite matrix. In this way, A-site (Ca,Sr) and B-site (Ti) cations in neighborouring perovskite blocks are brought into an antiphase position across the fault. The density of the observed planar faults was proportional to the added amount of SrO.

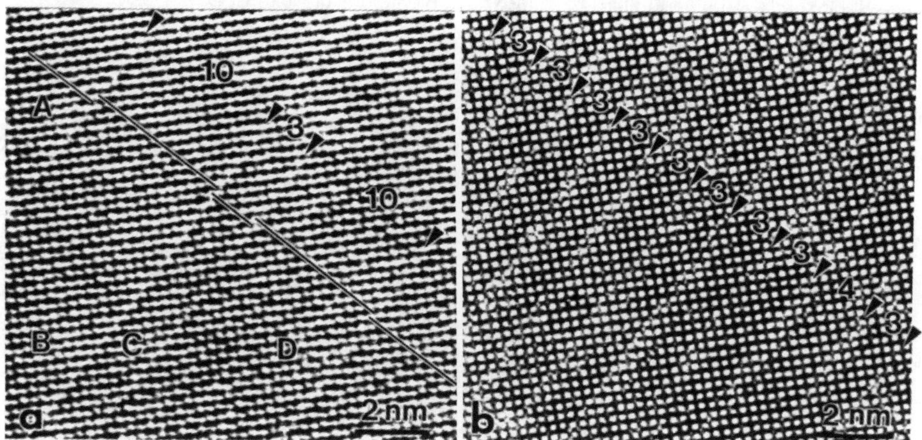

Fig. 4: Bright field HRTEM images of (a) isolated planar fault in [110] zone axis (sample sintered at 1350 °C). (b) ordered polytypoidic structure in [100] zone axis (sample sintered at 1550 °C).

This situation is being changed when the material is sintered at higher temperatures (T=1550 °C). Here we observe parallel planar faults (Fig. 4a) which are in some regions ordered into polytypoidic structures (Fig. 4b). Their composition can be represented with a general formula $A_4B_3O_{10}$ (or AO + 3 ABO_3), where A = (Ca,Sr) and B = Ti. Polytypoids of $A_4B_3O_{10}$ type posses a periodic sequence of parallel faults, where every AO fault is followed by three ABO_3 layers. Nonperiodic sequences having four or more perovskite layers are also observed, however these never form ordered polytypoids. Bothe single faults and polytypoids are like single faults coherently intergrown within the host perovskite.

CHEMISTRY OF INTERNAL INTERFACES IN SrO-DOPED CaTiO$_3$

Figure 5 shows a bright field TEM image of SrO-doped crystal that contains parallel single faults which are in some parts ordered into a $A_4B_3O_{10}$-type polytypoid EEL analyses were conducted in this very same region in the dedicated STEM. This was accomplished by recording images of this region in TEM at different magnifications, which were then used as a map to locate the same region in the STEM. The direction of the EEL linescans is marked in the TEM image (Fig. 5a,b). The results of EEL linescan across a few slabs of parallel faults indicate a significant increase in Ca concentration at the faults associated with a corresponding decrease in Ti concentration (Fig. 5c). Planar faults within the analysed region can be distinguished in the profile. The spacing between faults within the polytypoid is 1.36 nm and the width of each fault is 0.22 nm. The EEL spectra obtained from the region with no faults were subtracted from the EEL spectra from the region with faults, yielding the spectrum of the fault only. In the spectrum from the fault, the Ca/O atomic ratio is 0.94±0.09, which agrees with the CaO standard where Ca/O is 1. This indicates that the composition of parallel AO planar faults in the $A_4B_3O_{10}$-type polytypoid is close to CaO.

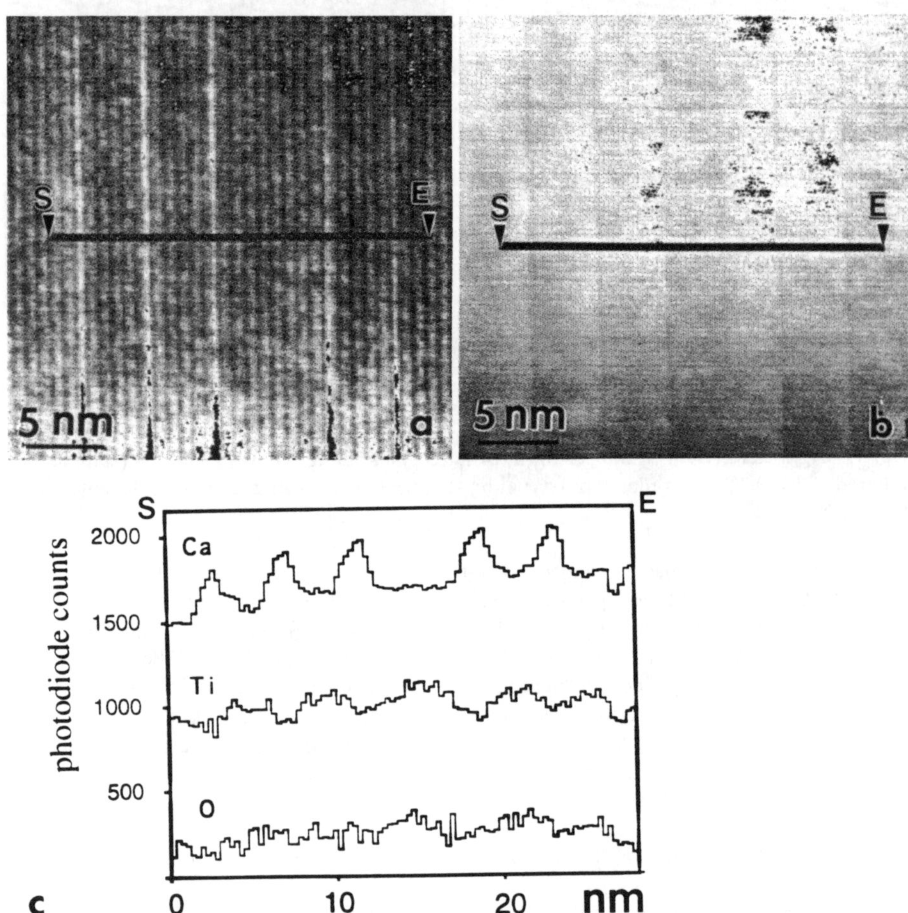

Fig. 5: (a) Bright field image and (b) the corresponding hgih angle annular darkfield image of analysed planar faults. (c) Elemental profiles for Ca, Ti and O. The line of scan is marked in the dark field image.

CONCLUSIONS

Advanced TEM techniques allow the structural and chemical characterisation of materials to the (nearly) atomistic scale. The characterisation comprises the structure. the chemistry and the bonding on the atomic level. In order to gain an optimum in information, constant advancement of the techniques is also required. In particular, the development of methods for the quantitative evaluation of the different signals obtainable on a modern TEM is an important aim of research.

The increasing resolution with which structural and compositional information can be obtained also means that the information is extracted from smaller and smaller volumes. Hence, a good correlation between macroscopic properties and the underlying structures on an atomistic level can only be obtained for model systems, like planar bimaterial interfaces or layered structures with different components. As an example, we have discussed the results obtained on the Cu/Al_2O_3 interface and the planar faults in SrO-doped $CaTiO_3$. We have shown that information can not only be obtained on the local arrangement of the atoms at the interface, but also on the interfacial chemistry and the type of chemical bonding at the interface. The inherent problem of TEM is that for real materials with complex microstructures it may be very critical to infer predictions on macroscopic properties from only a few local areas with nm-size dimensions. Another critical question is as to how much the preparation of the thin TEM specimens may induce artefacts or give non-representative sections through the overall microstructure. Both questions have to be kept in mind in any careful interpretation of TEM results.

ACKNOWLEDGEMENT

The authors are indebted to the Internationale Büro in Bonn (BMBF) for supporting the collaboration between the J. Stefan Institute at the University of Ljubljana (Slovenia) and the Max Planck Institut für Metallforschung in Stuttgart (Germany).

REFERENCES

1.   A.P. Sutton and R.W. Balluffi, Interfaces in Crystalline Materi-als, Clarendon Press, Oxford (1995).
2.   Interfacial Segregation, edited by W.C. Johnson and J.M. Blakely (American Society for Metals, Metals Park, Ohio, 1978).
3.   J. Maier, J. Electrochem. Soc. **134**, 1524 (1987).
4.   I. Majid, P.D. Bristowe and R.W. Balluffi, Phys. Rev. **B40**, 2779-2792 (1989).
5.   J.M. Cowley, Diffraction Physics, 3rd ed. (North Holland, Amsterdam, 1995)
6.   High-Voltage and High-Resolution Electron Microscopy, edited by M. Rühle, F. Phillipp, A. Seeger, J. Heydenreich, Ultramicroscopy **56**, 1-232 (1994).
7.   Proceedings of the Second International Workshop on Electron Energy Loss Spectroscopy and Imaging (EELSI), edited by O.L. Krivanek, Microsc. Microanal. Microstruct. **6**, 1-157 (1995).
8.   French-Japanese Seminar on In Situ Electron Microscopy, edited by F. Louchet, H. Saka, Microsc. Microanal. Microstruct. **4**, 101-346 (1993).
9.   S.J. Pennycook, D.E. Jesson, Ultramicroscopy **37**, 14-38 (1991).

10. P. Hirsch, A. Howie, R. Nicholson, D.W. Pashley, M.J. Whelan, Electron Microscopy of Thin Crystals (Butterworths, London 1965).

11. E.P. Butler, K.F. Hale, Dynamic Experiments in Electron Microscopy (North Holland, Amsterdam, 1981)

12. Lake Tahoe Workshop on Electron Energy Loss Spectroscopy, edited by O.L. Krivanek, Microsc. Microanal. Microstruct. 2, 143-411 (1990).

13. M.H. Loretto, Electron Beam Analysis of Materials, Chapman and Hall, London (1984).

14. C. Deininger, J. Mayer, M. Rühle, Optik 99, 135-140 (1995).

15. H.J. Kleebe, J. Bruley, M. Rühle, J. of the European Ceramic Society 14, 1-11 (1994).

16. N.D. Browning, M.F. Chisholm, S.J. Pennycook, Nature 366 143-146 (1993).

17. C. Gatts, G. Duscher, H. Müllejans and M. Rühle, Ultramicroscopy 59, 229-239 1995).

18. H. Müllejans in Proc. 1st Slowenian- German Seminar on Joint Projects in Materials Science and Technology, edited by D. Kolar, D. Suvorov (Jülich GmbH, Jülich 1995) 9-14.

19. R. Brydson, J. Bruley, H. Müllejans, C. Scheu, M. Rühle, Ultramicroscopy 59, 81-92 (1995).

20. R. Brydson, H. Sauer, W. Engel, in Transmission Electron Energy Loss Spectrometry in Materials Science, edited by M. M. Disko, C. C. Ahn, B. Fultz (TMS Monograph Series, Warrendale, USA 1992) 131-154.

21.   J. Bihr, G. Benner, D. Krahl, A. Rilk, E. Weimer, in Proc. 49th Ann. Meeting EMSA, edited by G.W. Bailey, (San Francisco Press, San Francisco 1991) 354-355.

22.   J. Mayer, D.V. Szabo, M. Rühle, M. Seher, R. Riedel, J. Eur. Cer. Soc. 15, 717-727 (1995).

23. O.L. Krivanek, C. Mory, M. Tence, C. Colliex, Microsc. Microanal. Microstruct. 2, 257-267 (1991).

24.   A. Berger, J. Mayer, H. Kohl, Ultramicroscopy 55, 101-112 (1994).

25. W. Jäger, J. Mayer, Ultramicroscopy 59, 33-45 (1995).

26. G. Möbus and M. Rühle, Ultramicroscopy 56, 54-70 (1994).

27. G. Möbus and G. Dehm, Ultramicroscopy 60, 205 (1996).

28. F. Ernst, M.W. Finnis, A. Koch, C. Schmidt, B. Straumal, W. Gust, Z. Metallkde 87, 911 (1996).

29. Joining of Ceramics, edited by M.G. Nicholas (Chapman and Hall, London 1990).

30. G. Dehm, M. Rühle, G. Ding and R. Raj, Phil. Mag. B 71, 1111-1124 (1995).

31. G. Dehm, F. Ernst, J. Mayer, G. Möbus, H. Müllejans, F. Phillipp, C. Scheu and M. Rühle, Z. Metallkde 87, 898-910 (1996).

32. M. Ceh, H. Gu, H. Müllejans, A.Recnik, to be published.

33. M.M. Elcombe, E.H. Kisi, K.D. Hawkins, T.J. White, P. Goodman and S. Matheson, Acta Cryst. B47, 305 (1991).

34. M. Ceh, V. Krasevec and D. Kolar, J. Solid State Chem. 103, 263 (1993).

THE PREPARATION OF KINETICALLY STABLE CRYSTALLINE COMPOUNDS FROM MODULATED ELEMENTAL REACTANTS

Marc D. Hornbostel, Myungkeun Noh, Christopher D. Johnson and David C. Johnson
Materials Science Institute & Department of Chemistry, University of Oregon, Eugene, Oregon
97403

ABSTRACT

Diffusion distances within elementally modulated reactants can be controlled on an
Ångstrom length scale. Below a critical repeat thickness, elementally modulated reactants
interdiffuse without nucleating crystalline compounds. Using this amorphous intermediate, we
can prepare metastable compounds by controlling nucleation. Above this critical repeat
thickness, crystalline compounds nucleate at the reacting interfaces. Using the architecture of the
initial reactant to control the diffusion distances in the initial reactant, we have found that we can
prepare crystalline superlattices. Crystalline superlattice compounds containing integral numbers
of inter grown transition metal dichalcogenide layers and alternating layers of transition metal
carbides have been prepared through controlled crystallization of superlattice reactants with
designed compositional modulation. High quality c-axis oriented dichalcogenide crystalline
superlattices result from extended annealing at relatively low temperatures. A large number of
$[00\ell]$ diffraction orders and off-axis $[10\ell]$ diffraction peaks are observed indicating that these
compounds are crystalline in three dimensions. Similar annealing conditions were used to
prepare carbide superlattices, however the limited low temperature diffusion rates of the carbides
limit the crystallite size to approximately 300Å. The rational synthesis of these intergrowth
compounds from superlattice reactants permits the exploratory synthesis of a new class of
compounds and the tailoring of physical properties as a function of compositional layer
thicknesses and native properties of the parent compounds.

INTRODUCTION

The designed synthesis of new molecular compounds depends upon the ability to tailor the
initial system to favor the kinetics of the desired product. By adjusting the structure of.the
reacting molecules and judicious choice of solvents and catalysts, the system is set up so the
easiest (lowest activation energy, smallest rearrangement, etc.) decrease in free energy occurs
towards the desired products. Solid state chemists have increasingly tried to apply similar
strategies in the development of new synthetic approaches to new solid state compounds [1].
Two different approaches have been used. One approach involves intimately mixing the
reactants in a "solution" at low temperatures. The utilization of low temperature flux and
hydrothermal techniques has resulted in a host of new compounds, often with high structural
complexity. The richness of the structures produced combined with the complexity of reaction
paths has limited the ability to control or predict the resulting structures. The second approach
involves making solid state precursors and then controlling the transformation to a desired
product. This "soft" chemical approach has been utilized with great insight to transform
precursors into desired metastable products [2]. This approach is often limited by the availability
of suitable precursors.

The difficulties in controlling kinetics to achieve a desired product is readily apparent in the
synthesis of superlattice structures. A limited number of superlattice structures have been
prepared via high temperature synthesis. This traditional synthesis approach is unable to prepare
an extended series of such compounds, however, because there is too little difference in free
energy between the various members of the series to control product formation using the overall

Mat. Res. Soc. Symp. Proc. Vol. 453 ©1997 Materials Research Society

composition and annealing conditions [3,4]. Historically, this limitation has been overcome using epitaxial routes which have had spectacular success in preparing superstructures containing two latticed matched materials. In particular, molecular beam epitaxy (MBE) controls the kinetics of growth using layer by layer deposition resulting in two dimensional growth. The kinetics are controlled on the growth surface by adjusting the temperature and the relative fluxes of the sources. MBE has given access to a new class of materials with accurately tailored superstructures as well as designed physical properties [5].

This paper discusses the development of an alternative approach to prepare crystalline compounds and crystalline superlattices using elemental superlattices as reactants. Elementally modulated reactants permit us to vary the starting structure on an Ångstrom scale to control the subsequent reactivity. The ability to control composition at an atomic level allows us to set the initial diffusion distances within these elementally modulated reactants. The ability to vary these distances provides an experimental variable to understand, control and predict the crystalline structures of the product compounds based upon the initial architecture of the reactant.

EXPERIMENTAL

The multilayer reactants were prepared in a high-vacuum evaporation system which has been described in detail elsewhere [6]. Briefly, the elements were sequentially evaporated under high vacuum (approx. 3×10^{-7} torr) under the control of a personal computer. Elements were deposited at rates of 0.5 Å/sec using either electron-beam guns or Knudsen cells. Deposition rates of each element were monitored using independent quartz crystal thickness monitors. Deposition was done simultaneously on several different substrates.

Different substrates were used for different applications. A polished silicon wafer was used to characterize the deposited sample and investigate interdiffusion of the reactants using low angle x-ray diffraction [7]. Multilayer composites deposited on PMMA coated wafers are floated from the substrate by dissolving the PMMA with acetone, then collected for differential scanning calorimetry (DSC) study. The total mass of each sample collected from the PMMA coated wafers was approximately 2-3 mg.

The stoichiometry of the samples studied were determined either by thermal-gravimetric analysis or electron microprobe analysis. Samples were annealed in a differential scanning calorimeter under a nitrogen atmosphere. Measured exotherms were correlated with x-ray results to identify and track the interdiffusion of the elements and the crystallization of any compounds. Samples on native oxide coated silicon substrates were annealed in a nitrogen dry box with less than a tenth of a part per million of oxygen.

RESULTS AND DISCUSSION

In the early eighties, W. L. Johnson's group discovered that amorphous metal alloys could be prepared by solid state amorphization reactions [8-11]. In this process, thin films consisting of alternate layers of crystalline elemental constituents are transformed into an amorphous alloy during a solid state reaction. When the amorphous alloy reaches a critical thickness, an equilibrium crystalline phase was found to nucleate. This upper limit to the thickness of amorphous alloy severely limits the ability to form bulk amorphous materials via this mechanism using bulk thicknesses of the reacting elements. Solid state amorphization reactions can be used to form bulk amorphous alloys if a multilayered reactant is used which is modulated on a length scale shorter than the critical thickness. The 'critical thickness' in a system depends upon many factors including the relevant diffusion coefficients, the density of the deposited elements, and the driving force for crystallization. In early transition metal-late transition metal systems such as Ni-Zr, it has been found that critical thicknesses of the repeating unit are from several hundred to a thousand Ångstroms while in transition metal - amorphous silicon systems critical thicknesses are from 50 to 250Å [12]. In transition metal-selenium systems such as molybdenum-selenium [13] or niobium-selenium [14] the critical thicknesses of the repeating bilayer are from 10 to 100Å in thickness.

Elementally modulated reactants which are layered below the critical thickness provide a kinetic pathway to metastable compounds. The evolution of a Hf-Fe-Sb multilayer film shown in Figures 1 and 2 as a function of annealing clearly illustrates this ability. The Hf-Fe-Sb film was prepared with a composition of 1:4:12 - Hf:Fe:Sb in an attempt to prepare a "filled skutterudite" structure containing hafnium as the ternary cation. The multilayer repeat distance (approximately 21Å) was chosen to be small enough that the elements could be reasonably expected to interdiffuse to a homogeneous amorphous intermediate before crystallizing. The sample was annealed in a differential scanning calorimeter at a scan rate of 10°C/min. and the results are shown in Figure 1. The data show three irreversible exotherms on heating, one at 230°C, one at 260°C and one at 350°C. Diffraction data collected after each of these exotherms as shown in Figure 2 indicate that the first exotherm results from the nucleation of the desired skutterudite compound, $Hf_{1-x}Fe_4Sb_{12}$, the second exotherm results from growth of this compound while the third exotherm results from the decomposition of the skutterudite compound into a mixture of binary compounds and elemental components. After this third exotherm, the sample contains the distribution of products observed from conventional high temperature synthesis. This data clearly demonstrates that the crystalline skutterudite compound is only kinetically stable, however, being metastable relative to a mix of binary compounds by the integrated heat of the third exotherm [15]. This metastability is presumably related to the decreased ionic radius of the hafnium cation 0.71Å compared with 1.03 Å for lanthanum [16,17] which forms a filled skutterudite compound using conventional high temperature synthesis.

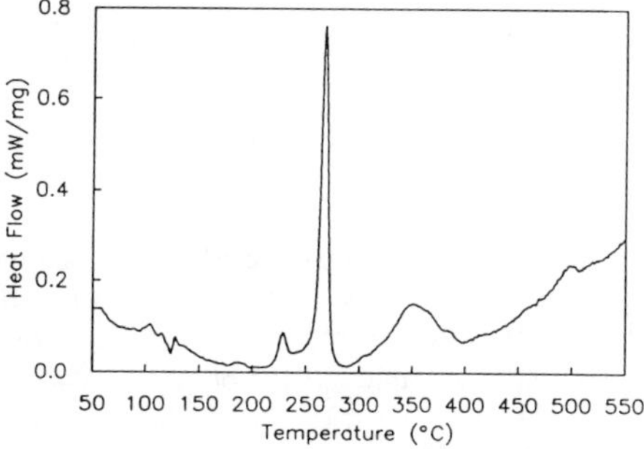

Figure 1: Differential scanning calorimetry data for a representative Hf-Fe-Sb sample.

The results presented on the synthesis of $Hf_{1-x}Fe_4Sb_{12}$ illustrate how short diffusion lengths can be used to form metastable crystalline compounds. With respect to crystalline superlattices, this initial success lead us to explore the fate of superlattice reactants containing different diffusion lengths. In particular, what were the kinetic constraints imposed upon a system containing short diffusion lengths for each component of a desired superlattice but a considerably longer diffusion length for the intermixing of the components with each other? The modulated elemental reactant shown in Figure 3 contains alternating layers of component A and B interleaved with component C. Given our results controlling nucleation and subsequent product formation in binary systems, we proposed three different length scale regimes and three potential mechanisms which can result in the formation of nanostructured composites and crystalline superlattices.

Case 1 - Interfacial Nucleation and Growth

This reaction pathway starts with interfacial nucleation of the compounds A_xC_y and B_uC_v at the A-C and B-C interfaces respectively. Following nucleation, rapid growth of these compounds along the plane of the interface is expected, since a plane at the interface is the only region in the initial reactant with the correct composition for these compounds to grow. Subsequent low temperature annealing will lead to growth of A_xC_y and B_uC_v perpendicular to the interfaces. Since diffusion rates can be constrained by low annealing temperatures and diffusion distances can be optimized in the initial structure of the modulated reactant, the growth process can be optimized to result in a modulated final structure with minimal interdiffusion of the components.

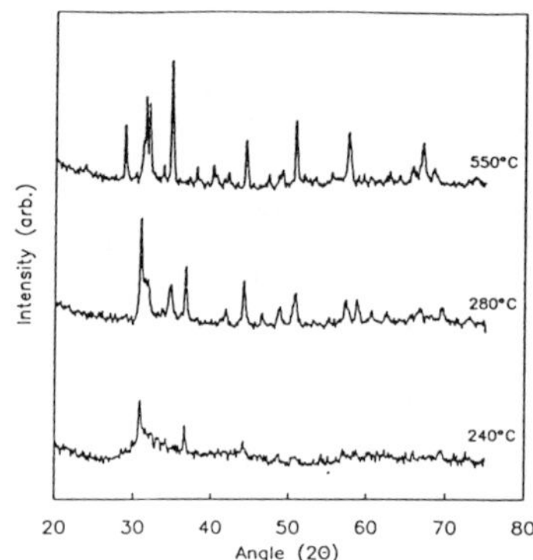

Figure 2: X-ray diffraction scans taken on the Hf-Fe-Sb sample shown in Figure 1 after heating to 240°C (bottom), 280°C (middle) and 550°C (top). The middle diffraction pattern is that of $Hf_{1-x}Fe_4Sb_{12}$ with a small amount of $FeSb_2$ impurity. The top diffraction pattern is that of a mixture of Sb, and $FeSb_2$ and other yet unidentified phases.

Case 2 - Nucleation From a Modulated Amorphous Intermediate

This reaction pathway starts with the low temperature interdiffusion of the elemental layers of the initial reactant forming a modulated amorphous intermediate containing amorphous A-C and amorphous B-C layers. We expect the interdiffusion rates between the amorphous layers to be much slower than the interdiffusion rates during the initial mixing of the elemental layers, since the chemical driving force should be substantially reduced. On subsequent, higher temperature annealing, one of the compounds, for example A_xC_y nucleates and grows. The remaining B-C amorphous layer may crystallize forming a compound B_uC_v in response to the nucleation event, especially if there are epitaxial relationships between the structure of B_uC_y and A_xC_y. It is also possible that the remaining B-C amorphous regions may only crystallize at a higher annealing temperatures. We expect some alloying of the crystalline compounds in the interfacial regions as a result of interdiffusion during the formation of the amorphous intermediates. This may limit the abruptness of composition profiles at the interface as well as the determine the lower limit for the layer thicknesses in the heterostructures produced using this reaction pathway

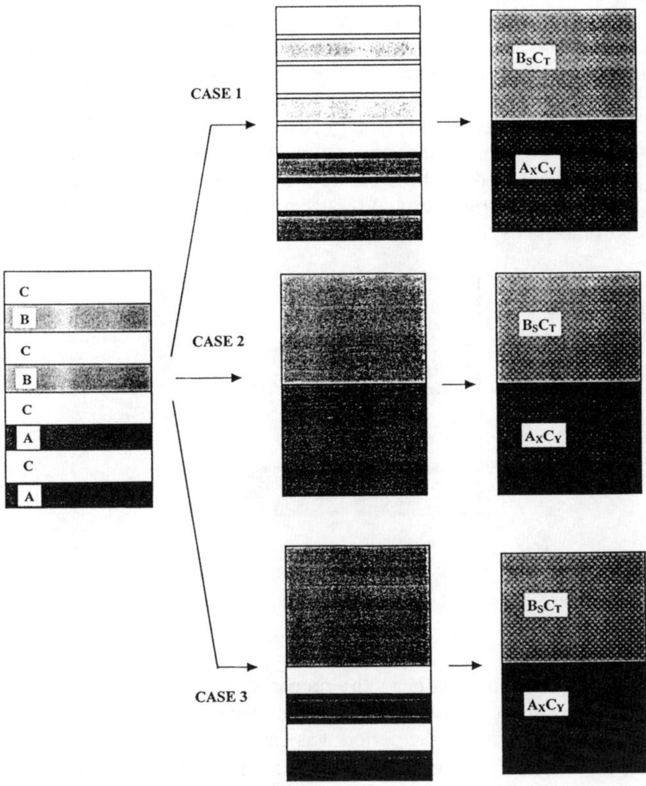

Figure 3: Schematic of a modulated elemental reactant designed to evolve into a crystalline superlattice or heterostructure, containing alternating layers of component A and B interleaved with component C. Possible reaction pathways resulting in a superlattice product are described in the text.

Case 3 - Interfacial Nucleation and Growth of One Component - Amorphous Intermediate Formation and Crystallization of the Other.

This reaction pathway starts with interfacial nucleation of one of the compounds, say A_xC_y at the A-C interface while the B-C layers interdiffuse forming an amorphous alloy. Again, we expect the rapid growth of A_xC_y along the A-C interface since a plane at the interface is the only region in the initial reactant with the correct composition for these compounds to grow. Subsequent low temperature annealing will lead to growth of A_xC_y perpendicular to the interfaces. On subsequent, higher temperature annealing, the remaining compound, B_uC_v, nucleates and grows. Since diffusion rates can be constrained by low annealing temperatures and diffusion distances can be optimized in the initial structure of the modulated reactant, growth process in this reaction pathway can be optimized to result in a modulated final structure with minimal interdiffusion of the components.

We first explored the formation of crystalline superlattice composites using the mechanism proposed in case 1, interfacial nucleation and growth. During our studies using modulated reactants to prepare amorphous intermediate in binary systems, several binary transition metal-selenium modulated reactants were observed to interfacially nucleate binary dichalcogenide compounds even when the initially modulation length was a small as 5Å. Diffraction data collected in these systems suggested that crystal growth first occurs along the plane of the interface. Subsequent annealing at higher temperatures was required for growth perpendicular to the substrate [14]. From these studies, these binary systems seemed ideal test candidates for the preparation of crystalline heterostructures and superlattices using the case 1, interfacial nucleation and growth mechanism.

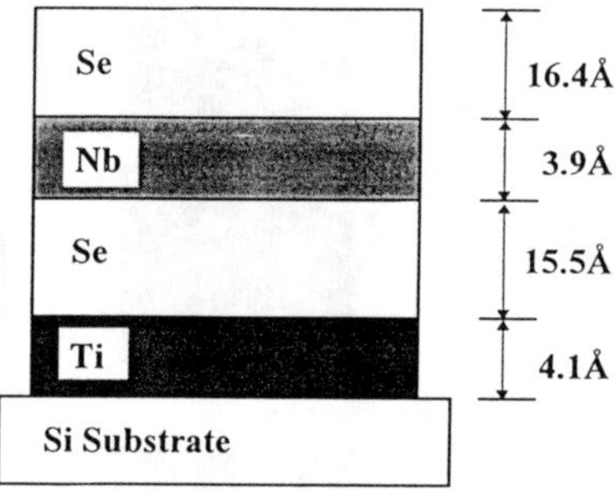

Figure 4: Schematic of a modulated elemental reactant designed to evolve into a crystalline $[NbSe_2]_3[TiSe_2]_3$ superlattice.

We tested the proposed reaction sequence by preparing a series of superlattice reactants with the overall structure shown in Figure 4 with varying thicknesses and number of Nb and Ti layers. The composition of each Nb-Se and Ti-Se period was chosen to be that of the desired MSe_2 compounds, and the thicknesses we chosen to be integral multiples of the known crystallographic unit cells so that there would be no elemental reactants left in the superlattice after extended annealing. Figure 5 shows the development of the [0 0 23] - [0 0 26] diffraction maxima of a $[NbSe_2]_6[TiSe_2]_6$ crystalline superlattice as a function of annealing temperature. The broad x-ray diffraction maxima evident at high angles upon low-temperature annealing results from the nucleation of the dichalcogenide layers. The diffraction patterns shown the continuous evolution of the high angle $[00\ell]$ Bragg diffraction peaks as the dichalcogenide nuclei grow. Rocking curve diffraction scans of the [0 0 24] superlattice diffraction peak taken as a function of annealing temperature are shown in Figure 5. The rocking curve peak shows a narrowing with increasing annealing time and temperature, with the half widths decreasing from greater than 3 to 1.0 degrees. This indicates that the superlattice is gradually evolving into the desired, kinetically trapped crystalline superlattice with the $[00\ell]$ diffraction planes aligned parallel with the initial interfaces present in the superlattice reactant. This diffraction data as well as additional data published previously is in qualitative agreement with the schematic for interfacial nucleation and growth as shown in Figure 3. [18-20].

The superlattice structure contained in the $[NbSe_2]_6[TiSe_2]_6$ product results from the structure of the initial elementally modulated reactant. By adjusting the thicknesses of the individual component layers on an Ångstrom level, one can control the superlattice period and structure on a unit cell basis. Figure 6 contains the diffraction patterns resulting from the annealing of a series of elementally modulated reactants designed to evolve into the indicated crystalline superlattices. The diffraction patterns contain many well resolved diffraction maxima, all of which can be indexed as [00ℓ] diffraction maxima resulting in a c-axis lattice parameters consistent with the desired number of $NbSe_2$ layers and $TiSe_2$ layers in the repeating unit of the superlattice sample. The intensity of the diffraction peaks in the diffraction patterns are also in qualitative agreement with the intended structure. One observes a convolution of the super-cell diffraction on top of that expected for the individual dichalcogenides. The high quality of the diffraction patterns as well as the observation of [10ℓ] diffraction peaks observed by off-axis scans in reciprocal space indicate the highly crystalline nature of the superlattice products. This is surprising, given the amorphous nature of the initial superlattice reactant. More detailed diffraction measurements as well as TEM studies are underway to determine the density of stacking faults and inclusions within these samples.

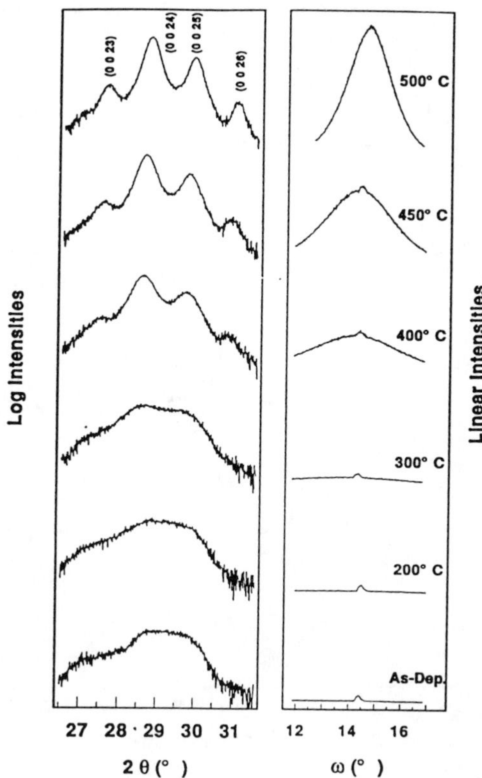

Figure 5: Evolution of the [0 0 23] - [0 0 26] diffraction maxima and rocking curve data for the [0 0 24] diffraction peak of a $[NbSe_2]_6[TiSe_2]_6$ crystalline superlattice as a function of annealing temperature.

As an extension of the superlattice systems containing dichalcogenides, we have begun to explore the application of this approach to other systems. Our studies of transition metal carbides suggest that they also should follow the interfacial nucleation reaction pathway shown in Figure 3. Figure 7 shows the evolution of the low angle diffraction pattern of a $(Mo_2C)_x(W_2C)_x$ as a function of annealing temperature. Figure 8 shows the evolution of the high angle diffraction pattern of this sample as a function of annealing temperature. The relative intensity of the [002] and [101] lines of the M_2C phase and rocking curve data indicate that the crystallites are preferentially oriented with respect to the substrate. Using the diffraction linewidths and the Scherrer equation , we calculate crystallite sizes of approximately 150 Å . This is twice the thickness of the superlattice period determined from the low-angle diffraction data, suggesting the crystallites have grown through several periods of the initial elementally modulated reactant [21]. Qualitatively, this diffraction data is similar to that found in early stages of annealing the dichalcogenide superlattices discussed earlier. This suggests that extended low temperature annealing is required to increase the crystallite size. These studies are currently underway.

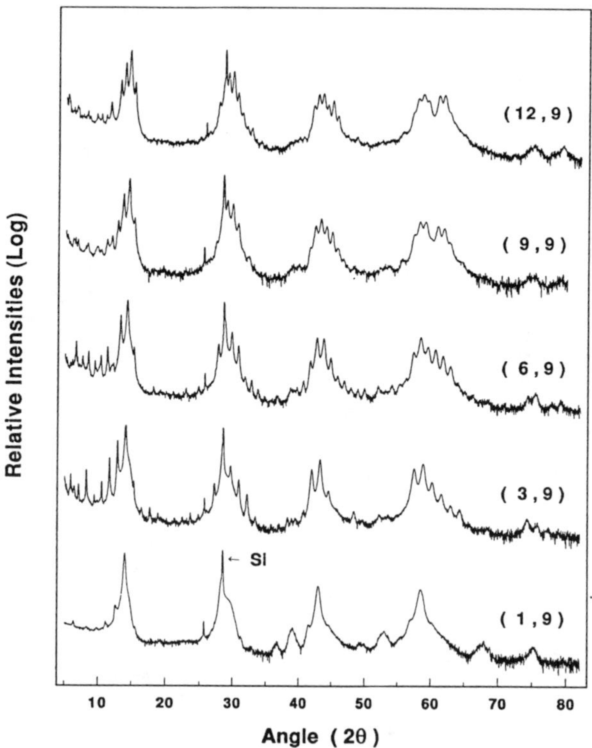

Figure 6: Diffraction patterns of a family of crystalline $[NbSe_2]_9[TiSe_2]_m$ superlattices.

We have concentrated our efforts on the mechanism described in case 1. Further investigation is needed to explore the other cases described. With respect to the mechanism described in case 2, nucleation from a modulated amorphous intermediate, the high interdiffusion rates at low temperature combined with the observed low nucleation temperature for the filled skutterudite, $Hf_{1-x}Fe_4Sb_{12}$, makes these skutterudite compounds an attractive candidate. These experiments are being initiated.

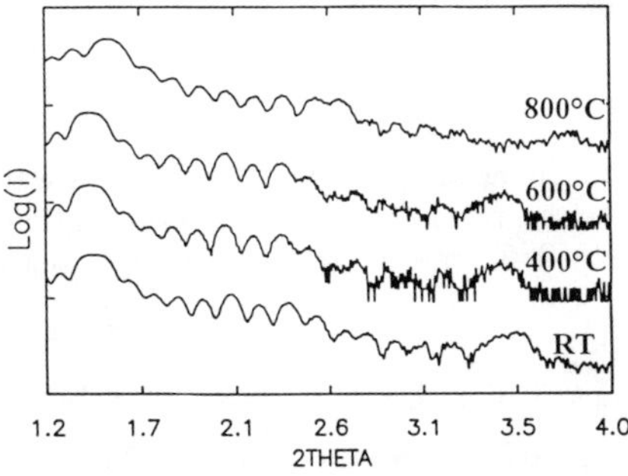

Figure 7: Evolution of the low angle diffraction pattern of a $(Mo_2C)_x(W_2C)_x$ superlattice as a function of annealing temperature.

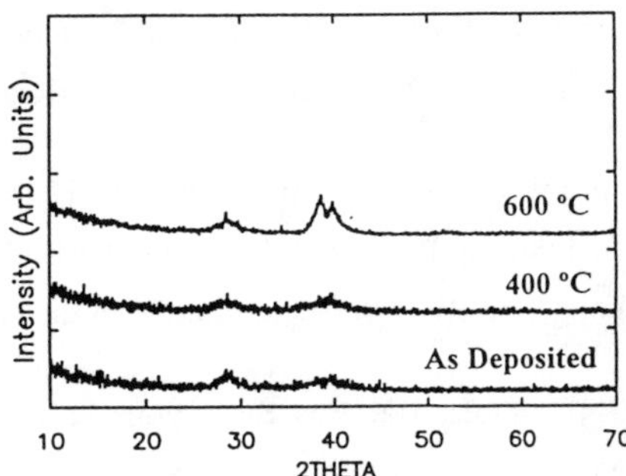

Figure 8: Evolution of the high angle diffraction pattern of a $(Mo_2C)_x(W_2C)_x$ superlattice as a function of annealing temperature.

CONCLUSIONS

Elementally modulated reactants provide a new tool for solid state chemists to rationally control the synthesis of kinetically stable compounds. Its strength is the ability to quickly explore composition space and rationally optimize the synthesis of new compounds. The low temperature synthesis of $Hf_{1-x}Fe_4Sb_{12}$ illustrates this ability. This synthetic approach also occupies a potentially valuable niche due to its ability to prepare crystalline superlattice materials with a variety of component materials with control of the superstructure on a unit cell basis. This approach does not compete with molecular beam epitaxy, as the defect and impurity levels are orders of magnitude higher than what can be accomplished with this more developed technique. Elementally modulated reactants are complementary, useful as an exploratory technique to probe a variety of systems for unusual properties which may result from the superlattice structure. In particular, it will permit the exploration of the transition from composite behavior to that of a new compound as the length scale of the compositional modulation decreases. Perhaps most importantly, the diffraction data collected as the initial reactants evolve into the kinetically trapped products provide a basis for understanding this transformation. This has resulted in a primitive understanding of the process, and the ability to control and predict the structures of the resulting products at the atomic level.

ACKNOWLEDGEMENTS:

Acknowledgment is made to the donors of the Petroleum Research Fund, administered by the ACS for partial support of this research. The support of the National Science Foundation (DMR-9308854 and DMR-9510562) is also gratefully acknowledged.

REFERENCES

1. A. Stein, S. W. Keller, and T. E. Mallouk, Science **259**, 1558-1564 (1993).

2. R. Marchand, L. Brohan, and M. Tournoux, Mat. Res. Bull. **15**, 1129-1133 (1980).

3. G. A. Weigers, A. Meelsma, R. J. Haange et al., Mater. Res. Bull. **23**, 1551 (1988).

4. C. Auriel, A. Meerschaut, R. Roesky et al., Eur. J. Solid State Inorg. Chem. **29**, 1079 (1992).

5. K. Ploog, Angewandte Chemie **27 (5)**, 593-621 (1988).

6. L. Fister, X. M. Li, Thomas Novet et al., J. Vac. Sci. & Technol. **A 11**, 3014-3019 (1993).

7. T. Novet, John M. McConnell, and D. C. Johnson, Materials Research Society **238**, 581-586 (1992).

8. E.J. Cotts, W.J. Meng, and W.L. Johnson, Phys. Rev. Lett. **57 (18)**, 2295-2298 (1986).

9. R. B. Schwarz and W. L. Johnson, Phys. Rev. Lett. **51 (5)**, 415-418 (1983).

10. R. B. Schwarz, K. L. Wong, and W. L. Johnson, Journal of the Non-Crystalline Solids **61 & 62**, 129-134 (1984).

11. B. M. Clemens and J. C. Buchholz, Mat. Res. Soc. Symp. Proc. **37**, 559-564 (1985).

12. B. M. Clemens and R. Sinclair, MRS Bulletin , 19-28 (1990).

13. L. Fister and D. C. Johnson, J. Am. Chem. Soc. **114**, 4639-4644 (1992).

14. M. Fukuto, M. D. Hornbostel, and D. C. Johnson, J. Am. Chem. Soc. **116**, 9136-9140 (1994).

15. M. D. Hornbostel, E. J. Hyer, J. Thiel et al., **submitted** J. Am. Chem. Soc. (1996).

16. R. D. Shannon, Acta Crystallogr. **A32**, 751 (1976).

17. R. D. Shannon and C. T. Prewitt, Acta Crystallogr. **B25**, 925 (1969).

18. M. Noh and D. C. Johnson, J. Am. Chem. Soc. **118**, 9117-9122 (1996).

19. M. Noh, J. Thiel, and D. C. Johnson, Science **270**, 1181-1184 (1995).

20. M. Noh and D. C. Johnson, Angewandte Chemie **accepted** (1996).

21. C. D. Johnson and D. C. Johnson, Mat. Res. Soc. Symp. Proc. **434**, 75-80 (1996).

SOLID-STATE REACTIONS IN MODEL OXIDE SYSTEMS

MATTHEW T. JOHNSON, PAUL G. KOTULA[†], RYAN S. THOMPSON AND C. BARRY CARTER

Department of Chemical Engineering and Materials Science, University of Minnesota, 421 Washington Ave. S.E., Minneapolis, MN 55455, [†]Current address: Center for Materials Science, MS K765, Los Alamos National Laboratory, Los Alamos, NM 87545, USA

ABSTRACT

The kinetics of thin-film solid-state reactions have been investigated in two model spinel forming oxide systems, NiO/Al_2O_3 and MgO/Fe_2O_3. In the NiO/Al_2O_3 system, thin-films of epitactic NiO were reacted with $(000\bar{1})$, $\{1\bar{1}00\}$, $\{11\bar{2}0\}$, and $\{1\bar{1}02\}$ orientated Al_2O_3 (corundum). The kinetics of the spinel forming reaction for this system were found to be linear-parabolic in nature. Additionally, it was found that the kinetics of the spinel-forming reaction varied by nearly two orders of magnitude between the fastest and slowest diffusion couples. The substrate determines the orientation of the overlayers and thereby the structure of the phase boundaries. In the MgO/Fe-oxide system, thin films of epitactic Fe oxide were reacted with $\{001\}$ MgO. The kinetics of this spinel forming reaction were parabolic in nature, indicative of diffusion control. In contrast to the NiO/Al_2O_3 system, the movement of phase boundaries are not the step controlling the reaction rate, but rather the diffusion of one of the cations across the reaction layer. In comparing the reaction rates for the two systems the activation energy for the formation of the spinel product in the MgO/Fe-oxide system was found to be almost a factor of 4 lower in comparison to the NiO/Al_2O_3 system.

INTRODUCTION

A thin-film on bulk substrate provides an ideal geometry for studying some of the fundamental processes occurring in solid-state reactions. These processes may include the effect that crystal orientation (or interfacial structure), grain boundaries, reaction temperature, surfaces and material system can have on the reaction products, kinetics, and interface morphology. Through the use of pulsed-laser deposition (PLD) a controlled geometry can be fabricated in which two high-quality oxides are in intimate contact, thus giving known starting conditions for the reaction. With this geometry the analysis of the reaction product is simplified because it is localized at the original interface of the two starting reactants.

Two classic spinel forming reactions were used to investigate these processes.[1,2,3,4,5] In the first, thin films of NiO were deposited on four orientations of monocrystalline Al_2O_3 (corundum). In these studies the resulting growth kinetics of the spinel reaction product have been studied using high-resolution low-voltage scanning electron microscopy (LVSEM). Additionally, a transmission electron microscopy (TEM) study of the interfaces and orientations in each of the diffusion couples was performed to gain an understanding of the interfaces that determine the reaction rates for this system. In a different spinel-forming system, thin films of Fe-oxide were deposited on monocrystalline $\{001\}$ magnesium oxide (MgO). Again the evolution of the spinel layer was followed through a series of heat treatments. The reaction couples were studied once again using LVSEM and the orientation of the reaction layers was investigated using TEM.

EXPERIMENTAL

In each of the two systems monocrystalline substrates were prepared prior to the thin-film deposition. In the reactions involving α-alumina, the substrates were acid cleaned and annealed at 1400°C for 8h prior to film deposition.[6] This preparation technique has been found to remove surface damage which occurred during polishing and in the case of the (0001) substrates, produced large (0001) terraces separated by surface steps.[7] In the reactions that involve MgO, bulk monocrystalline MgO was cleaved along <001>. The cleaved section was then immediately used for deposition.

PLD is a versatile thin-film deposition technique capable of producing high-quality oxide epilayers.[8] Details of the PLD apparatus are given elsewhere.[9,10,11] In brief, a pulsed KrF (λ=248nm) excimer laser was used to deposit various materials from a solid target source in a chamber filled with ~10mTorr oxygen. In the NiO/Al_2O_3 system, thin-films of NiO were deposited on Al_2O_3 substrates. Four orientation of Al_2O_3 were used to investigate the reaction kinetics: (0001), {1$\bar{1}$00}, {11$\bar{2}$0}, and {1$\bar{1}$02}. In order to avoid complications due to the nucleation of the reaction product, a buffer-layer of the spinel was deposited between the Al_2O_3 substrate and NiO thin-film.[12] The spinel buffer layer was produced from a polycrystalline $NiAl_2O_4$ target. 2000 laser pulses at 200mJ/pulse and 1Hz with a substrate temperature of 750°C in 6mtorr of oxygen was utilized to produce a 20nm film. A NiO film was deposited on the spinel/Al_2O_3 substrate. The parameters for the deposition were 50,000 pulses at 200mJ/pulse and 5Hz, with a substrate temperature of 750°C. In the Fe-oxide/MgO system thin-films of Fe-oxide were deposited onto {001} MgO. In producing the thin-film 90,000 laser pulses at 10Hz and 200mJ/pulse with a substrate temperature of 500°C were utilized to produce a 600nm Fe oxide film. All thin-film/substrates were then heat-treated in air for various times and temperatures, which were specific to the system.

The as-deposited and reacted thin films were characterized by scanning electron microscopy (SEM) and transmission electron microscopy (TEM) techniques. SEM and TEM samples were prepared in cross-section. The preparation of the Al_2O_3 and MgO SEM cross-section samples required the use of diamond lapping films followed by Syton©, to produce a polish which allowed the thin films (20-400nm thick) to be imaged. The SEM samples were then coated with approximately 0.5nm of Pt to reduce charging and then imaged in a Hitachi S-900 field-emission SEM operating at 5 kV. Cross-section and plan-view TEM was performed on selected samples in a Philips CM30 operating at 300 kV.

RESULTS AND DISCUSSION

Spinel formation in the NiO/alumina system

In studying the reaction kinetics for the NiO/Al_2O_3 the epitactic thin-film reaction couples were heat treated and then the thickness of the spinel layer was measured. Details of this procedure are given elsewhere.[13] For the (0001)- and {11$\bar{2}$0}-thin-film reaction couples the spinel layer was found to have linear growth kinetics, indicative of kinetic control by interfacial reaction, while for the {1$\bar{1}$00}- and {1$\bar{1}$02}-thin film reaction couples the kinetics were found to be linear-parabolic. Furthermore, the kinetics were found to vary by nearly two orders of magnitude between the fastest ({1$\bar{1}$00} and {1$\bar{1}$02} reaction couples) and the slowest ((0001) and {11$\bar{2}$0} reaction couples) at 1100°C. Figure 1 is a plot of the reaction-layer thickness as a function of time for the four orientations of alumina substrate.

The change in reaction rates can be attributed to the orientation of the monocrystalline substrate. This substrate dictates the orientation of the overlayers (NiO and spinel) and thereby the structure of the interfaces. It is the structure of the interfaces which controls the very earliest stages of growth, where the rate limiting step may be a phase boundary or interfacial reaction. The main difference between the (0001) and {1$\bar{1}$02} reaction couples and the {1$\bar{1}$00} and {1$\bar{1}$02} reaction couples is the coherency of the interfaces. The former group of reaction couples has {111} parallel to the substrate surface plane, while the latter group have a more complicated orientation relationship. Figure 2 is a cross-section TEM bright-field (BF) image of the as-deposited {1$\bar{1}$02} reaction couple taken 8° from the [$\bar{1}$102] pole of alumina. In this image there are two different orientations of NiO (and spinel) growing on this particular orientation of alumina. Under these imaging conditions one of the variants of the NiO is strongly diffracting in the BF image. Figure 3 is another TEM BF image of the {1$\bar{1}$02} reaction couple which has been heat treated at 1100°C for 16 minutes. This image is taken near the [010] pole of one of the NiO variants. It is clear from Figure 3 that one of the variants in the spinel preferentially grows into the alumina. In this way the original location of the spinel buffer-layer can be inferred (acting as a 'marker') as being approximately one quarter of the reaction layer thickness from the NiO/spinel phase boundary. This is consistent with the counter diffusion of cations being the reaction mechanism.

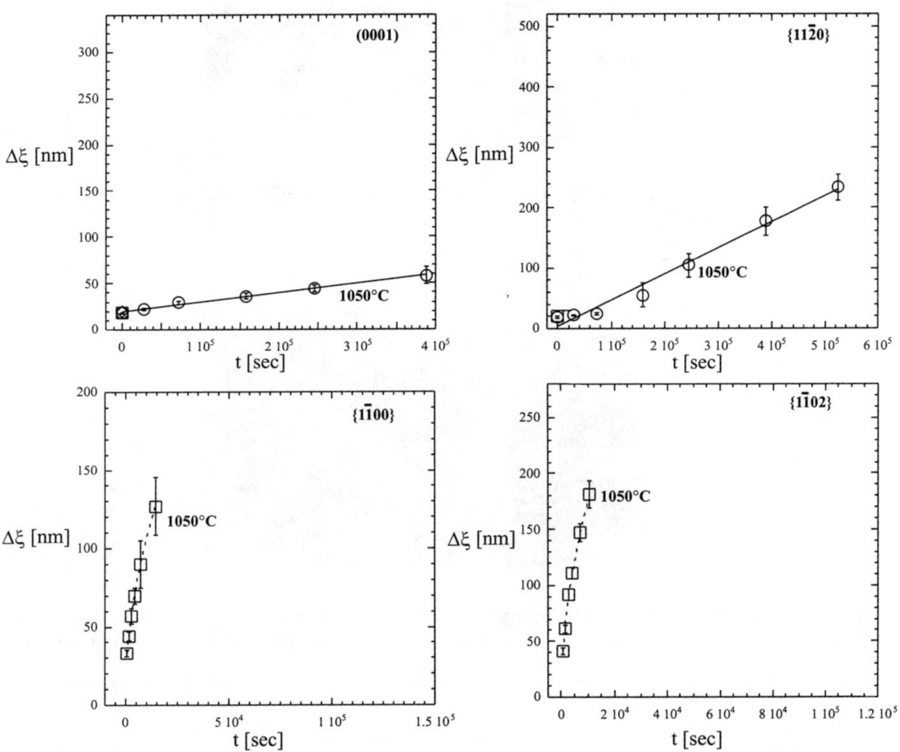

Figure 1. Plots showing the difference in reaction rate for the NiO/Al_2O_3 system for different substrate orientations. Notice the different scales. The slower reactions, in the upper two graphs give linear kinetics. While the reactions in the lower two graphs represent the faster reactions and give linear-parabolic kinetics.

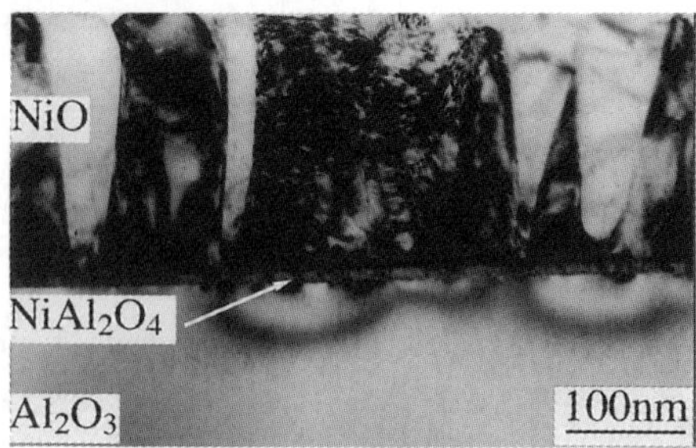

Figure 2. Cross-section TEM bright-field (BF) image of the as-deposited {1$\bar{1}$02} reaction couple taken 8° from the [1$\bar{1}$02] pole of Al$_2$O$_3$

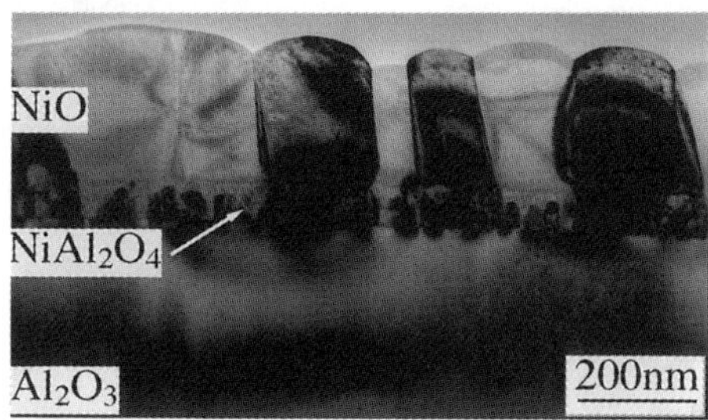

Figure 3. Cross-section TEM BF image of the {1$\bar{1}$02} reaction couple which has been heat treated to 1100°C for 16 minutes. This image is taken near the [010] pole of one of the NiO variants.

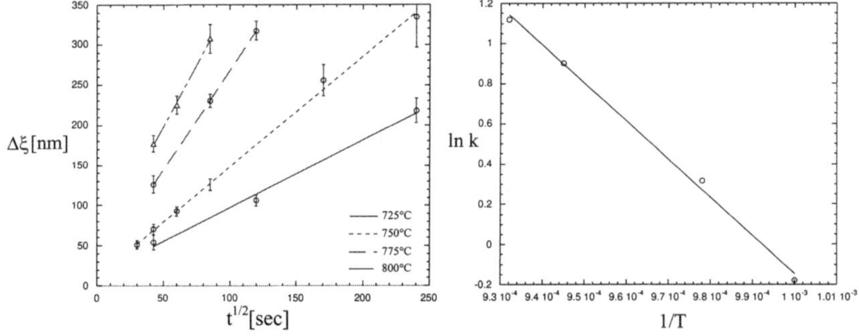

Figure 4. Left: Plot showing reaction rates for the Fe-oxide/MgO reaction couples for 4 different temperatures. Right: Arrhenius plot calculating the activation energy for the formation of the spinel reaction product, 157kJ/mol (1.6eV).

Spinel formation in the Fe_2O_3 /MgO system

Interfacial reactions between epitactic thin-films of Fe-oxide and {001} MgO were investigated in order to contrast the reaction mechanism and kinetic rates with the NiO/Al$_2$O$_3$. Thin-films of Fe-oxide were deposited onto {001} MgO. The structure of the as-deposited Fe oxide film is still under investigation, however preliminary studies indicate that the Fe oxide film is epitactic with the MgO substrate and is one of the spinel phases of iron oxide, either magnetite or maghemite. Previous studies on this material system have shown similar results.[14]

The as-deposited diffusion couples were then reacted in air for various temperatures and times to induce the spinel forming reaction. The method for studying the progression of the spinel layer was the same as used for the NiO/Al$_2$O$_4$ outlined in the previous section. Figure 4 is a graph showing the reaction rates for this system for 4 different temperatures. The reaction kinetics for all temperatures were found to be parabolic for measurable thicknesses. Calculations yielded an activation energy for this system of 1.6eV. This value represents the activation energy for the diffusion of the slower diffusing cation through the spinel layer.

TEM analysis was performed on selected samples both in cross-section and plan-view. Figure 5 is a cross-section TEM dark-field (DF) image of a Fe-oxide/MgO diffusion couple that was reacted at 800°C for 2.0 hours. The spinel layer that formed during the reaction is epitactic with the bulk MgO crystal. The {001} planes of the spinel are parallel to the {001} planes of the MgO and the <001> direction of the spinel is parallel to the <001> direction of the MgO. In addition, a remnant defect layer can be seen in the spinel reaction product. This remnant layer is located at a distance of one quarter of the total thickness of the spinel layer and is nearest to the MgO/spinel interface. This position is consistent with counter-diffusion of cations being the reaction mechanism. Figure 6 is a plan-view TEM BF image taken from a diffusion couple that was reacted at 750°C for 1.0 hour. As seen in this image the unreacted Fe oxide is heavily twinned and preliminary investigations into the structure of this layer indicate that the Fe oxide has transformed from the cubic spinel phase, in the as-deposited film, to that of hematite which has the corundum structure. This layer does not appear to be related crystallagraphically to the substrate in a simple manner.

CONCLUSION

Thin-film diffusion couples produced by PLD provide an ideal geometry for the study of solid-state reactions in oxide systems. In each of the spinel systems mentioned above a different kinetic mechanism controls the reaction rate. For the NiO/Al$_2$O$_3$ system, the effects of substrate orientation on the kinetics in the earliest stages of the reactions have been demonstrated. The substrate determines the orientation of the overlayers (NiO and spinel) and thereby the structure of the interfaces. It is the structure of the interfaces which controls the very earliest stages of growth, when the rate-limiting factor is a phase boundary or interfacial reaction. For the Fe-oxide/MgO system, the rate controlling step is the diffusion of the reactants through the spinel layer. The most striking difference between the two systems is that the activation energy for the formation of the spinel is a factor of 4 lower in the Fe-oxide/MgO system.

ACKNOWLEDGMENTS

The authors acknowledge the research support of the Center for Interfacial Engineering which is a National Science Foundation Engineering Research Center. They thank Prof. Stan Erlandsen for access to the Hitachi S-900 SEM and Chris Frethem for technical assistance.

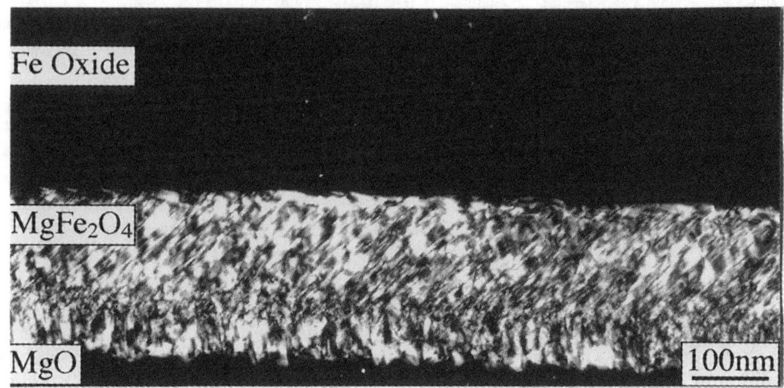

Figure 5. Cross-section TEM dark-field (DF) image taken from a Fe oxide/MgO diffusion couple that was reacted at 800°C for 2.0 hours. The image was formed with a [220] reflection of the spinel.

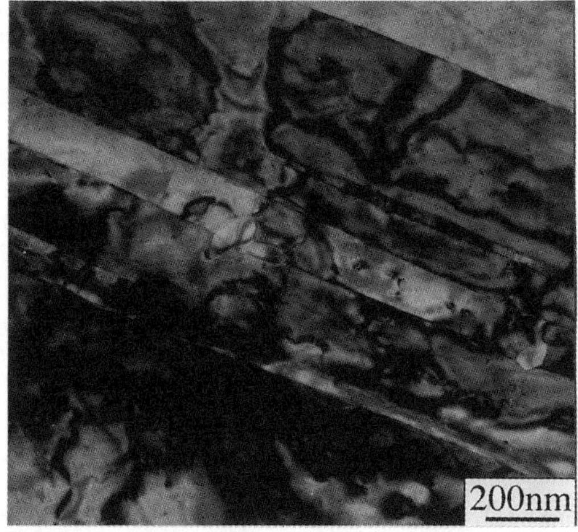

Figure 6. Plan-view TEM bright-field (BF) taken from a Fe oxide/MgO diffusion couple that was reacted at 750°C for 1.0 hours.

REFERENCES

1. R. E. Carter, J. Am. Ceram. Soc. **44**, 116-120 (1961).
2. G. de Roos, J. W. Geus, J. M. Fluit and J. H. de Wit, Nucl. Instrum. Methods **168**, 485-489 (1980).
3. F. S. Pettit, E. H. Randklev and E. J. Felten, J. Am. Ceram. Soc. **49**, 199-203 (1966).
4. A. E. Paladino, J. Am. Ceram. Sos. **43**, 183-191 (1960).
5. R. C. Rossi and R. M. Fulrath, J. Am. Ceram. Soc. **46**, 145-149 (1963).
6. D. W. Susnitsky and C. B. Carter, J. Am. Ceram. Soc. **75**, 2463-2478 (1992).
7. D. W. Susnitzky and C. B. Carter, J. Amer. Cer. Soc. **69**, C217-C220 (1986).
8. H. Sankur and J. T. Cheung, Appl. Phys. A **47**, 271-84 (1988).
9. J. R. Heffelfinger and C. B. Carter, Mat. Res. Soc. Proc. **317**, 553-558 (1993).
10. P. G. Kotula and C. B. Carter, Mat. Res. Soc. Proc. **285**, 373-379 (1993).
11. P. G. Kotula and C. B. Carter, **2**nd International Conference on Laser Ablation **288**, 231-236 (1993).
12. P. G. Kotula and C. B. Carter, Phys. Rev. Lett. **77**, 3367-3370 (1996).
13. P. G. Kotula and C. B. Carter, Proc. 53rd Annual Mtg. MSA 330-331 (1995).
14. I. M. Anderson, L. A. Tietz and C. B. Carter, Mat. Res. Soc. Symp. Proc. **238**, 807-814 (1992).

QUANTITATIVE INTERFACIAL CHEMISTRY AND BONDING IN CERAMIC MATERIALS

Hui Gu

Japan Science and Technology Corporation, "Ceramics Superplasticity" project, JFCC 2F, 2-4-1 Mutsuno, Atsuta, Nagoya 456, Japan, gu@ngo.jst-c.go.jp

ABSTRACT

Internal interfaces, between the same phase (grain boundary) or two different phases, often play an essential role in controlling various properties in ceramic materials. Analysis of such interfaces can achieve various newly available quantitative information on a sub-nanometer scale, as well as bonding picture associated with these interfaces. This analysis combines EELS spectrum profiling and advanced near-edge structure (ELNES) data processing, taking advantage from both the high spatial and energy resolutions provided by a dedicated STEM instrument. Demonstrated by chemical and structural study of grain boundary and inter-phase interfaces in Si_3N_4 ceramics where ~1 nm thick amorphous phases cover every crystalline boundaries, it reveals that the amorphous boundary films are substantially different to bulk amorphous phase and to the interface between crystalline and amorphous phases. Moreover, the boundaries between two different crystalline phases were found covered with two thin amorphous layers of different composition and bonding. These observations can shed further light on the influence of interfaces on the macroscopic properties of these ceramic materials. This interfacial analuysis method can be extended to broader research area in solid state physics and chemistry.

INTRODUCTION

Internal interfaces, such as a grain boundary of a same crystalline phase and phase boundary between two different crystalline grains or between a crystalline and an amorphous phase, can play important role in influencing or controlling various properties in high performance ceramics and composites. The mechanical properties of high-temperature structural ceramic Si_3N_4 are mainly controlled by ~1 nm thick amorphous films covering every grain boundaries, which in one side improve the sinterability and oxidation resistance of this ceramic material while deteriorate the strength and toughness at high temperature [1]. Knowledge of the chemical and physical properties of such grain boundary phases will be very valuable for understanding the microstructure-properties relations, the key factor in optimizing the performance of these and other ceramic materials. Recent development of microanalysis made it possible to investigate these interfacial regions in much great details. Not only an interface can be analyzed on a sub-nanometer scale, but also the bonding at interfacial region and various quantitative information can be obtained. This is largely due to the fully utilizing of dedicated analytical TEM, the scanning transmission electron microscope, which provides the best spatial and energy resolutions. In this paper, several types of interfaces in undoped and La-doped Si_3N_4 are studied by using EELS profiling and a combination of ELNES and quantification method [2-4], which reveal new interfacial phenomena and upgrade our understanding of this material.

Mat. Res. Soc. Symp. Proc. Vol. 453 © 1997 Materials Research Society

MATERIALS AND EXPERIMENT

High-purity ß-Si$_3$N$_4$ were fully-dense and sintered by HIPing at 1950°C for 1 h under 170 Mpa [5]. They contain 2-6 mol% of SiO$_2$ as secondary phase observed at grain boundaries and triple pockets. La-doped (5 mol%) Si$_3$N$_4$ were liquid phase sintered at high temperature and with a slow cooling rate [6]. TEM specimens were prepared by mechanical polishing and dimpling, finished with Ar-ion milling till perforation appeared. A dedicated STEM (VG-HB501), situated in Max-Planck-Institut für Metallforschung in Stuttgart, has been used in this work. This microscope has been fitted a cold field-emission tip and parallel EELS spectrometer (Gatan 666); it can provide an electron probe well less than 1 nm in diameter. The energy resolution achieved in this system was better than 0.7 eV measured at the zero-loss peak. This combined high performance makes it possible to fully use the EELS "spectrum imaging" facility incorporated into this STEM [7]. EELS spectra were recorded either from two scanning boxes *on* and *off* an interested interface or by using EELS profiling, 1D version of "spectrum imaging", to generate a large number of spectra, each corresponding to a position pixel, in a total length of 10-50 nm across an interface.

RESULTS AND DISCUSSIONS

EELS measurement for an interfacial region, either from a rectangular area in box mode or from a much finer electron probe in profiling, collects always considerable amount of energy-loss signals generated from the nearby bulk regions. This is because of the delocalization effect of the inelastic scattering and beam intensity from the extended tails of an electron probe even its size, usually measured as the width of half maximum, can be smaller than the width of a boundary region. To obtain signals belonging only to the boundary, it is necessary to process ELNES from the as-recorded spectra.

ELNES contains information of bonding, coordination and electronic structures of the probed atoms, and each resolved ELNES feature reflects and associates to a certain type of relation or interaction between the probed atom and its neighboring atoms. In ceramic materials, the local bonding and coordination for an element are usually different at the interface and from the bulk, this is likely to generate different ELNES belonging to the two regions for this element. The different ELNES can be separated due to the incoherent nature of energy-loss events. A systematic spectrum

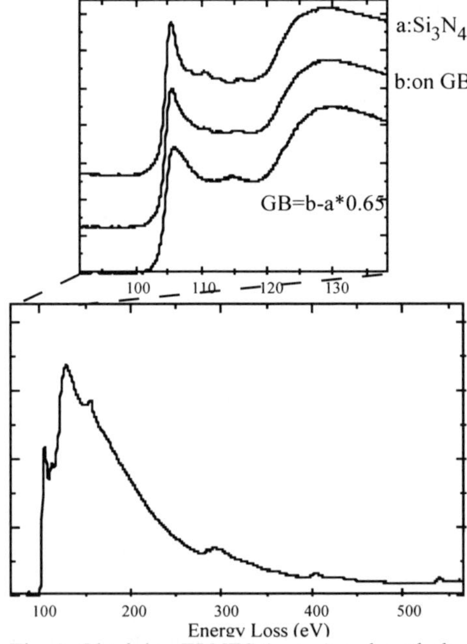

Fig. 1 Obtaining ELNES (upper) and a whole EELS spectrum (lower) for a grain boundary in undoped Si$_3$N$_4$ by bulk subtraction.

subtraction to achieve ELNES dedicated to grain boundaries has been demonstrated in [2]. This is in fact a procedure to find two independent vectors in a multiple-dimensional space of EELS spectra; any distinguishing ELNES feature belonging to the bulk should disappear in order to achieve ELNES belonging to a boundary. Moreover, when other core-loss edges have been acquired together, such ELNES processing can obtain an EELS spectrum for all these elements from the boundary region since the energy-loss edges of the other elements are just more coordinates of the two independent vectors corresponding to the bulk and the boundary regions respectively, an example is shown in Fig. 1 for ~1 nm thick grain boundary film in undoped Si_3N_4. The composition of the grain boundary region is quantified directly to be $SiN_{0.62}O_{1.39}$. Besides ELNES and composition of the boundary, quantification of the intensities of the separated signals from the two regions can give an ELNES-related chemical with of the boundary (d_{ELNES}) and the concentration (N^{GB}) of the involved element in the boundary region, as follows:

$$d_{ELNES} = w \left(1 - I^{on}/I^{off} + I^{GB}/I^{off}\right) \qquad (1)$$

$$N^{GB} = N^{bulk}/\left(1 + I^{off}/I^{GB} - I^{on}/I^{GB}\right) \qquad (2)$$

which are illustrated schematically in Fig. 2. I^{on} and I^{off} are intensities from the two boxes; I^{GB} is the signal of grain boundary separated from I^{on}; w is the common width of the two boxes; N^{bulk} is concentration of the same element in the bulk. From the spectra associated with the grain boundary film in the undoped Si_3N_4 (Fig. 1), we get a film width of 1.13 nm and silicon concentration of 28.5 nm^{-3}, or 69.2% of the bulk concentration. All these information can also be obtained from EELS profiling following the same procedure [4].

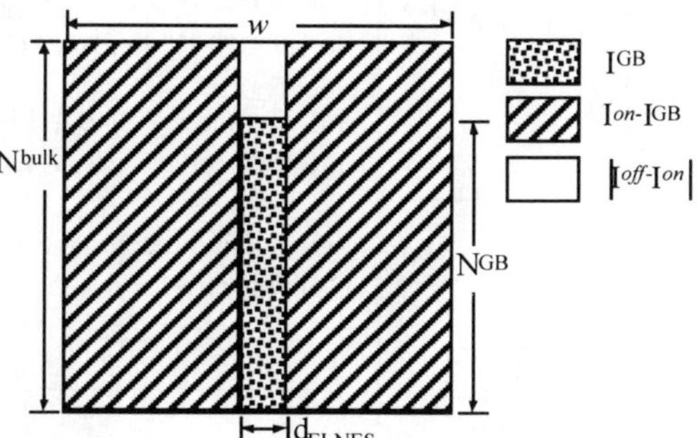

Fig. 2   Schematic for quantification of concentration and width of grain boundary.

Using EELS profiling elemental and ELNES profiles across a boundary can be obtained. These profiles are also convoluted with the beam profile. ELNES profiles for bulk and interface signals of a same element are processed by least-square fitting of every spectra with the corresponding ELNES [8]. Si, N and O profiles as well as Si_3N_4 and boundary ELNES profiles from Si-$L_{2,3}$ signals across the same grain boundary film are shown in Fig. 3. It is clearly shown a mainly oxide film containing significant amount of nitrogen. Silicon is considerably lower in the film, which agrees well with previous data. The Si_3N_4 ELNES profile demonstrates the important contribution from the tail intensity of the beam; there was still about 30% of Si-$L_{2,3}$ signal coming from the bulk even the probe size was about 0.5 nm and the film width is 1.1 nm.

Similar ELNES processing and profiling can be applied to the interface between Si_3N_4 grain and SiO_2 amorphous pocket. The Si-$L_{2,3}$ ELNES of the interface is obtained by removing both Si_3N_4 and SiO_2 signals, as shown in Fig. 4. This interface ELNES is not only remarkably different from those of the bulk phases, but also rather different from that of the amorphous grain boundary film in the same sample (Fig. 1). Fitting spectra in EELS profiling with the three ELNES in Fig. 4 gave corresponding ELNES profiles across this interface (Fig. 5). The interface ELNES profile not only reveals a wider interface than the grain boundary film, but its asymmetric shape shows also evidence for a hetero-phase boundary.

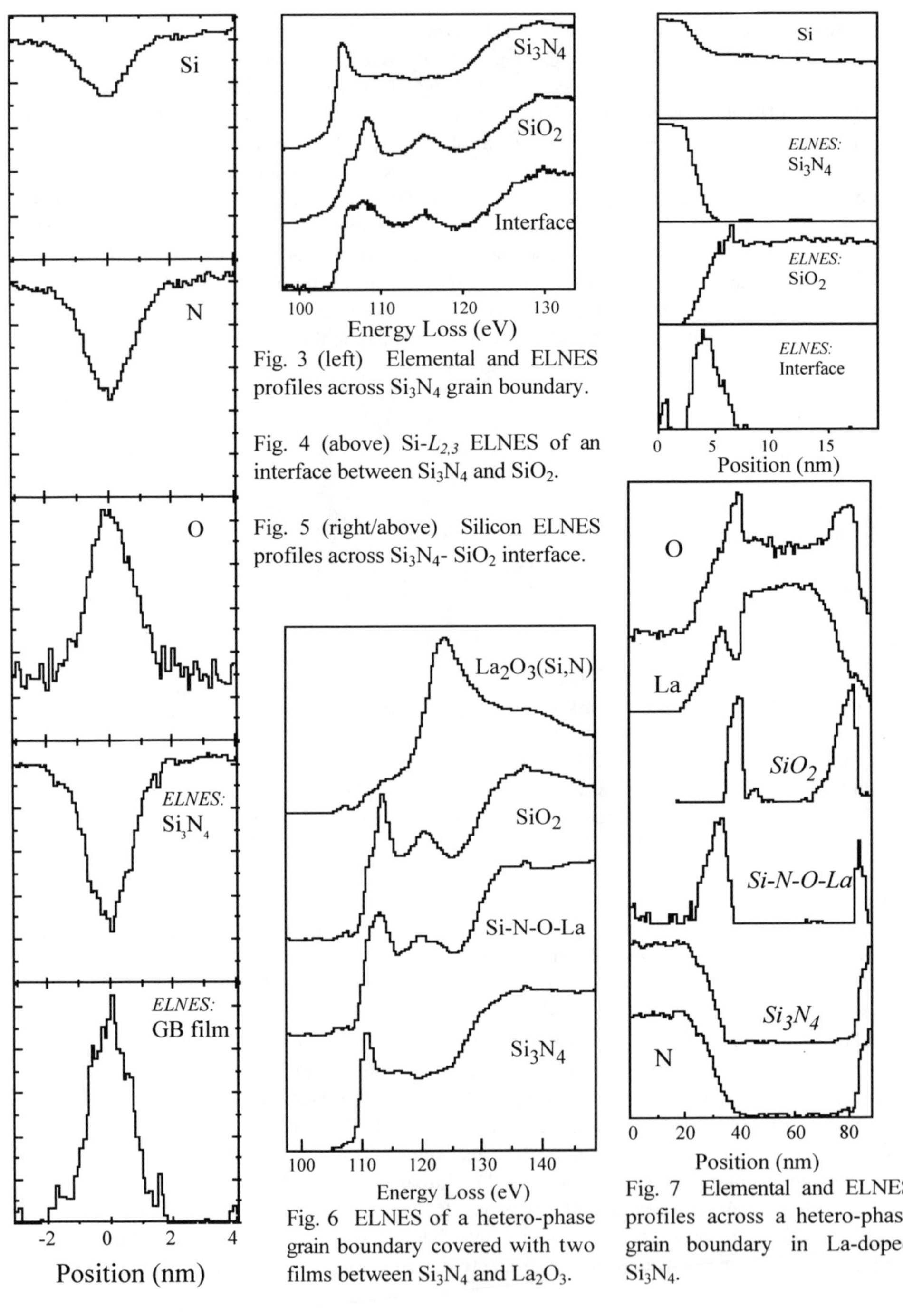

Fig. 3 (left) Elemental and ELNES profiles across Si_3N_4 grain boundary.

Fig. 4 (above) Si-$L_{2,3}$ ELNES of an interface between Si_3N_4 and SiO_2.

Fig. 5 (right/above) Silicon ELNES profiles across Si_3N_4- SiO_2 interface.

Fig. 6 ELNES of a hetero-phase grain boundary covered with two films between Si_3N_4 and La_2O_3.

Fig. 7 Elemental and ELNES profiles across a hetero-phase grain boundary in La-doped Si_3N_4.

The Si_3N_4 ceramic doped with 5 mol% of La_2O_3 contains La-Si-O-N amorphous films at grain boundaries, but the larger triple pockets have been successfully devitrificated into a crystalline secondary phase which is mainly La_2O_3 with small amount of Si incorporated into the structure [6]. However, the hetero-grain boundaries between Si_3N_4 and La_2O_3 phases were still found to be covered with amorphous films. EELS profiling across hetero-phase grain boundary films revealed that oxygen and lanthanum intensities peaked at different positions across the boundaries (Fig. 7). La-$N_{4,5}$ edge, from which the lathanium signal was measured, is heavily overlapped with Si-$L_{2,3}$ (Fig. 6). However, ELNES of La-$N_{4,5}$ is not expected to differ significantly from the second phase and from the amorphous films since ELNES features from this edge are dominated by atomic effect and much less sensitive to the change of local structures. Therefore the Si-$L_{2,3}$ ELNES from the two parts in the grain boundary films can be obtained using the similar ELNES processing procedure described in the previous examples to subtract the two bulk ELNES signals, one of them is dominately La-$N_{4,5}$ signal. This ELNES separation gives two significantly different Si-$L_{2,3}$ ELNES patterns in each hetero-grain boundary film corresponding to the peak positions for O and for La signals respectively. This indicates that the hetero-phase boundaries are covered by two distinctly different amorphous films. The film adjacent to Si_3N_4 grain shows ELNES rather similar to the ELNES of the amorphous grain boundary films in the undoped Si_3N_4 sample (Fig. 1), indicating a similar bonding configuration around the silicon atoms in this film. This signifies that there is significant amount of nitrogen in this film. This oxinitride amorphous film contains also some lanthanum, which is revealed by the peaked La concentration at the location of this film near Si_3N_4 grain in La profile (Fig. 7). ELNES from the other amorphous film, which is next to the grain of the La_2O_3-based secondary phase, exhibits a same Si-$L_{2,3}$ edge as for the amorphous SiO_2 (Fig. 4). This fact indicates that little or no nitrogen were dissolved into this oxide film. There is also less lanthanum in this film compared to the film next to Si_3N_4 grain, signifying the depletion of rare-earth cations from this typical silica phase. La might be completely expelled from this film, however, this is difficult to confirm since La-$N_{4,5}$ ELNES from the secondary phase and from the amorphous films are almost the same, which made it impossible to separate the contribution from the different parts and hence to know each of the concentrations.

This is the first evidence of two amorphous films existing at hetero-phase grain boundaries under thermodynamic equilibrium. This phenomanon indicates that the large difference of chemical potentials between the two phases has been balanced in a more delicate way. This fact will certainly provoke new thought about the interfacial wetting mechanism and its relation with segregation behavior. It is clear that each grain surface requires specific amorphous overlayer to lower its surface energy. For Si_3N_4 grain surfaces oxinitride films are always required, the same is true for either hetero-phase or homo-phase grain boundaries; contrarily, for La_2O_3 grains more or less pure oxide amorphous layers are enough to wet their surfaces. The segregation of La to the oxinitride films near Si_3N_4 grains, but not or less to the silica films near La_2O_3 grains, is then quite understandable since the co-solubility of nitrogen with rare-earth or alkaline-earth cations in the bulk silica amorphous phase is a well known phenomenon.

CONCLUSIONS

Analyzing internal interfaces in ceramic materials can reveals various important information using a newly established quantitative method based on ELNES information in a STEM environment as demonstrated by homo-phase and hetero-phase boundaries in two Si_3N_4 samples. In undoped sample, not only the composition, concentration and width can be obtained for amorphous films covering the grain boundaries, but the bonding in these films was quite different from that of the interfaces between Si_3N_4 grain and the amorphous SiO_2. In La-doped Si_3N_4 sample with crystalline second phase, the hetero-phase grain boundaries were found being covered by two different amorphous layers, an oxinitride film containing La segregates covers each Si_3N_4 grain surface and a typical amorphous SiO_2 layer with less or no La covers each La_2O_3 grain surface. These phenomena can significantly improve our understanding of various properties and behaviors in this group of structural ceramic materials, and from these observations further insight can be gained for wetting and segregation behaviors at interfaces under thermodynamic considerations in general materials.

ACKNOWLEDGMENT

The author acknowledges Max-Planck Society for the financial support during the period when the experiments were performed at Max-Planck-Institut für Metallforschung in Stuttgart. He thanks Prof. M. Rühle, Drs. R. Cannon, X. -Q. Pan, L. -T. Ma, I. Tanaka, H. Müllejans, G. Duscher for useful discussions as well as technical supports. The samples were provided by I. Tanaka and L. -T. Ma, the TEM specimens were prepared by U. Salzberger, and the "spectrum imaging" software package was programmed by Dr. M. Tencé, those efforts are all heartily appreciated by the author.

REFERENCES

1. F. F. Lange, Am. Ceram. Soc. Bull. **62**, 1368-99 (1983).

2. H. Gu, M. Ceh, S. Stemmer, H. Müllejans and M. Rühle, Ultramicroscopy **59**, p. 215-27 (1994).

3. H. Gu, R. M. Cannon and M. Rühle, J. Mater. Res., submitted.

4. H. Gu, Materials Transactions of Japan Institute of Metals **37** (12), in press (1996).

5. I. Tanaka, H.J. Kleebe, M.K. Cinibulk, J. Bruley, D.R. Clarke, and M. Rühle, J. Am. Cera. Soc. **77**, p. 911-17 (1994).

6. L. T. Ma, private communication.

7. C. Colliex, M. Tencé, E. Lefèvre, C. Mory, H. Gu, D. Bouchet and C. Jeanguillaume, Microchim. Acta **114/115**, p. 71-87 (1994).

8. M. Tencé, M. Quartuccio and C. Colliex, Ultramicroscopy **58**, p. 42-54 (1996).

CHEMICAL REACTIVITY BETWEEN SIALON CERAMICS
AND IRON-BASED ALLOYS

J. VLEUGELS, O. VAN DER BIEST
Department of Metallurgy and Materials Engineering (MTM), Katholieke Universiteit Leuven,
De Croylaan 2, B-3001 Leuven, Belgium.

ABSTRACT

The chemical interaction of a number of SiAlON ceramic - iron-based alloy combinations is studied by means of static interaction couple experiments at 1200 °C. During the interaction, silicon and nitrogen from the ceramic dissolve and diffuse into the iron alloy, whereas the remaining aluminium and oxygen form Al_2O_3 particles. The rate controlling step of the interaction was identified and the influence of alloying elements in the iron-based alloy as well as the chemical composition of the ceramic on the reactivity was investigated. The relative reactivity measured in the interaction couples was compared with the measured wear of the ceramics in turning of steels. The predictive capabilities of the results of the interaction couple experiments on the actual cutting tool performances in turning operations is highlighted.

INTRODUCTION

The universal trend in metal machining technology is high speed machining, i.e. metal removing by turning, milling and drilling operations at very high speed and feed. In Europe, environmental concern increases the demand for dry and semi-dry machining operations. During high speed as well as dry machining operations, high mechanical stresses and temperatures are generated at the contact surface between cutting tool and workpiece material. The essential requirements for new cutting tool materials in addition to low cost are: a high hardness and toughness at elevated temperatures, a good thermal and mechanical shock resistance, a high oxidation resistance and a high chemical stability with respect to the workpiece material.

In general, the total wear of a cutting tool can be seen as a combination of mechanical and chemical wear [1]. At lower cutting speeds, ceramic cutting tools will wear predominantly by abrasion. At high cutting speeds or under dry-machining conditions, chemical wear, i.e. the dissolution of tool material into the workpiece or chip, becomes the predominant mode of tool wear [2]. It is clear that in the development of tool materials for high speed machining, special attention should be given towards the chemical matching of tool and workpiece material at elevated temperatures. Therefore, our research group is investigating the ceramic-metal interaction in general, and especially the chemical compatibility of ceramic cutting tools with iron-based alloys.

EXPERIMENTAL PROCEDURE

The chemical compatibility experiments were performed by pressing together polished slices of sialon ceramic and iron-based alloy under a small load (2.5 MPa) in vacuum ($\approx$ 0.1 Pa) at 1200 °C. The chemical reactivity was studied by assessing the extent of interdiffusion of species between the two materials. The thickness of the interaction layers in the interaction couples was measured with an optical microscope. The values given are the mean of at least ten measurements in the central part of the interaction couple.

Mat. Res. Soc. Symp. Proc. Vol. 453 © 1997 Materials Research Society

Two sialon ceramics were prepared by conventional uniaxial hot pressing of Si_3N_4-Al_2O_3 powder mixtures under vacuum during 1 hour at 1650 °C with a mechanical load of 30 MPa. More details on the preparation and characterisation of these ceramics can be found elsewhere [3]. The composition of the starting powder mixture and the amount and chemical composition of the different phases in the hot pressed ceramics are given in Table 1. The third ceramic is a commercial YSiAlON cutting tool material (Céramétal, Luxemburg), obtained by sintering of Si_3N_4 - Al_2O_3 (4 wt. %) - Y_2O_3 (5 wt. %) powder mixtures.

The continuous turning tests were performed with ISO SNGN 120412 type inserts, with a feed of 0.268 mm, a depth of cut of 2 mm and a cutting speed of 100 m/min. The turning tests were performed dry on a high rigidity 18.5 kW TOS SUS63 lathe and the inserts were clamped in a Sandvik R 174.1 25-25-12 tool holder with an approach angle of 75° and a negative rake angle of 6°. The flank wear on the worn inserts was measured with an optical microscope. The maximum crater depth on the rake face was measured with a profilometer, after dissolving the adhering steel.

The chemical composition of the used steels are given in Table 2.

Table 1. Chemical composition of starting powder mixtures and hot pressed ceramics

sialon	starting powder mixture			sintered ceramic composition				
	Si_3N_4	Al_2O_3	Y_2O_3	β-sialon $Si_{6-z}Al_zO_zN_{8-z}$		O-sialon $Si_{2-x}Al_xO_{1+x}N_{2-x}$		X-sialon $Si_{12}Al_{18}O_{39}N_8$
	wt.%	wt.%	wt.%	z	wt.%	x	wt.%	wt.%
1	93.72	6.28	0.00	0.65	77	0.12	23	0
2	79.55	20.45	0.00	1.22	79	0.16	17	4
3	91.00	4.00	5.00	0.14	10	with ≈ 10 % YSiAlON glass		

Table 2. Nominal composition (wt. %) of the iron-based alloys

AISI	DIN	C	Si	Mn	Cr	Mo	Ni
/	50NiCr13	0.6	0.3	0.5	1.0	0.3	3.0
4140	42CrMo4	0.4	0.0	0.6	1.0	0.3	0.0

RESULTS AND DISCUSSION

The chemical reactivity was investigated by means of interaction couples at 1200 °C. As a typical example of the interaction between sialon ceramics and iron-based alloys, the interaction between sialon 2 and a carbon steel (AISI W110) after 3 hours at 1200 °C is shown in figure 1. Four regions can be clearly distinguished. Region A refers to the sialon ceramic; region B to the interaction zone on the ceramic side of the interaction couple; region C to the interaction zone on the steel side of the couple and region D to the steel. Experiments with inert markers showed that the interface between region B and C also marks the position of the original contact plane.

More information could be obtained from microprobe analysis (EPMA) in the four regions. The interaction zone on the ceramic side of the interaction couple (region B) consists of an aluminium and oxygen rich reaction product, embedded in an iron rich matrix with the same chemical composition as the interaction zone on the steel side (region C). The presence of alumina could be confirmed by X-ray diffraction. The interaction layer on the steel side of the interaction couple (region C) is α-Fe with 6-8 at. % silicon in solid solution. The thickness of this zone corresponds with the silicon diffusion depth into the steel. Aluminium and oxygen could not be detected by EPMA analysis in this zone. Diffusion couples with pure iron revealed also the presence of iron nitrides on the steel side.

Based on these observations, the following mechanism for the interaction between sialon ceramic and iron alloy is proposed. Silicon and nitrogen from the dissociating β-sialon and O-sialon grains in the ceramic dissolve and diffuse into the alloy, whereas aluminium and oxygen from the dissociation form Al_2O_3 particles that remain on the ceramic side of the interaction couple. At the same time, iron and alloying elements diffuse in between the Al_2O_3 particles in the opposite direction towards the reaction front in the ceramic. The dissolution of nitrogen is thought to proceed via the gas phase and during the interaction, a steady state nitrogen pressure will be built up at the metal-ceramic interface. As the chemical interaction proceeds, the reaction front advances into the ceramic material and the silicon diffusion front moves further into the steel. For the yttrium containing ceramic (sialon 3), an YSiAlON reaction product instead of Al_2O_3 was found [4].

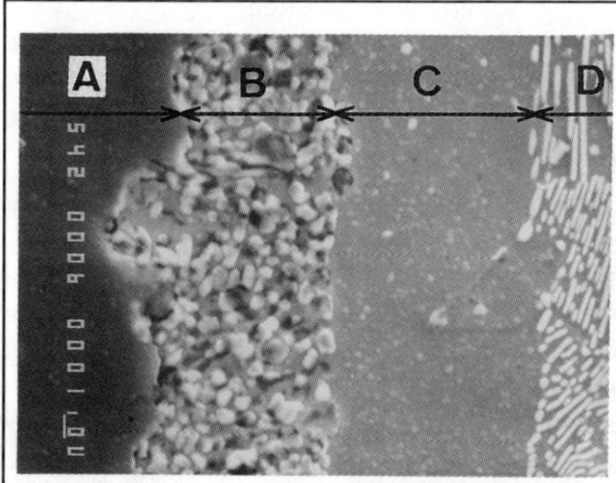

Figure 1: Sialon 2 - carbon steel interaction after 3 hours at 1200°C

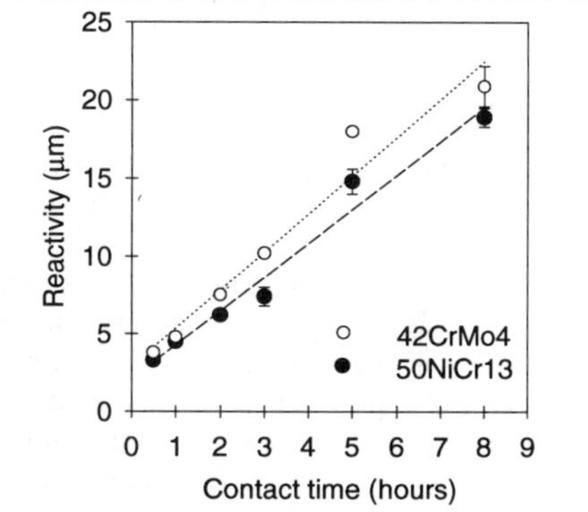

Figure 2: Reactivity of sialon 2 as a function of contact time at 1200°C

The reactivity of a metal-ceramic combination can be defined as the amount of ceramic that is dissolved at a given temperature after a predefined time. In this work, the reactivity is defined as the width of the interaction layer on the ceramic side of the interaction couple after 5 hours of contact at 1200°C. The reactivity is measured in the central part of the couple since an increased reactivity is always noticed at the edges, i.e. where the interaction zone is in contact with the furnace atmosphere (vacuum).

The reactivity of the sialon ceramics with 42CrMo4 and 50NiCr13 steel was found to increase linearly with time, indicating a reaction controlled interaction mechanism, as shown in figure 2 for the interaction with sialon 2. The intercepts of the least square fitted lines indicate that the initial interaction mechanism is faster.

Therefore, two regimes can be distinguished during the metal-ceramic interaction. The first regime is characterised by the possibility of N_2 gas removal from the interaction zone towards the furnace atmosphere. In the beginning of the interaction experiment, N_2 gas can be removed through the open porosity, initially existing between metal and ceramic. At the edges of the couple, N_2 gas can be easily removed all the time. If no nitrogen pressure is built up at the interface, the dissociation of the ceramic is not inhibited and the reaction rate will be maximum.

This explains the intercept of the curves in figure 2 and the higher reactivity at the edges of the interaction couples.

As the interaction proceeds and the porosity between metal and ceramic is closed, a nitrogen pressure will be built up at the metal-ceramic interface, limiting the dissociation of the ceramic. In this second regime, the solution of nitrogen or silicon into the metal or the formation of reaction products will become reaction rate controlling.

The observed reaction

$$SiAlON \leftrightarrow [Si]_{Fe} + [N]_{Fe} + Al_2O_3 + N_2 \tag{1}$$

in which the solution of nitrogen is thought to proceed via the gas phase, can be divided in the following reactions

$$
\begin{array}{lclcr}
SiAlON & \leftrightarrow & Si + Al + O_2 + N_2 & \text{(dissociation)} & (2) \\
Si & \leftrightarrow & [Si]_{Fe} & \text{(solution)} & (3) \\
1/2\,N_2 & \leftrightarrow & [N]_{Fe} & \text{(solution)} & (4) \\
2\,Al + 3/2\,O_2 & \leftrightarrow & Al_2O_3 & \text{(recombination)} & (5)
\end{array}
$$

with $[Si]_{Fe}$ and $[N]_{Fe}$ the Si and N dissolved in Fe.

Since the metal-ceramic interaction is reaction controlled, the rate controlling step in the interaction has to be part of the dissolution process (reaction 2-5).

When the metal-ceramic interaction is not limited by the formation of reaction products and the solution rate of nitrogen or silicon into the metal, the interaction rate will be maximum and controlled by the dissociation rate of the ceramic. This is the case at the edges of the interaction couple and in the initial stage of the interaction (regime 1).

In the situation where a nitrogen pressure is established, the reaction rate is controlled by the dissolution of N or Si, or the formation of Al_2O_3. Since Al_2O_3 does not form a continuous layer that could interfere as a diffusion barrier, the formation of Al_2O_3 will not be reaction rate limiting. It is also unlikely that the solution rate of silicon will be the rate controlling step, since the solubility of silicon in pure iron (18 at. % at 1200 °C) is much higher than that of nitrogen (0.089 at. % at 1200 °C) and the solubility limit of silicon was not reached in the interaction layer on the iron-based alloy side of the interaction couples. Therefore, the solution rate of nitrogen into the alloy is most likely the reaction rate controlling step.

Experimental results in which the reactivity of sialon ceramics is studied as a function of the type and amount of alloying elements in the iron-based alloy indeed show a qualitative relation between the measured reactivity and the calculated equilibrium nitrogen solubility in the iron-based alloys. This qualitative relation is illustrated in figure 3, where the reactivity of sialon 2 with pure iron, a number of binary alloys and steels after 5 hours at 1200°C is plotted together with the calculated nitrogen solubility at 1200°C. More details on the nitrogen solubility calculations are given in reference [5].

When the measured reactivity and the calculated nitrogen solubility of the investigated iron-based alloys are compared with that of pure iron, it is clear that alloying with elements that decrease the nitrogen solubility (Si and Ni) will reduce the reactivity with sialon ceramics, whereas alloying with elements that increase the nitrogen solubility (Cr and Mo) results in an increased reactivity. The lower reactivity of 50NiCr13 steel with respect to 42CrMo4 steel can be explained by the reactivity decreasing effect of Ni in the 50NiCr13 steel.

The influence of the sialon composition on the reactivity is shown in figure 4. The overall trend is that the amount of dissolved ceramic decreases with increasing nitrogen content in the ceramic, what can be explained by the nitrogen solubility controlled interaction mechanism.

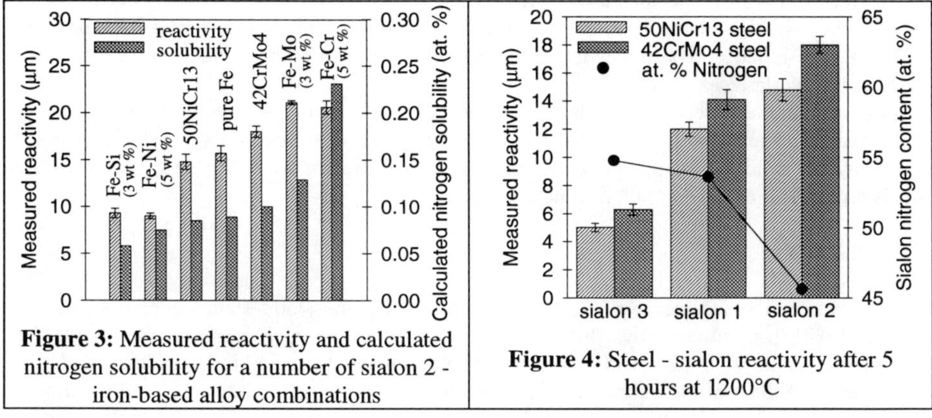

Figure 3: Measured reactivity and calculated nitrogen solubility for a number of sialon 2 - iron-based alloy combinations

Figure 4: Steel - sialon reactivity after 5 hours at 1200°C

The flank wear and crater depth on the sialon inserts after turning 50NiCr13 and 42CrMo4 steel for 2 minutes at 100 m/min are shown in figure 5 and 6. The predicted abrasive wear, calculated as $(K_{IC}^{3/4} \times H_V^{1/2})^{-1}$, with K_{IC} and H_V respectively the fracture toughness and hardness of the cutting tool material [6], and the calculated equilibrium solubility of the ceramics in pure iron at 1200°C are superimposed on these figures. More information on the calculation of the ceramic solubility in pure iron, that can be seen as a guideline in predicting the chemical wear of the different ceramics, can be found elsewhere [7]. Although the predicted abrasive and chemical wear are given on the same scale, they can not be added up since the chemical wear is expressed in (cm^3 ceramic / mole iron) and the abrasive wear in $((MPa\ m^{1/2})^{-3/4} \times GPa^{-1/2})$.

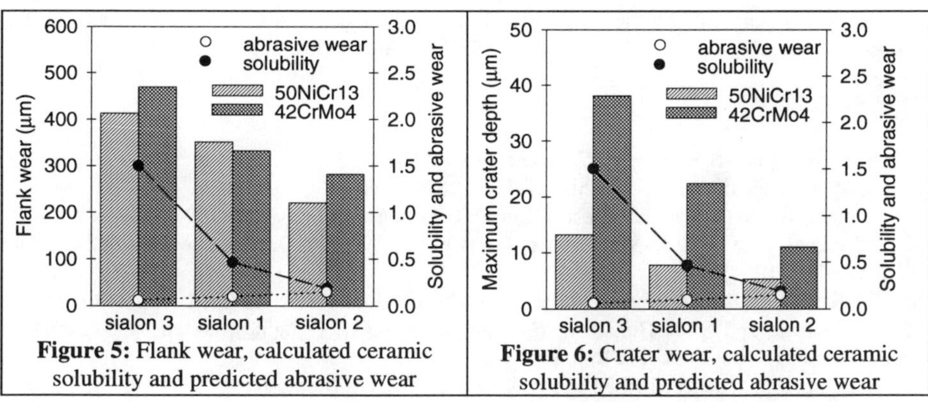

Figure 5: Flank wear, calculated ceramic solubility and predicted abrasive wear

Figure 6: Crater wear, calculated ceramic solubility and predicted abrasive wear

With respect to the sialon composition, it is clear from figures 5 and 6 that the tendency in flank and crater wear corresponds with the trend in calculated equilibrium solubility of the ceramics (chemical wear). The relative wear behaviour of the ceramics however, is in complete disagreement with their relative reactivity measured in the interaction couples (figure 4).

With respect to the influence of the workpiece composition however, the higher relative flank and crater wear when turning 42CrMo4 compared to 50NiCr13 steel with the three sialon ceramics, correlates with the higher reactivity of 42CrMo4 steel noticed in the reaction couples.

From this, it can be speculated that the results of the interaction couple study can be used to predict the relative chemical wear behaviour of a given sialon cutting tool when machining different iron-based alloys, under the assumption that the chip-tool contact temperature is the same for all workpieces.

Although the chemical wear behaviour during cutting tests is controlled by the solubility of the ceramic in the workpiece and not by the solution rate of nitrogen, as it is the case for the reactivity in static interaction couples, differences in nitrogen solubility among the different steels will influence the solubility of the ceramic.

CONCLUSIONS

Two interaction regimes can be distinguished during the static state interaction of sialon ceramics with iron-based alloys. In the initial stage of the interaction and at the edges of the interaction couple (regime 1), the reactivity is maximum and the reaction rate is controlled by the dissociation of the ceramic. After a nitrogen pressure is built up at the interface in the central part of the couple (regime 2), the nitrogen solution rate becomes interaction rate controlling and the extent of interaction is strongly influenced by the nitrogen solubility in the iron alloy.

Alloying elements that lower the nitrogen solubility (Si and Ni) were found to lower the reactivity with sialon ceramics, whereas alloying elements that increase the nitrogen solubility (Cr and Mo) also increased the reactivity.

The results of the interaction couple study can be used to predict the relative chemical wear behaviour of a given sialon cutting tool when machining different iron-based alloys, under the assumption that the chip-tool contact temperature is the same for all workpieces.

ACKNOWLEDGMENTS

This work was supported by the Ministry of Education under GOA Contract No. 92/96-01. J.V. thanks the Belgian I. W. O. N. L.-I.W.T. fund for a research fellowship.

REFERENCES

1.   S. T. Buljan and S.F. Wayne, Wear **133**, 309 (1989).
2.   B. M. Kramer, J. Vac. Sci. Technol. **A4** (6), 2870 (1986).
3.   J. Vleugels and O. Van Der Biest, J. Europ. Ceram. Soc. **13** (6), 529 (1994).
4.   J. Vleugels, T. Laoui, K. Vercammen, J. P. Celis and O. Van Der Biest, Mater. Sci. and Eng. **A187** (2), 177 (1994).
5.   J. Vleugels, L. Vandeperre and O. Van Der Biest, J. Mater. Res. **11** (5), 1265 (1996).
6.   S.F. Wayne and S. T. Buljan, in Friction and Wear of Ceramics, edited by S. Jahanmir, (Marcel Dekker, New York, 1994), p. 261-285.
7.   J. Vleugels, P. Jacobs, J.P. Kruth, P. Vanherck, W. Du Mong and O. Van Der Biest, Wear **189**, 32 (1995).

SYNCHROTRON X-RAY DIFFRACTION STUDY OF THE SURFACE LAYER IN POLED CERAMIC BaTiO$_3$[#]

D. BALZAR *,[†], P.W. STEPHENS **,[††], H. LEDBETTER *, J. LI ***, M.L. DUNN ***
*Materials Science and Engineering Laboratory, National Institute of Standards and Technology, 325 Broadway, Boulder, CO 80303, balzar@boulder.nist.gov
**Department of Physics, State University of New York, Stony Brook, NY 11794
***Department of Mechanical Engineering, University of Colorado, Boulder, CO 80309
[†]Division of Materials Science and Electronics, Ruđer Bošković Institute, P.O. Box 1016, 10001 Zagreb, Croatia
[††]National Synchrotron Light Source, Brookhaven National Laboratory, Upton, NY 11973

ABSTRACT

Both unpoled and poled BaTiO$_3$ were studied by laboratory and synchrotron x-ray sources. The shorter-wavelength synchrotron radiation was used to probe deeper below the specimen surface. Diffraction patterns revealed a distinctive surface layer of the same tetragonal structure as the main BaTiO$_3$ fraction. A $\sin^2 \Psi$ analysis of peak shift confirmed that no measurable change of elastic strain occurs upon poling. This indicates that poling may induce excessive strain that was relieved by microcracking or that a majority of domains reverse orientation. However, domain switching and possible microcracking may induce inhomogeneous strain (microstrain) and alter domain-size distribution. Line-broadening analysis showed large anisotropy of both coherently diffracting domain size and microstrain. The poled specimen shows a larger microstrain and smaller average domain size, which indicates possible effects of microcracking and additional defects created during poling.

INTRODUCTION

Among ferroelectric materials, BaTiO$_3$ is particularly interesting for its simple crystalline structure, good ferroelectric properties at ambient temperature, as well as chemical and mechanical stability, which facilitate easy fabrication of ceramic samples. BaTiO$_3$ polycrystals were studied much less than single crystals and there is a recent interest in the influence of BaTiO$_3$ particle size on structure and properties [1]. The aim here is to study the effect of poling on the near-surface region in BaTiO$_3$ ceramic by diffraction methods and to quantify microstructural changes, in particular the coherently diffracting domain size and microstrain.

EXPERIMENT

A cold pressed and sintered rectangular pellet of nominally 1.1 µm grain-size BaTiO$_3$ powder was poled parallel to the pressing direction by an electric field of 1 kV/mm for 1 h at 80°C. Both unpoled and poled pellets were fine polished (0.3 µm diamond paste) and surface stress was relieved at 250°C for 4 h before diffraction measurements. Powder-diffraction measurements were collected with a laboratory x-ray source using Cu$K\alpha_{1,2}$ radiation and at the X3B1 powder-diffraction beamline (1.14912 Å and 0.61988 Å) at the National Synchrotron Light Source (Brookhaven National Laboratory) with triple-axis geometry.

Mat. Res. Soc. Symp. Proc. Vol. 453 © 1997 Materials Research Society

RESULTS

<u>Surface-layer structure</u>

Although the observed strong [001] texture was corrected for, initial fits of laboratory x-ray data were not satisfactory. Closer investigation of peaks showed the probable reason was an anomaly at the lower-angle side, especially of 00l peaks. It caused a poor fit of 002 and 004 peaks, as depicted in Figure 1. It was often reported in the literature that $BaTiO_3$ exhibits a distinct surface layer. The grazing-incidence scan at 2° incidence confirms this (Figure 2). At this incident angle, the penetration depth is more than 90% reduced even for the first peak (001/100). Rietveld refinements of the whole pattern of the surface layer will be published elsewhere [2]. Although the pattern could be indexed as the cubic $BaTiO_3$ phase ($Pm\bar{3}m$) and no line splitting

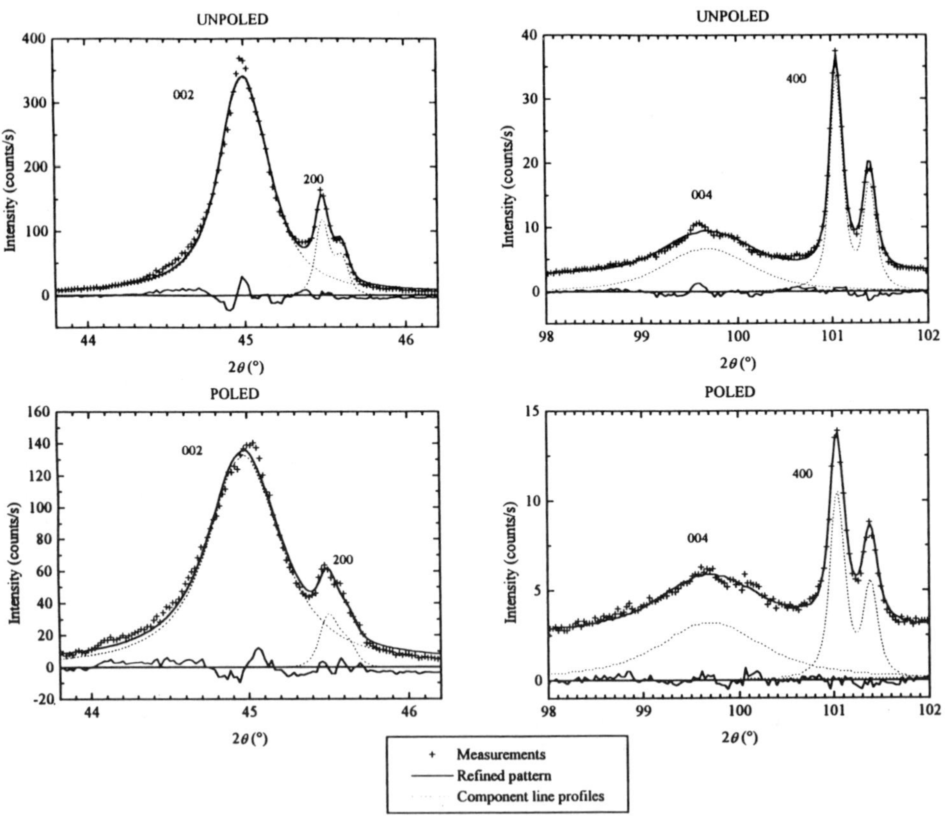

Figure 1. Unsatisfactory fits of diffraction 002/200 and 004/400 pairs of reflections (CuK$\alpha_{1,2}$ laboratory data).

could be seen because of broad peaks (thus indicating a significantly strained layer), a strongly [001]-textured tetragonal phase isostructural with the main fraction (*P4mm*) is more likely. Lattice parameters of the main fraction were calculated from all the available reflections by Cohen's method [3]. For both unpoled and poled specimens, measured at faces parallel and normal to the poling direction, lattice-parameters variations were within one standard deviation, the average value being $a = 3.992(1)$ Å, $c = 4.037(2)$ Å, $c/a = 1.01$.

Because the attenuation length in $BaTiO_3$ for $CuK\alpha_{1,2}$ radiation is only about 7 μm, we conducted measurements with higher-energy synchrotron radiation. The superior resolution at 1.14912 Å wavelength (attenuation length is about 15 μm) clearly revealed additional peaks (Figure 3a). At yet shorter wavelength (0.61988 Å, attenuation length is about 82 μm), the additional peaks are almost not visible (Figure 3b). The relative integrated intensities of 002 and 200 are now approximately in the 1:2 ratio, as expected for random-orientation powder.

<u>Poling process and microstructure</u>

In the poling process, reorientation and growth of ferroelectric domains occurs. In the tetragonal phase, both 90° and 180° domain switches are possible. Because the polarization vector is along [001] in tetragonal phase, these changes are theoretically trackable by monitoring the relative ratio of 00*l* and *h*00 diffraction-line intensities. However, because of Friedel's rule, the 180° domain switching will not affect relative intensities, although the 90° switching will (see, for

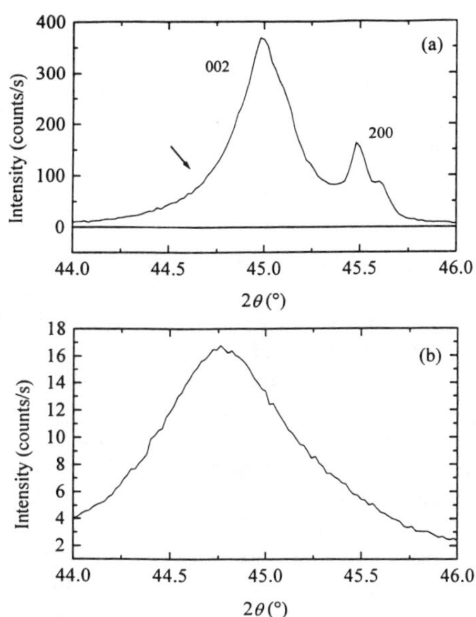

Figure 2. (a) 002 and 200 diffraction-line profiles of unpoled BaTiO₃ (CuKα₁,₂ laboratory data, coupled θ/2θ scan); (b) scanned at grazing 2 ° incidence angle.

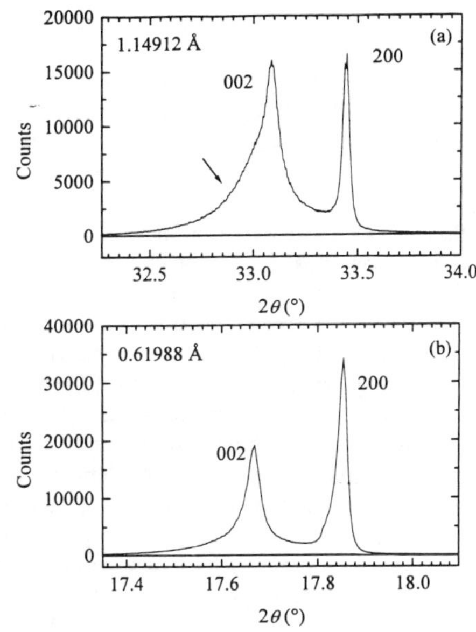

Figure 3. 002 and 200 diffraction-line profiles of unpoled BaTiO₃ at two synchrotron-radiation wavelengths: (a) 1.14912 Å; (b) 0.61988 Å.

instance, discussion in [4]). Similarly, while 90° domain switching upon poling changes the state of strain, the 180° domain switching does not, although it may involve microstructural change during poling. We monitored the lattice spacing d as a function of the specimen tilt angle ($\sin^2 \Psi$ method [5]), to detect a possible directional change, that is, the elastic strain. Because laboratory x-ray scans did not reveal any peak shift, synchrotron scans (which give better resolution) were also conducted, thus giving a higher sensitivity to the strain. However, Figure 4 shows that any possible strain-caused peak shift is under the detection capabilities of this method. This indicates that poling may have induced excessive strain that was relieved by microcracking or that most domains switched 180°. Although the former cannot be excluded, the latter is confirmed by a comparison of relative intensities of 00l/h00 pairs of reflections; the 004/400 intensity ratio before and after poling indicates that only about 4% of domains with polarization vector parallel to the surface align with the poling direction, thus switching for 90°.

Although the elastic strains may not be detectable after poling, during the reorientation and growth process defect concentration may change, which would be detected as an inhomogeneous strain (microstrain or strain of the third kind), as well as a domain-size distribution, thus resulting in change of diffraction coherent length (coherently diffracting domains). Both effects are detectable by analyzing diffraction line broadening. Diffraction-line profiles were corrected for instrumental broadening in a convolution-fitting procedure where the physically broadened profile is modeled by an exact Voigt function [6]. Table I shows the results for the major phase (synchrotron 0.61988 Å data), where the influence of the surface phase is least pronounced, but was accounted for (see Figure 5). There is a large anisotropy of both the coherently diffracting domain size and microstrain for the unpoled specimen. In absolute terms, however, this amount of line broadening is relatively small, which is reflected in relatively large errors of both domain size and microstrain. After poling, domain size is somewhat smaller and microstrain becomes significantly larger, which indicates that additional defects may have been created during poling. Figure 6 shows domain-size distributions that are more instructive than mere average values given in Table I. There is little change along [100], but orthogonal to the a-b plane ([001]) distribution narrows and shifts toward smaller sizes, thus implying that poling equalizes domain sizes. The coherently diffracting domain sizes (0.1-0.25 µm) are smaller 3-6 times than the observed and expected values for ferroelectric domains for 10 µm and larger grain size [7]. This is a known phenomenon: x-ray coherent length reflects subdomain and subgrain structure, such as small-angle boundaries and lattice-stacking faults.

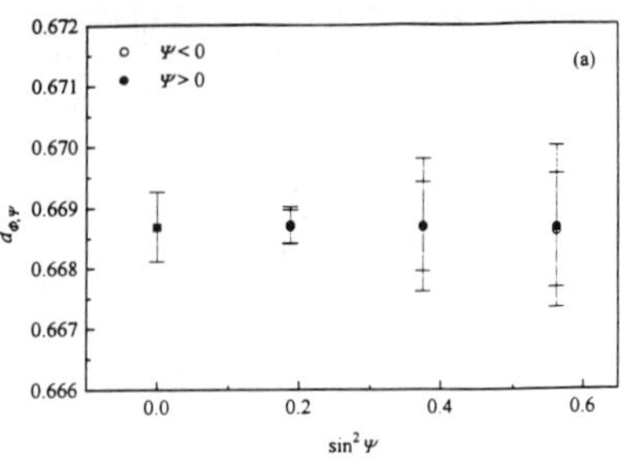
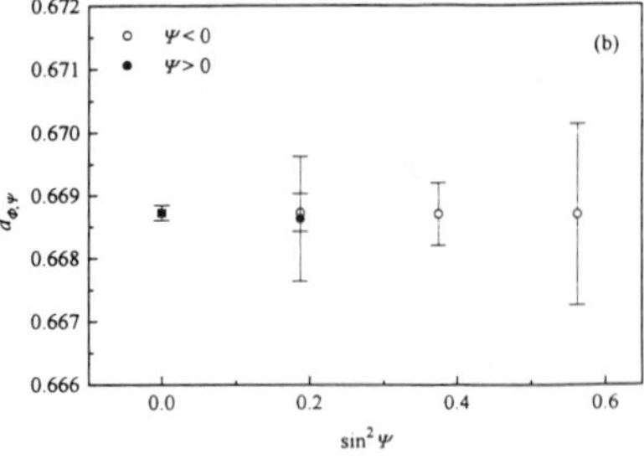

Figure 4. Lattice spacing d of 424 BaTiO$_3$ reflection as a function of specimen tilt angle Ψ (synchrotron 1.14912 Å data): (a) unpoled specimen; (b) poled specimen. The analogous measurements on different diffraction lines and with laboratory sources gave identical results.

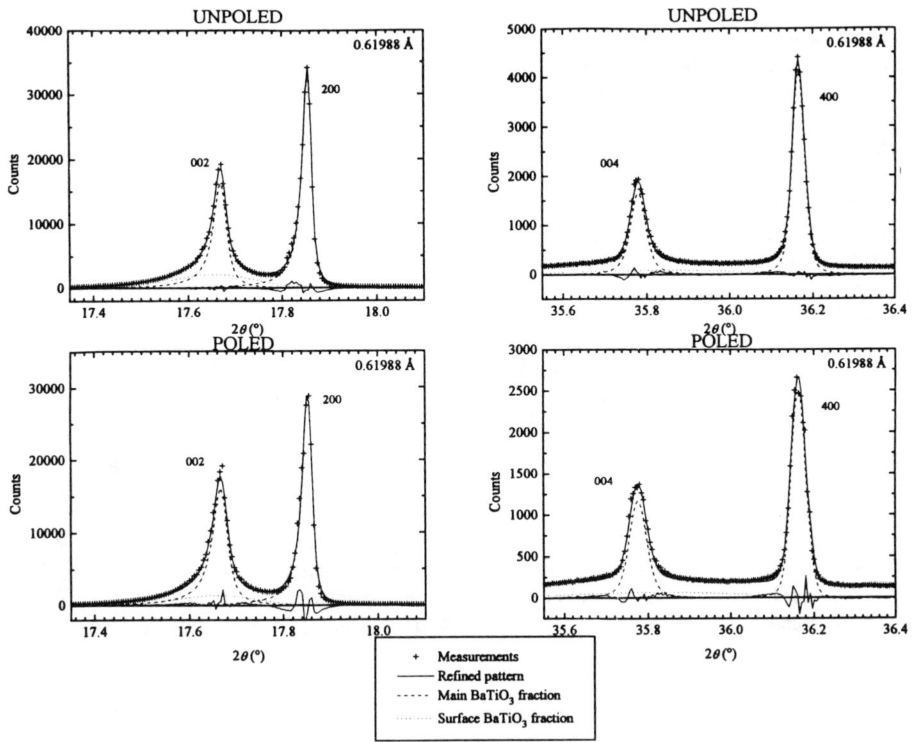

Figure 5. *Fits of synchrotron 0.61988 Å data. The refined pattern is a convolution of pre-determined instrumental-broadening profile and an assumed Voigt function that models the physical broadening.*

Table I. Results of line-broadening analysis (volume-weighted coherent domain size $<D>_v$ and root-mean-square microstrain $<\epsilon^2>^{1/2}$ averaged over $<D>_v$) for unpoled and poled $BaTiO_3$ specimens (synchrotron data at 0.61988 Å).

	Direction	$<D>_v$ (Å)	$<\epsilon^2(<D>_v/2)>^{1/2}$ (10^{-4})
Unpoled	[001]	1300(60)	1.2(8)
	[100]	2520(240)	0.0(2)
Poled	[001]	920(130)	3.1(11)
	[100]	2440(260)	3.0(2)

CONCLUSIONS

Microstructural analysis of $BaTiO_3$ ceramics proves the existence of a distinct severely textured and strained surface layer. Most likely, this layer has the same tetragonal structure as the main fraction. Upon poling, most of domain switching involves simple reversal in orientation, although there is an estimated 4% of 90° switching. Line-broadening analysis shows large microstructural changes: Coherently diffracting domain sizes decrease somewhat and microstrains increase significantly, while [001] domain-size distribution homogenizes. This may be explained by defect multiplication and possible microcracking that largely relieves elastic stresses that could not be detected by diffraction.

ACKNOWLEDGMENTS

We thank E. T. Park, J. L. Routbort, P. Nash, and Z. Li (Argonne National Laboratory) for the specimens and useful discussions.

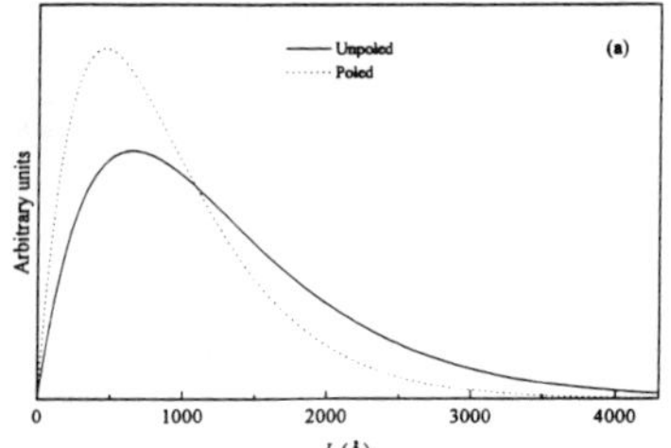

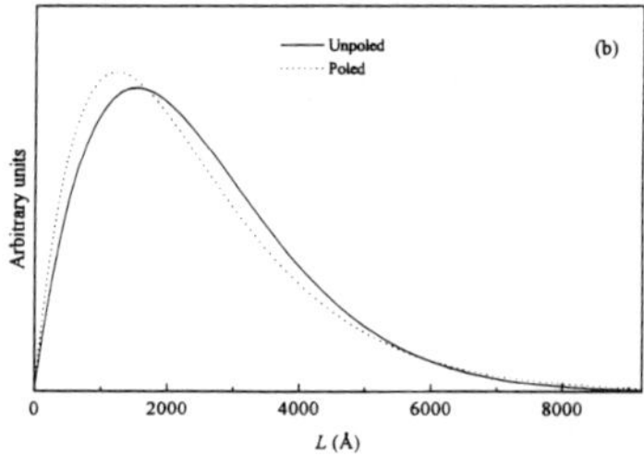

Figure 6. *Volume-weighted domain-size-distribution functions for unpoled and poled $BaTiO_3$ specimen (synchrotron 0.61988 Å data): (a) along [001]; (b) along [100].*

REFERENCES

1. M.H. Frey and D.A. Payne, Phys. Rev. B, **54**, 3158 (1996).
2. D.Balzar, P.W. Stephens, and H.Ledbetter (in preparation).
3. M. Cohen, Rev. Sci. Instr. **6**, 68 (1935).
4. E.C. Subarrao, M.C. McQuarrie, and W.R. Buessem, J. Appl. Phys. **28**, 1194 (1957).
5. H. Döle and V. Hauk, Z. Metallkd. **68**, 728 (1977).
6. D. Balzar, J. Res. Natl. Inst. Stand. Technol. **98**, 321 (1993).
7. G. Arlt, D. Hennings, and G. de With, J. Appl. Phys. **58**, 1619 (1985).

ATOMIC AND ELECTRONIC STRUCTURE OF THE
CORUNDUM (0001) SURFACE

V. E. PUCHIN [*], E. A. KOTOMIN [*], A. L. SHLUGER [**]
[*] University of Latvia, Rainis blvd. 19, Riga, LV-1586, Latvia
[**] Centre for Materials Research, Department of Physics, University College London,
Gower St., London WC1E 6BT, U.K.

ABSTRACT

The electronic structure and geometry of the Al terminated corundum (0001) surface were studied using a slab model within the *ab-initio* Hartree-Fock technique. The distance between the top Al plane and the next O basal plane is found to be considerably reduced on relaxation (by 0.57 Å, i.e. by 68% of the corresponding interlayer distance in the bulk). An interpretation of experimental photoelectron spectra (UPS HeI) and metastable impact electron spectra (MIES) is given. Calculated projected densities of states exhibit a strong dependence on the relaxation of surface atoms.

INTRODUCTION

Alumina surfaces have been studied extensively because of their technological importance [1]. Theoretically, the simplest nonpolar Al terminated (0001) corundum surface was predicted to be the most stable ideal termination of the bulk structure [2,3]. Pioneering *ab initio* Hartree-Fock (HF) calculations using the minimal (STO–3G) basis set of atomic orbitals (AO's) have predicted a large inward displacement (relaxation) of the terminating layer of Al atoms (0.4 Å, or 48% of Al − O interlayer spacing) [4]. Further calculations performed in the local density approximation (LDA) with pseudopotentials and a plane wave basis set predicted an even larger relaxation of about 0.7 Å [5]. The calculations using pair potentials based on the empirical shell-model [3] predict the surface relaxation to be intermediate between the HF [4] and LDA [5] data.

The incentive for this study comes from our recent combined experimental and theoretical study of the MgO (001) surface [6]. It has demonstrated that the UPS (He I) and MIES (He* $1s2s$) spectra which have quite different structures in the valence band energy region, can be well approximated by the total density of states (DOS) of a thin crystalline slab and the DOS projected to the surface O ($2p_z$) AO's, respectively (PDOS) [6]. Moreover, the results of these calculations demonstrated that the shape of both DOS and PDOS is very sensitive to the surface relaxation. This leads to the possibility of extracting information on surface atomic structure through the comparison of DOS and PDOS calculated for different surface relaxations with the experimental spectra.

In this paper we explore the above possibility in more detail for the case of the alumina surface. UPS and MIES spectra of the corundum monocrystalline (0001) surface and corundum films on Al (111) and W (110) substrates have recently been published [7-9]. As follows from LEED and ISS data, the films grown on Ta(110) [10] and W(110) [9] are Al terminated (0001) corundum or (111) γ-Al$_2$O$_3$ structures. Therefore Al electronic states must contribute to the valence band DOS [4], and the straightforward interpretation of electron spectra given for the simple ionic rocksalt-type insulators like MgO may not be valid for corundum.

We study the relationship between the atomic and electronic structure of the relaxed corundum (0001) surface and the UPS and MIES spectra. A novel method is proposed for the comparison of MIES spectra with theory: in order to simulate in a direct way the projectile–

Mat. Res. Soc. Symp. Proc. Vol. 453 © 1997 Materials Research Society

surface interaction (via Auger deexcitation connected with electron emission), the DOS projected onto the floating $1s$ orbital of He atom above the surface is calculated and analysed as a function of the He coordinates.

THEORETICAL METHOD AND RESULTS

To model the experimental spectra we have used the same method as discussed in ref. [6] for MgO. The electronic structure and the total energy of the corundum slab containing three O and 6 Al atomic layers parallel to (0001) plane terminated by an Al layer were calculated within the Hartree-Fock approximation as implemented in the CRYSTAL computer code [11]. An 85-11G/8-411G basis set augmented with polarization functions was used for Al and O, respectively. [12].

The atomic structure of the surface corresponding to the minimum of the slab total energy with respect to the 7 independent atomic coordinates was found using the least square fit of the energy surface to the quadratic form. The major displacements from the perfect bulk positions are that of Al atoms in the top surface layer. The inward relaxation value of 68% of the Al – O bulk interlayer distance have been found, which is 20% larger than the previous HF result [4]. This difference can be attributed to the more extended basis set and to the larger number of degrees of freedom taken into account in the minimization procedure. Further extension of the basis set with d polarization functions on Al increases this relaxation by 5%. However, our relaxation is still smaller by 17% than that obtained in the LDA calculations [5]. The surface formation energy have been caculated as a difference between the bulk and slab total Hartree-Fock energies. Before relaxation it was found to be 4.92 J/m^2. The surface relaxation energy, defined as the difference in total energies of relaxed and unrelaxed slabs, was obtained to be 3.12 J/m^2. Three quarters of the relaxation energy gain is due to the inward displacement of the surface Al. The final surface energy after the relaxation turns out to be 1.81 J/m^2.

In the MIES experiment the interaction of the He* probe atom with the surface takes place via the Auger deexcitation process along the entire trajectory in front of the surface. The projectile moves with thermal kinetic energy and the electron emission occurs on the incoming path of the trajectory at comparatively large distances (several atomic units, typically); the transition energy is not strongly influenced by the actual interaction potentials of the initial and final states. We have shown on several occasions [7, 6] that under these conditions the MIES spectra is proportional to the surface density of states relevant for the interaction with He*.

The spectrum of the emitted electrons may depend on the incident He* coordinate along the surface. In order to simulate this dependence, we have calculated the DOS projected to the floating He $1s$ orbital as a function of its coordinates. We find that the shape of this DOS does not depend on the distance from the surface (vertical z coordinate) while its magnitude decreases exponentially in the range of z varied between 2 and 4 Å reflecting the average behavior of the overlap integral between the floating AO and surface wave function.

In Fig. 1 we show the He (1s) projected DOS averaged along the surface at a distance $z =$ 2.1 Å which is close to the typical distance where the Auger deexcitation of He* projectile takes place. The dotted curve in Fig. 1 was calculated for the slab geometry predicted by the LDA method.

In this paper we do not develope the regorious theory for the UPS spectra, we merely assume that the structure in the valence band UPS (HeI) spectra is proportional to the valence band DOS of the near-surface region. It was proved to be a good approximation for the case of MgO (100) surface [6] and we believe that this is also the case for Al$_2$O$_3$ (0001). The DOS projected to the surface O atom is given in Fig. 1 along with UPS data.

COMPARISON WITH EXPERIMENT

The spectra represented by light lines in Fig. 1 were obtained from corundum films grown by coadsorbing Al and O_2 molecules onto a W (110) substrate held at 750K [8]; the film thickness was estimated as 13 Å. Electron energies are given with respect to the vacuum level. The value of 3.8 eV for the surface workfunction of the film was determined from the UPS spectra.

The MIES spectra show very little intensity above the top of the valence band (VB) ($E >$ 9.3 eV). We attribute the "soft tail" of the valence band emission to the ionization of such surface oxygen species which have a lower coordination than in the bulk, located at kinks, steps, etc. The VB emission consists of a pronounced peak ($E = 9.8$ eV) and a shoulder ($E \approx 12.3$ eV) towards higher binding energies. The UPS spectra show the same two VB features ($E = 9.8$; 12.3 eV). The weak intensity above the top of the VB is not well understood at present: it may be due to emission from the underlying substrate and/or from point defects located underneath the surface. The high energy part of both spectra above 16 eV is known to be affected by the emission of secondary electrons. Thus this energy region (denoted as A) will not be interpreted here.

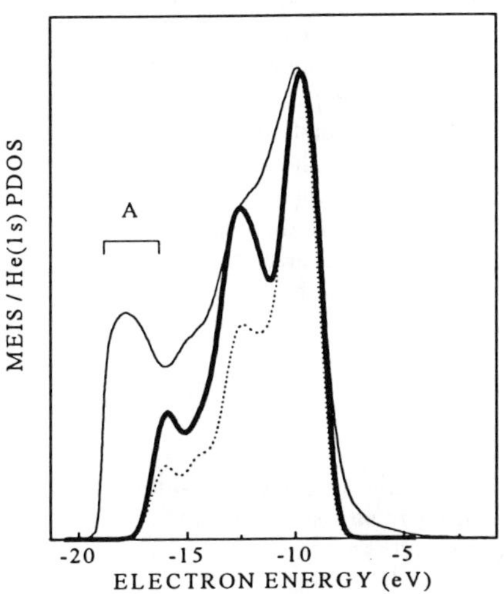
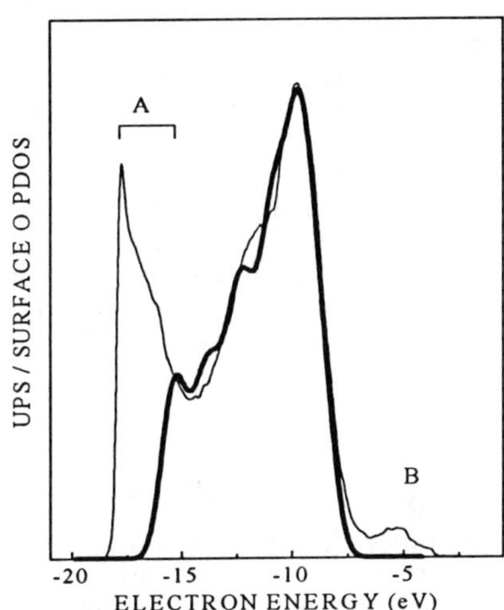

Fig. 1. The experimental MIES (left panel) and UPS (right panel) spectra for corundum films produced by coadsorption of Al and O_2 on W (110) substrate [8] (thin curves). Heavy curves show the result of theoretical simulations of MIES and UPS. The energy scale corresponds to the experimental electron binding energy. Theoretical curves are shifted by 1.7 eV in order to match the experimental top of the VB. Dotted curve is theoretical MIES spectrum for the surface structure predisted by LDA method [5].

The calculated surface DOS is in surprisingly good agreement with UPS spectra (Fig. 1). This confirms that the photoemission spectrum at HeI incident photon energy is mainly dependent on the initial state electron wavefunction. The final state effects including the dispersion of the conduction band and photoelectron scattering at the surface potential barrier might be important for larger photon energies and angle resolved UPS interpretation, but are probably less important in our case.

The result of the MIES spectra simulation with He $1s$ orbital as a probe of the surface DOS is close to the experiment. However, the peaks in the experimental spectra are broader than in the simulated one as it was also the case in MgO. Three possible explanations have been suggested for this discrepancy [6]. As it is seen from Fig. 1, the second peak in the simulated MIES spectrum at the LDA geometry has significantly lower amplitude relatively to the HF geometry, thus giving less satisfactory agreement with experimental spectrum.

From the comparison of the surface DOS and that projected to $1s$ orbital of the He^* projectile with the relevant UPS and MIES spectra, respectively, it is seen that the agreement between theory and experiment is satisfactory; the positions of the peaks and their relative magnitudes are similar. The comparison of the MIES spectrum with theoretical simulations allows us to discriminate between the HF and LDA optimized slab geometries in favor of that found in this paper using the HF technique.

CONCLUSION

The results of present calculations demonstrate that in corundum, like in MgO, the shape, relative positions and intensities of the peaks in the surface density of electronic states (DOS) are in satisfactory agreement with the experimental HeI photoelectron spectra (UPS). The metastable impact electron spectrum (MIES), arising from the Auger deexcitation of the He^* metastable atom near the surface can be approximated by the DOS projected to a floating $1s$ atomic orbital (AO) which simulates the probing He atom. This method appears to be more accurate for the simulation of the MIES spectra than just the comparison with DOS projected to the p_z AO of surface atoms used for MgO [6]. We also found that both DOS strongly depend on the surface relaxation. This allows us to study details of the surface atomic structure by means of a comparison of the projected DOS calculated using *ab initio* methods with experimental MIES and UPS spectra. The present Hartree-Fock calculations predict an inward relaxation of the top Al layer as large as 68% of the Al - O interlayer distance. This slab geometry gives the surface projected DOS which is in the best agreement with the experimental MIES and UPS spectra. More detailed study of corundum (0001) surface will be published elsewhere [13].

ACKNOWLEDGMENTS

The authors are greatly indebted to Prof. V. Kempter for the suppluing us with experimental data, numerous fruiful discussions and hospitality during their visits to Clausthal University. We are grateful to N. E. Christensen for providing us access to the CONVEX computer at Aarhus University, to A. Svane for supplying us with the code for the least square fitting to the quadratic form and to M. J. Gillan for providing the equilibrium atomic configuration of the corundum slab, obtained with LDA method.

REFERENCES.

1.   Science of Alumina (R. H. French, A. H. Heuer. Eds.) *J. Am. Ceram. Soc.* 77, 292-453 (1994)
2.   P. W. Tasker, *Adv. Ceram.* 10, 176-89 (1988)
3.   W. C. Mackrodt, *J. Chem. Soc. Faraday Trans. 2,* 85, 541 (1989); W. C. Mackrodt, *Phil. Trans. R. Soc. A,* 341, 301 (1992)
4.   C. Pisani, M. Causà, R. Dovesi, C. Roetti, *Progr. Surf. Sci.* 25, 119 (1987); M. Causà, R. Dovesi, C. Pisani, C. Roetti, *Surf. Sci.* 215, 259 (1989); L. Salasco, R. Dovesi, R. Orlando, M. Causà, V. R. Saunders, *Mol. Phys.* 72, 267 (1991)

5. I. Manassidis, A. De Vita, M. J. Gillan, *Surf. Sci. Lett.* **285**, L517 (1993); I. Manassidis, M. J. Gillan, *J. Am. Ceram. Soc.* **77**, 335 (1994)

6. D. Ochs, M. Brause, J. Günster, W. Maus-Friedrichs, V. Kempter, V. E. Puchin, L. N. Kantorovich, A. L. Shluger, *Surf. Sci.* (1996) in press.

7. A. Hitzke, J. Günster, J. Kolaczkiewicz, V. Kempter, *Surf. Sci.* **318**, 139 (1994);

8. J. Günster, A. Hitzke, M. Brause, Th. Mayer, V. Kempter, *Nucl. Instr. Meth. B*, **100**, 411 (1995);

9. J. Günster, Dissertation, TU Clausthal (1996), ISBN 3-931443-62-0

10. P. J. Chen, D. W. Goodman, *Surf. Sci.* **312**, L767 (1994)

11. C. Pisani, R. Dovesi, C. Roetti, *Hartree-Fock Ab-initio Treatment of Crystalline Systems, Lecture Notes in Chemistry*, Vol. **48**, Springer Verlag, Heidelberg, 1988; R. Dovesi, V. R. Saunders, C. Roetti, *CRYSTAL-92 User Manual*, QCPE program n.577, Bloomington, Indiana, 1989.

12. W. C. Mackrodt (to be published); J. D. Gale, C. R. A. Catlow and W. C. Mackrodt, *Modelling Simul. Mater. Sci. Eng.* **1**, 73 (1992)

13. V. E. Puchin, J. D. Gale A. L. Shluger E. A. Kotomin, J. Günster, M. Brause, V. Kempter, *Surf. Sci.* (in press).

SURFACE STABILITY OF SIDERITE UNDER ACIDIC ATMOSPHERE: AN ATOMIC FORCE MICROSCOPY STUDY

R.M. WEAVER, D.E.Grandstaff, G.H. Myer
Geology Dept., Temple University, 306 Beury Hall, Norris St., Philadelphia, PA 19122, rweave00@ nimbus.ocis.temple, grand@vm.temple.edu, gmyer@nimbus.ocis.temple.edu

ABSTRACT

The reactivity of siderite ($FeCO_3$) was investigated under ambient atmosphere and acidic aerosols using an atomic force microscope (AFM). Scans of freshly cleaved siderite crystals showed a relatively irregular microtopography, consisting of high-density kink-steps, compared to other isostructural carbonates (e.g. Iceland-spar calcite). Under ambient conditions siderite is inert with no spontaneous surface reconstruction as is reported for calcite. Under controlled conditions of pH and humidity, siderite was exposed to mists of HCl and H_2CO_3 and the reaction was imaged with submicron resolution. The kinetics of the reaction varied with pH, humidity levels, and also as a function of the initial microtopography. Preferential growth and dissolution were observed as a function of initial topography and due to crystallographic anisotropies. Although extreme reaction conditions were excluded by certain aspects of AFM (high capillary forces, tip corrosion, charged surface), this technique has allowed real-time, *in situ* observations of surface reactions.

INTRODUCTION

Acidic gases and aerosols originating from natural and anthropogenic sources can extensively alter the chemistry of natural waters and promote water-rock reactions. Carbonate minerals are common and are, in general, highly unstable under acidic conditions. Because carbonates are common rock-forming minerals and are common in statuary and building materials, the influence of acidic waters on the rate and extent of the chemical weathering of carbonates is of environmental significance. Many authors have recently reported on carbonate-system dissolution, precipitation and sorption reactions taking place in acidic solutions [1-4].

A distinct subset of these solution reactions includes those reactions occurring on (1) minimally, but continually wetted surfaces, (2) intermittently wetted surfaces and, (3) surfaces without a visible contacting solution. These reaction environments are relevant because most reactions taking place at the Earth's surface and in aquifers involve a high quantity of mineral surface area per unit volume of contacting solution. Minerals exposed to air may still have a significant amount of water sorbed on their surfaces. For example, at least 20 layers of water molecules are adsorbed on calcite surfaces exposed to ambient air [5]. Reactions occurring within this invisible water layer have been observed with AFM on calcite and are similar to those expected for a calcite-saturated solution and in equilibrium with the partial pressures of the contacting gases [6].

This paper reports initial findings of the reaction of the siderite ($FeCO_3$) surface during exposure to acidic mists/vapors of HCl and H_2CO_3. Experiments involving H_2CO_3 were performed in the absence of a visible contacting solution (at 300x) and also during rapid wetting-drying cycles. Experiments involving reaction with HCl (vapor) were performed in the absence of a visible contacting solution.

MATERIALS AND METHODS

A Burleigh ARIS-3300 AFM equipped with 70μm and 5μm scanners was used to acquire

Mat. Res. Soc. Symp. Proc. Vol. 453 © 1997 Materials Research Society

images in force and topographic modes in <1 to ~6 minutes for large area scans. Images contain 256 data points/scan line and Force$_{tip-sample}$ was typically between 20-95nN (F$_{tip-sample}$ was not calculated for each approach). Ambient condition scans acquired at >100nN force did not show detectable surface relaxation or erosion. Cantilevers used were 442µm, rectangular, Si nanosensors with high aspect (10:1) pyramidal tip and stiffness rating of ~0.045N/m. Coarse and fine scale standards were used to calibrate image extents and to characterize tip morphology and sharpness. Images have been tilt removed and scan line heights normalized. During exposure, when imaging was not taking place the AFM probe was retracted to prevent it from obstructing the vapor/mist transport. During constant flow the force exerted on the piezo scanner by the vapor stream had little to no influence on the image quality.

Water vapor saturated with ~1atm CO_2 and having a pH of *ca.* 3.95 was obtained by $CO_2(s)$ subliming in distilled water. This mist was released at a controlled rate from 2-3mm above the siderite surface at a slightly glancing angle. Prior to delivery the mist was carried vertically through a distance of 30cm to reduce the humidity. Reactions took place within an optically invisible fluid layer or in the presence of microdroplets that would briefly inhabit the surface before evaporating.

HCl (vapor) was obtained by bubbling N_2 through a 0.12M HCl solution with a pH of ~1.0 This acidic vapor was carried to the surface via tygon and glass at room temperature. A flow rate of 4.0 ±0.1ml/sec was maintained so that reactions took place within the time required for imaging and so that an optically invisible fluid layer was maintained.

To produce fresh (1014) surfaces, cleavage was assisted on euhedral siderite rhombs by repeatedly scoring the surface coincident with the cleavage directions and applying a small force. This method is known to produce broad terraces bounded by steps of up to several tens of nanometers on Iceland-spar calcite [7]. All observations pertain to the (1014) cleavage surface unless otherwise indicated.   X-Ray powder diffraction using a Rigaku DMAX-B diffractometer showed all of the expected siderite peaks.

RESULTS

<u>Optical and AFM Scale Surface Characterization</u>

After cleaving, the surface was examined under reflected light at 1800x to: (1) insure that preferential breakage had not occurred, (2) identify prominent surface features and, (3) locate areas displaying optical flatness ideal for AFM imaging. Common microtopographic features included linear and curved steps, ridges, ramps, pits and optically flat terraces (figure 1). Steps/ridges occurred frequently in three dominant directions, coincident with the [481] and [441] rhombohedral cleavage directions and also roughly parallel to (010) the bisector of

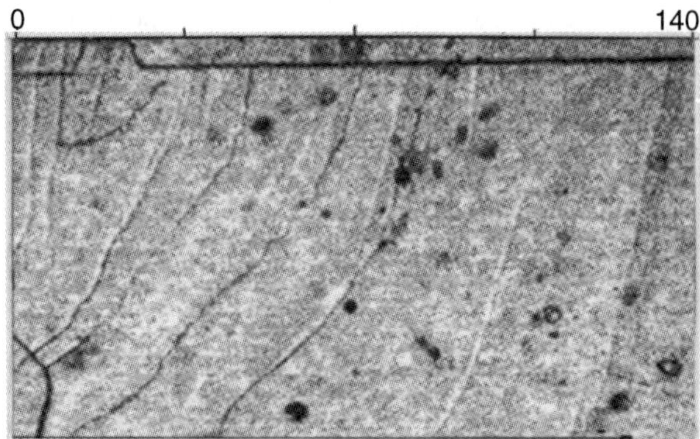

Figure 1. Photomicrograph of the siderite cleavage surface. Note the acute rhombohedral angle of ~78°, and the curved steps. The dark spots are areas where siderite has precipitated due to wetting/drying cycles. Scale is in micrometers.

the acute rhombohedral angle (*i.e.,* the long diagonal). High-density kink-steps/ridges traversing the surface in other directions were also common. This irregular rhombohedral topography was unexpected from the apparently well-formed nature of these crystals and has not been reported on other euhedral isostructural carbonates.

AFM images (figure 2) resolved a uniform to random arrangement of variously sized, ellipsoidal-triangular crystallites ranging in size from ~735 nm^3 to 1.6×10^6 nm^3. This volume was calculated on ellipsoidal particles and only includes the portion of the crystallite exposed at the surface. The irregular rhombohedral topography results from cleavage taking place along the crystallite boundaries. The three dominant macroscale step directions are in good agreement with the interfacial angles on the crystallites. In regions not displaying these crystallites near atomic flatness is observed. The ellipsoidal-polyhedral shape results from anisotropy of the kinetics of growth in nonequivalent crystallographic directions [8,9].

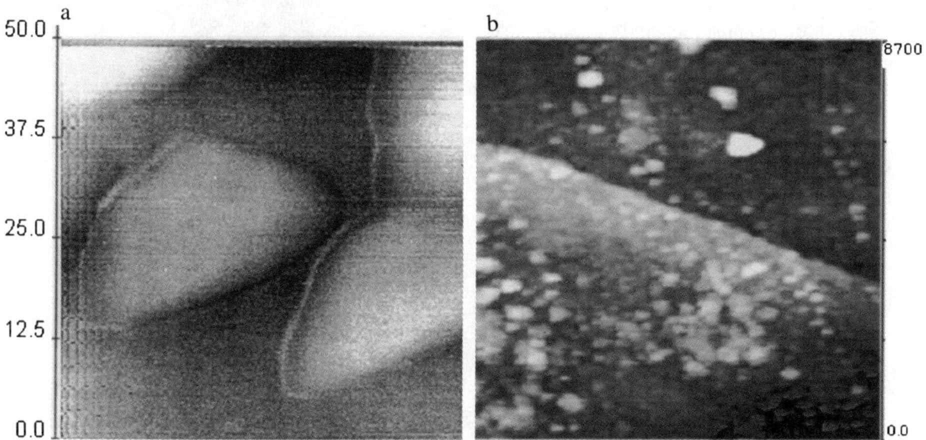

Figure **2**. AFM images. a) Small polyhedral crystallites. Image was obtained in force mode and vertical scale is ~4.1nm. b) The step present is parallel to the [481] cleavage trace. The aligned boundaries of the ellipsoidal-trigonal crystallites produce macroscale fracture ridges/steps. Vertical scale is ~150nm. Lateral scales are in nanometers.

The boundaries and long-range alignment of the crystallites control the fracture directions by causing steps/ridges to propagate parallel to their boundaries. Linear step segments result when the edges of neighboring crystallites are coincidentally aligned. Curved steps/ridges typically result from cleavage in less ordered areas and in areas where ellipsoidal particles dominate. A second phase occurring at the grain boundaries and porosity may also affect fracture.

The coarse-scale aggregate/granular habit of some siderite's has been acknowledged by optical scattering methods on iridescent surfaces; however, to the authors' knowledge, the identification of submicron crystallites on euhedral siderite has not been previously documented. A uniform arrangement of crystallites could cause the iridescent luster. Work is continuing to characterize the nature of the iridescence and to relate it to physical and/or chemical properties of the sample.

Although fluids were optically visible on fresh surfaces resulting from the rupture of fluid inclusions, under N_2, no spontaneous surface reconstruction was observed over a period of 6 days. Under controlled humidity (100-10%) existing optically visible fluids could be encouraged to advance out or retreat to their original locations during high and low humidity cycles, respectively. Samples left exposed to ~100% humidity atmosphere (noncondensing) for periods of up to 65 days showed regions where preferential dissolution had occurred due to crystallographic anisotropies and existing microtopography (figure 3).

These observations (optical microscopy and AFM) show that at ambient humidity, siderite is a relatively unreactive carbonate. However, under conditions of high humidity, siderite may undergo dissolution.

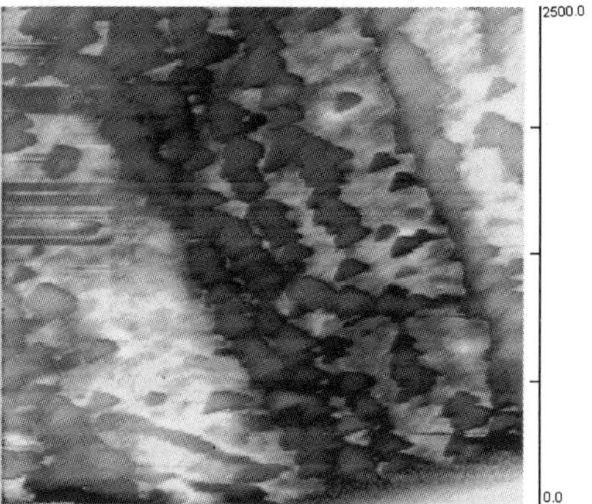

Figure 3. AFM image of siderite affected by dissolution in 100% humidity (noncondensing) environment for 65 days. Vertical scale is ~59nm. Horizontal lines and speckles (SE corner) are artifacts. Scale is in nanometers.

<u>Exposure to H_2CO_3 Vapor</u>

Experiments involving rapid wetting-drying cycles were performed by spraying microdroplets of H_2CO_3 onto the surface, where they would almost immediately evaporate. The extent of sample coverage by these microdroplets was controlled by adjusting the flow rate such that the individual microdroplets evaporated within *ca.* 5 seconds. After repeated exposure to these wetting/drying cycles, the topography as imaged with AFM showed areas of local dissolution and precipitation. These observations (growth and dissolution) could be quantified by profiling reacted verses unreacted surfaces; however, the heterogeneous nature of this reaction did not allow for a direct correlation between exposure time and growth/dissolution. The precipitation products of several of these microdroplets are visible in figures 4a,b and 1.

New growth occurred on existing crystallites and the ellipsoidal-polygonal shape was maintained. Proximal to these areas of local precipitation the crystallite size was seen to increase due to Ostwald ripening and agglomeration. This agglomeration probably results from surface tension forces in the rapidly evaporating microdroplets to cause $FeCO_3$ to be locally concentrated and precipitated.

Images obtained during exposure to H_2CO_3 in the absence of a visible contacting solution also showed that dissolution/reprecipitation reactions occur, although at slower rates and to lesser extent (figure 4c,d). Observations varied from infilling or widening of topographic lows to erosion or nucleation of highs. The heterogeneous nature of the dissolution/ precipitation reactions may be due to turbulent motion of the saturated solution at the interface.

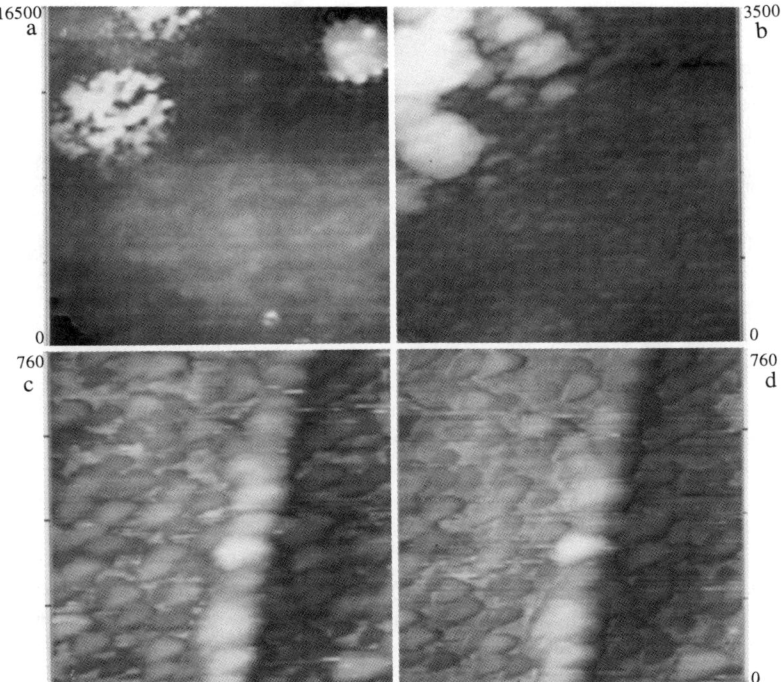

Figure 4. a) Image after exposure to repeated wetting and drying cycles over a 12 min. period. Three isolated regions where precipitation has taken place are seen (vertical scale is 280nm). b) Closeup of (a). It can be seen that new growth has occurred on the existing crystallites while maintaining their shape. Away from the island of precipitated crystallites there is less overgrowth. c) Initial unreacted surface (vertical scale is ~25.7nm). d) Image © after exposure to an invisible fluid layer for 6 minutes. Between images c and d, subtle differences can be seen in the shapes and heights of surface features (vertical scale is 26.5nm).

<u>Exposure to HCl Vapor</u>

During exposure to HCl vapor, the surface initially underwent dissolution. However, after prolonged exposure, imaging artifacts resulted from either, charging between the tip and surface or frictional effects of a partially wetted surface. These artifacts hindered attempts to obtain reliable topographic data owing to an apparent inversion of the topography (figure 5). The trigonally shaped topographic lows that existed under the influence of the HCl vapor now appear as topographic highs.

This surface-charge-dependent artifact may result from imaging taking place while the surface pH is of some intermediate value between that of the point of zero charge (PZC) of the Si tip and the $siderite_{PZC}$. The light areas (highs) that appear in figure 5a, roughly coincide with the boundary of the crystallites. This contrast may reflect lateral differences in the $siderite_{PZC}$ and may indicate that a cementing phase exists at the crystallite boundaries. Preferential resistance of this interstital material to dissolution was observed in other experiments.

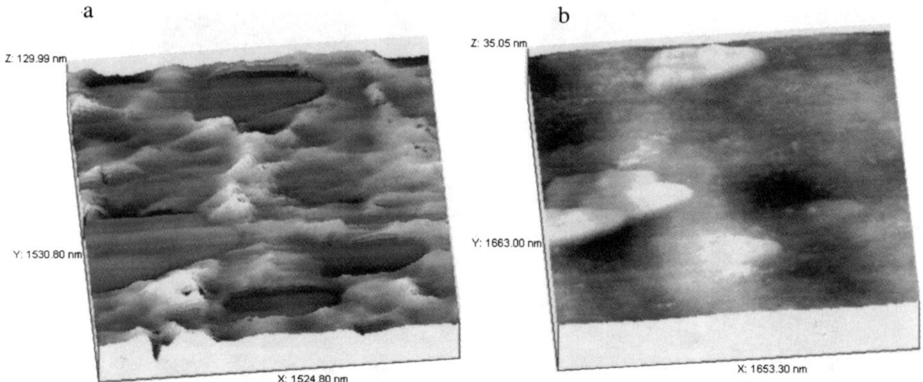

Figure **5**. AFM images during (a) and after (b) exposure to HCl (vapor). a) Image obtained after 72 minutes of exposure to HCl. b) Surface (a), 3-8 minutes after HCl vapor had been turned off (scan lasted 5 minutes).

CONCLUSIONS

These observations show that the siderite (1014) surface is relatively stable under ambient atmosphere (30-60% relative humidity), but after prolonged exposure to 100% noncondensing humidity the surface undergoes reconstruction. Reactions performed under acidic atmosphere involving H_2CO_3 show that material may be mobilized and the surface may show local signs of dissolution or precipitation. Surfaces reacted with HCl vapor show more rapid reorganization, primarily the development of dissolution features. However, in the presence of HCl a false topography may result from a differentially charged surface. These reactions may be aided (or inhibited) at least in part by microscale lateral chemical heterogeneities (cementing phase and included fluids), existing microtopography and a very high surface area/ volume.

These results show that the siderite-atmosphere interface is an active environment for geochemical reactions involving high humidity and/or the presence of acidic aerosols.

REFERENCES

1. W. Dreybrodt, J. Lauckner, L. Zaihua, U. Svensson and D. Buhmann, Geochim.Cosmochim Acta **60**, 3375 (1996).
2. P. Wersin, L. Charlet, R. Karthein and W.Stumm, Geochim.Cosmochim Acta **53**, 2787 (1989).
3. L. Charlet, P. Wersin, W. Stumm, Geochim.Cosmochim Acta **54**, 2329 (1990).
4. P. Van Cappellen, L. Charlet, W. Stumm and P. Wersin, Geochim.Cosmochim Acta **57**, 3505 (1993).
5. R.P. Chiarello, R.A. Wogelius and N.C. Sturchio, Geochim.Cosmochim Acta **57**, 4103 (1993).
6. S.L.S. Stipp, W. Gutemannsbauer and T. Lehmann, American Mineralogist, **81**, 1 (1996).
7. S.L.S. Stipp and M.F. Hochella, Geochim.Cosmochim Acta **55**, 1723 (1991).
8. S.L.S. Stipp, C.M. Eggleston and B.S. Neilsen, Geochim.Cosmochim Acta **58**, 3023 (1994).
9. P.E. Hillner, A.J. Gratz, S. Manne and P.K. Hansma, Geology, **20**, 359 (1992).

SYNTHESIS OF INORGANIC AROMATIC HYDROCARBON POLYMERS ON METAL SURFACES

DALE L. PERRY, CRYSTAL LEWIS, AND MISHI PUGH
Lawrence Berkeley National Laboratory, University of California, Berkeley, CA 94720.

ABSTRACT

Films of copper(II) imidazolate, an aromatic, heterocyclic hydrocarbon polymer, have been synthesized *in situ* at room temperature on copper surfaces. The films were made using a copper(II) salt formed as a thin film on an metallic copper substrate with subsequent complexation and deprotonation of the imidazole molecule. The *in-situ* formed copper(II) imidazolate polymer films have been studied by elemental analyses, scanning electron microscopy, and x-ray photoelectron and Auger spectroscopy. The copper Auger parameter has been derived for the films.

INTRODUCTION

Imidazole, an aromatic five-membered, heterocyclic two-nitrogen hydrocarbon, exhibits an extensive chemistry. The two ring nitrogen atoms are of two different types, a "pyrrole" nitrogen with a hydrogen atom attached to it and a "pyridine" nitrogen atom. Both nitrogen atoms possess a lone pair of electrons for donation to metal ion acceptors. The molecule and its substituted compounds easily form coordination complexes with transition metal ions [1-3], serve as a constituent in many biological molecules [4], and may be de-protonated at elevated pH values to form polymers with metal ions [5], a process which involves the removal of the "pyrrole" nitrogen hydrogen atom of the parent molecule (Fig. 1) to form a molecular anion.

The present work describes the formation of imidazolate films formed on metallic copper surfaces. Scanning electron microscopy, x-ray photoelectron, and Auger spectroscopy have been used to characterize the copper(II) imidazolate complex films formed on the elemental copper surface.

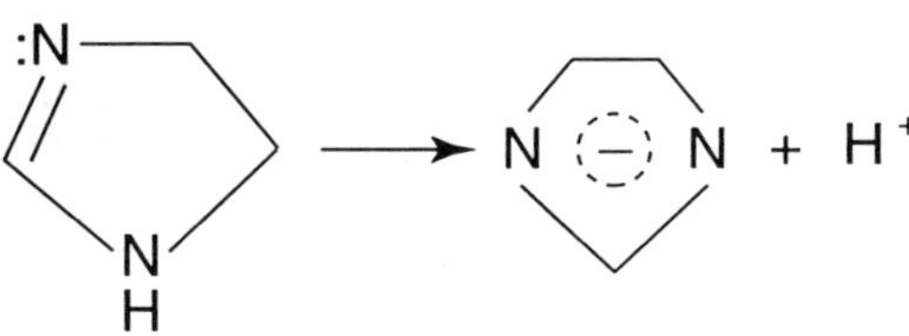

Figure 1. Imidazole and its deprotonated, anionic form.

EXPERIMENTAL

Materials and Chemicals

All chemicals used were reagent grade and used as received. Imidazole was purchased commercially from Aldrich Chemical Company, NaOH from Mallinckrodt, methanol from

Aldrich, and HNO₃ from Baker Chemical Company. Copper metal strips were purchased from
Copper and Brass Sales, Inc. All experiments were conducted using distilled, de-ionized water.

Instrumentation

The scanning electron microscope used was an ISI Model DS-130 dual stage scanning
electron microscope, with all micrographs taken at 20 kVwith a magnification of 3.53 kX. A
Polaron Model E5100 gold sputtering/coating unit was used to gold-coat the samples. X-ray
photoelectron and Auger spectra were obtained using a Physical Electronics Model 5300
spectrometer. Spectra were calibrated for charging using the adventitious carbon $1s$ photoelectron
line at 248.8 eV.

Synthesis of Copper(II) Imidazolate Films

The *in situ* method for forming the copper(II) imidazolate film on the metallic copper
substrate was accomplished by the following method. Strips of metallic copper were sequentially
immersed in dilute HNO₃ followed by immersion of the copper strip into an aqueous imidazole
solution, and, finally, into a 0.1 to 0.5 M aqueous solution of NaOH. The concentrations of HNO₃
varied in concentration, ranging from 20% to 50%, while the imidazole concentration ranged from
0.1 to 0.5 M. The films were air-dried overnight at room temperature after a final rinse with
methanol.

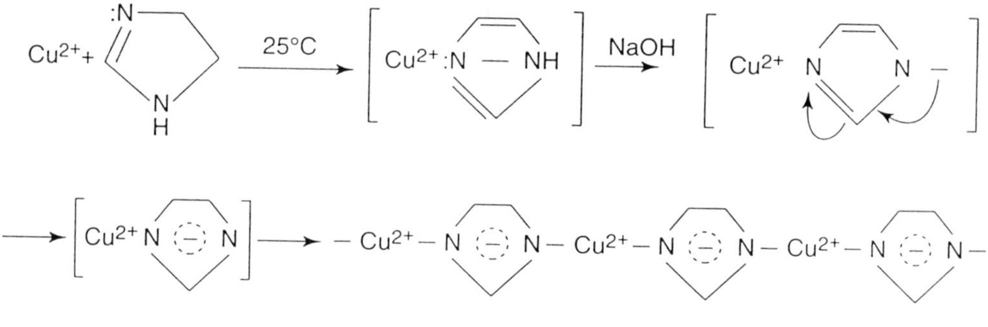

Figure 2. Reaction scheme for the preparation of copper(II) imidazolate polymeric films on copper
substrates. The scheme does not reflect details involving structurally inequivalent copper atoms
[Ref. 6] or nitrogen bonding modes.

RESULTS AND DISCUSSION

The formation of the copper(II) imidazolate films includes several discrete, individual
reaction steps. In the first one, the elemental copper surface is transformed into a layer of
copper(II) nitrate by the reaction of the copper with HNO₃. The subsequent immersion of the

altered copper strip into the imidazole solution forms the copper(II) nitrate coordination complex with the imidazole molecule. Finally, the introduction of the complexed strip with the NaOH solution results in the deprotonation of the complexed imidazole molecule to form the final copper(II) imidazolate polymer on the surface of the bulk elemental copper (Fig. 2). The formed films display a variety of morphological forms, with a typical one shown in Fig.3.

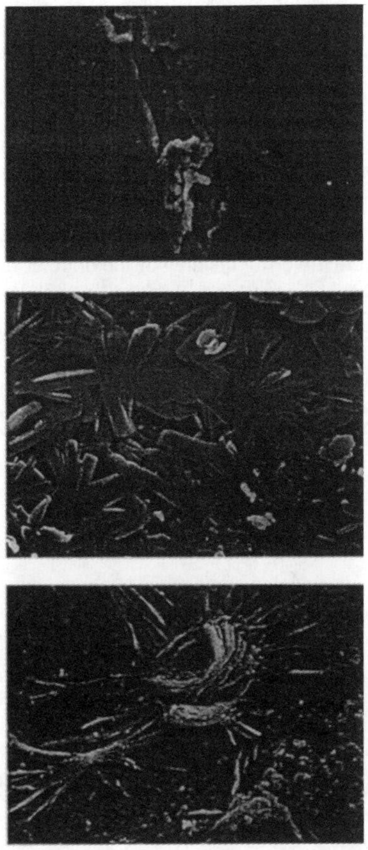

Figure 3. Scanning electron micrographs of metallic copper (top), metallic copper with a reactive surface layer of copper(II) nitrate (center), and a film of copper(II) imidazolate on metallic copper (bottom). The magnification for the three samples is 3.53 kx.

X-ray photoelectron data are in excellent agreement with the reported structure of the bulk copper(II) imidazolate [6] and also with x-ray photoelectron data for the imidazolate ion chemisorbed on silver [7].   Structurally, the copper(II) imidazolate molecule contains two structurally different types of copper. The copper $2p_{3/2,1/2}$ photoelectron spectra shown in Fig. 4 are in good agreement with the structure, including a slightly split doublet for the copper $2p_{3/2}$ line centered at 935.2 eV; the two copper $2p_{1/2}$ lines exhibit themselves as a broadened singlet centered at 955.0 eV. These lines are separated by 19.8 eV, the range of separation observed for copper(II) compounds [8-10]. Additionally, strong "shake-up" satellite structure appears ~ 9 eV to the high binding energy sides of both peaks, indicative of the paramagnetic copper(II) $[Ar]$ $3d^9$ electronic structure. The "shake-up" satellites are broad and asymmetric, consistent with two different copper atom environments This spectral signature is quite different from that of copper(I) with its diamagnetic $[Ar]$ $3d^{10}$ closed shell electronic structure, yielding a $2p_{3/2,1/2}$ spectrum in which there is only very weak "shake-up" satellite structure.

Other aspects of the x-ray photoelectron spectrum also agree with data previously published for the imidazolate ion.   The carbon $1s$ consists of a broad doublet that can be assigned values of 284.8 (the adventitious carbon line) and 285.9 eV, values that are in excellent agreement with those previously published [7]. The third line at 287.8 eV reported in the same paper appears as a broadened shoulder on the low binding energy side of the 285.9 eV peak.

The nitrogen $1s$ line is centered at 399.2 eV as a doublet, in very good agreement with previous work [7]. In that work, a very broad, unresolved singlet was reported being centered at 400.0 eV, with the peak width covering the area of the doublet reported here.

The broadened copper L_3VV Auger line was located at 338.4 eV. Taken in conjunction with the averaged of the slightly split, broad copper $2p_{3/2}$ line value of 935.2 eV , the Auger parameter [11] for copper could be calculated to be 1850.4.

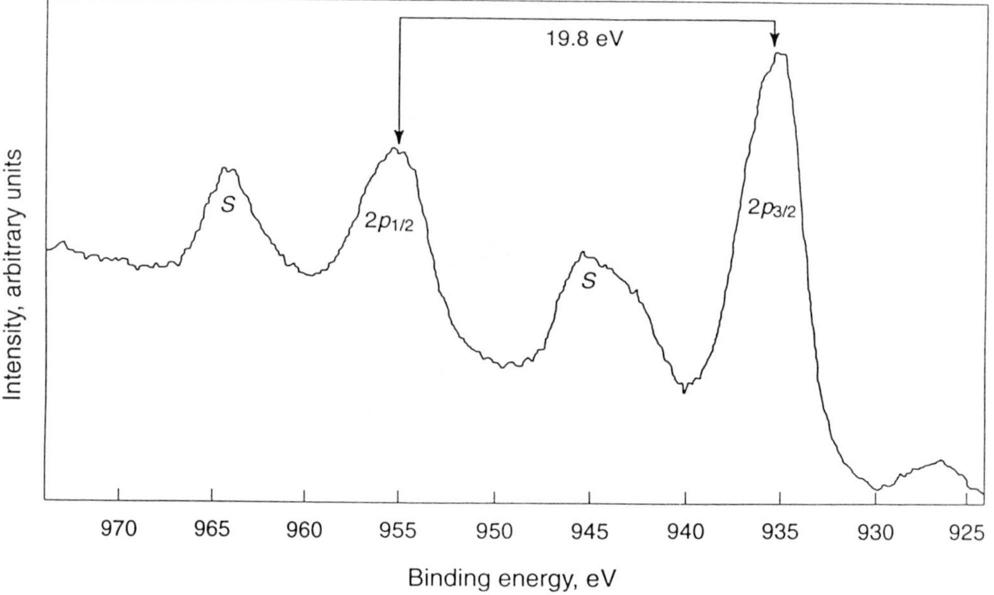

Figure 4. The copper $2p_{3/2,1/2}$ x-ray photoelectron spectrum of copper(II) imidazolate on copper. The S designates satellite structures related to the main photoelectron peaks.

ACKNOWLEDGMENTS

This work was performed under the Center for Science and Engineering Education (CSEE) at the Lawrence Berkeley National Laboratory with support from the Director, Office of Energy Research, Division of University and Science Education Programs, of the U. S. Department of Energy under Contract Number DE-AC03-76SF00098.

REFERENCES

1. W. J. Eilbeck, F. Holmes, and A. E. Underhill, J. Chem. Soc. A 757(1967).

2. B. Chiswell, F. Lions, and B. S. Ghosh, Inorg. Chem. **3**, 110(1964).

3. M. Goodgame and F. A. Cotton, J. Amer. Chem. Soc. **84**, 1543(1962).

4. R. J. Sundberg and R. B. Martin, Chem. Rev. **74**, 471(1974).

5. J. E. Bauman, Jr. and J. C. Wang, Inorg. Chem. **3**, 368(1964).

6. J. A. J. Jarvis, and A. F. Wells, Acta Cryst. **13**, 1027(1960).

7. Q. Xue, S.-Y. Liu, X.-Y. Huang, B. Sun, P.-X. Zhang, X.-Y. Li, and W.-L. Guo, Chinese Sci. Bull. **34**, 1090(1989).

8. G. Schon, Surf. Sci. **35**, 96(1973).

9. S. W. Gaarenstroom and N. Winograd, J. Chem. Phys. **67**, 3500(1977).

10. N. S. McIntyre, T. E. Rummery, M. G. Cook, and D. Owen, J. Electrochem. Soc. **123**, 1165(1976).

11. C. D. Wagner and A. Joshi, J. Electron Spectrosc. Relat. Phenom. **47**, 283(1988).

AUTHOR INDEX